Student's Solutions Manual

Peggy Irish

Teri Lovelace

Elementary and Intermediate Algebra

Third Edition

George Woodbury

College of the Sequoias

Addison-Wesley
is an imprint of

PEARSON

www.pearsonhighered.com

CONTENTS

1.1   QUICK CHECK

1. ⊢—┼—┼—●—┼—┼—┼—┼—┼—┼—▶
0 1 2 3 4 5 6 7 8 9 10

2. a) $8 > 3$

   b) $19 < 23$

3. a) 13

   b) -8

4. a) $-14 < -11$

   b) $6 > -20$

5. $|-9| = 9$

1.1   INTEGERS, OPPOSITES, AND ABSOLUTE VALUES

1. empty set

3. infinity

5. ⊢—┼—┼—┼—┼—┼—┼—●—┼—┼—▶
0 1 2 3 4 5 6 7 8 9 10

7. ⊢—┼—┼—┼—┼—┼—●—┼—┼—┼—▶
0 1 2 3 4 5 6 7 8 9 10

9. ⊢—┼—●—┼—┼—┼—┼—┼—┼—┼—▶
0 1 2 3 4 5 6 7 8 9 10

11. $3 < 13$

13. $8 > 6$

15. $45 > 42$

17. ◀—┼—┼—┼—┼—┼—●—┼—┼—┼—▶
−10   −8   −6   −4   −2   0

19. ◀—┼—┼—┼—┼—●—┼—┼—┼—▶
0   2   4   6   8   10

21. ◀—┼—┼—●—┼—┼—┼—┼—┼—┼—┼—┼—▶
−15 −12 −9 −6 −3   0

23. 7

25. -22

27. 0

29. $-7 > -9$

31. $-13 < -11$

33. $-16 < 0$

35. $9 > -14$

37. $|-15| = 15$

39. $|0| = 0$

41. $|-7| = 7$

43. $-|7| = -7$

45. $-|-29| = -29$

47. Since $|-7| = 7, |-7| > 4.$

49. Since $|-16| = 16, -16 < |-16|.$

51. Since $-|-24| = -24$ and $|-47| = 47,$
$-|-24| < |-47|$

53. A, B, C, D

55. B, C, D

57. C, D

59. $|-5| = 5$ and $|5| = 5,$ so $-5, 5$

61. No number possible.

63. $|-14| - 8 = 14 - 8 = 6,$ and
$|14| - 8 = 14 - 8 = 6,$ so $-14, 14$

65. No, the absolute value of 0 is not positive.

67. Negative. Explanations will vary. Example:
The opposite of a negative number is positive.
Also, the absolute value of a negative number
is positive.

1.2 QUICK CHECK

1. $14 + (-6) = 14 - 6 = 8$

2. $4 + (-17) = 4 - 17 = -13$

3. $-2 + (-9) = -2 - 9 = -11$

4. $11 - (-7) = 11 + 7 = 18$

5. $14 - 9 - (-22) - 6 + (-30) + 5$
$= 14 - 9 + 22 - 6 - 30 + 5$
$= 41 - 45$
$= -4$

6. $10(-6) = -60$

7. $(-7)(-9) = 63$

8. $-4(-10)(5)(-2) = 40(5)(-2)$
$= 200(-2)$
$= -400$

9. $72 \div (-8) = -9$

1.2 OPERATIONS WITH INTEGERS

1. larger

3. positive

5. positive

7. divisor

9. $8 + (-13) = 8 - 13 = -5$

11. $6 + (-33) = 6 - 33 = -27$

13. $(-4) + 5 = -4 + 5 = 1$

15. $-14 + 22 = 8$

17. $-5 + (-6) = -5 - 6 = -11$

19. $8 - 6 = 2$

21. $5 - 11 = -6$

23. $(-5) - 3 = -5 - 3 = -8$

25. $-9 - 13 = -22$

27. $36 - (-25) = 36 + 25 = 61$

29. $-42 - (-33) = -42 + 33 = -9$

31. $8 - 13 - 6 = 8 - 19 = -11$

33. $-9 + 7 - 4 = 7 - 13 = -6$

35. $6 - (-16) + 5 = 6 + 16 + 5 = 27$

37. $4 + (-15) - 13 - (-25) = 4 - 15 - 13 + 25$
$\qquad\qquad\qquad\qquad\qquad = 29 - 28$
$\qquad\qquad\qquad\qquad\qquad = 1$

39. $\$30 - \$22 = \$8$

41. $-8°C + 12°C = 4°C$

43. $1750 - (-400) = 1750 + 400 = 2150$ feet

45. $7(-6) = -42$

47. $-8 \cdot 5 = -40$

49. $-15(-12) = 180$

51. $82(-1) = -82$

53. $-6 \cdot 0 = 0$

55. $-6(-3)(5) = 18(5) = 90$

57. $5 \cdot 3(-2)(-6) = 15(-2)(-6) = (-30)(-6) = 180$

59. $45 \div (-5) = -9$

61. $-36 \div 6 = -6$

63. $-32 \div (-8) = 4$

65. $126 \div (-9) = -14$

67. $0 \div (-13) = 0$

69. $29 \div 0$ undefined

71. $-11(-12) = 132$

73. $5 - 13 = -8$

75. $17-(-11)-49 = 17+11-49 = 28-49 = -21$

77. $-432 \div 3 = -144$

79. $9(-24) = -216$

81. $5 \cdot 3(-17)(-29)(0) = 0$

83. $4(\$23) = \$92$

85. $400(-\$3) + 500(\$2) = -\$1200 + \$1000$
$$= -\$200$$

87. $? + (-34) = -15$
$? - 34 = -15$
Since $19 - 34 = 15$, 19 is the integer.

89. $? \div (-8) = 16$
Since $-128 \div (-8) = 16$, -128 is the integer.

91. True

93. True

95. Explanations will vary. Example:
Subtracting a negative integer results in a
"double negative" and is the same as adding the
opposite of the integer.

97. Explanations will vary. Example:
Multiplying a negative times a negative is like
taking the opposite of a negative, which results
in a positive.

1.3 **QUICK CHECK**

1. $1 \cdot 36 = 36$
$2 \cdot 18 = 36$
$3 \cdot 12 = 36$
$4 \cdot 9 = 36$
$6 \cdot 6 = 36$
$\{1, 2, 3, 4, 6, 9, 12, 18, 36\}$

2. **a)** Composite, since $3 \cdot 19 = 57$.

b) Prime

c) Composite, since $4 \cdot 12 = 48$.

3. $3 \cdot 3 \cdot 7 = 63$

4. a) $\dfrac{45}{210} = \dfrac{\not{3} \cdot 3 \cdot \not{5}}{2 \cdot \not{3} \cdot \not{5} \cdot 7} = \dfrac{3}{14}$

b) $\dfrac{24}{384} = \dfrac{\not{2} \cdot \not{2} \cdot \not{2} \cdot \not{3}}{\not{2} \cdot \not{2} \cdot \not{2} \cdot 2 \cdot 2 \cdot 2 \cdot 2 \cdot \not{3}} = \dfrac{1}{16}$

5. $\dfrac{121}{13}$

$$13\overline{)121} \quad \begin{array}{r} 9 \\ \hline 121 \\ -117 \\ \hline 4 \end{array}$$

$$\frac{121}{13} = 9\frac{4}{13}$$

6. $8\dfrac{1}{6} = \dfrac{49}{6}$

1.3 **FRACTIONS**

1. factor set

3. composite

5. top

7. common factors

9. improper

11. No, since there is no number x such that
$7 \cdot x = 247$.

13. Yes, since $6 \cdot 806 = 4836$.

15. Yes, since $15 \cdot 189 = 2835$.

17. $1 \cdot 48 = 48$
$2 \cdot 24 = 48$
$3 \cdot 16 = 48$
$4 \cdot 12 = 48$
$6 \cdot 8 = 48$
$\{1, 2, 3, 4, 6, 8, 12, 16, 24, 48\}$

19. $1 \cdot 27 = 27$
$3 \cdot 9 = 27$
$\{1, 3, 9, 27\}$

21. $1 \cdot 20 = 20$
$2 \cdot 10 = 20$
$4 \cdot 5 = 20$
$\{1, 2, 4, 5, 10, 20\}$

23. $1 \cdot 81 = 81$
$3 \cdot 27 = 81$
$9 \cdot 9 = 81$
$\{1, 3, 9, 27, 81\}$

25. $1 \cdot 31 = 31$
$\{1, 31\}$

27. $1 \cdot 91 = 91$
$7 \cdot 13 = 91$
$\{1, 7, 13, 91\}$

29. 18
$6 \cdot 3$
$2 \cdot 3$
$2 \cdot 3 \cdot 3 = 18$

31. 42
$6 \cdot 7$
$2 \cdot 3$
$2 \cdot 3 \cdot 7 = 42$

33. 39
$3 \cdot 13$
$3 \cdot 13 = 39$

35. 27
$9 \cdot 3$
$3 \cdot 3$
$3 \cdot 3 \cdot 3 = 27$

37. 125
$5 \cdot 25$
$5 \cdot 5$
$5 \cdot 5 \cdot 5 = 125$

39. 29 is prime.

41. 99
$9 \cdot 11$
$3 \cdot 3$
$3 \cdot 3 \cdot 11 = 99$

43. 31 is prime.

45. 120
$12 \cdot 10$
$6 \cdot 2 \quad 2 \cdot 5$
$3 \cdot 2$
$2 \cdot 2 \cdot 2 \cdot 3 \cdot 5 = 120$

47. $\dfrac{10}{16} = \dfrac{\cancel{2}^{1} \cdot 5}{\cancel{2}_{1} \cdot 2 \cdot 2 \cdot 2} = \dfrac{5}{8}$

49. $\dfrac{9}{45} = \dfrac{\cancel{3}^{1} \cdot \cancel{3}^{1}}{\cancel{3}_{1} \cdot \cancel{3}_{1} \cdot 5} = \dfrac{1}{5}$

51. $\dfrac{168}{378} = \dfrac{\cancel{2}^{1} \cdot 2 \cdot 2 \cdot \cancel{3}^{1} \cdot \cancel{7}^{1}}{\cancel{2}_{1} \cdot 3 \cdot 3 \cdot \cancel{3}_{1} \cdot \cancel{7}_{1}} = \dfrac{4}{9}$

53. $\dfrac{27}{64}$ does not simplify further.

55. $\dfrac{160}{176} = \dfrac{\cancel{2}^{1} \cdot \cancel{2}^{1} \cdot \cancel{2}^{1} \cdot \cancel{2}^{1} \cdot 2 \cdot 5}{\cancel{2}_{1} \cdot \cancel{2}_{1} \cdot \cancel{2}_{1} \cdot \cancel{2}_{1} \cdot 11} = \dfrac{10}{11}$

57. $\dfrac{49}{91} = \dfrac{\cancel{7}^{1} \cdot 7}{\cancel{7}_{1} \cdot 13} = \dfrac{7}{13}$

59. $3\dfrac{4}{5}$
Multiply $3 \cdot 5 = 15$.
Add $15 + 4 = 19$.
$3\dfrac{4}{5} = \dfrac{19}{5}$

61. $2\dfrac{16}{17}$
Multiply $2 \cdot 17 = 34$.
Add $34 + 16 = 50$.
$2\dfrac{16}{17} = \dfrac{50}{17}$

63. $13\dfrac{8}{11}$
Multiply $13 \cdot 11 = 143$.
Add $143 + 8 = 151$.
$13\dfrac{8}{11} = \dfrac{151}{11}$

65. $\dfrac{39}{5}$ $5\overline{\smash{)}\,39}$ $\dfrac{39}{5}=7\dfrac{4}{5}$
$\phantom{5\overline{)}}\;\underline{-35}$
$\phantom{5\overline{)39}}\;4$

67. $\dfrac{101}{7}$ $7\overline{\smash{)}\,101}$ $\dfrac{101}{7}=14\dfrac{3}{7}$
$\underline{-7}$
31
$\underline{-28}$
3

69. $\dfrac{141}{19}$ $19\overline{\smash{)}\,141}$ $\dfrac{141}{19}=7\dfrac{8}{19}$
$\underline{-133}$
8

71. Answers will vary. Example:
$$\frac{3}{4}\cdot\frac{2}{2}=\frac{6}{8}$$
$$\frac{3}{4}\cdot\frac{3}{3}=\frac{9}{12}$$
$$\frac{3}{4}\cdot\frac{4}{4}=\frac{12}{16}$$
$$\frac{3}{4}\cdot\frac{5}{5}=\frac{15}{20}$$
$$\frac{6}{8},\frac{9}{12},\frac{12}{16},\frac{15}{20}$$

73. Answers will vary. Example: $30, 42, 70, 105$

75. Answers will vary. Example:
A real-world situation involving a fraction is cooking with $\frac{1}{3}$ cup of oil. A real-world situation involving a mixed number is sawing $4\frac{1}{2}$ inches from the length of a board.

1.4   QUICK CHECK

1. $\dfrac{\cancel{10}^{\,5}}{\cancel{63}_{\,7}}\cdot\dfrac{\cancel{9}^{\,1}}{\cancel{16}_{\,8}}=\dfrac{5}{7}\cdot\dfrac{1}{8}=\dfrac{5}{56}$

2. $2\dfrac{2}{3}\cdot 8\dfrac{5}{8}=\dfrac{8}{3}\cdot\dfrac{69}{8}=\dfrac{\cancel{8}^{\,1}}{\cancel{3}_{\,1}}\cdot\dfrac{\cancel{69}^{\,23}}{\cancel{8}_{\,1}}=23$

3. $\dfrac{12}{25}\div\dfrac{63}{10}=\dfrac{12}{25}\cdot\dfrac{10}{63}=\dfrac{\cancel{12}^{\,4}}{\cancel{25}_{\,5}}\cdot\dfrac{\cancel{10}^{\,2}}{\cancel{63}_{\,21}}=\dfrac{8}{105}$

4. $\dfrac{20}{21}\div 2\dfrac{2}{3}=\dfrac{20}{21}\div\dfrac{8}{3}=\dfrac{20}{21}\cdot\dfrac{3}{8}=\dfrac{\cancel{20}^{\,5}}{\cancel{21}_{\,7}}\cdot\dfrac{\cancel{3}^{\,1}}{\cancel{8}_{\,2}}=\dfrac{5}{14}$

5. $\dfrac{17}{20}-\dfrac{5}{20}=\dfrac{12}{20}=\dfrac{3}{5}$

6. $6\dfrac{1}{8}+5\dfrac{5}{8}=\dfrac{49}{8}+\dfrac{45}{8}=\dfrac{94}{8}=11\dfrac{6}{8}=11\dfrac{3}{4}$

7. $18=2\cdot3\cdot3$
$42=2\cdot3\cdot7$
$\text{LCM}=2\cdot3\cdot3\cdot7=126$

8. $\dfrac{4}{9}+\dfrac{2}{15}=\dfrac{4}{9}\cdot\dfrac{5}{5}+\dfrac{2}{15}\cdot\dfrac{3}{3}=\dfrac{20}{45}+\dfrac{6}{45}=\dfrac{26}{45}$

9. $5\dfrac{1}{5}-3\dfrac{5}{6}=\dfrac{26}{5}-\dfrac{23}{6}$
$$=\dfrac{26}{5}\cdot\dfrac{6}{6}-\dfrac{23}{6}\cdot\dfrac{5}{5}$$
$$=\dfrac{156}{30}-\dfrac{115}{30}$$
$$=\dfrac{41}{30}$$
$$=1\dfrac{11}{30}$$

1.4   OPERATIONS WITH FRACTIONS

1. improper fraction

3. To divide a number by a fraction, invert the divisor and then multiply.

5. least common multiple

7. $\dfrac{\cancel{6}^{\,1}}{\cancel{8}_{\,2}}\cdot\dfrac{\cancel{4}^{\,1}}{\cancel{27}_{\,9}}=\dfrac{1}{18}$

9. $\dfrac{\cancel{20}^{\,2}}{\cancel{21}_{\,3}}\cdot\left(-\dfrac{\cancel{33}^{\,11}}{\cancel{90}_{\,9}}\right)=-\dfrac{22}{27}$

11. $4\dfrac{2}{7}\cdot\dfrac{14}{25}=\dfrac{\cancel{30}^{\,6}}{\cancel{7}_{\,1}}\cdot\dfrac{\cancel{14}^{\,2}}{\cancel{25}_{\,5}}=\dfrac{6}{1}\cdot\dfrac{2}{5}=\dfrac{12}{5}=2\dfrac{2}{5}$

13. $5\cdot6\dfrac{3}{10}=\dfrac{\cancel{5}^{\,1}}{1}\cdot\dfrac{63}{\cancel{10}_{\,2}}=\dfrac{1}{1}\cdot\dfrac{63}{2}=\dfrac{63}{2}=31\dfrac{1}{2}$

15. $\dfrac{2}{3}\cdot\dfrac{8}{9}=\dfrac{16}{27}$

17. $\dfrac{6}{25}\div\dfrac{8}{45}=\dfrac{\cancel{6}^{\,3}}{\cancel{25}_{\,5}}\cdot\dfrac{\cancel{45}^{\,9}}{\cancel{8}_{\,4}}=\dfrac{27}{20}$

19. $-\dfrac{22}{56} \div \dfrac{33}{147} = -\dfrac{\overset{2}{\cancel{22}}}{\underset{8}{\cancel{56}}} \cdot \dfrac{\overset{21}{\cancel{147}}}{\underset{3}{\cancel{33}}}$

$\quad\quad = -\dfrac{\overset{1}{\cancel{2}}}{\cancel{8}_4} \cdot \dfrac{\overset{7}{\cancel{21}}}{\cancel{3}_1}$

$\quad\quad = -\dfrac{7}{4}$

21. $\dfrac{17}{40} \div \dfrac{1}{2} = \dfrac{17}{\cancel{40}_{20}} \cdot \dfrac{\overset{1}{\cancel{2}}}{1} = \dfrac{17}{20}$

23. $\dfrac{4}{11} \div 3\dfrac{1}{5} = \dfrac{4}{11} \div \dfrac{16}{5} = \dfrac{\overset{1}{\cancel{4}}}{11} \cdot \dfrac{5}{\cancel{16}_4} = \dfrac{5}{44}$

25. $-2\dfrac{4}{5} \div 6\dfrac{2}{3} = -\dfrac{14}{5} \div \dfrac{20}{3} = -\dfrac{\overset{7}{\cancel{14}}}{5} \cdot \dfrac{3}{\cancel{20}_{10}} = -\dfrac{21}{50}$

27. $\dfrac{7}{15} + \dfrac{4}{15} = \dfrac{11}{15}$

29. $\dfrac{5}{9} + \dfrac{4}{9} = \dfrac{9}{9} = 1$

31. $\dfrac{2}{5} - \dfrac{4}{5} = -\dfrac{2}{5}$

33. $\dfrac{9}{16} - \dfrac{3}{16} = \dfrac{6}{16} = \dfrac{3}{8}$

35. $10 = 2 \cdot 5$
$15 = 3 \cdot 5$
$\text{LCM} = 2 \cdot 3 \cdot 5 = 30$

37. $12 = 2 \cdot 2 \cdot 3$
$42 = 2 \cdot 3 \cdot 7$
$\text{LCM} = 2 \cdot 2 \cdot 3 \cdot 7 = 84$

39. $16 = 2 \cdot 2 \cdot 2 \cdot 2$
$80 = 2 \cdot 2 \cdot 2 \cdot 2 \cdot 5$
$\text{LCM} = 2 \cdot 2 \cdot 2 \cdot 2 \cdot 5 = 80$

41. $8 = 2 \cdot 2 \cdot 2$
$10 = 2 \cdot 5$
$14 = 2 \cdot 7$
$\text{LCM} = 2 \cdot 2 \cdot 2 \cdot 5 \cdot 7 = 280$

43. $\dfrac{4}{5} + \dfrac{3}{4}$ $\text{LCM} = 20$

$\dfrac{4}{5} + \dfrac{3}{4} = \dfrac{4}{5} \cdot \dfrac{4}{4} + \dfrac{3}{4} \cdot \dfrac{5}{5} = \dfrac{16}{20} + \dfrac{15}{20} = \dfrac{31}{20}$

45. $\dfrac{7}{10} + \dfrac{5}{8}$ $\text{LCM} = 40$

$\dfrac{7}{10} + \dfrac{5}{8} = \dfrac{7}{10} \cdot \dfrac{4}{4} + \dfrac{5}{8} \cdot \dfrac{5}{5} = \dfrac{28}{40} + \dfrac{25}{40} = \dfrac{53}{40}$

47. $6\dfrac{1}{5} + 5 = 11\dfrac{1}{5}$

49. $6\dfrac{2}{3} + 5\dfrac{1}{6} = \dfrac{20}{3} + \dfrac{31}{6}$ $\text{LCM} = 6$

$\dfrac{20}{3} + \dfrac{31}{6} = \dfrac{20}{3} \cdot \dfrac{2}{2} + \dfrac{31}{6} = \dfrac{40}{6} + \dfrac{31}{6} = \dfrac{71}{6} = 11\dfrac{5}{6}$

51. $\dfrac{2}{3} - \dfrac{7}{15}$ $\text{LCM} = 15$

$\dfrac{2}{3} - \dfrac{7}{15} = \dfrac{2}{3} \cdot \dfrac{5}{5} - \dfrac{7}{15} = \dfrac{10}{15} - \dfrac{7}{15} = \dfrac{3}{15} = \dfrac{1}{5}$

53. $\dfrac{3}{4} - \dfrac{2}{7}$ $\text{LCM} = 28$

$\dfrac{3}{4} - \dfrac{2}{7} = \dfrac{3}{4} \cdot \dfrac{7}{7} - \dfrac{2}{7} \cdot \dfrac{4}{4} = \dfrac{21}{28} - \dfrac{8}{28} = \dfrac{13}{28}$

55. $7\dfrac{1}{2} - 3\dfrac{1}{4} = \dfrac{15}{2} - \dfrac{13}{4}$ $\text{LCM} = 4$

$\dfrac{15}{2} - \dfrac{13}{4} = \dfrac{15}{2} \cdot \dfrac{2}{2} - \dfrac{13}{4} = \dfrac{30}{4} - \dfrac{13}{4} = \dfrac{17}{4} = 4\dfrac{1}{4}$

57. $12\dfrac{3}{10} - 9 = \dfrac{123}{10} - 9$ $\text{LCM} = 10$

$\dfrac{123}{10} - 9 = \dfrac{123}{10} - \dfrac{9}{1} \cdot \dfrac{10}{10} = \dfrac{123}{10} - \dfrac{90}{10} = \dfrac{33}{10} = 3\dfrac{3}{10}$

59. $-\dfrac{5}{9} - \dfrac{7}{12} = -\dfrac{20}{36} - \dfrac{21}{36} = -\dfrac{41}{36}$

61. $-\dfrac{9}{16} + \dfrac{5}{24} = -\dfrac{27}{48} + \dfrac{10}{48} = -\dfrac{17}{48}$

63. $\dfrac{6}{7} - \left(-\dfrac{8}{15}\right) = \dfrac{6}{7} + \dfrac{8}{15} = \dfrac{90}{105} + \dfrac{56}{105} = \dfrac{146}{105}$

65. $-\dfrac{4}{15} + \left(-\dfrac{13}{18}\right) = -\dfrac{4}{15} - \dfrac{13}{18} = -\dfrac{24}{90} - \dfrac{65}{90} = -\dfrac{89}{90}$

67. $\dfrac{3}{16}+\dfrac{9}{20}-\dfrac{11}{12}=\dfrac{45}{240}+\dfrac{108}{240}-\dfrac{220}{240}$

$\qquad\qquad\quad =\dfrac{153}{240}-\dfrac{220}{240}$

$\qquad\qquad\quad =-\dfrac{67}{240}$

69. $\dfrac{8}{\cancel{9}_3}\cdot\dfrac{\cancel{3}^{\,1}}{5}=\dfrac{8}{15}$

71. $\dfrac{7}{30}\div\dfrac{35}{48}=\dfrac{\cancel{7}^{\,1}}{\cancel{30}_5}\cdot\dfrac{\cancel{48}^{\,8}}{\cancel{35}_5}=\dfrac{8}{25}$

73. $\dfrac{1}{6}-\dfrac{7}{8}\quad$ LCM = 24

$\dfrac{1}{6}-\dfrac{7}{8}=\dfrac{1}{6}\cdot\dfrac{4}{4}-\dfrac{7}{8}\cdot\dfrac{3}{3}=\dfrac{4}{24}-\dfrac{21}{24}=-\dfrac{17}{24}$

75. $\dfrac{19}{30}+\dfrac{11}{18}\quad$ LCM = 90

$\dfrac{19}{30}+\dfrac{11}{18}=\dfrac{19}{30}\cdot\dfrac{3}{3}+\dfrac{11}{18}\cdot\dfrac{5}{5}=\dfrac{57}{90}+\dfrac{55}{90}=\dfrac{112}{90}=\dfrac{56}{45}$

77. $13\div\dfrac{1}{8}=\dfrac{13}{1}\cdot\dfrac{8}{1}=13\cdot8=104$

79. $3\dfrac{4}{7}+6\dfrac{3}{5}-8=\dfrac{25}{7}+\dfrac{33}{5}-\dfrac{8}{1}\quad$ LCM = 35

$\dfrac{25}{7}+\dfrac{33}{5}-\dfrac{8}{1}=\dfrac{25}{7}\cdot\dfrac{5}{5}+\dfrac{33}{5}\cdot\dfrac{7}{7}-\dfrac{8}{1}\cdot\dfrac{35}{35}$

$\qquad\qquad\qquad =\dfrac{125}{35}+\dfrac{231}{35}-\dfrac{280}{35}$

$\qquad\qquad\qquad =\dfrac{356}{35}-\dfrac{280}{35}$

$\qquad\qquad\qquad =\dfrac{76}{35}$

$\qquad\qquad\qquad =2\dfrac{6}{35}$

81. $\dfrac{15}{56}+\left(-\dfrac{16}{21}\right)=\dfrac{15}{56}-\dfrac{16}{21}\quad$ LCM = 168

$\dfrac{15}{56}-\dfrac{16}{21}=\dfrac{15}{56}\cdot\dfrac{3}{3}-\dfrac{16}{21}\cdot\dfrac{8}{8}=\dfrac{45}{168}-\dfrac{128}{168}=-\dfrac{83}{168}$

83. $-\dfrac{9}{13}+\dfrac{19}{36}\quad$ LCM $=13\cdot36=468$

$-\dfrac{9}{13}+\dfrac{19}{36}=-\dfrac{9}{13}\cdot\dfrac{36}{36}+\dfrac{19}{36}\cdot\dfrac{13}{13}$

$\qquad\qquad\quad =-\dfrac{324}{468}+\dfrac{247}{468}$

$\qquad\qquad\quad =-\dfrac{77}{468}$

85. $\dfrac{11}{24}+\dfrac{5}{?}=\dfrac{13}{12}\cdot\dfrac{2}{2}$

$\dfrac{11}{24}+\dfrac{5}{?}\cdot\dfrac{3}{3}=\dfrac{26}{24}$

Since $?\cdot3=24$, 8 is the missing number.

87. $\dfrac{\cancel{10}^{\,2}}{21}\cdot\dfrac{?}{\cancel{75}_{15}}=\dfrac{4}{45}$

$\dfrac{2}{21}\cdot\dfrac{?}{15}=\dfrac{4}{45}$

$\dfrac{2\cdot?}{315}=\dfrac{4}{45}\cdot\dfrac{7}{7}$

$\dfrac{2\cdot?}{315}=\dfrac{28}{315}$

Since $2\cdot?=28$, 14 is the missing number.

89. $\dfrac{1}{2}+\dfrac{3}{4}+\dfrac{1}{3}\quad$ LCM = 12

$\dfrac{1}{2}+\dfrac{3}{4}+\dfrac{1}{3}=\dfrac{1}{2}\cdot\dfrac{6}{6}+\dfrac{3}{4}\cdot\dfrac{3}{3}+\dfrac{1}{3}\cdot\dfrac{4}{4}$

$\qquad\qquad\quad =\dfrac{6}{12}+\dfrac{9}{12}+\dfrac{4}{12}$

$\qquad\qquad\quad =\dfrac{19}{12}$

$\qquad\qquad\quad =1\dfrac{7}{12}$ cups

91. $\dfrac{23}{40}-\dfrac{1}{8}\quad$ LCM = 40

$\dfrac{23}{40}-\dfrac{1}{8}=\dfrac{23}{40}-\dfrac{1}{8}\cdot\dfrac{5}{5}$

$\qquad\qquad =\dfrac{23}{40}-\dfrac{5}{40}$

$\qquad\qquad =\dfrac{18}{40}$

$\qquad\qquad =\dfrac{9}{20}$ fluid ounce

93. $4\dfrac{1}{5} \div 6 = \dfrac{21}{5} \div \dfrac{6}{1} = \dfrac{\overset{7}{\cancel{21}}}{5} \cdot \dfrac{1}{\underset{2}{\cancel{6}}} = \dfrac{7}{10}$ of a foot

95. $2 \cdot \dfrac{5}{6} + 2 \cdot \dfrac{2}{3} = \dfrac{\overset{1}{\cancel{2}}}{1} \cdot \dfrac{5}{\underset{3}{\cancel{6}}} + \dfrac{2}{1} \cdot \dfrac{2}{3}$

$\qquad = \dfrac{1}{1} \cdot \dfrac{5}{3} + \dfrac{2}{1} \cdot \dfrac{2}{3}$

$\qquad = \dfrac{5}{3} + \dfrac{4}{3}$

$\qquad = \dfrac{9}{3}$

$\qquad = 3$

The total length of framing needed is 3 feet. Therefore, there is not enough material to make the frame.

97. You are only making half the recipe.

$\dfrac{1}{2} \cdot 1\dfrac{1}{3} = \dfrac{1}{\underset{1}{\cancel{2}}} \cdot \dfrac{\overset{2}{\cancel{4}}}{3} = \dfrac{2}{3}$ cup

99. Explanations will vary. Example:
You cannot divide by 2 from 4 and 6 in the

expression $\dfrac{4}{7} \cdot \dfrac{6}{11}$ because 4 and 6 are both in

the numerator.

1.5   QUICK CHECK

1. $\dfrac{3}{4}$

$$4\overline{)3.00} \quad \begin{array}{r} .75 \\ \hline \end{array}$$
$$\begin{array}{r} -28 \\ \hline 20 \\ -20 \\ \hline 0 \end{array}$$

$\dfrac{3}{4} = 0.75$

2. $\dfrac{5}{18}$

$$18\overline{)5.000} \quad \begin{array}{r} .277 \\ \hline \end{array}$$
$$\begin{array}{r} -36 \\ \hline 140 \\ -126 \\ \hline 140 \end{array}$$

$\dfrac{5}{18} = 0.2\overline{7}$

3. $0.425 = \dfrac{425}{1000} = \dfrac{17}{40}$

4. a) $\dfrac{7}{10} \cdot 100\% = \dfrac{7}{\underset{1}{\cancel{10}}} \cdot \dfrac{\overset{10}{\cancel{100}}}{1}\% = 70\%$

b) $\dfrac{21}{40} \cdot 100\% = \dfrac{21}{\underset{2}{\cancel{40}}} \cdot \dfrac{\overset{5}{\cancel{100}}}{1}\%$

$\qquad = \dfrac{105}{2}\%$

$\qquad = 52\dfrac{1}{2}\%$

5. $0.42 \cdot 100\% = 42\%$

6. a) $35\% = \dfrac{\overset{7}{\cancel{35}}}{1} \cdot \dfrac{1}{\underset{20}{\cancel{100}}} = \dfrac{7}{20}$

b) $11\dfrac{2}{3}\% = \dfrac{35}{3}\%$

$\qquad = \dfrac{35}{3} \cdot \dfrac{1}{100}$

$\qquad = \dfrac{\overset{7}{\cancel{35}}}{3} \cdot \dfrac{1}{\underset{20}{\cancel{100}}}$

$\qquad = \dfrac{7}{60}$

7. a) $8\% = 8 \div 100 = 0.08$

b) $240\% = 240 \div 100 = 2.4$

1.5   DECIMALS AND PERCENTS

1. tenths

3. numerator; denominator

5. divide

7.
$$\begin{array}{r} 4.23 \\ +3.62 \\ \hline 7.85 \end{array}$$

9.
$$\begin{array}{r} 5.2 \\ \times \;\; 7 \\ \hline 36.4 \end{array}$$
so, $-7(5.2) = -36.4$

11.
$$\begin{array}{r} 8.4 \\ -3.7 \\ \hline 4.7 \end{array}$$

13. $0.\underline{4}\overline{)13.568}$

$$\begin{array}{r} 33.92 \\ 4\overline{)135.68} \\ -12 \\ \hline 15 \\ -12 \\ \hline 36 \\ -36 \\ \hline 08 \\ -08 \\ \hline 0 \end{array}$$

33.92

15.
$$\begin{array}{r} 3.65 \\ \times -2.2 \\ \hline 730 \\ 7300 \\ \hline -8030 \end{array} = -8.03$$

17.
$$\begin{array}{r} 13.470 \\ +21.562 \\ \hline 35.032 \end{array}$$

19. Multiplying an even number of negative values gives a positive answer.

$$\begin{array}{r} 6.3 \\ \times 3.9 \\ \hline 567 \\ 1890 \\ \hline 24.57 \end{array}$$

$$\begin{array}{r} 24.57 \\ \times 2.25 \\ \hline 12285 \\ 49140 \\ 491400 \\ \hline 552825 \end{array}$$

55.2825

21.
$$\begin{array}{r} 37.278 \\ +56.722 \\ \hline 94.000 \end{array} = 94$$

23.
$$\begin{array}{r} 0.9 \\ 10\overline{)9.0} \\ -90 \\ \hline 0 \end{array}$$

$$\frac{9}{10} = 0.9$$

25.
$$\begin{array}{r} 2.875 \\ 8\overline{)23.000} \\ -16 \\ \hline 70 \\ -64 \\ \hline 60 \\ -56 \\ \hline 40 \\ -40 \\ \hline 0 \end{array}$$

$$-\frac{23}{8} = -2.875$$

27.
$$\begin{array}{r} 0.52 \\ 25\overline{)13.00} \\ -125 \\ \hline 50 \\ -50 \\ \hline 0 \end{array}$$

$$\frac{13}{25} = 0.52$$

29. $24\dfrac{29}{50}$

$$= \frac{1229}{50}$$

$$\begin{array}{r} 24.58 \\ 50\overline{)1229.00} \\ -100 \\ \hline 229 \\ -200 \\ \hline 290 \\ -250 \\ \hline 400 \\ -400 \\ \hline 0 \end{array}$$

$$24\frac{29}{50} = 24.58$$

31. $0.2 = \dfrac{2}{10} = \dfrac{1}{5}$

33. $0.85 = \dfrac{85}{100} = \dfrac{17}{20}$

35. $-0.74 = -\dfrac{74}{100} = -\dfrac{37}{50}$

37. $0.375 = \dfrac{375}{1000} = \dfrac{3}{8}$

39. $98.6 + 2.8$

$$\begin{array}{r} 98.6 \\ +\ 2.8 \\ \hline 101.4 \end{array}$$

$101.4°F$

41. Subtract the checks from the balance.

$$\begin{array}{r} 427.36 \\ -\ 19.95 \\ \hline 407.41 \\ -\ 34.40 \\ \hline 373.01 \\ -\ 148.68 \\ \hline 224.33 \end{array}$$

$\$224.33$

43.
$$\begin{array}{r} 21.47 \\ \times\ \ \ 12 \\ \hline 4294 \\ 21470 \\ \hline 25764 \end{array}$$

$\$257.64$

45. $\dfrac{3}{4} \cdot 100\% = \dfrac{3}{\cancel{4}_1} \cdot \dfrac{\cancel{100}^{25}}{1}\% = 75\%$

47. $\dfrac{4}{5} \cdot 100\% = \dfrac{4}{\cancel{5}_1} \cdot \dfrac{\cancel{100}^{20}}{1}\% = 80\%$

49. $\dfrac{7}{8} \cdot 100\% = \dfrac{7}{\cancel{8}_2} \cdot \dfrac{\cancel{100}^{25}}{1}\% = \dfrac{175}{2}\% = 87\dfrac{1}{2}\%$

51. $\dfrac{27}{4} \cdot 100\% = \dfrac{27}{\cancel{4}_1} \cdot \dfrac{\cancel{100}^{25}}{1}\% = 675\%$

53. $0.4 \cdot 100\% = 40\%$

55. $0.15 \cdot 100\% = 15\%$

57. $0.09 \cdot 100\% = 9\%$

59. $3.2 \cdot 100\% = 320\%$

61. $\dfrac{84}{1} \cdot \dfrac{1}{100} = \dfrac{\cancel{84}^{21}}{1} \cdot \dfrac{1}{\cancel{100}_{25}} = \dfrac{21}{25}$

63. $\dfrac{7}{1} \cdot \dfrac{1}{100} = \dfrac{7}{100}$

65. $11\dfrac{1}{9} \cdot \dfrac{1}{100} = \dfrac{100}{9} \cdot \dfrac{1}{100} = \dfrac{\cancel{100}^{1}}{9} \cdot \dfrac{1}{\cancel{100}_{1}} = \dfrac{1}{9}$

67. $\dfrac{520}{1} \cdot \dfrac{1}{100} = \dfrac{\cancel{520}^{26}}{1} \cdot \dfrac{1}{\cancel{100}_{5}} = \dfrac{26}{5} = 5\dfrac{1}{5}$

69. $54 \div 100 = 0.54$

71. $16 \div 100 = 0.16$

73. $7 \div 100 = 0.07$

75. $0.3 \div 100 = 0.003$

77. $400 \div 100 = 4$

79. Answers will vary. Example:

$$0.375 = \dfrac{375 \div 125}{1000 \div 125} = \dfrac{3}{8}$$

$$\dfrac{3}{8} \cdot \dfrac{2}{2} = \dfrac{6}{16}$$

$$\dfrac{3}{8} \cdot \dfrac{3}{3} = \dfrac{9}{24}$$

$$\dfrac{3}{8}, \dfrac{6}{16}, \dfrac{9}{24}$$

	Fraction	Decimal	Percent
81.	$\dfrac{1}{5}$	0.2	20%
83.	$\dfrac{8}{25}$	0.32	32%
85.	$\dfrac{13}{8}$	1.625	162.5%

87. Answers will vary. Example:
This was a good idea because they now match
U.S. currency – dollars and cents.

1.6   **QUICK CHECK**

1. $850, $1020, $970, $635, $795
The total of these values is $4270.

$$\text{Mean} = \frac{4270}{5} = \$854$$

2. a) 54, 21, 39, 16, 7, 75
Write the values in ascending order.

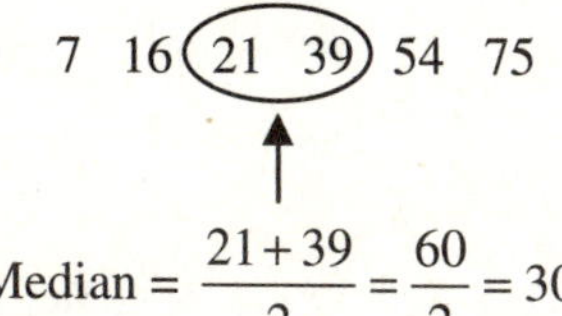

$$\text{Median} = \frac{21+39}{2} = \frac{60}{2} = 30$$

b) 51, 3, 29, 60, 62, 25, 43, 102, 14
Write the values in ascending order.

3 14 25 29 43 51 60 62 102

Median = 43

3. 56, 65, 66, 66, 65, 66, 65, 60, 61, 65
The value 65 is repeated the most times.
The mode is 65.
The largest value is 66 and the smallest value
is 56.

$$\text{Midrange} = \frac{66+56}{2} = \frac{122}{2} = 61$$

4. 165, 168, 174, 179, 182, 159, 171, 180
The largest value is 182 centimeters and the
smallest value is 159 centimeters.
Range = $182 - 159 = 23$ centimeters

5.

Ages	Frequency
25 to 34	3
35 to 44	7
45 to 54	6
55 to 64	15
65 to 74	35
75 to 84	20

On the horizontal axis, begin the labels at 25
and increase by 10 until reaching 85. On the
vertical axis, labels must reach at least 35,
which is the largest frequency.

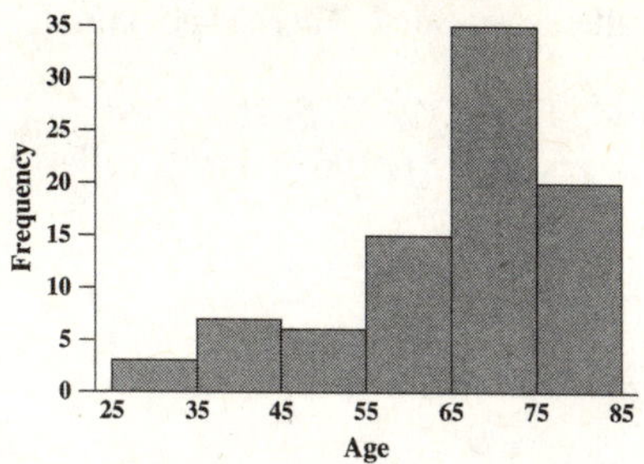

Age of Cruise Passengers

1.6   **BASIC STATISTICS**

1. center

3. mean

5. mode

7. range

9. 63, 98, 21, 42, 71
The total of these values is 295.

$$\text{Mean} = \frac{295}{5} = 59$$

11. 5, 17, 21, 35, 42, 59, 89, 106
The total of these values is 374.

$$\text{Mean} = \frac{374}{8} = 46.75$$

13. 97, 76, 22, 103, 80, 45, 66
Write the values in ascending order.

22 45 66 76 80 97 103

Median = 76

15. 68, 47, 32, 90, 85, 40, 83, 39, 50, 77
Write the values in ascending order.

32 39 40 47 50 68 77 83 85 90

$$\text{Median} = \frac{50+68}{2} = \frac{118}{2} = 59$$

17. 70, 56, 63, 35, 56, 63, 36, 56, 19
The value 56 is repeated the most times. The
mode is 56.

19. 7, 41, 32, 56, 41, 19, 8, 32, 25
The values 32 and 41 are repeated. The modes
are 32 and 41.

21. 5, 35, 89, 106, 42, 17, 59, 21
No value is repeated. There is no mode.

23. 70, 140, 87, 62, 196, 125, 155
The largest value is 196 and the smallest value is 62.

$$\text{Midrange} = \frac{196 + 62}{2} = \frac{258}{2} = 129$$

Range $= 196 - 62 = 134$

25. 406, 354, 509, 427, 516, 379
The largest value is 516 and the smallest value is 354.

$$\text{Midrange} = \frac{516 + 354}{2} = \frac{870}{2} = 435$$

Range $= 516 - 354 = 162$

27. 22, 13, 16, 30, 32, 19, 24, 30, 21

a) The total of these values is 207.
$$\text{Mean} = \frac{207}{9} = 23$$

b) Write the values in ascending order.

13 16 19 21 22 24 30 30 32

Median $= 22$

c) The value 30 is repeated. The mode is 30.

d) The largest value is 32 and the smallest value is 13.
$$\text{Midrange} = \frac{32 + 13}{2} = \frac{45}{2} = 22.5$$

e) Range $= 32 - 13 = 19$

29. 48, 45, 63, 36, 50, 38, 73, 63

a) The total of these values is 416.
$$\text{Mean} = \frac{416}{8} = 52$$

b) Write the values in ascending order.

36 38 45 48 50 63 63 73

Median $= \dfrac{48 + 50}{2} = \dfrac{98}{2} = 49$

c) The value 63 is repeated. The mode is 63.

d) The largest value is 73 and the smallest value is 36.
$$\text{Midrange} = \frac{73 + 36}{2} = \frac{109}{2} = 54.5$$

e) Range $= 73 - 36 = 37$

31. 63, 86, 76, 85, 59, 71, 34, 44, 30, 67, 44, 77, 50, 83, 76

a) The total of these values is 945.
$$\text{Mean} = \frac{945}{15} = 63$$

b) Write the values in ascending order.

30 34 44 44 50 59 63 67 71 76 76 77 83 85 86

Median $= 67$

c) The values 44 and 76 are repeated. The modes are 44 and 76.

d) The largest value is 86 and the smallest value is 30.
$$\text{Midrange} = \frac{86 + 30}{2} = \frac{116}{2} = 58$$

e) Range $= 86 - 30 = 56$

33. 196, 295, 213, 69, 371, 77, 253, 210, 298, 210, 426, 327, 270, 323, 262, 70, 459, 481, 278, 192

a) The total of these values is 5280.
$$\text{Mean} = \frac{5280}{20} = 264$$

b) Write the values in ascending order.

69 70 77 192 196 210 210 213 253 262 270
278 295 298 323 327 371 426 459 481

$$\text{Median} = \frac{262 + 270}{2}$$
$$= \frac{532}{2}$$
$$= 266$$

c) The value 210 is repeated. The mode is 210.

Continued on next page.

33. Continued.

   d) The largest value is 481 and the smallest value is 69.

$$\text{Midrange} = \frac{481+69}{2} = \frac{550}{2} = 275$$

   e) Range $= 481-69 = 412$

35. 95, 82, 104, 119, 118, 126, 82, 96, 116, 85, 90, 90
The total of these values is 1203.

$$\text{Mean} = \frac{1203}{12} = 100.25$$

37. 133, 112, 142, 154, 102, 139, 149
The total of these values is 931.

$$\text{Mean} = \frac{931}{7} = 133$$

39. 101, 110, 125, 120, 106, 113, 102, 108, 132, 135
The total of these values is 1152.

$$\text{Mean} = \frac{1152}{10} = 115.2$$

41. 243, 18, 21, 152, 93, 125
Write the values in ascending order.

18   21   93   125   152   243

$$\text{Median} = \frac{93+125}{2} = \frac{218}{2} = 109$$

43. 80, 96, 100, 89, 74, 96, 95, 98, 87
Write the values in ascending order.

74 80 87 89 95 96 96 98 100

Median = 95

45. 25, 12, 17, 3, 20, 20, 16, 34, 1, 9
Write the values in ascending order.

1 3 9 12 16 17 20 20 25 34

$$\text{Median} = \frac{16+17}{2} = \frac{33}{2} = 16.5$$

47. $219, $259, $127, $199, $259, $169, $219, $229, $299, $199

   a) The total of these values is $2178.

$$\text{Mean} = \frac{\$2178}{10} = \$217.80$$

   b) Write the values in ascending order.

$127 $169 $199 $199 $219 $219 $229 $259 $259 $299

$$\text{Median} = \frac{\$219+\$219}{2}$$
$$= \frac{\$438}{2}$$
$$= \$219$$

   c) The values $199, $219 and $259 are repeated. The modes are $199, $219 and $259.

   d) The largest value is $299 and the smallest value is $127.

$$\text{Midrange} = \frac{\$299+\$127}{2} = \frac{\$426}{2} = \$213$$

   e) Range $= \$299 - \$127 = \$172$

49. 289, 191, 180, 390, 271, 270, 198, 574

   a) The total of these values is 2363.

$$\text{Mean} = \frac{2363}{8} = 295.375$$

   b) Write the values in ascending order.

180 191 198 270 271 289 390 574

$$\text{Median} = \frac{270+271}{2} = \frac{541}{2} = 270.5$$

   c) No value is repeated. There is no mode.

   d) The largest value is 574 and the smallest value is 180.

$$\text{Midrange} = \frac{574+180}{2} = \frac{754}{2} = 377$$

   e) Range $= 574-180 = 394$

51. 636, 754, 662, 884, 1346, 659, 1006, 1357, 1129, 904, 1747, 1336, 1234, 388

a) The total of these values is 14,042.

$$\text{Mean} = \frac{14{,}042}{14} = 1003$$

b) Write the values in ascending order.

388 636 659 662 754 884 904 1006 1129 1234 1336 1346 1357 1747

$$\text{Median} = \frac{904 + 1006}{2}$$
$$= \frac{1910}{2}$$
$$= 955$$

c) No value is repeated. There is no mode.

d) The largest value is 1747 and the smallest value is 388.

$$\text{Midrange} = \frac{1747 + 388}{2} = \frac{2135}{2} = 1067.5$$

e) Range $= 1747 - 388 = 1359$

53. 80, 86, 100, 81, 30, 57, 90, x
The total of these values is $524 + x$.

$$\text{Mean} = \frac{524 + x}{8} = 75$$
$$524 + x = 600$$
$$x = 76$$

55. 112, 98, 121, 72, x, 65
Since there are an even number of values, the median is the average of the two middle values. Write the values in ascending order. If the median is 101.5, the missing value must follow 98.

65   72   98 x 112   121

$$\text{Median} = \frac{98 + x}{2} = 101.5$$
$$98 + x = 203$$
$$x = 105$$

57. 66, 25, 94, 37, x, 85, 42
Examine the known data values. The largest value is 94 and the smallest value is 25.

$$\text{Range} = 94 - 25 = 69$$

The actual range is 84, so the missing value must be either the largest or the smallest value. Try using x for the smallest value:

$$\text{Range} = 94 - x = 84$$
$$x = 10$$

$$\text{Midrange} = \frac{94 + 10}{2} = \frac{104}{2} = 52$$

Both range and midrange are correct. The missing value is 10.

59.

Average Score	Frequency
460 to 479	2
480 to 499	6
500 to 519	14
520 to 539	6
540 to 559	6
560 to 579	8
580 to 599	5
600 to 619	4

On the horizontal axis, begin the labels at 460 and increase by 20 until reaching 620. On the vertical axis, labels must reach at least 14, which is the largest frequency.

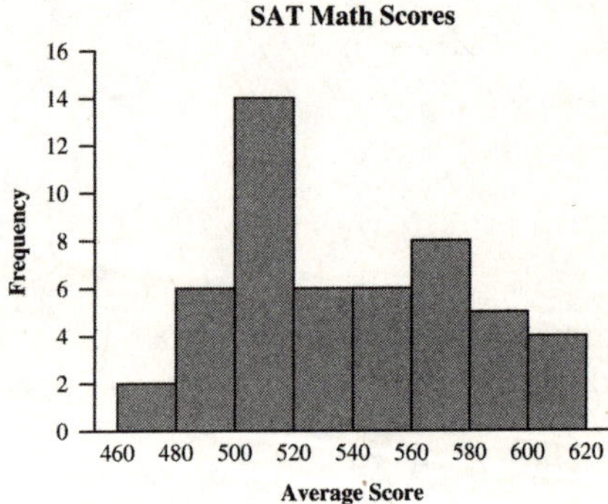

61.

Graduation Rate	Frequency
45.0% to 49.9%	1
50.0% to 54.9%	2
55.0% to 59.9%	2
60.0% to 64.9%	4
65.0% to 69.9%	7
70.0% to 74.9%	16
75.0% to 79.9%	13
80.0% to 84.9%	5

On the horizontal axis, begin the labels at 45 and increase by 5 until reaching 85. On the vertical axis, labels must reach at least 16, which is the largest frequency.

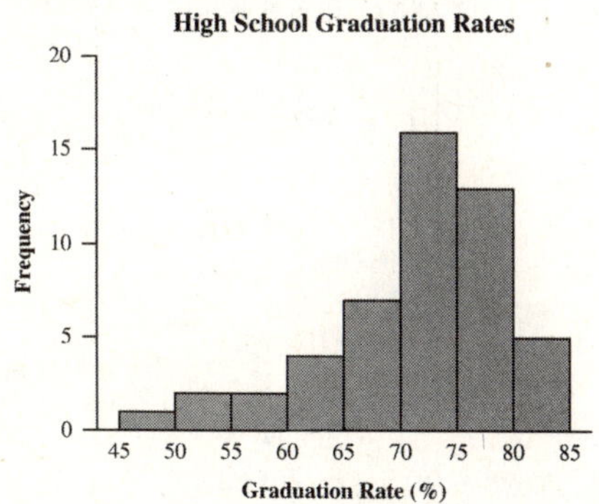

63. 471 505 515 503 506 507 471 522 548 478
514 490 511 463 515 514 490 485 467 531
487 482 500 506 504 492 522 497 508 499
515 499 516 495 499 496 510 520 509 500
488 512 501 506 488 497 498 503 496 488
522 505 517 497 500 502 472 525 477 506

Number of Heads	Frequency
450 to 469	2
470 to 489	11
490 to 509	30
510 to 529	15
530 to 549	2

On the horizontal axis, begin the labels at 450 and increase by 20 until reaching 550. On the vertical axis, labels must reach at least 30, which is the largest frequency.

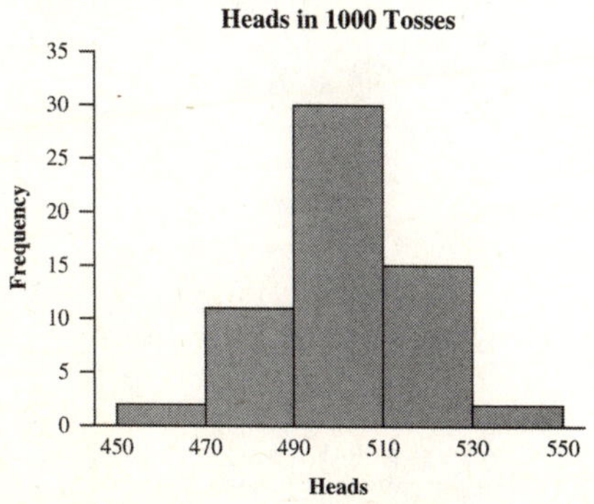

65. 104 114 88 96 105 120 92 107 134
110 95 133 132 119 115 116 129 114
100 95 105 140 110 131 97 113 104
120 95 128 128 106 96 114 127 133

IQ	Frequency
85 to 94	2
95 to 104	9
105 to 114	10
115 to 124	5
125 to 134	9
135 to 144	1

On the horizontal axis, begin the labels at 85 and increase by 10 until reaching 145. On the vertical axis, labels must reach at least 10, which is the largest frequency.

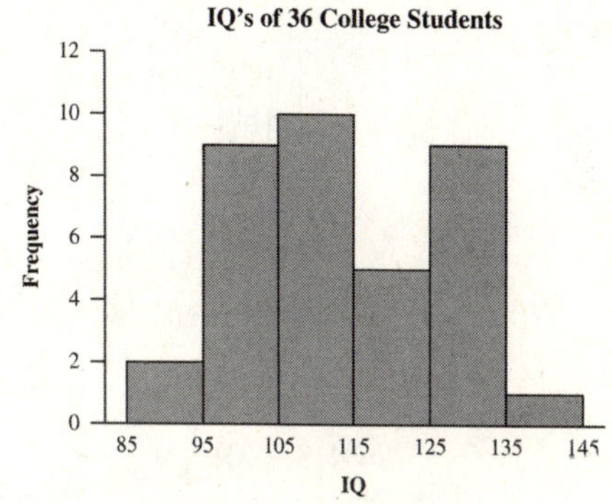

67.

Score	Frequency
50 to 59	9
60 to 69	8
70 to 79	11
80 to 89	8
90 to 99	9

69.

Height	Frequency
150 to 154	3
155 to 159	5
160 to 164	7
165 to 169	7
170 to 174	8
175 to 179	4

1.7   QUICK CHECK

1. $4^3 = 4 \cdot 4 \cdot 4 = 64$

2. $\left(\dfrac{1}{8}\right)^4 = \dfrac{1}{8} \cdot \dfrac{1}{8} \cdot \dfrac{1}{8} \cdot \dfrac{1}{8} = \dfrac{1}{4096}$

3. $\begin{aligned} -2 \cdot 7 + 11 - 3^4 &= -2 \cdot 7 + 11 - 81 \\ &= -14 + 11 - 81 \\ &= 11 - 95 \\ &= -84 \end{aligned}$

4. $\begin{aligned} 9^2 - 4(-2)(-10) &= 81 - 4(-2)(-10) \\ &= 81 + 8(-10) \\ &= 81 - 80 \\ &= 1 \end{aligned}$

5. $\begin{aligned} 20 \div 5 \cdot 10(3 \cdot 6 - 9) &= 20 \div 5 \cdot 10(18 - 9) \\ &= 20 \div 5 \cdot 10(9) \\ &= 4 \cdot 10(9) \\ &= 40(9) \\ &= 360 \end{aligned}$

6. $\begin{aligned} 4\left[3^2 + 5(2 - 8)\right] &= 4\left[3^2 + 5(-6)\right] \\ &= 4\left[9 + 5(-6)\right] \\ &= 4\left[9 - 30\right] \\ &= 4\left[-21\right] \\ &= -84 \end{aligned}$

1.7   EXPONENTS AND ORDER OF OPERATIONS

1. base

3. squared

5. grouping

7. $2 \cdot 2 \cdot 2 = 2^3$

9. $(-2)(-2)(-2)(-2) = (-2)^4$

11. $-3 \cdot 3 \cdot 3 \cdot 3 \cdot 3 = -3^5$

13. 5^3

15. $3^4 = 3 \cdot 3 \cdot 3 \cdot 3 = 81$

17. $7^5 = 7 \cdot 7 \cdot 7 \cdot 7 \cdot 7 = 16{,}807$

19. $10^6 = 10 \cdot 10 \cdot 10 \cdot 10 \cdot 10 \cdot 10 = 1{,}000{,}000$

21. $1^{723} = 1$

23. $\left(\dfrac{3}{4}\right)^3 = \dfrac{3}{4} \cdot \dfrac{3}{4} \cdot \dfrac{3}{4} = \dfrac{27}{64}$

25. $0.2^5 \, (0.2)(0.2)(0.2)(0.2)(0.2) = 0.00032$

27. $(-3)^4 = (-3)(-3)(-3)(-3) = 81$

29. $-2^7 = -2 \cdot 2 \cdot 2 \cdot 2 \cdot 2 \cdot 2 \cdot 2 = -128$

31. $2^3 \cdot 5^2 = 2 \cdot 2 \cdot 2 \cdot 5 \cdot 5 = 8 \cdot 25 = 200$

33. $\begin{aligned} -6^2 \cdot (-2)^2 &= -6 \cdot 6 \cdot (-2)(-2) \\ &= -36 \cdot (4) \\ &= -144 \end{aligned}$

35. $9 + 3 \cdot 4 = 9 + 12 = 21$

37. $20 \div 5 \cdot 2^2 = 20 \div 5 \cdot 4 = 4 \cdot 4 = 16$

39. $8 \cdot 5 - 9 \cdot 7 = 40 - 63 = -23$

41. $3.2 + 2.8(-6.3) = 3.2 - 17.64 = -14.44$

43. $17.1 - 8.58 \div 3.9 = 17.1 - 2.2 = 14.9$

45. $\begin{aligned} (-3)^2 - 4(-5)(-4) &= (-3)(-3) - 4(-5)(-4) \\ &= 9 - 80 \\ &= -71 \end{aligned}$

47. $3^2 + 4^2 = 9 + 16 = 25$

49. $\begin{aligned} -4 - 5(7 - 3 \cdot 6) &= -4 - 5(7 - 18) \\ &= -4 - 5(-11) \\ &= -4 + 55 \\ &= 51 \end{aligned}$

51. $\begin{aligned} \dfrac{1}{2} + \dfrac{1}{2} \cdot \dfrac{4}{7} &= \dfrac{1}{2} + \dfrac{1}{\cancel{2}_1} \cdot \dfrac{\cancel{4}^2}{7} \\ &= \dfrac{1}{2} + \dfrac{2}{7} \\ &= \dfrac{7}{14} + \dfrac{4}{14} \\ &= \dfrac{11}{14} \end{aligned}$

53. $\dfrac{2}{9} \div \dfrac{5}{3} \cdot \dfrac{33}{50} = \dfrac{2}{9} \cdot \dfrac{3}{5} \cdot \dfrac{33}{50}$

$$= \dfrac{\cancel{2}^{\,1}}{\cancel{9}_3} \cdot \dfrac{\cancel{3}^{\,1}}{5} \cdot \dfrac{33}{\cancel{50}_{25}}$$

$$= \dfrac{1}{\cancel{3}_1} \cdot \dfrac{1}{5} \cdot \dfrac{\cancel{33}^{\,11}}{25}$$

$$= \dfrac{11}{125}$$

55. $\dfrac{1}{9}\left(\dfrac{19}{28} - \dfrac{1}{4}\right) = \dfrac{1}{9}\left(\dfrac{19}{28} - \dfrac{7}{28}\right)$

$$= \dfrac{1}{9}\left(\dfrac{12}{28}\right)$$

$$= \dfrac{1}{9}\left(\dfrac{3}{7}\right)$$

$$= \dfrac{1}{\cancel{9}_3} \cdot \dfrac{\cancel{3}^{\,1}}{7}$$

$$= \dfrac{1}{21}$$

57. $\dfrac{2^2 - 9}{3^3 - 7} = \dfrac{4 - 9}{27 - 7} = \dfrac{-5}{20} = -\dfrac{1}{4}$

59. $\dfrac{(3+5)\cdot 6 - 8}{1 + 3^2 + 2} = \dfrac{(8)\cdot 6 - 8}{1 + 3^2 + 2}$

$$= \dfrac{(8)\cdot 6 - 8}{1 + 9 + 2}$$

$$= \dfrac{48 - 8}{1 + 9 + 2}$$

$$= \dfrac{40}{12}$$

$$= \dfrac{10}{3}$$

61. $3 - \left[4(5 - 6\cdot 7)\right] = 3 - \left[4(5 - 42)\right]$

$$= 3 - \left[4(-37)\right]$$

$$= 3 - (-148)$$

$$= 151$$

63. $-4\cdot 9 - \left|-7(3+5)\right| - 2^3$

$$= -4\cdot 9 - \left|-7(3+5)\right| - 8$$

$$= -4\cdot 9 - \left|-7(8)\right| - 8$$

$$= -36 - \left|-56\right| - 8$$

$$= -36 - 56 - 8$$

$$= -100$$

65. $9 - \left|3^2 + 2^3 - 10\cdot 9\right| = 9 - \left|9 + 8 - 10\cdot 9\right|$

$$= 9 - \left|9 + 8 - 90\right|$$

$$= 9 - \left|17 - 90\right|$$

$$= 9 - \left|-73\right|$$

$$= 9 - 73$$

$$= -64$$

67. $-21\left(\left|4 - 3\cdot 9\right| + \left|8(13 - 4)\right|\right)$

$$= -21\left(\left|4 - 27\right| + \left|8(9)\right|\right)$$

$$= -21\left(\left|-23\right| + \left|72\right|\right)$$

$$= -21(23 + 72)$$

$$= -21(95)$$

$$= -1995$$

69. Answers will vary. Example:
$8\cdot 3 + 8\cdot 2 + 1 = 41$

71. Answers will vary. Example:
$-4^2 + 9 - 2\cdot 3 = -13$

73. 1 kilobyte $= 2^{10}$ bytes $= 1024$ bytes

75. 1 gigabyte $= 2^{30}$ bytes $= 1{,}073{,}741{,}824$ bytes

77. Since $5\cdot 5\cdot 5 = 125$, the missing number is 3.

79. Since $9^3 = 729$, the missing number is 9.

81. Since $\left(\dfrac{3}{8}\right)^4 = \dfrac{81}{4096}$, the missing number is 8.

83. $3\cdot 5 - 9 + 8 = 14$

85. $(3 + 7\cdot 9) \div 2 = 33$

87. Explanations will vary. Example:

The base of $(-2)^6$ is (-2), while the base of -2^6 is just 2. Therefore, the former would be positive and the latter would be negative.

1.8   QUICK CHECK

1. $x+9$

2. $x-25$

3. $2x$

4. $\dfrac{x}{20}$

5. $5(11)+2=55+2=57$

6. $(-8)^2-13(-8)-40=64-13(-8)-40$
    $=64+104-40$
    $=168-40$
    $=128$

7. $(5)^2-4(-1)(18)=25-4(-1)(18)$
    $=25+4(18)$
    $=25+72$
    $=97$

8. $7(5x-4)=7\cdot5x-7\cdot4=35x-28$

9. $12(x-2y+3z)=12\cdot x-12\cdot2y+12\cdot3z$
    $=12x-24y+36z$

10. $-6(4x+11)=-6\cdot4x+(-6)\cdot11=-24x-66$

11. Four terms.
    $x^3,-x^2,23x,$ and -59
    $1,-1,23,-59$

12. $3x+7y+y-5x=-2x+8y$

13. $5(2x+3)+9x-8=10x+15+9x-8=19x+7$

14. $3(2x-7)-8(x-9)=6x-21-8x+72$
    $=-2x+51$

1.8   INTRODUCTION TO ALGEBRA

1. variable

3. evaluate

5. associative

7. term

9. like terms

11. $x+15$

13. $x-24$

15. $3x$

17. $2x+19$

19. $x+y$

21. $7(x-y)$

23. $325c$

25. $25,000+22a$

27. Nine less than a number

29. Seven times a number

31. Ten less than eight times a number

33. $8x+31$ for $x=6$
    $8(6)+31=48+31=79$

35. $7a-13b$ for $a=8$ and $b=11$
    $7(8)-13(11)=56-143=-87$

37. $2(3x+8)$ for $x=5$
    $2\left[3(5)+8\right]=2(15+8)=2(23)=46$

39. $x^2+9x+18$ for $x=5$
    $(5)^2+9(5)+18=25+9(5)+18$
    $=25+45+18$
    $=88$

41. $y^2+4y-17$ for $y=-3$
    $(-3)^2+4(-3)-17=9+4(-3)-17$
    $=9-12-17$
    $=9-29$
    $=-20$

43. m^2-9 for $m=3$
    $(3)^2-9=9-9=0$

45. b^2-4ac for $a=2$, $b=5$, and $c=-3$
    $(5)^2-4(2)(-3)=25-4(2)(-3)=25+24=49$

47. b^2-4ac for $a=-1$, $b=-2$, and $c=10$
    $(-2)^2-4(-1)(10)=4-4(-1)(10)$
    $=4+40$
    $=44$

49. $5(x+h)-17$ for $x=-3$ and $h=0.01$
$$5((-3)+(0.01))-17=5(-2.99)-17$$
$$=-14.95-17$$
$$=-31.95$$

51. $(x+h)^2-5(x+h)-14$
for $x=-3$ and $h=0.1$
$$(-3+0.1)^2-5(-3+0.1)-14$$
$$=(-2.9)^2-5(-2.9)-14$$
$$=8.41-5(-2.9)-14$$
$$=8.41+14.5-14$$
$$=8.91$$

53. **a)** $3-5\neq5-3$

   b) Yes, a and b must be equal, such as
   $7-7=7-7$.

55. $3(x-9)=3x-27$

57. $5(3-7x)=15-35x$

59. $-2(5x+7)=-10x-14$

61. $-6(9x-5)=-54x+30$

63. $7x+9x=16x$

65. $3x-8x=-5x$

67. $3x-7+4x+11=3x+4x-7+11=7x+4$

69. $5x-3y-7x-19y=5x-7x-3y-19y$
$$=-2x-22y$$

71. $3(2x-5)+4x+7=6x-15+4x+7$
$$=6x+4x-15+7$$
$$=10x-8$$

73. $2(4x-9)-11=8x-18-11=8x-29$

75. $6y-5(3y-17)=6y-15y+85=-9y+85$

77. $3(4z-7)+9(2z+3)=12z-21+18z+27$
$$=12z+18z-21+27$$
$$=30z+6$$

79. $-2(5a+4b-13c)+3b=-10a-8b+26c+3b$
$$=-10a-8b+3b+26c$$
$$=-10a-5b+26c$$

81. $5x^3+3x^2-7x-15$

   a) 4

   b) $5x^3,3x^2,-7x,-15$

   c) $5,3,-7,-15$

83. $3x-17$

   a) 2

   b) $3x,-17$

   c) $3,-17$

85. $5(3x^2-7x+11)-6x=15x^2-35x+55-6x$
$$=15x^2-35x-6x+55$$
$$=15x^2-41x+55$$

   a) 3

   b) $15x^2,-41x,55$

   c) $15,-41,55$

87. $5(-7a-3b+5c)-3(6b-9c-23)$
$$=-35a-15b+25c-18b+27c+69$$
$$=-35a-15b-18b+25c+27c+69$$
$$=-35a-33b+52c+69$$

   a) 4

   b) $-35a,-33b,52c,69$

   c) $-35,-33,52,69$

89. Answers will vary. Example:
A store sells DVD players for \$60 each and x is the number of DVD players sold. Then $60x$ represents the amount of money earned from the sale of DVD players.

91. Answers will vary.

CHAPTER 1 REVIEW

1. $-10 < -7$

2. $3 > -12$

3. $|8| = 8$

4. $|-13| = 13$

5. 6

6. -12

7. $9 + (-16) = 9 - 16 = -7$

8. $-10 + 7 = -3$

9. $-22 - 19 = -41$

10. $4 - (-23) = 4 + 23 = 27$

11. $3 - 16 - 24 = 3 - 40 = -37$

12. $8 - (-19) - 7 = 8 + 19 - 7 = 27 - 7 = 20$

13. $9(-6) = -54$

14. $-144 \div (-9) = 16$

15. $1 \cdot 42 = 42$
$2 \cdot 21 = 42$
$3 \cdot 14 = 42$
$6 \cdot 7 = 42$
$\{1, 2, 3, 6, 7, 14, 21, 42\}$

16. $1 \cdot 108 = 108$
$2 \cdot 54 = 108$
$3 \cdot 36 = 108$
$4 \cdot 27 = 108$
$6 \cdot 18 = 108$
$9 \cdot 12 = 108$
$\{1, 2, 3, 4, 6, 9, 12, 18, 27, 36, 54, 108\}$

17. $32 = 2 \cdot 2 \cdot 2 \cdot 2 \cdot 2$

18. $60 = 2 \cdot 2 \cdot 3 \cdot 5$

19. $\dfrac{24}{42} = \dfrac{2 \cdot 2 \cdot 2 \cdot 3}{2 \cdot 3 \cdot 7} = \dfrac{4}{7}$

20. $\dfrac{9}{72} = \dfrac{3 \cdot 3}{2 \cdot 2 \cdot 2 \cdot 3 \cdot 3} = \dfrac{1}{8}$

21. $\dfrac{40}{234} = \dfrac{2 \cdot 2 \cdot 2 \cdot 5}{2 \cdot 3 \cdot 3 \cdot 13} = \dfrac{20}{117}$

22. $\dfrac{98}{7} = \dfrac{2 \cdot 7 \cdot 7}{7} = 14$

23. $5\dfrac{2}{3}$

Multiply $5 \cdot 3 = 15$.
Add $15 + 2 = 17$.

$5\dfrac{2}{3} = \dfrac{17}{3}$

24. $12\dfrac{7}{25}$

Multiply $12 \cdot 25 = 300$.
Add $300 + 7 = 307$.

$12\dfrac{7}{25} = \dfrac{307}{25}$

25. $\dfrac{38}{10}$

$$10\overline{)38} \quad \begin{array}{r} 3 \\ \hline -30 \\ \hline 8 \end{array}$$

$\dfrac{38}{10} = 3\dfrac{8}{10} = 3\dfrac{4}{5}$

26. $\dfrac{55}{6}$

$$6\overline{)55} \quad \begin{array}{r} 9 \\ \end{array}$$
$$\begin{array}{r} -54 \\ \hline 1 \end{array}$$

$$\dfrac{55}{6} = 9\dfrac{1}{6}$$

27. $\dfrac{\cancel{9}^{3}}{\cancel{16}_{4}} \cdot \dfrac{\cancel{28}^{7}}{\cancel{75}_{25}} = \dfrac{21}{100}$

28. $5\dfrac{1}{2} \cdot \dfrac{14}{33} = \dfrac{\cancel{11}^{1}}{\cancel{2}_{1}} \cdot \dfrac{\cancel{14}^{7}}{\cancel{33}_{3}} = \dfrac{7}{3} = 2\dfrac{1}{3}$

29. $\dfrac{8}{15} \div \dfrac{20}{21} = \dfrac{8}{15} \cdot \dfrac{21}{20} = \dfrac{\cancel{8}^{2}}{\cancel{15}_{5}} \cdot \dfrac{\cancel{21}^{7}}{\cancel{20}_{5}} = \dfrac{14}{25}$

30. $3\dfrac{1}{5} \div 7\dfrac{1}{2} = \dfrac{16}{5} \div \dfrac{15}{2} = \dfrac{16}{5} \cdot \dfrac{2}{15} = \dfrac{32}{75}$

31. $\dfrac{5}{8} + \dfrac{7}{12}$ LCM = 24

$$\dfrac{5}{8} + \dfrac{7}{12} = \dfrac{5}{8} \cdot \dfrac{3}{3} + \dfrac{7}{12} \cdot \dfrac{2}{2} = \dfrac{15}{24} + \dfrac{14}{24} = \dfrac{29}{24}$$

32. $\dfrac{13}{20} + \dfrac{5}{6}$ LCM = 60

$$\dfrac{13}{20} + \dfrac{5}{6} = \dfrac{13}{20} \cdot \dfrac{3}{3} + \dfrac{5}{6} \cdot \dfrac{10}{10} = \dfrac{39}{60} + \dfrac{50}{60} = \dfrac{89}{60}$$

33. $\dfrac{11}{18} - \dfrac{4}{9}$ LCM = 18

$$\dfrac{11}{18} - \dfrac{4}{9} = \dfrac{11}{18} - \dfrac{4}{9} \cdot \dfrac{2}{2} = \dfrac{11}{18} - \dfrac{8}{18} = \dfrac{3}{18} = \dfrac{1}{6}$$

34. $\dfrac{7}{36} - \dfrac{13}{42}$ LCM = 252

$$\dfrac{7}{36} - \dfrac{13}{42} = \dfrac{7}{36} \cdot \dfrac{7}{7} - \dfrac{13}{42} \cdot \dfrac{6}{6} = \dfrac{49}{252} - \dfrac{78}{252} = -\dfrac{29}{252}$$

35. $\dfrac{\cancel{30}^{5}}{\cancel{49}_{7}} \cdot \dfrac{\cancel{35}^{5}}{\cancel{66}_{11}} = \dfrac{5}{7} \cdot \dfrac{5}{11} = \dfrac{25}{77}$

36. $\dfrac{7}{16} - \dfrac{11}{48}$ LCM = 48

$$\dfrac{7}{16} - \dfrac{11}{48} = \dfrac{7}{16} \cdot \dfrac{3}{3} - \dfrac{11}{48} = \dfrac{21}{48} - \dfrac{11}{48} = \dfrac{10}{48} = \dfrac{5}{24}$$

37. $\dfrac{9}{40} \div \dfrac{39}{35} = \dfrac{9}{40} \cdot \dfrac{35}{39} = \dfrac{\cancel{9}^{3}}{\cancel{40}_{8}} \cdot \dfrac{\cancel{35}^{7}}{\cancel{39}_{13}} = \dfrac{21}{104}$

38. $\dfrac{7}{20} + \dfrac{1}{15}$ LCM = 60

$$\dfrac{7}{20} + \dfrac{1}{15} = \dfrac{7}{20} \cdot \dfrac{3}{3} + \dfrac{1}{15} \cdot \dfrac{4}{4} = \dfrac{21}{60} + \dfrac{4}{60} = \dfrac{25}{60} = \dfrac{5}{12}$$

39.
$$\begin{array}{r} 8.70 \\ +3.92 \\ \hline 12.62 \end{array}$$

40.
$$\begin{array}{r} 24.308 \\ -15.490 \\ \hline 8.818 \end{array}$$

41.
$$\begin{array}{r} 8.4 \\ \times 3.6 \\ \hline 504 \\ +2520 \\ \hline 3024 \end{array} = 30.24$$

42. $4.65\overline{)40.92}$

$$\begin{array}{r} 8.8 \\ 465\overline{)4092.0} \\ -3720 \\ \hline 3720 \\ -3720 \\ \hline 0 \end{array}$$

8.8

43. $25\overline{)8.00}$
$$\begin{array}{r} .32 \\ 25\overline{)8.00} \\ -75 \\ \hline 50 \\ -50 \\ \hline 0 \end{array}$$

$$\dfrac{8}{25} = 0.32$$

44. $16\overline{)15.0000}$
$$\begin{array}{r} .9375 \\ 16\overline{)15.0000} \\ -144 \\ \hline 60 \\ -48 \\ \hline 120 \\ -112 \\ \hline 80 \\ -80 \\ \hline 0 \end{array}$$

$$\dfrac{15}{16} = 0.9375$$

45. $0.75 = \dfrac{75}{100} = \dfrac{3}{4}$

46. $0.28 = \dfrac{28}{100} = \dfrac{7}{25}$

47. $\dfrac{2}{5} \cdot 100\% = \dfrac{2}{\cancel{5}_1} \cdot \dfrac{\cancel{100}^{20}}{1}\% = 40\%$

48. $\dfrac{7}{25} \cdot 100\% = \dfrac{7}{\cancel{25}_1} \cdot \dfrac{\cancel{100}^{4}}{1}\% = 28\%$

49. $0.9 \cdot 100\% = 90\%$

50. $0.45 \cdot 100\% = 45\%$

51. $30\% = \dfrac{30}{100} = \dfrac{3}{10}$

52. $55\% = \dfrac{55}{100} = \dfrac{11}{20}$

53. $90\% = 90 \div 100 = 0.9$

54. $4\% = 4 \div 100 = 0.04$

55. $40 - 23 = 17$
$17

56. $78 - 125 = -47$
$-$47

57.
$$
\begin{array}{r}
12600 \\
3\overline{)37800} \\
\underline{-3} \\
07 \\
\underline{-6} \\
18 \\
\underline{-18} \\
00 \\
\underline{-\ 00} \\
0
\end{array}
$$
$12,600

58. $1\dfrac{1}{2} + 2\dfrac{2}{3} = \dfrac{3}{2} + \dfrac{8}{3}$ LCM $= 6$

$\dfrac{3}{2} + \dfrac{8}{3} = \dfrac{3}{2} \cdot \dfrac{3}{3} + \dfrac{8}{3} \cdot \dfrac{2}{2} = \dfrac{9}{6} + \dfrac{16}{6} = \dfrac{25}{6} = 4\dfrac{1}{6}$ cups

59. 77, 71, 83, 80, 73, 74, 73, 81
The total of these values is 612 inches.
Mean $= \dfrac{612}{8} = 76.5$ inches

60. 113, 127, 133, 128, 108, 126, 117, 129, 128, 121
Write the values in ascending order.

108 113 117 121 (126 127) 128 128 129 133

Median $= \dfrac{126 + 127}{2} = \dfrac{253}{2} = 126.5$

61. 96, 88, 95, 79, 88, 99
The value 88 is repeated. The mode is 88.
The largest value is 99 and the smallest value is 79.

Midrange $= \dfrac{99 + 79}{2} = \dfrac{178}{2} = 89$

Range $= 99 - 79 = 20$

62.

Blood Pressure (mmHg)	Frequency
80 to 89	2
90 to 99	8
100 to 109	12
110 to 119	19
120 to 129	6
130 to 139	3

On the horizontal axis, begin the labels at 80 and increase by 10 until reaching 140. On the vertical axis, labels must reach at least 19, which is the largest frequency.

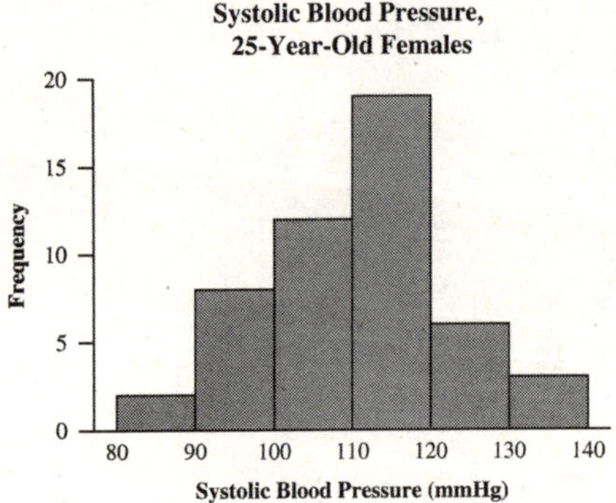

63. $4^3 = 4 \cdot 4 \cdot 4 = 64$

64. $\left(\dfrac{2}{5}\right)^3 = \left(\dfrac{2}{5}\right)\left(\dfrac{2}{5}\right)\left(\dfrac{2}{5}\right) = \dfrac{8}{125}$

65. $-2^6 = -2 \cdot 2 \cdot 2 \cdot 2 \cdot 2 \cdot 2 = -64$

66. $3^5 \cdot 5^2 = (3 \cdot 3 \cdot 3 \cdot 3 \cdot 3)(5 \cdot 5) = 243 \cdot 25 = 6075$

67. $5 + 8 \cdot 4 = 5 + 32 = 37$

68. $25 - 15 \div 5 = 25 - 3 = 22$

69. $3 + 13 \cdot 5 - 20 = 3 + 65 - 20 = 68 - 20 = 48$

70. $(3 + 13) \cdot 5 - 20 = (16) \cdot 5 - 20 = 80 - 20 = 60$

71. $54 - 27 \div 3^2 = 54 - 27 \div 9 = 54 - 3 = 51$

72. $\dfrac{3}{4} + \dfrac{1}{4} \cdot \dfrac{12}{25} = \dfrac{3}{4} + \dfrac{1}{\cancel{4}_1} \cdot \dfrac{\cancel{12}^3}{25} = \dfrac{3}{4} + \dfrac{3}{25}$
LCM $= 100$
$\dfrac{3}{4} + \dfrac{3}{25} = \dfrac{3}{4} \cdot \dfrac{25}{25} + \dfrac{3}{25} \cdot \dfrac{4}{4} = \dfrac{75}{100} + \dfrac{12}{100} = \dfrac{87}{100}$

73. $x + 14$

74. $x - 20$

75. $2x - 8$

76. $6x + 9$

77. $3.55c$

78. $20 + 0.15m$

79. $3x + 17$ for $x = 9$
$3(9) + 17 = 27 + 17 = 44$

80. $9 - 8x$ for $x = -2$
$9 - 8(-2) = 9 + 16 = 25$

81. $10a - 4b$ for $a = 2$ and $b = -9$
$10(2) - 4(-9) = 20 + 36 = 56$

82. $(8x - 9)(2x - 11)$ for $x = 4$
$(8(4) - 9)(2(4) - 11) = (32 - 9)(8 - 11)$
$\qquad\qquad\qquad\quad = (23)(-3)$
$\qquad\qquad\qquad\quad = -69$

83. $x^2 - 7x - 30$ for $x = -3$
$(-3)^2 - 7(-3) - 30 = 9 - 7(-3) - 30$
$\qquad\qquad\qquad\quad = 9 + 21 - 30$
$\qquad\qquad\qquad\quad = 30 - 30$
$\qquad\qquad\qquad\quad = 0$

84. $b^2 - 4ac$ for $a = -1, b = -8,$ and $c = 5$
$(-8)^2 - 4(-1)(5) = 64 - 4(-1)(5)$
$\qquad\qquad\qquad\quad = 64 + 4(5)$
$\qquad\qquad\qquad\quad = 64 + 20$
$\qquad\qquad\qquad\quad = 84$

85. $5(x + 7) = 5x + 35$

86. $6x + 21x = 27x$

87. $8x - 25 - 3x + 17 = 8x - 3x - 25 + 17 = 5x - 8$

88. $8y - 6(4y - 21) = 8y - 24y + 126 = -16y + 126$

89. $15 - 23k + 7(4k - 9) = 15 - 23k + 28k - 63$
$\qquad\qquad\qquad\qquad\quad = -23k + 28k + 15 - 63$
$\qquad\qquad\qquad\qquad\quad = 5k - 48$

90. $-8(2x + 25) - (103 - 19x)$
$\quad = -16x - 200 - 103 + 19x$
$\quad = -16x + 19x - 200 - 103$
$\quad = 3x - 303$

91. $x^3 - 4x^2 - 10x + 41$

a) 4

b) $x^3, -4x^2 - 10x, 41$

c) $1, -4, -10, 41$

92. $-x^2 + 5x - 30$

a) 3

b) $-x^2, 5x, -30$

c) $-1, 5, -30$

CHAPTER 1 TEST

1. $-15 > -18$

2. $|-17| = 17$

3. $7 + (-13) = 7 - 13 = -6$

4. $-7(-9) = 63$

5. $1 \cdot 45 = 45$
$3 \cdot 15 = 45$
$5 \cdot 9 = 45$
$\{1, 3, 5, 9, 15, 45\}$

6.
$$108$$
$$\boxed{2} \cdot 54$$
$$9 \cdot 6$$
$$\boxed{3} \cdot \boxed{3}\ \boxed{2} \cdot \boxed{3}$$
$$2 \cdot 2 \cdot 3 \cdot 3 \cdot 3 = 108$$

7. $\dfrac{60}{84} = \dfrac{\cancel{2}^{1} \cdot \cancel{2}^{1} \cdot \cancel{3}^{1} \cdot 5}{\cancel{2}_{1} \cdot \cancel{2}_{1} \cdot \cancel{3}_{1} \cdot 7} = \dfrac{5}{7}$

8. $\dfrac{67}{18} = 3\dfrac{13}{18}$

9. $\dfrac{\cancel{11}^{1}}{\cancel{63}_{21}} \cdot \dfrac{\cancel{15}^{5}}{\cancel{44}_{4}} = \dfrac{5}{84}$

10. $\dfrac{2}{9} \div \dfrac{8}{21} = \dfrac{2}{9} \cdot \dfrac{21}{8} = \dfrac{\cancel{2}^{1}}{\cancel{9}_{3}} \cdot \dfrac{\cancel{21}^{7}}{\cancel{8}_{4}} = \dfrac{7}{12}$

11. $\dfrac{3}{5} + \dfrac{11}{12} \quad \text{LCM} = 60$

$\dfrac{3}{5} + \dfrac{11}{12} = \dfrac{3}{5} \cdot \dfrac{12}{12} + \dfrac{11}{12} \cdot \dfrac{5}{5} = \dfrac{36}{60} + \dfrac{55}{60} = \dfrac{91}{60}$

12. $\dfrac{5}{24} - \dfrac{4}{9} \quad \text{LCM} = 72$

$\dfrac{5}{24} - \dfrac{4}{9} = \dfrac{5}{24} \cdot \dfrac{3}{3} - \dfrac{4}{9} \cdot \dfrac{8}{8} = \dfrac{15}{72} - \dfrac{32}{72} = -\dfrac{17}{72}$

13.
$$\begin{array}{r} 8.05 \\ \times 2.27 \\ \hline 5635 \\ 16100 \\ 161000 \\ \hline 182735 \end{array}$$
18.2735

14. $0.36 = \dfrac{36}{100} = \dfrac{9}{25}$

15.
$$\begin{array}{r} 499 \\ \times 13 \\ \hline 1497 \\ + 4990 \\ \hline 6487 \end{array}$$
$\$6487$

16.
$$\begin{array}{r} 1203.34 \\ - 407.83 \\ \hline 795.51 \end{array}$$
$\$795.51$

17. $72\% = \dfrac{72}{100} = \dfrac{18}{25}$

18. $6\% = 6 \div 100 = 0.06$

19. 43, 46, 71, 95, 85, 27, 37, 8, 44, 26, 34, 85, 91, 79, 89, 20

The total of these values is 880.

$\text{Mean} = \dfrac{880}{16} = 55$

Write the values in ascending order.

8 20 26 27 34 37 43 ⟨44 46⟩ 71 79 85 85 89 91 95

$\text{Median} = \dfrac{44 + 46}{2} = \dfrac{90}{2} = 45$

20. $16 - 8 \cdot 5 = 16 - 40 = -24$

21. $-9 + 4 \cdot 13 - 6 \cdot 3 = -9 + 52 - 18 = 52 - 27 = 25$

22. $\dfrac{4^{2} + 3^{2}}{3 + 4 \cdot 13} = \dfrac{16 + 9}{3 + 4 \cdot 13} = \dfrac{16 + 9}{3 + 52} = \dfrac{25}{55} = \dfrac{5}{11}$

23. $4n - 7$

24. $50 + 20h$

25. $16 - 5x$ for $x = -9$
$16 - 5(-9) = 16 + 45 = 61$

26. $x^{2} + 6x - 17$ for $x = -8$
$(-8)^{2} + 6(-8) - 17 = 64 + 6(-8) - 17$
$= 64 - 48 - 17$
$= -1$

27. $5(2x - 13) = 10x - 65$

28. $7y - 8(2y - 30) = 7y - 16y + 240 = -9y + 240$

CHAPTER 2   LINEAR EQUATIONS

2.1   QUICK CHECK

1. $4(-7)+23=-5$
$-28+23=-5$
$-5=-5$
Yes.

2. $\dfrac{x}{8}=-2$

$8\cdot\dfrac{x}{8}=8\cdot(-2)$

$x=-16$
$\{-16\}$

3. $4a=56$

$\dfrac{4a}{4}=\dfrac{56}{4}$

$a=14$
$\{14\}$

4. $-9a=144$

$\dfrac{-9a}{-9}=\dfrac{144}{-9}$

$a=-16$
$\{-16\}$

5. $\dfrac{9}{16}x=\dfrac{21}{8}$

$\dfrac{16}{9}\cdot\dfrac{9}{16}x=\dfrac{16}{9}\cdot\dfrac{21}{8}$

$x=\dfrac{\cancel{16}^{2}}{\cancel{9}_{3}}\cdot\dfrac{\cancel{21}^{7}}{\cancel{8}_{1}}$

$x=\dfrac{2}{3}\cdot\dfrac{7}{1}$

$x=\dfrac{14}{3}$

$\left\{\dfrac{14}{3}\right\}$

6. $a+22=-8$
$a+22-22=-8-22$
$a=-30$
$\{-30\}$

7. $-13=x-28$
$-13+28=x-28+28$
$15=x$
$\{15\}$

8. $x-26.50=38.50$
$x-26.50+26.50=38.50+26.50$
$x=65$
Josh had $65.

2.1   INTRODUCTION TO LINEAR EQUATIONS

1. equation

3. solution set

5. For any algebraic expressions A and B, and any number n, if $A=B$ then $A+n=B+n$ and $A-n=B-n$.

7. No, there is a term containing x^2.

9. Yes.

11. Yes.

13. Yes.

15. No, there is a variable in a denominator.

17. $5(7)-9=26$
$35-9=26$
$26=26$
Yes, $x=7$ is a solution to $5x-9=26$.

19. $3-2(8)=(8)+2(8)-11$
$3-16=8+16-11$
$-13=24-11$
$-13=13$
No, $a=8$ is not a solution.

21. $\dfrac{\cancel{2}^{1}}{3}\left(\dfrac{1}{\cancel{4}_{2}}\right)+\dfrac{11}{6}=2$

$\dfrac{1}{3}\cdot\dfrac{1}{2}+\dfrac{11}{6}=2$

$\dfrac{1}{6}+\dfrac{11}{6}=2$

$\dfrac{12}{6}=2$

$2=2$

Yes, $z=\dfrac{1}{4}$ is a solution to $\dfrac{2}{3}z+\dfrac{11}{6}=2$.

23. $3(3.4)-2=2(3.4)+0.4$
$10.2-2=6.8+0.4$
$8.2=7.2$
No, $m=3.4$ is not a solution.

25. $7x = -91$

$$\frac{7x}{7} = \frac{-91}{7}$$

$$x = -13$$

$$\{-13\}$$

27. $6y = 84$

$$\frac{6y}{6} = \frac{84}{6}$$

$$y = 14$$

$$\{14\}$$

29. $8b = 22$

$$\frac{8b}{8} = \frac{22}{8}$$

$$b = \frac{22}{8}$$

$$b = \frac{11}{4}$$

$$\left\{\frac{11}{4}\right\}$$

31. $5a = 0$

$$\frac{5a}{5} = \frac{0}{5}$$

$$a = 0$$

$$\{0\}$$

33. $-5t = 35$

$$\frac{-5t}{-5} = \frac{35}{-5}$$

$$a = -7$$

$$\{-7\}$$

35. $-2x = -28$

$$\frac{-2x}{-2} = \frac{-28}{-2}$$

$$x = 14$$

$$\{14\}$$

37. $-t = 45$

$$-t \cdot (-1) = 45 \cdot (-1)$$

$$t = -45$$

$$\{-45\}$$

39. $\dfrac{x}{3} = 7$

$$\frac{x}{3} \cdot 3 = 7 \cdot 3$$

$$x = 21$$

$$\{21\}$$

41. $-\dfrac{t}{8} = 12$

$$-\frac{t}{8} \cdot (-8) = 12 \cdot (-8)$$

$$t = -96$$

$$\{-96\}$$

43. $\dfrac{7}{12}x = \dfrac{14}{3}$

$$\frac{12}{7} \cdot \frac{7}{12}x = \frac{14}{3} \cdot \frac{12}{7}$$

$$x = 8$$

$$\{8\}$$

45. $-\dfrac{2}{5}x = 4$

$$\left(-\frac{5}{2}\right) \cdot \left(-\frac{2}{5}\right)x = \left(-\frac{5}{2}\right) \cdot 4$$

$$x = -10$$

$$\{-10\}$$

47. $3.2x = 6.944$

$$\frac{3.2x}{3.2} = \frac{6.944}{3.2}$$

$$x = 2.17$$

$$\{2.17\}$$

49. $a + 9 = 16$

$$a + 9 - 9 = 16 - 9$$

$$a = 7$$

$$\{7\}$$

51. $x + 11 = 3$

$$x + 11 - 11 = 3 - 11$$

$$x = -8$$

$$\{-8\}$$

53. $n - 13 = 30$

$$n - 13 + 13 = 30 + 13$$

$$n = 43$$

$$\{43\}$$

55.
$$a + 3.2 = 5.7$$
$$a + 3.2 - 3.2 = 5.7 - 3.2$$
$$a = 2.5$$
$$\{2.5\}$$

57.
$$12 = x + 3$$
$$12 - 3 = x + 3 - 3$$
$$9 = x$$
$$\{9\}$$

59.
$$b - 7 = 13$$
$$b - 7 + 7 = 13 + 7$$
$$b = 20$$
$$\{20\}$$

61.
$$t - 7 = -4$$
$$t - 7 + 7 = -4 + 7$$
$$t = 3$$
$$\{3\}$$

63.
$$-4 + x = 19$$
$$-4 + 4 + x = 19 + 4$$
$$x = 23$$
$$\{23\}$$

65.
$$x + 9 = 0$$
$$x + 9 - 9 = 0 - 9$$
$$x = -9$$
$$\{-9\}$$

67.
$$9 + a = 5$$
$$9 - 9 + a = 5 - 9$$
$$a = -4$$
$$\{-4\}$$

69.
$$a + 5 + 6 = 7$$
$$a + 11 = 7$$
$$a + 11 - 11 = 7 - 11$$
$$a = -4$$
$$\{-4\}$$

71.
$$60 = 5a$$
$$\frac{60}{5} = \frac{5a}{5}$$
$$12 = a$$
$$\{12\}$$

73.
$$x - 27 = -11$$
$$x - 27 + 27 = -11 + 27$$
$$x = 16$$
$$\{16\}$$

75.
$$-\frac{b}{10} = -3$$
$$-10 \cdot \left(-\frac{b}{10}\right) = -10 \cdot (-3)$$
$$b = 30$$
$$\{30\}$$

77.
$$x + 24 = -17$$
$$x + 24 - 24 = -17 - 24$$
$$x = -41$$
$$\{-41\}$$

79.
$$11 = -\frac{n}{7}$$
$$-7 \cdot 11 = -7 \cdot \left(-\frac{n}{7}\right)$$
$$-77 = n$$
$$\{-77\}$$

81.
$$0 = x + 56$$
$$0 - 56 = x + 56 - 56$$
$$-56 = x$$
$$\{-56\}$$

83.
$$-t = 18$$
$$\frac{-t}{-1} = \frac{18}{-1}$$
$$t = -18$$
$$\{-18\}$$

85.
$$\frac{4}{9}x = -\frac{14}{15}$$
$$\frac{9}{4} \cdot \frac{4}{9}x = \frac{9^3}{4_2}\left(-\frac{14^7}{15_5}\right)$$
$$x = \frac{3}{2}\left(-\frac{7}{5}\right)$$
$$x = -\frac{21}{10}$$
$$\left\{-\frac{21}{10}\right\}$$

87.
$$0 = 45m$$
$$\frac{0}{45} = \frac{45m}{45}$$
$$0 = m$$
$$\{0\}$$

89. Answers will vary. Example:
$$x - 3 = 4$$

91. Answers will vary. Example:
$2n = 5$

93. $n =$ number of nickels Zoe has.
$$0.05n = 1.35$$
$$\frac{0.05n}{0.05} = \frac{1.35}{0.05}$$
$$n = 27$$
27 nickels

95. $p =$ number of people who went to the party.
Proceeds are $\$425 + \$52 = \$477.$
$$3p = 477$$
$$\frac{3p}{3} = \frac{477}{3}$$
$$p = 159$$
159 people

97. $n =$ number of employees before 8 people were hired.
$$n + 8 = 174$$
$$n + 8 - 8 = 174 - 8$$
$$n = 166$$
166 employees

99. $T =$ temperature before the cold front.
$$T - 19 = 37$$
$$T - 19 + 19 = 37 + 19$$
$$T = 56$$
$56°\text{F}$

101. Explanations will vary. Example:
There is no number, x, which, when multiplied by 0, yields a product of 15. Therefore, $0x = 15$ has no solution.

2.2   QUICK CHECK

1.
$$5x - 2 = 33$$
$$5x - 2 + 2 = 33 + 2$$
$$5x = 35$$
$$\frac{5x}{5} = \frac{35}{5}$$
$$x = 7$$
$\{7\}$

2.
$$6 - 4x = 38$$
$$6 - 6 - 4x = 38 - 6$$
$$-4x = 32$$
$$\frac{-4x}{-4} = \frac{32}{-4}$$
$$x = -8$$
$\{-8\}$

3.
$$\frac{2}{7}x + \frac{1}{2} = \frac{4}{3}$$
$$42 \cdot \frac{2}{7}x + 42 \cdot \frac{1}{2} = 42 \cdot \frac{4}{3}$$
$$12x + 21 = 56$$
$$12x + 21 - 21 = 56 - 21$$
$$12x = 35$$
$$\frac{12x}{12} = \frac{35}{12}$$
$$x = \frac{35}{12}$$
$\left\{\dfrac{35}{12}\right\}$

4.
$$6x + 19 = 3x - 8$$
$$6x - 3x + 19 = 3x - 3x - 8$$
$$3x + 19 = -8$$
$$3x + 19 - 19 = -8 - 19$$
$$3x = -27$$
$$\frac{3x}{3} = \frac{-27}{3}$$
$$x = -9$$
$\{-9\}$

5.
$$6x - 9 + 4x = 3x + 13 - 8$$
$$10x - 9 = 3x + 5$$
$$10x - 3x - 9 = 3x - 3x + 5$$
$$7x - 9 = 5$$
$$7x - 9 + 9 = 5 + 9$$
$$7x = 14$$
$$\frac{7x}{7} = \frac{14}{7}$$
$$x = 2$$
$\{2\}$

6.
$$3(2x-7)+x=2(x-9)-18$$
$$6x-21+x=2x-18-18$$
$$7x-21=2x-36$$
$$7x-2x-21=2x-2x-36$$
$$5x-21=-36$$
$$5x-21+21=-36+21$$
$$5x=-15$$
$$\frac{5x}{5}=\frac{-15}{5}$$
$$x=-3$$
$$\{-3\}$$

7.
$$\frac{1}{10}x-\frac{2}{5}=\frac{1}{20}x-\frac{7}{10}$$
$$20\cdot\frac{1}{10}x-20\cdot\frac{2}{5}=20\cdot\frac{1}{20}x-20\cdot\frac{7}{10}$$
$$2x-8=x-14$$
$$2x-x-8=x-x-14$$
$$x-8=-14$$
$$x-8+8=-14+8$$
$$x=-6$$
$$\{-6\}$$

8.
$$3x-4=3x+4$$
$$3x-3x-4=3x-3x+4$$
$$-4=4$$
False.
$$\varnothing$$

9.
$$5a+4=4+5a$$
$$5a-5a+4=4+5a-5a$$
$$4=4$$
True.
$$\mathbb{R}$$

10.
$$x+2y=5$$
$$x-x+2y=5-x$$
$$2y=5-x$$
$$\frac{2y}{2}=\frac{5-x}{2}$$
$$y=\frac{5-x}{2}$$

11.
$$\frac{1}{5}xy=z$$
$$5\cdot\frac{1}{5}xy=5\cdot z$$
$$xy=5z$$
$$\frac{xy}{y}=\frac{5z}{y}$$
$$x=\frac{5z}{y}$$

2.2 *SOLVING LINEAR EQUATIONS: A GENERAL STRATEGY*

1. LCM

3. the empty set

5. b

7.
$$5x+31=16$$
$$5x+31-31=16-31$$
$$5x=-15$$
$$\frac{5x}{5}=\frac{-15}{5}$$
$$x=-3$$
$$\{-3\}$$

9.
$$23-4x=9$$
$$23-23-4x=9-23$$
$$-4x=-14$$
$$\frac{-4x}{-4}=\frac{-14}{-4}$$
$$x=\frac{7}{2}$$
$$\left\{\frac{7}{2}\right\}$$

11.
$$6=2a+20$$
$$6-20=2a+20-20$$
$$-14=2a$$
$$\frac{-14}{2}=\frac{2a}{2}$$
$$-7=a$$
$$\{-7\}$$

13.
$$9x+24=24$$
$$9x+24-24=24-24$$
$$9x=0$$
$$\frac{9x}{9}=\frac{0}{9}$$
$$x=0$$
$$\{0\}$$

15.
$$16.2x-43.8=48.54$$
$$16.2x-43.8+43.8=48.54+43.8$$
$$16.2x=92.34$$
$$x=5.7$$
$$\{5.7\}$$

17.
$$7a + 11 = 5a - 9$$
$$7a - 5a + 11 = 5a - 5a - 9$$
$$2a + 11 = -9$$
$$2a + 11 - 11 = -9 - 11$$
$$2a = -20$$
$$\frac{2a}{2} = \frac{-20}{2}$$
$$a = -10$$
$$\{-10\}$$

19.
$$16 - 4x = 2x + 61$$
$$16 - 4x + 4x = 2x + 4x + 61$$
$$16 = 6x + 61$$
$$16 - 61 = 6x + 61 - 61$$
$$-45 = 6x$$
$$\frac{-45}{6} = \frac{6x}{6}$$
$$-\frac{15}{2} = x$$
$$\left\{-\frac{15}{2}\right\}$$

21.
$$5x - 9 = -9 + 5x$$
$$5x - 5x - 9 = -9 + 5x - 5x$$
$$-9 = -9$$
True.
$$\mathbb{R}$$

23.
$$16 - 5x = 2x - 5$$
$$16 - 5x + 5x = 2x + 5x - 5$$
$$16 = 7x - 5$$
$$16 + 5 = 7x - 5 + 5$$
$$21 = 7x$$
$$\frac{21}{7} = \frac{7x}{7}$$
$$3 = x$$
$$\{3\}$$

25.
$$3.2x + 8.3 = 1.3x + 19.7$$
$$3.2x - 1.3x + 8.3 = 1.3x - 1.3x + 19.7$$
$$1.9x + 8.3 = 19.7$$
$$1.9x + 8.3 - 8.3 = 19.7 - 8.3$$
$$1.9x = 11.4$$
$$\frac{1.9x}{1.9} = \frac{11.4}{1.9}$$
$$x = 6$$
$$\{6\}$$

27.
$$3x + 8 + x = 2x - 9 + 13$$
$$4x + 8 = 2x + 4$$
$$4x - 2x + 8 = 2x - 2x + 4$$
$$2x + 8 = 4$$
$$2x + 8 - 8 = 4 - 8$$
$$2x = -4$$
$$\frac{2x}{2} = \frac{-4}{2}$$
$$x = -2$$
$$\{-2\}$$

29.
$$x + (x + 1) + (x + 2) = 378$$
$$3x + 3 = 378$$
$$3x + 3 - 3 = 378 - 3$$
$$3x = 375$$
$$\frac{3x}{3} = \frac{375}{3}$$
$$x = 125$$
$$\{125\}$$

31.
$$2L + 2(L - 5) = 94$$
$$2L + 2L - 10 = 94$$
$$4L - 10 = 94$$
$$4L - 10 + 10 = 94 + 10$$
$$4L = 104$$
$$\frac{4L}{4} = \frac{104}{4}$$
$$L = 26$$
$$\{26\}$$

33.
$$2(2w - 3) + 2w = 102$$
$$4w - 6 + 2w = 102$$
$$6w - 6 = 102$$
$$6w - 6 + 6 = 102 + 6$$
$$6w = 108$$
$$\frac{6w}{6} = \frac{108}{6}$$
$$w = 18$$
$$\{18\}$$

35.
$$4(2x - 3) - 5x = 3(x + 4)$$
$$8x - 12 - 5x = 3x + 12$$
$$3x - 12 = 3x + 12$$
$$3x - 3x - 12 = 3x - 3x + 12$$
$$-12 = 12$$
False.
$$\varnothing$$

37.
$$13(3x+4)-7(2x-5)=5x-13$$
$$39x+52-14x+35=5x-13$$
$$25x+87=5x-13$$
$$25x-5x+87=5x-5x-13$$
$$20x+87=-13$$
$$20x+87-87=-13-87$$
$$20x=-100$$
$$\frac{20x}{20}=\frac{-100}{20}$$
$$x=-5$$

$\{-5\}$

39.
$$0.06x+0.03(4000-x)=156$$
$$0.06x+120-0.03x=156$$
$$0.03x+120=156$$
$$0.03x+120-120=156-120$$
$$0.03x=36$$
$$\frac{0.03x}{0.03}=\frac{36}{0.03}$$
$$x=1200$$

$\{1200\}$

41.
$$0.17x-0.4(x+1700)=-2566$$
$$0.17x-0.4x-680=-2566$$
$$-0.23x-680=-2566$$
$$-0.23x-680+680=-2566+680$$
$$-0.23x=-1886$$
$$\frac{-0.23x}{-0.23}=\frac{-1886}{-0.23}$$
$$x=8200$$

$\{8200\}$

43.
$$0.48x+0.72(120-x)=0.66(120)$$
$$0.48x+86.4-0.72x=79.2$$
$$-0.24x+86.4=79.2$$
$$-0.24x+86.4-86.4=79.2-86.4$$
$$-0.24x=-7.2$$
$$\frac{-0.24x}{-0.24}=\frac{-7.2}{-0.24}$$
$$x=30$$

$\{30\}$

45.
$$\frac{1}{4}x-\frac{1}{3}=\frac{5}{12}\quad\text{The LCM is 12.}$$
$$12\cdot\left(\frac{1}{4}x-\frac{1}{3}\right)=12\cdot\frac{5}{12}$$
$$12^3\cdot\frac{1}{4_1}x-12^4\cdot\frac{1}{3_1}=12^1\cdot\frac{5}{12_1}$$
$$3x-4=5$$
$$3x-4+4=5+4$$
$$3x=9$$
$$\frac{3x}{3}=\frac{9}{3}$$
$$x=3$$

$\{3\}$

47.
$$\frac{x}{12}-\frac{11}{6}=\frac{5}{4}\quad\text{The LCM is 12.}$$
$$12\cdot\left(\frac{x}{12}-\frac{11}{6}\right)=12\cdot\frac{5}{4}$$
$$12^1\cdot\frac{x}{12_1}-12^2\cdot\frac{11}{6_1}=12^3\cdot\frac{5}{4_1}$$
$$x-22=15$$
$$x-22+22=15+22$$
$$x=37$$

$\{37\}$

49.
$$\frac{1}{2}x-3=\frac{11}{5}-\frac{3}{4}x\quad\text{The LCM is 20.}$$
$$20\cdot\left(\frac{1}{2}x-3\right)=20\cdot\left(\frac{11}{5}-\frac{3}{4}x\right)$$
$$20^{10}\cdot\frac{1}{2_1}x-20\cdot3=20^4\cdot\frac{11}{5_1}-20^5\cdot\frac{3}{4_1}x$$
$$10x-60=44-15x$$
$$10x+15x-60=44-15x+15x$$
$$25x-60=44$$
$$25x-60+60=44+60$$
$$25x=104$$
$$\frac{25x}{25}=\frac{104}{25}$$
$$x=\frac{104}{25}$$

$\left\{\dfrac{104}{25}\right\}$

51.
$$\frac{2}{9}x + \frac{3}{4} = \frac{1}{6}x - \frac{5}{3} \quad \text{The LCM is 36.}$$

$$36 \cdot \left(\frac{2}{9}x + \frac{3}{4}\right) = 36 \cdot \left(\frac{1}{6}x - \frac{5}{3}\right)$$

$$36^4 \cdot \frac{2}{9_1}x + 36^9 \cdot \frac{3}{4_1} = 36^6 \cdot \frac{1}{6_1}x - 36^{12} \cdot \frac{5}{3_1}$$

$$8x + 27 = 6x - 60$$
$$8x - 6x + 27 = 6x - 6x - 60$$
$$2x + 27 = -60$$
$$2x + 27 - 27 = -60 - 27$$
$$2x = -87$$
$$\frac{2x}{2} = \frac{-87}{2}$$
$$x = \frac{-87}{2}$$

$$\left\{-\frac{87}{2}\right\}$$

53. Answers will vary. Example:
$$5x - 6 = 5x - 7$$

55. Answers will vary. Example:
$$2(x - 4) - 3 = 2x - 11$$

57. Answers will vary. Example:
$$x + 3 = -8$$

59.
$$5x + y = -2$$
$$5x - 5x + y = -5x - 2$$
$$y = -5x - 2$$

61.
$$7x + 2y = 4$$
$$7x - 7x + 2y = -7x + 4$$
$$2y = -7x + 4$$
$$\frac{2y}{2} = \frac{-7x + 4}{2}$$
$$y = \frac{-7x + 4}{2}$$

63.
$$-4x + 3y = 10$$
$$-4x + 4x + 3y = 10 + 4x$$
$$3y = 10 + 4x$$
$$\frac{3y}{3} = \frac{10 + 4x}{3}$$
$$y = \frac{10 + 4x}{3}$$

65.
$$P = a + b + c$$
$$P - a = a - a + b + c$$
$$P - a = b + c$$
$$P - a - c = b + c - c$$
$$P - a - c = b$$
$$b = P - a - c$$

67.
$$d = r \cdot t$$
$$\frac{d}{r} = \frac{r \cdot t}{r}$$
$$\frac{d}{r} = t$$
$$t = \frac{d}{r}$$

69.
$$C = 2\pi r$$
$$\frac{C}{2\pi} = \frac{2\pi r}{2\pi}$$
$$\frac{C}{2\pi} = r$$
$$r = \frac{C}{2\pi}$$

71.
$$F = \frac{9}{5}C + 32$$
$$68 = \frac{9}{5}C + 32$$
$$68 - 32 = \frac{9}{5}C + 32 - 32$$
$$36 = \frac{9}{5}C$$
$$\frac{5}{9_1} \cdot 36^4 = \frac{5}{9} \cdot \frac{9}{5}C$$
$$20 = C$$
$$20°\text{C}$$

73.
$$3x + 64 = 325$$
$$3x + 64 - 64 = 325 - 64$$
$$3x = 261$$
$$\frac{3x}{3} = \frac{261}{3}$$
$$x = 87$$

75. a) Let x represent the number of kites to sell.
$$200 + 4x = 520$$
$$200 - 200 + 4x = 520 - 200$$
$$4x = 320$$
$$\frac{4x}{4} = \frac{320}{4}$$
$$x = 80$$
80 kites

b) Let y represent the amount charged for each kite.
$$80y = 520$$
$$\frac{80y}{80} = \frac{520}{80}$$
$$y = 6.5$$
$6.50

c) Let z represent the price per kite.
$$80z = 1020$$
$$\frac{80z}{80} = \frac{1020}{80}$$
$$z = 12.75$$
$12.75

77.
$$5x - ? = 11$$
$$5(3) - ? = 11$$
$$15 - ? = 11$$
$$? = 4$$
4

79.
$$2x + 9 = 6x + ?$$
$$2\left(-\frac{3}{2}\right) + 9 = 6\left(-\frac{3}{2}\right) + ?$$
$$-3 + 9 = -9 + ?$$
$$6 = -9 + ?$$
$$6 + 9 = -9 + 9 + ?$$
$$15 = ?$$
15

81.
$$14x = -105$$
$$\frac{14x}{14} = \frac{-105}{14}$$
$$x = -\frac{15}{2}$$
$$\left\{-\frac{15}{2}\right\}$$

83.
$$27x - 16 - 6x = 14 + 21x - 30$$
$$21x - 16 = 21x - 16$$
$$21x - 21x - 16 = 21x - 21x - 16$$
$$-16 = -16$$
True.
$\mathbb{R}$

85.
$$-5n + 11 - 4n + 9 = n - 45$$
$$-9n + 20 = n - 45$$
$$-9n + 9n + 20 = n + 9n - 45$$
$$20 = 10n - 45$$
$$20 + 45 = 10n - 45 + 45$$
$$65 = 10n$$
$$\frac{65}{10} = \frac{10n}{10}$$
$$\frac{13}{2} = n$$
$$\left\{\frac{13}{2}\right\}$$

87.
$$11x - 8y = 34$$
$$11x - 8y + 8y = 34 + 8y$$
$$11x = 34 + 8y$$
$$\frac{11x}{11} = \frac{34 + 8y}{11}$$
$$x = \frac{8y + 34}{11}$$

89.
$$\frac{2}{5}x - \frac{3}{8} = 3x - \frac{7}{10} \quad \text{The LCM is 40.}$$
$$40 \cdot \left(\frac{2}{5}x - \frac{3}{8}\right) = 40 \cdot \left(3x - \frac{7}{10}\right)$$
$$\cancel{40}^{8} \cdot \frac{2}{\cancel{5}_1}x - \cancel{40}^{5} \cdot \frac{3}{\cancel{8}_1} = 40 \cdot 3x - \cancel{40}^{4} \cdot \frac{7}{\cancel{10}_1}$$
$$16x - 15 = 120x - 28$$
$$16x - 16x - 15 = 120x - 16x - 28$$
$$-15 = 104x - 28$$
$$-15 + 28 = 104x - 28 + 28$$
$$13 = 104x$$
$$\frac{13}{104} = \frac{104x}{104}$$
$$\frac{1}{8} = x$$
$$\left\{\frac{1}{8}\right\}$$

91.
$$0.16x + 0.07(3000 - x) = 372$$
$$0.16x + 210 - 0.07x = 372$$
$$0.09x + 210 = 372$$
$$0.09x + 210 - 210 = 372 - 210$$
$$0.09x = 162$$
$$\frac{0.09x}{0.09} = \frac{162}{0.09}$$
$$x = 1800$$
$$\{1800\}$$

93. Answers will vary. Example:

$$14 = 2x - 9 \qquad \text{and} \qquad y = 2x - 9$$
$$14 + 9 = 2x - 9 + 9 \qquad y + 9 = 2x - 9 + 9$$
$$23 = 2x \qquad y + 9 = 2x$$
$$\frac{23}{2} = \frac{2x}{2} \qquad \frac{y+9}{2} = \frac{2x}{2}$$
$$\frac{23}{2} = x \qquad \frac{y+9}{2} = x$$

The only difference in the process of solving these two equations is that the $y + 9$ cannot be simplified into a single term like the $14 + 9$.

95. Answers will vary.

2.3   QUICK CHECK

1.
$$4x - 3 = 29$$
$$4x - 3 + 3 = 29 + 3$$
$$4x = 32$$
$$\frac{4x}{4} = \frac{32}{4}$$
$$x = 8$$

2. Length: $l = 2w - 8$
Width: w
$$P = 2l + 2w$$
$$80 = 2(2w - 8) + 2w$$
$$80 = 4w - 16 + 2w$$
$$80 = 6w - 16$$
$$80 + 16 = 6w - 16 + 16$$
$$96 = 6w$$
$$\frac{96}{6} = \frac{6w}{6}$$
$$16 = w$$
$$l = 2w - 8$$
$$l = 2(16) - 8$$
$$l = 32 - 8$$
$$l = 24$$

Length is 24 feet and width is 16 feet.

3. Longest side: $3s + 8$
Shortest side: s
Other side: $2s + 20$
$$112 = 3s + 8 + s + 2s + 20$$
$$112 = 6s + 28$$
$$112 - 28 = 6s + 28 - 28$$
$$84 = 6s$$
$$\frac{84}{6} = \frac{6s}{6}$$
$$14 = s$$
$$3s + 8 = 3(14) + 8$$
$$= 42 + 8$$
$$= 50$$
$$2s + 20 = 2(14) + 20$$
$$= 28 + 20$$
$$= 48$$

The lengths of the sides of the triangle are 14 inches, 48 inches and 50 inches.

4. #1: x
#2: $x + 1$
#3: $x + 2$
$$x + (x + 1) + (x + 2) = 213$$
$$3x + 3 = 213$$
$$3x + 3 - 3 = 213 - 3$$
$$3x = 210$$
$$\frac{3x}{3} = \frac{210}{3}$$
$$x = 70$$
$$x + 1 = 70 + 1 = 71$$
$$x + 2 = 70 + 2 = 72$$
$$70, 71, 72$$

5.

	Rate	Time	Distance
Jake	85	t	$85t$
Elwood	90	$t + 3$	$90(t + 3)$

$$85t + 90(t + 3) = 970$$
$$85t + 90t + 270 = 970$$
$$175t + 270 = 970$$
$$175t + 270 - 270 = 970 - 270$$
$$175t = 700$$
$$\frac{175t}{175} = \frac{700}{175}$$
$$t = 4$$
$$t + 3 = 4 + 3 = 7$$

Elwood drove for 7 hours.

6. #1: x

#2: $5x+1$

$$x+(5x+1)=103$$
$$6x+1=103$$
$$6x+1-1=103-1$$
$$6x=102$$
$$\frac{6x}{6}=\frac{102}{6}$$
$$x=17$$

$5x+1=5(17)+1=85+1=86$

17 and 86

7.

	Number	Value	Amount
Nickels	$n=q+13$	0.05	$0.05(q+13)$
Quarters	q	0.25	$0.25q$

$$0.05(q+13)+0.25q=3.35$$
$$0.05q+0.65+0.25q=3.35$$
$$0.30q+0.65=3.35$$
$$0.30q+0.65-0.65=3.35-0.65$$
$$0.30q=2.70$$
$$\frac{0.30q}{0.30}=\frac{2.70}{0.30}$$
$$q=9$$

$n=q+13=9+13=22$

Bert has 22 nickels and 9 quarters.

2.3 *PROBLEM SOLVING; APPLICATIONS OF LINEAR EQUATIONS*

1. b

3. b

5. $x+1;\ x+2$

7. $x+2;\ x+4$

9.
$$2x+5=97$$
$$2x+5-5=97-5$$
$$2x=92$$
$$\frac{2x}{2}=\frac{92}{2}$$
$$x=46$$

11.
$$5x-8=717$$
$$5x-8+8=717+8$$
$$5x=725$$
$$\frac{5x}{5}=\frac{725}{5}$$
$$x=145$$

13.
$$6x+3.2=56.6$$
$$6x+3.2-3.2=56.6-3.2$$
$$6x=53.4$$
$$\frac{6x}{6}=\frac{53.4}{6}$$
$$x=8.9$$

15. Unknowns

Length: l

Width: $w=l-5$

Knowns

Perimeter: 26 meters
$$P=2l+2w$$
$$26=2l+2(l-5)$$
$$26=2l+2l-10$$
$$26=4l-10$$
$$26+10=4l-10+10$$
$$36=4l$$
$$\frac{36}{4}=\frac{4l}{4}$$
$$9=l$$

Length: $l=9$ meters

Width: $l-5=9-5=4$ meters

17. Unknowns

Length: $l=2w-2$

Width: w

Knowns

Perimeter: 56 feet
$$P=2l+2w$$
$$56=2(2w-2)+2w$$
$$56=4w-4+2w$$
$$56=6w-4$$
$$56+4=6w-4+4$$
$$60=6w$$
$$\frac{60}{6}=\frac{6w}{6}$$
$$10=w$$

Length: $2w-2=2(10)-2=20-2=18$ feet

Width: $w=10$ feet

19. Unknowns
Length: $l = 4w$
Width: w

Knowns
Perimeter: 130 centimeters
$P = 2l + 2w$
$130 = 2(4w) + 2w$
$130 = 8w + 2w$
$130 = 10w$
$$\frac{130}{10} = \frac{10w}{10}$$
$13 = w$

Length: $4w = 4(13) = 52$ centimeters
Width: 13 centimeters

21. Unknown
Side: s

Known
Perimeter: 220 feet
$P = 4s$
$220 = 4s$
$$\frac{220}{4} = \frac{4s}{4}$$
$55 = s$

Side: $s = 55$ feet

23. Unknown
Side: s

Known
Perimeter: 135 centimeters
$P = 3s$
$135 = 3s$
$$\frac{135}{3} = \frac{3s}{3}$$
$45 = s$

Side: $s = 45$ centimeters

25. Unknowns
s_1: x
s_2: $x + 5$
s_3: $x + 7$

Known
Perimeter: 36 inches
$P = s_1 + s_2 + s_3$
$36 = x + (x + 5) + (x + 7)$
$36 = 3x + 12$
$36 - 12 = 3x + 12 - 12$
$24 = 3x$
$$\frac{24}{3} = \frac{3x}{3}$$
$8 = x$

s_1: $x = 8$ inches
s_2: $x + 5 = 8 + 5 = 13$ inches
s_3: $x + 7 = 8 + 7 = 15$ inches

27. Unknown
Radius: r

Known
Circumference: 69.08 inches
$C = 2\pi r$
$69.08 = 2(3.14)r$
$69.08 = 6.28r$
$$\frac{69.08}{6.28} = \frac{6.28r}{6.28}$$
$11 = r$

Radius: $r = 11$ inches

29. Angle $A = x + 30$
Angle $B = x$
$90° = $ Angle $A + $ Angle B
$90 = (x + 30) + x$
$90 = 2x + 30$
$90 - 30 = 2x + 30 - 30$
$60 = 2x$
$$\frac{60}{2} = \frac{2x}{2}$$
$30 = x$
Angle A: $x + 30 = 30 + 30 = 60°$
Angle B: $x = 30°$

31. Angle A: x
Angle B: $3x + 10$
$90° = $ Angle $A + $ Angle B
$90 = x + (3x + 10)$
$90 = 4x + 10$
$90 - 10 = 4x + 10 - 10$
$80 = 4x$
$$\frac{80}{4} = \frac{4x}{4}$$
$20 = x$
Angle A: $x = 20°$
Angle B: $3x + 10 = 3(20) + 10 = 70°$

33. Angle A: x
Angle B: $3x + 12$
$180° = $ Angle $A + $ Angle B
$180 = x + (3x + 12)$
$180 = 4x + 12$
$180 - 12 = 4x + 12 - 12$
$168 = 4x$
$$\frac{168}{4} = \frac{4x}{4}$$
$42 = x$
Angle A: $x = 42°$
Angle B: $3x + 12 = 3(42) + 12 = 138°$

35. Angle A: x
Angle B: $x + 10$
Angle C: $x + 20$

$$180° = \text{Angle } A + \text{Angle } B + \text{Angle } C$$
$$180 = x + (x + 10) + (x + 20)$$
$$180 = 3x + 30$$
$$180 - 30 = 3x + 30 - 30$$
$$150 = 3x$$
$$\frac{150}{3} = \frac{3x}{3}$$
$$50 = x$$

Angle A: $x = 50°$
Angle B: $x + 10 = 50 + 10 = 60°$
Angle C: $x + 20 = 50 + 20 = 70°$

37. Unknowns
Length: $l = w + 6$
Width: w
Area: $A = lw$

Known
Perimeter: 104 inches
$$P = 2l + 2w$$
$$104 = 2(w + 6) + 2w$$
$$104 = 2w + 12 + 2w$$
$$104 = 4w + 12$$
$$104 - 12 = 4w + 12 - 12$$
$$92 = 4w$$
$$\frac{92}{4} = \frac{4w}{4}$$
$$23 = w$$
Length: $w + 6 = 23 + 6 = 29$ inches
Width: $w = 23$ inches
The area of the rectangle is
$A = lw = 29 \cdot 23 = 667$ square inches.

39. Unknowns
Width of rectangle: w
Length of rectangle: $l = 2w - 7$
Side of square: $s = l$
Perimeter of square: $P = 4s = 4l$

Known
Perimeter of rectangle: 46 inches
$$P = 2l + 2w$$
$$46 = 2(2w - 7) + 2w$$
$$46 = 4w - 14 + 2w$$
$$46 = 6w - 14$$
$$46 + 14 = 6w - 14 + 14$$
$$60 = 6w$$
$$\frac{60}{6} = \frac{6w}{6}$$
$$10 = w$$
Length: $2w - 7 = 2(10) - 7 = 13$ inches
The perimeter of the square is
$P = 4 \cdot l = 4 \cdot 13 = 52$ inches.

41. Unknowns
#1: x
#2: $x + 1$
#3: $x + 2$
$$x + (x + 1) + (x + 2) = 459$$
$$3x + 3 = 459$$
$$3x + 3 - 3 = 459 - 3$$
$$3x = 456$$
$$\frac{3x}{3} = \frac{456}{3}$$
$$x = 152$$
$x + 1 = 152 + 1 = 153$
$x + 2 = 152 + 2 = 154$
The three consecutive numbers
are 152, 153, and 154.

43. Unknowns
#1: x
#2: $x + 2$
#3: $x + 4$
$$x + (x + 2) + (x + 4) = 246$$
$$3x + 6 = 246$$
$$3x + 6 - 6 = 246 - 6$$
$$3x = 240$$
$$\frac{3x}{3} = \frac{240}{3}$$
$$x = 80$$
$x + 2 = 80 + 2 = 82$
$x + 4 = 80 + 4 = 84$
The three consecutive even numbers
are 80, 82, and 84.

45. Unknowns
#1: x
#2: $x + 2$
#3: $x + 4$
#4: $x + 6$
$$x + (x + 2) + (x + 4) + (x + 6) = 304$$
$$4x + 12 = 304$$
$$4x + 12 - 12 = 304 - 12$$
$$4x = 292$$
$$\frac{4x}{4} = \frac{292}{4}$$
$$x = 73$$
$x + 2 = 73 + 2 = 75$
$x + 4 = 73 + 4 = 77$
$x + 6 = 73 + 6 = 79$
The four consecutive odd integers
are 73, 75, 77, and 79.

47. Unknowns
#1: x
#2: $x+1$
#3: $x+2$

$$x = (x+1)+(x+2)-18$$
$$x = 2x-15$$
$$x-2x = 2x-2x-15$$
$$-x = -15$$
$$(-1)\cdot -x = (-1)\cdot(-15)$$
$$x = 15$$

$x+1 = 15+1 = 16$
$x+2 = 15+2 = 17$
The three consecutive integers
are 15, 16, and 17.

49. Unknown
Distance: d

Known
Rate: 125 miles per hour
Time: 4 hours
$d = rt$
$d = 125\cdot 4$
$d = 500$
A.J. drove his race car 500 miles.

51. Unknown
Time: t

Known
Distance: 374 miles
Rate: 68 miles per hour

$$d = rt$$
$$374 = 68\cdot t$$
$$\frac{374}{68} = \frac{68\cdot t}{68}$$
$$5\frac{1}{2} = t$$

Mario drives for $5\frac{1}{2}$ hours.

53. Unknown
Rate: r

Known
Distance: 50 meters
Time: 28 seconds
$$d = rt$$
$$50 = r\cdot 28$$
$$\frac{50}{28} = \frac{r\cdot 28}{28}$$
$$r = \frac{50}{28} = 1\frac{22}{28} = 1\frac{11}{14}$$

Janet swam at a rate of $1\frac{11}{14}$ meters per second.

55.

	Rate	·	Time	=	Distance
To Phoenix	80		t		$80t$
To Home	70		$t+1$		$70(t+1)$

a)
$$80t = 70(t+1)$$
$$80t = 70t+70$$
$$80t-70t = 70t-70t+70$$
$$10t = 70$$
$$\frac{10t}{10} = \frac{70}{10}$$
$$t = 7$$
$t+1 = 7+1 = 8$
Total driving time $= 7+8 = 15$
Susan's total driving time was 15 hours.

b) Distance $= 80t = 80\cdot 7 = 560$
Susan lives 560 miles from Phoenix.

57.

	Rate	·	Time	=	Distance
Don	60		t		$60t$
Dennis	80		$t-2$		$80(t-2)$

Don's Distance = Dennis' Distance
$$60t = 80(t-2)$$
$$60t = 80t-160$$
$$60t-80t = 80t-80t-160$$
$$-20t = -160$$
$$\frac{-20t}{-20} = \frac{-160}{-20}$$
$$t = 8$$
$t-2 = 8-2 = 6$
It would take Dennis 6 hours to catch up
to Don.

59.

	Rate	·	Time	=	Distance
Lance	25		t		$25t$
Levi	20		$t-1$		$20(t-1)$

Lance's Distance + Levi's Distance
$\qquad\qquad$ = Distance apart
$$25t+20(t-1) = 250$$
$$25t+20t-20 = 250$$
$$45t-20 = 250$$
$$45t-20+20 = 250+20$$
$$45t = 270$$
$$\frac{45t}{45} = \frac{270}{45}$$
$$t = 6$$
$t-1 = 6-1 = 5$
The cyclists will be 250 miles apart 5 hours
after Levi leaves.

61. Unknowns
#1: x
#2: $x + 17$

$$x + (x + 17) = 71$$
$$2x + 17 = 71$$
$$2x + 17 - 17 = 71 - 17$$
$$2x = 54$$
$$\frac{2x}{2} = \frac{54}{2}$$
$$x = 27$$
$$x + 17 = 27 + 17 = 44$$

The two numbers are 27 and 44.

63. Unknowns
#1: x
#2: $x - 8$

$$x + (x - 8) = 96$$
$$2x - 8 = 96$$
$$2x - 8 + 8 = 96 + 8$$
$$2x = 104$$
$$\frac{2x}{2} = \frac{104}{2}$$
$$x = 52$$
$$x - 8 = 52 - 8 = 44$$

The two numbers are 44 and 52.

65. Unknowns
#1: x
#2: $x + 5$

$$2x + 3(x + 5) = 80$$
$$2x + 3x + 15 = 80$$
$$5x + 15 = 80$$
$$5x + 15 - 15 = 80 - 15$$
$$5x = 65$$
$$\frac{5x}{5} = \frac{65}{5}$$
$$x = 13$$
$$x + 5 = 13 + 5 = 18$$

The two numbers are 13 and 18.

67.

	Number $\cdot$	Value $=$	Amount
Dimes	x	0.10	$0.10x$
Quarters	$x + 7$	0.25	$0.25(x + 7)$

$$0.10x + 0.25(x + 7) = 5.60$$
$$100 \cdot \left[0.10x + 0.25(x + 7)\right] = 100 \cdot 5.60$$
$$100 \cdot 0.10x + 100 \cdot 0.25(x + 7) = 100 \cdot 5.60$$
$$10x + 25(x + 7) = 560$$
$$10x + 25x + 175 = 560$$
$$35x + 175 = 560$$
$$35x + 175 - 175 = 560 - 175$$
$$35x = 385$$
$$\frac{35x}{35} = \frac{385}{35}$$
$$x = 11$$
$$x + 7 = 11 + 7 = 18$$

There are 18 quarters in the jar.

69.

	Number $\cdot$	Value $=$	Amount
Students	$4x + 10$	\$4	$4(4x + 10)$
Nonstudents	x	\$7	$7x$

$$4(4x + 10) + 7x = 500$$
$$16x + 40 + 7x = 500$$
$$23x + 40 = 500$$
$$23x + 40 - 40 = 500 - 40$$
$$23x = 460$$
$$\frac{23x}{23} = \frac{460}{23}$$
$$x = 20$$
$$4x + 10 = 4(20) + 10 = 90$$

90 students attended the movie.

71. Answers will vary. Example:
A rectangular room has a perimeter of 140 feet. Its width is 30 feet less than its length. What is the area of the room?

2.4   QUICK CHECK

1. $25\% \cdot 44 = n$
$0.25 \cdot 44 = n$
$11 = n$

2. $39 = 60\% \cdot n$
$39 = 0.60 \cdot n$
$$\frac{39}{0.60} = \frac{0.60n}{0.60}$$
$65 = n$

3. $56 = p \cdot 80$
$$\frac{56}{80} = \frac{p \cdot 80}{80}$$
$0.7 = p$
$70\% = p$

4. $n = 28\% \cdot 500$
$n = 0.28 \cdot 500$
$n = 140$
140 children buy their lunch at school.

5. $30{,}000 - 13{,}800 = 16{,}200$
$16{,}200 = p \cdot 30{,}000$
$$\frac{16{,}200}{30{,}000} = \frac{p \cdot 30{,}000}{30{,}000}$$
$0.54 = p$
$54\% = p$

There is a 54% decrease in the value
of the car.

6. $n = 45\% \cdot 42$
$n = 0.45 \cdot 42$
$n = 18.90$
$42 + 18.90 = 60.90$
The selling price of an MP3 player is $60.90.

7.

	P	r	t	$I = Prt$
3%	x	0.03	1	$0.03x$
5%	$x + 2000$	0.05	1	$0.05(x + 2000)$

$0.03x + 0.05(x + 2000) = 500$
$100 \cdot \left[0.03x + 0.05(x + 2000) \right] = 100 \cdot 500$
$3x + 5(x + 2000) = 50{,}000$
$3x + 5x + 10{,}000 = 50{,}000$
$8x + 10{,}000 = 50{,}000$
$8x + 10{,}000 - 10{,}000 = 50{,}000 - 10{,}000$
$8x = 40{,}000$
$$\frac{8x}{8} = \frac{40{,}000}{8}$$
$x = 5000$

$x + 2000 = 5000 + 2000 = 7000$
Mark invested $5000 at 3% and $7000 at 5%.

8.

	P	r	t	$I = Prt$
4%	x	0.04	1	$0.04x$
5%	$1900 - x$	0.05	1	$0.05(1900 - x)$

$.05(1900 - x) + .04x = 86.50$
$95 - .05x + .04x = 86.50$
$95 - .01x = 86.50$
$95 - 95 - .01x = 86.50 - 95$
$-.01x = -8.5$
$$\frac{-.01x}{-.01} = \frac{-8.5}{-.01}$$
$x = 850$

$1900 - x = 1900 - 850 = 1050$
Patsy invested $850 at 4% and $1050 at 5%.

9.

	Solution	% Alcohol	Amount
Sol. 1 (30%)	x	0.30	$0.30x$
Sol. 2 (42%)	$80 - x$	0.42	$0.42(80 - x)$
Mixture	80	0.39	$0.39(80) = 31.2$

$0.30x + 0.42(80 - x) = 31.2$
$100 \cdot \left[0.30x + 0.42(80 - x) \right] = 100 \cdot 31.2$
$30x + 42(80 - x) = 3120$
$30x + 3360 - 42x = 3120$
$-12x + 3360 = 3120$
$-12x + 3360 - 3360 = 3120 - 3360$
$-12x = -240$
$$\frac{-12x}{-12} = \frac{-240}{-12}$$
$x = 20$

$80 - x = 80 - 20 = 60$
Marie uses 20 mL of 30% and 60 mL of
42% alcohol.

10. $$\frac{3}{17} = \frac{n}{187}$$
$3 \cdot 187 = 17 \cdot n$
$561 = 17n$
$$\frac{561}{17} = \frac{17n}{17}$$
$33 = n$
$\{33\}$

11. $$\frac{4}{5} = \frac{n}{85}$$
$4 \cdot 85 = 5 \cdot n$
$340 = 5n$
$$\frac{340}{5} = \frac{5n}{5}$$
$68 = n$
68 dentists would recommend sugarless gum to
their patients.

12. $\dfrac{1}{0.454} = \dfrac{175}{n}$

$n \cdot 1 = 0.454 \cdot 175$

$n = 79.45 \text{ kg}$

There are 79.45 kilograms in 175 pounds.

2.4 APPLICATIONS INVOLVING PERCENTAGES; RATIO AND PROPORTION

1. Amount = Percent · Base

3. original value

5. $I = P \cdot r \cdot t$

7. proportion

9. n is 45% of 80.
$n = 0.45 \cdot 80$
$n = 36$
45% of 80 is 36.

11. 92 is 40% of n.
$92 = 0.40 \cdot n$
$\dfrac{92}{0.40} = \dfrac{0.40n}{0.40}$
$230 = n$
40% of 230 is 92.

13. 224 is p% of 256.
$224 = p \cdot 256$
$\dfrac{224}{256} = \dfrac{256p}{256}$
$0.875 = p$
$87.5\% = p$
87.5% of 256 is 224.

15. n is $37\frac{1}{2}$% of 104.

$n = 0.375 \cdot 104$
$n = 39$

39 is $37\frac{1}{2}$% of 104.

17. n is 240% of 68.
$n = 2.4 \cdot 68$
$n = 163.2$
163.2 is 240% of 68.

19. 55 is 125% of n.
$55 = 1.25 \cdot n$
$\dfrac{55}{1.25} = \dfrac{1.25n}{1.25}$
$44 = n$
55 is 125% of 44.

21. Let n represent the number of brown M&M's.
n is 30% of 280.
$n = 0.30 \cdot 280$
$n = 84$
There are 84 brown M&M's in the bowl.

23. Let n represent the number of registered
voters in the precinct.
410 is 40% of n.
$410 = 0.4 \cdot n$
$\dfrac{410}{0.4} = \dfrac{0.4n}{0.4}$
$1025 = n$
There are 1025 registered voters in
the precinct.

25. Let n represent the number of milliliters
of alcohol.
n is 40% of 750.
$n = 0.40 \cdot 750$
$n = 300$
There are 300 milliliters of alcohol in a
750-milliliter bottle of rum.

27. Let n represent the amount of increase
in value.
n is 17% of 3500.
$n = 0.17 \cdot 3500$
$n = 595$
$3500 + 595 = 4095$
The new value of her portfolio is $4095.

29. Let n represent the orginal price of the milk.
0.26 is 8% of n.
$0.26 = 0.08 \cdot n$
$\dfrac{0.26}{0.08} = \dfrac{0.08n}{0.08}$
$3.25 = n$
The original price of the milk was $3.25.

31. Let p represent the percent decrease
in the number of parking spots.
120 is p% of 1600.
$120 = p \cdot 1600$
$\dfrac{120}{1600} = \dfrac{1600p}{1600}$
$0.075 = p$
$7.5\% = p$
The number of available parking spots
decreased by 7.5%.

33. a) Let n represent the amount of decrease last year.

n is 20% of 500.
$$n = 0.2 \cdot 500$$
$$n = 100$$
$$500 - 100 = 400$$
The new average score is 400.

b) 100 points

c) Let p represent percent increase next year.
100 is p% of 400.
$$100 = p \cdot 400$$
$$\frac{100}{400} = \frac{400p}{400}$$
$$0.25 = p$$
$$25\% = p$$
The percentage increase next year must be 25% to bring the average back to 500.

35. Let x represent the increase in value the first year.

$$x = 0.10 \cdot 30,000$$
$$x = 3000$$
$$30,000 + 3000 = 33,000$$
After the first year, the stock's value is $33,000.

Let y represent the decrease in value the second year.

$$y = 0.20 \cdot 33,000$$
$$y = 6600$$
$$33,000 - 6600 = 26,400$$
After the second year, the stock's value is $26,400.

Let p represent the increase in value required in the third year.

$$p \cdot 26,400 = 30,000 - 26,400$$
$$p \cdot 26,400 = 3600$$
$$\frac{p \cdot 26,400}{26,400} = \frac{3600}{26,400}$$
$$p \approx 0.136$$
$$p \approx 13.6\%$$
The stock must increase 13.6% in value to return to its original value.

37. Let v represent the original value of the stock portfolio.

$0.25 \cdot v$ is the decrease in value in 2007.

At the end of 2007, the stock portfolio's value is $v - 0.25v = 0.75v$.

$0.40 \cdot 0.75v$ is the decrease in value in 2008.

At the end of 2008, the stock portfolio's value is $0.75v - 0.40 \cdot 0.75v = 0.75v - 0.3v = 0.45v$.

Let p represent the gain in value required in 2009.

$$p \cdot 0.45v = v - 0.45v$$
$$p \cdot 0.45v = 0.55v$$
$$\frac{p \cdot 0.45v}{0.45v} = \frac{0.55v}{0.45v}$$
$$p \approx 1.222$$
$$p \approx 122.2\%$$
The stock portfolio must gain 122.2% in value to return to its pre-2007 value.

39.

	P	r	t	$I = Prt$
2%	x	0.02	1	$0.02x$
5%	$2x$	0.05	1	$0.05(2x)$
Total				84

$$0.02x + 0.05(2x) = 84$$
$$0.02x + 0.1x = 84$$
$$0.12x = 84$$
$$\frac{0.12x}{0.12} = \frac{8400}{0.12}$$
$$x = 700$$

$$2x = 2(700) = 1400$$
Janet invested $700 at 2% and $1400 at 5%.

41.

	P	r	t	$I = Prt$
4%	x	0.04	1	$0.04x$
5%	$6500 - x$	0.05	1	$0.05(6500 - x)$
Total	6500			300

$$0.04x + 0.05(6500 - x) = 300$$
$$0.04x + 325 - 0.05x = 300$$
$$-0.01x + 325 = 300$$
$$-0.01x + 325 - 325 = 300 - 325$$
$$-0.01x = -25$$
$$\frac{-0.01x}{-0.01} = \frac{-25}{-0.01}$$
$$x = 2500$$

$$6500 - x = 6500 - 2500 = 4000$$
Kamiran deposited $2500 at 4% and $4000 at 5%.

43.

	P	r	t	$I = Prt$
-8%	x	-0.08	1	$-0.08x$
-20%	$8000 - x$	-0.20	1	$-0.20(8000 - x)$
Total	8000			-1000

$$-0.08x - 0.20(8000 - x) = -1000$$
$$-0.08x - 1600 + 0.20x = -1000$$
$$0.12x - 1600 = -1000$$
$$0.12x - 1600 + 1600 = -1000 + 1600$$
$$0.12x = 600$$
$$\frac{0.12x}{0.12} = \frac{600}{0.12}$$
$$x = 5000$$

Aurora invested \$5000 at 8% .

45.

	P	r	t	$I = Prt$
6%	x	0.06	1	$0.06x$
-5%	$4800 - x$	-0.05	1	$-0.05(4800 - x)$
Total	4800			90

$$0.06x - 0.05(4800 - x) = 90$$
$$0.06x - 240 + 0.05x = 90$$
$$0.11x - 240 = 90$$
$$0.11x - 240 + 240 = 90 + 240$$
$$0.11x = 330$$
$$\frac{0.11x}{0.11} = \frac{330}{0.11}$$
$$x = 3000$$

Marquis invested \$3000 in the fund that earned a 6% profit.

47.

	Solution	Percent	Amount
Sol. 1 (70%)	x	0.70	$0.7x$
Sol. 2 (40%)	$60 - x$	0.40	$0.4(60 - x)$
Mixture	60	0.52	$0.52(60) = 31.2$

$$0.7x + 0.4(60 - x) = 31.2$$
$$0.7x + 24 - 0.4x = 31.2$$
$$0.3x + 24 = 31.2$$
$$0.3x + 24 - 24 = 31.2 - 24$$
$$0.3x = 7.2$$
$$\frac{0.3x}{0.3} = \frac{7.2}{0.3}$$
$$x = 24$$

$$60 - x = 60 - 24 = 36$$

The mechanic should mix 24 gallons of the 70% antifreeze with 36 gallons of the 40% antifreeze.

49.

	Solution	Percent	Amount
Sol. 1 (5%)	x	0.05	$0.05x$
Sol. 2 (2%)	$1000 - x$	0.02	$0.02(1000 - x)$
Mixture	1000	0.032	$0.032(1000) = 32$

$$0.05x + 0.02(1000 - x) = 32$$
$$0.05x + 20 - 0.02x = 32$$
$$0.03x + 20 = 32$$
$$0.03x + 20 - 20 = 32 - 20$$
$$0.03x = 12$$
$$\frac{0.03x}{0.03} = \frac{12}{0.03}$$
$$x = 400$$

$$1000 - x = 1000 - 400 = 600$$

The dairyman should mix 400 gallons of 5% butterfat and 600 gallons of 2% butterfat.

51.

	Solution	Percent	Amount
Sol. 1 (27%)	x	0.27	$0.27x$
Sol. 2 (43%)	$44 - x$	0.43	$0.43(44 - x)$
Mixture	44	0.39	$0.39(44) = 17.16$

$$0.27x + 0.43(44 - x) = 17.16$$
$$0.27x + 18.92 - 0.43x = 17.16$$
$$-0.16x + 18.92 = 17.16$$
$$-0.16x + 18.92 - 18.92 = 17.16 - 18.92$$
$$-0.16x = -1.76$$
$$\frac{-0.16x}{-0.16} = \frac{-1.76}{-0.16}$$
$$x = 11$$

$$44 - x = 44 - 11 = 33$$

The chemist should mix 11 milliliters of 27% solution with 33 milliliters of 43% solution.

53.

	Solution	Percent	Amount
Sol. 1 (40%)	x	0.4	$0.4x$
Sol. 2 (100%)	$1.6 - x$	1	$1(1.6 - x)$
Mixture	1.6	0.52	$0.52(1.6) = 0.832$

$$0.4x + 1(1.6 - x) = 0.832$$
$$0.4x + 1.6 - x = 0.832$$
$$-0.6x + 1.6 = 0.832$$
$$-0.6x + 1.6 - 1.6 = 0.832 - 1.6$$
$$-0.6x = -0.768$$
$$\frac{-0.6x}{-0.6} = \frac{-0.768}{-0.6}$$
$$x = 1.28$$

$$1.6 - x = 1.6 - 1.28 = 0.32$$

The chemist should mix 1.28 liters of 40% alcohol with 0.32 liters of 100% alcohol.

55.

	Solution	% Alcohol	Amount
Sol. 1 (80%)	400	0.80	400(0.80) = 320
Sol. 2 (0%)	x	0.00	$0(x) = 0$
Mixture	$400 + x$	0.50	$0.50(400 + x)$

$$320 = 0.5(400 + x)$$
$$320 = 200 + 0.5x$$
$$320 - 200 = 200 + 0.5x - 200$$
$$120 = 0.5x$$
$$\frac{120}{0.5} = \frac{0.5x}{0.5}$$
$$240 = x$$

Paula should add 240 millilters of water.

57.
$$\frac{2}{3} = \frac{n}{48}$$
$$2 \cdot 48 = 3 \cdot n$$
$$96 = 3n$$
$$\frac{96}{3} = \frac{3n}{3}$$
$$32 = n$$
$$\{32\}$$

59.
$$\frac{20}{n} = \frac{45}{81}$$
$$20 \cdot 81 = n \cdot 45$$
$$1620 = 45n$$
$$\frac{1620}{45} = \frac{45n}{45}$$
$$36 = n$$
$$\{36\}$$

61.
$$\frac{3}{4} = \frac{n}{109}$$
$$3 \cdot 109 = 4 \cdot n$$
$$327 = 4n$$
$$\frac{327}{4} = \frac{4n}{4}$$
$$81.75 = n$$
$$\{81.75\}$$

63.
$$\frac{13.8}{n} = \frac{2}{9}$$
$$9 \cdot 13.8 = n \cdot 2$$
$$124.2 = 2n$$
$$\frac{124.2}{2} = \frac{2n}{2}$$
$$62.1 = n$$
$$\{62.1\}$$

65.
$$\frac{3}{4} = \frac{n + 12}{28}$$
$$3 \cdot 28 = 4 \cdot (n + 12)$$
$$84 = 4n + 48$$
$$84 - 48 = 4n + 48 - 48$$
$$36 = 4n$$
$$\frac{36}{4} = \frac{4n}{4}$$
$$9 = n$$
$$\{9\}$$

67.
$$\frac{2n + 15}{18} = \frac{5n + 13}{24}$$
$$(2n + 15) \cdot 24 = 18 \cdot (5n + 13)$$
$$48n + 360 = 90n + 234$$
$$48n - 48n + 360 = 90n + 234 - 48n$$
$$360 - 234 = 42n + 234 - 234$$
$$126 = 42n$$
$$\frac{126}{42} = \frac{42n}{42}$$
$$3 = n$$
$$\{3\}$$

69. Let n represent the number of females.
$$\frac{3}{5} = \frac{n}{9200}$$
$$3 \cdot 9200 = 5 \cdot n$$
$$27,600 = 5n$$
$$\frac{27,600}{5} = \frac{5n}{5}$$
$$5520 = n$$

There are 5520 females.

71. Let n represent the number of tablespoons of plant food necessary.
$$\frac{3}{2} = \frac{n}{15}$$
$$3 \cdot 15 = 2 \cdot n$$
$$45 = 2n$$
$$\frac{45}{2} = \frac{2n}{2}$$
$$22.5 = n$$

You need 22.5 tablespoons of plant food to mix with 15 gallons of water.

73. Let n represent the number of qualified teachers.

$$\frac{5}{6} = \frac{n}{864}$$
$$5 \cdot 864 = 6 \cdot n$$
$$4320 = 6n$$
$$\frac{4320}{6} = \frac{6n}{6}$$
$$720 = n$$
$$864 - 720 = 144$$

The city has 144 teachers who do not meet minimum qualifications.

75. $1 \text{ in.} \approx 2.54 \text{ cm}$

$$\frac{1}{2.54} = \frac{n}{60}$$
$$1 \cdot 60 = 2.54 \cdot n$$
$$\frac{60}{2.54} = \frac{2.54n}{2.54}$$
$$23.6 \approx n$$
$$60 \text{ cm} \approx 23.6 \text{ in.}$$

77. $1 \text{ L} \approx 0.264 \text{ gal}$

$$\frac{1}{0.264} = \frac{15.2}{n}$$
$$1 \cdot n = 0.264 \cdot 15.2$$
$$n \approx 4.0$$
$$15.2 \text{ L} \approx 4.0 \text{ gal}$$

79. $1 \text{ kg} \approx 2.2 \text{ lb}$

$$\frac{1}{2.2} = \frac{80}{n}$$
$$1 \cdot n = 2.2 \cdot 80$$
$$n \approx 176$$
$$80 \text{ kg} \approx 176 \text{ lb}$$

81. Answers will vary. Example:
Joy invested in a 3% annual CD and a 5% annual CD. She invested $10,000 more in the 5% CD than in the 3% CD. If the two CD's made $2100 in interest in one year, how much did Joy invest in each CD?

2.5   *QUICK CHECK*

1. $8 < x$

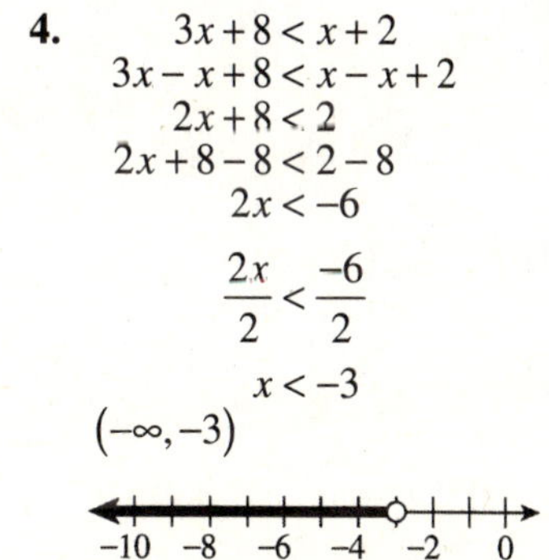

2.
$$4x + 3 < 31$$
$$4x + 3 - 3 < 31 - 3$$
$$4x < 28$$
$$\frac{4x}{4} < \frac{28}{4}$$
$$x < 7$$
$$(-\infty, 7)$$

3.
$$7 - 4x \le -13$$
$$7 - 7 - 4x \le -13 - 7$$
$$-4x \le -20$$
$$\frac{-4x}{-4} \ge \frac{-20}{-4}$$
$$x \ge 5$$
$$[5, \infty)$$

4.
$$3x + 8 < x + 2$$
$$3x - x + 8 < x - x + 2$$
$$2x + 8 < 2$$
$$2x + 8 - 8 < 2 - 8$$
$$2x < -6$$
$$\frac{2x}{2} < \frac{-6}{2}$$
$$x < -3$$
$$(-\infty, -3)$$

5.
$$(11x - 3) - (2x - 13) \ge 4(2x + 1)$$
$$11x - 3 - 2x + 13 \ge 8x + 4$$
$$9x + 10 \ge 8x + 4$$
$$9x - 8x + 10 \ge 8x - 8x + 4$$
$$x + 10 \ge 4$$
$$x + 10 - 10 \ge 4 - 10$$
$$x \ge -6$$
$$[-6, \infty)$$

6.
$$3x - 2 \leq 10 \quad \text{or} \quad 4x - 13 \geq 15$$
$$3x - 2 + 2 \leq 10 + 2 \quad \text{or} \quad 4x - 13 + 13 \geq 15 + 13$$
$$3x \leq 12 \quad \text{or} \quad 4x \geq 28$$
$$\frac{3x}{3} \leq \frac{12}{3} \quad \text{or} \quad \frac{4x}{4} \geq \frac{28}{4}$$
$$x \leq 4 \quad \text{or} \quad x \geq 7$$
$$(-\infty, 4] \cup [7, \infty)$$

7.
$$-11 < 4x + 1 < 7$$
$$-11 - 1 < 4x + 1 - 1 < 7 - 1$$
$$-12 < 4x < 6$$
$$\frac{-12}{4} < \frac{4x}{4} < \frac{6}{4}$$
$$-3 < x < \frac{3}{2}$$
$$\left(-3, \frac{3}{2}\right)$$

8. $x < 130$

9.
$$\frac{74 + 78 + 80 + x}{4} < 70$$
$$4 \cdot \frac{74 + 78 + 80 + x}{4} < 4 \cdot 70$$
$$74 + 78 + 80 + x < 280$$
$$232 + x < 280$$
$$232 - 232 + x < 280 - 232$$
$$x < 48$$
lower than 48

2.5   LINEAR INEQUALITIES

1. linear inequality

3. interval

5. compound

7. $x < 3$
$(-\infty, 3)$

9. $x \geq -1$
$[-1, \infty)$

11. $-2 < x < 8$
$(-2, 8)$

13. $x > \dfrac{9}{2}$
$\left(\dfrac{9}{2}, \infty\right)$

15. $x > 8$ or $x \leq 2$
$(-\infty, 2] \cup (8, \infty)$

17. $x > 5$

19. $-4 \leq x \leq 7$

21. $x < 2$ or $x > 9$

23. $x < -4$

25.
$$x + 9 < 5$$
$$x + 9 - 9 < 5 - 9$$
$$x < -4$$
$$(-\infty, -4)$$

27.
$$2x > 12$$
$$\frac{2x}{2} > \frac{12}{2}$$
$$x > 6$$
$$(6, \infty)$$

29.
$$-5x > 15$$
$$\frac{-5x}{-5} < \frac{15}{-5}$$
$$x < -3$$
$$(-\infty, -3)$$

31. $3x \geq -10.5$

$$\frac{3x}{3} \geq \frac{-10.5}{3}$$

$$x \geq -3.5$$

$$[-3.5, \infty)$$

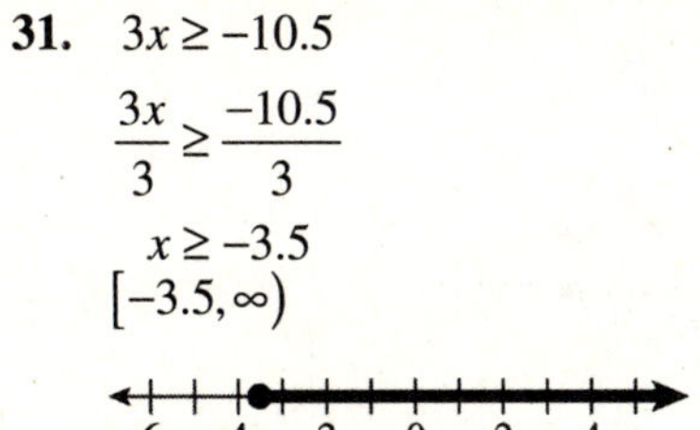

33. $3x + 2 < 14$
$3x + 2 - 2 < 14 - 2$
$3x < 12$

$$\frac{3x}{3} < \frac{12}{3}$$

$$x < 4$$

$$(-\infty, 4)$$

35. $-2x + 9 \leq 29$
$-2x + 9 - 9 \leq 29 - 9$
$-2x \leq 20$

$$\frac{-2x}{-2} \geq \frac{20}{-2}$$

$$x \geq -10$$

$$[-10, \infty)$$

37. $8 < 3x - 13$
$8 + 13 < 3x - 13 + 13$
$21 < 3x$

$$\frac{21}{3} < \frac{3x}{3}$$

$$7 < x$$
$$x > 7$$
$$(7, \infty)$$

39. $\dfrac{2}{3}x > \dfrac{10}{9}$

$$\frac{3}{2} \cdot \frac{2}{3}x > \frac{3}{2} \cdot \frac{10}{9}$$

$$x > \frac{5}{3}$$

$$\left(\frac{5}{3}, \infty\right)$$

41. $5x + 13 > 2x - 11$
$5x - 2x + 13 > 2x - 2x - 11$
$3x + 13 > -11$
$3x + 13 - 13 > -11 - 13$
$3x > -24$

$$\frac{3x}{3} > \frac{-24}{3}$$

$$x > -8$$

$$(-8, \infty)$$

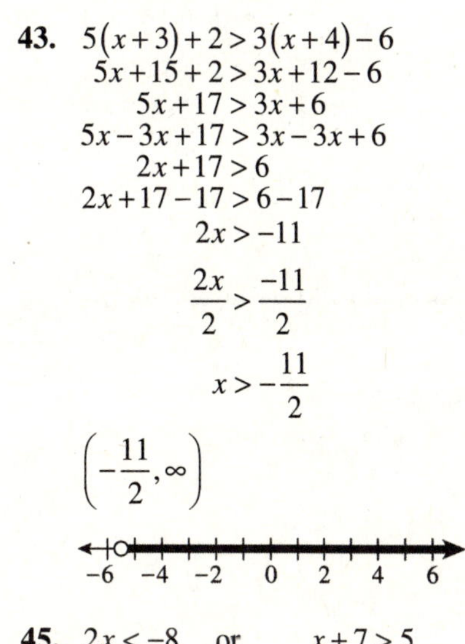

43. $5(x + 3) + 2 > 3(x + 4) - 6$
$5x + 15 + 2 > 3x + 12 - 6$
$5x + 17 > 3x + 6$
$5x - 3x + 17 > 3x - 3x + 6$
$2x + 17 > 6$
$2x + 17 - 17 > 6 - 17$
$2x > -11$

$$\frac{2x}{2} > \frac{-11}{2}$$

$$x > -\frac{11}{2}$$

$$\left(-\frac{11}{2}, \infty\right)$$

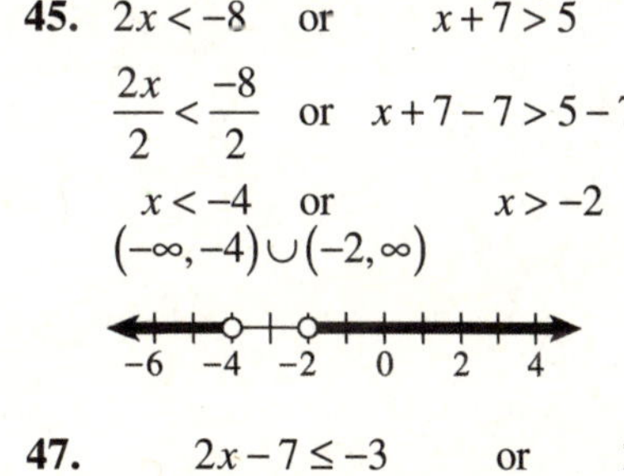

45. $2x < -8 \quad$ or $\quad x + 7 > 5$

$$\frac{2x}{2} < \frac{-8}{2} \quad \text{or} \quad x + 7 - 7 > 5 - 7$$

$$x < -4 \quad \text{or} \quad x > -2$$
$$(-\infty, -4) \cup (-2, \infty)$$

47. $2x - 7 \leq -3 \qquad$ or $\qquad 2x - 7 \geq 3$

$2x - 7 + 7 \leq -3 + 7 \quad$ or $\quad 2x - 7 + 7 \geq 3 + 7$

$$2x \leq 4 \qquad \text{or} \qquad 2x \geq 10$$

$$\frac{2x}{2} \leq \frac{4}{2} \qquad \text{or} \qquad \frac{2x}{2} \geq \frac{10}{2}$$

$$x \leq 2 \qquad \text{or} \qquad x \geq 5$$

$$(-\infty, 2] \cup [5, \infty)$$

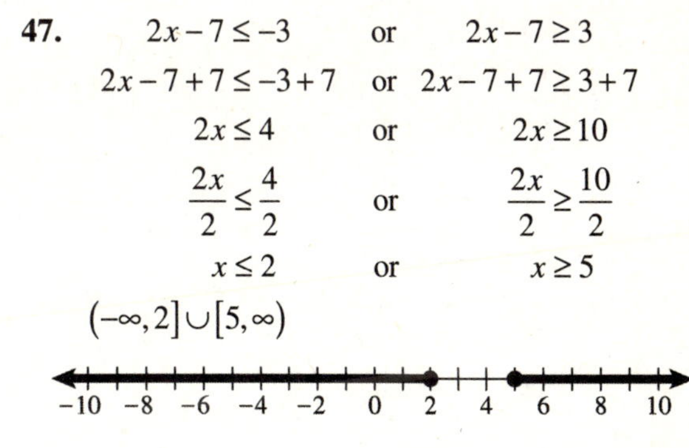

49.
$$-2x+17 < 5 \quad \text{or} \quad 10-x > 7$$
$$-2x+17-17 < 5-17 \quad \text{or} \quad 10-10-x > 7-10$$
$$-2x < -12 \quad \text{or} \quad -x > -3$$
$$\frac{-2x}{-2} > \frac{-12}{-2} \quad \text{or} \quad \frac{-x}{-1} < \frac{-3}{-1}$$
$$x > 6 \quad \text{or} \quad x < 3$$
$$(-\infty, 3) \cup (6, \infty)$$

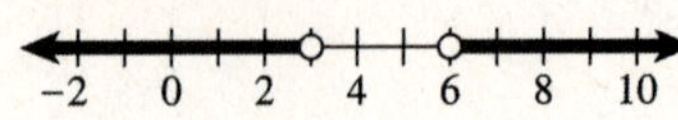

51.
$$3x+5 < 17 \quad \text{or} \quad 2x-9 < 5$$
$$3x+5-5 < 17-5 \quad \text{or} \quad 2x-9+9 < 5+9$$
$$3x < 12 \quad \text{or} \quad 2x < 14$$
$$\frac{3x}{3} < \frac{12}{3} \quad \text{or} \quad \frac{2x}{2} < \frac{14}{2}$$
$$x < 4 \quad \text{or} \quad x < 7$$
$$x < 7; \ (-\infty, 7)$$

53. First inequality:
$$\frac{1}{4}x + \frac{7}{24} \le \frac{2}{3}$$
$$24\left(\frac{1}{4}x + \frac{7}{24}\right) \le 24\left(\frac{2}{3}\right)$$
$$6x + 7 \le 16$$
$$6x + 7 - 7 \le 16 - 7$$
$$6x \le 9$$
$$\frac{6x}{6} \le \frac{9}{6}$$
$$x \le \frac{3}{2}$$

Second inequality:
$$\frac{1}{4}x + \frac{9}{20} \ge \frac{6}{5}$$
$$20\left(\frac{1}{4}x + \frac{9}{20}\right) \ge 20\left(\frac{6}{5}\right)$$
$$5x + 9 \ge 24$$
$$5x + 9 - 9 \ge 24 - 9$$
$$5x \ge 15$$
$$\frac{5x}{5} \ge \frac{15}{5}$$
$$x \ge 3$$
$$\left(-\infty, \frac{3}{2}\right] \cup [3, \infty)$$

55.
$$-4 < x - 3 < 1$$
$$-4 + 3 < x - 3 + 3 < 1 + 3$$
$$-1 < x < 4$$
$$(-1, 4)$$

57.
$$-2 < 5x - 7 < 28$$
$$-2 + 7 < 5x - 7 + 7 < 28 + 7$$
$$5 < 5x < 35$$
$$\frac{5}{5} < \frac{5x}{5} < \frac{35}{5}$$
$$1 < x < 7$$
$$(1, 7)$$

59.
$$6 < 3x + 15 \le 33$$
$$6 - 15 < 3x + 15 - 15 \le 33 - 15$$
$$-9 < 3x \le 18$$
$$\frac{-9}{3} < \frac{3x}{3} \le \frac{18}{3}$$
$$-3 < x \le 6$$
$$(-3, 6]$$

61.
$$\frac{1}{5} \le \frac{1}{2}x - \frac{1}{3} \le \frac{7}{4}$$
$$60 \cdot \left(\frac{1}{5}\right) \le 60 \cdot \left(\frac{1}{2}x - \frac{1}{3}\right) \le 60 \cdot \left(\frac{7}{4}\right)$$
$$12 \le 30x - 20 \le 105$$
$$12 + 20 \le 30x - 20 + 20 \le 105 + 20$$
$$32 \le 30x \le 125$$
$$\frac{32}{30} \le \frac{30x}{30} \le \frac{125}{30}$$
$$\frac{16}{15} \le x \le \frac{25}{6}$$
$$\left[\frac{16}{15}, \frac{25}{6}\right]$$

63. Answers will vary. Example:
Since $8 < 9x + 7$ and $9x + 7 < 3$ this implies that $8 < 3$, which is impossible. Therefore, the inequality $8 < 9x + 7 < 3$ has no solutions.

65. $x \ge 2$

67. $x < 30$

69. $0.10x \geq 30$

$$\frac{0.10x}{0.10} \geq \frac{30}{0.10}$$
$$x \geq 300$$
At least 300

71.
$$19.95 + 0.07x \leq 85$$
$$100(19.95 + 0.07x) \leq 100(85)$$
$$1995 + 7x \leq 8500$$
$$1995 - 1995 + 7x \leq 8500 - 1995$$
$$7x \leq 6505$$
$$\frac{7x}{7} \leq \frac{6505}{7}$$
$$x \leq 929.29$$
At most 929

73. $x \geq 0.15 \cdot 37{,}500$
$x \geq 5625$
At least \$5625

75.
$$\frac{92 + 93 + 85 + 96 + x}{5} \geq 80$$
$$5 \cdot \frac{92 + 93 + 85 + 96 + x}{5} \geq 5 \cdot 80$$
$$92 + 93 + 85 + 96 + x \geq 400$$
$$366 + x \geq 400$$
$$366 - 366 + x \geq 400$$
$$x \geq 34$$
Jacqui must score at least 34.

77. Explanations will vary. Example:
$2 < 4$ is a true statement. If we multiply both sides by -2 and want to keep the validity of the inequality, we must change the direction of the inequality, i.e. $-4 > -8$.

CHAPTER 2 REVIEW

1. $4(7) + 9 = 19$
$28 + 9 = 19$
$37 = 19$
No.

2.
$$(4(-9) - 15) - (2(-9) + 7) = -2(11 - (-9))$$
$$(-36 - 15) - (-18 + 7) = -2(11 + 9)$$
$$-51 - (-11) = -2(20)$$
$$-40 = -40$$
Yes.

3. $2x = 32$
$$\frac{2x}{2} = \frac{32}{2}$$
$$x = 16$$
$\{16\}$

4. $-9x = 54$
$$\frac{-9x}{-9} = \frac{54}{-9}$$
$$x = -6$$
$\{-6\}$

5. $x - 3 = -9$
$x - 3 + 3 = -9 + 3$
$x = -6$
$\{-6\}$

6. $a + 14 = 8$
$a + 14 - 14 = 8 - 14$
$a = -6$
$\{-6\}$

7. Let x be the number of people who attended.
$$8x - 250 = 694$$
$$8x - 250 + 250 = 694 + 250$$
$$8x = 944$$
$$\frac{8x}{8} = \frac{944}{8}$$
$$x = 118$$
118 people came to see "Battle of the Bands."

8. Let x be Randy's cholesterol two months ago.
$$x - 23 = 189$$
$$x - 23 + 23 = 189 + 23$$
$$x = 212$$
Randy's cholesterol was 212.

9. $5x - 8 = 27$
$5x - 8 + 8 = 27 + 8$
$$5x = 35$$
$$\frac{5x}{5} = \frac{35}{5}$$
$$x = 7$$
$\{7\}$

10. $-11 = 3a + 16$
$-11 - 16 = 3a + 16 - 16$
$-27 = 3a$
$$\frac{-27}{3} = \frac{3a}{3}$$
$$-9 = a$$
$\{-9\}$

11. $-4x + 15 = 7$
$-4x + 15 - 15 = 7 - 15$
$-4x = -8$
$$\frac{-4x}{-4} = \frac{-8}{-4}$$
$$x = 2$$
$\{2\}$

12.
$$51 - 2x = -21$$
$$51 - 51 - 2x = -21 - 51$$
$$-2x = -72$$
$$\frac{-2x}{-2} = \frac{-72}{-2}$$
$$x = 36$$
$$\{36\}$$

13.
$$\frac{x}{6} - \frac{5}{12} = \frac{5}{4} \quad \text{The LCM is 12.}$$
$$12 \cdot \frac{x}{6} - 12 \cdot \frac{5}{12} = 12 \cdot \frac{5}{4}$$
$$2x - 5 = 15$$
$$2x - 5 + 5 = 15 + 5$$
$$2x = 20$$
$$\frac{2x}{2} = \frac{20}{2}$$
$$x = 10$$
$$\{10\}$$

14.
$$\frac{1}{5}x - \frac{1}{2} = \frac{5}{4} \quad \text{The LCM is 20.}$$
$$20 \cdot \frac{1}{5}x - 20 \cdot \frac{1}{2} = 20 \cdot \frac{5}{4}$$
$$4x - 10 = 25$$
$$4x - 10 + 10 = 25 + 10$$
$$4x = 35$$
$$\frac{4x}{4} = \frac{35}{4}$$
$$x = \frac{35}{4}$$
$$\left\{ \frac{35}{4} \right\}$$

15.
$$4x - 19 = 5x + 14$$
$$4x - 4x - 19 = 5x - 4x + 14$$
$$-19 = x + 14$$
$$-19 - 14 = x + 14 - 14$$
$$-33 = x$$
$$\{-33\}$$

16.
$$2m + 23 = 17 - 8m$$
$$2m + 8m + 23 = 17 - 8m + 8m$$
$$10m + 23 = 17$$
$$10m + 23 - 23 = 17 - 23$$
$$10m = -6$$
$$\frac{10m}{10} = \frac{-6}{10}$$
$$m = -\frac{3}{5}$$
$$\left\{ -\frac{3}{5} \right\}$$

17.
$$2n - 11 + 3n + 4 = 7n - 39$$
$$5n - 7 = 7n - 39$$
$$5n - 5n - 7 = 7n - 5n - 39$$
$$-7 = 2n - 39$$
$$-7 + 39 = 2n - 39 + 39$$
$$32 = 2n$$
$$\frac{32}{2} = \frac{2n}{2}$$
$$16 = n$$
$$\{16\}$$

18.
$$3(2x - 7) - (x - 9) = 2x - 33$$
$$6x - 21 - x + 9 = 2x - 33$$
$$5x - 12 = 2x - 33$$
$$5x - 2x - 12 = 2x - 2x - 33$$
$$3x - 12 = -33$$
$$3x - 12 + 12 = -33 + 12$$
$$3x = -21$$
$$\frac{3x}{3} = \frac{-21}{3}$$
$$x = -7$$
$$\{-7\}$$

19.
$$7x + y = 8$$
$$7x - 7x + y = 8 - 7x$$
$$y = 8 - 7x$$

20.
$$15x + 7y = 30$$
$$15x - 15x + 7y = 30 - 15x$$
$$7y = 30 - 15x$$
$$\frac{7y}{7} = \frac{30 - 15x}{7}$$
$$y = \frac{30 - 15x}{7}$$

21.
$$C = \pi \cdot d$$
$$\frac{C}{\pi} = \frac{\pi d}{\pi}$$
$$\frac{C}{\pi} = d$$
$$d = \frac{C}{\pi}$$

22.
$$P = 3a + b + 2c$$
$$P - b = 3a + b - b + 2c$$
$$P - b = 3a + 2c$$
$$P - b - 2c = 3a + 2c - 2c$$
$$P - b - 2c = 3a$$
$$\frac{P - b - 2c}{3} = \frac{3a}{3}$$
$$\frac{P - b - 2c}{3} = a$$
$$a = \frac{P - b - 2c}{3}$$

23. #1: x
#2: $26 + x$
$$x + (26 + x) = 98$$
$$2x + 26 = 98$$
$$2x + 26 - 26 = 98 - 26$$
$$2x = 72$$
$$\frac{2x}{2} = \frac{72}{2}$$
$$x = 36$$
$$26 + x = 26 + 36 = 62$$

24. Length: l
Width: $w = l - 7$
$$P = 2l + 2w$$
$$50 = 2l + 2(l - 7)$$
$$50 = 2l + 2l - 14$$
$$50 = 4l - 14$$
$$50 + 14 = 4l - 14 + 14$$
$$64 = 4l$$
$$\frac{64}{4} = \frac{4l}{4}$$
$$16 = l$$
$$w = 16 - 7 = 9$$
The length is 16 feet and width is 9 feet.

25. #1: x
#2: $x + 2$
#3: $x + 4$
$$x + (x + 2) + (x + 4) = 204$$
$$3x + 6 = 204$$
$$3x + 6 - 6 = 204 - 6$$
$$3x = 198$$
$$\frac{3x}{3} = \frac{198}{3}$$
$$x = 66$$
$$x + 2 = 68$$
$$x + 4 = 70$$
$$66, 68, 70$$

26.
$$d = r \cdot t$$
$$342 = 72 \cdot t$$
$$\frac{342}{72} = \frac{72t}{72}$$
$$4\frac{3}{4} = t$$

It takes Jason $4\frac{3}{4}$ hours.

27.

	Number	Value	Money
Dimes	$3q - 4$	0.10	$0.10(3q - 4)$
Quarters	q	0.25	$0.25q$

$$0.10(3q - 4) + 0.25q = 7.85$$
$$0.3q - 0.4 + 0.25q = 7.85$$
$$0.55q - 0.4 = 7.85$$
$$0.55q - 0.4 + 0.4 = 7.85 + 0.4$$
$$0.55q = 8.25$$
$$\frac{0.55q}{0.55} = \frac{8.25}{0.55}$$
$$q = 15$$
$$3q - 4 = 3(15) - 4 = 41$$
Alycia has 41 dimes.

28. n is 20% of 95.
$$n = 0.20 \cdot 95$$
$$n = 19$$
20% of 95 is 19.

29. 51 is $p\%$ of 136.
$$51 = p \cdot 136$$
$$\frac{51}{136} = \frac{136p}{136}$$
$$0.375 = p$$
$$37.5\% = p$$
37.5% of 136 is 51.

30. 1360 is 32% of n.
$$1360 = 0.32 \cdot n$$
$$\frac{1360}{0.32} = \frac{0.32n}{0.32}$$
$$4250 = n$$
32% of 4250 is 1360.

31. Let n be the number of female students.
n is 55% of 10,520.
$$n = 0.55 \cdot 10,520$$
$$n = 5786$$
There are 5786 female students.

32. Let p be the percentage of students to pursue a Masters' degree.

204 is p% of 320.
$$204 = p \cdot 320$$
$$\frac{204}{320} = \frac{320p}{320}$$
$$0.6375 = p$$
$$63.75\% = p$$

$63\frac{3}{4}\%$ decided to pursue a Master's degree.

33.
$$I = P \cdot r \cdot t$$
$$553.50 = P \cdot 0.045 \cdot 1$$
$$553.50 = 0.045P$$
$$\frac{553.50}{0.045} = \frac{0.045P}{0.045}$$
$$12,300 = P$$
$$\$12,300$$

34. Let n represent the amount of the discount.

n is 20% of 59.95.
$$n = 0.20 \cdot 59.95$$
$$n = 11.99$$
$$59.95 - 11.99 = 47.96$$
The price after the discount is \$47.96.

35.

	P	r	t	I
6%	x	0.06	1	$0.06x$
5%	$2000 - x$	0.05	1	$0.05(2000 - x)$

$$0.06x + 0.05(2000 - x) = 117$$
$$0.06x + 100 - 0.05x = 117$$
$$0.01x + 100 = 117$$
$$0.01x + 100 - 100 = 117 - 100$$
$$0.01x = 17$$
$$\frac{0.01x}{0.01} = \frac{17}{0.01}$$
$$x = 1700$$

$2000 - x = 2000 - 1700 = 300$
Karl invested \$1700 at 6% and \$300 at 5%.

36.

	Solution	Percent	Amount
Sol. 1 (17%)	x	0.17	$0.17x$
Sol. 2 (42%)	$80 - x$	0.42	$0.42(80 - x)$
Mixture	80	0.32	$0.32(80) = 25.6$

$$0.17x + 0.42(80 - x) = 25.6$$
$$0.17x + 33.6 - 0.42x = 25.6$$
$$-0.25x + 33.6 = 25.6$$
$$-0.25x + 33.6 - 33.6 = 25.6 - 33.6$$
$$-0.25x = -8.0$$
$$\frac{-0.25x}{-0.25} = \frac{-8.0}{-0.25}$$
$$x = 32$$

$80 - x = 80 - 32 = 48$
32 milliliters of the 17% alcohol should be mixed with 48 milliliters of the 42% alcohol.

37.
$$\frac{n}{6} = \frac{18}{27}$$
$$n \cdot 27 = 6 \cdot 18$$
$$27n = 108$$
$$\frac{27n}{27} = \frac{108}{27}$$
$$n = 4$$
$$\{4\}$$

38.
$$\frac{4}{9} = \frac{n}{72}$$
$$4 \cdot 72 = 9 \cdot n$$
$$288 = 9n$$
$$\frac{288}{9} = \frac{9n}{9}$$
$$32 = n$$
$$\{32\}$$

39.
$$\frac{32}{13} = \frac{192}{n}$$
$$32 \cdot n = 13 \cdot 192$$
$$32n = 2496$$
$$\frac{32n}{32} = \frac{2496}{32}$$
$$n = 78$$
$$\{78\}$$

40.
$$\frac{38}{n} = \frac{57}{75}$$
$$38 \cdot 75 = n \cdot 57$$
$$2850 = 57n$$
$$\frac{2850}{57} = \frac{57n}{57}$$
$$50 = n$$
$$\{50\}$$

41. Let n be the number of students who plan to transfer.
$$\frac{3}{7} = \frac{n}{14,952}$$
$$3 \cdot 14,952 = 7 \cdot n$$
$$44,856 = 7n$$
$$\frac{44,856}{7} = \frac{7n}{7}$$
$$6408 = n$$
6408 students plan to transfer.

42. Let n be the number of tablespoons of weed killer needed.
$$\frac{6}{1} = \frac{n}{2\frac{1}{2}}$$
$$6 \cdot 2\frac{1}{2} = 1 \cdot n$$
$$\frac{6}{1} \cdot \frac{5}{2} = n$$
$$\frac{\cancel{6}^{3}}{1} \cdot \frac{5}{\cancel{2}_{1}} = n$$
$$15 = n$$
You need 15 tablespoons of weed killer.

43.
$$x + 5 < 2$$
$$x + 5 - 5 < 2 - 5$$
$$x < -3$$
$$(-\infty, -3)$$

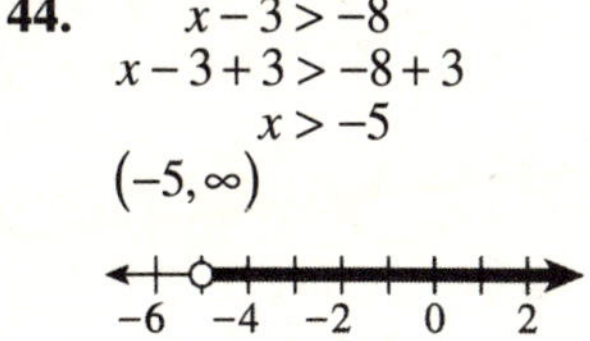

44.
$$x - 3 > -8$$
$$x - 3 + 3 > -8 + 3$$
$$x > -5$$
$$(-5, \infty)$$

45.
$$2x - 7 \geq 3$$
$$2x - 7 + 7 \geq 3 + 7$$
$$2x \geq 10$$
$$\frac{2x}{2} \geq \frac{10}{2}$$
$$x \geq 5$$
$$[5, \infty)$$

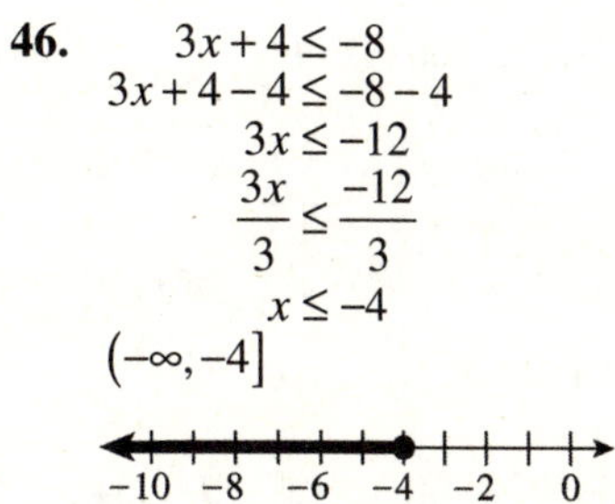

46.
$$3x + 4 \leq -8$$
$$3x + 4 - 4 \leq -8 - 4$$
$$3x \leq -12$$
$$\frac{3x}{3} \leq \frac{-12}{3}$$
$$x \leq -4$$
$$(-\infty, -4]$$

47.
$$3x < -18 \quad \text{or} \quad x + 3 > 10$$
$$\frac{3x}{3} < \frac{-18}{3} \quad \text{or} \quad x + 3 - 3 > 10 - 3$$
$$x < -6 \quad \text{or} \quad x > 7$$
$$(-\infty, -6) \cup (7, \infty)$$

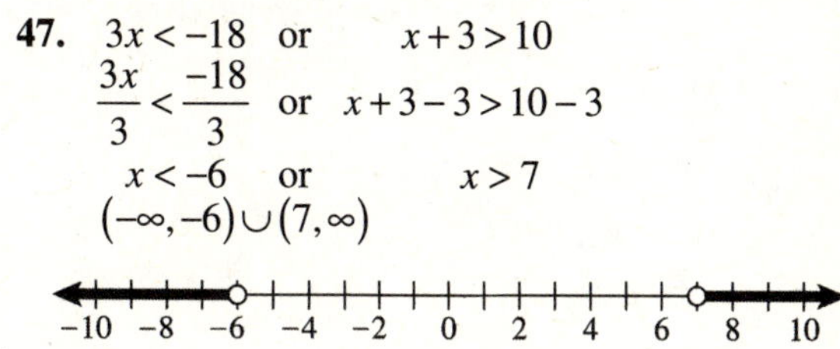

48.
$$x + 5 \leq -3 \quad \text{or} \quad 4x \geq -12$$
$$x + 5 - 5 \leq -3 - 5 \quad \text{or} \quad \frac{4x}{4} \geq \frac{-12}{4}$$
$$x \leq -8 \quad \text{or} \quad x \geq -3$$
$$(-\infty, -8] \cup [-3, \infty)$$

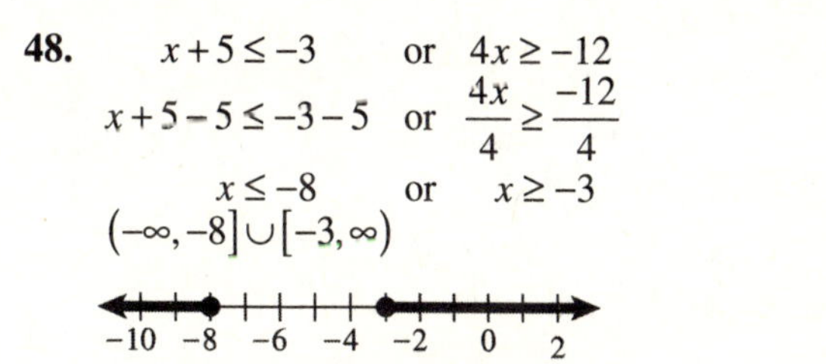

49.
$$-21 \leq 3x - 6 \leq 24$$
$$-21 + 6 \leq 3x - 6 + 6 \leq 24 + 6$$
$$-15 \leq 3x \leq 30$$
$$\frac{-15}{3} \leq \frac{3x}{3} \leq \frac{30}{3}$$
$$-5 \leq x \leq 10$$
$$[-5, 10]$$

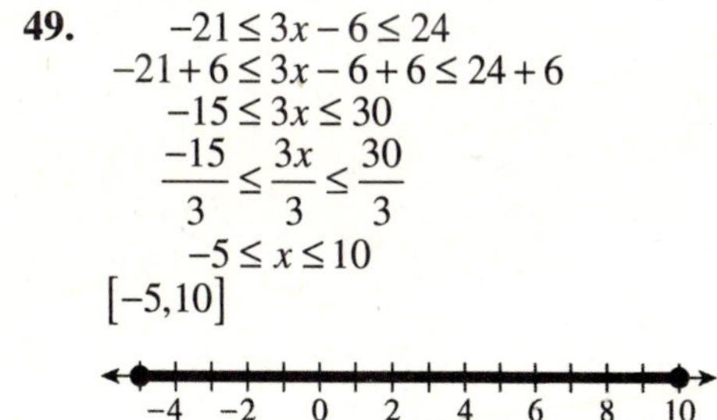

50.
$$-11 < 2x - 9 < 11$$
$$-11 + 9 < 2x - 9 + 9 < 11 + 9$$
$$-2 < 2x < 20$$
$$\frac{-2}{2} < \frac{2x}{2} < \frac{20}{2}$$
$$-1 < x < 10$$
$$(-1, 10)$$

51. $x \geq 7$

52. Let x be price of the stock.
$$2,000,000x \geq 69,500,000$$
$$\frac{2,000,000x}{2,000,000} \geq \frac{69,500,000}{2,000,000}$$
$$x \geq 34.75$$
The company needs to receive at least $34.75 per share.

53.
$$70 \leq \frac{58 + 72 + 66 + 75 + x}{5} < 80$$
$$5 \cdot 70 \leq 5 \cdot \frac{58 + 72 + 66 + 75 + x}{5} < 5 \cdot 80$$
$$350 \leq 58 + 72 + 66 + 75 + x < 400$$
$$350 \leq 271 + x < 400$$
$$350 - 271 \leq 271 - 271 + x < 400 - 271$$
$$79 \leq x < 129$$
Since the highest possible score is 100, a score from 79 to 100 will give him a C for the class.

CHAPTER 2 TEST

1.
$$-8x = 56$$
$$\frac{-8x}{-8} = \frac{56}{-8}$$
$$x = -7$$
$$\{-7\}$$

2.
$$x + 17 = -22$$
$$x + 17 - 17 = -22 - 17$$
$$x = -39$$
$$\{-39\}$$

3.
$$7x - 20 = 36$$
$$7x - 20 + 20 = 36 + 20$$
$$7x = 56$$
$$\frac{7x}{7} = \frac{56}{7}$$
$$x = 8$$
$$\{8\}$$

4.
$$72 = 30 - 8a$$
$$72 - 30 = 30 - 30 - 8a$$
$$42 = -8a$$
$$\frac{42}{-8} = \frac{-8a}{-8}$$
$$-\frac{21}{4} = a$$
$$\left\{-\frac{21}{4}\right\}$$

5.
$$3x + 34 = 5x - 21$$
$$3x - 3x + 34 = 5x - 3x - 21$$
$$34 = 2x - 21$$
$$34 + 21 = 2x - 21 + 21$$
$$55 = 2x$$
$$\frac{55}{2} = \frac{2x}{2}$$
$$\frac{55}{2} = x$$
$$\left\{\frac{55}{2}\right\}$$

6.
$$3(5x - 16) - 9x = x - 63$$
$$15x - 48 - 9x = x - 63$$
$$6x - 48 = x - 63$$
$$6x - x - 48 = x - x - 63$$
$$5x - 48 = -63$$
$$5x - 48 + 48 = -63 + 48$$
$$5x = -15$$
$$\frac{5x}{5} = \frac{-15}{5}$$
$$x = -3$$
$$\{-3\}$$

7. $\dfrac{1}{8}x - \dfrac{2}{3} = \dfrac{5}{6}$ The LCM is 24.
$$24 \cdot \frac{1}{8}x - 24 \cdot \frac{2}{3} = 24 \cdot \frac{5}{6}$$
$$3x - 16 = 20$$
$$3x - 16 + 16 = 20 + 16$$
$$3x = 36$$
$$\frac{3x}{3} = \frac{36}{3}$$
$$x = 12$$
$$\{12\}$$

8.
$$\frac{n}{100} = \frac{22}{40}$$
$$n \cdot 40 = 100 \cdot 22$$
$$40n = 2200$$
$$\frac{40n}{40} = \frac{2200}{40}$$
$$n = 55$$
$$\{55\}$$

9.
$$4x + 3y = 28$$
$$4x - 4x + 3y = 28 - 4x$$
$$3y = 28 - 4x$$
$$\frac{3y}{3} = \frac{28 - 4x}{3}$$
$$y = \frac{28 - 4x}{3}$$

10. n is 16% of 175.
$n = 0.16 \cdot 175$
$n = 28$
28 is 16% of 175.

11. 99 is p% of 660.
$99 = p \cdot 660$

$$\frac{99}{660} = \frac{660\,p}{660}$$

$0.15 = p$
$15\% = p$
15% of 660 is 99.

12.
$$6x + 8 \le -7$$
$$6x + 8 - 8 \le -7 - 8$$
$$6x \le -15$$
$$\frac{6x}{6} \le \frac{-15}{6}$$
$$x \le -\frac{5}{2}$$

$$\left(-\infty, -\frac{5}{2}\right]$$

13. First inequality:
$$x + 4 \le -5$$
$$x + 4 - 4 \le -5 - 4$$
$$x \le -9$$
Second inequality:
$$2x + 3 \ge 0$$
$$2x + 3 - 3 \ge 0 - 3$$
$$2x \ge -3$$
$$\frac{2x}{2} \ge \frac{-3}{2}$$
$$x \ge -\frac{3}{2}$$

$$(-\infty, -9] \cup \left[-\frac{3}{2}, \infty\right)$$

14.
$$-23 < 3x + 7 < 23$$
$$-23 - 7 < 3x + 7 - 7 < 23 - 7$$
$$-30 < 3x < 16$$
$$\frac{-30}{3} < \frac{3x}{3} < \frac{16}{3}$$
$$-10 < x < \frac{16}{3}$$

$$\left(-10, \frac{16}{3}\right)$$

15. Length: $l = 2w - 3$
Width: w
$$P = 2l + 2w$$
$$72 = 2(2w - 3) + 2w$$
$$72 = 4w - 6 + 2w$$
$$72 = 6w - 6$$
$$72 + 6 = 6w - 6 + 6$$
$$78 = 6w$$
$$\frac{78}{6} = \frac{6w}{6}$$
$$13 = w$$
$$l = 2w - 3 = 2(13) - 3 = 23$$
The length is 23 feet and the width is 13 feet.

16.

	Number	Value	Amount
\$5	$x + 7$	5	$5(x + 7)$
\$10	x	10	$10x$

$$5(x + 7) + 10x = 185$$
$$5x + 35 + 10x = 185$$
$$15x + 35 = 185$$
$$15x + 35 - 35 = 185 - 35$$
$$15x = 150$$
$$\frac{15x}{15} = \frac{150}{15}$$
$$x = 10$$
$$x + 7 = 10 + 7 = 17$$
Barry has 17 \$5 bills in his wallet.

17.

	P	r	t	$I = Prt$
11%	x	0.11	1	$0.11x$
4%	$10{,}000 - x$	0.04	1	$0.04(10{,}000 - x)$
Total	10,000			862

$$0.11x + 0.04(10{,}000 - x) = 862$$
$$0.11x + 400 - 0.04x = 862$$
$$0.07x + 400 = 862$$
$$0.07x + 400 - 400 = 862 - 400$$
$$0.07x = 462$$
$$\frac{0.07x}{0.07} = \frac{462}{0.07}$$
$$x = 6600$$
Ernesto invested \$6600 in the fund that made an 11% profit.

18.

	Solution	Percent	Amount
Sol. 1 (13%)	x	0.13	$0.13x$
Sol. 2 (21%)	$100 - x$	0.21	$0.21(100 - x)$
Mixture	100	0.18	$0.18(100) = 18$

$$0.13x + 0.21(100 - x) = 18$$
$$0.13x + 21 - 0.21x = 18$$
$$-0.08x + 21 = 18$$
$$-0.08x + 21 - 21 = 18 - 21$$
$$-0.08x = -3$$
$$\frac{-0.08x}{-0.08} = \frac{-3}{-0.08}$$
$$x = 37.5$$

$100 - x = 100 - 37.5 = 62.5$

37.5 milliliters of the 13% solution should be mixed with 62.5 milliliters of the 21% solution.

19. Let n represent the number of professors who earned their degrees outside of the U.S.

$$\frac{2}{9} = \frac{n}{396}$$
$$2 \cdot 396 = 9 \cdot n$$
$$792 = 9n$$
$$\frac{792}{9} = \frac{9n}{9}$$
$$88 = n$$

88 professors earned their degrees outside of the U.S.

20. Let n represent the number of cars Wayne sells.

$$500 \cdot n \geq 17,000$$
$$\frac{500n}{500} \geq \frac{17,000}{500}$$
$$n \geq 34$$

Wayne needs to sell at least 34 cars.

CHAPTER 3   GRAPHING LINEAR EQUATIONS

3.1   QUICK CHECK

1. $3(0) + 2(-6) = 12$
$$0 - 12 = 12$$
$$-12 = 12$$
No.

2.

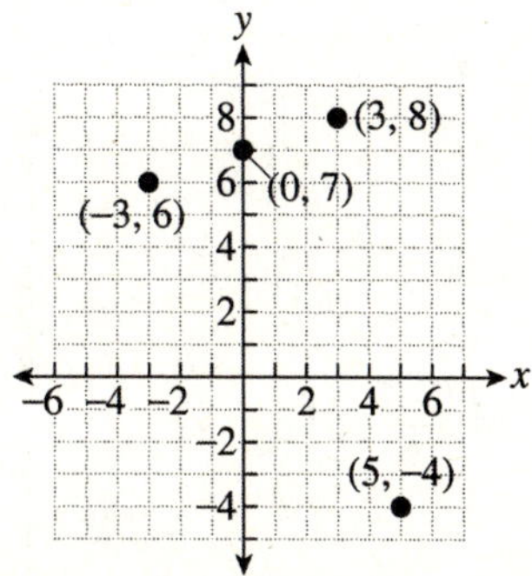

3. $A(-6, 2)$
$B(-2, 6)$
$C(0, -5)$
$D(-4, -2)$
$E(8, 1)$

4.

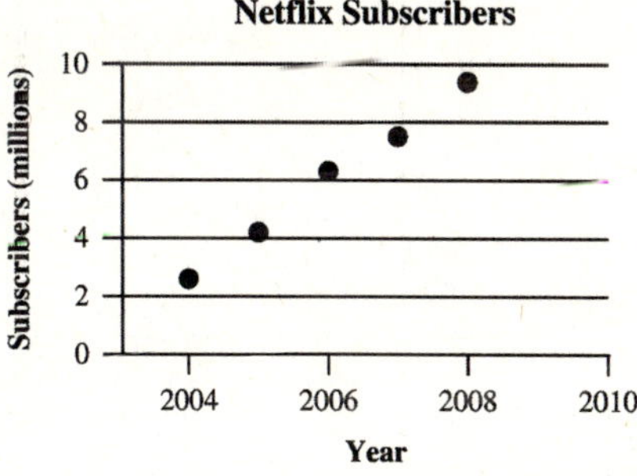

5. $y = -3(3) + 5$
$$y = -9 + 5$$
$$y = -4$$

6. $x - y = 6$

When $x = 4$,
$$4 - y = 6$$
$$4 - 4 - y = 6 - 4$$
$$-y = 2$$
$$-1 \cdot (-y) = -1 \cdot 2$$
$$y = -2$$

When $y = 3$,
$$x - 3 = 6$$
$$x - 3 + 3 = 6 + 3$$
$$x = 9$$

When $x = -1$,
$$-1 - y = 6$$
$$-1 + 1 - y = 6 + 1$$
$$-y = 7$$
$$-1 \cdot (-y) = -1 \cdot 7$$
$$y = -7$$

x	y
4	−2
9	3
−1	−7

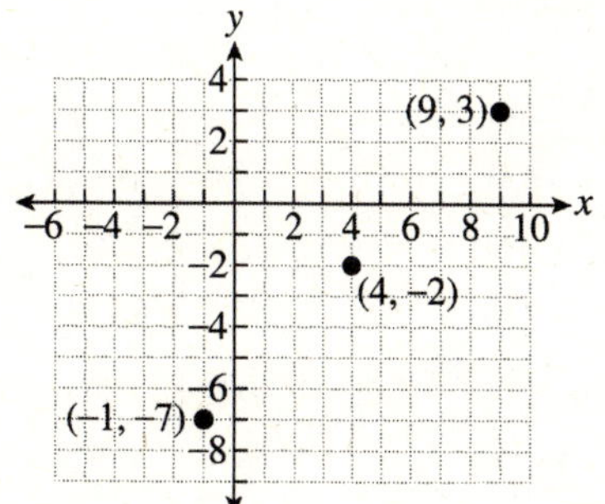

7. $y = 2x + 1$

$x = 0$	$x = 1$	$x = 2$
$y = 2(0) + 1$	$y = 2(1) + 1$	$y = 2(2) + 1$
$y = 0 + 1$	$y = 2 + 1$	$y = 4 + 1$
$y = 1$	$y = 3$	$y = 5$
$(0, 1)$	$(1, 3)$	$(2, 5)$

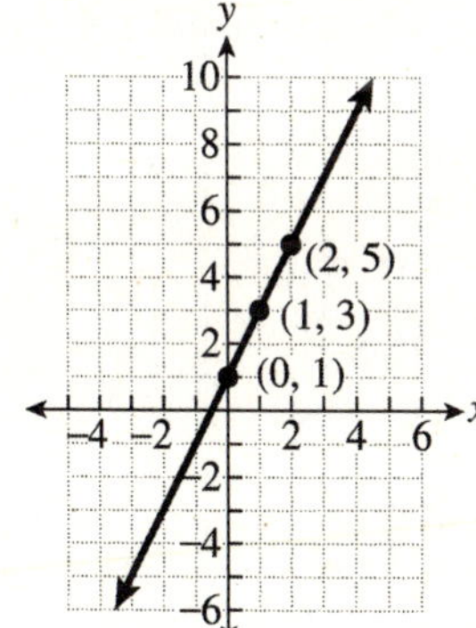

3.1   THE RECTANGULAR COORDINATE SYSTEM; EQUATIONS IN TWO VARIABLES

1. $Ax + By = C$

3. x-axis

5. origin

7. $5(7) + 3(6) = 53$
$\quad\quad 35 + 18 = 53$
$\quad\quad\quad\quad 53 = 53$
Yes.

9. $6(4) - 2(-3) = 18$
$\quad\quad\quad 24 + 6 = 18$
$\quad\quad\quad\quad\quad 30 = 18$
No.

11. $9 = \dfrac{3}{2}(10) - 6$
$9 = 15 - 6$
$9 = 9$
Yes.

13. $-2(3) + 5(2) = 4$
$\quad\quad -6 + 10 = 4$
$\quad\quad\quad\quad\quad 4 = 4$
Yes.

15. $-4(-3) + (0) = 12$
$\quad\quad\quad 12 + 0 = 12$
$\quad\quad\quad\quad\quad 12 = 12$
Yes.

17.

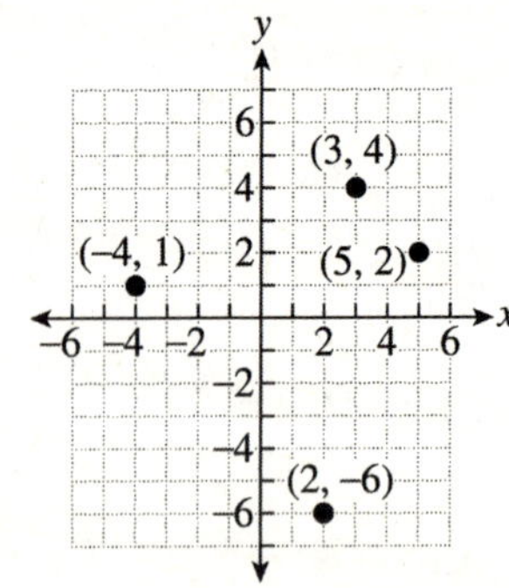

19.

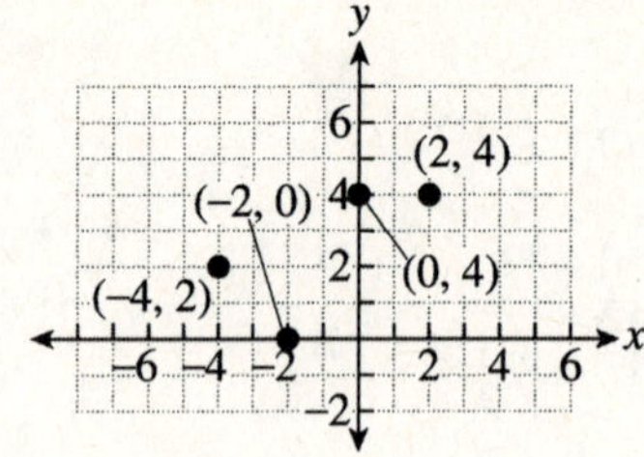

21.

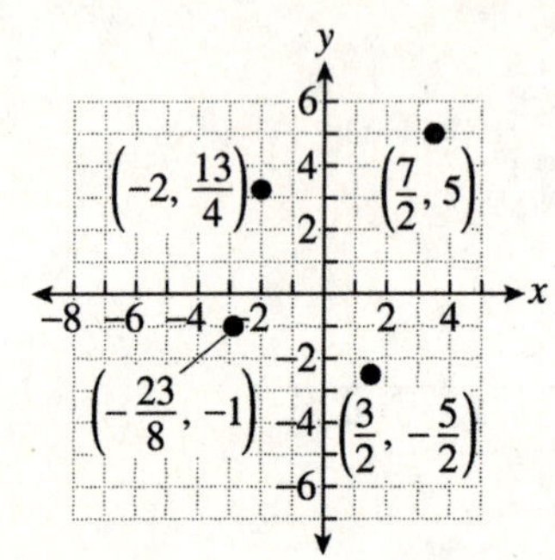

23. $A(-3,1)$, $B(2,-5)$, $C(-6,-4)$, $D(0,3)$

25. $A(30,-15)$, $B(-40,0)$, $C(-10,-30)$, $D(-15,15)$

27. IV

29. I

31. I

33. IV

35.

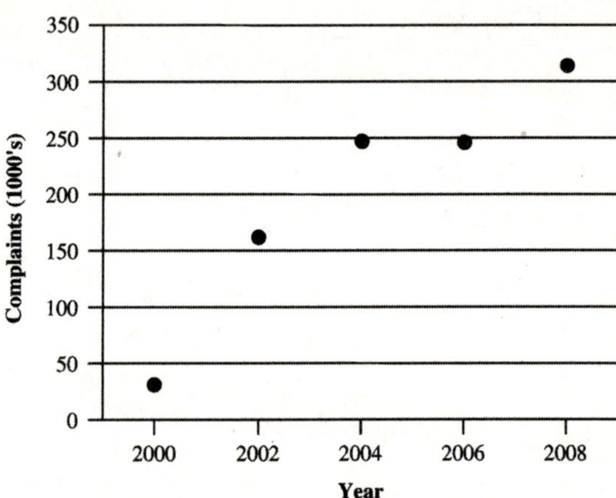

37.

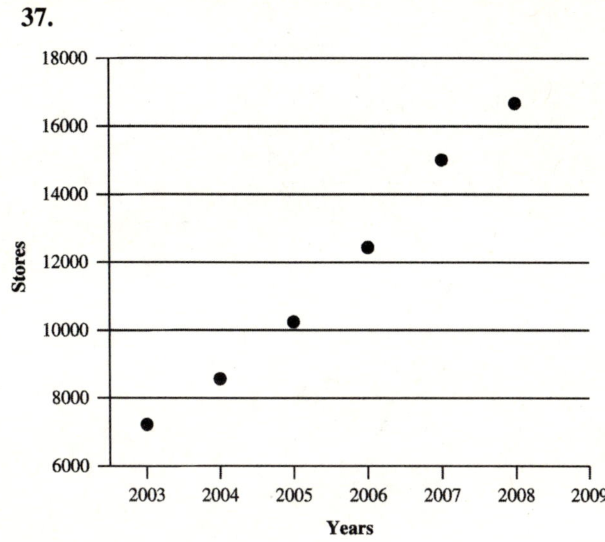

39.

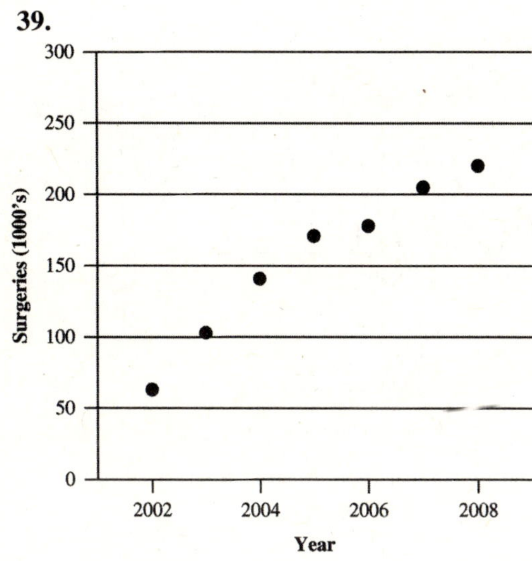

41.
$$3(2) - 2y = 8$$
$$6 - 2y = 8$$
$$6 - 6 - 2y = 8 - 6$$
$$-2y = 2$$
$$\frac{-2y}{-2} = \frac{2}{-2}$$
$$y = -1$$

43.
$$7x - 4(-6) = -11$$
$$7x + 24 = -11$$
$$7x + 24 - 24 = -11 - 24$$
$$7x = -35$$
$$\frac{7x}{7} = \frac{-35}{7}$$
$$x = -5$$

45.
$$y = 2(9) - 7$$
$$y = 18 - 7$$
$$y = 11$$

47.
$$7(0) - 11y = -22$$
$$0 - 11y = -22$$
$$\frac{-11y}{-11} = \frac{-22}{-11}$$
$$y = 2$$

49.
$$3x + \frac{1}{2}(8) = -2$$
$$3x + 4 = -2$$
$$3x + 4 - 4 = -2 - 4$$
$$3x = -6$$
$$\frac{3x}{3} = \frac{-6}{3}$$
$$x = -2$$

51.
$$6\left(\frac{2}{3}\right) + 2y = 9$$
$$4 + 2y = 9$$
$$4 - 4 + 2y = 9 - 4$$
$$2y = 5$$
$$\frac{2y}{2} = \frac{5}{2}$$
$$y = \frac{5}{2}$$

53.

When $x = -3$
$$5(-3) - 3y = -12$$
$$-15 - 3y = -12$$
$$-3y = 3$$
$$y = -1$$

When $x = 0$
$$5(0) - 3y = -12$$
$$-3y = -12$$
$$y = 4$$

When $x = 3$
$$5(3) - 3y = -12$$
$$15 - 3y = -12$$
$$-3y = -27$$
$$y = 9$$

x	y
-3	-1
0	4
3	9

55.

When $x = 0$
$$y = 4(0) - 9$$
$$y = -9$$

When $x = 1$
$$y = 4(1) - 9$$
$$y = -5$$

When $x = \frac{3}{2}$
$$y = 4\left(\frac{3}{2}\right) - 9$$
$$y = 6 - 9$$
$$y = -3$$

x	y
0	-9
1	-5
$\frac{3}{2}$	-3

57.

When $x = -3$

$y = \dfrac{2}{3}(-3) - 2$

$y = -2 - 2$

$y = -4$

When $x = 0$

$y = \dfrac{2}{3}(0) - 2$

$y = -2$

When $x = 6$

$y = \dfrac{2}{3}(6) - 2$

$y = 4 - 2$

$y = 2$

x	y
-3	-4
0	-2
6	2

59. Answers will vary. Example:
$x + y = 7,\ 2x - y = 8,\ y = 3x - 13$

61. Answers will vary. Example:
$5x + 4y = 24,\ x + y = 6,\ y = 3x + 6$

63. $y = 4x + 1$

$x = 0$	$x = 1$	$x = 2$
$y = 4(0) + 1$ $y = 1$	$y = 4(1) + 1$ $y = 5$	$y = 4(2) + 1$ $y = 9$
$(0, 1)$	$(1, 5)$	$(2, 9)$

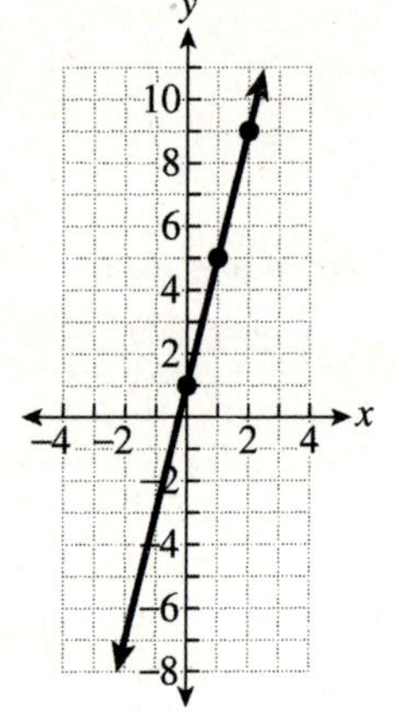

65. $y = -2x + 7$

$x = 0$	$x = 1$	$x = 2$
$y = -2(0) + 7$ $y = 7$	$y = -2(1) + 7$ $y = -2 + 7$ $y = 5$	$y = -2(2) + 7$ $y = -4 + 7$ $y = 3$
$(0, 7)$	$(1, 5)$	$(2, 3)$

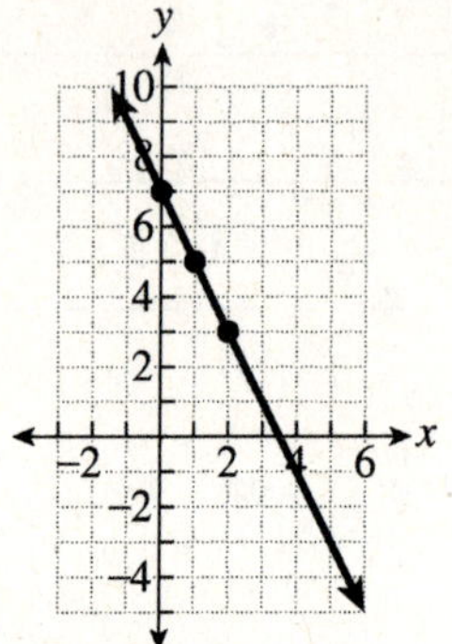

67. $y = -2x$

$x = 0$	$x = 1$	$x = 2$
$y = -2(0)$ $y = 0$	$y = -2(1)$ $y = -2$	$y = -2(2)$ $y = -4$
$(0, 0)$	$(1, -2)$	$(2, -4)$

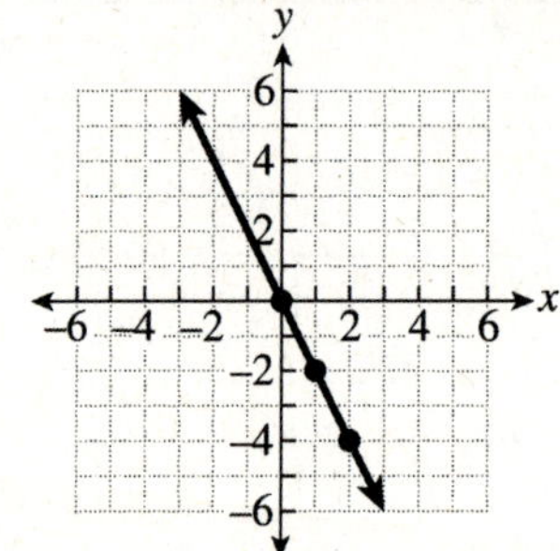

69. $y = \dfrac{1}{4}x + 3$

$x = 0$	$x = 1$	$x = 2$
$y = \dfrac{1}{4}(0) + 3$ $y = 3$	$y = \dfrac{1}{4}(1) + 3$ $y = \dfrac{1}{4} + \dfrac{3}{1}$ $y = \dfrac{1}{4} + \dfrac{12}{4}$ $y = \dfrac{13}{4}$	$y = \dfrac{1}{4}(2) + 3$ $y = \dfrac{1}{2} + \dfrac{3}{1}$ $y = \dfrac{1}{2} + \dfrac{6}{2}$ $y = \dfrac{7}{2}$
$(0,3)$	$\left(1, \dfrac{13}{4}\right)$	$\left(2, \dfrac{7}{2}\right)$

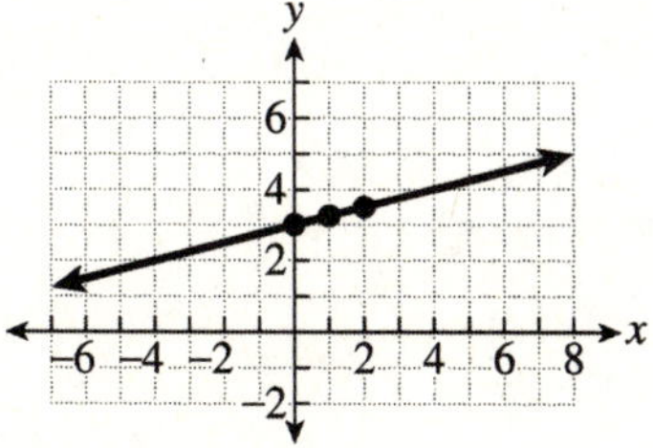

71. $y = -\dfrac{3}{5}x + 6$

$x = 0$	$x = 1$	$x = 2$
$y = -\dfrac{3}{5}(0) + 6$ $y = 6$	$y = -\dfrac{3}{5}(1) + 6$ $y = -\dfrac{3}{5} + \dfrac{30}{5}$ $y = \dfrac{27}{5}$	$y = -\dfrac{3}{5}(2) + 6$ $y = -\dfrac{6}{5} + \dfrac{30}{5}$ $y = \dfrac{24}{5}$
$(0,6)$	$\left(1, \dfrac{27}{5}\right)$	$\left(2, \dfrac{24}{5}\right)$

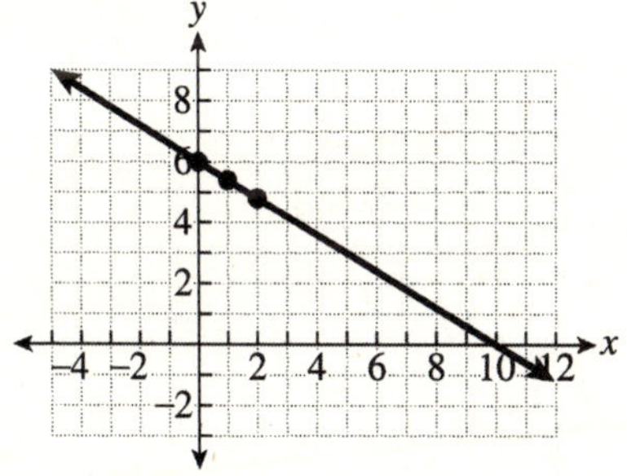

73. $3x + y = 9$

$x = 0$	$x = 1$	$x = 2$
$3(0) + y = 9$ $y = 9$	$3(1) + y = 9$ $3 + y = 9$ $y = 6$	$3(2) + y = 9$ $6 + y = 9$ $y = 3$
$(0,9)$	$(1,6)$	$(2,3)$

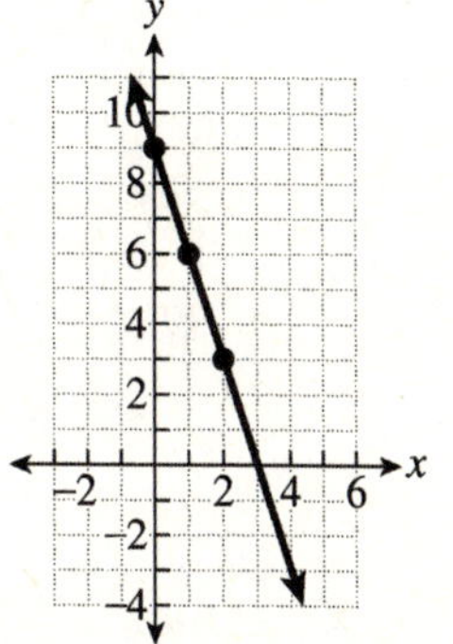

75. Explanations will vary. Example: Order is important so that the reader will know which is the x-coordinate and which is the y-coordinate. There is a difference between the points (a,b) and (b,a).

3.2   *QUICK CHECK*

1. $(-4,0), (0,5)$

2. $-3x + 4y = -12$

x-intercept:
$$-3x + 4(0) = -12$$
$$-3x = -12$$
$$\frac{-3x}{-3} = \frac{-12}{-3}$$
$$x = 4$$
$$(4,0)$$

y-intercept:
$$-3(0) + 4y = -12$$
$$4y = -12$$
$$\frac{4y}{4} = \frac{-12}{4}$$
$$y = -3$$
$$(0,-3)$$

3. $y = 2x - 9$

x-intercept:
$$0 = 2x - 9$$
$$0 + 9 = 2x - 9 + 9$$
$$9 = 2x$$
$$\frac{9}{2} = \frac{2x}{2}$$
$$\frac{9}{2} = x$$
$$\left(\frac{9}{2}, 0\right)$$

y-intercept:
$$y = 2(0) - 9$$
$$y = -9$$
$$(0,-9)$$

4. $2x - 3y = 12$

x-intercept:
$$2x - 3(0) = 12$$
$$2x = 12$$
$$\frac{2x}{2} = \frac{12}{2}$$
$$x = 6$$
$$(6, 0)$$

y-intercept:
$$2(0) - 3y = 12$$
$$\frac{-3y}{-3} = \frac{12}{-3}$$
$$y = -4$$
$$(0, -4)$$

3rd point at $x = 1$:
$$2(1) - 3y = 12$$
$$2 - 2 - 3y = 12 - 2$$
$$-3y = 10$$
$$\frac{-3y}{-3} = \frac{10}{-3}$$
$$y = -\frac{10}{3}$$
$$\left(1, -\frac{10}{3}\right)$$

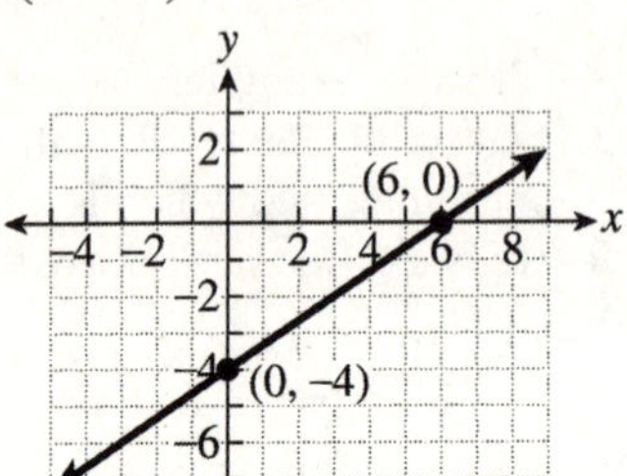

5. $y = -3x - 9$

x-intercept:
$$0 = -3x - 9$$
$$0 + 9 = -3x - 9 + 9$$
$$9 = -3x$$
$$\frac{9}{-3} = \frac{-3x}{-3}$$
$$-3 = x$$
$$(-3, 0)$$

y-intercept:
$$y = -3(0) - 9$$
$$y = -9$$
$$(0, -9)$$

3rd point at $x = -1$:
$$y = -3(-1) - 9$$
$$y = 3 - 9$$
$$y = -6$$
$$(-1, -6)$$

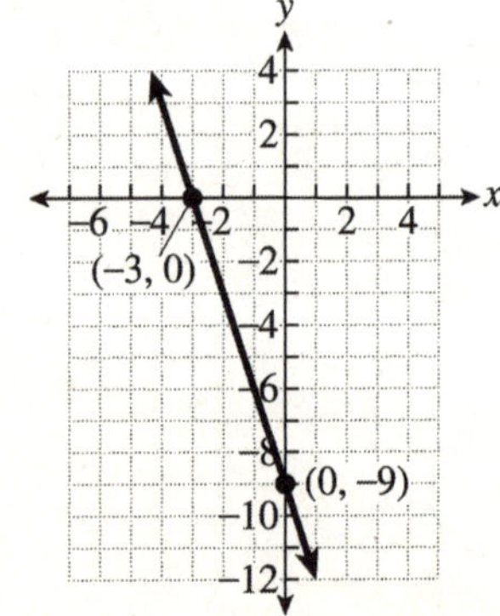

6. $y = 3x$
y-intercept:
$$y = 3(0)$$
$$y = 0$$
$$(0, 0)$$

2nd point at $x = 1$:
$$y = 3(1)$$
$$y = 3$$
$$(1, 3)$$

3rd point at $x = -1$:
$$y = 3(-1)$$
$$y = -3$$
$$(-1, -3)$$

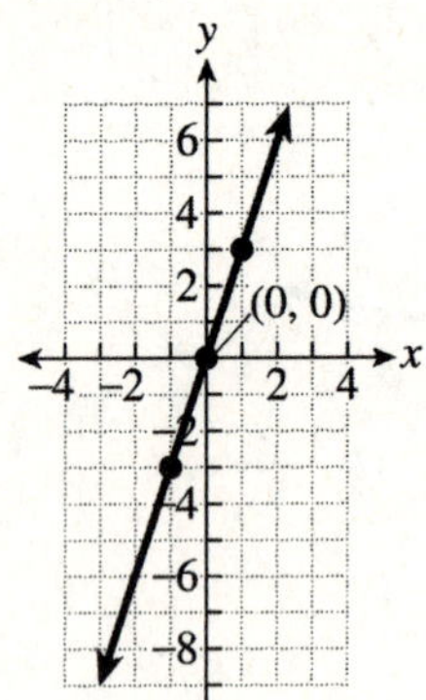

7. $y = 7$
Horizontal Line

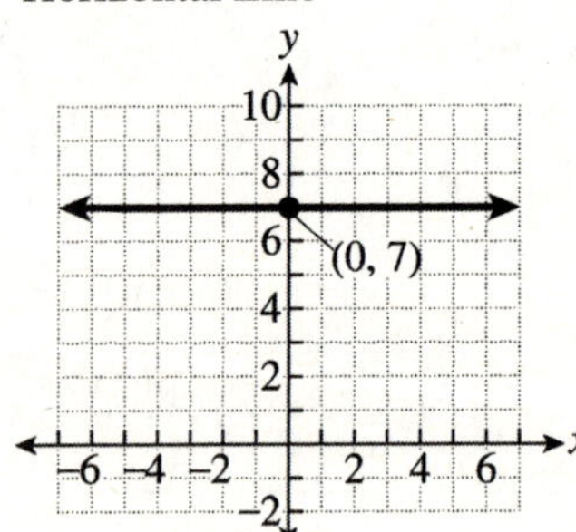

8. $x = -15$
Vertical Line

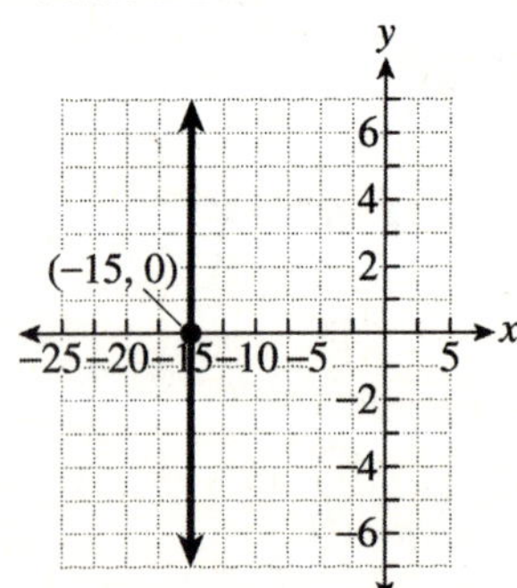

9. a) Initially, Nancy was \$10,000 in debt.

b) It will take five months until Nancy breaks even.

3.2 GRAPHING LINEAR EQUATIONS AND THEIR INTERCEPTS

1. x-intercept

3. vertical

5. Substitute 0 for y and solve for x.

7. x-intercept: $(-3,0)$ y-intercept: $(0,6)$

9. x-intercept: $(1,0)$ y-intercept: $(0,5)$

11. x-intercept: none y-intercept: $(0,2)$

13. x-intercept:
$$x+0=6$$
$$x=6$$
$$(6,0)$$
y-intercept:
$$0+y=6$$
$$y=6$$
$$(0,6)$$

15. x-intercept:
$$x-0=-8$$
$$x=-8$$
$$(-8,0)$$
y-intercept:
$$0-y=-8$$
$$-y=-8$$
$$y=8$$
$$(0,8)$$

17. x-intercept:
$$3x+2(0)=6$$
$$3x=6$$
$$x-2$$
$$(2,0)$$
y-intercept:
$$3(0)+2y=6$$
$$2y=6$$
$$y=3$$
$$(0,3)$$

19. x-intercept:
$$-3x+5(0)=-30$$
$$-3x=-30$$
$$x=10$$
$$(10,0)$$
y-intercept:
$$-3(0)+5y=-30$$
$$5y=-30$$
$$y=-6$$
$$(0,-6)$$

21. x-intercept:
$$\frac{3}{5}x+\frac{1}{2}(0)=3$$
$$\frac{3}{5}x=3$$
$$x=\frac{3}{1}\cdot\frac{5}{3}$$
$$x=5$$
$$(5,0)$$
y-intercept:
$$\frac{3}{5}(0)+\frac{1}{2}y=3$$
$$\frac{1}{2}y=3$$
$$y=6$$
$$(0,6)$$

23. x-intercept:
$$0=-x+8$$
$$x=8$$
$$(8,0)$$
y-intercept:
$$y=-(0)+8$$
$$y=8$$
$$(0,8)$$

25. x-intercept:
$$0=3x-10$$
$$10=3x$$
$$\frac{10}{3}=x$$
$$\left(\frac{10}{3},0\right)$$
y-intercept:
$$y=3(0)-10$$
$$y=-10$$
$$(0,-10)$$

27. x-intercept:
$$0=-2x$$
$$0=x$$
$$(0,0)$$
y-intercept:
$$y=-2(0)$$
$$y=0$$
$$(0,0)$$

29. x-intercept: none y-intercept: $(0,2)$

31. Answers will vary. Example:
$$x+y=4,\ 5x-y=20,\ y=2x-8$$

33. Answers will vary. Example:
$$-2x+5y=-50,\ x+y=-10,\ y=4x-10$$

35. x-intercept:
$$x+0=4$$
$$x=4$$
$$(4,0)$$
y-intercept:
$$0+y=4$$
$$y=4$$
$$(0,4)$$

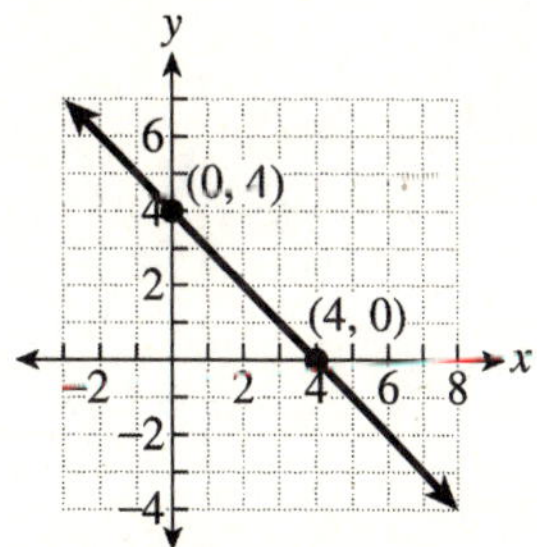

37. x-intercept:
$$x-0=-3$$
$$x=-3$$
$$(-3,0)$$
y-intercept:
$$0-y=-3$$
$$y=3$$
$$(0,3)$$

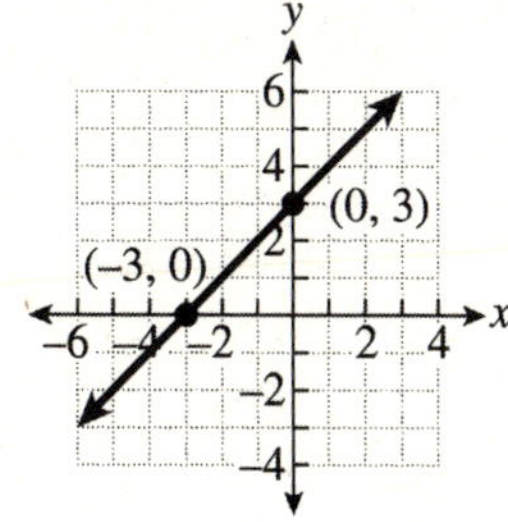

39.

x-intercept:	y-intercept:
$2x - 5(0) = 10$	$2(0) - 5y = 10$
$2x = 10$	$-5y = 10$
$x = 5$	$y = -2$
$(5, 0)$	$(0, -2)$

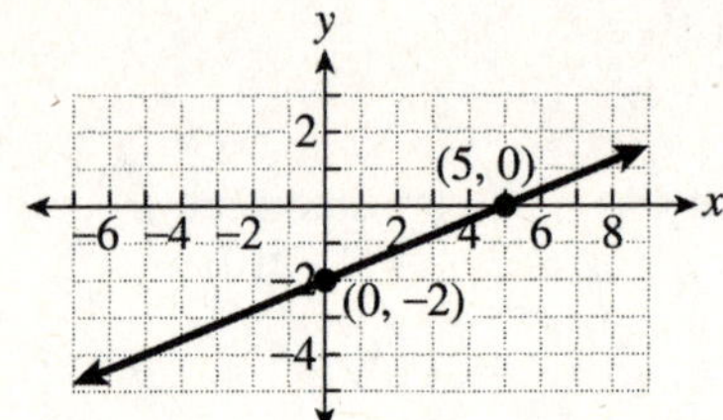

41.

x-intercept:	y-intercept:
$3x + 4(0) = -24$	$3(0) + 4y = -24$
$3x = -24$	$4y = -24$
$x = -8$	$y = -6$
$(-8, 0)$	$(0, -6)$

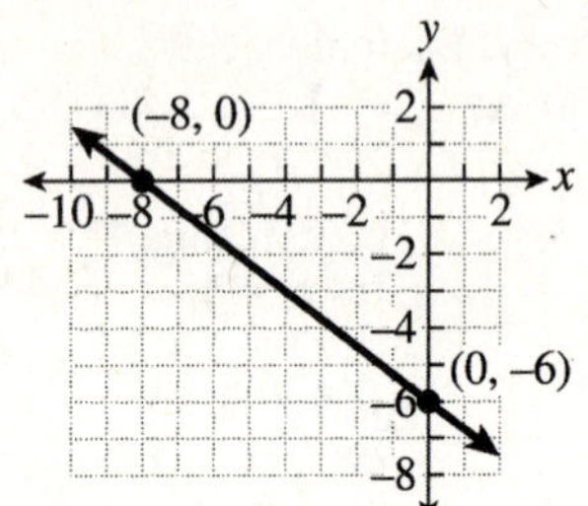

43.

x-intercept:	y-intercept:
$-4x + 7(0) = 14$	$-4(0) + 7y = 14$
$-4x = 14$	$7y = 14$
$x = -\dfrac{14}{4}$	$y = 2$
$x = -\dfrac{7}{2}$	$(0, 2)$
$\left(-\dfrac{7}{2}, 0\right)$	

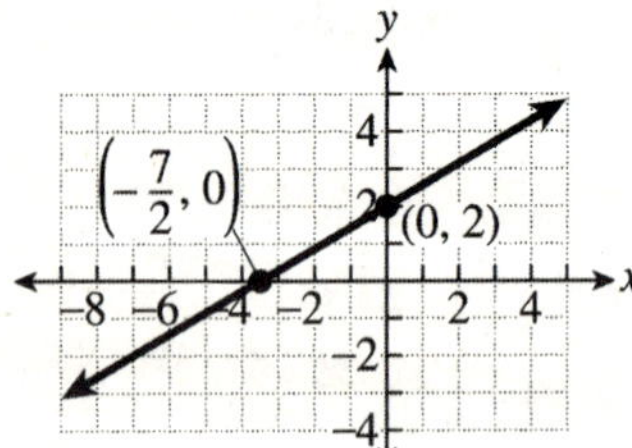

45.

x-intercept:	y-intercept:
$0 = x - 4$	$y = 0 - 4$
$4 = x$	$y = -4$
$(4, 0)$	$(0, -4)$

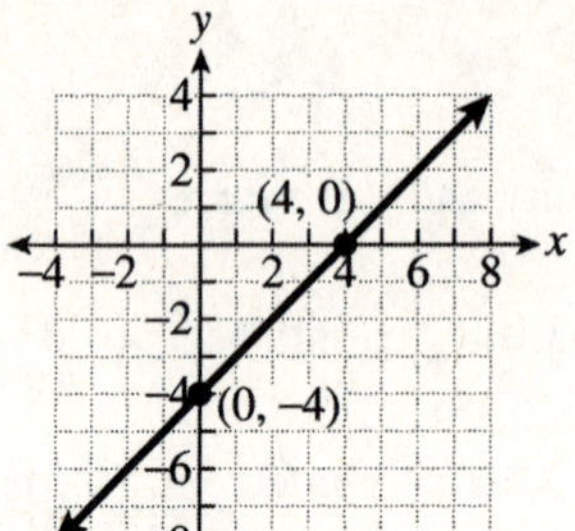

47.

x-intercept:	y-intercept:
$0 = -2x + 7$	$y = -2(0) + 7$
$2x = 7$	$y = 7$
$x = \dfrac{7}{2}$	$(0, 7)$
$\left(\dfrac{7}{2}, 0\right)$	

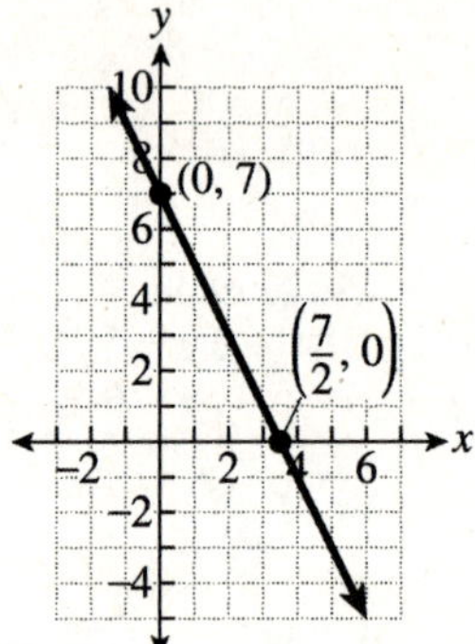

49.

x-intercept:	y-intercept:
$0 = 5x$	$y = 5(0)$
$0 = x$	$y = 0$
$(0, 0)$	$(0, 0)$

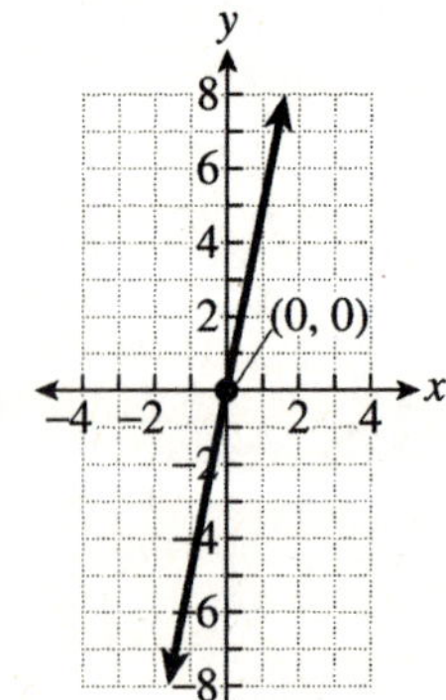

51. x-intercept: y-intercept:

$2x - 3(0) = 0$ $2(0) - 3y = 0$

$2x = 0$ $-3y = 0$

$x = 0$ $y = 0$

$(0, 0)$ $(0, 0)$

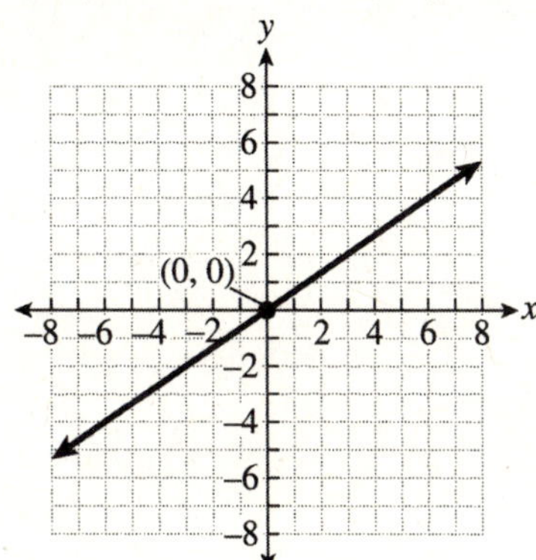

53. x-intercept: none y-intercept: $(0, 4)$

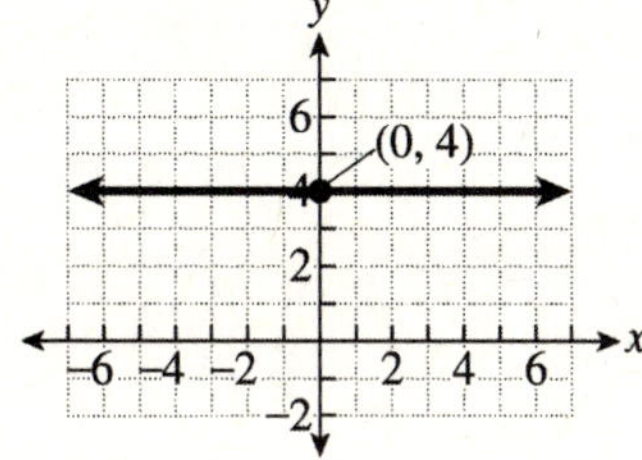

55. x-intercept: $(3, 0)$ y-intercept: none

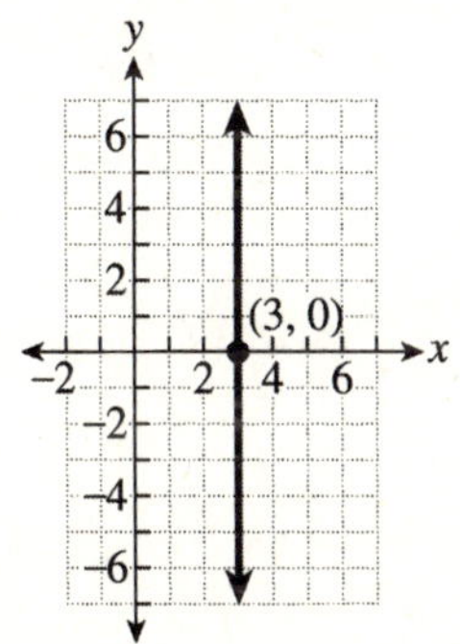

57. a) $y = 12,000 - 1500(0)$

$y = 12,000$

$(0, 12,000)$.

The original value of the copy machine is $12,000.

b) $0 = 12,000 - 1500x$

$1500x = 12,000$

$$x = \frac{12,000}{1500}$$

$x = 8$

$(8, 0)$

The copy machine has no value after 8 years.

c) $y = 12,000 - 1500(3)$

$y = 12,000 - 4500$

$y = 7500$

$7500

59. a) $y = 300 + 24(0)$

$y = 300$

$(0, 300)$

The cost to belong to the club is $300.

b) Because the minimum cost is $300, the cost to a member cannot be $0.

c) $y = 300 + 24(50)$

$y = 300 + 1200$

$y = 1500$

$1500

61. Answers will vary.

3.3   *QUICK CHECK*

1. Counting down from a y coordinate of 5 to a y coordinate of -7 is 12 units down. Counting right from an x coordinate of -2 to an x coordinate of 4 is 6 units right.

$$\text{slope} = \frac{-12}{6} = -2$$

2. $m = \dfrac{14 - 2}{5 - 1} = \dfrac{12}{4} = 3$

3. $m = \dfrac{-5 - 5}{1 - (-3)} = \dfrac{-10}{4} = -\dfrac{5}{2}$

4. The slope is undefined.

5. $m = 0$

6. $y = \dfrac{5}{2}x - 6$

$m = \dfrac{5}{2}$

y-intercept: $(0, -6)$

7. $6x + 2y = 10$
$$2y = -6x + 10$$
$$y = -3x + 5$$
$$m = -3$$
y-intercept: $(0, 5)$

8. $y = mx + b$

$$y = \frac{1}{2}x - \frac{4}{7}$$

9. $y = 2x + 6$
$m = 2$, y-intercept: $(0, 6)$
From the y-intercept use the slope and go up
2 and right 1 to find the next point $(1, 8)$.

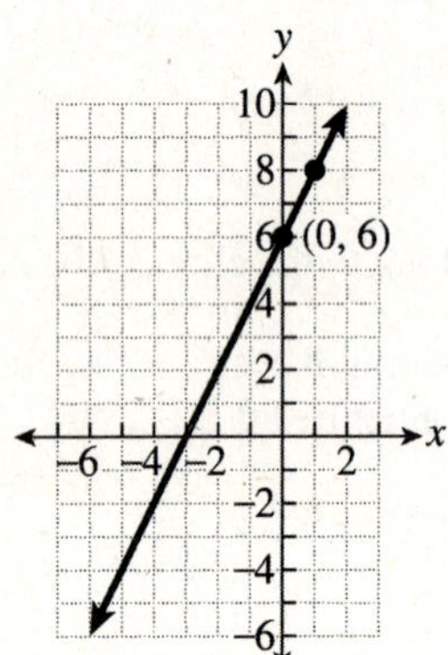

10. $y = -\dfrac{2}{3}x + 4$

$m = -\dfrac{2}{3}$, y-intercept: $(0, 4)$

From the y-intercept use the slope and go

down 2 and right 3 to find the next point $(3, 2)$.

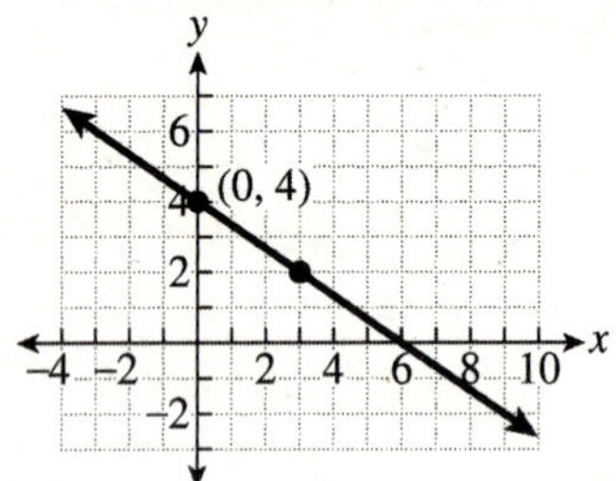

11. $y = 218x + 7485$

The slope is 218. The number of women
accepted to medical school increases by 218
per year.
The y-intercept is $(0, 7485)$. Approximately
7485 women were accepted to medical school
in 1997.

3.3   SLOPE OF A LINE

1. slope

3. negative

5. horizontal

7. slope-intercept

9. negative

11. positive

13. $-\dfrac{2}{4} = -\dfrac{1}{2}$

15. $-\dfrac{3}{4}$

17. 0

19. $m = \dfrac{7 - 5}{4 - 3} = \dfrac{2}{1} = 2$

21. $m = \dfrac{-6 - 3}{-5 - (-2)} = \dfrac{-9}{-3} = 3$

23. $m = \dfrac{2 - (-6)}{7 - 13} = \dfrac{8}{-6} = -\dfrac{4}{3}$

25. $m = \dfrac{-3 - 15}{-24 - (-32)} = \dfrac{-18}{8} = -\dfrac{9}{4}$

27. $m = \dfrac{6 - 6}{-4 - 3} = \dfrac{0}{-7} = 0$

29. $m = \dfrac{7 - 4}{0 - 0} = \dfrac{3}{0}$
The slope is undefined.

31. y-intercept: $(0, 4), m = 0$

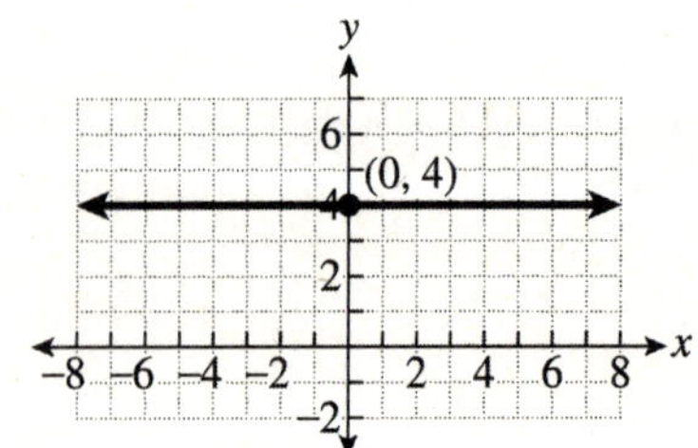

33. x-intercept: $(3,0)$, m is undefined

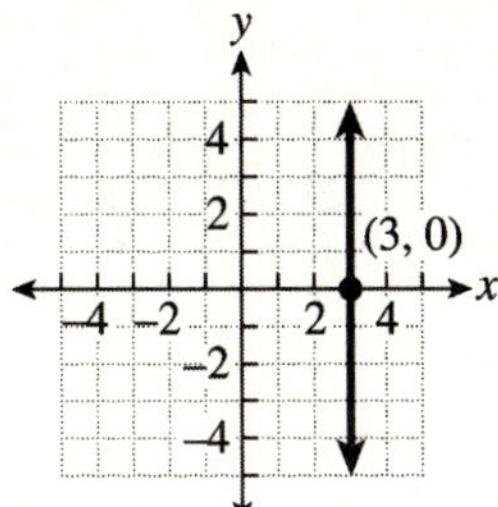

35. x-intercept: $(8,0)$, m is undefined

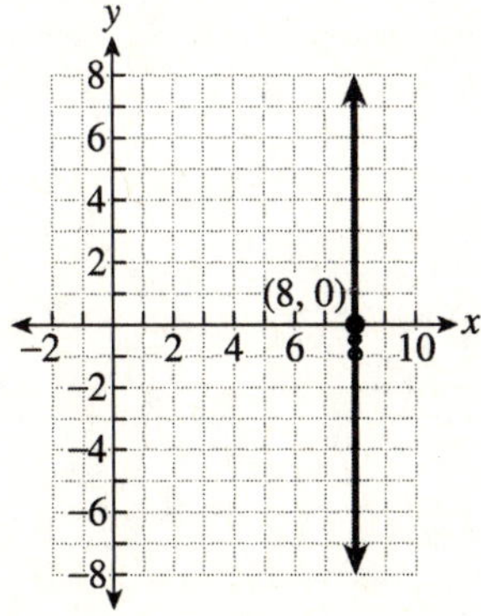

37. $x = 6$, m is undefined

39. $y = 2$, $m = 0$

41. $y = 6x - 7$
$m = 6$, y-intercept: $(0,-7)$

43. $y = -2x + 3$
$m = -2$, y-intercept: $(0,3)$

45. $6x + 4y = -10$
$\qquad 4y = -6x - 10$

$$y = -\frac{6}{4}x - \frac{10}{4}$$

$$y = -\frac{3}{2}x - \frac{5}{2}$$

$$m = -\frac{3}{2}, \ y\text{-intercept: } \left(0, -\frac{5}{2}\right)$$

47. $5x - 8y = 10$
$\qquad -8y = -5x + 10$

$$y = \frac{-5}{-8}x + \frac{10}{-8}$$

$$y = \frac{5}{8}x - \frac{5}{4}$$

$$m = \frac{5}{8}, \ y\text{-intercept: } \left(0, -\frac{5}{4}\right)$$

49. $x - 5y = 8$
$\qquad -5y = -x + 8$

$$y = \frac{-1}{-5}x + \frac{8}{-5}$$

$$y = \frac{1}{5}x - \frac{8}{5}$$

$$m = \frac{1}{5}, \ y\text{-intercept: } \left(0, -\frac{8}{5}\right)$$

51. $m = -2$, $b = 5$
$y = -2x + 5$

53. $m = 3$, $b = -6$
$y = 3x - 6$

55. $m = 0$, $b = -4$
$y = -4$

57. $y = 3x + 6$
$m = 3$, y-intercept: $(0,6)$
From the y-intercept count up 3 and right 1
to find the next point $(1,9)$.

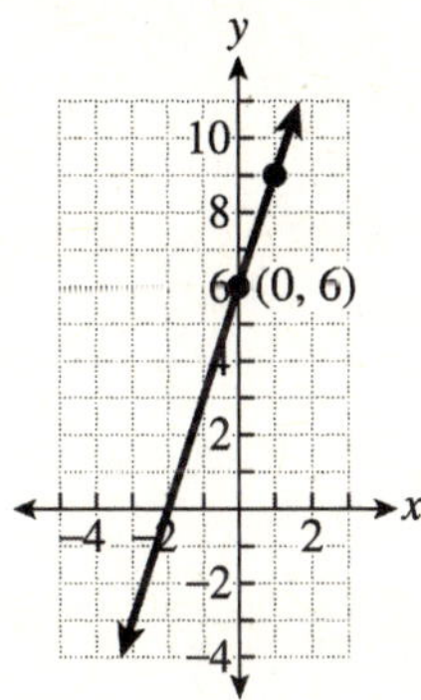

59. $y = -2x - 6$
$m = -2$, y-intercept: $(0,-6)$
From the y-intercept count down 2 and right 1
to find the next point $(1,-8)$.

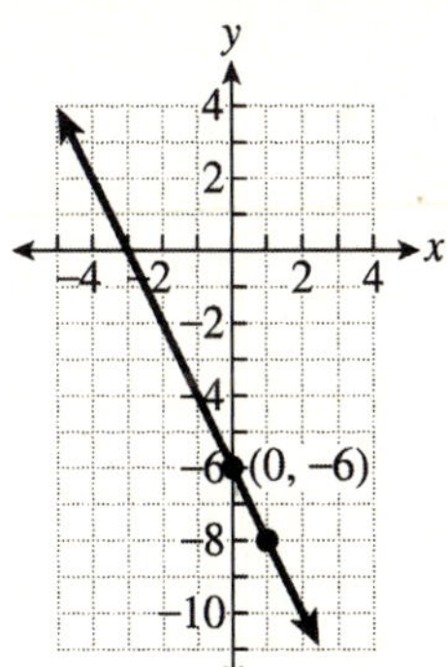

61. $y = x - 3$
$m = 1$, y-intercept: $(0, -3)$
From the y-intercept count up 1 and right 1
to find the next point $(1, -2)$.

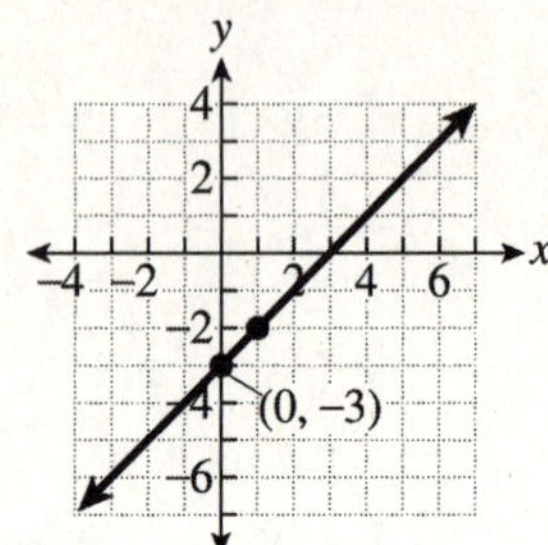

63. $y = 4x$
$m = 4$, y-intercept: $(0, 0)$
From the y-intercept count up 4 and right 1
to find the next point $(1, 4)$.

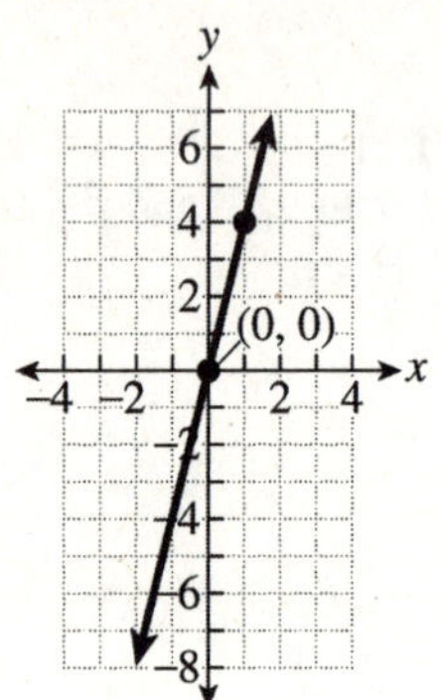

65. $y = \dfrac{7}{2}x - 7$

$m = \dfrac{7}{2}$, y-intercept: $(0, -7)$

From the y-intercept count up 7 and right 2
to find the next point $(2, 0)$.

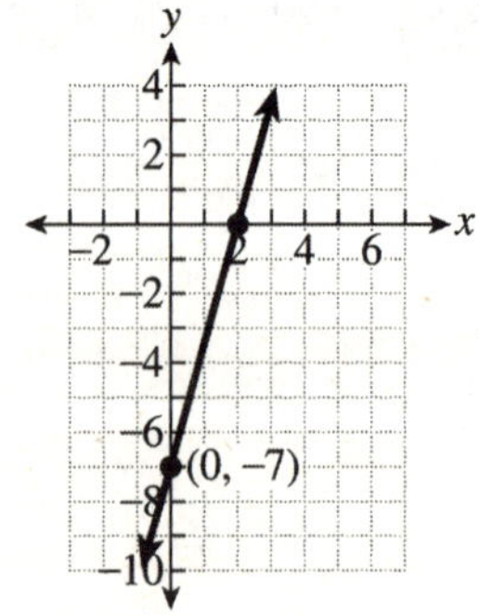

67. $3y = -9x - 6$
$$y = -\frac{9}{3}x - \frac{6}{3}$$
$$y = -3x - 2$$
$m = -3$, y-intercept: $(0, -2)$
From the y-intercept count down 3 and right 1
to find the next point $(1, -5)$.

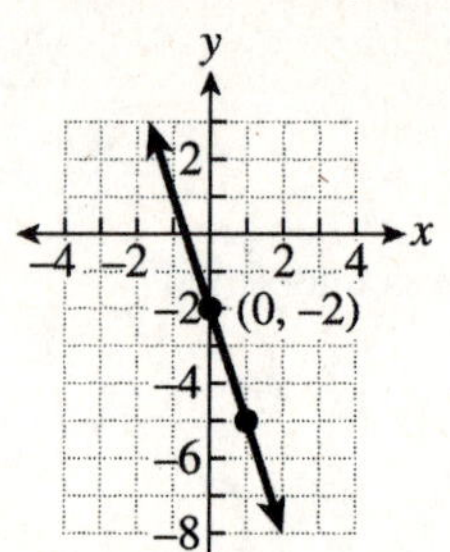

69. $-5x + 4y = 8$
$$4y = 5x + 8$$
$$y = \frac{5}{4}x + \frac{8}{4}$$
$$y = \frac{5}{4}x + 2$$

$m = \dfrac{5}{4}$, y-intercept: $(0, 2)$

From the y-intercept count up 5 and right 4
to find the next point $(4, 7)$.

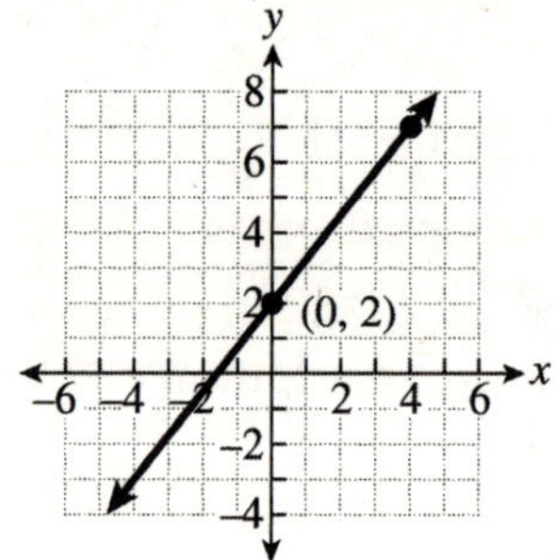

71. $y = -x + 2$
$m = -1$, y-intercept: $(0, 2)$
From the y-intercept count down 1 and right 1
to find the next point $(1, 1)$.

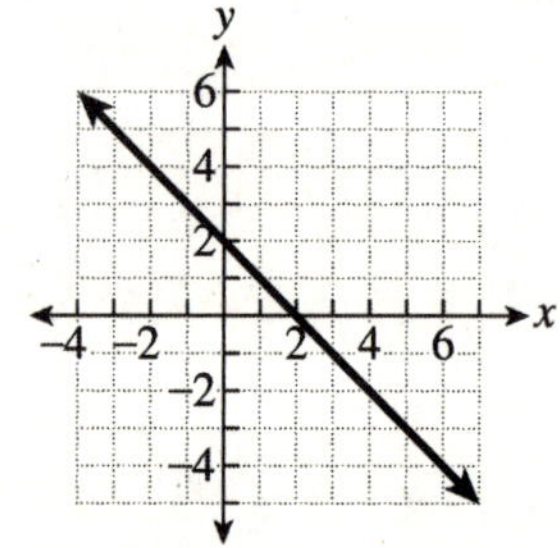

73. $y = -2x + 8$
$m = -2$, y-intercept: $(0,8)$
From the y-intercept count down 2 and right 1
to find the next point $(1,6)$.

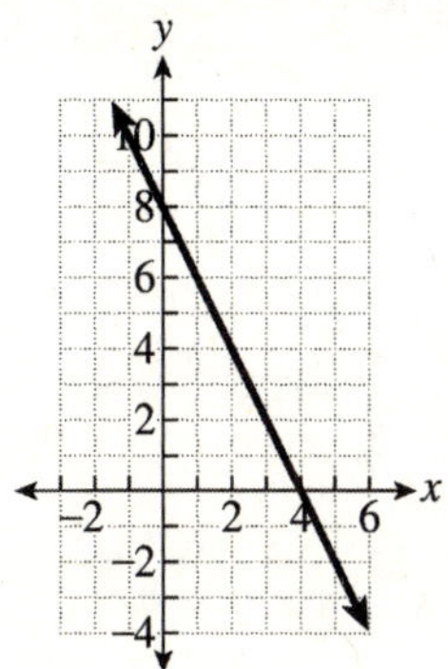

75. $y = 6x - 9$
$m = 6$, y-intercept: $(0,-9)$
From the y-intercept count up 6 and right 1
to find the next point $(1,-3)$.

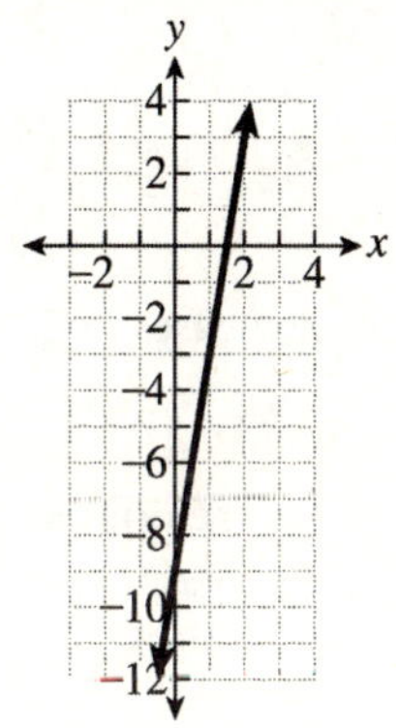

77. $y = -x - 5$
$m = -1$, y-intercept: $(0,-5)$
From the y-intercept count down 1 and right 1
to find the next point $(1,-6)$.

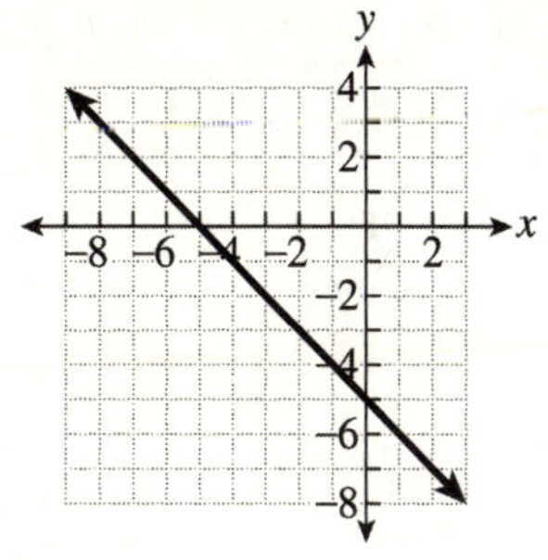

79. $3x - 2y = -12$

x-intercept:
$3x - 2(0) = -12$
$3x = -12$
$x = -4$
$(-4,0)$

y-intercept:
$3(0) - 2y = -12$
$-2y = -12$
$y = 6$
$(0,6)$

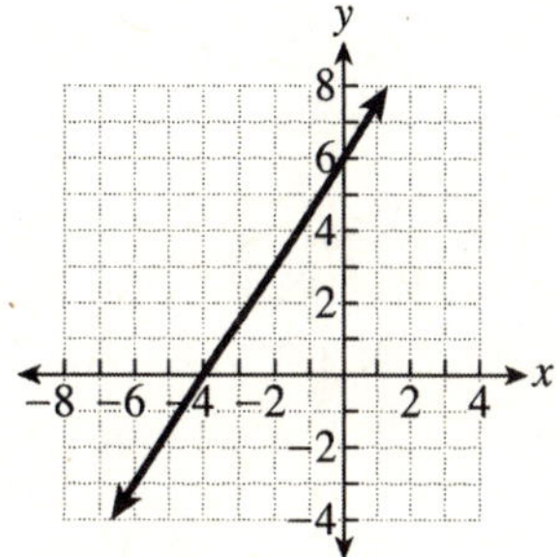

81. $y = \dfrac{4}{5}x$

$m = \dfrac{4}{5}$, y-intercept: $(0,0)$

From the y-intercept count up 4 and right 5
to find the next point $(5,4)$.

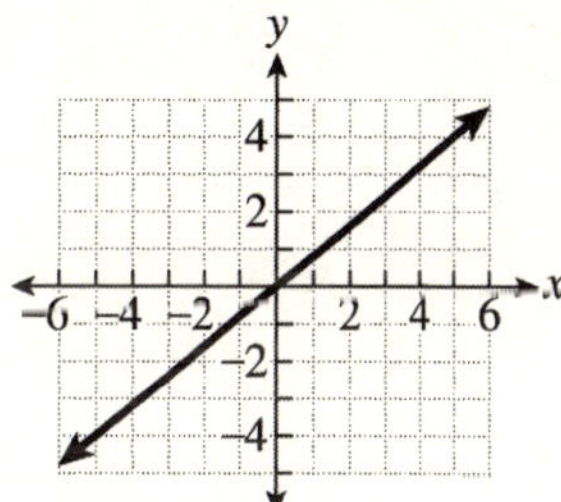

83. $y = 4$
x-intercept: None y-intercept: $(0,4)$

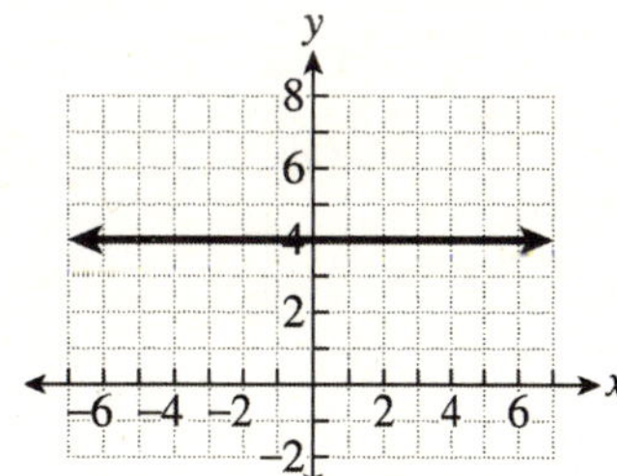

85. $3x + y = 1$

x-intercept:

$3x + (0) = 1$

$3x = 1$

$x = \dfrac{1}{3}$

$\left(\dfrac{1}{3}, 0\right)$

y-intercept:

$3(0) + y = 1$

$y = 1$

$(0, 1)$

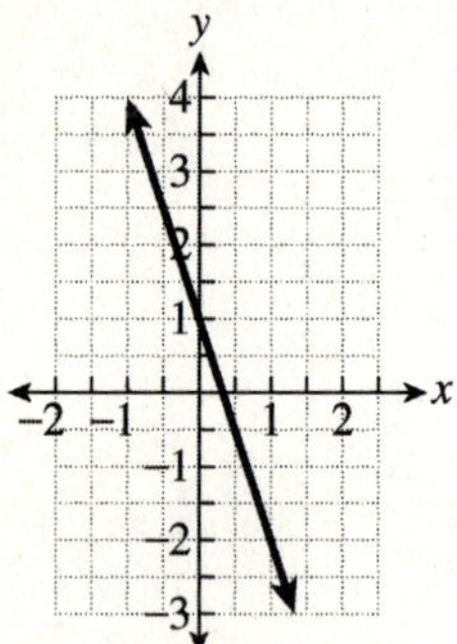

87. $5x - y = -5$

x-intercept:

$5x - (0) = -5$

$5x = -5$

$x = -1$

$(-1, 0)$

y-intercept:

$5(0) - y = -5$

$-y = -5$

$y = 5$

$(0, 5)$

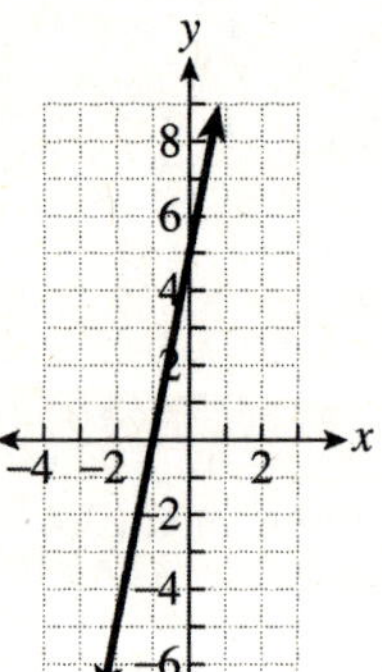

89. $y = 3x - 6$

y-intercept: $(0, -6)$

other points: $(1, -3)$ and $(2, 0)$

since $m = \dfrac{3}{1}$ up 3, right 1

$y = -x - 2$

y-intercept: $(0, -2)$

other points: $(1, -3)$ and $(-2, 0)$

since $m = \dfrac{-1}{1}$ down 1, right 1

point of intersection: $(1, -3)$

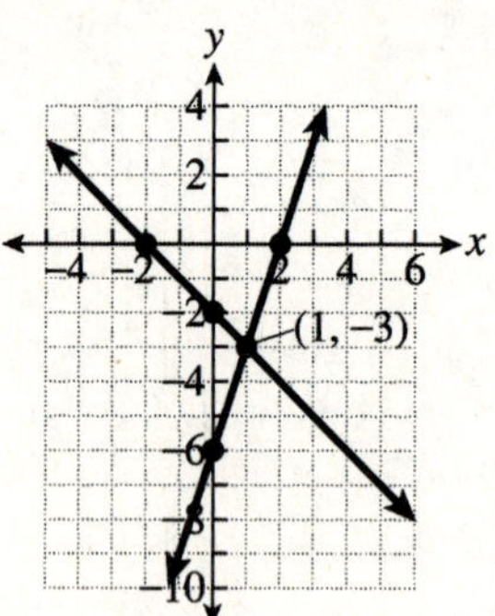

91. $y = -3000x + 25,000$

a) $m = -3000$, The value decreases by \$3000 per year.

b) y-intercept: $(0, 25,000)$, The original value of the car in 2008 was \$25,000.

c) In 2014, $x = 6$.

$y = -3000(6) + 25,000$

$y = -18,000 + 25,000$

$y = 7000$

\$7000

93. a) $m = \dfrac{\text{vertical change}}{\text{horizontal change}} = \dfrac{9}{27} = \dfrac{1}{3}$

b) $m = \dfrac{\text{vertical change}}{\text{horizontal change}} = \dfrac{36}{24} = \dfrac{3}{2}$

95. $m = \dfrac{\text{vertical change}}{\text{horizontal change}} = \dfrac{50}{1000} = \dfrac{1}{20}$

97. Positive. The unemployment rate (y) increases as the time after the statement (x) increases.

99. Answers will vary.

1. Cost $= 19.95 + 0.15x$, where x is the number of miles traveled.
   Domain: Set of all possible miles.
   Range: Set of all possible costs.

2. $g(-6) = 2(-6) + 9 = -12 + 9 = -3$

3. $g(a+8) = 7(a+8) - 12$
   $\qquad\quad = 7a + 56 - 12$
   $\qquad\quad = 7a + 44$

4. $f(x) = \dfrac{3}{4}x - 6$

   y-intercept: $(0, -6), m = \dfrac{3}{4}$

   since $m = \dfrac{3}{4}$, 3 up, 4 right

   other point: $(4, -3)$

   x-intercept: $(8, 0)$

   $0 = \dfrac{3}{4}x - 6$

   $6 = \dfrac{3}{4}x$

   $\dfrac{4}{3} \cdot \dfrac{6}{1} = x$

   $8 = x$

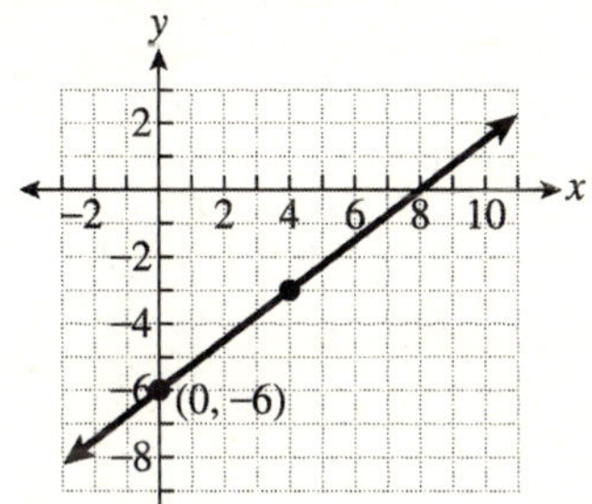

5. $f(x) = 4$
   y-intercept: $(0, 4)$
   Since $m = 0$, this is a horizontal line.

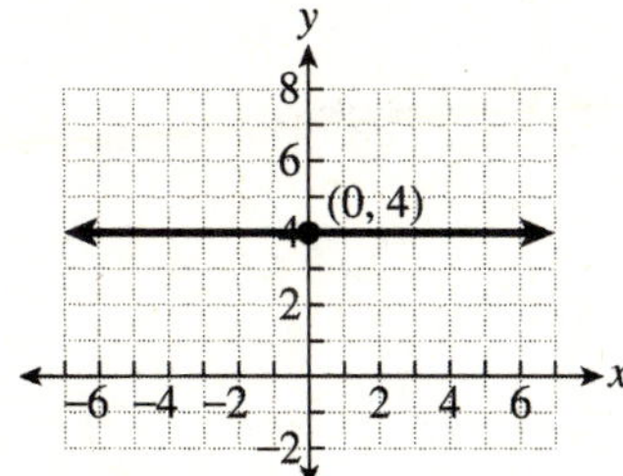

6. **a)** $f(1) = -6$

   b) If $f(x) = -2$, then $x = 3$.

1. function

3. range

5. linear

7. Yes, each player is listed with only one team.

9. Yes, each person has only one mother.

11. **a)** Yes. **b)** No.

13. **a)** Yes. **b)** No.

15. Yes.

17. No, some x-coordinates are associated with more than one y-coordinate.
    Example: $(5, 3)$ and $(5, -2)$

19. No, some x-coordinates are associated with more than one y-coordinate.
    Example: $(2, -5)$ and $(2, -1)$

21. **a)** $F(x) = \dfrac{9}{5}x + 32$

    b) $F(0) = \dfrac{9}{5}(0) + 32 = 0 + 32 = 32°\text{F}$

    $F(100) = \dfrac{9}{5}(100) + 32 = 180 + 32 = 212°\text{F}$

    $F(30) = \dfrac{9}{5}(30) + 32 = 54 + 32 = 86°\text{F}$

    $F(-10) = \dfrac{9}{5}(-10) + 32 = -18 + 32 = 14°\text{F}$

    $F(-40) = \dfrac{9}{5}(-40) + 32 = -72 + 32 = -40°\text{F}$

23. **a)** $f(x) = 7x + 36$

    b) $f(12) = 7(12) + 36 = 84 + 36 = 120$
    $120

25. $f(x) = 4x + 3$

27. $f(x) = -3x - 4$

29. $f(x) = 6$

31. $g(-13) = (-13) - 9 = -22$

33. $f(10) = -8(10) + 3 = -80 + 3 = -77$

35. $f(-10) = -\dfrac{2}{5}(-10) + 9 = 4 + 9 = 13$

37. $f(0) = 9(0) - 25 = 0 - 25 = -25$

39. $g\left(\dfrac{2}{3}\right) = 3\left(\dfrac{2}{3}\right) - 1 = 2 - 1 = 1$

41. $f(a) = 3(a) + 4 = 3a + 4$

43. $f(a+3) = 7(a+3) - 2 = 7a + 21 - 2 = 7a + 19$

45. $f(2a-5) = 16 - 3(2a-5)$
$\qquad\qquad = 16 - 6a + 15$
$\qquad\qquad = -6a + 31$

47. $f(x+h) = 6(x+h) + 4 = 6x + 6h + 4$

49. $f(x) = 6x - 6$
$m = 6$, y-intercept: $(0, -6)$
From the y-intercept count up 6 and right 1
to find the next point $(1, 0)$.

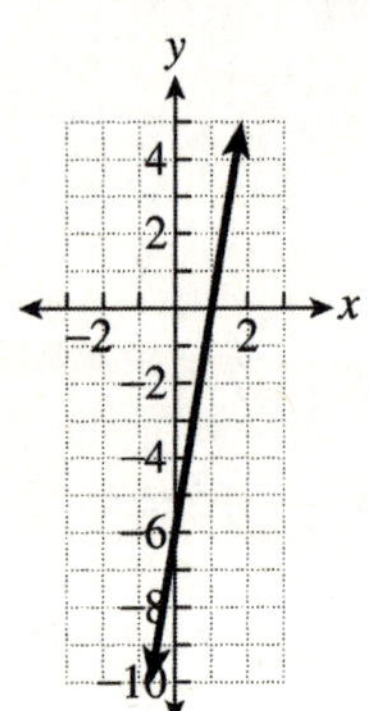

51. $f(x) = -3x + 3$
$m = -3$, y-intercept: $(0, 3)$
From the y-intercept count down 3 and right 1
to find the next point $(1, 0)$.

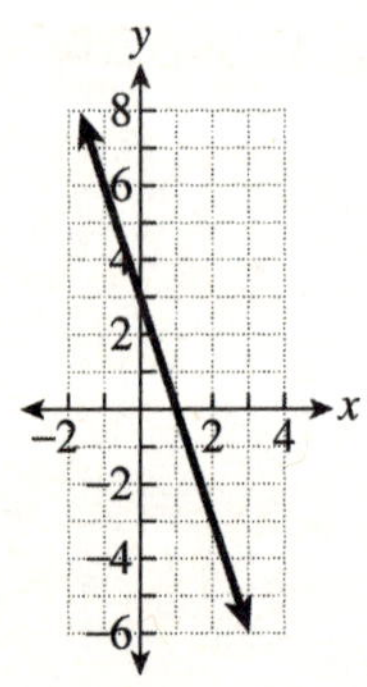

53. $f(x) = \dfrac{4}{3}x + 4$

$m = \dfrac{4}{3}$, y-intercept: $(0, 4)$

From the y-intercept count up 4 and right 3
to find the next point $(3, 8)$.

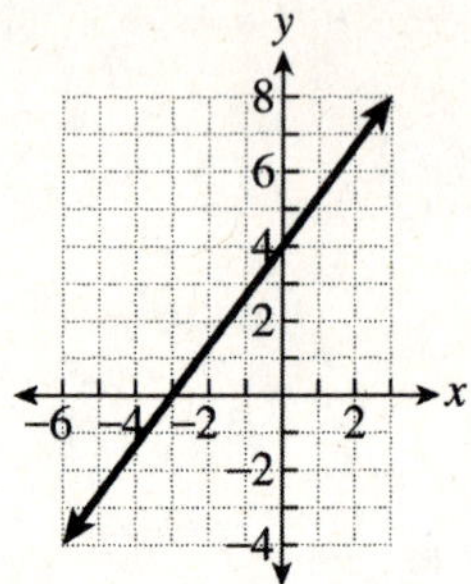

55. $f(x) = \dfrac{8}{3}x$

$m = \dfrac{8}{3}$, y-intercept: $(0, 0)$

From the y-intercept count up 8 and right 3
to find the next point $(3, 8)$.

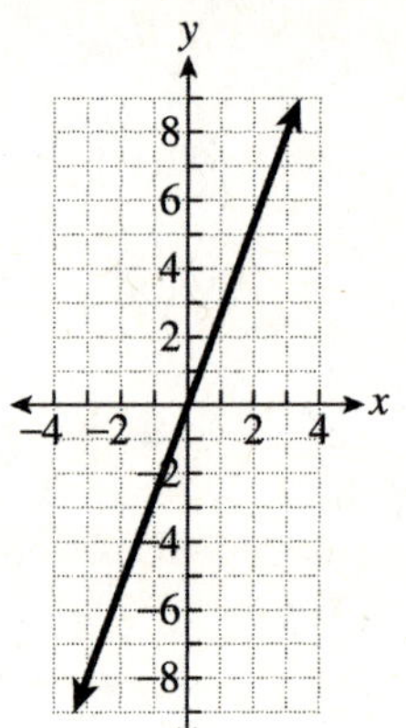

57. $f(x) = 4$
$m = 0$, y-intercept: $(0, 4)$

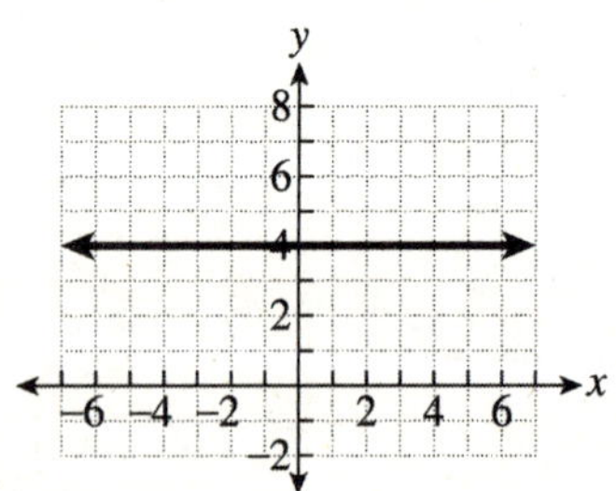

59. a) $f(5) = 4$ **b)** $f(-1) = -8$
                                      -1

  c) Domain: $(-\infty, \infty)$; range: $(-\infty, \infty)$

61. a) $f(2) = 3$ **b)** $f(-2) = 9$
                                        -2

  c) Domain: $(-\infty, \infty)$; range: $(-\infty, \infty)$

63. a) $f(4) = 5$

  b) Domain: $(-\infty, \infty)$; range: $\{5\}$

65. Answers will vary. Example:
A: Set of universities in the United States.
B: Set of states in the United States.
The rule that takes a university as its input
value and gives the state where the university is
located as its output value is a function; each
university is in one state. The rule that takes the
state as its input value and gives a university in
that state as its output value is not a function; a
state may have more than one university.

3.5 QUICK CHECK

1. $y = 4x + 3$ $y = -4x$
   $m = 4$ $m = -4$
No, the lines are not parallel.

2. $8x + 6y = 36$ $y = -\dfrac{4}{3}x + 5$
    $6y = -8x + 36$

$$y = \frac{-8}{6}x + \frac{36}{6} \qquad m = -\frac{4}{3}$$

$$y = -\frac{4}{3}x + 6$$

$$m = -\frac{4}{3}$$

Yes, the lines are parallel

3. $2x - 9y = 36$
    $-9y = -2x + 36$

$$y = \frac{-2}{-9}x + \frac{36}{-9}$$

$$y = \frac{2}{9}x - 4$$

$$m = \frac{2}{9}$$

4. Yes, the lines are perpendicular, since the
slopes are 1 and -1 respectively and these are
negative reciprocals of one another.

5. $6x + 4y = 12$ $y = \dfrac{2}{3}x + 5$
    $4y = -6x + 12$

$$y = \frac{-6}{4}x + \frac{12}{4} \qquad m = \frac{2}{3}$$

$$y = -\frac{3}{2}x + 3$$

$$m = -\frac{3}{2}$$

Yes, the lines are perpendicular.

6. $7x - 3y = 21$
    $-3y = -7x + 21$

$$y = \frac{-7}{-3}x + \frac{21}{-3}$$

$$y = \frac{7}{3}x - 7$$

$$m = \frac{7}{3}$$

The slope of the perpendicular line is $-\dfrac{3}{7}$.

7. $y = 2x - 5$ $y = -2x$
   $m = 2$ $m = -2$
The lines are neither parallel nor perpendicular.

8. $8x - 2y = 12$ $y = 4x + 3$
    $-2y = -8x + 12$ $m = 4$

$$y = \frac{-8}{-2}x + \frac{12}{-2}$$

$$y = 4x - 6$$

$$m = 4$$

The lines are parallel.

9. $5x - 2y = -8$ $y = -\dfrac{2}{5}x + 9$
    $-2y = -5x - 8$

$$y = \frac{-5}{-2}x - \frac{8}{-2} \qquad m = -\frac{2}{5}$$

$$y = \frac{5}{2}x + 4$$

$$m = \frac{5}{2}$$

The lines are perpendicular.

3.5   PARALLEL AND PERPENDICULAR LINES

1. parallel

3. vertical

5. parallel

7. $y = 3x + 5$ $y = 3x - 2$
$m = 3$ $m = 3$
Yes, the lines are parallel.

9. $4x + 2y = 9$ $3y = 6x + 7$
$\quad\quad 2y = -4x + 9$
$\quad\quad y = \dfrac{-4}{2}x + \dfrac{9}{2}$ $y = \dfrac{6}{3}x + \dfrac{7}{3}$
$\quad\quad y = -2x + \dfrac{9}{2}$ $y = 2x + \dfrac{7}{3}$
$\quad\quad\quad\quad\quad\quad\quad m = 2$
$m = -2$
No, the lines are not parallel.

11. $y = 6$ $y = -6$
$m = 0$ $m = 0$
Yes, the lines are parallel.

13. $y = 4x$ $y = \dfrac{1}{4}x - 3$
$m = 4$
$\quad\quad\quad\quad\quad\quad m = \dfrac{1}{4}$
No, the lines are not perpendicular.

15. $15x + 3y = 11$ $x - 5y = -4$
$\quad\quad 3y = -15x + 11$ $-5y = -x - 4$
$\quad\quad y = \dfrac{-15}{3}x + \dfrac{11}{3}$ $y = \dfrac{-1}{-5}x - \dfrac{4}{-5}$
$\quad\quad y = -5x + \dfrac{11}{3}$ $y = \dfrac{1}{5}x + \dfrac{4}{5}$
$m = -5$
$\quad\quad\quad\quad\quad\quad m = \dfrac{1}{5}$
Yes, the lines are perpendicular.

17. $x = 3$ $y = 4$
m is undefined $m = 0$
Yes, the lines are perpendicular.

19. $y = 6x - 2$ $y = -6x + 5$
$m = 6$ $m = -6$
The lines are neither parallel nor perpendicular.

21. $y = 4x + 3$ $y = -\dfrac{1}{4}x - 6$
$m = 4$
$\quad\quad\quad\quad\quad\quad m = -\dfrac{1}{4}$
The lines are perpendicular.

23. $8x + 6y = 12$ $12x + 9y = -27$
$\quad\quad 6y = -8x + 12$ $9y = -12x - 27$
$\quad\quad y = \dfrac{-8}{6}x + \dfrac{12}{6}$ $y = -\dfrac{12}{9}x - \dfrac{27}{9}$
$\quad\quad y = -\dfrac{4}{3}x + 2$ $y = -\dfrac{4}{3}x - 3$
$m = -\dfrac{4}{3}$ $m = -\dfrac{4}{3}$
The lines are parallel.

25. $5x - 4y = 44$ $10x + 8y = -32$
$\quad\quad -4y = -5x + 44$ $8y = -10x - 32$
$\quad\quad y = \dfrac{-5}{-4}x + \dfrac{44}{-4}$ $y = \dfrac{-10}{8}x - \dfrac{32}{8}$
$\quad\quad y = \dfrac{5}{4}x - 11$ $y = -\dfrac{5}{4}x - 4$
$m = \dfrac{1}{4}$ $m = -\dfrac{5}{4}$
The lines are neither parallel nor perpendicular.

27. $y = -\dfrac{8}{7}$ $8y - 7 = 0$
$\quad\quad\quad\quad\quad\quad\quad 8y = 7$
$y = 0 \cdot x - \dfrac{8}{7}$ $y = \dfrac{7}{8}$
$m = 0$
$\quad\quad\quad\quad\quad\quad\quad y = 0 \cdot x + \dfrac{7}{8}$
$\quad\quad\quad\quad\quad\quad\quad m = 0$
The lines are parallel.

29. $x = 5$ $y = -2$
m is undefined $m = 0$
The lines are perpendicular.

31. Slopes are 7 and 7.
Parallel.

33. Slopes are 0 and 6.
Neither.

35. Slopes are 1 and 1.
Parallel.

37. -6

39. 5

41. $12x + 4y = 8$
$\quad\quad 4y = -12x + 8$
$\quad\quad y = \dfrac{-12}{4}x + \dfrac{8}{4}$
$\quad\quad y = -3x + 2$
$m = -3$

43. $m = -\dfrac{1}{8}$

45. $10x + 6y = 30$
$6y = -10x + 30$
$y = -\dfrac{10}{6}x + \dfrac{30}{6}$
$y = -\dfrac{5}{3}x + 5$
$m = -\dfrac{1}{-\dfrac{5}{3}} = \dfrac{3}{5}$

47. $12x + 21y = 33$
$21y = -12x + 33$
$y = \dfrac{-12}{21}x + \dfrac{33}{21}$
$y = -\dfrac{4}{7}x + \dfrac{11}{7}$
$m = -\dfrac{1}{-\dfrac{4}{7}} = \dfrac{7}{4}$

49. $Ax + By = C$
$By = -Ax + C$
$y = \dfrac{-A}{B}x + \dfrac{C}{B}$
$m = -\dfrac{A}{B}$

51. $m = \dfrac{-1-7}{5-3}$ $y = -4x + 11$
$m = -4$
$m = \dfrac{-8}{2}$
$m = -4$
Yes, the lines are parallel.

53. $m = \dfrac{9-7}{-1-4}$ $-10x + 4y = 8$
$4y = 10x + 8$
$m = -\dfrac{2}{5}$ $y = \dfrac{10}{4}x + \dfrac{8}{4}$
$y = \dfrac{5}{2}x + 2$
$m = \dfrac{5}{2}$
Yes, the lines are perpendicular.

55. Answers will vary. Example:
1. Two streets that run east to west are parallel.
2. Telephone wires strung from the same poles are parallel.
3. Two mountains with the same steepness both going downhill are parallel.

57. Explanations will vary. Example:
Put each equation in slope-intercept form to find the slope. If the slopes are equal, then the lines are parallel.

3.6 *QUICK CHECK*

1. $y - 5 = 2(x - 1)$
$y - 5 = 2x - 2$
$y = 2x + 3$

2. $y - (-6) = -3(x - 4)$
$y + 6 = -3x + 12$
$y = -3x + 6$

3. $f(x) = -4x + b$
$f(-2) = 13$
$-4(-2) + b = 13$
$8 + b = 13$
$b = 5$
$f(x) = -4x + 5$

4. $m = \dfrac{-5 - 7}{6 - (-2)} = \dfrac{-12}{8} = -\dfrac{3}{2}$
$y - (-5) = -\dfrac{3}{2}(x - 6)$
$y + 5 = -\dfrac{3}{2}x + 9$
$y = -\dfrac{3}{2}x + 4$

5. $m = \dfrac{4 - 4}{3 - (-6)} = \dfrac{0}{9} = 0$
$y = 4$

6. Let x represent the number of years after 2003, then $x = 0$ in the year 2003 and $x = 4$ in the year 2007.
$m = \dfrac{4884 - 3620}{4 - 0} = \dfrac{1264}{4} = 316$
$y = 316x + 3620$

7. $y = \dfrac{2}{5}x - 9$

$m = \dfrac{2}{5}$

$y - 4 = \dfrac{2}{5}(x - 5)$

$y - 4 = \dfrac{2}{5}x - 2$

$y = \dfrac{2}{5}x + 2$

8. $9x - 12y = 4$
$-12y = -9x + 4$

$y = \dfrac{-9}{-12}x + \dfrac{4}{-12}$

$y = \dfrac{3}{4}x - \dfrac{1}{3}$

The slope of the perpendicular line is $-\dfrac{4}{3}$.

$y - (-2) = -\dfrac{4}{3}(x - 6)$

$y + 2 = -\dfrac{4}{3}x + 8$

$y = -\dfrac{4}{3}x + 6$

3.6 EQUATIONS OF LINES

1. $y - y_1 = m(x - x_1)$

3. $-6x + 2y = 10$
$2y = 6x + 10$

$y = \dfrac{6}{2}x + \dfrac{10}{2}$

$y = 3x + 5$

5. $x - 5y = 10$
$-5y = -x + 10$

$y = \dfrac{-1}{-5}x + \dfrac{10}{-5}$

$y = \dfrac{1}{5}x - 2$

7. $7x - y = 3$
$-y = -7x + 3$
$y = 7x - 3$

9. $y = -3x + 5$

11. $y = \dfrac{2}{3}x - 3$

13. $y = 5x + 8$

15. $y - 4 = 3(x - 7)$
$y - 4 = 3x - 21$
$y = 3x - 17$

17. $y - (-2) = -4(x - 6)$
$y + 2 = -4x + 24$
$y = -4x + 22$

19. $y - (-9) = \dfrac{3}{2}(x - (-6))$

$y + 9 = \dfrac{3}{2}(x + 6)$

$y + 9 = \dfrac{3}{2}x + 9$

$y = \dfrac{3}{2}x$

21. Since slope is 0, this is a horizontal line.
$y = 5$

23. $f(x) = -2x + b$
$f(-4) = 23$
$-2(-4) + b = 23$
$8 + b = 23$
$b = 15$
$f(x) = -2x + 15$

25. $f(x) = \dfrac{3}{5}x + b$

$f(-15) = -17$

$\dfrac{3}{5}(-15) + b = -17$

$-9 + b = -17$
$b = -8$

$f(x) = \dfrac{3}{5}x - 8$

27. $m = \dfrac{2 - (-3)}{7 - 2} = \dfrac{5}{5} = 1$

$y - 2 = 1(x - 7)$
$y - 2 = x - 7$
$y = x - 5$

29. $m = \dfrac{9 - (-3)}{1 - (-3)} = \dfrac{12}{4} = 3$

$y - 9 = 3(x - 1)$
$y - 9 = 3x - 3$
$y = 3x + 6$

31. $m = \dfrac{18 - (-12)}{15 - (-10)} = \dfrac{30}{25} = \dfrac{6}{5}$

$y - 18 = \dfrac{6}{5}(x - 15)$

$y - 18 = \dfrac{6}{5}x - 18$

$y = \dfrac{6}{5}x$

33. $m = \dfrac{-3 - (-9)}{-2 - (-2)} = \dfrac{6}{0}$

m is undefined.

$x = -2$

35. $m = \dfrac{2 - 0}{0 - 6} = \dfrac{2}{-6} = -\dfrac{1}{3}$

$y = -\dfrac{1}{3}x + 2$

37. $m = \dfrac{-4 - 0}{0 - 4} = \dfrac{-4}{-4} = 1$

$y = x - 4$

39. $m = \dfrac{-6 - 0}{0 - 6} = \dfrac{-6}{-6} = 1$

$y = x - 6$

41. a) $m = \dfrac{116 - 66}{8 - 3} = \dfrac{50}{5} = 10$

$y - 66 = 10(x - 3)$
$y - 66 = 10x - 30$
$y = 10x + 36$

b) $y = 10(0) + 36 = 36$
$\$36$

43. a) $m = \dfrac{1126 - 613.50}{76 - 35} = \dfrac{512.5}{41} = 12.50$

$y - 1126 = 12.50(x - 76)$
$y - 1126 = 12.50x - 950$
$y = 12.50x + 176$

b) $y = 12.50(0) + 176 = 176$
$\$176$

c) $m = 12.50$
$\$12.50$

45. a) $m = \dfrac{5750 - 4100}{175 - 120} = \dfrac{1650}{55} = 30$

$y - 4100 = 30(x - 120)$
$y - 4100 = 30x - 3600$
$y = 30x + 500$

b) $y = 30(0) + 500 = 500$
$\$500$

c) $m = 30$
$\$30$

d) $y = 30(200) + 500 = 6000 + 500 = 6500$
$\$6500$

47. $m = -2$
$y - (-3) = -2(x - (-6))$
$y + 3 = -2(x + 6)$
$y + 3 = -2x - 12$
$y = -2x - 15$

49. $-3x + y = 9$
$\qquad y = 3x + 9$
$m = 3$

$y - (-6) = 3(x - 4)$
$\quad y + 6 = 3x - 12$
$\qquad y = 3x - 18$

51. $y = 7$

53. $y = 3x - 7$ has a slope of 3.

Perpendicular line has a slope of $-\dfrac{1}{3}$.

$y - 4 = -\dfrac{1}{3}(x - 9)$

$y - 4 = -\dfrac{1}{3}x + 3$

$y = -\dfrac{1}{3}x + 7$

55. $4x + 5y = 9$
$$5y = -4x + 9$$
$$y = -\frac{4}{5}x + \frac{9}{5}$$

Perpendicular line has a slope of $\frac{5}{4}$:

$$y - 2 = \frac{5}{4}\left(x - (-4)\right)$$
$$y - 2 = \frac{5}{4}(x + 4)$$
$$y - 2 = \frac{5}{4}x + 5$$
$$y = \frac{5}{4}x + 7$$

57. $x = -5$

59. Slope of the graphed line is 1. Parallel line has a slope of 1 also and passes through $(-4, -1)$.
$$y - (-1) = 1\left(x - (-4)\right)$$
$$y + 1 = x + 4$$
$$y = x + 3$$

61. Slope of the graphed line is 4. Perpendicular line has a slope of $-\frac{1}{4}$ and passes through $(-4, 3)$.
$$y - 3 = -\frac{1}{4}\left(x - (-4)\right)$$
$$y - 3 = -\frac{1}{4}(x + 4)$$
$$y - 3 = -\frac{1}{4}x - 1$$
$$y = -\frac{1}{4}x + 2$$

63. $y - (-8) = -3(x - 4)$
$$y + 8 = -3x + 12$$
$$y = -3x + 4$$

65. $5x - 3y = -3$
$$-3y = -5x - 3$$
$$y = \frac{-5}{-3}x - \frac{3}{-3}$$
$$y = \frac{5}{3}x + 1$$
$$m = \frac{5}{3}$$

A line parallel to this line has slope $\frac{5}{3}$.

$$y - 8 = \frac{5}{3}\left(x - (-9)\right)$$
$$y - 8 = \frac{5}{3}(x + 9)$$
$$y - 8 = \frac{5}{3}x + 15$$
$$y = \frac{5}{3}x + 23$$

67. $m = 0$
$$y - 11 = 0(x - 13)$$
$$y - 11 = 0$$
$$y = 11$$

69. $x + 4y = 8$
$$4y = -x + 8$$
$$y = \frac{-1}{4}x + \frac{8}{4}$$
$$y = -\frac{1}{4}x + 2$$
$$m = -\frac{1}{4}$$

A line perpendicular to this line has slope
$$-\frac{1}{-\frac{1}{4}} = 4.$$

$$y - 13 = 4\left(x - (-2)\right)$$
$$y - 13 = 4(x + 2)$$
$$y - 13 = 4x + 8$$
$$y = 4x + 21$$

71. $m = \dfrac{33 - (-23)}{5 - 1} = \dfrac{56}{4} = 14$
$$y - 33 = 14(x - 5)$$
$$y - 33 = 14x - 70$$
$$y = 14x - 37$$

73. Explanations will vary. Example: Every point on a vertical line has the same x-coordinate. The equation is $x = a$. Similarly, every point on a horizontal line has the same y-coordinate. The equation is $y = b$.

3.7   QUICK CHECK

1. $y = 2x - 6$

$m = 2$, y-intercept: $(0, -6)$

From the y-intercept count up 2 and right 1 to find the next point $(1, -4)$.

Test point: $(0, 0)$
$0 \geq 2(0) - 6$
$0 \geq -6$

True. Shade on the side of the line that contains the test point.

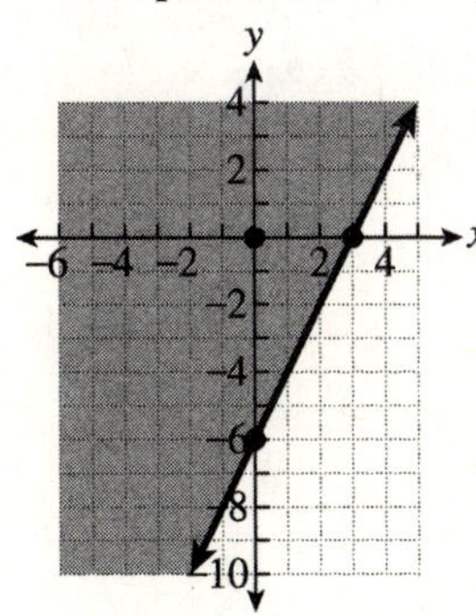

2. $x + 2y = 6$

x-intercept:
$x + 2(0) = 6$
$x = 6$
$(6, 0)$

y-intercept:
$(0) + 2y = 6$
$2y = 6$
$y = 3$
$(0, 3)$

Test Point: $(0, 0)$
$0 + 2(0) < 6$
$0 < 6$

True. Shade on the side of the line that contains the test point.

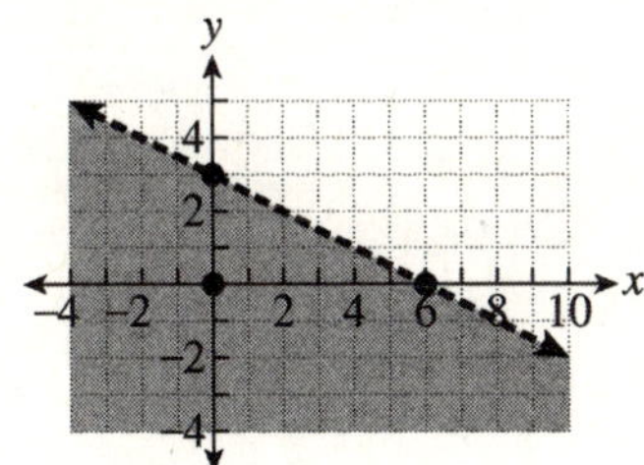

3. $y = -\dfrac{4}{3}x$

$m = -\dfrac{4}{3}$, y-intercept: $(0, 0)$

From the y-intercept count down 4 and right 3 to find the next point $(3, -4)$.

Test Point: $(1, 1)$

$1 \leq -\dfrac{4}{3}(1)$

$1 \leq -\dfrac{4}{3}$

False. Shade on the side of the line that does not contain the test point.

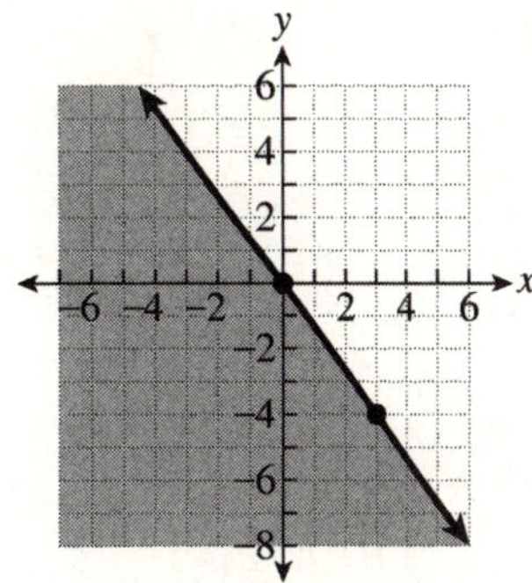

4. x represents the number of apples.
y represents the number of pears.
$x + y \geq 60$
Graph $x + y = 60$.

x-intercept:
$x + (0) = 60$
$x = 60$
$(60, 0)$

y-intercept:
$(0) + y = 60$
$y = 60$
$(0, 60)$

Test Point: $(0, 0)$
$0 + 0 \geq 60$
$0 \geq 60$

False. Shade on the side of the line that does not contain the test point.

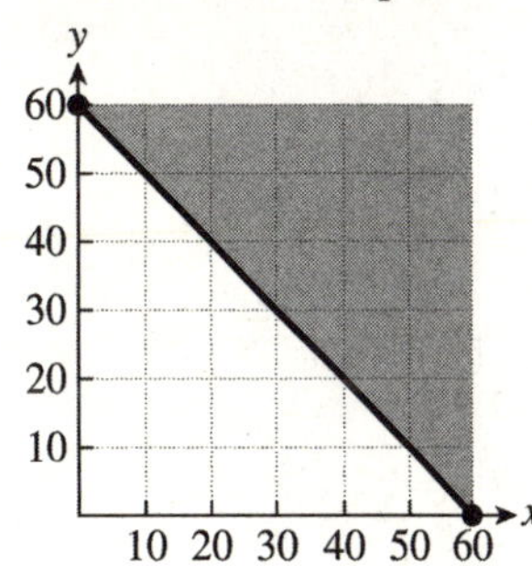

3.7   LINEAR INEQUALITIES

1. solution

3. dashed

5. $5x + 3y \le 22$

a) $5(0) + 3(0) \le 22$
$0 + 0 \le 22$
$0 \le 22$
Yes.

b) $5(8) + 3(-4) \le 22$
$40 - 12 \le 22$
$28 \le 22$
No.

c) $5(2) + 3(4) \le 22$
$10 + 12 \le 22$
$22 \le 22$
Yes.

d) $5(-3) + 3(9) \le 22$
$-15 + 27 \le 22$
$12 \le 22$
Yes.

7. $y < 6x - 11$

a) $8 < 6(5) - 11$
$8 < 30 - 11$
$8 < 19$
Yes.

b) $0 < 6(0) - 11$
$0 < 0 - 11$
$0 < -11$
No.

c) $1 < 6(2) - 11$
$1 < 12 - 11$
$1 < 1$
No.

d) $-13 < 6(-4) - 11$
$-13 < -24 - 11$
$-13 < -35$
No.

9. a) $0 < 7$ Yes.

b) $6 < 7$ Yes.

c) $10 < 7$ No.

d) $7 < 7$ No.

11. Test Point: $(0,0)$
$4(0) + 0 \ge 7$
$0 \ge 7$
False. Shade on the side of the line that does not contain the test point.

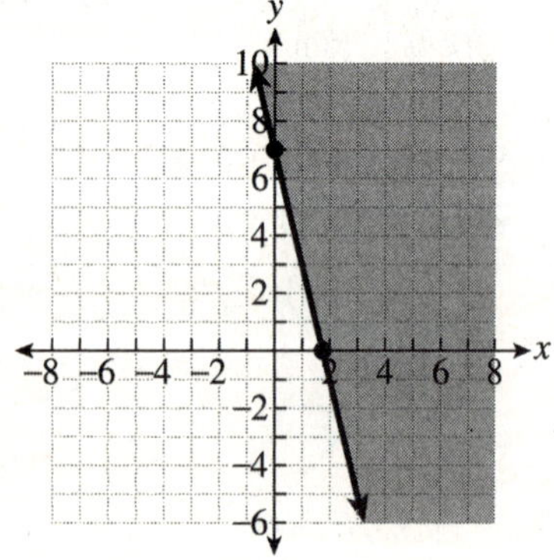

13. Test Point: $(0,0)$
$2(0) + 8(0) < -4$
$0 < -4$
False. Shade on the side of the line that does not contain the test point.

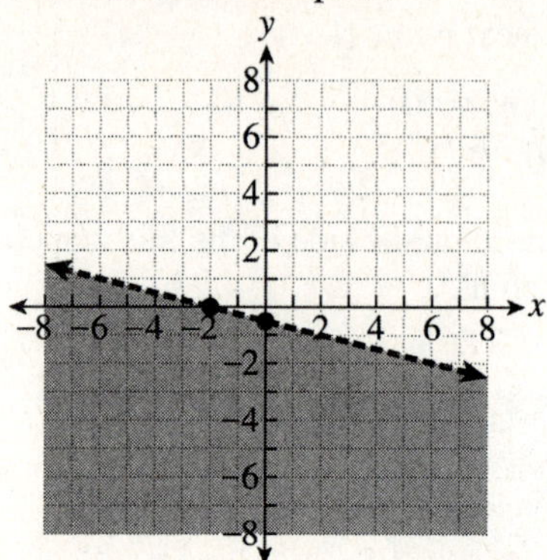

15. Test Point: $(0,0)$
$3(0) + 4(0) \ge 24$
$0 \ge 24$
False.
A

17. Since the inequality is $<$, the line must be dashed.
A

19. Test Point: $(0,0)$
$8(0) - 3(0) \_\_\_ 24$
$0 \_\_\_ 24$
$\le$

21. Test Point: $(1,0)$
$-5(1) + 8(0) \_\_\_ 0$
$-5 \_\_\_ 0$
$<$

23. $6x - 5y = -30$

x-intercept: $\qquad$ y-intercept:
$$6x - 5(0) = -30 \qquad 6(0) - 5y = -30$$
$$6x = -30 \qquad\quad -5y = -30$$
$$x = -5 \qquad\qquad y = 6$$
$(-5, 0)$ $\qquad\qquad$ $(0, 6)$

Test Point: $(0, 0)$
$$6(0) - 5(0) < -30$$
$$0 < -30$$
False. Shade on the side of the line that does not contain the test point.

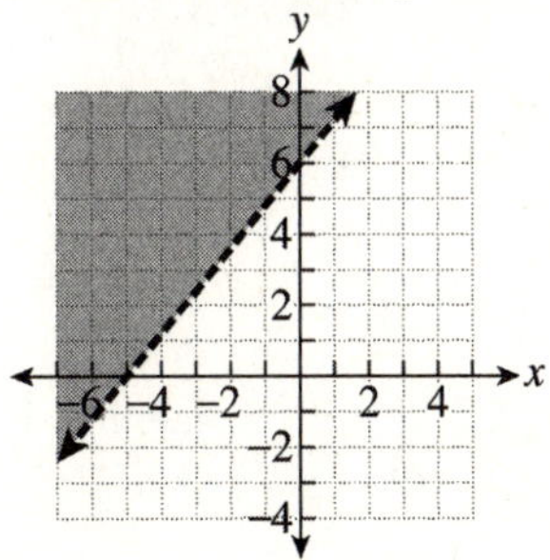

25. $-9x + 4y = 0$

x-intercept: $(0, 0)$ $\qquad$ y-intercept: $(0, 0)$

Test Point: $(1, 0)$
$$-9(1) + 4(0) \le 0$$
$$-9 \le 0$$
True. Shade on the side of the line that contains the test point.

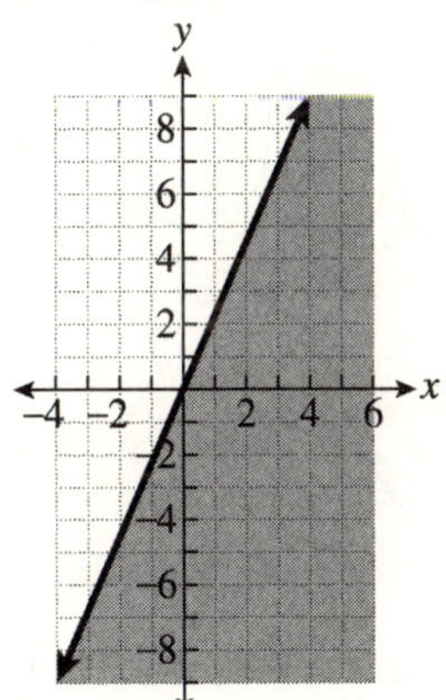

27. $y = 3$
Test Point: $(0, 0)$
$$0 > 3$$
False. Shade on the side of the line that does not contain the test point.

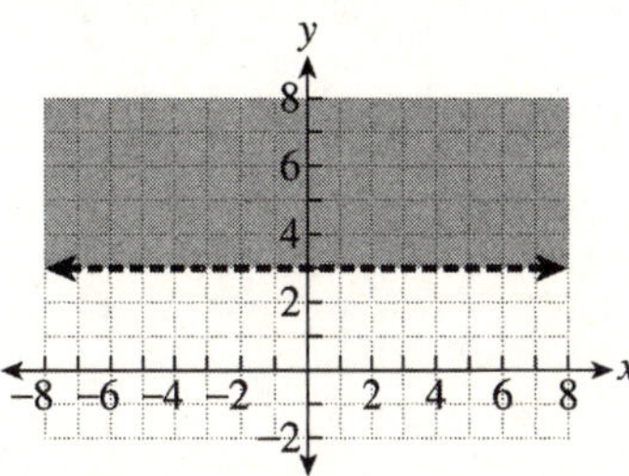

29. $y = \dfrac{2}{3}x - 6$

$m = \dfrac{2}{3}$, y-intercept: $(0, -6)$

From the y-intercept count up 2 and right 3 to find the next point $(3, -4)$.
Test Point: $(0, 0)$
$$0 \ge \frac{2}{3}(0) - 6$$
$$0 \ge -6$$
True. Shade on the side of the line that contains the test point.

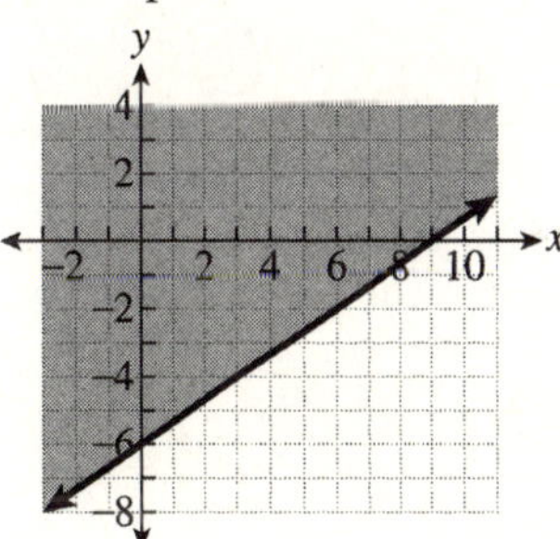

31. $y = -2x - 4$
$m = -2$, y-intercept: $(0, -4)$
From the y-intercept count down 2 and right 1 to find the next point $(1, -6)$.
Test Point: $(0, 0)$
$$0 \le -2(0) - 4$$
$$0 \le -4$$
False. Shade on the side of the line that does not contain the test point.

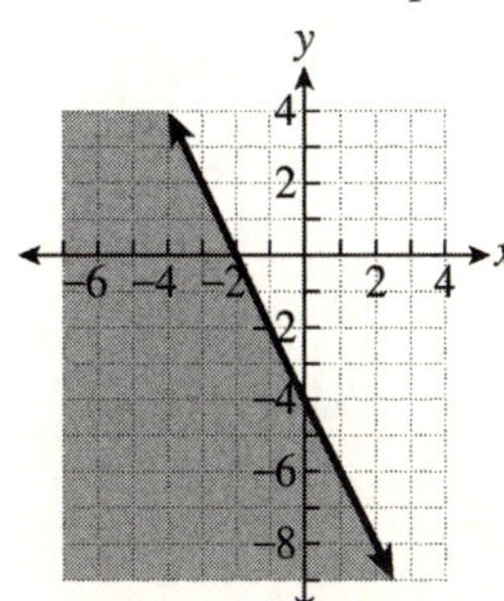

33. $y = -\dfrac{3}{2}x - \dfrac{3}{2}$

$m = -\dfrac{3}{2},\ \ y\text{-intercept: } \left(0, -\dfrac{3}{2}\right)$

From the y-intercept count down 3 and right 2

to find the next point $\left(2, -\dfrac{9}{2}\right)$.

Test Point: $(0, 0)$

$0 > -\dfrac{3}{2}(0) - \dfrac{3}{2}$

$0 > -\dfrac{3}{2}$

True. Shade on the side of the line that contains the test point.

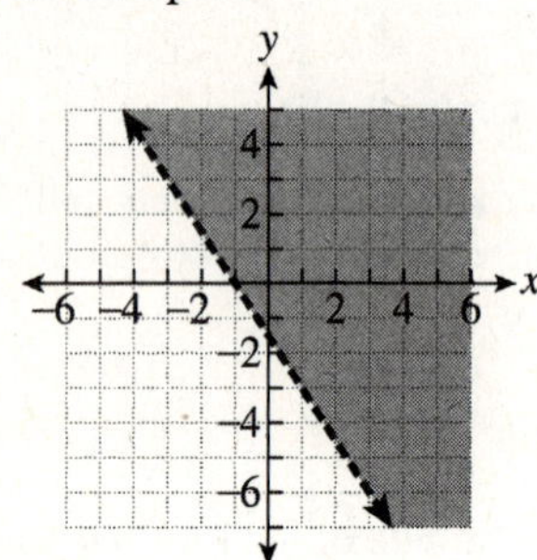

35. $x = -2$
Test Point: $(0, 0)$
$0 \le -2$
False. Shade on the side of the line that does not contain the test point.

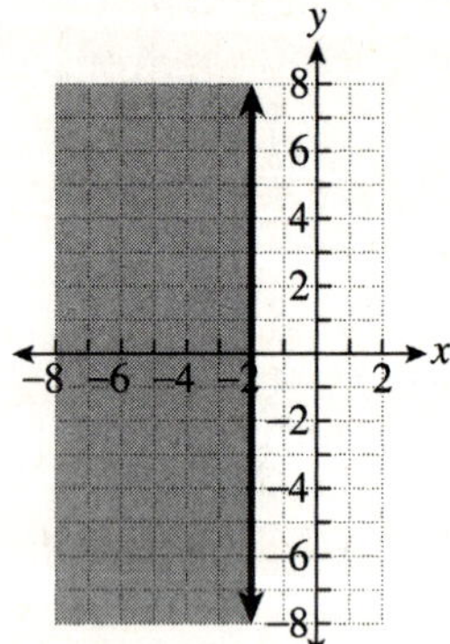

37. $y = -x - 5$
$m = -1$
y-intercept: $(0, -5)$
Test Point: $(-6, 0)$
$0 \underline{\ \ \ } -(-6) - 5$

$0 \underline{\ \ \ } 6 - 5$

$0 \ \underline{<}\ 1$

$y \le -x - 5$

39. $y = 8x$
$m = 8$
y-intercept: $(0, 0)$
Test Point: $(0, 2)$
$2 \underline{\ \ \ } 8(0)$

$2 \ \underline{>}\ 0$

$y > 8x$

41. $-x + 3y = 9$

x-intercept:	y-intercept:
$-x + 3(0) = 9$	$-(0) + 3y = 9$
$-x = 9$	$3y = 9$
$x = -9$	$y = 3$
$(-9, 0)$	$(0, 3)$

Test Point: $(0, 0)$
$-(0) + 3(0) \ge 9$
$0 \ge 9$

False. Shade on the side of the line that does not contain the test point.

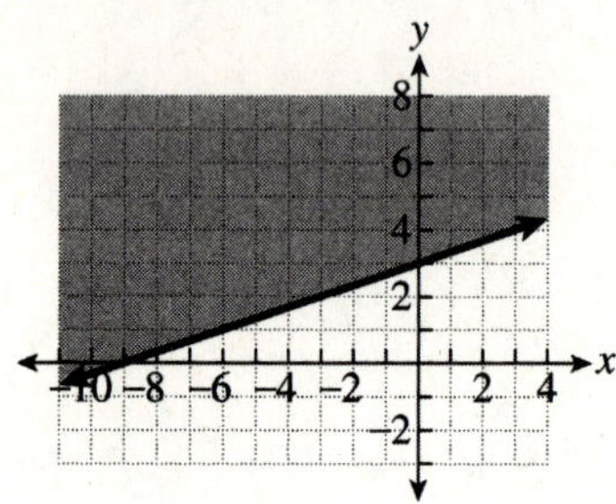

43. $4x - 3y = 24$

x-intercept:	y-intercept:
$4x - 3(0) = 24$	$4(0) - 3y = 24$
$4x = 24$	$-3y = 24$
$x = 6$	$y = -8$
$(6, 0)$	$(0, -8)$

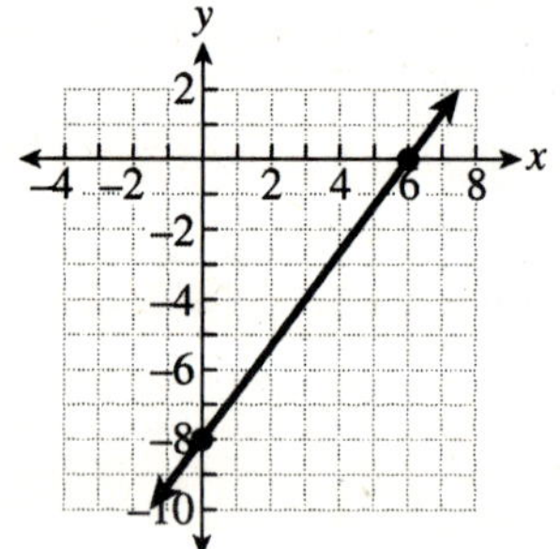

45. $5x + 4y = 0$
$\qquad 4y = -5x$
$$y = -\frac{5}{4}x$$

x-intercept and y-intercept are both $(0,0)$.

From the origin count down 5 and right 4 to find the next point $(4,-5)$.

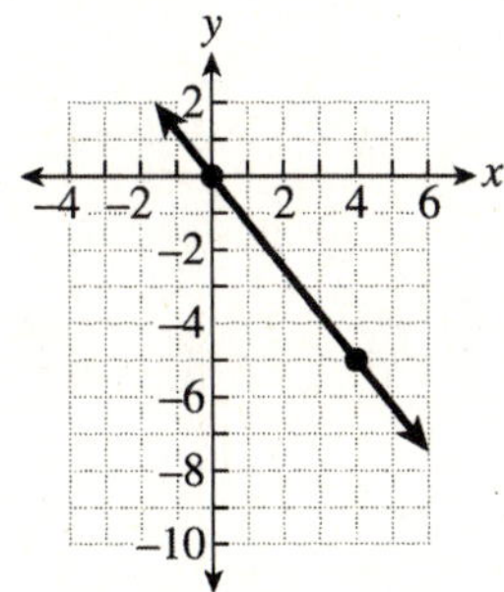

47. $y = -3x + 8$
$m = -3$, y-intercept: $(0,8)$

From the y-intercept count down 3 and right 1 to find the next point $(1,5)$.

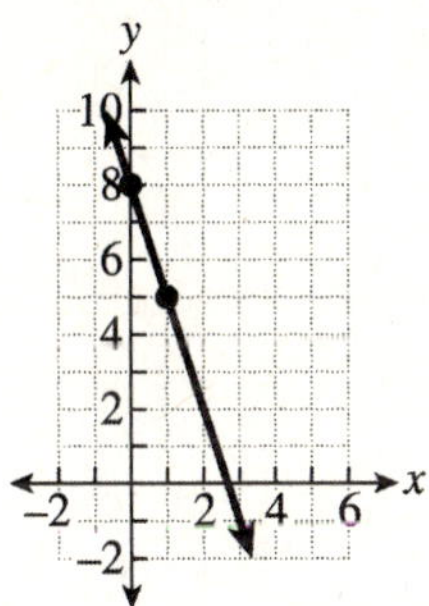

49. $y = 2x - 30$
$m = 2$, y-intercept: $(0,-30)$
x-intercept: $\quad 0 = 2x - 30$
$$\qquad\qquad -2x = -30$$
$$\qquad\qquad\quad x = 15$$
$(15,0)$
Test Point: $(0,0)$
$0 > 2(0) - 30$
$0 > -30$
True. Shade on the side of the line that contains the test point.

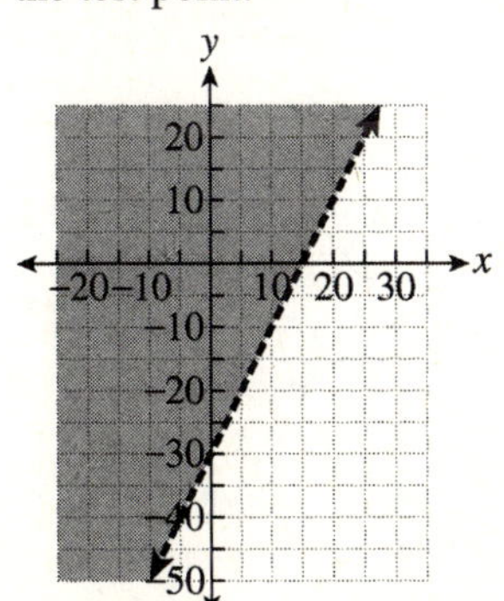

51. Let x represent the number of faculty in attendance.
Let y represent the number of students in attendance.
$x + y > 50$
Test Point: $(0,0)$
$0 + 0 > 50$
$\quad 0 > 50$
False. Shade on the side of the line that does not contain the test point.

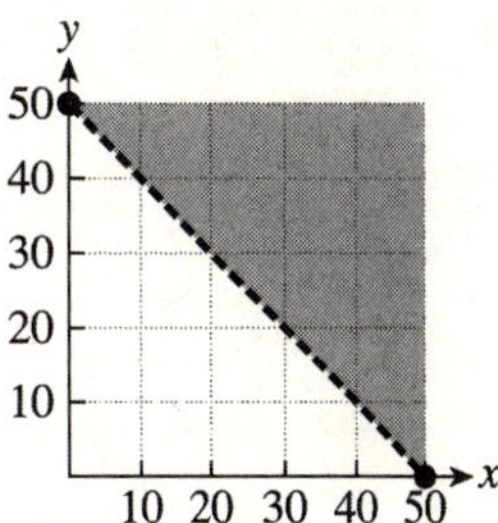

53. $x + y = 3$ $\qquad\qquad$ $y = x - 7$
$\quad y = -x + 3$ $\qquad$ y-intercept: $(0,-7)$
y-intercept: $(0,3)$ $\qquad$ x-intercept: $(7,0)$
x-intercept: $(3,0)$ $\qquad$ Test Point: $(0,0)$
Test Point: $(0,0)$ $\qquad$ $0 > 0 - 7$
$0 + 0 < 3$ $\qquad\qquad$ $0 > -7$
$\quad 0 < 3$ $\qquad\qquad$ True. Shade on the
True. Shade on the $\qquad$ side of the line that
side of the line that $\qquad$ contains the test point.
contains the test point.

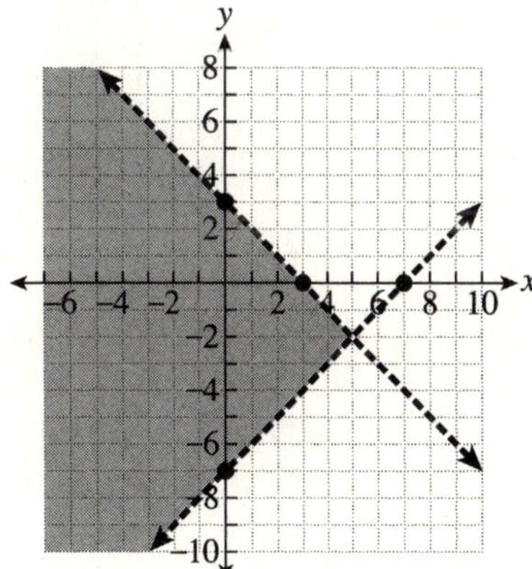

55. Explanations will vary. Example:
Since points on the line are not solutions when $<$ or $>$ are involved, we use a dashed line when graphing.
Since points on the line are solutions if $\leq$ or $\geq$ are involved, we use a solid line when graphing.

CHAPTER 3 REVIEW

1. $A(7,-2)$
  $B(-5,-1)$
  $C(-1,-5)$
  $D(0,8)$

2. III

3. II

4. $4(5)-2(-2)=16$
  $20+4=16$
  $24=16$
  No.

5. $4\left(\dfrac{7}{2}\right)+5\left(\dfrac{3}{5}\right)=17$
  $14+3=17$
  $17=17$
  Yes.

6. $(-6)-3(-4)=6$
  $-6+12=6$
  $6=6$
  Yes.

7. x-intercept:
  $2x-8(0)=-40$
  $2x=-40$
  $x=-20$
  $(-20,0)$

  y-intercept:
  $2(0)-8y=-40$
  $-8y=-40$
  $y=5$
  $(0,5)$

8. x-intercept:
  $5x-(0)=-15$
  $5x=-15$
  $x=-3$
  $(-3,0)$

  y-intercept:
  $5(0)-y=-15$
  $-y=-15$
  $y=15$
  $(0,15)$

9. x-intercept:
  $3x-7(0)=0$
  $3x=0$
  $x=0$
  $(0,0)$

  y-intercept:
  $3(0)-7y=0$
  $-7y=0$
  $y=0$
  $(0,0)$

10. x-intercept:
  $4x-7(0)=14$
  $4x=14$
  $x=\dfrac{7}{2}$
  $\left(\dfrac{7}{2},0\right)$

  y-intercept:
  $4(0)-7y=14$
  $-7y=14$
  $y=-2$
  $(0,-2)$

11. x-intercept:
  $-3x+2(0)=12$
  $-3x=12$
  $x=-4$
  $(-4,0)$

  y-intercept:
  $-3(0)+2y=12$
  $2y=12$
  $y=6$
  $(0,6)$

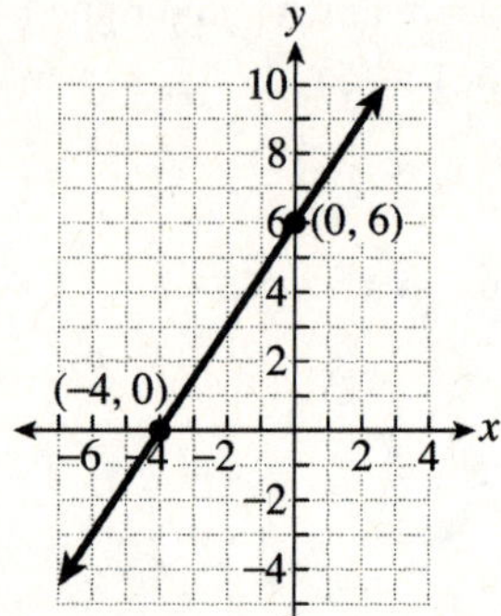

12. x-intercept:
  $x-4(0)=-8$
  $x=-8$
  $(-8,0)$

  y-intercept:
  $(0)-4y=-8$
  $-4y=-8$
  $y=2$
  $(0,2)$

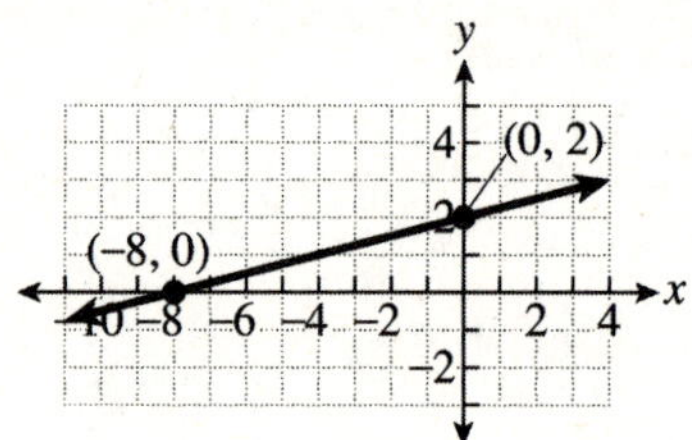

13. x-intercept:
  $2x+3(0)=0$
  $2x=0$
  $x=0$
  $(0,0)$

  y-intercept:
  $2(0)+3y=0$
  $3y=0$
  $y=0$
  $(0,0)$

  When $x=1$: $\left(1,-\dfrac{2}{3}\right)$
  $2(1)+3y=0$
  $2+3y=0$
  $3y=-2$
  $y=-\dfrac{2}{3}$

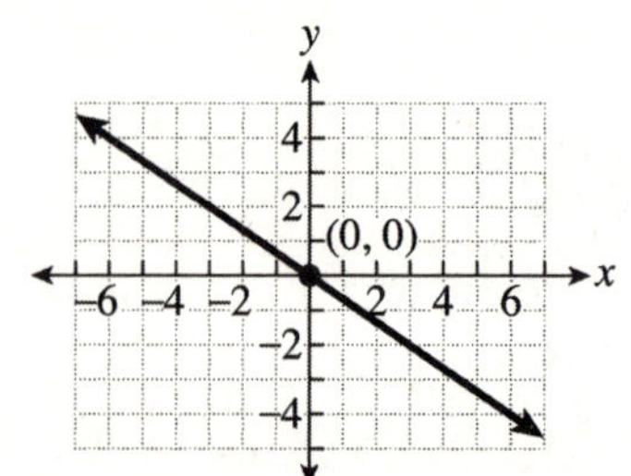

14. x-intercept:

$4x + 3(0) = -6$

$4x = -6$

$x = -\dfrac{3}{2}$

$\left(-\dfrac{3}{2}, 0\right)$

y-intercept:

$4(0) + 3y = -6$

$3y = -6$

$y = -2$

$(0, -2)$

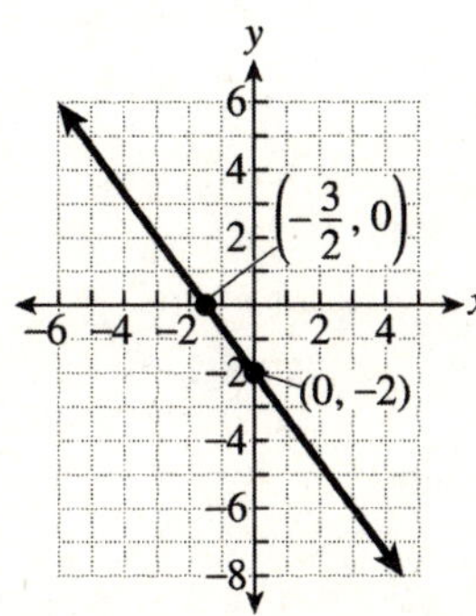

15. x-intercept:

$0 = -\dfrac{3}{2}x + 6$

$\dfrac{3}{2}x = 6$

$x = 4$

$(4, 0)$

y-intercept:

By slope-intercept form, $(0, 6)$.

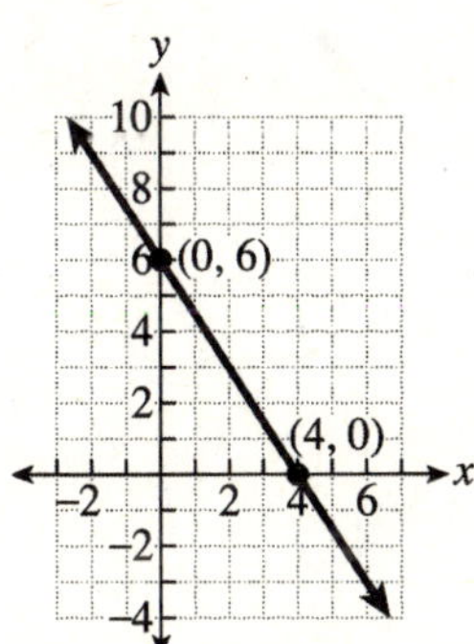

16. x-intercept:

$0 = 2x - 4$

$4 = 2x$

$2 = x$

$(2, 0)$

y-intercept:

By slope-intercept form, $(0, -4)$.

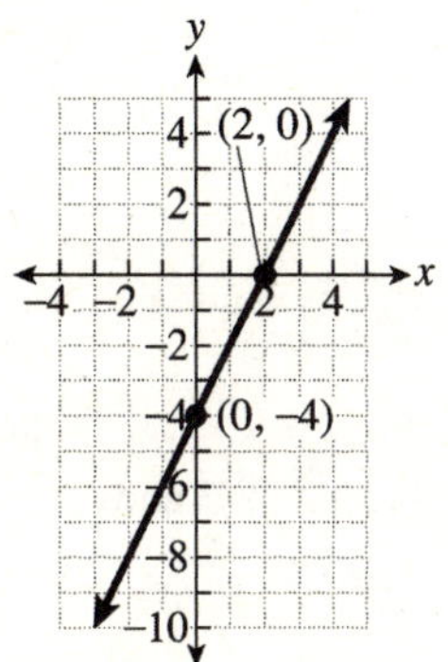

17. $m = \dfrac{9 - 3}{-3 - (-5)} = \dfrac{6}{2} = 3$

18. $m = \dfrac{10 - (-2)}{3 - 6} = \dfrac{12}{-3} = -4$

19. $m = \dfrac{-3 - (-8)}{8 - 3} = \dfrac{5}{5} = 1$

20. $m = \dfrac{-7 - (-7)}{7 - 2} = \dfrac{0}{5} = 0$

21. $m = -2$, y-intercept: $(0, 7)$

22. $m = 4$, y-intercept: $(0, -6)$

23. $m = -\dfrac{2}{3}$, y-intercept: $(0, 0)$

24. $4x - 2y = 9$

$-2y = -4x + 9$

$y = \dfrac{-4}{-2}x + \dfrac{9}{-2}$

$y = 2x - \dfrac{9}{2}$

$m = 2$, y-intercept: $\left(0, -\dfrac{9}{2}\right)$

25. $5x + 3y = 18$
$$3y = -5x + 18$$
$$y = \frac{-5}{3}x + \frac{18}{3}$$
$$y = -\frac{5}{3}x + 6$$
$m = -\frac{5}{3}$, y-intercept: $(0, 6)$

26. $m = 0$, y-intercept: $(0, -7)$

27. $m = 3$, y-intercept: $(0, -6)$
From the y-intercept count up 3 and right 1 to find the next point $(1, -3)$.
x-intercept: $0 = 3x - 6$
$$6 = 3x$$
$$2 = x$$
$(2, 0)$

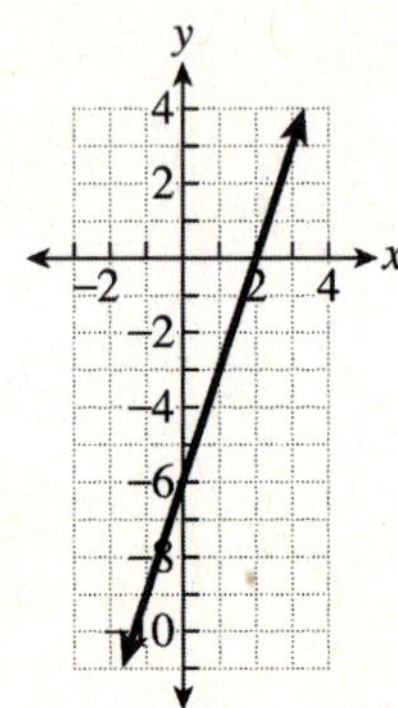

28. $m = -5$, y-intercept: $(0, -5)$
From the y-intercept count 5 down and 1 right to find the next point $(1, -10)$.
x-intercept: $0 = -5x - 5$
$$5x = -5$$
$$x = -1$$
$(-1, 0)$

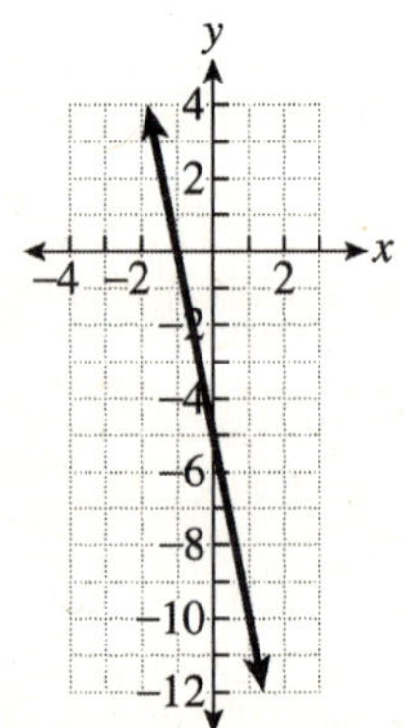

29. $m = \frac{2}{5}$, y-intercept: $(0, 2)$

From the y-intercept count up 2 and right 5 to find the next point $(5, 4)$.

x-intercept: $0 = \frac{2}{5}x + 2$
$$-2 = \frac{2}{5}x$$
$$-5 = x$$
$(-5, 0)$

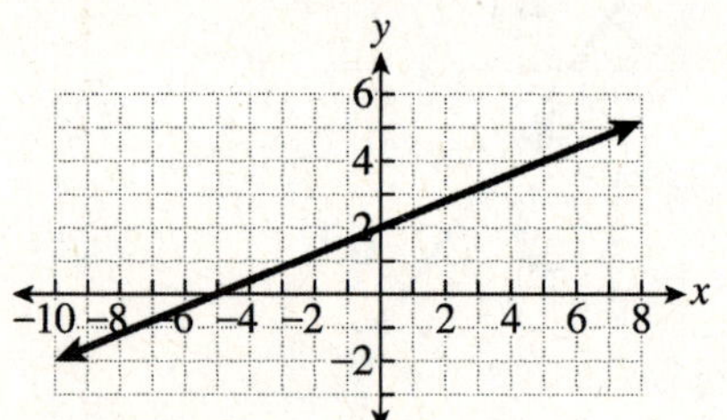

30. $m = 1$, y-intercept: $(0, 0)$
From the y-intercept count up 1 and right 1 to find the next point $(1, 1)$.
x-intercept: $(0, 0)$

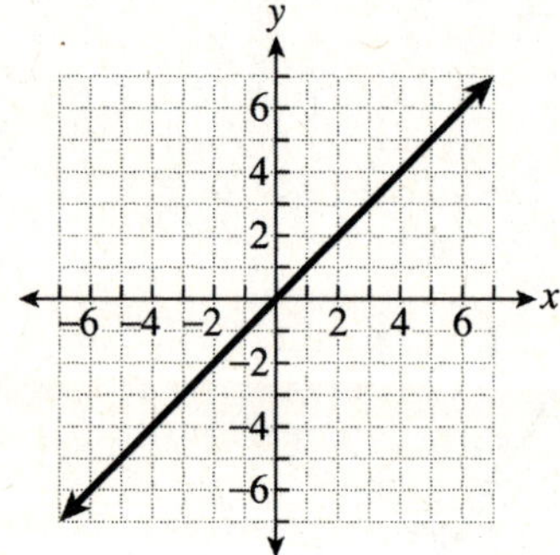

31. $m = -\dfrac{7}{3}$, y-intercept: $(0,3)$

From the y-intercept count down 7 and right 3 to find the next point $(3,-4)$.

x-intercept: $0 = -\dfrac{7}{3}x + 3$

$$\dfrac{7}{3}x = 3$$

$$x = \dfrac{9}{7}$$

$\left(\dfrac{9}{7}, 0\right)$

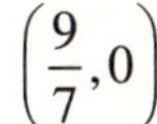

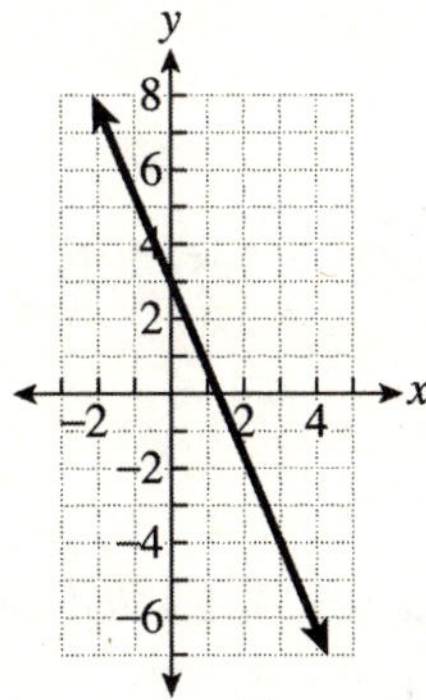

32. $m = \dfrac{1}{2}$, y-intercept: $\left(0, \dfrac{3}{2}\right)$

From the y-intercept count up 1 and right 2 to find the next point $\left(2, \dfrac{5}{2}\right)$.

x-intercept: $0 = \dfrac{1}{2}x + \dfrac{3}{2}$

$$-\dfrac{3}{2} = \dfrac{1}{2}x$$

$$-3 = x$$

$(-3,0)$

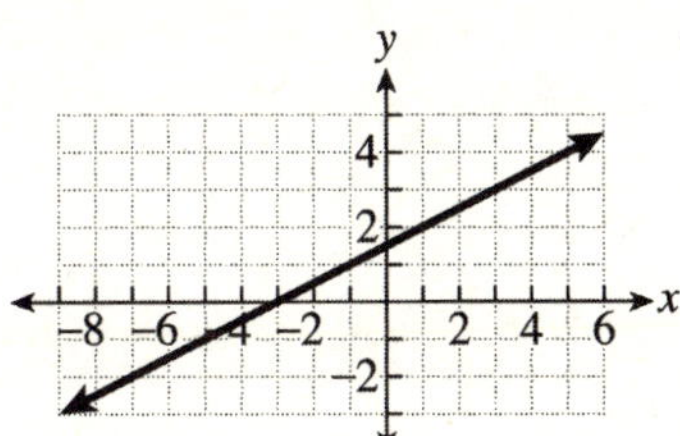

33. $y = 2x - 9$ $y = -2x + 9$
$m = 2$ $m = -2$
The lines are neither parallel or perpendicular.

34. $4x + y = 11$ $y = \dfrac{1}{4}x + \dfrac{5}{2}$
$\quad\quad y = -4x + 11$
$\quad m = -4$ $m = \dfrac{1}{4}$

The lines are perpendicular.

35. $12x - 9y = 17$ $-8x + 6y = 10$
$\quad\quad -9y = -12x + 17$ $6y = 8x + 10$

$$y = \dfrac{-12}{-9}x + \dfrac{17}{-9} \qquad\qquad y = \dfrac{8}{6}x + \dfrac{10}{6}$$

$$y = \dfrac{4}{3}x - \dfrac{17}{9} \qquad\qquad y = \dfrac{4}{3}x + \dfrac{5}{3}$$

$$m = \dfrac{4}{3} \qquad\qquad\qquad m = \dfrac{4}{3}$$

The lines are parallel.

36. $y = -4x - 2$

37. $y = \dfrac{2}{5}x - 6$

38. **a)** $f(x) = 720 + 50x$

b) $f(11) = 720 + 50(11) = 720 + 550 = 1270$
$\$1270$

c) $2750 = 720 + 50x$
$2030 = 50x$
$40.6 = x$
$41 \approx x$
41 months

39. $f(-2) = 9(-2) + 7 = -18 + 7 = -11$

40. $g(-5) = 3 - 8(-5) = 3 + 40 = 43$

41. $f(5a - 1) = 3(5a - 1) + 7$
$\quad\quad\quad\quad\quad = 15a - 3 + 7$
$\quad\quad\quad\quad\quad = 15a + 4$

42. **a)** $f(-2) = -7$ **b)** $x = 6$

c) $(-\infty, \infty)$ **d)** $(-\infty, \infty)$

43. $y - (-2) = 1(x - 4)$
$\quad\quad y + 2 = x - 4$
$\quad\quad\quad y = x - 6$

44. $y - 3 = -5(x - 1)$
$\quad y - 3 = -5x + 5$
$\quad\quad\quad y = -5x + 8$

45.
$$y - 9 = -\frac{3}{2}\big(x - (-4)\big)$$
$$y - 9 = -\frac{3}{2}(x + 4)$$
$$y - 9 = -\frac{3}{2}x - 6$$
$$y = -\frac{3}{2}x + 3$$

46.
$$m = \frac{-7 - 1}{2 - (-2)} = \frac{-8}{4} = -2$$
$$y - (-7) = -2(x - 2)$$
$$y + 7 = -2x + 4$$
$$y = -2x - 3$$

47.
$$m = \frac{7 - (-5)}{2 - (-6)} = \frac{12}{8} = \frac{3}{2}$$
$$y - 7 = \frac{3}{2}(x - 2)$$
$$y - 7 = \frac{3}{2}x - 3$$
$$y = \frac{3}{2}x + 4$$

48.
$$m = \frac{3 - 3}{-9 - 5} = \frac{0}{-14} = 0$$
$$y - 3 = 0(x - 3)$$
$$y - 3 = 0$$
$$y = 3$$

49.
$$m = \frac{3}{2}$$
y-intercept: $(0, 6)$
$$y = \frac{3}{2}x + 6$$

50. $m = 3$
y-intercept: $(0, 0)$
$$y = 3x$$

51. $m = 4$, y-intercept: $(0, 8)$
From the y-intercept count up 4 and right 1
to find the next point $(1, 12)$.
x-intercept: $0 = 4x + 8$
$$-8 = 4x$$
$$-2 = x$$
$(-2, 0)$

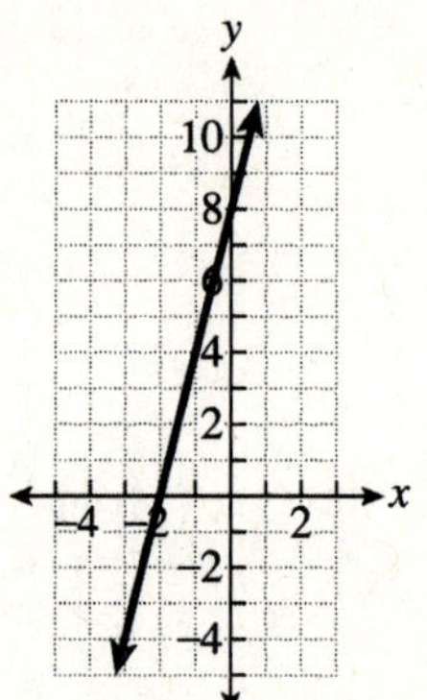

52. $m = -\frac{1}{2}$, y-intercept: $(0, 3)$

From the y-intercept count down 1 and right 2
to find the next point $(2, 2)$.

x-intercept: $0 = -\frac{1}{2}x + 3$
$$\frac{1}{2}x = 3$$
$$x = 6$$
$(6, 0)$

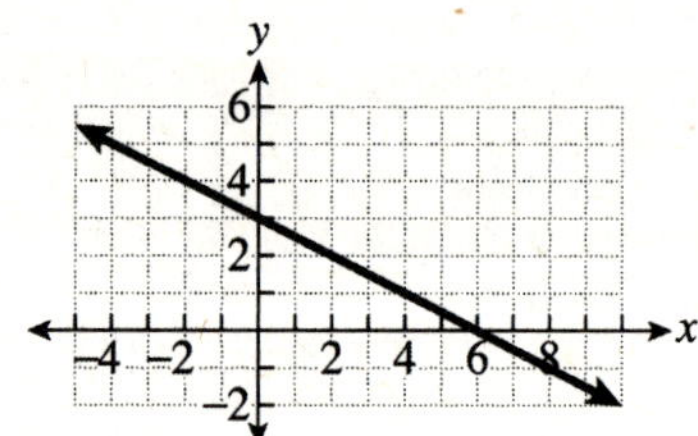

53. $m = -5$, y-intercept: $(0,2)$
From the y-intercept count down 5 and right 1
to find the next point $(1,-3)$.
x-intercept: $0 = -5x + 2$
$$5x = 2$$
$$x = \frac{2}{5}$$
$\left(\frac{2}{5}, 0\right)$

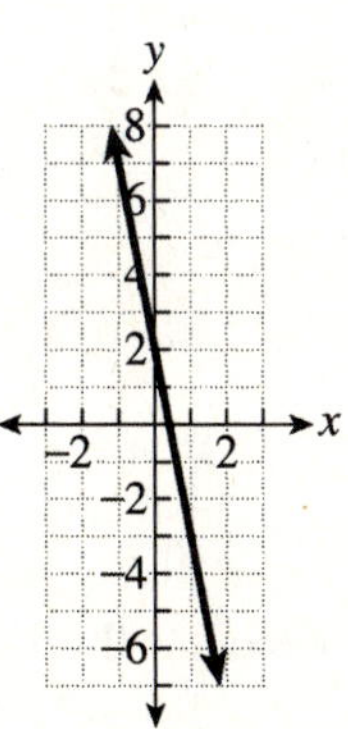

54. x-intercept: $(5,0)$ y-intercept: None

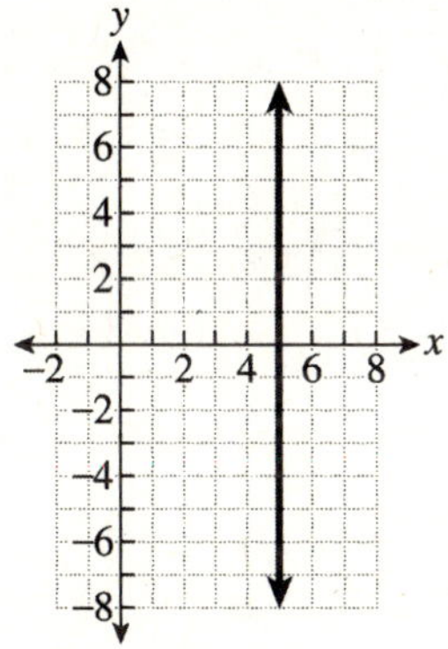

55. x-intercept: y-intercept:
$3x - 4(0) = 24$ $3(0) - 4y = 24$
$$3x = 24$$ $$-4y = 24$$
$$x = 8$$ $$y = -6$$
$(8,0)$ $(0,-6)$

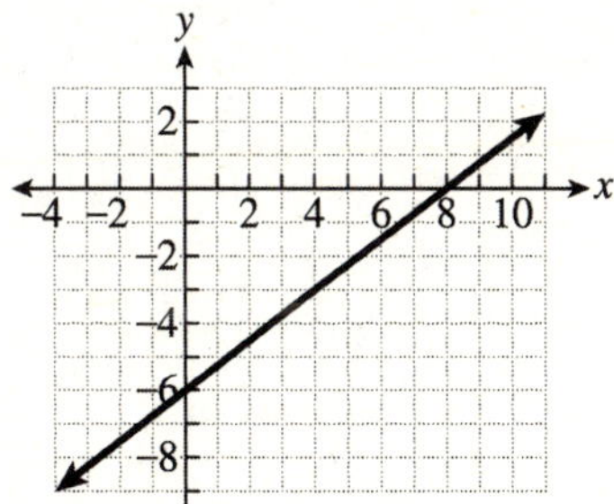

56. x-intercept: none y-intercept: $(0,-6)$

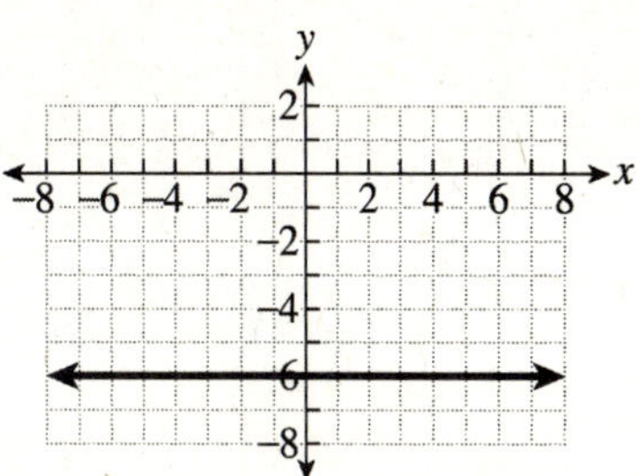

57. $3x + y = -9$
$$y = -3x - 9$$

y-intercept: $(0,-9)$
From the y-intercept count down 3 and right 1
to find the next point $(1,-12)$.
x-intercept: $(-3,0)$
Test Point: $(0,0)$
$$3(0) + 0 < -9$$
$$0 < -9$$
False. Shade on the side of the line that does
not contain the test point.

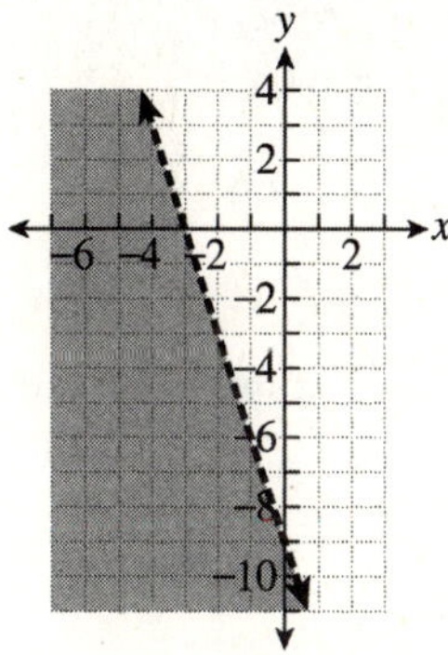

58. $2x + 7y = 14$
$$7y = -2x + 14$$
$$y = \frac{-2}{7}x + \frac{14}{7}$$
$$y = -\frac{2}{7}x + 2$$

x-intercept: y-intercept:
$2x + 7(0) = 14$ $(0,2)$
$$2x = 14$$
$$x = 7$$
$(7,0)$

Test Point: $(0,0)$
$2(0) + 7(0) \geq 14$
$$0 \geq 14$$

False. Shade on the side of the line that does not contain the test point.

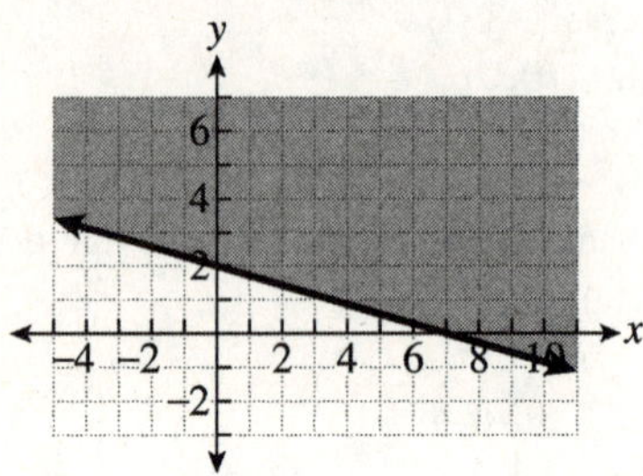

59. $y = -4$
Test Point: $(0,0)$
$0 \leq -4$
False. Shade on the side of the line that does not contain the test point.

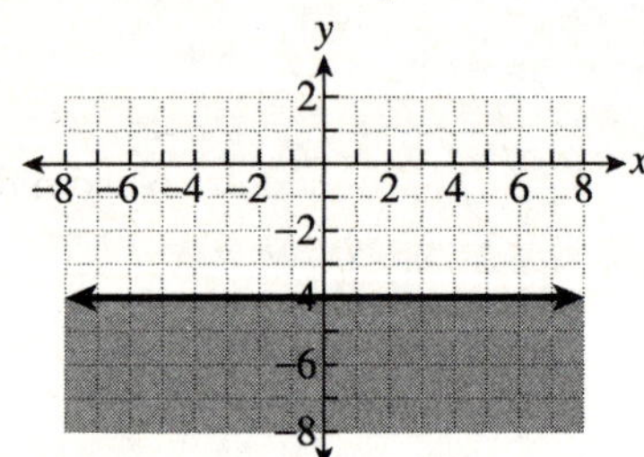

60. $y = \frac{3}{2}x + 5$

$m = \frac{3}{2}$, y-intercept: $(0,5)$
From the y-intercept count up 3 and right 2 to find the next point $(2,8)$.

x-intercept: $0 = \frac{3}{2}x + 5$
$$-5 = \frac{3}{2}x$$
$$-\frac{10}{3} = x$$
$\left(-\frac{10}{3}, 0\right)$

Test Point: $(0,0)$

$0 > \frac{3}{2}(0) + 5$
$$0 > 5$$
False. Shade on the side of the line that does not contain the test point.

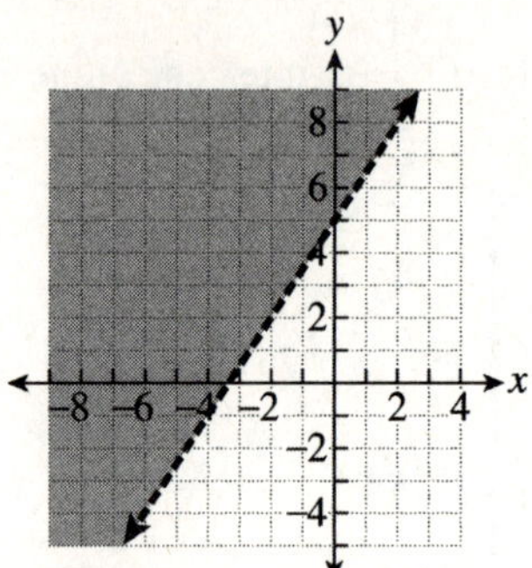

CHAPTER 3 TEST

1. II

2. $-3(4) - 7(-2) = 2$
$$-12 + 14 = 2$$
$$2 = 2$$
Yes.

3. x-intercept: y-intercept:
$5x - 6(0) = -15$ $5(0) - 6y = -15$
$$5x = -15$$ $$-6y = -15$$
$$x = -3$$
$(-3, 0)$ $$y = \frac{5}{2}$$
$$\left(0, \frac{5}{2}\right)$$

4. x-intercept: y-intercept:

$$0 = -\frac{7}{2}x + 21 \qquad y = -\frac{7}{2}(0) + 21$$

$$\frac{7}{2}x = 21 \qquad\qquad y = 21$$

$$\qquad\qquad\qquad\qquad (0, 21)$$

$$x = 6$$

$$(6, 0)$$

5. x-intercept: y-intercept:

$$4x - (0) = 6 \qquad 4(0) - y = 6$$

$$4x = 6 \qquad\qquad -y = 6$$

$$\qquad\qquad\qquad\qquad y = -6$$

$$x = \frac{3}{2} \qquad\qquad (0, -6)$$

$$\left(\frac{3}{2}, 0\right)$$

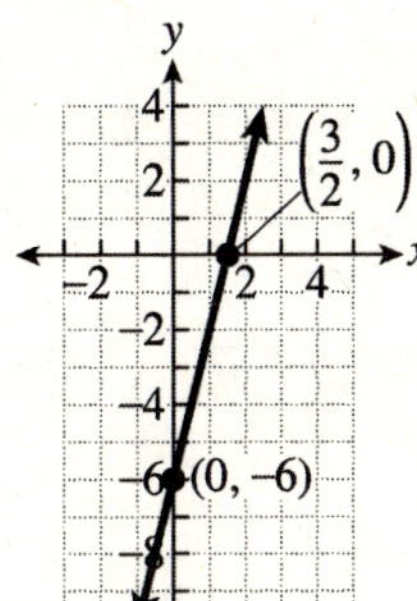

6. x-intercept: y-intercept:

$$0 = \frac{2}{3}x - 4 \qquad y = \frac{2}{3}(0) - 4$$

$$4 = \frac{2}{3}x \qquad\qquad y = -4$$

$$\qquad\qquad\qquad\qquad (0, -4)$$

$$6 = x$$

$$(6, 0)$$

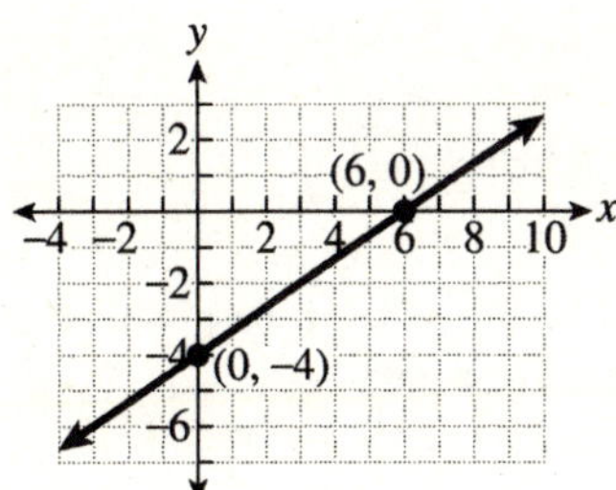

7. $m = \dfrac{9 - 8}{-3 - 2} = -\dfrac{1}{5}$

8. $m = 7$, y-intercept: $(0, -8)$

9. $3x + 5y = 45$

$$5y = -3x + 45$$

$$y = \frac{-3}{5}x + \frac{45}{5}$$

$$y = -\frac{3}{5}x + 9$$

$$m = -\frac{3}{5}, \ y\text{-intercept: } (0, 9)$$

10. $m = -2$, y-intercept: $(0, 3)$
From the y-intercept count down 2 and right 1
to find the next point $(1, 1)$.
x-intercept: $0 = -2x + 3$

$$2x = 3$$

$$x = \frac{3}{2}$$

$$\left(\frac{3}{2}, 0\right)$$

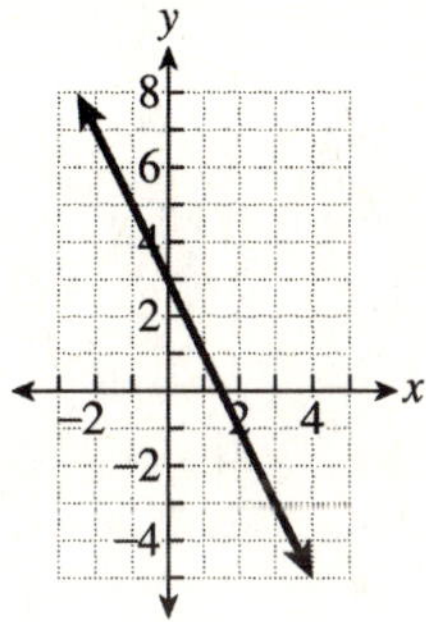

11. $m = -\dfrac{6}{5}$, y-intercept: $(0, 5)$

From the y-intercept count down 6 and right 5
to find the next point $(5, -1)$.

x-intercept: $0 = -\dfrac{6}{5}x + 5$

$$\frac{6}{5}x = 5$$

$$x = \frac{25}{6}$$

$$\left(\frac{25}{6}, 0\right)$$

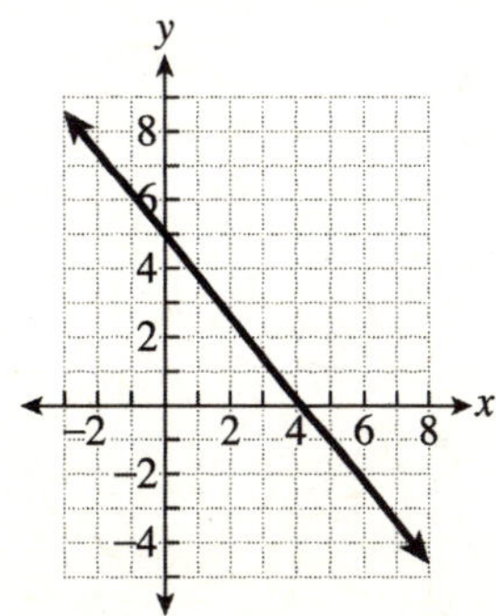

12. $4y = 8x - 11$

$$y = \frac{8}{4}x - \frac{11}{4}$$

$$y = 2x - \frac{11}{4}$$

$m = 2$

$2x - y = 44$

$-y = -2x + 44$

$y = 2x - 44$

$m = 2$

The lines are parallel.

13. $y = -6x + 5$

14. $f(-3) = 7 - 2(-3)$
$= 7 + 6$
$= 13$

15. $y - 9 = -\frac{5}{2}(x - (-2))$

$$y - 9 = -\frac{5}{2}(x + 2)$$

$$y - 9 = -\frac{5}{2}x - 5$$

$$y = -\frac{5}{2}x + 4$$

16. $m = \dfrac{-6 - 4}{-1 - (-6)} = \dfrac{-10}{5} = -2$

$y - 4 = -2(x - (-6))$
$y - 4 = -2(x + 6)$
$y - 4 = -2x - 12$
$y = -2x - 8$

17. $m = -4$, y-intercept: $(0, 6)$
From the y-intercept count down 4 andn right 1
to find the next point $(1, 2)$.
x-intercept: $0 = -4x + 6$
$4x = 6$
$$x = \frac{3}{2}$$

$\left(\dfrac{3}{2}, 0\right)$

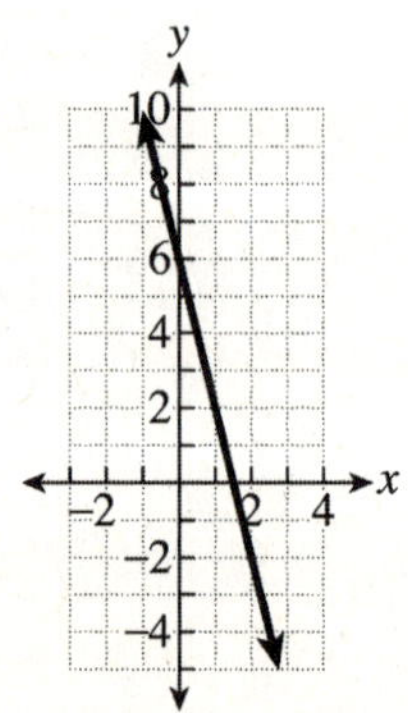

18. x-intercept:
$-4x + 5(0) = 20$
$-4x = 20$
$x = -5$
$(-5, 0)$

y-intercept:
$-4(0) + 5y = 20$
$5y = 20$
$y = 4$
$(0, 4)$

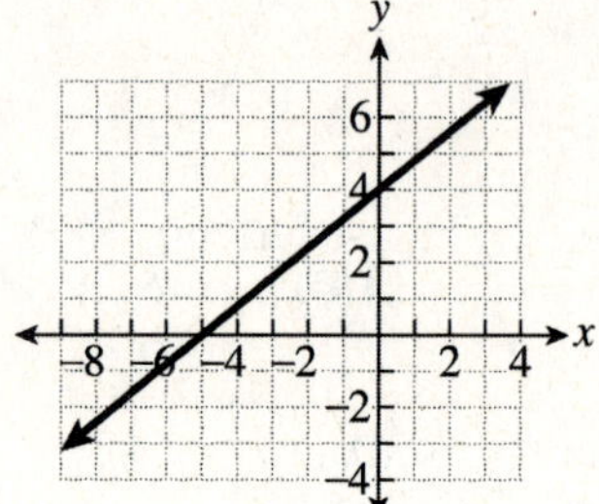

19. $m = 0$, y-intercept: $(0, 2)$

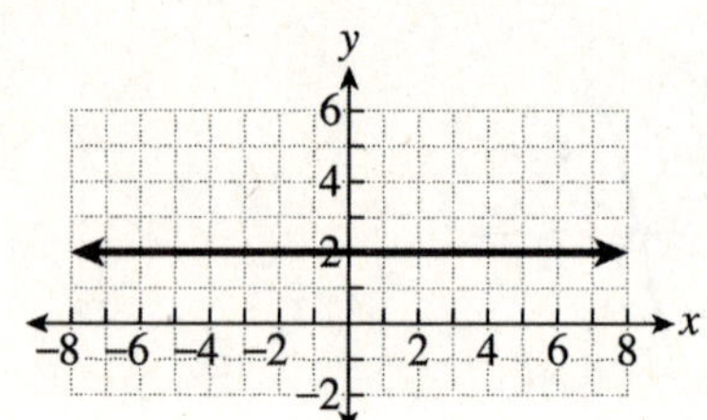

20. $x - 6y = -3$
$-6y = -x - 3$
$$y = \frac{-1}{-6}x - \frac{3}{-6}$$
$$y = \frac{1}{6}x + \frac{1}{2}$$

x-intercept:
$x - 6(0) = -3$
$x - 0 = -3$
$x = -3$
$(-3, 0)$

y-intercept:
$0 - 6y = -3$
$-6y = -3$
$$y = \frac{1}{2}$$
$\left(0, \dfrac{1}{2}\right)$

Test Point: $(0, 0)$
$0 - 6(0) < -3$
$0 < -3$
False. Shade on the side of the line that does
not contain the test point.

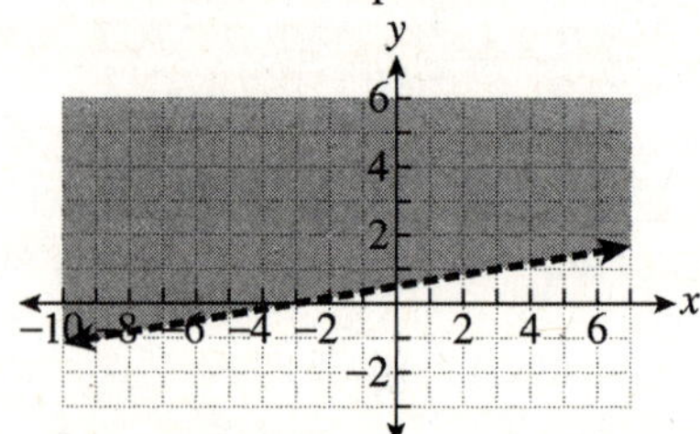

4.1   QUICK CHECK

1. $5(-6) - 2(3) = -36$
$$-30 - 6 = -36$$
$$-36 = -36$$
True.
$$-(-6) + 7(3) = 27$$
$$6 + 21 = 27$$
$$27 = 27$$
True.
Yes, $(-6, 3)$ is a solution.

2. $(5, 0)$

3. Graph the line $y = 3x - 2$.

$m = 3$, y-intercept: $(0, -2)$

Count from the y-intercept 3 up and 1 right to find the next point $(1, 1)$.

Graph the line $y = 5x - 6$.

$m = 5$, y-intercept: $(0, -6)$

Count from the y-intercept 5 up and 1 right to find the next point $(1, -1)$.

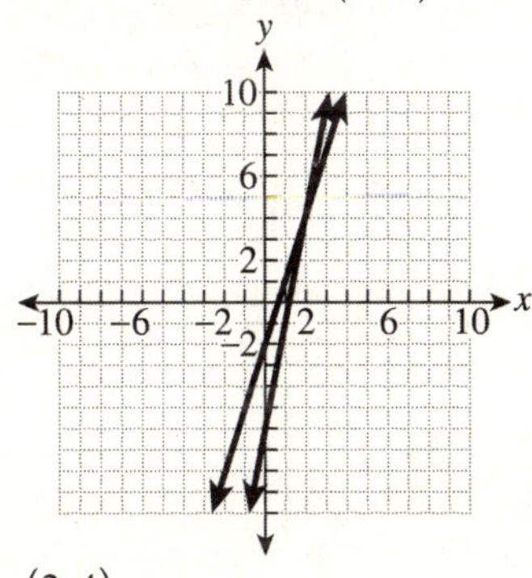

$(2, 4)$

4. $x + 5y = 5$

x-intercept:	y-intercept:
$x + 5(0) = 5$	$0 + 5y = 5$
$x = 5$	$y = 1$
$(5, 0)$	$(0, 1)$

$x - y = -7$

x-intercept:	y-intercept:
$x - 0 = -7$	$0 - y = -7$
$x = -7$	$y = 7$
$(-7, 0)$	$(0, 7)$

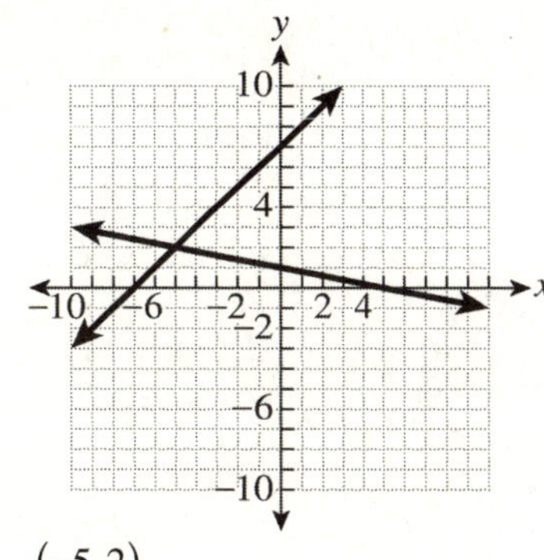

$(-5, 2)$

5. $y = -4x + 6$ $12x + 3y = 15$
$m = -4$ $3y = -12x + 15$
$$y = -\frac{12}{3}x + \frac{15}{3}$$
$$y = -4x + 5$$
$$m = -4$$

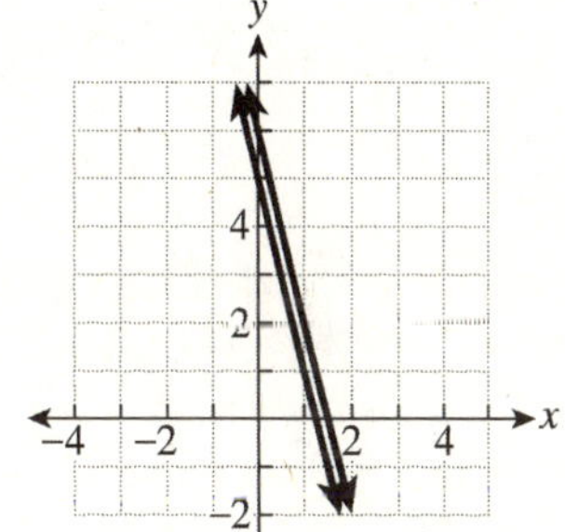

The lines are parallel, so no solution. $\varnothing$

6. $4x + 6y = 12$ $y = -\frac{2}{3}x + 2$
$$6y = -4x + 12$$
$$y = -\frac{4}{6}x + \frac{12}{6}$$
$$y = -\frac{2}{3}x + 2$$

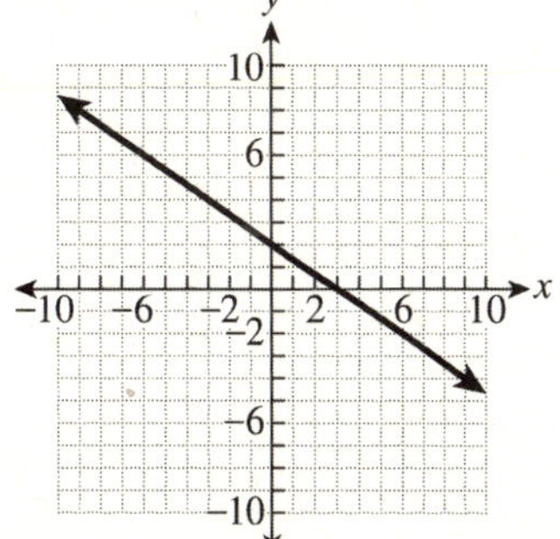

Both equations give the same line.
$$\left(x, -\frac{2}{3}x + 2\right)$$

4.1 **SYSTEMS OF LINEAR EQUATIONS; SOLVING SYSTEMS BY GRAPHING**

1. system of linear equations

3. solution

5. inconsistent

7.
$$2 = 2(5) - 8 \qquad 3(5) + 4(2) = 23$$
$$2 = 10 - 8 \qquad 15 + 8 = 23$$
$$2 = 2 \qquad 23 = 23$$
Yes.

9.
$$3(-6) + 2(1) = -16 \qquad 2(-6) - 5(1) = -7$$
$$-18 + 2 = -16 \qquad -12 - 5 = -7$$
$$-16 = -16 \qquad -17 = -7$$
No.

11.
$$3(-4) - 7(-10) = 58 \qquad -(-4) + 4(-10) = -36$$
$$-12 + 70 = 58 \qquad 4 - 40 = -36$$
$$58 = 58 \qquad -36 = -36$$
Yes.

13.
$$\frac{3}{4}(8) - \frac{2}{3}(-6) = 2 \qquad \frac{1}{8}(8) + 2(-6) = -11$$
$$6 + 4 = 2 \qquad 1 - 12 = -11$$
$$10 = 2 \qquad -11 = -11$$
No.

15.
$$8\left(\frac{3}{2}\right) - 12\left(\frac{5}{6}\right) = 2 \qquad 5\left(\frac{3}{2}\right) + 9\left(\frac{5}{6}\right) = 15$$
$$12 - 10 = 2 \qquad \frac{15}{2} + \frac{45}{6} = 15$$
$$2 = 2 \qquad \frac{45}{6} + \frac{45}{6} = 15$$
$$\frac{90}{6} = 15$$
$$15 = 15$$
Yes.

17. $(2, 4)$

19. $(-3, 6)$

21. Graph the line $y = x - 1$.

$m = 1$, y-intercept: $(0, -1)$

Count from the y-intercept 1 up and 1 right to find the next point $(1, 0)$.

$$3x - y = 9$$

x-intercept: $\qquad\qquad$ y-intercept:
$$3x - (0) = 9 \qquad\qquad 3(0) - y = 9$$
$$x = 3 \qquad\qquad\qquad y = -9$$
$$(3, 0) \qquad\qquad\qquad (0, -9)$$

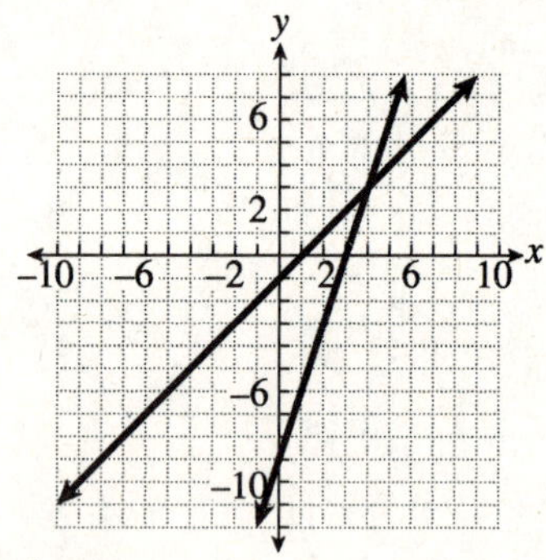

$(4, 3)$

23. Graph the line $y = -2x + 8$.

$m = -2$, y-intercept: $(0, 8)$

Count from the y-intercept 2 down and 1 right to find the next point $(1, 6)$.

Graph the line $y = \frac{3}{5}x - 5$.

$m = \frac{3}{5}$, y-intercept: $(0, -5)$

Count from the y-intercept 3 up and 5 right to find the next point $(5, -2)$.

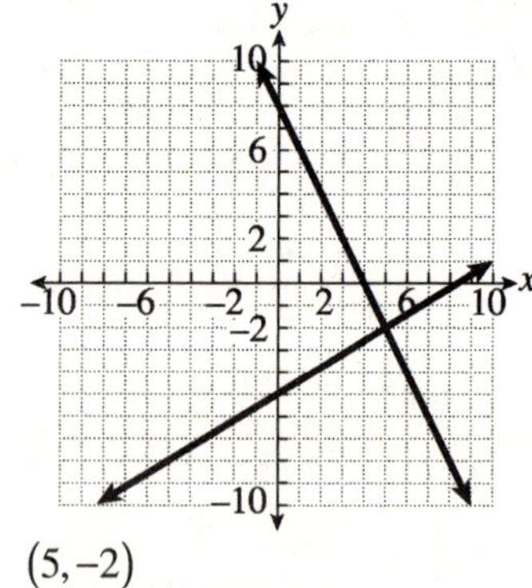

$(5, -2)$

25. Graph the line $y = 2x + 4$.

$m = 2$, y-intercept: $(0, 4)$

Count from the y-intercept 2 up and 1 right to find the next point $(1, 6)$.

$6x - 3y = -12$

x-intercept: $\quad$ y-intercept:
$6x - 3(0) = -12$ $\quad$ $6(0) - 3y = -12$
$\qquad x = -2$ $\qquad\qquad y = 4$
$(-2, 0)$ $\qquad\qquad$ $(0, 4)$

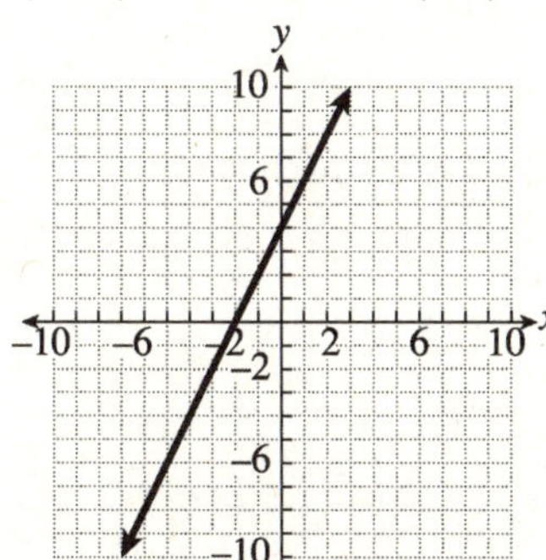

Dependent, $(x, 2x + 4)$

27. Graph the line $y = \dfrac{3}{2}x + 3$.

$m = \dfrac{3}{2}$, y-intercept: $(0, 3)$

Count from the y-intercept 3 up and 2 right to find the next point $(2, 6)$.

$7x - 2y = -14$

x-intercept: $\quad$ y-intercept:
$7x - 2(0) = -14$ $\quad$ $7(0) - 2y = -14$
$\qquad x = -2$ $\qquad\qquad y = 7$
$(-2, 0)$ $\qquad\qquad$ $(0, 7)$

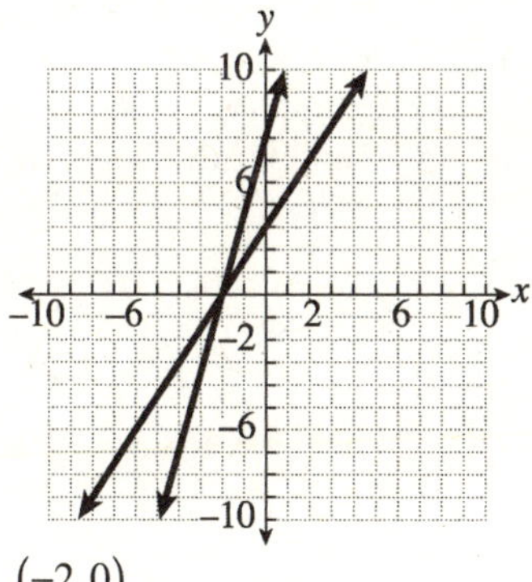

$(-2, 0)$

29. $5x - 3y = 15$

x-intercept: $\quad$ y-intercept:
$5x - 3(0) = 15$ $\quad$ $5(0) - 3y = 15$
$\qquad x = 3$ $\qquad\qquad y = -5$
$(3, 0)$ $\qquad\qquad$ $(0, -5)$

$2x - y = 4$

x-intercept: $\quad$ y-intercept:
$2x - (0) = 4$ $\quad$ $2(0) - y = 4$
$\qquad x = 2$ $\qquad\qquad y = -4$
$(2, 0)$ $\qquad\qquad$ $(0, -4)$

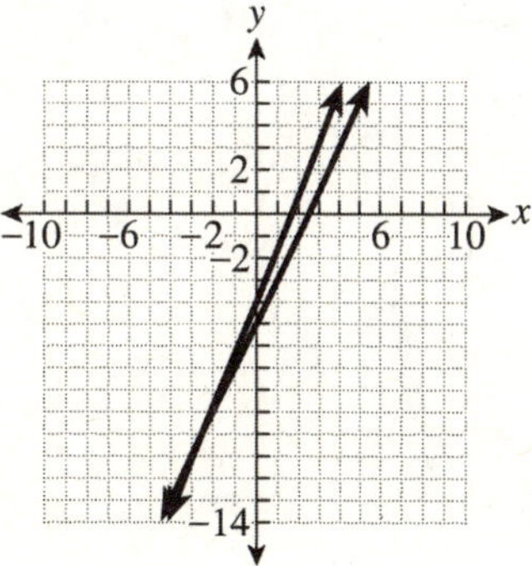

$(-3, -10)$

31. $4x - 5y = -20$

x-intercept: $\quad$ y-intercept:
$4x - 5(0) = -20$ $\quad$ $4(0) - 5y = -20$
$\qquad x = -5$ $\qquad\qquad y = 4$
$(-5, 0)$ $\qquad\qquad$ $(0, 4)$

$-4x + 7y = 28$

x-intercept: $\quad$ y-intercept:
$-4x + 7(0) = 28$ $\quad$ $-4(0) + 7y = 28$
$\qquad x = -7$ $\qquad\qquad y = 4$
$(-7, 0)$ $\qquad\qquad$ $(0, 4)$

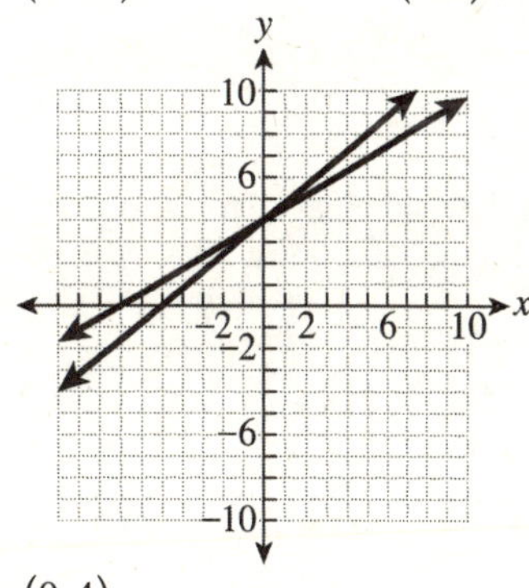

$(0, 4)$

33. Graph the line $y = 5x - 4$.
$m = 5$, y-intercept: $(0, -4)$
Count from the y-intercept 5 up and
1 right to find the next point $(1,1)$.

$10x - 2y = 6$
x-intercept: $\qquad$ y-intercept:
$10x - 2(0) = 6$ $\qquad$ $10(0) - 2y = 6$
$$x = \frac{6}{10} \qquad\qquad y = -3$$
$$x = \frac{3}{5} \qquad (0, -3)$$
$$\left(\frac{3}{5}, 0\right)$$

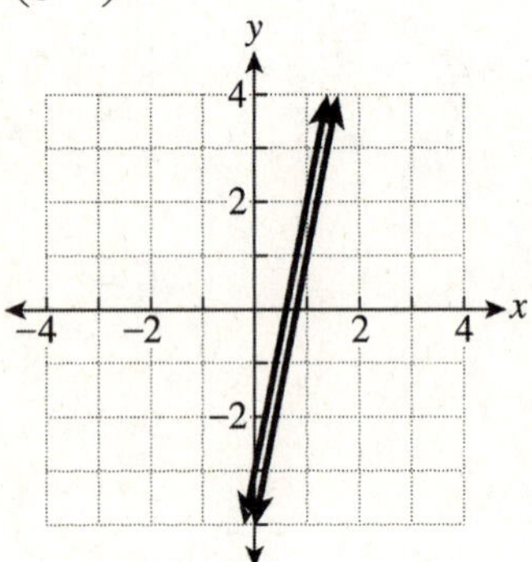

Inconsistent, $\varnothing$

35. $9x - 2y = -18$
x-intercept: $\qquad$ y-intercept:
$9x - 2(0) = -18$ $\qquad$ $9(0) - 2y = -18$
$\qquad x = -2$ $\qquad\qquad y = 9$
$(-2, 0)$ $\qquad\qquad (0, 9)$

Graph the line $y = \dfrac{1}{2}x - 7$.

$m = \dfrac{1}{2}$, y-intercept: $(0, -7)$

Count from the y-intercept 1 up and
2 right to find the next point $(2, -6)$.

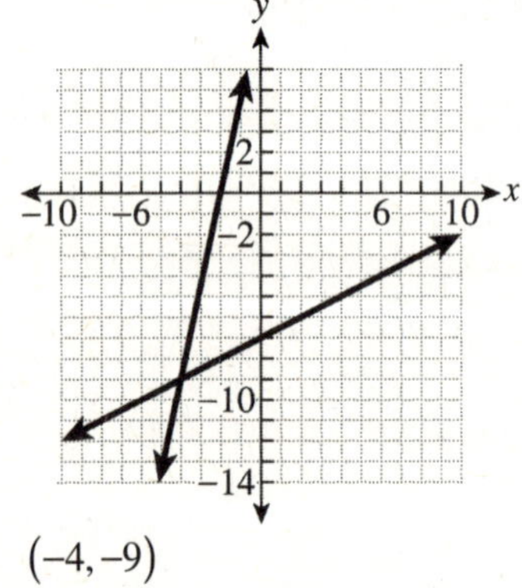

$(-4, -9)$

37. Answers will vary. Example:

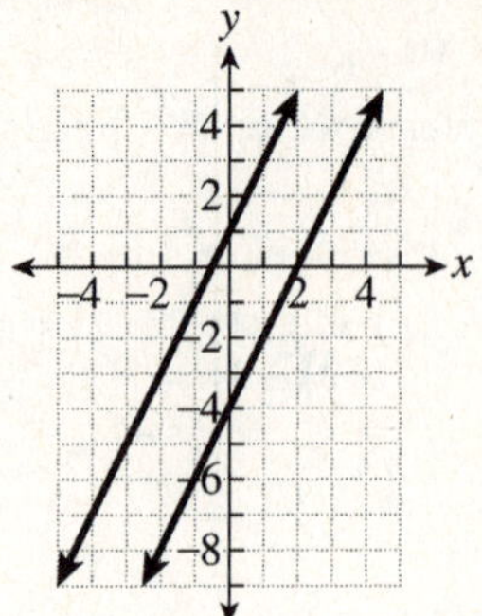

39. Answers will vary. Example:

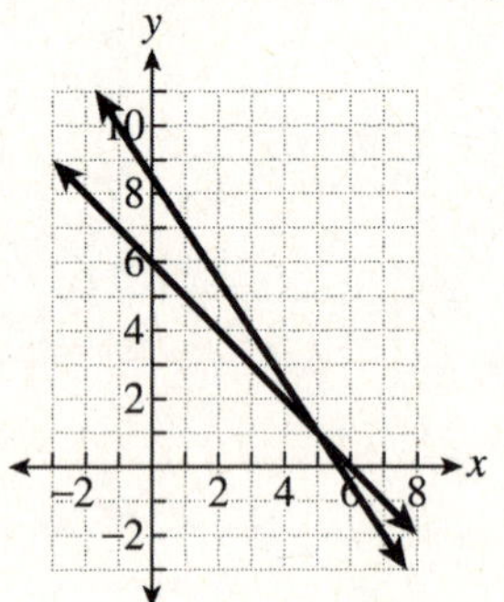

41. Answers will vary. Example:

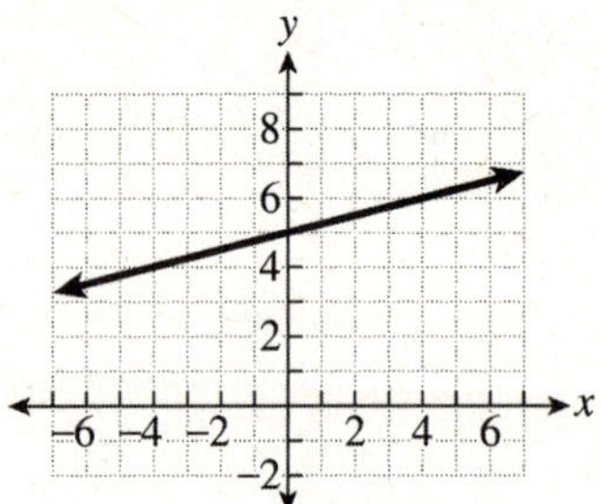

43. Explanations will vary. Example:
Since all points on a graph are solutions to its
given equation, the point of intersection is the
solution to both equations.

1.
$$7x - 4(x - 3) = 27$$
$$7x - 4x + 12 = 27$$
$$3x + 12 = 27$$
$$3x = 15$$
$$x = 5$$
$$y = x - 3$$
$$y = (5) - 3 = 2$$
$$(5, 2)$$

2.
$$x - y = -4$$
$$x = y - 4$$

$$3(y - 4) + 5y = 36$$
$$3y - 12 + 5y = 36$$
$$8y - 12 = 36$$
$$8y = 48$$
$$y = 6$$
$$x = y - 4$$
$$x = 6 - 4 = 2$$
$$(2, 6)$$

3.
$$-2x + y = 3$$
$$y = 2x + 3$$

$$8x - 4(2x + 3) = 10$$
$$8x - 8x - 12 = 10$$
$$-12 = 10$$
Inconsistent, $\varnothing$

4.
$$x - 6y = 4$$
$$x = 6y + 4$$

$$3(6y + 4) - 18y = 12$$
$$18y + 12 - 18y = 12$$
$$12 = 12$$
Dependent, $\left(x, \dfrac{1}{6}x - \dfrac{2}{3} \right)$

5.
$$2x + 4y = -14$$
$$2x = -4y - 14$$
$$x = -2y - 7$$

$$5(-2y - 7) - 3y = 17$$
$$-10y - 35 - 3y = 17$$
$$-13y - 35 = 17$$
$$-13y = 52$$
$$y = -4$$
$$x = -2y - 7$$
$$x = -2(-4) - 7 = 8 - 7 = 1$$
$$(1, -4)$$

6. Number of desktop computers: D
   Number of laptop computers: L

	Computers	Cost	Paid
Desktop	D	$700	$700D$
Laptop	L	$1000	$1000L$
Total	15		11,700

$$D + L = 15$$
$$D = 15 - L$$
$$700D + 1000L = 11,700$$
$$700(15 - L) + 1000L = 11,700$$
$$10,500 - 700L + 1000L = 11,700$$
$$10,500 + 300L = 11,700$$
$$300L = 1200$$
$$L = 4$$
$$D = 15 - L = 15 - 4 = 11$$
He bought 11 desktop and 4 laptop computers.

4.2   *SOLVING SYSTEMS OF EQUATIONS BY USING THE SUBSTITUTION METHOD*

1. one of the equations for one of the variables

3. dependent

5.
$$x + 3(3x - 5) = 15$$
$$x + 9x - 15 = 15$$
$$10x - 15 = 15$$
$$10x = 30$$
$$x = 3$$
$$y = 3(3) - 5 = 9 - 5 = 4$$
$$(3, 4)$$

7.
$$7(2 - 5y) + 10y = 89$$
$$14 - 35y + 10y = 89$$
$$14 - 25y = 89$$
$$-25y = 75$$
$$y = -3$$
$$x = 2 - 5(-3) = 2 + 15 = 17$$
$$(17, -3)$$

9.
$$-10x + 2(5x - 10) = 20$$
$$-10x + 10x - 20 = 20$$
$$-20 = 20$$
Inconsistent, $\varnothing$

11.
$$18x + 6(-3x - 2) = -12$$
$$18x - 18x - 12 = -12$$
$$-12 = -12$$
Dependent, $(x, -3x - 2)$

13. $-7x - 2(11 - 5x) = 14$
$-7x - 22 + 10x = 14$
$3x - 22 = 14$
$3x = 36$
$x = 12$
$y = 11 - 5(12) = 11 - 60 = -49$
$(12, -49)$

15. $3x + 2(2.2x - 6.8) = 16$
$3x + 4.4x - 13.6 = 16$
$7.4x = 29.6$
$x = 4$
$y = 2.2(4) - 6.8 = 8.8 - 6.8 = 2$
$(4, 2)$

17. $6\left(\dfrac{2}{3}y - 4\right) + 7y = 75$
$4y - 24 + 7y = 75$
$11y - 24 = 75$
$11y = 99$
$y = 9$
$x = \dfrac{2}{3}(9) - 4 = 6 - 4 = 2$
$(2, 9)$

19. $4x - 3 = -2x - 21$
$6x = -18$
$x = -3$
$y = 4x - 3 = 4(-3) - 3 = -12 - 3 = -15$
$(-3, -15)$

21. $\dfrac{5}{6}x + 3 = 3x - \dfrac{7}{2}$
$5x + 18 = 18x - 21$
$-13x = -39$
$x = 3$
$y = 3x - \dfrac{7}{2} = 3(3) - \dfrac{7}{2} = 9 - \dfrac{7}{2} = \dfrac{11}{2}$
$\left(3, \dfrac{11}{2}\right)$

23. $x + 3y = 19$
$x = -3y + 19$
$3(-3y + 19) - 4y = -21$
$-9y + 57 - 4y = -21$
$-13y = -78$
$y = 6$
$x + 3(6) = 19$
$x + 18 = 19$
$x = 1$
$(1, 6)$

25. $x + y = -6$
$x = -y - 6$
$4(-y - 6) - 3y = 46$
$-4y - 24 - 3y = 46$
$-7y = 70$
$y = -10$
$x + (-10) = -6$
$x = 4$
$(4, -10)$

27. $3x + y = -26$
$y = -3x - 26$
$5x + 2(-3x - 26) = -44$
$5x - 6x - 52 = -44$
$-x = 8$
$x = -8$
$y = -3(-8) - 26 = 24 - 26 = -2$
$(-8, -2)$

29. $2x - y = -1$
$-y = -2x - 1$
$y = 2x + 1$
$8x + 2(2x + 1) = 17$
$8x + 4x + 2 = 17$
$12x = 15$
$x = \dfrac{15}{12}$
$x = \dfrac{5}{4}$
$y = 2\left(\dfrac{5}{4}\right) + 1 = \dfrac{5}{2} + 1 = \dfrac{7}{2}$
$\left(\dfrac{5}{4}, \dfrac{7}{2}\right)$

31. $2(4y - 5) + 2y = 90$
$8y - 10 + 2y = 90$
$10y = 100$
$y = 10$
$x = 4(10) - 5x = 40 - 5 = 35$
$(35, 10)$

33. $x + y = 40$
$x = -y + 40$
$0.10(-y + 40) + 0.25y = 8.50$
$-0.10y + 4 + 0.25y = 8.50$
$0.15y = 4.50$
$y = 30$
$x = -(30) + 40 = 10$
$(10, 30)$

35. $x + y = 5000$
$x = -y + 5000$

$0.04(-y + 5000) + 0.07y = 260$
$-0.04y + 200 + 0.07y = 260$
$0.03y = 60$
$y = 2000$
$x = -(2000) + 5000 = 3000$
$(3000, 2000)$

37. $0.02x + 0.025(x + 100) = 22.75$
$0.02x + 0.025x + 2.5 = 22.75$
$0.045x = 20.25$
$x = 450$
$y = 450 + 100 = 550$
$(450, 550)$

39. $x + y = 80$
$y = -x + 80$

$0.18x + 0.33(-x + 80) = 0.24(80)$
$0.18x - 0.33x + 26.4 = 19.2$
$-0.15x = -7.2$
$x = 48$
$y = -(48) + 80 = 32$
$(48, 32)$

41. $6x + y = 21$
$y = -6x + 21$

$-18x - 3(-6x + 21) = -63$
$-18x + 18x - 63 = -63$
$-63 = -63$
Dependent, $(x, -6x + 21)$

43. $x + 2.8y = 3.4$
$x = -2.8y + 3.4$

$2(-2.8y + 3.4) - 9y = -37$
$-5.6y + 6.8 - 9y = -37$
$-14.6y + 6.8 = -37$
$-14.6y = -43.8$
$y = 3$
$x = -2.8(3) + 3.4 = -8.4 + 3.4 = -5$
$(-5, 3)$

45. $\dfrac{3}{2}x + y = -7$

$y = -\dfrac{3}{2}x - 7$

$-3x + 2\left(-\dfrac{3}{2}x - 7\right) = 34$

$-3x - 3x - 14 = 34$
$-6x - 14 = 34$
$-6x = 48$
$x = -8$
$y = -\dfrac{3}{2}(-8) - 7 = 12 - 7 = 5$
$(-8, 5)$

47. $4x + 2y = 10$
$2y = -4x + 10$
$y = -2x + 5$

$5x - 4(-2x + 5) = 6$
$5x + 8x - 20 = 6$
$13x - 20 = 6$
$13x = 26$
$x = 2$
$y = -2(2) + 5 = -4 + 5 = 1$
$(2, 1)$

49. $3x + 5y = 14$
$5y = -3x + 14$
$y = -\dfrac{3}{5}x + \dfrac{14}{5}$

$6x - 5\left(-\dfrac{3}{5}x + \dfrac{14}{5}\right) = -32$

$6x + 3x - 14 = -32$
$9x - 14 = -32$
$9x = -18$
$x = -2$
$y = -\dfrac{3}{5}(-2) + \dfrac{14}{5} = \dfrac{6}{5} + \dfrac{14}{5} = \dfrac{20}{5} = 4$
$(-2, 4)$

51.

	Attendees	Cost	Purchased
General	x	$5	$5x$
Students	y	$2	$2y$
Total	1200		$5250

$x + y = 1200$
$5x + 2y = 5250$
$x = 1200 - y$

$5(1200 - y) + 2y = 5250$
$6000 - 5y + 2y = 5250$
$6000 - 3y = 5250$
$-3y = -750$
$y = 250$
$x = 1200 - (250) = 950$

250 students, 950 nonstudents

53.

	Dinners	Cost	Profit
Chicken	x	$8	$8x$
Steak	y	$10	$10y$
Total	142		$1326

$x + y = 142$
$8x + 10y = 1326$
$x = 142 - y$

$8(142 - y) + 10y = 1326$
$1136 - 8y + 10y = 1326$
$1136 + 2y = 1326$
$2y = 190$
$y = 95$
$x = 142 - (95) = 47$

47 chicken dinners and 95 steak dinners

55. $y = 0.05x + 26.7$
$y = 0.10x + 24.6$

$0.05x + 26.7 = 0.10x + 24.6$
$26.7 = 0.05x + 24.6$
$2.1 = 0.05x$
$42 = x$

$1995 + 42 = 2037$
The year is 2037.

57. Explanations will vary. Example:
Solve for one variable in one equation.
Substitute the result into the other equation.
Solve for the remaining variable and back-
substitute to find the constant value of the other
variable.

59. Answers will vary.

4.3 QUICK CHECK

1. $4x + 3y = 31$
$-4x + 5y = -23$
$\overline{\qquad 8y = 8\qquad}$
$y = 1$

$4x + 3(1) = 31$
$4x + 3 = 31$
$4x = 28$
$x = 7$

$(7, 1)$

2. $2x - 9y = 33$
$3 \cdot (4x + 3y) = 3 \cdot (3)$

$2x - 9y = 33$
$12x + 9y = 9$
$\overline{\qquad 14x = 42\qquad}$
$x = 3$

$2(3) - 9y = 33$
$6 - 9y = 33$
$-9y = 27$
$y = -3$

$(3, -3)$

3. $2 \cdot (3x + 2y) = 2 \cdot (-26)$
$3 \cdot (-2x + 5y) = 3 \cdot (11)$

$6x + 4y = -52$
$-6x + 15y = 33$
$\overline{\qquad 19y = -19\qquad}$
$y = -1$

$3x + 2(-1) = -26$
$3x - 2 = -26$
$3x = -24$
$x = -8$

$(-8, -1)$

4. $2 \cdot (2x - 3y) = 2 \cdot (12)$
$-4x + 6y = -24$

$4x - 6y = 24$
$-4x + 6y = -24$
$\overline{\qquad 0 = 0\qquad}$

Dependent, $\left(x, \dfrac{2}{3}x - 4 \right)$

5. $-3 \cdot (3x - 4y) = -3 \cdot (11)$
$9x - 12y = 30$

$-9x + 12y = -33$
$9x - 12y = 30$
$\overline{\qquad 0 = -3\qquad}$

Inconsistent, $\varnothing$

6. $6 \cdot \left(-\dfrac{2}{3}x + \dfrac{5}{2}y \right) = 6 \cdot (14)$

$12 \cdot \left(\dfrac{5}{12}x + \dfrac{3}{4}y \right) = 12 \cdot \left(\dfrac{1}{2} \right)$

$5 \cdot (-4x + 15y) = 5 \cdot (84)$
$4 \cdot (5x + 9y) = 4 \cdot (6)$

$-20x + 75y = 420$
$20x + 36y = 24$
$\overline{\qquad 111y = 444\qquad}$
$y = 4$

$5x + 9(4) = 6$
$5x + 36 = 6$
$5x = -30$
$x = -6$

$(-6, 4)$

7. $x + y = 700$
$0.03x + 0.05y = 29.5$

$-5 \cdot (x + y) = -5 \cdot (700)$
$100 \cdot (0.03x + 0.05y) = 100 \cdot (29.5)$

$-5x - 5y = -3500$
$3x + 5y = 2950$
$\overline{\qquad -2x = -550\qquad}$
$x = 275$

$(275) + y = 700$
$y = 425$

$(275, 425)$

8.

	Coins	Value	Money
Nickels	x	$0.05	$0.05x$
Dimes	y	$0.10	$0.10y$
Total	75		$5.35

$$x + y = 75$$
$$0.05x + 0.10y = 5.35$$

$$-5 \cdot (x + y) = -5 \cdot (75)$$
$$100 \cdot (0.05x + 0.10y) = 100 \cdot (5.35)$$

$$-5x - 5y = -375$$
$$\underline{5x + 10y = 535}$$
$$5y = 160$$
$$y = 32$$

$$x + (32) = 75$$
$$x = 43$$

Erica has 43 nickels and 32 dimes.

4.3 **SOLVING SYSTEMS OF EQUATIONS BY USING THE ADDITION METHOD**

1. LCM

3. dependent

5. $\begin{aligned} 5x + 4y &= -56 \\ \underline{3x - 4y = -8} \\ 8x &= -64 \\ x &= -8 \end{aligned}$

$$5(-8) + 4y = -56$$
$$-40 + 4y = -56$$
$$4y = -16$$
$$y = -4$$
$$(-8, -4)$$

7. $\begin{aligned} -3x + 6y &= 36 \\ \underline{3x - 11y = -51} \\ -5y &= -15 \\ y &= 3 \end{aligned}$

$$-3x + 6(3) = 36$$
$$-3x + 18 = 36$$
$$-3x = 18$$
$$x = -6$$
$$(-6, 3)$$

9. $2 \cdot (4x + 3y) = 2 \cdot (1)$

$$\begin{aligned} 8x + 6y &= 2 \\ \underline{13x - 6y = 82} \\ 21x &= 84 \\ x &= 4 \end{aligned}$$

$$4(4) + 3y = 1$$
$$16 + 3y = 1$$
$$3y = -15$$
$$y = -5$$
$$(4, -5)$$

11. $2 \cdot (2x - 4y) = 2 \cdot (11)$

$$\begin{aligned} 4x - 8y &= 22 \\ \underline{-4x + 8y = -22} \\ 0 &= 0 \end{aligned}$$

Dependent $\left(x, \dfrac{1}{2}x - \dfrac{11}{4} \right)$

13. $-3 \cdot (x + 4y) = -3 \cdot (11)$

$$\begin{aligned} -3x - 12y &= -33 \\ \underline{3x + 16y = 37} \\ 4y &= 4 \\ y &= 1 \end{aligned}$$

$$x + 4(1) = 11$$
$$x + 4 = 11$$
$$x = 7$$
$$(7, 1)$$

15. $-4 \cdot (3x + 5y) = -4 \cdot (-31)$

$$\begin{aligned} -12x - 20y &= 124 \\ \underline{12x + 9y = -69} \\ -11y &= 55 \\ y &= -5 \end{aligned}$$

$$3x + 5(-5) = -31$$
$$3x - 25 = -31$$
$$3x = -6$$
$$x = -2$$
$$(-2, -5)$$

17. $5 \cdot (-x + 9y) = 5 \cdot (-39)$

$$\begin{aligned} 5x + 6y &= -9 \\ \underline{-5x + 45y = -195} \\ 51y &= -204 \\ y &= -4 \end{aligned}$$

$$5x + 6(-4) = -9$$
$$5x - 24 = -9$$
$$5x = 15$$
$$x = 3$$
$$(3, -4)$$

19. $3 \cdot (7x + 4y) = 3 \cdot (85)$
$2 \cdot (11x - 6y) = 2 \cdot (-63)$

$$\begin{aligned} 21x + 12y &= 255 \\ \underline{22x - 12y = -126} \\ 43x &= 129 \\ x &= 3 \end{aligned}$$

$$7(3) + 4y = 85$$
$$21 + 4y = 85$$
$$4y = 64$$
$$y = 16$$
$$(3, 16)$$

21.
$$3\cdot(-8x+5y)=3\cdot(-28)$$
$$8\cdot(3x+8y)=8\cdot(-29)$$
$$\begin{array}{r}-24x+15y=-84\\ 24x+64y=-232\\ \hline 79y=-316\\ y=-4\end{array}$$
$$3x+8(-4)=-29$$
$$3x-32=-29$$
$$3x=3$$
$$x=1$$
$$(1,-4)$$

23.
$$5\cdot(4x-3y)=5\cdot(-6)$$
$$3\cdot(6x+5y)=3\cdot(29)$$
$$\begin{array}{r}20x-15y=-30\\ 18x+15y=87\\ \hline 38x=57\end{array}$$
$$x=\frac{57}{38}$$
$$x=\frac{3}{2}$$
$$4\left(\frac{3}{2}\right)-3y=-6$$
$$6-3y=-6$$
$$-3y=-12$$
$$y=4$$
$$\left(\frac{3}{2},4\right)$$

25.
$$2\cdot(-6x+9y)=2\cdot(17)$$
$$-3\cdot(-4x+6y)=-3\cdot(12)$$
$$\begin{array}{r}-12x+18y=34\\ 12x-18y=-36\\ \hline 0=-2\end{array}$$
Inconsistent, $\varnothing$

27.
$$-4\cdot(5x+7y)=-4\cdot(16)$$
$$5\cdot(4x+8y)=5\cdot(14)$$
$$\begin{array}{r}-20x-28y=-64\\ 20x+40y=70\\ \hline 12y=6\end{array}$$
$$y=\frac{6}{12}$$
$$y=\frac{1}{2}$$

$$4x+8\left(\frac{1}{2}\right)=14$$
$$4x+4=14$$
$$4x=10$$
$$x=\frac{10}{4}$$
$$x=\frac{5}{2}$$
$$\left(\frac{5}{2},\frac{1}{2}\right)$$

29.
$$2\cdot(-3x+4y)=2\cdot(-13)$$
$$\begin{array}{r}-6x+8y=-26\\ 6x+5y=52\\ \hline 13y=26\\ y=2\end{array}$$
$$6x+5(2)=52$$
$$6x+10=52$$
$$6x=42$$
$$x=7$$
$$(7,2)$$

31.
$$\begin{array}{r}5x-3y=21\\ -5x+4y=-23\\ \hline y=-2\end{array}$$
$$5x=3(-2)+21$$
$$5x=-6+21$$
$$5x=15$$
$$x=3$$
$$(3,-2)$$

33.
$$10\cdot\left(\frac{1}{2}x+\frac{2}{5}y\right)=10\cdot\left(\frac{7}{5}\right)$$
$$12\cdot\left(\frac{1}{4}x-\frac{1}{6}y\right)=12\cdot\left(\frac{1}{3}\right)$$
$$5x+4y=14$$
$$3x-2y=4$$
$$2\cdot(3x-2y)=2\cdot(4)$$
$$\begin{array}{r}5x+4y=14\\ 6x-4y=8\\ \hline 11x=22\\ x=2\end{array}$$
$$5(2)+4y=14$$
$$10+4y=14$$
$$4y=4$$
$$y=1$$
$$(2,1)$$

35.
$$6 \cdot \left(\frac{2}{3}x + \frac{1}{6}y \right) = 6 \cdot (4)$$
$$3 \cdot \left(\frac{5}{3}x - y \right) = 3 \cdot (-7)$$
$$4x + y = 24$$
$$5x - 3y = -21$$
$$3 \cdot (4x + y) = 3 \cdot (24)$$
$$12x + 3y = 72$$
$$\underline{5x - 3y = -21}$$
$$17x = 51$$
$$x = 3$$

$$\frac{5}{3}(3) - y = -7$$
$$5 - y = -7$$
$$-y = -12$$
$$y = 12$$
$$(3, 12)$$

37.
$$(100)(0.03x + 0.07y) = (100)144$$
$$3x + 7y = 14,400$$
$$-3(x + y) = -3(3600)$$
$$-3x - 3y = -10,800$$
$$3x + 7y = 14,400$$
$$\underline{-3x - 3y = -10,800}$$
$$4y = 3600$$
$$y = 900$$
$$x + 900 = 3600$$
$$x = 2700$$
$$(2700, 900)$$

39.
$$(100)(8.50x + 5.75y) = (100)827.50$$
$$850x + 575y = 82,750$$
$$-850(x + y) = -850(120)$$
$$-850x - 850y = -102,000$$
$$850x + 575y = 82,750$$
$$\underline{-850x - 850y = -102,000}$$
$$-275y = -19,250$$
$$y = 70$$
$$x + 70 = 120$$
$$x = 50$$
$$(50, 70)$$

41.
$$(100)(0.36x + 0.48y) = (100)(0.45)(100)$$
$$36x + 48y = 4500$$
$$-36(x + y) = -36(100)$$
$$-36x - 36y = -3600$$
$$36x + 48y = 4500$$
$$\underline{-36x - 36y = -3600}$$
$$12y = 900$$
$$y = 75$$
$$x + 75 = 100$$
$$x = 25$$
$$(25, 75)$$

43.

	Flowers	Cost	Money
Rose	x	$6	$6x
Carnation	y	$4	$4y
Total	15		$68

$$x + y = 15$$
$$6x + 4y = 68$$
$$-4 \cdot (x + y) = -4 \cdot (15)$$
$$-4x - 4y = -60$$
$$\underline{6x + 4y = 68}$$
$$2x = 8$$
$$x = 4$$
$$4 + y = 15$$
$$y = 11$$

4 roses and 11 carnations

45. First family:

	Plants	Cost	Money
Trees	7	x	$7x$
Shrubs	4	y	$4y$
Total			$312

Second family:

	Plants	Cost	Money
Trees	4	x	$4x$
Shrubs	15	y	$15y$
Total			$280

$$7x + 4y = 312$$
$$4x + 15y = 280$$
$$-4(7x + 4y) = -4(312)$$
$$7(4x + 15y) = 7(280)$$
$$-28x - 16y = -1248$$
$$\underline{28x + 105y = 1960}$$
$$89y = 712$$
$$y = 8$$
$$7x + 4(8) = 312$$
$$7x + 32 = 312$$
$$7x = 280$$
$$x = 40$$

$40 for a tree and $8 for a shrub

47.

	Coins	Value	Money
Dimes	x	$0.10	$0.10x$
Nickels	y	$0.05	$0.05y$
Total	81		$6.50

$$x + y = 81$$
$$0.10x + 0.05y = 6.50$$

$$-5 \cdot (x + y) = -5 \cdot (81)$$
$$100 \cdot (0.10x + 0.05y) = 100 \cdot (6.50)$$

$$-5x - 5y = -405$$
$$\underline{10x + 5y = 650}$$
$$5x = 245$$
$$x = 49$$
$$49 + y = 81$$
$$y = 32$$

49 dimes and 32 nickels

49.

	Coins	Value	Money
Nickels	x	$0.05	$0.05x$
Quarters	y	$0.25	$0.25y$
Total			$6.95

$$x = y + 7$$
$$x - y = 7$$
$$0.05x + 0.25y = 6.95$$

$$25 \cdot (x - y) = 25 \cdot (7)$$
$$100 \cdot (0.05x + 0.25y) = 100 \cdot (6.95)$$

$$25x - 25y = 175$$
$$\underline{5x + 25y = 695}$$
$$30x = 870$$
$$x = 29$$
$$29 - y = 7$$
$$-y = -22$$
$$y = 22$$

29 nickels and 22 quarters

51. $2 \cdot (5x - 2y) = 2 \cdot (47)$

$$10x - 4y = 94$$
$$\underline{3x + 4y = 23}$$
$$13x = 117$$
$$x = 9$$

$$3(9) + 4y = 23$$
$$27 + 4y = 23$$
$$4y = -4$$
$$y = -1$$
$$(9, -1)$$

53. $8x - 5(2x + 20) = -86$

$$8x - 10x - 100 = -86$$
$$-2x - 100 = -86$$
$$-2x = 14$$
$$x = -7$$

$$y = 2(-7) + 20 = -14 + 20 = 6$$
$$(-7, 6)$$

55. $3x + 8\left(\dfrac{3}{4}x + 1\right) = 116$

$$3x + 6x + 8 = 116$$
$$9x + 8 = 116$$
$$9x = 108$$
$$x = 12$$

$$y = \frac{3}{4}(12) + 1 = 9 + 1 = 10$$
$$(12, 10)$$

57. $-12x + 3(4x - 8) = -24$

$$-12x + 12x - 24 = -24$$
$$-24 = -24$$

Dependent, $(x, 4x - 8)$

59. $5 \cdot (13x - 8y) = 5 \cdot (20)$
$4 \cdot (9x + 10y) = 4 \cdot (-126)$

$$65x - 40y = 100$$
$$\underline{36x + 40y = -504}$$
$$101x = -404$$
$$x = -4$$

$$13(-4) - 8y = 20$$
$$-52 - 8y = 20$$
$$-8y = 72$$
$$y = -9$$
$$(-4, -9)$$

61. $4x + 5(6x - 14) = -19$

$$4x + 30x - 70 = -19$$
$$34x = 51$$
$$x = \frac{51}{34}$$
$$x = \frac{3}{2}$$

$$y = 6\left(\frac{3}{2}\right) - 14 = 9 - 14 = -5$$
$$\left(\frac{3}{2}, -5\right)$$

63. Answers will vary.

1. Number of math instructors: x
Number of English instructors: y
$x + y = 68$
$y = x + 12$
Substitution method
$x + (x + 12) = 68$
$2x + 12 = 68$
$2x = 56$
$x = 28$
$y = 28 + 12 = 40$

40 English instructors

2. Length: ℓ
Width: w
Perimeter $= 2\ell + 2w$
$230 = 2\ell + 2w$
$\ell = w + 35$
Substitution method
$230 = 2(w + 35) + 2w$
$230 = 2w + 70 + 2w$
$230 = 4w + 70$
$160 = 4w$
$40 = w$
$\ell = 40 + 35 = 75$

The concrete slab is 75 feet by 40 feet.

3.

Account	Principal	Rate	Earned
CD #1	x	0.05	$0.05x$
CD #2	y	0.04	$0.04y$
Total	$4200		$200

$x + y = 4200$
$0.05x + 0.04y = 200$

$$-5 \cdot (x + y) = -5 \cdot (4200)$$
$$100 \cdot (0.05x + 0.04y) = 100 \cdot (200)$$

$-5x - 5y = -21,000$
$\underline{5x + 4y = 20,000}$
$-y = -1000$
$y = 1000$

$x + 1000 = 4200$
$x = 3200$

$3200 at 5% and $1000 at 4%

4.

Nuts	Pounds	Cost	Total
Almonds	x	$5	$5x$
Cashews	y	$8	$8y$
Total	48		$48(\$6) = \$288$

$x + y = 48$
$5x + 8y = 288$

$$-5 \cdot (x + y) = -5 \cdot (48)$$

$-5x - 5y = -240$
$\underline{5x + 8y = 288}$
$3y = 48$
$y = 16$
$x + 16 = 48$
$x = 32$

32 pounds of almonds and 16 pounds of cashews

5.

Solution	Volume	Percent	Acid
60%	x	0.6	$0.6x$
50%	y	0.5	$0.5y$
Mix (54%)	400	0.54	$0.54(400) = 216$

$x + y = 400$
$0.6x + 0.5y = 216$

$$-5 \cdot (x + y) = -5 \cdot (400)$$
$$10 \cdot (0.6x + 0.5y) = 10 \cdot (216)$$

$-5x - 5y = -2000$
$\underline{6x + 5y = 2160}$
$x = 160$
$160 + y = 400$
$y = 240$

160 milliliters of 60% solution and 240 milliliters of 50% solution

6.

	Rate	Time	Distance
No Construction	70	x	$70x$
Construction	40	y	$40y$
Total		5	335

$x + y = 5$
$70x + 40y = 335$
$-40 \cdot (x + y) = -40 \cdot (5)$
$-40x - 40y = -200$
$\underline{70x + 40y = 335}$
$30x = 135$
$x = \dfrac{135}{30}$
$x = 4\dfrac{1}{2}$

$4\dfrac{1}{2} + y = 5$
$y = \dfrac{1}{2}$

$\dfrac{1}{2}$ hour

7.

	Rate	Time	Distance
With Wind	$x + y$	4.5	2250
Against Wind	$x - y$	5.0	2250

Where x is the speed of the plane in calm air.
Where y is the speed of the wind.

$$4.5(x + y) = 2250$$

$$5.0(x - y) = 2250$$

$$10 \cdot (4.5x + 4.5y) = 10 \cdot (2250)$$

$$9 \cdot (5x - 5y) = 9 \cdot (2250)$$

$$\begin{aligned} 45x + 45y &= 22,500 \\ 45x - 45y &= 20,250 \\ \hline 90x &= 42,750 \\ x &= 475 \end{aligned}$$

$$\begin{aligned} 5(475) - 5y &= 2250 \\ 2375 - 5y &= 2250 \\ -5y &= -125 \\ y &= 25 \end{aligned}$$

The plane's speed is 475 mph and the wind's
speed is 25 mph.

4.4   APPLICATIONS OF SYSTEMS OF EQUATIONS

1. $P = 2L + 2W$

3. $d = r \cdot t$

5.   Number of algebra students: x
Number of statistics students: y
$$x + y = 93$$
$$y = x + 17$$
Substitution method
$$\begin{aligned} x + (x + 17) &= 93 \\ 2x + 17 &= 93 \\ 2x &= 76 \\ x &= 38 \end{aligned}$$

$$y = 38 + 17 = 55$$
38 algebra students and 55 statistics students

7.   Andre's father's age: x
Andre's age: y
$$x + y = 98$$
$$x = y + 28$$
Substitution method
$$\begin{aligned} (y + 28) + y &= 98 \\ 2y + 28 &= 98 \\ 2y &= 70 \\ y &= 35 \end{aligned}$$

$$x = 35 + 28 = 63$$
Andre's age is 35 and his father's age is 63.

9.   Number of home runs (#1): x
Number of home runs (#2): y
$$x + y = 79$$
$$x = 2y + 7$$
Substitution method
$$\begin{aligned} (2y + 7) + y &= 79 \\ 3y + 7 &= 79 \\ 3y &= 72 \\ y &= 24 \end{aligned}$$

$$x = 2(24) + 7 = 48 + 7 = 55$$

The number of home runs for player #1 is 55
and for player #2 is 24.

11.   Number of field goals: x
Number of extra points: y
$$x + y = 47$$
$$3x + y = 77$$

$$-1 \cdot (x + y) = -1 \cdot (47)$$

$$\begin{aligned} -x - y &= -47 \\ 3x + y &= 77 \\ \hline 2x &= 30 \\ x &= 15 \end{aligned}$$

$$\begin{aligned} 15 + y &= 47 \\ y &= 32 \end{aligned}$$

15 field goals, 32 extra points

13.   Number of correct responses: x
Number of incorrect responses: y
$$x + y = 27$$
$$y = -x + 27$$

$$5x - 2y = 93$$

Substitution method
$$\begin{aligned} 5x - 2(-x + 27) &= 93 \\ 5x + 2x - 54 &= 93 \\ 7x - 54 &= 93 \\ 7x &= 147 \\ x &= 21 \end{aligned}$$

21 correct responses

15.

	Coins	Value	Money
Dimes	x	\$0.10	\$0.10x
Nickels	y	\$0.05	\$0.05y
Total	55		\$4.50

$$x + y = 55$$
$$0.10x + 0.05y = 4.50$$

$$-5 \cdot (x + y) = -5 \cdot (55)$$
$$100 \cdot (0.10x + 0.05y) = 100 \cdot (4.50)$$

$$\begin{aligned} -5x - 5y &= -275 \\ 10x + 5y &= 450 \\ \hline 5x &= 175 \\ x &= 35 \end{aligned}$$

35 dimes

17. Cost per pizza: x
Cost per pitcher of soda: y
$$3x + 2y = 41$$
$$5x + 3y = 67$$

$$-3 \cdot (3x + 2y) = -3 \cdot (41)$$
$$2 \cdot (5x + 3y) = 2 \cdot (67)$$

$$\begin{aligned} -9x - 6y &= -123 \\ 10x + 6y &= 134 \\ \hline x &= 11 \end{aligned}$$

$$3(11) + 2y = 41$$
$$33 + 2y = 41$$
$$2y = 8$$
$$y = 4$$

$11 per pizza and $4 per pitcher of soda

19. Length: ℓ
Width: w
$$2\ell + 2w = 92$$
$$w = \ell - 8$$
Substitution method
$$2\ell + 2(\ell - 8) = 92$$
$$2\ell + 2\ell - 16 = 92$$
$$4\ell - 16 = 92$$
$$4\ell = 108$$
$$\ell = 27$$

$$w = 27 - 8 = 19$$
Length is 27 inches and the width is 19 inches.

21. Length: ℓ
Width: w
$$2\ell + 2w = 330$$
$$\ell = 2w + 75$$
Substitution method
$$2(2w + 75) + 2w = 330$$
$$4w + 150 + 2w = 330$$
$$6w + 150 = 330$$
$$6w = 180$$
$$w = 30$$

$$\ell = 2(30) + 75 = 135$$

Length is 135 feet and the width is 30 feet.

23. Length: ℓ
Width: w
$$2\ell + 2w = 500$$
$$\ell = 2w + 25$$
Substitution method
$$2(2w + 25) + 2w = 500$$
$$4w + 50 + 2w = 500$$
$$6w + 50 = 500$$
$$6w = 450$$
$$w = 75$$

$$\ell = 2(75) + 25 = 175$$

Length is 175 centimeters and the width is 75 centimeters.

25. Length: ℓ
Width: w
$$2(\ell + 12) + 2(w + 12) = 178$$
$$\ell = w + 15$$
Substitution method
$$2[(w + 15) + 12] + 2(w + 12) = 178$$
$$2(w + 27) + 2(w + 12) = 178$$
$$2w + 54 + 2w + 24 = 178$$
$$4w + 78 = 178$$
$$4w = 100$$
$$w = 25$$

$$\ell = 25 + 15 = 40$$
Length is 40 feet and the width is 25 feet.

27.

Account	Principal	Rate	Earned
6% Fund	x	0.06	0.06x
3% Fund	y	0.03	0.03y
Total	2000		84

$$x + y = 2000$$
$$0.06x + 0.03y = 84$$

$$-6 \cdot (x + y) = -6 \cdot (2000)$$
$$100 \cdot (0.06x + 0.03y) = 100 \cdot (84)$$

$$\begin{aligned} -6x - 6y &= -12,000 \\ 6x + 3y &= 8,400 \\ \hline -3y &= -3,600 \\ y &= 1200 \end{aligned}$$

$$x + 1200 = 2000$$
$$x = 800$$
$800 at 6% and $1200 at 3%

29.

Account	Principal	Rate	Earned
4.25% CD	x	0.0425	0.0425x
3.5% CD	y	0.035	0.035y
Total	25,000		1025

$$x + y = 25,000$$
$$0.0425x + 0.035y = 1025$$

$$-350 \cdot (x + y) = -350 \cdot (25,000)$$
$$10,000 \cdot (0.0425x + 0.035y) = 10,000 \cdot (1025)$$

$$\begin{aligned} -350x - 350y &= -8,750,000 \\ 425x + 350y &= 10,250,000 \\ \hline 75x &= 1,500,000 \\ x &= 20,000 \end{aligned}$$

$$20,000 + y = 25,000$$
$$y = 5000$$

$20,000 at 4.25% and $5000 at 3.5%

31.

Account	Principal	Rate	Earned
4% Fund	x	0.04	$0.04x$
$-$ 13% Fund	y	-0.13	$-0.13y$
Total	7500		45

$x + y = 7500$
$0.04x - 0.13y = 45$

$13 \cdot (x + y) = 13 \cdot (7500)$
$100 \cdot (0.04x - 0.13y) = 100 \cdot (45)$

$13x + 13y = 97,500$
$\underline{4x - 13y = 4500}$
$\qquad 17x = 102,000$
$\qquad\quad x = 6000$

$6000 + y = 7500$
$\qquad\quad y = 1500$

$6000 at 4% profit and $1500 at 13% loss

33.

Granola	Pounds	Cost per Pound	Total Cost
Fruit	x	11.50	$11.50x$
No fruit	y	9.00	$9.00y$
Mix	40	10.00	400

$x + y = 40$
$11.50x + 9.00y = 400$

$-90 \cdot (x + y) = -90 \cdot (40)$
$10 \cdot (11.50x + 9.00) = 10 \cdot (400)$

$-90x - 90y = -3600$
$\underline{115x + 90y = 4000}$
$\qquad 25x = 400$
$\qquad\quad x = 16$

$16 + y = 40$
$\qquad y = 24$

16 pounds of granola with dried fruit and
24 pounds of granola without dried fruit

35.

Coffee	Pounds	Cost per Pound	Total Cost
Kona	x	45	$45x$
Regular	y	5	$5y$
Mix	80	20	1600

$x + y = 80$
$45x + 5y = 1600$

$-5 \cdot (x + y) = -5 \cdot (80)$

$-5x - 5y = -400$
$\underline{45x + 5y = 1600}$
$\qquad 40x = 1200$
$\qquad\quad x = 30$

$30 + y = 80$
$\qquad y = 50$

30 pounds of Kona beans and 50 pounds of
regular beans

37.

Solution	Volume	% Acid	Acid
80%	x	0.80	$0.80x$
44%	y	0.44	$0.44y$
Mix	320	0.53	$0.53(320)=169.6$

$x + y = 320$
$0.80x + 0.44y = 169.6$

$-80 \cdot (x + y) = -80 \cdot (320)$
$100 \cdot (0.80x + 0.44y) = 100 \cdot (169.6)$

$-80x - 80y = -25,600$
$\underline{80x + 44y = \quad 16,960}$
$\qquad -36y = -8,640$
$\qquad\quad y = 240$

$x + 240 = 320$
$\qquad\quad x = 80$

80 milliliters of the 80% acid solution and
240 milliliters of 44% acid solution

39.

Solution	Volume	% Alcohol	Alcohol
20%	x	0.20	$0.20x$
50%	y	0.50	$0.50y$
Mix	18	0.30	$0.30(18)=5.4$

$x + y = 18$
$0.20x + 0.50y = 5.4$

$-2 \cdot (x + y) = -2 \cdot (18)$
$10 \cdot (0.20x + 0.50y) = 10 \cdot (5.4)$

$-2x - 2y = -36$
$\underline{2x + 5y = 54}$
$\qquad 3y = 18$
$\qquad\quad y = 6$

$x + 6 = 18$
$\qquad x = 12$

12 ounces of 20% alcohol and 6 ounces of
50% alcohol

41. Amount of vodka: x
Amount of tonic: y
$x + y = 4$
$0.4x + 0y = 0.22(4) = 0.88$

$\qquad 0.4x = 0.88$
$\qquad\quad x = 2.2$
$2.2 + y = 4$
$\qquad y = 1.8$

2.2 liters of vodka and 1.8 liters of tonic water

43.

Driver	Rate	Time	Distance
George	70	x	$70x$
Tina	85	y	$85y$
Total		6	480

$x + y = 6$
$70x + 85y = 480$

$-70 \cdot (x + y) = -70 \cdot (6)$

$\begin{aligned} -70x - 70y &= -420 \\ 70x + 85y &= 480 \\ \hline 15y &= 60 \\ y &= 4 \end{aligned}$

$x + 4 = 6$
$\quad x = 2$

George drove 2 hours and Tina drove 4 hours.

45. Time without engine trouble: x
Time with engine trouble: y
$x + y = 3.5$
$160x + 130y = 500$

$-160 \cdot (x + y) = -160 \cdot (3.5)$

$\begin{aligned} -160x - 160y &= -560 \\ 160x + 130y &= 500 \\ \hline -30y &= -60 \\ y &= 2 \end{aligned}$

$x + 2 = 3.5$
$\quad x = 1.5$

after 1.5 hours

47.

	Rate	Time	Distance
With Current	$x + y$	3	18
Against Current	$x - y$	9	18

Speed of kayak in still water: x
Speed of current: y

$3(x + y) = 18$
$9(x - y) = 18$

$3x + 3y = 18$
$9x - 9y = 18$

$3 \cdot (3x + 3y) = 3 \cdot (18)$

$\begin{aligned} 9x + 9y &= 54 \\ 9x - 9y &= 18 \\ \hline 18x &= 72 \\ x &= 4 \end{aligned}$

$3(4) + 3y = 18$
$12 + 3y = 18$
$\quad 3y = 6$
$\quad y = 2$

The speed of the kayak is 4 mph and the speed
of the current is 2 mph.

49. Speed of plane in calm air: x
Speed of wind: y
$5(x + y) = 3000$
$6(x - y) = 3000$

$5x + 5y = 3000$
$6x - 6y = 3000$

$6 \cdot (5x + 5y) = 6 \cdot (3000)$
$5 \cdot (6x - 6y) = 5 \cdot (3000)$

$\begin{aligned} 30x + 30y &= 18,000 \\ 30x - 30y &= 15,000 \\ \hline 60x &= 33,000 \\ x &= 550 \end{aligned}$

$5(550) + 5y = 3000$
$2750 + 5y = 3000$
$\quad 5y = 250$
$\quad y = 50$

The speed of the plane is 550 mph and the
speed of the wind is 50 mph.

51. Answers will vary. Example:
The perimeter of a rectangle is 46 meters. If the
width is seven meters less than the length, find
the dimensions of the rectangle.

53. Answers will vary. Example:
Andrea mixes a 40% acid solution with a 58%
acid solution. Find the amount of each needed
to make 90 ml of a 42% acid solution.

4.5 QUICK CHECK

1. Graph a dashed line at $y = x + 3$.
Test Point: $(0, 0)$
$0 > 0 + 3$
$0 > 3$
False. Shade the half-plane that does not
contain the test point.

Graph a dashed line at $2x + 3y = 6$.
Test Point: $(0, 0)$
$2(0) + 3(0) < 6$
$\quad\quad 0 < 6$
True. Shade the half-plane that contains
the test point.

The solution of the system of linear inequalities
is the region where the two solutions intersect.

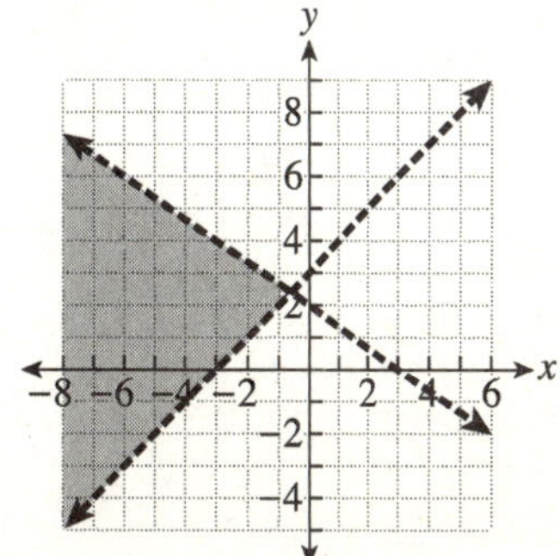

2. $-3x + 4y = 8$
$$4y = 3x + 8$$
$$y = \frac{3}{4}x + 2$$

Graph a solid line at $y = \frac{3}{4}.x + 2$.

Test Point: $(0,0)$
$$-3(0) + 4(0) \le 8$$
$$0 \le 8$$
True. Shade the half-plane that contains the test point.

Graph a solid line at $y = 0$.
Test Point: $(0,3)$
$$3 \ge 0$$
True. Shade the half-plane that contains the test point.

The solution of the system of linear inequalities is the region where the two solutions intersect.

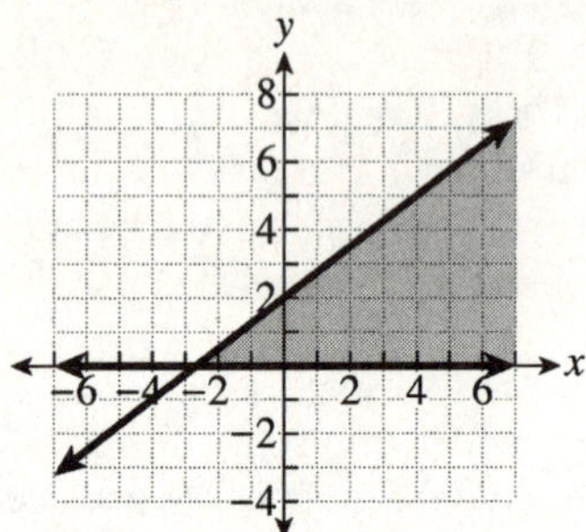

3. Graph a solid line at $x + y = 8$.
Test Point: $(0,0)$
$$0 + 0 \le 8$$
$$0 \le 8$$
True. Shade the half-plane that contains the test point.

Graph a solid line at $x = 1$.
Test Point: $(0,0)$
$$0 \ge 1$$
False. Shade the half-plane that does not contain the test point.

Graph a solid line at $y = -1$.
Test Point: $(0,0)$
$$0 \ge -1$$
True. Shade the half-plane that contains the test point.

The solution of the system of linear inequalities is the region where the three solutions intersect.

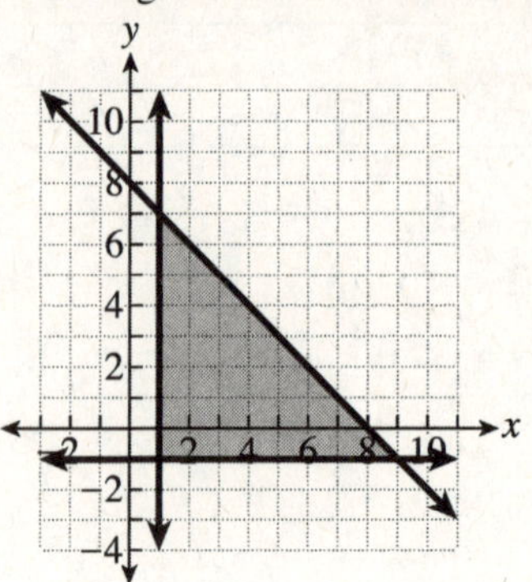

4.5 SYSTEMS OF LINEAR INEQUALITIES

1. system of linear inequalities

3. solid

5. Graph a solid line at $y = 5x - 2$.
Test Point: $(0,0)$
$$0 \ge 5(0) - 2$$
$$0 \ge -2$$
True. Shade the half-plane that contains the test point.

Graph a solid line at $y = -x + 1$.
Test Point: $(0,0)$
$$0 \le -(0) + 1$$
$$0 \le 1$$
True. Shade the half-plane that contains the test point.

The solution of the system of linear inequalities is the region where the two solutions intersect.

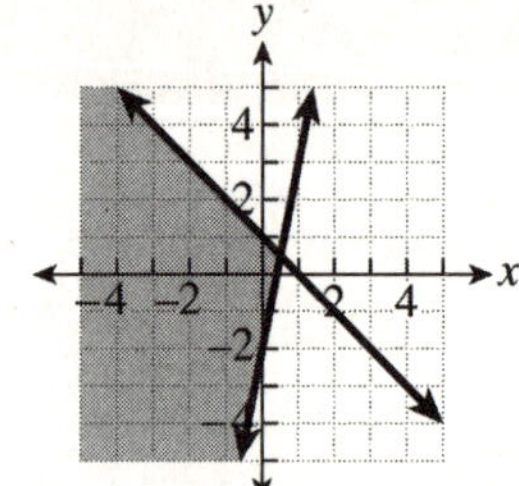

7. Graph a solid line at $y = 3x - 1$.
Test Point: $(0,0)$
$0 \le 3(0) - 1$
$0 \le -1$
False. Shade the half-plane that does not contain the test point.

Graph a dashed line at $y = -\dfrac{3}{4}x + 6$.

Test Point: $(0,0)$
$0 > -\dfrac{3}{4}(0) + 6$
$0 > 6$
False. Shade the half-plane that does not contain the test point.

The solution of the system of linear inequalities is the region where the two solutions intersect.

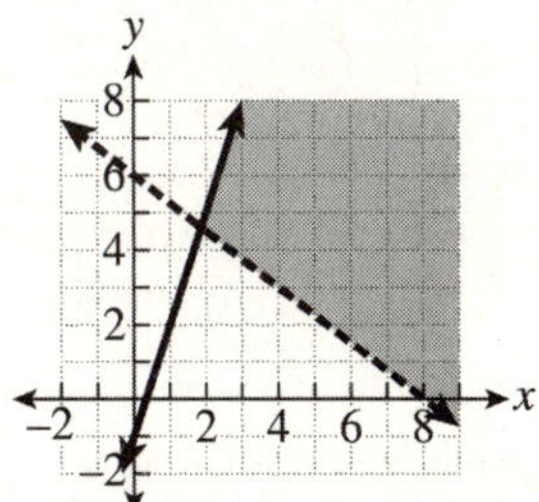

9. Graph a solid line at $x + y = 5$.
Test Point: $(0,0)$
$0 + 0 \ge 5$
$0 \ge 5$
False. Shade the half-plane that does not contain the test point.

Graph a solid line at $x - y = 3$.
Test Point: $(0,0)$
$0 - 0 \ge 3$
$0 \ge 3$
False. Shade the half-plane that does not contain the test point.

The solution of the system of linear inequalities is the region where the two solutions intersect.

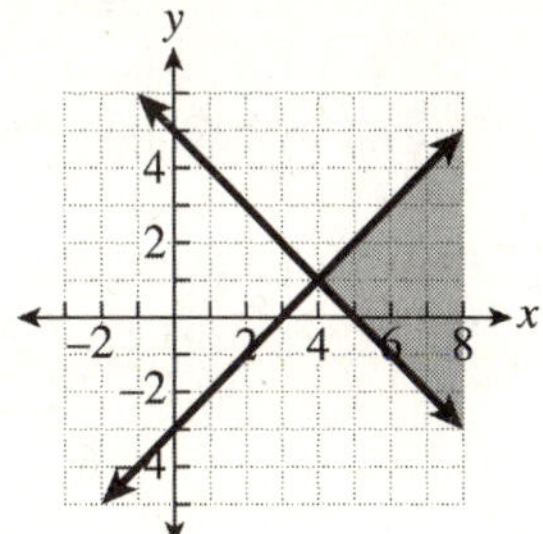

11. Graph a dashed line at $y = 5x + 5$.
Test Point: $(0,0)$
$0 < 5(0) + 5$
$0 < 5$
True. Shade the half-plane that contains the test point.

Graph a dashed line at $y = 5x - 5$.
Test Point: $(0,0)$
$0 > 5(0) - 5$
$0 > -5$
True. Shade the half-plane that contains the test point.

The solution of the system of linear inequalities is the region where the two solutions intersect.

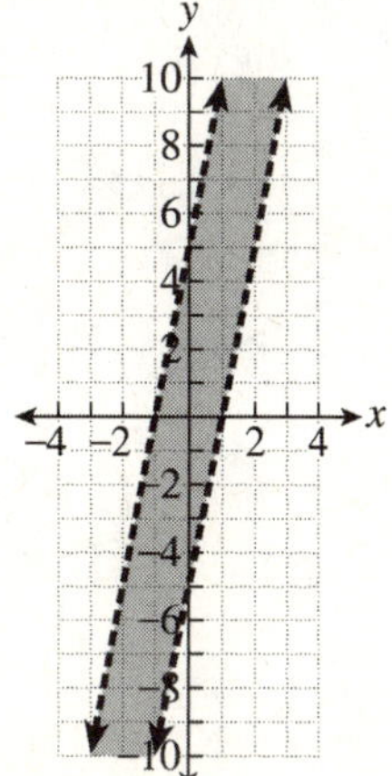

13. Graph a dashed line at $x + 4y = 8$.
Test Point: $(0,0)$
$0 + 4(0) < 8$
$0 < 8$
True. Shade the half-plane that contains the test point.

Graph a dashed line at $3x - 2y = 6$.
Test Point: $(0,0)$
$3(0) - 2(0) < 6$
$0 < 6$
True. Shade the half-plane that contains the test point.

The solution of the system of linear inequalities is the region where the two solutions intersect.

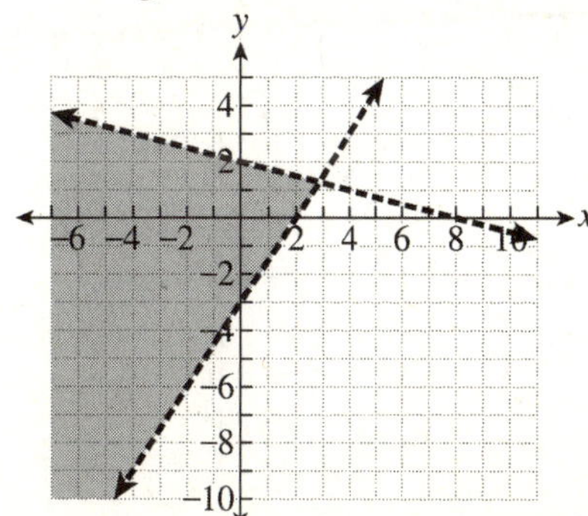

15. Graph a dashed line at $-7x+3y=21$.
Test Point: $(0,0)$
$-7(0)+3(0)>21$
$\qquad 0>21$
False. Shade the half-plane that does not contain the test point.

Graph a solid line at $4x+9y=36$.
Test Point: $(0,0)$
$4(0)+9(0)\le 36$
$\qquad 0\le 36$
True. Shade the half-plane that contains the test point.

The solution of the system of linear inequalities is the region where the two solutions intersect.

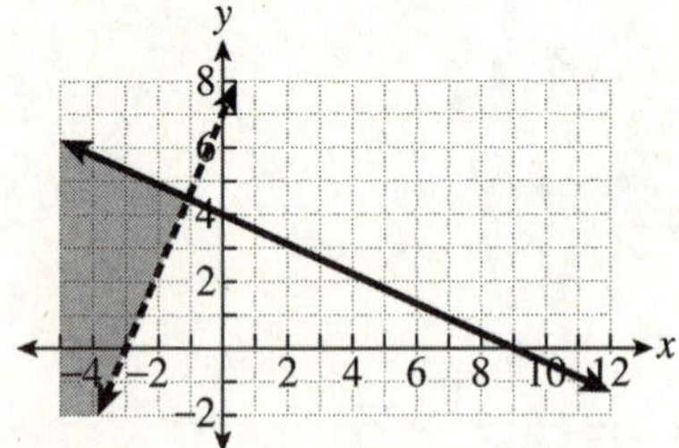

17. Graph a solid line at $y=5$
Test Point: $(0,0)$
$0\ge 5$
False. Shade the half-plane that does not contain the test point.

$5x+2y=0$
$\qquad 2y=-5x$
$\qquad y=-\dfrac{5}{2}x$

Graph a dashed line at $y=-\dfrac{5}{2}x$.

Test Point: $(1,1)$
$5(1)+2(1)<0$
$\qquad 7<0$
False. Shade the half-plane that does not contain the test point.

The solution of the system of linear inequalities is the region where the two solutions intersect.

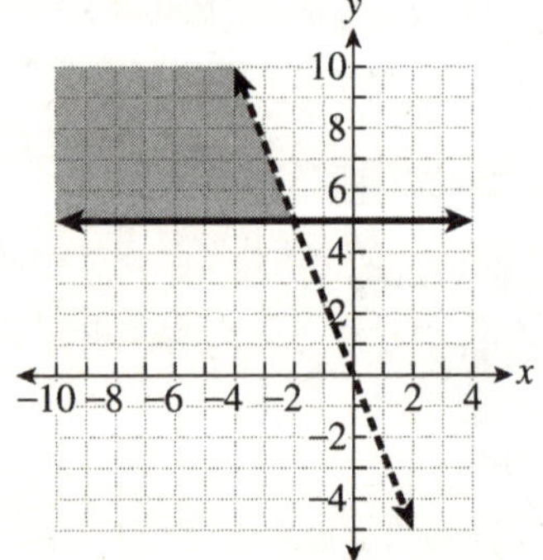

19. Graph a solid line at $y=3x$.
Test Point: $(1,1)$
$1\ge 3(1)$
$1\ge 3$
False. Shade the half-plane that does not contain the test point.

Graph a dashed line at $y=3$.
Test Point: $(0,0)$
$0>3$
False. Shade the half-plane that does not contain the test point.

The solution of the system of linear inequalities is the region where the two solutions intersect.

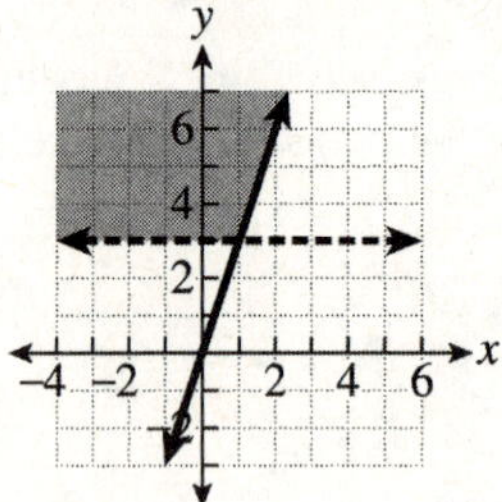

21. Graph a dashed line at $y=-\dfrac{3}{2}x+6$.

Test Point: $(0,0)$

$0>-\dfrac{3}{2}(0)+6$

$0>6$
False. Shade the half-plane that does not contain the test point.

$4x-3y=-6$
$\qquad -3y=-4x-6$
$\qquad y=\dfrac{4}{3}x+2$

Graph a solid line at $y=\dfrac{4}{3}x+2$

Test Point: $(0,0)$
$4(0)-3(0)\ge -6$
$\qquad 0\ge -6$
True. Shade the half-plane that contains the test point.

The solution of the system of linear inequalities is the region where the two solutions intersect.

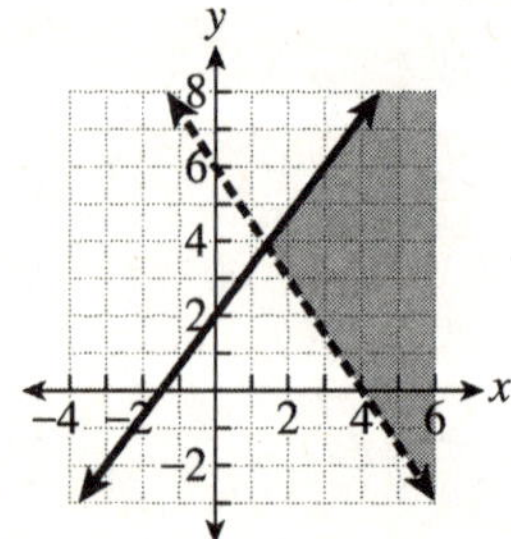

23. Graph a dashed line at $x = 1$.
Test Point: $(0,0)$
$0 > 1$
False. Shade the half-plane that does not contain the test point.

$$8x - 4y = 12$$
$$-4y = -8x + 12$$
$$y = 2x - 3$$

Graph a dashed line at $y = 2x - 3$.
Test Point: $(0,0)$
$8(0) - 4(0) > 12$
$0 > 12$
False. Shade the half-plane that does not contain the test point.

The solution of the system of linear inequalities is the region where the two solutions intersect.

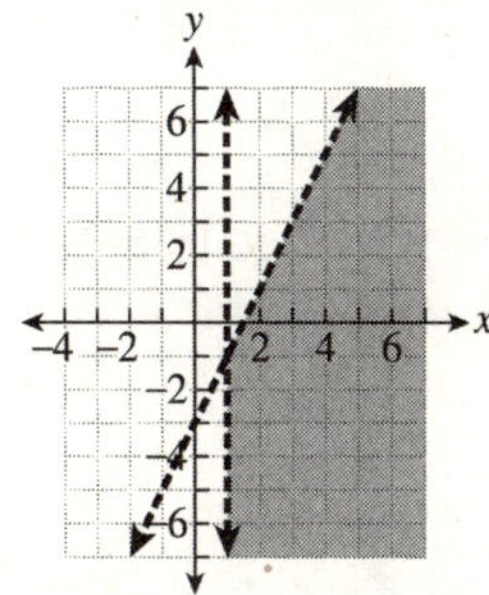

25. Graph a dashed line at $y = 2x$.
Test Point: $(1,1)$
$1 < 2(1)$
$1 < 2$
True. Shade the half-plane that contains the test point.

Graph a dashed line at $y = 6$.
Test Point: $(0,0)$
$0 < 6$
True. Shade the half-plane that contains the test point.

Graph a dashed line at $x = 2$.
Test Point: $(0,0)$
$0 > 2$
False. Shade the half-plane that does not contain the test point.

The solution of the system of linear inequalities is the region where the three solutions intersect.

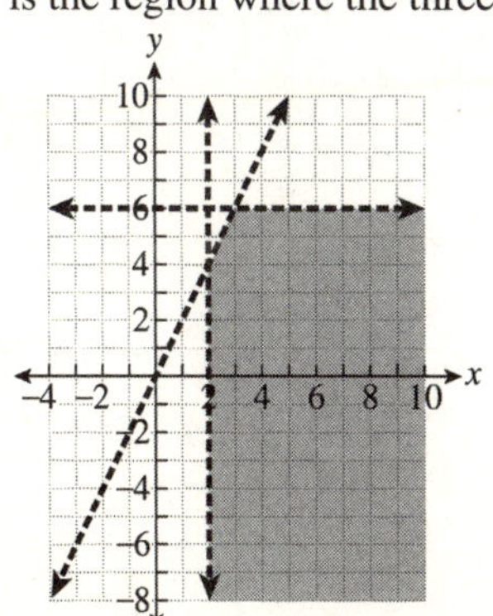

27. $3x - 2y = -10$
$$-2y = -3x - 10$$
$$y = \frac{3}{2}x + 5$$

Graph a solid line at $y = \frac{3}{2}x + 5$.

Test Point: $(0,0)$
$3(0) - 2(0) \le -10$
$0 \le -10$
False. Shade the half-plane that does not contain the test point.

Graph a dashed line at $y = -4x - 6$.
Test Point: $(0,0)$
$0 < -4(0) - 6$
$0 < -6$
False. Shade the half-plane that does not contain the test point.

$$-x + 3y = 12$$
$$3y = x + 12$$
$$y = \frac{1}{3}x + 4$$

Graph a dashed line at $y = \frac{1}{3}x + 4$.

Test Point: $(0,0)$
$-(0) + 3(0) < 12$
$0 < 12$
True. Shade the half-plane that contains the test point.

The solution of the system of linear inequalities is the region where the three solutions intersect.

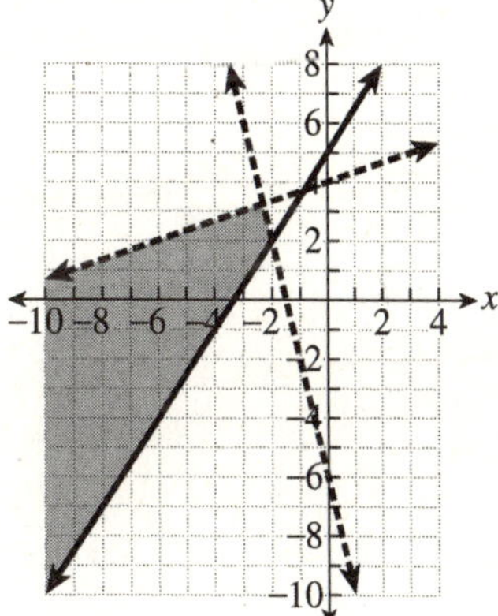

29. $x > 0 \qquad y > 0$

31. Test each inequality at $(0,0)$.

$x = 2$ $y = 1$ $y = -x + 9$
$0 \geq 2$ False. $0 \geq 1$ False. $0 \leq -(0) + 9$
                                      $0 \leq 9$ True.

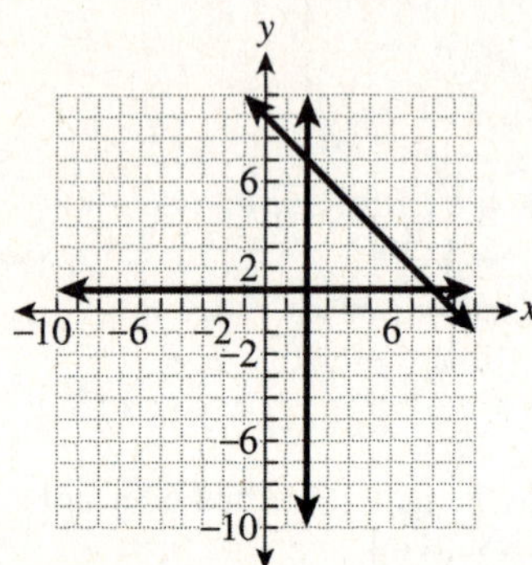

Find the area inside the triangle.
base $= 8 - 2 = 6$
height $= 7 - 1 = 6$

$$\text{Area of a Triangle} = \frac{1}{2}(\text{base})(\text{height})$$
$$= \frac{1}{2}(6)(6)$$
$$= 18 \text{ square units}$$

CHAPTER 4 REVIEW

1. $(-4, -5)$

2. $(6, -1)$

3. $(-3, 6)$

4. $(-7, 0)$

5. Graph the line $y = x - 5$.
$m = 1$, y-intercept: $(0, -5)$
Count from the y-intercept 1 up and
1 right to find the next point $(1, -4)$.
Graph the line $y = -\frac{1}{2}x + 7$.
$m = -\frac{1}{2}$, y-intercept: $(0, 7)$
Count from the y-intercept 1 down and
2 right to find the next point $(2, 6)$.

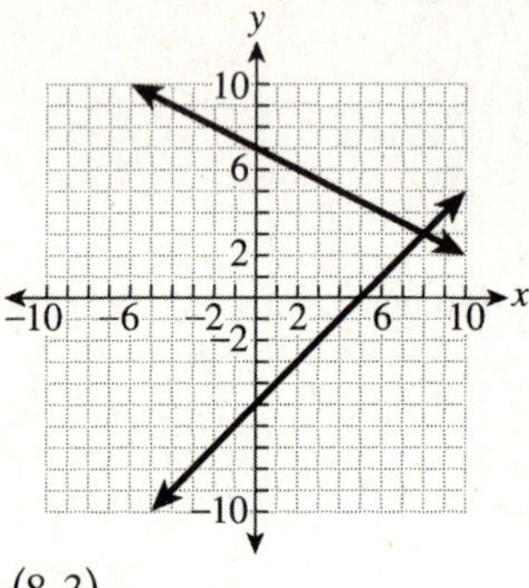

$(8, 3)$

6. Graph the line $y = \frac{1}{3}x - 4$.

$m = \frac{1}{3}$, y-intercept: $(0, -4)$
Count from the y-intercept 1 up and
3 right to find the next point $(3, -3)$.
Graph the line $y = -x$.
$m = -1$, y-intercept: $(0, 0)$
Count from the y-intercept 1 down and
1 right to find the next point $(1, -1)$.

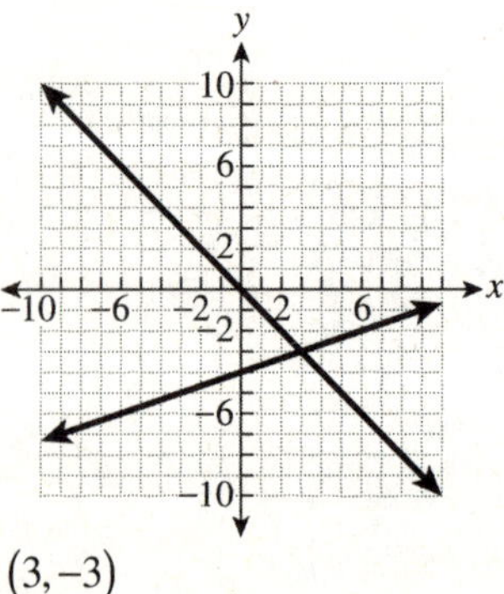

$(3, -3)$

7. Graph the line $y = -2x + 8$.
$m = -2$, y-intercept: $(0, 8)$

Count from the y-intercept 2 down and 1 right to find the next point $(1, 6)$.

Graph the line $y = \dfrac{1}{2}x + 3$.

$m = \dfrac{1}{2}$, y-intercept: $(0, 3)$

Count from the y-intercept 1 up and 2 right to find the next point $(2, 4)$.

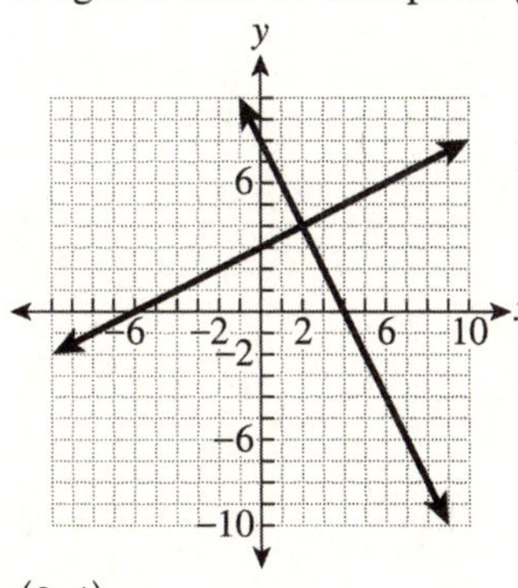

$(2, 4)$

8. $2x - y = -8$

x-intercept:
$2x - (0) = -8$
$x = -4$
$(-4, 0)$

y-intercept:
$2(0) - y = -8$
$y = 8$
$(0, -8)$

$-3x + 2y = 10$

x-intercept:
$-3x + 2(0) = 10$
$x = \dfrac{10}{-3}$
$\left(-\dfrac{10}{3}, 0\right)$

y-intercept:
$-3(0) + 2y = 10$
$y = 5$
$(0, 5)$

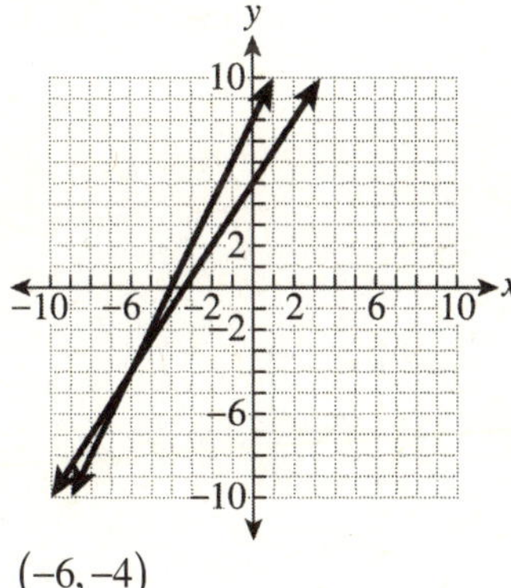

$(-6, -4)$

9.
$$2(2y + 1) - 5y = 4$$
$$4y + 2 - 5y = 4$$
$$2 - y = 4$$
$$-y = 2$$
$$y = -2$$

$x = 2(-2) + 1 = -4 + 1 = -3$
$(-3, -2)$

10.
$$3x + 4(-x - 3) = -5$$
$$3x - 4x - 12 = -5$$
$$-x - 12 = -5$$
$$-x = 7$$
$$x = -7$$

$y = -(-7) - 3 = 7 - 3 = 4$
$(-7, 4)$

11.
$$4x + y = -5$$
$$y = -4x - 5$$

$$5x - 3(-4x - 5) = 15$$
$$5x + 12x + 15 = 15$$
$$17x = 0$$
$$x = 0$$

$$4(0) + y = -5$$
$$y = -5$$
$(0, -5)$

12.
$$-4x + 2y = 18$$
$$2y = 4x + 18$$
$$y = 2x + 9$$

$$2x - (2x + 9) = 3$$
$$2x - 2x - 9 = 3$$
$$-9 = 3$$
Inconsistent, $\varnothing$

13. $3x - 4\left(\dfrac{2}{3}x - 5\right) = 23$

$3x - \dfrac{8}{3}x + 20 = 23$

$3x - \dfrac{8}{3}x = 3$

$3 \cdot \left(3x - \dfrac{8}{3}x\right) = 3 \cdot 3$

$9x - 8x = 9$
$x = 9$

$y = \dfrac{2}{3}(9) - 5 = 6 - 5 = 1$

$(9, 1)$

14.
$$3x + 2y = 15$$
$$\underline{5x - 2y = 17}$$
$$8x = 32$$
$$x = 4$$

$$3(4) + 2y = 15$$
$$12 + 2y = 15$$
$$2y = 3$$
$$y = \frac{3}{2}$$

$$\left(4, \frac{3}{2}\right)$$

15.
$$2 \cdot (-3x + 4y) = 2 \cdot (-27)$$
$$-6x + 8y = -54$$
$$\underline{6x + 7y = -66}$$
$$15y = -120$$
$$y = -8$$

$$6x + 7(-8) = -66$$
$$6x - 56 = -66$$
$$6x = -10$$
$$x = -\frac{10}{6}$$
$$x = -\frac{5}{3}$$

$$\left(-\frac{5}{3}, -8\right)$$

16.
$$2 \cdot (2x - y) = 2 \cdot (4)$$
$$4x - 2y = 8$$
$$\underline{-4x + 2y = -8}$$
$$0 = 0$$
Dependent, $(x, 2x - 4)$

17.
$$y = 4x + 5$$
$$-4x + y = 5$$

$$2 \cdot (-4x + y) = 2 \cdot (5)$$
$$-8x + 2y = 10$$
$$\underline{5x - 2y = -4}$$
$$-3x = 6$$
$$x = -2$$

$$y = 4(-2) + 5 = -8 + 5 = -3$$
$$(-2, -3)$$

18.
$$-6 \cdot \left(\frac{1}{2}x + \frac{2}{3}y\right) = 2 \cdot -6$$
$$-3x - 4y = -12$$
$$2 \cdot (-3x - 4y) = -12 \cdot 2$$
$$-6x - 8y = -24$$

$$2 \cdot \left(3x - \frac{1}{2}y\right) = -15 \cdot 2$$
$$6x - y = -30$$

$$-6x - 8y = -24$$
$$\underline{6x - y = -30}$$
$$-9y = -54$$
$$y = 6$$

$$6x - (6) = -30$$
$$6x = -24$$
$$x = -4$$
$$(-4, 6)$$

19.
$$4 \cdot (-x + 6y) = 4 \cdot (-26)$$
$$-4x + 24y = -104$$
$$\underline{4x - 3y = 20}$$
$$21y = -84$$
$$y = -4$$

$$4x - 3(-4) = 20$$
$$4x + 12 = 20$$
$$4x = 8$$
$$x = 2$$
$$(2, -4)$$

20.
$$5 \cdot (x + y) = 5 \cdot (4)$$
$$5x + 5y = 20$$
$$\underline{-5x + 4y = 43}$$
$$9y = 63$$
$$y = 7$$

$$x + (7) = 4$$
$$x = -3$$
$$(-3, 7)$$

21.
$$3x - 7(2x - 1) = 29$$
$$3x - 14x + 7 = 29$$
$$-11x + 7 = 29$$
$$-11x = 22$$
$$x = -2$$

$$y = 2(-2) - 1 = -4 - 1 = -5$$
$$(-2, -5)$$

22.
$$8(5) - 9y = 67$$
$$40 - 9y = 67$$
$$-9y = 27$$
$$y = -3$$
$$(5, -3)$$

23.
$$-9x + 3(3x - 2) = -6$$
$$-9x + 9x - 6 = -6$$
$$-6 = -6$$
Dependent, $(x, 3x - 2)$

24.
$$6 \cdot (x + y) = 6 \cdot (-4)$$
$$6x + 6y = -24$$
$$\underline{4x - 6y = -1}$$
$$10x = -25$$
$$x = \frac{-25}{10}$$
$$x = -\frac{5}{2}$$
$$\left(-\frac{5}{2}\right) + y = -4$$
$$y = -\frac{8}{2} + \frac{5}{2}$$
$$y = -\frac{3}{2}$$
$$\left(-\frac{5}{2}, -\frac{3}{2}\right)$$

25.
$$3x - 2\left(\frac{4}{5}x - 6\right) = 12$$
$$3x - \frac{8}{5}x + 12 = 12$$
$$\frac{15}{5}x - \frac{8}{5}x = 0$$
$$\frac{7}{5}x = 0$$
$$x = 0$$
$$y = \frac{4}{5}(0) - 6 = -6$$
$$(0, -6)$$

26.
$$-4 \cdot (x + 2y) = -4 \cdot (-5)$$
$$12 \cdot \left(\frac{1}{3}x - \frac{1}{4}y\right) = 12 \cdot (13)$$
$$-4x - 8y = 20$$
$$\underline{4x - 3y = 156}$$
$$-11y = 176$$
$$y = -16$$
$$x + 2(-16) = -5$$
$$x - 32 = -5$$
$$x = 27$$
$$(27, -16)$$

27.
$$4 \cdot \left(-\frac{1}{2}x + 2y\right) = 4 \cdot (3)$$
$$-2x + 8y = 12$$
$$\underline{2x - 8y = -9}$$
$$0 = 3$$
Inconsistent, $\varnothing$

28.
$$7 \cdot (7x - 5y) = 7 \cdot (-46)$$
$$5 \cdot (6x + 7y) = 5 \cdot (17)$$
$$49x - 35y = -322$$
$$\underline{30x + 35y = 85}$$
$$79x = -237$$
$$x = -3$$
$$7(-3) - 5y = -46$$
$$-21 - 5y = -46$$
$$-5y = -25$$
$$y = 5$$
$$(-3, 5)$$

29.
$$3x - 2(0.3x + 2.4) = 7.2$$
$$3x - 0.6x - 4.8 = 7.2$$
$$2.4x - 4.8 = 7.2$$
$$2.4x = 12$$
$$x = 5$$
$$y = 0.3(5) + 2.4 = 1.5 + 2.4 = 3.9$$
$$(5, 3.9)$$

30.
$$100 \cdot (0.04x + 0.05y) = 340 \cdot 100$$
$$4x + 5y = 34000$$
$$-4 \cdot (x + y) = 8000 \cdot -4$$
$$-4x - 4y = -32000$$
$$4x + 5y = 34000$$
$$\underline{-4x - 4y = -32000}$$
$$y = 2000$$
$$x + 2000 = 8000$$
$$x = 6000$$
$$(6000,\ 2000)$$

31.
$$x + y = 87$$
$$y = 2x + 18$$
$$x + (2x + 18) = 87$$
$$3x + 18 = 87$$
$$3x = 69$$
$$x = 23$$
$$y = 2(23) + 18 = 64$$
The numbers are 23 and 64.

32. Number of baseball cards Hal has: x
Number of baseball cards Mac has: y
$$x = y + 284$$
$$x + y = 1870$$
$$(y + 284) + y = 1870$$
$$2y + 284 = 1870$$
$$2y = 1586$$
$$y = 793$$
$$x = 793 + 284 = 1077$$
Hal has 1077 baseball cards.

33. Number of men's tickets: x
Number of women's tickets: y
$$x + y = 1764$$
$$10x + 5y = 13,470$$
$$-10(x + y) = -10(1764)$$
$$\begin{array}{r} -10x - 10y = -17,640 \\ 10x + 5y = 13,470 \\ \hline -5y = -4,170 \\ y = 834 \end{array}$$
There were 834 women at the game.

34. Number of general tickets: x
Number of senior tickets: y
$$x + y = 206$$
$$7.5x + 4y = 1384$$
$$-7.5 \cdot (x + y) = -7.5 \cdot (206)$$
$$\begin{array}{r} -7.5x - 7.5y = -1545 \\ 7.5x + 4y = 1384 \\ \hline -3.5y = -161 \\ y = 46 \end{array}$$
There were 46 senior citizens.

35. Length: ℓ
Width: w
$$2\ell + 2w = 1320$$
$$\ell = w + 140$$
Substitution method
$$2(w + 140) + 2w = 1320$$
$$2w + 280 + 2w = 1320$$
$$4w + 280 = 1320$$
$$4w = 1040$$
$$w = 260$$
$$\ell = 260 + 140 = 400$$
Length is 400 feet and the width is 260 feet.

36. Height: h Width: w
$$2h + 2w = 224$$
$$h = 2w + 4$$
Substitution method
$$2(2w + 4) + 2w = 224$$
$$4w + 8 + 2w = 224$$
$$6w + 8 = 224$$
$$6w = 216$$
$$w = 36$$
$$h = 2(36) + 4 = 76$$

The height of the door is 76 inches.

37. Amount invested at 5%: x
Amount invested at 4.25%: y
$$x + y = 4000$$
$$0.05x + 0.0425y = 179$$
$$-500 \cdot (x + y) = -500 \cdot (4000)$$
$$10,000 \cdot (0.05x + 0.0425y) = 10,000 \cdot (179)$$
$$\begin{array}{r} -500x - 500y = -2,000,000 \\ 500x + 425y = 1,790,000 \\ \hline -75y = -210,000 \\ y = 2800 \end{array}$$
$$x + 2800 = 4000$$
$$x = 1200$$
$1200 at 5% and $2800 at 4.25%

38. Amount invested at 20%: x
Amount invested at 8%: y
$$x + y = 2500$$
$$0.20x + 0.08y = 320$$
$$-20 \cdot (x + y) = -20 \cdot (2500)$$
$$100 \cdot (0.20x + 0.08y) = 100 \cdot (320)$$
$$\begin{array}{r} -20x - 20y = -50,000 \\ 20x + 8y = 32,000 \\ \hline -12y = -18,000 \\ y = 1500 \end{array}$$
$$x + 1500 = 2500$$
$$x = 1000$$
$1000 at 20% and $1500 at 8%

39. Amount of 4% protein feed: x
Amount of 13% protein feed: y
$$x + y = 45$$
$$0.04x + 0.13y = 0.10 \cdot (45) = 4.5$$
$$-4 \cdot (x + y) = -4 \cdot (45)$$
$$100 \cdot (0.04x + 0.13y) = 100 \cdot (4.5)$$
$$\begin{array}{r} -4x - 4y = -180 \\ 4x + 13y = 450 \\ \hline 9y = 270 \\ y = 30 \end{array}$$
$$x + 30 = 45$$
$$x = 15$$
15 pounds of 4% protein feed and 30 pounds of 13% protein feed

40. Speed of the kayak in still water: x
Speed of the current: y

	Rate	Time	Distance
Downstream	$x + y$	1.5	12
Upstream	$x - y$	6.0	12

$$1.5(x + y) = 12$$
$$6(x - y) = 12$$
$$4 \cdot (1.5x + 1.5y) = 4 \cdot (12)$$
$$6x + 6y = 48$$
$$6x - 6y = 12$$
$$\overline{12x = 60}$$
$$x = 5$$

The speed of the kayak in still water is 5 mph.

41. Graph a solid line at $y = 2x + 7$.

Test Point: $(0,0)$
$$0 \le 2(0) + 7$$
$$0 \le 7$$
True. Shade the half-plane that contains the test point.

Graph a dashed line at $y = x - 5$.

Test Point: $(0,0)$
$$0 > 0 - 5$$
$$0 > -5$$
True. Shade the half-plane that contains the test point.

The solution of the system of linear inequalities is the region where the two solutions intersect.

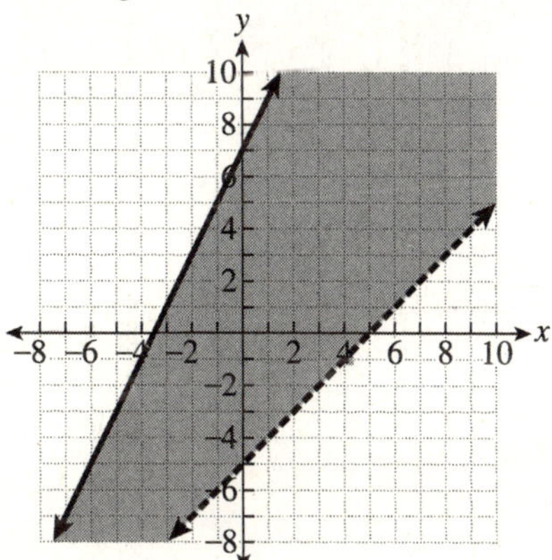

42. Graph a dashed line at $3x + 4y = 24$.
Test Point: $(0,0)$
$$3(0) + 4(0) < 24$$
$$0 < 24$$
True. Shade the half-plane that contains the test point.

Graph a solid line at $5x + y = 10$.
Test Point: $(0,0)$
$$5(0) + 0 \ge 10$$
$$0 \ge 10$$
False. Shade the half-plane that does not contain the test point.

The solution of the system of linear inequalities is the region where the two solutions intersect.

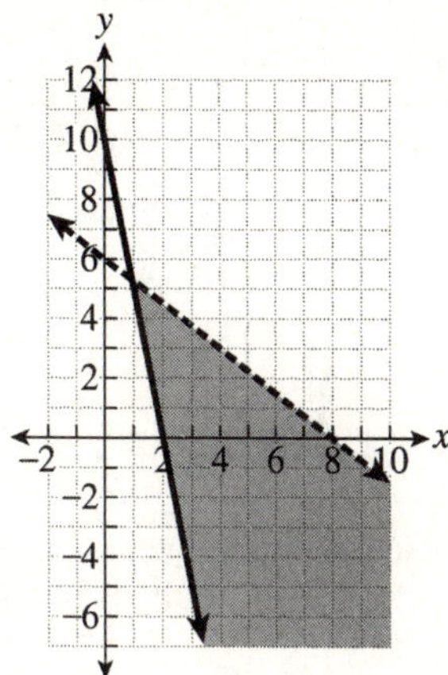

43. Graph a solid line at $y = -4x + 7$.
Test Point: $(0,0)$
$$0 \le -4(0) + 7$$
$$0 \le 7$$
True. Shade the half-plane that contains the test point.

Graph a solid line at $y = \dfrac{3}{2}x$.

Test Point: $(4,0)$

$$0 \ge \frac{3}{2}(4)$$

$$0 \ge 6$$
False. Shade the half-plane that does not contain the test point.

The solution of the system of linear inequalities is the region where the two solutions intersect.

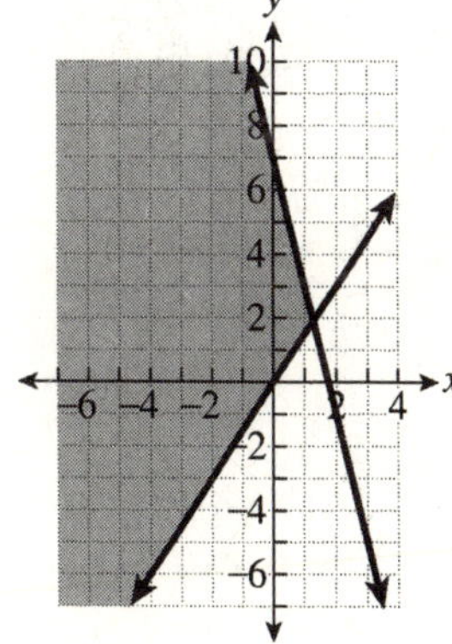

44. Graph a dashed line at $y = \dfrac{1}{2}x - 3$.

Test Point: $(0,0)$

$0 > \dfrac{1}{2}(0) - 3$

$0 > -3$

True. Shade the half-plane that contains the test point.

Draw a dashed line at $y = -2$.
Test Point: $(0,0)$
$0 > -2$
True. Shade the half-plane that contains the test point.

The solution of the system of linear inequalities is the region where the two solutions intersect.

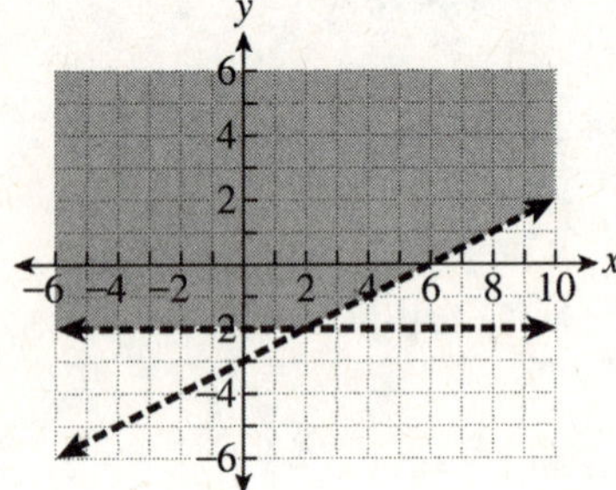

CHAPTER 4 TEST

1. $(-5, 2)$

2. Graph the line $y = -2x + 3$.

$m = -\dfrac{2}{1}$, y-intercept: $(0,3)$

Count from the y-intercept 2 down and 1 right to find the next point $(1,1)$.

Graph the line $y = -3x + 7$.

$m = \dfrac{-3}{1}$, y-intercept: $(0,7)$

Count from the y-intercept 3 down and 1 right to find the next point $(1,4)$.

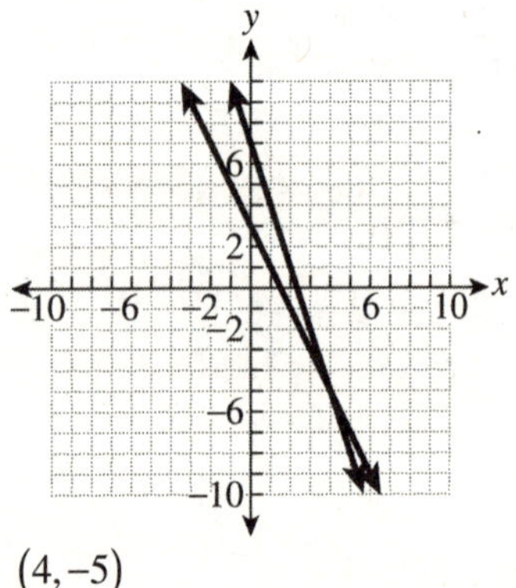

$(4, -5)$

3. $x + 4y = -9$
$x = -4y - 9$

$2(-4y - 9) - 5y = 21$
$-8y - 18 - 5y = 21$
$-13y - 18 = 21$
$-13y = 39$
$y = -3$

$x = -4(-3) - 9 = 3$
$(3, -3)$

4. $5 \cdot (5x + 2y) = 5 \cdot (-22)$
$2 \cdot (2x - 5y) = 2 \cdot (-3)$

$25x + 10y = -110$
$\underline{4x - 10y = -6}$
$29x = -116$
$x = -4$

$5(-4) + 2y = -22$
$-20 + 2y = -22$
$2y = -2$
$y = -1$
$(-4, -1)$

5. $-x + 4y = 11$
$\underline{3x - 4y = -17}$
$2x = -6$
$x = -3$

$-(-3) + 4y = 11$
$3 + 4y = 11$
$4y = 8$
$y = 2$
$(-3, 2)$

6. $4x + 2(-2x + 18) = 9$
$4x - 4x + 36 = 9$
$36 = 9$

Inconsistent, $\varnothing$

7. $2 \cdot (x + y) = 2 \cdot (11)$

$2x + 2y = 22$
$\underline{5x - 2y = 13}$
$7x = 35$
$x = 5$

$5 + y = 11$
$y = 6$
$(5, 6)$

8. $2 \cdot (7x - 11y) = 2 \cdot (-47)$
$11 \cdot (9x + 2y) = 11 \cdot (101)$

$14x - 22y = -94$
$99x + 22y = 1111$
$\overline{ 113x = 1017}$
$x = 9$

$7(9) - 11y = -47$
$63 - 11y = -47$
$-11y = -110$
$y = 10$

$(9, 10)$

9. $13(7y + 4) + 6y = 52$
$91y + 52 + 6y = 52$
$97y + 52 = 52$
$97y = 0$
$y = 0$

$x = 7(0) + 4 = 4$

$(4, 0)$

10. $10 \cdot (4x - 6y) = 10 \cdot (18)$
$4 \cdot (-10x + 15y) = 4 \cdot (-45)$

$40x - 60y = 180$
$-40x + 60y = -180$
$\overline{ 0 = 0}$

Dependent, $\left(x, \dfrac{2}{3}x - 3 \right)$

11. Number of walks: x
Number of home runs: y
$x = y + 123$
$x + y = 237$
$(y + 123) + y = 237$
$2y + 123 = 237$
$2y = 114$
$y = 57$

$x = 57 + 123 = 180$
The player walked 180 times.

12. Number of adults at the park: x
Number of children at the park: y
$x + y = 13,900$
$20x + 5y = 197,000$

$-20 \cdot (x + y) = -20 \cdot (13,900)$

$-20x - 20y = -278,000$
$20x + 5y = 197,000$
$\overline{ -15y = -81,000}$
$y = 5400$
5400 children under 8 years old

13. Length: ℓ
Width: w
$2\ell + 2w = 66$
$\ell = 3w + 1$
$2(3w + 1) + 2w = 66$
$6w + 2 + 2w = 66$
$8w = 64$
$w = 8$

$\ell = 3(8) + 1 = 25$

The length is 25 feet and the width is 8 feet.

14. Amount in 7% fund: x
Amount in 3% fund: y
$x + y = 32,000$
$0.07x + 0.03y = 1760$
$-3 \cdot (x + y) = -3 \cdot (32,000)$
$100 \cdot (0.07x + 0.03y) = 100 \cdot (1760)$
$-3x - 3y = -96,000$
$7x + 3y = 176,000$
$\overline{ 4x = 80,000}$
$x = 20,000$

$20,000 + y = 32,000$
$y = 12,000$

$\$20,000$ at 7% and $\$12,000$ at 3%

15. Draw a dashed line at $y = \dfrac{1}{2}x + 4$.

Test Point: $(0, 0)$

$0 < \dfrac{1}{2}(0) + 4$

$0 < 4$
True. Shade the half-plane that contains the test point.

$5x + 2y = -16$
$2y = -5x - 16$
$y = -\dfrac{5}{2}x - 8$

Draw a dashed line at $y = -\dfrac{5}{2}x - 8$.

Test Point: $(0, 0)$
$5(0) + 2(0) > -16$
$0 > -16$
True. Shade the half-plane that contains the test point.

Continued on next page.

15. Continued.

The solution of the system of linear inequalities is the region where the two solutions intersect.

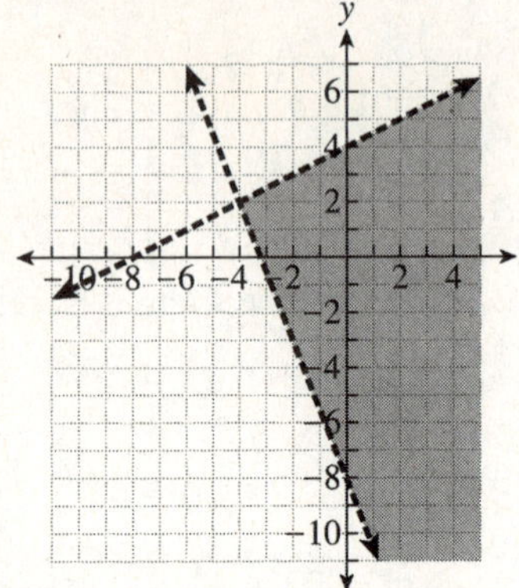

CUMULATIVE REVIEW CHAPTERS 1–4

1. $-3 > -8$

2. $-9 - 25 = -34$

3. $15(-11) = -165$

4. $-234 \div (-18) = 13$

5.

$$3 \cdot 3 \cdot 2 \cdot 2 \cdot 2 = 2^3 \cdot 3^2 = 72$$

6. $\dfrac{\cancel{6}^{\,2}}{\cancel{35}_{\,7}} \cdot \dfrac{\cancel{5}^{\,1}}{\cancel{21}_{\,7}} = \dfrac{2}{7} \cdot \dfrac{1}{7} = \dfrac{2}{49}$

7. $\dfrac{3}{4} + \dfrac{9}{10} = \dfrac{3 \cdot 5}{4 \cdot 5} + \dfrac{9 \cdot 2}{10 \cdot 2} = \dfrac{15}{20} + \dfrac{18}{20} = \dfrac{33}{20}$

8.
$$\begin{array}{r} 9.80 \\ -3.72 \\ \hline 6.08 \end{array}$$

9.
$$\begin{array}{r} 4.7 \\ \times\, 3.8 \\ \hline 376 \\ 1410 \\ \hline 1786 \end{array} = 17.86$$

10.
$$\begin{array}{r} 32 \\ \times\, 14 \\ \hline 128 \\ 320 \\ \hline 448 \end{array}$$
$448

11. $1\dfrac{1}{3} + 2\dfrac{3}{4} = \dfrac{4}{3} + \dfrac{11}{4}$

$$= \dfrac{4 \cdot 4}{3 \cdot 4} + \dfrac{11 \cdot 3}{4 \cdot 3}$$

$$= \dfrac{16}{12} + \dfrac{33}{12}$$

$$= \dfrac{49}{12}$$

$$= 4\dfrac{1}{12} \text{ cups}$$

12. $9 + 3 \cdot 4 - 2^6 = 9 + 3 \cdot 4 - 64$
$$= 9 + 12 - 64$$
$$= 21 - 64$$
$$= -43$$

13. $7 + 2(9 - 4 \cdot 5) - 6(-5) = 7 + 2(9 - 20) - 6(-5)$
$$= 7 + 2(-11) - 6(-5)$$
$$= 7 - 22 + 30$$
$$= 37 - 22$$
$$= 15$$

14. $5(-4) - 8 = -20 - 8 = -28$

15. $8x - 11 - 3x + 8 = 5x - 3$

16. $3(3x - 8) - 5(4x + 9) = 9x - 24 - 20x - 45$
$$= -11x - 69$$

17.
$$3x - 4 = -31$$
$$3x - 4 + 4 = -31 + 4$$
$$3x = -27$$
$$\dfrac{3x}{3} = \dfrac{-27}{3}$$
$$x = -9$$
$$\{-9\}$$

18.
$$-2x + 15 = 8$$
$$-2x + 15 - 15 = 8 - 15$$
$$-2x = -7$$
$$\dfrac{-2x}{-2} = \dfrac{-7}{-2}$$
$$x = \dfrac{7}{2}$$
$$\left\{\dfrac{7}{2}\right\}$$

19.
$$3x+17 = 6x-25$$
$$3x-6x+17 = 6x-6x-25$$
$$-3x+17 = -25$$
$$-3x+17-17 = -25-17$$
$$-3x = -42$$
$$\frac{-3x}{-3} = \frac{-42}{-3}$$
$$x = 14$$
$$\{14\}$$

20.
$$2(2x+5)-3(x-8) = 20$$
$$4x+10-3x+24 = 20$$
$$x+34 = 20$$
$$x+34-34 = 20-34$$
$$x = -14$$
$$\{-14\}$$

21. length: ℓ
width: w
$$2\ell+2w = 70$$
$$w = \ell-9$$
$$2\ell+2(\ell-9) = 70$$
$$2\ell+2\ell-18 = 70$$
$$4\ell-18 = 70$$
$$4\ell-18+18 = 70+18$$
$$4\ell = 88$$
$$\frac{4\ell}{4} = \frac{88}{4}$$
$$\ell = 22$$
$$w = 22-9 = 13$$
The length is 22 feet and the width is 13 feet.

22. Number of \$5 bills: $x+8$
Number of \$10 bills: x
$$5(x+8)+10x = 205$$
$$5x+40+10x = 205$$
$$15x+40 = 205$$
$$15x+40-40 = 205-40$$
$$15x = 165$$
$$\frac{15x}{15} = \frac{165}{15}$$
$$x = 11$$
$$x+8 = 11+8 = 19$$
19 \$5 bills

23. 20% of 125 is x
$$0.20 \cdot 125 = x$$
$$2500 = x$$
$$25 = x$$

24. p percent of 275 is 66.
$$p \cdot 275 = 66$$
$$p = \frac{66}{275}$$
$$p = 0.24$$
$$p = 24\%$$

25. The increase in value of the home: x
$$0.13(220,000) = x$$
$$28,600 = x$$
$$220,000+28,600 = 248,600$$
$$\$248,600$$

26.
$$\frac{n}{8} = \frac{27}{36}$$
$$36n = 8(27)$$
$$\frac{36n}{36} = \frac{216}{36}$$
$$n = 6$$
$$\{6\}$$

27. Number of students who plan to transfer: n
$$\frac{5}{9} = \frac{n}{8307}$$
$$5(8307) = 9n$$
$$\frac{41,535}{9} = \frac{9n}{9}$$
$$4615 = n$$

28.
$$2x-5 \geq 3x-8$$
$$2x-2x-5 \geq 3x-2x-8$$
$$-5 \geq x-8$$
$$-5+8 \geq x-8+8$$
$$3 \geq x$$
$$x \leq 3$$
$$(-\infty,3]$$

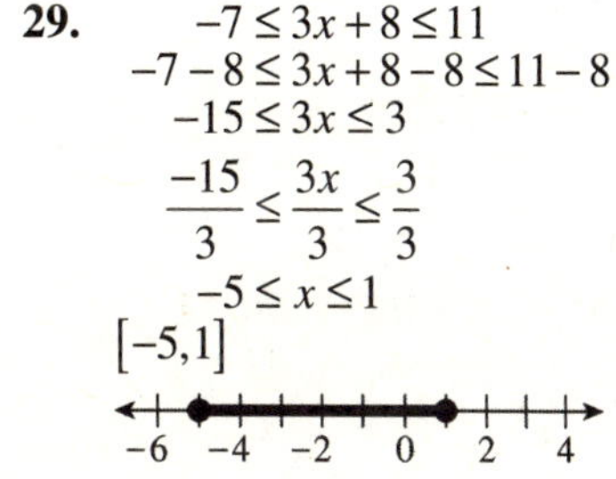

29.
$$-7 \leq 3x+8 \leq 11$$
$$-7-8 \leq 3x+8-8 \leq 11-8$$
$$-15 \leq 3x \leq 3$$
$$\frac{-15}{3} \leq \frac{3x}{3} \leq \frac{3}{3}$$
$$-5 \leq x \leq 1$$
$$[-5,1]$$

30. x-intercept:
$$2x - (0) = 6$$
$$2x = 6$$
$$x = 3$$
$$(3, 0)$$

y-intercept:
$$2(0) - y = 6$$
$$-y = 6$$
$$y = -6$$
$$(0, -6)$$

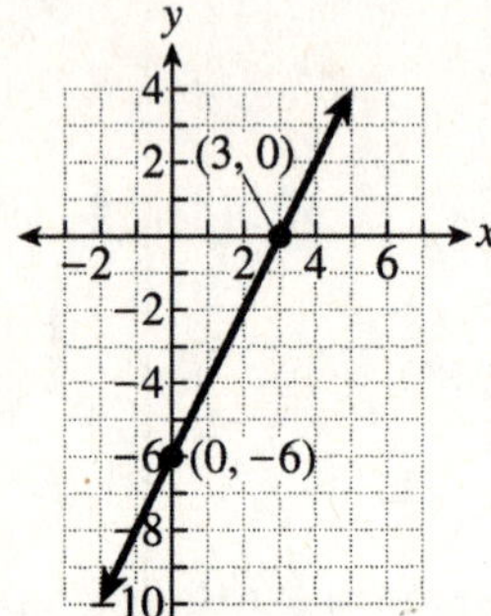

31. x-intercept:
$$-3x + 4(0) = -24$$
$$-3x = -24$$
$$x = 8$$
$$(8, 0)$$

y-intercept:
$$-3(0) + 4y = -24$$
$$4y = -24$$
$$y = -6$$
$$(0, -6)$$

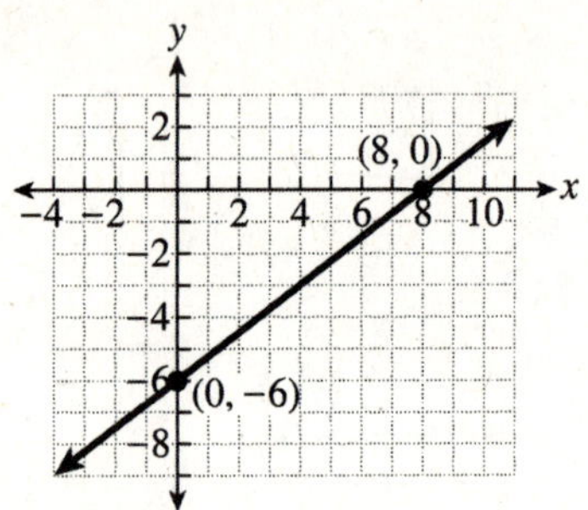

32. $m = \dfrac{5 - (-7)}{3 - (-3)} = \dfrac{5 + 7}{3 + 3} = \dfrac{12}{6} = 2$

33. $y = mx + b$ where m is the slope and b is the y-coordinate of the y-intercept.
$$2x + 3y = 21$$
$$3y = -2x + 21$$
$$y = -\frac{2}{3}x + 7$$
$$m = -\frac{2}{3}, \; y\text{-intercept: } (0, 7)$$

34. $y = 2x - 3$
$$m = 2$$
y-intercept: $(0, -3)$
x-intercept:
$$0 = 2x - 3$$
$$3 = 2x$$
$$\frac{3}{2} = x$$
$$\left(\frac{3}{2}, 0\right)$$

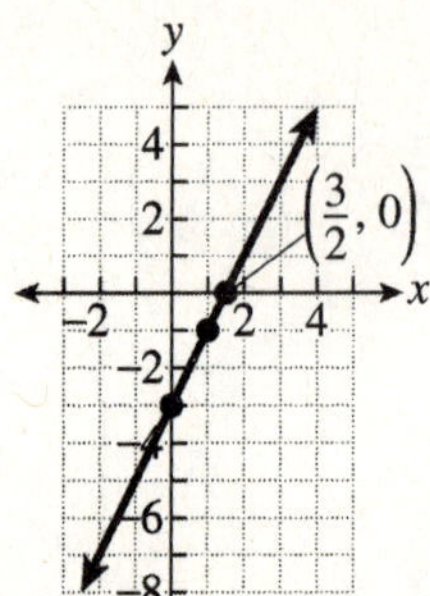

35. $y = \dfrac{3}{4}x - 3$
$$m = \frac{3}{4}$$
y-intercept: $(0, -3)$

x-intercept:
$$0 = \frac{3}{4}x - 3$$
$$3 = \frac{3}{4}x$$
$$\left(\frac{4}{3}\right) \cdot 3 = \left(\frac{4}{3}\right) \cdot \frac{3}{4}x$$
$$4 = x$$
$$(4, 0)$$

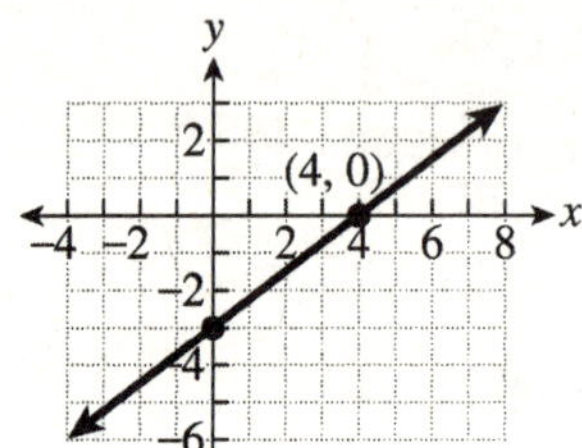

36. $5x + 3y = 9$
$$3y = -5x + 9$$
$$y = -\frac{5}{3}x + 3$$
$$m_1 = -\frac{5}{3}$$

$3x - 5y = -5$
$$-5y = -3x - 5$$
$$y = \frac{3}{5}x + 1$$
$$m_2 = \frac{3}{5}$$

The lines are perpendicular.

37. **a)** Number of months saved: x

$$f(x) = 220 + 40x$$

b) After 8 months: $x = 8$

$$f(8) = 220 + 40(8) = 220 + 320 = 540$$

$540

c) $f(x) = 800$

$$800 = 220 + 40x$$
$$800 - 220 = 220 - 220 + 40x$$
$$580 = 40x$$
$$\frac{580}{40} = \frac{40x}{40}$$
$$14.5 = x$$

15 months

38. $f(14) = 24(14) + 55 = 336 + 55 = 391$

39. $g(-32) = -7(-32) + 905 = 224 + 905 = 1129$

40. $y = mx + b$
$y = -2x + b$

Use $(3, -4)$ to find b.

$$-4 = -2(3) + b$$
$$-4 = -6 + b$$
$$-4 + 6 = -6 + 6 + b$$
$$2 = b$$
$$y = -2x + 2$$

41. $m = \dfrac{-5 - 1}{4 - 2} = \dfrac{-6}{2} = -3$

$y = -3x + b$

Use $(2, 1)$ to find b.

$$1 = -3(2) + b$$
$$1 = -6 + b$$
$$1 + 6 = -6 + 6 + b$$
$$7 = b$$
$$y = -3x + 7$$

42. Graph a dashed line at $3x - 2y = 12$.

Test Point: $(0, 0)$

$$3(0) - 2(0) > 12$$
$$0 > 12$$

False. Shade on the side of the line that does not contain the test point.

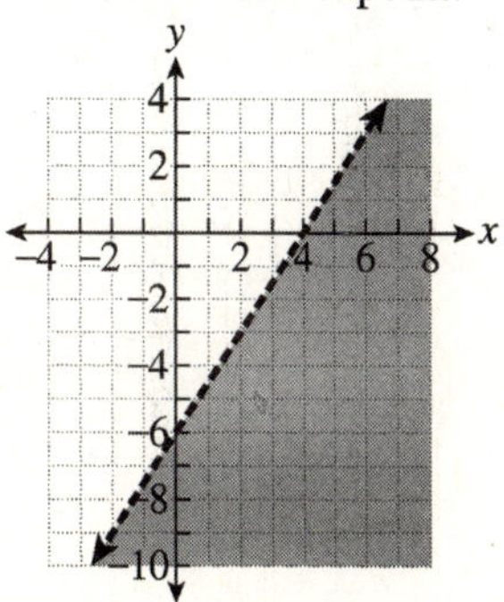

43. Addition Method

$$-8 \cdot (4x + y) = 8 \cdot (22)$$

$$\begin{array}{r} -32x - 8y = -176 \\ -3x + 8y = 1 \\ \hline -35x = -175 \\ x = 5 \end{array}$$

$$4(5) + y = 22$$
$$20 + y = 22$$
$$y = 2$$

$(5, 2)$

44. Substitution Method

$$2x - 3(3x - 2) = -15$$
$$2x - 9x + 6 = -15$$
$$-7x = -21$$
$$x = 3$$

$$y = 3(3) - 2 = 7$$
$(3, 7)$

45. Addition Method

$$3 \cdot (9x + 8y) = 3 \cdot (25)$$
$$4 \cdot (5x - 6y) = 4 \cdot (-7)$$

$$\begin{array}{r} 27x + 24y = 75 \\ 20x - 24y = -28 \\ \hline 47x = 47 \\ x = 1 \end{array}$$

$$9(1) + 8y = 25$$
$$9 + 8y = 25$$
$$8y = 16$$
$$y = 2$$

$(1, 2)$

46. $10 \cdot \left(\dfrac{3}{2}x + \dfrac{3}{5}y \right) = 10 \cdot \dfrac{9}{2}$

$$15x + 6y = 45$$
$$4x + y = 6$$
$$y = -4x + 6$$

Substitution Method

$$15x + 6(-4x + 6) = 45$$
$$15x - 24x + 36 = 45$$
$$-9x = 9$$
$$x = -1$$

$$y = -4(-1) + 6 = 4 + 6 = 10$$
$$(-1, 10)$$

47. Number of children at the game: x
Number of adults at the game: y

$$x + y = 1050$$
$$4x + 5y = 5000$$

$$-5 \cdot (x + y) = -5 \cdot (1050)$$

$$\begin{aligned} -5x - 5y &= -5250 \\ 4x + 5y &= 5000 \\ \hline -x &= -250 \\ x &= 250 \end{aligned}$$

250 children

48. length: ℓ
width: w

$$2\ell + 2w = 130$$
$$\ell = 2w - 10$$

Subsitution Method

$$2(2w - 10) + 2w = 130$$
$$4w - 20 + 2w = 130$$
$$6w - 20 = 130$$
$$6w = 150$$
$$w = 25$$

$$\ell = 2(25) - 10 = 40$$

The length is 40 feet and the width is 25 feet.

49. Amount in 3% CD: x
Amount in 2.5 % CD: y

$$x + y = 15,000$$
$$0.03x + 0.025y = 430$$

$$-30 \cdot (x + y) = -30 \cdot (15,000)$$
$$1000 \cdot (0.03x + 0.025y) = 1000 \cdot (430)$$

$$\begin{aligned} -30x - 30y &= -450,000 \\ 30x + 25y &= 430,000 \\ \hline -5y &= -20,000 \\ y &= 4000 \end{aligned}$$

$$x + 4000 = 15,000$$
$$x = 11,000$$

$11,000 at 3% and $4000 at 2.5%

50. Graph a dashed line at $y = x + 4$.

Test Point: $(0, 0)$
$0 < 0 + 4$
$0 < 4$
True. Shade the half-plane that contains the test point.

Graph a solid line at $y = 4x - 2$.

Test Point: $(0, 0)$
$0 \geq 4(0) - 2$
$0 \geq -2$
True. Shade the half-plane that contains the test point.

The solution of the system of linear inequalities is the region where the two solutions intersect.

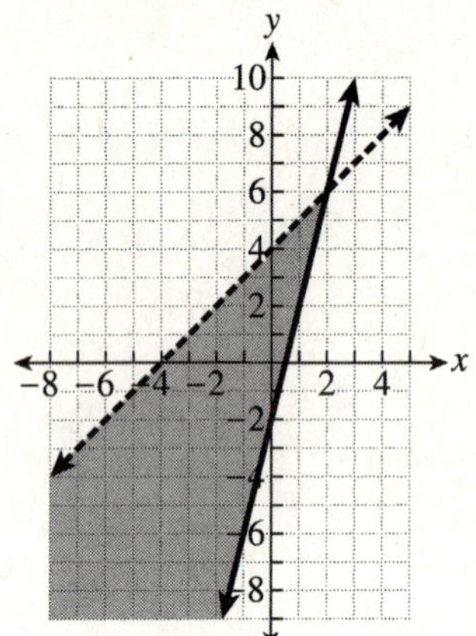

CHAPTER 5   EXPONENTS AND POLYNOMIALS

5.1   QUICK CHECK

1. a) $x^5 \cdot x^4 = x^{5+4} = x^9$

   b) $3^3 \cdot 3^2 \cdot 3 = 3^{3+2+1} = 3^6 = 729$

2. $(a-6)^{10} \cdot (a-6)^{17} = (a-6)^{10+17} = (a-6)^{27}$

3. $\left(a^6 b^5\right)\left(a^9 b^5\right) = a^6 b^5 a^9 b^5$
$$= a^6 a^9 b^5 b^5$$
$$= a^{6+9} b^{5+5}$$
$$= a^{15} b^{10}$$

4. $\left(x^9\right)^7 = x^{9 \cdot 7} = x^{63}$

5. $\left(x^4\right)^6 \left(x^5\right)^2 = x^{4 \cdot 6} x^{5 \cdot 2} = x^{24} x^{10} = x^{24+10} = x^{34}$

6. $\left(a^5 b^3\right)^8 = a^{5 \cdot 8} b^{3 \cdot 8} = a^{40} b^{24}$

7. $\left(4a^2 b^3\right)^3 \left(a^4 b^2 c^7\right)^5 = 4^3 a^{2 \cdot 3} b^{3 \cdot 3} a^{4 \cdot 5} b^{2 \cdot 5} c^{7 \cdot 5}$
$$= 64 a^6 b^9 a^{20} b^{10} c^{35}$$
$$= 64 a^{26} b^{19} c^{35}$$

8. a) $\dfrac{x^{20}}{x^4} = x^{20-4} = x^{16}$

   b) $x^{24} \div x^8 = \dfrac{x^{24}}{x^8} = x^{24-8} = x^{16}$

9. $\dfrac{20 a^{10} b^{12} c^{15}}{5 a^2 b^6 c^9} = 4 a^{10-2} b^{12-6} c^{15-9} = 4 a^8 b^6 c^6$

10. a) $124^0 = 1$

   b) $\left(3x^2\right)^0 = 1$

   c) $9x^0 = 9 \cdot 1 = 9$

11. $\left(\dfrac{ab^7}{c^3 d^4}\right)^5 = \dfrac{a^5 b^{7 \cdot 5}}{c^{3 \cdot 5} d^{4 \cdot 5}} = \dfrac{a^5 b^{35}}{c^{15} d^{20}}$

12. $f(-4) = (-4)^6 = 4096$

13. $f\left(a^8\right) = \left(a^8\right)^7 = a^{56}$

5.1   EXPONENTS

1. $x^m \cdot x^n = x^{m+n}$

3. $(xy)^n = x^n y^n$

5. $x^0 = 1$

7. $2^4 \cdot 2^2 = 2^{4+2} = 2^6 = 64$

9. $x^8 \cdot x^5 = x^{8+5} = x^{13}$

11. $m^{21} \cdot m^{19} = m^{21+19} = m^{40}$

13. $b^5 \cdot b^6 \cdot b^3 = b^{5+6+3} = b^{14}$

15. $(2x-3)^4 \cdot (2x-3)^{10} = (2x-3)^{4+10}$
$$= (2x-3)^{14}$$

17. $\left(x^6 y^5\right)\left(x^2 y^8\right) = x^8 y^{13}$

19. $x^4 \cdot x^5 = x^{4+5} = x^9$
Therefore, x^5 is the missing factor.

21. $b \cdot b^5 = b^{1+5} = b^6$
Therefore, b is the missing factor.

23. $\left(x^4\right)^6 = x^{4 \cdot 6} = x^{24}$

25. $\left(a^7\right)^9 = a^{7 \cdot 9} = a^{63}$

27. $\left(2^5\right)^2 = 2^{10} = 1024$

29. $\left(x^3\right)^7 \left(x^4\right)^2 = x^{21} \cdot x^8 = x^{29}$

31. $\left(x^6\right)^7 = x^{6 \cdot 7} = x^{42}$
Therefore, 7 is the missing exponent.

33. $\left(x^3\right)^8\left(x^5\right)^4 = x^{24}x^{20} = x^{44}$
Therefore, 5 is the missing exponent.

35. $(3x)^4 = 3^4 x^4 = 81x^4$

37. $\left(-8x^3\right)^2 = (-8)^2 x^6 = 64x^6$

39. $\left(m^8 n^5\right)^7 = m^{56}n^{35}$

41. $\left(3x^4 y\right)^4 = 3^4 x^{16} y^4 = 81x^{16} y^4$

43. $\left(x^2 y^5 z^3\right)^7\left(y^4 z^5\right)^8 = \left(x^{14}y^{35}z^{21}\right)\left(y^{32}z^{40}\right)$
$$= x^{14}y^{67}z^{61}$$

45. $\left(x^5 y^4\right)^7 = x^{35}y^{28}$
Therefore, 7 is the missing exponent.

47. $\left(2s^6 t^{13}\right)^4 = 2^4 s^{24} t^{52} = 16s^{24}t^{52}$
Therefore, 6 and 13 are the missing exponents.

49. $\dfrac{x^{12}}{x^3} = x^{12-3} = x^9$

51. $r^{25} \div r^{10} = \dfrac{r^{25}}{r^{10}} = r^{25-10} = r^{15}$

53. $\dfrac{16x^{16}}{2x^2} = 8x^{16-2} = 8x^{14}$

55. $\dfrac{(a+5b)^{13}}{(a+5b)^{11}} = (a+5b)^2$

57. $\dfrac{a^8 b^6}{a^3 b} = a^5 b^5$

59. $\dfrac{x^{14}}{x^{12}} = x^2$
Therefore, 12 is the missing exponent.

61. $m^{30} \div m^{11} = \dfrac{m^{30}}{m^{11}} = m^{19}$
Therefore, 30 is the missing exponent.

63. $9^0 = 1$

65. $(-16)^0 = 1$

67. $\dfrac{3}{7^0} = \dfrac{3}{1} = 3$

69. $(13x)^0 = 1$

71. $\left(\dfrac{3}{4}\right)^3 = \dfrac{3^3}{4^3} = \dfrac{27}{64}$

73. $\left(\dfrac{a}{b}\right)^8 = \dfrac{a^8}{b^8}$

75. $\left(\dfrac{x^4}{y^3}\right)^5 = \dfrac{x^{20}}{y^{15}}$

77. $\left(\dfrac{2a^6}{b^7}\right)^3 = \dfrac{2^3 a^{18}}{b^{21}} = \dfrac{8a^{18}}{b^{21}}$

79. $\left(\dfrac{a^5 b^4}{b^2 c^7}\right)^9 = \dfrac{a^{45}b^{36}}{b^{18}c^{63}} = \dfrac{a^{45}b^{18}}{c^{63}}$

81. $\left(\dfrac{3}{4}\right)^4 = \dfrac{3^4}{4^4} = \dfrac{81}{256}$
Therefore, 4 is the missing exponent.

83. $\left(\dfrac{a^3 b^6}{c^8}\right)^{12} = \dfrac{a^{36}b^{72}}{c^{96}}$
Therefore, 12 is the missing exponent.

85. $f(4) = (4)^2 = 16$

87. $g(-2) = (-2)^4 = 16$

89. $f\left(a^4\right) = \left(a^4\right)^6 = a^{24}$

91. $\left(a^7\right)^6 = a^{42}$

93. $\left(\dfrac{5x^4}{2y^7}\right)^3 = \dfrac{5^3 x^{12}}{2^3 y^{21}} = \dfrac{125x^{12}}{8y^{21}}$

95. $\dfrac{x^{13}}{x} = x^{12}$

97. $\left(\dfrac{a^8 b^9}{2c^2}\right)^5 = \dfrac{a^{40} b^{45}}{2^5 c^{10}} = \dfrac{a^{40} b^{45}}{32 c^{10}}$

99. $\left(b^{10}\right)^2 = b^{20}$

101. $\left(x^{15}\right)^6 = x^{90}$

103. $\left(\dfrac{3a^2 b^9}{7cd^4}\right)^4 = \dfrac{3^4 a^8 b^{36}}{7^4 c^4 d^{16}} = \dfrac{81 a^8 b^{36}}{2401 c^4 d^{16}}$

105. $f(3) = 16(3)^2 = 16(9) = 144$
144 feet

107. $A(90) = (90)^2 = 8100$
8100 square feet

109. $A(14) = \pi(14)^2 = 3.14(196)$
615.44 square feet

111. $x^5 \cdot x^4$; explanations will vary. Example:
When you multiply like variables you add
exponents, so $x^5 \cdot x^4 = x^{5+4} = x^9$.

5.2 *QUICK CHECK*

1. $4^{-3} = \dfrac{1}{4^3} = \dfrac{1}{64}$

2. $a^5 b^{-4} = \dfrac{a^5}{b^4}$

3. a) $\dfrac{x^{12}}{y^{-7}} = x^{12} y^7$

  b) $\dfrac{x^{-2} y^4}{z^{-1} w^{-9}} = \dfrac{y^4 z w^9}{x^2}$

4. $x^{-13} \cdot x^6 = x^{-13+6} = x^{-7} = \dfrac{1}{x^7}$

5. $\left(x^{-6}\right)^{-7} = x^{42}$

6. $\left(9ab^{-3} c^2\right)^{-2} = 9^{-2} a^{-2} b^6 c^{-4}$
$$= \dfrac{b^6}{9^2 a^2 c^4}$$
$$= \dfrac{b^6}{81 a^2 c^4}$$

7. $\dfrac{x^8}{x^{14}} = x^{8-14} = x^{-6} = \dfrac{1}{x^6}$

8. $\left(\dfrac{x^5}{y^{-4}}\right)^{-4} = \dfrac{x^{-20}}{y^{16}} = \dfrac{1}{x^{20} y^{16}}$

9. a) $0.0046 = 4.6 \times 10^{-3}$

  b) $3{,}570{,}000 = 3.57 \times 10^6$

10. a) $3.2 \times 10^6 = 3.2 \times 1{,}000{,}000$
$$= 3{,}200{,}000$$

  b) $7.21 \times 10^{-4} = 7.21 \times 0.0001 = 0.000721$

11. $\left(5.8 \times 10^4\right)\left(1.2 \times 10^9\right) = (5.8)(1.2)\left(10^4\right)\left(10^9\right)$
$$= 6.96 \times 10^{13}$$

12. $\left(9.9 \times 10^5\right) \div \left(3.3 \times 10^{-7}\right) = \dfrac{9.9 \times 10^5}{3.3 x 10^{-7}}$
$$= \dfrac{9.9}{3.3} \times 10^{5-(-7)}$$
$$= 3.0 \times 10^{12}$$

13. $1{,}488{,}000{,}000 = 1.488 \times 10^9$
$$\left(1.488 \times 10^9\right) \div \left(1.86 \times 10^5\right) = \dfrac{1.488 \times 10^9}{1.86 \times 10^5}$$
$$= \dfrac{1.488}{1.86} \times 10^{9-5}$$
$$= 0.8 \times 10^4$$
$$= 8.0 \times 10^3$$
$$= 8000 \text{ seconds}$$

5.2   NEGATIVE EXPONENTS; SCIENTIFIC NOTATION

1. $\dfrac{1}{x^n}$

3. $5^{-2} = \dfrac{1}{5^2} = \dfrac{1}{25}$

5. $4^{-3} = \dfrac{1}{4^3} = \dfrac{1}{64}$

7. $-13^{-2} = -\dfrac{1}{13^2} = -\dfrac{1}{169}$

9. $b^{-12} = \dfrac{1}{b^{12}}$

11. $12x^{-6} = \dfrac{12}{x^6}$

13. $-5m^{-19} = -\dfrac{5}{m^{19}}$

15. $\dfrac{1}{x^{-5}} = x^5$

17. $\dfrac{3}{y^{-4}} = 3y^4$

19. $\dfrac{x^{-11}}{y^{-2}} = \dfrac{y^2}{x^{11}}$

21. $\dfrac{-8a^{-6}b^5}{c^7} = -\dfrac{8b^5}{a^6 c^7}$

23. $\dfrac{5a^{-6}b^{-9}}{c^{-4}d^5} = \dfrac{5c^4}{a^6 b^9 d^5}$

25. $x^{-12} \cdot x^7 = x^{-5} = \dfrac{1}{x^5}$

27. $a^{16} \cdot a^{-10} = a^6$

29. $m^{-9} \cdot m^{-12} = m^{-21} = \dfrac{1}{m^{21}}$

31. $\left(x^6\right)^{-3} = x^{-18} = \dfrac{1}{x^{18}}$

33. $\left(x^{-6}\right)^{-7} = x^{42}$

35. $\left(x^5 y^4 z^{-6}\right)^{-2} = x^{-10} y^{-8} z^{12} = \dfrac{z^{12}}{x^{10} y^8}$

37. $\left(4a^{-5}b^{-8}z^2\right)^{-3} = 4^{-3} a^{15} b^{24} z^{-6}$

$$= \dfrac{a^{15}b^{24}}{4^3 z^6}$$

$$= \dfrac{a^{15}b^{24}}{64 z^6}$$

39. $\dfrac{x^{-5}}{x^{10}} = \dfrac{1}{x^5 x^{10}} = \dfrac{1}{x^{15}}$

41. $\dfrac{x^{11}}{x^{-6}} = x^{11} x^6 = x^{17}$

43. $\dfrac{a^{-8}}{a^{-5}} = \dfrac{a^5}{a^8} = a^{5-8} = a^{-3} = \dfrac{1}{a^3}$

45. $\dfrac{x^7}{x^{18}} = x^{7-18} = x^{-11} = \dfrac{1}{x^{11}}$

47. $\dfrac{x^5 \cdot x^{-11}}{x^{-6}} = \dfrac{x^{-6}}{x^{-6}} = x^{-6-(-6)} = x^0 = 1$

49. $\left(\dfrac{x^3}{y^4}\right)^{-5} = \dfrac{x^{-15}}{y^{-20}} = \dfrac{y^{20}}{x^{15}}$

51. $\left(\dfrac{a^4 b^7}{3c^5 d^{-8}}\right)^{-2} = \dfrac{a^{-8} b^{-14}}{3^{-2} c^{-10} d^{16}}$

$$= \dfrac{3^2 c^{10}}{a^8 b^{14} d^{16}}$$

$$= \dfrac{9 c^{10}}{a^8 b^{14} d^{16}}$$

53. $3.07 \times 10^{-7} = 3.07 \times 0.0000001$
$$= 0.000000307$$

55. $8.935 \times 10^9 = 8.935 \times 1{,}000{,}000{,}000$
$$= 8{,}935{,}000{,}000$$

57. $9.021 \times 10^4 = 90{,}210$

59. $0.00027 = 2.7 \times 10^{-4}$

61. $8,600,000 = 8.6 \times 10^{6}$

63. $420,000,000,000 = 4.2 \times 10^{11}$

65. $\left(4.1 \times 10^{8}\right)\left(2.3 \times 10^{11}\right) = (4.1)(2.3)\left(10^{8}\right)\left(10^{11}\right)$
$$= 9.43 \times 10^{19}$$

67. $\left(1.598 \times 10^{-12}\right) \div \left(4.7 \times 10^{9}\right) = \dfrac{1.598 \times 10^{-12}}{4.7 \times 10^{9}}$
$$= \frac{1.598}{4.7} \times 10^{-12-9}$$
$$= 0.34 \times 10^{-21}$$
$$= 3.4 \times 10^{-22}$$

69. $\left(5.32 \times 10^{-15}\right)\left(7.8 \times 10^{3}\right)$
$$= (5.32)(7.8)\left(10^{-15}\right)\left(10^{3}\right)$$
$$= 41.496 \times 10^{-12}$$
$$= 4.1496 \times 10^{-11}$$

71. $(87,000,000,000)(0.000002)$
$$= \left(8.7 \times 10^{10}\right)\left(2.0 \times 10^{-6}\right)$$
$$= (8.7)(2.0)\left(10^{10}\right)\left(10^{-6}\right)$$
$$= 17.4 \times 10^{4}$$
$$= 1.74 \times 10^{5}$$

73. $\left(8.0 \times 10^{-10}\right)\left(4.2 \times 10^{13}\right)$
$$= (8.0)(4.2)\left(10^{-10}\right)\left(10^{13}\right)$$
$$= 33.6 \times 10^{3}$$
$$= 3.36 \times 10^{4} \text{ seconds}$$
$$= 33,600 \text{ seconds}$$

75. $10 \text{ minutes} = 10 \cdot 60 = 600 \text{ seconds}$
$$\left(6.0 \times 10^{2}\right)\left(1.86 \times 10^{5}\right)$$
$$= (6.0)(1.86)\left(10^{2}\right)\left(10^{5}\right)$$
$$= 11.16 \times 10^{7}$$
$$= 1.116 \times 10^{8} \text{ miles}$$

77. $\left(5.0 \times 10^{11}\right)\left(2.2 \times 10^{33}\right)$
$$= (5.0)(2.2)\left(10^{11}\right)\left(10^{33}\right)$$
$$= 11.0 \times 10^{44}$$
$$= 1.1 \times 10^{45} \text{ grams}$$

79. $\left(6.118 \times 10^{10}\right) + \left(3.369 \times 10^{10}\right)$
$$= (6.118 + 3.369) \times 10^{10}$$
$$= 9.487 \times 10^{10}$$
$$= \$9.487 \times 10^{10}$$

81. $\left(3.3 \times 10^{6}\right)(0.7) = (3.3)(0.7) \times 10^{6}$
$$= 2.31 \times 10^{6}$$
$$= 2,310,000 \text{ students}$$

83. Answers will vary. Example:
A negative exponent means to take the
reciprocal of the base $\left(\text{i.e. } 4^{-2} = \dfrac{1}{4^{2}}\right)$. A
negative sign does not mean reciprocal. So
$(-4)^{2} = (-4)(-4) = 16$

5.3 *QUICK CHECK*

1. a) Trinomial, 2, 1, 0

   b) Binomial, 2, 0

   c) Monomial, 7

2. $1, -6, -11, 32$

3. $-x^{3} + 4x^{2} + x + 9$
Leading term: $-x^{3}$
Leading coefficient: -1
Degree: 3

4. $(-3)^{4} - 5(-3)^{3} - 6(-3)^{2} + 10(-3) - 21$
$$= 81 - 5(-27) - 6(9) + 10(-3) - 21$$
$$= 81 + 135 - 54 - 30 - 21$$
$$= 216 - 105$$
$$= 111$$

5. $f(-5) = (-5)^{3} + 12(-5)^{2} - 21(-5)$
$$= -125 + 12(25) - 21(-5)$$
$$= -125 + 300 + 105$$
$$= -125 + 405$$
$$= 280$$

6. $\left(x^3 - x^2 - 9x + 25\right) + \left(x^2 + 12x + 144\right)$

$= x^3 - x^2 - 9x + 25 + x^2 + 12x + 144$

$= x^3 + 3x + 169$

7. $\left(5x^2 + 6x + 27\right) - \left(3x^3 + 6x^2 - 9x + 22\right)$

$= 5x^2 + 6x + 27 - 3x^3 - 6x^2 + 9x - 22$

$= -3x^3 - x^2 + 15x + 5$

8. $f(x) - g(x)$
$= \left(4x^2 + 30x - 45\right) - \left(-2x^2 + 17x + 52\right)$

$= 4x^2 + 30x - 45 + 2x^2 - 17x - 52$

$= 6x^2 + 13x - 97$

9. $(-9)^2 - 4(1)(-52) = 81 - 4(1)(-52)$
$= 81 + 208$
$= 289$

10. Each term: 5, 8, 10
Polynomial: 10

11. $\left(5x^4 y^2 - 6x^3 y^3\right) + \left(8x^4 y^2 + 2x^3 y^3\right)$

$= 5x^4 y^2 - 6x^3 y^3 + 8x^4 y^2 + 2x^3 y^3$

$= 13x^4 y^2 - 4x^3 y^3$

5.3   POLYNOMIALS; ADDITION AND SUBTRACTION OF POLYNOMIALS

1. polynomial

3. monomial

5. trinomial

7. descending order

9. $4, 2, 1, 0$

11. $1, 7, 4$

13. $7, 1, -15$

15. $10, -17, 6, -1, 2$

17. trinomial

19. binomial

21. monomial

23. $3x^2 + 8x - 7$

Leading term: $3x^2$
Leading coefficient: 3
Degree: 2

25. $2x^4 + 6x^2 - 11x + 10$

Leading term: $2x^4$
Leading coefficient: 2
Degree: 4

27. $(5)^2 + 3(5) - 10 = 25 + 3(5) - 10$
$= 25 + 15 - 10$
$= 30$

29. $(-2)^4 - 8(-2)^3 - 48 = 16 - 8(-8) - 48$
$= 16 + 64 - 48$
$= 32$

31. $-(10)^2 + 7(10) + 22 = -(100) + 7(10) + 22$
$= -100 + 70 + 22$
$= -8$

33. $f(8) = (8)^2 - 7(8) - 10$
$= 64 - 7(8) - 10$
$= 64 - 56 - 10$
$= 64 - 66$
$= -2$

35. $g(-5) = 3(-5)^3 - 8(-5)^2 + 5(-5) + 9$
$= 3(-125) - 8(25) + 5(-5) + 9$
$= -375 - 200 - 25 + 9$
$= -600 + 9$
$= -591$

37. $\left(5x^2 + 8x - 11\right) + \left(3x^2 - 14x + 14\right)$

$= 5x^2 + 8x - 11 + 3x^2 - 14x + 14$

$= 8x^2 - 6x + 3$

39. $\left(4x^2 - 7x + 30\right) - \left(2x^2 + 10x - 50\right)$

$= 4x^2 - 7x + 30 - 2x^2 - 10x + 50$

$= 2x^2 - 17x + 80$

41. $\left(2x^3 + 7x^2 - 19x\right) + \left(x^2 + 5x - 11\right)$

$= 2x^3 + 7x^2 - 19x + x^2 + 5x - 11$

$= 2x^3 + 8x^2 - 14x - 11$

43. $\left(x^3 + 5x^2 - 16\right) - \left(6x^2 - 5x - 19\right)$

$= x^3 + 5x^2 - 16 - 6x^2 + 5x + 19$

$= x^3 - x^2 + 5x + 3$

45. $\left(2x^9 - 5x^4 + 7x^2\right) - \left(-7x^6 + 4x^5 - 12\right)$

$= 2x^9 - 5x^4 + 7x^2 + 7x^6 - 4x^5 + 12$

$= 2x^9 + 7x^6 - 4x^5 - 5x^4 + 7x^2 + 12$

47. $3x^2 + 4x^2 = 7x^2$
$8x + (-3x) = 5x$
$11 + (-13) = -2$
The missing polynomial is
$4x^2 - 3x - 13.$

49. $3x^4 - 2x^4 = x^4$
$-5x^3 - \left(-3x^3\right) = -2x^3$
$0 - x^2 = -x^2$
$6x - 11x = -5x$
$12 - 11 = 1$
The missing polynomial is
$2x^4 - 3x^3 + x^2 + 11x + 11.$

51. $f(x) + g(x) = \left(7x^2 + 10x + 3\right) + \left(5x^2 - 9x + 6\right)$

$\qquad = 7x^2 + 10x + 3 + 5x^2 - 9x + 6$

$\qquad = 12x^2 + x + 9$

$f(x) - g(x) = \left(7x^2 + 10x + 3\right) - \left(5x^2 - 9x + 6\right)$

$\qquad = 7x^2 + 10x + 3 - 5x^2 + 9x - 6$

$\qquad = 2x^2 + 19x - 3$

53. $f(x) + g(x)$

$= \left(x^3 - 8x - 31\right) + \left(4x^3 + x^2 + 3x + 25\right)$

$= x^3 - 8x - 31 + 4x^3 + x^2 + 3x + 25$

$= 5x^3 + x^2 - 5x - 6$

$f(x) - g(x)$

$= \left(x^3 - 8x - 31\right) - \left(4x^3 + x^2 + 3x + 25\right)$

$= x^3 - 8x - 31 - 4x^3 - x^2 - 3x - 25$

$= -3x^3 - x^2 - 11x - 56$

55. $(2)^2 - 7(2)(5) + 10(5)^2 = 4 - 7(2)(5) + 10(25)$
$\qquad\qquad = 4 - 70 + 250$
$\qquad\qquad = 254 - 70$
$\qquad\qquad = 184$

57. $(-2)^2 - 4(-9)(4) = 4 - 4(-9)(4)$
$\qquad\qquad = 4 + 144$
$\qquad\qquad = 148$

59. $3(4)^2(5)(-2)^3 - 4(4)(5)^2(-2)$
$\quad - 6(4)^4(5)^2(-2)$
$= 3(16)(5)(-8) - 4(4)(25)(-2)$
$\quad - 6(256)(25)(-2)$
$= -1920 + 800 + 76,800$
$= -1920 + 77,600$
$= 75,680$

61. 10, 6, 3, Polynomial: 10

63. 8, 9, 11, Polynomial: 11

65. $\left(15x^2 y + 8xy^2 - 7x^2 y^3\right)$

$\quad + \left(-8x^2 y + 3xy^2 + 4x^2 y^3\right)$

$= 15x^2 y + 8xy^2 - 7x^2 y^3$

$\quad - 8x^2 y + 3xy^2 + 4x^2 y^3$

$= 7x^2 y + 11xy^2 - 3x^2 y^3$

67. $\left(a^3 b^2 + 11a^5 b + 24a^2 b^3\right)$

$\quad - \left(19a^2 b^3 - 2a^5 b + 15a^3 b^2\right)$

$= a^3 b^2 + 11a^5 b + 24a^2 b^3$

$\quad - 19a^2 b^3 + 2a^5 b - 15a^3 b^2$

$= -14a^3 b^2 + 13a^5 b + 5a^2 b^3$

69. $\left(2x^3 yz^2 - xy^4 z^3 - 10x^2 y^2 z^5\right)$

$\quad - \left(4x^2 yz^3 + 14xy^4 z^3 - 3x^2 y^2 z^5\right)$

$= 2x^3 yz^2 - xy^4 z^3 - 10x^2 y^2 z^5$

$\quad - 4x^2 yz^3 - 14xy^4 z^3 + 3x^2 y^2 z^5$

$= 2x^3 yz^2 - 15xy^4 z^3 - 7x^2 y^2 z^5 - 4x^2 yz^3$

71. $R(x) = -400x^2 + 13,600x + 1,230,000$
$R(6) = -400(6)^2 + 13,600(6) + 1,230,000$
$\qquad = -400(36) + 13,600(6) + 1,230,000$
$\qquad = -14,400 + 81,600 + 1,230,000$
$\qquad = 1,297,200$
$\$1,297,200$

73. $g(x) = 6x^2 - 224x + 4941$

$g(39) = 6(39)^2 - 224(39) + 4941$

$\qquad = 6(1521) - 224(39) + 4941$

$\qquad = 9126 - 8736 + 4941$

$\qquad = 5331$ students

75. Explanations will vary. Example:
Using parentheses is a good idea so that you
don't "lose" a negative sign.

5.4   QUICK CHECK

1. $7x \cdot 8x = 56x^2$

2. $5x^6 yz^7 \cdot 12x^8 z^6 = 60x^{14} yz^{13}$

3. $4x\left(7x^3 - 6x^2 + 5x - 8\right)$

$= 4x \cdot 7x^3 - 4x \cdot 6x^2 + 4x \cdot 5x - 4x \cdot 8$

$= 28x^4 - 24x^3 + 20x^2 - 32x$

4. $-3x^5\left(-2x^4 + x^3 - x^2 - 15x + 21\right)$

$= \left(-3x^5\right)\left(-2x^4\right) + \left(-3x^5\right)\left(x^3\right) - \left(-3x^5\right)\left(x^2\right)$

$\quad - \left(-3x^5\right)(15x) + \left(-3x^5\right)(21)$

$= 6x^9 + \left(-3x^8\right) - \left(-3x^7\right) - \left(-45x^6\right) + \left(-63x^5\right)$

$= 6x^9 - 3x^8 + 3x^7 + 45x^6 - 63x^5$

5. $(x+11)(x+9) = x \cdot x + x \cdot 9 + 11 \cdot x + 11 \cdot 9$

$\qquad\qquad\qquad = x^2 + 9x + 11x + 99$

$\qquad\qquad\qquad = x^2 + 20x + 99$

6. $(5x-2)(2x-9)$

$= 5x \cdot 2x - 5x \cdot 9 - 2 \cdot 2x - 2(-9)$

$= 10x^2 - 45x - 4x + 18$

$= 10x^2 - 49x + 18$

7. $f(x) \cdot g(x) = (x+8)\left(3x^2 + 6x - 2\right)$

$= x \cdot 3x^2 + x \cdot 6x - x \cdot 2 + 8 \cdot 3x^2 + 8 \cdot 6x - 8 \cdot 2$

$= 3x^3 + 6x^2 - 2x + 24x^2 + 48x - 16$

$= 3x^3 + 30x^2 + 46x - 16$

8. $(x+10)(x-10) = x^2 - 10^2 = x^2 - 100$

9. $(2x+7)(2x-7) = (2x)^2 - 7^2 = 4x^2 - 49$

10. $(5x-8)^2 = (5x)^2 - 2 \cdot (5x)(8) + 8^2$

$\qquad\qquad = 25x^2 - 80x + 64$

11. $(x+6)^2 = x^2 + 2 \cdot x \cdot 6 + 6^2 = x^2 + 12x + 36$

5.4   MULTIPLYING POLYNOMIALS

1. add

3. multiply

5. $4x \cdot 9x^4 = 36x^5$

7. $9m^7\left(-6m^{11}\right) = -54m^{18}$

9. $13a^7 b^{10} \cdot 7a^5 b = 91a^{12} b^{11}$

11. $-5x^3 y^2 z^4 \cdot 14xz^5 w^4 = -70x^4 y^2 z^9 w^4$

13. $2x^5 \cdot 6x^3 \cdot 7x^2 = 12x^8 \cdot 7x^2 = 84x^{10}$

15. $5 \cdot 6 = 30$

$x^5 \cdot x^{25} = x^{30}$

The missing monomial is $6x^{25}$.

17. $2(-12) = -24$

$x^9 \cdot x^9 = x^{18}$

The missing monomial is $-12x^9$.

19. $5(3x-4) = 5 \cdot 3x - 5 \cdot 4 = 15x - 20$

21. $-2(6x-9) = -2 \cdot 6x - (-2)9 = -12x + 18$

23. $6x(3x+5) = 6x \cdot 3x + 6x \cdot 5 = 18x^2 + 30x$

25. $x^3\left(3x^2 - 4x + 7\right) = x^3 \cdot 3x^2 - x^3 \cdot 4x + x^3 \cdot 7$

$\qquad\qquad\qquad\qquad = 3x^5 - 4x^4 + 7x^3$

27. $2xy^2\left(3x^2 - 6xy + 7y^2\right)$

$= 2xy^2 \cdot 3x^2 - 2xy^2 \cdot 6xy + 2xy^2 \cdot 7y^2$

$= 6x^3 y^2 - 12x^2 y^3 + 14xy^4$

29. $(x+7)(x-9) = x \cdot x + x(-9) + 7 \cdot x + 7(-9)$
$$= x^2 - 9x + 7x - 63$$
$$= x^2 - 2x - 63$$

31. $(x-3)(x-9) = x \cdot x + x(-9) - 3 \cdot x - 3(-9)$
$$= x^2 - 9x - 3x + 27$$
$$= x^2 - 12x + 27$$

33. $(4x+3)(x-6) = 4x \cdot x + 4x(-6) + 3 \cdot x + 3(-6)$
$$= 4x^2 - 24x + 3x - 18$$
$$= 4x^2 - 21x - 18$$

35. $(x-8)(x+9) = x \cdot x + x \cdot 9 - 8 \cdot x - 8 \cdot 9$
$$= x^2 + 9x - 8x - 72$$
$$= x^2 + x - 72$$

37. $(2x+3)(2x-13)$
$$= 2x \cdot 2x + 2x(-13) + 3 \cdot 2x + 3(-13)$$
$$= 4x^2 - 26x + 6x - 39$$
$$= 4x^2 - 20x - 39$$

39. $(x+6)(3x-5) = x \cdot 3x + x(-5) + 6 \cdot 3x + 6(-5)$
$$= 3x^2 - 5x + 18x - 30$$
$$= 3x^2 + 13x - 30$$

41. $(3x+2)\left(x^2 - 5x - 9\right)$
$$= 3x \cdot x^2 - 3x \cdot 5x - 3x \cdot 9 + 2 \cdot x^2 - 2 \cdot 5x - 2 \cdot 9$$
$$= 3x^3 - 15x^2 - 27x + 2x^2 - 10x - 18$$
$$= 3x^3 - 13x^2 - 37x - 18$$

43. $\left(x^2 - 7x + 10\right)\left(x^2 + 3x - 40\right)$
$$= x^2 \cdot x^2 + x^2 \cdot 3x - x^2 \cdot 40 - 7x \cdot x^2 - 7x \cdot 3x$$
$$\quad - 7x \cdot (-40) + 10 \cdot x^2 + 10 \cdot 3x - 10 \cdot 40$$
$$= x^4 + 3x^3 - 40x^2 - 7x^3 - 21x^2 + 280x + 10x^2$$
$$\quad + 30x - 400$$
$$= x^4 - 4x^3 - 51x^2 + 310x - 400$$

45. $(x+2y)(x-4y)$
$$= x \cdot x - x \cdot 4y + 2y \cdot x - 2y \cdot 4y$$
$$= x^2 - 4xy + 2xy - 8y^2$$
$$= x^2 - 2xy - 8y^2$$

47. $(4xy+3)(5xy+6)$
$$= 4xy \cdot 5xy + 4xy \cdot 6 + 3 \cdot 5xy + 3 \cdot 6$$
$$= 20x^2 y^2 + 24xy + 15xy + 18$$
$$= 20x^2 y^2 + 39xy + 18$$

49. $3x^3 \cdot 2x^2 = 6x^5$
$$3x^3 \cdot (-7x) = -21x^4$$
$$3x^3 \cdot (-10) = -30x^3$$
The missing factor is $3x^3$.

51. $4x^4\left(3x^3\right) = 12x^7$
$$4x^4\left(-5x^2\right) = -20x^6$$
$$4x^4(-12) = -48x^4$$
The missing factor is
$$\left(3x^3 - 5x^2 - 12\right).$$

53. $(x+3)(x+5) = x^2 + 5x + 3x + 15$
$$= x^2 + 8x + 15$$
The missing term is 3.

55. $(x+3)(x+6) = x^2 + 6x + 3x + 18$
$$= x^2 + 9x + 18$$
The missing terms are 3 and 6.

57. $f(x) \cdot g(x) = (x-9)(x+2)$
$$= x^2 + 2x - 9x - 18$$
$$= x^2 - 7x - 18$$

59. $f(x) \cdot g(x) = \left(6x^5\right)\left(-3x^4\right) = -18x^9$

61. $f(x) \cdot g(x) = \left(x^3 - 5x^2 + 8x + 3\right)\left(-4x^5\right)$
$$= -4x^8 + 20x^7 - 32x^6 - 12x^5$$

63. $(x+9)(x-9) = x^2 - 9^2 = x^2 - 81$

65. $(3x-7)(3x+7) = (3x)^2 - 7^2 = 9x^2 - 49$

67. $(x+7)^2 = x^2 + 2 \cdot x \cdot 7 + 7^2 = x^2 + 14x + 49$

69. $(4x-3)^2 = (4x)^2 - 2\cdot(4x)\cdot(3)+3^2$
$\qquad = 16x^2 - 24x + 9$

71. $(x+9)(x-9) = x^2 - 9^2 = x^2 - 81$
The missing factor is $(x-9)$.

73. $(x-6)^2 = x^2 - 2\cdot x\cdot 6 + 6^2 = x^2 - 12x + 36$
The missing factor is $(x-6)$.

75. $(x+7)(x-1) = x^2 - x + 7x - 7 = x^2 + 6x - 7$

77. $5x(x^2 - 8x - 9) = 5x^3 - 40x^2 - 45x$

79. $-7x^2 y^3 \cdot 4x^4 y^6 = -28x^6 y^9$

81. $-5x^6(-4x^8) = 20x^{14}$

83. $(5x+3)(5x-3) = (5x)^2 - 3^2 = 25x^2 - 9$

85. $(x+13)^2 = x^2 + 2\cdot x\cdot 13 + 13^2 = x^2 + 26x + 169$

87. $2x^2 y(3x^2 - x^4 y^3 - 7y)$
$\quad = 6x^4 y - 2x^6 y^4 - 14x^2 y^2$

89. $(4x-9)^2 = (4x)^2 - 2\cdot(4x)\cdot(9) + 9^2$
$\qquad = 16x^2 - 72x + 81$

91. $(x^2 + 3x + 4)(x^2 - 5x + 4)$
$\quad = x^4 - 5x^3 + 4x^2 + 3x^3 - 15x^2 + 12x + 4x^2$
$\qquad - 20x + 16$
$\quad = x^4 - 2x^3 - 7x^2 - 8x + 16$

93. Answers will vary. Example:
$8x^2 + 3x^2$ involves addition. Add the coefficients. The variable part remains unchanged. $8x^2 + 3x^2 = (8+3)x^2 = 11x^2$

$(8x^2)(3x^2)$ involves multiplication. Multiply the coefficients and add the exponents of the variable part. $(8x^2)(3x^2) = 8\cdot 3x^{2+2} = 24x^4$

95. Answers will vary.

5.5 **QUICK CHECK**

1. $\dfrac{32x^9}{4x^4} = 8x^{9-4} = 8x^5$

2. $\dfrac{40x^3 y^{11}}{8xy^2} = 5x^{3-1} y^{11-2} = 5x^2 y^9$

3. $\dfrac{7x^2 - 21x - 49}{7} = \dfrac{7x^2}{7} - \dfrac{21x}{7} - \dfrac{49}{7}$
$\qquad = x^2 - 3x - 7$

4. $\dfrac{48x^{10} + 12x^7 + 30x^5}{6x^2}$
$= \dfrac{48x^{10}}{6x^2} + \dfrac{12x^7}{6x^2} + \dfrac{30x^5}{6x^2}$
$= 8x^{10-2} + 2x^{7-2} + 5x^{5-2}$
$= 8x^8 + 2x^5 + 5x^3$

5.
$$\begin{array}{r} x+9 \\ x+4\overline{\smash{\big)}x^2 + 13x + 36} \\ \underline{-x^2 \mp 4x} \\ 9x + 36 \\ \underline{-9x \mp 36} \\ 0 \end{array}$$
$x + 9$

6.
$$\begin{array}{r} x-10 \\ x+7\overline{\smash{\big)}x^2 - 3x - 8} \\ \underline{-x^2 \mp 7x} \\ 10x - 8 \\ \underline{\overset{+}{}10x \overset{+}{} 70} \\ 62 \end{array}$$
$x - 10 + \dfrac{62}{x+7}$

7.
$$\begin{array}{r} 4x+3 \\ 3x-4\overline{\smash{\big)}12x^2 - 7x + 4} \\ \underline{-12x^2 \overset{+}{} 16x} \\ 9x + 4 \\ \underline{-9x \overset{+}{} 12} \\ 16 \end{array}$$
$4x + 3 + \dfrac{16}{3x-4}$

8.

$$\require{enclose}\begin{array}{r} x^2 - x - 2 \\ x-2 \enclose{longdiv}{x^3 - 3x^2 + 0x - 9} \\ \underline{-x^3 + 2x^2} \\ -x^2 + 0x \\ \underline{+x^2 - 2x} \\ -2x - 9 \\ \underline{+2x - 4} \\ -13 \end{array}$$

$$x^2 - x - 2 - \frac{13}{x-2}$$

5.5 DIVIDING POLYNOMIALS

1. term

3. factor

5. $\dfrac{24x^{24}}{3x^3} = 8x^{24-3} = 8x^{21}$

7. $\dfrac{-30n^{11}}{6n^4} = -5n^{11-4} = -5n^7$

9. $\dfrac{26a^7 b^5}{2ab^3} = 13a^{7-1}b^{5-3} = 13a^6 b^2$

11. $\left(15x^6\right) \div \left(3x^2\right) = \dfrac{15x^6}{3x^2} = 5x^{6-2} = 5x^4$

13. $\dfrac{12x^6}{8x^4} = \dfrac{3x^{6-4}}{2} = \dfrac{3x^2}{2}$

15. $\dfrac{7x^{12}}{21x^4} = \dfrac{x^{12-4}}{3} = \dfrac{x^8}{3}$

17. Since $\dfrac{10x^{11}}{5x^4} = 2x^7$,

the missing monomial is $10x^{11}$.

19. Since $\dfrac{24x^9}{-4x^5} = -6x^4$,

the missing monomial is $-4x^5$.

21. $\dfrac{15x^2 - 25x - 40}{5} = \dfrac{15x^2}{5} - \dfrac{25x}{5} - \dfrac{40}{5}$

$$= 3x^2 - 5x - 8$$

23. $\dfrac{24x^4 + 30x^2 - 27x}{3x} = \dfrac{24x^4}{3x} + \dfrac{30x^2}{3x} - \dfrac{27x}{3x}$

$$= 8x^3 + 10x - 9$$

25. $\dfrac{6x^7 + 9x^6 + 15x^5}{3x} = \dfrac{6x^7}{3x} + \dfrac{9x^6}{3x} + \dfrac{15x^5}{3x}$

$$= 2x^6 + 3x^5 + 5x^4$$

27. $\dfrac{20x^6 - 30x^4}{-10x^2} = \dfrac{20x^6}{-10x^2} - \dfrac{30x^4}{-10x^2} = -2x^4 + 3x^2$

29. $\dfrac{x^6 y^6 - x^4 y^5 + x^2 y^4}{xy^2} = \dfrac{x^6 y^6}{xy^2} - \dfrac{x^4 y^5}{xy^2} + \dfrac{x^2 y^4}{xy^2}$

$$= x^5 y^4 - x^3 y^3 + xy^2$$

31. $\dfrac{24x^3 - 48x^2 + 36x}{6x} = 4x^2 - 8x + 6$

The missing divisor is $6x$.

33. $\dfrac{6x^7 - 8x^5 - 18x^3}{2x^3} = 3x^4 - 4x^2 - 9$

The missing dividend $6x^7 - 8x^5 - 18x^3$.

35.

$$\require{enclose}\begin{array}{r} x+5 \\ x+8 \enclose{longdiv}{x^2 + 13x + 40} \\ \underline{-x^2 - 8x} \\ 5x + 40 \\ \underline{-5x - 40} \\ 0 \end{array}$$

$x + 5$

37.

$$\require{enclose}\begin{array}{r} x+6 \\ x-14 \enclose{longdiv}{x^2 - 8x - 84} \\ \underline{-x^2 + 14x} \\ 6x - 84 \\ \underline{-6x + 84} \\ 0 \end{array}$$

$x + 6$

39.

$$\require{enclose}\begin{array}{r} x-9 \\ x-4 \enclose{longdiv}{x^2 - 13x + 36} \\ \underline{-x^2 + 4x} \\ -9x + 36 \\ \underline{+9x - 36} \\ 0 \end{array}$$

$x - 9$

41.

$$x - 8 \overline{)\, x^2 - 4x - 29} \quad \begin{array}{c} x + 4 \end{array}$$

$$\underline{-x^2 \overset{+}{-} 8x}$$
$$4x - 29$$
$$\underline{-4x \overset{+}{-} 32}$$
$$3$$

$$x + 4 + \frac{3}{x - 8}$$

43.

$$x + 1 \overline{)\, x^3 - 11x^2 - 37x + 14} \quad \begin{array}{c} x^2 - 12x - 25 \end{array}$$

$$\underline{-x^3 \overset{-}{\mp} x^2}$$
$$-12x^2 - 37x$$
$$\underline{\overset{+}{-}12x^2 \overset{+}{-} 12x}$$
$$-25x + 14$$
$$\underline{\overset{+}{-} 25x \overset{+}{-} 25}$$
$$39$$

$$x^2 - 12x - 25 + \frac{39}{x + 1}$$

45.

$$x + 5 \overline{)\, 2x^2 + 3x - 32} \quad \begin{array}{c} 2x - 7 \end{array}$$

$$\underline{-2x^2 \overset{-}{\mp} 10x}$$
$$-7x - 32$$
$$\underline{\overset{+}{-} 7x \overset{+}{-} 35}$$
$$3$$

$$2x - 7 + \frac{3}{x + 5}$$

47.

$$2x + 7 \overline{)\, 6x^2 + 25x + 10} \quad \begin{array}{c} 3x + 2 \end{array}$$

$$\underline{-6x^2 \overset{-}{\mp} 21x}$$
$$4x + 10$$
$$\underline{-4x \overset{-}{\mp} 14}$$
$$-4$$

$$3x + 2 - \frac{4}{2x + 7}$$

49.

$$x + 13 \overline{)\, x^2 + 0x - 169} \quad \begin{array}{c} x - 13 \end{array}$$

$$\underline{-x^2 \overset{-}{\mp} 13x}$$
$$-13x - 169$$
$$\underline{\overset{+}{-} 13x \overset{+}{-} 169}$$
$$0$$

$$x - 13$$

51.

$$x - 3 \overline{)\, x^4 + 0x^3 + 2x^2 - 15x + 32} \quad \begin{array}{c} x^3 + 3x^2 + 11x + 18 \end{array}$$

$$\underline{-x^4 \overset{+}{-} 3x^3}$$
$$3x^3 + 2x^2$$
$$\underline{-3x^3 \overset{+}{-} 9x^2}$$
$$11x^2 - 15x$$
$$\underline{-11x^2 \overset{+}{-} 33x}$$
$$18x + 32$$
$$\underline{-18x \overset{+}{-} 54}$$
$$86$$

$$x^3 + 3x^2 + 11x + 18 + \frac{86}{x - 3}$$

53.

$$x - 5 \overline{)\, x^3 + 0x^2 + 0x - 125} \quad \begin{array}{c} x^2 + 5x + 25 \end{array}$$

$$\underline{-x^3 \overset{+}{-} 5x^2}$$
$$5x^2 + 0x$$
$$\underline{-5x^2 \overset{+}{-} 25x}$$
$$25x - 125$$
$$\underline{-25x \overset{+}{-} 125}$$
$$0$$

$$x^2 + 5x + 25$$

55. $(x + 8)(x - 3)$

$$= x^2 - 3x + 8x - 24$$
$$= x^2 + 5x - 24$$

The missing dividend is $x^2 + 5x - 24$.

57.

$$x - 3 \overline{)\, x^2 + 10x - 39} \quad \begin{array}{c} x + 13 \end{array}$$

$$\underline{-x^2 \overset{+}{-} 3x}$$
$$13x - 39$$
$$\underline{-13x \overset{+}{-} 39}$$
$$0$$

The missing divisor is $x + 13$.

59.

$$x + 9 \overline{)\, x^2 + 28x + 171} \quad \begin{array}{c} x + 19 \end{array}$$

$$\underline{-x^2 \overset{-}{\mp} 9x}$$
$$19x + 171$$
$$\underline{-19x \overset{-}{\mp} 171}$$
$$0$$

Yes.

61.
$$\begin{array}{r} 2x+1 \\ 2x-5\overline{)4x^2-8x-15} \\ \underline{-4x^2 \overset{+}{} 10x} \\ 2x-15 \\ \underline{-2x \overset{+}{} 5} \\ -10 \end{array}$$

No.

63.
$$\begin{array}{r} x^2+9x-27 \\ x+3\overline{)x^3+12x^2+\ 0x-9} \\ \underline{-x^3 \overset{-}{} 3x^2} \\ 9x^2+\ 0x \\ \underline{-9x^2 \overset{-}{} 27x} \\ -27x-9 \\ \underline{\overset{+}{} 27x \overset{+}{} 81} \\ 72 \end{array}$$

$$x^2+9x-27+\frac{72}{x+3}$$

65.
$$\begin{array}{r} x^2-4x+24 \\ x+4\overline{)x^3+0x^2+8x-19} \\ \underline{-x^3 \overset{-}{} 4x^2} \\ -4x^2+8x \\ \underline{\overset{+}{} 4x^2 \overset{+}{} 16x} \\ 24x-19 \\ \underline{-24x \overset{-}{} 96} \\ -115 \end{array}$$

$$x^2-4x+24-\frac{115}{x+4}$$

67.
$$\frac{12x^6-8x^4+20x^3-4x^2}{4x^2}$$

$$=\frac{12x^6}{4x^2}-\frac{8x^4}{4x^2}+\frac{20x^3}{4x^2}-\frac{4x^2}{4x^2}$$

$$3x^4-2x^2+5x-1$$

69.
$$\begin{array}{r} 8x-6 \\ x+6\overline{)8x^2+42x-25} \\ \underline{-8x^2 \overset{-}{} 48x} \\ -6x-25 \\ \underline{\overset{+}{} 6x \overset{+}{} 36} \\ 11 \end{array}$$

$$8x-6+\frac{11}{x+6}$$

71.
$$\begin{array}{r} 5x-9 \\ 4x-3\overline{)20x^2-51x-6} \\ \underline{-20x^2 \overset{+}{} 15x} \\ -36x-6 \\ \underline{\overset{+}{} 36x \overset{-}{} 27} \\ -33 \end{array}$$

$$5x-9-\frac{33}{4x-3}$$

73.
$$\frac{27x^9-18x^7-6x^6-3x^3}{-3x^2}$$

$$=\frac{27x^9}{-3x^2}-\frac{18x^7}{-3x^2}-\frac{6x^6}{-3x^2}-\frac{3x^3}{-3x^2}$$

$$=-9x^7+6x^5+2x^4+x$$

75.
$$\begin{array}{r} 4x^2-6x-10 \\ 2x+3\overline{)8x^3+0x^2-38x-39} \\ \underline{-8x^3 \overset{-}{} 12x^2} \\ -12x^2-38x \\ \underline{\overset{+}{} 12x^2 \overset{+}{} 18x} \\ -20x-39 \\ \underline{\overset{+}{} 20x \overset{+}{} 30} \\ -9 \end{array}$$

$$4x^2-6x-10-\frac{9}{2x+3}$$

77.
$$\begin{array}{r} 3x^2-5x-17 \\ 2x-9\overline{)6x^3-37x^2+11x+153} \\ \underline{-6x^3 \overset{+}{} 27x^2} \\ -10x^2+11x \\ \underline{\overset{+}{} 10x^2 \overset{-}{} 45x} \\ -34x+153 \\ \underline{\overset{+}{} 34x \overset{-}{} 153} \\ 0 \end{array}$$

$$3x^2-5x-17$$

79. $\dfrac{21x^9y^6z^{17}}{-7x^7y^5z^{12}}=-3x^2yz^5$

81. Answers will vary.

CHAPTER 5 REVIEW

1. $\dfrac{x^9}{x^3} = x^{9-3} = x^6$

2. $\left(4x^7\right)^3 = 4^3 x^{21} = 64x^{21}$

3. $7x^0 = 7 \cdot 1 = 7$

4. $\left(\dfrac{5x^9}{2y^2}\right)^4 = \dfrac{5^4 x^{36}}{2^4 y^8} = \dfrac{625x^{36}}{16y^8}$

5. $7x^7 \cdot 4x^{12} = 28x^{19}$

6. $3x^8 y^5 \cdot 8x^8 y^3 = 24x^{16} y^8$

7. $\left(-6a^5 b^3 c^6\right)^2 = (-6)^2 a^{10} b^6 c^{12} = 36a^{10} b^6 c^{12}$

8. $15^0 - 3x^0 = 1 - 3 \cdot 1 = -2$

9. $\left(x^{10} y^{13} z^4\right)^4 = x^{40} y^{52} z^{16}$

10. $\dfrac{x^{16}}{x^6} = x^{10}$

11. $5^{-2} = \dfrac{1}{5^2} = \dfrac{1}{25}$

12. $10^{-3} = \dfrac{1}{10^3} = \dfrac{1}{1000}$

13. $7x^{-4} = \dfrac{7}{x^4}$

14. $-6x^{-6} = -\dfrac{6}{x^6}$

15. $\dfrac{-8}{y^{-5}} = -8y^5$

16. $\left(3x^{-4}\right)^3 = 3^3 x^{-12} = \dfrac{27}{x^{12}}$

17. $\left(2x^{-5}\right)^{-2} = 2^{-2} x^{10} = \dfrac{x^{10}}{2^2} = \dfrac{x^{10}}{4}$

18. $\dfrac{x^{10}}{x^{-5}} = x^{10-(-5)} = x^{15}$

19. $\left(\dfrac{a^{-4}}{b^{-7}}\right)^3 = \dfrac{a^{-12}}{b^{-21}} = \dfrac{b^{21}}{a^{12}}$

20. $m^{13} \cdot m^{-19} = m^{-6} = \dfrac{1}{m^6}$

21. $\left(4x^{-5} y^4 z^{-7}\right)^{-3} = 4^{-3} x^{15} y^{-12} z^{21}$

$$= \dfrac{x^{15} z^{21}}{4^3 y^{12}}$$

$$= \dfrac{x^{15} z^{21}}{64 y^{12}}$$

22. $\left(\dfrac{x^{-11}}{y^{-6}}\right)^6 = \dfrac{x^{-66}}{y^{-36}} = \dfrac{y^{36}}{x^{66}}$

23. $a^{-15} \cdot a^{-5} = a^{-15+(-5)} = a^{-20} = \dfrac{1}{a^{20}}$

24. $\dfrac{x^{-20}}{x^{-9}} = x^{-20-(-9)} = x^{-20+9} = x^{-11} = \dfrac{1}{x^{11}}$

25. $1,400,000,000 = 1.4 \times 10^9$

26. $0.0000000000021 = 2.1 \times 10^{-12}$

27. $0.000005002 = 5.002 \times 10^{-6}$

28. $1.23 \times 10^{-4} = 0.000123$

29. $4.075 \times 10^7 = 40,750,000$

30. $6.1275 \times 10^{12} = 6,127,500,000,000$

31. $\left(2.5 \times 10^7\right)\left(6.0 \times 10^6\right) = (2.5)(6.0)\left(10^7\right)\left(10^6\right)$

$$= 15.0 \times 10^{13}$$

$$= 1.5 \times 10^{14}$$

32. $\left(3.0 \times 10^{-13}\right) \div \left(1.2 \times 10^5\right) = \dfrac{3.0}{1.2} \times 10^{-13-5}$

$$= 2.5 \times 10^{-18}$$

33. $1,500,000,000,000 = 1.5 \times 10^{12}$

$\left(1.66 \times 10^{-24}\right)\left(1.5 \times 10^{12}\right)$

$= (1.66)(1.5)\left(10^{-24}\right)\left(10^{12}\right)$

$= 2.49 \times 10^{-12}$ grams

34. $\left(5.0 \times 10^{-12}\right)\left(3.0 \times 10^{15}\right)$

$= (5.0)(3.0)\left(10^{-12}\right)\left(10^{15}\right)$

$= 15 \times 10^{3}$

$= 1.5 \times 10^{4}$ seconds

35. $(-5)^2 - 8(-5) + 15 = 25 - 8(-5) + 15$
$= 25 + 40 + 15$
$= 80$

36. $(2)^2 - 11(2) - 29 = 4 - 11(2) - 29$
$= 4 - 22 - 29$
$= 4 - 51$
$= -47$

37. $-(6)^2 + 16(6) - 30 = -36 + 16(6) - 30$
$= -36 + 96 - 30$
$= -66 + 96$
$= 30$

38. $-3(-7)^2 - 4(-7) + 7 = -3(49) - 4(-7) + 7$
$= -147 + 28 + 7$
$= -147 + 35$
$= -112$

39. $5(-3)^2 + 10(-3) + 12 = 5(9) + 10(-3) + 12$
$= 45 - 30 + 12$
$= 57 - 30$
$= 27$

40. $7(8)^2 - 14(8) + 35 = 7(64) - 14(8) + 35$
$= 448 - 112 + 35$
$= 483 - 112$
$= 371$

41. $\left(x^2 + 3x - 15\right) + \left(x^2 - 9x + 6\right)$

$= x^2 + 3x - 15 + x^2 - 9x + 6$

$= 2x^2 - 6x - 9$

42. $\left(3x^2 - 31\right) + \left(5x^2 - 19x - 19\right)$

$= 3x^2 - 31 + 5x^2 - 19x - 19$

$= 8x^2 - 19x - 50$

43. $\left(x^2 - 6x - 13\right) - \left(x^2 - 14x + 8\right)$

$= x^2 - 6x - 13 - x^2 + 14x - 8$

$= 8x - 21$

44. $\left(2x^2 + 8x - 7\right) - \left(x^2 - x + 15\right)$

$= 2x^2 + 8x - 7 - x^2 + x - 15$

$= x^2 + 9x - 22$

45. $\left(3x^3 + x^2 - 10\right) - \left(6x^2 - 13x + 20\right)$

$= 3x^3 + x^2 - 10 - 6x^2 + 13x - 20$

$= 3x^3 - 5x^2 + 13x - 30$

46. $\left(10x^2 - 21x - 35\right) - \left(4x^3 + 6x^2 - 15x + 80\right)$

$= 10x^2 - 21x - 35 - 4x^3 - 6x^2 + 15x - 80$

$= -4x^3 + 4x^2 - 6x - 115$

47. $f(-4) = (-4)^2 - 25 = 16 - 25 = -9$

48. $f(5) = 2(5)^2 - 6(5) + 17$
$= 2(25) - 6(5) + 17$
$= 50 - 30 + 17$
$= 67 - 30$
$= 37$

49. $f\left(a^4\right) = \left(a^4\right)^2 - 7\left(a^4\right) + 10 = a^8 - 7a^4 + 10$

50. $f\left(2a^3\right) = \left(2a^3\right)^2 + 3\left(2a^3\right) - 30$

$= 4a^6 + 3\left(2a^3\right) - 30$

$= 4a^6 + 6a^3 - 30$

51. $f(x) + g(x) = \left(3x^2 - 8x - 9\right) + \left(7x^2 + 6x + 40\right)$

$= 3x^2 - 8x - 9 + 7x^2 + 6x + 40$

$= 10x^2 - 2x + 31$

$f(x) - g(x) = \left(3x^2 - 8x - 9\right) - \left(7x^2 + 6x + 40\right)$

$= 3x^2 - 8x - 9 - 7x^2 - 6x - 40$

$= -4x^2 - 14x - 49$

52.
$$f(x) + g(x) = \left(x^2 + 5x - 30\right) + \left(-8x^2 - 15x + 6\right)$$
$$= x^2 + 5x - 30 - 8x^2 - 15x + 6$$
$$= -7x^2 - 10x - 24$$

$$f(x) - g(x) = \left(x^2 + 5x - 30\right) - \left(-8x^2 - 15x + 6\right)$$
$$= x^2 + 5x - 30 + 8x^2 + 15x - 6$$
$$= 9x^2 + 20x - 36$$

53. $(x-5)^2 = x^2 - 2 \cdot x \cdot 5 + 5^2 = x^2 - 10x + 25$

54. $(x-4)(x+13) = x^2 + 13x - 4x - 52$
$$= x^2 + 9x - 52$$

55. $4x\left(3x^2 - 7x - 16\right) = 12x^3 - 28x^2 - 64x$

56. $(2x-7)^2 = (2x)^2 - 2 \cdot (2x) \cdot 7 + 7^2$
$$= 4x^2 - 28x + 49$$

57. $5x^7 \cdot 3x^6 = 15x^{13}$

58. $(x-8)(x+8) = x^2 - 8^2 = x^2 - 64$

59. $(3x+10)(2x-13) = 6x^2 - 39x + 20x - 130$
$$= 6x^2 - 19x - 130$$

60. $-9x^5 \cdot 2x = -18x^6$

61. $(6x+5)(6x-5) - (6x)^2 \quad 5^2 - 36x^2 - 25$

62. $(x+10)^2 = x^2 + 2 \cdot x \cdot 10 + 10^2$
$$= x^2 + 20x + 100$$

63. $(4x+7)^2 = (4x)^2 + 2 \cdot (4x) \cdot 7 + 7^2$
$$= 16x^2 + 56x + 49$$

64. $-8x^2\left(3x^5 - 7x^4 - 6x^3\right) = -24x^7 + 56x^6 + 48x^5$

65. $(x+9)\left(x^2 - 6x + 12\right)$
$$= x^3 - 6x^2 + 12x + 9x^2 - 54x + 108$$
$$= x^3 + 3x^2 - 42x + 108$$

66. $(x-2)\left(3x^2 - 5x - 9\right)$
$$= 3x^3 - 5x^2 - 9x - 6x^2 + 10x + 18$$
$$= 3x^3 - 11x^2 + x + 18$$

67. $\left(10x^5\right)\left(-3x^7\right)\left(4x^6\right) = \left(-30x^{12}\right)\left(4x^6\right)$
$$= -120x^{18}$$

68. $-2x^4\left(-5x^2 - 11x + 12\right) = 10x^6 + 22x^5 - 24x^4$

69. $f(x) \cdot g(x) = (x+10)(x-7)$
$$= x^2 - 7x + 10x - 70$$
$$= x^2 + 3x - 70$$

70. $f(x) \cdot g(x) = -6x^6\left(-4x^2 - 9x + 20\right)$
$$= 24x^8 + 54x^7 - 120x^6$$

71. $\dfrac{6x^4 - 8x^3 + 20x^2}{2x^2} = \dfrac{6x^4}{2x^2} - \dfrac{8x^3}{2x^2} + \dfrac{20x^2}{2x^2}$
$$= 3x^2 - 4x + 10$$

72.
$$x - 4 \overline{)\,3x^2 - 8x - 25}$$
$$\underline{-3x^2 \overset{+}{-} 12x}$$
$$4x - 25$$
$$\underline{-4x \overset{+}{-} 16}$$
$$-9$$
$$3x + 4 - \frac{9}{x-4}$$

73. $\dfrac{4x^5 y^7}{-2x^5 y} = -2y^6$

74.
$$x + 5 \overline{)\,x^3 + 3x^2 + 0x - 15}$$
$$\underline{-x^3 \overset{-}{\ne} 5x^2}$$
$$-2x^2 + 0x$$
$$\underline{\overset{+}{-} 2x^2 \overset{+}{-} 10x}$$
$$10x - 15$$
$$\underline{-10x \overset{-}{\ne} 50}$$
$$-65$$
$$x^2 - 2x + 10 - \frac{65}{x+5}$$

75.

$$x+3\overline{)x^4+0x^3+0x^2+0x+81} \quad\to\quad x^3-3x^2+9x-27$$

$$-x^4 \mp 3x^3$$
$$-3x^3+0x^2$$
$$\pm 3x^3 \pm 9x^2$$
$$9x^2+0x$$
$$-9x^2 \mp 27x$$
$$-27x+81$$
$$\pm 27x \pm 81$$
$$162$$

$$x^3-3x^2+9x-27+\frac{162}{x+3}$$

76. $\dfrac{18x^3y^7z^6}{3x^3y^6z}=6yz^5$

77.

$$x+9\overline{)8x^2+87x+171} \quad\to\quad 8x+15$$

$$-8x^2 \mp 72x$$
$$15x+171$$
$$-15x \mp 135$$
$$36$$

$$8x+15+\frac{36}{x+9}$$

78.

$$x-7\overline{)6x^2-30x-84} \quad\to\quad 6x+12$$

$$-6x^2 \pm 42x$$
$$12x-84$$
$$-12x \pm 84$$
$$0$$

$$6x+12$$

79.

$$5x+14\overline{)15x^2-43x-238} \quad\to\quad 3x-17$$

$$-15x^2 \mp 42x$$
$$-85x-238$$
$$\pm 85x \pm 238$$
$$0$$

$$3x-17$$

80.

$$8x-3\overline{)16x^2+2x+20} \quad\to\quad 2x+1$$

$$-16x^2 \pm 6x$$
$$8x+20$$
$$-8x \pm 3$$
$$23$$

$$2x+1+\frac{23}{8x-3}$$

CHAPTER 5 TEST

1. $\dfrac{x^{10}}{x^2}=x^{10-2}=x^8$

2. $\left(\dfrac{x^5}{3y^4}\right)^6=\dfrac{x^{30}}{3^6y^{24}}=\dfrac{x^{30}}{729y^{24}}$

3. $\left(x^9y^5z^8\right)^8=x^{72}y^{40}z^{64}$

4. $4^{-4}=\dfrac{1}{4^4}=\dfrac{1}{256}$

5. $\left(2x^{-7}\right)^5=2^5x^{-35}=\dfrac{2^5}{x^{35}}=\dfrac{32}{x^{35}}$

6. $\dfrac{x^4}{x^{-13}}=x^{4-(-13)}=x^{17}$

7. $x^{22}\cdot x^{-15}=x^{22+(-15)}=x^7$

8. $\left(3xy^{-6}z^5\right)^{-2}=3^{-2}x^{-2}y^{12}z^{-10}$

$$=\frac{y^{12}}{3^2x^2z^{10}}$$

$$=\frac{y^{12}}{9x^2z^{10}}$$

9. $23{,}500{,}000=2.35\times10^7$

10. $4.7\times10^{-8}=0.000000047$

11. $\left(4.3\times10^{13}\right)\left(1.8\times10^{-9}\right)$

$$=(4.3)(1.8)\left(10^{13}\right)\left(10^{-9}\right)$$

$$=7.74\times10^4$$

12. $(-8)^2 + 12(-8) - 38 = 64 + 12(-8) - 38$
$$= 64 - 96 - 38$$
$$= 64 - 134$$
$$= -70$$

13. $\left(x^2 - 6x - 32\right) - \left(5x^2 + 8x - 33\right)$

$= x^2 - 6x - 32 - 5x^2 - 8x + 33$

$= -4x^2 - 14x + 1$

14. $\left(4x^2 - 3x - 15\right) + \left(-2x^2 + 17x - 49\right)$

$= 4x^2 - 3x - 15 - 2x^2 + 17x - 49$

$= 2x^2 + 14x - 64$

15. $f(-7) = (-7)^2 + 3(-7) - 14$
$$= 49 + 3(-7) - 14$$
$$= 49 - 21 - 14$$
$$= 49 - 35$$
$$= 14$$

16. $5x^3\left(6x^2 + 8x - 17\right) = 30x^5 + 40x^4 - 85x^3$

17. $(x+6)(x-6) = x^2 - 6^2 = x^2 - 36$

18. $(5x-9)(4x+7) = 20x^2 + 35x - 36x - 63$
$$= 20x^2 - x - 63$$

19. $\dfrac{21x^7 + 33x^6 - 15x^5}{3x^2} = \dfrac{21x^7}{3x^2} + \dfrac{33x^6}{3x^2} - \dfrac{15x^5}{3x^2}$
$$= 7x^5 + 11x^4 - 5x^3$$

20.
$$
\begin{array}{r}
6x - 29 \\
x+3 \overline{\smash{)}\ 6x^2 - 11x + 38} \\
\underline{-6x^2 \mp 18x} \\
-29x + 38 \\
\underline{\pm 29x \pm 87} \\
125
\end{array}
$$

$6x - 29 + \dfrac{125}{x+3}$

CHAPTER 6   FACTORING AND QUADRATIC EQUATIONS

6.1   QUICK CHECK

1. $16 = 2^4$

$20 = 2^2 \cdot 5$

$\text{GCF} = 2^2 = 4$

2. $48 = 2^4 \cdot 3$

$120 = 2^3 \cdot 3 \cdot 5$

$156 = 2^2 \cdot 3 \cdot 13$

$\text{GCF} = 2^2 \cdot 3 = 12$

3. a) $12x^5 = 2^2 \cdot 3 \cdot x^5$

$28x^3 = 2^2 \cdot 7 \cdot x^3$

$\text{GCF} = 2^2 \cdot x^3 = 4x^3$

b) $x^3 y^4 z^9 = x^3 \cdot y^4 \cdot z^9$

$x^6 y^2 z^{10} = x^6 \cdot y^2 \cdot z^{10}$

$x^7 z^4 = x^7 \cdot z^4$

$\text{GCF} = x^3 \cdot z^4 = x^3 z^4$

4. $\text{GCF} = 6x^2$

$6x^4 - 42x^3 - 90x^2 = 6x^2 \left(x^2 - 7x - 15 \right)$

5. $\text{GCF} = 3x^4$

$15x^7 - 30x^5 + 3x^4 = 3x^4 \left(5x^3 - 10x + 1 \right)$

6. a) $\text{GCF} = (x - 9)$

$5x(x - 9) + 14(x - 9) = (x - 9)(5x + 14)$

b) $\text{GCF} = (x - 8)$

$7x(x - 8) - 6(x - 8) = (x - 8)(7x - 6)$

7. $x^3 + 4x^2 + 7x + 28 = x^2 (x + 4) + 7x + 28$

$\qquad = x^2 (x + 4) + 7(x + 4)$

$\qquad = (x + 4)\left(x^2 + 7 \right)$

8. $2x^2 - 10x - 9x + 45 = 2x(x - 5) - 9x + 45$

$\qquad = 2x(x - 5) - 9(x - 5)$

$\qquad = (x - 5)(2x - 9)$

9. $4x^3 - 36x^2 + x - 9 = 4x^2 (x - 9) + x - 9$

$\qquad = 4x^2 (x - 9) + 1(x - 9)$

$\qquad = (x - 9)\left(4x^2 + 1 \right)$

10. $3x^2 + 24x - 12x - 96 = 3\left(x^2 + 8x - 4x - 32 \right)$

$\qquad = 3\left[x(x + 8) - 4(x + 8) \right]$

$\qquad = 3(x + 8)(x - 4)$

6.1   AN INTRODUCTION TO FACTORING; THE GREATEST COMMON FACTOR; FACTORING BY GROUPING

1. factored

3. smallest

5. $6 = 2 \cdot 3$

$8 = 2^3$

$\text{GCF} = 2$

7. $30 = 2 \cdot 3 \cdot 5$

$42 = 2 \cdot 3 \cdot 7$

$\text{GCF} = 2 \cdot 3 = 6$

9. $16 = 2^4$

$40 = 2^3 \cdot 5$

$60 = 2^2 \cdot 3 \cdot 5$

$\text{GCF} = 2^2 = 4$

11. x^3

x^7

$\text{GCF} = x^3$

13. $a^2 b^3 = a^2 \cdot b^3$

$a^5 b^2 = a^5 \cdot b^2$

$\text{GCF} = a^2 \cdot b^2 = a^2 b^2$

15. $4x^4 = 2^2 \cdot x^4$

$6x^3 = 2 \cdot 3 \cdot x^3$

$\text{GCF} = 2 \cdot x^3 = 2x^3$

17. $15a^5 b^2 c = 3 \cdot 5 \cdot a^5 \cdot b^2 \cdot c$

$25a^2 c^3 = 5^2 \cdot a^2 \cdot c^3$

$5a^3 bc^2 = 5 \cdot a^3 \cdot b \cdot c^2$

$\text{GCF} = 5 \cdot a^2 \cdot c = 5a^2 c$

19. $\text{GCF} = 7$

$7x - 14 = 7(x - 2)$

21. GCF $= x$
$$5x^2 + 4x = x(5x + 4)$$

23. GCF $= 4x$
$$8x^3 + 20x = 4x(2x^2 + 5)$$

25. GCF $= 6$
$$60x^2 + 36x + 6 = 6(10x^2 + 6x + 1)$$

27. GCF $= 5x^3$
$$20x^6 - 35x^4 - 50x^3 = 5x^3(4x^3 - 7x - 10)$$

29. GCF $= m^5 n^3$
$$m^7 n^3 - m^5 n^4 + m^6 n^6 = m^5 n^3(m^2 - n + mn^3)$$

31. GCF $= 10x^2 y^4$
$$10x^5 y^5 - 30x^7 y^4 + 80x^2 y^{10}$$
$$= 10x^2 y^4(x^3 y - 3x^5 + 8y^6)$$

33. GCF $= 4x$
$$-8x^3 + 12x^2 - 16x = 4x(-2x^2 + 3x - 4)$$

35. GCF $= (2x - 7)$
$$5x(2x - 7) + 8(2x - 7) = (2x - 7)(5x + 8)$$

37. GCF $= (3x - 4)$
$$x(3x - 4) - 9(3x - 4) = (3x - 4)(x - 9)$$

39. GCF $= (4x + 7)$
$$5x(4x + 7) - (4x + 7) = 5x(4x + 7) - 1(4x + 7)$$
$$= (4x + 7)(5x - 1)$$

41. $x^2 + 10x + 3x + 30 = x(x + 10) + 3x + 30$
$$= x(x + 10) + 3(x + 10)$$
$$= (x + 10)(x + 3)$$

43. $x^3 - 9x^2 + 6x - 54 = x^2(x - 9) + 6x - 54$
$$= x^2(x - 9) + 6(x - 9)$$
$$= (x - 9)(x^2 + 6)$$

45. $x^2 - 5x - 12x + 60 = x(x - 5) - 12x + 60$
$$= x(x - 5) - 12(x - 5)$$
$$= (x - 5)(x - 12)$$

47. $3x^2 + 15x + 4x + 20 = 3x(x + 5) + 4x + 20$
$$= 3x(x + 5) + 4(x + 5)$$
$$= (x + 5)(3x + 4)$$

49. $7x^2 + 28x - 6x - 24 = 7x(x + 4) - 6x - 24$
$$= 7x(x + 4) - 6(x + 4)$$
$$= (x + 4)(7x - 6)$$

51. $3x^2 + 21x + x + 7 = 3x(x + 7) + 1(x + 7)$
$$= (x + 7)(3x + 1)$$

53. $2x^2 - 10x - x + 5 = 2x(x - 5) - 1(x - 5)$
$$= (x - 5)(2x - 1)$$

55. $x^3 + 8x^2 + 6x + 48 = x^2(x + 8) + 6(x + 8)$
$$= (x + 8)(x^2 + 6)$$

57. $4x^3 + 12x + 3x^2 + 9 = 4x(x^2 + 3) + 3(x^2 + 3)$
$$= (x^2 + 3)(4x + 3)$$

59. $2x^2 + 10x + 6x + 30 = 2(x^2 + 5x + 3x + 15)$
$$= 2[x(x + 5) + 3(x + 5)]$$
$$= 2(x + 5)(x + 3)$$

61. Answers will vary. Example:
$$(x - 5)(x + 8) = x \cdot x + x \cdot 8 - 5 \cdot x - 5 \cdot 8$$
$$= x^2 + 8x - 5x - 40$$
$$= x^2 + 3x - 40$$
Therefore $x^2 + 3x - 40$ has a factor of $x - 5$.

63. Answers will vary. Example:
$$(2x + 9)(x - 4)$$
$$= 2x \cdot x + 2x \cdot (-4) + 9 \cdot x + 9 \cdot (-4)$$
$$= 2x^2 - 8x + 9x - 36$$
$$= 2x^2 + x - 36$$
Therefore $2x^2 + x - 36$ has a factor of $2x + 9$.

65. Answers will vary. Example:
Multiply the factored form using the Distributive Property. The product should be equal to the expanded form.

6.2 QUICK CHECK

1. a) Factors of 10: $(1)(10),(2)(5)$

$$x^2 + 7x + 10 = (x+2)(x+5)$$

b) Factors of 30:

$$(-1)(-30),(-3)(-10),(-5)(-6)$$
$$x^2 - 11x + 30 = (x-5)(x-6)$$

2. a) Factors of -36:

$$(-1)(36),(-2)(18),(-3)(12),(-4)(9),(-6)(6)$$
$$(1)(-36),(2)(-18),(3)(-12),(4)(-9),(6)(-6)$$
$$x^2 + 9x - 36 = (x+12)(x-3)$$

b) Factors of -42:

$$(-1)(42),(-2)(21),(-3)(14),(-6)(7)$$
$$(1)(-42),(2)(-21),(3)(-14),(6)(-7)$$
$$x^2 - x - 42 = (x+6)(x-7)$$

3. a) $x^2 + 5x - 6 = (x+6)(x-1)$

b) $x^2 + 5x + 6 = (x+2)(x+3)$

c) $x^2 - 5x + 6 = (x-2)(x-3)$

d) $x^2 - 5x - 6 = (x-6)(x+1)$

4. $x^2 + 10x + 25 = (x+5)(x+5) = (x+5)^2$

5. Prime

6. $5x^2 + 40x + 60 = 5\left(x^2 + 8x + 12\right)$
$$= 5(x+2)(x+6)$$

7. $-x^2 + 14x - 48 = -\left(x^2 - 14x + 48\right)$
$$= -(x-6)(x-8)$$

8. $x^2 - 4xy - 32y^2 = (x-8y)(x+4y)$

9. $x^2 y^2 + 10xy - 24 = (xy+12)(xy-2)$

6.2 FACTORING TRINOMIALS OF THE FORM
$x^2 + bx + c$

1. leading coefficient

3. prime

5. Factors of -20:

$$(-1)(20),(-2)(10),(-4)(5)$$
$$(1)(-20),(2)(-10),(4)(-5)$$
$$x^2 - 8x - 20 = (x-10)(x+2)$$

7. Factors of -36:

$$(-1)(36),(-2)(18),(-3)(12),(-4)(9),(-6)(6)$$
$$(1)(-36),(2)(-18),(3)(-12),(4)(-9),(6)(-6)$$
$$x^2 + 5x - 36 = (x+9)(x-4)$$

9. $x^2 + 12x + 24$ is prime.

11. $x^2 + 14x + 48 = (x+6)(x+8)$

13. $x^2 + 13x - 30 = (x+15)(x-2)$

15. $x^2 - 5x + 36$ is prime.

17. $x^2 - 17x - 60 = (x-20)(x+3)$

19. $x^2 + 11x - 12 = (x+12)(x-1)$

21. $x^2 + 14x + 49 = (x+7)(x+7) = (x+7)^2$

23. $x^2 + 20x + 91 = (x+7)(x+13)$

25. $x^2 - 24x + 144 = (x-12)(x-12) = (x-12)^2$

27. $x^2 - 13x + 40 = (x-5)(x-8)$

29. $x^2 - 13x - 30 = (x-15)(x+2)$

31. $x^2 - 9x + 20 = (x-4)(x-5)$

33. $x^2 - 16x + 60 = (x-6)(x-10)$

35. $6x^2 - 54x + 120 = 6\left(x^2 - 9x + 20\right)$
$$= 6(x-4)(x-5)$$

37. $5x^2 + 35x - 150 = 5\left(x^2 + 7x - 30\right)$
$\qquad\qquad\qquad = 5(x - 3)(x + 10)$

39. $-x^2 + 3x + 70 = -\left(x^2 - 3x - 70\right)$
$\qquad\qquad\qquad = -(x + 7)(x - 10)$

41. $x^3 + 8x^2 + 16x = x\left(x^2 + 8x + 16\right)$
$\qquad\qquad\qquad = x(x + 4)(x + 4)$
$\qquad\qquad\qquad = x(x + 4)^2$

43. $-3x^5 - 6x^4 + 240x^3 = -3x^3\left(x^2 + 2x - 80\right)$
$\qquad\qquad\qquad\qquad = -3x^3(x + 10)(x - 8)$

45. $10x^8 + 20x^7 - 990x^6 = 10x^6\left(x^2 + 2x - 99\right)$
$\qquad\qquad\qquad\qquad = 10x^6(x + 11)(x - 9)$

47. $x^2 + xy - 42y^2 = (x + 7y)(x - 6y)$

49. $x^2 - 5xy - 36y^2 = (x + 4y)(x - 9y)$

51. $x^2 y^2 + 17xy + 72 = (xy + 8)(xy + 9)$

53. $5x^5 + 50x^4 y + 105x^3 y^2$
$\quad = 5x^3\left(x^2 + 10xy + 21y^2\right)$
$\quad = 5x^3(x + 3y)(x + 7y)$

55. $x^4 + 9x^3 + 5x + 45 = x^3(x + 9) + 5(x + 9)$
$\qquad\qquad\qquad\qquad = (x + 9)\left(x^3 + 5\right)$

57. $x^2 + x + 30$ is prime.

59. $x^2 + 22x + 40 = (x + 2)(x + 20)$

61. $x^2 + 40x + 400 = (x + 20)(x + 20) = (x + 20)^2$

63. $4x^2 + 8x - 96 = 4\left(x^2 + 2x - 24\right)$
$\qquad\qquad\qquad = 4(x + 6)(x - 4)$

65. $x^4 - 15x^3 = x^3(x - 15)$

67. $x^2 - 16x + 63 = (x - 7)(x - 9)$

69. $x^2 + 10x - 25$ is prime.

71. $x^2 - 21x - 100 = (x - 25)(x + 4)$

73. Since $(x + 2)(x + 8) = x^2 + 8x + 2x + 16$
$\qquad\qquad\qquad\quad = x^2 + 10x + 16$
the missing value is 16.

75. Since $(x - 4)(x + 15) = x^2 + 15x - 4x - 60$
$\qquad\qquad\qquad\qquad = x^2 + 11x - 60$
the missing value is 60.

6.3   QUICK CHECK

1. Find $m \cdot n = 3 \cdot 12 = 36$ and $m + n = 13$.
$\quad m = 4, n = 9$

$\quad 3x^2 + 13x + 12 = 3x^2 + 4x + 9x + 12$
$\qquad\qquad\qquad\quad = x(3x + 4) + 3(3x + 4)$
$\qquad\qquad\qquad\quad = (3x + 4)(x + 3)$

2. Find $m \cdot n = 4 \cdot 36 = 144$ and $m + n = -25$.
$\quad m = -9, n = -16$

$\quad 4x^2 - 25x + 36 = 4x^2 - 9x - 16x + 36$
$\qquad\qquad\qquad\quad = x(4x - 9) - 4(4x - 9)$
$\qquad\qquad\qquad\quad = (4x - 9)(x - 4)$

3. Find $m \cdot n = 16 \cdot 21 = 336$ and $m + n = -38$.
$\quad m = -24, n = -14$

$\quad 16x^2 - 38x + 21 = 16x^2 - 24x - 14x + 21$
$\qquad\qquad\qquad\qquad = 8x(2x - 3) - 7(2x - 3)$
$\qquad\qquad\qquad\qquad = (2x - 3)(8x - 7)$

4. $2x^2 + x - 28 = (x + 4)(2x - 7)$

5. $20x^2 + 17x - 24 = (4x - 3)(5x + 8)$

6. $9x^2 + 27x - 162 = 9\left(x^2 + 3x - 18\right)$
$\qquad\qquad\qquad\quad = 9(x + 6)(x - 3)$

6.3   FACTORING TRINOMIALS OF THE FORM
$\quad$ ***ax*** ^{2} ***+ bx + c*, WHERE *a* $\neq$ 1**

1. $ax^2 + bx + c$, where $a \neq 1$

3. Find $m \cdot n = 5 \cdot 3 = 15$ and $m + n = 16$.
$\quad m = 15, n = 1$

$\quad 5x^2 + 16x + 3 = 5x^2 + 15x + x + 3$
$\qquad\qquad\qquad\quad = 5x(x + 3) + (x + 3)$
$\qquad\qquad\qquad\quad = (x + 3)(5x + 1)$

5. Find $m \cdot n = 4 \cdot 35 = 140$ and $m + n = 24$.
$m = 14, n = 10$

$$4x^2 + 24x + 35 = 4x^2 + 14x + 10x + 35$$
$$= 2x(2x + 7) + 5(2x + 7)$$
$$= (2x + 5)(2x + 7)$$

7. Find $m \cdot n = 6 \cdot 10 = 60$ and $m + n = -19$.
$m = -4, n = -15$

$$6x^2 - 19x + 10 = 6x^2 - 4x - 15x + 10$$
$$= 2x(3x - 2) - 5(3x - 2)$$
$$= (3x - 2)(2x - 5)$$

9. Find $m \cdot n = 14 \cdot (-9) = -126$ and $m + n = 15$.
$m = 21, n = -6$

$$14x^2 + 15x - 9 = 14x^2 + 21x - 6x - 9$$
$$= 7x(2x + 3) - 3(2x + 3)$$
$$= (7x - 3)(2x + 3)$$

11. Find $m \cdot n = 12 \cdot (-15) = -180$ and $m + n = 8$.
$m = 18, n = -10$

$$12x^2 + 8x - 15 = 12x^2 + 18x - 10x - 15$$
$$= 6x(2x + 3) - 5(2x + 3)$$
$$= (2x + 3)(6x - 5)$$

13. Find $m \cdot n = 12 \cdot 12 = 144$ and $m + n = -25$.
$m = -9, n = -16$

$$12x^2 - 25x + 12 = 12x^2 - 9x - 16x + 12$$
$$= 3x(4x - 3) - 4(4x - 3)$$
$$= (3x - 4)(4x - 3)$$

15. Find $m \cdot n = 2 \cdot 7 = 14$ and $m + n = 13$.
We cannot find two numbers with a product of
14 and a sum of 13. $2x^2 + 13x + 7$ is prime.

17. $3x^2 + 13x - 10 = (3x - 2)(x + 5)$

19. Find $m \cdot n = 12 \cdot (-56) = -672$ and $m + n = 11$.
$m = 32, n = -21$

$$12x^2 + 11x - 56 = 12x^2 + 32x - 21x - 56$$
$$= 4x(3x + 8) - 7(3x + 8)$$
$$= (3x + 8)(4x - 7)$$

21. $16x^2 - 24x + 9 = (4x - 3)(4x - 3) = (4x - 3)^2$

23. $15x^2 - 24x - 12 = 3(5x^2 - 8x - 4)$
$$= 3(5x + 2)(x - 2)$$

25. $16x^2 + 24x - 40 = 8(2x^2 + 3x - 5)$
$$= 8(2x + 5)(x - 1)$$

27. $16x^2 - 64x + 64 = 16(x^2 - 4x + 4)$
$$= 16(x - 2)(x - 2)$$
$$= 16(x - 2)^2$$

29. $15x^2 - 25x - 560 = 5(3x^2 - 5x - 112)$
$$= 5(3x + 16)(x - 7)$$

31. $x^2 + 14x + 13 = (x + 1)(x + 13)$

33. $18x - 9 = 9(2x - 1)$

35. $2x^2 - 7x + 6 = (2x - 3)(x - 2)$

37. Find $m \cdot n = 3 \cdot (-4) = -12$ and $m + n = 2$.
We cannot find two numbers with a product of
-12 and a sum of 2. $3x^2 + 2x - 4$ is prime.

39. $x^5 - 6x^4 = x^4(x - 6)$

41. $6x^2 - 48x = 6x(x - 8)$

43. $-x^2 - 4x + 77 = -\left(x^2 + 4x - 77\right)$
$$= -(x + 11)(x - 7)$$

45. Find $m \cdot n = 4 \cdot (-10) = -40$ and $m + n = -7$.
We cannot find two numbers with a product of
-40 and a sum of -7. $4x^2 - 7x - 10$ is prime.

47. $x^5 - 7x^3 + 4x^2 - 28 = x^3\left(x^2 - 7\right) + 4\left(x^2 - 7\right)$
$$= \left(x^2 - 7\right)\left(x^3 + 4\right)$$

49. $x^3 + 6x^2 - 15x - 90 = x^2(x + 6) - 15(x + 6)$
$$= (x + 6)\left(x^2 - 15\right)$$

51. $x^2 - 24x + 144 = (x - 12)(x - 12) = (x - 12)^2$

53. $9x^2 - 45x - 54 = 9\left(x^2 - 5x - 6\right)$
$$= 9(x - 6)(x + 1)$$

55. $-5x^2 + 28x + 12 = -\left(5x^2 - 28x - 12\right)$
$= -(5x+2)(x-6)$

57. $2x^2 - 15x - 27 = (x-9)(2x+3)$

59. $x^2 + 12x - 45 = (x+15)(x-3)$

61. $-3x^2 - 12x + 96 = -3\left(x^2 + 4x - 32\right)$
$= -3(x+8)(x-4)$

63. $15x^2 + 16x - 15 = (5x-3)(3x+5)$

65. $2x^2 + 9x + 4 = (2x+1)(x+4)$

67. Answers will vary.

6.4 QUICK CHECK

1. $x^2 - 25 = (x)^2 - (5)^2 = (x+5)(x-5)$

2. $81x^{12} - 100y^{16} = \left(9x^6\right)^2 - \left(10y^8\right)^2$
$= \left(9x^6 + 10y^8\right)\left(9x^6 - 10y^8\right)$

3. $7x^2 - 175 = 7\left(x^2 - 25\right)$
$= 7\left[(x)^2 - (5)^2\right]$
$= 7(x+5)(x-5)$

4. $x^4 - y^4 = \left(x^2\right)^2 - \left(y^2\right)^2$
$= \left(x^2 + y^2\right)\left(x^2 - y^2\right)$
$= \left(x^2 + y^2\right)\left[(x)^2 - (y)^2\right]$
$= \left(x^2 + y^2\right)(x+y)(x-y)$

5. $x^3 - 125 = (x)^3 - (5)^3$
$= (x-5)(x \cdot x + x \cdot 5 + 5 \cdot 5)$
$= (x-5)\left(x^2 + 5x + 25\right)$

6. $x^{12} - 8 = \left(x^4\right)^3 - (2)^3$
$= \left(x^4 - 2\right)\left(x^4 \cdot x^4 + x^4 \cdot 2 + 2 \cdot 2\right)$
$= \left(x^4 - 2\right)\left(x^8 + 2x^4 + 4\right)$

7. $x^3 + 216 = (x)^3 + (6)^3$
$= (x+6)(x \cdot x - x \cdot 6 + 6 \cdot 6)$
$= (x+6)\left(x^2 - 6x + 36\right)$

6.4 FACTORING SPECIAL BINOMIALS

1. difference of squares

3. multiples of 2

5. difference of cubes

7. $x^2 - 64 = (x)^2 - (8)^2 = (x+8)(x-8)$

9. $x^2 - 100 = (x)^2 - (10)^2 = (x+10)(x-10)$

11. $25a^2 - 49 = (5a)^2 - (7)^2 = (5a+7)(5a-7)$

13. $x^2 - 20$ is prime.

15. $a^2 - 16b^2 = (a)^2 - (4b)^2 = (a+4b)(a-4b)$

17. $25x^2 - 64y^2 = (5x)^2 - (8y)^2$
$= (5x+8y)(5x-8y)$

19. $16 - x^2 = (4)^2 - (x)^2 = (4+x)(4-x)$

21. $4x^2 - 196 = 4\left(x^2 - 49\right)$
$= 4\left[(x)^2 - (7)^2\right]$
$= 4(x+7)(x-7)$

23. $x^2 + 25$ is prime.

25. $5x^2 + 20 = 5\left(x^2 + 4\right)$

27. $x^4 - 16 = \left(x^2\right)^2 - (4)^2$
$= \left(x^2 + 4\right)\left(x^2 - 4\right)$
$= \left(x^2 + 4\right)\left[(x)^2 - (2)^2\right]$
$= \left(x^2 + 4\right)(x+2)(x-2)$

29. $x^4 - 81y^2 = \left(x^2\right)^2 - (9y)^2$
$= \left(x^2 + 9y\right)\left(x^2 - 9y\right)$

31. $x^3 - 1 = (x)^3 - (1)^3$
$\quad\quad = (x-1)(x \cdot x + x \cdot 1 + 1 \cdot 1)$
$\quad\quad = (x-1)(x^2 + x + 1)$

33. $x^3 - 8 = (x)^3 - (2)^3$
$\quad\quad = (x-2)(x \cdot x + x \cdot 2 + 2 \cdot 2)$
$\quad\quad = (x-2)(x^2 + 2x + 4)$

35. $1000 - y^3 = (10)^3 - (y)^3$
$\quad\quad = (10-y)(10 \cdot 10 + 10 \cdot y + y \cdot y)$
$\quad\quad = (10-y)(100 + 10y + y^2)$

37. $a^3 - 64b^3 = (a)^3 - (4b)^3$
$\quad\quad = (a-4b)(a \cdot a + a \cdot 4b + 4b \cdot 4b)$
$\quad\quad = (a-4b)(a^2 + 4ab + 16b^2)$

39. $125x^3 - 8y^3$
$\quad = (5x)^3 - (2y)^3$
$\quad = (5x-2y)(5x \cdot 5x + 5x \cdot 2y + 2y \cdot 2y)$
$\quad = (5x-2y)(25x^2 + 10xy + 4y^2)$

41. $x^3 + y^3 = (x)^3 + (y)^3$
$\quad\quad = (x+y)(x \cdot x - x \cdot y + y \cdot y)$
$\quad\quad = (x+y)(x^2 - xy + y^2)$

43. $x^3 + 8 = (x)^3 + (2)^3$
$\quad\quad = (x+2)(x \cdot x - x \cdot 2 + 2 \cdot 2)$
$\quad\quad = (x+2)(x^2 - 2x + 4)$

45. $x^6 + 27 = (x^2)^3 + (3)^3$
$\quad\quad = (x^2+3)(x^2 \cdot x^2 - x^2 \cdot 3 + 3 \cdot 3)$
$\quad\quad = (x^2+3)(x^4 - 3x^2 + 9)$

47. $64m^3 + n^3 = (4m)^3 + (n)^3$
$\quad\quad = (4m+n)(4m \cdot 4m - 4m \cdot n + n \cdot n)$
$\quad\quad = (4m+n)(16m^2 - 4mn + n^2)$

49. $3x^3 - 375 = 3(x^3 - 125)$
$\quad\quad = 3[(x)^3 - (5)^3]$
$\quad\quad = 3(x-5)(x \cdot x + x \cdot 5 + 5 \cdot 5)$
$\quad\quad = 3(x-5)(x^2 + 5x + 25)$

51. $320a^3 + 135b^3$
$\quad = 5(64a^3 + 27b^3)$
$\quad = 5[(4a)^3 + (3b)^3]$
$\quad = 5[(4a+3b)(4a \cdot 4a - 4a \cdot 3b + 3b \cdot 3b)]$
$\quad = 5[(4a+3b)(16a^2 - 12ab + 9b^2)]$

53. $x^2 - 16x + 64 = (x-8)(x-8) = (x-8)^2$

55. $x^2 + 16x + 28 = (x+2)(x+14)$

57. $22x^4 - 55x^2 = 11x^2(2x^2 - 5)$

59. $6x^2 - 5x - 25 = (2x-5)(3x+5)$

61. $x^5 - 16x^4 + 60x^3 = x^3(x^2 - 16x + 60)$
$\quad\quad\quad\quad\quad\quad\quad = x^3(x-10)(x-6)$

63. $-9x^2 - 225 = -9(x^2 + 25)$

65. $-6x^2 + 60x - 144 = -6(x^2 - 10x + 24)$
$\quad\quad\quad\quad\quad\quad\quad = -6(x-4)(x-6)$

67. $x^2 - 10x - 39 = (x-13)(x+3)$

69. $25x^2 - 36y^8 = (5x)^2 - (6y^4)^2$
$\quad\quad\quad\quad\quad = (5x+6y^4)(5x-6y^4)$

71. $2x^2 + x - 13$ is prime.

73. $x^2 - 64y^6 = (x)^2 - (8y^3)^2$
$\quad\quad\quad\quad = (x+8y^3)(x-8y^3)$

75. $8x^5 - 20x^4 - 52x^2 = 4x^2(2x^3 - 5x^2 - 13)$

77. $5x^2 + 45 = 5(x^2 + 9)$

79. $x^2 + 12x + 35 = (x+5)(x+7)$

81. $3x^2 - 13x - 38 = (x+2)(3x-19)$

83. $\begin{aligned} x^3 + 12x^2 - 6x - 72 &= x^2(x+12) - 6(x+12) \\ &= (x+12)(x^2 - 6) \end{aligned}$

85. $\begin{aligned} x^3 + 125y^3 &= (x)^3 + (5y)^3 \\ &= (x+5y)(x \cdot x - x \cdot 5y + 5y \cdot 5y) \\ &= (x+5y)(x^2 - 5xy + 25y^2) \end{aligned}$

87. $\begin{aligned} 4x^7 + 12x^6 + 9x^5 &= x^5(4x^2 + 12x + 9) \\ &= x^5(2x+3)(2x+3) \\ &= x^5(2x+3)^2 \end{aligned}$

89. d

91. f

93. g

95. b

97. Answers will vary. Example:
If there is subtraction between the two terms and each term can be written as a perfect square, then the binomial is a difference of squares.

6.5　QUICK CHECK

1. $\begin{aligned} -6x^2 + 54x + 60 &= -6(x^2 - 9x - 10) \\ &= -6(x-10)(x+1) \end{aligned}$

2. $\begin{aligned} 216x^{12}y^{21} + 125 \\ = (6x^4y^7)^3 + (5)^3 \\ = (6x^4y^7 + 5)(36x^8y^{14} - 30x^4y^7 + 25) \end{aligned}$

3. $\begin{aligned} x^4 + 5x^3 - 8x - 40 \\ = x^3(x+5) - 8(x+5) \\ = (x+5)(x^3 - 8) \\ = (x+5)\left[(x)^3 - (2)^3\right] \\ = (x+5)(x-2)(x^2 + 2x + 4) \end{aligned}$

4. $\begin{aligned} x^{12} - y^6 \\ = (x^6)^2 - (y^3)^2 \\ = (x^6 + y^3)(x^6 - y^3) \\ = \left[(x^2)^3 + (y)^3\right]\left[(x^2)^3 - (y)^3\right] \\ = (x^2 + y)(x^4 - x^2y + y^2)(x^2 - y)(x^4 + x^2y + y^2) \end{aligned}$

5. $\begin{aligned} x^6 + 64 &= (x^2)^3 + (4)^3 \\ &= (x^2 + 4)(x^4 - 4x^2 + 16) \end{aligned}$

6. $\begin{aligned} 8x^2 + 42x - 36 &= 2(4x^2 + 21x - 18) \\ &= 2(4x - 3)(x + 6) \end{aligned}$

6.5　FACTORING POLYNOMIALS: A GENERAL STRATEGY

1. common factors

3. sum of squares

5. sum of cubes

7. $x^7 + 18x^5 - 36x = x(x^6 + 18x^4 - 36)$

9. $84x^2 + 4x - 20 = 4(21x^2 + x - 5)$

11. $\begin{aligned} a^5b^7 - 3a^4b^6 + 8a^2b^9 \\ = a^2b^6(a^3b - 3a^2 + 8b^3) \end{aligned}$

13. $\begin{aligned} x^3 + 6x^2 + 5x + 30 &= x^2(x+6) + 5(x+6) \\ &= (x+6)(x^2 + 5) \end{aligned}$

15. $\begin{aligned} x^3 - 7x^2 - 3x + 21 &= x^2(x-7) - 3(x-7) \\ &= (x-7)(x^2 - 3) \end{aligned}$

17. $x^2 - 8x - 65 = (x - 13)(x + 5)$

19. $x^2 + 13x + 42 = (x + 6)(x + 7)$

21. $x^2 + 4x - 165 = (x + 15)(x - 11)$

23. $x^2 + 17x - 30$ is prime.

25. $x^2 - 23x + 42 = (x - 2)(x - 21)$

27. $x^2 + 20x + 100 = (x + 10)(x + 10) = (x + 10)^2$

29. $x^2 + 7xy - 60y^2 = (x + 12y)(x - 5y)$

31. $x^2y^2 + 17xy + 16 = (xy + 1)(xy + 16)$

33. $4x^2 + 56x + 192 = 4\left(x^2 + 14x + 48\right)$
$\qquad\qquad\qquad = 4(x+6)(x+8)$

35. $3x^2 - 42x + 135 = 3\left(x^2 - 14x + 45\right)$
$\qquad\qquad\qquad = 3(x-9)(x-5)$

37. $-10x^2 + 70x + 180 = -10\left(x^2 - 7x - 18\right)$
$\qquad\qquad\qquad\quad = -10(x-9)(x+2)$

39. $3x^2 + 10x - 48 = (3x-8)(x+6)$

41. $21x^2 + 31x + 4 = (3x+4)(7x+1)$

43. $2x^2 + 7x + 14$ is prime.

45. $12x^2 - 41x + 22 = (4x-11)(3x-2)$

47. $9x^2 - 30x + 25 = (3x-5)(3x-5) = (3x-5)^2$

49. $18x^2 - 60x + 42 = 6\left(3x^2 - 10x + 7\right)$
$\qquad\qquad\qquad = 6(3x-7)(x-1)$

51. $x^2 - 64 = (x)^2 - (8)^2 = (x+8)(x-8)$

53. $x^8 - 16y^6 = \left(x^4\right)^2 - \left(4y^3\right)^2$
$\qquad\qquad = \left(x^4 + 4y^3\right)\left(x^4 - 4y^3\right)$

55. $x^4 - 36x^2 = x^2\left(x^2 - 36\right)$
$\qquad\qquad = x^2\left[(x)^2 - (6)^2\right]$
$\qquad\qquad = x^2(x+6)(x-6)$

57. $x^4 + 81y^2$ is prime.

59. $x^3 - 1000 = (x)^3 - (10)^3$
$\qquad\qquad = (x-10)\left(x^2 + 10x + 100\right)$

61. $27x^3 - 8y^3 = (3x)^3 - (2y)^3$
$\qquad\qquad\quad = (3x-2y)\left(9x^2 + 6xy + 4y^2\right)$

63. $x^3 + 216 = (x)^3 + (6)^3 = (x+6)\left(x^2 - 6x + 36\right)$

65. $x^6 + 729y^3 = \left(x^2\right)^3 + (9y)^3$
$\qquad\qquad\quad = \left(x^2 + 9y\right)\left(x^4 - 9x^2y + 81y^2\right)$

67. $7x^3 + 189 = 7\left(x^3 + 27\right)$
$\qquad\qquad = 7\left[(x)^3 + (3)^3\right]$
$\qquad\qquad = 7(x+3)\left(x^2 - 3x + 9\right)$

69. $x^3 - 9x^2 - 4x + 36 = x^2(x-9) - 4(x-9)$
$\qquad\qquad\qquad\quad = (x-9)\left(x^2 - 4\right)$
$\qquad\qquad\qquad\quad = (x-9)\left[(x)^2 - (2)^2\right]$
$\qquad\qquad\qquad\quad = (x-9)(x+2)(x-2)$

71. $4x^3 - 28x^2 - x + 7 = 4x^2(x-7) - 1(x-7)$
$\qquad\qquad\qquad\quad = (x-7)\left(4x^2 - 1\right)$
$\qquad\qquad\qquad\quad = (x-7)\left[(2x)^2 - (1)^2\right]$
$\qquad\qquad\qquad\quad = (x-7)(2x+1)(2x-1)$

73. $x^4 + 16x^3 + x + 16 = x^3(x+16) + (x+16)$
$\qquad\qquad\qquad\quad = (x+16)\left(x^3 + 1\right)$
$\qquad\qquad\qquad\quad = (x+16)\left[(x)^3 + (1)^3\right]$
$\qquad\qquad\qquad\quad = (x+16)(x+1)\left(x^2 - x + 1\right)$

75. $x^8 - 1 = \left(x^4\right)^2 - (1)^2$
$\qquad\quad = \left(x^4 + 1\right)\left(x^4 - 1\right)$
$\qquad\quad = \left(x^4 + 1\right)\left[\left(x^2\right)^2 - (1)^2\right]$
$\qquad\quad = \left(x^4 + 1\right)\left(x^2 + 1\right)\left(x^2 - 1\right)$
$\qquad\quad = \left(x^4 + 1\right)\left(x^2 + 1\right)\left[(x)^2 - (1)^2\right]$
$\qquad\quad = \left(x^4 + 1\right)\left(x^2 + 1\right)(x+1)(x-1)$

77. $x^6 - y^6$
$\quad = \left(x^3\right)^2 - \left(y^3\right)^2$
$\quad = \left(x^3 + y^3\right)\left(x^3 - y^3\right)$
$\quad = \left[(x)^3 + (y)^3\right]\left[(x)^3 - (y)^3\right]$
$\quad = (x+y)\left(x^2 - xy + y^2\right)(x-y)\left(x^2 + xy + y^2\right)$

79. $(x+4)^2 - y^2 = (x+4+y)(x+4-y)$

81. $\left(x^2 - 10x + 25\right) - y^2 = (x-5)^2 - y^2$
$$= (x-5+y)(x-5-y)$$

83. $x^2 + 6x + 9 - y^2 = (x+3)^2 - y^2$
$$= (x+3+y)(x+3-y)$$

85. $4x^2 + 8x - 5 = (2x+5)(2x-1)$

87. $x^3 - 6x^2 + 8x - 48 = x^2(x-6) + 8(x-6)$
$$= (x-6)\left(x^2 + 8\right)$$

89. $x^2 + 11x - 60 = (x+15)(x-4)$

91. $x^2 + 13x - 36$ is prime.

93. $8x^3 + 343 = (2x)^3 + (7)^3$
$$= (2x+7)\left(4x^2 - 14x + 49\right)$$

95. $x^2 + 21x + 98 = (x+7)(x+14)$

97. $16x^4 - 81 = \left(4x^2\right)^2 - (9)^2$
$$= \left(4x^2 + 9\right)\left(4x^2 - 9\right)$$
$$= \left(4x^2 + 9\right)\left[(2x)^2 - (3)^2\right]$$
$$= \left(4x^2 + 9\right)(2x+3)(2x-3)$$

99. $6x^2 + 78x + 240 = 6\left(x^2 + 13x + 40\right)$
$$= 6(x+5)(x+8)$$

101. $9x^2 - 121y^6 = (3x)^2 - \left(11y^3\right)^2$
$$= \left(3x + 11y^3\right)\left(3x - 11y^3\right)$$

103. $x^3 + 4x^2 - 25x - 100 = x^2(x+4) - 25(x+4)$
$$= (x+4)\left(x^2 - 25\right)$$
$$= (x+4)\left[(x)^2 - (5)^2\right]$$
$$= (x+4)(x+5)(x-5)$$

105. $729x^3 - 512y^3$
$$= (9x)^3 - (8y)^3$$
$$= (9x - 8y)\left(81x^2 + 72xy + 64y^2\right)$$

107. $x^2 y^2 - 15xy + 54 = (xy)^2 - 15xy + 54$
$$= (xy - 6)(xy - 9)$$

109. Find $m \cdot n = 3 \cdot (-20) = -60$ and $m + n = -28$.
$m = -30, n = 2$
$$3x^2 - 28x - 20 = 3x^2 - 30x + 2x - 20$$
$$= 3x(x-10) + 2(x-10)$$
$$= (3x+2)(x-10)$$

111. $2x^2 + 19x + 25$ is prime.

113. $5x^5 + 20x^3 - 35x^2 = 5x^2\left(x^3 + 4x - 7\right)$

115. $64x^2 - 169 = (8x)^2 - (13)^2 = (8x+13)(8x-13)$

117. Answers will vary.

6.6   QUICK CHECK

1. a) $(x-2)(x+8) = 0$
$x - 2 = 0$ or $x + 8 = 0$
$x = 2$ or $x = -8$
$\{2, -8\}$

b) $x(4x+9) = 0$
$x = 0$ or $4x + 9 = 0$
$x = 0$ or $4x = -9$
$x = 0$ or $x = -\dfrac{9}{4}$
$\left\{0, -\dfrac{9}{4}\right\}$

2. $x^2 - x - 20 = 0$
$(x-5)(x+4) = 0$
$x - 5 = 0$ or $x + 4 = 0$
$x = 5$ or $x = -4$
$\{5, -4\}$

3. $4x^2 + 60x + 224 = 0$
$4\left(x^2 + 15x + 56\right) = 0$
$4(x+7)(x+8) = 0$
$x + 7 = 0$ or $x + 8 = 0$
$x = -7$ or $x = -8$
$\{-7, -8\}$

4. $6x^2 - 23x + 7 = 0$
$(2x-7)(3x-1) = 0$
$2x - 7 = 0$ or $3x - 1 = 0$
$2x = 7$ or $3x = 1$
$x = \dfrac{7}{2}$ or $x = \dfrac{1}{3}$
$\left\{\dfrac{7}{2}, \dfrac{1}{3}\right\}$

5.
$$x^2 + 4x = 45$$
$$x^2 + 4x - 45 = 0$$
$$(x+9)(x-5) = 0$$
$$x + 9 = 0 \quad \text{or} \quad x - 5 = 0$$
$$x = -9 \quad \text{or} \quad x = 5$$
$$\{-9, 5\}$$

6.
$$x(x-3) = 70$$
$$x^2 - 3x = 70$$
$$x^2 - 3x - 70 = 0$$
$$(x+7)(x-10) = 0$$
$$x + 7 = 0 \quad \text{or} \quad x - 10 = 0$$
$$x = -7 \quad \text{or} \quad x = 10$$
$$\{-7, 10\}$$

7.
$$\frac{2}{45}x^2 + \frac{4}{15}x - \frac{6}{5} = 0$$
$$45 \cdot \left(\frac{2}{45}x^2 + \frac{4}{15}x - \frac{6}{5} \right) = 45 \cdot 0$$
$$2x^2 + 12x - 54 = 0$$
$$2\left(x^2 + 6x - 27\right) = 0$$
$$2(x-3)(x+9) = 0$$
$$x - 3 = 0 \quad \text{or} \quad x + 9 = 0$$
$$x = 3 \quad \text{or} \quad x = -9$$
$$\{3, -9\}$$

8.
$$(x+6)(x-8) = 0$$
$$x^2 - 8x + 6x - 48 = 0$$
$$x^2 - 2x - 48 = 0$$

6.6 SOLVING QUADRATIC EQUATIONS BY FACTORING

1. quadratic equation

3. zero-factor

5.
$$(x+7)(x-100) = 0$$
$$x + 7 = 0 \quad \text{or} \quad x - 100 = 0$$
$$x = -7 \quad \text{or} \quad x = 100$$
$$\{-7, 100\}$$

7.
$$x(x-12) = 0$$
$$x = 0 \quad \text{or} \quad x - 12 = 0$$
$$x = 0 \quad \text{or} \quad x = 12$$
$$\{0, 12\}$$

9.
$$7(x+8)(x-6) = 0$$
$$x + 8 = 0 \quad \text{or} \quad x - 6 = 0$$
$$x = -8 \quad \text{or} \quad x = 6$$
$$\{-8, 6\}$$

11.
$$(x+7)(3x+4) = 0$$
$$x + 7 = 0 \quad \text{or} \quad 3x + 4 = 0$$
$$x = -7 \quad \text{or} \quad 3x = -4$$
$$x = -7 \quad \text{or} \quad x = -\frac{4}{3}$$
$$\left\{ -7, -\frac{4}{3} \right\}$$

13.
$$x^2 - 11x + 18 = 0$$
$$(x-2)(x-9) = 0$$
$$x - 2 = 0 \quad \text{or} \quad x - 9 = 0$$
$$x = 2 \quad \text{or} \quad x = 9$$
$$\{2, 9\}$$

15.
$$x^2 + 15x + 44 = 0$$
$$(x+11)(x+4) = 0$$
$$x + 11 = 0 \quad \text{or} \quad x + 4 = 0$$
$$x = -11 \quad \text{or} \quad x = -4$$
$$\{-11, -4\}$$

17.
$$x^2 + 6x - 40 = 0$$
$$(x+10)(x-4) = 0$$
$$x + 10 = 0 \quad \text{or} \quad x - 4 = 0$$
$$x = -10 \quad \text{or} \quad x = 4$$
$$\{-10, 4\}$$

19.
$$x^2 - 12x - 45 = 0$$
$$(x-15)(x+3) = 0$$
$$x - 15 = 0 \quad \text{or} \quad x + 3 = 0$$
$$x = 15 \quad \text{or} \quad x = -3$$
$$\{15, -3\}$$

21.
$$x^2 - 16x + 64 = 0$$
$$(x-8)(x-8) = 0$$
$$x - 8 = 0 \quad \text{or} \quad x - 8 = 0$$
$$x = 8 \quad \text{or} \quad x = 8$$
$$\{8\}$$

23.
$$x^2 - 81 = 0$$
$$(x+9)(x-9) = 0$$
$$x + 9 = 0 \quad \text{or} \quad x - 9 = 0$$
$$x = -9 \quad \text{or} \quad x = 9$$
$$\{-9, 9\}$$

25.
$$x^2 + 7x = 0$$
$$x(x+7) = 0$$
$$x = 0 \quad \text{or} \quad x + 7 = 0$$
$$x = 0 \quad \text{or} \quad x = -7$$
$$\{0, -7\}$$

27. $5x^2 - 22x = 0$
$x(5x - 22) = 0$

$x = 0$ or $5x - 22 = 0$
$x = 0$ or $5x = 22$
$x = 0$ or $x = \dfrac{22}{5}$
$\left\{0, \dfrac{22}{5}\right\}$

29. $3x^2 - 3x - 270 = 0$
$3\left(x^2 - x - 90\right) = 0$
$3(x + 9)(x - 10) = 0$

$x + 9 = 0$ or $x - 10 = 0$
$x = -9$ or $x = 10$
$\{-9, 10\}$

31. $-x^2 + 12x - 35 = 0$
$-\left(x^2 - 12x + 35\right) = 0$
$-(x - 5)(x - 7) = 0$

$x - 5 = 0$ or $x - 7 = 0$
$x = 5$ or $x = 7$
$\{5, 7\}$

33. $3x^2 + 22x - 16 = 0$
$(3x - 2)(x + 8) = 0$

$3x - 2 = 0$ or $x + 8 = 0$
$3x = 2$ or $x = -8$
$x = \dfrac{2}{3}$ or $x = -8$
$\left\{-8, \dfrac{2}{3}\right\}$

35. $27x^2 - 3x - 2 = 0$
$(9x + 2)(3x - 1) = 0$

$9x + 2 = 0$ or $3x - 1 = 0$
$9x = -2$ or $3x = 1$
$x = -\dfrac{2}{9}$ or $x = \dfrac{1}{3}$
$\left\{-\dfrac{2}{9}, \dfrac{1}{3}\right\}$

37. $x^2 - 2x = 35$
$x^2 - 2x - 35 = 0$
$(x - 7)(x + 5) = 0$

$x - 7 = 0$ or $x + 5 = 0$
$x = 7$ or $x = -5$
$\{7, -5\}$

39. $x^2 = 8x - 16$
$x^2 - 8x = -16$
$x^2 - 8x + 16 = 0$
$(x - 4)(x - 4) = 0$

$x - 4 = 0$ or $x - 4 = 0$
$x = 4$ or $x = 4$
$\{4\}$

41. $x^2 + 12x = 3x - 18$
$x^2 + 9x = -18$
$x^2 + 9x + 18 = 0$
$(x + 6)(x + 3) = 0$

$x + 6 = 0$ or $x + 3 = 0$
$x = -6$ or $x = -3$
$\{-6, -3\}$

43. $x^2 - 3x - 7 = 4x + 11$
$x^2 - 7x - 7 = 11$
$x^2 - 7x - 18 = 0$
$(x + 2)(x - 9) = 0$

$x + 2 = 0$ or $x - 9 = 0$
$x = -2$ or $x = 9$
$\{-2, 9\}$

45. $x^2 = 9$
$x^2 - 9 = 0$
$(x + 3)(x - 3) = 0$

$x + 3 = 0$ or $x - 3 = 0$
$x = -3$ or $x = 3$
$\{-3, 3\}$

47. $x^2 + 11x = 11x + 36$
$x^2 = 36$
$x^2 - 36 = 0$
$(x + 6)(x - 6) = 0$

$x + 6 = 0$ or $x - 6 = 0$
$x = -6$ or $x = 6$
$\{-6, 6\}$

49. $x(x + 4) = 4(x + 16)$
$x^2 + 4x = 4x + 64$
$x^2 = 64$
$x^2 - 64 = 0$
$(x + 8)(x - 8) = 0$

$x + 8 = 0$ or $x - 8 = 0$
$x = -8$ or $x = 8$
$\{-8, 8\}$

51.
$$x(x-7)=30$$
$$x^2-7x=30$$
$$x^2-7x-30=0$$
$$(x+3)(x-10)=0$$
$$x+3=0 \quad \text{or} \quad x-10=0$$
$$x=-3 \quad \text{or} \quad x=10$$
$$\{-3,10\}$$

53.
$$(x-5)(x+6)=-18$$
$$x^2+6x-5x-30=-18$$
$$x^2+x-30=-18$$
$$x^2+x-12=0$$
$$(x+4)(x-3)=0$$
$$x+4=0 \quad \text{or} \quad x-3=0$$
$$x=-4 \quad \text{or} \quad x=3$$
$$\{-4,3\}$$

55.
$$(x+2)(x+3)=(x+7)(x-4)$$
$$x^2+3x+2x+6=x^2-4x+7x-28$$
$$x^2+5x+6=x^2+3x-28$$
$$5x+6=3x-28$$
$$2x+6=-28$$
$$2x=-34$$
$$x=-17$$
$$\{-17\}$$

57.
$$\frac{1}{6}x^2-\frac{3}{2}x+3=0$$
$$6\cdot\left(\frac{1}{6}x^2-\frac{3}{2}x+3\right)=6\cdot0$$
$$x^2-9x+18=0$$
$$(x-3)(x-6)=0$$
$$x-3=0 \quad \text{or} \quad x-6=0$$
$$x=3 \quad \text{or} \quad x=6$$
$$\{3,6\}$$

59.
$$\frac{1}{6}x^2-\frac{5}{4}x-\frac{9}{4}=0$$
$$12\cdot\left(\frac{1}{6}x^2-\frac{5}{4}x-\frac{9}{4}\right)=12\cdot0$$
$$2x^2-15x-27=0$$
$$(2x+3)(x-9)=0$$
$$2x+3=0 \quad \text{or} \quad x-9=0$$
$$2x=-3 \quad \text{or} \quad x=9$$
$$x=-\frac{3}{2} \quad \text{or} \quad x=9$$
$$\left\{-\frac{3}{2},9\right\}$$

61.
$$\frac{2}{15}x^2-\frac{7}{10}x+\frac{2}{3}=0$$
$$30\cdot\left(\frac{2}{15}x^2-\frac{7}{10}x+\frac{2}{3}\right)=30\cdot0$$
$$4x^2-21x+20=0$$
$$(4x-5)(x-4)=0$$
$$4x-5=0 \quad \text{or} \quad x-4=0$$
$$4x=5 \quad \text{or} \quad x=4$$
$$x=\frac{5}{4} \quad \text{or} \quad x=4$$
$$\left\{\frac{5}{4},4\right\}$$

63.
$$x=4 \quad \text{or} \quad x=5$$
$$x-4=0 \quad \text{or} \quad x-5=0$$
$$(x-4)(x-5)=0$$
$$x^2-5x-4x+20=0$$
$$x^2-9x+20=0$$

65.
$$x=0 \quad \text{or} \quad x=6$$
$$x=0 \quad \text{or} \quad x-6=0$$
$$x(x-6)=0$$
$$x^2-6x=0$$

67.
$$x=-5 \quad \text{or} \quad x=5$$
$$x+5=0 \quad \text{or} \quad x-5=0$$
$$(x+5)(x-5)=0$$
$$x^2-5x+5x-25=0$$
$$x^2-25=0$$

69.
$$x=-\frac{2}{5} \quad \text{or} \quad x=\frac{1}{2}$$
$$5x=-2 \quad \text{or} \quad 2x=1$$
$$5x+2=0 \quad \text{or} \quad 2x-1=0$$
$$(5x+2)(2x-1)=0$$
$$10x^2-5x+4x-2=0$$
$$10x^2-x-2=0$$

71.
$$x=\frac{7}{2}$$
$$2x=7$$
$$2x-7=0$$
$$\text{so}\,(2x-7)(3x+14)=6x^2+7x-98$$
$$(3x+14)=0$$
$$3x=-14$$
$$x=-\frac{14}{3}$$

73.
$$x=23$$
$$x-23=0$$
$$\text{and }(x-23)(x+29)=x^2+6x-667$$
$$\text{so, }x+29=0$$
$$x=-29$$

75.
$$x = -\frac{19}{3}$$
$$3x = -19$$
$$3x + 19 = 0$$
and $(3x+19)(5x-28) = 15x^2 + 11x - 532$
so, $5x - 28 = 0$
$$5x = 28$$
$$x = \frac{28}{5}$$

77. Answers will vary. Example:
If the product of two numbers is zero then one or the other is equal to zero. When solving a quadratic equation, you need to factor and then apply the zero-factor property.

79. Answers will vary.

6.7   QUICK CHECK

1. $f(-6) = (-6)^2 - 9(-6) + 405$
$ = 36 + 54 + 405$
$ = 495$

2. $f(x) = x^2 - 17x + 72 = 0$
$f(x) = (x-8)(x-9) = 0$
$x - 8 = 0$ or $x - 9 = 0$
$ x = 8$ or $ x = 9$
$\{8, 9\}$

3. $f(x) = x^2 - 6x + 20 = 92$
$f(x) = x^2 - 6x - 72 = 0$
$f(x) = (x+6)(x-12) = 0$
$x + 6 = 0$ or $x - 12 = 0$
$ x = -6$ or $ x = 12$
$\{-6, 12\}$

4. a) 48 feet

   b) 3 seconds

   c) 1 second

   d) 64 feet

6.7   QUADRATIC FUNCTIONS

1. quadratic function

3. $f(6) = (6)^2 + 8(6) + 20 = 36 + 48 + 20 = 104$

5. $f(-3) = (-3)^2 + 7(-3) - 33$
$ = 9 - 21 + -33$
$ = -45$

7. $f(5) = (5)^2 - 7(5) + 10 = 25 - 35 + 10 = 0$

9. $g(-15) = (-15)^2 + 12(-15) - 45$
$ = 225 - 180 - 45$
$ = 0$

11. $g(7) = 3(7)^2 + 2(7) - 14$
$ = 3(49) + 14 - 14$
$ = 147$

13. $f(-3) = -2(-3)^2 - 9(-3) + 8$
$ = -2(9) - 9(-3) + 8$
$ = -18 + 27 + 8$
$ = -18 + 35$
$ = 17$

15. $h(4) = -16(4)^2 + 96(4) + 32$
$ = -16(16) + 96(4) + 32$
$ = -256 + 384 + 32$
$ = -256 + 416$
$ = 160$

17. $f(4a) = (4a)^2 + 4(4a) - 20$
$ = 16a^2 + 4(4a) - 20$
$ = 16a^2 + 16a - 20$

19. $f(a+5) = (a+5)^2 - 7(a+5) - 25$
$ = (a+5)(a+5) - 7(a+5) - 25$
$ = a^2 + 5a + 5a + 25 - 7a - 35 - 25$
$ = a^2 + 10a - 7a + 25 - 60$
$ = a^2 + 3a - 35$

21. $f(x) = x^2 + x - 42 = 0$
$ (x+7)(x-6) = 0$
$x + 7 = 0$ or $x - 6 = 0$
$ x = -7$ or $ x = 6$

23. $f(x) = x^2 - 9x + 20 = 0$
$ (x-4)(x-5) = 0$
$x - 4 = 0$ or $x - 5 = 0$
$ x = 4$ or $ x = 5$

25. $f(x) = x^2 - 23x + 144 = 18$
$$x^2 - 23x + 126 = 0$$
$$(x-9)(x-14) = 0$$
$$x - 9 = 0 \quad \text{or} \quad x - 14 = 0$$
$$x = 9 \quad \text{or} \quad x = 14$$

27. $f(x) = x^2 + 11 = 60$
$$x^2 - 49 = 0$$
$$(x+7)(x-7) = 0$$
$$x + 7 = 0 \quad \text{or} \quad x - 7 = 0$$
$$x = -7 \quad \text{or} \quad x = 7$$

29. $x^2 + 7x - 16 = x$
$$x^2 + 6x - 16 = 0$$
$$(x+8)(x-2) = 0$$
$$x + 8 = 0 \quad \text{or} \quad x - 2 = 0$$
$$x = -8 \quad \text{or} \quad x = 2$$

31. $x^2 + 14x + 40 = x$
$$x^2 + 13x + 40 = 0$$
$$(x+5)(x+8) = 0$$
$$x + 5 = 0 \quad \text{or} \quad x + 8 = 0$$
$$x = -5 \quad \text{or} \quad x = -8$$

33. a) $h(t) = 400 - 16t^2 = 0$
$$h(t) = -16(t^2 - 25) = 0$$
$$h(t) = -16(t+5)(t-5) = 0$$
$$t + 5 = 0 \quad \text{or} \quad t - 5 = 0$$
$$t = -5 \quad \text{or} \quad t = 5$$
5 seconds

b) $h(2) = 400 - 16(2)^2$
$$= 400 - 16(4)$$
$$= 400 - 64$$
$$= 336$$
336 feet

35. a) $t = 3$
$$h(3) = -16(3)^2 + 32(3) + 240$$
$$= -16(9) + 32(3) + 240$$
$$= -144 + 96 + 240$$
$$= -144 + 336$$
$$= 192$$
192 feet

b) $h(t) = -16t^2 + 32t + 240 = 0$
$$h(t) = -16(t^2 - 2t - 15) = 0$$
$$h(t) = -16(t-5)(t+3) = 0$$
$$t - 5 = 0 \quad \text{or} \quad t + 3 = 0$$
$$t = 5 \quad \text{or} \quad t = -3$$
5 seconds

c) 1 second

d) $t = 1$
$$h(1) = -16(1)^2 + 32(1) + 240$$
$$= -16 + 32 + 240$$
$$= -16 + 272$$
$$= 256$$
256 feet

37. a) When $x = 0, f(0) = -70$.
Start up costs are $70.

b) When $f(x) = 0, x = 5$.
5 mugs

c) 20 mugs give the maximum profit.

d) When $x = 25, f(25) = 80$.
$80 profit

6.8   QUICK CHECK

1. <u>Unknowns</u> #1: x
 #2: $x + 1$
$$x(x+1) = 132$$
$$x^2 + x = 132$$
$$x^2 + x - 132 = 0$$
$$(x+12)(x-11) = 0$$
$$x + 12 = 0 \quad \text{or} \quad x - 11 = 0$$
$$x = -12 \quad \text{or} \quad x = 11$$
#1: $x = 11$
#2: $x + 1 = 11 + 1 = 12$
11 and 12

2. <u>Unknowns</u> #1: x
 #2: $21 - x$
$$x(21 - x) = 108$$
$$21x - x^2 = 108$$
$$21x = x^2 + 108$$
$$0 = x^2 - 21x + 108$$
$$0 = (x-9)(x-12)$$
$$0 = x - 9 \quad \text{or} \quad 0 = x - 12$$
$$9 = x \quad \text{or} \quad 12 = x$$
#1: $x = 9$
#2: $21 - x = 21 - 9 = 12$
9 and 12

3. <u>Unknowns</u> length: x
width: $x - 8$

$$x(x-8) = 105$$
$$x^2 - 8x = 105$$
$$x^2 - 8x - 105 = 0$$
$$(x-15)(x+7) = 0$$
$$x - 15 = 0 \quad \text{or} \quad x + 7 = 0$$
$$x = 15 \quad \text{or} \quad x = -7$$

length: $x = 15$ feet
width: $x - 8 = 15 - 8 = 7$ feet

4. <u>Unknowns</u> length: $3 + x$
width: x

$$(x+2+2)(3+x+2+2) = 270$$
$$(x+4)(x+7) = 270$$
$$x^2 + 7x + 4x + 28 = 270$$
$$x^2 + 11x - 242 = 0$$
$$(x+22)(x-11) = 0$$
$$x + 22 = 0 \quad \text{or} \quad x - 11 = 0$$
$$x = -22 \quad \text{or} \quad x = 11$$

width: $x = 11$ feet
length: $x + 3 = 11 + 3 = 14$ feet

5. a) $v_o = 144, s = 0$
$$h(t) = -16t^2 + 144t$$

b) $h(t) = -16t^2 + 144 = 0$
$$h(t) = -16t(t-9) = 0$$
$$t = 0 \quad \text{or} \quad t - 9 = 0$$
$$t = 0 \quad \text{or} \quad t = 9$$
9 seconds

c) $[0, 9]$

6. a) $f(n) = \dfrac{1}{2}n^2 + \dfrac{1}{2}n$
$$n = 100$$
$$f(100) = \frac{1}{2}(100)^2 + \frac{1}{2}(100)$$
$$= \frac{1}{2}(10,000) + \frac{1}{2}(100)$$
$$= 5000 + 50$$
$$= 5050$$

b) $f(n) = \dfrac{1}{2}n^2 + \dfrac{1}{2}n$
$$f(n) = 45$$
$$\frac{1}{2}n^2 + \frac{1}{2}n = 45$$
$$2 \cdot \left(\frac{1}{2}n^2 + \frac{1}{2}n\right) = 2 \cdot 45$$
$$n^2 + n = 90$$
$$n^2 + n - 90 = 0$$
$$(n+10)(n-9) = 0$$
$$n + 10 = 0 \quad \text{or} \quad n - 9 = 0$$
$$n = -10 \quad \text{or} \quad n = 9$$
9

6.8   APPLICATIONS OF QUADRATIC EQUATIONS AND QUADRATIC FUNCTIONS

1. $x;\ x+1;\ x+2$

3. Area $=$ Length $\cdot$ Width

5. <u>Unknowns</u> #1: x
#2: $x + 1$

$$x(x+1) = 132$$
$$x^2 + x = 132$$
$$x^2 + x - 132 = 0$$
$$(x-11)(x+12) = 0$$
$$x - 11 = 0 \quad \text{or} \quad x + 12 = 0$$
$$x = 11 \quad \text{or} \quad x = -12$$

#1: $x = 11$
#2: $x + 1 = 11 + 1 = 12$

7. <u>Unknowns</u> #1: x
#2: $x + 2$

$$x(x+2) = 528$$
$$x^2 + 2x = 528$$
$$x^2 + 2x - 528 = 0$$
$$(x-22)(x+24) = 0$$
$$x - 22 = 0 \quad \text{or} \quad x + 24 = 0$$
$$x = 22 \quad \text{or} \quad x = -24$$

#1: $x = 22$
#2: $x + 2 = 22 + 2 = 24$

9. Unknowns #1: x
#2: $x + 2$

$$x(x + 2) = 255$$
$$x^2 + 2x = 255$$
$$x^2 + 2x - 255 = 0$$
$$(x + 17)(x - 15) = 0$$
$$x + 17 = 0 \quad \text{or} \quad x - 15 = 0$$
$$x = -17 \quad \text{or} \quad x = 15$$

#1: $x = 15$
#2: $x + 2 = 15 + 2 = 17$

11. Unknowns #1: x
#2: $x + 1$

$$x(x + 1) = 27 + 5(x + 1)$$
$$x^2 + x = 27 + 5x + 5$$
$$x^2 + x = 32 + 5x$$
$$x^2 - 4x = 32$$
$$x^2 - 4x - 32 = 0$$
$$(x - 8)(x + 4) = 0$$
$$x - 8 = 0 \quad \text{or} \quad x + 4 = 0$$
$$x = 8 \quad \text{or} \quad x = -4$$

#1: $x = 8$
#2: $x + 1 = 8 + 1 = 9$

13. Unknowns #1: x
#2: $x + 2$

$$x(x + 2) = 398 + (x + x + 2)$$
$$x^2 + 2x = 398 + 2x + 2$$
$$x^2 + 2x = 400 + 2x$$
$$x^2 = 400$$
$$x^2 - 400 = 0$$
$$(x + 20)(x - 20) = 0$$
$$x + 20 = 0 \quad \text{or} \quad x - 20 = 0$$
$$x = -20 \quad \text{or} \quad x = 20$$

#1: $x = 20$
#2: $x + 2 = 20 + 2 = 22$

15. Unknowns #1: x
#2: $x + 5$

$$x(x + 5) = 66$$
$$x^2 + 5x = 66$$
$$x^2 + 5x - 66 = 0$$
$$(x + 11)(x - 6) = 0$$
$$x + 11 = 0 \quad \text{or} \quad x - 6 = 0$$
$$x = -11 \quad \text{or} \quad x = 6$$

#1: $x = 6$
#2: $x + 5 = 6 + 5 = 11$

17. Unknowns #1: x
#2: $3x - 4$

$$x(3x - 4) = 84$$
$$3x^2 - 4x = 84$$
$$3x^2 - 4x - 84 = 0$$
$$(3x + 14)(x - 6) = 0$$
$$3x + 14 = 0 \quad \text{or} \quad x - 6 = 0$$
$$3x = -14 \quad \text{or} \quad x = 6$$
$$x = -\frac{14}{3} \quad \text{or} \quad x = 6$$

#1: $x = 6$
#2: $3x - 4 = 3(6) - 4 = 14$

19. Unknowns #1: x
#2: $29 - x$

$$x(29 - x) = 210$$
$$29x - x^2 = 210$$
$$29x = x^2 + 210$$
$$0 = x^2 - 29x + 210$$
$$0 = (x - 14)(x - 15)$$
$$0 = x - 14 \quad \text{or} \quad 0 = x - 15$$
$$14 = x \quad \text{or} \quad 15 = x$$

#1: $x = 14$
#2: $29 - x = 29 - 14 = 15$

21. Unknowns #1: x
#2: $25 - x$

$$x(25 - x) = 66$$
$$25x - x^2 = 66$$
$$25x = x^2 + 66$$
$$0 = x^2 - 25x + 66$$
$$0 = (x - 3)(x - 22)$$
$$0 = x - 3 \quad \text{or} \quad 0 = x - 22$$
$$3 = x \quad \text{or} \quad 22 = x$$

#1: $x = 3$
#2: $25 - x = 25 - 3 = 22$

23. Unknowns #1: x
#2: $x + 3$

$$x(x + 3) = 88$$
$$x^2 + 3x = 88$$
$$x^2 + 3x - 88 = 0$$
$$(x + 11)(x - 8) = 0$$
$$x + 11 = 0 \quad \text{or} \quad x - 8 = 0$$
$$x = -11 \quad \text{or} \quad x = 8$$

#1: $x = 8$
#2: $x + 3 = 8 + 3 = 11$

25. Zorayda's age: x
Gabriela's age: $x + 5$

$$x(x+5) = 126$$
$$x^2 + 5x = 126$$
$$x^2 + 5x - 126 = 0$$
$$(x-9)(x+14) = 0$$
$$x - 9 = 0 \quad \text{or} \quad x + 14 = 0$$
$$x = 9 \quad \text{or} \quad x = -14$$

Gabriela's age: $x + 5 = 9 + 5 = 14$

27. Clyde's age: $2x + 5$
Bonnie's age: x

$$x(2x+5) = 168$$
$$2x^2 + 5x = 168$$
$$2x^2 + 5x - 168 = 0$$
$$(2x+21)(x-8) = 0$$
$$2x + 21 = 0 \quad \text{or} \quad x - 8 = 0$$
$$2x = -21 \quad \text{or} \quad x = 8$$
$$x = -\frac{21}{2} \quad \text{or} \quad x = 8$$

Clyde's age: $2x + 5 = 2(8) + 5 = 21$

29. Unknowns width: x
length: $x + 7$

$$x(x+7) = 120$$
$$x^2 + 7x = 120$$
$$x^2 + 7x - 120 = 0$$
$$(x+15)(x-8) = 0$$
$$x + 15 = 0 \quad \text{or} \quad x - 8 = 0$$
$$x = -15 \quad \text{or} \quad x = 8$$

width: $x = 8$ inches
length: $x + 7 = 8 + 7 = 15$ inches

31. Unknowns width: x
length: $2x + 2$

$$x(2x+2) = 420$$
$$2x^2 + 2x = 420$$
$$2x^2 + 2x - 420 = 0$$
$$2(x^2 + x - 210) = 0$$
$$2(x+15)(x-14) = 0$$
$$x + 15 = 0 \quad \text{or} \quad x - 14 = 0$$
$$x = -15 \quad \text{or} \quad x = 14$$

width: $x = 14$ centimeters
length: $2x + 2 = 2(14) + 2$
$$= 28 + 2$$
$$= 30 \text{ centimeters}$$

33. Unknowns length: $2x + 1$
width: x

$$x(2x+1) = 105$$
$$2x^2 + x = 105$$
$$2x^2 + x - 105 = 0$$
$$(2x+15)(x-7) = 0$$
$$2x + 15 = 0 \quad \text{or} \quad x - 7 = 0$$
$$2x = -15 \quad \text{or} \quad x = 7$$
$$x = -\frac{15}{2} \quad \text{or} \quad x = 7$$

length: $2x + 1 = 2(7) + 1 = 15$ meters
width: $x = 7$ meters

35. Unknowns length: $2x - 8$
width: x

$$x(2x-8) = 960$$
$$2x^2 - 8x = 960$$
$$2x^2 - 8x - 960 = 0$$
$$2(x^2 - 4x - 480) = 0$$
$$2(x+20)(x-24) = 0$$
$$x + 20 = 0 \quad \text{or} \quad x - 24 = 0$$
$$x = -20 \quad \text{or} \quad x = 24$$

length: $2x - 8 = 2(24) - 8 = 40$ feet
width: $x = 24$ feet

37. Unknowns length: x
width: $x - 3$

$$(x+1+1)(x-3+1+1) = 70$$
$$(x+2)(x-1) = 70$$
$$x^2 - x + 2x - 2 = 70$$
$$x^2 + x - 72 = 0$$
$$(x-8)(x+9) = 0$$
$$x - 8 = 0 \quad \text{or} \quad x + 9 = 0$$
$$x = 8 \quad \text{or} \quad x = -9$$

length: $x = 8$ inches
width: $x - 3 = 8 - 3 = 5$ inches

39. Unknowns length: $2x - 1$
width: x

$$(2x-1+2+2)(x+2+2) = 117$$
$$(2x+3)(x+4) = 117$$
$$2x^2 + 8x + 3x + 12 = 117$$
$$2x^2 + 11x - 105 = 0$$
$$(2x+21)(x-5) = 0$$
$$2x + 21 = 0 \quad \text{or} \quad x - 5 = 0$$
$$2x = -21 \quad \text{or} \quad x = 5$$
$$x = -\frac{21}{2} \quad \text{or} \quad x = 5$$

length: $2x - 1 = 2(5) - 1 = 9$ feet
width: $x = 5$ feet

41. Unknown length: x

$$x^2 = 64$$
$$x^2 - 64 = 0$$
$$(x+8)(x-8) = 0$$
$$x+8 = 0 \quad \text{or} \quad x-8 = 0$$
$$x = -8 \quad \text{or} \quad x = 8$$

length: $x = 8$ meters

43. Unknowns base: x
height: $2x+5$

$$\frac{1}{2}(x)(2x+5) = 84$$
$$2 \cdot \frac{1}{2}(x)(2x+5) = 2 \cdot 84$$
$$(x)(2x+5) = 168$$
$$2x^2 + 5x = 168$$
$$2x^2 + 5x - 168 = 0$$
$$(2x+21)(x-8) = 0$$
$$2x+21 = 0 \quad \text{or} \quad x-8 = 0$$
$$2x = -21 \quad \text{or} \quad x = 8$$
$$x = -\frac{21}{2} \quad \text{or} \quad x = 8$$

base: $x = 8$ feet
height: $2x+5 = 2(8)+5 = 21$ feet

45. $v_o = 128, s = 320, h(t) = 0$
$$-16t^2 + 128t + 320 = 0$$
$$-16\left(t^2 - 8t - 20\right) = 0$$
$$-16(t-10)(t+2) = 0$$
$$t-10 = 0 \quad \text{or} \quad t+2 = 0$$
$$t = 10 \quad \text{or} \quad t = -2$$
10 seconds

47. $v_o = 64, s = 80, h(t) = 0$
$$-16t^2 + 64t + 80 = 0$$
$$-16\left(t^2 - 4t - 5\right) = 0$$
$$-16(t-5)(t+1) = 0$$
$$t-5 = 0 \quad \text{or} \quad t+1 = 0$$
$$t = 5 \quad \text{or} \quad t = -1$$
5 seconds

49. $v_o = 64, s = 40, h(t) = 88$
$$-16t^2 + 64t + 40 = 88$$
$$-16t^2 + 64t - 48 = 0$$
$$-16\left(t^2 - 4t + 3\right) = 0$$
$$-16(t-3)(t-1) = 0$$
$$t-3 = 0 \quad \text{or} \quad t-1 = 0$$
$$t = 3 \quad \text{or} \quad t = 1$$
1 second, 3 seconds

51. $v_o = 128, s = 0, h(t) = 0$
$$-16t^2 + 128t + 0 = 0$$
$$-16t(t-8) = 0$$
$$t = 0 \quad \text{or} \quad t-8 = 0$$
$$t = 0 \quad \text{or} \quad t = 8$$
8 seconds

53. $s = 0, t = 5, h(5) = 0$
$$-16(5)^2 + v_o(5) + 0 = 0$$
$$-16(25) + 5v_o = 0$$
$$-400 + 5v_o = 0$$
$$5v_o = 400$$
$$v_o = 80$$
80 feet/second

55. $s = 144, h(t) = 0$
$$-16t^2 + 144 = 0$$
$$-16\left(t^2 - 9\right) = 0$$
$$-16(t+3)(t-3) = 0$$
$$t+3 = 0 \quad \text{or} \quad t-3 = 0$$
$$t = -3 \quad \text{or} \quad t = 3$$
3 seconds

57. $f(n) = \frac{1}{2}n^2 + \frac{1}{2}n$

$f(n) = 36$

$$\frac{1}{2}n^2 + \frac{1}{2}n = 36$$
$$2 \cdot \left(\frac{1}{2}n^2 + \frac{1}{2}n\right) = 2 \cdot 36$$
$$n^2 + n = 72$$
$$n^2 + n - 72 = 0$$
$$(n+9)(n-8) = 0$$
$$n+9 = 0 \quad \text{or} \quad n-8 = 0$$
$$n = -9 \quad \text{or} \quad n = 8$$
8

59. $f(n) = \frac{1}{2}n^2 + \frac{1}{2}n$

$f(n) = 210$

$$\frac{1}{2}n^2 + \frac{1}{2}n = 210$$
$$2 \cdot \left(\frac{1}{2}n^2 + \frac{1}{2}n\right) = 2 \cdot 210$$
$$n^2 + n = 420$$
$$n^2 + n - 420 = 0$$
$$(n+21)(n-20) = 0$$
$$n+21 = 0 \quad \text{or} \quad n-20 = 0$$
$$n = -21 \quad \text{or} \quad n = 20$$
20

61. Answers will vary. Example:
George has a square shaped garden. If he widens the width by 5 feet and extends the length by 9 feet, the area of the garden is now 192 square feet. Find the original length of the garden.

CHAPTER 6 REVIEW

1. $3x - 21 = 3(x - 7)$

2. $x^2 - 20x = x(x - 20)$

3. $5x^7 + 15x^4 - 20x^3 = 5x^3\left(x^4 + 3x - 4\right)$

4. $6a^5b^2 - 10a^3b^3 + 15a^4b$
$= a^3b\left(6a^2b - 10b^2 + 15a\right)$

5. $x^3 + 6x^2 + 4x + 24 = x^2(x + 6) + 4(x + 6)$
$= (x + 6)\left(x^2 + 4\right)$

6. $x^3 + 9x^2 - 7x - 63 = x^2(x + 9) - 7(x + 9)$
$= (x + 9)\left(x^2 - 7\right)$

7. $x^3 - 3x^2 - 11x + 33 = x^2(x - 3) - 11(x - 3)$
$= (x - 3)\left(x^2 - 11\right)$

8. $x^3 - 3x^2 - 9x + 27 = x^2(x - 3) - 9(x - 3)$
$= (x - 3)\left(x^2 - 9\right)$
$= (x - 3)(x + 3)(x - 3)$
$= (x - 3)^2(x + 3)$

9. $x^2 - 8x + 16 = (x - 4)(x - 4) = (x - 4)^2$

10. $x^2 + 3x - 28 = (x + 7)(x - 4)$

11. $x^2 - 11x + 24 = (x - 3)(x - 8)$

12. $-5x^2 - 20x + 160 = -5\left(x^2 + 4x - 32\right)$
$= -5(x + 8)(x - 4)$

13. $x^2 + 14xy + 45y^2 = (x + 5y)(x + 9y)$

14. $x^2 + 9x - 23$ is prime.

15. $x^2 + 17x + 30 = (x + 2)(x + 15)$

16. $x^2 - x - 42 = (x - 7)(x + 6)$

17. $10x^2 - 83x + 24 = (x - 8)(10x - 3)$

18. $4x^2 - 3x - 10 = (4x + 5)(x - 2)$

19. $18x^2 + 12x - 6 = 6\left(3x^2 + 2x - 1\right)$
$= 6(3x - 1)(x + 1)$

20. $4x^2 - 12x + 9 = (2x - 3)(2x - 3) = (2x - 3)^2$

21. $3x^2 - 75 = 3\left(x^2 - 25\right)$
$= 3\left[(x)^2 - (5)^2\right]$
$= 3(x + 5)(x - 5)$

22. $4x^2 - 25y^2 = (2x)^2 - (5y)^2$
$= (2x + 5y)(2x - 5y)$

23. $x^3 + 8y^3 = (x)^3 + (2y)^3$
$= (x + 2y)\left(x^2 - 2xy + 4y^2\right)$

24. $x^2 + 121$ is prime.

25. $x^3 - 216 = (x)^3 - (6)^3$
$= (x - 6)\left(x^2 + 6x + 36\right)$

26. $7x^3 - 189 = 7\left(x^3 - 27\right)$
$= 7\left[(x)^3 - (3)^3\right]$
$= 7(x - 3)\left(x^2 + 3x + 9\right)$

27. $(x - 5)(x + 7) = 0$
$x - 5 = 0$ or $x + 7 = 0$
$x = 5$ or $x = -7$
$\{5, -7\}$

28. $(3x - 8)(2x + 1) = 0$
$3x - 8 = 0$ or $2x + 1 = 0$
$3x = 8$ or $2x = -1$
$x = \dfrac{8}{3}$ or $x = -\dfrac{1}{2}$
$\left\{\dfrac{8}{3}, -\dfrac{1}{2}\right\}$

29. $x^2 - 25 = 0$
$(x+5)(x-5) = 0$
$x+5 = 0$ or $x-5 = 0$
$x = -5$ or $x = 5$
$\{-5, 5\}$

30. $x^2 - 14x + 40 = 0$
$(x-4)(x-10) = 0$
$x-4 = 0$ or $x-10 = 0$
$x = 4$ or $x = 10$
$\{4, 10\}$

31. $x^2 + 16x + 64 = 0$
$(x+8)(x+8) = 0$
$x+8 = 0$ or $x+8 = 0$
$x = -8$ or $x = -8$
$\{-8\}$

32. $x^2 + 3x - 108 = 0$
$(x+12)(x-9) = 0$
$x+12 = 0$ or $x-9 = 0$
$x = -12$ or $x = 9$
$\{-12, 9\}$

33. $x^2 - 9x - 70 = 0$
$(x-14)(x+5) = 0$
$x-14 = 0$ or $x+5 = 0$
$x = 14$ or $x = -5$
$\{14, -5\}$

34. $x^2 - 16x = 0$
$x(x-16) = 0$
$x = 0$ or $x-16 = 0$
$x = 0$ or $x = 16$
$\{0, 16\}$

35. $x^2 + 19x + 60 = 0$
$(x+4)(x+15) = 0$
$x+4 = 0$ or $x+15 = 0$
$x = -4$ or $x = -15$
$\{-4, -15\}$

36. $x^2 - 18x + 80 = 0$
$(x-8)(x-10) = 0$
$x-8 = 0$ or $x-10 = 0$
$x = 8$ or $x = 10$
$\{8, 10\}$

37. $x^2 - 15x - 16 = 0$
$(x-16)(x+1) = 0$
$x-16 = 0$ or $x+1 = 0$
$x = 16$ or $x = -1$
$\{16, -1\}$

38. $2x^2 - 7x - 60 = 0$
$(2x-15)(x+4) = 0$
$2x-15 = 0$ or $x+4 = 0$
$2x = 15$ or $x = -4$
$x = \dfrac{15}{2}$ or $x = -4$
$\left\{\dfrac{15}{2}, -4\right\}$

39. $7x^2 - 252 = 0$
$7\left(x^2 - 36\right) = 0$
$7(x+6)(x-6) = 0$
$x+6 = 0$ or $x-6 = 0$
$x = -6$ or $x = 6$
$\{-6, 6\}$

40. $4x^2 - 60x + 144 = 0$
$4\left(x^2 - 15x + 36\right) = 0$
$4(x-3)(x-12) = 0$
$x-3 = 0$ or $x-12 = 0$
$x = 3$ or $x = 12$
$\{3, 12\}$

41. $x(x+10) = -21$
$x^2 + 10x = -21$
$x^2 + 10x + 21 = 0$
$(x+3)(x+7) = 0$
$x+3 = 0$ or $x+7 = 0$
$x = -3$ or $x = -7$
$\{-3, -7\}$

42. $(x+4)(x+5) = 72$
$x^2 + 5x + 4x + 20 = 72$
$x^2 + 9x + 20 = 72$
$x^2 + 9x - 52 = 0$
$(x+13)(x-4) = 0$
$x+13 = 0$ or $x-4 = 0$
$x = -13$ or $x = 4$
$\{-13, 4\}$

43. $x = 6$ or $x = 7$
$x-6 = 0$ or $x-7 = 0$
$(x-6)(x-7) = 0$
$x^2 - 7x - 6x + 42 = 0$
$x^2 - 13x + 42 = 0$

44.
$$x = -4 \quad \text{or} \quad x = 4$$
$$x + 4 = 0 \quad \text{or} \quad x - 4 = 0$$
$$(x + 4)(x - 4) = 0$$
$$x^2 - 4x + 4x - 16 = 0$$
$$x^2 - 16 = 0$$

45.
$$x = 2 \quad \text{or} \quad x = -\frac{3}{5}$$
$$x - 2 = 0 \quad \text{or} \quad 5x = -3$$
$$x - 2 = 0 \quad \text{or} \quad 5x + 3 = 0$$
$$(x - 2)(5x + 3) = 0$$
$$5x^2 + 3x - 10x - 6 = 0$$
$$5x^2 - 7x - 6 = 0$$

46.
$$x = -\frac{5}{4} \quad \text{or} \quad x = \frac{2}{7}$$
$$4x = -5 \quad \text{or} \quad 7x = 2$$
$$4x + 5 = 0 \quad \text{or} \quad 7x - 2 = 0$$
$$(4x + 5)(7x - 2) = 0$$
$$28x^2 - 8x + 35x - 10 = 0$$
$$28x^2 + 27x - 10 = 0$$

47. $f(-3) = (-3)^2 = 9$

48. $f(4) = (4)^2 - 8(4) = 16 - 32 = -16$

49.
$$f(-8) = (-8)^2 + 10(-8) + 24$$
$$= 64 - 80 + 24$$
$$= 88 - 80$$
$$= 8$$

50.
$$f(5) = 3(5)^2 + 8(5) - 19$$
$$= 3(25) + 8(5) - 19$$
$$= 75 + 40 - 19$$
$$= 115 - 19$$
$$= 96$$

51.
$$x^2 - 36 = 0$$
$$(x + 6)(x - 6) = 0$$
$$x + 6 = 0 \quad \text{or} \quad x - 6 = 0$$
$$x = -6 \quad \text{or} \quad x = 6$$
$$\{-6, 6\}$$

52.
$$x^2 + 3x - 54 = 0$$
$$(x + 9)(x - 6) = 0$$
$$x + 9 = 0 \quad \text{or} \quad x - 6 = 0$$
$$x = -9 \quad \text{or} \quad x = 6$$
$$\{-9, 6\}$$

53.
$$x^2 + 12x + 20 = 0$$
$$(x + 2)(x + 10) = 0$$
$$x + 2 = 0 \quad \text{or} \quad x + 10 = 0$$
$$x = -2 \quad \text{or} \quad x = -10$$
$$\{-2, -10\}$$

54.
$$x^2 - 7x - 44 = 0$$
$$(x - 11)(x + 4) = 0$$
$$x - 11 = 0 \quad \text{or} \quad x + 4 = 0$$
$$x = 11 \quad \text{or} \quad x = -4$$
$$\{11, -4\}$$

55. a) 80 feet

b) 4 seconds

c) 1 second

d)
$$f(1) = -16(1)^2 + 32(1) + 128$$
$$= -16 + 32 + 128$$
$$= -16 + 160$$
$$= 144$$
144 feet

56. a) $f(800)$
$$= 0.0001(800)^2 - 0.08(800) + 20.85$$
$$= 0.0001(640,000) - 0.08(800) + 20.85$$
$$= 64 - 64 + 20.85$$
$$= 20.85$$
$20.85

b) 400

c) $f(400)$
$$= 0.0001(400)^2 - 0.08(400) + 20.85$$
$$= 0.0001(160,000) - 0.08(400) + 20.85$$
$$= 16 - 32 + 20.85$$
$$= 36.85 - 32$$
$$= 4.85$$
$4.85

57. <u>Unknowns</u> #1: x
#2: $x + 2$
$$x(x + 2) = 288$$
$$x^2 + 2x = 288$$
$$x^2 + 2x - 288 = 0$$
$$(x + 18)(x - 16) = 0$$
$$x + 18 = 0 \quad \text{or} \quad x - 16 = 0$$
$$x = -18 \quad \text{or} \quad x = 16$$
#1: $x = 16$
#2: $x + 2 = 16 + 2 = 18$

58. Rosa's age: $x - 8$
Dale's age: x

$$x(x-8) = 105$$
$$x^2 - 8x = 105$$
$$x^2 - 8x - 105 = 0$$
$$(x+7)(x-15) = 0$$
$$x + 7 = 0 \quad \text{or} \quad x - 15 = 0$$
$$x = -7 \quad \text{or} \quad x = 15$$

Dale's age: $x = 15$
Rosa's age: $x - 8 = 15 - 8 = 7$

59. Unknowns length: $2x - 3$
width: x

$$x(2x-3) = 104$$
$$2x^2 - 3x = 104$$
$$2x^2 - 3x - 104 = 0$$
$$(2x+13)(x-8) = 0$$
$$2x + 13 = 0 \quad \text{or} \quad x - 8 = 0$$
$$2x = -13 \quad \text{or} \quad x = 8$$
$$x = -\frac{13}{2} \quad \text{or} \quad x = 8$$

length: $2x - 3 = 2(8) - 3 = 13$ meters
width: $x = 8$ meters

60. a) $v_o = 80, s = 384, h(t) = 0$
$$-16t^2 + 80t + 384 = 0$$
$$-16(t^2 - 5t - 24) = 0$$
$$-16(t-8)(t+3) = 0$$
$$t - 8 = 0 \quad \text{or} \quad t + 3 = 0$$
$$t = 8 \quad \text{or} \quad t = -3$$
8 seconds

b) $v_o = 80, s = 384, h(t) = 480$
$$-16t^2 + 80t + 384 = 480$$
$$-16t^2 + 80t - 96 = 0$$
$$-16(t^2 - 5t + 6) = 0$$
$$-16(t-2)(t-3) = 0$$
$$t - 2 = 0 \quad \text{or} \quad t - 3 = 0$$
$$t = 2 \quad \text{or} \quad t = 3$$
2 seconds, 3 seconds

CHAPTER 6 TEST

1. $x^2 - 25x = x(x - 25)$

2. $x^3 - 3x^2 - 5x + 15 = x^2(x-3) - 5(x-3)$
$$= (x-3)(x^2 - 5)$$

3. $x^2 - 9x + 20 = (x-4)(x-5)$

4. $x^2 - 14x - 72 = (x-18)(x+4)$

5. $3x^2 + 60x + 57 = 3(x^2 + 20x + 19)$
$$= 3(x+1)(x+19)$$

6. $6x^2 - x - 5 = (6x+5)(x-1)$

7. $x^2 - 25y^2 = (x)^2 - (5y)^2 = (x+5y)(x-5y)$

8. $x^3 - 216y^3 = (x)^3 - (6y)^3$
$$= (x-6y)(x^2 + 6xy + 36y^2)$$

9.
$$x^2 - 100 = 0$$
$$(x+10)(x-10) = 0$$
$$x + 10 = 0 \quad \text{or} \quad x - 10 = 0$$
$$x = -10 \quad \text{or} \quad x = 10$$
$$\{-10, 10\}$$

10. $x^2 - 13x + 36 = 0$
$$(x-4)(x-9) = 0$$
$$x - 4 = 0 \quad \text{or} \quad x - 9 = 0$$
$$x = 4 \quad \text{or} \quad x = 9$$
$$\{4, 9\}$$

11. $x^2 + 7x - 60 = 0$
$$(x+12)(x-5) = 0$$
$$x + 12 = 0 \quad \text{or} \quad x - 5 = 0$$
$$x = -12 \quad \text{or} \quad x = 5$$
$$\{-12, 5\}$$

12.
$$(x+2)(x-9) = 60$$
$$x^2 - 9x + 2x - 18 = 60$$
$$x^2 - 7x - 78 = 0$$
$$(x+6)(x-13) = 0$$
$$x + 6 = 0 \quad \text{or} \quad x - 13 = 0$$
$$x = -6 \quad \text{or} \quad x = 13$$
$$\{-6, 13\}$$

13.
$$x = -2 \quad \text{or} \quad x = 2$$
$$x + 2 = 0 \quad \text{or} \quad x - 2 = 0$$
$$(x+2)(x-2) = 0$$
$$x^2 - 2x + 2x - 4 = 0$$
$$x^2 - 4 = 0$$

14.
$$x = -6 \quad \text{or} \quad x = \frac{9}{5}$$
$$x + 6 = 0 \quad \text{or} \quad 5x = 9$$
$$x + 6 = 0 \quad \text{or} \quad 5x - 9 = 0$$
$$(x+6)(5x-9) = 0$$
$$5x^2 - 9x + 30x - 54 = 0$$
$$5x^2 + 21x - 54 = 0$$

15.
$$f(-5) = (-5)^2 - 12(-5) + 35$$
$$= 25 + 60 + 35$$
$$= 120$$

16.
$$f(7) = -(7)^2 + 2(7) + 17$$
$$= -49 + 14 + 17$$
$$= -49 + 31$$
$$= -18$$

17.
$$x^2 + 14x + 48 = 0$$
$$(x+6)(x+8) = 0$$
$$x + 6 = 0 \quad \text{or} \quad x + 8 = 0$$
$$x = -6 \quad \text{or} \quad x = -8$$
$$\{-6, -8\}$$

18. **a)**
$$f(6500) = 0.00000006(6500)^2$$
$$-0.00096(6500) + 5.95$$
$$= 2.535 - 6.24 + 5.95$$
$$= 8.485 - 6.24$$
$$= 2.245$$
$2.245

b) 8000

c)
$$f(8000) = 0.00000006(8000)^2$$
$$-0.00096(8000) + 5.95$$
$$= 3.84 - 7.68 + 5.95$$
$$= 9.79 - 7.68$$
$$= 2.11$$
$2.11

19. <u>Unknowns</u> length: $3x + 5$
width: x
$$x(3x+5) = 100$$
$$3x^2 + 5x = 100$$
$$3x^2 + 5x - 100 = 0$$
$$(3x+20)(x-5) = 0$$
$$3x + 20 = 0 \quad \text{or} \quad x - 5 = 0$$
$$3x = -20 \quad \text{or} \quad x = 5$$
$$x = -\frac{20}{3} \quad \text{or} \quad x = 5$$
length: $3x + 5 = 3(5) + 5 = 20$ feet
width: $x = 5$ feet

20. **a)** $v_o = 32, s = 240, h(t) = 0$
$$-16t^2 + 32t + 240 = 0$$
$$-16\left(t^2 - 2t - 15\right) = 0$$
$$-16(t-5)(t+3) = 0$$
$$t - 5 = 0 \quad \text{or} \quad t + 3 = 0$$
$$t = 5 \quad \text{or} \quad t = -3$$
5 seconds

b) $v_o = 32, s = 240, h(t) = 192$
$$-16t^2 + 32t + 240 = 192$$
$$-16t^2 + 32t + 48 = 0$$
$$-16\left(t^2 - 2t - 3\right) = 0$$
$$-16(t-3)(t+1) = 0$$
$$t - 3 = 0 \quad \text{or} \quad t + 1 = 0$$
$$t = 3 \quad \text{or} \quad t = -1$$
3 seconds

CHAPTER 7   RATIONAL EXPRESSIONS AND EQUATIONS

7.1   QUICK CHECK

1. $\dfrac{(-4)^2 - 3(-4) - 18}{(-4)^2 - 5(-4) - 6} = \dfrac{16 + 12 - 18}{16 + 20 - 6} = \dfrac{10}{30} = \dfrac{1}{3}$

2. $2x - 7 = 0$
$\quad\ 2x = 7$
$\quad\ \ x = \dfrac{7}{2}$

3. $x^2 - 8x - 9 = 0$
$(x - 9)(x + 1) = 0$
$x - 9 = 0 \ \text{ or } \ x + 1 = 0$
$\quad x = 9 \ \text{ or } \quad\ x = -1$

4. $r(-6) = \dfrac{(-6)^2 - 3(-6) - 12}{(-6)^2 + 7(-6) - 15}$
$\qquad = \dfrac{36 + 18 - 12}{36 - 42 - 15}$
$\qquad = \dfrac{42}{-21}$
$\qquad = -2$

5. $x^2 - 4x - 45 = 0$
$(x - 9)(x + 5) = 0$
$x - 9 = 0 \ \text{ or } \ x + 5 = 0$
$\quad x = 9 \ \text{ or } \quad\ x = -5$
All real numbers except 9 and -5.
$(-\infty, -5) \cup (-5, 9) \cup (9, \infty)$

6. $\dfrac{15x^4}{21x^3} = \dfrac{\cancel{15}^{\,5}\, x^4}{\cancel{21}_{\,7}\, x^3} = \dfrac{5x^4}{7x^3} = \dfrac{5x}{7}$

7. $\dfrac{x^2 + 10x + 24}{x^2 - 2x - 48} = \dfrac{(x + 6)(x + 4)}{(x - 8)(x + 6)}$
$\qquad = \dfrac{\cancel{(x+6)}^{\,1}(x+4)}{(x-8)\,\cancel{(x+6)}_{\,1}}$
$\qquad = \dfrac{x+4}{x-8}$

8. $\dfrac{3x^2 + 16x + 5}{x^2 + 8x + 15} = \dfrac{(x+5)(3x+1)}{(x+3)(x+5)}$
$\qquad = \dfrac{\cancel{(x+5)}^{\,1}(3x+1)}{(x+3)\,\cancel{(x+5)}_{\,1}}$
$\qquad = \dfrac{3x+1}{x+3}$

9. $\dfrac{x^2 + 8x - 9}{1 - x^2} = \dfrac{(x+9)(x-1)}{(1+x)(1-x)}$
$\qquad = \dfrac{(x+9)\,\cancel{(x-1)}^{\,1}}{(1+x)\,\cancel{(1-x)}_{\,-1}}$
$\qquad = -\dfrac{x+9}{1+x}$

7.1   RATIONAL EXPRESSIONS AND FUNCTIONS

1. rational expression

3. rational function

5. lowest terms

7. $\dfrac{6}{(4) + 4} = \dfrac{6}{8} = \dfrac{3}{4}$

9. $\dfrac{(-25) + 3}{(-25) - 8} = \dfrac{-22}{-33} = \dfrac{2}{3}$

11. $\dfrac{(3)^2 - 13(3) - 48}{(3)^2 - 6(3) - 12} = \dfrac{9 - 39 - 48}{9 - 18 - 12}$
$\qquad = \dfrac{-78}{-21}$
$\qquad = \dfrac{26}{7}$

13. $\dfrac{(-9)^2 - 6(-9) + 15}{(-9)^2 + 14(-9) - 7} = \dfrac{81 + 54 + 15}{81 - 126 - 7}$
$\qquad = \dfrac{150}{-52}$
$\qquad = -\dfrac{75}{26}$

15. $x - 5 = 0$
$x = 5$

17. $x^2 + x - 56 = 0$
$(x + 8)(x - 7) = 0$
$x + 8 = 0 \quad$ or $\quad x - 7 = 0$
$x = -8 \quad$ or $\quad x = 7$

19. $3x^2 + 13x - 10 = 0$
$(3x - 2)(x + 5) = 0$
$3x - 2 = 0 \quad$ or $\quad x + 5 = 0$
$3x = 2 \quad$ or $\quad x = -5$
$x = \dfrac{2}{3} \quad$ or $\quad x = -5$

21. $x^2 - 36 = 0$
$(x + 6)(x - 6) = 0$
$x + 6 = 0 \quad$ or $\quad x - 6 = 0$
$x = -6 \quad$ or $\quad x = 6$

23. $r(-5) = \dfrac{20}{(-5)^2 - 5(-5) + 10}$
$= \dfrac{20}{25 + 25 + 10}$
$= \dfrac{20}{60}$
$= \dfrac{1}{3}$

25. $r(4) = \dfrac{(4)^2 + 10(4) + 24}{(4)^2 - 5(4) - 66}$
$= \dfrac{16 + 40 + 24}{16 - 20 - 66}$
$= \dfrac{80}{-70}$
$= -\dfrac{8}{7}$

27. $r(10) = \dfrac{(10)^3 - 7(10)^2 - 11(10) + 20}{(10)^2 + 8(10) - 20}$
$= \dfrac{1000 - 7(100) - 11(10) + 20}{100 + 8(10) - 20}$
$= \dfrac{1000 - 700 - 110 + 20}{100 + 80 - 20}$
$= \dfrac{210}{160}$
$= \dfrac{21}{16}$

29. $x^2 + 10x = 0$
$x(x + 10) = 0$
$x = 0 \quad$ or $\quad x + 10 = 0$
$x = 0 \quad$ or $\quad x = -10$
All real numbers except 0 and -10.
$(-\infty, -10) \cup (-10, 0) \cup (0, \infty)$

31. $x^2 - 2x - 15 = 0$
$(x - 5)(x + 3) = 0$
$x - 5 = 0 \quad$ or $\quad x + 3 = 0$
$x = 5 \quad$ or $\quad x = -3$
All real numbers except 5 and -3.
$(-\infty, -3) \cup (-3, 5) \cup (5, \infty)$

33. $x^2 + 3x - 18 = 0$
$(x - 3)(x + 6) = 0$
$x - 3 = 0 \quad$ or $\quad x + 6 = 0$
$x = 3 \quad$ or $\quad x = -6$
All real numbers except 3 and -6.
$(-\infty, -6) \cup (-6, 3) \cup (3, \infty)$

35. Quadratic.

37. Rational.

39. Linear.

41. $\dfrac{3x^5}{9x^8} = \dfrac{\cancel{3}^1 x^5}{\cancel{9}_3 x^8} = \dfrac{x^5}{3x^8} = \dfrac{1}{3x^3}$

43. $\dfrac{x + 5}{x^2 + 11x + 30} = \dfrac{\cancel{x + 5}^1}{(x + 6)\cancel{(x + 5)}_1} = \dfrac{1}{(x + 6)}$

45. $\dfrac{x^2 + 11x + 28}{x^2 + 4x - 21} = \dfrac{(x + 4)\cancel{(x + 7)}^1}{(x - 3)\cancel{(x + 7)}_1}$
$= \dfrac{x + 4}{x - 3}$

47. $\dfrac{x^2 - 4x}{x^2 - 14x + 40} = \dfrac{x\cancel{(x - 4)}^1}{\cancel{(x - 4)}_1 (x - 10)} = \dfrac{x}{x - 10}$

49. $\dfrac{x^2 - 4x - 45}{x^3 + 125} = \dfrac{(x - 9)\cancel{(x + 5)}^1}{\cancel{(x + 5)}_1 (x^2 - 5x + 25)}$
$= \dfrac{x - 9}{x^2 - 5x + 25}$

51. $\dfrac{2x^2+13x+6}{2x^2-7x-4}=\dfrac{\cancel{(2x+1)}^{1}(x+6)}{(x-4)\cancel{(2x+1)}_{1}}=\dfrac{x+6}{x-4}$

53. $\dfrac{3x^2-25x-50}{x^2-3x-70}=\dfrac{\cancel{(x-10)}^{1}(3x+5)}{\cancel{(x-10)}_{1}(x+7)}=\dfrac{3x+5}{x+7}$

55. $\dfrac{3x^2-6x-105}{x^2-11x+28}=\dfrac{3\left(x^2-2x-35\right)}{x^2-11x+28}$

$\qquad =\dfrac{3\cancel{(x-7)}^{1}(x+5)}{\cancel{(x-7)}_{1}(x-4)}$

$\qquad =\dfrac{3(x+5)}{x-4}$

57. Not opposites.

59. Opposites.

61. Opposites.

63. $\dfrac{49-x^2}{x^2-12x+35}=\dfrac{-\left(x^2-49\right)}{x^2-12x+35}$

$\qquad =\dfrac{-(x+7)\cancel{(x-7)}^{1}}{(x-5)\cancel{(x-7)}_{1}}$

$\qquad =-\dfrac{x+7}{x-5}$

65. $\dfrac{8x-x^2}{x^2-19x+88}=\dfrac{-\left(x^2-8x\right)}{x^2-19x+88}$

$\qquad =\dfrac{-x\cancel{(x-8)}^{1}}{(x-11)\cancel{(x-8)}_{1}}$

$\qquad =-\dfrac{x}{x-11}$

67. $\dfrac{x^3-8}{4-x^2}=\dfrac{x^3-8}{-\left(x^2-4\right)}$

$\qquad =\dfrac{\cancel{(x-2)}^{1}\left(x^2+2x+4\right)}{-(x+2)\cancel{(x-2)}_{1}}$

$\qquad =-\dfrac{x^2+2x+4}{x+2}$

69. a) $r(1)=4$

b) $r(-4)=-1$

c) $r(4)=1;\ 4$

71. Answers will vary. Example:
Set denominator equal to zero and solve.

73. Answers will vary. Example:
Two factors are opposites if they differ by a multiple of -1. $x-2$ is the opposite of $2-x$ since $-1(x-2)=2-x$ and $-1(2-x)=x-2$.

7.2 QUICK CHECK

1. $\dfrac{x^2-3x-18}{x^2-9x+20}\cdot\dfrac{x^2-3x-10}{x^2+13x+30}$

$=\dfrac{(x-6)(x+3)}{(x-5)(x-4)}\cdot\dfrac{(x-5)(x+2)}{(x+3)(x+10)}$

$=\dfrac{(x-6)\cancel{(x+3)}^{1}}{\cancel{(x-5)}_{1}(x-4)}\cdot\dfrac{\cancel{(x-5)}^{1}(x+2)}{\cancel{(x+3)}_{1}(x+10)}$

$=\dfrac{(x-6)(x+2)}{(x-4)(x+10)}$

2. $\dfrac{36-x^2}{x^2+9x+14}\cdot\dfrac{x^2-7x-18}{x^2-15x+54}$

$=\dfrac{(6-x)(6+x)}{(x+7)(x+2)}\cdot\dfrac{(x-9)(x+2)}{(x-9)(x-6)}$

$=\dfrac{\cancel{(6-x)}^{-1}(6+x)}{(x+7)\cancel{(x+2)}_{1}}\cdot\dfrac{\cancel{(x-9)}^{1}\cancel{(x+2)}^{1}}{\cancel{(x-9)}_{1}\cancel{(x-6)}_{1}}$

$=-\dfrac{6+x}{x+7}$

3. $f(x)\cdot g(x)$

$=\dfrac{x+4}{49-x^2}\cdot\dfrac{x^2-10x+21}{x^2+x-12}$

$=\dfrac{x+4}{(7+x)(7-x)}\cdot\dfrac{(x-7)(x-3)}{(x+4)(x-3)}$

$=\dfrac{\cancel{x+4}^{1}}{(7+x)\cancel{(7-x)}_{-1}}\cdot\dfrac{\cancel{(x-7)}^{1}\cancel{(x-3)}^{1}}{\cancel{(x+4)}_{1}\cancel{(x-3)}_{1}}$

$=-\dfrac{1}{7+x}$

4. $\dfrac{x^2-4}{x^2-5x+4} \div \dfrac{x^2-6x-16}{x^2+3x-28}$

$= \dfrac{x^2-4}{x^2-5x+4} \cdot \dfrac{x^2+3x-28}{x^2-6x-16}$

$= \dfrac{(x+2)(x-2)}{(x-4)(x-1)} \cdot \dfrac{(x+7)(x-4)}{(x-8)(x+2)}$

$= \dfrac{\cancel{(x+2)}^{\,1}(x-2)}{\cancel{(x-4)}_{\,1}(x-1)} \cdot \dfrac{(x+7)\cancel{(x-4)}^{\,1}}{(x-8)\cancel{(x+2)}_{\,1}}$

$= \dfrac{(x-2)(x+7)}{(x-1)(x-8)}$

5. $\dfrac{x^2-13x+36}{x^2-x-12} \div \dfrac{9-x}{x^2+8x+15}$

$= \dfrac{x^2-13x+36}{x^2-x-12} \cdot \dfrac{x^2+8x+15}{9-x}$

$= \dfrac{(x-9)(x-4)}{(x-4)(x+3)} \cdot \dfrac{(x+3)(x+5)}{9-x}$

$= \dfrac{\cancel{(x-9)}^{\,1}\cancel{(x-4)}^{\,1}}{\cancel{(x-4)}_{\,1}\cancel{(x+3)}_{\,1}} \cdot \dfrac{\cancel{(x+3)}^{\,1}(x+5)}{\cancel{9-x}_{\,-1}}$

$= -(x+5)$

6. $f(x) \div g(x)$

$= \dfrac{x^2-12x+36}{x^2-11x+10} \div \dfrac{x^2-8x+12}{1}$

$= \dfrac{x^2-12x+36}{x^2-11x+10} \cdot \dfrac{1}{x^2-8x+12}$

$= \dfrac{(x-6)(x-6)}{(x-1)(x-10)} \cdot \dfrac{1}{(x-6)(x-2)}$

$= \dfrac{(x-6)\cancel{(x-6)}^{\,1}}{(x-1)(x-10)} \cdot \dfrac{1}{\cancel{(x-6)}_{\,1}(x-2)}$

$= \dfrac{x-6}{(x-1)(x-10)(x-2)}$

7.2 MULTIPLICATION AND DIVISION OF RATIONAL EXPRESSIONS

1. factoring

3. factored

5. $\dfrac{x^2+15x+54}{x^2+2x-15} \cdot \dfrac{x^2-25}{x^2+4x-45}$

$= \dfrac{(x+6)(x+9)}{(x+5)(x-3)} \cdot \dfrac{(x+5)(x-5)}{(x+9)(x-5)}$

$= \dfrac{(x+6)\cancel{(x+9)}^{\,1}}{\cancel{(x+5)}_{\,1}(x-3)} \cdot \dfrac{\cancel{(x+5)}^{\,1}\cancel{(x-5)}^{\,1}}{\cancel{(x+9)}_{\,1}\cancel{(x-5)}_{\,1}}$

$= \dfrac{x+6}{x-3}$

7. $\dfrac{121-x^2}{x^2-22x+120} \cdot \dfrac{x^2-24x+140}{x^2-18x+77}$

$= \dfrac{(11+x)(11-x)}{(x-10)(x-12)} \cdot \dfrac{(x-10)(x-14)}{(x-7)(x-11)}$

$= \dfrac{(11+x)\cancel{(11-x)}^{\,1}}{\cancel{(x-10)}_{\,1}(x-12)} \cdot \dfrac{\cancel{(x-10)}^{\,1}(x-14)}{(x-7)\cancel{(x-11)}_{\,-1}}$

$= -\dfrac{(11+x)(x-14)}{(x-12)(x-7)}$

9. $\dfrac{x^2-8x-33}{x^2-8x+16} \cdot \dfrac{2x^2-7x-4}{x^2+5x+6}$

$= \dfrac{(x-11)(x+3)}{(x-4)(x-4)} \cdot \dfrac{(2x+1)(x-4)}{(x+2)(x+3)}$

$= \dfrac{(x-11)\cancel{(x+3)}^{\,1}}{(x-4)\cancel{(x-4)}_{\,1}} \cdot \dfrac{(2x+1)\cancel{(x-4)}^{\,1}}{(x+2)\cancel{(x+3)}_{\,1}}$

$= \dfrac{(x-11)(2x+1)}{(x-4)(x+2)}$

11. $\dfrac{x^2-11x}{x^2+5x+4}\cdot\dfrac{x^2+10x+9}{x^2+6x}$

$=\dfrac{x(x-11)}{(x+4)(x+1)}\cdot\dfrac{(x+9)(x+1)}{x(x+6)}$

$=\dfrac{\cancel{x}^{\,1}(x-11)}{(x+4)\cancel{(x+1)}}\cdot\dfrac{(x+9)\cancel{(x+1)}^{\,1}}{\cancel{x}_{\,1}(x+6)}$

$=\dfrac{(x-11)(x+9)}{(x+4)(x+6)}$

13. $\dfrac{x^3-8}{x^2+x-42}\cdot\dfrac{x^2+9x+14}{x^2-4}$

$=\dfrac{(x-2)(x^2+2x+4)}{(x+7)(x-6)}\cdot\dfrac{(x+7)(x+2)}{(x+2)(x-2)}$

$=\dfrac{\cancel{(x-2)}^{\,1}(x^2+2x+4)}{\cancel{(x+7)}_{\,1}(x-6)}\cdot\dfrac{\cancel{(x+7)}^{\,1}\cancel{(x+2)}^{\,1}}{\cancel{(x+2)}_{\,1}\cancel{(x-2)}_{\,1}}$

$=\dfrac{x^2+2x+4}{x-6}$

15. $\dfrac{9-x^2}{x^2-x-90}\cdot\dfrac{x^2+13x+36}{x^2+10x+21}$

$=\dfrac{(3+x)(3-x)}{(x-10)(x+9)}\cdot\dfrac{(x+4)(x+9)}{(x+7)(x+3)}$

$=\dfrac{\cancel{(3+x)}^{\,1}(3-x)}{(x-10)\cancel{(x+9)}^{\,1}}\cdot\dfrac{(x+4)\cancel{(x+9)}^{\,1}}{(x+7)\cancel{(x+3)}^{\,1}}$

$=\dfrac{(3-x)(x+4)}{(x-10)(x+7)}$

17. $\dfrac{x^2-11x+24}{x^2-7x+12}\cdot\dfrac{x^2+4x-32}{x^2-64}$

$=\dfrac{(x-8)(x-3)}{(x-4)(x-3)}\cdot\dfrac{(x+8)(x-4)}{(x+8)(x-8)}$

$=\dfrac{\cancel{(x-8)}^{\,1}\cancel{(x-3)}^{\,1}}{\cancel{(x-4)}_{\,1}\cancel{(x-3)}_{\,1}}\cdot\dfrac{\cancel{(x+8)}^{\,1}\cancel{(x-4)}^{\,1}}{\cancel{(x+8)}_{\,1}\cancel{(x-8)}_{\,1}}$

$=1$

19. $\dfrac{2x^2-x-36}{x^2+x-90}\cdot\dfrac{9x-x^2}{x^2-2x-24}$

$=\dfrac{(2x-9)(x+4)}{(x+10)(x-9)}\cdot\dfrac{x(9-x)}{(x-6)(x+4)}$

$=\dfrac{(2x-9)\cancel{(x+4)}^{\,1}}{(x+10)\cancel{(x-9)}_{\,1}}\cdot\dfrac{x\cancel{(9-x)}^{-1}}{(x-6)\cancel{(x+4)}_{\,1}}$

$=-\dfrac{x(2x-9)}{(x+10)(x-6)}$

21. $f(x)\cdot g(x)$

$=\dfrac{2}{x-3}\cdot\dfrac{x^2-12x+27}{x^2-5x-36}$

$=\dfrac{2}{(x-3)}\cdot\dfrac{(x-9)(x-3)}{(x-9)(x+4)}$

$=\dfrac{2}{\cancel{(x-3)}_{\,1}}\cdot\dfrac{\cancel{(x-9)}^{\,1}\cancel{(x-3)}^{\,1}}{\cancel{(x-9)}_{\,1}(x+4)}$

$=\dfrac{2}{x+4}$

23. $f(x)\cdot g(x)$

$=\dfrac{x^2-13x+22}{x^2+11x-12}\cdot\dfrac{x^2+4x-5}{x^2-6x+8}$

$=\dfrac{(x-11)(x-2)}{(x+12)(x-1)}\cdot\dfrac{(x+5)(x-1)}{(x-4)(x-2)}$

$=\dfrac{(x-11)\cancel{(x-2)}^{\,1}}{(x+12)\cancel{(x-1)}_{\,1}}\cdot\dfrac{(x+5)\cancel{(x-1)}^{\,1}}{(x-4)\cancel{(x-2)}_{\,1}}$

$=\dfrac{(x-11)(x+5)}{(x+12)(x-4)}$

25. $\dfrac{x^2+3x-28}{x^2-36}\div\dfrac{x^2-13x+36}{x^2+11x+30}$

$=\dfrac{x^2+3x-28}{x^2-36}\cdot\dfrac{x^2+11x+30}{x^2-13x+36}$

$=\dfrac{(x+7)(x-4)}{(x+6)(x-6)}\cdot\dfrac{(x+6)(x+5)}{(x-9)(x-4)}$

$=\dfrac{(x+7)\cancel{(x-4)}^{\,1}}{\cancel{(x+6)}_{\,1}(x-6)}\cdot\dfrac{\cancel{(x+6)}^{\,1}(x+5)}{(x-9)\cancel{(x-4)}_{\,1}}$

$=\dfrac{(x+7)(x+5)}{(x-6)(x-9)}$

27.
$$\frac{x^2-6x+9}{x^2-8x-9} \div \frac{x^2-9}{x^2+13x+12}$$

$$= \frac{x^2-6x+9}{x^2-8x-9} \cdot \frac{x^2+13x+12}{x^2-9}$$

$$= \frac{(x-3)(x-3)}{(x-9)(x+1)} \cdot \frac{(x+12)(x+1)}{(x+3)(x-3)}$$

$$= \frac{\cancel{(x-3)}^{1}(x-3)}{(x-9)\,\cancel{(x+1)}_{1}} \cdot \frac{(x+12)\,\cancel{(x+1)}^{1}}{(x+3)\,\cancel{(x-3)}_{1}}$$

$$= \frac{(x-3)(x+12)}{(x-9)(x+3)}$$

29.
$$\frac{2x^2-13x+20}{x^2+x-6} \div \frac{5x-2x^2}{x^2+14x+33}$$

$$= \frac{2x^2-13x+20}{x^2+x-6} \cdot \frac{x^2+14x+33}{5x-2x^2}$$

$$= \frac{(2x-5)(x-4)}{(x+3)(x-2)} \cdot \frac{(x+11)(x+3)}{x(5-2x)}$$

$$= \frac{\cancel{(2x-5)}^{1}(x-4)}{\cancel{(x+3)}_{1}(x-2)} \cdot \frac{(x+11)\,\cancel{(x+3)}^{1}}{x\,\cancel{(5-2x)}_{-1}}$$

$$= -\frac{(x-4)(x+11)}{x(x-2)}$$

31.
$$\frac{x^2+6x-7}{x^2+16x+55} \div \frac{x^2+13x+42}{x^2+7x-44}$$

$$= \frac{x^2+6x-7}{x^2+16x+55} \cdot \frac{x^2+7x-44}{x^2+13x+42}$$

$$= \frac{(x+7)(x-1)}{(x+5)(x+11)} \cdot \frac{(x+11)(x-4)}{(x+7)(x+6)}$$

$$= \frac{\cancel{(x+7)}^{1}(x-1)}{(x+5)\,\cancel{(x+11)}_{1}} \cdot \frac{\cancel{(x+11)}^{1}(x-4)}{\cancel{(x+7)}_{1}(x+6)}$$

$$= \frac{(x-1)(x-4)}{(x+5)(x+6)}$$

33.
$$\frac{x^2-4x-5}{x^2+x-30} \div \frac{x^2-x-2}{x^2+4x-12}$$

$$= \frac{x^2-4x-5}{x^2+x-30} \cdot \frac{x^2+4x-12}{x^2-x-2}$$

$$= \frac{(x-5)(x+1)}{(x+6)(x-5)} \cdot \frac{(x+6)(x-2)}{(x-2)(x+1)}$$

$$= \frac{\cancel{(x-5)}^{1}\,\cancel{(x+1)}^{1}}{\cancel{(x+6)}_{1}\,\cancel{(x-5)}_{1}} \cdot \frac{\cancel{(x+6)}^{1}\,\cancel{(x-2)}^{1}}{\cancel{(x-2)}_{1}\,\cancel{(x+1)}_{1}}$$

$$= 1$$

35.
$$\frac{x^2-2x-63}{x^2-4x-21} \div \frac{81-x^2}{x^2-7x-30}$$

$$= \frac{x^2-2x-63}{x^2-4x-21} \cdot \frac{x^2-7x-30}{81-x^2}$$

$$= \frac{(x-9)(x+7)}{(x-7)(x+3)} \cdot \frac{(x-10)(x+3)}{(9+x)(9-x)}$$

$$= \frac{\cancel{(x-9)}^{1}(x+7)}{(x-7)\,\cancel{(x+3)}_{1}} \cdot \frac{(x-10)\,\cancel{(x+3)}^{1}}{(9+x)\,\cancel{(9-x)}_{-1}}$$

$$= -\frac{(x+7)(x-10)}{(x-7)(9+x)}$$

37.
$$\frac{x^2+10x+24}{4x^2-19x-5} \div \frac{x^2+16x+60}{x^2+5x-50}$$

$$= \frac{x^2+10x+24}{4x^2-19x-5} \cdot \frac{x^2+5x-50}{x^2+16x+60}$$

$$= \frac{(x+6)(x+4)}{(x-5)(4x+1)} \cdot \frac{(x+10)(x-5)}{(x+10)(x+6)}$$

$$= \frac{\cancel{(x+6)}^{1}(x+4)}{\cancel{(x-5)}_{1}(4x+1)} \cdot \frac{\cancel{(x+10)}^{1}\,\cancel{(x-5)}^{1}}{\cancel{(x+10)}_{1}\,\cancel{(x+6)}_{1}}$$

$$= \frac{x+4}{4x+1}$$

39. $\dfrac{x^2+16x+63}{2x-x^2} \div \dfrac{x^2+x-72}{x^2-5x+6}$

$= \dfrac{x^2+16x+63}{2x-x^2} \cdot \dfrac{x^2-5x+6}{x^2+x-72}$

$= \dfrac{(x+9)(x+7)}{x(2-x)} \cdot \dfrac{(x-3)(x-2)}{(x+9)(x-8)}$

$= \dfrac{(x+9)^1(x+7)}{x(2-x)_{-1}} \cdot \dfrac{(x-3)(x-2)^1}{(x+9)_1(x-8)}$

$= -\dfrac{(x+7)(x-3)}{x(x-8)}$

41. $f(x) \div g(x) = \dfrac{x^2+8x-9}{x^2-10x+25} \div (x+9)$

$= \dfrac{x^2+8x-9}{x^2-10x+25} \cdot \dfrac{1}{(x+9)}$

$= \dfrac{(x+9)(x-1)}{(x-5)(x-5)} \cdot \dfrac{1}{(x+9)}$

$= \dfrac{(x+9)^1(x-1)}{(x-5)(x-5)} \cdot \dfrac{1}{(x+9)_1}$

$= \dfrac{x-1}{(x-5)^2}$

43. $f(x) \div g(x) = \dfrac{2x^2-5x}{x^2-1} \div \dfrac{2x^2+x-15}{x^2-8x-9}$

$= \dfrac{2x^2-5x}{x^2-1} \cdot \dfrac{x^2-8x-9}{2x^2+x-15}$

$= \dfrac{x(2x-5)}{(x+1)(x-1)} \cdot \dfrac{(x-9)(x+1)}{(2x-5)(x+3)}$

$= \dfrac{x(2x-5)^1}{(x+1)_1(x-1)} \cdot \dfrac{(x-9)(x+1)^1}{(2x-5)_1(x+3)}$

$= \dfrac{x(x-9)}{(x-1)(x+3)}$

45. $\dfrac{x^2-10x+16}{x^2+4x-77} \cdot \dfrac{(x-3)(x+11)}{(2-x)(x+2)}$

$= \dfrac{(x-8)(x-2)}{(x+11)(x-7)} \cdot \dfrac{(x-3)(x+11)}{(2-x)(x+2)}$

$= \dfrac{(x-8)(x-2)^1}{(x+11)_1(x-7)} \cdot \dfrac{(x-3)(x+11)^1}{(2-x)_{-1}(x+2)}$

$= -\dfrac{(x-8)(x-3)}{(x-7)(x+2)}$

$= -\dfrac{x^2-3x-8x+24}{x^2+2x-7x-14}$

$= -\dfrac{x^2-11x+24}{x^2-5x-14}$

The missing factor is $\dfrac{(x-3)(x+11)}{(2-x)(x+2)}$.

47. $\dfrac{x^2-10x+9}{x^2+7x} \div \dfrac{(x-9)(x+4)}{x(x-12)}$

$= \dfrac{(x-9)(x-1)}{x(x+7)} \cdot \dfrac{x(x-12)}{(x-9)(x+4)}$

$= \dfrac{(x-9)^1(x-1)}{x_1(x+7)} \cdot \dfrac{x^1(x-12)}{(x-9)_1(x+4)}$

$= \dfrac{(x-1)(x-12)}{(x+7)(x+4)}$

$= \dfrac{x^2-12x-x+12}{x^2+4x+7x+28}$

$= \dfrac{x^2-13x+12}{x^2+11x+28}$

The missing factor is $\dfrac{(x-9)(x+4)}{x(x-12)}$.

49. $\dfrac{x^2-9x-10}{100-x^2} \cdot \dfrac{x^2+4x-60}{x^2-7x+6}$

$= \dfrac{(x-10)(x+1)}{(10+x)(10-x)} \cdot \dfrac{(x+10)(x-6)}{(x-6)(x-1)}$

$= \dfrac{(x-10)^1(x+1)}{(10+x)_1(10-x)_{-1}} \cdot \dfrac{(x+10)^1(x-6)^1}{(x-6)_1(x-1)}$

$= -\dfrac{x+1}{x-1}$

51. $\dfrac{3x^2-13x-10}{2x^2-17x+35} = \dfrac{(3x+2)(x-5)}{(2x-7)(x-5)}$

$\qquad = \dfrac{(3x+2)\cancel{(x-5)}^{\,1}}{(2x-7)\cancel{(x-5)}_{\,1}}$

$\qquad = \dfrac{3x+2}{2x-7}$

53. $\dfrac{x^2-14x+24}{x^2+x-6} \div \dfrac{x^2+2x-80}{x^2+13x+30}$

$\quad = \dfrac{x^2-14x+24}{x^2+x-6} \cdot \dfrac{x^2+13x+30}{x^2+2x-80}$

$\quad = \dfrac{(x-12)(x-2)}{(x+3)(x-2)} \cdot \dfrac{(x+10)(x+3)}{(x+10)(x-8)}$

$\quad = \dfrac{(x-12)\cancel{(x-2)}^{\,1}}{\cancel{(x+3)}_{\,1}\cancel{(x-2)}_{\,1}} \cdot \dfrac{\cancel{(x+10)}^{\,1}\cancel{(x+3)}^{\,1}}{\cancel{(x+10)}_{\,1}(x-8)}$

$\quad = \dfrac{x-12}{x-8}$

55. $\dfrac{x^2-20x+96}{144-x^2} = \dfrac{(x-12)(x-8)}{(12+x)(12-x)}$

$\qquad = \dfrac{\cancel{(x-12)}^{\,1}(x-8)}{(12+x)\cancel{(12-x)}_{\,-1}}$

$\qquad = -\dfrac{x-8}{12+x}$

57. $\dfrac{x^2+5x-84}{x^2-x-12} \cdot \dfrac{x^2+11x+24}{2x^2-15x+7}$

$\quad = \dfrac{(x+12)(x-7)}{(x-4)(x+3)} \cdot \dfrac{(x+8)(x+3)}{(2x-1)(x-7)}$

$\quad = \dfrac{(x+12)\cancel{(x-7)}^{\,1}}{(x-4)\cancel{(x+3)}_{\,1}} \cdot \dfrac{(x+8)\cancel{(x+3)}^{\,1}}{(2x-1)\cancel{(x-7)}_{\,1}}$

$\quad = \dfrac{(x+12)(x+8)}{(x-4)(2x-1)}$

59. $\dfrac{x^2+3x-54}{x^2-2x-3} \div \dfrac{x^2+9x}{x^2-7x-8}$

$\quad = \dfrac{x^2+3x-54}{x^2-2x-3} \cdot \dfrac{x^2-7x-8}{x^2+9x}$

$\quad = \dfrac{(x+9)(x-6)}{(x-3)(x+1)} \cdot \dfrac{(x-8)(x+1)}{x(x+9)}$

$\quad = \dfrac{\cancel{(x+9)}^{\,1}(x-6)}{(x-3)\cancel{(x+1)}_{\,1}} \cdot \dfrac{(x-8)\cancel{(x+1)}^{\,1}}{x\cancel{(x+9)}_{\,1}}$

$\quad = \dfrac{(x-6)(x-8)}{x(x-3)}$

61. Answers will vary. Example:
A similarity in both is that we can divide by taking the reciprocal of the divisor and multiplying.

7.3   QUICK CHECK

1. $\dfrac{9}{2x+7}+\dfrac{12}{2x+7}=\dfrac{9+12}{2x+7}=\dfrac{21}{2x+7}$

2. $\dfrac{2x+54}{x^2+4x-32}+\dfrac{3x-14}{x^2+4x-32}$

$\quad = \dfrac{(2x+54)+(3x-14)}{x^2+4x-32}$

$\quad = \dfrac{5x+40}{x^2+4x-32}$

$\quad = \dfrac{5\cancel{(x+8)}^{\,1}}{\cancel{(x+8)}_{\,1}(x-4)}$

$\quad = \dfrac{5}{x-4}$

3. $\dfrac{x^2+6x+4}{x^2+4x-21}+\dfrac{5x+24}{x^2+4x-21}$

$\quad = \dfrac{\left(x^2+6x+4\right)+(5x+24)}{x^2+4x-21}$

$\quad = \dfrac{x^2+11x+28}{x^2+4x-21}$

$\quad = \dfrac{\cancel{(x+7)}^{\,1}(x+4)}{\cancel{(x+7)}_{\,1}(x-3)}$

$\quad = \dfrac{x+4}{x-3}$

4. $\dfrac{25}{6x+36} - \dfrac{17}{6x+36} = \dfrac{8}{6x+36}$

$$= \dfrac{8}{6(x+6)}$$

$$= \dfrac{\cancel{8}^{4}}{\cancel{6}_{3}(x+6)}$$

$$= \dfrac{4}{3(x+6)}$$

5. $\dfrac{3x^2+6x-37}{64-x^2} - \dfrac{2x^2+17x-61}{64-x^2}$

$$= \dfrac{\left(3x^2+6x-37\right)-\left(2x^2+17x-61\right)}{64-x^2}$$

$$= \dfrac{3x^2+6x-37-2x^2-17x+61}{64-x^2}$$

$$= \dfrac{x^2-11x+24}{64-x^2}$$

$$= \dfrac{\cancel{(x-8)}^{1}(x-3)}{\cancel{(8-x)}_{-1}(8+x)}$$

$$= -\dfrac{x-3}{8+x}$$

6. $\dfrac{x}{5x-20} + \dfrac{4}{20-5x} = \dfrac{x}{5x-20} - \dfrac{4}{5x-20}$

$$= \dfrac{x-4}{5x-20}$$

$$= \dfrac{\cancel{(x-4)}^{1}}{5\cancel{(x-4)}_{1}}$$

$$= \dfrac{1}{5}$$

7. $\dfrac{x^2-2x-11}{x^2-16} - \dfrac{7x-25}{16-x^2}$

$$= \dfrac{x^2-2x-11}{x^2-16} + \dfrac{7x-25}{x^2-16}$$

$$= \dfrac{\left(x^2-2x-11\right)+\left(7x-25\right)}{x^2-16}$$

$$= \dfrac{x^2+5x-36}{x^2-16}$$

$$= \dfrac{(x+9)\cancel{(x-4)}^{1}}{(x+4)\cancel{(x-4)}_{1}}$$

$$= \dfrac{x+9}{x+4}$$

7.3   ADDITION AND SUBTRACTION OF RATIONAL EXPRESSIONS THAT HAVE THE SAME DENOMINATOR

1. numerators

3. first term

5. $\dfrac{5}{x+3} + \dfrac{8}{x+3} = \dfrac{13}{x+3}$

7. $\dfrac{x}{x-5} + \dfrac{10}{x-5} = \dfrac{x+10}{x-5}$

9. $\dfrac{x}{x^2+5x-36} + \dfrac{9}{x^2+5x-36} = \dfrac{x+9}{x^2+5x-36}$

$$= \dfrac{\cancel{(x+9)}^{1}}{\cancel{(x+9)}_{1}(x-4)}$$

$$= \dfrac{1}{x-4}$$

11. $\dfrac{x^2-5x+9}{x^2-8x+15}+\dfrac{x-6}{x^2-8x+15}$

$$=\dfrac{\left(x^2-5x+9\right)+(x-6)}{x^2-8x+15}$$

$$=\dfrac{x^2-4x+3}{x^2-8x+15}$$

$$=\dfrac{\cancel{(x-3)}^{\,1}(x-1)}{(x-5)\cancel{(x-3)}_{\,1}}$$

$$=\dfrac{x-1}{x-5}$$

13. $\dfrac{x^2+3x+11}{x^2-2x-48}+\dfrac{12x+43}{x^2-2x-48}=\dfrac{x^2+15x+54}{x^2-2x-48}$

$$=\dfrac{(x+9)\cancel{(x+6)}^{\,1}}{(x-8)\cancel{(x+6)}_{\,1}}$$

$$=\dfrac{x+9}{x-8}$$

15. $\dfrac{x^2-16x-45}{x^2-3x-18}+\dfrac{x^2-2x-27}{x^2-3x-18}=\dfrac{2x^2-18x-72}{x^2-3x-18}$

$$=\dfrac{2\left(x^2-9x-36\right)}{(x-6)(x+3)}$$

$$=\dfrac{2(x-12)\cancel{(x+3)}^{\,1}}{(x-6)\cancel{(x+3)}_{\,1}}$$

$$=\dfrac{2(x-12)}{x-6}$$

17. $f(x)+g(x)=\dfrac{3}{x}+\dfrac{5}{x}=\dfrac{8}{x}$

19. $f(x)+g(x)=\dfrac{x^2+3x-15}{x^2-4x-45}+\dfrac{4x+25}{x^2-4x-45}$

$$=\dfrac{x^2+7x+10}{x^2-4x-45}$$

$$=\dfrac{\cancel{(x+5)}^{\,1}(x+2)}{\cancel{(x+5)}_{\,1}(x-9)}$$

$$=\dfrac{x+2}{x-9}$$

21. $\dfrac{13}{x+1}-\dfrac{9}{x+1}=\dfrac{4}{x+1}$

23. $\dfrac{x}{x+7}-\dfrac{7}{x+7}=\dfrac{x-7}{x+7}$

25. $\dfrac{3x-7}{x-8}-\dfrac{x+9}{x-8}=\dfrac{(3x-7)-(x+9)}{x-8}$

$$=\dfrac{3x-7-x-9}{x-8}$$

$$=\dfrac{2x-16}{x-8}$$

$$=\dfrac{2\cancel{(x-8)}^{\,1}}{\cancel{(x-8)}_{\,1}}$$

$$=2$$

27. $\dfrac{x}{x^2-13x+30}-\dfrac{3}{x^2-13x+30}=\dfrac{x-3}{x^2-13x+30}$

$$=\dfrac{\cancel{(x-3)}^{\,1}}{(x-10)\cancel{(x-3)}_{\,1}}$$

$$=\dfrac{1}{x-10}$$

29. $\dfrac{4x-7}{x^2-16}-\dfrac{2x-15}{x^2-16}=\dfrac{(4x-7)-(2x-15)}{x^2-16}$

$$=\dfrac{4x-7-2x+15}{x^2-16}$$

$$=\dfrac{2x+8}{x^2-16}$$

$$=\dfrac{2\cancel{(x+4)}^{\,1}}{\cancel{(x+4)}_{\,1}(x-4)}$$

$$=\dfrac{2}{x-4}$$

31. $\dfrac{x^2-2x-3}{x^2-x-72}-\dfrac{5x+15}{x^2-x-72}$

$=\dfrac{\left(x^2-2x-3\right)-\left(5x+15\right)}{x^2-x-72}$

$=\dfrac{x^2-2x-3-5x-15}{x^2-x-72}$

$=\dfrac{x^2-7x-18}{x^2-x-72}$

$=\dfrac{\cancel{(x-9)}^{1}(x+2)}{\cancel{(x-9)}_{1}(x+8)}$

$=\dfrac{x+2}{x+8}$

33. $\dfrac{x^2-3x+67}{3x^2+17x+10}-\dfrac{7-20x}{3x^2+17x+10}$

$=\dfrac{\left(x^2-3x+67\right)-\left(7-20x\right)}{3x^2+17x+10}$

$=\dfrac{x^2-3x+67-7+20x}{3x^2+17x+10}$

$=\dfrac{x^2+17+60}{3x^2+17x+10}$

$=\dfrac{(x+12)\cancel{(x+5)}^{1}}{(3x+2)\cancel{(x+5)}_{1}}$

$=\dfrac{x+12}{3x+2}$

35. $\dfrac{(x-6)(x+3)}{x^2-2x-15}-\dfrac{9(x-5)}{x^2-2x-15}$

$=\dfrac{(x-6)(x+3)-9(x-5)}{x^2-2x-15}$

$=\dfrac{x^2+3x-6x-18-9x+45}{x^2-2x-15}$

$=\dfrac{x^2-12x+27}{x^2-2x-15}$

$=\dfrac{(x-3)(x-9)}{(x-5)(x+3)}$

37. $f(x)-g(x)=\dfrac{12}{x+8}-\dfrac{23}{x+8}=-\dfrac{11}{x+8}$

39. $f(x)-g(x)=\dfrac{x^2+3x-5}{x^2-11x+18}-\dfrac{8x+9}{x^2-11x+18}$

$=\dfrac{x^2+3x-5-8x-9}{x^2-11x+18}$

$=\dfrac{x^2-5x-14}{x^2-11x+18}$

$=\dfrac{(x-7)(x+2)}{(x-2)(x-9)}$

41. $\dfrac{2x}{x-5}+\dfrac{10}{5-x}=\dfrac{2x}{x-5}-\dfrac{10}{x-5}$

$=\dfrac{2x-10}{x-5}$

$=\dfrac{2\cancel{(x-5)}^{1}}{\cancel{(x-5)}_{1}}$

$=2$

43. $\dfrac{x^2+8x}{x-1}-\dfrac{2x-11}{1-x}=\dfrac{x^2+8x}{x-1}+\dfrac{2x-11}{x-1}$

$=\dfrac{x^2+10x-11}{x-1}$

$=\dfrac{\cancel{(x-1)}^{1}(x+11)}{\cancel{(x-1)}_{1}}$

$=x+11$

45. $\dfrac{x^2-3x-9}{x-9}+\dfrac{x^2-7x+27}{9-x}$

$=\dfrac{x^2-3x-9}{x-9}-\dfrac{x^2-7x+27}{x-9}$

$=\dfrac{x^2-3x-9-x^2+7x-27}{x-9}$

$=\dfrac{4x-36}{x-9}$

$=\dfrac{4\cancel{(x-9)}^{1}}{\cancel{(x-9)}_{1}}$

$=4$

47. $\dfrac{x^2-3x+5}{2x-20}+\dfrac{4x+35}{20-2x}=\dfrac{x^2-3x+5}{2x-20}-\dfrac{4x+35}{2x-20}$

$$=\dfrac{x^2-7x-30}{2x-20}$$

$$=\dfrac{\cancel{(x-10)}^{\,1}(x+3)}{2\cancel{(x-10)}_{\,1}}$$

$$=\dfrac{x+3}{2}$$

49. $\dfrac{x^2-14x+49}{x^2+x-30}\div\dfrac{3x^2-9x-84}{x^2-6x+5}$

$$=\dfrac{x^2-14x+49}{x^2+x-30}\cdot\dfrac{x^2-6x+5}{3\left(x^2-3x-28\right)}$$

$$=\dfrac{\cancel{(x-7)}^{\,1}(x-7)}{(x+6)\cancel{(x-5)}_{\,1}}\cdot\dfrac{\cancel{(x-5)}^{\,1}(x-1)}{3(x+4)\cancel{(x-7)}_{\,1}}$$

$$=\dfrac{(x-7)(x-1)}{3(x+6)(x+4)}$$

51. $\dfrac{x^2+2x-360}{x^2+30x+200}=\dfrac{\cancel{(x+20)}^{\,1}(x-18)}{\cancel{(x+20)}_{\,1}(x+10)}$

$$=\dfrac{x-18}{x+10}$$

53. $\dfrac{x^2-3x+45}{x^2-81}+\dfrac{x^2+29x+27}{x^2-81}$

$$=\dfrac{2x^2+26x+72}{x^2-81}$$

$$=\dfrac{2\left(x^2+13x+36\right)}{x^2-81}$$

$$=\dfrac{2\cancel{(x+9)}^{\,1}(x+4)}{\cancel{(x+9)}_{\,1}(x-9)}$$

$$=\dfrac{2(x+4)}{x-9}$$

55. $\dfrac{121-x^2}{x^2-3x-4}\cdot\dfrac{x^2+4x+3}{x^2-3x-88}$

$$=\dfrac{(11+x)\cancel{(11-x)}^{\,-1}}{(x-4)\cancel{(x+1)}_{\,1}}\cdot\dfrac{(x+3)\cancel{(x+1)}^{\,1}}{\cancel{(x-11)}_{\,1}(x+8)}$$

$$=-\dfrac{(11+x)(x+3)}{(x-4)(x+8)}$$

57. $\dfrac{2x^2-15x+39}{x^2+3x-4}+\dfrac{x^2-18x-9}{x^2+3x-4}=\dfrac{3x^2-33x+30}{x^2+3x-4}$

$$=\dfrac{3\left(x^2-11x+10\right)}{x^2+3x-4}$$

$$=\dfrac{3(x-10)\cancel{(x-1)}^{\,1}}{(x+4)\cancel{(x-1)}_{\,1}}$$

$$=\dfrac{3(x-10)}{x+4}$$

59. $\dfrac{6x}{3x-2}-\dfrac{4}{2-3x}=\dfrac{6x}{3x-2}+\dfrac{4}{3x-2}$

$$=\dfrac{6x+4}{3x-2}$$

$$=\dfrac{2(3x+2)}{3x-2}$$

61. $\dfrac{5x^2+3x-13}{x^2-8x-9}-\dfrac{4x^2-7x-22}{x^2-8x-9}$

$$=\dfrac{5x^2+3x-13-4x^2+7x+22}{x^2-8x-9}$$

$$=\dfrac{x^2+10x+9}{x^2-8x-9}$$

$$=\dfrac{(x+9)\cancel{(x+1)}^{\,1}}{(x-9)\cancel{(x+1)}_{\,1}}$$

$$=\dfrac{x+9}{x-9}$$

63. $\dfrac{x^2+x-90}{3x^2+2x-16} \div \dfrac{x^2-9x}{3x^2+17x+24}$

$$= \dfrac{x^2+x-90}{3x^2+2x-16} \cdot \dfrac{3x^2+17x+24}{x^2-9x}$$

$$= \dfrac{(x+10)\,\cancel{(x-9)}^{\,1}}{\cancel{(3x+8)}_{\,1}\,(x-2)} \cdot \dfrac{\cancel{(3x+8)}^{\,1}\,(x+3)}{x\,\cancel{(x-9)}_{\,1}}$$

$$= \dfrac{(x+10)(x+3)}{x(x-2)}$$

65. $\dfrac{2x^2+5x+20}{x^2+5x+6} + \dfrac{3x^2-7x-4}{x^2+5x+6} - \dfrac{4x^2-2x+25}{x^2+5x+6}$

$$= \dfrac{5x^2-2x+16}{x^2+5x+6} - \dfrac{4x^2-2x+25}{x^2+5x+6}$$

$$= \dfrac{5x^2-2x+16-4x^2+2x-25}{x^2+5x+6}$$

$$= \dfrac{x^2-9}{x^2+5x+6}$$

$$= \dfrac{\cancel{(x+3)}^{\,1}(x-3)}{\cancel{(x+3)}_{\,1}(x+2)}$$

$$= \dfrac{x-3}{x+2}$$

67. $\dfrac{3x}{x+6} + \dfrac{18}{x+6} = \dfrac{3x+18}{x+6} = \dfrac{3\,\cancel{(x+6)}^{\,1}}{\cancel{(x+6)}_{\,1}} = 3$

The missing numerator is $3x$.

69. $\dfrac{x^2+7x+17}{(x+8)(x+5)} + \dfrac{4x+7}{(x+8)(x+5)} = \dfrac{x^2+11x+24}{(x+8)(x+5)}$

$$= \dfrac{\cancel{(x+8)}^{\,1}(x+3)}{\cancel{(x+8)}_{\,1}(x+5)}$$

$$= \dfrac{x+3}{x+5}$$

The missing numerator is $4x+7$.

71. Answers will vary. Example:
If each term in the first denominator differs by a sign compared to each term in the second denominator, then the denominators are opposites.

7.4   QUICK CHECK

1. LCD $= 36r^2s^6$

2. $x^2-13x+40=(x-8)(x-5)$
$x^2-4x-32=(x-8)(x+4)$
LCD $=(x-8)(x-5)(x+4)$

3. $x^2-81=(x+9)(x-9)$
$x^2+18x+81=(x+9)(x+9)=(x+9)^2$
LCD $=(x+9)^2(x-9)$

4. $\dfrac{5}{x-2} + \dfrac{6}{x+6} = \dfrac{5}{x-2}\cdot\dfrac{x+6}{x+6} + \dfrac{6}{x+6}\cdot\dfrac{x-2}{x-2}$

$$= \dfrac{5(x+6)}{(x-2)(x+6)} + \dfrac{6(x-2)}{(x-2)(x+6)}$$

$$= \dfrac{5x+30+6x-12}{(x-2)(x+6)}$$

$$= \dfrac{11x+18}{(x-2)(x+6)}$$

5. $\dfrac{x}{x^2-6x+5} + \dfrac{1}{x^2+2x-3}$

$$= \dfrac{x}{(x-5)(x-1)} + \dfrac{1}{(x+3)(x-1)}$$

$$= \dfrac{x}{(x-5)(x-1)}\cdot\dfrac{x+3}{x+3} + \dfrac{1}{(x+3)(x-1)}\cdot\dfrac{x-5}{x-5}$$

$$= \dfrac{x(x+3)}{(x-5)(x-1)(x+3)} + \dfrac{1(x-5)}{(x-5)(x-1)(x+3)}$$

$$= \dfrac{x^2+3x+x-5}{(x-5)(x-1)(x+3)}$$

$$= \dfrac{x^2+4x-5}{(x-5)(x-1)(x+3)}$$

$$= \dfrac{(x+5)\,\cancel{(x-1)}^{\,1}}{(x-5)\,\cancel{(x-1)}_{\,1}(x+3)}$$

$$= \dfrac{(x+5)}{(x-5)(x+3)}$$

6. $\dfrac{x+5}{x^2+8x+12}+\dfrac{x+9}{x^2-36}$

$=\dfrac{x+5}{(x+6)(x+2)}+\dfrac{x+9}{(x+6)(x-6)}$

$=\dfrac{(x+5)}{(x+6)(x+2)}\cdot\dfrac{x-6}{x-6}+\dfrac{(x+9)}{(x+6)(x-6)}\cdot\dfrac{x+2}{x+2}$

$=\dfrac{(x+5)(x-6)}{(x+6)(x+2)(x-6)}+\dfrac{(x+9)(x+2)}{(x+6)(x+2)(x-6)}$

$=\dfrac{x^2-6x+5x-30+x^2+2x+9x+18}{(x+6)(x+2)(x-6)}$

$=\dfrac{2x^2+10x-12}{(x+6)(x+2)(x-6)}$

$=\dfrac{2\left(x^2+5x-6\right)}{(x+6)(x+2)(x-6)}$

$=\dfrac{2\,\cancel{(x+6)}^{\,1}(x-1)}{\cancel{(x+6)}_{\,1}(x+2)(x-6)}$

$=\dfrac{2(x-1)}{(x+2)(x-6)}$

7.4   ADDITION AND SUBTRACTION OF RATIONAL EXPRESSIONS THAT HAVE DIFFERENT DENOMINATORS

1. LCD

3. LCD $=6a$

5. LCD $=(x-9)(x+5)$

7. $x^2-9=(x+3)(x-3)$
$x^2-9x+18=(x-3)(x-6)$
LCD $=(x+3)(x-3)(x-6)$

9. $x^2+7x+10=(x+2)(x+5)$
$x^2+4x+4=(x+2)(x+2)=(x+2)^2$
LCD $=(x+2)^2(x+5)$

11. $\dfrac{3x}{8}+\dfrac{5x}{12}=\dfrac{3x}{8}\cdot\dfrac{3}{3}+\dfrac{5x}{12}\cdot\dfrac{2}{2}=\dfrac{9x}{24}+\dfrac{10x}{24}=\dfrac{19x}{24}$

13. $\dfrac{9}{5n}-\dfrac{7}{4n}=\dfrac{9}{5n}\cdot\dfrac{4}{4}-\dfrac{7}{4n}\cdot\dfrac{5}{5}=\dfrac{36}{20n}-\dfrac{35}{20n}=\dfrac{1}{20n}$

15. $\dfrac{10}{m^7n^2}+\dfrac{13}{m^5n^3}=\dfrac{10}{m^7n^2}\cdot\dfrac{n}{n}+\dfrac{13}{m^5n^3}\cdot\dfrac{m^2}{m^2}$

$=\dfrac{10n}{m^7n^3}+\dfrac{13m^2}{m^7n^3}$

$=\dfrac{10n+13m^2}{m^7n^3}$

17. $\dfrac{5}{x+2}+\dfrac{3}{x+4}=\dfrac{5}{x+2}\cdot\dfrac{x+4}{x+4}+\dfrac{3}{x+4}\cdot\dfrac{x+2}{x+2}$

$=\dfrac{5(x+4)}{(x+2)(x+4)}+\dfrac{3(x+2)}{(x+2)(x+4)}$

$=\dfrac{5x+20+3x+6}{(x+2)(x+4)}$

$=\dfrac{8x+26}{(x+2)(x+4)}$

$=\dfrac{2(4x+13)}{(x+2)(x+4)}$

19. $\dfrac{10}{x+3}-\dfrac{4}{x-3}=\dfrac{10}{x+3}\cdot\dfrac{x-3}{x-3}-\dfrac{4}{x-3}\cdot\dfrac{x+3}{x+3}$

$=\dfrac{10(x-3)}{(x+3)(x-3)}-\dfrac{4(x+3)}{(x+3)(x-3)}$

$=\dfrac{10x-30-4x-12}{(x+3)(x-3)}$

$=\dfrac{6x-42}{(x+3)(x-3)}$

$=\dfrac{6(x-7)}{(x+3)(x-3)}$

21. $\dfrac{3}{x^2+7x+10}+\dfrac{8}{x^2-x-6}$

$=\dfrac{3}{(x+2)(x+5)}+\dfrac{8}{(x+2)(x-3)}$

$=\dfrac{3}{(x+2)(x+5)}\cdot\dfrac{x-3}{x-3}+\dfrac{8}{(x+2)(x-3)}\cdot\dfrac{x+5}{x+5}$

$=\dfrac{3(x-3)}{(x+2)(x+5)(x-3)}+\dfrac{8(x+5)}{(x+2)(x+5)(x-3)}$

$=\dfrac{3x-9+8x+40}{(x+2)(x+5)(x-3)}$

$=\dfrac{11x+31}{(x+2)(x+5)(x-3)}$

23. $\dfrac{1}{x^2-5x-6}-\dfrac{5}{x^2+8x+7}$

$=\dfrac{1}{(x-6)(x+1)}-\dfrac{5}{(x+1)(x+7)}$

$=\dfrac{1}{(x-6)(x+1)}\cdot\dfrac{x+7}{x+7}-\dfrac{5}{(x+1)(x+7)}\cdot\dfrac{x-6}{x-6}$

$=\dfrac{1(x+7)}{(x-6)(x+1)(x+7)}-\dfrac{5(x-6)}{(x-6)(x+1)(x+7)}$

$=\dfrac{x+7-5x+30}{(x-6)(x+1)(x+7)}$

$=\dfrac{-4x+37}{(x-6)(x+1)(x+7)}$

25. $\dfrac{5}{x^2-15x+50}+\dfrac{9}{x^2-x-20}$

$=\dfrac{5}{(x-10)(x-5)}+\dfrac{9}{(x+4)(x-5)}$

$=\dfrac{5}{(x-10)(x-5)}\cdot\dfrac{x+4}{x+4}+\dfrac{9}{(x+4)(x-5)}\cdot\dfrac{x-10}{x-10}$

$=\dfrac{5(x+4)}{(x-10)(x-5)(x+4)}+\dfrac{9(x-10)}{(x-10)(x-5)(x+4)}$

$=\dfrac{5x+20+9x-90}{(x-10)(x-5)(x+4)}$

$=\dfrac{14x-70}{(x-10)(x-5)(x+4)}$

$=\dfrac{14\,\overset{1}{\cancel{(x-5)}}}{(x-10)\,\underset{1}{\cancel{(x-5)}}(x+4)}$

$=\dfrac{14}{(x-10)(x+4)}$

27. $\dfrac{3}{x^2-13x+40}-\dfrac{4}{x^2-12x+32}$

$=\dfrac{3}{(x-8)(x-5)}-\dfrac{4}{(x-8)(x-4)}$

$=\dfrac{3}{(x-8)(x-5)}\cdot\dfrac{x-4}{x-4}-\dfrac{4}{(x-8)(x-4)}\cdot\dfrac{x-5}{x-5}$

$=\dfrac{3(x-4)}{(x-8)(x-5)(x-4)}-\dfrac{4(x-5)}{(x-8)(x-5)(x-4)}$

$=\dfrac{3x-12-4x+20}{(x-8)(x-5)(x-4)}$

$=\dfrac{-x+8}{(x-8)(x-5)(x-4)}$

$=\dfrac{\overset{-1}{\cancel{-x+8}}}{\underset{1}{\cancel{(x-8)}}(x-5)(x-4)}$

$=-\dfrac{1}{(x-5)(x-4)}$

29. $\dfrac{5}{2x^2+3x-2}+\dfrac{7}{2x^2-9x+4}$

$=\dfrac{5}{(2x-1)(x+2)}+\dfrac{7}{(2x-1)(x-4)}$

$=\dfrac{5}{(2x-1)(x+2)}\cdot\dfrac{x-4}{x-4}+\dfrac{7}{(2x-1)(x-4)}\cdot\dfrac{x+2}{x+2}$

$=\dfrac{5(x-4)}{(2x-1)(x+2)(x-4)}+\dfrac{7(x+2)}{(2x-1)(x+2)(x-4)}$

$=\dfrac{5x-20+7x+14}{(2x-1)(x+2)(x-4)}$

$=\dfrac{12x-6}{(2x-1)(x+2)(x-4)}$

$=\dfrac{6\,\overset{1}{\cancel{(2x-1)}}}{\underset{1}{\cancel{(2x-1)}}(x+2)(x-4)}$

$=\dfrac{6}{(x+2)(x-4)}$

31.
$$\frac{x+9}{x^2+8x+15}-\frac{2}{x^2+9x+20}$$
$$=\frac{x+9}{(x+3)(x+5)}-\frac{2}{(x+4)(x+5)}$$
$$=\frac{x+9}{(x+3)(x+5)}\cdot\frac{x+4}{x+4}-\frac{2}{(x+4)(x+5)}\cdot\frac{x+3}{x+3}$$
$$=\frac{(x+9)(x+4)}{(x+3)(x+5)(x+4)}-\frac{2(x+3)}{(x+3)(x+5)(x+4)}$$
$$=\frac{x^2+4x+9x+36-2x-6}{(x+3)(x+5)(x+4)}$$
$$=\frac{x^2+11x+30}{(x+3)(x+5)(x+4)}$$
$$=\frac{(x+6)\cancel{(x+5)}^{1}}{(x+3)\cancel{(x+5)}_{1}(x+4)}$$
$$=\frac{x+6}{(x+3)(x+4)}$$

33.
$$\frac{x-3}{x^2-x-20}+\frac{6}{x^2-2x-24}$$
$$=\frac{x-3}{(x-5)(x+4)}+\frac{6}{(x-6)(x+4)}$$
$$=\frac{x-3}{(x-5)(x+4)}\cdot\frac{x-6}{x-6}+\frac{6}{(x-6)(x+4)}\cdot\frac{x-5}{x-5}$$
$$=\frac{(x-3)(x-6)}{(x-5)(x+4)(x-6)}+\frac{6(x-5)}{(x-5)(x+4)(x-6)}$$
$$=\frac{x^2-6x-3x+18+6x-30}{(x-5)(x+4)(x-6)}$$
$$=\frac{x^2-3x-12}{(x-5)(x+4)(x-6)}$$

35.
$$\frac{x+14}{x^2+5x-50}-\frac{8}{x^2+15x+50}$$
$$=\frac{x+14}{(x+10)(x-5)}-\frac{8}{(x+10)(x+5)}$$
$$=\frac{x+14}{(x+10)(x-5)}\cdot\frac{x+5}{x+5}-\frac{8}{(x+10)(x+5)}\cdot\frac{x-5}{x-5}$$
$$=\frac{(x+14)(x+5)}{(x+10)(x-5)(x+5)}-\frac{8(x-5)}{(x+10)(x-5)(x+5)}$$
$$=\frac{x^2+5x+14x+70-8x+40}{(x+10)(x-5)(x+5)}$$
$$=\frac{x^2+11x+110}{(x+10)(x-5)(x+5)}$$

37.
$$\frac{x-9}{x^2-49}+\frac{1}{x^2-7x}$$
$$=\frac{x-9}{(x+7)(x-7)}+\frac{1}{x(x-7)}$$
$$=\frac{x-9}{(x+7)(x-7)}\cdot\frac{x}{x}+\frac{1}{x(x-7)}\cdot\frac{x+7}{x+7}$$
$$=\frac{x(x-9)}{x(x+7)(x-7)}+\frac{1(x+7)}{x(x+7)(x-7)}$$
$$=\frac{x^2-9x+x+7}{x(x+7)(x-7)}$$
$$=\frac{x^2-8x+7}{x(x+7)(x-7)}$$
$$=\frac{(x-1)\cancel{(x-7)}^{1}}{x(x+7)\cancel{(x-7)}_{1}}$$
$$=\frac{x-1}{x(x+7)}$$

39. $\dfrac{3}{x^2+11x+28}+\dfrac{x-4}{x^2-x-56}$

$=\dfrac{3}{(x+7)(x+4)}+\dfrac{x-4}{(x+7)(x-8)}$

$=\dfrac{3}{(x+7)(x+4)}\cdot\dfrac{x-8}{x-8}+\dfrac{x-4}{(x+7)(x-8)}\cdot\dfrac{x+4}{x+4}$

$=\dfrac{3(x-8)}{(x+7)(x+4)(x-8)}+\dfrac{(x-4)(x+4)}{(x+7)(x+4)(x-8)}$

$=\dfrac{3x-24+x^2-16}{(x+7)(x+4)(x-8)}$

$=\dfrac{x^2+3x-40}{(x+7)(x+4)(x-8)}$

$=\dfrac{(x-5)(x+8)}{(x+7)(x+4)(x-8)}$

41. $\dfrac{x+1}{x^2+2x-8}+\dfrac{x+3}{x^2+10x+24}$

$=\dfrac{x+1}{(x-2)(x+4)}+\dfrac{x+3}{(x+4)(x+6)}$

$=\dfrac{x+1}{(x-2)(x+4)}\cdot\dfrac{x+6}{x+6}+\dfrac{x+3}{(x+4)(x+6)}\cdot\dfrac{x-2}{x-2}$

$=\dfrac{(x+1)(x+6)}{(x-2)(x+4)(x+6)}+\dfrac{(x+3)(x-2)}{(x-2)(x+4)(x+6)}$

$=\dfrac{x^2+6x+x+6+x^2-2x+3x-6}{(x-2)(x+4)(x+6)}$

$=\dfrac{2x^2+8x}{(x-2)(x+4)(x+6)}$

$=\dfrac{2x\cancel{(x+4)}^{1}}{(x-2)\cancel{(x+4)}_{1}(x+6)}$

$=\dfrac{2x}{(x-2)(x+6)}$

43. $\dfrac{x+9}{2x^2-9x-18}-\dfrac{x-4}{2x^2+17x+21}$

$=\dfrac{x+9}{(2x+3)(x-6)}-\dfrac{x-4}{(2x+3)(x+7)}$

$=\dfrac{x+9}{(2x+3)(x-6)}\cdot\dfrac{x+7}{x+7}-\dfrac{x-4}{(2x+3)(x+7)}\cdot\dfrac{x-6}{x-6}$

$=\dfrac{(x+9)(x+7)}{(2x+3)(x-6)(x+7)}-\dfrac{(x-4)(x-6)}{(2x+3)(x-6)(x+7)}$

$=\dfrac{x^2+7x+9x+63-\left(x^2-6x-4x+24\right)}{(2x+3)(x-6)(x+7)}$

$=\dfrac{x^2+7x+9x+63-x^2+6x+4x-24}{(2x+3)(x-6)(x+7)}$

$=\dfrac{26x+39}{(2x+3)(x-6)(x+7)}$

$=\dfrac{13\cancel{(2x+3)}^{1}}{\cancel{(2x+3)}_{1}(x-6)(x+7)}$

$=\dfrac{13}{(x-6)(x+7)}$

45. $\dfrac{9}{x^2-11x+28}+\dfrac{3}{x^2-7x+12}$

$=\dfrac{9}{(x-7)(x-4)}+\dfrac{3}{(x-3)(x-4)}$

$=\dfrac{9}{(x-7)(x-4)}\cdot\dfrac{x-3}{x-3}+\dfrac{3}{(x-3)(x-4)}\cdot\dfrac{x-7}{x-7}$

$=\dfrac{9(x-3)}{(x-7)(x-4)(x-3)}+\dfrac{3(x-7)}{(x-7)(x-4)(x-3)}$

$=\dfrac{9x-27+3x-21}{(x-7)(x-4)(x-3)}$

$=\dfrac{12x-48}{(x-7)(x-4)(x-3)}$

$=\dfrac{12\cancel{(x-4)}^{1}}{(x-7)\cancel{(x-4)}_{1}(x-3)}$

$=\dfrac{12}{(x-7)(x-3)}$

47. $\dfrac{x+10}{x^2+10x+16}+\dfrac{x+7}{x^2+13x+40}$

$=\dfrac{x+10}{(x+8)(x+2)}+\dfrac{x+7}{(x+8)(x+5)}$

$=\dfrac{x+10}{(x+8)(x+2)}\cdot\dfrac{x+5}{x+5}+\dfrac{x+7}{(x+8)(x+5)}\cdot\dfrac{x+2}{x+2}$

$=\dfrac{(x+10)(x+5)}{(x+8)(x+2)(x+5)}+\dfrac{(x+7)(x+2)}{(x+8)(x+2)(x+5)}$

$=\dfrac{x^2+5x+10x+50+x^2+2x+7x+14}{(x+8)(x+2)(x+5)}$

$=\dfrac{2x^2+24x+64}{(x+8)(x+2)(x+5)}$

$=\dfrac{2\left(x^2+12x+32\right)}{(x+8)(x+2)(x+5)}$

$=\dfrac{2(x+8)^{1}(x+4)}{(x+8)_{1}(x+2)(x+5)}$

$=\dfrac{2(x+4)}{(x+2)(x+5)}$

49. $\dfrac{x^2-2x-24}{x^2+9x+20}\cdot\dfrac{2x^2+x-45}{x^2-7x+6}$

$=\dfrac{(x-6)^{1}(x+4)^{1}}{(x+5)_{1}(x+4)_{1}}\cdot\dfrac{(2x-9)(x+5)^{1}}{(x-6)_{1}(x-1)}$

$=\dfrac{2x-9}{x-1}$

51. $\dfrac{x+2}{x^2+4x-32}-\dfrac{1}{x^2+18x+80}$

$=\dfrac{x+2}{(x+8)(x-4)}-\dfrac{1}{(x+10)(x+8)}$

$=\dfrac{x+2}{(x+8)(x-4)}\cdot\dfrac{x+10}{x+10}-\dfrac{1}{(x+10)(x+8)}\cdot\dfrac{x-4}{x-4}$

$=\dfrac{(x+2)(x+10)}{(x+8)(x-4)(x+10)}-\dfrac{1(x-4)}{(x+8)(x-4)(x+10)}$

$=\dfrac{x^2+10x+2x+20-x+4}{(x+8)(x-4)(x+10)}$

$=\dfrac{x^2+11x+24}{(x+8)(x-4)(x+10)}$

$=\dfrac{(x+3)(x+8)^{1}}{(x+8)_{1}(x-4)(x+10)}$

$=\dfrac{x+3}{(x-4)(x+10)}$

53. $\dfrac{3x+17}{x^2-5x}-\dfrac{x^2-15x+18}{5x-x^2}$

$=\dfrac{3x+17}{x(x-5)}-\dfrac{x^2-15x+18}{-x(x-5)}$

$=\dfrac{3x+17}{x(x-5)}+\dfrac{x^2-15x+18}{x(x-5)}$

$=\dfrac{3x+17+x^2-15x+18}{x(x-5)}$

$=\dfrac{x^2-12x+35}{x(x-5)}$

$=\dfrac{(x-7)(x-5)^{1}}{x(x-5)_{1}}$

$=\dfrac{x-7}{x}$

55. $\dfrac{2x^2+19x+42}{36-x^2}=\dfrac{(2x+7)(x+6)}{(6-x)(6+x)}$

$=\dfrac{(2x+7)(x+6)^{1}}{(6-x)(6+x)_{1}}$

$=\dfrac{2x+7}{6-x}$

57. $\dfrac{3x^2+2x-8}{x^2-9x} \div \dfrac{x^3+8}{x^2-16x+63}$

$= \dfrac{3x^2+2x-8}{x^2-9x} \cdot \dfrac{x^2-16x+63}{x^3+8}$

$= \dfrac{(3x-4)(x+2)}{x(x-9)} \cdot \dfrac{(x-9)(x-7)}{(x+2)(x^2-2x+4)}$

$= \dfrac{(3x-4)\,\cancel{(x+2)}^{1}}{x\,\cancel{(x-9)}_{1}} \cdot \dfrac{\cancel{(x-9)}^{1}(x-7)}{\cancel{(x+2)}_{1}(x^2-2x+4)}$

$= \dfrac{(3x-4)(x-7)}{x(x^2-2x+4)}$

59. $\dfrac{x^2-7x+23}{x^2-81} + \dfrac{12x-67}{81-x^2}$

$= \dfrac{x^2-7x+23}{x^2-81} + \dfrac{12x-67}{-(x^2-81)}$

$= \dfrac{x^2-7x+23-12x+67}{x^2-81}$

$= \dfrac{x^2-19x+90}{x^2-81}$

$= \dfrac{(x-10)\,\cancel{(x-9)}^{1}}{(x+9)\,\cancel{(x-9)}_{1}}$

$= \dfrac{x-10}{x+9}$

61. Answers will vary. Example:
Factor the denominators first. Select the individual factors and their respective greatest exponent in the denominator to find the LCD.

63. Answers will vary.

1. $\dfrac{\dfrac{2}{5}+\dfrac{3}{8}}{\dfrac{1}{4}+\dfrac{7}{10}} = \dfrac{40}{40} \cdot \dfrac{\dfrac{2}{5}+\dfrac{3}{8}}{\dfrac{1}{4}+\dfrac{7}{10}}$

$= \dfrac{40\cdot\dfrac{2}{5}+40\cdot\dfrac{3}{8}}{40\cdot\dfrac{1}{4}+40\cdot\dfrac{7}{10}}$

$= \dfrac{16+15}{10+28}$

$= \dfrac{31}{38}$

2. $\dfrac{1+\dfrac{3}{x}-\dfrac{10}{x^2}}{1-\dfrac{2}{x}} = \dfrac{x^2}{x^2} \cdot \dfrac{1+\dfrac{3}{x}-\dfrac{10}{x^2}}{1-\dfrac{2}{x}}$

$= \dfrac{x^2\cdot 1+x^2\cdot\dfrac{3}{x}-x^2\cdot\dfrac{10}{x^2}}{x^2\cdot 1-x^2\cdot\dfrac{2}{x}}$

$= \dfrac{x^2+3x-10}{x^2-2x}$

$= \dfrac{(x+5)\,\cancel{(x-2)}^{1}}{x\,\cancel{(x-2)}_{1}}$

$= \dfrac{x+5}{x}$

3. $\dfrac{\dfrac{1}{64}-\dfrac{1}{x^2}}{\dfrac{1}{8}-\dfrac{1}{x}} = \dfrac{64x^2}{64x^2} \cdot \dfrac{\dfrac{1}{64}-\dfrac{1}{x^2}}{\dfrac{1}{8}-\dfrac{1}{x}}$

$= \dfrac{\cancel{64}^{1}x^2\cdot\dfrac{1}{\cancel{64}_{1}}-64\,\cancel{x^2}^{1}\cdot\dfrac{1}{\cancel{x^2}^{1}}}{\cancel{64}^{8}x^2\cdot\dfrac{1}{\cancel{8}_{1}}-64\,\cancel{x^2}^{x}\cdot\dfrac{1}{\cancel{x}_{1}}}$

$= \dfrac{x^2-64}{8x^2-64x}$

$= \dfrac{(x+8)\,\cancel{(x-8)}^{1}}{8x\,\cancel{(x-8)}_{1}}$

$= \dfrac{x+8}{8x}$

4. $\dfrac{\dfrac{6}{x-4}+\dfrac{5}{x+7}}{\dfrac{x+2}{x-4}}$

$$=\frac{(x-4)(x+7)}{(x-4)(x+7)}\cdot\frac{\dfrac{6}{x-4}+\dfrac{5}{x+7}}{\dfrac{x+2}{x-4}}$$

$$=\frac{(x-4)^1(x+7)\cdot\dfrac{6}{(x-4)_1}+(x-4)(x+7)^1\cdot\dfrac{5}{(x+7)_1}}{(x-4)^1(x+7)\cdot\dfrac{x+2}{(x-4)_1}}$$

$$=\frac{6(x+7)+5(x-4)}{(x+7)(x+2)}$$

$$=\frac{6x+42+5x-20}{(x+7)(x+2)}$$

$$=\frac{11x+22}{(x+7)(x+2)}$$

$$=\frac{11(x+2)^1}{(x+7)(x+2)_1}$$

$$=\frac{11}{x+7}$$

5. $\dfrac{\dfrac{x^2-2x-48}{x^2+7x-30}}{\dfrac{x^2+7x+6}{x^2-5x+6}}=\dfrac{x^2-2x-48}{x^2+7x-30}\div\dfrac{x^2+7x+6}{x^2-5x+6}$

$$=\frac{x^2-2x-48}{x^2+7x-30}\cdot\frac{x^2-5x+6}{x^2+7x+6}$$

$$=\frac{(x-8)(x+6)}{(x+10)(x-3)}\cdot\frac{(x-2)(x-3)}{(x+6)(x+1)}$$

$$=\frac{(x-8)(x+6)^1}{(x+10)(x-3)_1}\cdot\frac{(x-2)(x-3)^1}{(x+6)_1(x+1)}$$

$$=\frac{(x-8)(x-2)}{(x+10)(x+1)}$$

7.5 COMPLEX FRACTIONS

1. complex fraction

3. $\dfrac{\dfrac{2}{5}-\dfrac{1}{4}}{\dfrac{9}{10}+\dfrac{5}{2}}=\dfrac{20}{20}\cdot\dfrac{\dfrac{2}{5}-\dfrac{1}{4}}{\dfrac{9}{10}+\dfrac{5}{2}}$

$$=\frac{20\cdot\dfrac{2}{5}-20\cdot\dfrac{1}{4}}{20\cdot\dfrac{9}{10}+20\cdot\dfrac{5}{2}}$$

$$=\frac{8-5}{18+50}$$

$$=\frac{3}{68}$$

5. $\dfrac{2-\dfrac{3}{8}}{\dfrac{5}{4}+\dfrac{1}{3}}=\dfrac{24}{24}\cdot\dfrac{2-\dfrac{3}{8}}{\dfrac{5}{4}+\dfrac{1}{3}}$

$$=\frac{24\cdot2-24\cdot\dfrac{3}{8}}{24\cdot\dfrac{5}{4}+24\cdot\dfrac{1}{3}}$$

$$=\frac{48-9}{30+8}$$

$$=\frac{39}{38}$$

7. $\dfrac{x+\dfrac{3}{5}}{x+\dfrac{4}{7}}=\dfrac{35}{35}\cdot\dfrac{x+\dfrac{3}{5}}{x+\dfrac{4}{7}}$

$$=\frac{35\cdot x+35\cdot\dfrac{3}{5}}{35\cdot x+35\cdot\dfrac{4}{7}}$$

$$=\frac{35x+21}{35x+20}$$

$$=\frac{7(5x+3)}{5(7x+4)}$$

9. $\dfrac{6+\dfrac{15}{x}}{x+\dfrac{5}{2}} = \dfrac{2x}{2x}\cdot\dfrac{6+\dfrac{15}{x}}{x+\dfrac{5}{2}}$

$\qquad = \dfrac{2x\cdot 6 + 2x\cdot\dfrac{15}{x}}{2x\cdot x + 2x\cdot\dfrac{5}{2}}$

$\qquad = \dfrac{12x+30}{2x^2+5x}$

$\qquad = \dfrac{6\,(2x+5)^1}{x\,(2x+5)_1}$

$\qquad = \dfrac{6}{x}$

11. $\dfrac{14-\dfrac{4}{x}}{21-\dfrac{6}{x}} = \dfrac{x}{x}\cdot\dfrac{14-\dfrac{4}{x}}{21-\dfrac{6}{x}}$

$\qquad = \dfrac{x\cdot 14 - x\cdot\dfrac{4}{x}}{x\cdot 21 - x\cdot\dfrac{6}{x}}$

$\qquad = \dfrac{14x-4}{21x-6}$

$\qquad = \dfrac{2\,(7x-2)^1}{3\,(7x-2)_1}$

$\qquad = \dfrac{2}{3}$

13. $\dfrac{3+\dfrac{15}{x}}{1-\dfrac{25}{x^2}} = \dfrac{x^2}{x^2}\cdot\dfrac{3+\dfrac{15}{x}}{1-\dfrac{25}{x^2}}$

$\qquad = \dfrac{x^2\cdot 3 + x^2\cdot\dfrac{15}{x}}{x^2\cdot 1 - x^2\cdot\dfrac{25}{x^2}}$

$\qquad = \dfrac{3x^2+15x}{x^2-25}$

$\qquad = \dfrac{3x\,(x+5)^1}{(x+5)_1(x-5)}$

$\qquad = \dfrac{3x}{x-5}$

15. $\dfrac{\dfrac{4}{x+3}+\dfrac{2}{x+6}}{\dfrac{x+5}{x+3}}$

$\qquad = \dfrac{(x+3)(x+6)}{(x+3)(x+6)}\cdot\dfrac{\dfrac{4}{x+3}+\dfrac{2}{x+6}}{\dfrac{x+5}{x+3}}$

$\qquad = \dfrac{(x+3)^1(x+6)\cdot\dfrac{4}{x+3_1}+(x+3)\,(x+6)^1\cdot\dfrac{2}{x+6_1}}{(x+3)^1(x+6)\cdot\dfrac{x+5}{x+3_1}}$

$\qquad = \dfrac{4(x+6)+2(x+3)}{(x+6)(x+5)}$

$\qquad = \dfrac{4x+24+2x+6}{(x+6)(x+5)}$

$\qquad = \dfrac{6x+30}{(x+6)(x+5)}$

$\qquad = \dfrac{6\,(x+5)^1}{(x+6)\,(x+5)_1}$

$\qquad = \dfrac{6}{x+6}$

17. $\dfrac{\dfrac{12}{x}+\dfrac{3}{x-5}}{\dfrac{x+4}{x}+\dfrac{2}{x-5}} = \dfrac{x(x-5)}{x(x-5)}\cdot\dfrac{\dfrac{12}{x}+\dfrac{3}{x-5}}{\dfrac{x+4}{x}+\dfrac{2}{x-5}}$

$\qquad = \dfrac{x^1(x-5)\cdot\dfrac{12}{x_1}+x\,(x-5)^1\cdot\dfrac{3}{x-5_1}}{x^1(x-5)\cdot\dfrac{x+4}{x_1}+x\,(x-5)^1\cdot\dfrac{2}{x-5_1}}$

$\qquad = \dfrac{12(x-5)+3x}{(x+4)(x-5)+2x}$

$\qquad = \dfrac{12x-60+3x}{x^2-5x+4x-20+2x}$

$\qquad = \dfrac{15x-60}{x^2+x-20}$

$\qquad = \dfrac{15\,(x-4)^1}{(x-4)_1(x+5)}$

$\qquad = \dfrac{15}{x+5}$

19.
$$\dfrac{1+\dfrac{4}{x}-\dfrac{32}{x^2}}{1+\dfrac{13}{x}+\dfrac{40}{x^2}}=\dfrac{x^2}{x^2}\cdot\dfrac{1+\dfrac{4}{x}-\dfrac{32}{x^2}}{1+\dfrac{13}{x}+\dfrac{40}{x^2}}$$

$$=\dfrac{x^2\cdot1+x^2\cdot\dfrac{4}{x}-x^2\cdot\dfrac{32}{x^2}}{x^2\cdot1+x^2\cdot\dfrac{13}{x}+x^2\cdot\dfrac{40}{x^2}}$$

$$=\dfrac{x^2+4x-32}{x^2+13x+40}$$

$$=\dfrac{(x+8)(x-4)}{(x+8)(x+5)}$$

$$=\dfrac{x-4}{x+5}$$

21.
$$\dfrac{\dfrac{8}{x^2}-\dfrac{8}{x}+2}{\dfrac{4}{x^2}-1}=\dfrac{x^2}{x^2}\cdot\dfrac{\dfrac{8}{x^2}-\dfrac{8}{x}+2}{\dfrac{4}{x^2}-1}$$

$$=\dfrac{x^2\cdot\dfrac{8}{x^2}-x^2\cdot\dfrac{8}{x}+x^2\cdot2}{x^2\cdot\dfrac{4}{x^2}-x^2\cdot1}$$

$$=\dfrac{8-8x+2x^2}{4-x^2}$$

$$=\dfrac{2x^2-8x+8}{4-x^2}$$

$$=\dfrac{2\left(x^2-4x+4\right)}{(2+x)(2-x)}$$

$$=\dfrac{2(x-2)(x-2)}{(2+x)(2-x)}$$

$$=-\dfrac{2(x-2)}{2+x}$$

23.
$$\dfrac{\dfrac{x^2+14x+48}{x^2+3x-40}}{\dfrac{x^2-3x-54}{x^2-25}}$$

$$=\dfrac{x^2+14x+48}{x^2+3x-40}\div\dfrac{x^2-3x-54}{x^2-25}$$

$$=\dfrac{x^2+14x+48}{x^2+3x-40}\cdot\dfrac{x^2-25}{x^2-3x-54}$$

$$=\dfrac{(x+8)(x+6)}{(x-5)(x+8)}\cdot\dfrac{(x+5)(x-5)}{(x+6)(x-9)}$$

$$=\dfrac{x+5}{x-9}$$

25.
$$\dfrac{\dfrac{4x^2-16x-9}{8x-x^2}}{\dfrac{4x^2-1}{x^2-x-56}}$$

$$=\dfrac{4x^2-16x-9}{8x-x^2}\div\dfrac{4x^2-1}{x^2-x-56}$$

$$=\dfrac{4x^2-16x-9}{8x-x^2}\cdot\dfrac{x^2-x-56}{4x^2-1}$$

$$=\dfrac{(2x-9)(2x+1)}{x(8-x)}\cdot\dfrac{(x-8)(x+7)}{(2x+1)(2x-1)}$$

$$=-\dfrac{(2x-9)(x+7)}{x(2x-1)}$$

27.
$$\dfrac{x^2+4x}{x^2-x-42}+\dfrac{x-6}{x^2-x-42}=\dfrac{x^2+5x-6}{x^2-x-42}$$

$$=\dfrac{(x+6)(x-1)}{(x+6)(x-7)}$$

$$=\dfrac{(x+6)(x-1)}{(x+6)(x-7)}$$

$$=\dfrac{x-1}{x-7}$$

29. $\dfrac{x+3}{x^2-2x-24}+\dfrac{5}{x^2-8x+12}$

$$=\dfrac{x+3}{(x-6)(x+4)}+\dfrac{5}{(x-6)(x-2)}$$

$$=\dfrac{x+3}{(x-6)(x+4)}\cdot\dfrac{x-2}{x-2}+\dfrac{5}{(x-6)(x-2)}\cdot\dfrac{x+4}{x+4}$$

$$=\dfrac{x^2-2x+3x-6+5x+20}{(x-6)(x+4)(x-2)}$$

$$=\dfrac{x^2+6x+14}{(x-6)(x+4)(x-2)}$$

31. $\dfrac{x^2+7x}{3x^2-19x+20}\div\dfrac{x^2+12x+35}{x^2-3x-10}$

$$=\dfrac{x^2+7x}{3x^2-19x+20}\cdot\dfrac{x^2-3x-10}{x^2+12x+35}$$

$$=\dfrac{x(x+7)}{(3x-4)(x-5)}\cdot\dfrac{(x-5)(x+2)}{(x+5)(x+7)}$$

$$=\dfrac{x\,\cancel{(x+7)}^{\,1}}{(3x-4)\,\cancel{(x-5)}_{1}}\cdot\dfrac{\cancel{(x-5)}^{\,1}(x+2)}{(x+5)\,\cancel{(x+7)}_{1}}$$

$$=\dfrac{x(x+2)}{(3x-4)(x+5)}$$

33. $\dfrac{x+5}{x^2-49}-\dfrac{6}{x^2-7x}$

$$=\dfrac{x+5}{(x+7)(x-7)}-\dfrac{6}{x(x-7)}$$

$$=\dfrac{x+5}{(x+7)(x-7)}\cdot\dfrac{x}{x}-\dfrac{6}{x(x-7)}\cdot\dfrac{x+7}{x+7}$$

$$=\dfrac{x^2+5x-6x-42}{x(x+7)(x-7)}$$

$$=\dfrac{x^2-x-42}{x(x+7)(x-7)}$$

$$=\dfrac{(x-7)(x+6)}{x(x+7)(x-7)}$$

$$=\dfrac{\cancel{(x-7)}^{\,1}(x+6)}{x(x+7)\,\cancel{(x-7)}_{1}}$$

$$=\dfrac{x+6}{x(x+7)}$$

35. $\dfrac{5x}{x-2}+\dfrac{10}{2-x}=\dfrac{5x}{x-2}-\dfrac{10}{x-2}$

$$=\dfrac{5x-10}{(x-2)}$$

$$=\dfrac{5\,\cancel{(x-2)}^{\,1}}{\cancel{(x-2)}_{1}}$$

$$=5$$

37. $\dfrac{\dfrac{7}{x+2}-\dfrac{2}{x-3}}{\dfrac{x-5}{x+2}}$

$$=\dfrac{(x+2)(x-3)}{(x+2)(x-3)}\cdot\dfrac{\dfrac{7}{x+2}-\dfrac{2}{x-3}}{\dfrac{x-5}{x+2}}$$

$$=\dfrac{\cancel{(x+2)}(x-3)\cdot\dfrac{7}{\cancel{x+2}}-(x+2)\,\cancel{(x-3)}\cdot\dfrac{2}{\cancel{x-3}}}{\cancel{(x+2)}(x-3)\cdot\dfrac{(x-5)}{\cancel{(x+2)}}}$$

$$=\dfrac{7(x-3)-2(x+2)}{(x-3)(x-5)}$$

$$=\dfrac{7x-21-2x-4}{(x-3)(x-5)}$$

$$=\dfrac{5x-25}{(x-3)(x-5)}$$

$$=\dfrac{5\,\cancel{(x-5)}^{\,1}}{(x-3)\,\cancel{(x-5)}_{1}}$$

$$=\dfrac{5}{x-3}$$

39.
$$\frac{x^2-13x+14}{x^2-6x}-\frac{x^2+4x-32}{6x-x^2}$$
$$=\frac{x^2-13x+14}{x(x-6)}-\frac{x^2+4x-32}{x(6-x)}$$
$$=\frac{x^2-13x+14}{x(x-6)}-\frac{x^2+4x-32}{-x(x-6)}$$
$$=\frac{x^2-13x+14}{x(x-6)}+\frac{x^2+4x-32}{x(x-6)}$$
$$=\frac{x^2-13x+14+x^2+4x-32}{x(x-6)}$$
$$=\frac{2x^2-9x-18}{x(x-6)}$$
$$=\frac{(2x+3)(x-6)^1}{x(x-6)_1}$$
$$=\frac{2x+3}{x}$$

41.
$$\frac{\frac{1}{8}-\frac{1}{x}}{\frac{1}{x^2}-\frac{1}{64}}=\frac{64x^2}{64x^2}\cdot\frac{\frac{1}{8}-\frac{1}{x}}{\frac{1}{x^2}-\frac{1}{64}}$$
$$=\frac{64^8x^2\cdot\frac{1}{8_1}-64x^{2\,x}\cdot\frac{1}{x_1}}{64x^{2\,1}\cdot\frac{1}{x^2_1}-64^1x^2\cdot\frac{1}{64_1}}$$
$$=\frac{8x^2-64x}{64-x^2}$$
$$=\frac{8x(x-8)^1}{(8+x)(8-x)_{-1}}$$
$$=-\frac{8x}{8+x}$$

43.
$$\frac{8}{x^2+7x+12}+\frac{4}{x^2+10x+24}$$
$$=\frac{8}{(x+3)(x+4)}+\frac{4}{(x+4)(x+6)}$$
$$=\frac{8}{(x+3)(x+4)}\cdot\frac{x+6}{x+6}+\frac{4}{(x+4)(x+6)}\cdot\frac{x+3}{x+3}$$
$$=\frac{8x+48+4x+12}{(x+3)(x+4)(x+6)}$$
$$=\frac{12x+60}{(x+3)(x+4)(x+6)}$$
$$=\frac{12(x+5)}{(x+3)(x+4)(x+6)}$$

45.
$$\frac{x^3+64}{x^2+14x+49}\cdot\frac{x^2+3x-28}{x^2-16}$$
$$=\frac{(x+4)(x^2-4x+16)}{(x+7)(x+7)}\cdot\frac{(x+7)(x-4)}{(x+4)(x-4)}$$
$$=\frac{(x+4)^1(x^2-4x+16)}{(x+7)_1(x+7)}\cdot\frac{(x+7)^1(x-4)^1}{(x+4)_1(x-4)_1}$$
$$=\frac{x^2-4x+16}{x+7}$$

47.
$$\frac{x^2+18x+77}{x^2-4x-32}\cdot\frac{x^2+6x+8}{x^2+9x+14}$$
$$=\frac{(x+11)(x+7)}{(x-8)(x+4)}\cdot\frac{(x+4)(x+2)}{(x+7)(x+2)}$$
$$=\frac{(x+11)(x+7)^1}{(x-8)(x+4)_1}\cdot\frac{(x+4)^1(x+2)^1}{(x+7)_1(x+2)_1}$$
$$=\frac{x+11}{x-8}$$

49.
$$\frac{x^2+7x}{x^2-2x-3}-\frac{4x+18}{x^2-2x-3}=\frac{x^2+7x-4x-18}{x^2-2x-3}$$
$$=\frac{x^2+3x-18}{x^2-2x-3}$$
$$=\frac{(x+6)(x-3)^1}{(x+1)(x-3)_1}$$
$$=\frac{x+6}{x+1}$$

51.
$$\frac{x^3-1000}{x^2+10x-11}\div\frac{3x^2-37x+70}{x^2+4x-5}$$
$$=\frac{x^3-1000}{x^2+10x-11}\cdot\frac{x^2+4x-5}{3x^2-37x+70}$$
$$=\frac{(x-10)^1(x^2+10x+100)}{(x+11)(x-1)_1}\cdot\frac{(x+5)(x-1)^1}{(x-10)_1(3x-7)}$$
$$=\frac{(x^2+10x+100)(x+5)}{(x+11)(3x-7)}$$

53. Answers will vary. Example:
Complex fractions are fractions whose numerator and/or denominator are fractions. In contrast, the numerators and denominators of the rational expressions in Section 7.1 were polynomials.

7.6 QUICK CHECK

1.

$$\frac{6}{x} - \frac{1}{8} = \frac{7}{40}$$

$$40x \cdot \left(\frac{6}{x} - \frac{1}{8}\right) = 40x \cdot \frac{7}{40}$$

$$240 - 5x = 7x$$
$$240 = 12x$$
$$20 = x$$

Check:

$$\frac{6}{20} - \frac{1}{8} = \frac{7}{40}$$

$$\frac{12}{40} - \frac{5}{40} = \frac{7}{40}$$

$$\frac{7}{40} = \frac{7}{40}$$

True.

$$\{20\}$$

2.

$$1 = \frac{5}{x} + \frac{24}{x^2}$$

$$x^2 \cdot 1 = x^2\left(\frac{5}{x} + \frac{24}{x^2}\right)$$

$$x^2 = 5x + 24$$
$$x^2 - 5x = 24$$
$$x^2 - 5x - 24 = 0$$
$$(x+3)(x-8) = 0$$

$$x + 3 = 0 \quad \text{or} \quad x - 8 = 0$$
$$x = -3 \quad \text{or} \quad x = 8$$

$$\{-3, 8\}$$

3.

$$\frac{7}{x-4} + 3 = \frac{2x-1}{x-4}$$

$$(x-4) \cdot \frac{7}{x-4} + (x-4) \cdot 3 = (x-4) \cdot \frac{2x-1}{x-4}$$

$$7 + 3(x-4) = 2x - 1$$
$$7 + 3x - 12 = 2x - 1$$
$$3x - 5 = 2x - 1$$
$$x - 5 = -1$$
$$x = 4$$

$\varnothing$ since $x = 4$ makes the original equation undefined.

4.

$$\frac{x+2}{x^2 - 3x - 54} = \frac{2}{x^2 - 12x + 27}$$

$$\frac{x+2}{(x-9)(x+6)} = \frac{2}{(x-3)(x-9)}$$

$$(x-3)(x-9)(x+6) \cdot \frac{x+2}{(x-9)(x+6)} = (x-3)(x-9)(x+6) \cdot \frac{2}{(x-3)(x-9)}$$

$$(x-3)(x+2) = 2(x+6)$$
$$x^2 + 2x - 3x - 6 = 2x + 12$$
$$x^2 - x - 6 = 2x + 12$$
$$x^2 - 3x - 18 = 0$$
$$(x+3)(x-6) = 0$$

$$x + 3 = 0 \quad \text{or} \quad x - 6 = 0$$
$$x = -3 \quad \text{or} \quad x = 6$$

$$\{-3, 6\}$$

5.

$$\frac{x+3}{x^2 + x - 12} - \frac{3}{x^2 - 2x - 3} = \frac{x+7}{x^2 + 5x + 4}$$

$$\frac{x+3}{(x+4)(x-3)} - \frac{3}{(x-3)(x+1)} = \frac{x+7}{(x+4)(x+1)}$$

$$(x+1)(x+4)(x-3) \cdot \left(\frac{x+3}{(x+4)(x-3)} - \frac{3}{(x-3)(x+1)}\right) = (x+1)(x+4)(x-3) \cdot \frac{x+7}{(x+4)(x+1)}$$

$$(x+1)(x+3) - 3(x+4) = (x-3)(x+7)$$
$$x^2 + 3x + 1x + 3 - 3x - 12 = x^2 + 7x - 3x - 21$$
$$x^2 + x - 9 = x^2 + 4x - 21$$
$$x - 9 = 4x - 21$$
$$12 = 3x$$
$$4 = x$$

$$\{4\}$$

6.
$$\frac{2}{x}+\frac{3}{y}=\frac{4}{z}$$

$$xyz\cdot\left(\frac{2}{x}+\frac{3}{y}\right)=xyz\cdot\frac{4}{z}$$

$$2yz+3xz=4xy$$
$$2yz=4xy-3xz$$
$$2yz=x(4y-3z)$$

$$\frac{2yz}{4y-3z}=x$$

7.
$$y=\frac{5x}{4x-3}$$

$$y(4x-3)=\frac{5x}{4x-3_{1}}\cdot(4x-3)^{1}$$

$$4xy-3y=5x$$
$$-3y=5x-4xy$$
$$-3y=x(5-4y)$$

$$\frac{-3y}{5-4y}=\frac{x(5-4y)^{1}}{5-4y_{1}}$$

$$\frac{-3y}{5-4y}=x$$

$$x=\frac{3y}{4y-5}$$

7.6 RATIONAL EQUATIONS

1. rational equation

3. extraneous solution

5.
$$\frac{x}{9}+\frac{11}{18}=\frac{7}{6}$$

$$18\cdot\left(\frac{x}{9}+\frac{11}{18}\right)=\frac{18}{1}\cdot\frac{7}{6}$$

$$2x+11=21$$
$$2x=10$$
$$x=5$$
$$\{5\}$$

7.
$$\frac{5}{6}+\frac{11}{x}=\frac{7}{4}$$

$$12x\cdot\left(\frac{5}{6}+\frac{11}{x}\right)=12x\cdot\frac{7}{4}$$

$$10x+132=21x$$
$$132=11x$$
$$12=x$$
$$\{12\}$$

9.
$$\frac{25}{21}-\frac{17}{x}=\frac{7}{12}$$

$$84x\cdot\left(\frac{25}{21}-\frac{17}{x}\right)=84x\cdot\frac{7}{12}$$

$$100x-1428=49x$$
$$51x=1428$$
$$x=28$$
$$\{28\}$$

11.
$$x-5+\frac{12}{x}=2$$

$$x\cdot\left(x-5+\frac{12}{x}\right)=x\cdot2$$

$$x^2-5x+12=2x$$
$$x^2-7x+12=0$$
$$(x-3)(x-4)=0$$
$$x-3=0\quad\text{or}\quad x-4=0$$
$$x=3\quad\text{or}\qquad x=4$$
$$\{3,4\}$$

13.
$$\frac{x}{2}+\frac{6}{x}=4$$

$$2x\cdot\left(\frac{x}{2}+\frac{6}{x}\right)=2x\cdot4$$

$$x^2+12=8x$$

$$x^2-8x+12=0$$
$$(x-2)(x-6)=0$$
$$x-2=0\quad\text{or}\quad x-6=0$$
$$x=2\quad\text{or}\qquad x=6$$
$$\{2,6\}$$

15.
$$1-\frac{7}{x}+\frac{10}{x^2}=0$$

$$x^2\cdot\left(1-\frac{7}{x}+\frac{10}{x^2}\right)=x^2\cdot0$$

$$x^2-7x+10=0$$
$$(x-2)(x-5)=0$$
$$x-2=0\quad\text{or}\quad x-5=0$$
$$x=2\quad\text{or}\qquad x=5$$
$$\{2,5\}$$

17.
$$1 - \frac{25}{x^2} = 0$$
$$x^2 \cdot \left(1 - \frac{25}{x^2}\right) = x^2 \cdot 0$$
$$x^2 - 25 = 0$$
$$(x+5)(x-5) = 0$$
$$x+5 = 0 \quad \text{or} \quad x-5 = 0$$
$$x = -5 \quad \text{or} \quad x = 5$$
$$\{-5, 5\}$$

19.
$$\frac{8x-3}{x+7} = \frac{2x+15}{x+7}$$
$$(x+7) \cdot \frac{8x-3}{x+7} = (x+7) \cdot \frac{2x+15}{x+7}$$
$$8x - 3 = 2x + 15$$
$$6x - 3 = 15$$
$$6x = 18$$
$$x = 3$$
$$\{3\}$$

21.
$$\frac{7x-11}{5x-2} = \frac{2x-9}{5x-2}$$
$$(5x-2) \cdot \frac{7x-11}{5x-2} = (5x-2) \cdot \frac{2x-9}{5x-2}$$
$$7x - 11 = 2x - 9$$
$$5x = 2$$
$$x = \frac{2}{5}$$

$\varnothing$ since $x = \frac{2}{5}$ makes the original equation undefined.

23.
$$8 + \frac{6}{x-4} = \frac{x-12}{x-4}$$
$$(x-4) \cdot \left(8 + \frac{6}{x-4}\right) = (x-4) \cdot \frac{x-12}{x-4}$$
$$8(x-4) + 6 = x - 12$$
$$8x - 32 + 6 = x - 12$$
$$8x - 26 = x - 12$$
$$7x - 26 = -12$$
$$7x = 14$$
$$x = 2$$
$$\{2\}$$

25.
$$x + \frac{x+11}{x-6} = \frac{8x-31}{x-6}$$
$$(x-6) \cdot \left(x + \frac{x+11}{x-6}\right) = (x-6) \cdot \frac{8x-31}{x-6}$$
$$x(x-6) + x + 11 = 8x - 31$$
$$x^2 - 6x + x + 11 = 8x - 31$$
$$x^2 - 5x + 11 = 8x - 31$$
$$x^2 - 13x + 42 = 0$$
$$(x-6)(x-7) = 0$$
$$x - 6 = 0 \quad \text{or} \quad x - 7 = 0$$
$$x = 6 \quad \text{or} \quad x = 7$$

$\{7\}$ since $x = 6$ makes the original equation undefined.

27.
$$\frac{3}{x-7} = \frac{7}{x+5}$$
$$(x-7)(x+5) \cdot \frac{3}{x-7} = (x-7)(x+5) \cdot \frac{7}{x+5}$$
$$3(x+5) = 7(x-7)$$
$$3x + 15 = 7x - 49$$
$$-4x + 15 = -49$$
$$-4x = -64$$
$$x = 16$$

$$\{16\}$$

29.
$$\frac{x-3}{x+9} = \frac{6}{x+2}$$
$$(x+9)(x+2) \cdot \frac{x-3}{x+9} = (x+9)(x+2) \cdot \frac{6}{x+2}$$
$$(x+2)(x-3) = 6(x+9)$$
$$x^2 - 3x + 2x - 6 = 6x + 54$$
$$x^2 - x - 6 = 6x + 54$$
$$x^2 - 7x - 60 = 0$$
$$(x+5)(x-12) = 0$$
$$x + 5 = 0 \quad \text{or} \quad x - 12 = 0$$
$$x = -5 \quad \text{or} \quad x = 12$$
$$\{-5, 12\}$$

31.
$$\frac{3}{x+2}-\frac{1}{x+1}=\frac{x+3}{x^2+3x+2}$$
$$\frac{3}{x+2}-\frac{1}{x+1}=\frac{x+3}{(x+2)(x+1)}$$
$$(x+2)(x+1)\cdot\left(\frac{3}{x+2}-\frac{1}{x+1}\right)=(x+2)(x+1)\cdot\frac{x+3}{(x+2)(x+1)}$$
$$3(x+1)-(x+2)=x+3$$
$$3x+3-x-2=x+3$$
$$2x+1=x+3$$
$$x=2$$

$\{2\}$

33.
$$\frac{4}{x+4}+\frac{3}{x-4}=\frac{24}{x^2-16}$$
$$\frac{4}{x+4}+\frac{3}{x-4}=\frac{24}{(x+4)(x-4)}$$
$$(x+4)(x-4)\cdot\left(\frac{4}{x+4}+\frac{3}{x-4}\right)=(x+4)(x-4)\cdot\frac{24}{(x+4)(x-4)}$$
$$4(x-4)+3(x+4)=24$$
$$4x-16+3x+12=24$$
$$7x-4=24$$
$$7x=28$$
$$x=4$$

$\varnothing$ since $x=4$ makes the original equation undefined.

35.
$$\frac{2}{x^2-x-2}+\frac{10}{x^2-2x-3}=\frac{x+12}{x^2-x-2}$$
$$\frac{2}{(x-2)(x+1)}+\frac{10}{(x-3)(x+1)}=\frac{x+12}{(x-2)(x+1)}$$
$$(x-2)(x+1)(x-3)\cdot\left(\frac{2}{(x-2)(x+1)}+\frac{10}{(x-3)(x+1)}\right)=(x-2)(x+1)(x-3)\cdot\frac{x+12}{(x-2)(x+1)}$$
$$2(x-3)+10(x-2)=(x-3)(x+12)$$
$$2x-6+10x-20=x^2+12x-3x-36$$
$$12x-26=x^2+9x-36$$
$$0=x^2-3x-10$$
$$0=(x+2)(x-5)$$
$$x+2=0 \quad \text{or} \quad x-5=0$$
$$x=-2 \quad \text{or} \quad x=5$$

$\{-2,5\}$

37.
$$\frac{x-8}{x-5}+\frac{x-9}{x-4}=\frac{x+7}{x^2-9x+20}$$
$$\frac{x-8}{x-5}+\frac{x-9}{x-4}=\frac{x+7}{(x-5)(x-4)}$$
$$(x-5)(x-4)\cdot\left(\frac{x-8}{x-5}+\frac{x-9}{x-4}\right)=(x-5)(x-4)\cdot\frac{x+7}{(x-5)(x-4)}$$
$$(x-4)(x-8)+(x-5)(x-9)=x+7$$
$$x^2-8x-4x+32+x^2-9x-5x+45=x+7$$
$$2x^2-26x+77=x+7$$
$$2x^2-27x+70=0$$
$$(2x-7)(x-10)=0$$
$$\begin{array}{ccc} 2x-7=0 & \text{or} & x-10=0 \\ 2x=7 & \text{or} & x=10 \\ x=\dfrac{7}{2} & \text{or} & x=10 \end{array}$$
$$\left\{\frac{7}{2},10\right\}$$

39.
$$\frac{x+3}{x^2-4x-12}+\frac{x-11}{x^2-2x-24}=\frac{x+1}{x^2+6x+8}$$
$$\frac{x+3}{(x-6)(x+2)}+\frac{x-11}{(x-6)(x+4)}=\frac{x+1}{(x+2)(x+4)}$$
$$(x-6)(x+2)(x+4)\cdot\left(\frac{x+3}{(x-6)(x+2)}+\frac{x-11}{(x-6)(x+4)}\right)=(x-6)(x+2)(x+4)\cdot\frac{x+1}{(x+2)(x+4)}$$
$$(x+4)(x+3)+(x+2)(x-11)=(x-6)(x+1)$$
$$x^2+3x+4x+12+x^2-11x+2x-22=x^2+x-6x-6$$
$$2x^2-2x-10=x^2-5x-6$$
$$x^2+3x-4=0$$
$$(x+4)(x-1)=0$$
$$\begin{array}{ccc} x+4=0 & \text{or} & x-1=0 \\ x=-4 & \text{or} & x=1 \end{array}$$
$\{1\}$ since $x=-4$ makes the original equation undefined.

41.
$$\frac{3x-4}{x^2-10x+21}-\frac{x-8}{x^2-18x+77}=\frac{x-5}{x^2-14x+33}$$
$$\frac{3x-4}{(x-3)(x-7)}-\frac{x-8}{(x-11)(x-7)}=\frac{x-5}{(x-11)(x-3)}$$
$$(x-3)(x-7)(x-11)\cdot\left(\frac{3x-4}{(x-3)(x-7)}-\frac{x-8}{(x-11)(x-7)}\right)=(x-3)(x-7)(x-11)\cdot\frac{x-5}{(x-11)(x-3)}$$
$$(x-11)(3x-4)-(x-3)(x-8)=(x-7)(x-5)$$
$$3x^2-4x-33x+44-\left(x^2-8x-3x+24\right)=x^2-5x-7x+35$$
$$3x^2-4x-33x+44-x^2+8x+3x-24=x^2-5x-7x+35$$
$$2x^2-26x+20=x^2-12x+35$$
$$x^2-14x-15=0$$
$$(x+1)(x-15)=0$$
$$\begin{array}{ccc} x+1=0 & \text{or} & x-15=0 \\ x=-1 & \text{or} & x=15 \end{array}$$
$$\{-1,15\}$$

43.
$$1+\frac{13}{x}+\frac{42}{x^2}=0$$
$$x^2\cdot\left(1+\frac{13}{x}+\frac{42}{x^2}\right)=x^2\cdot 0$$
$$x^2+13x+42=0$$
$$(x+7)(x+6)=0$$
$$x+7=0 \quad\text{or}\quad x+6=0$$
$$x=-7 \quad\text{or}\quad x=-6$$
$$\{-7,-6\}$$

45.
$$x^2-14x-120=0$$
$$(x+6)(x-20)=0$$
$$x+6=0 \quad\text{or}\quad x-20=0$$
$$x=-6 \quad\text{or}\quad x=20$$
$$\{-6,20\}$$

47.
$$2(3x+4)-19=4x-1$$
$$6x+8-19=4x-1$$
$$2x=10$$
$$x=5$$
$$\{5\}$$

49.
$$\frac{x}{x+3}+\frac{x-4}{x-3}=\frac{9x-5}{x^2-9}$$
$$\frac{x}{x+3}+\frac{x-4}{x-3}=\frac{9x-5}{(x+3)(x-3)}$$
$$(x+3)(x-3)\cdot\left(\frac{x}{x+3}+\frac{x-4}{x-3}\right)=(x+3)(x-3)\cdot\frac{9x-5}{(x+3)(x-3)}$$
$$x(x-3)+(x+3)(x-4)=9x-5$$
$$x^2-3x+x^2-4x+3x-12=9x-5$$
$$2x^2-4x-12=9x-5$$
$$2x^2-13x-7=0$$
$$(2x+1)(x-7)=0$$
$$2x+1=0 \quad\text{or}\quad x-7=0$$
$$2x=-1 \quad\text{or}\quad x=7$$
$$x=-\frac{1}{2} \quad\text{or}\quad x=7$$
$$\left\{-\frac{1}{2},7\right\}$$

51.
$$6x^2+29x-5=0$$
$$(x+5)(6x-1)=0$$
$$x+5=0 \quad\text{or}\quad 6x-1=0$$
$$x=-5 \quad\text{or}\quad 6x=1$$
$$x=-5 \quad\text{or}\quad x=\frac{1}{6}$$
$$\left\{-5,\frac{1}{6}\right\}$$

53.
$$\frac{6}{x+2}=\frac{25}{3x+13}$$
$$(x+2)(3x+13)\cdot\frac{6}{x+2}=(x+2)(3x+13)\cdot\frac{25}{3x+13}$$
$$6(3x+13)=25(x+2)$$
$$18x+78=25x+50$$
$$-7x=-28$$
$$x=4$$
$$\{4\}$$

55.
$$\frac{x}{x+6}+\frac{3}{x+4}=\frac{8}{x^2+10x+24}$$
$$\frac{x}{x+6}+\frac{3}{x+4}=\frac{8}{(x+6)(x+4)}$$
$$(x+6)(x+4)\cdot\left(\frac{x}{x+6}+\frac{3}{x+4}\right)=(x+6)(x+4)\cdot\frac{8}{(x+6)(x+4)}$$
$$x(x+4)+3(x+6)=8$$
$$x^2+4x+3x+18=8$$
$$x^2+7x+10=0$$
$$(x+5)(x+2)=0$$
$$x+5=0 \quad\text{or}\quad x+2=0$$
$$x=-5 \quad\text{or}\quad x=-2$$
$$\{-5,-2\}$$

57.
$$5x-17=8x+13$$
$$-3x=30$$
$$x=-10$$
$$\{-10\}$$

59.
$$L=\frac{A}{W}$$
$$W\cdot L=W\cdot\frac{A}{W}$$
$$W\cdot L=A$$
$$W=\frac{A}{L}$$

61.
$$y=\frac{x}{2x+5}$$
$$(2x+5)\cdot y=(2x+5)\cdot\frac{x}{2x+5}$$
$$2xy+5y=x$$
$$5y=x-2xy$$
$$5y=-x(2y-1)$$
$$\frac{5y}{2y-1}=-x$$
$$x=-\frac{5y}{2y-1}$$

63.
$$y=\frac{2x-9}{3x-8}$$
$$(3x-8)\cdot y=(3x-8)\cdot\frac{2x-9}{3x-8}$$
$$3xy-8y=2x-9$$
$$3xy-2x=8y-9$$
$$x(3y-2)=8y-9$$
$$x=\frac{8y-9}{3y-2}$$

65.
$$\frac{x}{r}+\frac{y}{2r}=1$$
$$2r\cdot\frac{x}{r}+2r\cdot\frac{y}{2r}=2r\cdot1$$
$$2x+y=2r$$
$$r=\frac{2x+y}{2}$$

67.
$$m=\frac{y-y_1}{x-x_1}$$
$$(x-x_1)m=(x-x_1)\frac{y-y_1}{x-x_1}$$
$$mx-mx_1=y-y_1$$
$$mx=y-y_1+mx_1$$
$$x=\frac{y-y_1+mx_1}{m}$$

69. a) Substitue $x = 9$ into the given equation.

$$\frac{9-4}{9-1} + \frac{7}{9+3} = \frac{?}{9^2 + 2 \cdot 9 - 3}$$

$$\frac{5}{8} + \frac{7}{12} = \frac{?}{81 + 18 - 3}$$

$$\frac{15}{24} + \frac{14}{24} = \frac{?}{96}$$

$$\frac{29}{24} = \frac{?}{96}$$

$$\frac{29}{24} \cdot \frac{4}{4} = \frac{?}{96}$$

$$\frac{116}{96} = \frac{?}{96}$$

The missing numerator is 116.

b)

$$\frac{x-4}{x-1} + \frac{7}{x+3} = \frac{116}{x^2 + 2x - 3}$$

$$\frac{x-4}{x-1} + \frac{7}{x+3} = \frac{116}{(x+3)(x-1)}$$

$$(x+3)(x-1) \cdot \left(\frac{x-4}{x-1} + \frac{7}{x+3} \right) = \frac{116}{(x+3)(x-1)} \cdot (x+3)(x-1)$$

$$(x+3)(x-4) + 7(x-1) = 116$$

$$x^2 - 4x + 3x - 12 + 7x - 7 = 116$$

$$x^2 + 6x - 19 = 116$$

$$x^2 + 6x - 135 = 0$$

$$(x-9)(x+15) = 0$$

$$x - 9 = 0 \quad \text{or} \quad x + 15 = 0$$
$$x = 9 \quad \text{or} \quad x = -15$$

The other solution is $x = -15$.

71. Answers will vary. Example:
Evaluate the original equation with the specified value.
If any of the denominators become zero, then the value
is an extraneous solution.

73. Answers will vary.

7.7   QUICK CHECK

1.   Unknown: x

$$\frac{1}{x}+\frac{3}{8}=\frac{19}{40}$$

$$40x\cdot\left(\frac{1}{x}+\frac{3}{8}\right)=40x\cdot\frac{19}{40}$$

$$40+15x=19x$$
$$40=4x$$
$$10=x$$

2.   Smaller number: x
Larger number: $x+9$

$$2\cdot\frac{1}{x}+4\cdot\frac{1}{x+9}=1$$

$$\frac{2}{x}+\frac{4}{x+9}=1$$

$$x(x+9)\cdot\left(\frac{2}{x}+\frac{4}{x+9}\right)=x(x+9)\cdot1$$

$$2(x+9)+4x=x(x+9)$$
$$2x+18+4x=x^2+9x$$
$$6x+18=x^2+9x$$
$$0=x^2+3x-18$$
$$0=(x+6)(x-3)$$

$$x+6=0 \quad\text{or}\quad x-3=0$$
$$x=-6 \quad\text{or}\quad x=3$$

Smaller number: $x=3$
Larger number: $x+9=3+9=12$

3.

Person	Job Time	Rate	Time	Portion
New	20	$\frac{1}{20}$	t	$\frac{t}{20}$
Old	35	$\frac{1}{35}$	t	$\frac{t}{35}$

$$\frac{t}{20}+\frac{t}{35}=1$$

$$140\cdot\left(\frac{t}{20}+\frac{t}{35}\right)=140\cdot1$$

$$7t+4t=140$$
$$11t=140$$

$$t=\frac{140}{11}$$

$$t=12\frac{8}{11}\ \text{minutes}$$

4.

Pipes	Job Time	Rate	Time	Portion
Small	$t+16$	$\frac{1}{t+16}$	6	$\frac{6}{t+16}$
Larger	t	$\frac{1}{t}$	6	$\frac{6}{t}$

$$\frac{6}{t+16}+\frac{6}{t}=1$$

$$t(t+16)\cdot\left(\frac{6}{t+16}+\frac{6}{t}\right)=t(t+16)\cdot1$$

$$6t+6(t+16)=t(t+16)$$
$$6t+6t+96=t^2+16t$$
$$12t+96=t^2+16t$$
$$0=t^2+4t-96$$
$$0=(t+12)(t-8)$$

$$t+12=0 \quad\text{or}\quad t-8=0$$
$$t=-12 \quad\text{or}\quad t=8$$

$$t+16=8+16=24\ \text{hours}$$

5.

	Distance	Rate	Time
To	300 miles	r	$\frac{300}{r}$
From	300 miles	$r+15$	$\frac{300}{r+15}$

$$\frac{300}{r}+\frac{300}{r+15}=9$$

$$r(r+15)\cdot\left(\frac{300}{r}+\frac{300}{r+15}\right)=r(r+15)\cdot9$$

$$300(r+15)+300r=9r(r+15)$$
$$300r+4500+300r=9r^2+135r$$
$$600r+4500=9r^2+135r$$
$$0=9r^2-465r-4500$$
$$0=3\left(3r^2-155r-1500\right)$$
$$0=3(3r+25)(r-60)$$

$$3r+25=0 \quad\text{or}\quad r-60=0$$
$$3r=-25 \quad\text{or}\quad r=60$$

$$r=-\frac{25}{3} \quad\text{or}\quad r=60$$

Rehema's Speed: $r+15=60+15=75\ \text{mph}$

6.

	Distance	Rate	Time
Upstream	8 miles	$r-2$	$\dfrac{8}{r-2}$
Downstream	8 miles	$r+2$	$\dfrac{8}{r+2}$

$$\frac{8}{r-2}+\frac{8}{r+2}=3$$

$$(r-2)(r+2)\cdot\left(\frac{8}{r-2}+\frac{8}{r+2}\right)=(r-2)(r+2)\cdot 3$$

$$8(r+2)+8(r-2)=3\left(r^2-4\right)$$

$$8r+16+8r-16=3r^2-12$$

$$16r=3r^2-12$$

$$0=3r^2-16r-12$$

$$0=(3r+2)(r-6)$$

$$3r+2=0 \quad \text{or} \quad r-6=0$$

$$3r=-2 \quad \text{or} \quad r=6$$

$$r=-\frac{2}{3} \quad \text{or} \quad r=6$$

6 mph

7. Let y represent the tuition.
Let x represent the number of units.
$$y=kx$$
$$288=k\cdot 12$$
$$24=k$$
$$y=24\cdot 15=360$$
$360

8. Let y represent the time.
Let x represent the average speed.
$$y=\frac{k}{x}$$
$$131=\frac{k}{5}$$
$$5\cdot 131=5\cdot\frac{k}{5}$$
$$655=k$$
$$y=\frac{655}{4}=163\frac{3}{4}$$
$163\frac{3}{4}$ minutes

9.
$$y=kx\cdot\sqrt{z}$$
$$1000=k(50)\cdot\sqrt{25}$$
$$1000=k(50)\cdot 5$$
$$1000=250k$$
$$4=k$$
$$y=4\cdot 72\cdot\sqrt{9}=288\cdot 3=864$$

7.7   APPLICATIONS OF RATIONAL EQUATIONS

1. reciprocal

3. divided

5. vary directly

7. Unknown: x
$$\frac{1}{x}+\frac{23}{30}=\frac{5}{6}$$
$$30x\cdot\left(\frac{1}{x}+\frac{23}{30}\right)=30x\cdot\frac{5}{6}$$
$$30+23x=25x$$
$$30=2x$$
$$15=x$$

9. Unknown: x
$$3\cdot\frac{1}{x}+\frac{13}{30}=\frac{7}{12}$$
$$\frac{3}{x}+\frac{13}{30}=\frac{7}{12}$$
$$60x\cdot\left(\frac{3}{x}+\frac{13}{30}\right)=60x\cdot\frac{7}{12}$$
$$180+26x=35x$$
$$180=9x$$
$$20=x$$

11. Unknown: x
$$\frac{1}{x}-\frac{5}{18}=\frac{2}{9}$$
$$18x\cdot\left(\frac{1}{x}-\frac{5}{18}\right)=18x\cdot\frac{2}{9}$$
$$18-5x=4x$$
$$18=9x$$
$$2=x$$

13. Smaller number: x
Larger number: $x + 3$

$$4 \cdot \frac{1}{x} + 3 \cdot \frac{1}{x+3} = 1$$

$$\frac{4}{x} + \frac{3}{x+3} = 1$$

$$x(x+3) \cdot \left(\frac{4}{x} + \frac{3}{x+3} \right) = x(x+3) \cdot 1$$

$$4(x+3) + 3x = x(x+3)$$

$$4x + 12 + 3x = x^2 + 3x$$

$$7x + 12 = x^2 + 3x$$

$$0 = x^2 - 4x - 12$$

$$0 = (x-6)(x+2)$$

$$x - 6 = 0 \quad \text{or} \quad x + 2 = 0$$

$$x = 6 \quad \text{or} \quad x = -2$$

Smaller number: $x = 6$
Larger number: $x + 3 = 6 + 3 = 9$

15. Smaller number: $x - 10$
Larger number: x

$$\frac{1}{x-10} + 3 \cdot \frac{1}{x} = \frac{1}{4}$$

$$\frac{1}{x-10} + \frac{3}{x} = \frac{1}{4}$$

$$4x(x-10) \cdot \left(\frac{1}{x-10} + \frac{3}{x} \right) = 4x(x-10) \cdot \frac{1}{4}$$

$$4x + 12(x-10) = x(x-10)$$

$$4x + 12x - 120 = x^2 - 10x$$

$$16x - 120 = x^2 - 10x$$

$$0 = x^2 - 26x + 120$$

$$0 = (x-20)(x-6)$$

$$x - 20 = 0 \quad \text{or} \quad x - 6 = 0$$

$$x = 20 \quad \text{or} \quad x = 6$$

$x \neq 6$ since $x - 10 = 6 - 10 = -4$,
which is not a positive number.

Smaller number: $x - 10 = 20 - 10 = 10$
Larger number: $x = 20$

17.

Copiers	Job Time	Rate	Time	Portion
Older	15 min	$\frac{1}{15}$	t	$\frac{t}{15}$
Newer	10 min	$\frac{1}{10}$	t	$\frac{t}{10}$

$$\frac{t}{15} + \frac{t}{10} = 1$$

$$30 \cdot \left(\frac{t}{15} + \frac{t}{10} \right) = 30 \cdot 1$$

$$2t + 3t = 30$$

$$5t = 30$$

$$t = 6 \text{ minutes}$$

19.

Hose	Job Time	Rate	Time	Portion
Hose 1	18 hrs	$\frac{1}{18}$	t	$\frac{t}{18}$
Hose 2	24 hrs	$\frac{1}{24}$	t	$\frac{t}{24}$

$$\frac{t}{18} + \frac{t}{24} = 1$$

$$72 \cdot \left(\frac{t}{18} + \frac{t}{24} \right) = 72 \cdot 1$$

$$4t + 3t = 72$$

$$7t = 72$$

$$t = \frac{72}{7}$$

$$t = 10\frac{2}{7} \text{ hours}$$

21.

Person	Job Time	Rate	Time	Portion
Tina	30 min	$\frac{1}{30}$	t	$\frac{t}{30}$
Marisa	40 min	$\frac{1}{40}$	t	$\frac{t}{40}$

$$\frac{t}{30} + \frac{t}{40} = 1$$

$$120 \cdot \left(\frac{t}{30} + \frac{t}{40} \right) = 120 \cdot 1$$

$$4t + 3t = 120$$

$$7t = 120$$

$$t = \frac{120}{7}$$

$$t = 17\frac{1}{7} \text{ minutes}$$

23.

Pipe	Job Time	Rate	Time	Portion
Small	$x+7$	$\dfrac{1}{x+7}$	12	$\dfrac{12}{x+7}$
Large	x	$\dfrac{1}{x}$	12	$\dfrac{12}{x}$

$$\frac{12}{x+7}+\frac{12}{x}=1$$

$$x(x+7)\cdot\left(\frac{12}{x+7}+\frac{12}{x}\right)=x(x+7)\cdot 1$$

$$12x+12(x+7)=x(x+7)$$
$$12x+12x+84=x^2+7x$$
$$24x+84=x^2+7x$$
$$0=x^2-17x-84$$
$$0=(x-21)(x+4)$$

$$x-21=0 \quad\text{or}\quad x+4=0$$
$$x=21 \quad\text{or}\quad x=-4$$

Smaller pipe: $x+7=21+7=28$ hours

25.

Person	Job Time	Rate	Time	Portion
Steve	x	$\dfrac{1}{x}$	6	$\dfrac{6}{x}$
Ross	$x+9$	$\dfrac{1}{x+9}$	6	$\dfrac{6}{x+9}$

$$\frac{6}{x}+\frac{6}{x+9}=1$$

$$x(x+9)\cdot\left(\frac{6}{x}+\frac{6}{x+9}\right)=x(x+9)\cdot 1$$

$$6(x+9)+6x=x(x+9)$$
$$6x+54+6x=x^2+9x$$
$$12x+54=x^2+9x$$
$$0=x^2-3x-54$$
$$0=(x-9)(x+6)$$

$$x-9=0 \quad\text{or}\quad x+6=0$$
$$x=9 \quad\text{or}\quad x=-6$$

Ross' time: $x+9=9+9=18$ hours

27.

Person	Job Time	Rate	Time	Portion
Sarah	$2x$	$\dfrac{1}{2x}$	2	$\dfrac{2}{2x}=\dfrac{1}{x}$
Jeff	x	$\dfrac{1}{x}$	2	$\dfrac{2}{x}$

$$\frac{1}{x}+\frac{2}{x}=1$$

$$\frac{3}{x}=1$$

$$x\cdot\frac{3}{x}=x\cdot 1$$

$$3=x$$

Sarah's time: $2x=2(3)=6$ hours

29.

Person	Job Time	Rate	Time	Portion
Brent	$x+2$	$\dfrac{1}{x+2}$	3	$\dfrac{3}{x+2}$
Tracy	x	$\dfrac{1}{x}$	2	$\dfrac{2}{x}$

$$\frac{3}{x+2}+\frac{2}{x}=1$$

$$x(x+2)\cdot\left(\frac{3}{x+2}+\frac{2}{x}\right)=x(x+2)\cdot 1$$

$$3x+2(x+2)=x(x+2)$$
$$3x+2x+4=x^2+2x$$
$$5x+4=x^2+2x$$
$$0=x^2-3x-4$$
$$0=(x-4)(x+1)$$

$$x-4=0 \quad\text{or}\quad x+1=0$$
$$x=4 \quad\text{or}\quad x=-1$$

Brent's time: $x+2=4+2=6$ hours

31.

	Distance	Rate	Time
Bicycle	40	$x+15$	$\dfrac{40}{x+15}$
Running	16	x	$\dfrac{16}{x}$

$$\frac{40}{x+15}=\frac{16}{x}$$

$$x(x+15)\cdot\frac{40}{x+15}=x(x+15)\cdot\frac{16}{x}$$

$$40x=16(x+15)$$
$$40x=16x+240$$
$$24x=240$$

Running speed: $x=10$ mph

33.

	Distance	Rate	Time
First part	9 miles	x	$\dfrac{9}{x}$
Second part	5 miles	$x+4$	$\dfrac{5}{x+4}$

$$\frac{9}{x}+\frac{5}{x+4}=2$$

$$x(x+4)\cdot\left(\frac{9}{x}+\frac{5}{x+4}\right)=x(x+4)\cdot 2$$

$$9(x+4)+5x=2x(x+4)$$

$$9x+36+5x=2x^2+8x$$

$$14x+36=2x^2+8x$$

$$0=2x^2-6x-36$$

$$0=2\left(x^2-3x-18\right)$$

$$0=2(x-6)(x+3)$$

$$x-6=0 \quad \text{or} \quad x+3=0$$

$$x=6 \quad \text{or} \quad x=-3$$

Speed first 9 miles: $x=6$ mph

35.

	Distance	Rate	Time
Upstream	30 km	$x-10$	$\dfrac{30}{x-10}$
Downstream	50 km	$x+10$	$\dfrac{50}{x+10}$

$$\frac{30}{x-10}=\frac{50}{x+10}$$

$$(x-10)(x+10)\cdot\frac{30}{x-10}=(x-10)(x+10)\cdot\frac{50}{x+10}$$

$$30(x+10)=50(x-10)$$

$$30x+300=50x-500$$

$$300=20x-500$$

$$800=20x$$

$$40=x$$

40 kilometers/hour

37.

	Distance	Rate	Time
With Wind	72	$x+30$	$\dfrac{72}{x+30}$
Against Wind	72	$x-30$	$\dfrac{72}{x-30}$

$$\frac{72}{x+30}+\frac{72}{x-30}=1$$

$$(x+30)(x-30)\cdot\left(\frac{72}{(x+30)}+\frac{72}{(x-30)}\right)=(x+30)(x-30)\cdot 1$$

$$72(x-30)+72(x+30)=(x+30)(x-30)$$

$$72x-2160+72x+2160=x^2-900$$

$$144x=x^2-900$$

$$0=x^2-144x-900$$

$$0=(x-150)(x+6)$$

$$x-150=0 \quad \text{or} \quad x+6=0$$

$$x=150 \quad \text{or} \quad x=-6$$

150 mph

39. $y=kx$
$30=k\cdot 3$
$10=k$

$y=10\cdot 8=80$

41.
$$y = \frac{k}{x}$$
$$8 = \frac{k}{7}$$
$$56 = k$$
$$y = \frac{56}{4} = 14$$

43.
$$y = kx^2$$
$$150 = k \cdot 5^2$$
$$150 = 25k$$
$$6 = k$$
$$y = 6 \cdot (7)^2 = 6 \cdot 49 = 294$$

45.
$$y = kxz$$
$$270 = k(10)(6)$$
$$270 = 60k$$
$$4.5 = k$$
$$y = 4.5(3)(8) = 108$$

47. Let y represent Sam's gross pay.
Let x represent number of hours worked.
$$y = kx$$
$$304 = k \cdot 32$$
$$9.5 = k$$
$$y = 9.5 \cdot 20 = 190$$
$190

49. Let y represent the electric current.
Let x represent the voltage.
$$y = kx$$
$$8 = k \cdot 24$$
$$\frac{8}{24} = k$$
$$\frac{1}{3} = k$$
$$y = \frac{1}{3} \cdot 9 = 3$$
3 amperes

51. Let y represent the money contributed by each person.
Let x represent the number of people contributing.
$$y = \frac{k}{x}$$
$$24 = \frac{k}{10}$$
$$240 = k$$
$$y = \frac{240}{16} = 15$$
$15

53. Let y represent the electric current.
Let x represent the resistance.
$$y = \frac{k}{x}$$
$$30 = \frac{k}{3}$$
$$90 = k$$
$$y = \frac{90}{5} = 18$$
18 amperes

55. Let y represent the illumination of an object.
Let x represent the distance from a light source.
$$y = \frac{k}{x^2}$$
$$30 = \frac{k}{10^2}$$
$$30 = \frac{k}{100}$$
$$3000 = k$$
$$y = \frac{3000}{20^2} = \frac{3000}{400} = 7\frac{1}{2}$$
$7\frac{1}{2}$ foot-candles

57. Answers will vary. Example:
If it takes Joan 8 hours to paint a house and Jim 10 hours to paint the same house, how long would it take them together to paint the house?

CHAPTER 7 REVIEW

1. $\dfrac{3}{(-6)-9} = \dfrac{3}{-15} = -\dfrac{1}{5}$

2. $\dfrac{(19)-7}{(19)+13} = \dfrac{12}{32} = \dfrac{3}{8}$

3. $\dfrac{(3)^2+9(3)-11}{(3)^2+2(3)-5} = \dfrac{9+27-11}{9+6-5}$
$= \dfrac{36-11}{15-5}$
$= \dfrac{25}{10}$
$= \dfrac{5}{2}$

4. $\dfrac{(-4)^2-5(-4)-15}{(-4)^2+17(-4)+66} = \dfrac{16+20-15}{16-68+66}$
$= \dfrac{36-15}{82-68}$
$= \dfrac{21}{14}$
$= \dfrac{3}{2}$

5. $x+9=0$
$x=-9$

6. $x^2-3x-54=0$
$(x+6)(x-9)=0$
$x+6=0 \quad$ or $\quad x-9=0$
$x=-6 \quad$ or $\quad x=9$

7. $\dfrac{x+7}{x^2+4x-21} = \dfrac{(x+7)^1}{(x+7)_1(x-3)} = \dfrac{1}{x-3}$

8. $\dfrac{x^2+6x-40}{x^2-9x+20} = \dfrac{(x+10)(x-4)^1}{(x-5)(x-4)_1} = \dfrac{x+10}{x-5}$

9. $\dfrac{36-x^2}{x^2-7x+6} = \dfrac{(6+x)(6-x)^{-1}}{(x-6)_1(x-1)} = -\dfrac{6+x}{x-1}$

10. $\dfrac{x^2+6x+9}{4x^2+15x+9} = \dfrac{(x+3)^1(x+3)}{(x+3)_1(4x+3)} = \dfrac{x+3}{4x+3}$

11. $r(3) = \dfrac{3+7}{(3)^2+8(3)-15}$
$= \dfrac{10}{9+24-15}$
$= \dfrac{10}{33-15}$
$= \dfrac{10}{18}$
$= \dfrac{5}{9}$

12. $r(-2) = \dfrac{(-2)^2-15(-2)}{(-2)^2-11(-2)+18}$
$= \dfrac{4+30}{4+22+18}$
$= \dfrac{34}{44}$
$= \dfrac{17}{22}$

13. $x^2+6x=0$
$x(x+6)=0$
$x=0 \quad$ or $\quad x+6=0$
$x=0 \quad$ or $\qquad x=-6$
All real numbers except -6 and 0.
$(-\infty,-6)\cup(-6,0)\cup(0,\infty)$

14. $x^2-5x+4=0$
$(x-1)(x-4)=0$
$x-1=0 \quad$ or $\quad x-4=0$
$x=1 \quad$ or $\qquad x=4$
All real numbers except 1 and 4.
$(-\infty,1)\cup(1,4)\cup(4,\infty)$

15. $\dfrac{x+9}{x+5} \cdot \dfrac{x^2-7x}{x^2+2x-63} = \dfrac{(x+9)^1}{x+5} \cdot \dfrac{x(x-7)^1}{(x+9)_1(x-7)_1}$
$= \dfrac{x}{x+5}$

16. $\dfrac{x^2+14x+45}{x^2-6x-7} \cdot \dfrac{x^2+9x+8}{x^2+3x-10}$
$= \dfrac{(x+9)(x+5)^1}{(x-7)(x+1)_1} \cdot \dfrac{(x+8)(x+1)^1}{(x+5)_1(x-2)}$
$= \dfrac{(x+9)(x+8)}{(x-7)(x-2)}$

17. $\dfrac{x^2-4x}{x^2-15x+54}\cdot\dfrac{x^2-x-72}{x^2+8x}$

$=\dfrac{\cancel{x}^{1}(x-4)}{\cancel{(x-9)}_{1}(x-6)}\cdot\dfrac{\cancel{(x-9)}^{1}\cancel{(x+8)}^{1}}{\cancel{x}_{1}\cancel{(x+8)}_{1}}$

$=\dfrac{x-4}{x-6}$

18. $\dfrac{16-x^2}{x+2}\cdot\dfrac{7x^2+13x-2}{x^2+2x-24}$

$=\dfrac{(4+x)\cancel{(4-x)}^{-1}}{\cancel{x+2}_{1}}\cdot\dfrac{\cancel{(x+2)}^{1}(7x-1)}{(x+6)\cancel{(x-4)}_{1}}$

$=-\dfrac{(4+x)(7x-1)}{x+6}$

19. $f(x)\cdot g(x)=\dfrac{x-6}{x-1}\cdot\dfrac{x^2+6x+5}{x^2-3x-18}$

$=\dfrac{\cancel{x-6}^{1}}{x-1}\cdot\dfrac{(x+1)(x+5)}{\cancel{(x-6)}_{1}(x+3)}$

$=\dfrac{(x+1)(x+5)}{(x-1)(x+3)}$

20. $f(x)\cdot g(x)=\dfrac{x^2+16x+64}{x^2+x-42}\cdot\dfrac{49-x^2}{x^2+11x+24}$

$=\dfrac{(x+8)\cancel{(x+8)}^{1}}{\cancel{(x+7)}_{1}(x-6)}\cdot\dfrac{\cancel{(7+x)}^{1}(7-x)}{(x+3)\cancel{(x+8)}_{1}}$

$=\dfrac{(x+8)(7-x)}{(x-6)(x+3)}$

21. $\dfrac{x^2+9x+18}{x^2-7x+12}\div\dfrac{x+3}{x-3}=\dfrac{x^2+9x+18}{x^2-7x+12}\cdot\dfrac{x-3}{x+3}$

$=\dfrac{(x+6)(x+3)}{(x-4)(x-3)}\cdot\dfrac{x-3}{x+3}$

$=\dfrac{(x+6)\cancel{(x+3)}^{1}}{(x-4)\cancel{(x-3)}_{1}}\cdot\dfrac{\cancel{x-3}^{1}}{\cancel{x+3}_{1}}$

$=\dfrac{x+6}{x-4}$

22. $\dfrac{x^2-x-6}{x^2-19x+88}\div\dfrac{x^2+10x+16}{x^2+4x-96}$

$=\dfrac{x^2-x-6}{x^2-19x+88}\cdot\dfrac{x^2+4x-96}{x^2+10x+16}$

$=\dfrac{(x-3)(x+2)}{(x-11)(x-8)}\cdot\dfrac{(x+12)(x-8)}{(x+8)(x+2)}$

$=\dfrac{(x-3)\cancel{(x+2)}^{1}}{(x-11)\cancel{(x-8)}_{1}}\cdot\dfrac{(x+12)\cancel{(x-8)}^{1}}{(x+8)\cancel{(x+2)}_{1}}$

$=\dfrac{(x-3)(x+12)}{(x-11)(x+8)}$

23. $\dfrac{x^2-3x-4}{x^2+x}\div\dfrac{3x^2-10x-8}{x^2-1}$

$=\dfrac{x^2-3x-4}{x^2+x}\cdot\dfrac{x^2-1}{3x^2-10x-8}$

$=\dfrac{(x-4)(x+1)}{x(x+1)}\cdot\dfrac{(x+1)(x-1)}{(x-4)(3x+2)}$

$=\dfrac{\cancel{(x-4)}^{1}(x+1)}{x\cancel{(x+1)}_{1}}\cdot\dfrac{\cancel{(x+1)}^{1}(x-1)}{\cancel{(x-4)}_{1}(3x+2)}$

$=\dfrac{(x+1)(x-1)}{x(3x+2)}$

24. $\dfrac{9-x^2}{x-7}\div\dfrac{x^2-8x+15}{x^2-5x-14}$

$=\dfrac{9-x^2}{x-7}\cdot\dfrac{x^2-5x-14}{x^2-8x+15}$

$=\dfrac{(3+x)(3-x)}{x-7}\cdot\dfrac{(x-7)(x+2)}{(x-5)(x-3)}$

$=\dfrac{(3+x)\cancel{(3-x)}^{-1}}{\cancel{x-7}_{1}}\cdot\dfrac{\cancel{(x-7)}^{1}(x+2)}{(x-5)\cancel{(x-3)}_{1}}$

$=-\dfrac{(3+x)(x+2)}{x-5}$

25. $f(x) \div g(x) = \dfrac{x^2+3x}{x^2+10x+25} \div \dfrac{x^2-6x-27}{x^2-x-30}$

$\qquad = \dfrac{x^2+3x}{x^2+10x+25} \cdot \dfrac{x^2-x-30}{x^2-6x-27}$

$\qquad = \dfrac{x(x+3)}{(x+5)(x+5)} \cdot \dfrac{(x-6)(x+5)}{(x-9)(x+3)}$

$\qquad = \dfrac{x\,\cancel{(x+3)}^{\,1}}{(x+5)\,\cancel{(x+5)}_{\,1}} \cdot \dfrac{(x-6)\,\cancel{(x+5)}^{\,1}}{(x-9)\,\cancel{(x+3)}_{\,1}}$

$\qquad = \dfrac{x(x-6)}{(x+5)(x-9)}$

26. $f(x) \div g(x) = \dfrac{x^2+8x+12}{x^2+5x+4} \div \dfrac{x^2-4x-60}{x^2+7x+12}$

$\qquad = \dfrac{x^2+8x+12}{x^2+5x+4} \cdot \dfrac{x^2+7x+12}{x^2-4x-60}$

$\qquad = \dfrac{(x+6)(x+2)}{(x+4)(x+1)} \cdot \dfrac{(x+4)(x+3)}{(x+6)(x-10)}$

$\qquad = \dfrac{\cancel{(x+6)}^{\,1}(x+2)}{\cancel{(x+4)}_{\,1}(x+1)} \cdot \dfrac{\cancel{(x+4)}^{\,1}(x+3)}{\cancel{(x+6)}_{\,1}(x-10)}$

$\qquad = \dfrac{(x+2)(x+3)}{(x+1)(x-10)}$

27. $\dfrac{5}{x+6} + \dfrac{7}{x+6} = \dfrac{12}{x+6}$

28. $\dfrac{x^2+10x}{x^2+15x+56} - \dfrac{5x+24}{x^2+15x+56}$

$\qquad = \dfrac{x^2+10x-(5x+24)}{x^2+15x+56}$

$\qquad = \dfrac{x^2+10x-5x-24}{x^2+15x+56}$

$\qquad = \dfrac{x^2+5x-24}{x^2+15x+56}$

$\qquad = \dfrac{\cancel{(x+8)}^{\,1}(x-3)}{\cancel{(x+8)}_{\,1}(x+7)}$

$\qquad = \dfrac{x-3}{x+7}$

29. $\dfrac{x^2-6x-13}{x^2-x-20} + \dfrac{10x-32}{x^2-x-20}$

$\qquad = \dfrac{x^2-6x-13+10x-32}{x^2-x-20}$

$\qquad = \dfrac{x^2+4x-45}{x^2-x-20}$

$\qquad = \dfrac{(x+9)\,\cancel{(x-5)}^{\,1}}{(x+4)\,\cancel{(x-5)}_{\,1}}$

$\qquad = \dfrac{x+9}{x+4}$

30. $\dfrac{5x-34}{x-7} - \dfrac{x-8}{7-x} = \dfrac{5x-34}{x-7} + \dfrac{x-8}{x-7}$

$\qquad = \dfrac{5x-34+x-8}{x-7}$

$\qquad = \dfrac{6x-42}{x-7}$

$\qquad = \dfrac{6\,\cancel{(x-7)}^{\,1}}{\cancel{x-7}_{\,1}}$

$\qquad = 6$

31. $\dfrac{2x^2+7x-3}{x-2} + \dfrac{x^2+3x+9}{2-x}$

$\qquad = \dfrac{2x^2+7x-3}{x-2} - \dfrac{x^2+3x+9}{x-2}$

$\qquad = \dfrac{2x^2+7x-3-\left(x^2+3x+9\right)}{x-2}$

$\qquad = \dfrac{2x^2+7x-3-x^2-3x-9}{x-2}$

$\qquad = \dfrac{x^2+4x-12}{x-2}$

$\qquad = \dfrac{(x+6)\,\cancel{(x-2)}^{\,1}}{\cancel{x-2}_{\,1}}$

$\qquad = x+6$

32. $\dfrac{x^2-3x-12}{x^2-25}-\dfrac{2x-18}{25-x^2}$

$$=\dfrac{x^2-3x-12}{x^2-25}+\dfrac{2x-18}{x^2-25}$$

$$=\dfrac{x^2-3x-12+2x-18}{x^2-25}$$

$$=\dfrac{x^2-x-30}{x^2-25}$$

$$=\dfrac{\cancel{(x+5)}^{\,1}(x-6)}{\cancel{(x+5)}_{\,1}(x-5)}$$

$$=\dfrac{x-6}{x-5}$$

33. $\dfrac{5}{x^2+3x-4}+\dfrac{2}{x^2-4x+3}$

$$=\dfrac{5}{(x+4)(x-1)}+\dfrac{2}{(x-3)(x-1)}$$

$$=\dfrac{5}{(x+4)(x-1)}\cdot\dfrac{x-3}{x-3}+\dfrac{2}{(x-1)(x-3)}\cdot\dfrac{x+4}{x+4}$$

$$=\dfrac{5(x-3)+2(x+4)}{(x-1)(x-3)(x+4)}$$

$$=\dfrac{5x-15+2x+8}{(x-1)(x-3)(x+4)}$$

$$=\dfrac{7x-7}{(x-1)(x-3)(x+4)}$$

$$=\dfrac{7\cancel{(x-1)}^{\,1}}{\cancel{(x-1)}_{\,1}(x-3)(x+4)}$$

$$=\dfrac{7}{(x+4)(x-3)}$$

34. $\dfrac{6}{x^2-10x+16}-\dfrac{1}{x^2-15x+56}$

$$=\dfrac{6}{(x-8)(x-2)}-\dfrac{1}{(x-8)(x-7)}$$

$$=\dfrac{6}{(x-8)(x-2)}\cdot\dfrac{x-7}{x-7}-\dfrac{1}{(x-8)(x-7)}\cdot\dfrac{x-2}{x-2}$$

$$=\dfrac{6(x-7)-1(x-2)}{(x-8)(x-7)(x-2)}$$

$$=\dfrac{6x-42-x+2}{(x-8)(x-7)(x-2)}$$

$$=\dfrac{5x-40}{(x-8)(x-7)(x-2)}$$

$$=\dfrac{5\cancel{(x-8)}^{\,1}}{\cancel{(x-8)}_{\,1}(x-7)(x-2)}$$

$$=\dfrac{5}{(x-2)(x-7)}$$

35. $\dfrac{x}{x^2-5x-50}-\dfrac{4}{x^2-14x+40}$

$$=\dfrac{x}{(x-10)(x+5)}-\dfrac{4}{(x-10)(x-4)}$$

$$=\dfrac{x}{(x-10)(x+5)}\cdot\dfrac{x-4}{x-4}-\dfrac{4}{(x-10)(x-4)}\cdot\dfrac{x+5}{x+5}$$

$$=\dfrac{x(x-4)-4(x+5)}{(x-10)(x-4)(x+5)}$$

$$=\dfrac{x^2-4x-4x-20}{(x-10)(x-4)(x+5)}$$

$$=\dfrac{x^2-8x-20}{(x-10)(x-4)(x+5)}$$

$$=\dfrac{\cancel{(x-10)}^{\,1}(x+2)}{\cancel{(x-10)}_{\,1}(x-4)(x+5)}$$

$$=\dfrac{x+2}{(x-4)(x+5)}$$

36. $\dfrac{x-5}{x^2+14x+33}+\dfrac{4}{x^2+20x+99}$

$=\dfrac{x-5}{(x+11)(x+3)}+\dfrac{4}{(x+11)(x+9)}$

$=\dfrac{x-5}{(x+11)(x+3)}\cdot\dfrac{x+9}{x+9}+\dfrac{4}{(x+11)(x+9)}\cdot\dfrac{x+3}{x+3}$

$=\dfrac{(x-5)(x+9)+4(x+3)}{(x+11)(x+3)(x+9)}$

$=\dfrac{x^2+9x-5x-45+4x+12}{(x+11)(x+3)(x+9)}$

$=\dfrac{x^2+8x-33}{(x+11)(x+3)(x+9)}$

$=\dfrac{\cancel{(x+11)}^{1}(x-3)}{\cancel{(x+11)}_{1}(x+3)(x+9)}$

$=\dfrac{x-3}{(x+3)(x+9)}$

37. $\dfrac{x+1}{x^2+2x-24}+\dfrac{x-5}{x^2-6x+8}$

$=\dfrac{x+1}{(x+6)(x-4)}+\dfrac{x-5}{(x-4)(x-2)}$

$=\dfrac{x+1}{(x+6)(x-4)}\cdot\dfrac{x-2}{x-2}+\dfrac{x-5}{(x-4)(x-2)}\cdot\dfrac{x+6}{x+6}$

$=\dfrac{(x+1)(x-2)+(x-5)(x+6)}{(x+6)(x-4)(x-2)}$

$=\dfrac{x^2-2x+x-2+x^2+6x-5x-30}{(x+6)(x-4)(x-2)}$

$=\dfrac{2x^2-32}{(x+6)(x-4)(x-2)}$

$=\dfrac{2\left(x^2-16\right)}{(x+6)(x-4)(x-2)}$

$=\dfrac{2(x+4)\cancel{(x-4)}^{1}}{(x+6)\cancel{(x-4)}_{1}(x-2)}$

$=\dfrac{2(x+4)}{(x+6)(x-2)}$

38. $f(x)+g(x)$

$=\dfrac{x+3}{x^2-x-2}+\dfrac{5}{x^2-7x+10}$

$=\dfrac{x+3}{(x-2)(x+1)}+\dfrac{5}{(x-2)(x-5)}$

$=\dfrac{x+3}{(x-2)(x+1)}\cdot\dfrac{x-5}{x-5}+\dfrac{5}{(x-2)(x-5)}\cdot\dfrac{x+1}{x+1}$

$=\dfrac{(x+3)(x-5)+5(x+1)}{(x-2)(x-5)(x+1)}$

$=\dfrac{x^2-5x+3x-15+5x+5}{(x-2)(x-5)(x+1)}$

$=\dfrac{x^2+3x-10}{(x-2)(x-5)(x+1)}$

$=\dfrac{(x+5)\cancel{(x-2)}^{1}}{\cancel{(x-2)}_{1}(x-5)(x+1)}$

$=\dfrac{x+5}{(x-5)(x+1)}$

39. $f(x)-g(x)$

$=\dfrac{x+2}{x^2+4x-5}-\dfrac{1}{x^2+12x+35}$

$=\dfrac{x+2}{(x+5)(x-1)}-\dfrac{1}{(x+5)(x+7)}$

$=\dfrac{x+2}{(x+5)(x-1)}\cdot\dfrac{x+7}{x+7}-\dfrac{1}{(x+5)(x+7)}\cdot\dfrac{x-1}{x-1}$

$=\dfrac{(x+2)(x+7)-(x-1)}{(x+5)(x-1)(x+7)}$

$=\dfrac{x^2+7x+2x+14-x+1}{(x+5)(x-1)(x+7)}$

$=\dfrac{x^2+8x+15}{(x+5)(x-1)(x+7)}$

$=\dfrac{\cancel{(x+5)}^{1}(x+3)}{\cancel{(x+5)}_{1}(x-1)(x+7)}$

$=\dfrac{x+3}{(x-1)(x+7)}$

40.
$$\dfrac{1-\dfrac{9}{x}}{1-\dfrac{81}{x^2}}=\dfrac{x^2}{x^2}\cdot\dfrac{1-\dfrac{9}{x}}{1-\dfrac{81}{x^2}}$$

$$=\dfrac{x^2\cdot 1-x^2\cdot\dfrac{9}{x}}{x^2\cdot 1-x^2\cdot\dfrac{81}{x^2}}$$

$$=\dfrac{x^2-9x}{x^2-81}$$

$$=\dfrac{x(x-9)}{(x+9)(x-9)}$$

$$=\dfrac{x\,\overset{1}{\cancel{(x-9)}}}{(x+9)\,\underset{1}{\cancel{(x-9)}}}$$

$$=\dfrac{x}{x+9}$$

41.
$$\dfrac{\dfrac{x-5}{x-3}-\dfrac{3}{x-7}}{\dfrac{x^2+7x-44}{x^2-10x+21}}=\dfrac{\dfrac{x-5}{x-3}-\dfrac{3}{x-7}}{\dfrac{(x+11)(x-4)}{(x-3)(x-7)}}$$

$$=\dfrac{(x-3)(x-7)}{(x-3)(x-7)}\cdot\dfrac{\dfrac{x-5}{x-3}-\dfrac{3}{x-7}}{\dfrac{(x+11)(x-4)}{(x-3)(x-7)}}$$

$$=\dfrac{(x-3)(x-7)\cdot\left(\dfrac{x-5}{x-3}-\dfrac{3}{x-7}\right)}{(x-3)(x-7)\cdot\dfrac{(x+11)(x-4)}{(x-3)(x-7)}}$$

$$=\dfrac{(x-7)(x-5)-3(x-3)}{(x+11)(x-4)}$$

$$=\dfrac{x^2-5x-7x+35-3x+9}{(x+11)(x-4)}$$

$$=\dfrac{x^2-15x+44}{(x+11)(x-4)}$$

$$=\dfrac{(x-11)\,\overset{1}{\cancel{(x-4)}}}{(x+11)\,\underset{1}{\cancel{(x-4)}}}$$

$$=\dfrac{x-11}{x+11}$$

42.
$$\dfrac{1+\dfrac{2}{x}-\dfrac{35}{x^2}}{1-\dfrac{7}{x}+\dfrac{10}{x^2}}=\dfrac{x^2}{x^2}\cdot\dfrac{1+\dfrac{2}{x}-\dfrac{35}{x^2}}{1-\dfrac{7}{x}+\dfrac{10}{x^2}}$$

$$=\dfrac{x^2\cdot 1+\cancel{x^2}^{\,x}\cdot\dfrac{2}{\cancel{x}_1}-\cancel{x^2}^{1}\cdot\dfrac{35}{\cancel{x^2}_1}}{x^2\cdot 1-\cancel{x^2}^{\,x}\cdot\dfrac{7}{\cancel{x}_1}+\cancel{x^2}^{1}\cdot\dfrac{10}{\cancel{x^2}_1}}$$

$$=\dfrac{x^2+2x-35}{x^2-7x+10}$$

$$=\dfrac{(x+7)\,\overset{1}{\cancel{(x-5)}}}{(x-2)\,\underset{1}{\cancel{(x-5)}}}$$

$$=\dfrac{x+7}{x-2}$$

43.
$$\dfrac{\dfrac{x^2-8x+15}{x^2-11x+18}}{\dfrac{x^2+x-30}{x^2-6x-27}}=\dfrac{x^2-8x+15}{x^2-11x+18}\div\dfrac{x^2+x-30}{x^2-6x-27}$$

$$=\dfrac{x^2-8x+15}{x^2-11x+18}\cdot\dfrac{x^2-6x-27}{x^2+x-30}$$

$$=\dfrac{(x-3)\,\overset{1}{\cancel{(x-5)}}}{(x-2)\,\underset{1}{\cancel{(x-9)}}}\cdot\dfrac{\overset{1}{\cancel{(x-9)}}(x+3)}{(x+6)\,\underset{1}{\cancel{(x-5)}}}$$

$$=\dfrac{(x-3)(x+3)}{(x-2)(x+6)}$$

44.
$$\dfrac{9}{x}-\dfrac{3}{4}=\dfrac{7}{8}$$

$$8x\cdot\left(\dfrac{9}{x}-\dfrac{3}{4}\right)=8x\cdot\dfrac{7}{8}$$

$$72-6x=7x$$
$$72=13x$$

$$\dfrac{72}{13}=x$$

$$\left\{\dfrac{72}{13}\right\}$$

45.
$$x - 5 + \frac{12}{x} = 2$$
$$x \cdot \left(x - 5 + \frac{12}{x}\right) = x \cdot 2$$
$$x^2 - 5x + 12 = 2x$$
$$x^2 - 7x + 12 = 0$$
$$(x - 3)(x - 4) = 0$$
$$x - 3 = 0 \quad \text{or} \quad x - 4 = 0$$
$$x = 3 \quad \text{or} \quad x = 4$$
$$\{3, 4\}$$

46.
$$\frac{3}{x - 7} = \frac{7}{x + 5}$$
$$(x - 7)(x + 5) \cdot \frac{3}{x - 7} = (x - 7)(x + 5) \cdot \frac{7}{x + 5}$$
$$3(x + 5) = 7(x - 7)$$
$$3x + 15 = 7x - 49$$
$$15 = 4x - 49$$
$$64 = 4x$$
$$16 = x$$
$$\{16\}$$

47.
$$\frac{2x - 11}{x - 1} = \frac{x - 7}{x + 3}$$
$$(x - 1)(x + 3) \cdot \frac{2x - 11}{x - 1} = (x - 1)(x + 3) \cdot \frac{x - 7}{x + 3}$$
$$(x + 3)(2x - 11) = (x - 1)(x - 7)$$
$$2x^2 - 11x + 6x - 33 = x^2 - 7x - x + 7$$
$$2x^2 - 5x - 33 = x^2 - 8x + 7$$
$$x^2 + 3x - 40 = 0$$
$$(x + 8)(x - 5) = 0$$
$$x + 8 = 0 \quad \text{or} \quad x - 5 = 0$$
$$x = -8 \quad \text{or} \quad x = 5$$
$$\{-8, 5\}$$

48.
$$\frac{3}{x + 2} - \frac{1}{x + 1} = \frac{x + 3}{x^2 + 3x + 2}$$
$$\frac{3}{x + 2} - \frac{1}{x + 1} = \frac{x + 3}{(x + 2)(x + 1)}$$
$$(x + 2)(x + 1) \cdot \left(\frac{3}{x + 2} - \frac{1}{x + 1}\right) = (x + 2)(x + 1) \cdot \frac{x + 3}{(x + 2)(x + 1)}$$
$$3(x + 1) - 1(x + 2) = x + 3$$
$$3x + 3 - x - 2 = x + 3$$
$$2x + 1 = x + 3$$
$$x = 2$$
$$\{2\}$$

49.

$$\frac{x}{x+5}+\frac{3}{x-7}=\frac{36}{x^2-2x-35}$$

$$\frac{x}{x+5}+\frac{3}{x-7}=\frac{36}{(x+5)(x-7)}$$

$$(x+5)(x-7)\cdot\left(\frac{x}{x+5}+\frac{3}{x-7}\right)=(x+5)(x-7)\cdot\frac{36}{(x+5)(x-7)}$$

$$x(x-7)+3(x+5)=36$$
$$x^2-7x+3x+15=36$$
$$x^2-4x+15=36$$
$$x^2-4x-21=0$$
$$(x-7)(x+3)=0$$

$$x-7=0 \quad\text{or}\quad x+3=0$$
$$x=7 \quad\text{or}\quad x=-3$$

$\{-3\}$ since $x=7$ makes the original equation undefined.

50.

$$\frac{2}{x^2-x-2}+\frac{10}{x^2-2x-3}=\frac{x+12}{x^2-x-2}$$

$$\frac{2}{(x-2)(x+1)}+\frac{10}{(x-3)(x+1)}=\frac{x+12}{(x-2)(x+1)}$$

$$(x-2)(x+1)(x-3)\cdot\left(\frac{2}{(x-2)(x+1)}+\frac{10}{(x-3)(x+1)}\right)=(x-2)(x+1)(x-3)\cdot\frac{x+12}{(x-2)(x+1)}$$

$$2(x-3)+10(x-2)=(x-3)(x+12)$$
$$2x-6+10x-20=x^2+12x-3x-36$$
$$12x-26=x^2+9x-36$$
$$0=x^2-3x-10$$
$$0=(x+2)(x-5)$$

$$x+2=0 \quad\text{or}\quad x-5=0$$
$$x=-2 \quad\text{or}\quad x=5$$

$\{-2,5\}$

51.

$$\frac{20}{x^2+12x+27}-\frac{4}{x^2+14x+45}=\frac{x+10}{x^2+8x+15}$$

$$\frac{20}{(x+3)(x+9)}-\frac{4}{(x+9)(x+5)}=\frac{x+10}{(x+3)(x+5)}$$

$$(x+3)(x+9)(x+5)\cdot\left(\frac{20}{(x+3)(x+9)}-\frac{4}{(x+9)(x+5)}\right)=(x+3)(x+9)(x+5)\cdot\frac{x+10}{(x+3)(x+5)}$$

$$20(x+5)-4(x+3)=(x+9)(x+10)$$
$$20x+100-4x-12=x^2+10x+9x+90$$
$$16x+88=x^2+19x+90$$
$$0=x^2+3x+2$$
$$0=(x+2)(x+1)$$

$$x+2=0 \quad\text{or}\quad x+1=0$$
$$x=-2 \quad\text{or}\quad x=-1$$

$\{-2,-1\}$

52.
$$\frac{1}{2} = \frac{A}{bh}$$
$$bh \cdot \frac{1}{2} = bh \cdot \frac{A}{bh}$$
$$\frac{bh}{2} = A$$
$$\frac{2}{b} \cdot \frac{bh}{2} = \frac{2}{b} \cdot A$$
$$h = \frac{2A}{b}$$

53.
$$y = \frac{3x}{4x-7}$$
$$y(4x-7) = \frac{3x}{4x-7} \cdot (4x-7)$$
$$4xy - 7y = 3x$$
$$4xy - 3x - 7y = 0$$
$$4xy - 3x = 7y$$
$$x(4y-3) = 7y$$
$$x = \frac{7y}{4y-3}$$

54.
$$\frac{x}{2r} - \frac{y}{3r} = \frac{1}{5}$$
$$30r \cdot \left(\frac{x}{2r} - \frac{y}{3r}\right) = 30r \cdot \frac{1}{5}$$
$$15x - 10y = 6r$$
$$r = \frac{15x - 10y}{6}$$

55. Unknown: x
$$\frac{1}{x} + \frac{5}{6} = \frac{23}{24}$$
$$24x \cdot \left(\frac{1}{x} + \frac{5}{6}\right) = 24x \cdot \frac{23}{24}$$
$$24 + 20x = 23x$$
$$24 = 3x$$
$$8 = x$$

56. Smaller number: x
Larger number: $x+5$
$$3 \cdot \frac{1}{x} + 4 \cdot \frac{1}{x+5} = 1$$
$$\frac{3}{x} + \frac{4}{x+5} = 1$$
$$x(x+5) \cdot \left(\frac{3}{x} + \frac{4}{x+5}\right) = x(x+5) \cdot 1$$
$$3(x+5) + 4x = x(x+5)$$
$$3x + 15 + 4x = x^2 + 5x$$
$$7x + 15 = x^2 + 5x$$
$$0 = x^2 - 2x - 15$$
$$0 = (x-5)(x+3)$$
$$x - 5 = 0 \quad \text{or} \quad x + 3 = 0$$
$$x = 5 \quad \text{or} \quad x = -3$$
Smaller number: $x = 5$
Larger number: $x + 5 = 5 + 5 = 10$

57.

Person	Job Time	Rate	Time	Portion
Rob	30 min	$\frac{1}{30}$	t	$\frac{t}{30}$
Sean	60 min	$\frac{1}{60}$	t	$\frac{t}{60}$

$$\frac{t}{30} + \frac{t}{60} = 1$$
$$60 \cdot \left(\frac{t}{30} + \frac{t}{60}\right) = 60 \cdot 1$$
$$2t + t = 60$$
$$3t = 60$$
$$t = 20$$
20 minutes

58.

Pipe	Job Time	Rate	Time	Portion
Small	$x+42$	$\frac{1}{x+42}$	28	$\frac{28}{x+42}$
Large	x	$\frac{1}{x}$	28	$\frac{28}{x}$

$$\frac{28}{x+42} + \frac{28}{x} = 1$$
$$x(x+42) \cdot \left(\frac{28}{x+42} + \frac{28}{x}\right) = x(x+42) \cdot 1$$
$$28x + 28(x+42) = x(x+42)$$
$$28x + 28x + 1176 = x^2 + 42x$$
$$56x + 1176 = x^2 + 42x$$
$$0 = x^2 - 14x - 1176$$
$$0 = (x-42)(x+28)$$
$$x - 42 = 0 \quad \text{or} \quad x + 28 = 0$$
$$x = 42 \quad \text{or} \quad x = -28$$
Small pipe: $x + 42 = 42 + 42 = 84$ minutes

59.

	Distance	Rate	Time
1st part	100 mi	x	$\dfrac{100}{x}$
2nd part	175 mi	$x+15$	$\dfrac{175}{x+15}$

$$\frac{100}{x}+\frac{175}{x+15}=4$$

$$x(x+15)\cdot\left(\frac{100}{x}+\frac{175}{x+15}\right)=x(x+15)\cdot 4$$

$$100(x+15)+175x=4x(x+15)$$
$$100x+1500+175x=4x^2+60x$$
$$275x+1500=4x^2+60x$$
$$0=4x^2-215x-1500$$
$$0=(4x+25)(x-60)$$

$$4x+25=0 \quad \text{or} \quad x-60=0$$
$$4x=-25 \quad \text{or} \quad x=60$$
$$x=-\frac{25}{4} \quad \text{or} \quad x=60$$

60 mph

60.

	Distance	Rate	Time
Upstream	5 mi	$x-2$	$\dfrac{5}{x-2}$
Downstream	10 mi	$x+2$	$\dfrac{10}{x+2}$

$$\frac{5}{x-2}=\frac{10}{x+2}$$

$$(x-2)(x+2)\cdot\frac{5}{x-2}=(x-2)(x+2)\cdot\frac{10}{x+2}$$

$$5(x+2)=10(x-2)$$
$$5x+10=10x-20$$
$$-5x+10=-20$$
$$-5x=-30$$
$$x=6$$

Dominique paddles in still water at 6 mph.

61. Let y represent the number of calories.
Let x represent the amount of milk.

$$y=kx$$
$$195=k\cdot(12)$$

$$\frac{195}{12}=k$$

$$y=\frac{195}{12}\cdot\frac{8}{1}=\frac{195}{\cancel{12}_3}\cdot\frac{\cancel{8}^2}{1}=\frac{390}{3}=130$$

130 calories

62. Let y represent the distance to brake.
Let x represent the speed of the car.

$$y=kx^2$$
$$100=k\cdot 40^2$$
$$100=k\cdot 1600$$

$$\frac{100}{1600}=k$$

$$\frac{1}{16}=k$$

$$y=\frac{1}{16}\cdot 60^2=\frac{1}{16}\cdot 3600=225$$

225 ft

63. Let y represent the maximum load.
Let x represent the length.

$$y=\frac{k}{x}$$

$$900=\frac{k}{6}$$

$$5400=k$$

$$y=\frac{5400}{8}=675$$

675 lbs

64. Let y represent the illumination.
Let x represent the distance from
the source of the light.

$$y=\frac{k}{x^2}$$

$$20=\frac{k}{12^2}$$

$$20=\frac{k}{144}$$

$$2880=k$$

$$y=\frac{2880}{6^2}=\frac{2880}{36}=80$$

80 foot-candles

CHAPTER 7 TEST

1. $\dfrac{10}{(-7)-5}=\dfrac{10}{-12}=-\dfrac{5}{6}$

2. $x^2+10x+21=0$
$(x+7)(x+3)=0$
$\quad x+7=0 \quad$ or $\quad x+3=0$
$\quad\quad x=-7 \quad$ or $\quad\quad x=-3$

3. $\dfrac{x^2-8x+15}{x^2+7x-30}=\dfrac{(x-5)(x-3)}{(x+10)(x-3)}$
$\qquad =\dfrac{(x-5)\,\cancel{(x-3)}^{\,1}}{(x+10)\,\cancel{(x-3)}_{\,1}}$
$\qquad =\dfrac{x-5}{x+10}$

4. $r(-3)=\dfrac{(-3)^2+3(-3)-13}{(-3)^2-4(-3)-20}$
$\qquad =\dfrac{9-9-13}{9+12-20}$
$\qquad =\dfrac{-13}{21-20}$
$\qquad =\dfrac{-13}{1}$
$\qquad =-13$

5. $\dfrac{x^2+x-42}{x^2+6x-7}\cdot\dfrac{x^2-2x+1}{x^2-9x+18}$
$\qquad =\dfrac{(x+7)(x-6)}{(x+7)(x-1)}\cdot\dfrac{(x-1)(x-1)}{(x-6)(x-3)}$
$\qquad =\dfrac{\cancel{(x+7)}^{\,1}\cancel{(x-6)}^{\,1}}{\cancel{(x+7)}_{\,1}\cancel{(x-1)}_{\,1}}\cdot\dfrac{\cancel{(x-1)}^{\,1}(x-1)}{\cancel{(x-6)}_{\,1}(x-3)}$
$\qquad =\dfrac{x-1}{x-3}$

6. $\dfrac{x^2-10x+25}{x^2+11x+24}\div\dfrac{25-x^2}{x^2+3x-40}$
$\qquad =\dfrac{x^2-10x+25}{x^2+11x+24}\cdot\dfrac{x^2+3x-40}{25-x^2}$
$\qquad =\dfrac{(x-5)(x-5)}{(x+3)(x+8)}\cdot\dfrac{(x+8)(x-5)}{(5+x)(5-x)}$
$\qquad =\dfrac{(x-5)(x-5)}{(x+3)\cancel{(x+8)}_{\,1}}\cdot\dfrac{\cancel{(x+8)}^{\,1}\cancel{(x-5)}^{\,1}}{(5+x)\cancel{(5-x)}_{\,-1}}$
$\qquad =-\dfrac{(x-5)(x-5)}{(x+3)(5+x)}$
$\qquad =-\dfrac{(x-5)^2}{(x+3)(5+x)}$

7. $\dfrac{5x+3}{x-6}-\dfrac{2x+21}{x-6}=\dfrac{5x+3-(2x+21)}{x-6}$
$\qquad =\dfrac{5x+3-2x-21}{x-6}$
$\qquad =\dfrac{3x-18}{x-6}$
$\qquad =\dfrac{3\cancel{(x-6)}^{\,1}}{\cancel{x-6}_{\,1}}$
$\qquad =3$

8. $\dfrac{2x^2+6x-11}{x-4}+\dfrac{x^2+2x+21}{4-x}$
$\qquad =\dfrac{2x^2+6x-11}{x-4}-\dfrac{x^2+2x+21}{x-4}$
$\qquad =\dfrac{2x^2+6x-11-(x^2+2x+21)}{x-4}$
$\qquad =\dfrac{2x^2+6x-11-x^2-2x-21}{x-4}$
$\qquad =\dfrac{x^2+4x-32}{x-4}$
$\qquad =\dfrac{(x+8)\cancel{(x-4)}^{\,1}}{\cancel{x-4}_{\,1}}$
$\qquad =x+8$

9. $\dfrac{6}{x^2+8x+7}+\dfrac{5}{x^2-3x-4}$

$=\dfrac{6}{(x+7)(x+1)}+\dfrac{5}{(x-4)(x+1)}$

$=\dfrac{6}{(x+7)(x+1)}\cdot\dfrac{x-4}{x-4}+\dfrac{5}{(x-4)(x+1)}\cdot\dfrac{x+7}{x+7}$

$=\dfrac{6(x-4)+5(x+7)}{(x+7)(x+1)(x-4)}$

$=\dfrac{6x-24+5x+35}{(x+7)(x+1)(x-4)}$

$=\dfrac{11x+11}{(x+7)(x+1)(x-4)}$

$=\dfrac{11\,\cancel{(x+1)}^{\,1}}{(x+7)\,\cancel{(x+1)}_{\,1}(x-4)}$

$=\dfrac{11}{(x+7)(x-4)}$

10. $\dfrac{x-4}{x^2-13x+40}-\dfrac{8}{x^2-10x+16}$

$=\dfrac{x-4}{(x-8)(x-5)}-\dfrac{8}{(x-8)(x-2)}$

$=\dfrac{x-4}{(x-8)(x-5)}\cdot\dfrac{x-2}{x-2}-\dfrac{8}{(x-8)(x-2)}\cdot\dfrac{x-5}{x-5}$

$=\dfrac{(x-4)(x-2)-8(x-5)}{(x-8)(x-5)(x-2)}$

$=\dfrac{x^2-2x-4x+8-8x+40}{(x-8)(x-5)(x-2)}$

$=\dfrac{x^2-14x+48}{(x-8)(x-5)(x-2)}$

$=\dfrac{(x-6)\,\cancel{(x-8)}^{\,1}}{\cancel{(x-8)}_{\,1}(x-5)(x-2)}$

$=\dfrac{x-6}{(x-5)(x-2)}$

11. $\dfrac{1+\dfrac{4}{x}-\dfrac{45}{x^2}}{1+\dfrac{2}{x}-\dfrac{35}{x^2}}=\dfrac{x^2}{x^2}\cdot\dfrac{1+\dfrac{4}{x}-\dfrac{45}{x^2}}{1+\dfrac{2}{x}-\dfrac{35}{x^2}}$

$=\dfrac{x^2\cdot1+\cancel{x^2}^{\,x}\cdot\dfrac{4}{\cancel{x}_1}-\cancel{x^2}^{\,1}\cdot\dfrac{45}{\cancel{x^2}_{\,1}}}{x^2\cdot1+\cancel{x^2}^{\,x}\cdot\dfrac{2}{\cancel{x}_1}-\cancel{x^2}^{\,1}\cdot\dfrac{35}{\cancel{x^2}_{\,1}}}$

$=\dfrac{x^2+4x-45}{x^2+2x-35}$

$=\dfrac{(x+9)\,\cancel{(x-5)}^{\,1}}{(x+7)\,\cancel{(x-5)}_{\,1}}$

$=\dfrac{x+9}{x+7}$

12. $\dfrac{x+5}{x^2+3x-54}\cdot\dfrac{x^2-11x+30}{x^2+16x+55}$

$=\dfrac{x+5}{(x+9)(x-6)}\cdot\dfrac{(x-6)(x-5)}{(x+5)(x+11)}$

$=\dfrac{\cancel{x+5}^{\,1}}{(x+9)\,\cancel{(x-6)}_{\,1}}\cdot\dfrac{\cancel{(x-6)}^{\,1}(x-5)}{\cancel{(x+5)}_{\,1}(x+11)}$

$=\dfrac{x-5}{(x+9)(x+11)}$

13. $\dfrac{x^2-3x-40}{x^2+7x+6}\div\dfrac{x^2-8x}{x^2+10x+9}$

$=\dfrac{x^2-3x-40}{x^2+7x+6}\cdot\dfrac{x^2+10x+9}{x^2-8x}$

$=\dfrac{(x-8)(x+5)}{(x+6)(x+1)}\cdot\dfrac{(x+9)(x+1)}{x(x-8)}$

$=\dfrac{\cancel{(x-8)}^{\,1}(x+5)}{(x+6)\,\cancel{(x+1)}_{\,1}}\cdot\dfrac{(x+9)\,\cancel{(x+1)}^{\,1}}{x\,\cancel{(x-8)}_{\,1}}$

$=\dfrac{(x+5)(x+9)}{x(x+6)}$

14. $\dfrac{x^2+7x-14}{x^2+6x+8}+\dfrac{3x+38}{x^2+6x+8}$

$=\dfrac{x^2+7x-14+3x+38}{x^2+6x+8}$

$=\dfrac{x^2+10x+24}{x^2+6x+8}$

$=\dfrac{\overset{1}{\cancel{(x+4)}}(x+6)}{\underset{1}{\cancel{(x+4)}}(x+2)}$

$=\dfrac{x+6}{x+2}$

15. $\dfrac{x-7}{x^2+7x-8}-\dfrac{2}{x^2+x-2}$

$=\dfrac{x-7}{(x+8)(x-1)}-\dfrac{2}{(x+2)(x-1)}$

$=\dfrac{x-7}{(x+8)(x-1)}\cdot\dfrac{x+2}{x+2}-\dfrac{2}{(x+2)(x-1)}\cdot\dfrac{x+8}{x+8}$

$=\dfrac{(x-7)(x+2)-2(x+8)}{(x-1)(x+8)(x+2)}$

$=\dfrac{x^2+2x-7x-14-2x-16}{(x-1)(x+8)(x+2)}$

$=\dfrac{x^2-7x-30}{(x-1)(x+8)(x+2)}$

$=\dfrac{(x-10)(x+3)}{(x-1)(x+8)(x+2)}$

16. $\dfrac{15}{x}+\dfrac{3}{8}=\dfrac{19}{24}$

$24x\cdot\left(\dfrac{15}{x}+\dfrac{3}{8}\right)=24x\cdot\dfrac{19}{24}$

$360+9x=19x$

$360=10x$

$36=x$

$\{36\}$

17. $\dfrac{x+2}{x^2+5x+4}-\dfrac{4}{x^2+x-12}=\dfrac{2}{x^2-2x-3}$

$\dfrac{x+2}{(x+4)(x+1)}-\dfrac{4}{(x+4)(x-3)}=\dfrac{2}{(x-3)(x+1)}$

$(x-3)(x+1)(x+4)\cdot\left(\dfrac{x+2}{(x+4)(x+1)}-\dfrac{4}{(x+4)(x-3)}\right)=(x-3)(x+1)(x+4)\cdot\dfrac{2}{(x-3)(x+1)}$

$(x-3)(x+2)-4(x+1)=2(x+4)$

$x^2+2x-3x-6-4x-4=2x+8$

$x^2-5x-10=2x+8$

$x^2-7x-18=0$

$(x+2)(x-9)=0$

$x+2=0 \quad \text{or} \quad x-9=0$

$x=-2 \quad \text{or} \quad x=9$

$\{-2,9\}$

18.
$$x = \frac{2y}{7y-5}$$

$$(7y-5)\cdot x = (7y-5)\cdot \frac{2y}{7y-5}$$

$$7xy - 5x = 2y$$
$$-5x = 2y - 7xy$$
$$-5x = y(2 - 7x)$$

$$\frac{-5x}{2-7x} = y$$

$$\frac{5x}{7x-2} = y$$

19.

Pipe	Job Time	Rate	Time	Portion
Small	$x+18$	$\dfrac{1}{x+18}$	$40\,\text{min}$	$\dfrac{40}{x+18}$
Large	x	$\dfrac{1}{x}$	$40\,\text{min}$	$\dfrac{40}{x}$

$$\frac{40}{x+18} + \frac{40}{x} = 1$$

$$x(x+18)\cdot\left(\frac{40}{x+18} + \frac{40}{x}\right) = x(x+18)\cdot 1$$

$$40x + 40(x+18) = x(x+18)$$
$$40x + 40x + 720 = x^2 + 18x$$
$$80x + 720 = x^2 + 18x$$
$$0 = x^2 - 62x - 720$$
$$0 = (x-72)(x+10)$$

$$x - 72 = 0 \quad \text{or} \quad x + 10 = 0$$
$$x = 72 \quad \text{or} \qquad x = -10$$

Small pipe: $x + 18 = 72 + 18 = 90$ minutes

20.

	Distance	Rate	Time
1st part	10 mi	x	$\dfrac{10}{x}$
2nd part	15 mi	$x+10$	$\dfrac{15}{x+10}$

$$\frac{10}{x} + \frac{15}{x+10} = 1$$

$$x(x+10)\cdot\left(\frac{10}{x} + \frac{15}{x+10}\right) = x(x+10)\cdot 1$$

$$10(x+10) + 15x = x(x+10)$$
$$10x + 100 + 15x = x^2 + 10x$$
$$25x + 100 = x^2 + 10x$$
$$0 = x^2 - 15x - 100$$
$$0 = (x-20)(x+5)$$

$$x - 20 = 0 \quad \text{or} \quad x + 5 = 0$$
$$x = 20 \quad \text{or} \qquad x = -5$$

Speed for first 10 minutes: 20 mph

1. $\dfrac{x^{11}}{x^4} = x^{11-4} = x^7$

2. $12m^{10}n^7 \cdot 9m^6 n^{14} = 108m^{10+6}n^{7+14}$
$$= 108m^{16}n^{21}$$

3. $\left(\dfrac{5a^6 b^{11}}{c^7}\right)^4 = \dfrac{5^4 a^{6\cdot 4} b^{11\cdot 4}}{c^{7\cdot 4}} = \dfrac{625a^{24}b^{44}}{c^{28}}$

4. $\dfrac{t^{12}}{t^{-9}} = t^{12}\cdot t^9 = t^{21}$

5. $\left(a^{-6}b^7\right)^{-8} = a^{48}b^{-56} = \dfrac{a^{48}}{b^{56}}$

6. $x^{-16}\cdot x^{-19} = x^{-35} = \dfrac{1}{x^{35}}$

7. $330{,}000{,}000{,}000 = 3.3\times 10^{11}$

8. $0.000000000000714 = 7.14\times 10^{-13}$

9. $(0.000000000005)(200{,}000{,}000)$
$$= \left(5.0\times 10^{-12}\right)\left(2.0\times 10^8\right)$$
$$= 10\times 10^{-4}$$
$$= 1.0\times 10^{-3}$$
$$= 0.001 \text{ second}$$

10. $(-7)^2 + 11(-7) - 16 = 49 - 77 - 16$
$$= 49 - 93$$
$$= -44$$

11. $\left(x^2 + 6x - 20\right) + \left(x^2 - 13x - 39\right)$
$$= x^2 + 6x - 20 + x^2 - 13x - 39$$
$$= 2x^2 - 7x - 59$$

12. $\left(x^2 - 15x - 9\right) - \left(2x^2 - 3x - 31\right)$
$$= x^2 - 15x - 9 - 2x^2 + 3x + 31$$
$$= -x^2 - 12x + 22$$

13. $f(8) = (8)^2 + 5(8) - 37$
$$= 64 + 40 - 37$$
$$= 104 - 37$$
$$= 67$$

14. $3x\left(5x^2 - 8x + 12\right) = 3x \cdot 5x^2 - 3x \cdot 8x + 3x \cdot 12$
$$= 15x^3 - 24x^2 + 36x$$

15. $(4x - 7)(3x - 8) = 12x^2 - 32x - 21x + 56$
$$= 12x^2 - 53x + 56$$

16. $\dfrac{10x^{12} + 24x^9 - 18x^6}{2x^5} = \dfrac{10x^{12}}{2x^5} + \dfrac{24x^9}{2x^5} - \dfrac{18x^6}{2x^5}$
$$= 5x^7 + 12x^4 - 9x$$

17.
$$\begin{array}{r}
x + 7 \\
x + 6 \overline{)\, x^2 + 13x + 39} \\
\underline{-x^2 \mp 6x} \\
7x + 39 \\
\underline{-7x \mp 42} \\
-3
\end{array}$$
$$x + 7 - \dfrac{3}{x + 6}$$

18. $x^3 - 4x^2 + 6x - 24 = x^2(x - 4) + 6(x - 4)$
$$= (x - 4)\left(x^2 + 6\right)$$

19. $2x^2 - 26x + 80 = 2\left(x^2 - 13x + 40\right)$
$$= 2(x - 5)(x - 8)$$

20. $x^2 + 5x - 36 = (x + 9)(x - 4)$

21. $2x^2 - x - 10 = (2x - 5)(x + 2)$

22. $x^3 - 125 = (x)^3 - (5)^3$
$$= (x - 5)(x \cdot x + x \cdot 5 + 5 \cdot 5)$$
$$= (x - 5)\left(x^2 + 5x + 25\right)$$

23. $x^2 - 81 = (x)^2 - (9)^2 = (x + 9)(x - 9)$

24. $x^2 - 49 = 0$
$(x + 7)(x - 7) = 0$
$x + 7 = 0 \quad$ or $\quad x - 7 = 0$
$\quad x = -7 \quad$ or $\quad\quad x = 7$
$\{-7, 7\}$

25. $x^2 - 11x + 24 = 0$
$(x - 3)(x - 8) = 0$
$x - 3 = 0 \quad$ or $\quad x - 8 = 0$
$\quad x = 3 \quad$ or $\quad\quad x = 8$
$\{3, 8\}$

26. $x^2 + 14x + 49 = 0$
$(x + 7)(x + 7) = 0$
$x + 7 = 0 \quad$ or $\quad x + 7 = 0$
$\quad x = -7 \quad$ or $\quad\quad x = -7$
$\{-7\}$

27. $x(x + 8) = 20$
$x^2 + 8x = 20$
$x^2 + 8x - 20 = 0$
$(x + 10)(x - 2) = 0$
$x + 10 = 0 \quad$ or $\quad x - 2 = 0$
$\quad x = -10 \quad$ or $\quad\quad x = 2$
$\{-10, 2\}$

28. $x = -2 \quad$ or $\quad x = 7$
$x + 2 = 0 \quad$ or $\quad x - 7 = 0$
$(x + 2)(x - 7) = 0$
$x^2 - 7x + 2x - 14 = 0$
$x^2 - 5x - 14 = 0$

29. $f(x) = 0$
$x^2 - 10 = 0$
$(x + 10)(x - 10) = 0$
$x + 10 = 0 \quad$ or $\quad x - 10 = 0$
$\quad x = -10 \quad$ or $\quad\quad x = 10$
$\{-10, 10\}$

30. $f(x) = 0$
$x^2 + 2x - 35 = 0$
$(x + 7)(x - 5) = 0$
$x + 7 = 0 \quad$ or $\quad x - 5 = 0$
$\quad x = -7 \quad$ or $\quad\quad x = 5$
$\{-7, 5\}$

31. a) $t = 3$
$h(3) = -16(3)^2 + 64(3) + 192$
$\quad\quad = -16(9) + 64(3) + 192$
$\quad\quad = -144 + 192 + 192$
$\quad\quad = -144 + 384$
$\quad\quad = 240$
240 feet

b) 6 seconds

c) 2 seconds

d) $h(2) = -16(2)^2 + 64(2) + 192$
$\quad\quad = -16(4) + 64(2) + 192$
$\quad\quad = -64 + 128 + 192$
$\quad\quad = -64 + 320$
$\quad\quad = 256$
256 feet

32. 1st even positive integer: x
2nd even positive integer: $x + 2$

$$x(x+2) = 168$$
$$x^2 + 2x = 168$$
$$x^2 + 2x - 168 = 0$$
$$(x+14)(x-12) = 0$$

$$x + 14 = 0 \quad \text{or} \quad x - 12 = 0$$
$$x = -14 \quad \text{or} \quad x = 12$$

1st even positive integer: $x = 12$
2nd even positive integer: $x + 2 = 12 + 2 = 14$

33. Length: $2x + 5$
Width: x

$$(\text{length}) \cdot (\text{width}) = A$$
$$(2x+5)(x) = 133$$
$$2x^2 + 5x = 133$$
$$2x^2 + 5x - 133 = 0$$
$$(2x+19)(x-7) = 0$$

$$2x + 19 = 0 \quad \text{or} \quad x - 7 = 0$$
$$x = -\frac{19}{2} \quad \text{or} \quad x = 7$$

Width: $x = 7$ meters
Length: $2x + 5 = 2(7) + 5 = 19$ meters

34. $x^2 - 14x + 45 = 0$
$(x-5)(x-9) = 0$

$$x - 5 = 0 \quad \text{or} \quad x - 9 = 0$$
$$x = 5 \quad \text{or} \quad x = 9$$

35. $\dfrac{x^2 + 3x - 54}{x^2 + 16x + 63} = \dfrac{(x+9)(x-6)}{(x+9)(x+7)}$

$$= \frac{\cancel{(x+9)}^{1}(x-6)}{\cancel{(x+9)}_{1}(x+7)}$$

$$= \frac{x-6}{x+7}$$

36. $r(3) = \dfrac{(3)+12}{(3)^2 + 7(3) + 5} = \dfrac{3+12}{9+21+5} = \dfrac{15}{35} = \dfrac{3}{7}$

37. $\dfrac{x^2 + 3x - 18}{x^2 + 15x + 44} \cdot \dfrac{x^2 + 2x - 8}{x^2 + 8x + 12}$

$$= \frac{(x+6)(x-3)}{(x+4)(x+11)} \cdot \frac{(x+4)(x-2)}{(x+6)(x+2)}$$

$$= \frac{\cancel{(x+6)}^{1}(x-3)}{\cancel{(x+4)}_{1}(x+11)} \cdot \frac{\cancel{(x+4)}^{1}(x-2)}{\cancel{(x+6)}_{1}(x+2)}$$

$$= \frac{(x-3)(x-2)}{(x+11)(x+2)}$$

38. $\dfrac{x^2 + x - 90}{x^2 + 14x + 49} \div \dfrac{x^2 - 8x - 9}{x^2 + 5x - 14}$

$$= \frac{x^2 + x - 90}{x^2 + 14x + 49} \cdot \frac{x^2 + 5x - 14}{x^2 - 8x - 9}$$

$$= \frac{(x+10)(x-9)}{(x+7)(x+7)} \cdot \frac{(x+7)(x-2)}{(x-9)(x+1)}$$

$$= \frac{(x+10)\cancel{(x-9)}^{1}}{\cancel{(x+7)}_{1}(x+7)} \cdot \frac{\cancel{(x+7)}^{1}(x-2)}{\cancel{(x-9)}_{1}(x+1)}$$

$$= \frac{(x+10)(x-2)}{(x+7)(x+1)}$$

39. $\dfrac{5x+7}{x^2 - 2x - 8} - \dfrac{2x+19}{x^2 - 2x - 8} = \dfrac{5x + 7 - (2x+19)}{x^2 - 2x - 8}$

$$= \frac{5x + 7 - 2x - 19}{x^2 - 2x - 8}$$

$$= \frac{3x - 12}{x^2 - 2x - 8}$$

$$= \frac{3\cancel{(x-4)}^{1}}{\cancel{(x-4)}_{1}(x+2)}$$

$$= \frac{3}{x+2}$$

40. $\dfrac{6}{x^2 + x - 2} + \dfrac{2}{x^2 - 6x + 5}$

$$= \frac{6}{(x+2)(x-1)} + \frac{2}{(x-5)(x-1)}$$

$$= \frac{6}{(x+2)(x-1)} \cdot \frac{x-5}{x-5} + \frac{2}{(x-5)(x-1)} \cdot \frac{x+2}{x+2}$$

$$= \frac{6(x-5) + 2(x+2)}{(x+2)(x-1)(x-5)}$$

$$= \frac{6x - 30 + 2x + 4}{(x+2)(x-1)(x-5)}$$

$$= \frac{8x - 26}{(x+2)(x-1)(x-5)}$$

$$= \frac{2(4x - 13)}{(x+2)(x-1)(x-5)}$$

41. $\dfrac{x-5}{x^2-10x+21}-\dfrac{3}{x^2-9}$

$$=\dfrac{x-5}{(x-7)(x-3)}-\dfrac{3}{(x+3)(x-3)}$$

$$=\dfrac{x-5}{(x-7)(x-3)}\cdot\dfrac{x+3}{x+3}-\dfrac{3}{(x+3)(x-3)}\cdot\dfrac{x-7}{x-7}$$

$$=\dfrac{(x-5)(x+3)-3(x-7)}{(x-7)(x-3)(x+3)}$$

$$=\dfrac{x^2+3x-5x-15-3x+21}{(x-7)(x-3)(x+3)}$$

$$=\dfrac{x^2-5x+6}{(x-7)(x-3)(x+3)}$$

$$=\dfrac{(x-2)\cancel{(x-3)}^{1}}{(x-7)\cancel{(x-3)}_{1}(x+3)}$$

$$=\dfrac{x-2}{(x-7)(x+3)}$$

42. $\dfrac{x+4}{x^2+13x+42}+\dfrac{6}{x^2+12x+35}$

$$=\dfrac{x+4}{(x+7)(x+6)}+\dfrac{6}{(x+7)(x+5)}$$

$$=\dfrac{x+4}{(x+7)(x+6)}\cdot\dfrac{x+5}{x+5}+\dfrac{6}{(x+7)(x+5)}\cdot\dfrac{x+6}{x+6}$$

$$=\dfrac{(x+4)(x+5)+6(x+6)}{(x+7)(x+6)(x+5)}$$

$$=\dfrac{x^2+5x+4x+20+6x+36}{(x+7)(x+6)(x+5)}$$

$$=\dfrac{x^2+15x+56}{(x+7)(x+6)(x+5)}$$

$$=\dfrac{(x+8)\cancel{(x+7)}^{1}}{\cancel{(x+7)}_{1}(x+6)(x+5)}$$

$$=\dfrac{x+8}{(x+6)(x+5)}$$

43. $\dfrac{\dfrac{x-5}{x+3}-\dfrac{1}{x-6}}{\dfrac{x^2+5x-24}{x^2-3x-18}}$

$$=\dfrac{\dfrac{x-5}{x+3}-\dfrac{1}{x-6}}{\dfrac{(x+8)(x-3)}{(x-6)(x+3)}}$$

$$=\dfrac{(x-6)(x+3)}{(x-6)(x+3)}\cdot\left(\dfrac{\dfrac{x-5}{x+3}-\dfrac{1}{x-6}}{\dfrac{(x+8)(x-3)}{(x-6)(x+3)}}\right)$$

$$=\dfrac{(x-6)(x+3)\cdot\dfrac{x-5}{x+3}-(x-6)(x+3)\cdot\dfrac{1}{x-6}}{\cancel{(x-6)}^{1}\cancel{(x+3)}^{1}\cdot\dfrac{(x+8)(x-3)}{\cancel{(x-6)}_{1}\cancel{(x+3)}_{1}}}$$

$$=\dfrac{(x-6)(x-5)-1(x+3)}{(x+8)(x-3)}$$

$$=\dfrac{x^2-5x-6x+30-x-3}{(x+8)(x-3)}$$

$$=\dfrac{x^2-12x+27}{(x+8)(x-3)}$$

$$=\dfrac{(x-9)\cancel{(x-3)}^{1}}{(x+8)\cancel{(x-3)}_{1}}$$

$$=\dfrac{x-9}{x+8}$$

44. $\dfrac{1+\dfrac{1}{x}-\dfrac{20}{x^2}}{1+\dfrac{13}{x}+\dfrac{40}{x^2}}=\dfrac{x^2}{x^2}\cdot\dfrac{1+\dfrac{1}{x}-\dfrac{20}{x^2}}{1+\dfrac{13}{x}+\dfrac{40}{x^2}}$

$$=\dfrac{x^2\cdot1+\cancel{x^2}\,^{x}\cdot\dfrac{1}{\cancel{x}_{1}}-\cancel{x^2}\,^{1}\cdot\dfrac{20}{\cancel{x^2}_{1}}}{x^2\cdot1+\cancel{x^2}\,^{x}\cdot\dfrac{13}{\cancel{x}_{1}}+\cancel{x^2}\,^{1}\cdot\dfrac{40}{\cancel{x^2}_{1}}}$$

$$=\dfrac{x^2+x-20}{x^2+13x+40}$$

$$=\dfrac{\cancel{(x+5)}^{1}(x-4)}{\cancel{(x+5)}_{1}(x+8)}$$

$$=\dfrac{x-4}{x+8}$$

45.
$$6 + \frac{4}{x-3} = \frac{3x+1}{x-3}$$
$$(x-3) \cdot \left(6 + \frac{4}{x-3}\right) = (x-3) \cdot \frac{3x+1}{x-3}$$
$$6(x-3) + 4 = 3x+1$$
$$6x - 18 + 4 = 3x+1$$
$$6x - 14 = 3x+1$$
$$3x - 14 = 1$$
$$3x = 15$$
$$x = 5$$

$\{5\}$

46.
$$\frac{2x-3}{x+4} = \frac{x+3}{x-4}$$
$$(x+4)(x-4) \cdot \frac{2x-3}{x+4} = (x+4)(x-4) \cdot \frac{x+3}{x-4}$$
$$(x-4)(2x-3) = (x+4)(x+3)$$
$$2x^2 - 3x - 8x + 12 = x^2 + 3x + 4x + 12$$
$$2x^2 - 11x + 12 = x^2 + 7x + 12$$
$$x^2 - 18x = 0$$
$$x(x-18) = 0$$
$$x = 0 \quad \text{or} \quad x - 18 = 0$$
$$x = 0 \quad \text{or} \quad x = 18$$

$\{0, 18\}$

47.
$$\frac{5}{x+7} + \frac{4}{x-3} = \frac{x-11}{x^2 + 4x - 21}$$
$$\frac{5}{x+7} + \frac{4}{x-3} = \frac{x-11}{(x+7)(x-3)}$$
$$(x+7)(x-3) \cdot \left(\frac{5}{x+7} + \frac{4}{x-3}\right) = (x+7)(x-3) \cdot \frac{x-11}{(x+7)(x-3)}$$
$$5(x-3) + 4(x+7) = x-11$$
$$5x - 15 + 4x + 28 = x-11$$
$$9x + 13 = x-11$$
$$8x + 13 = -11$$
$$8x = -24$$
$$x = -3$$

$\{-3\}$

48.
$$\frac{1}{x} + \frac{1}{y} = 2$$
$$xy \cdot \frac{1}{x} + xy \cdot \frac{1}{y} = xy \cdot 2$$
$$y + x = 2xy$$
$$y = 2xy - x$$
$$y = x(2y - 1)$$
$$x = \frac{y}{2y - 1}$$

49. Unknown: x
$$\frac{1}{x} + \frac{4}{15} = \frac{11}{30}$$
$$30x \cdot \left(\frac{1}{x} + \frac{4}{15}\right) = 30x \cdot \frac{11}{30}$$
$$30 + 8x = 11x$$
$$30 = 3x$$
$$10 = x$$

50.

Pipe	Job Time	Rate	Time	Portion
Small	$x+8$	$\dfrac{1}{x+8}$	3	$\dfrac{3}{x+8}$
Large	x	$\dfrac{1}{x}$	3	$\dfrac{3}{x}$

$$\frac{3}{x+8} + \frac{3}{x} = 1$$
$$x(x+8) \cdot \left(\frac{3}{x+8} + \frac{3}{x}\right) = x(x+8) \cdot 1$$
$$x(x+8) \cdot \frac{3}{x+8} + x(x+8) \cdot \frac{3}{x} = x(x+8) \cdot 1$$
$$3x + 3(x+8) = x(x+8)$$
$$3x + 3x + 24 = x^2 + 8x$$
$$6x + 24 = x^2 + 8x$$
$$0 = x^2 + 2x - 24$$
$$0 = (x+6)(x-4)$$
$$x + 6 = 0 \quad \text{or} \quad x - 4 = 0$$
$$x = -6 \quad \text{or} \quad x = 4$$

Small: $x + 8 = 4 + 8 = 12$ hours

CHAPTER 8   A TRANSITION

8.1   QUICK CHECK

1. $9x + 13 = -14$
$9x = -27$
$x = -3$
$\{-3\}$

2. $3x - 2(4x - 5) = 2x - 11$
$3x - 8x + 10 = 2x - 11$
$-5x + 10 = 2x - 11$
$-7x + 10 = -11$
$-7x = -21$
$x = 3$
$\{3\}$

3.
$$\frac{2}{3}x - \frac{1}{5} = \frac{1}{2}x + \frac{5}{6}$$
$$30\left(\frac{2}{3}x - \frac{1}{5}\right) = 30\left(\frac{1}{2}x + \frac{5}{6}\right)$$
$$\overset{10}{\cancel{30}}\cdot\frac{2}{\cancel{3}}x - \overset{6}{\cancel{30}}\cdot\frac{1}{\cancel{5}} = \overset{15}{\cancel{30}}\cdot\frac{1}{\cancel{2}}x + \overset{5}{\cancel{30}}\cdot\frac{5}{\cancel{6}}$$
$$20x - 6 = 15x + 25$$
$$5x - 6 = 25$$
$$5x = 31$$
$$x = \frac{31}{5}$$
$\left\{\dfrac{31}{5}\right\}$

4. $(2x + 1) - (5 - 3x) = 2(3x - 2) - x$
$2x + 1 - 5 + 3x = 6x - 4 - x$
$5x - 4 = 5x - 4$
$-4 = -4$
True.
$\mathbb{R}$

5. $5x - 7 = 6(x + 2) - x$
$5x - 7 = 6x + 12 - x$
$5x - 7 = 5x + 12$
$-7 = 12$
False.
$\varnothing$

6. $|x| = 7$
$x = 7$ or $x = -7$
$\{-7, 7\}$

7. $|3x + 8| = 5$
$3x + 8 = -5$ or $3x + 8 = 5$
$3x = -13$ or $3x = -3$
$x = \dfrac{-13}{3}$ or $x = -1$
$\left\{-1, -\dfrac{13}{3}\right\}$

8. $4|2x - 5| - 9 = 7$
$4|2x - 5| = 16$
$|2x - 5| = 4$
$2x - 5 = -4$ or $2x - 5 = 4$
$2x = 1$ or $2x = 9$
$x = \dfrac{1}{2}$ or $x = \dfrac{9}{2}$
$\left\{\dfrac{9}{2}, \dfrac{1}{2}\right\}$

9. $|2x + 3| = -6$
$\varnothing$ Since an absolute value set equal to a negative number has no solution.

10. $|x - 9| = |2x + 13|$
$x - 9 = 2x + 13$ or $x - 9 = -(2x + 13)$
$-x - 9 = 13$ or $x - 9 = -2x - 13$
$-x = 22$ or $3x - 9 = -13$
$x = -22$ or $3x = -4$
$x = -22$ or $x = -\dfrac{4}{3}$
$\left\{-22, -\dfrac{4}{3}\right\}$

8.1   LINEAR EQUATIONS AND ABSOLUTE VALUE EQUATIONS

1. equation

3. LCM

5. the empty set

7. $X = a,\ X = -a$

9. $m - 5 = -9$
$m = -4$
$\{-4\}$

11. $5t = -8$

$$t = -\frac{8}{5}$$

$$\left\{ -\frac{8}{5} \right\}$$

13. $7n - 11 = -39$
$7n = -28$
$n = -4$
$\{-4\}$

15. $-2x + 9 = 31$
$-2x = 22$
$x = -11$
$\{-11\}$

17. $9m - 30 = 39$
$9m = 69$

$$m = \frac{69}{9}$$

$$m = \frac{23}{3}$$

$$\left\{ \frac{23}{3} \right\}$$

19. $4x + 17 = 2x - 19$
$2x = -36$
$x = -18$
$\{-18\}$

21. $3x - 7 = -x + 15$
$4x = 22$

$$x = \frac{22}{4}$$

$$x = \frac{11}{2}$$

$$\left\{ \frac{11}{2} \right\}$$

23. $-9r + 22 = 3r + 67$
$-12r = 45$

$$r = \frac{45}{-12}$$

$$r = -\frac{15}{4}$$

$$\left\{ -\frac{15}{4} \right\}$$

25. $2(2x - 4) - 7 = 6x - 11$
$4x - 8 - 7 = 6x - 11$
$4x - 15 = 6x - 11$
$-2x - 15 = -11$
$-2x = 4$
$x = -2$
$\{-2\}$

27. $5x - 3(2x - 9) = 4(3x - 8) + 7x$
$5x - 6x + 27 = 12x - 32 + 7x$
$-x + 27 = 19x - 32$
$-20x + 27 = -32$
$-20x = -59$

$$x = \frac{59}{20}$$

$$\left\{ \frac{59}{20} \right\}$$

29. $2(3x - 1) + 5(x + 4) = 3(x + 7) - (x + 6)$
$6x - 2 + 5x + 20 = 3x + 21 - x - 6$
$11x + 18 = 2x + 15$
$9x = -3$

$$x = \frac{-3}{9}$$

$$x = -\frac{1}{3}$$

$$\left\{ -\frac{1}{3} \right\}$$

31. $5x - 3(4x + 1) = 3 + 4(3x + 8)$
$5x - 12x - 3 = 3 + 12x + 32$
$-7x - 3 = 12x + 35$
$-19x = 38$
$x = -2$
$\{-2\}$

33.
$$\frac{3}{4}t - 6 = \frac{1}{3}t - 1$$

$$12\left(\frac{3}{4}t - 6 \right) = 12\left(\frac{1}{3}t - 1 \right)$$

$$\overset{3}{\cancel{12}} \cdot \frac{3}{\underset{1}{\cancel{4}}}t - 12 \cdot 6 = \overset{4}{\cancel{12}} \cdot \frac{1}{\underset{1}{\cancel{3}}}t - 12 \cdot 1$$

$9t - 72 = 4t - 12$
$5t - 72 = -12$
$5t = 60$
$t = 12$
$\{12\}$

35.
$$\frac{1}{2}(2x-7)-\frac{1}{3}x=\frac{4}{3}x-5$$
$$6\left(\frac{1}{2}(2x-7)-\frac{1}{3}x\right)=6\left(\frac{4}{3}x-5\right)$$
$$\cancel{6}^{3}\cdot\frac{1}{\cancel{2}_{1}}(2x-7)-\cancel{6}^{2}\cdot\frac{1}{\cancel{3}_{1}}x=\cancel{6}^{2}\cdot\frac{4}{\cancel{3}_{1}}x-6\cdot5$$
$$3(2x-7)-2x=8x-30$$
$$6x-21-2x=8x-30$$
$$4x-21=8x-30$$
$$-4x-21=-30$$
$$-4x=-9$$
$$x=\frac{9}{4}$$
$$\left\{\frac{9}{4}\right\}$$

37. $2b-9=2b-9$
$\quad -9=-9$
True.
$\mathbb{R}$

39. $3(2n-1)+5=2(3n+2)$
$\quad 6n-3+5=6n+4$
$\quad\quad 6n+2=6n+4$
$\quad\quad\quad 2=4$
False.
$\varnothing$

41. $5x-2(3-2x)=3(3x-2)$
$\quad 5x-6+4x=9x-6$
$\quad\quad 9x-6=9x-6$
$\quad\quad\quad -6=-6$
True.
$\mathbb{R}$

43. $|x|=2$
$x=-2$ or $x=2$
$\{-2,2\}$

45. $|x+6|=13$
$x+6=-13$ or $x+6=13$
$\quad x=-19$ or $\quad x=7$
$\{-19,7\}$

47. $|3x+2|=20$
$3x+2=-20$ or $3x+2=20$
$\quad 3x=-22$ or $\quad 3x=18$
$$x=-\frac{22}{3}\quad\text{or}\quad x=6$$
$$\left\{-\frac{22}{3},6\right\}$$

49. $|x|+7=3$
$\quad |x|=-4$
$\varnothing$ since an absolute value cannot be
equal to a negative number.

51. $|x+4|-12=6$
$\quad |x+4|=18$
$x+4=-18$ or $x+4=18$
$\quad x=-22$ or $\quad x=14$
$\{-22,14\}$

53. $|x-3|+7=12$
$\quad |x-3|=5$
$x-3=-5$ or $x-3=5$
$\quad x=-2$ or $\quad x=8$
$\{-2,8\}$

55. $|4x-11|-19=-8$
$\quad |4x-11|=11$
$4x-11=-11$ or $4x-11=11$
$\quad 4x=0$ or $\quad 4x=22$
$$x=0\quad\text{or}\quad x=\frac{22}{4}$$
$$x=0\quad\text{or}\quad x=\frac{11}{2}$$
$$\left\{0,\frac{11}{2}\right\}$$

57. $2|3x+2|-9=17$
$\quad 2|3x+2|=26$
$\quad\quad |3x+2|=13$
$3x+2=-13$ or $3x+2=13$
$\quad 3x=-15$ or $\quad 3x=11$
$$x=-5\quad\text{or}\quad x=\frac{11}{3}$$
$$\left\{-5,\frac{11}{3}\right\}$$

59. $-|x+3|-10=-16$
$\quad -|x+3|=-6$
$\quad\quad |x+3|=6$
$x+3=-6$ or $x+3=6$
$\quad x=-9$ or $\quad x=3$
$\{-9,3\}$

61. $|4x-5| = |3x-23|$

$4x-5 = -(3x-23)$ or $4x-5 = 3x-23$
$4x-5 = -3x+23$ or $x-5 = -23$
$7x-5 = 23$ or $x = -18$
$7x = 28$
$x = 4$
$\{-18, 4\}$

63. $|x+12| = |2x-9|$

$x+12 = -(2x-9)$ or $x+12 = 2x-9$
$x+12 = -2x+9$ or $-x+12 = -9$
$3x+12 = 9$ or $-x = -21$
$3x = -3$ or $x = 21$
$x = -1$
$\{-1, 21\}$

65. $|x+13| - 12 = -7$

$|x+13| = 5$

$x+13 = -5$ or $x+13 = 5$
$x = -18$ or $x = -8$
$\{-18, -8\}$

67. $|x-6| + 14 = 9$

$|x-6| = -5$

$\varnothing$ since an absolute value cannot be equal to a negative number.

69. $25 = 4 - 6x$
$21 = -6x$

$\dfrac{21}{-6} = x$

$-\dfrac{7}{2} = x$

$\left\{ -\dfrac{7}{2} \right\}$

71. $10x - 17 = 6x - 53$
$4x = -36$
$x = -9$
$\{-9\}$

73.
$$\frac{3}{5}x - \frac{1}{2} = \frac{2}{3}x + \frac{3}{4}$$

$$60\left(\frac{3}{5}x - \frac{1}{2}\right) = 60\left(\frac{2}{3}x + \frac{3}{4}\right)$$

$$\overset{12}{60}\cdot\frac{3}{\underset{1}{5}}x - \overset{30}{60}\cdot\frac{1}{\underset{1}{2}} = \overset{20}{60}\cdot\frac{2}{\underset{1}{3}}x + \overset{15}{60}\cdot\frac{3}{\underset{1}{4}}$$

$36x - 30 = 40x + 45$
$-4x = 75$

$$x = -\frac{75}{4}$$

$\left\{ -\dfrac{75}{4} \right\}$

75. $|7x-11| = |4x+6|$

$7x-11 = -(4x+6)$ or $7x-11 = 4x+6$
$7x-11 = -4x-6$ or $3x = 17$

$11x = 5$ or $x = \dfrac{17}{3}$

$x = \dfrac{5}{11}$

$\left\{ \dfrac{5}{11}, \dfrac{17}{3} \right\}$

77. $|3x+16| - 4 = 9$

$|3x+16| = 13$

$3x+16 = -13$ or $3x+16 = 13$
$3x = -29$ or $3x = -3$

$x = -\dfrac{29}{3}$ or $x = -1$

$\left\{ -\dfrac{29}{3}, -1 \right\}$

79. $3(4x-5) + 7x = 5(2x-9) + 12$
$12x - 15 + 7x = 10x - 45 + 12$
$19x - 15 = 10x - 33$
$9x = -18$
$x = -2$
$\{-2\}$

81. Answers will vary. Example: $|x| = 2$

83. Answers will vary. Example: $|x+1| = 4$

85. Answers will vary. Example:
Isolating the absolute value yields
$|2x-7| = -2$. Since an absolute value cannot be equal to a negative number, there is no solution.

8.2   QUICK CHECK

1. $2x + 5 \le 9$
$2x \le 4$
$x \le 2$
$(-\infty, 2]$

2. $3(2x - 1) + 4 \ge 2x - 7$
$6x - 3 + 4 \ge 2x - 7$
$6x + 1 \ge 2x - 7$
$4x + 1 \ge -7$
$4x \ge -8$
$x \ge -2$
$[-2, \infty)$

3. $5 - 2x > -5$
$-2x > -10$
$x < 5$
$(-\infty, 5)$

4. $-8 \le 3x + 7 \le 1$
$-15 \le 3x \le -6$
$-5 \le x \le -2$
$[-5, -2]$

5. $x + 3 < -3 \quad$ or $\quad 2x + 5 > 14$
$x < -6 \quad$ or $\quad 2x > 9$

$x < -6 \quad$ or $\quad x > \dfrac{9}{2}$

$(-\infty, -6) \cup \left(\dfrac{9}{2}, \infty \right)$

6. $|x + 3| \le 1$
$-1 \le x + 3 \le 1$
$-4 \le x \le -2$
$[-4, -2]$

7. $|2x - 1| + 4 < 9$

$|2x - 1| < 5$
$-5 < 2x - 1 < 5$
$-4 < 2x < 6$
$-2 < x < 3$
$(-2, 3)$

8. $|x + 1| - 3 < -8$

$|x + 1| < -5$
$\varnothing$ since an absolute value cannot be less than a negative number.

9. $|x - 2| > 4$
$x - 2 < -4 \quad$ or $\quad x - 2 > 4$
$x < -2 \quad$ or $\quad x > 6$
$(-\infty, -2) \cup (6, \infty)$

10. $|2x + 7| + 5 > 10$

$|2x + 7| > 5$

$2x + 7 < -5 \quad$ or $\quad 2x + 7 > 5$
$2x < -12 \quad$ or $\quad 2x > -2$
$x < -6 \quad$ or $\quad x > -1$
$(-\infty, -6) \cup (-1, \infty)$

11. $|7x - 13| \ge -1$
$\mathbb{R}$ since an absolute value is always greater than a negative number.
$(-\infty, \infty)$

8.2   LINEAR INEQUALITIES AND ABSOLUTE VALUE INEQUALITIES

1. linear inequality

3. interval

5. compound

7. $x - 6 \le 2$
$x \le 8$
$(-\infty, 8]$

9. $3x + 2 \ge -10$
$3x \ge -12$
$x \ge -4$
$[-4, \infty)$

11. $-3x + 2 < 17$
$-3x < 15$
$x > -5$
$(-5, \infty)$

13.
$$\frac{2}{3}x - \frac{5}{6} > -2$$
$$6 \cdot \left(\frac{2}{3}x - \frac{5}{6}\right) > 6 \cdot (-2)$$
$$\cancel{6}^2 \cdot \frac{2}{\cancel{3}_1}x - \cancel{6}^1 \cdot \frac{5}{\cancel{6}_1} > -12$$
$$4x - 5 > -12$$
$$4x > -7$$
$$x > -\frac{7}{4}$$
$$\left(-\frac{7}{4}, \infty\right)$$

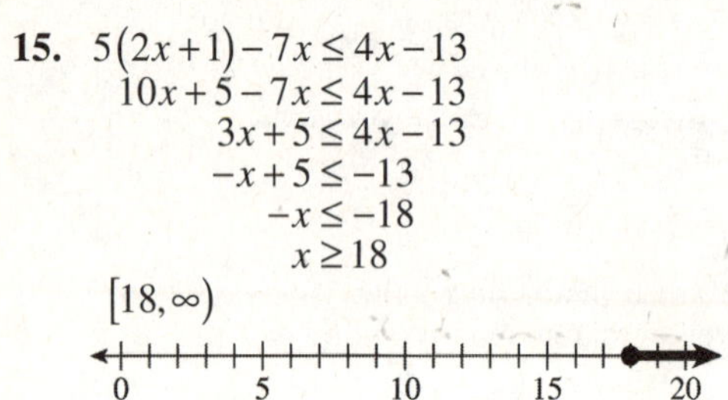

15.
$$5(2x+1) - 7x \le 4x - 13$$
$$10x + 5 - 7x \le 4x - 13$$
$$3x + 5 \le 4x - 13$$
$$-x + 5 \le -13$$
$$-x \le -18$$
$$x \ge 18$$
$$[18, \infty)$$

17.
$$5 \le x - 3 \le 11$$
$$8 \le x \le 14$$
$$[8, 14]$$

19.
$$-10 \le 7x + 4 \le 53$$
$$-14 \le 7x \le 49$$
$$-2 \le x \le 7$$
$$[-2, 7]$$

21.
$$-9 < 2x + 5 < -4$$
$$-14 < 2x < -9$$
$$-7 < x < -\frac{9}{2}$$
$$\left(-7, -\frac{9}{2}\right)$$

23.
$$-6 < \frac{3}{5}x + \frac{2}{5} < -2$$
$$5 \cdot (-6) < 5 \cdot \frac{3}{5}x + 5 \cdot \frac{2}{5} < 5 \cdot (-2)$$
$$5 \cdot (-6) < \cancel{5}^1 \cdot \frac{3}{\cancel{5}_1}x + \cancel{5}^1 \cdot \frac{2}{\cancel{5}_1} < 5 \cdot (-2)$$
$$-30 < 3x + 2 < -10$$
$$-32 < 3x < -12$$
$$-\frac{32}{3} < x < -4$$
$$\left(-\frac{32}{3}, -4\right)$$

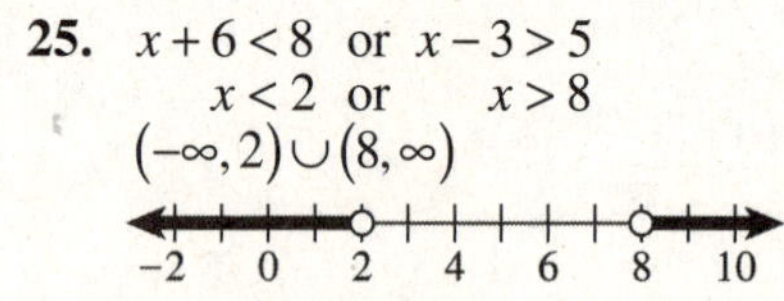

25.
$$x + 6 < 8 \quad \text{or} \quad x - 3 > 5$$
$$x < 2 \quad \text{or} \quad x > 8$$
$$(-\infty, 2) \cup (8, \infty)$$

27.
$$3x + 4 \le -17 \quad \text{or} \quad 2x - 9 \ge 9$$
$$3x \le -21 \quad \text{or} \quad 2x \ge 18$$
$$x \le -7 \quad \text{or} \quad x \ge 9$$
$$(-\infty, -7] \cup [9, \infty)$$

29.
$$3x - 11 > 5x - 3 \quad \text{or} \quad 2x + 31 < 6x + 7$$
$$-2x - 11 > -3 \quad \text{or} \quad -4x + 31 < 7$$
$$-2x > 8 \quad \text{or} \quad -4x < -24$$
$$x < -4 \quad \text{or} \quad x > 6$$
$$(-\infty, -4) \cup (6, \infty)$$

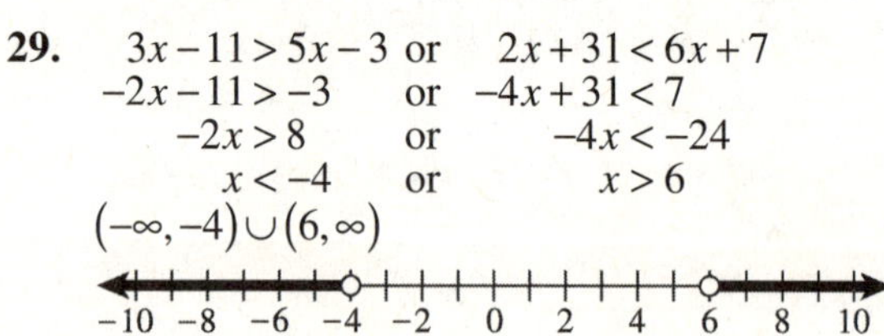

31.
$$|x + 2| < 3$$
$$-3 < x + 2 < 3$$
$$-5 < x < 1$$
$$(-5, 1)$$

33.
$$|x - 5| + 7 \le 13$$
$$|x - 5| \le 6$$
$$-6 \le x - 5 \le 6$$
$$-1 \le x \le 11$$
$$[-1, 11]$$

35. $|4x+9|-8 \le -3$

$|4x+9| \le 5$

$-5 \le 4x+9 \le 5$

$-14 \le 4x \le -4$

$-\dfrac{14}{4} \le x \le -1$

$-\dfrac{7}{2} \le x \le -1$

$\left[-\dfrac{7}{2}, -1\right]$

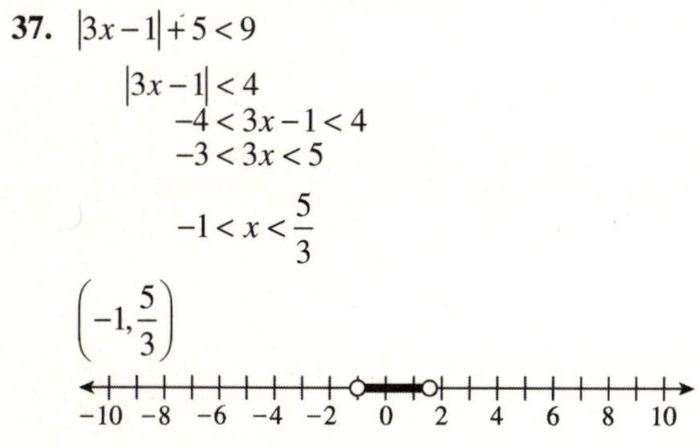

37. $|3x-1|+5 < 9$

$|3x-1| < 4$

$-4 < 3x-1 < 4$

$-3 < 3x < 5$

$-1 < x < \dfrac{5}{3}$

$\left(-1, \dfrac{5}{3}\right)$

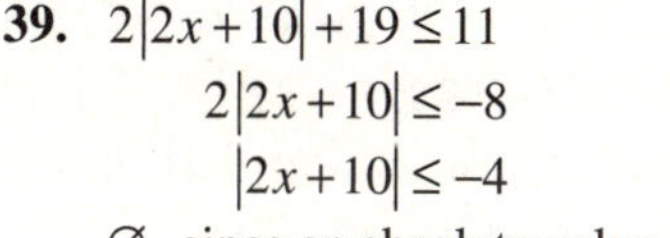

39. $2|2x+10|+19 \le 11$

$2|2x+10| \le -8$

$|2x+10| \le -4$

$\varnothing$, since an absolute value cannot be less than or equal to a negative number.

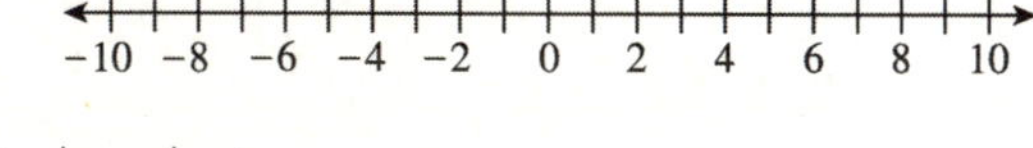

41. $|x-4| > 1$

$x-4 < -1$ or $x-4 > 1$

$x < 3$ or $x > 5$

$(-\infty, 3) \cup (5, \infty)$

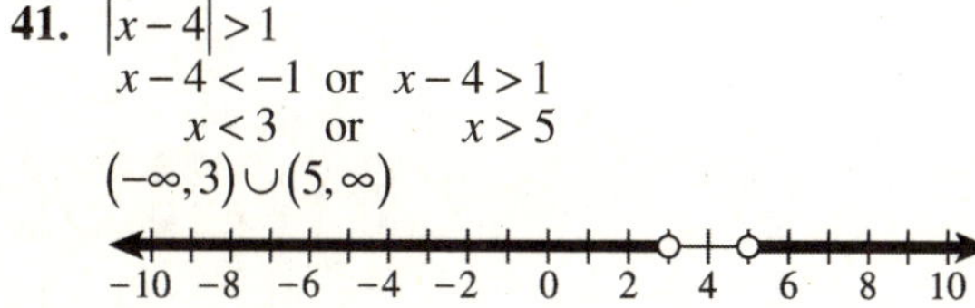

43. $|x+6|+5 \ge 8$

$|x+6| \ge 3$

$x+6 \le -3$ or $x+6 \ge 3$

$x \le -9$ or $x \ge -3$

$(-\infty, -9] \cup [-3, \infty)$

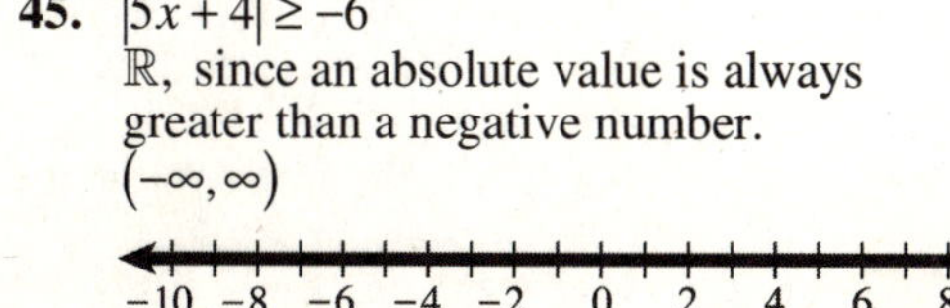

45. $|5x+4| \ge -6$

$\mathbb{R}$, since an absolute value is always greater than a negative number.

$(-\infty, \infty)$

47. $|4x-17|-5 > 6$

$|4x-17| > 11$

$4x-17 < -11$ or $4x-17 > 11$

$4x < 6$ or $4x > 28$

$x < \dfrac{6}{4}$ or $x > 7$

$x < \dfrac{3}{2}$

$\left(-\infty, \dfrac{3}{2}\right) \cup (7, \infty)$

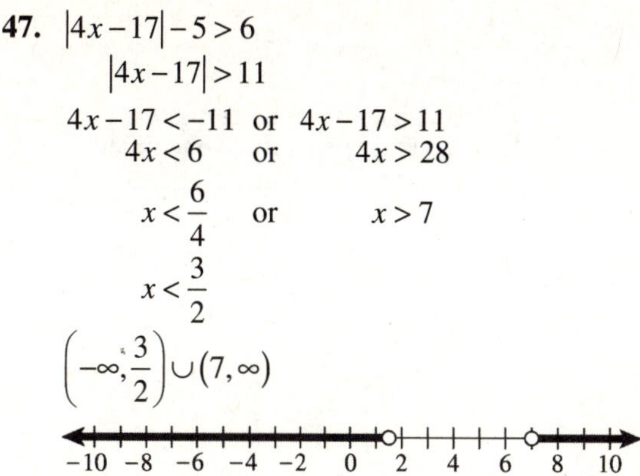

49. $2|2x+3|+5 > 19$

$2|2x+3| > 14$

$|2x+3| > 7$

$2x+3 < -7$ or $2x+3 > 7$

$2x < -10$ or $2x > 4$

$x < -5$ or $x > 2$

$(-\infty, -5) \cup (2, \infty)$

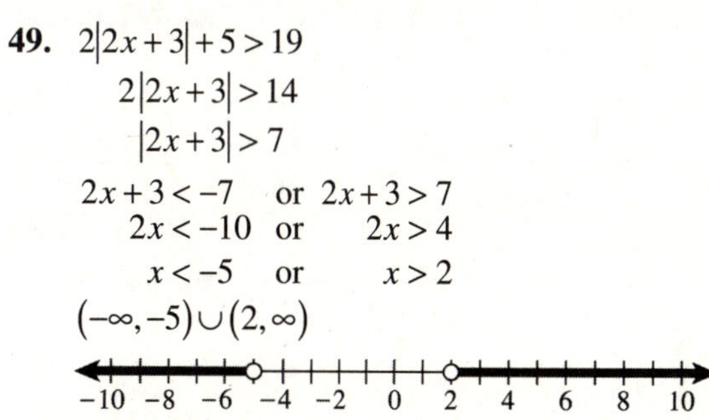

51. $|x+4| \le 6$

$-6 \le x+4 \le 6$

$-10 \le x \le 2$

$[-10, 2]$

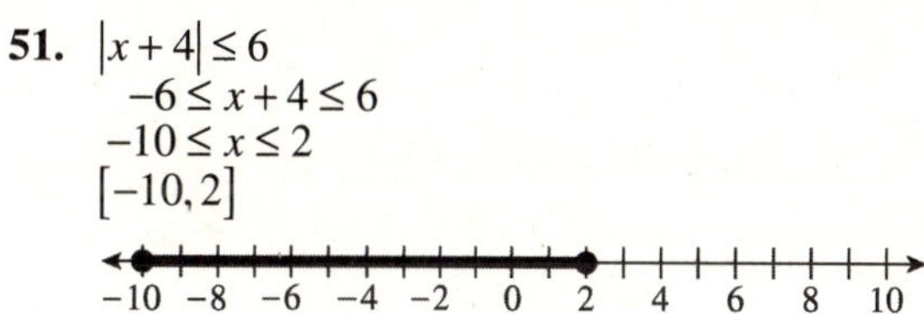

53. $-3x+10 \ge 34$ or $6x-9 \ge 33$

$-3x \ge 24$ or $6x \ge 42$

$x \le -8$ or $x \ge 7$

$(-\infty, -8] \cup [7, \infty)$

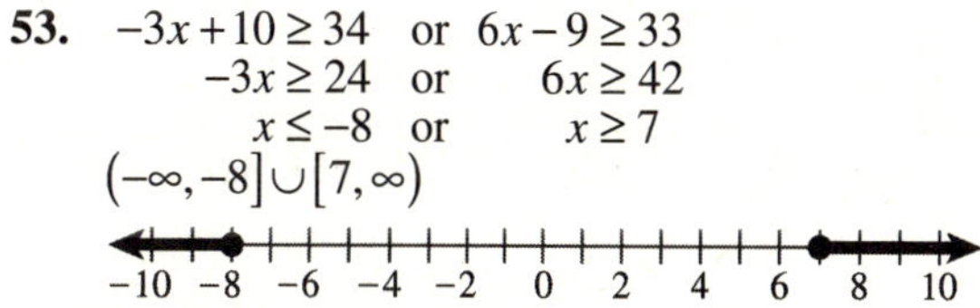

55. $|x-5|+9 > 4$

$|x-5| > -5$

$\mathbb{R}$, since an absolute value is always greater than a negative number.

$(-\infty, \infty)$

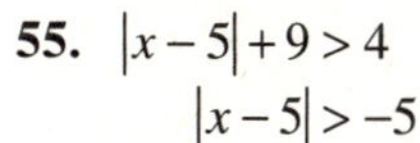
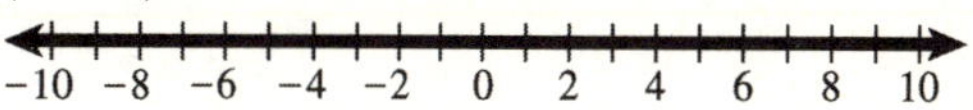

57. $2x+7 \ge -9$

$2x \ge -16$

$x \ge -8$

$[-8, \infty)$

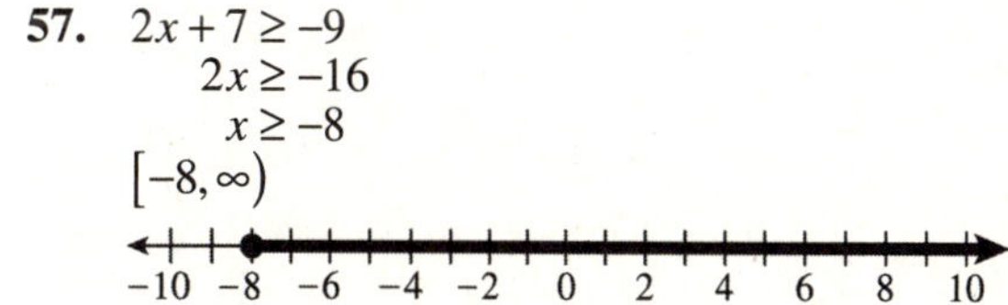

59. $|x-9|+8 \le 3$

$\qquad |x-9| \le -5$

$\varnothing$, since an absolute value cannot be less than or equal to a negative number.

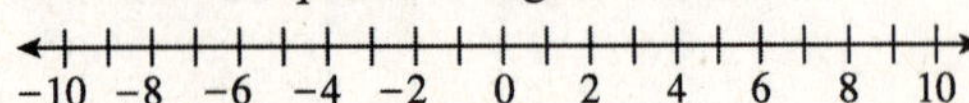

61. $|2x+5|-9 \ge 4$

$\qquad |2x+5| \ge 13$

$2x+5 \le -13 \quad \text{or} \quad 2x+5 \ge 13$

$\qquad 2x \le -18 \quad \text{or} \qquad 2x \ge 8$

$\qquad\quad x \le -9 \quad \text{or} \qquad\quad x \ge 4$

$(-\infty, -9] \cup [4, \infty)$

63. $17 \le 5x-3 < 37$

$20 \le 5x < 40$

$\quad 4 \le x < 8$

$[4, 8)$

65. $|x| < 2$

67. $|x-5| > 4$

69. Answers will vary. Example:

The inequality $|x+5| < -3$ has no solution since an absolute value is never less than a negative number.

The inequality $|x+5| > -3$ has every real number as a solution since an absolute value is always greater than a negative number.

71. Answers will vary.

1. $5x+2y = -10$

x-intercept: Let $y = 0$.

$5x+2(0) = -10$

$\qquad 5x = -10$

$\qquad\quad x = -2$

$(-2, 0)$

y-intercept: Let $x = 0$.

$5(0)+2y = -10$

$\qquad 2y = -10$

$\qquad\quad y = -5$

$(0, -5)$

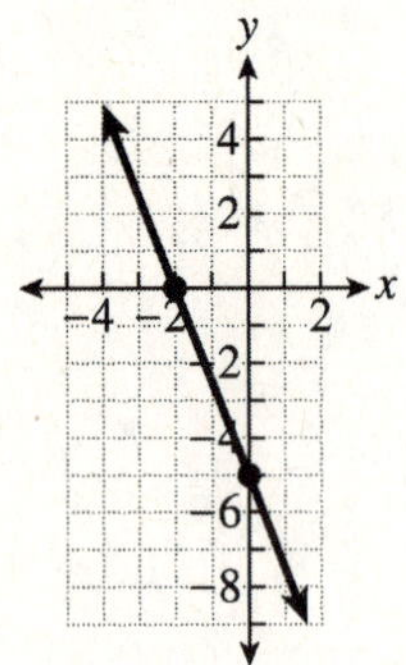

2. $y = \dfrac{7}{2}x - 4$

y-intercept: $(0, -4)$ using $y = mx+b$

2nd point: $(2, 3)$ using slope $= \dfrac{7}{2}$

(up 7, right 2)

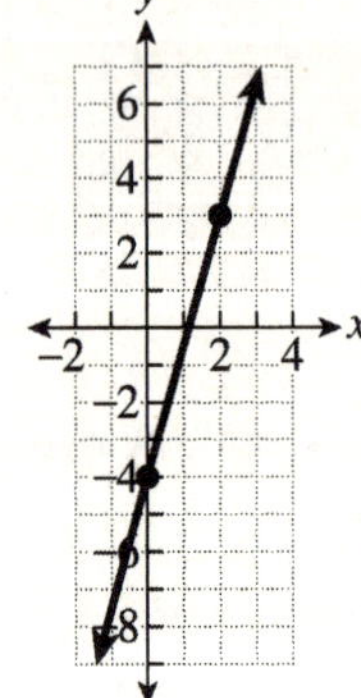

3. $f(x) = -x + 6$

y-intercept: $(0,6)$ using $f(x) = mx + b$
2nd point: $(1,5)$ using slope $= -1$
(down 1, right 1)

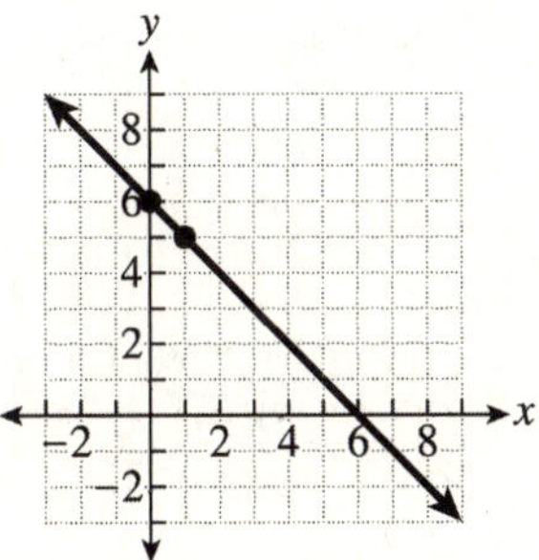

4. $x + 1 = 0$
$\quad\quad x = -1$

x	$f(x) = \lvert x+1\rvert + 3$	$(x, f(x))$
-3	$f(-3) = \lvert -3+1\rvert + 3 = 2 + 3 = 5$	$(-3,5)$
-2	$f(-2) = \lvert -2+1\rvert + 3 = 1 + 3 = 4$	$(-2,4)$
-1	$f(-1) = \lvert -1+1\rvert + 3 = 0 + 3 = 3$	$(-1,3)$
0	$f(0) = \lvert 0+1\rvert + 3 = 1 + 3 = 4$	$(0,4)$
1	$f(1) = \lvert 1+1\rvert + 3 = 2 + 3 = 5$	$(1,5)$

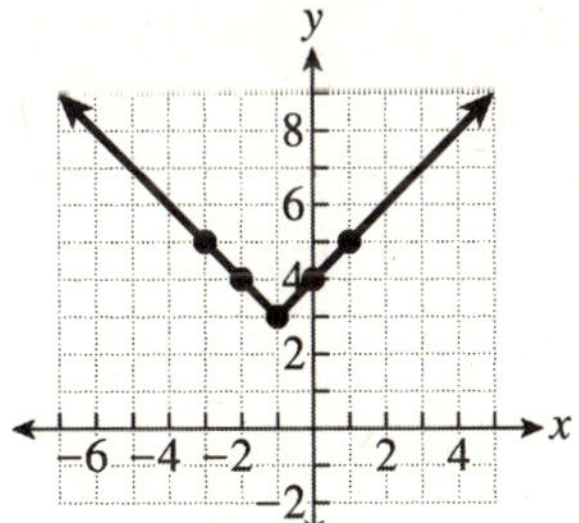

5. $x - 5 = 0$
$\quad\quad x = 5$

x	$f(x) = \lvert x-5\rvert - 4$	$(x, f(x))$
3	$f(3) = \lvert 3-5\rvert - 4 = 2 - 4 = -2$	$(3,-2)$
4	$f(4) = \lvert 4-5\rvert - 4 = 1 - 4 = -3$	$(4,-3)$
5	$f(5) = \lvert 5-5\rvert - 4 = 0 - 4 = -4$	$(5,-4)$
6	$f(6) = \lvert 6-5\rvert - 4 = 1 - 4 = -3$	$(6,-3)$
7	$f(7) = \lvert 7-5\rvert - 4 = 2 - 4 = -2$	$(7,-2)$

Domain: $(-\infty, \infty)$
Range: $[-4, \infty)$

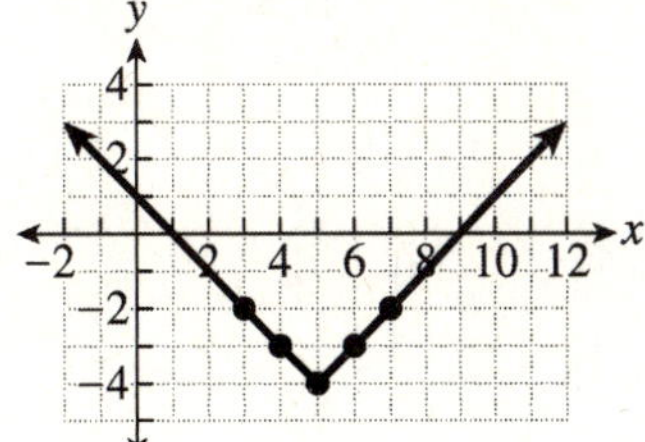

6. $x - 4 = 0$
$\quad\quad x = 4$

x	$f(x) = -\lvert x-4\rvert - 5$	$(x, f(x))$
2	$f(2) = -\lvert 2-4\rvert - 5 = -2 - 5 = -7$	$(2,-7)$
3	$f(3) = -\lvert 3-4\rvert - 5 = -1 - 5 = -6$	$(3,-6)$
4	$f(4) = -\lvert 4-4\rvert - 5 = 0 - 5 = -5$	$(4,-5)$
5	$f(5) = -\lvert 5-4\rvert - 5 = -1 - 5 = -6$	$(5,-6)$
6	$f(6) = -\lvert 6-4\rvert - 5 = -2 - 5 = -7$	$(6,-7)$

Domain: $(-\infty, \infty)$
Range: $(-\infty, -5]$

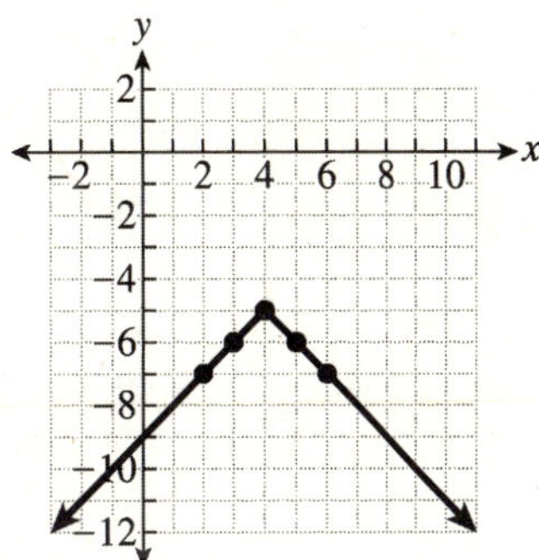

7. $f(x) = \lvert x+6\rvert - 3$

8.3   GRAPHING LINEAR EQUATIONS AND FUNCTIONS; GRAPHING ABSOLUTE VALUE FUNCTIONS

1. absolute value function

3. y-intercept

5. $4x + 3y = -24$

x-intercept: Let $y = 0$.
$4x + 3(0) = -24$
$4x = -24$
$x = -6$
$(-6, 0)$

y-intercept: Let $x = 0$.
$4(0) + 3y = -24$
$3y = -24$
$y = -8$
$(0, -8)$

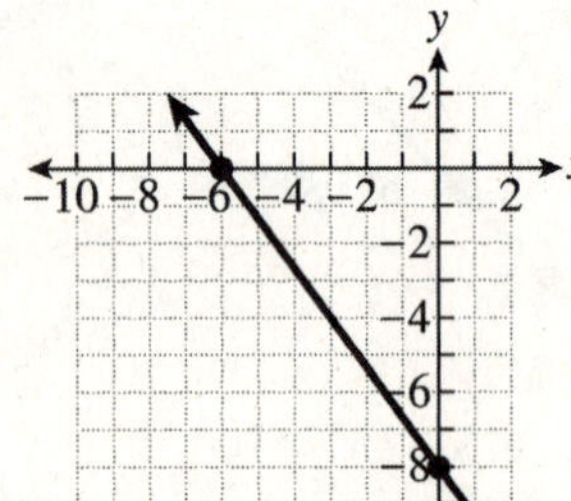

7. $2x - 5y = 10$

x-intercept: Let $y = 0$.
$2x - 5(0) = 10$
$2x = 10$
$x = 5$
$(5, 0)$

y-intercept: Let $x = 0$.
$2(0) - 5y = 10$
$-5y = 10$
$y = -2$
$(0, -2)$

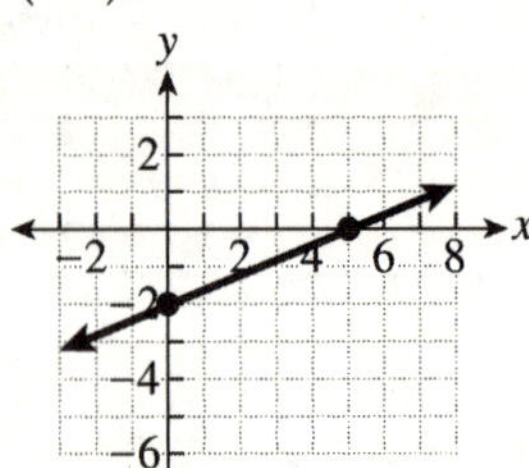

9. $\dfrac{2}{9}x + \dfrac{1}{3}y = 2$

x-intercept: Let $y = 0$.
$\dfrac{2}{9}x + \dfrac{1}{3}(0) = 2$
$\dfrac{2}{9}x = 2$
$x = 9$
$(9, 0)$

y-intercept: Let $x = 0$.
$\dfrac{2}{9}(0) + \dfrac{1}{3}y = 2$
$\dfrac{1}{3}y = 2$
$y = 6$
$(0, 6)$

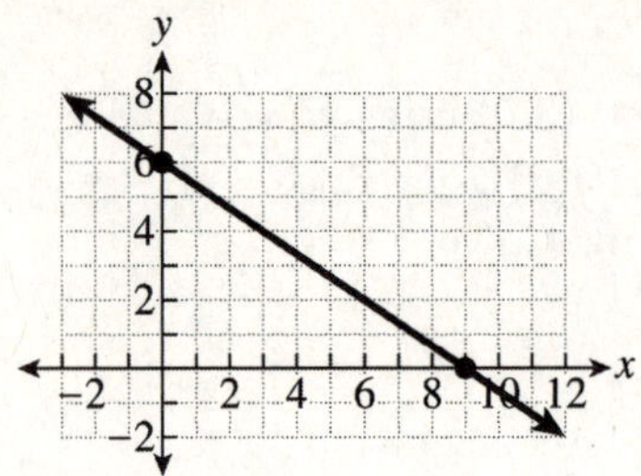

11. $8x - 3y = 12$

x-intercept: Let $y = 0$.
$8x - 3(0) = 12$
$8x = 12$
$x = \dfrac{12}{8}$
$x = \dfrac{3}{2}$
$\left(\dfrac{3}{2}, 0\right)$

y-intercept: Let $x = 0$.
$8(0) - 3y = 12$
$-3y = 12$
$y = -4$
$(0, -4)$

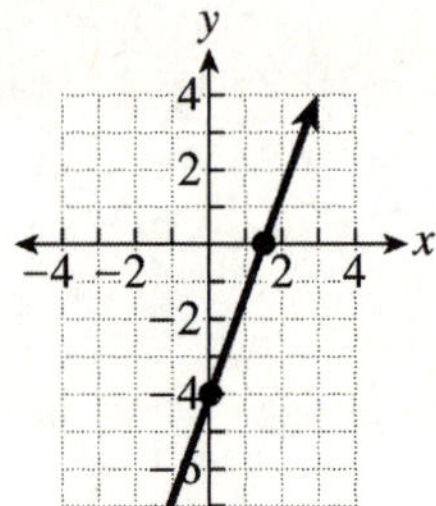

13. $y = -3x + 4$

x-intercept: Let $y = 0$.
$0 = -3x + 4$
$3x = 4$
$x = \dfrac{4}{3}$
$\left(\dfrac{4}{3}, 0\right)$

y-intercept: $(0, 4)$ using $y = mx + b$

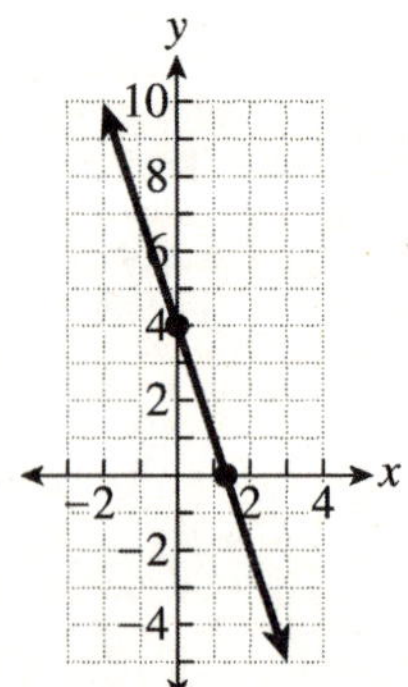

15. $y = 3x + 2$

y-intercept: $(0, 2)$ using $y = mx + b$
2nd point: $(1, 5)$ using slope $= 3$
(up 3, right 1)

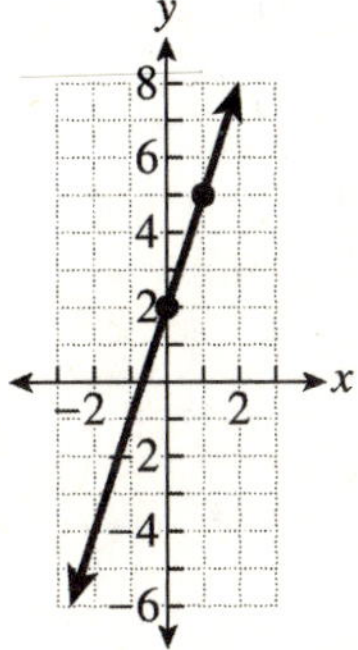

17. $y = -4x + 8$

y-intercept: $(0, 8)$ using $y = mx + b$
2nd point: $(1, 4)$ using slope $= -4$
(down 4, right 1)

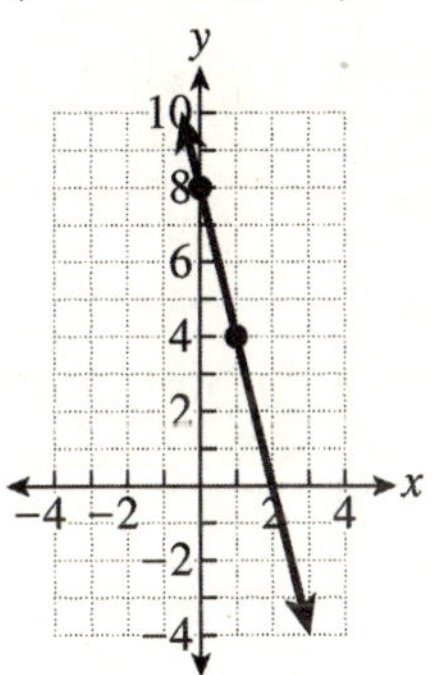

19. $y = \dfrac{2}{3}x - 6$

y-intercept: $(0, -6)$ using $y = mx + b$
2nd point: $(3, -4)$ using slope $= \dfrac{2}{3}$
(up 2, right 3)

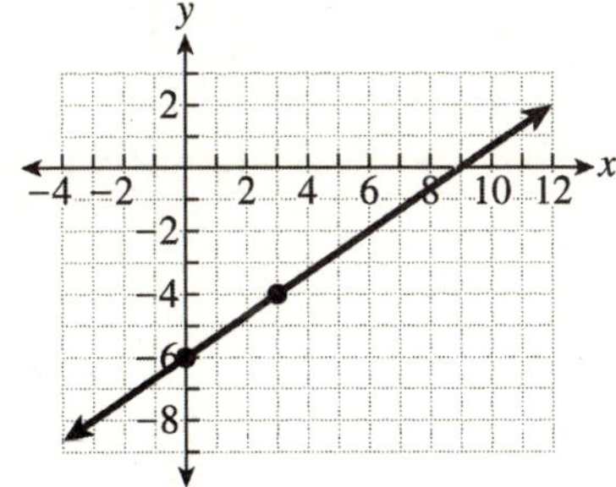

21. $y = 4x$

y-intercept: $(0, 0)$ using $y = mx + b$
2nd point: $(1, 4)$ using slope $= 4$
(up 4, right 1)

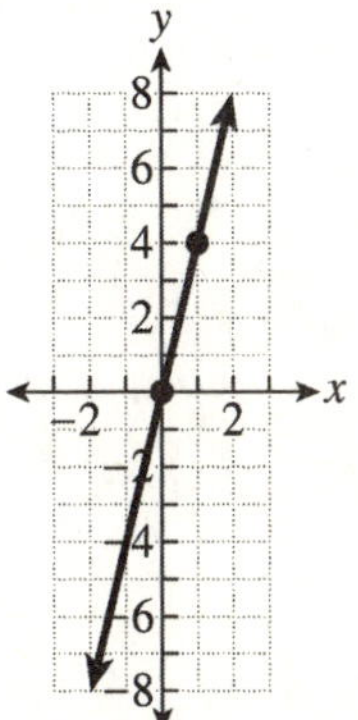

23. $f(x) = x + 7$

y-intercept: $(0, 7)$
2nd point: $(1, 8)$ using slope $= 1$
(up 1, right 1)

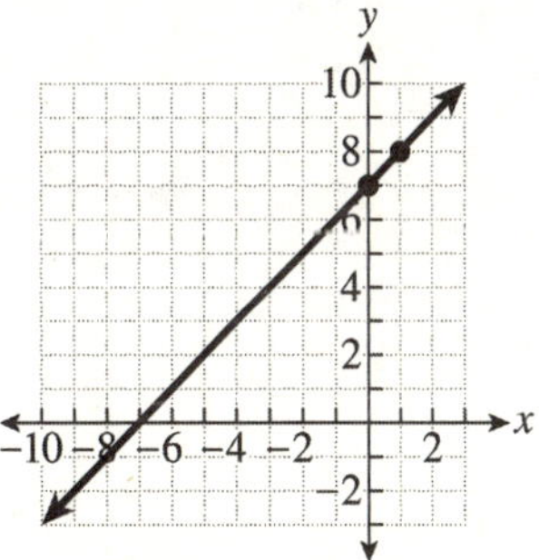

25. $f(x) = 4x - 5$

y-intercept: $(0, -5)$
2nd point: $(1, -1)$ using slope $= 4$
(up 4, right 1)

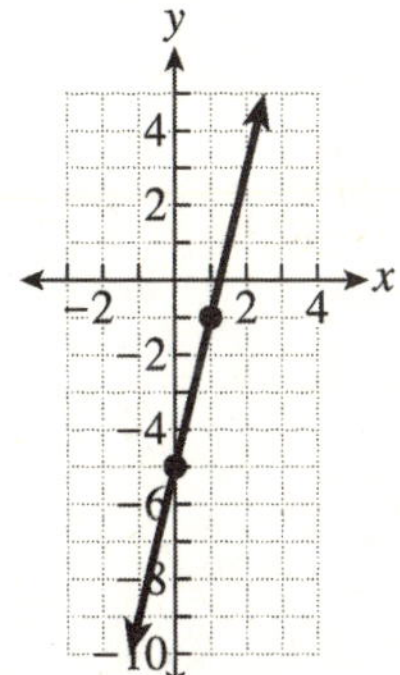

27. $f(x) = -5x + 8$

y-intercept: $(0,8)$

2nd point: $(1,3)$ using slope $= -5$
(down 5, right 1)

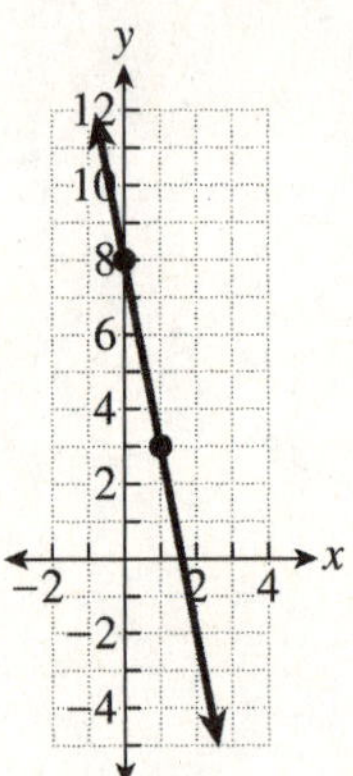

29. $f(x) = \dfrac{3}{4}x - 6$

y-intercept: $(0,-6)$

2nd point: $(4,-3)$ using slope $= \dfrac{3}{4}$
(up 3, right 4)

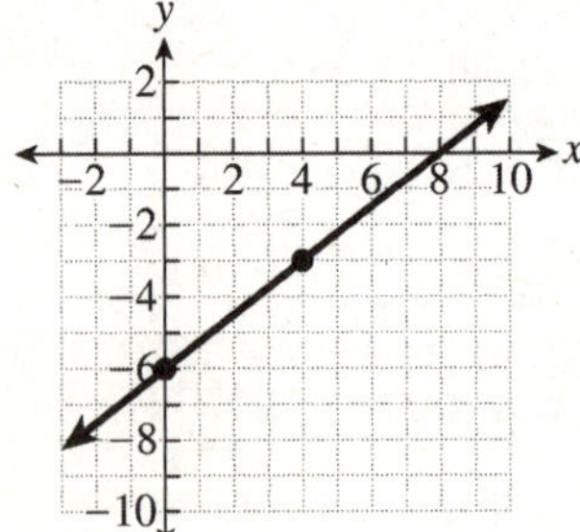

31. $f(x) = 2$
Since slope is 0, the function is a
horizontal line at $y = 2$.

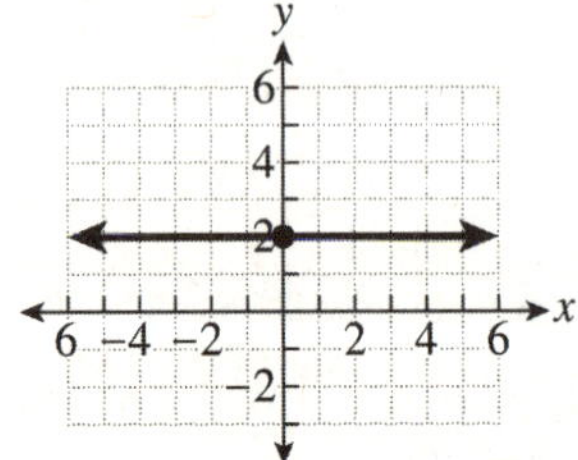

33. $x - 7 = 0$
  $x = 7$

x	$f(x) = |x-7|$	$(x, f(x))$
5	$f(5) = |5-7| = 2$	$(5,2)$
6	$f(6) = |6-7| = 1$	$(6,1)$
7	$f(7) = |7-7| = 0$	$(7,0)$
8	$f(8) = |8-7| = 1$	$(8,1)$
9	$f(9) = |9-7| = 2$	$(9,2)$

Domain: $(-\infty, \infty)$ Range: $[0, \infty)$

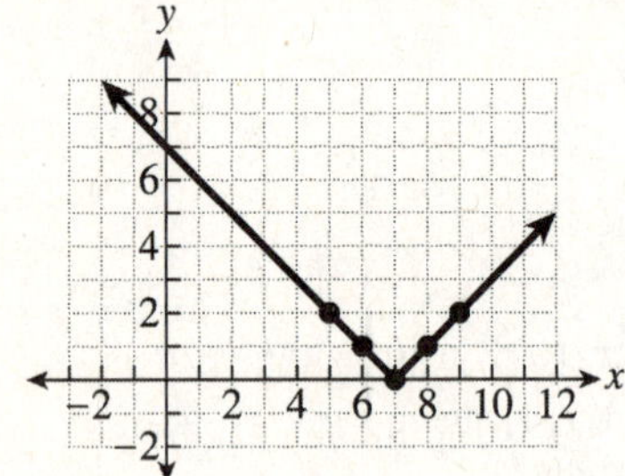

35.

x	$f(x) = |x| + 3$	$(x, f(x))$
-2	$f(-2) = |-2| + 3 = 5$	$(-2,5)$
-1	$f(-1) = |-1| + 3 = 4$	$(-1,4)$
0	$f(0) = |0| + 3 = 3$	$(0,3)$
1	$f(1) = |1| + 3 = 4$	$(1,4)$
2	$f(2) = |2| + 3 = 5$	$(2,5)$

Domain: $(-\infty, \infty)$ Range: $[3, \infty)$

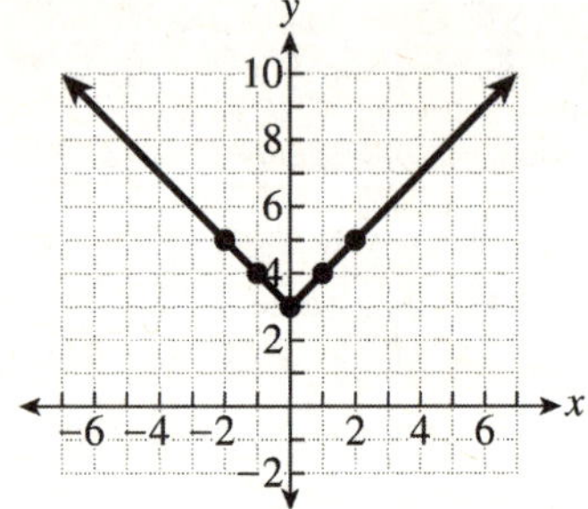

37. $x + 3 = 0$
   $x = -3$

x	$f(x) = \lvert x+3 \rvert - 4$	$(x, f(x))$
-5	$f(-5) = \lvert -5+3 \rvert - 4 = -2$	$(-5, -2)$
-4	$f(-4) = \lvert -4+3 \rvert - 4 = -3$	$(-4, -3)$
-3	$f(-3) = \lvert -3+3 \rvert - 4 = -4$	$(-3, -4)$
-2	$f(-2) = \lvert -2+3 \rvert - 4 = -3$	$(-2, -3)$
-1	$f(-1) = \lvert -1+3 \rvert - 4 = -2$	$(-1, -2)$

Domain: $(-\infty, \infty)$ Range: $[-4, \infty)$

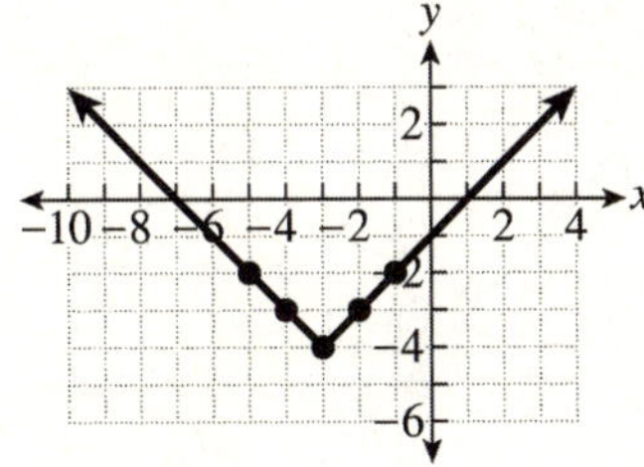

41. $x + 4 = 0$
   $x = -4$

x	$f(x) = \lvert x+4 \rvert - 2$	$(x, f(x))$
-6	$f(-6) = \lvert -6+4 \rvert - 2 = 0$	$(-6, 0)$
-5	$f(-5) = \lvert -5+4 \rvert - 2 = -1$	$(-5, -1)$
-4	$f(-4) = \lvert -4+4 \rvert - 2 = -2$	$(-4, -2)$
-3	$f(-3) = \lvert -3+4 \rvert - 2 = -1$	$(-3, -1)$
-2	$f(-2) = \lvert -2+4 \rvert - 2 = 0$	$(-2, 0)$

Domain: $(-\infty, \infty)$ Range: $[-2, \infty)$

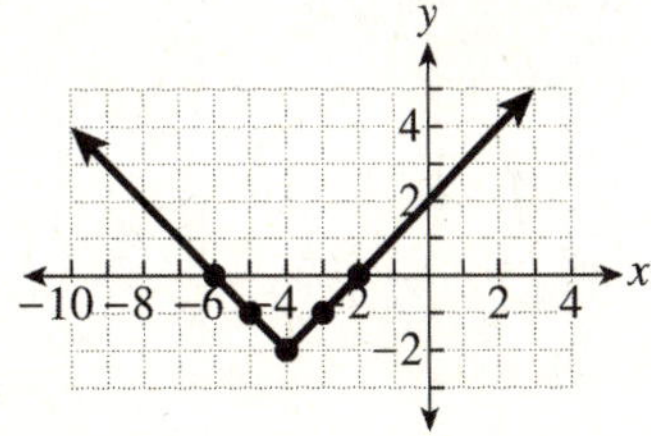

39. $x - 6 = 0$
   $x = 6$

x	$f(x) = \lvert x-6 \rvert + 3$	$(x, f(x))$
4	$f(4) = \lvert 4-6 \rvert + 3 = 5$	$(4, 5)$
5	$f(5) = \lvert 5-6 \rvert + 3 = 4$	$(5, 4)$
6	$f(6) = \lvert 6-6 \rvert + 3 = 3$	$(6, 3)$
7	$f(7) = \lvert 7-6 \rvert + 3 = 4$	$(7, 4)$
8	$f(8) = \lvert 8-6 \rvert + 3 = 5$	$(8, 5)$

Domain: $(-\infty, \infty)$ Range: $[3, \infty)$

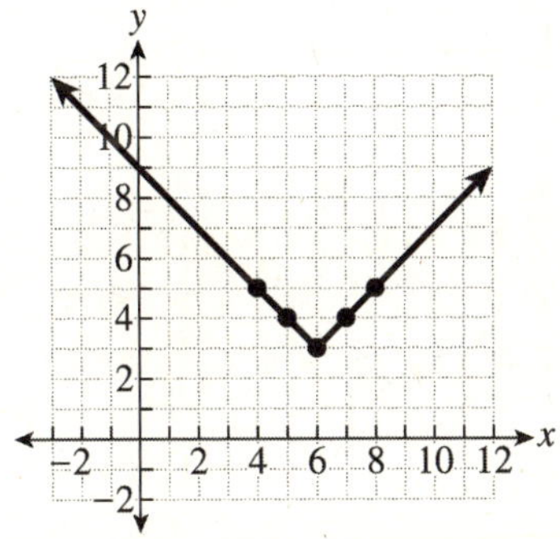

43. $2x - 6 = 0$
   $2x = 6$
   $x = 3$

x	$f(x) = \lvert 2x-6 \rvert$	$(x, f(x))$
1	$f(1) = \lvert 2(1)-6 \rvert = 4$	$(1, 4)$
2	$f(2) = \lvert 2(2)-6 \rvert = 2$	$(2, 2)$
3	$f(3) = \lvert 2(3)-6 \rvert = 0$	$(3, 0)$
4	$f(4) = \lvert 2(4)-6 \rvert = 2$	$(4, 2)$
5	$f(5) = \lvert 2(5)-6 \rvert = 4$	$(5, 4)$

Domain: $(-\infty, \infty)$ Range: $[0, \infty)$

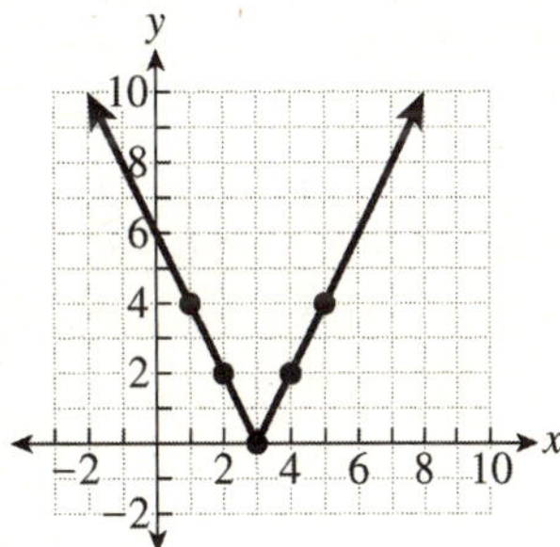

45. $4x + 8 = 0$
$\quad\quad 4x = -8$
$\quad\quad\; x = -2$

x	$f(x) = \lvert 4x + 8 \rvert - 5$	$(x, f(x))$
-4	$f(-4) = \lvert 4(-4) + 8 \rvert - 5 = 3$	$(-4, 3)$
-3	$f(-3) = \lvert 4(-3) + 8 \rvert - 5 = -1$	$(-3, -1)$
-2	$f(-2) = \lvert 4(-2) + 8 \rvert - 5 = -5$	$(-2, -5)$
-1	$f(-1) = \lvert 4(-1) + 8 \rvert - 5 = -1$	$(-1, -1)$
0	$f(0) = \lvert 4(0) + 8 \rvert - 5 = 3$	$(0, 3)$

Domain: $(-\infty, \infty)$ Range: $[-5, \infty)$

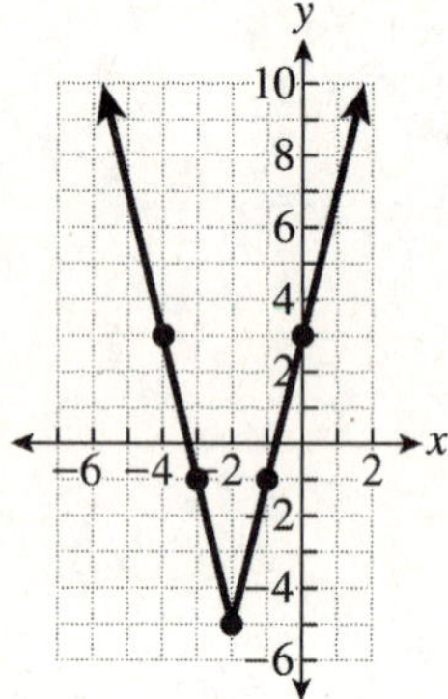

47. $x - 1 = 0$
$\quad\quad x = 1$

x	$f(x) = 2\lvert x - 1 \rvert + 3$	$(x, f(x))$
-1	$f(-1) = 2\lvert -1 - 1 \rvert + 3 = 7$	$(-1, 7)$
0	$f(0) = 2\lvert 0 - 1 \rvert + 3 = 5$	$(0, 5)$
1	$f(1) = 2\lvert 1 - 1 \rvert + 3 = 3$	$(1, 3)$
2	$f(2) = 2\lvert 2 - 1 \rvert + 3 = 5$	$(2, 5)$
3	$f(3) = 2\lvert 3 - 1 \rvert + 3 = 7$	$(3, 7)$

Domain: $(-\infty, \infty)$ Range: $[3, \infty)$

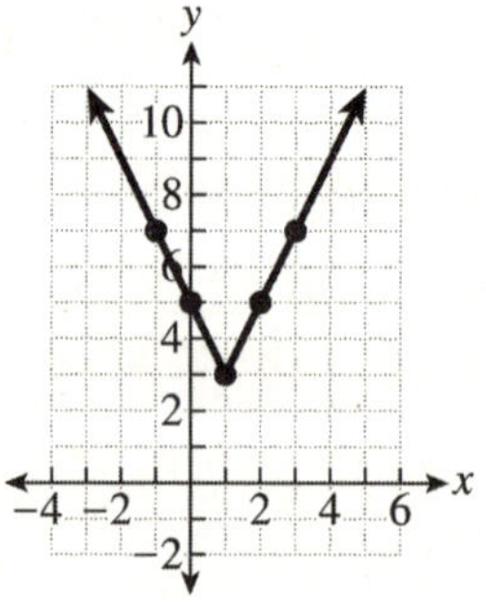

49. $x + 6 = 0$
$\quad\quad x = -6$

x	$f(x) = -\lvert x + 6 \rvert$	$(x, f(x))$
-8	$f(-8) = -\lvert -8 + 6 \rvert = -2$	$(-8, -2)$
-7	$f(-7) = -\lvert -7 + 6 \rvert = -1$	$(-7, -1)$
-6	$f(-6) = -\lvert -6 + 6 \rvert = 0$	$(-6, 0)$
-5	$f(-5) = -\lvert -5 + 6 \rvert = -1$	$(-5, -1)$
-4	$f(-4) = -\lvert -4 + 6 \rvert = -2$	$(-4, -2)$

Domain: $(-\infty, \infty)$ Range: $(-\infty, 0]$

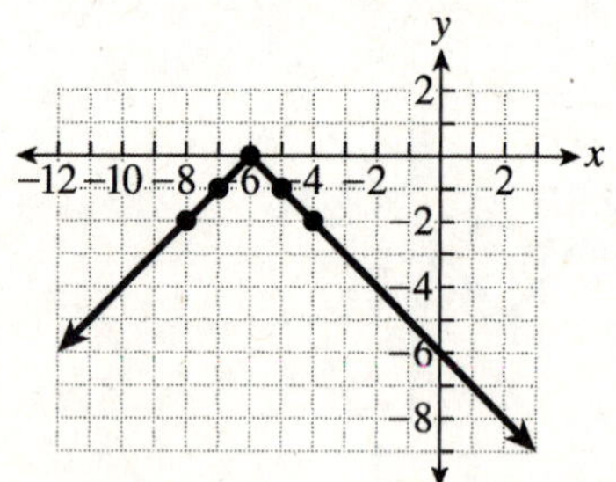

51. $x - 3 = 0$
$\quad\quad x = 3$

x	$f(x) = -\lvert x - 3 \rvert - 4$	$(x, f(x))$
1	$f(1) = -\lvert 1 - 3 \rvert - 4 = -6$	$(1, -6)$
2	$f(2) = -\lvert 2 - 3 \rvert - 4 = -5$	$(2, -5)$
3	$f(3) = -\lvert 3 - 3 \rvert - 4 = -4$	$(3, -4)$
4	$f(4) = -\lvert 4 - 3 \rvert - 4 = -5$	$(4, -5)$
5	$f(5) = -\lvert 5 - 3 \rvert - 4 = -6$	$(5, -6)$

Domain: $(-\infty, \infty)$ Range: $(-\infty, -4]$

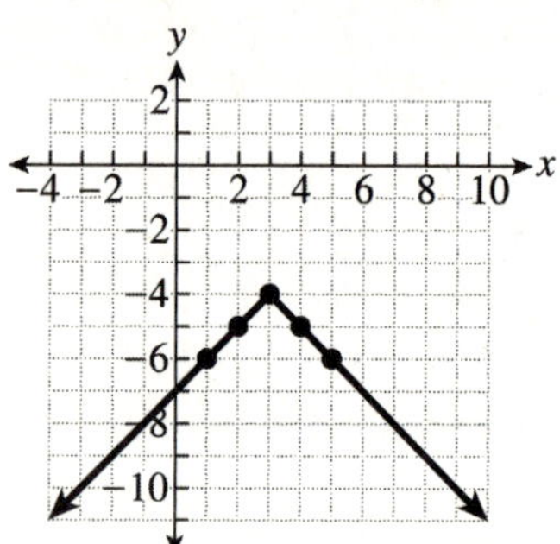

53. $x + 2 = 0$
   $x = -2$

x	$f(x) = -3\lvert x + 2\rvert + 9$	$(x, f(x))$
-4	$f(-4) = -3\lvert -4 + 2\rvert + 9 = 3$	$(-4, 3)$
-3	$f(-3) = -3\lvert -3 + 2\rvert + 9 = 6$	$(-3, 6)$
-2	$f(-2) = -3\lvert -2 + 2\rvert + 9 = 9$	$(-2, 9)$
-1	$f(-1) = -3\lvert -1 + 2\rvert + 9 = 6$	$(-1, 6)$
0	$f(0) = -3\lvert 0 + 2\rvert + 9 = 3$	$(0, 3)$

Domain: $(-\infty, \infty)$ Range: $(-\infty, 9]$

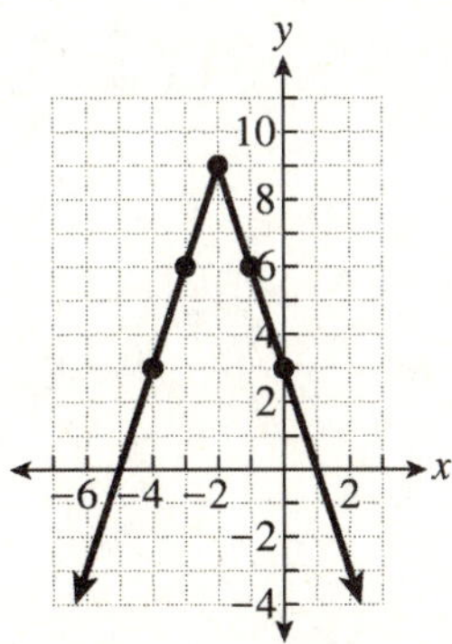

55. x-intercept: Let $f(x) = 0$.
$\lvert x - 3\rvert - 7 = 0$
$\lvert x - 3\rvert = 7$

$x - 3 = -7$ or $x - 3 = 7$
   $x = -4$ or $x = 10$
$(-4, 0), (10, 0)$

y-intercept: Let $x = 0$.
$f(0) = \lvert 0 - 3\rvert - 7 = -4$
$(0, -4)$

57. x-intercept: Let $f(x) = 0$.
$\lvert x\rvert - 6 = 0$
$\lvert x\rvert = 6$
$(-6, 0), (6, 0)$

y-intercept: Let $x = 0$.
$f(0) = \lvert 0\rvert - 6 = -6$
$(0, -6)$

59. x-intercept: Let $f(x) = 0$.
$\lvert x + 2\rvert + 4 = 0$
$\lvert x + 2\rvert = -4$
No solution. No x-intercept.

y-intercept: Let $x = 0$.
$f(0) = \lvert 0 + 2\rvert + 4 = 6$
$(0, 6)$

61. x-intercept: Let $f(x) = 0$.
$-\lvert x - 4\rvert + 3 = 0$
$-\lvert x - 4\rvert = -3$
$\lvert x - 4\rvert = 3$

$x - 4 = -3$ or $x - 4 = 3$
   $x = 1$ or $x = 7$
$(1, 0), (7, 0)$

y-intercept: Let $x = 0$.
$f(0) = -\lvert 0 - 4\rvert + 3 = -4 + 3 = -1$
$(0, -1)$

63. $f(x) = \lvert x + 6\rvert$

65. $f(x) = \lvert x + 1\rvert + 3$

67. $f(x) = -\lvert x + 2\rvert + 5$

69. $y = -\dfrac{3}{2}x + 7$

x-intercept: Let $y = 0$.

$0 = -\dfrac{3}{2}x + 7$

$\dfrac{3}{2}x = 7$

$3x = 14$

$x = \dfrac{14}{3}$

$\left(\dfrac{14}{3}, 0\right)$

y-intercept: $(0, 7)$ using $y = mx + b$

3rd point: $(2, 4)$ using slope $= -\dfrac{3}{2}$

(down 3, right 2)

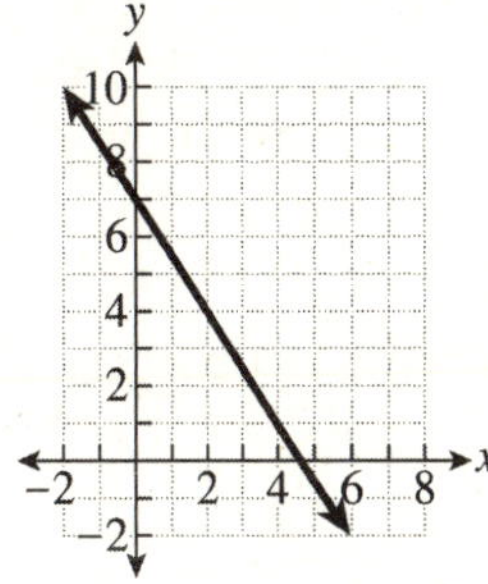

71. $x - 3 = 0$
$\quad\quad x = 3$

x	$f(x) = |x-3| - 2$	$(x, f(x))$
1	$f(1) = |1-3| - 2 = 0$	$(1,0)$
2	$f(2) = |2-3| - 2 = -1$	$(2,-1)$
3	$f(3) = |3-3| - 2 = -2$	$(3,-2)$
4	$f(4) = |4-3| - 2 = -1$	$(4,-1)$
5	$f(5) = |5-3| - 2 = 0$	$(5,0)$

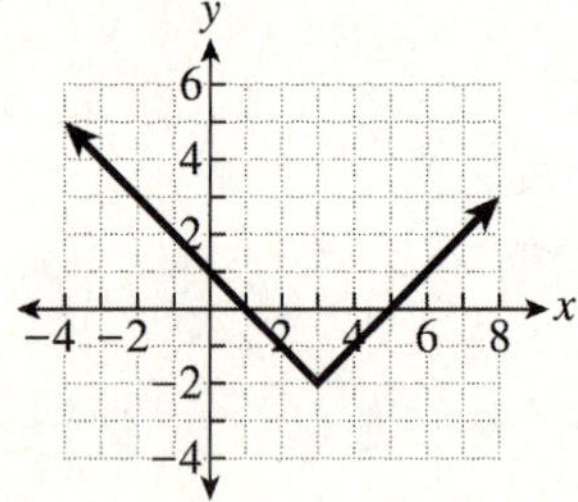

73. $y = x + 3$

x-intercept: Let $y = 0$.
$\quad 0 = x + 3$
$-3 = x$
$(-3, 0)$

y-intercept: $(0,3)$ using $y = mx + b$
3rd point: $(1,4)$ using slope $= 1$
(up 1, right 1)

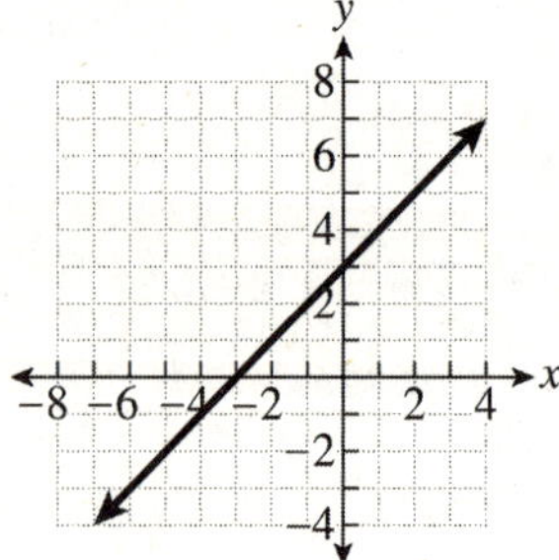

75. $x + 4 = 0$
$\quad\quad x = -4$

x	$f(x) = |x+4|$	$(x, f(x))$
-6	$f(-6) = |-6+4| = 2$	$(-6,2)$
-5	$f(-5) = |-5+4| = 1$	$(-5,1)$
-4	$f(-4) = |-4+4| = 0$	$(-4,0)$
-3	$f(-3) = |-3+4| = 1$	$(-3,1)$
-2	$f(-2) = |-2+4| = 2$	$(-2,2)$

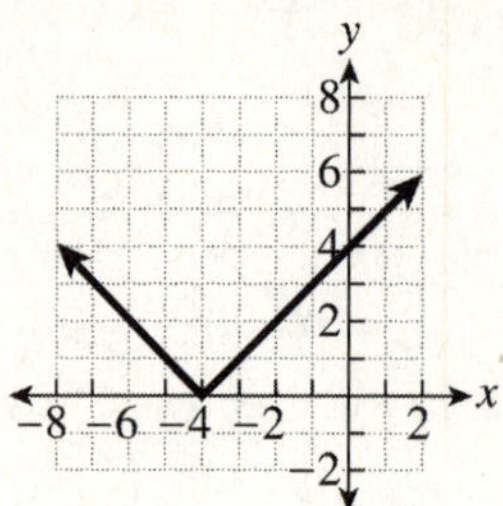

77. $x + 2 = 0$
$\quad\quad x = -2$

x	$f(x) = |x+2| - 7$	$(x, f(x))$
-4	$f(-4) = |-4+2| - 7 = -5$	$(-4,-5)$
-3	$f(-3) = |-3+2| - 7 = -6$	$(-3,-6)$
-2	$f(-2) = |-2+2| - 7 = -7$	$(-2,-7)$
-1	$f(-1) = |-1+2| - 7 = -6$	$(-1,-6)$
0	$f(0) = |0+2| - 7 = -5$	$(0,-5)$

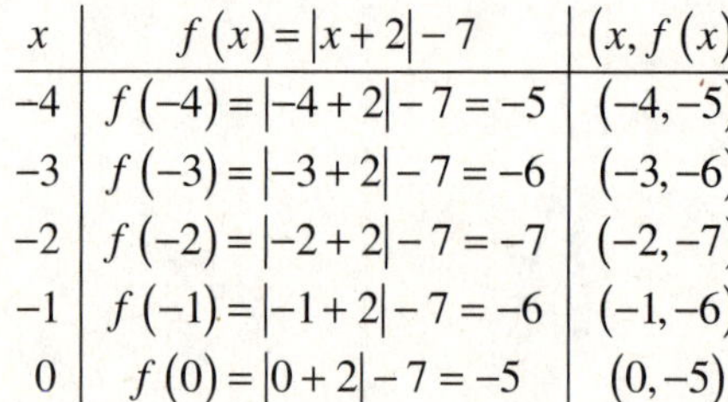

79. $y = \dfrac{3}{7}x$

x-intercept: Let $y = 0$.
$0 = \dfrac{3}{7}x$
$0 = x$
$(0,0)$

y-intercept: $(0,0)$ using $y = mx + b$
2nd point: $(7,3)$ using slope $= \dfrac{3}{7}$
(up 3, right 7)

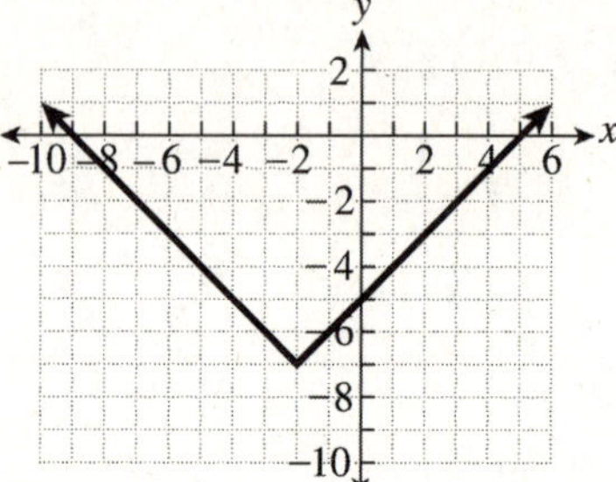

81. No; explanations will vary. Example:
To generate a complete accurate graph, the student should evaluate the function at values to the left and to the right of the turning point, $x = 5$.

8.4   QUICK CHECK

1. $36x^6 + 60x^5 - 12x^3 = 12x^3\left(3x^3 + 5x^2 - 1\right)$

2. $x^3 - 8x^2 + 4x - 32 = x^2(x-8) + 4(x-8)$
$$= (x-8)\left(x^2 + 4\right)$$

3. a) $x^2 + 11x + 28 = (x+4)(x+7)$

b) $x^2 - 13x + 30 = (x-3)(x-10)$

4. $2x^2 - 14x - 36 = 2\left(x^2 - 7x - 18\right)$
$$= 2(x-9)(x+2)$$

5. $-x^2 + 17x - 70 = -\left(x^2 - 17x + 70\right)$
$$= -(x-7)(x-10)$$

6. $6x^2 - 7x - 20 = (2x-5)(3x+4)$

7. $9x^2 - 100 = (3x)^2 - (10)^2 = (3x+10)(3x-10)$

8. $27x^3 - 64y^3$
$$= (3x)^3 - (4y)^3$$
$$= (3x-4y)(3x\cdot 3x + 3x\cdot 4y + 4y\cdot 4y)$$
$$= (3x-4y)\left(9x^2 + 12xy + 16y^2\right)$$

9.
$$x(x+3) = 7x + 32$$
$$x^2 + 3x = 7x + 32$$
$$x^2 - 4x - 32 = 0$$
$$(x+4)(x-8) = 0$$
$$x+4 = 0 \quad \text{or} \quad x-8 = 0$$
$$x = -4 \quad \text{or} \quad x = 8$$
$$\{-4, 8\}$$

10.
$$\frac{1}{5}x^2 - \frac{1}{10}x - 1 = 0$$
$$10\cdot\left(\frac{1}{5}x^2 - \frac{1}{10}x - 1\right) = 10\cdot 0$$
$$\overset{2}{\cancel{10}}\cdot\frac{1}{\cancel{5}_{1}}x^2 - \overset{1}{\cancel{10}}\cdot\frac{1}{\cancel{10}_{1}}x - 10\cdot 1 = 0$$
$$2x^2 - x - 10 = 0$$
$$(x+2)(2x-5) = 0$$
$$x+2 = 0 \quad \text{or} \quad 2x-5 = 0$$
$$x = -2 \quad \text{or} \quad 2x = 5$$
$$x = -2 \quad \text{or} \quad x = \frac{5}{2}$$
$$\left\{-2, \frac{5}{2}\right\}$$

11.
$$\frac{x}{x-4} - \frac{8}{x-3} = 1$$
$$(x-4)(x-3)\left(\frac{x}{x-4} - \frac{8}{x-3}\right) = (x-4)(x-3)\cdot 1$$
$$\overset{1}{\cancel{(x-4)}}(x-3)\cdot\frac{x}{\cancel{x-4}_{1}} - (x-4)\overset{1}{\cancel{(x-3)}}\cdot\frac{8}{\cancel{x-3}_{1}} = (x-4)(x-3)\cdot 1$$
$$x(x-3) - 8(x-4) = (x-4)(x-3)$$
$$x^2 - 3x - 8x + 32 = x^2 - 3x - 4x + 12$$
$$x^2 - 11x + 32 = x^2 - 7x + 12$$
$$-11x + 32 = -7x + 12$$
$$-4x = -20$$
$$x = 5$$

$$\{5\}$$

12.

$$\frac{x}{x+3} + \frac{1}{x+2} = \frac{7x}{x^2+5x+6}$$

$$(x+3)(x+2)\cdot\left(\frac{x}{x+3} + \frac{1}{x+2}\right) = (x+3)(x+2)\cdot\frac{7x}{(x+3)(x+2)}$$

$$\frac{\cancel{(x+3)}^{\,1}(x+2)\cdot x}{\cancel{(x+3)}_{\,1}} + (x+3)\cancel{(x+2)}^{\,1}\cdot\frac{1}{\cancel{(x+2)}_{\,1}} = \cancel{(x+3)}^{\,1}\cancel{(x+2)}^{\,1}\cdot\frac{7x}{\cancel{(x+3)}_{\,1}\cancel{(x+2)}_{\,1}}$$

$$x(x+2) + 1(x+3) = 7x$$
$$x^2 + 2x + x + 3 = 7x$$
$$x^2 + 3x + 3 = 7x$$
$$x^2 - 4x + 3 = 0$$
$$(x-1)(x-3) = 0$$
$$x - 1 = 0 \quad \text{or} \quad x - 3 = 0$$
$$x = 1 \quad \text{or} \quad x = 3$$

$$\{1, 3\}$$

**8.4   REVIEW OF FACTORING; QUADRATIC
EQUATIONS AND RATIONAL EQUATIONS**

1. factored

3. common factors

5. difference of cubes

7. zero-factor

9. $36x^8 - 16x^4 + 50x^3 = 2x^3\left(18x^5 - 8x + 25\right)$

11. $20a^3b - 28a^2b^4 - 16a^5b^2$
$= 4a^2b\left(5a - 7b^3 - 4a^3b\right)$

13. $n^3 + 4n^2 + 5n + 20 = n^2(n+4) + 5(n+4)$
$\qquad\qquad\qquad\quad = (n+4)\left(n^2+5\right)$

15. $x^3 + 7x^2 - 3x - 21 = x^2(x+7) - 3(x+7)$
$\qquad\qquad\qquad\quad = (x+7)\left(x^2-3\right)$

17. $x^2 - 13x + 40 = (x-5)(x-8)$

19. $2x^2 - 8x - 64 = 2\left(x^2 - 4x - 32\right)$
$\qquad\qquad\quad = 2(x-8)(x+4)$

21. $a^2 + 13ab + 36b^2 = (a+4b)(a+9b)$

23. $4x^2 - 20x - 56 = 4\left(x^2 - 5x - 14\right)$
$\qquad\qquad\quad = 4(x+2)(x-7)$

25. $2x^2 + x - 36 = (2x+9)(x-4)$

27. $6x^2 + 25x + 24 = (2x+3)(3x+8)$

29. $x^2 - 36 = (x)^2 - (6)^2 = (x+6)(x-6)$

31. $16b^2 - 9 = (4b)^2 - (3)^2 = (4b+3)(4b-3)$

33. $x^2 + 16$ is prime. It cannot be factored.

35. $x^3 - 8 = (x)^3 - (2)^3$
$\qquad = (x-2)(x\cdot x + x\cdot 2 + 2\cdot 2)$
$\qquad = (x-2)\left(x^2 + 2x + 4\right)$

37. $27x^3 + 125 = (3x)^3 + (5)^3$
$\qquad\qquad = (3x+5)(3x\cdot 3x - 3x\cdot 5 + 5\cdot 5)$
$\qquad\qquad = (3x+5)\left(9x^2 - 15x + 25\right)$

39. $x^4 - 81 = \left(x^2\right)^2 - (9)^2$
$\qquad\quad = \left(x^2 + 9\right)\left(x^2 - 9\right)$
$\qquad\quad = \left(x^2 + 9\right)\left[(x)^2 - (3)^2\right]$
$\qquad\quad = \left(x^2 + 9\right)(x+3)(x-3)$

41.
$$\begin{aligned}
x^3 - 7x^2 - 9x + 63 &= \left(x^3 - 7x^2\right) - (9x - 63) \\
&= x^2(x-7) - 9(x-7) \\
&= (x-7)\left(x^2 - 9\right) \\
&= (x-7)\left[(x)^2 - (3)^2\right] \\
&= (x-7)(x+3)(x-3)
\end{aligned}$$

43.
$$\begin{aligned}
n^4 &+ 4n^3 + 8n + 32 \\
&= \left(n^4 + 4n^3\right) + (8n + 32) \\
&= n^3(n+4) + 8(n+4) \\
&= (n+4)\left(n^3 + 8\right) \\
&= (n+4)\left[(n)^3 + (2)^3\right] \\
&= (n+4)(n+2)(n \cdot n - 2 \cdot n + 2 \cdot 2) \\
&= (n+4)(n+2)\left(n^2 - 2n + 4\right)
\end{aligned}$$

45.
$$\begin{aligned}
x^2 + 3x - 4 &= 0 \\
(x+4)(x-1) &= 0
\end{aligned}$$
$$\begin{array}{lll}
x + 4 = 0 & \text{or} & x - 1 = 0 \\
x = -4 & \text{or} & x = 1
\end{array}$$
$$\{-4, 1\}$$

47.
$$\begin{aligned}
t^2 + 17t + 72 &= 0 \\
(t+9)(t+8) &= 0
\end{aligned}$$
$$\begin{array}{lll}
t + 9 = 0 & \text{or} & t + 8 = 0 \\
t = -9 & \text{or} & t = -8
\end{array}$$
$$\{-9, -8\}$$

49.
$$\begin{aligned}
x^2 - 2x - 48 &= 0 \\
(x+6)(x-8) &= 0
\end{aligned}$$
$$\begin{array}{lll}
x + 6 = 0 & \text{or} & x - 8 = 0 \\
x = -6 & \text{or} & x = 8
\end{array}$$
$$\{-6, 8\}$$

51.
$$\begin{aligned}
4x^2 - 44x + 72 &= 0 \\
4\left(x^2 - 11x + 18\right) &= 0 \\
4(x-2)(x-9) &= 0
\end{aligned}$$
$$\begin{array}{lll}
x - 2 = 0 & \text{or} & x - 9 = 0 \\
x = 2 & \text{or} & x = 9
\end{array}$$
$$\{2, 9\}$$

53.
$$\begin{aligned}
2x^2 + 7x + 6 &= 0 \\
(x+2)(2x+3) &= 0
\end{aligned}$$
$$\begin{array}{lll}
x + 2 = 0 & \text{or} & 2x + 3 = 0 \\
x = -2 & \text{or} & 2x = -3 \\
x = -2 & \text{or} & x = -\dfrac{3}{2}
\end{array}$$
$$\left\{-2, -\dfrac{3}{2}\right\}$$

55.
$$\begin{aligned}
x^2 - 25 &= 0 \\
(x+5)(x-5) &= 0
\end{aligned}$$
$$\begin{array}{lll}
x + 5 = 0 & \text{or} & x - 5 = 0 \\
x = -5 & \text{or} & x = 5
\end{array}$$
$$\{-5, 5\}$$

57.
$$\begin{aligned}
9x^2 - 4 &= 0 \\
(3x)^2 - (2)^2 &= 0 \\
(3x+2)(3x-2) &= 0
\end{aligned}$$
$$\begin{array}{lll}
3x + 2 = 0 & \text{or} & 3x - 2 = 0 \\
3x = -2 & \text{or} & 3x = 2 \\
x = -\dfrac{2}{3} & \text{or} & x = \dfrac{2}{3}
\end{array}$$
$$\left\{-\dfrac{2}{3}, \dfrac{2}{3}\right\}$$

59.
$$\begin{aligned}
x^2 - 25x &= 0 \\
x(x - 25) &= 0
\end{aligned}$$
$$\begin{array}{lll}
x = 0 & \text{or} & x - 25 = 0 \\
x = 0 & \text{or} & x = 25
\end{array}$$
$$\{0, 25\}$$

61.
$$\begin{aligned}
x^2 + 4x &= 21 \\
x^2 + 4x - 21 &= 0 \\
(x+7)(x-3) &= 0
\end{aligned}$$
$$\begin{array}{lll}
x + 7 = 0 & \text{or} & x - 3 = 0 \\
x = -7 & \text{or} & x = 3
\end{array}$$
$$\{-7, 3\}$$

63.
$$\begin{aligned}
x^2 + 11x &= 4x - 6 \\
x^2 + 7x + 6 &= 0 \\
(x+6)(x+1) &= 0
\end{aligned}$$
$$\begin{array}{lll}
x + 6 = 0 & \text{or} & x + 1 = 0 \\
x = -6 & \text{or} & x = -1
\end{array}$$
$$\{-6, -1\}$$

65.
$$\frac{4}{3}x^2 + \frac{14}{3}x + 2 = 0$$
$$3 \cdot \left(\frac{4}{3}x^2 + \frac{14}{3}x + 2\right) = 3 \cdot 0$$
$$\cancel{3}^1 \cdot \frac{4}{\cancel{3}_1}x^2 + \cancel{3}^1 \cdot \frac{14}{\cancel{3}_1}x + 3 \cdot 2 = 0$$
$$4x^2 + 14x + 6 = 0$$
$$2\left(2x^2 + 7x + 3\right) = 0$$
$$2(x+3)(2x+1) = 0$$
$$x + 3 = 0 \quad \text{or} \quad 2x + 1 = 0$$
$$x = -3 \quad \text{or} \quad 2x = -1$$
$$x = -3 \quad \text{or} \quad x = -\frac{1}{2}$$
$$\left\{-3, -\frac{1}{2}\right\}$$

67. $\left|x^2 - 13x\right| = 30$
$$x^2 - 13x = -30 \quad \text{or} \quad x^2 - 13x = 30$$
$$x^2 - 13x + 30 = 0 \quad \text{or} \quad x^2 - 13x - 30 = 0$$
$$(x-3)(x-10) = 0 \quad \text{or} \quad (x+2)(x-15) = 0$$
$$(x-3)(x-10) = 0$$
$$x - 3 = 0 \quad \text{or} \quad x - 10 = 0$$
$$x = 3 \quad \text{or} \quad x = 10$$
$$(x+2)(x-15) = 0$$
$$x + 2 = 0 \quad \text{or} \quad x - 15 = 0$$
$$x = -2 \quad \text{or} \quad x = 15$$
$$\{-2, 3, 10, 15\}$$

69. $\left|x^2 + 10x\right| = 24$
$$x^2 + 10x = -24 \quad \text{or} \quad x^2 + 10x = 24$$
$$x^2 + 10x + 24 = 0 \quad \text{or} \quad x^2 + 10x - 24 = 0$$
$$(x+4)(x+6) = 0 \quad \text{or} \quad (x+12)(x-2) = 0$$
$$(x+4)(x+6) = 0$$
$$x + 4 = 0 \quad \text{or} \quad x + 6 = 0$$
$$x = -4 \quad \text{or} \quad x = -6$$
$$(x+12)(x-2) = 0$$
$$x + 12 = 0 \quad \text{or} \quad x - 2 = 0$$
$$x = -12 \quad \text{or} \quad x = 2$$
$$\{-12, -6, -4, 2\}$$

71.
$$x = -9 \quad \text{or} \quad x = 4$$
$$x + 9 = 0 \quad \text{or} \quad x - 4 = 0$$
$$(x+9)(x-4) = 0$$
$$x^2 - 4x + 9x - 36 = 0$$
$$x^2 + 5x - 36 = 0$$

73.
$$x = \frac{1}{3} \quad \text{or} \quad x = 5$$
$$3x = 1 \quad \text{or} \quad x - 5 = 0$$
$$3x - 1 = 0 \quad \text{or} \quad x - 5 = 0$$
$$(3x-1)(x-5) = 0$$
$$3x^2 - 15x - x + 5 = 0$$
$$3x^2 - 16x + 5 = 0$$

75.
$$x = -12 \quad \text{or} \quad x = 12$$
$$x + 12 = 0 \quad \text{or} \quad x - 12 = 0$$
$$(x+12)(x-12) = 0$$
$$x^2 - 12x + 12x - 144 = 0$$
$$x^2 - 144 = 0$$

77.
$$\frac{8}{x} = \frac{7}{x-2}$$
$$\cancel{x}^1(x-2) \cdot \frac{8}{\cancel{x}_1} = x\cancel{(x-2)}^1 \cdot \frac{7}{\cancel{x-2}_1}$$
$$8(x-2) = 7x$$
$$8x - 16 = 7x$$
$$x - 16 = 0$$
$$x = 16$$
$$\{16\}$$

79.
$$x - 3 = \frac{10}{x}$$
$$x \cdot (x-3) = x \cdot \frac{10}{x}$$
$$x^2 - 3x = 10$$
$$x^2 - 3x - 10 = 0$$
$$(x+2)(x-5) = 0$$
$$x + 2 = 0 \quad \text{or} \quad x - 5 = 0$$
$$x = -2 \quad \text{or} \quad x = 5$$
$$\{-2, 5\}$$

81.
$$\frac{x^2 - 23}{x - 3} + 7 = 0$$
$$(x-3) \cdot \left(\frac{x^2 - 23}{x-3} + 7\right) = (x-3) \cdot 0$$
$$\cancel{(x-3)}^1 \cdot \frac{x^2 - 23}{\cancel{(x-3)}_1} + (x-3) \cdot 7 = (x-3) \cdot 0$$
$$x^2 - 23 + 7(x-3) = 0$$
$$x^2 - 23 + 7x - 21 = 0$$
$$x^2 + 7x - 44 = 0$$
$$(x+11)(x-4) = 0$$
$$x + 11 = 0 \quad \text{or} \quad x - 4 = 0$$
$$x = -11 \quad \text{or} \quad x = 4$$
$$\{-11, 4\}$$

83.
$$\frac{x+1}{x+2} = \frac{6}{x+6}$$

$$(x+2)^1(x+6)\cdot\frac{x+1}{(x+2)_1} = (x+2)\,(x+6)^1\cdot\frac{6}{(x+6)_1}$$

$$(x+6)(x+1) = 6(x+2)$$
$$x^2 + x + 6x + 6 = 6x + 12$$
$$x^2 + 7x + 6 = 6x + 12$$
$$x^2 + x - 6 = 0$$
$$(x+3)(x-2) = 0$$

$$x + 3 = 0 \quad \text{or} \quad x - 2 = 0$$
$$x = -3 \quad \text{or} \quad x = 2$$

$$\{-3, 2\}$$

85.
$$\frac{x}{x-3} + \frac{4}{x+5} = \frac{8x}{(x-3)(x+5)}$$

$$(x-3)(x+5)\cdot\left(\frac{x}{x-3} + \frac{4}{x+5}\right) = (x-3)(x+5)\cdot\frac{8x}{(x-3)(x+5)}$$

$$(x-3)^1(x+5)\cdot\frac{x}{(x-3)_1} + (x-3)\,(x+5)^1\cdot\frac{4}{(x+5)_1} = (x-3)^1(x+5)^1\cdot\frac{8x}{(x-3)_1(x+5)_1}$$

$$x(x+5) + 4(x-3) = 8x$$
$$x^2 + 5x + 4x - 12 = 8x$$
$$x^2 + 9x - 12 = 8x$$
$$x^2 + x - 12 = 0$$
$$(x+4)(x-3) = 0$$

$$x + 4 = 0 \quad \text{or} \quad x - 3 = 0$$
$$x = -4 \quad \text{or} \quad x = 3$$

$\{-4\}$, since $x = 3$ makes the original equation undefined.

87.
$$\frac{x-4}{x-1} + \frac{4}{x+1} = \frac{x+17}{x^2-1}$$

$$\frac{x-4}{x-1} + \frac{4}{x+1} = \frac{x+17}{(x+1)(x-1)}$$

$$(x+1)(x-1)\cdot\left(\frac{x-4}{x-1} + \frac{4}{x+1}\right) = (x+1)(x+1)\cdot\frac{x+17}{(x+1)(x-1)}$$

$$(x+1)\,(x-1)^1\cdot\frac{x-4}{(x-1)_1} + (x+1)^1(x-1)\cdot\frac{4}{(x+1)_1} = (x+1)^1(x-1)^1\cdot\frac{x+17}{(x+1)_1(x-1)_1}$$

$$(x+1)(x-4) + 4(x-1) = x + 17$$
$$x^2 - 4x + x - 4 + 4x - 4 = x + 17$$
$$x^2 + x - 8 = x + 17$$
$$x^2 - 25 = 0$$
$$(x+5)(x-5) = 0$$

$$x + 5 = 0 \quad \text{or} \quad x - 5 = 0$$
$$x = -5 \quad \text{or} \quad x = 5$$

$$\{-5, 5\}$$

89.

$$\frac{3}{x-1}+\frac{7}{x+2}=\frac{9}{x^2+x-2}$$

$$\frac{3}{x-1}+\frac{7}{x+2}=\frac{9}{(x+2)(x-1)}$$

$$(x-1)(x+2)\cdot\left(\frac{3}{x-1}+\frac{7}{x+2}\right)=(x-1)(x+2)\cdot\frac{9}{(x+2)(x-1)}$$

$$(x-1)^1(x+2)\cdot\frac{3}{(x-1)_1}+(x-1)(x+2)^1\cdot\frac{7}{(x+2)_1}=(x-1)^1(x+2)^1\cdot\frac{9}{(x+2)_1(x-1)_1}$$

$$3(x+2)+7(x-1)=9$$
$$3x+6+7x-7=9$$
$$10x-1=9$$
$$10x=10$$
$$x=1$$

$\varnothing$, since $x=1$ makes the original equation undefined.

91. $\left|\dfrac{2x-15}{x}\right|=3$

$$\frac{2x-15}{x}=-3 \quad\text{or}\quad \frac{2x-15}{x}=3$$
$$2x-15=-3x \quad\text{or}\quad 2x-15=3x$$
$$-15=-5x \quad\text{or}\quad -15=x$$
$$3=x \quad\text{or}\quad -15=x$$
$$\{-15,3\}$$

93. $\left|\dfrac{x^2+5x-48}{x}\right|=3$

$$\frac{x^2+5x-48}{x}=-3 \quad\text{or}\quad \frac{x^2+5x-48}{x}=3$$
$$x^2+5x-48=-3x \quad\text{or}\quad x^2+5x-48=3x$$
$$x^2+8x-48=0 \quad\text{or}\quad x^2+2x-48=0$$
$$(x+12)(x-4)=0 \quad\text{or}\quad (x+8)(x-6)=0$$

$$(x+12)(x-4)=0$$
$$x+12=0 \quad\text{or}\quad x-4=0$$
$$x=-12 \quad\text{or}\quad x=4$$

$$(x+8)(x-6)=0$$
$$x+8=0 \quad\text{or}\quad x-6=0$$
$$x=-8 \quad\text{or}\quad x=6$$

$$\{-12,-8,4,6\}$$

95. Answers will vary. Example:
An extraneous solution is a value that appears to be a solution, but which makes the original equation undefined. We omit it from the solution set since it is not part of the domain.

8.5   QUICK CHECK

1.
$$2x-5y=-24 \quad (\text{Eq. 1})$$
$$4x+y=-4 \quad (\text{Eq. 2})$$
Solving for y in Eq. 2
$$y=-4x-4$$
Substitute $y=-4x-4$ into Eq. 1.
$$2x-5(-4x-4)=-24$$
$$2x+20x+20=-24$$
$$22x=-44$$
$$x=-2$$
Substitute $x=-2$ into Eq. 2.
$$y=-4(-2)-4=8-4=4$$
$$(-2,4)$$

2.
$$3x+2y=13$$
$$5x-3y=9$$

$$9x+6y=39 \quad (3\cdot\text{Eq. 1})$$
$$\underline{10x-6y=18 \quad (2\cdot\text{Eq. 2})}$$
$$19x=57$$
$$x=3$$
Substitute $x=3$ into Eq. 1.
$$3(3)+2y=13$$
$$9+2y=13$$
$$2y=4$$
$$y=2$$
$$(3,2)$$

3. Unknowns:
x = volume of 60% solution
y = volume of 50% solution

Solution	Volume of Solution	% Acid	Volume of Acid
60%	x	0.6	$0.6x$
50%	y	0.5	$0.5y$
Mixture (52%)	400	0.52	$400 \cdot 0.52 = 208$

The system to solve is:
$$x + y = 400$$
$$0.6x + 0.5y = 208$$
Multiply the second equation by 10 to clear the decimals:
$$10(0.6x + 0.5y) = 10(208)$$
$$6x + 5y = 2080$$
Multiply the first equation by -5 so the coefficients of the y-terms are opposites.
$$-5(x + y) = -5(400)$$
$$-5x - 5y = -2000$$
Add the new equations:
$$\begin{array}{r} -5x - 5y = -2000 \\ 6x + 5y = 2080 \\ \hline x \phantom{{}+5y} = 80 \end{array}$$
Substitute 80 for x in $x + y = 400$:
$$x + y = 400$$
$$80 + y = 400$$
$$y - 320$$
The chemist should mix 80 milliliters of the 60% solution and 320 milliliters of the 50% solution.

4. Substitute $x = 4, y = -2, z = 5$.
$$x - 2y - z = 3$$
$$4 - 2(-2) - 5 = 3$$
$$4 + 4 - 5 = 3$$
$$3 = 3$$
True.
$$3x + 2y + 2z = 26$$
$$3(4) + 2(-2) + 2(5) = 26$$
$$12 - 4 + 10 = 26$$
$$18 \neq 26$$
False.
$$5x - 3y - 4z = 6$$
$$5(4) - 3(-2) - 4(5) = 6$$
$$20 + 6 - 20 = 6$$
$$6 = 6$$
True.
No, since $(4, -2, 5)$ is not a solution to
$3x + 2y + 2z = 26$.

5.
$$x + y + z = 0 \quad \text{(Eq. 1)}$$
$$2x - y + 3z = -9 \quad \text{(Eq. 2)}$$
$$-3x + 2y - 4z = 13 \quad \text{(Eq. 3)}$$

$$\begin{array}{r} x + y + z = 0 \quad \text{(Eq. 1)} \\ + \ 2x - y + 3z = -9 \quad \text{(Eq. 2)} \\ \hline 3x + 4z = -9 \quad \text{(Eq. 4)} \end{array}$$

$$\begin{array}{r} -2x - 2y - 2z = 0 \quad (-2 \cdot \text{Eq. 1}) \\ + \ -3x + 2y - 4z = 13 \quad \text{(Eq. 3)} \\ \hline -5x - 6z = 13 \quad \text{(Eq. 5)} \end{array}$$

$$3x + 4z = -9 \quad \text{(Eq. 4)}$$
$$-5x - 6z = 13 \quad \text{(Eq. 5)}$$

$$\begin{array}{r} 15x + 20z = -45 \quad (5 \cdot \text{Eq. 4}) \\ + \ -15x - 18z = 39 \quad (3 \cdot \text{Eq. 5}) \\ \hline 2z = -6 \\ z = -3 \end{array}$$
Substitute $z = -3$ into Eq. 4.
$$3x + 4(-3) = -9$$
$$3x - 12 = -9$$
$$3x = 3$$
$$x = 1$$
Substitute $x = 1, z = -3$ into Eq. 1.
$$1 + y - 3 = 0$$
$$y - 2 = 0$$
$$y = 2$$
$$(1, 2, -3)$$

6.
$$2x + y - z = -8 \quad \text{(Eq. 1)}$$
$$-4x - 3y + 3z = 25 \quad \text{(Eq. 2)}$$
$$6x + 5y + 4z = 3 \quad \text{(Eq. 3)}$$

$$\begin{array}{r} 4x + 2y - 2z = -16 \quad (2 \cdot \text{Eq. 1}) \\ + \ -4x - 3y + 3z = 25 \quad \text{(Eq. 2)} \\ \hline -y + z = 9 \quad \text{(Eq. 4)} \end{array}$$

$$\begin{array}{r} -6x - 3y + 3z = 24 \quad (-3 \cdot \text{Eq. 1}) \\ + \ 6x + 5y + 4z = 3 \quad \text{(Eq. 3)} \\ \hline 2y + 7z = 27 \quad \text{(Eq. 5)} \end{array}$$

$$\begin{array}{r} -2y + 2z = 18 \quad (2 \cdot \text{Eq. 4}) \\ + \ 2y + 7z = 27 \quad \text{(Eq. 5)} \\ \hline 9z = 45 \\ z = 5 \end{array}$$
Substitute $z = 5$ into Eq. 4.
$$-y + 5 = 9$$
$$-y = 4$$
$$y = -4$$
Substitute $y = -4, z = 5$ into Eq. 1.
$$2x - 4 - 5 = -8$$
$$2x - 9 = -8$$
$$2x = 1$$
$$x = \frac{1}{2}$$
$$\left(\frac{1}{2}, -4, 5 \right)$$

7. Unknowns:
x = number of $1 bills
y = number of $5 bills
z = number of $10 bills

$$\begin{aligned}
x + y + z &= 30 \quad &&\text{(Eq. 1)}\\
1x + 5y + 10z &= 111 \quad &&\text{(Eq. 2)}\\
y &= z + 4 \quad &&\text{(Eq. 3)}
\end{aligned}$$

$$\begin{aligned}
-x - y - z &= -30 \quad &&(-1 \cdot \text{Eq. 1})\\
+\quad x + 5y + 10z &= 111 \quad &&\text{(Eq. 2)}\\
\hline
4y + 9z &= 81 \quad &&\text{(Eq. 4)}
\end{aligned}$$

Substitute Eq. 3 into Eq. 4.
$$\begin{aligned}
4(z+4) + 9z &= 81\\
4z + 16 + 9z &= 81\\
13z &= 65\\
z &= 5
\end{aligned}$$

Substitute $z = 5$ into Eq. 3.
$$y = 5 + 4 = 9$$

Substitute $y = 9$, $z = 5$ into Eq. 1.
$$\begin{aligned}
x + 9 + 5 &= 30\\
x + 14 &= 30\\
x &= 16
\end{aligned}$$
16 $1 bills, 9 $5 bills, 5 $10 bills

8.5 SYSTEMS OF EQUATIONS (TWO EQUATIONS IN TWO UNKNOWNS, THREE EQUATIONS IN THREE UNKNOWNS)

1. system of linear equations

3. independent

5. dependent

7. ordered triple

9. $y = 2x - 16$
$3x + 2y = 17$

Substitute $y = 2x - 16$ into Eq. 2.
$$\begin{aligned}
3x + 2(2x - 16) &= 17\\
3x + 4x - 32 &= 17\\
7x &= 49\\
x &= 7
\end{aligned}$$
Substitute $x = 7$ into Eq. 1.
$$y = 2(7) - 16 = 14 - 16 = -2$$
$(7, -2)$

11. $2x - 7y = 41$
$x = 4y + 23$

Substitute $x = 4y + 23$ into Eq. 1.
$$\begin{aligned}
2(4y + 23) - 7y &= 41\\
8y + 46 - 7y &= 41\\
y &= -5
\end{aligned}$$
Substitute $y = -5$ into Eq. 2.
$$x = 4(-5) + 23 = -20 + 23 = 3$$
$(3, -5)$

13. $y = \dfrac{2}{3}x + 11$
$4x - 3y = -45$

Substitute $y = \dfrac{2}{3}x + 11$ into Eq. 2.

$$\begin{aligned}
4x - 3\left(\frac{2}{3}x + 11\right) &= -45\\
4x - 2x - 33 &= -45\\
2x &= -12\\
x &= -6
\end{aligned}$$
Substitute $x = -6$ into Eq. 1.
$$y = \frac{2}{3}(-6) + 11 = -4 + 11 = 7$$
$(-6, 7)$

15. $y = 3x + 7$
$6x - 2y = -14$

Substitute $y = 3x + 7$ into Eq. 2.
$$\begin{aligned}
6x - 2(3x + 7) &= -14\\
6x - 6x - 14 &= -14\\
-14 &= -14
\end{aligned}$$
True.
This is a dependent system.
All points $(x, 3x + 7)$ are solutions.

17.
$$\begin{aligned}
-4x + 6y &= -20 \quad &&(-2 \cdot \text{Eq. 1})\\
+\quad 4x - 2y &= 28 \quad &&\text{(Eq. 2)}\\
\hline
4y &= 8\\
y &= 2
\end{aligned}$$

Substitute $y = 2$ into Eq. 2.
$$\begin{aligned}
4x - 2(2) &= 28\\
4x - 4 &= 28\\
4x &= 32\\
x &= 8
\end{aligned}$$
$(8, 2)$

19. $x - 4y = 7$
$\quad\quad x = 4y + 7$

Substitute $x = 4y + 7$ into Eq. 2.
$$\begin{aligned}
3(4y + 7) + 2y &= 14\\
12y + 21 + 2y &= 14\\
14y &= -7
\end{aligned}$$
$$y = \frac{-7}{14}$$
$$y = -\frac{1}{2}$$

Substitute $y = -\dfrac{1}{2}$ into Eq. 1.
$$x = 4\left(-\frac{1}{2}\right) + 7 = -2 + 7 = 5$$
$\left(5, -\dfrac{1}{2}\right)$

21.
$$8x + 6y = -56 \quad (2 \cdot \text{Eq. 1})$$
$$+\ \underline{9x - 6y = -12} \quad (3 \cdot \text{Eq. 2})$$
$$17x = -68$$
$$x = -4$$

Substitute $x = -4$ into Eq. 1.
$$4(-4) + 3y = -28$$
$$-16 + 3y = -28$$
$$3y = -12$$
$$y = -4$$
$$(-4, -4)$$

23.
$$18x - 12y = 33 \quad (3 \cdot \text{Eq. 1})$$
$$+\ \underline{-18x + 12y = -34} \quad (-2 \cdot \text{Eq. 2})$$
$$0 = -1$$

False.
This is an inconsistent system.
$\varnothing$

25.
$$12 \cdot \left(\frac{3}{4}x + \frac{1}{3}y\right) = 12 \cdot (1) \quad (12 \cdot \text{Eq. 1})$$
$$20 \cdot \left(\frac{2}{5}x - \frac{1}{4}y\right) = 20 \cdot \left(\frac{31}{10}\right) \quad (20 \cdot \text{Eq. 2})$$
$$9x + 4y = 12 \quad (\text{Eq. 3})$$
$$8x - 5y = 62 \quad (\text{Eq. 4})$$
$$45x + 20y = 60 \quad (5 \cdot \text{Eq. 3})$$
$$+\ \underline{32x - 20y = 248} \quad (4 \cdot \text{Eq. 4})$$
$$77x = 308$$
$$x = 4$$

Substitute $x = 4$ into Eq. 3.
$$9(4) + 4y = 12$$
$$36 + 4y = 12$$
$$4y = -24$$
$$y = -6$$
$$(4, -6)$$

27. Unknowns:
$x =$ length of rectangle
$y =$ width of rectangle

length is 3 more than twice the width:
$$x = 2y + 3 \quad (\text{Eq. 1})$$

perimeter is 54:
$$2x + 2y = 54 \quad (\text{Eq. 2})$$

Substitute $x = 2y + 3$ into Eq. 2.
$$2(2y + 3) + 2y = 54$$
$$4y + 6 + 2y = 54$$
$$6y = 48$$
$$y = 8$$

Substitute $y = 8$ into Eq. 1.
$$x = 2(8) + 3 = 16 + 3 = 19$$

The length is 19 inches and the width is 8 inches.

29. Unknowns:
$x =$ number of dimes
$y =$ number of quarters

65 coins, some are dimes, the rest are quarters:
$$x + y = 65 \quad (\text{Eq. 1})$$

value of coins is \$13.10:
$$0.10x + 0.25y = 13.10 \quad (\text{Eq. 2})$$

Solve Eq. 1 for y.
$$y = -x + 65 \quad (\text{Eq. 3})$$

Multiply Eq. 2 by 100 to clear the decimals.
$$100(0.10x + 0.25y) = 100(13.10)$$
$$10x + 25y = 1310 \quad (\text{Eq. 4})$$

Substitute $y = -x + 65$ into Eq. 4.
$$10x + 25(-x + 65) = 1310$$
$$10x - 25x + 1625 = 1310$$
$$-15x = -315$$
$$x = 21$$

Substitute $x = 21$ into Eq. 3.
$$y = -(21) + 65 = 44$$

The jar contains 21 dimes and 44 quarters.

31. Unknowns:
$x =$ amount in account paying 6%
$y =$ amount in account paying 3%

\$7500 deposited in the two accounts:
$$x + y = 7500 \quad (\text{Eq. 1})$$

interest earned was \$387:
$$0.06x + 0.03y = 387 \quad (\text{Eq. 2})$$

Solve Eq. 1 for y.
$$y = -x + 7500 \quad (\text{Eq. 3})$$

Multiply Eq. 2 by 100 to clear the decimals.
$$100(0.06x + 0.03y) = 100(387)$$
$$6x + 3y = 38,700 \quad (\text{Eq. 4})$$

Substitute $y = -x + 7500$ into Eq. 4.
$$6x + 3(-x + 7500) = 38,700$$
$$6x - 3x + 22,500 = 38,700$$
$$3x = 16,200$$
$$x = 5400$$

Substitute $x = 5400$ into Eq. 3.
$$y = -5400 + 7500 = 2100$$

Faustino deposited \$5400 in the account earning 6% and \$2100 in the account earning 3%.

33. Unknowns:
x = volume of 15% solution
y = volume of 30% solution

Solution	Volume of Solution	% Acid	Volume of Acid
15%	x	0.15	$0.15x$
30%	y	0.3	$0.3y$
Mixture (25%)	60	0.25	$60 \cdot 0.25 = 15$

The system to solve is:
$x + y = 60$
$0.15x + 0.3y = 15$
Multiply the second equation by 100 to clear the decimals:
$100(0.15x + 0.3y) = 100(15)$
$15x + 30y = 1500$

Multiply the first equation by -15 so the coefficients of the
x-terms are opposites.
$-15(x + y) = -15(60)$
$-15x - 15y = -900$

Add the new equations:
$-15x - 15y = -900$
$\underline{15x + 30y = 1500}$
$15y = 600$
$y = 40$

Substitute 40 for y in $x + y = 60$:
$x + y = 60$
$x + 40 = 60$
$x = 20$
The chemist should mix 20 milliliters of the
15% solution and 40 milliliters of the
30% solution.

35. Substitute $x = 3$, $y = 2$, $z = 6$.

$x + y + z = 11$	$x - y + 2z = 17$	$4x - 3y - 2z = -6$
$3 + 2 + 6 = 11$	$3 - 2 + 2(6) = 17$	$4(3) - 3(2) - 2(6) = -6$
$11 = 11$	$3 - 2 + 12 = 17$	$12 - 6 - 12 = -6$
	$13 \neq 17$	$-6 = -6$

No, since $(3, 2, 6)$ is not a solution of $x - y + 2z = 17$.

37. Substitute $x = -5$, $y = 2$, $z = 0$.

$x + 2y + 3z = -1$	$3x - y - 6z = -17$	$4x - 3y + z = -26$
$-5 + 2(2) + 3(0) = -1$	$3(-5) - 2 - 6(0) = -17$	$4(-5) - 3(2) + 0 = -26$
$-5 + 4 + 0 = -1$	$-15 - 2 - 0 = -17$	$-20 - 6 + 0 = -26$
$-1 = -1$	$-17 = -17$	$-26 = -26$

Yes, $(-5, 2, 0)$ is a solution.

39.
$$x + 2y + 3z = 23 \quad \text{(Eq. 1)}$$
$$-2x + 3y + z = 17 \quad \text{(Eq. 2)}$$
$$-3x + 5y - 4z = 12 \quad \text{(Eq. 3)}$$

$$\begin{array}{rl} 2x + 4y + 6z = 46 & (2 \cdot \text{Eq. 1}) \\ + \quad -2x + 3y + z = 17 & (\text{Eq. 2}) \\ \hline 7y + 7z = 63 & \\ y + z = 9 & (\text{Eq. 4}) \end{array}$$

$$\begin{array}{rl} 3x + 6y + 9z = 69 & (3 \cdot \text{Eq. 1}) \\ + \quad -3x + 5y - 4z = 12 & (\text{Eq. 3}) \\ \hline 11y + 5z = 81 & (\text{Eq. 5}) \end{array}$$

$$\begin{array}{rl} -5y - 5z = -45 & (-5 \cdot \text{Eq. 4}) \\ + \quad 11y + 5z = 81 & (\text{Eq. 5}) \\ \hline 6y = 36 & \\ y = 6 & \end{array}$$

Substitute $y = 6$ into Eq. 4.
$$6 + z = 9$$
$$z = 3$$

Substitute $y = 6, z = 3$ into Eq. 1.
$$x + 2(6) + 3(3) = 23$$
$$x + 12 + 9 = 23$$
$$x = 2$$
$$(2, 6, 3)$$

41.
$$x + y + z = 2 \quad \text{(Eq. 1)}$$
$$2x + 3y - z = 5 \quad \text{(Eq. 2)}$$
$$7x + 4y + 2z = 25 \quad \text{(Eq. 3)}$$

$$\begin{array}{rl} x + y + z = 2 & (\text{Eq. 1}) \\ + \quad 2x + 3y - z = 5 & (\text{Eq. 2}) \\ \hline 3x + 4y = 7 & (\text{Eq. 4}) \end{array}$$

$$\begin{array}{rl} -2x - 2y - 2z = -4 & (-2 \cdot \text{Eq. 1}) \\ + \quad 7x + 4y + 2z = 25 & (\text{Eq. 3}) \\ \hline 5x + 2y = 21 & (\text{Eq. 5}) \end{array}$$

$$\begin{array}{rl} -10x - 4y = -42 & (-2 \cdot \text{Eq. 5}) \\ + \quad 3x + 4y = 7 & (\text{Eq. 4}) \\ \hline -7x = -35 & \\ x = 5 & \end{array}$$

Substitute $x = 5$ into Eq. 5.
$$5(5) + 2y = 21$$
$$25 + 2y = 21$$
$$2y = -4$$
$$y = -2$$

Substitute $x = 5, y = -2$ into Eq. 1.
$$5 - 2 + z = 2$$
$$3 + z = 2$$
$$z = -1$$
$$(5, -2, -1)$$

43.
$$3x + 2y + 5z = -8 \quad \text{(Eq. 1)}$$
$$9x - 4y + 6z = -38 \quad \text{(Eq. 2)}$$
$$-6x - 6y + 7z = -62 \quad \text{(Eq. 3)}$$

$$\begin{array}{rl} -9x - 6y - 15z = 24 & (-3 \cdot \text{Eq. 1}) \\ + \quad 9x - 4y + 6z = -38 & (\text{Eq. 2}) \\ \hline -10y - 9z = -14 & (\text{Eq. 4}) \end{array}$$

$$\begin{array}{rl} 6x + 4y + 10z = -16 & (2 \cdot \text{Eq. 1}) \\ + \quad -6x - 6y + 7z = -62 & (\text{Eq. 3}) \\ \hline -2y + 17z = -78 & (\text{Eq. 5}) \end{array}$$

$$\begin{array}{rl} -10y - 9z = -14 & (\text{Eq. 4}) \\ + \quad 10y - 85z = 390 & (-5 \cdot \text{Eq. 5}) \\ \hline -94z = 376 & \\ z = -4 & \end{array}$$

Substitute $z = -4$ into Eq. 4.
$$-10y - 9(-4) = -14$$
$$-10y + 36 = -14$$
$$-10y = -50$$
$$y = 5$$

Substitute $y = 5, z = -4$ into Eq. 1.
$$3x + 2(5) + 5(-4) = -8$$
$$3x + 10 - 20 = -8$$
$$3x - 10 = -8$$
$$3x = 2$$
$$x = \frac{2}{3}$$
$$\left(\frac{2}{3}, 5, -4 \right)$$

45.
$$2x + 3y + z = 5 \quad \text{(Eq. 1)}$$
$$-2x - y + 4z = 60 \quad \text{(Eq. 2)}$$
$$x + 2y - 8z = -30 \quad \text{(Eq. 3)}$$

$$\begin{array}{rl} 2x + 3y + z = 5 & (\text{Eq. 1}) \\ + \quad -2x - y + 4z = 60 & (\text{Eq. 2}) \\ \hline 2y + 5z = 65 & (\text{Eq. 4}) \end{array}$$

$$\begin{array}{rl} -2x - y + 4z = 60 & (\text{Eq. 2}) \\ + \quad 2x + 4y - 16z = -60 & (2 \cdot \text{Eq. 3}) \\ \hline 3y - 12z = 0 & (\text{Eq. 5}) \end{array}$$

$$\begin{array}{rl} 6y + 15z = 195 & (3 \cdot \text{Eq. 4}) \\ + \quad -6y + 24z = 0 & (-2 \cdot \text{Eq. 5}) \\ \hline 39z = 195 & \\ z = 5 & \end{array}$$

Substitute $z = 5$ into Eq. 5.
$$3y - 12(5) = 0$$
$$3y - 60 = 0$$
$$3y = 60$$
$$y = 20$$

Substitute $y = 20, z = 5$ into Eq. 3.
$$x + 2(20) - 8(5) = -30$$
$$x + 40 - 40 = -30$$
$$x = -30$$
$$(-30, 20, 5)$$

252 **CHAPTER 8   A TRANSITION**

47.
$$2x - 3y - z = -6 \quad \text{(Eq. 1)}$$
$$-3x + 2y + 4z = 24 \quad \text{(Eq. 2)}$$
$$5x - 4y + 3z = -18 \quad \text{(Eq. 3)}$$

$$\begin{aligned} 8x - 12y - 4z &= -24 \quad (4 \cdot \text{Eq. 1}) \\ + \; -3x + 2y + 4z &= 24 \quad \;\;\text{(Eq. 2)} \\ \hline 5x - 10y &= 0 \quad \;\;\;\text{(Eq. 4)} \end{aligned}$$

$$\begin{aligned} 6x - 9y - 3z &= -18 \quad (3 \cdot \text{Eq. 1}) \\ + \; 5x - 4y + 3z &= -18 \quad \;\;\text{(Eq. 3)} \\ \hline 11x - 13y &= -36 \quad \;\;\text{(Eq. 5)} \end{aligned}$$

$$\begin{aligned} -55x + 110y &= 0 \quad\;\; (-11 \cdot \text{Eq. 4}) \\ + \; 55x - 65y &= -180 \quad (5 \cdot \text{Eq. 5}) \\ \hline 45y &= -180 \\ y &= -4 \end{aligned}$$

Substitute $y = -4$ into Eq. 4.
$$5x - 10(-4) = 0$$
$$5x + 40 = 0$$
$$5x = -40$$
$$x = -8$$
Substitute $x = -8$, $y = -4$ into Eq. 1.
$$2(-8) - 3(-4) - z = -6$$
$$-16 + 12 - z = -6$$
$$-4 - z = -6$$
$$-z = -2$$
$$z = 2$$
$$(-8, -4, 2)$$

49.
$$4x + 2y + z = -5 \quad \text{(Eq. 1)}$$
$$2x - y - 6z = 0 \quad \text{(Eq. 2)}$$
$$-8x + 6y + 3z = -30 \quad \text{(Eq. 3)}$$

$$\begin{aligned} 24x + 12y + 6z &= -30 \quad (6 \cdot \text{Eq. 1}) \\ + \; 2x - y - 6z &= 0 \quad \;\;\;\text{(Eq. 2)} \\ \hline 26x + 11y &= -30 \quad \text{(Eq. 4)} \end{aligned}$$

$$\begin{aligned} -12x - 6y - 3z &= 15 \quad (-3 \cdot \text{Eq. 1}) \\ + \; -8x + 6y + 3z &= -30 \quad \;\;\text{(Eq. 3)} \\ \hline -20x &= -15 \\ x &= \frac{-15}{-20} \\ x &= \frac{3}{4} \end{aligned}$$

Substitute $x = \dfrac{3}{4}$ into Eq. 4.
$$26\left(\frac{3}{4}\right) + 11y = -30$$
$$\frac{78}{4} + 11y = -30$$
$$4 \cdot \left(\frac{78}{4} + 11y\right) = 4 \cdot -30$$
$$78 + 44y = -120$$
$$44y = -198$$
$$y = -\frac{198}{44}$$
$$y = -\frac{9}{2}$$

Substitute $x = \dfrac{3}{4}$, $y = -\dfrac{9}{2}$ into Eq. 1.
$$4\left(\frac{3}{4}\right) + 2\left(-\frac{9}{2}\right) + z = -5$$
$$3 - 9 + z = -5$$
$$-6 + z = -5$$
$$z = 1$$
$$\left(\frac{3}{4}, -\frac{9}{2}, 1\right)$$

51. Multiply by the LCD to eliminate the fractions.
$$6 \cdot \left(\frac{1}{2}x + \frac{1}{3}y + z\right) = 6 \cdot 3$$
$$3 \cdot \left(\frac{2}{3}x - 4y - 3z\right) = 3 \cdot 13$$

$$3x + 2y + 6z = 18 \quad \text{(Eq. 1)}$$
$$2x - 12y - 9z = 39 \quad \text{(Eq. 2)}$$
$$3x + 4y + 5z = 11 \quad \text{(Eq. 3)}$$

$$\begin{aligned} 18x + 12y + 36z &= 108 \quad (6 \cdot \text{Eq. 1}) \\ + \; 2x - 12y - 9z &= 39 \quad \;\;\text{(Eq. 2)} \\ \hline 20x + 27z &= 147 \quad \text{(Eq. 4)} \end{aligned}$$

$$\begin{aligned} -6x - 4y - 12z &= -36 \quad (-2 \cdot \text{Eq. 1}) \\ + \; 3x + 4y + 5z &= 11 \quad \;\;\text{(Eq. 3)} \\ \hline -3x - 7z &= -25 \quad \text{(Eq. 5)} \end{aligned}$$

$$\begin{aligned} 60x + 81z &= 441 \quad (3 \cdot \text{Eq. 4}) \\ + \; -60x - 140z &= -500 \quad (20 \cdot \text{Eq. 5}) \\ \hline -59z &= -59 \\ z &= 1 \end{aligned}$$

Continued on next page.

51. Continued.
Substitute $z = 1$ into Eq. 4.
$$20x + 27(1) = 147$$
$$20x + 27 = 147$$
$$20x = 120$$
$$x = 6$$
Substitute $x = 6$, $z = 1$ into Eq. 1.
$$3(6) + 2y + 6(1) = 18$$
$$18 + 2y + 6 = 18$$
$$2y + 24 = 18$$
$$2y = -6$$
$$y = -3$$
$(6, -3, 1)$

53.
$$x + 2y + z = 8 \quad (\text{Eq. 1})$$
$$3x + 4y + 2z = 9 \quad (\text{Eq. 2})$$
$$3x - 4y = -37 \quad (\text{Eq. 3})$$

$$
\begin{array}{ll}
-2x - 4y - 2z = -16 & (-2 \cdot \text{Eq. 1}) \\
+ \quad 3x + 4y + 2z = 9 & (\text{Eq. 2}) \\
\hline
x = -7 &
\end{array}
$$

Substitute $x = -7$ into Eq. 3.
$$3(-7) - 4y = -37$$
$$-21 - 4y = -37$$
$$-4y = -16$$
$$y = 4$$
Substitute $x = -7$, $y = 4$ into Eq. 1.
$$-7 + 2(4) + z = 8$$
$$-7 + 8 + z = 8$$
$$1 + z = 8$$
$$z = 7$$
$(-7, 4, 7)$

55. Unknowns:
x = number of \$1 bills
y = number of \$5 bills
z = number of \$10 bills

$$x + y + z = 35 \quad (\text{Eq. 1})$$
$$1x + 5y + 10z = 246 \quad (\text{Eq. 2})$$
$$-x + y = 4 \quad (\text{Eq. 3})$$

$$
\begin{array}{ll}
-10x - 10y - 10z = -350 & (-10 \cdot \text{Eq. 1}) \\
+ \quad x + 5y + 10z = 246 & (\text{Eq. 2}) \\
\hline
-9x - 5y = -104 & (\text{Eq. 4})
\end{array}
$$

$$
\begin{array}{ll}
-5x + 5y = 20 & (5 \cdot \text{Eq. 3}) \\
+ \quad -9x - 5y = -104 & (\text{Eq. 4}) \\
\hline
-14x = -84 & \\
x = 6 &
\end{array}
$$

Substitute $x = 6$ into Eq. 3.
$$-6 + y = 4$$
$$y = 10$$

Substitute $x = 6$, $y = 10$ into Eq. 1.
$$6 + 10 + z = 35$$
$$16 + z = 35$$
$$z = 19$$
Jonah has 19 \$10 bills.

57. Unknowns:
x = number of cups of coffee
y = number of muffins
z = number of bacon-and-egg sandwiches

$$x + y + z = 50 \quad (\text{Eq. 1})$$
$$1x + 2y + 3z = 85 \quad (\text{Eq. 2})$$
$$y = x + 5 \quad (\text{Eq. 3})$$

Substitute Eq. 3 into Eq. 1.
$$x + (x + 5) + z = 50$$
$$2x + z = 45 \quad (\text{Eq. 4})$$
Substitute Eq. 3 into Eq. 2.
$$x + 2(x + 5) + 3z = 85$$
$$x + 2x + 10 + 3z = 85$$
$$3x + 3z = 75 \quad (\text{Eq. 5})$$

$$
\begin{array}{ll}
-6x - 3z = -135 & (-3 \cdot \text{Eq. 4}) \\
+ \quad 6x + 6z = 150 & (2 \cdot \text{Eq. 5}) \\
\hline
3z = 15 & \\
z = 5 &
\end{array}
$$

5 bacon-and-egg sandwiches were sold.

59. Unknowns:
x = Abraham's age
y = Belen's age
z = Celeste's age

$$x = y + 8 \quad (\text{Eq. 1})$$
$$y + z = 2x + 3$$
$$-2x + y + z = 3 \quad (\text{Eq. 2})$$
$$x + y + z = 138 \quad (\text{Eq. 3})$$

Substitute Eq. 1 into Eq. 2.
$$-2(y + 8) + y + z = 3$$
$$-2y - 16 + y + z = 3$$
$$-y + z = 19 \quad (\text{Eq. 4})$$
Substitute Eq. 1 into Eq. 3.
$$(y + 8) + y + z = 138$$
$$2y + z = 130 \quad (\text{Eq. 5})$$

$$
\begin{array}{ll}
y - z = -19 & (-1 \cdot \text{Eq. 4}) \\
+ \quad 2y + z = 130 & (\text{Eq. 5}) \\
\hline
3y = 111 & \\
y = 37 &
\end{array}
$$

Substitute $y = 37$ into Eq. 1.
$$x = 37 + 8$$
$$x = 45$$
Substitute $x = 45$, $y = 37$ into Eq. 3.
$$45 + 37 + z = 138$$
$$82 + z = 138$$
$$z = 56$$
Abraham is 45, Belen is 37, and Celeste is 56.

61. Unknowns:
A = measure of angle A
B = measure of angle B
C = measure of angle C

$$\begin{aligned}
A + B + C &= 180 \quad &\text{(Eq. 1)} \\
A &= C - 20 \quad &\text{(Eq. 2)} \\
C &= 2B \quad &\text{(Eq. 3)}
\end{aligned}$$

Substitute Eq. 2 into Eq. 1.
$$(C - 20) + B + C = 180$$
$$B + 2C = 200 \quad \text{(Eq. 4)}$$

Substitute Eq. 3 into Eq. 4.
$$B + 2(2B) = 200$$
$$B + 4B = 200$$
$$5B = 200$$
$$B = 40$$

Substitute $B = 40$ into Eq. 3.
$$C = 2(40) = 80$$

Substitute $C = 80$ into Eq. 2.
$$A = 80 - 20 = 60$$

Measure of angle $A = 60°$, $B = 40°$, $C = 80°$.

63. Unknowns:
x = amount invested at 3%
y = amount invested at 5%
z = amount invested at 6%

$$\begin{aligned}
x + y + z &= 40{,}000 \quad &\text{(Eq. 1)} \\
0.03x + 0.05y + 0.06z &= 1960 \quad &\text{(Eq. 2)} \\
y &= z + 8000 \quad &\text{(Eq. 3)}
\end{aligned}$$

Substitute Eq. 3 into Eq. 1.
$$x + (z + 8000) + z = 40{,}000$$
$$x + 2z = 32{,}000 \quad \text{(Eq. 4)}$$
Multiply Eq. 2 by 100 to eliminate decimals.
$$3x + 5y + 6z = 196{,}000 \quad \text{(Eq. 5)}$$

Substitute Eq. 3 into Eq. 5.
$$3x + 5(z + 8000) + 6z = 196{,}000$$
$$3x + 5z + 40{,}000 + 6z = 196{,}000$$
$$3x + 11z = 156{,}000 \quad \text{(Eq. 6)}$$

$$\begin{aligned}
-3x - 6z &= -96{,}000 \quad &(-3 \cdot \text{Eq. 4}) \\
+ \quad 3x + 11z &= 156{,}000 \quad &\text{(Eq. 6)} \\
\hline
5z &= 60{,}000 \\
z &= 12{,}000
\end{aligned}$$

Substitute $z = 12{,}000$ into Eq. 3.
$$y = 12{,}000 + 8000 = 20{,}000$$

Substitute $z = 12{,}000$ into Eq. 4.
$$x + 2(12{,}000) = 32{,}000$$
$$x + 24{,}000 = 32{,}000$$
$$x = 8000$$
Mary invested $8000 at 3%, $20,000 at 5%, and $12,000 at 6%.

65. Unknowns:
x = amount in fund that increased 2%
y = amount in fund that increased 10%
z = amount in fund that increased 25%

$$\begin{aligned}
x + y + z &= 8000 \quad &\text{(Eq. 1)} \\
0.02x + 0.10y + 0.25z &= 550 \quad &\text{(Eq. 2)} \\
z &= y - 1000 \quad &\text{(Eq. 3)}
\end{aligned}$$

Substitute Eq. 3 into Eq. 1.
$$x + y + (y - 1000) = 8000$$
$$x + 2y = 9000 \quad \text{(Eq. 4)}$$
Multiply Eq. 2 by 100 to eliminate decimals.
$$2x + 10y + 25z = 55{,}000 \quad \text{(Eq. 5)}$$

Substitute Eq. 3 into Eq. 5.
$$2x + 10y + 25(y - 1000) = 55{,}000$$
$$2x + 10y + 25y - 25{,}000 = 55{,}000$$
$$2x + 35y = 80{,}000 \quad \text{(Eq. 6)}$$

$$\begin{aligned}
-2x - 4y &= -18{,}000 \quad &(-2 \cdot \text{Eq. 4}) \\
+ \quad 2x + 35y &= 80{,}000 \quad &\text{(Eq. 6)} \\
\hline
31y &= 62{,}000 \\
y &= 2000
\end{aligned}$$

Substitute $y = 2000$ into Eq. 3.
$$z = 2000 - 1000 = 1000$$

Substitute $y = 2000$, $z = 1000$ into Eq. 1.
$$x + 2000 + 1000 = 8000$$
$$x + 3000 = 8000$$
$$x = 5000$$
Pilar invested $5000 in the 2% fund, $2000 in the 10% fund, and $1000 in the 25% fund.

67. Unknowns:
x = amount in stock that increased by 5%
y = amount in stock that increased by 7%
z = amount in stock that decreased by 40%

$$\begin{aligned}
x + y + z &= 25{,}000 \quad &\text{(Eq. 1)} \\
0.05x + 0.07y - 0.40z &= -1330 \quad &\text{(Eq. 2)} \\
y &= z \quad &\text{(Eq. 3)}
\end{aligned}$$

Substitute Eq. 3 into Eq. 1.
$$x + (z) + z = 25{,}000$$
$$x + 2z = 25{,}000 \quad \text{(Eq. 4)}$$
Multiply Eq. 2 by 100 to eliminate decimals.
$$5x + 7y - 40z = -133{,}000 \quad \text{(Eq. 5)}$$

Substitute Eq. 3 into Eq. 5.
$$5x + 7(z) - 40z = -133{,}000$$
$$5x + 7z - 40z = -133{,}000$$
$$5x - 33z = -133{,}000 \quad \text{(Eq. 6)}$$

$$\begin{aligned}
-5x - 10z &= -125{,}000 \quad &(-5 \cdot \text{Eq. 4}) \\
+ \quad 5x - 33z &= -133{,}000 \quad &\text{(Eq. 6)} \\
\hline
-43z &= -258{,}000 \\
z &= 6000
\end{aligned}$$

Continued on next page.

67. Continued

Substitute $z = 6000$ into Eq. 3.
$y = 6000$

Substitute $y = 6000$, $z = 6000$ into Eq. 1.
$$x + 6000 + 6000 = 25,000$$
$$x + 12,000 = 25,000$$
$$x = 13,000$$
Salim invested \$13,000 in the 5% stock, \$6000 in the 7% stock, and \$6000 in the −40% stock.

69. Unknowns:
x = amount of Solution A (10% alcohol)
y = amount of Solution B (15% alcohol)
z = amount of Solution C (30% alcohol)

$$x + \quad y + \quad z = 100 \qquad (\text{Eq. 1})$$
$$0.10x + 0.15y + 0.30z = 0.135(100)$$
$$0.10x + 0.15y + 0.30z = 13.5 \qquad (\text{Eq. 2})$$
$$x = 2y \qquad (\text{Eq. 3})$$

Substitute Eq. 3 into Eq. 1.
$$(2y) + y + z = 100$$
$$3y + z = 100 \qquad (\text{Eq. 4})$$
Multiply Eq. 2 by 100 to eliminate decimals.
$$10x + 15y + 30z = 1350 \qquad (\text{Eq. 5})$$

Substitute Eq. 3 into Eq. 5.
$$10(2y) + 15y + 30z = 1350$$
$$20y + 15y + 30z = 1350$$
$$35y + 30z = 1350$$
$$7y + 6z = 270 \qquad (\text{Eq. 6})$$

$$\begin{aligned} -18y - 6z &= -600 \quad (-6 \cdot \text{Eq. 4}) \\ + \quad 7y + 6z &= 270 \qquad (\text{Eq. 6}) \\ \hline -11y &= -330 \\ y &= 30 \end{aligned}$$

Substitute $y = 30$ into Eq. 3.
$$x = 2(30) = 60$$

Substitute $x = 60$, $y = 30$ into Eq. 1.
$$60 + 30 + z = 100$$
$$90 + z = 100$$
$$z = 10$$
The chemist should combine 60 milliliters of the 10% solution, 30 milliliters of the 15% solution, and 10 milliliters of the 30% solution.

71. Answers will vary. Example:
The local high school sold 300 total tickets to their "End of the Year" play. They sold 20 more senior citizen tickets than children tickets. They also sold 160 fewer children tickets than adult tickets. How many of each type of ticket did they sell?

1. $2x - 17 = -5$
$$2x = 12$$
$$x = 6$$
$\{6\}$

2. $3x - 9 = 17 - 5x$
$$8x - 9 = 17$$
$$8x = 26$$
$$x = \frac{26}{8}$$
$$x = \frac{13}{4}$$
$\left\{\dfrac{13}{4}\right\}$

3.
$$\frac{1}{3}x - \frac{1}{6} = \frac{2}{15}x + \frac{1}{3}$$
$$30 \cdot \left(\frac{1}{3}x - \frac{1}{6}\right) = 30 \cdot \left(\frac{2}{15}x + \frac{1}{3}\right)$$
$$\cancel{30}^{10} \cdot \frac{1}{\cancel{3}_1}x - \cancel{30}^{5} \cdot \frac{1}{\cancel{6}_1} = \cancel{30}^{2} \cdot \frac{2}{\cancel{15}_1}x + \cancel{30}^{10} \cdot \frac{1}{\cancel{3}_1}$$
$$10x - 5 = 4x + 10$$
$$6x - 5 = 10$$
$$6x = 15$$
$$x = \frac{15}{6}$$
$$x = \frac{5}{2}$$
$\left\{\dfrac{5}{2}\right\}$

4. $4x + 2(3x - 1) = 17x - 3(4x - 11)$
$$4x + 6x - 2 = 17x - 12x + 33$$
$$10x - 2 = 5x + 33$$
$$5x - 2 = 33$$
$$5x = 35$$
$$x = 7$$
$\{7\}$

5. $|x| = 6$
$$x = -6 \quad \text{or} \quad x = 6$$
$\{-6, 6\}$

6. $|x| = -6$
$\varnothing$, since an absolute value cannot be equal to a negative number.

7. $|x+7|-14=-6$
$|x+7|=8$
$x+7=-8$ or $x+7=8$
$x=-15$ or $x=1$
$\{-15,1\}$

8. $|3x+13|+20=28$
$|3x+13|=8$
$3x+13=-8$ or $3x+13=8$
$3x=-21$ or $3x=-5$
$x=-7$ or $x=-\dfrac{5}{3}$
$\left\{-7,-\dfrac{5}{3}\right\}$

9. $|6x+25|=|x+10|$
$6x+25=-(x+10)$ or $6x+25=x+10$
$6x+25=-x-10$ or $5x+25=10$
$7x+25=-10$ or $5x=-15$
$7x=-35$ or $x=-3$
$x=-5$ or
$\{-5,-3\}$

10. $|4x-3|=|2x-15|$
$4x-3=-(2x-15)$ or $4x-3=2x-15$
$4x-3=-2x+15$ or $2x-3=-15$
$6x-3=15$ or $2x=-12$
$6x=18$ or $x=-6$
$x=3$
$\{3,-6\}$

11. $3x+16<10$
$3x<-6$
$x<-2$
$(-\infty,-2)$

12. $-5x+19\ge-16$
$-5x\ge-35$
$x\le 7$
$(-\infty,7]$

13. $2x-13\le 4x-31$
$-2x-13\le-31$
$-2x\le-18$
$x\ge 9$
$[9,\infty)$

14. $13\le 2x-21\le 18$
$34\le 2x\le 39$
$17\le x\le\dfrac{39}{2}$
$\left[17,\dfrac{39}{2}\right]$

15. $-14<3x+10<37$
$-24<3x<27$
$-8<x<9$
$(-8,9)$

16. $6x+13<25$ or $2x-19\ge-7$
$6x<12$ or $2x\ge 12$
$x<2$ or $x\ge 6$
$(-\infty,2)\cup[6,\infty)$

17. $|x|\le 10$
$-10\le x\le 10$
$[-10,10]$

18. $|3x-11|<7$
$-7<3x-11<7$
$4<3x<18$
$\dfrac{4}{3}<x<6$
$\left(\dfrac{4}{3},6\right)$

19. $|2x+1|-9<6$
$|2x+1|<15$
$-15<2x+1<15$
$-16<2x<14$
$-8<x<7$
$(-8,7)$

20. $|x|+4>13$
$|x|>9$
$x<-9$ or $x>9$
$(-\infty,-9)\cup(9,\infty)$

21. $|6x+21|-10 \geq 5$

$\qquad |6x+21| \geq 15$

$6x+21 \leq -15 \quad$ or $\quad 6x+21 \geq 15$

$\qquad 6x \leq -36 \quad$ or $\qquad 6x \geq -6$

$\qquad x \leq -6 \quad$ or $\qquad x \geq -1$

$(-\infty,-6] \cup [-1,\infty)$

22. $|x-8|+13 \geq 8$

$\qquad |x-8| \geq -5$

$\mathbb{R}$, since an absolute value is always greater than a negative number.

$(-\infty,\infty)$

23. $5x+6y=30$

x-intercept: Let $y=0$. $\qquad$ y-intercept: Let $x=0$.

$5x+6(0)=30 \qquad\qquad 5(0)+6y=30$

$\qquad 5x=30 \qquad\qquad\qquad 6y=30$

$\qquad x=6 \qquad\qquad\qquad y=5$

$(6,0) \qquad\qquad\qquad\qquad (0,5)$

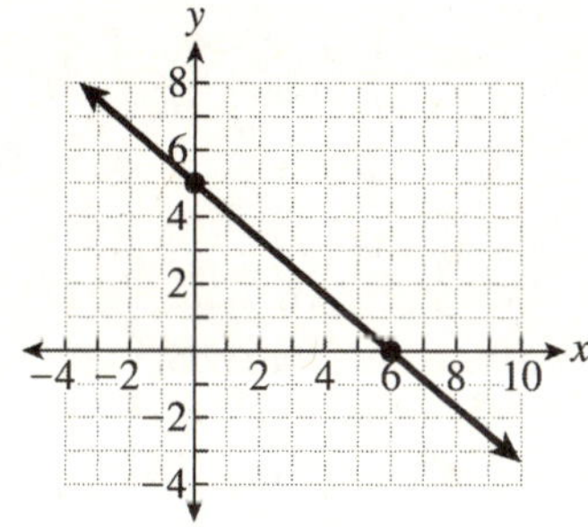

24. $8x-3y=-12$

x-intercept: Let $y=0$. $\qquad$ y-intercept: Let $x=0$.

$8x-3(0)=-12 \qquad\qquad 8(0)-3y=-12$

$\qquad 8x=-12 \qquad\qquad\qquad -3y=-12$

$\qquad\qquad\qquad\qquad\qquad\qquad y=4$

$\qquad x=\dfrac{-12}{8} \qquad\qquad (0,4)$

$\qquad x=-\dfrac{3}{2}$

$\left(-\dfrac{3}{2},0\right)$

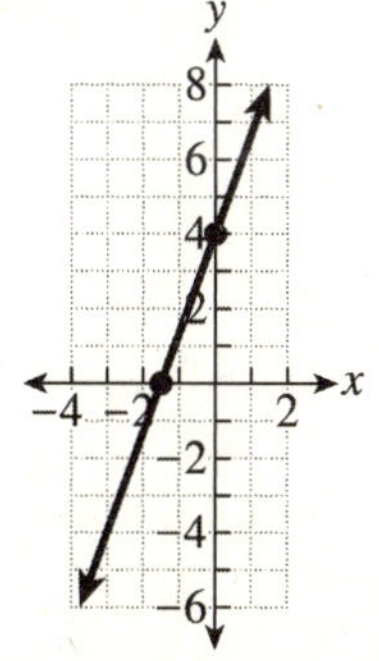

25. $y=-2x+9$

y-intercept: $(0,9)$ using $y=mx+b$

2nd point: $(1,7)$ using slope $=-2$

(down 2, right 1)

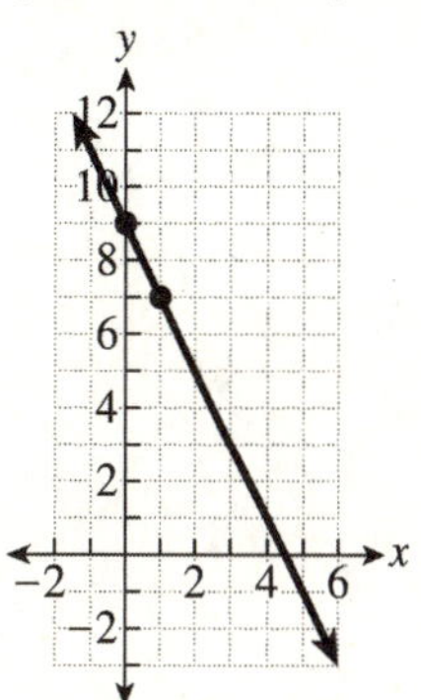

26. $y=\dfrac{2}{3}x+6$

y-intercept: $(0,6)$ using $y=mx+b$

2nd point: $(3,8)$ using slope $=\dfrac{2}{3}$

(up 2, right 3)

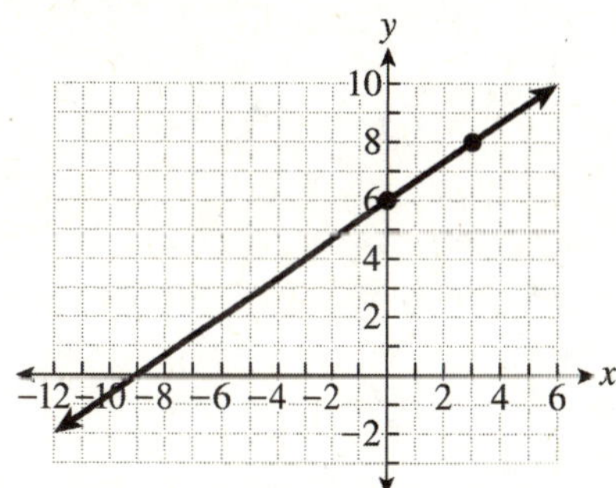

27. $f(x)=-x+5$

y-intercept: $(0,5)$ using $f(x)=mx+b$

2nd point: $(1,4)$ using slope $=-1$

(down 1, right 1)

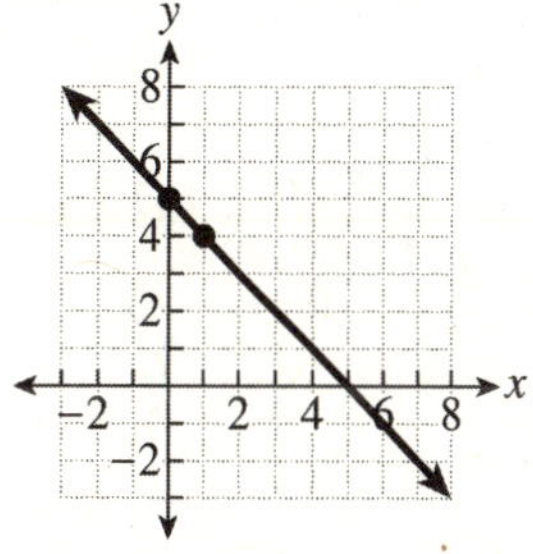

28. $f(x) = 3x$

y-intercept: $(0,0)$ using $f(x) = mx + b$
2nd point: $(1,3)$ using slope $= 3$
(up 3, right 1)

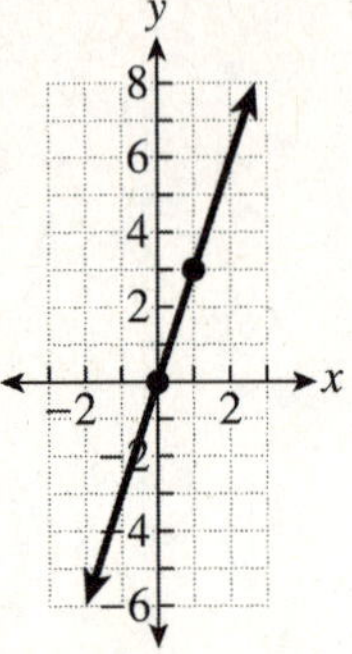

29. $f(x) = \dfrac{3}{5}x - 6$

y-intercept: $(0,-6)$ using $f(x) = mx + b$
2nd point: $(5,-3)$ using slope $= \dfrac{3}{5}$
(up 3, right 5)

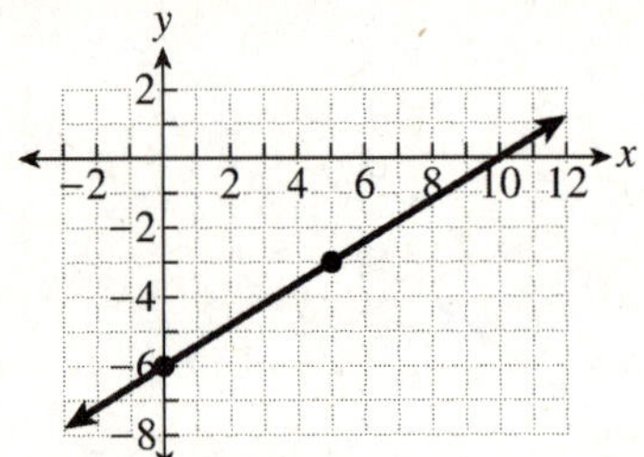

30. $f(x) = -\dfrac{1}{6}x + 1$

y-intercept: $(0,1)$ using $f(x) = mx + b$
2nd point: $(6,0)$ using slope $= -\dfrac{1}{6}$
(down 1, right 6)

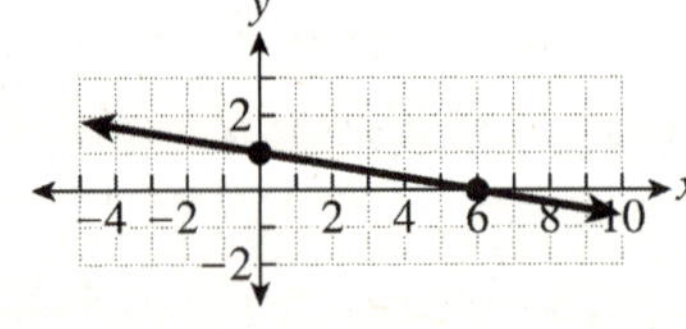

31.

x	$f(x) = |x| + 3$	$(x, f(x))$
-2	$f(-2) = |-2| + 3 = 5$	$(-2,5)$
-1	$f(-1) = |-1| + 3 = 4$	$(-1,4)$
0	$f(0) = |0| + 3 = 3$	$(0,3)$
1	$f(1) = |1| + 3 = 4$	$(1,4)$
2	$f(2) = |2| + 3 = 5$	$(2,5)$

Domain: $(-\infty, \infty)$, Range: $[3, \infty)$

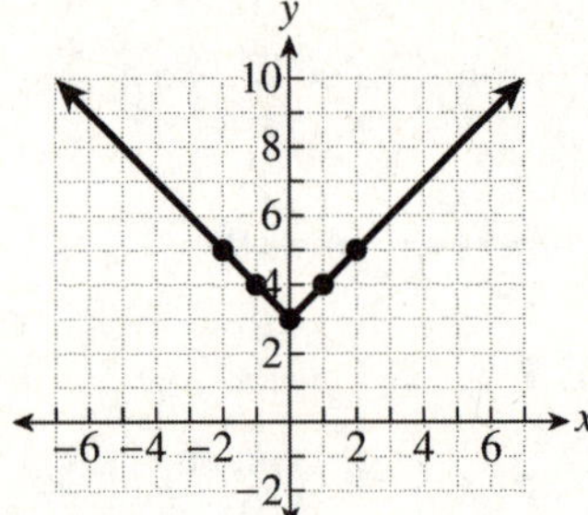

32. $x - 5 = 0$
$\quad\;\; x = 5$

x	$f(x) = |x - 5|$	$(x, f(x))$
3	$f(3) = |3 - 5| = 2$	$(3,2)$
4	$f(4) = |4 - 5| = 1$	$(4,1)$
5	$f(5) = |5 - 5| = 0$	$(5,0)$
6	$f(6) = |6 - 5| = 1$	$(6,1)$
7	$f(7) = |7 - 5| = 2$	$(7,2)$

Domain: $(-\infty, \infty)$, Range: $[0, \infty)$

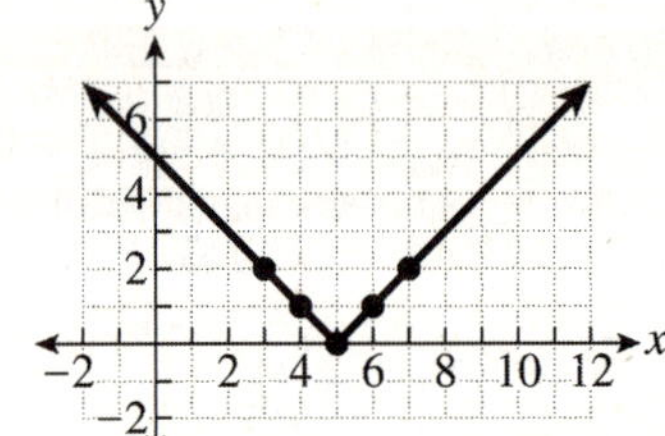

33. $x + 2 = 0$
  $x = -2$

x	$f(x) = |x+2| - 4$	$(x, f(x))$
-4	$f(-4) = |-4+2| - 4 = -2$	$(-4, -2)$
-3	$f(-3) = |-3+2| - 4 = -3$	$(-3, -3)$
-2	$f(-2) = |-2+2| - 4 = -4$	$(-2, -4)$
-1	$f(-1) = |-1+2| - 4 = -3$	$(-1, -3)$
0	$f(0) = |0+2| - 4 = -2$	$(0, -2)$

Domain: $(-\infty, \infty)$, Range: $[-4, \infty)$

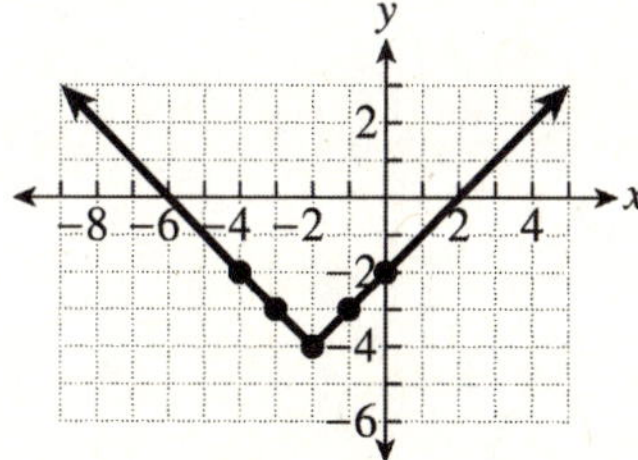

34. $x - 1 = 0$
  $x = 1$

x	$f(x) = -|x-1| + 2$	$(x, f(x))$
-1	$f(-1) = -|-1-1| + 2 = 0$	$(-1, 0)$
0	$f(0) = -|0-1| + 2 = 1$	$(0, 1)$
1	$f(1) = -|1-1| + 2 = 2$	$(1, 2)$
2	$f(2) = -|2-1| + 2 = 1$	$(2, 1)$
3	$f(3) = -|3-1| + 2 = 0$	$(3, 0)$

Domain: $(-\infty, \infty)$, Range: $(-\infty, 2]$

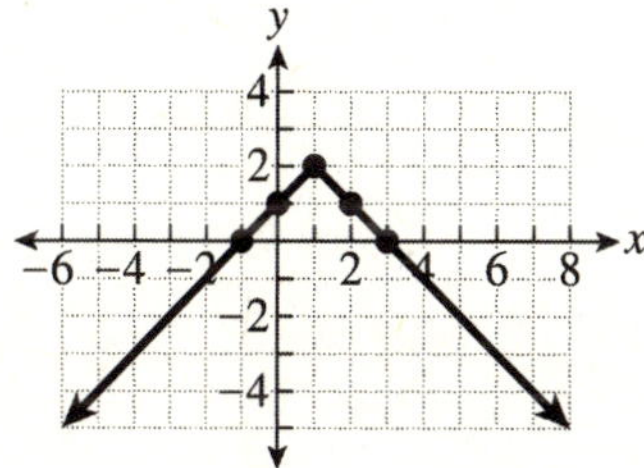

35. $6x^9 - 4x^7 + 10x^3 - 18x^2 + 2x$
$= 2x\left(3x^8 - 2x^6 + 5x^2 - 9x + 1\right)$

36. $5a^4b^3 - 10a^2b^5 - 25a^3b$
$= 5a^2b\left(a^2b^2 - 2b^4 - 5a\right)$

37. $x^3 - 8x^2 + 11x - 88 = x^2(x-8) + 11(x-8)$
$= (x-8)\left(x^2 + 11\right)$

38. $x^3 + 9x^2 - 4x - 36 = x^2(x+9) - 4(x+9)$
$= (x+9)\left(x^2 - 4\right)$
$= (x+9)(x+2)(x-2)$

39. $x^2 - 5x - 14 = (x-7)(x+2)$

40. $3x^2 - 33x + 72 = 3\left(x^2 - 11x + 24\right)$
$= 3(x-3)(x-8)$

41. $x^2 + 18xy + 80y^2 = (x+8y)(x+10y)$

42. $x^2 + 13x - 30 = (x+15)(x-2)$

43. $4x^2 + 20x - 24 = 4\left(x^2 + 5x - 6\right)$
$= 4(x+6)(x-1)$

44. $x^2 - 3x - 54 = (x-9)(x+6)$

45. $6x^2 - 29x + 9 = (2x-9)(3x-1)$

46. $4x^2 - 8x - 21 = (2x+3)(2x-7)$

47. $x^2 - 36 = (x)^2 - (6)^2 = (x+6)(x-6)$

48. $9x^2 - 1 = (3x)^2 - (1)^2 = (3x+1)(3x-1)$

49. $4x^2 - 49 = (2x)^2 - (7)^2 = (2x+7)(2x-7)$

50. $25x^2 - 64y^2 = (5x)^2 - (8y)^2$
$= (5x+8y)(5x-8y)$

51. $x^3 + 27 = (x)^3 + (3)^3$
$= (x+3)(x \cdot x - x \cdot 3 + 3 \cdot 3)$
$= (x+3)\left(x^2 - 3x + 9\right)$

52. $x^3 - 729 = (x)^3 - (9)^3$
$= (x-9)(x \cdot x + x \cdot 9 + 9 \cdot 9)$
$= (x-9)\left(x^2 + 9x + 81\right)$

53.
$$x^2 - x - 6 = 0$$
$$(x+2)(x-3) = 0$$
$$x+2 = 0 \quad \text{or} \quad x-3 = 0$$
$$x = -2 \quad \text{or} \quad x = 3$$
$$\{-2, 3\}$$

54.
$$x^2 + 10x + 24 = 0$$
$$(x+4)(x+6) = 0$$
$$x+4 = 0 \quad \text{or} \quad x+6 = 0$$
$$x = -4 \quad \text{or} \quad x = -6$$
$$\{-4, -6\}$$

55.
$$x^2 + 13x + 25 = 6x + 43$$
$$x^2 + 7x - 18 = 0$$
$$(x+9)(x-2) = 0$$
$$x+9 = 0 \quad \text{or} \quad x-2 = 0$$
$$x = -9 \quad \text{or} \quad x = 2$$
$$\{-9, 2\}$$

56.
$$x^2 = 144$$
$$x^2 - 144 = 0$$
$$(x+12)(x-12) = 0$$
$$x+12 = 0 \quad \text{or} \quad x-12 = 0$$
$$x = -12 \quad \text{or} \quad x = 12$$
$$\{-12, 12\}$$

57.
$$\frac{9}{x+2} = \frac{3}{x-8}$$
$$(x+2)(x-8) \cdot \frac{9}{(x+2)} = (x+2)(x-8) \cdot \frac{3}{(x-8)}$$
$$9(x-8) = 3(x+2)$$
$$9x - 72 = 3x + 6$$
$$6x - 72 = 6$$
$$6x = 78$$
$$x = 13$$
$$\{13\}$$

58.
$$x - 5 - \frac{24}{x} = 0$$
$$x \cdot \left(x - 5 - \frac{24}{x} \right) = x \cdot 0$$
$$x \cdot x - x \cdot 5 - x \cdot \frac{24}{x} = x \cdot 0$$
$$x^2 - 5x - 24 = 0$$
$$(x+3)(x-8) = 0$$
$$x+3 = 0 \quad \text{or} \quad x-8 = 0$$
$$x = -3 \quad \text{or} \quad x = 8$$
$$\{-3, 8\}$$

59.
$$\frac{4}{x+5} + \frac{1}{x-4} = \frac{9}{x^2 + x - 20}$$
$$\frac{4}{x+5} + \frac{1}{x-4} = \frac{9}{(x+5)(x-4)}$$
$$(x+5)(x-4) \cdot \left(\frac{4}{x+5} + \frac{1}{x-4} \right) = (x+5)(x-4) \cdot \frac{9}{(x+5)(x-4)}$$
$$(x+5)(x-4) \cdot \frac{4}{(x+5)} + (x+5)(x-4) \cdot \frac{1}{(x-4)} = (x+5)(x-4) \cdot \frac{9}{(x+5)(x-4)}$$
$$4(x-4) + 1(x+5) = 9$$
$$4x - 16 + x + 5 = 9$$
$$5x - 11 = 9$$
$$5x = 20$$
$$x = 4$$

$\varnothing$ since $x = 4$ makes the original equation undefined.

60.

$$\frac{x-2}{x+3} - \frac{3}{x+6} = \frac{9}{x^2+9x+18}$$

$$\frac{x-2}{x+3} - \frac{3}{x+6} = \frac{9}{(x+3)(x+6)}$$

$$(x+3)(x+6)\cdot\left(\frac{x-2}{x+3} - \frac{3}{x+6}\right) = (x+3)(x+6)\cdot\frac{9}{(x+3)(x+6)}$$

$$(x+3)^1(x+6)\cdot\frac{x-2}{(x+3)_1} - (x+3)(x+6)^1\cdot\frac{3}{(x+6)_1} = (x+3)^1(x+6)^1\cdot\frac{9}{(x+3)_1(x+6)_1}$$

$$(x+6)(x-2) - 3(x+3) = 9$$
$$x^2 - 2x + 6x - 12 - 3x - 9 = 9$$
$$x^2 + x - 21 = 9$$
$$x^2 + x - 30 = 0$$
$$(x+6)(x-5) = 0$$
$$x+6 = 0 \quad \text{or} \quad x-5 = 0$$
$$x = -6 \quad \text{or} \quad x = 5$$

$\{5\}$, since $x = -6$ makes the original equation undefined.

61. $2x + y = 7$
  $\quad\quad y = -2x + 7$

Substitute $y = -2x + 7$ into Eq. 2.
$3x + 4(-2x + 7) = 8$
$\quad 3x - 8x + 28 = 8$
$\quad\quad\quad -5x = -20$
$\quad\quad\quad\quad x = 4$

Substitute $x = 4$ into Eq. 1.
$y = -2(4) + 7 = -8 + 7 = -1$
$(4, -1)$

62. Substitute $x = 7 - 4y$ into Eq. 2.
$6(7 - 4y) - 5y = -45$
$\quad 42 - 24y - 5y = -45$
$\quad\quad 42 - 29y = -45$
$\quad\quad\quad -29y = -87$
$\quad\quad\quad\quad y = 3$

Substitute $y = 3$ into Eq. 1.
$x = 7 - 4(3) = 7 - 12 = -5$
$(-5, 3)$

63. $\quad 2x + 3y = 23 \quad$ (Eq. 1)
  $\quad -6x + 2y = 8 \quad$ (Eq. 2)

$\quad\quad 6x + 9y = 69 \quad (3\cdot\text{Eq. 1})$
$+ \quad -6x + 2y = 8 \quad\quad (\text{Eq. 2})$
$\quad\quad\quad\quad 11y = 77$
$\quad\quad\quad\quad\quad y = 7$

Substitute $y = 7$ into Eq. 1.
$2x + 3(7) = 23$
$\quad 2x + 21 = 23$
$\quad\quad 2x = 2$
$\quad\quad\quad x = 1$

$(1, 7)$

64. $3x + 5y = 19 \quad$ (Eq. 1)
  $\quad 5x - 4y = -67 \quad$ (Eq. 2)

$\quad\quad 12x + 20y = 76 \quad (4\cdot\text{Eq. 1})$
$+ \quad 25x - 20y = -335 \quad (5\cdot\text{Eq. 2})$
$\quad\quad\quad 37x = -259$
$\quad\quad\quad\quad x = -7$

Substitute $x = -7$ into Eq. 1.
$3(-7) + 5y = 19$
$\quad -21 + 5y = 19$
$\quad\quad\quad 5y = 40$
$\quad\quad\quad\quad y = 8$

$(-7, 8)$

65.
$$x - 4y + 2z = -15 \quad \text{(Eq. 1)}$$
$$2x + 3y = 22 \quad \text{(Eq. 2)}$$
$$y = 4 \quad \text{(Eq. 3)}$$
Substitute $y = 4$ into Eq. 2.
$$2x + 3(4) = 22$$
$$2x + 12 = 22$$
$$2x = 10$$
$$x = 5$$
Substitute $x = 5$, $y = 4$ into Eq. 1.
$$5 - 4(4) + 2z = -15$$
$$5 - 16 + 2z = -15$$
$$-11 + 2z = -15$$
$$2z = -4$$
$$z = -2$$
$$(5, 4, -2)$$

66.
$$x + y - z = -9 \quad \text{(Eq. 1)}$$
$$3x + 5y + 6z = -7 \quad \text{(Eq. 2)}$$
$$4y + 7z = -4 \quad \text{(Eq. 3)}$$

$$-3x - 3y + 3z = 27 \quad (-3 \cdot \text{Eq. 1})$$
$$+ \quad \underline{3x + 5y + 6z = -7 \quad \text{(Eq. 2)}}$$
$$2y + 9z = 20 \quad \text{(Eq. 4)}$$

$$-4y - 18z = -40 \quad (-2 \cdot \text{Eq. 4})$$
$$+ \quad \underline{4y + 7z = -4 \quad \text{(Eq. 3)}}$$
$$-11z = -44$$
$$z = 4$$
Substitute $z = 4$ into Eq. 3.
$$4y + 7(4) = -4$$
$$4y + 28 = -4$$
$$4y = -32$$
$$y = -8$$
Substitute $y = -8$, $z = 4$ into Eq. 1.
$$x + (-8) - 4 = -9$$
$$x - 12 = -9$$
$$x = 3$$
$$(3, -8, 4)$$

67.
$$x + y + z = 0 \quad \text{(Eq. 1)}$$
$$2x + 3y - z = 24 \quad \text{(Eq. 2)}$$
$$3x - 2y + 4z = -36 \quad \text{(Eq. 3)}$$

$$x + y + z = 0 \quad \text{(Eq. 1)}$$
$$+ \quad \underline{2x + 3y - z = 24 \quad \text{(Eq. 2)}}$$
$$3x + 4y = 24 \quad \text{(Eq. 4)}$$

$$-4x - 4y - 4z = 0 \quad (-4 \cdot \text{Eq. 1})$$
$$+ \quad \underline{3x - 2y + 4z = -36 \quad \text{(Eq. 3)}}$$
$$-x - 6y = -36 \quad \text{(Eq. 5)}$$

$$3x + 4y = 24 \quad \text{(Eq. 4)}$$
$$+ \quad \underline{-3x - 18y = -108 \quad (3 \cdot \text{Eq. 5})}$$
$$-14y = -84$$
$$y = 6$$

Substitute $y = 6$ into Eq. 4.
$$3x + 4(6) = 24$$
$$3x + 24 = 24$$
$$3x = 0$$
$$x = 0$$
Substitute $x = 0$, $y = 6$ into Eq. 1.
$$0 + 6 + z = 0$$
$$z = -6$$
$$(0, 6, -6)$$

68.
$$x + 3y - 4z = -13 \quad \text{(Eq. 1)}$$
$$-3x + 9y + 8z = -5 \quad \text{(Eq. 2)}$$
$$-5x + 6y + 6z = -19 \quad \text{(Eq. 3)}$$

$$3x + 9y - 12z = -39 \quad (3 \cdot \text{Eq. 1})$$
$$+ \quad \underline{-3x + 9y + 8z = -5 \quad \text{(Eq. 2)}}$$
$$18y - 4z = -44 \quad \text{(Eq. 4)}$$

$$5x + 15y - 20z = -65 \quad (5 \cdot \text{Eq. 1})$$
$$+ \quad \underline{-5x + 6y + 6z = -19 \quad \text{(Eq. 3)}}$$
$$21y - 14z = -84 \quad \text{(Eq. 5)}$$

$$126y - 28z = -308 \quad (7 \cdot \text{Eq. 4})$$
$$+ \quad \underline{-42y + 28z = 168 \quad (-2 \cdot \text{Eq. 5})}$$
$$84y = -140$$
$$y = \frac{-140}{84}$$
$$y = -\frac{5}{3}$$

Substitute $y = -\frac{5}{3}$ into Eq. 4.
$$18\left(-\frac{5}{3}\right) - 4z = -44$$
$$-30 - 4z = -44$$
$$-4z = -14$$
$$z = \frac{-14}{-4}$$
$$z = \frac{7}{2}$$

Substitute $y = -\frac{5}{3}$, $z = \frac{7}{2}$ into Eq. 1.
$$x + 3\left(-\frac{5}{3}\right) - 4\left(\frac{7}{2}\right) = -13$$
$$x - 5 - 14 = -13$$
$$x - 19 = -13$$
$$x = 6$$
$$\left(6, -\frac{5}{3}, \frac{7}{2}\right)$$

69. Unknowns:
x = number of children
y = number of seniors
z = number of general admission

$$\begin{aligned}
x + y + z &= 200 \quad &(\text{Eq. 1})\\
8x + 10y + 15z &= 2540 \quad &(\text{Eq. 2})\\
-x + y &= 20 \quad &(\text{Eq. 3})
\end{aligned}$$

$$\begin{aligned}
-15x - 15y - 15z &= -3000 \quad (-15 \cdot \text{Eq. 1})\\
+\quad 8x + 10y + 15z &= 2540 \quad\quad (\text{Eq. 2})\\
\hline
-7x - 5y &= -460 \quad\quad (\text{Eq. 4})
\end{aligned}$$

$$\begin{aligned}
7x - 7y &= -140 \quad (-7 \cdot \text{Eq. 3})\\
+\quad -7x - 5y &= -460 \quad\quad (\text{Eq. 4})\\
\hline
-12y &= -600\\
y &= 50
\end{aligned}$$

50 seniors played.

70. Unknowns:
A = measure of angle A
B = measure of angle B
C = measure of angle C

$$\begin{aligned}
A + B + C &= 180 \quad (\text{Eq. 1})\\
A - B &= 24 \quad (\text{Eq. 2})\\
A + B - 2C &= 0 \quad (\text{Eq. 3})
\end{aligned}$$

$$\begin{aligned}
2A + 2B + 2C &= 360 \quad (2 \cdot \text{Eq. 1})\\
+\quad A + B - 2C &= 0 \quad\quad (\text{Eq. 3})\\
\hline
3A + 3B &= 360 \quad (\text{Eq. 4})
\end{aligned}$$

$$\begin{aligned}
3A - 3B &= 72 \quad (3 \cdot \text{Eq. 2})\\
+\quad 3A + 3B &= 360 \quad (\text{Eq. 3})\\
\hline
6A &= 432\\
A &= 72
\end{aligned}$$

Substitute $A = 72$ into Eq. 2.
$$\begin{aligned}
72 - B &= 24\\
-B &= -48\\
B &= 48
\end{aligned}$$

Substitute $A = 72, B = 48$ into Eq. 1.
$$\begin{aligned}
72 + 48 + C &= 180\\
120 + C &= 180\\
C &= 60
\end{aligned}$$

Measure of angle $A = 72°$, $B = 48°$, $C = 60°$.

CHAPTER 8 TEST

1. $6(2x - 5) + 8 = 5(3x + 7) - 9$
$$\begin{aligned}
12x - 30 + 8 &= 15x + 35 - 9\\
12x - 22 &= 15x + 26\\
-3x - 22 &= 26\\
-3x &= 48\\
x &= -16
\end{aligned}$$
$\{-16\}$

2. $|x + 9| = 7$
$$\begin{aligned}
x + 9 = -7 \quad &\text{or} \quad x + 9 = 7\\
x = -16 \quad &\text{or} \quad x = -2
\end{aligned}$$
$\{-16, -2\}$

3. $|x - 5| - 11 = -3$
$$|x - 5| = 8$$
$$\begin{aligned}
x - 5 = -8 \quad &\text{or} \quad x - 5 = 8\\
x = -3 \quad &\text{or} \quad x = 13
\end{aligned}$$
$\{-3, 13\}$

4. $-5x < 3x + 32$
$$\begin{aligned}
-8x &< 32\\
x &> -4
\end{aligned}$$
$(-4, \infty)$

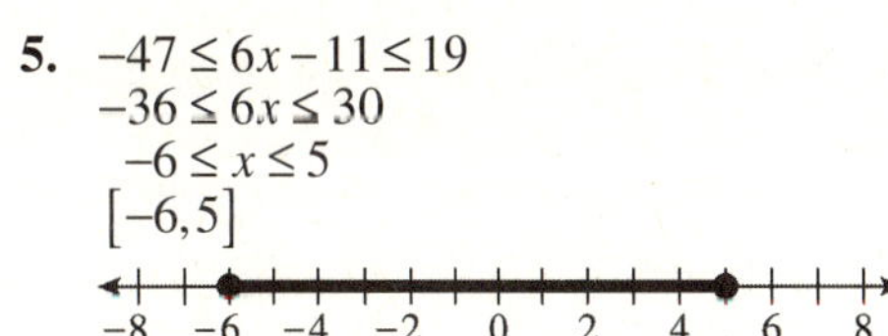

5. $-47 \le 6x - 11 \le 19$
$$\begin{aligned}
-36 &\le 6x \le 30\\
-6 &\le x \le 5
\end{aligned}$$
$[-6, 5]$

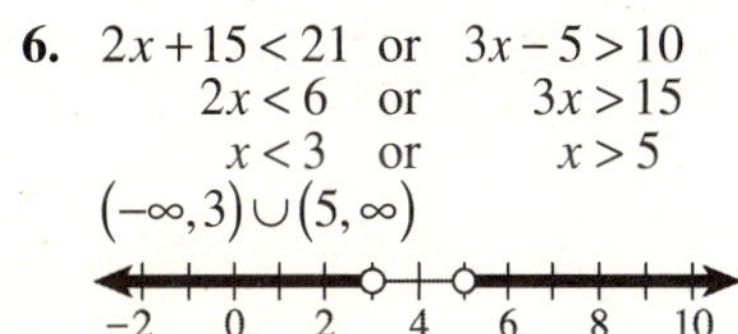

6. $2x + 15 < 21 \quad \text{or} \quad 3x - 5 > 10$
$$\begin{aligned}
2x < 6 \quad &\text{or} \quad 3x > 15\\
x < 3 \quad &\text{or} \quad x > 5
\end{aligned}$$
$(-\infty, 3) \cup (5, \infty)$

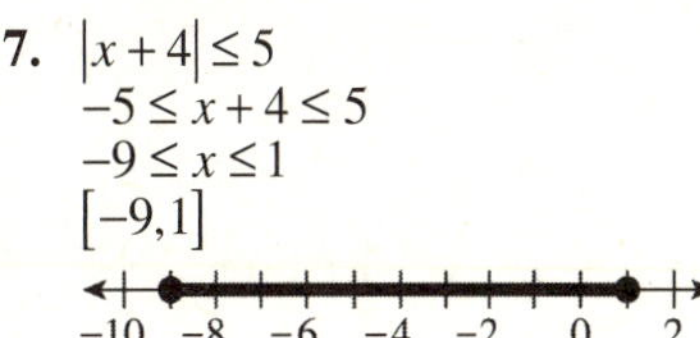

7. $|x + 4| \le 5$
$$\begin{aligned}
-5 &\le x + 4 \le 5\\
-9 &\le x \le 1
\end{aligned}$$
$[-9, 1]$

8. $|2x-5|+8>13$
$|2x-5|>5$

$2x-5<-5$ or $2x-5>5$
$2x<0$ or $2x>10$
$x<0$ or $x>5$
$(-\infty,0)\cup(5,\infty)$

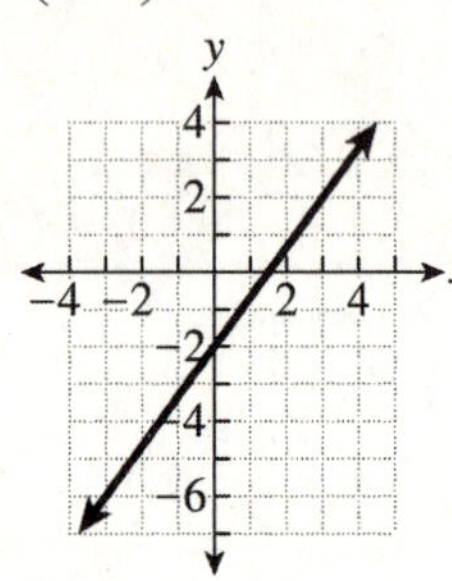

9. $4x-3y=6$

x-intercept: Let $y=0$. y-intercept: Let $x=0$.
$4x-3(0)=6$ $4(0)-3y=6$
$4x=6$ $-3y=6$
                                   $y=-2$
$x=\dfrac{6}{4}$ $(0,-2)$

$x=\dfrac{3}{2}$

$\left(\dfrac{3}{2},0\right)$

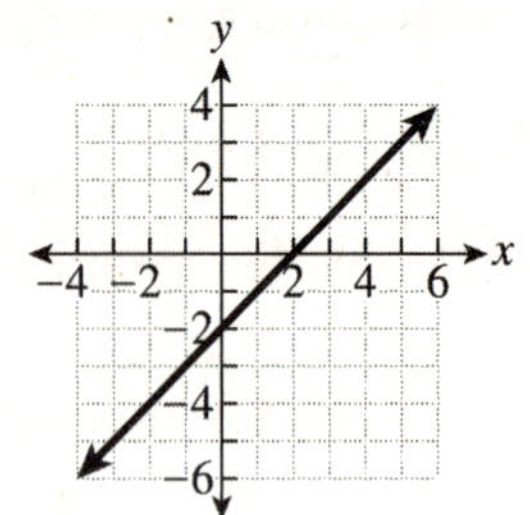

10. $f(x)=x-2$

y-intercept: $(0,-2)$ using $f(x)=mx+b$
2nd point: $(1,-1)$ using slope $=1$
(up 1, right 1)

11. $x-6=0$
$x=6$

x	$f(x)=|x-6|+3$	$(x,f(x))$
4	$f(4)=|4-6|+3=2+3=5$	$(4,5)$
5	$f(5)=|5-6|+3=1+3=4$	$(5,4)$
6	$f(6)=|6-6|+3=0+3=3$	$(6,3)$
7	$f(7)=|7-6|+3=1+3=4$	$(7,4)$
8	$f(8)=|8-6|+3=2+3=5$	$(8,5)$

Domain: $(-\infty,\infty)$, Range: $[3,\infty)$

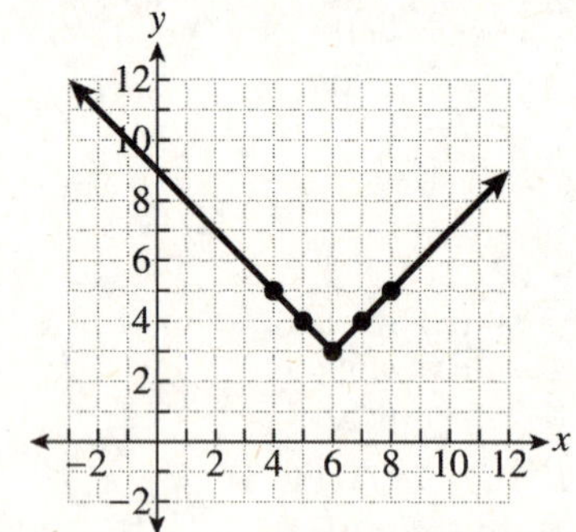

12. $x^2+3x-40=(x+8)(x-5)$

13. $4x^2-28x+40=4\left(x^2-7x+10\right)$
$=4(x-2)(x-5)$

14. $6x^2+19x+10=(2x+5)(3x+2)$

15. $9x^2-100=(3x)^2-(10)^2=(3x+10)(3x-10)$

16. $x^2+6x-27=0$
$(x+9)(x-3)=0$

$x+9=0$ or $x-3=0$
$x=-9$ or $x=3$
$\{-9,3\}$

17.

$$\frac{x+7}{x^2+5x+4}+\frac{6}{x^2-3x-4}=\frac{3}{x^2-16}$$

$$\frac{x+7}{(x+1)(x+4)}+\frac{6}{(x+1)(x-4)}=\frac{3}{(x+4)(x-4)}$$

$$(x+1)(x+4)(x-4)\cdot\left(\frac{x+7}{(x+1)(x+4)}+\frac{6}{(x+1)(x-4)}\right)=(x+1)(x+4)(x-4)\cdot\frac{3}{(x+4)(x-4)}$$

$$(x-4)(x+7)+6(x+4)=3(x+1)$$
$$x^2+7x-4x-28+6x+24=3x+3$$
$$x^2+9x-4=3x+3$$
$$x^2+6x-7=0$$
$$(x+7)(x-1)=0$$
$$x+7=0 \quad \text{or} \quad x-1=0$$
$$x=-7 \quad \text{or} \quad x=1$$

$\{-7,1\}$

18. $5x+\ y=-11$ (Eq. 1)
$4x+3y=0$ (Eq. 2)

Solve for y from Eq. 1.
$5x+y=-11$
$\quad y=-5x-11$ (Eq. 3)

Substitute $y=-5x-11$ into Eq. 2.
$4x+3(-5x-11)=0$
$\quad 4x-15x-33=0$
$\quad\quad -11x-33=0$
$\quad\quad\quad -11x=33$
$\quad\quad\quad\quad x=-3$

Substitute $x=-3$ into Eq. 3.
$y=-5(-3)-11=15-11=4$
$(-3,4)$

19. $x+\ y+\ z=11$ (Eq. 1)
$3x-2y+4z=24$ (Eq. 2)
$-2x+3y-5z=-15$ (Eq. 3)

$\quad -4x-4y-4z=-44$ $(-4\cdot$ Eq. 1)
$+\ \underline{\quad 3x-2y+4z=24}$ (Eq. 2)
$\quad\quad -x-6y=-20$ (Eq. 4)

$\quad 5x+5y+5z=55$ $(5\cdot$ Eq. 1)
$+\ \underline{-2x+3y-5z=-15}$ (Eq. 3)
$\quad\quad 3x+8y=40$ (Eq. 5)

$\quad -3x-18y=-60$ $(3\cdot$ Eq. 4)
$+\ \underline{\quad 3x+\ 8y=40}$ (Eq. 5)
$\quad\quad -10y=-20$
$\quad\quad\quad y=2$

Substitute $y=2$ into Eq. 4.
$-x-6(2)=-20$
$\quad -x-12=-20$
$\quad\quad -x=-8$
$\quad\quad\quad x=8$

Substitute $x=8$, $y=2$ into Eq. 1.
$8+2+z=11$
$\quad 10+z=11$
$\quad\quad z=1$
$(8,2,1)$

20. Unknowns:
A = measure of angle A
B = measure of angle B
C = measure of angle C

$$\begin{aligned}
A + B + C &= 180 \quad &(\text{Eq. 1})\\
B - C &= 25 \quad &(\text{Eq. 2})\\
A - 2B &= 25 \quad &(\text{Eq. 3})
\end{aligned}$$

$$\begin{aligned}
-A - B - C &= -180 \quad &(-1 \cdot \text{Eq. 1})\\
+ \quad A - 2B &= 25 \quad &(\text{Eq. 3})\\
\hline
-3B - C &= -155 \quad &(\text{Eq. 4})
\end{aligned}$$

$$\begin{aligned}
-B + C &= -25 \quad &(-1 \cdot \text{Eq. 2})\\
+ \quad -3B - C &= -155 \quad &(\text{Eq. 4})\\
\hline
-4B &= -180\\
B &= 45
\end{aligned}$$

Substitute $B = 45$ into Eq. 2.
$$\begin{aligned}
45 - C &= 25\\
-C &= -20\\
C &= 20
\end{aligned}$$

Substitute $B = 45, C = 20$ into Eq. 1.
$$\begin{aligned}
A + 45 + 20 &= 180\\
A + 65 &= 180\\
A &= 115
\end{aligned}$$

Measure of angle $A = 115°$, $B = 45°$, $C = 20°$.

CHAPTER 9   RADICAL EXPRESSIONS AND EQUATIONS

9.1   QUICK CHECK

1. a) $\sqrt{49} = 7$

b) $\sqrt{\dfrac{4}{25}} = \dfrac{2}{5}$

c) $-\sqrt{36} = -6$

2. a) $\sqrt{a^{14}} = \sqrt{\left(a^{7}\right)^{2}} = \left|a^{7}\right|$

b) $\sqrt{25b^{22}} = \sqrt{\left(5b^{11}\right)^{2}} = \left|5b^{11}\right| = 5\left|b^{11}\right|$

3. $\sqrt{64x^{24}} = 8x^{12}$

4. $\sqrt{109} \approx 10.440$

5. a) $\sqrt[3]{27} = 3$

b) $\sqrt[4]{256} = 4$

c) $\sqrt[3]{-343} = -7$

6. $\sqrt[6]{x^{42}} = \sqrt[6]{\left(x^{7}\right)^{6}} = x^{7}$

7. $\sqrt[5]{-243x^{35}y^{40}} = \sqrt[5]{\left(-3x^{7}y^{8}\right)^{5}} = -3x^{7}y^{8}$

8. $\sqrt{x^{2}+10x+25} = \sqrt{(x+5)^{2}} = x+5$

9. $\sqrt{6b^{5}} \cdot \sqrt{150b^{3}} = \sqrt{900b^{8}} = 30b^{4}$

10. $3\sqrt{6} \cdot 8\sqrt{6} = 24\sqrt{36} = 24 \cdot 6 = 144$

11. a) $\dfrac{\sqrt{350}}{\sqrt{14}} = \sqrt{\dfrac{350}{14}} = \sqrt{25} = 5$

b) $\dfrac{\sqrt[5]{2916a^{18}}}{\sqrt[5]{12a^{3}}} = \sqrt[5]{\dfrac{2916a^{18}}{12a^{3}}} = \sqrt[5]{243a^{15}} = 3a^{3}$

12. $\begin{aligned} f(-3) &= \sqrt{3(-3)+13} + 33 \\ &= \sqrt{-9+13} + 33 \\ &= \sqrt{4} + 33 \\ &= 2 + 33 \\ &= 35 \end{aligned}$

13. $x+18 \geq 0$
$\qquad x \geq -18$
$[-18, \infty)$

14. Since this radical function is an odd root, its domain is the set of all real numbers.
$(-\infty, \infty)$

9.1   SQUARE ROOTS; RADICAL NOTATION

1. square root

3. nth root

5. radicand

7. radical function

9. $\sqrt{36} = 6$

11. $\sqrt{4} = 2$

13. $\sqrt{\dfrac{1}{81}} = \dfrac{1}{9}$

15. $\sqrt{\dfrac{25}{36}} = \dfrac{5}{6}$

17. $\sqrt{-16}$ is not a real number.

19. $-\sqrt{49} = -7$

21. $\sqrt{a^{14}} = \sqrt{\left(a^{7}\right)^{2}} = \left|a^{7}\right|$

23. $\sqrt{x^{34}} = \sqrt{\left(x^{17}\right)^{2}} = \left|x^{17}\right|$

25. $\sqrt{9x^{6}} = \sqrt{\left(3x^{3}\right)^{2}} = \left|3x^{3}\right| = 3\left|x^{3}\right|$

27. $\sqrt{\dfrac{1}{49}x^4} = \sqrt{\left(\dfrac{1}{7}x^2\right)^2} = \left|\dfrac{1}{7}x^2\right| = \dfrac{1}{7}x^2$

29. $\sqrt{a^{18}b^{22}} = \sqrt{\left(a^9 b^{11}\right)^2} = \left|a^9 b^{11}\right|$

31. $\sqrt{x^8 y^{10} z^{14}} = \sqrt{\left(x^4 y^5 z^7\right)^2}$
$\qquad = \left|x^4 y^5 z^7\right|$
$\qquad = x^4 \left|y^5 z^7\right|$

33. Since $9^2 = 81$, the missing number is 81.

35. Since $(3x)^2 = 9x^2$, the missing expression is $9x^2$.

37. $\sqrt{55} \approx 7.416$

39. $\sqrt{326} \approx 18.055$

41. $\sqrt{0.53} \approx 0.728$

43. $\sqrt[3]{125} = 5$

45. $\sqrt[5]{-243} = -3$

47. $\sqrt[4]{x^{20}} = \sqrt[4]{\left(x^5\right)^4} = x^5$

49. $\sqrt[6]{a^{42}b^{24}} = \sqrt[6]{\left(a^7 b^4\right)^6} = a^7 b^4$

51. $\sqrt[3]{343 m^{21}} = \sqrt[3]{\left(7m^7\right)^3} = 7m^7$

53. $\sqrt[4]{81 x^{12} y^{20}} = \sqrt[4]{\left(3x^3 y^5\right)^4} = 3x^3 y^5$

55. $\sqrt{x^2 + 6x + 9} = \sqrt{(x+3)^2} = x+3$

57. $\sqrt{27} \cdot \sqrt{3} = \sqrt{81} = 9$

59. $\sqrt{10} \cdot \sqrt{90} = \sqrt{900} = 30$

61. $\sqrt{a^9} \cdot \sqrt{a^{17}} = \sqrt{a^{26}} = a^{13}$

63. $\sqrt{x^{19}} \cdot \sqrt{x^{19}} = \sqrt{x^{38}} = x^{19}$

65. $\dfrac{\sqrt{180}}{\sqrt{5}} = \sqrt{\dfrac{180}{5}} = \sqrt{36} = 6$

67. $\dfrac{\sqrt{800}}{\sqrt{8}} = \sqrt{\dfrac{800}{8}} = \sqrt{100} = 10$

69. $\dfrac{\sqrt{x^{23}}}{\sqrt{x^5}} = \sqrt{\dfrac{x^{23}}{x^5}} = \sqrt{x^{18}} = x^9$

71. $\dfrac{\sqrt{a^{13}}}{\sqrt{a}} = \sqrt{\dfrac{a^{13}}{a}} = \sqrt{a^{12}} = a^6$

73. $\sqrt[3]{6} \cdot \sqrt[3]{36} = \sqrt[3]{216} = 6$

75. $\sqrt[4]{48} \cdot \sqrt[4]{27} = \sqrt[4]{1296} = 6$

77. $\sqrt[4]{x^7} \cdot \sqrt[4]{x^{13}} = \sqrt[4]{x^{20}} = x^5$

79. $\sqrt[5]{a^{11}b^8} \cdot \sqrt[5]{a^4 b^2} = \sqrt[5]{a^{15}b^{10}} = a^3 b^2$

81. $\dfrac{\sqrt[4]{324}}{\sqrt[4]{4}} = \sqrt[4]{\dfrac{324}{4}} = \sqrt[4]{81} = 3$

83. $\dfrac{\sqrt[5]{x^{32}}}{\sqrt[5]{x^7}} = \sqrt[5]{\dfrac{x^{32}}{x^7}} = \sqrt[5]{x^{25}} = x^5$

85. $7\sqrt{6} \cdot 8\sqrt{6} = 7 \cdot 8 \cdot \sqrt{6 \cdot 6}$
$\qquad = 56 \cdot \sqrt{36}$
$\qquad = 56 \cdot 6$
$\qquad = 336$

87. $5\sqrt{12} \cdot 10\sqrt{27} = 5 \cdot 10 \cdot \sqrt{12 \cdot 27}$
$\qquad = 50\sqrt{324}$
$\qquad = 50 \cdot 18$
$\qquad = 900$

89. $9\sqrt{x} \cdot 12\sqrt{x} = 9 \cdot 12 \cdot \sqrt{x \cdot x}$
$\qquad = 108\sqrt{x^2}$
$\qquad = 108x$

91. Since $10^2 = 100$, $\sqrt{100} = 10$.
$\sqrt{100} = \sqrt{20 \cdot 5} = \sqrt{20} \cdot \sqrt{5}$,
so the missing number is 5.

93. Since $7^2 = 49$, $\sqrt{49} = 7$.

$$\sqrt{49} = \sqrt{\frac{392}{8}} = \frac{\sqrt{392}}{\sqrt{8}},$$

so the missing number is 392.

95. $f(21) = \sqrt{21-5} = \sqrt{16} = 4$

97. $\begin{aligned} f(3) &= \sqrt{2(3)+19} - 2 \\ &= \sqrt{6+19} - 2 \\ &= \sqrt{25} - 2 \\ &= 5 - 2 \\ &= 3 \end{aligned}$

99. $\begin{aligned} f(10) &= \sqrt{2(10)+9} + 7 \\ &= \sqrt{20+9} + 7 \\ &= \sqrt{29} + 7 \\ &\approx 5.385 + 7 \\ &\approx 12.385 \end{aligned}$

101. $\begin{aligned} f(4) &= \sqrt[3]{(4)^2 + 7(4) + 81} \\ &= \sqrt[3]{16 + 28 + 81} \\ &= \sqrt[3]{125} \\ &= 5 \end{aligned}$

103. $x - 4 \geq 0$
$x \geq 4$
$[4, \infty)$

105. $3x - 8 \geq 0$
$3x \geq 8$
$x \geq \dfrac{8}{3}$
$\left[\dfrac{8}{3}, \infty \right)$

107. $13 - 2x \geq 0$
$-2x \geq -13$
$x \leq \dfrac{13}{2}$
$\left(-\infty, \dfrac{13}{2} \right]$

109. $(-\infty, \infty)$ since the index is odd.

111. Explanations will vary. Example:
$-\sqrt{64}, -\sqrt[3]{64}$, and $\sqrt[3]{-64}$ are real.
The only value that is not real is $\sqrt{-64}$, since there is no real number a such that $a^2 = -64$.

9.2 *QUICK CHECK*

1. a) $36^{1/2} = \sqrt{36} = 6$

b) $\left(-32x^{15}\right)^{1/5} = \sqrt[5]{-32x^{15}} = \sqrt[5]{\left(-2x^3\right)^5} = -2x^3$

2. $\begin{aligned} \left(x^{32}y^4z^{24}\right)^{1/4} &= \sqrt[4]{x^{32}y^4z^{24}} \\ &= \sqrt[4]{\left(x^8yz^6\right)^4} \\ &= x^8yz^6 \end{aligned}$

3. $\sqrt[9]{a} = a^{1/9}$

4. $\begin{aligned} \left(4096x^{18}\right)^{5/6} &= \left(\sqrt[6]{4096x^{18}}\right)^5 \\ &= \left(4x^3\right)^5 \\ &= 1024x^{15} \end{aligned}$

5. $\sqrt[12]{x^5} = x^{5/12}$

6. $x^{5/8} \cdot x^{7/12} = x^{5/8 + 7/12} = x^{15/24 + 14/24} = x^{29/24}$

7. $\left(x^{7/10}\right)^{4/9} = x^{7/10 \cdot 4/9} = x^{14/45}$

8. $\left(\dfrac{x^9}{y^3 z^{12}}\right)^{5/3} = \dfrac{x^{9 \cdot 5/3}}{y^{3 \cdot 5/3} z^{12 \cdot 5/3}} = \dfrac{x^{15}}{y^5 z^{20}}$

9. $81^{-3/2} = \dfrac{1}{81^{3/2}} = \dfrac{1}{\left(\sqrt{81}\right)^3} = \dfrac{1}{9^3} = \dfrac{1}{729}$

10. $\begin{aligned} \sqrt[3]{x} \cdot \sqrt[4]{x} &= x^{1/3} \cdot x^{1/4} \\ &= x^{1/3 + 1/4} \\ &= x^{4/12 + 3/12} \\ &= x^{7/12} \\ &= \sqrt[12]{x^7} \end{aligned}$

9.2 **RATIONAL EXPONENTS**

1. $\sqrt[n]{a}$

3. rational

5. $\sqrt[4]{x} = x^{1/4}$

7. $\sqrt{7} = 7^{1/2}$

9. $9\sqrt[4]{d} = 9d^{1/4}$

11. $64^{1/2} = \sqrt{64} = 8$

13. $(-343)^{1/3} = \sqrt[3]{-343} = -7$

15. $\left(x^{30}\right)^{1/5} = \sqrt[5]{x^{30}} = \sqrt[5]{\left(x^6\right)^5} = x^6$

17. $\left(64a^{33}b^{21}\right)^{1/3} = \sqrt[3]{64a^{33}b^{21}}$
$$= \sqrt[3]{\left(4a^{11}b^7\right)^3}$$
$$4a^{11}b^7$$

19. $\left(-243x^{55}y^{25}z^5\right)^{1/5} = \sqrt[5]{-243x^{55}y^{25}z^5}$
$$= \sqrt[5]{\left(-3x^{11}y^5z\right)^5}$$
$$= -3x^{11}y^5z$$

21. $\left(\sqrt[5]{x}\right)^3 = \left(x^{1/5}\right)^3 = x^{3/5}$

23. $\left(\sqrt[9]{y}\right)^8 = \left(y^{1/9}\right)^8 = y^{8/9}$

25. $\left(\sqrt[3]{3x^2}\right)^7 = \left(3x^2\right)^{7/3}$

27. $\left(\sqrt[8]{10x^4y^5}\right)^3 = \left(10x^4y^5\right)^{3/8}$

29. $25^{3/2} = \left(\sqrt{25}\right)^3 = 5^3 = 125$

31. $32^{2/5} = \left(\sqrt[5]{32}\right)^2 = 2^2 = 4$

33. $\left(x^{12}\right)^{2/3} = \left(\sqrt[3]{x^{12}}\right)^2 = \left(x^4\right)^2 = x^8$

35. $\left(256a^8b^{24}\right)^{3/4} = \left(\sqrt[4]{256a^8b^{24}}\right)^3$
$$= \left(4a^2b^6\right)^3$$
$$= 64a^6b^{18}$$

37. $\left(-125x^9y^{15}z^3\right)^{4/3} = \left(\sqrt[3]{-125x^9y^{15}z^3}\right)^4$
$$= \left(-5x^3y^5z\right)^4$$
$$= 625x^{12}y^{20}z^4$$

39. $x^{7/10} \cdot x^{1/10} = x^{7/10 + 1/10} = x^{8/10} = x^{4/5}$

41. $x^{4/3} \cdot x^{1/2} = x^{4/3 + 1/2} = x^{8/6 + 3/6} = x^{11/6}$

43. $a^{5/3} \cdot a^{3/4} \cdot a^{1/6} = a^{5/3 + 3/4 + 1/6}$
$$= a^{20/12 + 9/12 + 2/12}$$
$$= a^{31/12}$$

45. $x^{1/3}y^{3/4} \cdot x^{1/6}y^{1/6} = x^{1/3}x^{1/6} \cdot y^{3/4}y^{1/6}$
$$= x^{1/3 + 1/6} \cdot y^{3/4 + 1/6}$$
$$= x^{2/6 + 1/6} \cdot y^{9/12 + 2/12}$$
$$= x^{3/6} \cdot y^{11/12}$$
$$= x^{1/2}y^{11/12}$$

47. $\left(x^{3/4}\right)^{2/7} = x^{3/4 \cdot 2/7} = x^{3/14}$

49. $\left(a^{3/5}\right)^{3/5} = a^{3/5 \cdot 3/5} = a^{9/25}$

51. $\left(x^{5/12}\right)^4 = x^{5/12 \cdot 4} = x^{5/3}$

53. $\dfrac{x^{4/5}}{x^{3/10}} = x^{4/5 - 3/10} = x^{8/10 - 3/10} = x^{5/10} = x^{1/2}$

55. $\dfrac{x^{5/8}}{x^{1/3}} = x^{5/8 - 1/3} = x^{15/24 - 8/24} = x^{7/24}$

57. $\left(x^{5/4}\right)^0 = x^{5/4 \cdot 0} = x^0 = 1$

59. $32^{-1/5} = \dfrac{1}{32^{1/5}} = \dfrac{1}{\sqrt[5]{32}} = \dfrac{1}{2}$

61. $27^{-4/3} = \dfrac{1}{27^{4/3}} = \dfrac{1}{\left(\sqrt[3]{27}\right)^4} = \dfrac{1}{3^4} = \dfrac{1}{81}$

63. $216^{-4/3} \cdot 216^{-1/3} = 216^{-4/3 + \left(-1/3\right)}$

$$= 216^{-5/3}$$
$$= \dfrac{1}{216^{5/3}}$$
$$= \dfrac{1}{\left(\sqrt[3]{216}\right)^5}$$
$$= \dfrac{1}{6^5}$$
$$= \dfrac{1}{7776}$$

65. $\dfrac{125^{2/3}}{125^{7/3}} = 125^{2/3 - 7/3}$

$$= 125^{-5/3}$$
$$= \dfrac{1}{125^{5/3}}$$
$$= \dfrac{1}{\left(\sqrt[3]{125}\right)^5}$$
$$= \dfrac{1}{5^5}$$
$$= \dfrac{1}{3125}$$

67. $\dfrac{216^{7/3}}{216^{8/3}} = 216^{7/3 - 8/3}$

$$= 216^{-1/3}$$
$$= \dfrac{1}{216^{1/3}}$$
$$= \dfrac{1}{\sqrt[3]{216}}$$
$$= \dfrac{1}{6}$$

69. $\sqrt[4]{x} \cdot \sqrt[5]{x} = x^{1/4} \cdot x^{1/5}$

$$= x^{1/4 + 1/5}$$
$$= x^{5/20 + 4/20}$$
$$= x^{9/20}$$
$$= \sqrt[20]{x^9}$$

71. $\sqrt[12]{a} \cdot \sqrt[4]{a} = a^{1/12} \cdot a^{1/4}$

$$= a^{1/12 + 3/12}$$
$$= a^{4/12}$$
$$= a^{1/3}$$
$$= \sqrt[3]{a}$$

73. $\dfrac{\sqrt[4]{m}}{\sqrt[12]{m}} = \dfrac{m^{1/4}}{m^{1/12}}$

$$= m^{1/4 - 1/12}$$
$$= m^{3/12 - 1/12}$$
$$= m^{2/12}$$
$$= m^{1/6}$$
$$= \sqrt[6]{m}$$

75. $\dfrac{\sqrt[30]{x}}{\sqrt[5]{x}} = \dfrac{x^{1/30}}{x^{1/5}}$

$$= x^{1/30 - 1/5}$$
$$= x^{1/30 - 6/30}$$
$$= x^{-5/30}$$
$$= x^{-1/6}$$
$$= \dfrac{1}{x^{1/6}}$$
$$= \dfrac{1}{\sqrt[6]{x}}$$

77. Yes; explanations will vary. Example:

The base is positive: $-16^{1/2} = -\sqrt{16} = -4$.

9.3   QUICK CHECK

1. $\sqrt[5]{x^{17}} = \sqrt[5]{x^{15} \cdot x^2} = x^3 \sqrt[5]{x^2}$

2. $\sqrt{a^8 b^{15} c^7 d} = \sqrt{a^8 \left(b^{14} \cdot b\right)\left(c^6 \cdot c\right)d}$
$= a^4 b^7 c^3 \sqrt{bcd}$

3. $\sqrt{90} = \sqrt{9 \cdot 10} = 3\sqrt{10}$

4. $\sqrt[3]{280} = \sqrt[3]{8 \cdot 35} = 2\sqrt[3]{35}$

5. $\sqrt[5]{x^{33} y^6 z^{50} w^{18}} = x^6 y z^{10} w^3 \sqrt[5]{x^3 y w^3}$

6. $9\sqrt{10} - 13\sqrt{5} + 6\sqrt{10} + 8\sqrt{5} = 15\sqrt{10} - 5\sqrt{5}$

7. $16\sqrt[5]{a^4 b^3} - 11\sqrt[7]{a^4 b^3} - 5\sqrt[5]{a^4 b^3}$
$= 11\sqrt[5]{a^4 b^3} - 11\sqrt[7]{a^4 b^3}$

8. $\sqrt{63} + \sqrt{7} = \sqrt{9 \cdot 7} + \sqrt{7} = 3\sqrt{7} + \sqrt{7} = 4\sqrt{7}$

9. $\sqrt{18} - \sqrt{32} + \sqrt{98} = \sqrt{9 \cdot 2} - \sqrt{16 \cdot 2} + \sqrt{49 \cdot 2}$
$= 3\sqrt{2} - 4\sqrt{2} + 7\sqrt{2}$
$= 6\sqrt{2}$

9.3   SIMPLIFYING, ADDING, AND SUBTRACTING RADICAL EXPRESSIONS

1. like radicals

3. $\sqrt{12} = \sqrt{4 \cdot 3} = 2\sqrt{3}$

5. $\sqrt{45} = \sqrt{9 \cdot 5} = 3\sqrt{5}$

7. $\sqrt{432} = \sqrt{144 \cdot 3} = 12\sqrt{3}$

9. $\sqrt[3]{192} = \sqrt[3]{64 \cdot 3} = 4\sqrt[3]{3}$

11. $\sqrt[3]{1296} = \sqrt[3]{216 \cdot 6} = 6\sqrt[3]{6}$

13. $\sqrt[4]{1200} = \sqrt[4]{16 \cdot 75} = 2\sqrt[4]{75}$

15. $\sqrt{x^9} = \sqrt{x^8 \cdot x} = x^4 \sqrt{x}$

17. $\sqrt[3]{m^{13}} = \sqrt[3]{m^{12} \cdot m} = m^4 \sqrt[3]{m}$

19. $\sqrt[5]{x^{74}} = \sqrt[5]{x^{70} \cdot x^4} = x^{14} \sqrt[5]{x^4}$

21. $\sqrt{x^{15} y^{12}} = \sqrt{\left(x^{14} \cdot x\right) y^{12}} = x^7 y^6 \sqrt{x}$

23. $\sqrt{x y^7} = \sqrt{x\left(y^6 \cdot y\right)} = y^3 \sqrt{xy}$

25. $\sqrt{x^{33} y^{17} z^{16}} = \sqrt{\left(x^{32} \cdot x\right)\left(y^{16} \cdot y\right) z^{16}}$
$= x^{16} y^8 z^8 \sqrt{xy}$

27. $\sqrt[3]{x^{13} y^{11} z} = \sqrt[3]{\left(x^{12} \cdot x\right)\left(y^9 \cdot y^2\right) z} = x^4 y^3 \sqrt[3]{x y^2 z}$

29. $\sqrt[4]{a^{14} b^{35} c^9} = \sqrt[4]{\left(a^{12} \cdot a^2\right)\left(b^{32} \cdot b^3\right)\left(c^8 \cdot c\right)}$
$= a^3 b^8 c^2 \sqrt[4]{a^2 b^3 c}$

31. $\sqrt{27 x^{11} y^{12}} = \sqrt{(9 \cdot 3)\left(x^{10} \cdot x\right) y^{12}}$
$= 3 x^5 y^6 \sqrt{3x}$

33. $\sqrt[4]{162 x^{23} y^6 z^{16}}$
$= \sqrt[4]{(81 \cdot 2)\left(x^{20} \cdot x^3\right)\left(y^4 \cdot y^2\right) z^{16}}$
$= 3 x^5 y z^4 \sqrt[4]{2 x^3 y^2}$

35. $10\sqrt{5} + 17\sqrt{5} = 27\sqrt{5}$

37. $9\sqrt[3]{4} - 15\sqrt[3]{4} = -6\sqrt[3]{4}$

39. $10\sqrt{15} - 3\sqrt{15} + 8\sqrt{15} = 15\sqrt{15}$

41. $10\sqrt{7} - 3\sqrt{3} - 18\sqrt{3} - 5\sqrt{7} = 5\sqrt{7} - 21\sqrt{3}$

43. $\left(8\sqrt{2} - 4\sqrt{3}\right) - \left(6\sqrt{2} + 9\sqrt{3}\right)$
$= 8\sqrt{2} - 4\sqrt{3} - 6\sqrt{2} - 9\sqrt{3}$
$= 2\sqrt{2} - 13\sqrt{3}$

45. $5\sqrt{x} - 7\sqrt{x} = -2\sqrt{x}$

47. $3\sqrt[5]{a} + 8\sqrt[5]{a} = 11\sqrt[5]{a}$

49. $16\sqrt{x} - 9\sqrt{x} + 15\sqrt{x} = 22\sqrt{x}$

51. $4\sqrt{x} - 11\sqrt{y} - 6\sqrt{y} + \sqrt{x} = 5\sqrt{x} - 17\sqrt{y}$

53. $8\sqrt[3]{x} + 7\sqrt[4]{x} - 10\sqrt[4]{x} + 13\sqrt[3]{x} = 21\sqrt[3]{x} - 3\sqrt[4]{x}$

55. $\sqrt{50} + \sqrt{128} = \sqrt{25 \cdot 2} + \sqrt{64 \cdot 2}$
$$= 5\sqrt{2} + 8\sqrt{2}$$
$$= 13\sqrt{2}$$

57. $7\sqrt{24} - 9\sqrt{150} = 7\sqrt{4 \cdot 6} - 9\sqrt{25 \cdot 6}$
$$= 7 \cdot 2\sqrt{6} - 9 \cdot 5\sqrt{6}$$
$$= 14\sqrt{6} - 45\sqrt{6}$$
$$= -31\sqrt{6}$$

59. $10\sqrt{3} + 2\sqrt{27} + 5\sqrt{108}$
$$= 10\sqrt{3} + 2\sqrt{9 \cdot 3} + 5\sqrt{36 \cdot 3}$$
$$= 10\sqrt{3} + 2 \cdot 3\sqrt{3} + 5 \cdot 6\sqrt{3}$$
$$= 10\sqrt{3} + 6\sqrt{3} + 30\sqrt{3}$$
$$= 46\sqrt{3}$$

61. $6\sqrt[4]{2} + 7\sqrt[4]{1250} - 2\sqrt[4]{162}$
$$= 6\sqrt[4]{2} + 7\sqrt[4]{625 \cdot 2} - 2\sqrt[4]{81 \cdot 2}$$
$$= 6\sqrt[4]{2} + 7 \cdot 5\sqrt[4]{2} - 2 \cdot 3\sqrt[4]{2}$$
$$= 6\sqrt[4]{2} + 35\sqrt[4]{2} - 6\sqrt[4]{2}$$
$$= 35\sqrt[4]{2}$$

63. $20\sqrt{12} - 9\sqrt{18} - 13\sqrt{147} - 8\sqrt{72}$
$$= 20\sqrt{4 \cdot 3} - 9\sqrt{9 \cdot 2} - 13\sqrt{49 \cdot 3} - 8\sqrt{36 \cdot 2}$$
$$= 20 \cdot 2\sqrt{3} - 9 \cdot 3\sqrt{2} - 13 \cdot 7\sqrt{3} - 8 \cdot 6\sqrt{2}$$
$$= 40\sqrt{3} - 27\sqrt{2} - 91\sqrt{3} - 48\sqrt{2}$$
$$= -51\sqrt{3} - 75\sqrt{2}$$

65. $\left(8\sqrt{3} + 7\sqrt{2}\right) + \left(7\sqrt{3} - 11\sqrt{2}\right) = 15\sqrt{3} - 4\sqrt{2}$

The missing expression is $7\sqrt{3} - 11\sqrt{2}$.

67. Since $3\sqrt{50} - \sqrt{405} = 3\sqrt{25 \cdot 2} - \sqrt{81 \cdot 5}$
$$= 15\sqrt{2} - 9\sqrt{5}$$
and $3\sqrt{98} - 13\sqrt{20} = 3\sqrt{49 \cdot 2} - 13\sqrt{4 \cdot 5}$
$$= 21\sqrt{2} - 26\sqrt{5},$$
then $\left(15\sqrt{2} - 9\sqrt{5}\right) - \left(-6\sqrt{2} + 17\sqrt{5}\right)$
$$= 15\sqrt{2} - 9\sqrt{5} + 6\sqrt{2} - 17\sqrt{5}$$
$$= 21\sqrt{2} - 26\sqrt{5},$$
the missing expression is $-6\sqrt{2} + 17\sqrt{5}$.

69. Answers will vary. Example:
Write the radicand as factors that are perfect squares, if possible. Then take the principal square root of each factor independently.
$$\sqrt{360} = \sqrt{36 \cdot 10} = \sqrt{36} \cdot \sqrt{10} = 6\sqrt{10}$$

9.4 QUICK CHECK

1. a) $\sqrt{45} \cdot \sqrt{80} = \sqrt{3600} = 60$

b) $\sqrt[3]{x^4 y^2 z} \cdot \sqrt[3]{x^8 y z^4} = \sqrt[3]{x^{12} y^3 z^5}$
$$= \sqrt[3]{x^{12} y^3 \left(z^3 \cdot z^2\right)}$$
$$= x^4 y z \sqrt[3]{z^2}$$

c) $4\sqrt{8} \cdot 9\sqrt{6} = 36\sqrt{48}$
$$= 36\sqrt{16 \cdot 3}$$
$$= 36 \cdot 4\sqrt{3}$$
$$= 144\sqrt{3}$$

2. $\sqrt{12}\left(\sqrt{3} + \sqrt{15}\right) = \sqrt{12} \cdot \sqrt{3} + \sqrt{12} \cdot \sqrt{15}$
$$= \sqrt{36} + \sqrt{180}$$
$$= 6 + \sqrt{36 \cdot 5}$$
$$= 6 + 6\sqrt{5}$$

3. $\sqrt{x^9 y^6}\left(\sqrt{x^3 y^6} - \sqrt{x^8 y^7}\right)$
$$= \sqrt{x^9 y^6} \cdot \sqrt{x^3 y^6} - \sqrt{x^9 y^6} \cdot \sqrt{x^8 y^7}$$
$$= \sqrt{x^{12} y^{12}} - \sqrt{x^{17} y^{13}}$$
$$= x^6 y^6 - x^8 y^6 \sqrt{xy}$$

4. a) $\left(5\sqrt{3} + 4\sqrt{2}\right)\left(7\sqrt{3} - 6\sqrt{2}\right)$
$$= 5\sqrt{3} \cdot 7\sqrt{3} - 5\sqrt{3} \cdot 6\sqrt{2}$$
$$+ 4\sqrt{2} \cdot 7\sqrt{3} - 4\sqrt{2} \cdot 6\sqrt{2}$$
$$= 35 \cdot 3 - 30\sqrt{6} + 28\sqrt{6} - 24 \cdot 2$$
$$= 105 - 30\sqrt{6} + 28\sqrt{6} - 48$$
$$= 57 - 2\sqrt{6}$$

b) $\left(\sqrt{7} - \sqrt{10}\right)^2$
$$= \left(\sqrt{7} - \sqrt{10}\right)\left(\sqrt{7} - \sqrt{10}\right)$$
$$= \sqrt{7} \cdot \sqrt{7} - \sqrt{7} \cdot \sqrt{10} - \sqrt{10} \cdot \sqrt{7} + \sqrt{10} \cdot \sqrt{10}$$
$$= 7 - \sqrt{70} - \sqrt{70} + 10$$
$$= 17 - 2\sqrt{70}$$

5. a) $\left(\sqrt{38}-\sqrt{29}\right)\left(\sqrt{38}+\sqrt{29}\right)$

$= \sqrt{38}\cdot\sqrt{38}-\sqrt{29}\cdot\sqrt{29}$

$= 38-29$

$= 9$

b) $\left(8\sqrt{11}-5\sqrt{7}\right)\left(8\sqrt{11}+5\sqrt{7}\right)$

$= 8\sqrt{11}\cdot 8\sqrt{11}-5\sqrt{7}\cdot 5\sqrt{7}$

$= 64\cdot 11-25\cdot 7$

$= 704-175$

$= 529$

6. $\dfrac{\sqrt{70}}{\sqrt{3}} = \dfrac{\sqrt{70}}{\sqrt{3}}\cdot\dfrac{\sqrt{3}}{\sqrt{3}} = \dfrac{\sqrt{210}}{3}$

7. $\dfrac{2}{\sqrt{20}} = \dfrac{2}{\sqrt{20}}\cdot\dfrac{\sqrt{5}}{\sqrt{5}} = \dfrac{2\sqrt{5}}{\sqrt{100}} = \dfrac{2\sqrt{5}}{10} = \dfrac{\sqrt{5}}{5}$

8. $\sqrt{\dfrac{12x^2y^8z^5}{75x^5y^3z^{15}}} = \sqrt{\dfrac{4y^5}{25x^3z^{10}}}$

$= \dfrac{\sqrt{4y^5}}{\sqrt{25x^3z^{10}}}$

$= \dfrac{2y^2\sqrt{y}}{5xz^5\sqrt{x}}$

$= \dfrac{2y^2\sqrt{y}}{5xz^5\sqrt{x}}\cdot\dfrac{\sqrt{x}}{\sqrt{x}}$

$= \dfrac{2y^2\sqrt{xy}}{5x^2z^5}$

9. $\dfrac{\sqrt{15}}{\sqrt{5}+\sqrt{3}} = \dfrac{\sqrt{15}}{\left(\sqrt{5}+\sqrt{3}\right)}\cdot\dfrac{\left(\sqrt{5}-\sqrt{3}\right)}{\left(\sqrt{5}-\sqrt{3}\right)}$

$= \dfrac{\sqrt{15}\cdot\sqrt{5}-\sqrt{15}\cdot\sqrt{3}}{\sqrt{5}\cdot\sqrt{5}-\sqrt{3}\cdot\sqrt{3}}$

$= \dfrac{\sqrt{75}-\sqrt{45}}{5-3}$

$= \dfrac{\sqrt{25\cdot 3}-\sqrt{9\cdot 5}}{2}$

$= \dfrac{5\sqrt{3}-3\sqrt{5}}{2}$

10. $\dfrac{2\sqrt{6}-7\sqrt{2}}{2\sqrt{6}+3\sqrt{2}}$

$= \dfrac{\left(2\sqrt{6}-7\sqrt{2}\right)}{\left(2\sqrt{6}+3\sqrt{2}\right)}\cdot\dfrac{\left(2\sqrt{6}-3\sqrt{2}\right)}{\left(2\sqrt{6}-3\sqrt{2}\right)}$

$= \dfrac{2\sqrt{6}\cdot 2\sqrt{6}-2\sqrt{6}\cdot 3\sqrt{2}-7\sqrt{2}\cdot 2\sqrt{6}+7\sqrt{2}\cdot 3\sqrt{2}}{2\sqrt{6}\cdot 2\sqrt{6}-3\sqrt{2}\cdot 3\sqrt{2}}$

$= \dfrac{4\cdot 6-6\sqrt{12}-14\sqrt{12}+21\cdot 2}{4\cdot 6-9\cdot 2}$

$= \dfrac{24-20\sqrt{12}+42}{24-18}$

$= \dfrac{66-20\sqrt{4\cdot 3}}{6}$

$= \dfrac{66-40\sqrt{3}}{6}$

$= \dfrac{\overset{1}{\cancel{2}}\left(33-20\sqrt{3}\right)}{\underset{3}{\cancel{6}}}$

$= \dfrac{33-20\sqrt{3}}{3}$

9.4 MULTIPLYING AND DIVIDING RADICAL EXPRESSIONS

1. x

3. rationalizing

5. $\sqrt{12}\cdot\sqrt{75} = \sqrt{900} = 30$

7. $4\sqrt{27}\cdot 6\sqrt{3} = 24\sqrt{81} = 24\cdot 9 = 216$

9. $7\sqrt{32}\cdot 5\sqrt{6} = 35\sqrt{192}$

$= 35\sqrt{64\cdot 3}$

$= 35\cdot 8\sqrt{3}$

$= 280\sqrt{3}$

11. $6\sqrt[3]{24}\cdot 2\sqrt[3]{18} = 12\sqrt[3]{432}$

$= 12\sqrt[3]{216\cdot 2}$

$= 12\cdot 6\sqrt[3]{2}$

$= 72\sqrt[3]{2}$

13. $\sqrt{x^{17}}\cdot\sqrt{x^5} = \sqrt{x^{22}} = x^{11}$

15. $\sqrt{6a^7b^6} \cdot \sqrt{10a^5b^9} = \sqrt{60a^{12}b^{15}}$

$$= \sqrt{(4 \cdot 15)a^{12}\left(b^{14} \cdot b\right)}$$

$$= 2a^6b^7\sqrt{15b}$$

17. $\sqrt[4]{x^{15}y^{10}z^5} \cdot \sqrt[4]{x^9y^2z^{18}} = \sqrt[4]{x^{24}y^{12}z^{23}}$

$$= \sqrt[4]{x^{24}y^{12}\left(z^{20} \cdot z^3\right)}$$

$$= x^6y^3z^5\sqrt[4]{z^3}$$

19. $\sqrt{2}\left(\sqrt{8} - \sqrt{6}\right) = \sqrt{2} \cdot \sqrt{8} - \sqrt{2} \cdot \sqrt{6}$

$$= \sqrt{16} - \sqrt{12}$$

$$= 4 - \sqrt{4 \cdot 3}$$

$$= 4 - 2\sqrt{3}$$

21. $\sqrt{98}\left(\sqrt{18} + \sqrt{10}\right) = \sqrt{49 \cdot 2}\left(\sqrt{9 \cdot 2} + \sqrt{10}\right)$

$$= 7\sqrt{2} \cdot 3\sqrt{2} + 7\sqrt{2} \cdot \sqrt{10}$$

$$= 21 \cdot 2 + 7\sqrt{20}$$

$$= 42 + 7\sqrt{4 \cdot 5}$$

$$= 42 + 7 \cdot 2\sqrt{5}$$

$$= 42 + 14\sqrt{5}$$

23. $6\sqrt{3}\left(9\sqrt{6} - 4\sqrt{15}\right) = 6\sqrt{3} \cdot 9\sqrt{6} - 6\sqrt{3} \cdot 4\sqrt{15}$

$$= 54\sqrt{18} - 24\sqrt{45}$$

$$= 54\sqrt{9 \cdot 2} - 24\sqrt{9 \cdot 5}$$

$$= 54 \cdot 3\sqrt{2} - 24 \cdot 3\sqrt{5}$$

$$= 162\sqrt{2} - 72\sqrt{5}$$

25. $\sqrt{m^5n^9}\left(\sqrt{m^3n^4} + \sqrt{m^8n^7}\right)$

$$= \sqrt{m^5n^9} \cdot \sqrt{m^3n^4} + \sqrt{m^5n^9} \cdot \sqrt{m^8n^7}$$

$$= \sqrt{m^8n^{13}} + \sqrt{m^{13}n^{16}}$$

$$= \sqrt{m^8\left(n^{12} \cdot n\right)} + \sqrt{\left(m^{12} \cdot m\right)n^{16}}$$

$$= m^4n^6\sqrt{n} + m^6n^8\sqrt{m}$$

27. $\sqrt[3]{a^7b^{10}c^{13}}\left(\sqrt[3]{a^{11}b^{35}} + \sqrt[3]{a^{23}c^8}\right)$

$$= \sqrt[3]{a^{18}b^{45}c^{13}} + \sqrt[3]{a^{30}b^{10}c^{21}}$$

$$= \sqrt[3]{a^{18}b^{45}\left(c^{12} \cdot c\right)} + \sqrt[3]{a^{30}\left(b^9 \cdot b\right)c^{21}}$$

$$= a^6b^{15}c^4\sqrt[3]{c} + a^{10}b^3c^7\sqrt[3]{b}$$

29. $\left(\sqrt{2} + \sqrt{5}\right)\left(\sqrt{2} - \sqrt{3}\right)$

$$= \sqrt{2 \cdot 2} - \sqrt{2 \cdot 3} + \sqrt{5 \cdot 2} - \sqrt{5 \cdot 3}$$

$$= 2 - \sqrt{6} + \sqrt{10} - \sqrt{15}$$

31. $\left(7\sqrt{2} + 4\sqrt{3}\right)\left(5\sqrt{2} - \sqrt{3}\right)$

$$= 7 \cdot 5\sqrt{2 \cdot 2} - 7\sqrt{2 \cdot 3} + 4 \cdot 5\sqrt{3 \cdot 2} - 4\sqrt{3 \cdot 3}$$

$$= 35 \cdot 2 - 7\sqrt{6} + 20\sqrt{6} - 4 \cdot 3$$

$$= 70 + 13\sqrt{6} - 12$$

$$= 58 + 13\sqrt{6}$$

33. $\left(2\sqrt{7} - 7\sqrt{3}\right)\left(2\sqrt{7} - 6\sqrt{3}\right)$

$$= 2 \cdot 2\sqrt{7 \cdot 7} - 2 \cdot 6\sqrt{7 \cdot 3} - 7 \cdot 2\sqrt{3 \cdot 7} + 7 \cdot 6\sqrt{3 \cdot 3}$$

$$= 4 \cdot 7 - 12\sqrt{21} - 14\sqrt{21} + 42 \cdot 3$$

$$= 28 - 26\sqrt{21} + 126$$

$$= 154 - 26\sqrt{21}$$

35. $\left(6\sqrt{3} + 4\sqrt{5}\right)\left(3\sqrt{5} - 2\sqrt{3}\right)$

$$= 6 \cdot 3\sqrt{3 \cdot 5} - 6 \cdot 2\sqrt{3 \cdot 3} + 4 \cdot 3\sqrt{5 \cdot 5} - 4 \cdot 2\sqrt{5 \cdot 3}$$

$$= 18\sqrt{15} - 12 \cdot 3 + 12 \cdot 5 - 8\sqrt{15}$$

$$= 10\sqrt{15} - 36 + 60$$

$$= 24 + 10\sqrt{15}$$

37. $\left(4\sqrt{5} - 3\sqrt{2}\right)^2$

$$= \left(4\sqrt{5} - 3\sqrt{2}\right)\left(4\sqrt{5} - 3\sqrt{2}\right)$$

$$= 4 \cdot 4\sqrt{5 \cdot 5} - 4 \cdot 3\sqrt{5 \cdot 2} - 3 \cdot 4\sqrt{2 \cdot 5} + 3 \cdot 3\sqrt{2 \cdot 2}$$

$$= 16 \cdot 5 - 12\sqrt{10} - 12\sqrt{10} + 9 \cdot 2$$

$$= 80 - 24\sqrt{10} + 18$$

$$= 98 - 24\sqrt{10}$$

39. $\left(\sqrt{7} - \sqrt{10}\right)\left(\sqrt{7} + \sqrt{10}\right) = \sqrt{7} \cdot \sqrt{7} - \sqrt{10} \cdot \sqrt{10}$

$$= 7 - 10$$

$$= -3$$

41. $\left(9\sqrt{6} + 3\sqrt{8}\right)\left(9\sqrt{6} - 3\sqrt{8}\right)$

$$= 9\sqrt{6} \cdot 9\sqrt{6} - 3\sqrt{8} \cdot 3\sqrt{8}$$

$$= 81 \cdot 6 - 9 \cdot 8$$

$$= 486 - 72$$

$$= 414$$

43. $\left(8-11\sqrt{2}\right)\left(8+11\sqrt{2}\right) = 8\cdot 8 - 11\sqrt{2}\cdot 11\sqrt{2}$
$$= 64 - 121\cdot 2$$
$$= 64 - 242$$
$$= -178$$

45. $\sqrt{\dfrac{12}{x^4 y^6}} = \dfrac{\sqrt{12}}{\sqrt{x^4 y^6}} = \dfrac{\sqrt{4\cdot 3}}{x^2 y^3} = \dfrac{2\sqrt{3}}{x^2 y^3}$

47. $\sqrt{\dfrac{126}{7}} = \sqrt{18} = \sqrt{9\cdot 2} = 3\sqrt{2}$

49. $\sqrt{\dfrac{135}{20}} = \sqrt{\dfrac{27}{4}} = \dfrac{\sqrt{9\cdot 3}}{\sqrt{4}} = \dfrac{3\sqrt{3}}{2}$

51. $\dfrac{\sqrt{8}}{\sqrt{3}} = \dfrac{\sqrt{8}}{\sqrt{3}}\cdot\dfrac{\sqrt{3}}{\sqrt{3}} = \dfrac{\sqrt{24}}{3} = \dfrac{\sqrt{4\cdot 6}}{3} = \dfrac{2\sqrt{6}}{3}$

53. $\dfrac{1}{\sqrt{2}} = \dfrac{1}{\sqrt{2}}\cdot\dfrac{\sqrt{2}}{\sqrt{2}} = \dfrac{\sqrt{2}}{2}$

55. $\sqrt{\dfrac{27}{11a^7}} = \dfrac{\sqrt{27}}{\sqrt{11a^7}}$
$$= \dfrac{\sqrt{9\cdot 3}}{\sqrt{11\left(a^6\cdot a\right)}}$$
$$= \dfrac{3\sqrt{3}}{a^3\sqrt{11a}}\cdot\dfrac{\sqrt{11a}}{\sqrt{11a}}$$
$$= \dfrac{3\sqrt{33a}}{a^3\cdot 11a}$$
$$= \dfrac{3\sqrt{33a}}{11a^4}$$

57. $\sqrt[3]{\dfrac{s^8}{2r^2 t}} = \dfrac{\sqrt[3]{s^8}}{\sqrt[3]{2r^2 t}}$
$$= \dfrac{\sqrt[3]{s^6 s^2}}{\sqrt[3]{2r^2 t}}$$
$$= \dfrac{s^2\sqrt[3]{s^2}}{\sqrt[3]{2r^2 t}}\cdot\dfrac{\sqrt[3]{4rt^2}}{\sqrt[3]{4rt^2}}$$
$$= \dfrac{s^2\sqrt[3]{4rs^2 t^2}}{2rt}$$

59. $\dfrac{14}{\sqrt{12}} = \dfrac{14}{\sqrt{4\cdot 3}} = \dfrac{14}{2\sqrt{3}} = \dfrac{7}{\sqrt{3}}\cdot\dfrac{\sqrt{3}}{\sqrt{3}} = \dfrac{7\sqrt{3}}{3}$

61. $\dfrac{8}{\sqrt{18a}} = \dfrac{8}{\sqrt{9\cdot 2a}}$
$$= \dfrac{8}{3\sqrt{2a}}\cdot\dfrac{\sqrt{2a}}{\sqrt{2a}}$$
$$= \dfrac{8\sqrt{2a}}{3\cdot 2a}$$
$$= \dfrac{4\sqrt{2a}}{3a}$$

63. $\dfrac{x^4 z^8}{\sqrt{x^3 y^5 z^{10}}} = \dfrac{x^4 z^8}{\sqrt{\left(x^2\cdot x\right)\left(y^4\cdot y\right)z^{10}}}$
$$= \dfrac{x^4 z^8}{xy^2 z^5\sqrt{xy}}$$
$$= \dfrac{x^3 z^3}{y^2\sqrt{xy}}\cdot\dfrac{\sqrt{xy}}{\sqrt{xy}}$$
$$= \dfrac{x^3 z^3\sqrt{xy}}{y^2\cdot xy}$$
$$= \dfrac{x^2 z^3\sqrt{xy}}{y^3}$$

65. $\dfrac{9}{\sqrt{13}-\sqrt{7}} = \dfrac{9}{\left(\sqrt{13}-\sqrt{7}\right)}\cdot\dfrac{\left(\sqrt{13}+\sqrt{7}\right)}{\left(\sqrt{13}+\sqrt{7}\right)}$
$$= \dfrac{9\sqrt{13}+9\sqrt{7}}{\sqrt{13}\cdot\sqrt{13}-\sqrt{7}\cdot\sqrt{7}}$$
$$= \dfrac{9\sqrt{13}+9\sqrt{7}}{13-7}$$
$$= \dfrac{\cancel{9}^{\,3}\left(\sqrt{13}+\sqrt{7}\right)}{\cancel{6}_{\,2}}$$
$$= \dfrac{3\left(\sqrt{13}+\sqrt{7}\right)}{2}$$
$$= \dfrac{3\sqrt{13}+3\sqrt{7}}{2}$$

67.
$$\frac{4\sqrt{3}}{\sqrt{3}+\sqrt{11}} = \frac{4\sqrt{3}}{\left(\sqrt{3}+\sqrt{11}\right)} \cdot \frac{\left(\sqrt{3}-\sqrt{11}\right)}{\left(\sqrt{3}-\sqrt{11}\right)}$$

$$= \frac{4\sqrt{3}\cdot\sqrt{3}-4\sqrt{3}\cdot\sqrt{11}}{\sqrt{3}\cdot\sqrt{3}-\sqrt{11}\cdot\sqrt{11}}$$

$$= \frac{4\cdot 3-4\sqrt{33}}{3-11}$$

$$= \frac{12-4\sqrt{33}}{-8}$$

$$= \frac{\overset{1}{\cancel{4}}\left(3-\sqrt{33}\right)}{\underset{-2}{\cancel{-8}}}$$

$$= -\frac{3-\sqrt{33}}{2}$$

69.
$$\frac{6\sqrt{3}}{\sqrt{13}-3} = \frac{6\sqrt{3}}{\left(\sqrt{13}-3\right)} \cdot \frac{\left(\sqrt{13}+3\right)}{\left(\sqrt{13}+3\right)}$$

$$= \frac{6\sqrt{3}\cdot\sqrt{13}+6\sqrt{3}\cdot 3}{\sqrt{13}\cdot\sqrt{13}-3\cdot 3}$$

$$= \frac{6\sqrt{39}+18\sqrt{3}}{13-9}$$

$$= \frac{6\sqrt{39}+18\sqrt{3}}{4}$$

$$= \frac{\overset{1}{\cancel{2}}\left(3\sqrt{39}+9\sqrt{3}\right)}{\underset{2}{\cancel{4}}}$$

$$= \frac{3\sqrt{39}+9\sqrt{3}}{2}$$

71.
$$\frac{5\sqrt{5}-3\sqrt{3}}{4\sqrt{5}+2\sqrt{3}}$$

$$= \frac{\left(5\sqrt{5}-3\sqrt{3}\right)}{\left(4\sqrt{5}+2\sqrt{3}\right)} \cdot \frac{\left(4\sqrt{5}-2\sqrt{3}\right)}{\left(4\sqrt{5}-2\sqrt{3}\right)}$$

$$= \frac{5\sqrt{5}\cdot 4\sqrt{5}-5\sqrt{5}\cdot 2\sqrt{3}-3\sqrt{3}\cdot 4\sqrt{5}+3\sqrt{3}\cdot 2\sqrt{3}}{4\sqrt{5}\cdot 4\sqrt{5}-2\sqrt{3}\cdot 2\sqrt{3}}$$

$$= \frac{20\cdot 5-10\sqrt{15}-12\sqrt{15}+6\cdot 3}{16\cdot 5-4\cdot 3}$$

$$= \frac{100-10\sqrt{15}-12\sqrt{15}+18}{80-12}$$

$$= \frac{118-22\sqrt{15}}{68}$$

$$= \frac{59-11\sqrt{15}}{34}$$

73.
$$\frac{7\sqrt{11}+2\sqrt{2}}{4\sqrt{11}-5\sqrt{2}}$$

$$= \frac{\left(7\sqrt{11}+2\sqrt{2}\right)}{\left(4\sqrt{11}-5\sqrt{2}\right)} \cdot \frac{\left(4\sqrt{11}+5\sqrt{2}\right)}{\left(4\sqrt{11}+5\sqrt{2}\right)}$$

$$= \frac{7\sqrt{11}\cdot 4\sqrt{11}+7\sqrt{11}\cdot 5\sqrt{2}+2\sqrt{2}\cdot 4\sqrt{11}+2\sqrt{2}\cdot 5\sqrt{2}}{4\sqrt{11}\cdot 4\sqrt{11}-5\sqrt{2}\cdot 5\sqrt{2}}$$

$$= \frac{28\cdot 11+35\sqrt{22}+8\sqrt{22}+10\cdot 2}{16\cdot 11-25\cdot 2}$$

$$= \frac{308+35\sqrt{22}+8\sqrt{22}+20}{176-50}$$

$$= \frac{328+43\sqrt{22}}{126}$$

75.
$$\frac{4\sqrt{2}-\sqrt{3}}{2\sqrt{2}-3\sqrt{3}}$$

$$= \frac{\left(4\sqrt{2}-\sqrt{3}\right)}{\left(2\sqrt{2}-3\sqrt{3}\right)} \cdot \frac{\left(2\sqrt{2}+3\sqrt{3}\right)}{\left(2\sqrt{2}+3\sqrt{3}\right)}$$

$$= \frac{4\sqrt{2}\cdot 2\sqrt{2}+4\sqrt{2}\cdot 3\sqrt{3}-\sqrt{3}\cdot 2\sqrt{2}-\sqrt{3}\cdot 3\sqrt{3}}{2\sqrt{2}\cdot 2\sqrt{2}-3\sqrt{3}\cdot 3\sqrt{3}}$$

$$= \frac{8\cdot 2+12\sqrt{6}-2\sqrt{6}-3\cdot 3}{4\cdot 2-9\cdot 3}$$

$$= \frac{16+12\sqrt{6}-2\sqrt{6}-9}{8-27}$$

$$= \frac{7+10\sqrt{6}}{-19}$$

$$= -\frac{7+10\sqrt{6}}{19}$$

77.
$$\sqrt{\frac{90}{98}} = \sqrt{\frac{45}{49}} = \frac{\sqrt{45}}{\sqrt{49}} = \frac{\sqrt{9\cdot 5}}{7} = \frac{3\sqrt{5}}{7}$$

79.
$$\left(8\sqrt{7}+5\sqrt{5}\right)^2$$

$$= \left(8\sqrt{7}+5\sqrt{5}\right)\left(8\sqrt{7}+5\sqrt{5}\right)$$

$$= 8\cdot 8\sqrt{7\cdot 7}+8\cdot 5\sqrt{7\cdot 5}+5\cdot 8\sqrt{5\cdot 7}+5\cdot 5\sqrt{5\cdot 5}$$

$$= 64\cdot 7+40\sqrt{35}+40\sqrt{35}+25\cdot 5$$

$$= 448+80\sqrt{35}+125$$

$$= 573+80\sqrt{35}$$

81. $\sqrt[3]{25a^8bc^7} \cdot \sqrt[3]{25ab^5c^7}$

$= \sqrt[3]{625a^9b^6c^{14}}$

$= \sqrt[3]{(125 \cdot 5)a^9b^6\left(c^{12} \cdot c^2\right)}$

$= 5a^3b^2c^4\sqrt[3]{5c^2}$

83. $\dfrac{10\sqrt{5} - 7\sqrt{7}}{\sqrt{5} + \sqrt{7}}$

$= \dfrac{\left(10\sqrt{5} - 7\sqrt{7}\right)}{\left(\sqrt{5} + \sqrt{7}\right)} \cdot \dfrac{\left(\sqrt{5} - \sqrt{7}\right)}{\left(\sqrt{5} - \sqrt{7}\right)}$

$= \dfrac{10\sqrt{5} \cdot \sqrt{5} - 10\sqrt{5} \cdot \sqrt{7} - 7\sqrt{7} \cdot \sqrt{5} + 7\sqrt{7} \cdot \sqrt{7}}{\sqrt{5} \cdot \sqrt{5} - \sqrt{7} \cdot \sqrt{7}}$

$= \dfrac{10 \cdot 5 - 10\sqrt{35} - 7\sqrt{35} + 7 \cdot 7}{5 - 7}$

$= \dfrac{50 - 17\sqrt{35} + 49}{-2}$

$= \dfrac{99 - 17\sqrt{35}}{-2}$

$= -\dfrac{99 - 17\sqrt{35}}{2}$

85. $\sqrt[3]{1080} = \sqrt[3]{216 \cdot 5} = 6\sqrt[3]{5}$

87. $\left(3\sqrt{10} - 7\sqrt{2}\right)\left(4\sqrt{10} + 9\sqrt{5}\right)$

$= 3 \cdot 4\sqrt{10 \cdot 10} + 3 \cdot 9\sqrt{10 \cdot 5} - 7 \cdot 4\sqrt{2 \cdot 10} - 7 \cdot 9\sqrt{2 \cdot 5}$

$= 12 \cdot 10 + 27\sqrt{50} - 28\sqrt{20} - 63\sqrt{10}$

$= 120 + 27\sqrt{25 \cdot 2} - 28\sqrt{4 \cdot 5} - 63\sqrt{10}$

$= 120 + 135\sqrt{2} - 56\sqrt{5} - 63\sqrt{10}$

89. $-\sqrt{7x^9}\left(2\sqrt{14x} - \sqrt{7x^{13}}\right)$

$= -\sqrt{7x^9} \cdot 2\sqrt{14x} + \sqrt{7x^9} \cdot \sqrt{7x^{13}}$

$= -2\sqrt{98x^{10}} + \sqrt{49x^{22}}$

$= -2\sqrt{49 \cdot 2x^{10}} + \sqrt{49x^{22}}$

$= -14x^5\sqrt{2} + 7x^{11}$

91. $\sqrt[4]{a^{17}b^{34}c^{44}} = \sqrt[4]{\left(a^{16} \cdot a\right)\left(b^{32} \cdot b^2\right)c^{44}}$

$\qquad = a^4b^8c^{11}\sqrt[4]{ab^2}$

93. $\sqrt{\dfrac{a^5b^8}{12c^9}} = \dfrac{\sqrt{\left(a^4 \cdot a\right)b^8}}{\sqrt{4 \cdot 3\left(c^8 \cdot c\right)}}$

$= \dfrac{a^2b^4\sqrt{a}}{2c^4\sqrt{3c}}$

$= \dfrac{a^2b^4\sqrt{a}}{2c^4\sqrt{3c}} \cdot \dfrac{\sqrt{3c}}{\sqrt{3c}}$

$= \dfrac{a^2b^4\sqrt{3ac}}{2c^4 \cdot 3c}$

$= \dfrac{a^2b^4\sqrt{3ac}}{6c^5}$

95. $\dfrac{10x^2}{\sqrt[3]{36x^8}} = \dfrac{10x^2}{\sqrt[3]{36x^8}} \cdot \dfrac{\sqrt[3]{6x}}{\sqrt[3]{6x}}$

$= \dfrac{10x^2\sqrt[3]{6x}}{\sqrt[3]{216x^9}}$

$= \dfrac{10x^2\sqrt[3]{6x}}{6x^3}$

$= \dfrac{5\sqrt[3]{6x}}{3x}$

97. $\left(18\sqrt{7} - \sqrt{11}\right)\left(18\sqrt{7} + \sqrt{11}\right)$

$= 18\sqrt{7} \cdot 18\sqrt{7} - \sqrt{11} \cdot \sqrt{11}$

$= 324 \cdot 7 - 11$

$= 2268 - 11$

$= 2257$

99. $\left(\sqrt{a} + \sqrt{b}\right)^2$

$= \left(\sqrt{a} + \sqrt{b}\right)\left(\sqrt{a} + \sqrt{b}\right)$

$= \sqrt{a} \cdot \sqrt{a} + \sqrt{a} \cdot \sqrt{b} + \sqrt{b} \cdot \sqrt{a} + \sqrt{b} \cdot \sqrt{b}$

$= a + \sqrt{ab} + \sqrt{ab} + b$

$= a + 2\sqrt{ab} + b$

101. Answers will vary. Example:
If the terms are the same except one of the terms has the opposite sign, then the expressions are conjugates.

9.5 QUICK CHECK

1. a) $\sqrt{x+2}=7$

$$\left(\sqrt{x+2}\right)^2=(7)^2$$
$$x+2=49$$
$$x=47$$
$$\{47\}$$

b) $\sqrt[3]{x+9}+10=6$

$$\sqrt[3]{x+9}=-4$$
$$\left(\sqrt[3]{x+9}\right)^3=(-4)^3$$
$$x+9=-64$$
$$x=-73$$
$$\{-73\}$$

2. $\sqrt{2x-9}-8=-11$

$$\sqrt{2x-9}=-3$$
$$\left(\sqrt{2x-9}\right)^2=(-3)^2$$
$$2x-9=9$$
$$2x=18$$
$$x=9$$

check:
$$\sqrt{2(9)-9}-8=-11$$
$$\sqrt{18-9}-8=-11$$
$$\sqrt{9}-8=-11$$
$$3-8=-11$$
$$-5=-11$$

False
$\varnothing$

3. Let $f(x)=\sqrt{3x-8}+4=9$.

$$\sqrt{3x-8}=5$$
$$\left(\sqrt{3x-8}\right)^2=(5)^2$$
$$3x-8=25$$
$$3x=33$$
$$x=11$$

4. a) $x^{1/2}-10=-7$

$$\sqrt{x}-10=-7$$
$$\sqrt{x}=3$$
$$\left(\sqrt{x}\right)^2=(3)^2$$
$$x=9$$
$$\{9\}$$

b) $(x+4)^{1/3}+8=2$

$$\sqrt[3]{x+4}+8=2$$
$$\sqrt[3]{x+4}=-6$$
$$\left(\sqrt[3]{x+4}\right)^3=(-6)^3$$
$$x+4=-216$$
$$x=-220$$
$$\{-220\}$$

5. $\sqrt{12x-20}=x$

$$\left(\sqrt{12x-20}\right)^2=(x)^2$$
$$12x-20=x^2$$
$$0=x^2-12x+20$$
$$0=(x-2)(x-10)$$
$$x-2=0 \quad\text{or}\quad x-10=0$$
$$x=2 \quad\text{or}\quad x=10$$
$$\{2,10\}$$

6. $\sqrt{2x}+4=x$

$$\sqrt{2x}=x-4$$
$$\left(\sqrt{2x}\right)^2=(x-4)^2$$
$$2x=(x-4)(x-4)$$
$$2x=x^2-8x+16$$
$$0=x^2-10x+16$$
$$0=(x-8)(x-2)$$
$$x-8=0 \quad\text{or}\quad x-2=0$$
$$x=8 \quad\text{or}\quad x=2$$

check: $x=2$
$$\sqrt{2(2)}+4=2$$
$$\sqrt{4}+4=2$$
$$2+4=2$$
$$6=2$$

False
$\{8\}$

7. $\sqrt[3]{5x-11}=\sqrt[3]{7x+33}$

$$\left(\sqrt[3]{5x-11}\right)^3=\left(\sqrt[3]{7x+33}\right)^3$$
$$5x-11=7x+33$$
$$-2x-11=33$$
$$-2x=44$$
$$x=-22$$
$$\{-22\}$$

8. $\sqrt{x+3} - \sqrt{x-2} = 1$

$$\sqrt{x+3} = 1 + \sqrt{x-2}$$
$$\left(\sqrt{x+3}\right)^2 = \left(1+\sqrt{x-2}\right)^2$$
$$x+3 = \left(1+\sqrt{x-2}\right)\left(1+\sqrt{x-2}\right)$$
$$x+3 = 1 + 2\sqrt{x-2} + x - 2$$
$$x+3 = x - 1 + 2\sqrt{x-2}$$
$$3 = -1 + 2\sqrt{x-2}$$
$$4 = 2\sqrt{x-2}$$
$$2 = \sqrt{x-2}$$
$$(2)^2 = \left(\sqrt{x-2}\right)^2$$
$$4 = x - 2$$
$$6 = x$$

$\{6\}$

9. $T = 2\pi\sqrt{\dfrac{L}{32}} = 2\pi\sqrt{\dfrac{6}{32}} = 2\pi\sqrt{\dfrac{3}{16}} \approx 2.72$

2.72 seconds

10.
$$T = 2\pi\sqrt{\dfrac{L}{32}}$$
$$3 = 2\pi\sqrt{\dfrac{L}{32}}$$
$$\dfrac{3}{2\pi} = \sqrt{\dfrac{L}{32}}$$
$$\left(\dfrac{3}{2\pi}\right)^2 = \left(\sqrt{\dfrac{L}{32}}\right)^2$$
$$\dfrac{9}{4\pi^2} = \dfrac{L}{32}$$
$$\overset{8}{\cancel{32}} \cdot \dfrac{9}{\cancel{4}_1\pi^2} = \overset{1}{\cancel{32}} \cdot \dfrac{L}{\cancel{32}_1}$$
$$\dfrac{72}{\pi^2} = L$$
$$L \approx 7.30 \text{ feet}$$

11. $s = \sqrt{30df} = \sqrt{30(215)(0.75)} = \sqrt{4837.5}s \approx 70$

70 miles per hour

9.5   RADICAL EQUATIONS AND APPLICATIONS OF RADICAL EQUATIONS

1. radical

3. isolate

5. pendulum

7.
$$\sqrt{x+3} = 10$$
$$\left(\sqrt{x+3}\right)^2 = (10)^2$$
$$x + 3 = 100$$
$$x = 97$$
check:
$$\sqrt{97+3} = 10$$
$$\sqrt{100} = 10$$
$$10 = 10$$
True
$\{97\}$

9.
$$\sqrt[3]{2x+9} = 5$$
$$\left(\sqrt[3]{2x+9}\right)^3 = (5)^3$$
$$2x + 9 = 125$$
$$2x = 116$$
$$x = 58$$
check:
$$\sqrt[3]{2(58)+9} = 5$$
$$\sqrt[3]{116+9} = 5$$
$$\sqrt[3]{125} = 5$$
$$5 = 5$$
True
$\{58\}$

11.
$$\sqrt{5x+17} = -6$$
$$\left(\sqrt{5x+17}\right)^2 = (-6)^2$$
$$5x + 17 = 36$$
$$5x = 19$$
$$x = \dfrac{19}{5}$$
check:
$$\sqrt{5\left(\dfrac{19}{5}\right)+17} = -6$$
$$\sqrt{19+17} = -6$$
$$\sqrt{36} = -6$$
$$6 = -6$$
False
$\varnothing$

13. $\sqrt{x+5}-7=-2$

$\sqrt{x+5}=5$

$\left(\sqrt{x+5}\right)^2=(5)^2$

$x+5=25$

$x=20$

check:

$\sqrt{20+5}-7=-2$

$\sqrt{25}-7=-2$

$5-7=-2$

$-2=-2$

True

$\{20\}$

15. $\sqrt[3]{2x+7}+3=6$

$\sqrt[3]{2x+7}=3$

$\left(\sqrt[3]{2x+7}\right)^3=(3)^3$

$2x+7=27$

$2x=20$

$x=10$

check:

$\sqrt[3]{2(10)+7}+3=6$

$\sqrt[3]{27}+3=6$

$3+3=6$

$6=6$

True

$\{10\}$

17. $\sqrt{x^2+5x-1}-2=5$

$\sqrt{x^2+5x-1}=7$

$\left(\sqrt{x^2+5x-1}\right)^2=(7)^2$

$x^2+5x-1=49$

$x^2+5x-50=0$

$(x+10)(x-5)=0$

$x+10=0 \quad$ or $\quad x-5=0$

$x=-10 \quad$ or $\qquad x=5$

check $x=-10$:

$\sqrt{(-10)^2+5(-10)-1}-2=5$

$\sqrt{49}-2=5$

$7-2=5$

$5=5$

True

check $x=5$:

$\sqrt{(5)^2+5(5)-1}-2=5$

$\sqrt{49}-2=5$

$7-2=5$

$5=5$

True

$\{-10,5\}$

19. Let $f(x)=12$.

$\sqrt{3x+9}=12$

$\left(\sqrt{3x+9}\right)^2=(12)^2$

$3x+9=144$

$3x=135$

$x=45$

21. Let $f(x)=9$.

$\sqrt[3]{5x-1}+5=9$

$\sqrt[3]{5x-1}=4$

$\left(\sqrt[3]{5x-1}\right)^3=(4)^3$

$5x-1=64$

$5x=65$

$x=13$

23. Let $f(x)=14$.

$\sqrt{x^2-5x+2}+10=14$

$\sqrt{x^2-5x+2}=4$

$\left(\sqrt{x^2-5x+2}\right)^2=(4)^2$

$x^2-5x+2=16$

$x^2-5x-14=0$

$(x+2)(x-7)=0$

$x+2=0 \quad$ or $\quad x-7=0$

$x=-2 \quad$ or $\qquad x=7$

25. $x^{1/2}+8=10$

$\sqrt{x}+8=10$

$\sqrt{x}=2$

$\left(\sqrt{x}\right)^2=(2)^2$

$x=4$

check:

$(4)^{1/2}+8=10$

$\sqrt{4}+8=10$

$2+8=10$

$10=10$

True

$\{4\}$

27.
$$(x+10)^{1/2} - 4 = 3$$
$$\sqrt{x+10} - 4 = 3$$
$$\sqrt{x+10} = 7$$
$$\left(\sqrt{x+10}\right)^2 = (7)^2$$
$$x + 10 = 49$$
$$x = 39$$

check:
$$(39+10)^{1/2} - 4 = 3$$
$$49^{1/2} - 4 = 3$$
$$\sqrt{49} - 4 = 3$$
$$7 - 4 = 3$$
$$3 = 3$$

True
$\{39\}$

29.
$$(5x+6)^{1/3} - 8 = -2$$
$$\sqrt[3]{5x+6} - 8 = -2$$
$$\sqrt[3]{5x+6} = 6$$
$$\left(\sqrt[3]{5x+6}\right)^3 = (6)^3$$
$$5x + 6 = 216$$
$$5x = 210$$
$$x = 42$$

check:
$$(5(42)+6)^{1/3} - 8 = -2$$
$$(210+6)^{1/3} - 8 = -2$$
$$216^{1/3} - 8 = -2$$
$$\sqrt[3]{216} - 8 = -2$$
$$6 - 8 = -2$$
$$-2 = -2$$

True
$\{42\}$

31.
$$x = \sqrt{2x+48}$$
$$(x)^2 = \left(\sqrt{2x+48}\right)^2$$
$$x^2 = 2x + 48$$
$$x^2 - 2x - 48 = 0$$
$$(x-8)(x+6) = 0$$
$$x - 8 = 0 \quad \text{or} \quad x + 6 = 0$$
$$x = 8 \quad \text{or} \quad x = -6$$

check:

$x = 8$	$x = -6$
$8 = \sqrt{2(8)+48}$	$-6 = \sqrt{2(-6)+48}$
$8 = \sqrt{16+48}$	$-6 = \sqrt{-12+48}$
$8 = \sqrt{64}$	$-6 = \sqrt{36}$
$8 = 8$	$-6 = 6$
True	False

$\{8\}$

33.
$$\sqrt{4x+13} = x - 2$$
$$\left(\sqrt{4x+13}\right)^2 = (x-2)^2$$
$$4x + 13 = (x-2)(x-2)$$
$$4x + 13 = x^2 - 4x + 4$$
$$0 = x^2 - 8x - 9$$
$$0 = (x-9)(x+1)$$
$$x - 9 = 0 \quad \text{or} \quad x + 1 = 0$$
$$x = 9 \quad \text{or} \quad x = -1$$

check:

$x = 9$	$x = -1$
$\sqrt{4(9)+13} = 9 - 2$	$\sqrt{4(-1)+13} = -1 - 2$
$\sqrt{49} = 7$	$\sqrt{9} = -3$
$7 = 7$	$3 = -3$
True	False

$\{9\}$

35.
$$\sqrt{2x-5} + 4 = x$$
$$\sqrt{2x-5} = x - 4$$
$$\left(\sqrt{2x-5}\right)^2 = (x-4)^2$$
$$2x - 5 = (x-4)(x-4)$$
$$2x - 5 = x^2 - 8x + 16$$
$$0 = x^2 - 10x + 21$$
$$0 = (x-3)(x-7)$$
$$x - 3 = 0 \quad \text{or} \quad x - 7 = 0$$
$$x = 3 \quad \text{or} \quad x = 7$$

check:

$x = 3$	$x = 7$
$\sqrt{2(3)-5} + 4 = 3$	$\sqrt{2(7)-5} + 4 = 7$
$\sqrt{1} + 4 = 3$	$\sqrt{9} + 4 = 7$
$1 + 4 = 3$	$3 + 4 = 7$
$5 = 3$	$7 = 7$
False	True

$\{7\}$

37.

$$x = \sqrt{49 - 8x} + 7$$
$$x - 7 = \sqrt{49 - 8x}$$
$$(x - 7)^2 = \left(\sqrt{49 - 8x}\right)^2$$
$$(x - 7)(x - 7) = 49 - 8x$$
$$x^2 - 14x + 49 = 49 - 8x$$
$$x^2 - 6x = 0$$
$$x(x - 6) = 0$$
$$x = 0 \quad \text{or} \quad x - 6 = 0$$
$$x = 0 \quad \text{or} \quad x = 6$$

check:

$x = 0$	$x = 6$
$0 = \sqrt{49 - 8(0)} + 7$	$6 = \sqrt{49 - 8(6)} + 7$
$0 = \sqrt{49} + 7$	$6 = \sqrt{1} + 7$
$0 = 14$	$6 = 8$
False	False

$\varnothing$

39.

$$2x - 3 = \sqrt{30 - 7x}$$
$$(2x - 3)^2 = \left(\sqrt{30 - 7x}\right)^2$$
$$(2x - 3)(2x - 3) = 30 - 7x$$
$$4x^2 - 12x + 9 = 30 - 7x$$
$$4x^2 - 5x - 21 = 0$$
$$(4x + 7)(x - 3) = 0$$
$$4x + 7 = 0 \quad \text{or} \quad x - 3 = 0$$
$$x = -\frac{7}{4} \quad \text{or} \quad x = 3$$

check:

$x = -\dfrac{7}{4}$	$x = 3$
$2\left(-\dfrac{7}{4}\right) - 3 = \sqrt{30 - 7\left(-\dfrac{7}{4}\right)}$	$2(3) - 3 = \sqrt{30 - 7(3)}$
$-\dfrac{7}{2} - 3 = \sqrt{30 + \dfrac{49}{4}}$	$6 - 3 = \sqrt{30 - 21}$
$-\dfrac{7}{2} - \dfrac{6}{2} = \sqrt{\dfrac{120}{4} + \dfrac{49}{4}}$	$3 = \sqrt{9}$
$-\dfrac{13}{2} = \sqrt{\dfrac{169}{4}}$	$3 = 3$
$-\dfrac{13}{2} = \dfrac{13}{2}$	
False	True

$\{3\}$

41.

$$3x = 1 + \sqrt{4x^2 + x + 7}$$
$$3x - 1 = \sqrt{4x^2 + x + 7}$$
$$(3x - 1)^2 = \left(\sqrt{4x^2 + x + 7}\right)^2$$
$$(3x - 1)(3x - 1) = 4x^2 + x + 7$$
$$9x^2 - 6x + 1 = 4x^2 + x + 7$$
$$5x^2 - 7x - 6 = 0$$
$$(5x + 3)(x - 2) = 0$$
$$5x + 3 = 0 \quad \text{or} \quad x - 2 = 0$$
$$x = -\frac{3}{5} \quad \text{or} \quad x = 2$$

check:

$x = -\dfrac{3}{5}$	$x = 2$
$3\left(-\dfrac{3}{5}\right) = 1 + \sqrt{4\left(-\dfrac{3}{5}\right)^2 + \left(-\dfrac{3}{5}\right) + 7}$	$3(2) = 1 + \sqrt{4(2)^2 + 2 + 7}$
$-\dfrac{9}{5} = 1 + \sqrt{\dfrac{36}{25} - \dfrac{3}{5} + 7}$	$6 = 1 + \sqrt{16 + 2 + 7}$
$-\dfrac{9}{5} = 1 + \sqrt{\dfrac{36}{25} - \dfrac{15}{25} + \dfrac{175}{25}}$	$6 = 1 + \sqrt{25}$
$-\dfrac{9}{5} = 1 + \sqrt{\dfrac{196}{25}}$	$6 = 1 + 5$
$-\dfrac{9}{5} = 1 + \dfrac{14}{5}$	$6 = 6$
$-\dfrac{9}{5} = \dfrac{19}{5}$	
False	True

$\{2\}$

43.

$$\sqrt{4x-15} = \sqrt{3x+11}$$
$$\left(\sqrt{4x-15}\right)^2 = \left(\sqrt{3x+11}\right)^2$$
$$4x-15 = 3x+11$$
$$x-15 = 11$$
$$x = 26$$

check:
$$\sqrt{4(26)-15} = \sqrt{3(26)+11}$$
$$\sqrt{104-15} = \sqrt{78+11}$$
$$\sqrt{89} = \sqrt{89}$$
True
$\{26\}$

45.

$$\sqrt[4]{x^2-8x+4} = \sqrt[4]{3x-14}$$
$$\left(\sqrt[4]{x^2-8x+4}\right)^4 = \left(\sqrt[4]{3x-14}\right)^4$$
$$x^2-8x+4 = 3x-14$$
$$x^2-11x+18 = 0$$
$$(x-9)(x-2) = 0$$
$$x-9=0 \quad \text{or} \quad x-2=0$$
$$x=9 \quad \text{or} \quad x=2$$

check:

$x=9$	$x=2$
$\sqrt[4]{(9)^2-8(9)+4} = \sqrt[4]{3(9)-14}$	$\sqrt[4]{(2)^2-8(2)+4} = \sqrt[4]{3(2)-14}$
$\sqrt[4]{81-72+4} = \sqrt[4]{27-14}$	$\sqrt[4]{4-16+4} = \sqrt[4]{6-14}$
$\sqrt[4]{13} = \sqrt[4]{13}$	$\sqrt[4]{-8} = \sqrt[4]{-8}$
True	undefined
$\{9\}$	

47.

$$\sqrt{x+4} = \sqrt{x-1}+1$$
$$\left(\sqrt{x+4}\right)^2 = \left(\sqrt{x-1}+1\right)^2$$
$$x+4 = \left(\sqrt{x-1}+1\right)\left(\sqrt{x-1}+1\right)$$
$$x+4 = x-1+2\sqrt{x-1}+1$$
$$x+4 = x+2\sqrt{x-1}$$
$$4 = 2\sqrt{x-1}$$
$$2 = \sqrt{x-1}$$
$$(2)^2 = \left(\sqrt{x-1}\right)^2$$
$$4 = x-1$$
$$5 = x$$

check:
$$\sqrt{5+4} = \sqrt{5-1}+1$$
$$\sqrt{9} = \sqrt{4}+1$$
$$3 = 2+1$$
$$3 = 3$$
True
$\{5\}$

49.

$$\sqrt{2x+3} = 1+\sqrt{x+1}$$
$$\left(\sqrt{2x+3}\right)^2 = \left(1+\sqrt{x+1}\right)^2$$
$$2x+3 = \left(1+\sqrt{x+1}\right)\left(1+\sqrt{x+1}\right)$$
$$2x+3 = 1\cdot1+1\cdot\sqrt{x+1}+1\cdot\sqrt{x+1}+\sqrt{x+1}\cdot\sqrt{x+1}$$
$$2x+3 = 1+2\sqrt{x+1}+x+1$$
$$x+1 = 2\sqrt{x+1}$$
$$(x+1)^2 = \left(2\sqrt{x+1}\right)^2$$
$$x^2+x+x+1 = 4(x+1)$$
$$x^2+2x+1 = 4x+4$$
$$x^2-2x-3 = 0$$
$$(x+1)(x-3) = 0$$
$$x+1=0 \quad \text{or} \quad x-3=0$$
$$x=-1 \quad \text{or} \quad x=3$$

check:

$x=-1$	$x=3$
$\sqrt{2(-1)+3} = 1+\sqrt{-1+1}$	$\sqrt{2(3)+3} = 1+\sqrt{3+1}$
$\sqrt{1} = 1+\sqrt{0}$	$\sqrt{9} = 1+\sqrt{4}$
$1 = 1$	$3 = 1+2$
	$3 = 3$
True	True
$\{-1,3\}$	

51.

$$\sqrt{2x+12} = 1+\sqrt{x+5}$$
$$\left(\sqrt{2x+12}\right)^2 = \left(1+\sqrt{x+5}\right)^2$$
$$2x+12 = \left(1+\sqrt{x+5}\right)\left(1+\sqrt{x+5}\right)$$
$$2x+12 = 1\cdot 1 + 1\cdot\sqrt{x+5} + 1\cdot\sqrt{x+5} + \sqrt{x+5}\cdot\sqrt{x+5}$$
$$2x+12 = 1+2\sqrt{x+5}+x+5$$
$$x+6 = 2\sqrt{x+5}$$
$$(x+6)^2 = \left(2\sqrt{x+5}\right)^2$$
$$x^2+6x+6x+36 = 4(x+5)$$
$$x^2+12x+36 = 4x+20$$
$$x^2+8x+16 = 0$$
$$(x+4)(x+4) = 0$$
$$x+4 = 0$$
$$x = -4$$

check:
$$\sqrt{2(-4)+12} = 1+\sqrt{-4+5}$$
$$\sqrt{4} = 1+\sqrt{1}$$
$$2 = 1+1$$
$$2 = 2$$
True
$$\{-4\}$$

53.

$$\sqrt{3x+1}-\sqrt{x+4} = 1$$
$$\sqrt{3x+1} = 1+\sqrt{x+4}$$
$$\left(\sqrt{3x+1}\right)^2 = \left(1+\sqrt{x+4}\right)^2$$
$$3x+1 = \left(1+\sqrt{x+4}\right)\left(1+\sqrt{x+4}\right)$$
$$3x+1 = 1\cdot 1 + 1\cdot\sqrt{x+4} + 1\cdot\sqrt{x+4} + \sqrt{x+4}\cdot\sqrt{x+4}$$
$$3x+1 = 1+2\sqrt{x+4}+x+4$$
$$2x-4 = 2\sqrt{x+4}$$
$$x-2 = \sqrt{x+4}$$
$$(x-2)^2 = \left(\sqrt{x+4}\right)^2$$
$$x^2-2x-2x+4 = x+4$$
$$x^2-4x+4 = x+4$$
$$x^2-5x = 0$$
$$x(x-5) = 0$$
$$x=0 \quad\text{or}\quad x-5=0$$
$$x=0 \quad\text{or}\quad x=5$$

check:

$x=0$	$x=5$
$\sqrt{3(0)+1}-\sqrt{0+4} = 1$	$\sqrt{3(5)+1}-\sqrt{5+4} = 1$
$\sqrt{1}-\sqrt{4} = 1$	$\sqrt{16}-\sqrt{9} = 1$
$1-2 = 1$	$4-3 = 1$
$-1 = 1$	$1 = 1$
False	True

$$\{5\}$$

55. $T = 2\pi\sqrt{\dfrac{L}{32}} = 2\pi\sqrt{\dfrac{5}{32}} \approx 2.48$
2.48 seconds

57. $T = 2\pi\sqrt{\dfrac{L}{32}} = 2\pi\sqrt{\dfrac{1.8}{32}} = 2\pi\sqrt{\dfrac{0.9}{16}} \approx 1.49$
1.49 seconds

59.
$$T = 2\pi\sqrt{\dfrac{L}{32}}$$
$$1.8 = 2\pi\sqrt{\dfrac{L}{32}}$$
$$\dfrac{1.8}{2\pi} = \sqrt{\dfrac{L}{32}}$$
$$\left(\dfrac{0.9}{\pi}\right)^2 = \left(\sqrt{\dfrac{L}{32}}\right)^2$$
$$\dfrac{0.81}{\pi^2} = \dfrac{L}{32}$$
$$32 \cdot \dfrac{0.81}{\pi^2} = \overset{1}{\cancel{32}} \cdot \dfrac{L}{\underset{1}{\cancel{32}}}$$
$$\dfrac{25.92}{\pi^2} = L$$
$$L \approx 2.6 \text{ feet}$$

61. $s = \sqrt{30df} = \sqrt{30(70)(0.75)} = \sqrt{1575} \approx 40$
40 mph

63. $s = \sqrt{30df} = \sqrt{30(185)(0.95)} = \sqrt{5272.5}s \approx 73$
73 mph

65. $BSA = \sqrt{\dfrac{hw}{3600}} = \sqrt{\dfrac{177 \cdot 89}{3600}} = \sqrt{\dfrac{15,753}{3600}} \approx 2.09$
2.09 square meters

67. $BSA = \sqrt{\dfrac{hw}{3600}} = \sqrt{\dfrac{151 \cdot 50}{3600}} = \sqrt{\dfrac{7550}{3600}} \approx 1.45$
1.45 square meters

69. $BSA = \sqrt{\dfrac{hw}{3600}} = \dfrac{\sqrt{hw}}{\sqrt{3600}} = \dfrac{\sqrt{hw}}{60}$
$$1.9 = \dfrac{\sqrt{190w}}{60}$$
$$60 \cdot 1.9 = \dfrac{\sqrt{190w}}{\cancel{60}} \cdot \cancel{60}$$
$$114 = \sqrt{190w}$$
$$114^2 = \left(\sqrt{190w}\right)^2$$
$$12,996 = 190w$$
$$w = \dfrac{12,996}{190} \approx 68 \text{ kilograms}$$

71. $r = 19.8\sqrt{d} = 19.8\sqrt{49} = 19.8(7) = 138.6$
138.6 gallons/minute

73. $r = 19.8\sqrt{d} = 19.8\sqrt{13} \approx 71.4$
71.4 gallons/minute

75.
$$r = 19.8\sqrt{d}$$
$$30 = 19.8\sqrt{d}$$
$$\dfrac{30}{19.8} = \sqrt{d}$$
$$\left(\dfrac{30}{19.8}\right)^2 = \left(\sqrt{d}\right)^2$$
$$d \approx 2.3 \text{ feet}$$

77. $d = \sqrt{1.5h} = \sqrt{1.5 \cdot 2} = \sqrt{3} \approx 1.7$
1.7 miles

79. $d = \sqrt{1.5h} = \sqrt{1.5 \cdot 100} = \sqrt{150} \approx 12.2$
12.2 miles

81. Answers will vary. Example:
A vehicle that was involved in an accident made 200 feet of skid marks on a concrete road before crashing. If the drag factor for concrete is 0.95, find the speed the car was traveling when it started skidding. Round to the nearest mile per hour.

$s = \sqrt{30df} = \sqrt{30(200)(0.95)} = \sqrt{5700} \approx 75$
75 mph

83. Answers will vary.

9.6   QUICK CHECK

1. a) $\sqrt{-36} = \sqrt{36(-1)} = 6i$

b) $\sqrt{-63} = \sqrt{9 \cdot 7 \cdot (-1)} = \sqrt{9 \cdot 7} \cdot \sqrt{-1} = 3i\sqrt{7}$

c) $-\sqrt{-54} = -\sqrt{9 \cdot 6 \cdot (-1)}$
$= -\sqrt{9 \cdot 6} \cdot \sqrt{-1} = -3i\sqrt{6}$

2. a) $(3 + 12i) + (-8 + 15i)$
$= 3 + 12i - 8 + 15i$
$= -5 + 27i$

b) $(14 - 6i) - (3 + 22i)$
$= 14 - 6i - 3 - 22i$
$= 11 - 28i$

3. $4i \cdot 15i = 60i^2 = 60(-1) = -60$

4. $\sqrt{-5} \cdot \sqrt{-120} = \sqrt{5}i \cdot \sqrt{120}i$
$= \sqrt{600}\, i^2$
$= \sqrt{100 \cdot 6}\,(-1)$
$= -10\sqrt{6}$

5. $-6i(7 + 8i) = -42i - 48i^2$
$= -42i - 48(-1)$
$= 48 - 42i$

6. $(3 - 2i)(8 + i) = 24 + 3i - 16i - 2i^2$
$= 24 - 13i - 2(-1)$
$= 24 - 13i + 2$
$= 26 - 13i$

7. $(9 - 4i)^2 = (9 - 4i)(9 - 4i)$
$= 81 - 36i - 36i + 16i^2$
$= 81 - 72i + 16(-1)$
$= 81 - 72i - 16$
$= 65 - 72i$

8. $(7 + 4i)(7 - 4i) = 49 - 28i + 28i - 16i^2$
$= 49 - 16(-1)$
$= 49 + 16$
$= 65$

9. Since $(a - bi)(a + bi) = a^2 + b^2$,
$(11 - 4i)(11 + 4i) = 11^2 + 4^2 = 121 + 16 = 137$

10. $\dfrac{15}{8 + 6i} = \dfrac{15}{(8 + 6i)} \cdot \dfrac{(8 - 6i)}{(8 - 6i)}$
$= \dfrac{120 - 90i}{8^2 + 6^2}$
$= \dfrac{120 - 90i}{64 + 36}$
$= \dfrac{120 - 90i}{100}$
$= \dfrac{12 - 9i}{10}$
$= \dfrac{6}{5} - \dfrac{9}{10}i$

11. $\dfrac{9 + 2i}{7 - 3i} = \dfrac{(9 + 2i)}{(7 - 3i)} \cdot \dfrac{(7 + 3i)}{(7 + 3i)}$
$= \dfrac{63 + 27i + 14i + 6i^2}{7^2 + 3^2}$
$= \dfrac{63 + 41i + 6(-1)}{49 + 9}$
$= \dfrac{57 + 41i}{58}$
$= \dfrac{57}{58} + \dfrac{41}{58}i$

12. $\dfrac{15 - 8i}{12i} = \dfrac{15 - 8i}{12i} \cdot \dfrac{i}{i}$
$= \dfrac{15i - 8i^2}{12i^2}$
$= \dfrac{15i - 8(-1)}{12(-1)}$
$= \dfrac{15i + 8}{-12}$
$= -\dfrac{8 + 15i}{12}$
$= -\dfrac{2}{3} - \dfrac{5}{4}i$

13. $i^{30} = i^{28} \cdot i^2 = \left(i^4\right)^7 \cdot i^2 = i^2 = -1$

9.6   THE COMPLEX NUMBERS

1. imaginary

3. -1

5. real

7. $\sqrt{-4} = \sqrt{4} \cdot \sqrt{-1} = 2i$

9. $\sqrt{-81} = \sqrt{81} \cdot \sqrt{-1} = 9i$

11. $-\sqrt{-169} = -\sqrt{169} \cdot \sqrt{-1} = -13i$

13. $\sqrt{-50} = \sqrt{25 \cdot 2} \cdot \sqrt{-1} = 5\sqrt{2} \cdot i = 5i\sqrt{2}$

15. $-\sqrt{-252} = -\sqrt{36 \cdot 7} \cdot \sqrt{-1} = -6\sqrt{7} \cdot i = -6i\sqrt{7}$

17. $(8 + 9i) + (3 + 5i) = 8 + 9i + 3 + 5i = 11 + 14i$

19. $(6 - 7i) - (2 + 10i) = 6 - 7i - 2 - 10i = 4 - 17i$

21. $(-3 + 8i) + (3 - 4i) = -3 + 8i + 3 - 4i$
$$= 0 + 4i$$
$$= 4i$$

23. $(12 + 13i) - (12 - 13i) = 12 + 13i - 12 + 13i = 26i$

25. $(1 - 12i) - (8 + 8i) + (5 - 4i)$
$$= 1 - 12i - 8 - 8i + 5 - 4i$$
$$= -2 - 24i$$

27. $(4 - 7i) + (6 + 9i) = 4 - 7i + 6 + 9i = 10 + 2i$
The missing number is $6 + 9i$.

29. $(3 + 4i) - (2 + 9i) = 3 + 4i - 2 - 9i = 1 - 5i$
The missing number is $2 + 9i$.

31. $7i \cdot 14i = 98i^2 = 98(-1) = -98$

33. $-4i \cdot 5i = -20i^2 = -20(-1) = 20$

35. $\sqrt{-49} \cdot \sqrt{-64} = \sqrt{49} \cdot \sqrt{-1} \cdot \sqrt{64} \cdot \sqrt{-1}$
$$= 7i \cdot 8i$$
$$= 56i^2$$
$$= 56(-1)$$
$$= -56$$

37. $\sqrt{-12} \cdot \sqrt{-18} = \sqrt{12} \cdot \sqrt{-1} \cdot \sqrt{18} \cdot \sqrt{-1}$
$$= \sqrt{216} \, i^2$$
$$= \sqrt{36 \cdot 6} \, (-1)$$
$$= -6\sqrt{6}$$

39. $\sqrt{-20} \cdot \sqrt{-45} = \sqrt{20} \cdot \sqrt{-1} \cdot \sqrt{45} \cdot \sqrt{-1}$
$$= \sqrt{900} \, i^2$$
$$= 30(-1)$$
$$= -30$$

41. $7i(8 + 5i) = 56i + 35i^2$
$$= 56i + 35(-1)$$
$$= -35 + 56i$$

43. $-6i(9 + 4i) = -54i - 24i^2$
$$= -54i - 24(-1)$$
$$= 24 - 54i$$

45. $(7 + 2i)(5 + i) = 35 + 7i + 10i + 2i^2$
$$= 35 + 17i + 2(-1)$$
$$= 35 + 17i - 2$$
$$= 33 + 17i$$

47. $(3 - 5i)(4 + 10i) = 12 + 30i - 20i - 50i^2$
$$= 12 + 10i - 50(-1)$$
$$= 12 + 10i + 50$$
$$= 62 + 10i$$

49. $(6 + 7i)^2 = (6 + 7i)(6 + 7i)$
$$= 36 + 42i + 42i + 49i^2$$
$$= 36 + 84i + 49(-1)$$
$$= 36 + 84i - 49$$
$$= -13 + 84i$$

51. $(1 - 5i)^2 = (1 - 5i)(1 - 5i)$
$$= 1 - 5i - 5i + 25i^2$$
$$= 1 - 10i + 25(-1)$$
$$= 1 - 10i - 25$$
$$= -24 - 10i$$

53. $(5 + 2i)(5 - 2i) = 5^2 + 2^2$
$$= 25 + 4$$
$$= 29$$

55. $(12 - 8i)(12 + 8i) = 12^2 + 8^2 = 144 + 64 = 208$

57. $(5 + 6i)(5 - 6i) = 5^2 + 6^2 = 25 + 36 = 61$

59. $\dfrac{6}{5+i} = \dfrac{6}{5+i} \cdot \dfrac{5-i}{5-i}$

$= \dfrac{30-6i}{5^2+1^2}$

$= \dfrac{30-6i}{26}$

$= \dfrac{\overset{1}{\cancel{2}}(15-3i)}{\underset{13}{\cancel{26}}}$

$= \dfrac{15-3i}{13}$

$= \dfrac{15}{13} - \dfrac{3}{13}i$

61. $\dfrac{25i}{9-7i} = \dfrac{25i}{9-7i} \cdot \dfrac{9+7i}{9+7i}$

$= \dfrac{225i+175i^2}{9^2+7^2}$

$= \dfrac{225i-175}{130}$

$= \dfrac{\overset{1}{\cancel{5}}(45i-35)}{\underset{26}{\cancel{130}}}$

$= \dfrac{45i-35}{26}$

$= -\dfrac{35}{26} + \dfrac{45}{26}i$

63. $\dfrac{4+7i}{7-2i} = \dfrac{4+7i}{7-2i} \cdot \dfrac{7+2i}{7+2i}$

$= \dfrac{28+8i+49i+14i^2}{7^2+2^2}$

$= \dfrac{28+57i+14i^2}{49+4}$

$= \dfrac{28+57i-14}{53}$

$= \dfrac{14+57i}{53}$

$= \dfrac{14}{53} + \dfrac{57}{53}i$

65. $\dfrac{1+6i}{7+6i} = \dfrac{1+6i}{7+6i} \cdot \dfrac{7-6i}{7-6i}$

$= \dfrac{7-6i+42i-36i^2}{7^2+6^2}$

$= \dfrac{7+36i-36i^2}{49+36}$

$= \dfrac{7+36i+36}{85}$

$= \dfrac{43+36i}{85}$

$= \dfrac{43}{85} + \dfrac{36}{85}i$

67. $\dfrac{8}{i} = \dfrac{8}{i} \cdot \dfrac{i}{i} = \dfrac{8i}{i^2} = \dfrac{8i}{-1} = -8i$

69. $\dfrac{5-7i}{3i} = \dfrac{(5-7i)}{3i} \cdot \dfrac{i}{i}$

$= \dfrac{5i-7i^2}{3i^2}$

$= \dfrac{5i-7(-1)}{3(-1)}$

$= \dfrac{7+5i}{-3}$

$= -\dfrac{7+5i}{3}$

$= -\dfrac{7}{3} - \dfrac{5}{3}i$

71. $\dfrac{6+i}{-4i} = \dfrac{(6+i)}{-4i} \cdot \dfrac{i}{i}$

$= \dfrac{6i+i^2}{-4i^2}$

$= \dfrac{6i-1}{-4(-1)}$

$= \dfrac{-1+6i}{4}$

$= -\dfrac{1}{4} + \dfrac{3}{2}i$

73.
$$\begin{aligned}
(7+3i) \div 2i &= \frac{7+3i}{2i} \\
&= \frac{7+3i}{2i} \cdot \frac{i}{i} \\
&= \frac{7i+3i^2}{2i^2} \\
&= \frac{7i-3}{-2} \\
&= -\frac{7i-3}{2} \\
&= \frac{-7i+3}{2} \\
&= \frac{3}{2} - \frac{7}{2}i
\end{aligned}$$

75.
$$\begin{aligned}
(8-7i) \div (1+3i) &= \frac{8-7i}{1+3i} \\
&= \frac{8-7i}{1+3i} \cdot \frac{1-3i}{1-3i} \\
&= \frac{8-24i-7i+21i^2}{1^2+3^2} \\
&= \frac{8-31i-21}{1+9} \\
&= \frac{-13-31i}{10} \\
&= -\frac{13}{10} - \frac{31}{10}i
\end{aligned}$$

77. To find the missing number, divide $4+35i$ by $4+i$.
$$\begin{aligned}
\frac{4+35i}{4+i} &= \frac{(4+35i)}{(4+i)} \cdot \frac{(4-i)}{(4-i)} \\
&= \frac{16-4i+140i-35i^2}{4^2+1^2} \\
&= \frac{16+136i-35(-1)}{16+1} \\
&= \frac{51+136i}{17} \\
&= \frac{17(3+8i)}{17} \\
&= 3+8i
\end{aligned}$$
The missing number is $3+8i$.

79. To find the missing number, multiply $1+9i$ and $4-3i$.
$$\begin{aligned}
(1+9i)(4-3i) &= 4-3i+36i-27i^2 \\
&= 4+33i-27(-1) \\
&= 31+33i
\end{aligned}$$
The missing number is $31+33i$.

81. $i^{45} = i^{44} \cdot i = \left(i^4\right)^{11} \cdot i = i$

83. $i^{42} = i^{40} \cdot i^2 = \left(i^4\right)^{10} \cdot i^2 = i^2 = -1$

85. $\dfrac{i^{37}}{i^{15}} = i^{22} = i^{20} \cdot i^2 = \left(i^4\right)^5 \cdot i^2 = i^2 = -1$

87.
$$\begin{aligned}
i^{21}+i^{53} &= i^{20} \cdot i + i^{52} \cdot i \\
&= \left(i^4\right)^5 \cdot i + \left(i^4\right)^{13} \cdot i \\
&= i+i \\
&= 2i
\end{aligned}$$

89.
$$\begin{aligned}
(7+2i)(6-3i) &= 42-21i+12i-6i^2 \\
&= 42-9i+6 \\
&= 48-9i
\end{aligned}$$

91.
$$\begin{aligned}
\frac{9+4i}{5i} &= \frac{(9+4i)}{5i} \cdot \frac{i}{i} \\
&= \frac{9i+4i^2}{5i^2} \\
&= \frac{-4+9i}{-5} \\
&= -\frac{-4+9i}{5} \\
&= \frac{4-9i}{5} \\
&= \frac{4}{5} - \frac{9}{5}i
\end{aligned}$$

93.
$$\begin{aligned}
-9i(12+5i) &= -108i-45i^2 \\
&= -108i+45 \\
&= 45-108i
\end{aligned}$$

95. $12i \div (4 - 2i) = \dfrac{12i}{4 - 2i}$

$$= \frac{12i}{(4-2i)} \cdot \frac{(4+2i)}{(4+2i)}$$
$$= \frac{48i + 24i^2}{4^2 + 2^2}$$
$$= \frac{-24 + 48i}{16 + 4}$$
$$= \frac{-24 + 48i}{20}$$
$$= \frac{-\overset{1}{\cancel{4}}(6 - 12i)}{\underset{5}{\cancel{20}}}$$
$$= \frac{-6 + 12i}{5}$$
$$= -\frac{6}{5} + \frac{12}{5}i$$

97. $(2 - 5i)(2 + 5i) = 2^2 + 5^2 = 4 + 25 = 29$

99. $\sqrt{-24} \cdot \sqrt{-27} = \sqrt{4 \cdot 6(-1)} \cdot \sqrt{9 \cdot 3(-1)}$

$$= 2\sqrt{6}\,i \cdot 3\sqrt{3}\,i$$
$$= 6\sqrt{18}\,i^2$$
$$= -6\sqrt{9 \cdot 2}$$
$$= -6 \cdot 3\sqrt{2}$$
$$= -18\sqrt{2}$$

101. $\dfrac{15i}{9 - 2i} = \dfrac{15i}{(9 - 2i)} \cdot \dfrac{(9 + 2i)}{(9 + 2i)}$

$$= \frac{135i + 30i^2}{9^2 + 2^2}$$
$$= \frac{135i - 30}{81 + 4}$$
$$= \frac{-30 + 135i}{85}$$
$$= \frac{-\overset{1}{\cancel{5}}(6 - 27i)}{\underset{17}{\cancel{85}}}$$
$$= \frac{-6 + 27i}{17}$$
$$= -\frac{6}{17} + \frac{27}{17}i$$

103. $\sqrt{-252} = \sqrt{36 \cdot 7(-1)} = 6i\sqrt{7}$

105. $(9 - 3i)^2 = (9 - 3i)(9 - 3i)$

$$= 81 - 27i - 27i + 9i^2$$
$$= 81 - 54i - 9$$
$$= 72 - 54i$$

107. $(8 - 11i) + (15 + 21i) = 8 - 11i + 15 + 21i$
$$= 23 + 10i$$

109. $-14i \cdot 9i = -126i^2 = 126$

111. $i^{323} = i^{320} \cdot i^3 = \left(i^4\right)^{80} \cdot i^3 = i^3 = -i$

113. $(30 - 17i) - (54 - 33i) = 30 - 17i - 54 + 33i$
$$= -24 + 16i$$

115. Answers will vary. Example:
Add two complex numbers by adding the real portions and the imaginary portions independently. This is similar to adding like terms when adding two variable expressions.

CHAPTER 9 REVIEW

1. $\sqrt{25} = 5$

2. $\sqrt{169} = 13$

3. $\sqrt[3]{-64} = -4$

4. $\sqrt[3]{x^9} = \sqrt[3]{\left(x^3\right)^3} = x^3$

5. $\sqrt{81x^{12}} = \sqrt{\left(9x^6\right)^2} = 9x^6$

6. $\sqrt[5]{243x^{15}y^{10}z^{25}} = \sqrt[5]{\left(3x^3y^2z^5\right)^5}$
$$= 3x^3y^2z^5$$

7. $\sqrt{17} \approx 4.123$

8. $\sqrt{41} \approx 6.403$

9. $f(91) = \sqrt{91 - 10} = \sqrt{81} = 9$

10. $f(3) = \sqrt{(3)^2 + 6(3) - 19}$
$$= \sqrt{9 + 18 - 19}$$
$$= \sqrt{8} \approx 2.828$$

11. $2x + 6 \geq 0$
$$2x \geq -6$$
$$x \geq -3$$
$$[-3, \infty)$$

12. $3x + 25 \geq 0$
$$3x \geq -25$$
$$x \geq -\frac{25}{3}$$
$$\left[-\frac{25}{3}, \infty\right)$$

13. $\sqrt[5]{a} = a^{1/5}$

14. $\left(\sqrt[4]{x}\right)^3 = x^{3/4}$

15. $\left(\sqrt{x}\right)^{13} = x^{13/2}$

16. $\sqrt[5]{n^8} = n^{8/5}$

17. $4^{1/2} = \sqrt{4} = 2$

18. $\left(343x^{21}\right)^{2/3} = \left(\sqrt[3]{343x^{21}}\right)^2$
$$= \left(\sqrt[3]{\left(7x^7\right)^3}\right)^2$$
$$= \left(7x^7\right)^2$$
$$= 49x^{14}$$

19. $x^{3/2} \cdot x^{1/6} = x^{3/2 + 1/6} = x^{9/6 + 1/6} = x^{10/6} = x^{5/3}$

20. $\left(x^{3/8}\right)^{7/6} = x^{3/8 \cdot 7/6} = x^{7/16}$

21. $\dfrac{x^{2/3}}{x^{5/12}} = x^{2/3 - 5/12} = x^{8/12 - 5/12} = x^{3/12} = x^{1/4}$

22. $49^{-1/2} = \dfrac{1}{49^{1/2}} = \dfrac{1}{\sqrt{49}} = \dfrac{1}{7}$

23. $64^{-5/3} = \dfrac{1}{64^{5/3}} = \dfrac{1}{\left(\sqrt[3]{64}\right)^5} = \dfrac{1}{4^5} = \dfrac{1}{1024}$

24. $\dfrac{27^{4/3}}{27^{8/3}} = 27^{4/3 - 8/3}$
$$= 27^{-4/3}$$
$$= \dfrac{1}{27^{4/3}}$$
$$= \dfrac{1}{\left(\sqrt[3]{27}\right)^4}$$
$$= \dfrac{1}{3^4}$$
$$= \dfrac{1}{81}$$

25. $\sqrt{28} = \sqrt{4 \cdot 7} = 2\sqrt{7}$

26. $\sqrt[3]{x^{10}} = \sqrt[3]{x^9 \cdot x} = x^3 \sqrt[3]{x}$

27. $\sqrt[5]{x^{11}y^8z^{15}} = \sqrt[5]{\left(x^{10} \cdot x\right)\left(y^5 \cdot y^3\right)z^{15}}$
$$= x^2 yz^3 \sqrt[5]{xy^3}$$

28. $\sqrt[3]{405a^{25}b^{12}c^{16}}$
$$= \sqrt[3]{(27 \cdot 15)\left(a^{24} \cdot a\right)b^{12}\left(c^{15} \cdot c\right)}$$
$$= 3a^8 b^4 c^5 \sqrt[3]{15ac}$$

29. $\sqrt{180r^3s^{11}t^{12}} = \sqrt{(36 \cdot 5)\left(r^2 \cdot r\right)\left(s^{10} \cdot s\right)t^{12}}$
$$= 6rs^5 t^6 \sqrt{5rs}$$

30. $\sqrt{175r^6s^8t^4} = \sqrt{(25 \cdot 7)r^6s^8t^4} = 5r^3 s^4 t^2 \sqrt{7}$

31. $4\sqrt{2} + 8\sqrt{2} = 12\sqrt{2}$

32. $7\sqrt[3]{x} - \sqrt[3]{x} = 6\sqrt[3]{x}$

33. $5\sqrt{20} + 3\sqrt{125} = 5\sqrt{4 \cdot 5} + 3\sqrt{25 \cdot 5}$
$$= 5 \cdot 2\sqrt{5} + 3 \cdot 5\sqrt{5}$$
$$= 10\sqrt{5} + 15\sqrt{5}$$
$$= 25\sqrt{5}$$

34. $6\sqrt{12} - 8\sqrt{50} + 9\sqrt{75}$
$$= 6\sqrt{4 \cdot 3} - 8\sqrt{25 \cdot 2} + 9\sqrt{25 \cdot 3}$$
$$= 6 \cdot 2\sqrt{3} - 8 \cdot 5\sqrt{2} + 9 \cdot 5\sqrt{3}$$
$$= 12\sqrt{3} - 40\sqrt{2} + 45\sqrt{3}$$
$$= 57\sqrt{3} - 40\sqrt{2}$$

35. $\sqrt[3]{4x^2} \cdot \sqrt[3]{18x^4} = \sqrt[3]{72x^6}$

$\qquad = \sqrt[3]{8 \cdot 9x^6}$

$\qquad = 2x^2 \sqrt[3]{9}$

36. $5\sqrt{3}\left(2\sqrt{6} - 4\sqrt{3}\right) = 10\sqrt{18} - 20 \cdot 3$

$\qquad = 10\sqrt{9 \cdot 2} - 60$

$\qquad = 10 \cdot 3\sqrt{2} - 60$

$\qquad = 30\sqrt{2} - 60$

37. $\left(\sqrt{3} + \sqrt{8}\right)\left(\sqrt{6} - \sqrt{2}\right) = \sqrt{18} - \sqrt{6} + \sqrt{48} - \sqrt{16}$

$\qquad = \sqrt{9 \cdot 2} - \sqrt{6} + \sqrt{16 \cdot 3} - 4$

$\qquad = 3\sqrt{2} - \sqrt{6} + 4\sqrt{3} - 4$

38. $\left(10\sqrt{5} - 7\sqrt{11}\right)\left(9\sqrt{5} + 2\sqrt{11}\right)$

$\qquad = 90 \cdot 5 + 20\sqrt{55} - 63\sqrt{55} - 14 \cdot 11$

$\qquad = 450 - 43\sqrt{55} - 154$

$\qquad = 296 - 43\sqrt{55}$

39. $\left(2\sqrt{15} - 9\sqrt{3}\right)^2$

$\qquad = \left(2\sqrt{15} - 9\sqrt{3}\right)\left(2\sqrt{15} - 9\sqrt{3}\right)$

$\qquad = 4 \cdot 15 - 18\sqrt{45} - 18\sqrt{45} + 81 \cdot 3$

$\qquad = 60 - 36\sqrt{45} + 243$

$\qquad = 303 - 36\sqrt{9 \cdot 5}$

$\qquad = 303 - 36 \cdot 3\sqrt{5}$

$\qquad = 303 - 108\sqrt{5}$

40. $\left(2\sqrt{13} - 10\sqrt{7}\right)\left(2\sqrt{13} + 10\sqrt{7}\right)$

$\qquad = 4 \cdot 13 + 20\sqrt{91} - 20\sqrt{91} - 100 \cdot 7$

$\qquad = 52 - 700$

$\qquad = -648$

41. $\dfrac{\sqrt{224x^5}}{\sqrt{2x}} = \sqrt{\dfrac{224x^5}{2x}}$

$\qquad = \sqrt{112x^4}$

$\qquad = \sqrt{16 \cdot 7 \cdot x^4}$

$\qquad = 4x^2\sqrt{7}$

42. $\sqrt{\dfrac{50a^9b^8c^3}{8a^3bc^{13}}} = \sqrt{\dfrac{25a^6b^7}{4c^{10}}}$

$\qquad = \dfrac{\sqrt{25a^6b^7}}{\sqrt{4c^{10}}}$

$\qquad = \dfrac{5a^3b^3\sqrt{b}}{2c^5}$

43. $\sqrt[3]{\dfrac{16}{5}} = \dfrac{\sqrt[3]{16}}{\sqrt[3]{5}} \cdot \dfrac{\sqrt[3]{25}}{\sqrt[3]{25}}$

$\qquad = \dfrac{\sqrt[3]{400}}{\sqrt[3]{125}}$

$\qquad = \dfrac{\sqrt[3]{8 \cdot 50}}{5}$

$\qquad = \dfrac{2\sqrt[3]{50}}{5}$

44. $\dfrac{8}{\sqrt{14}} = \dfrac{8}{\sqrt{14}} \cdot \dfrac{\sqrt{14}}{\sqrt{14}} = \dfrac{8\sqrt{14}}{14} = \dfrac{4\sqrt{14}}{7}$

45. $\sqrt{\dfrac{11}{18}} = \dfrac{\sqrt{11}}{\sqrt{18}}$

$\qquad = \dfrac{\sqrt{11}}{\sqrt{9 \cdot 2}}$

$\qquad = \dfrac{\sqrt{11}}{3\sqrt{2}} \cdot \dfrac{\sqrt{2}}{\sqrt{2}}$

$\qquad = \dfrac{\sqrt{22}}{3 \cdot 2}$

$\qquad = \dfrac{\sqrt{22}}{6}$

46. $\dfrac{a^5bc^3}{\sqrt{a^4b^5c}} = \dfrac{a^5bc^3}{a^2b^2\sqrt{bc}}$

$\qquad = \dfrac{a^3c^3}{b\sqrt{bc}} \cdot \dfrac{\sqrt{bc}}{\sqrt{bc}}$

$\qquad = \dfrac{a^3c^3\sqrt{bc}}{b^2c}$

$\qquad = \dfrac{a^3c^2\sqrt{bc}}{b^2}$

47.
$$\frac{8}{\sqrt{11}-\sqrt{5}} = \frac{8}{\sqrt{11}-\sqrt{5}}\cdot\frac{\sqrt{11}+\sqrt{5}}{\sqrt{11}+\sqrt{5}}$$
$$= \frac{8\left(\sqrt{11}+\sqrt{5}\right)}{11-5}$$
$$= \frac{8\left(\sqrt{11}+\sqrt{5}\right)}{6}$$
$$= \frac{4\left(\sqrt{11}+\sqrt{5}\right)}{3}$$

48.
$$\frac{2\sqrt{3}}{10\sqrt{15}-3\sqrt{12}} = \frac{2\sqrt{3}}{10\sqrt{15}-3\sqrt{12}}\cdot\frac{10\sqrt{15}+3\sqrt{12}}{10\sqrt{15}+3\sqrt{12}}$$
$$= \frac{20\sqrt{45}+6\sqrt{36}}{100\cdot15-9\cdot12}$$
$$= \frac{20\sqrt{9\cdot5}+6\sqrt{36}}{1500-108}$$
$$= \frac{20\cdot3\sqrt{5}+6\cdot6}{1392}$$
$$= \frac{60\sqrt{5}+36}{1392}$$
$$= \frac{12\left(5\sqrt{5}+3\right)}{1392}$$
$$= \frac{5\sqrt{5}+3}{116}$$

49.
$$\frac{\sqrt{7}+5\sqrt{2}}{2\sqrt{7}-3\sqrt{2}} = \frac{2\sqrt{3}}{10\sqrt{15}-3\sqrt{12}}\cdot\frac{10\sqrt{15}+3\sqrt{12}}{10\sqrt{15}+3\sqrt{12}}$$
$$= \frac{\left(\sqrt{7}+5\sqrt{2}\right)}{\left(2\sqrt{7}-3\sqrt{2}\right)}\cdot\frac{\left(2\sqrt{7}+3\sqrt{2}\right)}{\left(2\sqrt{7}+3\sqrt{2}\right)}$$
$$= \frac{2\cdot7+3\sqrt{14}+10\sqrt{14}+15\cdot2}{4\cdot7-9\cdot2}$$
$$= \frac{14+13\sqrt{14}+30}{28-18}$$
$$= \frac{44+13\sqrt{14}}{10}$$

50.
$$\frac{3-\sqrt{5}}{6+7\sqrt{5}} = \frac{\left(3-\sqrt{5}\right)}{\left(6+7\sqrt{5}\right)}\cdot\frac{\left(6-7\sqrt{5}\right)}{\left(6-7\sqrt{5}\right)}$$
$$= \frac{18-21\sqrt{5}-6\sqrt{5}+7\cdot5}{36-49\cdot5}$$
$$= \frac{18-27\sqrt{5}+35}{36-245}$$
$$= \frac{53-27\sqrt{5}}{-209}$$
$$= \frac{-53+27\sqrt{5}}{209}$$

51.
$$\sqrt{4x+1} = 7$$
$$\left(\sqrt{4x+1}\right)^2 = (7)^2$$
$$4x+1 = 49$$
$$4x = 48$$
$$x = 12$$
check:
$$\sqrt{4(12)+1} = 7$$
$$\sqrt{49} = 7$$
$$7 = 7$$
True
$\{12\}$

52.
$$\sqrt[3]{7x-15} = 5$$
$$\left(\sqrt[3]{7x-15}\right)^3 = (5)^3$$
$$7x-15 = 125$$
$$7x = 140$$
$$x = 20$$
check:
$$\sqrt[3]{7(20)-15} = 5$$
$$\sqrt[3]{125} = 5$$
$$5 = 5$$
True
$\{20\}$

53. $\sqrt{3x+22}+2=x$

$\sqrt{3x+22}=x-2$

$\left(\sqrt{3x+22}\right)^2=(x-2)^2$

$3x+22=(x-2)(x-2)$

$3x+22=x^2-4x+4$

$0=x^2-7x-18$

$0=(x-9)(x+2)$

$x-9=0$ or $x+2=0$

$x=9$ or $x=-2$

check:

$x=9$	$x=-2$
$\sqrt{3(9)+22}+2=9$	$\sqrt{3(-2)+22}+2=-2$
$\sqrt{49}+2=9$	$\sqrt{16}+2=-2$
$7+2=9$	$4+2=-2$
$9=9$	$6=-2$
True	False

$\{9\}$

54. $\sqrt{17-4x}-x=1$

$\sqrt{17-4x}=x+1$

$\left(\sqrt{17-4x}\right)^2=(x+1)^2$

$17-4x=(x+1)(x+1)$

$17-4x=x^2+2x+1$

$0=x^2+6x-16$

$0=(x+8)(x-2)$

$x+8=0$ or $x-2=0$

$x=-8$ or $x=2$

check:

$x=-8$	$x=2$
$\sqrt{17-4(-8)}-(-8)=1$	$\sqrt{17-4(2)}-(2)=1$
$\sqrt{49}+8=1$	$\sqrt{9}-2=1$
$7+8=1$	$3-2=1$
$15=1$	$1=1$
False	True

$\{2\}$

55. $\sqrt{x+5}=\sqrt{3x-13}$

$\left(\sqrt{x+5}\right)^2=\left(\sqrt{3x-13}\right)^2$

$x+5=3x-13$

$-2x+5=-13$

$-2x=-18$

$x=9$

check:

$\sqrt{(9)+5}=\sqrt{3(9)-13}$

$\sqrt{14}=\sqrt{14}$

True

$\{9\}$

56. $\sqrt{x+8}=2-\sqrt{x-12}$

$\left(\sqrt{x+8}\right)^2=\left(2-\sqrt{x-12}\right)^2$

$x+8=\left(2-\sqrt{x-12}\right)\left(2-\sqrt{x-12}\right)$

$x+8=4-4\sqrt{x-12}+x-12$

$x+8=x-8-4\sqrt{x-12}$

$8=-8-4\sqrt{x-12}$

$16=-4\sqrt{x-12}$

$-4=\sqrt{x-12}$

$(-4)^2=\left(\sqrt{x-12}\right)^2$

$16=x-12$

$28=x$

check:

$\sqrt{(28)+8}=2-\sqrt{(28)-12}$

$\sqrt{36}=2-\sqrt{16}$

$6=2-4$

$6=-2$

False

$\varnothing$

57. $f(x)=\sqrt{3x+4}=4$

$\left(\sqrt{3x+4}\right)^2=(4)^2$

$3x+4=16$

$3x=12$

$x=4$

58. Let $f(x)=11$.

$f(x)=\sqrt[3]{x^2-6x+11}+8=11$

$\sqrt[3]{x^2-6x+11}=3$

$\left(\sqrt[3]{x^2-6x+11}\right)^3=(3)^3$

$x^2-6x+11=27$

$x^2-6x-16=0$

$(x+2)(x-8)=0$

$x+2=0$ or $x-8=0$

$x=-2$ or $x=8$

59. $T=2\pi\sqrt{\dfrac{L}{32}}=2\pi\sqrt{\dfrac{5}{32}}\approx 2.48$

2.48 seconds

60. $s=\sqrt{30df}=\sqrt{30(200)(0.75)}\approx 67$

67 mph

61. $\sqrt{-49}=\sqrt{49\cdot(-1)}=\sqrt{49}\cdot\sqrt{-1}=7i$

62. $\sqrt{-40} = \sqrt{4 \cdot 10 \cdot (-1)} = \sqrt{4 \cdot 10} \cdot \sqrt{-1} = 2i\sqrt{10}$

63. $\sqrt{-252} = \sqrt{36 \cdot 7 \cdot (-1)} = \sqrt{36 \cdot 7} \cdot \sqrt{-1} = 6i\sqrt{7}$

64. $-\sqrt{-675} = -\sqrt{225 \cdot 3 \cdot (-1)}$
$$= -\sqrt{225 \cdot 3} \cdot \sqrt{-1}$$
$$= -15i\sqrt{3}$$

65. $(4 + 2i) + (7 + 3i) = 4 + 2i + 7 + 3i = 11 + 5i$

66. $(9 - 4i) + (7 + 2i) = 9 - 4i + 7 + 2i = 16 - 2i$

67. $(11 - 8i) - (10 - 13i) = 11 - 8i - 10 + 13i = 1 + 5i$

68. $(6 + 12i) - (5 + 5i) = 6 + 12i - 5 - 5i = 1 + 7i$

69. $2i \cdot 8i = 16i^2 = 16(-1) = -16$

70. $\sqrt{-6} \cdot \sqrt{-50} = \sqrt{6} \cdot \sqrt{-1} \cdot \sqrt{25 \cdot 2} \cdot \sqrt{-1}$
$$= \sqrt{6}\,i \cdot 5\sqrt{2}\,i$$
$$= 5\sqrt{12}\,i^2$$
$$= 5\sqrt{4 \cdot 3} \cdot (-1)$$
$$= -5 \cdot 2\sqrt{3}$$
$$= -10\sqrt{3}$$

71. $7i(3 - 4i) = 21i - 28i^2$
$$= 21i - 28(-1)$$
$$= 28 + 21i$$

72. $(10 + 2i)(13 + 8i) = 130 + 80i + 26i + 16i^2$
$$= 130 + 106i + 16(-1)$$
$$= 114 + 106i$$

73. $(9 - 3i)^2 = (9 - 3i)(9 - 3i)$
$$= 81 - 27i - 27i + 9i^2$$
$$= 81 - 54i + 9(-1)$$
$$= 72 - 54i$$

74. $(3 + 14i)(3 - 14i) = 3^2 + 14^2 = 9 + 196 = 205$

75. $\dfrac{6}{7 - 3i} = \dfrac{6}{7 - 3i} \cdot \dfrac{7 + 3i}{7 + 3i}$
$$= \frac{6(7 + 3i)}{7^2 + 3^2}$$
$$= \frac{42 + 18i}{49 + 9}$$
$$= \frac{\overset{1}{\cancel{2}}(21 + 9i)}{\underset{29}{\cancel{58}}}$$
$$= \frac{21 + 9i}{29}$$
$$= \frac{21}{29} + \frac{9}{29}i$$

76. $\dfrac{15i}{6 + 8i} = \dfrac{15i}{6 + 8i} \cdot \dfrac{6 - 8i}{6 - 8i}$
$$= \frac{90i - 120i^2}{6^2 + 8^2}$$
$$= \frac{90i + 120}{36 + 64}$$
$$= \frac{120 + 90i}{100}$$
$$= \frac{\overset{1}{\cancel{10}}(12 + 9i)}{\underset{10}{\cancel{100}}}$$
$$= \frac{12 + 9i}{10}$$
$$= \frac{6}{5} + \frac{9}{10}i$$

77. $\dfrac{2 + i}{9 - 5i} = \dfrac{2 + i}{9 - 5i} \cdot \dfrac{9 + 5i}{9 + 5i}$
$$= \frac{18 + 10i + 9i + 5i^2}{9^2 + 5^2}$$
$$= \frac{18 + 19i - 5}{81 + 25}$$
$$= \frac{13 + 19i}{106}$$
$$= \frac{13}{106} + \frac{19}{106}i$$

78. $\dfrac{16}{i} = \dfrac{16}{i} \cdot \dfrac{i}{i} = \dfrac{16i}{i^2} = \dfrac{16i}{-1} = -16i$

79. $i^{53} = i^{52} \cdot i = \left(i^4\right)^{13} \cdot i = i$

80. $i^{22} = i^{20} \cdot i^2 = \left(i^4\right)^5 \cdot i^2 = i^2 = -1$

CHAPTER 9 TEST

1. $\sqrt{25x^{10}y^{14}} = 5x^5 y^7$

2. $f(-5) = \sqrt{(-5)^2 - 9(-5) + 6}$
$= \sqrt{25 + 45 + 6}$
$= \sqrt{76}$
≈ 8.718

3. $a^{3/10} \cdot a^{1/5} = a^{3/10 + 1/5} = a^{3/10 + 2/10} = a^{5/10} = a^{1/2}$

4. $\dfrac{b^{7/8}}{b^{5/12}} = b^{7/8 - 5/12} = b^{21/24 - 10/24} = b^{11/24}$

5. $125^{-2/3} = \dfrac{1}{125^{2/3}} = \dfrac{1}{\left(\sqrt[3]{125}\right)^2} = \dfrac{1}{5^2} = \dfrac{1}{25}$

6. $\sqrt{72} = \sqrt{36 \cdot 2} = 6\sqrt{2}$

7. $\sqrt[4]{a^{21}b^{42}c^4} = \sqrt[4]{\left(a^{20} \cdot a\right)\left(b^{40} \cdot b^2\right)c^4}$
$= a^5 b^{10} c \sqrt[4]{ab^2}$

8. $8\sqrt{18} + 11\sqrt{2} - 3\sqrt{200}$
$= 8\sqrt{9 \cdot 2} + 11\sqrt{2} - 3\sqrt{100 \cdot 2}$
$= 8 \cdot 3\sqrt{2} + 11\sqrt{2} - 3 \cdot 10\sqrt{2}$
$= 24\sqrt{2} + 11\sqrt{2} - 30\sqrt{2}$
$= 5\sqrt{2}$

9. $4\sqrt{5}\left(7\sqrt{10} - 2\sqrt{35}\right) = 28\sqrt{50} - 8\sqrt{175}$
$= 28\sqrt{25 \cdot 2} - 8\sqrt{25 \cdot 7}$
$= 28 \cdot 5\sqrt{2} - 8 \cdot 5\sqrt{7}$
$= 140\sqrt{2} - 40\sqrt{7}$

10. $\left(3\sqrt{6} - 4\sqrt{8}\right)\left(9\sqrt{6} - 2\sqrt{8}\right)$
$= 27 \cdot 6 - 6\sqrt{48} - 36\sqrt{48} + 8 \cdot 8$
$= 162 - 42\sqrt{48} + 64$
$= 226 - 42\sqrt{16 \cdot 3}$
$= 226 - 42 \cdot 4\sqrt{3}$
$= 226 - 168\sqrt{3}$

11. $\dfrac{x^3 y^2}{\sqrt{x^3 y^{11}}} = \dfrac{x^3 y^2}{xy^5 \sqrt{xy}}$
$= \dfrac{x^2}{y^3 \sqrt{xy}} \cdot \dfrac{\sqrt{xy}}{\sqrt{xy}}$
$= \dfrac{x^2 \sqrt{xy}}{xy^4}$
$= \dfrac{x\sqrt{xy}}{y^4}$

12. $\dfrac{15\sqrt{7} - 8\sqrt{2}}{4\sqrt{7} + 9\sqrt{2}} = \dfrac{15\sqrt{7} - 8\sqrt{2}}{4\sqrt{7} + 9\sqrt{2}} \cdot \dfrac{4\sqrt{7} - 9\sqrt{2}}{4\sqrt{7} - 9\sqrt{2}}$
$= \dfrac{60 \cdot 7 - 135\sqrt{14} - 32\sqrt{14} + 72 \cdot 2}{16 \cdot 7 - 81 \cdot 2}$
$= \dfrac{420 - 167\sqrt{14} + 144}{112 - 162}$
$= \dfrac{564 - 167\sqrt{14}}{-50}$
$= \dfrac{-564 + 167\sqrt{14}}{50}$

13. $\sqrt[3]{6x + 5} = 5$
$\left(\sqrt[3]{6x + 5}\right)^3 = (5)^3$
$6x + 5 = 125$
$6x = 120$
$x = 20$
$\{20\}$

14. $\sqrt{x + 28} + 2 = x$
$\sqrt{x + 28} = x - 2$
$\left(\sqrt{x + 28}\right)^2 = (x - 2)^2$
$x + 28 = (x - 2)(x - 2)$
$x + 28 = x^2 - 4x + 4$
$0 = x^2 - 5x - 24$
$0 = (x - 8)(x + 3)$

$x - 8 = 0$ or $x + 3 = 0$
$x = 8$ or $x = -3$

check:

$x = 8$	$x = -3$
$\sqrt{(8) + 28} + 2 = 8$	$\sqrt{(-3) + 28} + 2 = -3$
$\sqrt{36} + 2 = 8$	$\sqrt{25} + 2 = -3$
$6 + 2 = 8$	$5 + 2 = -3$
$8 = 8$	$7 = -3$
True	False

$\{8\}$

15.
$$T = 2\pi\sqrt{\frac{L}{32}}$$

$$4 = 2\pi\sqrt{\frac{L}{32}}$$

$$\frac{4}{2\pi} = \sqrt{\frac{L}{32}}$$

$$\frac{2}{\pi} = \sqrt{\frac{L}{32}}$$

$$\left(\frac{2}{\pi}\right)^2 = \left(\sqrt{\frac{L}{32}}\right)^2$$

$$\frac{4}{\pi^2} = \frac{L}{32}$$

$$\frac{4 \cdot 32}{\pi^2} = L$$

$$\frac{128}{\pi^2} = L$$

$$L \approx 13.0 \text{ feet}$$

16. $\sqrt{-99} = \sqrt{9 \cdot 11(-1)} = \sqrt{9 \cdot 11} \cdot \sqrt{-1} = 3i\sqrt{11}$

17. $(2 - 3i) - (20 + 8i) = 2 - 3i - 20 - 8i = -18 - 11i$

18. $(6 + 7i)(6 - 7i) = 6^2 + 7^2 = 36 + 49 = 85$

19.
$$\frac{4 + 9i}{3 - 5i} = \frac{4 + 9i}{3 - 5i} \cdot \frac{3 + 5i}{3 + 5i}$$

$$= \frac{12 + 20i + 27i + 45i^2}{3^2 + 5^2}$$

$$= \frac{12 + 47i - 45}{9 + 25}$$

$$= \frac{-33 + 47i}{34}$$

$$= -\frac{33}{34} + \frac{47}{34}i$$

CHAPTER 10 QUADRATIC EQUATIONS

10.1 QUICK CHECK

1. a)
$$x^2 = 2x + 15$$
$$x^2 - 2x - 15 = 0$$
$$(x+3)(x-5) = 0$$
$$x+3 = 0 \quad \text{or} \quad x-5 = 0$$
$$x = -3 \quad \text{or} \quad x = 5$$
$$\{-3, 5\}$$

b)
$$x^2 = 121$$
$$x^2 - 121 = 0$$
$$(x+11)(x-11) = 0$$
$$x+11 = 0 \quad \text{or} \quad x-11 = 0$$
$$x = -11 \quad \text{or} \quad x = 11$$
$$\{-11, 11\}$$

2. a)
$$x^2 = 28$$
$$\sqrt{x^2} = \pm\sqrt{28}$$
$$x = \pm 2\sqrt{7}$$
$$\left\{-2\sqrt{7}, 2\sqrt{7}\right\}$$

b)
$$x^2 + 32 = 0$$
$$x^2 = -32$$
$$\sqrt{x^2} = \pm\sqrt{-32}$$
$$x = \pm 4\sqrt{2}i$$
$$\left\{-4i\sqrt{2}, 4i\sqrt{2}\right\}$$

3. a)
$$(3x+1)^2 = 100$$
$$\sqrt{(3x+1)^2} = \pm\sqrt{100}$$
$$3x+1 = \pm 10$$
$$3x = -1 \pm 10$$
$$x = \frac{-1 \pm 10}{3}$$
$$x = \frac{-1-10}{3} \quad \text{or} \quad x = \frac{-1+10}{3}$$
$$x = \frac{-11}{3} \quad \text{or} \quad x = \frac{9}{3}$$
$$x = -\frac{11}{3} \quad \text{or} \quad x = 3$$
$$\left\{-\frac{11}{3}, 3\right\}$$

b)
$$2(2x-3)^2 + 17 = -1$$
$$2(2x-3)^2 = -18$$
$$(2x-3)^2 = -9$$
$$\sqrt{(2x-3)^2} = \pm\sqrt{-9}$$
$$2x-3 = \pm 3i$$
$$2x = 3 \pm 3i$$
$$x = \frac{3 \pm 3i}{2}$$
$$\left\{\frac{3-3i}{2}, \frac{3+3i}{2}\right\}$$

4.
$$x^2 = 150$$
$$\sqrt{x^2} = \pm\sqrt{150}$$
$$x = \pm 5\sqrt{6}$$
$$x \approx \pm 12.2$$
12.2 inches

5.
$$\pi r^2 = 20$$
$$\frac{\pi r^2}{\pi} = \frac{20}{\pi}$$
$$r^2 = \frac{20}{\pi}$$
$$r = \pm\sqrt{\frac{20}{\pi}}$$
$$r \approx \pm 2.5$$
2.5 centimeters

6. a)
$$x^2 + 8x + 12 = 0$$
$$x^2 + 8x = -12$$
$$x^2 + 8x + \left(\frac{8}{2}\right)^2 = -12 + \left(\frac{8}{2}\right)^2$$
$$x^2 + 8x + 16 = -12 + 16$$
$$(x+4)^2 = 4$$
$$\sqrt{(x+4)^2} = \pm\sqrt{4}$$
$$x+4 = \pm 2$$
$$x = -4 \pm 2$$
$$x = -4-2 \quad \text{or} \quad x = -4+2$$
$$x = -6 \quad \text{or} \quad x = -2$$
$$\{-6, -2\}$$

Continued on next page.

6. Continued.

b)
$$x^2 - 6x + 10 = 0$$
$$x^2 - 6x = -10$$
$$x^2 - 6x + \left(-\frac{6}{2}\right)^2 = -10 + \left(-\frac{6}{2}\right)^2$$
$$x^2 - 6x + 9 = -10 + 9$$
$$(x-3)^2 = -1$$
$$\sqrt{(x-3)^2} = \pm\sqrt{-1}$$
$$x - 3 = \pm i$$
$$x = 3 \pm i$$
$$\{3 - i, 3 + i\}$$

7.
$$x^2 + 7x - 18 = 0$$
$$x^2 + 7x = 18$$
$$x^2 + 7x + \left(\frac{7}{2}\right)^2 = 18 + \left(\frac{7}{2}\right)^2$$
$$x^2 + 7x + \frac{49}{4} = \frac{72}{4} + \frac{49}{4}$$
$$\left(x + \frac{7}{2}\right)^2 = \frac{121}{4}$$
$$\sqrt{\left(x + \frac{7}{2}\right)^2} = \pm\sqrt{\frac{121}{4}}$$
$$x + \frac{7}{2} = \pm\frac{11}{2}$$
$$x = -\frac{7}{2} \pm \frac{11}{2}$$
$$x = -\frac{7}{2} - \frac{11}{2} \quad \text{or} \quad x = -\frac{7}{2} + \frac{11}{2}$$
$$x = -\frac{18}{2} \quad \text{or} \quad x = \frac{4}{2}$$
$$x = -9 \quad \text{or} \quad x = 2$$
$$\{-9, 2\}$$

8.
$$2x^2 - 32x + 92 = 0$$
$$\frac{2x^2 - 32x + 92}{2} = \frac{0}{2}$$
$$x^2 - 16x + 46 = 0$$
$$x^2 - 16x = -46$$
$$x^2 - 16x + \left(\frac{-16}{2}\right)^2 = -46 + \left(\frac{-16}{2}\right)^2$$
$$x^2 - 16x + 64 = -46 + 64$$
$$(x-8)^2 = 18$$
$$\sqrt{(x-8)^2} = \pm\sqrt{18}$$
$$x - 8 = \pm 3\sqrt{2}$$
$$x = 8 \pm 3\sqrt{2}$$
$$\{8 - 3\sqrt{2}, 8 + 3\sqrt{2}\}$$

9.
$$x = \frac{5}{3} \quad \text{or} \quad x = -2$$
$$3x = 5 \quad \text{or} \quad x + 2 = 0$$
$$3x - 5 = 0 \quad \text{or} \quad x + 2 = 0$$
$$(3x - 5)(x + 2) = 0$$
$$3x^2 + 6x - 5x - 10 = 0$$
$$3x^2 + x - 10 = 0$$

10.
$$x = 10i \quad \text{or} \quad x = -10i$$
$$x - 10i = 0 \quad \text{or} \quad x + 10i = 0$$
$$(x - 10i)(x + 10i) = 0$$
$$x^2 + 10xi - 10xi - 100i^2 = 0$$
$$x^2 - 100(-1) = 0$$
$$x^2 + 100 = 0$$

10.1 SOLVING QUADRATIC EQUATIONS BY EXTRACTING SQUARE ROOTS; COMPLETING THE SQUARE

1. extracting square roots

3. completing the square

5.
$$x^2 - 3x - 10 = 0$$
$$(x+2)(x-5) = 0$$
$$x + 2 = 0 \quad \text{or} \quad x - 5 = 0$$
$$x = -2 \quad \text{or} \quad x = 5$$
$$\{-2, 5\}$$

7.
$$x^2 + 7x = 0$$
$$x(x+7) = 0$$
$$x = 0 \quad \text{or} \quad x + 7 = 0$$
$$x = 0 \quad \text{or} \quad x = -7$$
$$\{-7, 0\}$$

9.
$$3x^2 - 13x + 4 = 0$$
$$(3x-1)(x-4) = 0$$
$$3x - 1 = 0 \quad \text{or} \quad x - 4 = 0$$
$$3x = 1 \quad \text{or} \quad x = 4$$
$$x = \frac{1}{3} \quad \text{or} \quad x = 4$$
$$\left\{\frac{1}{3}, 4\right\}$$

11.
$$x^2 - 2x = 24$$
$$x^2 - 2x - 24 = 0$$
$$(x+4)(x-6) = 0$$
$$x + 4 = 0 \quad \text{or} \quad x - 6 = 0$$
$$x = -4 \quad \text{or} \quad x = 6$$
$$\{-4, 6\}$$

13.
$$x^2 - 9 = 0$$
$$(x+3)(x-3) = 0$$
$$x + 3 = 0 \quad \text{or} \quad x - 3 = 0$$
$$x = -3 \quad \text{or} \quad x = 3$$
$$\{-3, 3\}$$

15.
$$x^2 - 4x = 7x + 26$$
$$x^2 - 11x = 26$$
$$x^2 - 11x - 26 = 0$$
$$(x+2)(x-13) = 0$$
$$x + 2 = 0 \quad \text{or} \quad x - 13 = 0$$
$$x = -2 \quad \text{or} \quad x = 13$$
$$\{-2, 13\}$$

17.
$$x^2 = 64$$
$$\sqrt{x^2} = \pm\sqrt{64}$$
$$x = \pm 8$$
$$\{-8, 8\}$$

19.
$$x^2 - 24 = 0$$
$$x^2 = 24$$
$$\sqrt{x^2} = \pm\sqrt{24}$$
$$x = \pm 2\sqrt{6}$$
$$\left\{-2\sqrt{6}, 2\sqrt{6}\right\}$$

21.
$$x^2 = -72$$
$$\sqrt{x^2} = \pm\sqrt{-72}$$
$$x = \pm 6i\sqrt{2}$$
$$\left\{-6i\sqrt{2}, 6i\sqrt{2}\right\}$$

23.
$$x^2 + 52 = 32$$
$$x^2 = -20$$
$$\sqrt{x^2} = \pm\sqrt{-20}$$
$$x = \pm 2i\sqrt{5}$$
$$\left\{-2i\sqrt{5}, 2i\sqrt{5}\right\}$$

25.
$$(x-7)^2 = 18$$
$$\sqrt{(x-7)^2} = \pm\sqrt{18}$$
$$x - 7 = \pm 3\sqrt{2}$$
$$x = 7 \pm 3\sqrt{2}$$
$$\left\{7 - 3\sqrt{2}, 7 + 3\sqrt{2}\right\}$$

27.
$$(x+6)^2 = -49$$
$$\sqrt{(x+6)^2} = \pm\sqrt{-49}$$
$$x + 6 = \pm 7i$$
$$x = -6 \pm 7i$$
$$\{-6 - 7i, -6 + 7i\}$$

29.
$$(x+5)^2 + 33 = 15$$
$$(x+5)^2 = -18$$
$$\sqrt{(x+5)^2} = \pm\sqrt{-18}$$
$$x + 5 = \pm 3i\sqrt{2}$$
$$x = -5 \pm 3i\sqrt{2}$$
$$\left\{-5 - 3i\sqrt{2}, -5 + 3i\sqrt{2}\right\}$$

31.
$$2(x-9)^2 - 17 = 83$$
$$2(x-9)^2 = 100$$
$$(x-9)^2 = 50$$
$$\sqrt{(x-9)^2} = \pm\sqrt{50}$$
$$x-9 = \pm 5\sqrt{2}$$
$$x = 9 \pm 5\sqrt{2}$$
$$\{9-5\sqrt{2},\, 9+5\sqrt{2}\}$$

33.
$$\left(x-\frac{2}{5}\right)^2 = \frac{49}{25}$$
$$\sqrt{\left(x-\frac{2}{5}\right)^2} = \pm\sqrt{\frac{49}{25}}$$
$$x-\frac{2}{5} = \pm\frac{7}{5}$$
$$x = \frac{2}{5} \pm \frac{7}{5}$$
$$x = \frac{2}{5} - \frac{7}{5} \quad \text{or} \quad x = \frac{2}{5} + \frac{7}{5}$$
$$x = -\frac{5}{5} \quad \text{or} \quad x = \frac{9}{5}$$
$$x = -1 \quad \text{or} \quad x = \frac{9}{5}$$
$$\left\{-1, \frac{9}{5}\right\}$$

35. $x^2 + 12x +$ _____
$$\left(\frac{12}{2}\right)^2 = 6^2 = 36, \text{ so } x^2 + 12x + 36 = (x+6)^2.$$

37. $x^2 - 5x +$ _____
$$\left(-\frac{5}{2}\right)^2 = \frac{25}{4}, \text{ so } x^2 - 5x + \frac{25}{4} = \left(x-\frac{5}{2}\right)^2.$$

39. $x^2 - 8x +$ _____
$$\left(-\frac{8}{2}\right)^2 = (-4)^2 = 16, \text{ so } x^2 - 8x + 16 = (x-4)^2.$$

41. $x^2 + \dfrac{3}{4}x +$ _____
$$\left(\frac{\frac{3}{4}}{2}\right)^2 = \left(\frac{3}{8}\right)^2 = \frac{9}{64},$$
$$\text{so } x^2 + \frac{3}{4}x + \frac{9}{64} = \left(x+\frac{3}{8}\right)^2.$$

43.
$$x^2 = 81$$
$$\sqrt{x^2} = \pm\sqrt{81}$$
$$x = \pm 9$$
9 meters

45.
$$x^2 = 128$$
$$\sqrt{x^2} = \pm\sqrt{128}$$
$$x = \pm 8\sqrt{2}$$
$$x \approx \pm 11.3$$
11.3 inches

47.
$$\pi r^2 = 24$$
$$\frac{\pi r^2}{\pi} = \frac{24}{\pi}$$
$$r^2 = \frac{24}{\pi}$$
$$r = \pm\sqrt{\frac{24}{\pi}}$$
$$r \approx \pm 2.8$$
2.8 inches

49.
$$\pi r^2 = 250$$
$$\frac{\pi r^2}{\pi} = \frac{250}{\pi}$$
$$r^2 = \frac{250}{\pi}$$
$$r = \pm\sqrt{\frac{250}{\pi}}$$
$$r \approx \pm 8.9$$
8.9 meters

51.
$$x^2 - 2x - 63 = 0$$
$$x^2 - 2x = 63$$
$$x^2 - 2x + \left(-\frac{2}{2}\right)^2 = 63 + \left(-\frac{2}{2}\right)^2$$
$$x^2 - 2x + 1 = 63 + 1$$
$$(x-1)^2 = 64$$
$$\sqrt{(x-1)^2} = \pm\sqrt{64}$$
$$x - 1 = \pm 8$$
$$x = 1 \pm 8$$
$$x = 1 - 8 \quad \text{or} \quad x = 1 + 8$$
$$x = -7 \quad \text{or} \quad x = 9$$
$$\{-7, 9\}$$

53.
$$x^2 + 6x - 11 = 0$$
$$x^2 + 6x = 11$$
$$x^2 + 6x + \left(\frac{6}{2}\right)^2 = 11 + \left(\frac{6}{2}\right)^2$$
$$x^2 + 6x + 9 = 11 + 9$$
$$(x+3)^2 = 20$$
$$\sqrt{(x+3)^2} = \pm\sqrt{20}$$
$$x + 3 = \pm 2\sqrt{5}$$
$$x = -3 \pm 2\sqrt{5}$$
$$x = -3 - 2\sqrt{5} \quad \text{or} \quad x = -3 + 2\sqrt{5}$$
$$\left\{-3 - 2\sqrt{5}, -3 + 2\sqrt{5}\right\}$$

55.
$$x^2 - 4x = -12$$
$$x^2 - 4x + \left(-\frac{4}{2}\right)^2 = -12 + \left(-\frac{4}{2}\right)^2$$
$$x^2 - 4x + 4 = -12 + 4$$
$$(x-2)^2 = -8$$
$$\sqrt{(x-2)^2} = \pm\sqrt{-8}$$
$$x - 2 = \pm 2i\sqrt{2}$$
$$x = 2 \pm 2i\sqrt{2}$$
$$\left\{2 - 2i\sqrt{2}, 2 + 2i\sqrt{2}\right\}$$

57.
$$x^2 - 14x + 30 = 0$$
$$x^2 - 14x = -30$$
$$x^2 - 14x + \left(-\frac{14}{2}\right)^2 = -30 + \left(-\frac{14}{2}\right)^2$$
$$x^2 - 14x + 49 = -30 + 49$$
$$(x-7)^2 = 19$$
$$\sqrt{(x-7)^2} = \pm\sqrt{19}$$
$$x - 7 = \pm\sqrt{19}$$
$$x = 7 \pm \sqrt{19}$$
$$\left\{7 - \sqrt{19}, 7 + \sqrt{19}\right\}$$

59.
$$x^2 + 8x + 44 = 0$$
$$x^2 + 8x = -44$$
$$x^2 + 8x + \left(\frac{8}{2}\right)^2 = -44 + \left(\frac{8}{2}\right)^2$$
$$x^2 + 8x + 16 = -44 + 16$$
$$(x+4)^2 = -28$$
$$\sqrt{(x+4)^2} = \pm\sqrt{-28}$$
$$x + 4 = \pm 2i\sqrt{7}$$
$$x = -4 \pm 2i\sqrt{7}$$
$$\left\{-4 - 2i\sqrt{7}, -4 + 2i\sqrt{7}\right\}$$

61.
$$x^2 - 16 = -6x$$
$$x^2 + 6x = 16$$
$$x^2 + 6x + \left(\frac{6}{2}\right)^2 = 16 + \left(\frac{6}{2}\right)^2$$
$$x^2 + 6x + 9 = 16 + 9$$
$$(x+3)^2 = 25$$
$$\sqrt{(x+3)^2} = \pm\sqrt{25}$$
$$x + 3 = \pm 5$$
$$x = -3 \pm 5$$
$$x = -3 - 5 \quad \text{or} \quad x = -3 + 5$$
$$x = -8 \quad \text{or} \quad x = 2$$
$$\{-8, 2\}$$

63.
$$x^2 - 3x - 54 = 0$$
$$x^2 - 3x = 54$$
$$x^2 - 3x + \left(-\frac{3}{2}\right)^2 = 54 + \left(-\frac{3}{2}\right)^2$$
$$x^2 - 3x + \frac{9}{4} = 54 + \frac{9}{4}$$
$$\left(x - \frac{3}{2}\right)^2 = \frac{225}{4}$$
$$\sqrt{\left(x - \frac{3}{2}\right)^2} = \pm\sqrt{\frac{225}{4}}$$
$$x - \frac{3}{2} = \pm\frac{15}{2}$$
$$x = \frac{3}{2} \pm \frac{15}{2}$$
$$x = \frac{3}{2} - \frac{15}{2} \quad \text{or} \quad x = \frac{3}{2} + \frac{15}{2}$$
$$x = -\frac{12}{2} \quad \text{or} \quad x = \frac{18}{2}$$
$$x = -6 \quad \text{or} \quad x = 9$$
$$\{-6, 9\}$$

65.
$$x^2 + 7x - 10 = 0$$
$$x^2 + 7x = 10$$
$$x^2 + 7x + \left(\frac{7}{2}\right)^2 = 10 + \left(\frac{7}{2}\right)^2$$
$$x^2 + 7x + \frac{49}{4} = 10 + \frac{49}{4}$$
$$\left(x + \frac{7}{2}\right)^2 = \frac{89}{4}$$
$$\sqrt{\left(x + \frac{7}{2}\right)^2} = \pm\sqrt{\frac{89}{4}}$$
$$x + \frac{7}{2} = \pm\frac{\sqrt{89}}{2}$$
$$x = -\frac{7}{2} \pm \frac{\sqrt{89}}{2}$$
$$\left\{\frac{-7 - \sqrt{89}}{2}, \frac{-7 + \sqrt{89}}{2}\right\}$$

67.
$$x^2 - 9x + 36 = 0$$
$$x^2 - 9x = -36$$
$$x^2 - 9x + \left(-\frac{9}{2}\right)^2 = -36 + \left(-\frac{9}{2}\right)^2$$
$$x^2 - 9x + \frac{81}{4} = -36 + \frac{81}{4}$$
$$\left(x - \frac{9}{2}\right)^2 = -\frac{63}{4}$$
$$\sqrt{\left(x - \frac{9}{2}\right)^2} = \pm\sqrt{-\frac{63}{4}}$$
$$x - \frac{9}{2} = \pm\frac{3i\sqrt{7}}{2}$$
$$x = \frac{9}{2} \pm \frac{3i\sqrt{7}}{2}$$
$$\left\{\frac{9 - 3i\sqrt{7}}{2}, \frac{9 + 3i\sqrt{7}}{2}\right\}$$

69.
$$x^2 + \frac{5}{2}x + 1 = 0$$
$$x^2 + \frac{5}{2}x = -1$$
$$x^2 + \frac{5}{2}x + \left(\frac{5}{4}\right)^2 = -1 + \left(\frac{5}{4}\right)^2$$
$$x^2 + \frac{5}{2}x + \frac{25}{16} = -\frac{16}{16} + \frac{25}{16}$$
$$\left(x + \frac{5}{4}\right)^2 = \frac{9}{16}$$
$$\sqrt{\left(x + \frac{5}{4}\right)^2} = \pm\sqrt{\frac{9}{16}}$$
$$x + \frac{5}{4} = \pm\frac{3}{4}$$
$$x = -\frac{5}{4} \pm \frac{3}{4}$$
$$x = -\frac{5}{4} - \frac{3}{4} \quad \text{or} \quad x = -\frac{5}{4} + \frac{3}{4}$$
$$x = -\frac{8}{4} \quad \text{or} \quad x = -\frac{2}{4}$$
$$x = -2 \quad \text{or} \quad x = -\frac{1}{2}$$
$$\left\{-2, -\frac{1}{2}\right\}$$

71.

$$2x^2 - 2x - 144 = 0$$

$$\frac{2x^2 - 2x - 144}{2} = \frac{0}{2}$$

$$x^2 - x - 72 = 0$$

$$x^2 - x = 72$$

$$x^2 - x + \left(\frac{-1}{2}\right)^2 = 72 + \left(\frac{-1}{2}\right)^2$$

$$x^2 - x + \frac{1}{4} = \frac{288}{4} + \frac{1}{4}$$

$$\left(x - \frac{1}{2}\right)^2 = \frac{289}{4}$$

$$\sqrt{\left(x - \frac{1}{2}\right)^2} = \pm\sqrt{\frac{289}{4}}$$

$$x - \frac{1}{2} = \pm\frac{17}{2}$$

$$x = \frac{1}{2} \pm \frac{17}{2}$$

$$x = \frac{1}{2} - \frac{17}{2} \quad \text{or} \quad x = \frac{1}{2} + \frac{17}{2}$$

$$x = \frac{-16}{2} \quad \text{or} \quad x = \frac{18}{2}$$

$$x = -8 \quad \text{or} \quad x = 9$$

$$\{-8, 9\}$$

73.

$$2x^2 - 9x + 4 = 0$$

$$\frac{2x^2 - 9x + 4}{2} = \frac{0}{2}$$

$$x^2 - \frac{9}{2}x + 2 = 0$$

$$x^2 - \frac{9}{2}x = -2$$

$$x^2 - \frac{9}{2}x + \left(-\frac{9}{4}\right)^2 = -2 + \left(-\frac{9}{4}\right)^2$$

$$x^2 - \frac{9}{2}x + \frac{81}{16} = -2 + \frac{81}{16}$$

$$\left(x - \frac{9}{4}\right)^2 = \frac{49}{16}$$

$$\sqrt{\left(x - \frac{9}{4}\right)^2} = \pm\sqrt{\frac{49}{16}}$$

$$\left(x - \frac{9}{4}\right) = \pm\frac{7}{4}$$

$$x = \frac{9}{4} \pm \frac{7}{4}$$

$$x = \frac{9}{4} - \frac{7}{4} \quad \text{or} \quad x = \frac{9}{4} + \frac{7}{4}$$

$$x = \frac{2}{4} \quad \text{or} \quad x = \frac{16}{4}$$

$$x = \frac{1}{2} \quad \text{or} \quad x = 4$$

$$\left\{\frac{1}{2}, 4\right\}$$

75.
$$2x^2 - 6x + 5 = 0$$
$$\frac{2x^2 - 6x + 5}{2} = \frac{0}{2}$$
$$x^2 - 3x + \frac{5}{2} = 0$$
$$x^2 - 3x = -\frac{5}{2}$$
$$x^2 - 3x + \left(-\frac{3}{2}\right)^2 = -\frac{5}{2} + \left(-\frac{3}{2}\right)^2$$
$$x^2 - 3x + \frac{9}{4} = -\frac{5}{2} + \frac{9}{4}$$
$$\left(x - \frac{3}{2}\right)^2 = -\frac{1}{4}$$
$$\sqrt{\left(x - \frac{3}{2}\right)^2} = \pm\sqrt{-\frac{1}{4}}$$
$$\left(x - \frac{3}{2}\right) = \pm\frac{i}{2}$$
$$x = \frac{3}{2} \pm \frac{i}{2}$$
$$\left\{\frac{3 - i}{2}, \frac{3 + i}{2}\right\}$$

77.
$$x = 2 \quad \text{or} \quad x = -5$$
$$x - 2 = 0 \quad \text{or} \quad x + 5 = 0$$
$$(x - 2)(x + 5) = 0$$
$$x^2 + 5x - 2x - 10 = 0$$
$$x^2 + 3x - 10 = 0$$

79.
$$x = 6 \quad \text{or} \quad x = 8$$
$$x - 6 = 0 \quad \text{or} \quad x - 8 = 0$$
$$(x - 6)(x - 8) = 0$$
$$x^2 - 8x - 6x + 48 = 0$$
$$x^2 - 14x + 48 = 0$$

81.
$$x = -2 \quad \text{or} \quad x = \frac{3}{4}$$
$$x + 2 = 0 \quad \text{or} \quad 4x = 3$$
$$x + 2 = 0 \quad \text{or} \quad 4x - 3 = 0$$
$$(x + 2)(4x - 3) = 0$$
$$4x^2 - 3x + 8x - 6 = 0$$
$$4x^2 + 5x - 6 = 0$$

83.
$$x = 3 \quad \text{or} \quad x = -3$$
$$x - 3 = 0 \quad \text{or} \quad x + 3 = 0$$
$$(x - 3)(x + 3) = 0$$
$$x^2 - 9 = 0$$

85.
$$x = 5i \quad \text{or} \quad x = -5i$$
$$x - 5i = 0 \quad \text{or} \quad x + 5i = 0$$
$$(x - 5i)(x + 5i) = 0$$
$$x^2 + 5xi - 5xi - 25i^2 = 0$$
$$x^2 - 25(-1) = 0$$
$$x^2 + 25 = 0$$

87.
$$x^2 + 13x + 30 = 0$$
$$(x + 3)(x + 10) = 0$$
$$x + 3 = 0 \quad \text{or} \quad x + 10 = 0$$
$$x = -3 \quad \text{or} \quad x = -10$$
$$\{-3, -10\}$$

89.
$$x^2 + 6x - 17 = 0$$
$$x^2 + 6x = 17$$
$$x^2 + 6x + \left(\frac{6}{2}\right)^2 = 17 + \left(\frac{6}{2}\right)^2$$
$$x^2 + 6x + 9 = 17 + 9$$
$$(x + 3)^2 = 26$$
$$\sqrt{(x + 3)^2} = \pm\sqrt{26}$$
$$x + 3 = \pm\sqrt{26}$$
$$x = -3 \pm \sqrt{26}$$
$$\left\{-3 - \sqrt{26}, -3 + \sqrt{26}\right\}$$

91.
$$x^2 - 6x - 7 = 0$$
$$(x + 1)(x - 7) = 0$$
$$x + 1 = 0 \quad \text{or} \quad x - 7 = 0$$
$$x = -1 \quad \text{or} \quad x = 7$$
$$\{-1, 7\}$$

93.
$$2x^2 - 15x + 25 = 0$$
$$\frac{2x^2 - 15x + 25}{2} = \frac{0}{2}$$
$$x^2 - \frac{15}{2}x + \frac{25}{2} = 0$$
$$x^2 - \frac{15}{2}x = -\frac{25}{2}$$
$$x^2 - \frac{15}{2}x + \left(-\frac{15}{4}\right)^2 = -\frac{25}{2} + \left(-\frac{15}{4}\right)^2$$
$$x^2 - \frac{15}{2}x + \frac{225}{16} = -\frac{25}{2} + \frac{225}{16}$$
$$\left(x - \frac{15}{4}\right)^2 = \frac{25}{16}$$
$$\sqrt{\left(x - \frac{15}{4}\right)^2} = \pm\sqrt{\frac{25}{16}}$$
$$\left(x - \frac{15}{4}\right) = \pm\frac{5}{4}$$
$$x = \frac{15}{4} \pm \frac{5}{4}$$
$$x = \frac{15}{4} - \frac{5}{4} \quad \text{or} \quad x = \frac{15}{4} + \frac{5}{4}$$
$$x = \frac{10}{4} \quad \text{or} \quad x = \frac{20}{4}$$
$$x = \frac{5}{2} \quad \text{or} \quad x = 5$$
$$\left\{\frac{5}{2}, 5\right\}$$

95.
$$(x - 4)^2 - 11 = 16$$
$$(x - 4)^2 = 27$$
$$\sqrt{(x - 4)^2} = \pm\sqrt{27}$$
$$x - 4 = \pm 3\sqrt{3}$$
$$x = 4 \pm 3\sqrt{3}$$
$$\left\{4 - 3\sqrt{3}, 4 + 3\sqrt{3}\right\}$$

97.
$$x^2 - 4x = 20$$
$$x^2 - 4x + \left(-\frac{4}{2}\right)^2 = 20 + \left(-\frac{4}{2}\right)^2$$
$$x^2 - 4x + 4 = 20 + 4$$
$$(x - 2)^2 = 24$$
$$\sqrt{(x - 2)^2} = \pm\sqrt{24}$$
$$x - 2 = \pm 2\sqrt{6}$$
$$x = 2 \pm 2\sqrt{6}$$
$$\left\{2 - 2\sqrt{6}, 2 + 2\sqrt{6}\right\}$$

99.
$$x^2 - 5x - 6 = 0$$
$$(x + 1)(x - 6) = 0$$
$$x + 1 = 0 \quad \text{or} \quad x - 6 = 0$$
$$x = -1 \quad \text{or} \quad x = 6$$
$$\{-1, 6\}$$

101.
$$x^2 - 8x + 19 = 0$$
$$x^2 - 8x = -19$$
$$x^2 - 8x + \left(-\frac{8}{2}\right)^2 = -19 + \left(-\frac{8}{2}\right)^2$$
$$x^2 - 8x + 16 = -19 + 16$$
$$(x - 4)^2 = -3$$
$$\sqrt{(x - 4)^2} = \pm\sqrt{-3}$$
$$x - 4 = \pm i\sqrt{3}$$
$$x = 4 \pm i\sqrt{3}$$
$$\left\{4 - i\sqrt{3}, 4 + i\sqrt{3}\right\}$$

103.
$$3(x + 3)^2 + 13 = -11$$
$$3(x + 3)^2 = -24$$
$$(x + 3)^2 = -8$$
$$\sqrt{(x + 3)^2} = \pm\sqrt{-8}$$
$$x + 3 = \pm 2i\sqrt{2}$$
$$x = -3 \pm 2i\sqrt{2}$$
$$\left\{-3 - 2i\sqrt{2}, -3 + 2i\sqrt{2}\right\}$$

105.
$$x^2 + 3x + 9 = 0$$
$$x^2 + 3x = -9$$
$$x^2 + 3x + \left(\frac{3}{2}\right)^2 = -9 + \left(\frac{3}{2}\right)^2$$
$$x^2 + 3x + \frac{9}{4} = -\frac{36}{4} + \frac{9}{4}$$
$$\left(x + \frac{3}{2}\right)^2 = -\frac{27}{4}$$
$$\sqrt{\left(x + \frac{3}{2}\right)^2} = \pm\sqrt{-\frac{27}{4}}$$
$$x + \frac{3}{2} = \pm\frac{3i\sqrt{3}}{2}$$
$$x = -\frac{3}{2} \pm \frac{3i\sqrt{3}}{2}$$
$$\left\{\frac{-3 - 3i\sqrt{3}}{2}, \frac{-3 + 3i\sqrt{3}}{2}\right\}$$

107. $x^2 + 16x = 0$

$x(x+16) = 0$

$x = 0$ or $x + 16 = 0$

$x = 0$ or $x = -16$

$\{-16, 0\}$

109. $x(x+5) - 7(x+5) = 0$

$(x+5)(x-7) = 0$

$x + 5 = 0$ or $x - 7 = 0$

$x = -5$ or $x = 7$

$\{-5, 7\}$

111. $-16x^2 + 64x + 80 = 0$

$$\frac{-16x^2 + 64x + 80}{-16} = \frac{0}{-16}$$

$x^2 - 4x - 5 = 0$

$(x+1)(x-5) = 0$

$x + 1 = 0$ or $x - 5 = 0$

$x = -1$ or $x = 5$

$\{-1, 5\}$

113. $\dfrac{2}{3}x^2 + \dfrac{8}{3}x - \dfrac{10}{3} = 0$

$$3 \cdot \left(\frac{2}{3}x^2 + \frac{8}{3}x - \frac{10}{3} \right) = 3 \cdot 0$$

$2x^2 + 8x - 10 = 0$

$$\frac{2x^2 + 8x - 10}{2} = \frac{0}{2}$$

$x^2 + 4x - 5 = 0$

$(x+5)(x-1) = 0$

$x + 5 = 0$ or $x - 1 = 0$

$x = -5$ or $x = 1$

$\{-5, 1\}$

115. Answers will vary. Example:
We use the $\pm$ symbol when taking the square root of a side to make sure to include all solutions.

10.2 QUICK CHECK

1. a) $x^2 + 11x - 13 = 0$

$a = 1,\ b = 11,\ c = -13$

b) $5x^2 - 11 = 9x$

$5x^2 - 9x - 11 = 0$

$a = 5,\ b = -9,\ c = -11$

c) $x^2 = 30$

$x^2 - 30 = 0$

$a = 1,\ b = 0,\ c = -30$

2. a) $x^2 + 7x - 30 = 0$

$a = 1,\ b = 7,\ c = -30$

$$x = \frac{-7 \pm \sqrt{7^2 - 4(1)(-30)}}{2(1)}$$

$$x = \frac{-7 \pm \sqrt{49 + 120}}{2}$$

$$x = \frac{-7 \pm \sqrt{169}}{2}$$

$$x = \frac{-7 \pm 13}{2}$$

$$x = -\frac{20}{2} \quad \text{or} \quad x = \frac{6}{2}$$

$x = -10$ or $x = 3$

$\{-10, 3\}$

b) $7x^2 + 21x = -9$

$7x^2 + 21x + 9 = 0$

$a = 7,\ b = 21,\ c = 9$

$$x = \frac{-21 \pm \sqrt{21^2 - 4(7)(9)}}{2(7)}$$

$$x = \frac{-21 \pm \sqrt{189}}{14}$$

$$x = \frac{-21 \pm 3\sqrt{21}}{14}$$

$$\left\{ \frac{-21 + 3\sqrt{21}}{14},\ \frac{-21 - 3\sqrt{21}}{14} \right\}$$

3.
$$x^2 - 7x = -19$$
$$x^2 - 7x + 19 = 0$$
$$a = 1,\ b = -7,\ c = 19$$

$$x = \frac{-(-7) \pm \sqrt{(-7)^2 - 4(1)(19)}}{2(1)}$$

$$x = \frac{7 \pm \sqrt{-27}}{2}$$

$$x = \frac{7 \pm 3i\sqrt{3}}{2}$$

$$\left\{ \frac{7 + 3i\sqrt{3}}{2},\ \frac{7 - 3i\sqrt{3}}{2} \right\}$$

4.
$$\frac{1}{4}x^2 + \frac{1}{3}x + \frac{1}{2} = 0$$

$$12 \cdot \left(\frac{1}{4}x^2 + \frac{1}{3}x + \frac{1}{2} \right) = 12 \cdot 0$$

$$3x^2 + 4x + 6 = 0$$

$$x = \frac{-4 \pm \sqrt{4^2 - 4(3)(6)}}{2(3)}$$

$$x = \frac{-4 \pm \sqrt{-56}}{6}$$

$$x = \frac{-4 \pm 2i\sqrt{14}}{6}$$

$$x = \frac{-2 \pm i\sqrt{14}}{3}$$

$$\left\{ \frac{-2 + i\sqrt{14}}{3},\ \frac{-2 - i\sqrt{14}}{3} \right\}$$

5. $-6x^2 - 7x + 20 = 0$
$$0 = 6x^2 + 7x - 20$$

$$x = \frac{-7 \pm \sqrt{7^2 - 4(6)(-20)}}{2(6)}$$

$$x = \frac{-7 \pm \sqrt{529}}{12}$$

$$x = \frac{-7 \pm 23}{12}$$

$$x = -\frac{30}{12} \quad \text{or} \quad x = \frac{16}{12}$$

$$x = -\frac{5}{2} \quad \text{or} \quad x = \frac{4}{3}$$

$$\left\{ -\frac{5}{2}, \frac{4}{3} \right\}$$

6. a) $x^2 + 18x - 63 = 0$
$$(18)^2 - 4(1)(-63) = 576 \text{ positive}$$
Two real solutions

b) $x^2 + 5x + 42 = 0$
$$(5)^2 - 4(1)(42) = -143 \text{ negative}$$
Two nonreal complex solutions

c) $x^2 - 20x + 100 = 0$
$$(-20)^2 - 4(1)(100) = 0$$
One real solution

7. a) $x^2 + 14x + 12$
$$(14)^2 - 4(1)(12) = 148$$
Not a perfect square
Not factorable

b) $5x^2 - 36x - 32$
$$(-36)^2 - 4(5)(-32) = 1936$$
$$\sqrt{1936} = 44$$
A perfect square
Factorable

8. <u>Unknown :</u>   <u>Knowns :</u>
time: t $v_o : 80$ ft/sec
                    $s : 0$ ft
                    $h(t) : 0$ ft

$$0 = -16t^2 + 80t$$
$$0 = -16t(t - 5)$$
$$-16t = 0 \quad \text{or} \quad t - 5 = 0$$
$$t = 0 \quad \text{or} \quad t = 5$$
5 seconds

9. <u>Unknown :</u>   <u>Knowns :</u>
time: t $v_o : 36$ ft/sec
                    $s : 40$ ft
                    $h(t) : 50$ ft

$$50 = -16t^2 + 36t + 40$$
$$16t^2 - 36t + 10 = 0$$

$$t = \frac{-(-36) \pm \sqrt{(-36)^2 - 4(16)(10)}}{2(16)}$$

$$t = \frac{36 \pm \sqrt{656}}{32}$$

$$t = \frac{36 \pm 4\sqrt{41}}{32}$$

$$t = \frac{9 \pm \sqrt{41}}{8}$$

$$t \approx 0.32 \text{ seconds or } 1.93 \text{ seconds}$$

10.
$$80 = -16t^2 + 36t + 40$$
$$16t^2 - 36t + 40 = 0$$
$$t = \frac{-(-36) \pm \sqrt{(-36)^2 - 4(16)(40)}}{2(16)}$$
$$t = \frac{36 \pm \sqrt{-1264}}{32}$$

No, the projectile will never reach a height of 80 feet.

10.2 THE QUADRATIC FORMULA

1. quadratic

3. zero

5. two

7. $x^2 - 5x - 36 = 0$
$$x = \frac{-(-5) \pm \sqrt{(-5)^2 - 4(1)(-36)}}{2(1)}$$
$$x = \frac{5 \pm \sqrt{169}}{2}$$
$$x = \frac{5 \pm 13}{2}$$
$$x = -\frac{8}{2} \quad \text{or} \quad x = \frac{18}{2}$$
$$x = -4 \quad \text{or} \quad x = 9$$
$$\{-4, 9\}$$

9. $x^2 - 4x + 2 = 0$
$$x = \frac{-(-4) \pm \sqrt{(-4)^2 - 4(1)(2)}}{2(1)}$$
$$x = \frac{4 \pm \sqrt{8}}{2}$$
$$x = \frac{4 \pm 2\sqrt{2}}{2}$$
$$x = 2 \pm \sqrt{2}$$
$$\left\{2 - \sqrt{2}, 2 + \sqrt{2}\right\}$$

11. $x^2 + x + 7 = 0$
$$x = \frac{-1 \pm \sqrt{1^2 - 4(1)(7)}}{2(1)}$$
$$x = \frac{-1 \pm \sqrt{-27}}{2}$$
$$x = \frac{-1 \pm 3i\sqrt{3}}{2}$$
$$\left\{\frac{-1 - 3i\sqrt{3}}{2}, \frac{-1 + 3i\sqrt{3}}{2}\right\}$$

13. $x^2 + 12 = 0$
$$x = \frac{-0 \pm \sqrt{0^2 - 4(1)(12)}}{2(1)}$$
$$x = \frac{\pm\sqrt{-48}}{2}$$
$$x = \frac{\pm 4i\sqrt{3}}{2}$$
$$x = \pm 2i\sqrt{3}$$
$$\left\{-2i\sqrt{3}, 2i\sqrt{3}\right\}$$

15. $x^2 + 7x + 11 = 0$
$$x = \frac{-7 \pm \sqrt{7^2 - 4(1)(11)}}{2(1)}$$
$$x = \frac{-7 \pm \sqrt{5}}{2}$$
$$\left\{\frac{-7 - \sqrt{5}}{2}, \frac{-7 + \sqrt{5}}{2}\right\}$$

17. $x^2 - 3x - 88 = 0$
$$x = \frac{-(-3) \pm \sqrt{(-3)^2 - 4(1)(-88)}}{2(1)}$$
$$x = \frac{3 \pm \sqrt{361}}{2}$$
$$x = \frac{3 \pm 19}{2}$$
$$x = \frac{3 - 19}{2} \quad \text{or} \quad x = \frac{3 + 19}{2}$$
$$x = -\frac{16}{2} \quad \text{or} \quad x = \frac{22}{2}$$
$$x = -8 \quad \text{or} \quad x = 11$$
$$\{-8, 11\}$$

19. $2x^2 - 5x - 12 = 0$

$$x = \frac{-(-5) \pm \sqrt{(-5)^2 - 4(2)(-12)}}{2(2)}$$

$$x = \frac{5 \pm \sqrt{121}}{4}$$

$$x = \frac{5 \pm 11}{4}$$

$$x = \frac{5 - 11}{4} \quad \text{or} \quad x = \frac{5 + 11}{4}$$

$$x = -\frac{6}{4} \quad \text{or} \quad x = \frac{16}{4}$$

$$x = -\frac{3}{2} \quad \text{or} \quad x = 4$$

$$\left\{-\frac{3}{2}, 4\right\}$$

21. $x^2 - 4x = -32$

$x^2 - 4x + 32 = 0$

$$x = \frac{-(-4) \pm \sqrt{(-4)^2 - 4(1)(32)}}{2(1)}$$

$$x = \frac{4 \pm \sqrt{-112}}{2}$$

$$x = \frac{4 \pm 4i\sqrt{7}}{2}$$

$$x = 2 \pm 2i\sqrt{7}$$

$$\left\{2 - 2i\sqrt{7}, 2 + 2i\sqrt{7}\right\}$$

23. $x^2 - 3x = 9$

$x^2 - 3x - 9 = 0$

$$x = \frac{-(-3) \pm \sqrt{(-3)^2 - 4(1)(-9)}}{2(1)}$$

$$x = \frac{3 \pm \sqrt{45}}{2}$$

$$x = \frac{3 \pm 3\sqrt{5}}{2}$$

$$\left\{\frac{3 - 3\sqrt{5}}{2}, \frac{3 + 3\sqrt{5}}{2}\right\}$$

25. $\qquad -4 = 19x - 5x^2$

$5x^2 - 19x - 4 = 0$

$$x = \frac{-(-19) \pm \sqrt{(-19)^2 - 4(5)(-4)}}{2(5)}$$

$$x = \frac{19 \pm \sqrt{441}}{10}$$

$$x = \frac{19 \pm 21}{10}$$

$$x = -\frac{2}{10} \quad \text{or} \quad x = \frac{40}{10}$$

$$x = -\frac{1}{5} \quad \text{or} \quad x = 4$$

$$\left\{-\frac{1}{5}, 4\right\}$$

27. $x^2 - 6x + 9 = 0$

$$x = \frac{-(-6) \pm \sqrt{(-6)^2 - 4(1)(9)}}{2(1)}$$

$$x = \frac{6 \pm \sqrt{0}}{2}$$

$$x = \frac{6}{2}$$

$$x = 3$$

$$\{3\}$$

29. $x^2 - 24 = 0$

$$x = \frac{-0 \pm \sqrt{0^2 - 4(1)(-24)}}{2(1)}$$

$$x = \frac{\pm\sqrt{96}}{2}$$

$$x = \frac{\pm 4\sqrt{6}}{2}$$

$$x = \pm 2\sqrt{6}$$

$$\left\{-2\sqrt{6}, 2\sqrt{6}\right\}$$

31.
$$x(x-4)+3x=20$$
$$x^2-4x+3x-20=0$$
$$x^2-x-20=0$$
$$x=\frac{-(-1)\pm\sqrt{(-1)^2-4(1)(-20)}}{2(1)}$$
$$x=\frac{1\pm\sqrt{81}}{2}$$
$$x=\frac{1\pm9}{2}$$
$$x=-\frac{8}{2}\quad\text{or}\quad x=\frac{10}{2}$$
$$x=-4\quad\text{or}\quad x=5$$
$$\{-4,5\}$$

33.
$$x^2-\frac{1}{5}x+\frac{3}{4}=0$$
$$20\cdot\left(x^2-\frac{1}{5}x+\frac{3}{4}\right)=20\cdot0$$
$$20x^2-4x+15=0$$
$$x=\frac{-(-4)\pm\sqrt{(-4)^2-4(20)(15)}}{2(20)}$$
$$x=\frac{4\pm\sqrt{-1184}}{40}$$
$$x=\frac{4\pm4i\sqrt{74}}{40}$$
$$x=\frac{1\pm i\sqrt{74}}{10}$$
$$\left\{\frac{1-i\sqrt{74}}{10},\frac{1+i\sqrt{74}}{10}\right\}$$

35. $-x^2+7x-12=0$
$$0=x^2-7x+12$$
$$x=\frac{-(-7)\pm\sqrt{(-7)^2-4(1)(12)}}{2(1)}$$
$$x=\frac{7\pm\sqrt{1}}{2}$$
$$x=\frac{7\pm1}{2}$$
$$x=\frac{6}{2}\quad\text{or}\quad x=\frac{8}{2}$$
$$x=3\quad\text{or}\quad x=4$$
$$\{3,4\}$$

37. $x^2+12x-30=0$
$$(12)^2-4(1)(-30)=264\text{ positive}$$
Two real solutions

39. $2x^2-3x+5=0$
$$(-3)^2-4(2)(5)=-31\text{ negative}$$
Two nonreal complex solutions

41.
$$x^2+\frac{2}{5}x+\frac{5}{6}=0$$
$$30\cdot\left(x^2+\frac{2}{5}x+\frac{5}{6}\right)=30\cdot0$$
$$30x^2+12x+25=0$$
$$(12)^2-4(30)(25)=-2856\text{ negative}$$
Two nonreal complex solutions

43. $9x^2-12x+4=0$
$$(-12)^2-4(9)(4)=0$$
One real solution

45. $x^2+12x-35$
$$(12)^2-4(1)(-35)=284$$
Not a perfect square
Prime

47. $x^2+52x+667$
$$(52)^2-4(1)(667)=36$$
$$\sqrt{36}=6$$
A perfect square
Factorable

49. $35x^2-116x+65$
$$(-116)^2-4(35)(65)=4356$$
$$\sqrt{4356}=66$$
A perfect square
Factorable

51. $5x^2-16x-18$
$$(-16)^2-4(5)(-18)=616$$
Not a perfect square
Prime

53. $x^2-5x-15=0$
$$x=\frac{-(-5)\pm\sqrt{(-5)^2-4(1)(-15)}}{2(1)}$$
$$x=\frac{5\pm\sqrt{85}}{2}$$
$$\left\{\frac{5-\sqrt{85}}{2},\frac{5+\sqrt{85}}{2}\right\}$$

55. $3x^2 + 2x - 1 = 0$
$(x+1)(3x-1) = 0$
$x + 1 = 0 \quad \text{or} \quad 3x - 1 = 0$

$x = -1 \quad \text{or} \quad x = \dfrac{1}{3}$

$\left\{ -1, \dfrac{1}{3} \right\}$

57. $(5x - 4)^2 = 36$
$\sqrt{(5x-4)^2} = \pm\sqrt{36}$
$5x - 4 = \pm 6$
$5x = 4 \pm 6$

$x = \dfrac{4 \pm 6}{5}$

$x = -\dfrac{2}{5} \quad \text{or} \quad x = \dfrac{10}{5}$

$x = -\dfrac{2}{5} \quad \text{or} \quad x = 2$

$\left\{ -\dfrac{2}{5}, 2 \right\}$

59. $2x^2 + 8x = -9$
$2x^2 + 8x + 9 = 0$

$x = \dfrac{-8 \pm \sqrt{8^2 - 4(2)(9)}}{2(2)}$

$x = \dfrac{-8 \pm \sqrt{-8}}{4}$

$x = \dfrac{-8 \pm 2i\sqrt{2}}{4}$

$x = \dfrac{-4 \pm i\sqrt{2}}{2}$

$\left\{ \dfrac{-4 - i\sqrt{2}}{2}, \dfrac{-4 + i\sqrt{2}}{2} \right\}$

61. $x^2 + 324 = 0$
$x^2 = -324$
$\sqrt{x^2} = \pm\sqrt{-324}$
$x = \pm 18i$
$\{ -18i, 18i \}$

63. $6x^2 - 29x + 28 = 0$
$(3x - 4)(2x - 7) = 0$
$3x - 4 = 0 \quad \text{or} \quad 2x - 7 = 0$
$3x = 4 \quad \text{or} \quad 2x = 7$
$x = \dfrac{4}{3} \quad \text{or} \quad x = \dfrac{7}{2}$

$\left\{ \dfrac{4}{3}, \dfrac{7}{2} \right\}$

65. $3(2x + 1)^2 - 7 = 23$

$3(2x + 1)^2 = 30$

$(2x + 1)^2 = 10$

$\sqrt{(2x+1)^2} = \pm\sqrt{10}$

$2x + 1 = \pm\sqrt{10}$

$2x = -1 \pm \sqrt{10}$

$x = \dfrac{-1 \pm \sqrt{10}}{2}$

$\left\{ \dfrac{-1 - \sqrt{10}}{2}, \dfrac{-1 + \sqrt{10}}{2} \right\}$

67. $x^2 - 9x - 21 = 0$

$x = \dfrac{-(-9) \pm \sqrt{(-9)^2 - 4(1)(-21)}}{2(1)}$

$x = \dfrac{9 \pm \sqrt{165}}{2}$

$\left\{ \dfrac{9 - \sqrt{165}}{2}, \dfrac{9 + \sqrt{165}}{2} \right\}$

69. $16x^2 - 24x + 9 = 0$
$(4x - 3)(4x - 3) = 0$
$4x - 3 = 0$
$4x = 3$

$x = \dfrac{3}{4}$

$\left\{ \dfrac{3}{4} \right\}$

71. $x^2 + x + 20 = 0$

$$x = \frac{-1 \pm \sqrt{1^2 - 4(1)(20)}}{2(1)}$$

$$x = \frac{-1 \pm \sqrt{-79}}{2}$$

$$x = \frac{-1 \pm i\sqrt{79}}{2}$$

$$\left\{ \frac{-1 - i\sqrt{79}}{2}, \frac{-1 + i\sqrt{79}}{2} \right\}$$

73. $x^2 - 4x - 2 = 0$

$$x = \frac{-(-4) \pm \sqrt{(-4)^2 - 4(1)(-2)}}{2(1)}$$

$$x = \frac{4 \pm \sqrt{24}}{2}$$

$$x = \frac{4 \pm 2\sqrt{6}}{2}$$

$$x = 2 \pm \sqrt{6}$$

$$\left\{ 2 - \sqrt{6}, 2 + \sqrt{6} \right\}$$

75. $x^2 - 6x + 10 = 0$

$$x = \frac{-(-6) \pm \sqrt{(-6)^2 - 4(1)(10)}}{2(1)}$$

$$x = \frac{6 \pm \sqrt{-4}}{2}$$

$$x = \frac{6 \pm 2i}{2}$$

$$x = 3 \pm i$$

$$\left\{ 3 - i, 3 + i \right\}$$

77. $\dfrac{3}{4}x^2 + \dfrac{2}{3}x - \dfrac{1}{2} = 0$

$$12 \cdot \left(\frac{3}{4}x^2 + \frac{2}{3}x - \frac{1}{2} \right) = 12 \cdot 0$$

$$9x^2 + 8x - 6 = 0$$

$$x = \frac{-8 \pm \sqrt{8^2 - 4(9)(-6)}}{2(9)}$$

$$x = \frac{-8 \pm \sqrt{280}}{18}$$

$$x = \frac{-8 \pm 2\sqrt{70}}{18}$$

$$x = \frac{-4 \pm \sqrt{70}}{9}$$

$$\left\{ \frac{-4 - \sqrt{70}}{9}, \frac{-4 + \sqrt{70}}{9} \right\}$$

79. $2x(3x + 7) - 5(3x + 7) = 0$

$$(3x + 7)(2x - 5) = 0$$

$$3x + 7 = 0 \quad \text{or} \quad 2x - 5 = 0$$

$$3x = -7 \quad \text{or} \quad 2x = 5$$

$$x = -\frac{7}{3} \quad \text{or} \quad x = \frac{5}{2}$$

$$\left\{ -\frac{7}{3}, \frac{5}{2} \right\}$$

81. $2(2x - 9)^2 + 13 = 77$

$$2(2x - 9)^2 = 64$$

$$(2x - 9)^2 = 32$$

$$\sqrt{(2x - 9)^2} = \pm\sqrt{32}$$

$$2x - 9 = \pm 4\sqrt{2}$$

$$2x = 9 \pm 4\sqrt{2}$$

$$x = \frac{9 \pm 4\sqrt{2}}{2}$$

$$\left\{ \frac{9 - 4\sqrt{2}}{2}, \frac{9 + 4\sqrt{2}}{2} \right\}$$

83. $x^2 + 3x - 4 = 0$

$$(x + 4)(x - 1) = 0$$

$$x + 4 = 0 \quad \text{or} \quad x - 1 = 0$$

$$x = -4 \quad \text{or} \quad x = 1$$

$$\left\{ -4, 1 \right\}$$

85. $x^2 - 10x + 18 = 0$

$$x = \frac{-(-10) \pm \sqrt{(-10)^2 - 4(1)(18)}}{2(1)}$$

$$x = \frac{10 \pm \sqrt{28}}{2}$$

$$x = \frac{10 \pm 2\sqrt{7}}{2}$$

$$x = 5 \pm \sqrt{7}$$

$$\left\{ 5 - \sqrt{7}, 5 + \sqrt{7} \right\}$$

87. $x^2 - 16x + 63 = 0$
$(x - 7)(x - 9) = 0$
$x - 7 = 0$ or $x - 9 = 0$
$\quad x = 7$ or $\quad x = 9$
$\{7, 9\}$

89. Unknown: Knowns:
time: t v_o: 128 ft/sec
   s: 0 ft
   $h(t)$: 0 ft

$0 = -16t^2 + 128t$
$0 = -16t(t - 8)$

$-16t = 0$ or $t - 8 = 0$
$\quad t = 0$ or $\quad t = 8$
time: 8 seconds

91. Unknown: Knowns:
time: t v_o: 70 ft/sec
   s: 90 ft
   $h(t)$: 0 ft

$0 = -16t^2 + 70t + 90$
$16t^2 - 70t - 90 = 0$

$$t = \frac{-(-70) \pm \sqrt{(-70)^2 - 4(16)(-90)}}{2(16)}$$

$$t = \frac{70 \pm \sqrt{10,660}}{32}$$

$t \approx 5.4$
time: about 5.4 seconds

93. Unknown: Knowns:
time: t v_o: 80 ft/sec
   s: 0 ft
   $h(t)$: 64 ft

$64 = -16t^2 + 80t$
$16t^2 - 80t + 64 = 0$
$16\left(t^2 - 5t + 4\right) = 0$
$16(t - 1)(t - 4) = 0$

$t - 1 = 0$ or $t - 4 = 0$
$\quad t = 1$ or $\quad t = 4$
time: 1 second and 4 seconds

95. Unknown: Knowns:
time: t v_o: 120 ft/sec
   s: 50 ft
   $h(t)$: 195 ft

$195 = -16t^2 + 120t + 50$
$16t^2 - 120t + 145 = 0$

$$t = \frac{-(-120) \pm \sqrt{(-120)^2 - 4(16)(145)}}{2(16)}$$

$$t = \frac{120 \pm \sqrt{5120}}{32}$$

$$t = \frac{120 \pm 32\sqrt{5}}{32}$$

$$t = \frac{15 \pm 4\sqrt{5}}{4}$$

$$t = \frac{15 - 4\sqrt{5}}{4} \quad \text{or} \quad t = \frac{15 + 4\sqrt{5}}{4}$$

$t \approx 1.5$ or $t \approx 6.0$
time: about 1.5 seconds and 6.0 seconds

97. Unknown: Knowns:
time: t v_o: 16 ft/sec
   s: 5 ft
   $h(t)$: 30 ft

$30 = -16t^2 + 16t + 5$
$16t^2 - 16t + 25 = 0$

$$t = \frac{-(-16) \pm \sqrt{(-16)^2 - 4(16)(25)}}{2(16)}$$

$$t = \frac{16 \pm \sqrt{-1344}}{32}$$

The discriminant is negative. This is not a real number. No; the rock never reachs a height of 30 feet.

99. Unknown: Knowns:
time: t v_o: 42 ft/sec
          s: 5 ft
          $h(t)$: 30 ft

$$30 = -16t^2 + 42t + 5$$
$$16t^2 - 42t + 25 = 0$$
$$t = \frac{-(-42) \pm \sqrt{(-42)^2 - 4(16)(25)}}{2(16)}$$
$$t = \frac{42 \pm \sqrt{164}}{32}$$
$$t = \frac{42 \pm 2\sqrt{41}}{32}$$
$$t = \frac{21 \pm \sqrt{41}}{16}$$
$$t = \frac{21 - \sqrt{41}}{16} \text{ or } t = \frac{21 + \sqrt{41}}{16}$$
$$t \approx 0.9 \qquad \text{or} \quad t \approx 1.7$$

Yes; the rock reaches the height of 30 feet in about 0.9 seconds.

101. Unknown: Knowns:
initial velocity: v s: 0 ft
          t: 6 sec
          $h(6)$: 0 ft

$$0 = -16(6)^2 + v(6)$$
$$0 = -576 + 6v$$
$$576 = 6v$$
$$96 = v$$

initial velocity: 96 feet per second

103. Answers will vary. Example:
In the quadratic formula, the discriminant is the radicand. So if the discriminate is positive, the radical yields two real numbers. Therefore, there are two real solutions.
If the discriminant is zero, the radical yields zero. Therefore, there is one real solution.
If the discriminant is negative, the radical yields two imaginary numbers. Therefore, there are two complex solutions.

10.3 QUICK CHECK

1. $x^4 - x^2 - 12 = 0$
Let $u = x^2$ and $u^2 = x^4$.
$$u^2 - u - 12 = 0$$
$$(u - 4)(u + 3) = 0$$
$$u - 4 = 0 \qquad \text{or} \quad u + 3 = 0$$
$$u = 4 \qquad \text{or} \qquad u = -3$$
$$x^2 = 4 \qquad \text{or} \qquad x^2 = -3$$
$$\sqrt{x^2} = \pm\sqrt{4} \quad \text{or} \quad \sqrt{x^2} = \pm\sqrt{-3}$$
$$x = \pm 2 \qquad \text{or} \qquad x = \pm i\sqrt{3}$$
$$\left\{ -2, 2, -i\sqrt{3}, i\sqrt{3} \right\}$$

2. a) $x + 11\sqrt{x} - 26 = 0$
Let $u = \sqrt{x}$.

b) $2x^{2/3} - 17x^{1/3} + 8 = 0$
Let $u = x^{1/3}$.

c) $\left(x^2 - 4x \right)^2 - 9\left(x^2 - 4x \right) - 36 = 0$
Let $u = x^2 - 4x$.

3. a) $x - 6x^{1/2} + 5 = 0$
Let $u = x^{1/2}$ and $u^2 = x$.
$$u^2 - 6u + 5 = 0$$
$$(u - 1)(u - 5) = 0$$
$$u - 1 = 0 \quad \text{or} \quad u - 5 = 0$$
$$u = 1 \quad \text{or} \qquad u = 5$$
$$x^{1/2} = 1 \quad \text{or} \qquad x^{1/2} = 5$$
$$\left(x^{1/2} \right)^2 = 1^2 \quad \text{or} \quad \left(x^{1/2} \right)^2 = 5^2$$
$$x = 1 \quad \text{or} \qquad x = 25$$
$$\left\{ 1, 25 \right\}$$

b) $x^{2/3} - 4x^{1/3} + 3 = 0$
Let $u = x^{1/3}$ and $u^2 = x^{2/3}$.
$$u^2 - 4u + 3 = 0$$
$$(u - 1)(u - 3) = 0$$
$$u - 1 = 0 \quad \text{or} \quad u - 3 = 0$$
$$u = 1 \quad \text{or} \qquad u = 3$$
$$x^{1/3} = 1 \quad \text{or} \qquad x^{1/3} = 3$$
$$\left(x^{1/3} \right)^3 = 1^3 \quad \text{or} \quad \left(x^{1/3} \right)^3 = 3^3$$
$$x = 1 \quad \text{or} \qquad x = 27$$
$$\left\{ 1, 27 \right\}$$

4.
$$x + 7 = \sqrt{x + 9}$$
$$(x + 7)^2 = \left(\sqrt{x + 9}\right)^2$$
$$(x + 7)(x + 7) = x + 9$$
$$x^2 + 7x + 7x + 49 = x + 9$$
$$x^2 + 13x + 40 = 0$$
$$(x + 8)(x + 5) = 0$$

$$x + 8 = 0 \quad \text{or} \quad x + 5 = 0$$
$$x = -8 \quad \text{or} \quad x = -5$$

$x = -8$	$x = -5$
$(-8) + 7 = \sqrt{-8 + 9}$	$(-5) + 7 = \sqrt{-5 + 9}$
$-1 = \sqrt{1}$	$2 = \sqrt{4}$
$-1 = 1$	$2 = 2$
False	True

$\{-5\}$

5.
$$\frac{2}{x + 1} + \frac{1}{x - 1} = 1$$
$$(x + 1)(x - 1) \cdot \left(\frac{2}{x + 1} + \frac{1}{x - 1}\right) = (x + 1)(x - 1) \cdot 1$$
$$2(x - 1) + 1(x + 1) = (x + 1)(x - 1)$$
$$2x - 2 + x + 1 = x^2 - 1$$
$$0 = x^2 - 3x$$
$$0 = x(x - 3)$$

$$x = 0 \quad \text{or} \quad x - 3 = 0$$
$$x = 0 \quad \text{or} \quad x = 3$$

Neither of these values makes a denominator of the original equation 0.

$\{0, 3\}$

6.

Worker	Time to Complete the Job Alone	Work-Rate	Time Working	Portion of the Job Completed
Gabe	$t - 50$ minutes	$\dfrac{1}{t - 50}$	45 minutes	$\dfrac{45}{t - 50}$
Rob	t minutes	$\dfrac{1}{t}$	45 minutes	$\dfrac{45}{t}$

$$\frac{45}{t - 50} + \frac{45}{t} = 1$$

$$t(t - 50) \cdot \left(\frac{45}{t - 50} + \frac{45}{t}\right) = t(t - 50) \cdot 1$$

$$45t + 45(t - 50) = t(t - 50)$$
$$45t + 45t - 2250 = t^2 - 50t$$
$$0 = t^2 - 140t + 2250$$

$$t = \frac{-(-140) \pm \sqrt{(-140)^2 - 4(1)(2250)}}{2(1)}$$

$$t = \frac{140 \pm \sqrt{10,600}}{2}$$

$t \approx 18.5$ or $t \approx 121.5$

The value $t \approx 18.5$ is omitted because it gives a negative solution for Gabe's time.

Gabe's time alone: $t - 50 \approx 121.5 - 50 = 71.5$ minutes

10.3 EQUATIONS THAT ARE QUADRATIC IN FORM

1. u-substitution

3. $x^4 - 5x^2 + 4 = 0$
Let $u = x^2$ and $u^2 = x^4$.
$$u^2 - 5u + 4 = 0$$
$$(u-1)(u-4) = 0$$
$$\begin{array}{lll} u-1=0 & \text{or} & u-4=0 \\ u=1 & \text{or} & u=4 \\ x^2=1 & \text{or} & x^2=4 \\ \sqrt{x^2}=\pm\sqrt{1} & \text{or} & \sqrt{x^2}=\pm\sqrt{4} \\ x=\pm 1 & \text{or} & x=\pm 2 \end{array}$$
$$\{1, -1, 2, -2\}$$

5. $x^4 - 6x^2 + 9 = 0$
Let $u = x^2$ and $u^2 = x^4$.
$$u^2 - 6u + 9 = 0$$
$$(u-3)(u-3) = 0$$
$$u - 3 = 0$$
$$u = 3$$
$$x^2 = 3$$
$$\sqrt{x^2} = \pm\sqrt{3}$$
$$x = \pm\sqrt{3}$$
$$\{-\sqrt{3}, \sqrt{3}\}$$

7. $x^4 + 7x^2 - 18 = 0$
Let $u = x^2$ and $u^2 = x^4$.
$$u^2 + 7u - 18 = 0$$
$$(u+9)(u-2) = 0$$
$$\begin{array}{lll} u+9=0 & \text{or} & u-2=0 \\ u=-9 & \text{or} & u=2 \\ x^2=-9 & \text{or} & x^2=2 \\ \sqrt{x^2}=\pm\sqrt{-9} & \text{or} & \sqrt{x^2}=\pm\sqrt{2} \\ x=\pm 3i & \text{or} & x=\pm\sqrt{2} \end{array}$$
$$\{-3i, 3i, -\sqrt{2}, \sqrt{2}\}$$

9. $x^4 - 13x^2 + 36 = 0$
Let $u = x^2$ and $u^2 = x^4$.
$$u^2 - 13u + 36 = 0$$
$$(u-4)(u-9) = 0$$
$$\begin{array}{lll} u-4=0 & \text{or} & u-9=0 \\ u=4 & \text{or} & u=9 \\ x^2=4 & \text{or} & x^2=9 \\ \sqrt{x^2}=\pm\sqrt{4} & \text{or} & \sqrt{x^2}=\pm\sqrt{9} \\ x=\pm 2 & \text{or} & x=\pm 3 \end{array}$$
$$\{-2, 2, -3, 3\}$$

11. $x - 13\sqrt{x} + 36 = 0$
Let $u = \sqrt{x}$ and $u^2 = x$.
$$u^2 - 13u + 36 = 0$$
$$(u-4)(u-9) = 0$$
$$\begin{array}{lll} u-4=0 & \text{or} & u-9=0 \\ u=4 & \text{or} & u=9 \\ \sqrt{x}=4 & \text{or} & \sqrt{x}=9 \\ \left(\sqrt{x}\right)^2=(4)^2 & \text{or} & \left(\sqrt{x}\right)^2=(9)^2 \\ x=16 & \text{or} & x=81 \end{array}$$
check:

$x = 16$	$x = 81$
$16 - 13\sqrt{16} + 36 = 0$	$81 - 13\sqrt{81} + 36 = 0$
$16 - 13(4) + 36 = 0$	$81 - 13(9) + 36 = 0$
$16 - 52 + 36 = 0$	$81 - 117 + 36 = 0$
$0 = 0$	$0 = 0$
True	True

$$\{16, 81\}$$

13. $x + 2\sqrt{x} - 48 = 0$
Let $u = \sqrt{x}$ and $u^2 = x$.
$$u^2 + 2u - 48 = 0$$
$$(u+8)(u-6) = 0$$
$$\begin{array}{lll} u+8=0 & \text{or} & u-6=0 \\ u=-8 & \text{or} & u=6 \\ \sqrt{x}=-8 & \text{or} & \sqrt{x}=6 \\ \left(\sqrt{x}\right)^2=(-8)^2 & \text{or} & \left(\sqrt{x}\right)^2=(6)^2 \\ x=64 & \text{or} & x=36 \end{array}$$
check:

$x = 64$	$x = 36$
$64 + 2\sqrt{64} - 48 = 0$	$36 + 2\sqrt{36} - 48 = 0$
$64 + 2(8) - 48 = 0$	$36 + 2(6) - 48 = 0$
$80 - 48 = 0$	$48 - 48 = 0$
$32 = 0$	$0 = 0$
False	True

$$\{36\}$$

15. $x + 9\sqrt{x} + 20 = 0$

Let $u = \sqrt{x}$ and $u^2 = x$.

$u^2 + 9u + 20 = 0$

$(u+4)(u+5) = 0$

$$\begin{array}{lll} u+4=0 & \text{or} & u+5=0 \\ u=-4 & \text{or} & u=-5 \\ \sqrt{x}=-4 & \text{or} & \sqrt{x}=-5 \\ \left(\sqrt{x}\right)^2=(-4)^2 & \text{or} & \left(\sqrt{x}\right)^2=(-5)^2 \\ x=16 & \text{or} & x=25 \end{array}$$

check:

$x = 16$	$x = 25$
$16 + 9\sqrt{16} + 20 = 0$	$25 + 9\sqrt{25} + 20 = 0$
$16 + 9(4) + 20 = 0$	$25 + 9(5) + 20 = 0$
$16 + 36 + 20 = 0$	$25 + 45 + 20 = 0$
$72 = 0$	$90 = 0$
False	False

$\varnothing$

17. $x - 8x^{1/2} + 7 = 0$

Let $u = x^{1/2}$ and $u^2 = x$.

$u^2 - 8u + 7 = 0$

$(u-1)(u-7) = 0$

$$\begin{array}{lll} u-1=0 & \text{or} & u-7=0 \\ u=1 & \text{or} & u=7 \\ x^{1/2}=1 & \text{or} & x^{1/2}=7 \\ \left(x^{1/2}\right)^2=(1)^2 & \text{or} & \left(x^{1/2}\right)^2=(7)^2 \\ x=1 & \text{or} & x=49 \end{array}$$

check:

$x = 1$	$x = 49$
$1 - 8(1)^{1/2} + 7 = 0$	$49 - 8(49)^{1/2} + 7 = 0$
$1 - 8(1) + 7 = 0$	$49 - 8(7) + 7 = 0$
$1 - 8 + 7 = 0$	$49 - 56 + 7 = 0$
$0 = 0$	$0 = 0$
True	True

$\{1, 49\}$

19. $x + 9x^{1/2} + 18 = 0$

Let $u = x^{1/2}$ and $u^2 = x$.

$u^2 + 9u + 18 = 0$

$(u+6)(u+3) = 0$

$$\begin{array}{lll} u+6=0 & \text{or} & u+3=0 \\ u=-6 & \text{or} & u=-3 \\ x^{1/2}=-6 & \text{or} & x^{1/2}=-3 \\ \left(x^{1/2}\right)^2=(-6)^2 & \text{or} & \left(x^{1/2}\right)^2=(-3)^2 \\ x=36 & \text{or} & x=9 \end{array}$$

check:

$x = 36$	$x = 9$
$36 + 9(36)^{1/2} + 18 = 0$	$9 + 9(9)^{1/2} + 18 = 0$
$36 + 9(6) + 18 = 0$	$9 + 9(3) + 18 = 0$
$36 + 54 + 18 = 0$	$9 + 27 + 18 = 0$
$108 = 0$	$54 = 0$
False	False

$\varnothing$

21. $x - 2x^{1/2} - 80 = 0$

Let $u = x^{1/2}$ and $u^2 = x$.

$u^2 - 2u - 80 = 0$

$(u+8)(u-10) = 0$

$$\begin{array}{lll} u+8=0 & \text{or} & u-10=0 \\ u=-8 & \text{or} & u=10 \\ x^{1/2}=-8 & \text{or} & x^{1/2}=10 \\ \left(x^{1/2}\right)^2=(-8)^2 & \text{or} & \left(x^{1/2}\right)^2=(10)^2 \\ x=64 & \text{or} & x=100 \end{array}$$

check:

$x = 64$	$x = 100$
$64 - 2(64)^{1/2} - 80 = 0$	$100 - 2(100)^{1/2} - 80 = 0$
$64 - 2(8) - 80 = 0$	$100 - 2(10) - 80 = 0$
$64 - 16 - 80 = 0$	$100 - 20 - 80 = 0$
$-32 = 0$	$0 = 0$
False	True

$\{100\}$

23. $x^{2/3} - 5x^{1/3} + 6 = 0$

Let $u = x^{1/3}$ and $u^2 = x^{2/3}$.

$$u^2 - 5u + 6 = 0$$
$$(u-3)(u-2) = 0$$

$$
\begin{array}{lll}
u - 3 = 0 & \text{or} & u - 2 = 0 \\
u = 3 & \text{or} & u = 2 \\
x^{1/3} = 3 & \text{or} & x^{1/3} = 2 \\
\left(x^{1/3}\right)^3 = (3)^3 & \text{or} & \left(x^{1/3}\right)^3 = (2)^3 \\
x = 27 & \text{or} & x = 8
\end{array}
$$

$\{8, 27\}$

25. $x^{2/3} + 5x^{1/3} - 6 = 0$

Let $u = x^{1/3}$ and $u^2 = x^{2/3}$.

$$u^2 + 5u - 6 = 0$$
$$(u+6)(u-1) = 0$$

$$
\begin{array}{lll}
u + 6 = 0 & \text{or} & u - 1 = 0 \\
u = -6 & \text{or} & u = 1 \\
x^{1/3} = -6 & \text{or} & x^{1/3} = 1 \\
\left(x^{1/3}\right)^3 = (-6)^3 & \text{or} & \left(x^{1/3}\right)^3 = (1)^3 \\
x = -216 & \text{or} & x = 1
\end{array}
$$

$\{-216, 1\}$

27. $x^{2/3} + 9x^{1/3} + 20 = 0$

Let $u = x^{1/3}$ and $u^2 = x^{2/3}$.

$$u^2 + 9u + 20 = 0$$
$$(u+4)(u+5) = 0$$

$$
\begin{array}{lll}
u + 4 = 0 & \text{or} & u + 5 = 0 \\
u = -4 & \text{or} & u = -5 \\
x^{1/3} = -4 & \text{or} & x^{1/3} = -5 \\
\left(x^{1/3}\right)^3 = (-4)^3 & \text{or} & \left(x^{1/3}\right)^3 = (-5)^3 \\
x = -64 & \text{or} & x = -125
\end{array}
$$

$\{-64, -125\}$

29. $(x-3)^2 + 5(x-3) + 4 = 0$

Let $u = x - 3$ and $u^2 = (x-3)^2$.

$$u^2 + 5u + 4 = 0$$
$$(u+4)(u+1) = 0$$

$$
\begin{array}{lll}
u + 4 = 0 & \text{or} & u + 1 = 0 \\
u = -4 & \text{or} & u = -1 \\
x - 3 = -4 & \text{or} & x - 3 = -1 \\
x = -1 & \text{or} & x = 2
\end{array}
$$

$\{-1, 2\}$

31. $(2x-9)^2 - 6(2x-9) - 27 = 0$

Let $u = 2x - 9$ and $u^2 = (2x-9)^2$.

$$u^2 - 6u - 27 = 0$$
$$(u+3)(u-9) = 0$$

$$
\begin{array}{lll}
u + 3 = 0 & \text{or} & u - 9 = 0 \\
u = -3 & \text{or} & u = 9 \\
2x - 9 = -3 & \text{or} & 2x - 9 = 9 \\
2x = 6 & \text{or} & 2x = 18 \\
x = 3 & \text{or} & x = 9
\end{array}
$$

$\{3, 9\}$

33.
$$\sqrt{x^2 + 6x} = 4$$
$$\left(\sqrt{x^2 + 6x}\right)^2 = (4)^2$$
$$x^2 + 6x = 16$$
$$x^2 + 6x - 16 = 0$$
$$(x+8)(x-2) = 0$$

$$
\begin{array}{lll}
x + 8 = 0 & \text{or} & x - 2 = 0 \\
x = -8 & \text{or} & x = 2
\end{array}
$$

check:

$x = -8$	$x = 2$
$\sqrt{(-8)^2 + 6(-8)} = 4$	$\sqrt{2^2 + 6(2)} = 4$
$\sqrt{64 - 48} = 4$	$\sqrt{4 + 12} = 4$
$\sqrt{16} = 4$	$\sqrt{16} = 4$
$4 = 4$	$4 = 4$
True	True

$\{-8, 2\}$

35.
$$\sqrt{3x - 6} = x - 2$$
$$\left(\sqrt{3x - 6}\right)^2 = (x-2)^2$$
$$3x - 6 = (x-2)(x-2)$$
$$3x - 6 = x^2 - 2x - 2x + 4$$
$$0 = x^2 - 7x + 10$$
$$0 = (x-2)(x-5)$$

$$
\begin{array}{lll}
x - 2 = 0 & \text{or} & x - 5 = 0 \\
x = 2 & \text{or} & x = 5
\end{array}
$$

check:

$x = 2$	$x = 5$
$\sqrt{3(2) - 6} = 2 - 2$	$\sqrt{3(5) - 6} = 5 - 2$
$\sqrt{6 - 6} = 0$	$\sqrt{15 - 6} = 3$
$\sqrt{0} = 0$	$\sqrt{9} = 3$
$0 = 0$	$3 = 3$
True	True

$\{2, 5\}$

37.
$$\sqrt{x+15} - x = 3$$
$$\sqrt{x+15} = x+3$$
$$\left(\sqrt{x+15}\right)^2 = (x+3)^2$$
$$x+15 = (x+3)(x+3)$$
$$x+15 = x^2 + 3x + 3x + 9$$
$$0 = x^2 + 5x - 6$$
$$0 = (x+6)(x-1)$$

$$x+6=0 \quad \text{or} \quad x-1=0$$
$$x=-6 \quad \text{or} \quad x=1$$

check:

$x=-6$	$x=1$
$\sqrt{-6+15} - (-6) = 3$	$\sqrt{1+15} - 1 = 3$
$\sqrt{9} + 6 = 3$	$\sqrt{16} - 1 = 3$
$3+6=3$	$4-1=3$
$9=3$	$3=3$
False	True

$\{1\}$

39.
$$\sqrt{x-1} + 2 = \sqrt{2x+5}$$
$$\left(\sqrt{x-1}+2\right)^2 = \left(\sqrt{2x+5}\right)^2$$
$$\left(\sqrt{x-1}+2\right)\left(\sqrt{x-1}+2\right) = 2x+5$$
$$x-1+2\sqrt{x-1}+2\sqrt{x-1}+4 = 2x+5$$
$$x+3+4\sqrt{x-1} = 2x+5$$
$$4\sqrt{x-1} = x+2$$
$$\left(4\sqrt{x-1}\right)^2 = (x+2)^2$$
$$16(x-1) = (x+2)(x+2)$$
$$16x-16 = x^2 + 2x + 2x + 4$$
$$0 = x^2 - 12x + 20$$
$$0 = (x-2)(x-10)$$

$$x-2=0 \quad \text{or} \quad x-10=0$$
$$x=2 \quad \text{or} \quad x=10$$

check:

$x=2$	$x=10$
$\sqrt{2-1}+2 = \sqrt{2(2)+5}$	$\sqrt{10-1}+2 = \sqrt{2(10)+5}$
$\sqrt{1}+2 = \sqrt{4+5}$	$\sqrt{9}+2 = \sqrt{20+5}$
$1+2 = \sqrt{9}$	$3+2 = \sqrt{25}$
$3=3$	$5=5$
True	True

$\{2,10\}$

41.
$$x = \frac{40}{x+6}$$
$$(x+6)\cdot x = (x+6)\cdot \frac{40}{(x+6)}$$
$$x^2 + 6x = 40$$
$$x^2 + 6x - 40 = 0$$
$$(x+10)(x-4) = 0$$

$$x+10=0 \quad \text{or} \quad x-4=0$$
$$x=-10 \quad \text{or} \quad x=4$$

Neither of these values makes the denominator of the original equation 0.

$\{-10,4\}$

43.
$$1 + \frac{7}{x} - \frac{60}{x^2} = 0$$
$$x^2 \cdot \left(1 + \frac{7}{x} - \frac{60}{x^2}\right) = x^2 \cdot 0$$
$$x^2 + 7x - 60 = 0$$
$$(x+12)(x-5) = 0$$

$$x+12=0 \quad \text{or} \quad x-5=0$$
$$x=-12 \quad \text{or} \quad x=5$$

Neither of these values makes a denominator of the original equation 0.

$\{-12,5\}$

45.
$$\frac{1}{x} + \frac{7}{x+2} = \frac{10}{x(x+2)}$$
$$x(x+2)\left(\frac{1}{x} + \frac{7}{x+2}\right) = x(x+2)\cdot \frac{10}{x(x+2)}$$
$$x+2+7x = 10$$
$$8x+2 = 10$$
$$8x = 8$$
$$x = 1$$

This value does not make a denominator of the original equation 0.

$\{1\}$

47.

$$\frac{2}{x+3}+\frac{x+7}{x+1}=\frac{9}{4}$$

$$4(x+3)(x+1)\cdot\left(\frac{2}{x+3}+\frac{x+7}{x+1}\right)=4(x+3)(x+1)\cdot\frac{9}{4}$$

$$8(x+1)+4(x+3)(x+7)=9(x+3)(x+1)$$
$$8x+8+4\left(x^2+7x+3x+21\right)=9\left(x^2+x+3x+3\right)$$
$$8x+8+4\left(x^2+10x+21\right)=9\left(x^2+4x+3\right)$$
$$8x+8+4x^2+40x+84=9x^2+36x+27$$
$$4x^2+48x+92=9x^2+36x+27$$
$$0=5x^2-12x-65$$
$$0=(5x+13)(x-5)$$

$$\begin{array}{lll}5x+13=0 & \text{or} & x-5=0\\ 5x=-13 & \text{or} & x=5\\ x=-\dfrac{13}{5} & \text{or} & x=5\end{array}$$

Neither of these values makes a denominator of the original equation 0.

$$\left\{-\frac{13}{5},5\right\}$$

49.

Pipe	Time to Complete the Job Alone	Work-Rate	Time Working	Portion of the Job Completed
Small	$2t$ minutes	$\dfrac{1}{2t}$	40 minutes	$\dfrac{40}{2t}$
Large	t minutes	$\dfrac{1}{t}$	40 minutes	$\dfrac{40}{t}$

$$\frac{40}{2t}+\frac{40}{t}=1$$

$$2t\cdot\left(\frac{40}{2t}+\frac{40}{t}\right)=2t\cdot1$$

$$40+80=2t$$
$$120=2t$$
$$t=60$$
$$2t=2(60)=120$$

Small: 120 minutes
Large: 60 minutes

51.

Hose	Time to Complete the Job Alone	Work-Rate	Time Working	Portion of the Job Completed
Small	$t+1$ hours	$\dfrac{1}{t+1}$	3 hours	$\dfrac{3}{t+1}$
Large	t hours	$\dfrac{1}{t}$	3 hours	$\dfrac{3}{t}$

$$\frac{3}{t+1}+\frac{3}{t}=1$$

$$t(t+1)\left(\frac{3}{t+1}+\frac{3}{t}\right)=t(t+1)\cdot 1$$

$$3t+3(t+1)=t(t+1)$$
$$3t+3t+3=t^2+t$$
$$6t+3=t^2+t$$
$$0=t^2-5t-3$$
$$t=\frac{-(-5)\pm\sqrt{(-5)^2-4(1)(-3)}}{2(1)}$$
$$t=\frac{5\pm\sqrt{37}}{2}$$

$t\approx-0.5$ or $t\approx 5.5$
Small hose: $t+1\approx 5.5+1=6.5$ hours

53.

Printer	Time to Complete the Job Alone	Work-Rate	Time Working	Portion of the Job Completed
New	$t-15$ minutes	$\dfrac{1}{t-15}$	40 minutes	$\dfrac{40}{t-15}$
Old	t minutes	$\dfrac{1}{t}$	40 minutes	$\dfrac{40}{t}$

$$\frac{40}{t-15}+\frac{40}{t}=1$$

$$t(t-15)\left(\frac{40}{t-15}+\frac{40}{t}\right)=t(t-15)\cdot 1$$

$$40t+40(t-15)=t(t-15)$$
$$40t+40t-600=t^2-15t$$
$$80t-600=t^2-15t$$
$$0=t^2-95t+600$$
$$t=\frac{-(-95)\pm\sqrt{(-95)^2-4(1)(600)}}{2(1)}$$
$$t=\frac{95\pm\sqrt{6625}}{2}$$

$t\approx 6.8$ or $t\approx 88.2$
The value $t\approx 6.8$ is omitted because it gives a negative solution for the new printer.
Old: about 88.2 minutes

55. $x^2 - 8x - 13 = 0$

$$x = \frac{-(-8) \pm \sqrt{(-8)^2 - 4(1)(-13)}}{2(1)}$$

$$x = \frac{8 \pm \sqrt{116}}{2}$$

$$x = \frac{8 \pm 2\sqrt{29}}{2}$$

$$x = 4 \pm \sqrt{29}$$

$$\left\{4 - \sqrt{29}, 4 + \sqrt{29}\right\}$$

57. $x - 2\sqrt{x} - 48 = 0$

Let $u = \sqrt{x}$ and $u^2 = x$.

$$u^2 - 2u - 48 = 0$$

$$(u - 8)(u + 6) = 0$$

$u - 8 = 0$	or	$u + 6 = 0$
$u = 8$	or	$u = -6$
$\sqrt{x} = 8$	or	$\sqrt{x} = -6$
$\left(\sqrt{x}\right)^2 = (8)^2$	or	$\left(\sqrt{x}\right)^2 = (-6)^2$
$x = 64$	or	$x = 36$

check:

$x = 64$	$x = 36$
$64 - 2\sqrt{64} - 48 = 0$	$36 - 2\sqrt{36} - 48 = 0$
$64 - 2(8) - 48 = 0$	$36 - 2(6) - 48 = 0$
$64 - 16 - 48 = 0$	$36 - 12 - 48 = 0$
$0 = 0$	$-24 = 0$
True	False

$\{64\}$

59. $(3x - 2)^2 = 32$

$$\sqrt{(3x - 2)^2} = \pm\sqrt{32}$$

$$3x - 2 = \pm 4\sqrt{2}$$

$$3x = 2 \pm 4\sqrt{2}$$

$$x = \frac{2 \pm 4\sqrt{2}}{3}$$

$$\left\{\frac{2 - 4\sqrt{2}}{3}, \frac{2 + 4\sqrt{2}}{3}\right\}$$

61. $x^2 - 5x + 14 = 0$

$$x = \frac{-(-5) \pm \sqrt{(-5)^2 - 4(1)(14)}}{2(1)}$$

$$x = \frac{5 \pm \sqrt{-31}}{2}$$

$$x = \frac{5 \pm i\sqrt{31}}{2}$$

$$\left\{\frac{5 - i\sqrt{31}}{2}, \frac{5 + i\sqrt{31}}{2}\right\}$$

63. $x^2 - 6x - 91 = 0$

$$(x + 7)(x - 13) = 0$$

$x + 7 = 0$ or $x - 13 = 0$

$x = -7$ or $x = 13$

$\{-7, 13\}$

65.

$$\frac{x + 1}{x + 3} + \frac{7}{x + 4} = \frac{5}{x^2 + 7x + 12}$$

$$\frac{x + 1}{x + 3} + \frac{7}{x + 4} = \frac{5}{(x + 3)(x + 4)}$$

$$(x + 3)(x + 4)\left(\frac{x + 1}{x + 3} + \frac{7}{x + 4}\right) = \frac{5}{(x + 3)(x + 4)} \cdot (x + 3)(x + 4)$$

$$(x + 4)(x + 1) + (x + 3) \cdot 7 = 5$$

$$x^2 + x + 4x + 4 + 7x + 21 = 5$$

$$x^2 + 12x + 20 = 0$$

$$(x + 10)(x + 2) = 0$$

$x + 10 = 0$ or $x + 2 = 0$

$x = -10$ or $x = -2$

Neither of these values makes a denominator of the original equation 0.

$\{-10, -2\}$

67.
$$\sqrt{3x-2} = x-2$$
$$\left(\sqrt{3x-2}\right)^2 = (x-2)^2$$
$$3x-2 = (x-2)(x-2)$$
$$3x-2 = x^2 - 2x - 2x + 4$$
$$0 = x^2 - 7x + 6$$
$$0 = (x-6)(x-1)$$

$$x - 6 = 0 \quad \text{or} \quad x - 1 = 0$$
$$x = 6 \quad \text{or} \quad x = 1$$

check:

$x = 6$	$x = 1$
$\sqrt{3(6)-2} = 6-2$	$\sqrt{3(1)-2} = 1-2$
$\sqrt{16} = 4$	$\sqrt{1} = -1$
$4 = 4$	$1 = -1$
True	False

$\{6\}$

69. $(5x+3)^2 + 2(5x+3) - 15 = 0$
Let $u = 5x+3$ and $u^2 = (5x+3)^2$.
$$u^2 + 2u - 15 = 0$$
$$(u+5)(u-3) = 0$$

$$u + 5 = 0 \quad \text{or} \quad u - 3 = 0$$
$$u = -5 \quad \text{or} \quad u = 3$$
$$5x + 3 = -5 \quad \text{or} \quad 5x + 3 = 3$$
$$5x = -8 \quad \text{or} \quad 5x = 0$$

$$x = -\frac{8}{5} \quad \text{or} \quad x = 0$$

$$\left\{-\frac{8}{5}, 0\right\}$$

71.
$$(x+8)^2 = -144$$
$$\sqrt{(x+8)^2} = \pm\sqrt{-144}$$
$$x + 8 = \pm 12i$$
$$x = -8 \pm 12i$$
$$\{-8 - 12i, -8 + 12i\}$$

73. $x^4 - 3x^2 - 4 = 0$
Let $u = x^2$ and $u^2 = x^4$.
$$u^2 - 3u - 4 = 0$$
$$(u+1)(u-4) = 0$$

$$u + 1 = 0 \quad \text{or} \quad u - 4 = 0$$
$$u = -1 \quad \text{or} \quad u = 4$$
$$x^2 = -1 \quad \text{or} \quad x^2 = 4$$
$$\sqrt{x^2} = \sqrt{-1} \quad \text{or} \quad \sqrt{x^2} = \sqrt{4}$$
$$x = \pm i \quad \text{or} \quad x = \pm 2$$
$$\{-i, i, -2, 2\}$$

75. $x^{2/3} + 12x^{1/3} + 35 = 0$
Let $u = x^{1/3}$ and $u^2 = x^{2/3}$.
$$u^2 + 12u + 35 = 0$$
$$(u+5)(u+7) = 0$$

$$u + 5 = 0 \quad \text{or} \quad u + 7 = 0$$
$$u = -5 \quad \text{or} \quad u = -7$$
$$x^{1/3} = -5 \quad \text{or} \quad x^{1/3} = -7$$
$$\left(x^{1/3}\right)^3 = (-5)^3 \quad \text{or} \quad \left(x^{1/3}\right)^3 = (-7)^3$$
$$x = -125 \quad \text{or} \quad x = -343$$
$$\{-125, -343\}$$

77. $x^2 + 10x + 25 = 0$
$$(x+5)(x+5) - 0$$
$$x + 5 = 0$$
$$x = -5$$
$$\{-5\}$$

79. After replacing u with x^2, take the square root of each side.

10.4 QUICK CHECK

1. $y = x^2 - 8x + 7$

y-intercept: Let $x = 0$.
$y = 0^2 - 8(0) + 7 = 7$
$(0, 7)$

x-intercepts: Let $y = 0$.
$0 = x^2 - 8x + 7$
$0 = (x - 1)(x - 7)$
$x - 1 = 0 \quad \text{or} \quad x - 7 = 0$
$\quad x = 1 \quad \text{or} \quad\quad x = 7$
$\quad\quad (1, 0) \quad\quad\quad\quad (7, 0)$

vertex:
$$x = \frac{-b}{2a} = \frac{-(-8)}{2(1)} = 4$$

$$y = 4^2 - 8(4) + 7 = 16 - 32 + 7 = -9$$

$(4, -9)$

axis of symmetry: $x = 4$
point symmetric to the y-intercept: $(8, 7)$

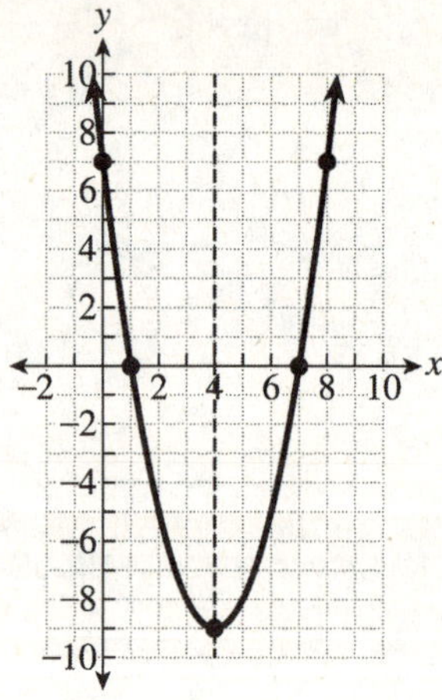

2. $y = x^2 + 6x - 9$

y-intercept: Let $x = 0$
$y = 0^2 + 6(0) - 9 = -9$
$(0, -9)$

x-intercepts: Let $y = 0$
$0 = x^2 + 6x - 9$

$$x = \frac{-6 \pm \sqrt{6^2 - 4(1)(-9)}}{2(1)}$$

$$x = \frac{-6 \pm \sqrt{72}}{2}$$

$$x = \frac{-6 \pm 6\sqrt{2}}{2}$$

$x = -3 \pm 3\sqrt{2}$
$(1.2, 0) \quad (-7.2, 0)$

vertex:
$$x = \frac{-b}{2a} = \frac{-6}{2(1)} = -3$$

$$y = (-3)^2 + 6(-3) - 9 = 9 - 18 - 9 = -18$$

$(-3, -18)$

axis of symmetry: $x = -3$
point symmetric to the y-intercept: $(-6, -9)$

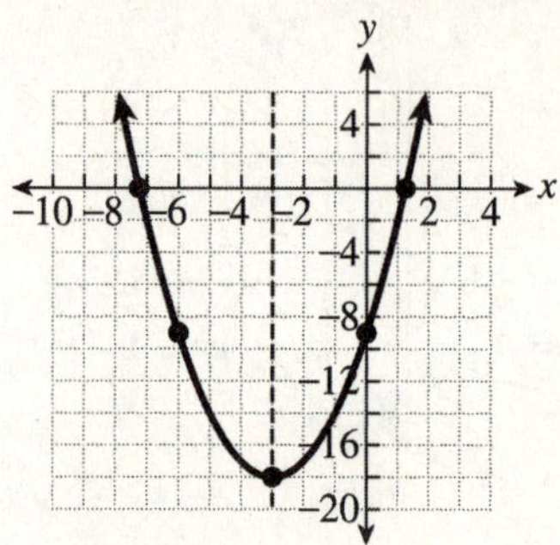

3. $y = x^2 + 6x + 12$

vertex:
$$x = \frac{-b}{2a} = \frac{-6}{2(1)} = -3$$

$$y = (-3)^2 + 6(-3) + 12 = 9 - 18 + 12 = 3$$

$(-3, 3)$

The equation is in standard form.

y-intercept: $(0, 12)$

x-intercepts: The vertex is above the
x-axis and the parabola opens upward.
No x-intercepts.

axis of symmetry: $x = -3$
point symmetric to the y-intercept: $(-6, 12)$

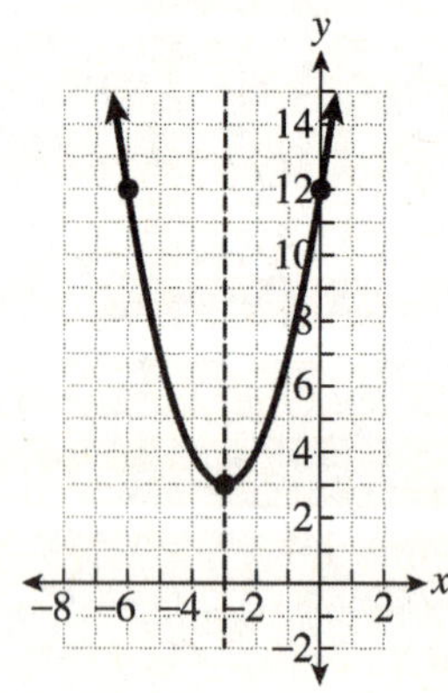

4. $y = -x^2 + 9x + 10$

vertex:

$$x = \frac{-b}{2a} = \frac{-9}{2(-1)} = \frac{9}{2}$$

$$y = -\left(\frac{9}{2}\right)^2 + 9\left(\frac{9}{2}\right) + 10$$

$$y = -\frac{81}{4} + \frac{81}{2} + \frac{10}{1}$$

$$y = -\frac{81}{4} + \frac{162}{4} + \frac{40}{4}$$

$$y = \frac{121}{4} \text{ or } 30.25$$

$$\left(\frac{9}{2}, \frac{121}{4}\right) \text{ or } (4.5,\ 30.25)$$

The equation is in standard form.

y-intercept: $(0,10)$

x-intercepts: Let $y = 0$.

$$0 = -x^2 + 9x + 10$$
$$x^2 - 9x - 10 = 0$$
$$(x+1)(x-10) = 0$$
$$x + 1 = 0 \quad \text{or} \quad x - 10 = 0$$
$$x = -1 \quad \text{or} \qquad x = 10$$
$$(-1, 0) \qquad\qquad (10, 0)$$

axis of symmetry: $x = \dfrac{9}{2}$

point symmetric to the y-intercept: $(9,10)$

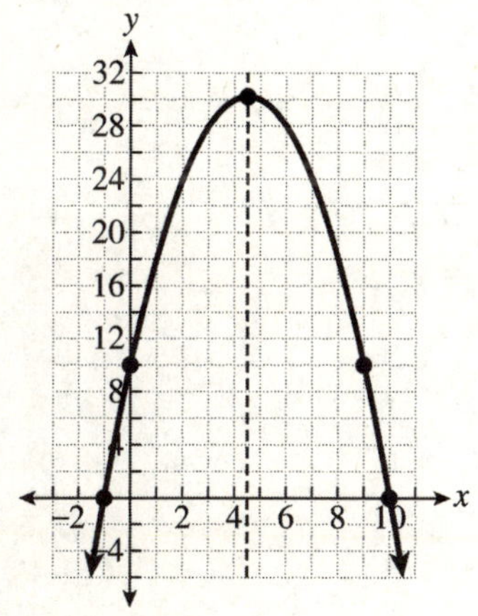

5. $y = -\dfrac{1}{4}x^2 - \dfrac{5}{2}x - 4$

vertex:

$$x = \frac{-b}{2a} = \frac{-\left(-\dfrac{5}{2}\right)}{2\left(-\dfrac{1}{4}\right)} = \frac{\dfrac{5}{2}}{-\dfrac{1}{2}} = -5$$

$$y = -\frac{1}{4}(-5)^2 - \frac{5}{2}(-5) - 4$$

$$y = -\frac{1}{4}(25) + \frac{25}{2} - \frac{4}{1}$$

$$y = -\frac{25}{4} + \frac{50}{4} - \frac{16}{4}$$

$$y = \frac{9}{4}$$

$$\left(-5, \frac{9}{4}\right)$$

The equation is in standard form.

y-intercept: $(0,-4)$

x-intercepts: Let $y = 0$.

$$0 = -\frac{1}{4}x^2 - \frac{5}{2}x - 4$$
$$-4(0) = -4\left(-\frac{1}{4}x^2 - \frac{5}{2}x - 4\right)$$
$$0 = x^2 + 10x + 16$$
$$0 = (x+2)(x+8)$$
$$x + 2 = 0 \quad \text{or} \quad x + 8 = 0$$
$$x = -2 \quad \text{or} \qquad x = -8$$
$$(-2, 0) \qquad\qquad (-8, 0)$$

axis of symmetry: $x = -5$

point symmetric to the y-intercept: $(-10,-4)$

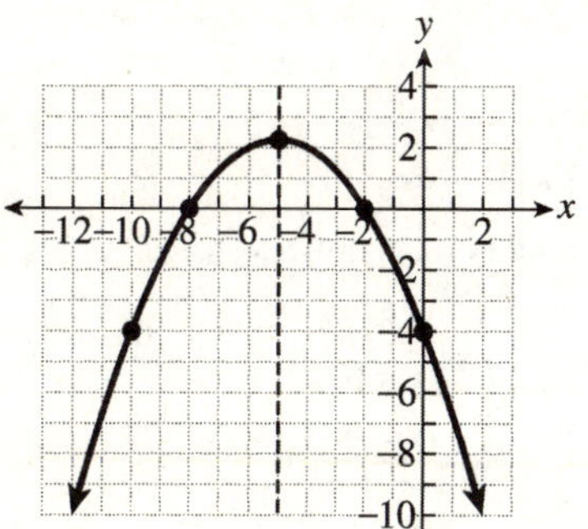

6. a) $y = (x+2)^2 - 8$

The equation is in the form
$y = a(x-h)^2 + k.$
vertex: $(h,k) = (-2,-8)$
axis of symmetry: $x = h,$ or $x = -2$

b) $y = -2(x-8)^2 + 7$

The equation is in the form
$y = a(x-h)^2 + k.$
vertex: $(h,k) = (8,7)$
axis of symmetry: $x = h,$ or $x = 8$

7. $y = (x+2)^2 + 3$

The parabola opens upward.

vertex: $(-2,3)$

y-intercept: Let $x = 0.$

$y = (0+2)^2 + 3 = 4 + 3 = 7$
$(0,7)$

x-intercepts: The vertex is above the
x-axis and the parabola opens upward.
No x-intercepts.

axis of symmetry: $x = -2$

point symmetric to the y-intercept: $(-4,7)$

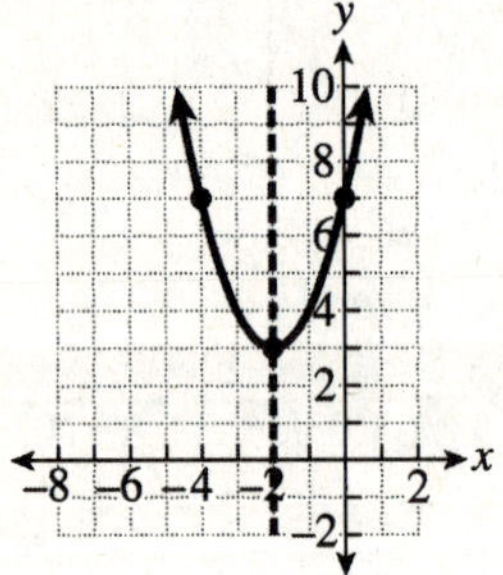

8. $y = -(x+1)^2 + 4$

The parabola opens downward.

vertex: $(-1,4)$

y-intercept: Let $x = 0.$

$y = -(0+1)^2 + 4 = -1 + 4 = 3$
$(0,3)$

x-intercepts: Let $y = 0.$

$$0 = -(x+1)^2 + 4$$
$$(x+1)^2 = 4$$
$$\sqrt{(x+1)^2} = \pm\sqrt{4}$$
$$x+1 = \pm 2$$
$$x = -1 \pm 2$$
$$x = -1 - 2 \quad \text{or} \quad x = -1 + 2$$
$$x = -3 \quad\quad \text{or} \quad x = 1$$
$$(-3,0), (1,0)$$

axis of symmetry: $x = -1$
point symmetric to the y-intercept: $(-2,3)$

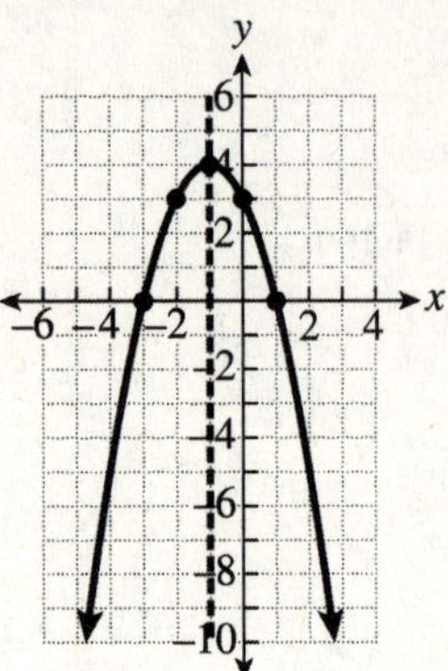

10.4   GRAPHING QUADRATIC EQUATIONS

1. parabola

3. negative

5. y-intercept

7. $y = x^2 + 8x - 22$

vertex:
$$x = \frac{-b}{2a} = \frac{-8}{2(1)} = -4$$
$$y = (-4)^2 + 8(-4) - 22 = 16 - 32 - 22 = -38$$
$$(-4,-38)$$

axis of symmetry: $x = -4$

9. $y = x^2 - 7x + 10$

vertex:
$$x = \frac{-b}{2a} = \frac{-(-7)}{2(1)} = \frac{7}{2}$$

$$y = \left(\frac{7}{2}\right)^2 - 7\left(\frac{7}{2}\right) + 10$$

$$= \frac{49}{4} - \frac{49}{2} + \frac{10}{1}$$

$$= \frac{49}{4} - \frac{98}{4} + \frac{40}{4}$$

$$= -\frac{9}{4}$$

$$\left(\frac{7}{2}, -\frac{9}{4}\right)$$

axis of symmetry: $x = \frac{7}{2}$

11. $y = -x^2 + 12x + 62$

vertex:

$x = \dfrac{-b}{2a} = \dfrac{-12}{2(-1)} = 6$

$y = -(6)^2 + 12(6) + 62 = -36 + 72 + 62 = 98$

$(6, 98)$

axis of symmetry: $x = 6$

13. $y = 2x^2 + 8x - 19$

vertex:

$x = \dfrac{-b}{2a} = \dfrac{-8}{2(2)} = -2$

$y = 2(-2)^2 + 8(-2) - 19 = 8 - 16 - 19 = -27$

$(-2, -27)$

axis of symmetry: $x = -2$

15. $y = -4x^2 + 24x - 11$

vertex:

$x = \dfrac{-b}{2a} = \dfrac{-24}{2(-4)} = 3$

$y = -4(3)^2 + 24(3) - 11 = -36 + 72 - 11 = 25$

$(3, 25)$

axis of symmetry: $x = 3$

17. $y = x^2 + \dfrac{3}{2}x + 8$

vertex:

$x = \dfrac{-b}{2a} = \dfrac{-\left(\dfrac{3}{2}\right)}{2(1)} = -\dfrac{3}{2} \cdot \dfrac{1}{2} = -\dfrac{3}{4}$

$y = \left(-\dfrac{3}{4}\right)^2 + \dfrac{3}{2}\left(-\dfrac{3}{4}\right) + 8$

$= \dfrac{9}{16} - \dfrac{9}{8} + 8$

$= \dfrac{9}{16} - \dfrac{18}{16} + \dfrac{128}{16}$

$= \dfrac{119}{16}$

$\left(-\dfrac{3}{4}, \dfrac{119}{16}\right)$

axis of symmetry: $x = -\dfrac{3}{4}$

19. $y = x^2 + 10x$

vertex:

$x = \dfrac{-b}{2a} = \dfrac{-10}{2(1)} = -5$

$y = (-5)^2 + 10(-5) = 25 - 50 = -25$

$(-5, -25)$

axis of symmetry: $x = -5$

21. $y = (x - 7)^2 + 12$

The equation is in the form $y = a(x - h)^2 + k$.
vertex: $(h, k) = (7, 12)$
axis of symmetry: $x = h$, or $x = 7$

23. $y = (x + 1)^2 - 5$

The equation is in the form $y = a(x - h)^2 + k$.
vertex: $(h, k) = (-1, -5)$
axis of symmetry: $x = h$, or $x = -1$

25. $y = -(x - 9)^2 - 7$

The equation is in the form $y = a(x - h)^2 + k$.
vertex: $(h, k) = (9, -7)$
axis of symmetry: $x = h$, or $x = 9$

27. $y = x^2 + 3x - 40$
The equation is in standard form.
y-intercept: $(0, -40)$

x-intercepts: Let $y = 0$.
$0 = x^2 + 3x - 40$
$0 = (x + 8)(x - 5)$
$x + 8 = 0 \quad$ or $\quad x - 5 = 0$
$\quad x = -8 \quad$ or $\qquad x = 5$
$(-8, 0), (5, 0)$

29. $y = x^2 - 5x + 3$
The equation is in standard form.
y-intercept: $(0, 3)$

x-intercepts: Let $y = 0$.
$0 = x^2 - 5x + 3$

$x = \dfrac{-(-5) \pm \sqrt{(-5)^2 - 4(1)(3)}}{2(1)}$

$x = \dfrac{5 \pm \sqrt{13}}{2}$

$x = \dfrac{5 - \sqrt{13}}{2} \quad$ or $\quad x = \dfrac{5 + \sqrt{13}}{2}$

$x \approx 0.7 \qquad$ or $\quad x \approx 4.3$
$(0.7, 0), (4.3, 0)$

31. $y = x^2 + 5x + 15$
The equation is in standard form.
y-intercept: $(0, 15)$

x-intercepts: Let $y = 0$.
$0 = x^2 + 5x + 15$

$$x = \frac{-5 \pm \sqrt{5^2 - 4(1)(15)}}{2(1)}$$

$$x = \frac{-5 \pm \sqrt{25 - 60}}{2}$$

$$x = \frac{-5 \pm \sqrt{-35}}{2}$$

The discriminant is negative; the equation does not have any real solutions.
No x-intercepts.

33. $y = 2x^2 + 6x - 15$
The equation is in standard form.
y-intercept: $(0, -15)$

x-intercepts: Let $y = 0$.
$0 = 2x^2 + 6x - 15$

$$x = \frac{-6 \pm \sqrt{6^2 - 4(2)(-15)}}{2(2)}$$

$$x = \frac{-6 \pm \sqrt{156}}{4}$$

$$x = \frac{-6 \pm 2\sqrt{39}}{4}$$

$$x = \frac{-3 \pm \sqrt{39}}{2}$$

$$x = \frac{-3 - \sqrt{39}}{2} \quad \text{or} \quad x = \frac{-3 + \sqrt{39}}{2}$$
$$x \approx -4.6 \qquad \text{or} \quad x \approx 1.6$$
$$(-4.6, 0), (1.6, 0)$$

35. $y = -x^2 - 5x + 6$
The equation is in standard form.
y-intercept: $(0, 6)$

x-intercepts: Let $y = 0$.
$0 = -x^2 - 5x + 6$
$0 = -\left(x^2 + 5x - 6\right)$
$0 = -(x + 6)(x - 1)$

$$x + 6 = 0 \quad \text{or} \quad x - 1 = 0$$
$$x = -6 \quad \text{or} \qquad x = 1$$
$$(-6, 0), (1, 0)$$

37. $y = -x^2 + 6x + 20$
The equation is in standard form.
y-intercept: $(0, 20)$

x-intercepts: Let $y = 0$.
$0 = -x^2 + 6x + 20$

$$x = \frac{-6 \pm \sqrt{6^2 - 4(-1)(20)}}{2(-1)}$$

$$x = \frac{-6 \pm \sqrt{116}}{-2}$$

$$x = \frac{-6 \pm 2\sqrt{29}}{-2}$$

$$x = 3 \pm \sqrt{29}$$

$$x = 3 - \sqrt{29} \quad \text{or} \quad x = 3 + \sqrt{29}$$
$$x \approx -2.4 \qquad \text{or} \quad x \approx 8.4$$
$$(-2.4, 0), (8.4, 0)$$

39. $y = -x^2 + 7x - 15$
The equation is in standard form.
y-intercept: $(0, -15)$

x-intercepts: Let $y = 0$.
$0 = -x^2 + 7x - 15$
$0 = x^2 - 7x + 15$

$$x = \frac{-(-7) \pm \sqrt{(-7)^2 - 4(1)(15)}}{2(1)}$$

$$x = \frac{7 \pm \sqrt{49 - 60}}{2}$$

$$x = \frac{7 \pm \sqrt{-11}}{2}$$

The discriminant is negative; the equation does not have any real solutions.
No x-intercepts.

41. $y = x^2 - 8x + 16$
The equation is in standard form.
y-intercept: $(0, 16)$

x-intercepts: Let $y = 0$.
$0 = x^2 - 8x + 16$
$0 = (x - 4)(x - 4)$
$x - 4 = 0$
$\quad x = 4$
$(4, 0)$

43. $y = (x-7)^2 - 9$
y-intercept: Let $x = 0$.
$y = (0-7)^2 - 9 = 49 - 9 = 40$
$(0, 40)$

x-intercepts: Let $y = 0$.
$$0 = (x-7)^2 - 9$$
$$(x-7)^2 = 9$$
$$\sqrt{(x-7)^2} = \pm\sqrt{9}$$
$$x - 7 = \pm 3$$
$$x = 7 \pm 3$$
$x = 7 - 3 \quad$ or $\quad x = 7 + 3$
$x = 4 \qquad$ or $\quad x = 10$
$(4, 0), (10, 0)$

45. $y = (x+5)^2 + 4$
y-intercept: Let $x = 0$.
$y = (0+5)^2 + 4 = 25 + 4 = 29$
$(0, 29)$

x-intercepts: The vertex is above the x-axis and the parabola opens upward. No x-intercepts.

47. $y = -(x-1)^2 + 12$
y-intercept: Let $x = 0$.
$y = -(0-1)^2 + 12 = -1 + 12 = 11$
$(0, 11)$

x-intercepts: Let $y = 0$.
$$0 = -(x-1)^2 + 12$$
$$(x-1)^2 = 12$$
$$\sqrt{(x-1)^2} = \pm\sqrt{12}$$
$$x - 1 = \pm 2\sqrt{3}$$
$$x = 1 \pm 2\sqrt{3}$$
$x = 1 - 2\sqrt{3} \quad$ or $\quad x = 1 + 2\sqrt{3}$
$x \approx -2.5 \qquad$ or $\quad x \approx 4.5$
$(-2.5, 0), (4.5, 0)$

49. $y = x^2 - 2x - 3$
vertex:
$$x = \frac{-b}{2a} = \frac{-(-2)}{2(1)} = 1$$
$$y = (1)^2 - 2(1) - 3 = -4$$
$(1, -4)$

The equation is in standard form.
y-intercept: $(0, -3)$

x-intercepts: Let $y = 0$.
$$0 = x^2 - 2x - 3$$
$$0 = (x+1)(x-3)$$
$x + 1 = 0 \quad$ or $\quad x - 3 = 0$
$\quad x = -1 \quad$ or $\qquad x = 3$
$(-1, 0), (3, 0)$

axis of symmetry: $x = 1$
point symmetric to the y-intercept: $(2, -3)$

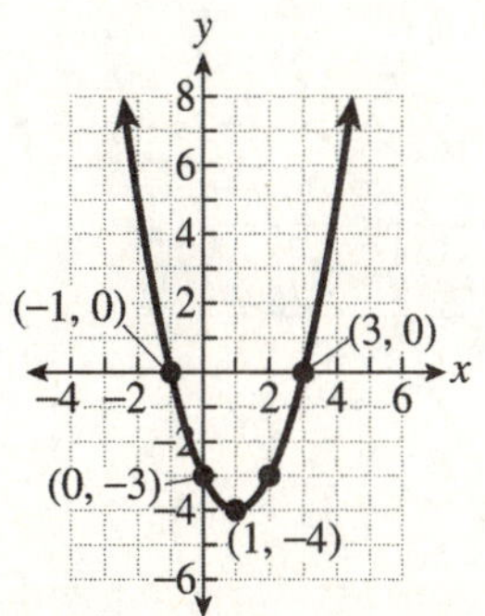

51. $y = x^2 - 5x$
vertex:
$$x = \frac{-b}{2a} = \frac{-(-5)}{2(1)} = \frac{5}{2}$$
$$y = \left(\frac{5}{2}\right)^2 - 5\left(\frac{5}{2}\right) = \frac{25}{4} - \frac{25}{2} = \frac{25}{4} - \frac{50}{4} = -\frac{25}{4}$$
$\left(\frac{5}{2}, -\frac{25}{4}\right)$

The equation is in standard form.
y-intercept: $(0, 0)$

x-intercepts: Let $y = 0$.
$$0 = x^2 - 5x$$
$$0 = x(x-5)$$
$x = 0 \quad$ or $\quad x - 5 = 0$
$x = 0 \quad$ or $\qquad x = 5$
$(0, 0), (5, 0)$

axis of symmetry: $x = \frac{5}{2}$

point symmetric to the y-intercept: $(5, 0)$

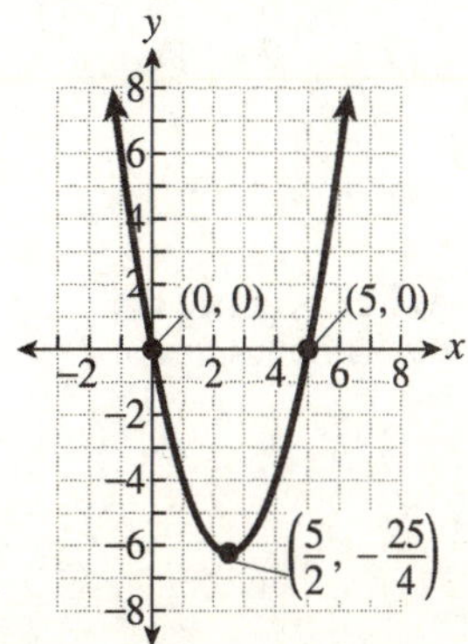

53. $y = x^2 - 2x - 1$

vertex:

$$x = \frac{-b}{2a} = \frac{-(-2)}{2(1)} = 1$$

$$y = (1)^2 - 2(1) - 1 = 1 - 2 - 1 = -2$$
$$(1, -2)$$

The equation is in standard form.
y-intercept: $(0, -1)$

x-intercepts: Let $y = 0$.

$$0 = x^2 - 2x - 1$$

$$x = \frac{-(-2) \pm \sqrt{(-2)^2 - 4(1)(-1)}}{2(1)}$$

$$x = \frac{2 \pm \sqrt{8}}{2}$$

$$x = \frac{2 \pm 2\sqrt{2}}{2}$$

$$x = 1 \pm \sqrt{2}$$
$$\left(1 - \sqrt{2}, 0\right), \left(1 + \sqrt{2}, 0\right)$$

axis of symmetry: $x = 1$
point symmetric to the y-intercept: $(2, -1)$

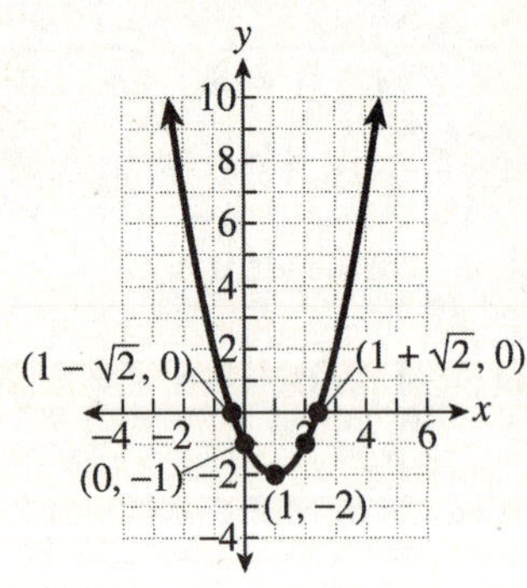

55. $y = x^2 - 6x + 10$

vertex:

$$x = \frac{-b}{2a} = \frac{-(-6)}{2(1)} = 3$$

$$y = (3)^2 - 6(3) + 10 = 9 - 18 + 10 = 1$$
$$(3, 1)$$

The equation is in standard form.
y-intercept: $(0, 10)$

x-intercepts: The vertex is above the
x-axis and the parabola opens upward.
No x-intercepts.

axis of symmetry: $x = 3$
point symmetric to the y-intercept: $(6, 10)$

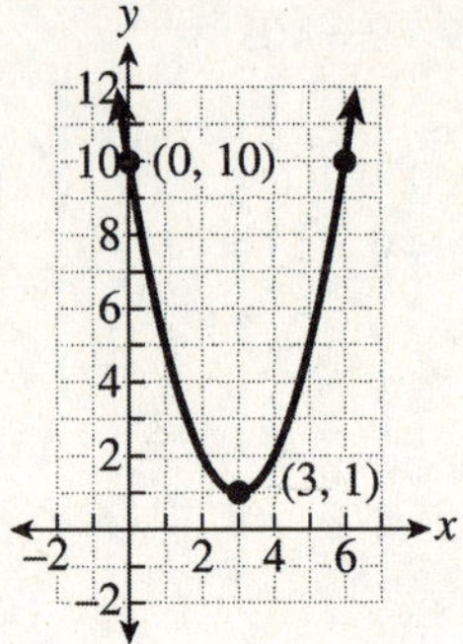

57. $y = -x^2 + 4x - 4$

vertex:

$$x = \frac{-b}{2a} = \frac{-4}{2(-1)} = 2$$

$$y = -(2)^2 + 4(2) - 4 = -4 + 8 - 4 = 0$$
$$(2, 0)$$

The equation is in standard form.
y-intercept: $(0, -4)$

x-intercepts: Let $y = 0$.
$$0 = -x^2 + 4x - 4$$
$$0 = x^2 - 4x + 4$$
$$0 = (x - 2)(x - 2)$$
$$x - 2 = 0$$
$$x = 2$$
$$(2, 0)$$

axis of symmetry: $x = 2$
point symmetric to the y-intercept: $(4, -4)$

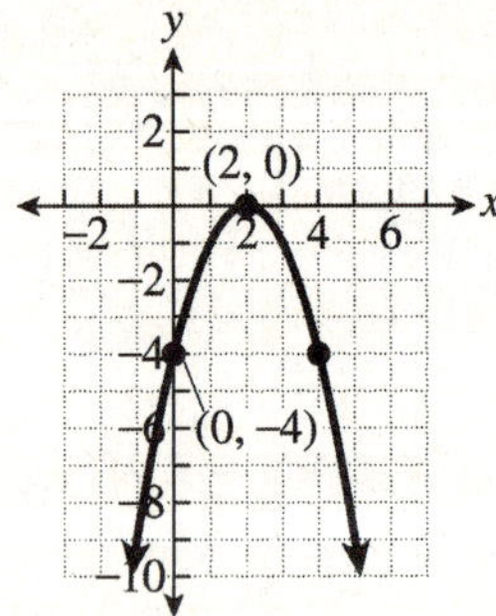

59. $y = -\dfrac{1}{2}x^2 + 3x - 3$

vertex:

$$x = \frac{-b}{2a} = \frac{-3}{2\left(-\dfrac{1}{2}\right)} = 3$$

$$y = -\frac{1}{2}(3)^2 + 3(3) - 3$$

$$= -\frac{9}{2} + 9 - 3$$

$$= -\frac{9}{2} + \frac{18}{2} - \frac{6}{2}$$

$$= \frac{3}{2}$$

$$\left(3, \frac{3}{2}\right)$$

The equation is in standard form.
y-intercept: $(0, -3)$

x-intercepts: Let $y = 0$.

$$0 = -\frac{1}{2}x^2 + 3x - 3$$

$$(-2)\cdot 0 = \left(-\frac{1}{2}x^2 + 3x - 3\right)(-2)$$

$$0 = x^2 - 6x + 6$$

$$x = \frac{-(-6) \pm \sqrt{(-6)^2 - 4(1)(6)}}{2(1)}$$

$$x = \frac{6 \pm \sqrt{12}}{2}$$

$$x = \frac{6 \pm 2\sqrt{3}}{2}$$

$$x = 3 \pm \sqrt{3}$$

$$\left(3 - \sqrt{3}, 0\right), \left(3 + \sqrt{3}, 0\right)$$

axis of symmetry: $x = 3$
point symmetric to the y-intercept: $(6, -3)$

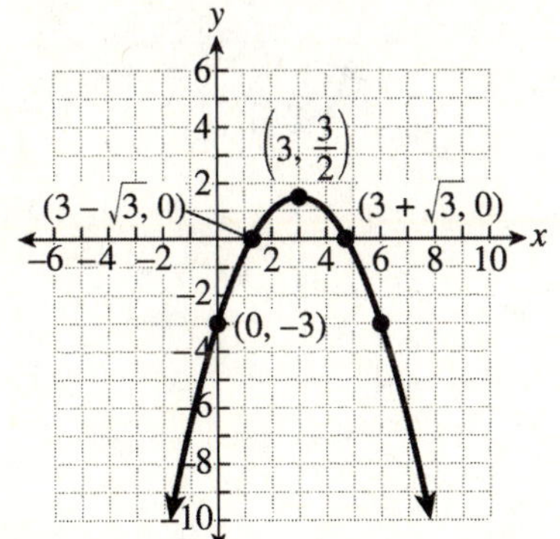

61. $y = -x^2 + 3x - 4$

vertex:

$$x = \frac{-b}{2a} = \frac{-3}{2(-1)} = \frac{3}{2}$$

$$y = -\left(\frac{3}{2}\right)^2 + 3\left(\frac{3}{2}\right) - 4$$

$$= -\frac{9}{4} + \frac{9}{2} - \frac{4}{1}$$

$$= -\frac{9}{4} + \frac{18}{4} - \frac{16}{4}$$

$$= -\frac{7}{4}$$

$$\left(\frac{3}{2}, -\frac{7}{4}\right)$$

The equation is in standard form.
y-intercept: $(0, -4)$

x-intercepts: The vertex is below the
x-axis and the parabola opens downward.
No x-intercepts.

axis of symmetry: $x = \dfrac{3}{2}$

point symmetric to the y-intercept: $(3, -4)$

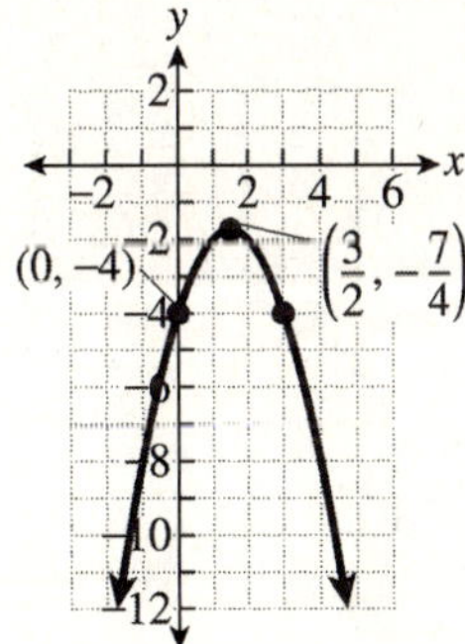

63. $y = (x-3)^2 - 4$

vertex: $(3,-4)$
y-intercept: Let $x = 0$.
$y = (0-3)^2 - 4 = 9 - 4 = 5$
$(0,5)$

x-intercepts: Let $y = 0$.
$$0 = (x-3)^2 - 4$$
$$(x-3)^2 = 4$$
$$\sqrt{(x-3)^2} = \pm\sqrt{4}$$
$$x - 3 = \pm 2$$
$$x = 3 \pm 2$$
$x = 3 - 2$ or $x = 3 + 2$
$x = 1$ or $x = 5$
$(1,0), (5,0)$

axis of symmetry: $x = 3$
point symmetric to the y-intercept: $(6,5)$

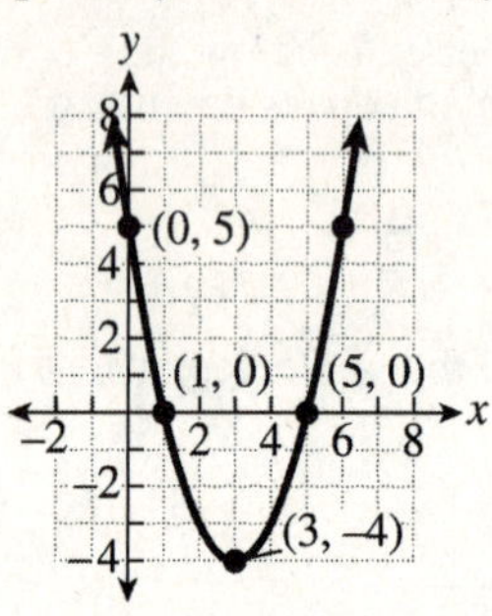

65. $y = -(x-2)^2 + 1$

vertex: $(2,1)$
y-intercept: Let $x = 0$.
$y = -(0-2)^2 + 1 = -4 + 1 = -3$
$(0,-3)$

x-intercepts: Let $y = 0$.
$$0 = -(x-2)^2 + 1$$
$$(x-2)^2 = 1$$
$$\sqrt{(x-2)^2} = \pm\sqrt{1}$$
$$x - 2 = \pm 1$$
$$x = 2 \pm 1$$
$x = 2 - 1$ or $x = 2 + 1$
$x = 1$ or $x = 3$
$(1,0), (3,0)$

axis of symmetry: $x = 2$
point symmetric to the y-intercept: $(4,-3)$

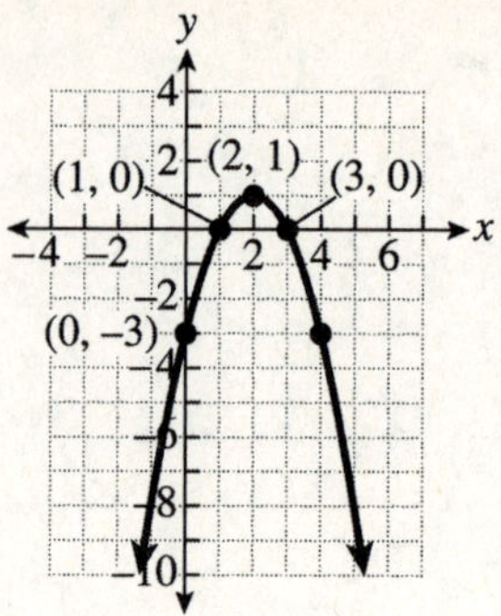

67. $y = (x-4)^2 - 2$

vertex: $(4,-2)$
y-intercept: Let $x = 0$.
$y = (0-4)^2 - 2 = 16 - 2 = 14$
$(0,14)$

x-intercepts: Let $y = 0$.
$$0 = (x-4)^2 - 2$$
$$(x-4)^2 = 2$$
$$\sqrt{(x-4)^2} = \pm\sqrt{2}$$
$$x - 4 = \pm\sqrt{2}$$
$$x = 4 \pm \sqrt{2}$$
$\left(4 - \sqrt{2}, 0\right), \left(4 + \sqrt{2}, 0\right)$

axis of symmetry: $x = 4$
point symmetric to the y-intercept: $(8,14)$

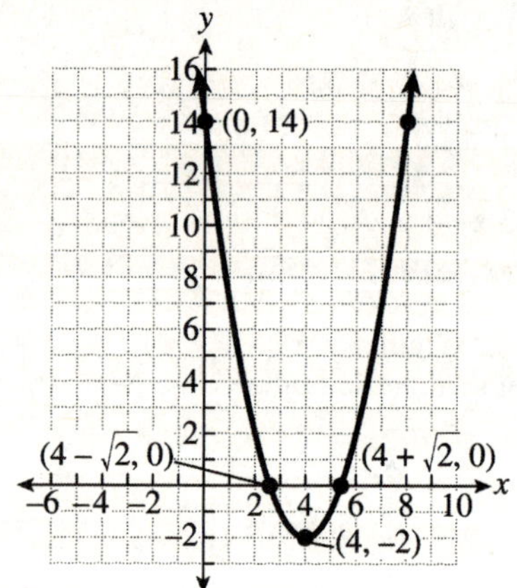

69. $y = -(x-2)^2 - 4$

vertex: $(2, -4)$
y-intercept: Let $x = 0$.
$y = -(0-2)^2 - 4 = -4 - 4 = -8$
$(0, -8)$

x-intercepts: The vertex is below the
x-axis and the parabola opens downward.
No x-intercepts.

axis of symmetry: $x = 2$
point symmetric to the y-intercept: $(4, -8)$

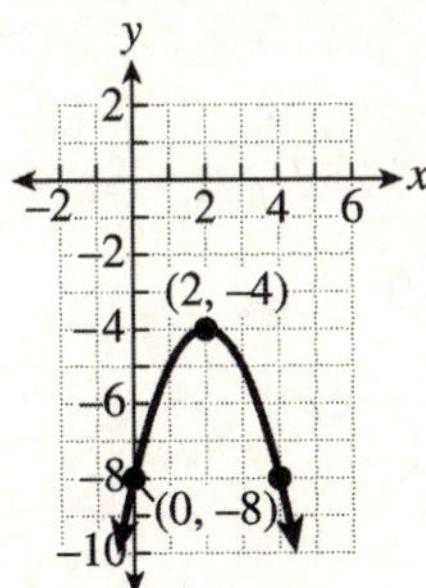

71. Answers will vary. Example:
If the parabola opens upward and the vertex is
above the x-axis, or the parabola opens
downward and the vertex is below the x-axis,
the parabola does not have any x-intercepts.

10.5 *Quick Check*

1. Unknowns: Known:
Base: $3x + 2$ Area: 60 in^2
Height: x

$$\text{Area} = \frac{1}{2}bh$$
$$60 = \frac{1}{2}(3x+2)(x)$$
$$2 \cdot 60 = 2 \cdot \frac{1}{2}(3x+2)(x)$$
$$120 = (3x+2)(x)$$
$$120 = 3x^2 + 2x$$
$$0 = 3x^2 + 2x - 120$$
$$0 = (3x+20)(x-6)$$
$$3x + 20 = 0 \quad \text{or} \quad x - 6 = 0$$
$$3x = -20 \quad \text{or} \quad x = 6$$
$$x = -\frac{20}{3} \quad \text{or} \quad x = 6$$

Height: $x = 6$ inches
Base: $3x + 2 = 3(6) + 2 = 20$ inches

2. Unknowns: Known:
Length: $x + 8$ Area: 350 in.^2
Width: x

$$(x+8)x = 350$$
$$x^2 + 8x = 350$$
$$x^2 + 8x - 350 = 0$$
$$x = \frac{-8 \pm \sqrt{8^2 - 4(1)(-350)}}{2(1)}$$
$$x = \frac{-8 \pm \sqrt{1464}}{2}$$
$$x = \frac{-8 \pm 2\sqrt{366}}{2}$$
$$x = -4 \pm \sqrt{366}$$
$$x = -4 - \sqrt{366} \quad \text{or} \quad x = -4 + \sqrt{366}$$
$$x \approx -23.1 \quad \text{or} \quad x \approx 15.1$$

Width: $x \approx 15.1$ inches
Length: $x + 8 \approx 15.1 + 8 \approx 23.1$ inches

3. Unknown: Knowns:
Leg: a Leg b: 5 in.
              Hypotenuse: 11 in.

$$a^2 + 5^2 = 11^2$$
$$a^2 + 25 = 121$$
$$a^2 = 96$$
$$\sqrt{a^2} = \pm\sqrt{96}$$
$$a = \pm\sqrt{96}$$
$$a \approx 9.80 \text{ inches}$$

4. Unknown: Knowns:
Height: h Base: 2 ft
             Ladder: 8 ft

$$h^2 + 2^2 = 8^2$$
$$h^2 + 4 = 64$$
$$h^2 = 60$$
$$\sqrt{h^2} = \pm\sqrt{60}$$
$$h \approx 7.7 \text{ feet}$$

5. Unknown: Knowns:
Hose: h Length: 240 ft
            Width: 70 ft

$$70^2 + 240^2 = h^2$$
$$4900 + 57600 = h^2$$
$$62500 = h^2$$
$$\pm\sqrt{62500} = \sqrt{h^2}$$
$$h = 250 \text{ feet}$$

10.5 APPLICATIONS USING QUADRATIC EQUATIONS

1. $A = \dfrac{1}{2}bh$

3. legs

5. Unknowns: Known:
Length: $x+5$ Area: 66 in.^{2}
Width: x

$$(x+5)x = 66$$
$$x^2 + 5x = 66$$
$$x^2 + 5x - 66 = 0$$
$$(x+11)(x-6) = 0$$
$$x+11 = 0 \quad \text{or} \quad x-6 = 0$$
$$x = -11 \quad \text{or} \quad x = 6$$

Width: $x = 6$ inches
Length: $x+5 = 6+5 = 11$ inches

7. Unknowns: Known:
Width: x Area: 120 m^2
Length: $2x$

$$(2x)x = 120$$
$$2x^2 = 120$$
$$x^2 = 60$$
$$\sqrt{x^2} = \pm\sqrt{60}$$
$$x = \pm 2\sqrt{15}$$
$$x \approx 7.7$$

Width: $x \approx 7.7$ meters
Length: $2x \approx 2(7.7) \approx 15.4$ meters

9. Unknowns: Known:
Width: $2x-13$ Area: 68 ft^2
Length: x

$$x(2x-13) = 68$$
$$2x^2 - 13x = 68$$
$$2x^2 - 13x - 68 = 0$$
$$x = \frac{-(-13) \pm \sqrt{(-13)^2 - 4(2)(-68)}}{2(2)}$$
$$x = \frac{13 \pm \sqrt{713}}{4}$$
$$x = \frac{13 - \sqrt{713}}{4} \quad \text{or} \quad x = \frac{13 + \sqrt{713}}{4}$$
$$x \approx -3.4 \quad \text{or} \quad x \approx 9.9$$

Length: $x \approx 9.9$ feet
Width: $2x-13 \approx 2(9.9)-13 \approx 6.8$ feet

11. Unknowns: Known:
Base: $x+5$ Area: 42 in.^{2}
Height: x

$$\frac{1}{2}(x+5)(x) = 42$$
$$x(x+5) = 84$$
$$x^2 + 5x = 84$$
$$x^2 + 5x - 84 = 0$$
$$(x+12)(x-7) = 0$$
$$x+12 = 0 \quad \text{or} \quad x-7 = 0$$
$$x = -12 \quad \text{or} \quad x = 7$$

Height: $x = 7$ inches
Base: $x+5 = 7+5 = 12$ inches

13. Unknowns: Known:
Base: x Area: 15 in^2
Height: $2x+1$

$$\frac{1}{2}(x)(2x+1) = 15$$
$$x(2x+1) = 30$$
$$2x^2 + x = 30$$
$$2x^2 + x - 30 = 0$$
$$x = \frac{-1 \pm \sqrt{1^2 - 4(2)(-30)}}{2(2)}$$
$$x = \frac{-1 \pm \sqrt{241}}{4}$$
$$x = \frac{-1 - \sqrt{241}}{4} \quad \text{or} \quad x = \frac{-1 + \sqrt{241}}{4}$$
$$x \approx -4.1 \quad \text{or} \quad x \approx 3.6$$

Base: $x \approx 3.6$ inches
Height: $2x+1 \approx 2(3.6)+1 = 8.2$ inches

15. Unknowns: Known:
Width: $x-2$ Area: 80 in.^{2}
Height: x

$$x(x-2) = 80$$
$$x^2 - 2x = 80$$
$$x^2 - 2x - 80 = 0$$
$$(x+8)(x-10) = 0$$
$$x+8 = 0 \quad \text{or} \quad x-10 = 0$$
$$x = -8 \quad \text{or} \quad x = 10$$

Height: $x = 10$ inches
Width: $x-2 = 10-2 = 8$ inches

17. Unknowns: Known:
Length: $2x - 10$ Area: 2160 in.^{2}
Width: x

$$(2x - 10)x = 2160$$
$$2x^2 - 10x = 2160$$
$$2x^2 - 10x - 2160 = 0$$
$$x = \frac{-(-10) \pm \sqrt{(-10)^2 - 4(2)(-2160)}}{2(2)}$$
$$x = \frac{10 \pm \sqrt{17,380}}{4}$$
$$x = \frac{10 - \sqrt{17,380}}{4} \quad \text{or} \quad x = \frac{10 + \sqrt{17,380}}{4}$$
$$x \approx -30.5 \quad \text{or} \quad x \approx 35.5$$

Width: $x \approx 35.5$ inches
Length: $2x - 10 \approx 2(35.5) - 10 \approx 61.0$ inches

19. Unknowns: Known:
Width: $x - 16$ Area: 540 in.^{2}
Length: x

$$x(x - 16) = 540$$
$$x^2 - 16x = 540$$
$$x^2 - 16x - 540 = 0$$
$$x = \frac{-(-16) \pm \sqrt{(-16)^2 - 4(1)(-540)}}{2(1)}$$
$$x = \frac{16 \pm \sqrt{2416}}{2}$$
$$x = \frac{16 - \sqrt{2416}}{2} \quad \text{or} \quad x = \frac{16 + \sqrt{2416}}{2}$$
$$x \approx -16.6 \quad \text{or} \quad x \approx 32.6$$

Length: $x \approx 32.6$ inches
Width: $x - 16 \approx 32.6 - 16 \approx 16.6$ inches

21. Unknowns: Known:
Height: $x + 8$ Area: 1920 in.^{2}
Height with border: $x + 16$
Width: x
Width with border: $x + 8$

$$(x + 16)(x + 8) = 1920$$
$$x^2 + 8x + 16x + 128 = 1920$$
$$x^2 + 24x - 1792 = 0$$
$$(x + 56)(x - 32) = 0$$
$$x + 56 = 0 \quad \text{or} \quad x - 32 = 0$$
$$x = -56 \quad \text{or} \quad x = 32$$

Width: $x = 32$ inches
Height: $x + 8 = 32 + 8 = 40$ inches

23. Unknowns: Known:
Base: $2x$ Area: 256 in.^{2}
Height: x

$$\frac{1}{2}(2x)x = 256$$
$$x^2 = 256$$
$$\sqrt{x^2} = \pm\sqrt{256}$$
$$x = \pm 16$$

Height: $x = 16$ inches
Base: $2x = 2(16) = 32$ inches

25. Unknowns: Known:
Base: x Area: 46 ft^2
Height: $x + 8$

$$\frac{1}{2}(x)(x + 8) = 46$$
$$x(x + 8) = 92$$
$$x^2 + 8x = 92$$
$$x^2 + 8x - 92 = 0$$
$$x = \frac{-8 \pm \sqrt{8^2 - 4(1)(-92)}}{2(1)}$$
$$x = \frac{-8 \pm \sqrt{432}}{2}$$
$$x \approx 6.4$$

Base: $x \approx 6.4$ feet
Height: $x + 8 \approx 6.4 + 8 \approx 14.4$ feet

27. Unknown: Knowns:
Hypotenuse: c Leg a: 10 in.
                     Leg b: 16 in.

$$(10)^2 + (16)^2 = c^2$$
$$100 + 256 = c^2$$
$$356 = c^2$$
$$\pm\sqrt{356} = \sqrt{c^2}$$
$$\pm 18.9 \approx c$$

Hypotenuse: $c \approx 18.9$ inches

29. Unknown: Knowns:
Leg: a Leg b: 9 feet
          Hypotenuse: 15 feet

$$a^2 + (9)^2 = (15)^2$$
$$a^2 + 81 = 225$$
$$a^2 = 144$$
$$\sqrt{a^2} = \pm\sqrt{144}$$
$$a = \pm 12$$

Leg $a = 12$ feet

31. Unknown: Knowns:
Leg: a Leg b: 12 meters
 Hypotenuse: 24 meters.

$$a^2 + (12)^2 = (24)^2$$
$$a^2 + 144 = 576$$
$$a^2 = 432$$
$$\sqrt{a^2} = \pm\sqrt{432}$$
$$a \approx \pm 20.8$$
Leg: $a \approx 20.8$ meters

33. $(80)^2 + (60)^2 = d^2$
$$6400 + 3600 = d^2$$
$$10,000 = d^2$$
$$\sqrt{10,000} = \sqrt{d^2}$$
$$\pm 100 = d$$
distance $d = 100$ miles

35.
$$x^2 + (x-2)^2 = 10^2$$
$$x^2 + (x-2)(x-2) = 100$$
$$x^2 + x^2 - 2x - 2x + 4 = 100$$
$$2x^2 - 4x + 4 = 100$$
$$2x^2 - 4x - 96 = 0$$
$$2(x^2 - 2x - 48) = 0$$
$$2(x-8)(x+6) = 0$$
$$x - 8 = 0 \quad \text{or} \quad x + 6 = 0$$
$$x = 8 \quad \text{or} \quad x = -6$$

George's height: $x - 2 = 8 - 2 = 6$ feet

37.
$$b^2 + (b+5)^2 = 40^2$$
$$b^2 + (b+5)(b+5) = 1600$$
$$b^2 + b^2 + 5b + 5b + 25 = 1600$$
$$2b^2 + 10b - 1575 = 0$$
$$b = \frac{-10 \pm \sqrt{10^2 - 4(2)(-1575)}}{2(2)}$$
$$b = \frac{-10 \pm \sqrt{12,700}}{4}$$
$$b \approx 25.7$$
Pole height: $b + 5 \approx 25.7 + 5 \approx 30.7$ feet

39. $(10)^2 + (13)^2 = d^2$
$$100 + 169 = d^2$$
$$269 = d^2$$
$$\pm\sqrt{269} = \sqrt{d^2}$$
$$16.4 \approx d$$
Length of diagonal: $d \approx 16.4$ inches

41. Unknowns: Known:
Length: $w+1$ Hypotenuse: 6.5 feet
Width: w

$$w^2 + (w+1)^2 = (6.5)^2$$
$$w^2 + (w+1)(w+1) = 42.25$$
$$w^2 + w^2 + w + w + 1 = 42.25$$
$$2w^2 + 2w - 41.25 = 0$$
$$w = \frac{-2 \pm \sqrt{2^2 - 4(2)(-41.25)}}{2(2)}$$
$$w = \frac{-2 \pm \sqrt{334}}{4}$$
$$w \approx 4.1$$
Width: $w \approx 4.1$ feet
Length: $w + 1 \approx 4.1 + 1 \approx 5.1$ feet

43. a) Unknowns: Known:
Base: x Area: 54 ft^2
Height: $x - 3$

$$\frac{1}{2}(x)(x-3) = 54$$
$$x(x-3) = 108$$
$$x^2 - 3x = 108$$
$$x^2 - 3x - 108 = 0$$
$$(x-12)(x+9) = 0$$
$$x - 12 = 0 \quad \text{or} \quad x + 9 = 0$$
$$x = 12 \quad \text{or} \quad x = -9$$
Base: $x = 12$ feet
Height: $x - 3 = 12 - 3 = 9$ feet

b) $(9)^2 + (12)^2 = c^2$
$$81 + 144 = c^2$$
$$225 = c^2$$
$$\pm\sqrt{225} = c^2$$
$$\pm 15 = c$$
Hypotenuse: $c = 15$ feet

45. a) $40^2 + h^2 = 80^2$
$$1600 + h^2 = 6400$$
$$h^2 = 4800$$
$$\sqrt{h^2} = \pm\sqrt{4800}$$
$$h \approx 69.3$$
height: $h \approx 69.3$ centimeters

b) $A \approx \dfrac{1}{2}(80)(69.3)$

$A \approx 2772$ square centimeters

47. Answers will vary. Example:
Yoshi is building a rectangular garden in his backyard. If the area in 126 square feet and the length is 5 feet more than the width, find the dimension of the garden.

10.6 QUICK CHECK

1. $x^2 - 14x + 48 \le 0$
$(x-6)(x-8) \le 0$

Test Point	$x-6$	$x-8$	Product
0	−	−	+
7	+	−	−
9	+	+	+

$[6,8]$

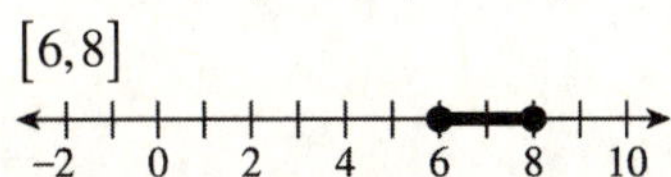

2. $x^2 + 3x - 15 \ge 0$

$$x = \frac{-3 \pm \sqrt{3^2 - 4(1)(-15)}}{2(1)}$$

$$x = \frac{-3 \pm \sqrt{69}}{2}$$

$x \approx -5.65$ or $x \approx 2.65$

Test Points	$x^2 + 3x - 15$	Sign
−6	$(-6)^2 + 3(-6) - 15 = 3$	+
0	$(0)^2 + 3(0) - 15 = -15$	−
3	$(3)^2 + 3(3) - 15 = 3$	+

$$\left(-\infty, \frac{-3-\sqrt{69}}{2}\right] \cup \left[\frac{-3+\sqrt{69}}{2}, \infty\right)$$

3. $x^2 - 5x + 20 > 0$

$$x = \frac{-(-5) \pm \sqrt{(-5)^2 - 4(1)(20)}}{2(1)}$$

$$x = \frac{5 \pm \sqrt{-55}}{2}$$

No real zeros.

Test Point	$x^2 - 5x + 20$	Sign
0	$(0)^2 - 5(0) + 20 = 20$	+

$(-\infty, \infty)$

4. $\dfrac{(x+2)(x-6)}{(x-8)(x+7)} \le 0$

Numerator is zero:
$(x+2)(x-6) = 0$
$x+2 = 0$ or $x-6 = 0$
$\qquad x = -2$ or $\qquad x = 6$

Denominator is zero:
$(x-8)(x+7) = 0$
$x-8 = 0$ or $x+7 = 0$
$\qquad x = 8$ or $\qquad x = -7$

Test Points	$x+2$	$x-6$	$x-8$	$x+7$	$\dfrac{(x+2)(x-6)}{(x-8)(x+7)}$
−8	−	−	−	−	+
−5	−	−	−	+	−
0	+	−	−	+	+
7	+	+	−	+	−
9	+	+	+	+	+

$(-7, -2] \cup [6, 8)$

5. $\dfrac{x^2 - 6x}{x^2 + 5x + 4} > 0$

$$\frac{x(x-6)}{(x+4)(x+1)} > 0$$

Numerator is zero:
$x(x-6) = 0$
$x = 0$ or $x-6 = 0$
$x = 0$ or $\qquad x = 6$

Denominator is zero:
$(x+4)(x+1) = 0$
$x+4 = 0$ or $x+1 = 0$
$\qquad x = -4$ or $\qquad x = -1$

Test Points	x	$x-6$	$x+4$	$x+1$	$\dfrac{x(x-6)}{(x+4)(x+1)}$
−5	−	−	−	−	+
−2	−	−	+	−	−
$-\frac{1}{2}$	−	−	+	+	+
2	+	−	+	+	−
8	+	+	+	+	+

$(-\infty, -4) \cup (-1, 0) \cup (6, \infty)$

6.
$$\frac{x^2 - 17}{x - 3} < 4$$

$$\frac{x^2 - 17}{x - 3} - 4 < 0$$

$$\frac{x^2 - 17}{x - 3} - 4 \cdot \frac{x - 3}{x - 3} < 0$$

$$\frac{x^2 - 17 - 4x + 12}{x - 3} < 0$$

$$\frac{x^2 - 4x - 5}{x - 3} < 0$$

$$\frac{(x - 5)(x + 1)}{x - 3} < 0$$

Numerator is zero:
$$(x - 5)(x + 1) = 0$$
$x - 5 = 0$ or $x + 1 = 0$
$x = 5$ or $x = -1$

Denominator is zero:
$x - 3 = 0$
$x = 3$

Test Points	$x - 5$	$x + 1$	$x - 3$	$\dfrac{(x-5)(x+1)}{x-3}$
-2	$-$	$-$	$-$	$-$
0	$-$	$+$	$-$	$+$
4	$-$	$+$	$+$	$-$
6	$+$	$+$	$+$	$+$

$(-\infty, -1) \cup (3, 5)$

7. $f(x) = x^2 + 10x + 35 < 14$
$$x^2 + 10x + 21 < 0$$
$$(x + 7)(x + 3) < 0$$

Find the zeros:
$(x + 7)(x + 3) = 0$
$x + 7 = 0$ or $x + 3 = 0$
$x = -7$ or $x = -3$

Test Points	$x + 7$	$x + 3$	Product
-8	$-$	$-$	$+$
-5	$+$	$-$	$-$
0	$+$	$+$	$+$

$(-7, -3)$

8. $h(t) = -16t^2 + 272t$
$h(t) \geq 960$

$$-16t^2 + 272t \geq 960$$
$$-16t^2 + 272t - 960 \geq 0$$
Multiply both sides by -1.
$$16t^2 - 272t + 960 \leq 0$$
$$16\left(t^2 - 17t + 60\right) \leq 0$$
$$16(t - 12)(t - 5) \leq 0$$

Find zeros:
$16(t - 12)(t - 5) = 0$
$t = 12$ or $t = 5$

$t = 1$	$t = 6$	$t = 13$
$h(1) = -16(1)^2 + 272(1)$	$h(6) = -16(6)^2 + 272(6)$	$h(1) = -16(13)^2 + 272(13)$
$= -16 + 272$	$= -576 + 1632$	$= -2704 + 3536$
$= 256$	$= 1056$	$= 832$

$h(t) \geq 960$ is true only in the interval containing $t = 6$.

From 5 seconds after launch until 12 seconds after launch

10.6 *QUADRATIC AND RATIONAL INEQUALITIES*

1. quadratic

3. strict

5. $(x-5)(x-1)<0$

Find the zeros:
$(x-5)(x-1)=0$
$x-5=0$ or $x-1=0$
$x=5$ or $x=1$

Test Points	$x-5$	$x-1$	Product
0	$-$	$-$	$+$
2	$-$	$+$	$-$
6	$+$	$+$	$+$

$(1,5)$

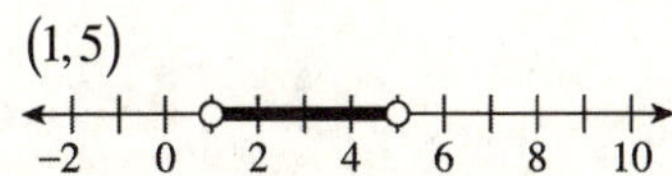

7. $-2(x-6)(x+3)\le 0$

Find the zeros:
$-2(x-6)(x+3)=0$
$x-6=0$ or $x+3=0$
$x=6$ or $x=-3$

Test Points	-2	$x-6$	$x+3$	Product
-4	$-$	$-$	$-$	$-$
0	$-$	$-$	$+$	$+$
8	$-$	$+$	$+$	$-$

$(-\infty,-3]\cup[6,\infty)$

9. $x^2-4x-21<0$
$(x+3)(x-7)<0$

Find the zeros:
$(x+3)(x-7)=0$
$x+3=0$ or $x-7=0$
$x=-3$ or $x=7$

Test Points	$x+3$	$x-7$	Product
-4	$-$	$-$	$+$
0	$+$	$-$	$-$
8	$+$	$+$	$+$

$(-3,7)$

11. $x^2-10x+16\ge 0$
$(x-2)(x-8)\ge 0$

Find the zeros:
$(x-2)(x-8)=0$
$x-2=0$ or $x-8=0$
$x=2$ or $x=8$

Test Points	$x-2$	$x-8$	Product
0	$-$	$-$	$+$
5	$+$	$-$	$-$
9	$+$	$+$	$+$

$(-\infty,2]\cup[8,\infty)$

13. $x^2-25\le 0$
$(x+5)(x-5)\le 0$

Find the zeros:
$(x+5)(x-5)=0$
$x+5=0$ or $x-5=0$
$x=-5$ or $x=5$

Test Points	$x+5$	$x-5$	Product
-6	$-$	$-$	$+$
0	$+$	$-$	$-$
6	$+$	$+$	$+$

$[-5,5]$

15. $x^2+6x+3>0$

Find the zeros:
$x^2+6x+3=0$

$$x=\frac{-6\pm\sqrt{6^2-4(1)(3)}}{2(1)}$$

$$x=\frac{-6\pm\sqrt{24}}{2}$$

$$x=\frac{-6\pm2\sqrt{6}}{2}$$

$x=-3\pm\sqrt{6}$

$x\approx-0.55$ or $x\approx-5.45$

Test Points	x^2+6x+3	Sign
-6	$(-6)^2+6(-6)+3=3$	$+$
-4	$(-4)^2+6(-4)+3=-5$	$-$
0	$(0)^2+6(0)+3=3$	$+$

$\left(-\infty,-3-\sqrt{6}\right)\cup\left(-3+\sqrt{6},\infty\right)$

17. $3x^2 - 5x - 1 \geq 0$

Find the zeros:

$3x^2 - 5x - 1 = 0$

$$x = \frac{-(-5) \pm \sqrt{(-5)^2 - 4(3)(-1)}}{2(3)}$$

$$x = \frac{5 \pm \sqrt{37}}{6}$$

$x \approx -0.18$ or $x \approx 1.85$

Test Points	$3x^2 - 5x - 1$	Sign
-1	$3(-1)^2 - 5(-1) - 1 = 7$	$+$
0	$3(0)^2 - 5(0) - 1 = -1$	$-$
3	$3(3)^2 - 5(3) - 1 = 11$	$+$

$$\left(-\infty, \frac{5 - \sqrt{37}}{6}\right] \cup \left[\frac{5 + \sqrt{37}}{6}, \infty\right)$$

19. $x^2 + 3x + 5 > 0$

Find the zeros:

$x^2 + 3x + 5 = 0$

$$x = \frac{-3 \pm \sqrt{3^2 - 4(1)(5)}}{2(1)}$$

$$x = \frac{-3 \pm \sqrt{-11}}{2}$$

No real zeros.

Test Point	$x^2 + 3x + 5$	Sign
0	$(0)^2 + 3(0) + 5 = 5$	$+$

$(-\infty, \infty)$

21. $x^2 - 2x + 24 < 0$

Find the zeros:

$x^2 - 2x + 24 = 0$

$$x = \frac{-(-2) \pm \sqrt{(-2)^2 - 4(1)(24)}}{2(1)}$$

$$x = \frac{2 \pm \sqrt{-92}}{2}$$

No real zeros.

Test Point	$x^2 - 2x + 24$	Sign
0	$(0)^2 - 2(0) + 24 = 24$	$+$

$\varnothing$

23. $x^2 - x \geq 56$

$x^2 - x - 56 \geq 0$

$(x + 7)(x - 8) \geq 0$

Find the zeros:

$(x + 7)(x - 8) = 0$

$x + 7 = 0$ or $x - 8 = 0$

$x = -7$ or $x = 8$

Test Points	$x + 7$	$x - 8$	Product
-8	$-$	$-$	$+$
0	$+$	$-$	$-$
9	$+$	$+$	$+$

$(-\infty, -7] \cup [8, \infty)$

25. $-x^2 + 10x - 5 > 0$

Multiply both sides by -1.

$x^2 - 10x + 5 < 0$

Find the zeros:

$x^2 - 10x + 5 = 0$

$$x = \frac{-(-10) \pm \sqrt{(-10)^2 - 4(1)(5)}}{2(1)}$$

$$x = \frac{10 \pm \sqrt{80}}{2}$$

$$x = \frac{10 \pm 4\sqrt{5}}{2}$$

$x = 5 \pm 2\sqrt{5}$

$x \approx 0.53$ or $x \approx 9.47$

Test Points	$x^2 - 10x + 5$	Sign
0	$(0)^2 - 10(0) + 5 = 5$	$+$
1	$(1)^2 - 10(1) + 5 = -4$	$-$
10	$(10)^2 - 10(10) + 5 = 5$	$+$

$\left(5 - 2\sqrt{5},\ 5 + 2\sqrt{5}\right)$

27. $(x-1)(x+2)(x-4) \geq 0$

Find the zeros:
$(x-1)(x+2)(x-4) = 0$
$x-1=0$ or $x+2=0$ or $x-4=0$
$x=1$ or $x=-2$ or $x=4$

Test Points	$x-1$	$x+2$	$x-4$	Product
-3	$-$	$-$	$-$	$-$
0	$-$	$+$	$-$	$+$
2	$+$	$+$	$-$	$-$
5	$+$	$+$	$+$	$+$

$[-2,1] \cup [4, \infty)$

29. $(x-7)^2 (x+5)(x+9) \leq 0$

Find the zeros:
$(x-7)^2 (x+5)(x+9) = 0$
$x-7=0$ or $x+5=0$ or $x+9=0$
$x=7$ or $x=-5$ or $x=-9$

Test Points	$(x-7)^2$	$x+5$	$x+9$	Product
-10	$+$	$-$	$-$	$+$
-6	$+$	$-$	$+$	$-$
0	$+$	$+$	$+$	$+$
8	$+$	$+$	$+$	$+$

$[-9,-5] \cup \{7\}$

31. $\dfrac{x-3}{x+4} < 0$

Numerator is zero:
$x-3=0$
$x=3$

Denominator is zero:
$x+4=0$
$x=-4$

Test Points	$x-3$	$x+4$	$\dfrac{x-3}{x+4}$
-6	$-$	$-$	$+$
0	$-$	$+$	$-$
4	$+$	$+$	$+$

$(-4,3)$

33. $\dfrac{(x+2)(x-2)}{x+1} \leq 0$

Numerator is zero:
$x+2=0$ or $x-2=0$
$x=-2$ or $x=2$

Denominator is zero:
$x+1=0$
$x=-1$

Test Points	$x+2$	$x-2$	$x+1$	$\dfrac{(x+2)(x-2)}{x+1}$
-4	$-$	$-$	$-$	$-$
$-1\dfrac{1}{2}$	$+$	$-$	$-$	$+$
0	$+$	$-$	$+$	$-$
4	$+$	$+$	$+$	$+$

$(-\infty,-2] \cup (-1,2]$

35. $\dfrac{x-4}{x^2+9x+18} < 0$

$\dfrac{x-4}{(x+6)(x+3)} < 0$

Numerator is zero:
$x-4=0$
$x=4$

Denominator is zero:
$x+6=0$ or $x+3=0$
$x=-6$ or $x=-3$

Test Points	$x-4$	$x+6$	$x+3$	$\dfrac{x-4}{(x+6)(x+3)}$
-7	$-$	$-$	$-$	$-$
-5	$-$	$+$	$-$	$+$
0	$-$	$+$	$+$	$-$
5	$+$	$+$	$+$	$+$

$(-\infty,-6) \cup (-3,4)$

37. $\dfrac{x^2 + x - 20}{x^2 - 4} \geq 0$

$\dfrac{(x+5)(x-4)}{(x+2)(x-2)} \geq 0$

Numerator is zero:
$x + 5 = 0 \quad$ or $\quad x - 4 = 0$
$\quad x = -5 \quad$ or $\qquad x = 4$

Denominator is zero:
$x + 2 = 0 \quad$ or $\quad x - 2 = 0$
$\quad x = -2 \quad$ or $\qquad x = 2$

Test Points	$x+5$	$x-4$	$x+2$	$x-2$	$\dfrac{(x+5)(x-4)}{(x+2)(x-2)}$
-6	$-$	$-$	$-$	$-$	$+$
-3	$+$	$-$	$-$	$-$	$-$
0	$+$	$-$	$+$	$-$	$+$
3	$+$	$-$	$+$	$+$	$-$
5	$+$	$+$	$+$	$+$	$+$

$(-\infty, -5] \cup (-2, 2) \cup [4, \infty)$

39. $\dfrac{x^2 - 2x - 48}{x^2 + 13x + 36} \leq 0$

$\dfrac{(x+6)(x-8)}{(x+9)(x+4)} \leq 0$

Numerator is zero:
$x + 6 = 0 \quad$ or $\quad x - 8 = 0$
$\quad x = -6 \quad$ or $\qquad x = 8$

Denominator is zero:
$x + 9 = 0 \quad$ or $\quad x + 4 = 0$
$\quad x = -9 \quad$ or $\qquad x = -4$

Test Points	$x+6$	$x-8$	$x+9$	$x+4$	$\dfrac{(x+6)(x-8)}{(x+9)(x+4)}$
-10	$-$	$-$	$-$	$-$	$+$
-7	$-$	$-$	$+$	$-$	$-$
-5	$+$	$-$	$+$	$-$	$+$
0	$+$	$-$	$+$	$+$	$-$
9	$+$	$+$	$+$	$+$	$+$

$(-9, -6] \cup (-4, 8]$

41. $\dfrac{x^2 - 4x + 3}{x^2 + 4x - 32} < 0$

$\dfrac{(x-3)(x-1)}{(x+8)(x-4)} < 0$

Numerator is zero:
$x - 3 = 0 \quad$ or $\quad x - 1 = 0$
$\quad x = 3 \quad$ or $\qquad x = 1$

Denominator is zero:
$x + 8 = 0 \quad$ or $\quad x - 4 = 0$
$\quad x = -8 \quad$ or $\qquad x = 4$

Test Points	$x-3$	$x-1$	$x+8$	$x-4$	$\dfrac{(x-3)(x-1)}{(x+8)(x-4)}$
-9	$-$	$-$	$-$	$-$	$+$
0	$-$	$-$	$+$	$-$	$-$
2	$-$	$+$	$+$	$-$	$+$
$3\frac{1}{2}$	$+$	$+$	$+$	$-$	$-$
6	$+$	$+$	$+$	$+$	$+$

$(-8, 1) \cup (3, 4)$

43. $\dfrac{4x - 7}{x - 4} > 3$

$\dfrac{4x - 7}{x - 4} - 3 > 0$

$\dfrac{4x - 7}{x - 4} - 3 \cdot \dfrac{x - 4}{x - 4} > 0$

$\dfrac{4x - 7 - 3x + 12}{x - 4} > 0$

$\dfrac{x + 5}{x - 4} > 0$

Numerator is zero:
$x + 5 = 0$
$\quad x = -5$

Denominator is zero:
$x - 4 = 0$
$\quad x = 4$

Test Points	$x+5$	$x-4$	$\dfrac{x+5}{x-4}$
-6	$-$	$-$	$+$
0	$+$	$-$	$-$
5	$+$	$+$	$+$

$(-\infty, -5) \cup (4, \infty)$

45.
$$\frac{x^2+9x+12}{x+3} \geq 2$$

$$\frac{x^2+9x+12}{x+3} - 2 \geq 0$$

$$\frac{x^2+9x+12}{x+3} - 2 \cdot \frac{x+3}{x+3} \geq 0$$

$$\frac{x^2+9x+12-2x-6}{x+3} \geq 0$$

$$\frac{x^2+7x+6}{x+3} \geq 0$$

$$\frac{(x+6)(x+1)}{x+3} \geq 0$$

Numerator is zero:
$$(x+6)(x+1) = 0$$
$$x+6 = 0 \quad \text{or} \quad x+1 = 0$$
$$x = -6 \quad \text{or} \quad x = -1$$

Denominator is zero:
$$x+3 = 0$$
$$x = -3$$

Test Points	$x+6$	$x+1$	$x+3$	$\dfrac{(x+6)(x+1)}{x+3}$
-7	$-$	$-$	$-$	$-$
-5	$+$	$-$	$-$	$+$
-2	$+$	$-$	$+$	$-$
0	$+$	$+$	$+$	$+$

$[-6,-3) \cup [-1, \infty)$

47.
$$\frac{x^2+3x+10}{x-2} \leq x$$

$$\frac{x^2+3x+10}{x-2} - x \leq 0$$

$$\frac{x^2+3x+10}{x-2} - x \cdot \frac{x-2}{x-2} \leq 0$$

$$\frac{x^2+3x+10-x^2+2x}{x-2} \leq 0$$

$$\frac{5x+10}{x-2} \leq 0$$

$$\frac{5(x+2)}{x-2} \leq 0$$

Numerator is zero:
$$x+2 = 0$$
$$x = -2$$

Denominator is zero:
$$x-2 = 0$$
$$x = 2$$

Test Points	$x+2$	$x-2$	$\dfrac{x+2}{x-2}$
-3	$-$	$-$	$+$
0	$+$	$-$	$-$
3	$+$	$+$	$+$

$[-2, 2)$

49. $h(t) = -16t^2 + 64t + 27 \geq 75$
$$-16t^2 + 64t - 48 \geq 0$$
Multiply both sides by -1.
$$16t^2 - 64t + 48 \leq 0$$
Find the zeros:
$$16\left(t^2 - 4t + 3\right) = 0$$
$$16(t-1)(t-3) = 0$$
$$t-1 = 0 \quad \text{or} \quad t-3 = 0$$
$$t = 1 \quad \text{or} \quad t = 3$$

Use test points $t = 0$, $t = 2$, $t = 4$.

$t = 0$	$t = 2$	$t = 4$
$h(0) = -16(0)^2 + 64(0) + 27$	$h(2) = -16(2)^2 + 64(2) + 27$	$h(4) = -16(4)^2 + 64(4) + 27$
$\quad = 0 + 0 + 27$	$\quad = -64 + 128 + 27$	$\quad = -256 + 256 + 27$
$\quad = 27$	$\quad = 91$	$\quad = 27$

$h(t) \geq 75$ only at the test point $t = 2$.

From 1 second to 3 seconds

51. $h(t) = -16t^2 + 32t \geq 12$

$-16t^2 + 32t - 12 \geq 0$

Multiply both sides by -1.

$16t^2 - 32t + 12 \leq 0$

Find the zeros:

$4\left(4t^2 - 8t + 3\right) = 0$

$4(2t - 1)(2t - 3) = 0$

$2t - 1 = 0$ or $2t - 3 = 0$

$\quad 2t = 1$ or $\quad 2t = 3$

$\quad t = \dfrac{1}{2}$ or $\quad t = \dfrac{3}{2}$

Use test points $t = 0$, $t = 1$, $t = 2$.

$t = 0$	$t = 1$	$t = 2$
$h(0) = -16(0)^2 + 32(0)$	$h(1) = -16(1)^2 + 32(1)$	$h(2) = -16(2)^2 + 32(2)$
$\quad = 0 + 0$	$\quad = -16 + 32$	$\quad = -64 + 64$
$\quad = 0$	$\quad = 16$	$\quad = 0$

$h(t) \geq 12$ only at the test point $t = 1$.

From 0.5 seconds to 1.5 seconds

53. a) $h(t) = -16t^2 + 125t + 48$

b) $-16t^2 + 125t + 48 \geq 200$

$-16t^2 + 125t - 152 \geq 0$

Multiply both sides by -1.

$16t^2 - 125t + 152 \leq 0$

Find the zeros:

$16t^2 - 125t + 152 = 0$

$t = \dfrac{-(-125) \pm \sqrt{(-125)^2 - 4(16)(152)}}{2(16)}$

$t = \dfrac{125 \pm \sqrt{5897}}{32}$

$t \approx 1.5$ or $t \approx 6.3$

Use test points $t = 0$, $t = 2$, $t = 7$.

$t = 0$	$t = 2$	$t = 7$
$h(0) = -16(0)^2 + 125(0) + 48$	$h(2) = -16(2)^2 + 125(2) + 48$	$h(7) = -16(7)^2 + 125(7) + 48$
$\quad = 0 + 0 + 48$	$\quad = -64 + 250 + 48$	$\quad = -784 + 875 + 48$
$\quad = 48$	$\quad = 234$	$\quad = 139$

$h(t) \geq 200$ only at the test point $t = 2$.

From 1.5 seconds to 6.3 seconds

55. $f(x) = x^2 - 8x + 12 \le 0$
$$(x-6)(x-2) \le 0$$

Find the zeros:
$(x-6)(x-2) = 0$
$x - 6 = 0$ or $x - 2 = 0$
$\quad x = 6$ or $\quad x = 2$

Test Points	$x-6$	$x-2$	Product
0	$-$	$-$	$+$
4	$-$	$+$	$-$
8	$+$	$+$	$+$

$[2, 6]$

57. $f(x) = x^2 - 5x + 13 \ge 7$
$$x^2 - 5x + 6 \ge 0$$
$$(x-3)(x-2) \ge 0$$

Find the zeros:
$(x-3)(x-2) = 0$
$x - 3 = 0$ or $x - 2 = 0$
$\quad x = 3$ or $\quad x = 2$

Test Points	$x-3$	$x-2$	Product
0	$-$	$-$	$+$
$2\frac{1}{2}$	$-$	$+$	$-$
5	$+$	$+$	$+$

$(-\infty, 2] \cup [3, \infty)$

59. $\dfrac{x+9}{x-6} \le 0$

Numerator is zero:
$x + 9 = 0$
$\quad x = -9$

Denominator is zero:
$x - 6 = 0$
$\quad x = 6$

Test Points	$x+9$	$x-6$	$\dfrac{x+9}{x-6}$
-10	$-$	$-$	$+$
0	$+$	$-$	$-$
8	$+$	$+$	$+$

$[-9, 6)$

61. $f(x) = \dfrac{x^2 - 8x - 33}{x^2 + 13x + 42} > 0$
$$\dfrac{(x-11)(x+3)}{(x+6)(x+7)} > 0$$

Numerator is zero:
$x - 11 = 0$ or $x + 3 = 0$
$\quad x = 11$ or $\quad x = -3$

Denominator is zero:
$x + 6 = 0$ or $x + 7 = 0$
$\quad x = -6$ or $\quad x = -7$

Test Points	$x-11$	$x+3$	$x+6$	$x+7$	$\dfrac{(x-11)(x+3)}{(x+6)(x+7)}$
-8	$-$	$-$	$-$	$-$	$+$
$-6\frac{1}{2}$	$-$	$-$	$-$	$+$	$-$
-4	$-$	$-$	$+$	$+$	$+$
0	$-$	$+$	$+$	$+$	$-$
12	$+$	$+$	$+$	$+$	$+$

$(-\infty, -7) \cup (-6, -3) \cup (11, \infty)$

63. Zeros: $x = -1$, $x = 3$

Test Points	$x+1$	$x-3$	Product
-2	$-$	$-$	$+$
0	$+$	$-$	$-$
4	$+$	$+$	$+$

$(x+1)(x-3) < 0$
$x^2 - 2x - 3 < 0$

65. Zeros: $x = -2$, $x = 2$

Test Points	$x+2$	$x-2$	Product
-3	$-$	$-$	$+$
0	$+$	$-$	$-$
3	$+$	$+$	$+$

$(x+2)(x-2) \ge 0$
$x^2 - 4 \ge 0$

67. Zeros in the numerator: $x = 5$
Zeros in the denominator: $x = -4$

Test Points	$x-5$	$x+4$	$\dfrac{x-5}{x+4}$
-6	$-$	$-$	$+$
0	$-$	$+$	$-$
7	$+$	$+$	$+$

$\dfrac{x-5}{x+4} \le 0$

69. Zeros in the numerator: $x = -1, x = 3$
Zeros in the denominator: $x = -6, x = -4$

Test Points	$x+1$	$x-3$	$x+6$	$x+4$	$\dfrac{(x+1)(x-3)}{(x+6)(x+4)}$
-7	$-$	$-$	$-$	$-$	$+$
-5	$-$	$-$	$+$	$-$	$-$
-2	$-$	$-$	$+$	$+$	$+$
0	$+$	$-$	$+$	$+$	$-$
4	$+$	$+$	$+$	$+$	$+$

$$\frac{(x+1)(x-3)}{(x+6)(x+4)} \le 0$$

$$\frac{x^2 - 2x - 3}{x^2 + 10x + 24} \le 0$$

71. Answers will vary.

CHAPTER 10 REVIEW

1. $x^2 + 11x + 28 = 0$
$(x+7)(x+4) = 0$
$x+7 = 0 \quad$ or $\quad x+4 = 0$
$\quad x = -7 \quad$ or $\quad\quad x = -4$
$\{-7, -4\}$

2. $x^2 + 7x - 30 = 0$
$(x+10)(x-3) = 0$
$x+10 = 0 \quad$ or $\quad x-3 = 0$
$\quad x = -10 \quad$ or $\quad\quad x = 3$
$\{-10, 3\}$

3. $x^2 + 4x = 0$
$x(x+4) = 0$
$x = 0 \quad$ or $\quad x+4 = 0$
$x = 0 \quad$ or $\quad\quad x = -4$
$\{0, -4\}$

4. $2x^2 - 13x + 15 = 0$
$(2x-3)(x-5) = 0$
$2x-3 = 0 \quad$ or $\quad x-5 = 0$
$\quad 2x = 3 \quad$ or $\quad\quad x = 5$
$\quad\quad x = \dfrac{3}{2} \quad$ or $\quad\quad x = 5$
$\left\{ \dfrac{3}{2}, 5 \right\}$

5. $x^2 + 7x - 12 = 3x - 7$
$x^2 + 4x - 12 = -7$
$x^2 + 4x - 5 = 0$
$(x+5)(x-1) = 0$
$x+5 = 0 \quad$ or $\quad x-1 = 0$
$\quad x = -5 \quad$ or $\quad\quad x = 1$
$\{-5, 1\}$

6. $x^2 - 9x + 23 = 9(x-6)$
$x^2 - 9x + 23 = 9x - 54$
$x^2 - 18x + 23 = -54$
$x^2 - 18x + 77 = 0$
$(x-7)(x-11) = 0$
$x-7 = 0 \quad$ or $\quad x-11 = 0$
$\quad x = 7 \quad$ or $\quad\quad x = 11$
$\{7, 11\}$

7. $x^2 = 49$
$\sqrt{x^2} = \pm\sqrt{49}$
$x = \pm 7$
$\{-7, 7\}$

8. $x^2 - 72 = 0$
$x^2 = 72$
$\sqrt{x^2} = \pm\sqrt{72}$
$x = \pm 6\sqrt{2}$
$\left\{ -6\sqrt{2}, 6\sqrt{2} \right\}$

9. $x^2 + 4 = 0$
$x^2 = -4$
$\sqrt{x^2} = \pm\sqrt{-4}$
$x = \pm 2i$
$\{-2i, 2i\}$

10. $(x-6)^2 = -36$
$\sqrt{(x-6)^2} = \pm\sqrt{-36}$
$x - 6 = \pm 6i$
$x = 6 \pm 6i$
$\{6 - 6i, 6 + 6i\}$

11. $(x-8)^2 + 30 = 21$
$(x-8)^2 = -9$
$\sqrt{(x-8)^2} = \pm\sqrt{-9}$
$x - 8 = \pm 3i$
$x = 8 \pm 3i$
$\{8 - 3i, 8 + 3i\}$

12.
$$(2x+7)^2 + 20 = 141$$
$$(2x+7)^2 = 121$$
$$\sqrt{(2x+7)^2} = \pm\sqrt{121}$$
$$2x+7 = \pm 11$$
$$2x+7 = -11 \quad \text{or} \quad 2x+7 = 11$$
$$2x = -18 \quad \text{or} \quad 2x = 4$$
$$x = -9 \quad \text{or} \quad x = 2$$
$$\{-9, 2\}$$

13.
$$x^2 - 2x - 8 = 0$$
$$x^2 - 2x = 8$$
$$x^2 - 2x + \left(-\frac{2}{2}\right)^2 = 8 + \left(-\frac{2}{2}\right)^2$$
$$x^2 - 2x + 1 = 8 + 1$$
$$(x-1)^2 = 9$$
$$\sqrt{(x-1)^2} = \pm\sqrt{9}$$
$$x - 1 = \pm 3$$
$$x - 1 = -3 \quad \text{or} \quad x - 1 = 3$$
$$x = -2 \quad \text{or} \quad x = 4$$
$$\{-2, 4\}$$

14.
$$x^2 - 6x + 3 = 0$$
$$x^2 - 6x = -3$$
$$x^2 - 6x + \left(-\frac{6}{2}\right)^2 = -3 + \left(-\frac{6}{2}\right)^2$$
$$x^2 - 6x + 9 = -3 + 9$$
$$(x-3)^2 = 6$$
$$\sqrt{(x-3)^2} = \pm\sqrt{6}$$
$$x - 3 = \pm\sqrt{6}$$
$$x = 3 \pm \sqrt{6}$$
$$\{3 - \sqrt{6}, 3 + \sqrt{6}\}$$

15.
$$x^2 + 14x + 47 = 0$$
$$x^2 + 14x = -47$$
$$x^2 + 14x + \left(\frac{14}{2}\right)^2 = -47 + \left(\frac{14}{2}\right)^2$$
$$x^2 + 14x + 49 = -47 + 49$$
$$(x+7)^2 = 2$$
$$\sqrt{(x+7)^2} = \pm\sqrt{2}$$
$$x + 7 = \pm\sqrt{2}$$
$$x = -7 \pm \sqrt{2}$$
$$\{-7 - \sqrt{2}, -7 + \sqrt{2}\}$$

16.
$$x^2 + 5x + 10 = 0$$
$$x^2 + 5x = -10$$
$$x^2 + 5x + \left(\frac{5}{2}\right)^2 = -10 + \left(\frac{5}{2}\right)^2$$
$$x^2 + 5x + \frac{25}{4} = -10 + \frac{25}{4}$$
$$\left(x + \frac{5}{2}\right)^2 = -\frac{40}{4} + \frac{25}{4}$$
$$\left(x + \frac{5}{2}\right)^2 = -\frac{15}{4}$$
$$\sqrt{\left(x + \frac{5}{2}\right)^2} = \pm\sqrt{-\frac{15}{4}}$$
$$x + \frac{5}{2} = \pm i\frac{\sqrt{15}}{2}$$
$$x = -\frac{5}{2} \pm i\frac{\sqrt{15}}{2}$$
$$x = \frac{-5 \pm i\sqrt{15}}{2}$$
$$\left\{\frac{-5 - i\sqrt{15}}{2}, \frac{-5 + i\sqrt{15}}{2}\right\}$$

17. $x^2 + 7x + 1 = 0$
$$x = \frac{-7 \pm \sqrt{7^2 - 4(1)(1)}}{2(1)}$$
$$x = \frac{-7 \pm \sqrt{45}}{2}$$
$$x = \frac{-7 \pm 3\sqrt{5}}{2}$$
$$\left\{\frac{-7 - 3\sqrt{5}}{2}, \frac{-7 + 3\sqrt{5}}{2}\right\}$$

18. $x^2 + 4x - 2 = 0$
$$x = \frac{-4 \pm \sqrt{4^2 - 4(1)(-2)}}{2(1)}$$
$$x = \frac{-4 \pm \sqrt{24}}{2}$$
$$x = \frac{-4 \pm 2\sqrt{6}}{2}$$
$$x = -2 \pm \sqrt{6}$$
$$\{-2 - \sqrt{6}, -2 + \sqrt{6}\}$$

19. $x^2 - 3x + 9 = 0$

$$x = \frac{-(-3) \pm \sqrt{(-3)^2 - 4(1)(9)}}{2(1)}$$

$$x = \frac{3 \pm \sqrt{-27}}{2}$$

$$x = \frac{3 \pm 3i\sqrt{3}}{2}$$

$$\left\{ \frac{3 - 3i\sqrt{3}}{2}, \frac{3 + 3i\sqrt{3}}{2} \right\}$$

20. $4x^2 - 20x + 29 = 0$

$$x = \frac{-(-20) \pm \sqrt{(-20)^2 - 4(4)(29)}}{2(4)}$$

$$x = \frac{20 \pm \sqrt{-64}}{8}$$

$$x = \frac{20 \pm 8i}{8}$$

$$x = \frac{5 \pm 2i}{2}$$

$$\left\{ \frac{5 - 2i}{2}, \frac{5 + 2i}{2} \right\}$$

21. $2x^2 - 5x + 9 = x^2 - 9x + 30$

$$x^2 - 5x + 9 = -9x + 30$$
$$x^2 + 4x + 9 = 30$$
$$x^2 + 4x - 21 = 0$$

$$x = \frac{-4 \pm \sqrt{4^2 - 4(1)(-21)}}{2(1)}$$

$$x = \frac{-4 \pm \sqrt{100}}{2}$$

$$x = \frac{-4 \pm 10}{2}$$

$$x = \frac{-4 - 10}{2} \quad \text{or} \quad x = \frac{-4 + 10}{2}$$

$$x = \frac{-14}{2} \quad \text{or} \quad x = \frac{6}{2}$$

$$x = -7 \quad \text{or} \quad x = 3$$

$$\{-7, 3\}$$

22.
$$x(x + 4) = 16x - 37$$
$$x^2 + 4x = 16x - 37$$
$$x^2 - 12x = -37$$
$$x^2 - 12x + 37 = 0$$

$$x = \frac{-(-12) \pm \sqrt{(-12)^2 - 4(1)(37)}}{2(1)}$$

$$x = \frac{12 \pm \sqrt{-4}}{2}$$

$$x = \frac{12 \pm 2i}{2}$$

$$x = 6 \pm i$$

$$\{6 - i, 6 + i\}$$

23. $x^4 - 5x^2 + 4 = 0$

Let $u = x^2$ and $u^2 = x^4$.

$$u^2 - 5u + 4 = 0$$
$$(u - 1)(u - 4) = 0$$

$$
\begin{array}{lll}
u - 1 = 0 & \text{or} & u - 4 = 0 \\
u = 1 & \text{or} & u = 4 \\
x^2 = 1 & \text{or} & x^2 = 4 \\
\sqrt{x^2} = \pm\sqrt{1} & \text{or} & \sqrt{x^2} = \pm\sqrt{4} \\
x = \pm 1 & \text{or} & x = \pm 2
\end{array}
$$

$$\{-1, 1, -2, 2\}$$

24. $x - 4\sqrt{x} - 32 = 0$

Let $u = \sqrt{x}$ and $u^2 = x$.

$$u^2 - 4u - 32 = 0$$
$$(u + 4)(u - 8) = 0$$

$$
\begin{array}{lll}
u + 4 = 0 & \text{or} & u - 8 = 0 \\
u = -4 & \text{or} & u = 8 \\
\sqrt{x} = -4 & \text{or} & \sqrt{x} = 8 \\
\left(\sqrt{x}\right)^2 = (-4)^2 & \text{or} & \left(\sqrt{x}\right)^2 = (8)^2 \\
x = 16 & \text{or} & x = 64
\end{array}
$$

check:

$x = 16$	$x = 64$
$16 - 4\sqrt{16} - 32 = 0$	$64 - 4\sqrt{64} - 32 = 0$
$16 - 4(4) - 32 = 0$	$64 - 4(8) - 32 = 0$
$16 - 16 - 32 = 0$	$64 - 32 - 32 = 0$
$-32 = 0$	$0 = 0$
False	True

$$\{64\}$$

25. $x + 7x^{1/2} - 8 = 0$

Let $u = \sqrt{x}$ and $u^2 = x$.

$$u^2 + 7u - 8 = 0$$
$$(u + 8)(u - 1) = 0$$

$$u + 8 = 0 \quad \text{or} \quad u - 1 = 0$$
$$u = -8 \quad \text{or} \quad u = 1$$
$$\sqrt{x} = -8 \quad \text{or} \quad \sqrt{x} = 1$$
$$\left(\sqrt{x}\right)^2 = (-8)^2 \quad \text{or} \quad \left(\sqrt{x}\right)^2 = (1)^2$$
$$x = 64 \quad \text{or} \quad x = 1$$

check:

$x = 64$	$x = 1$
$64 + 7(64)^{1/2} - 8 = 0$	$1 + 7(1)^{1/2} - 8 = 0$
$64 + 7(8) - 8 = 0$	$1 + 7 - 8 = 0$
$64 + 56 - 8 = 0$	$0 = 0$
$112 = 0$	True
False	

$\{1\}$

26. $x^{2/3} - 4x^{1/3} - 21 = 0$

Let $u = x^{1/3}$ and $u^2 = x^{2/3}$.

$$u^2 - 4u - 21 = 0$$
$$(u + 3)(u - 7) = 0$$

$$u + 3 = 0 \quad \text{or} \quad u - 7 = 0$$
$$u = -3 \quad \text{or} \quad u = 7$$
$$x^{1/3} = -3 \quad \text{or} \quad x^{1/3} = 7$$
$$\left(x^{1/3}\right)^3 = (-3)^3 \quad \text{or} \quad \left(x^{1/3}\right)^3 = (7)^3$$
$$x = -27 \quad \text{or} \quad x = 343$$

$\{-27, 343\}$

27. $\sqrt{x^2 + x - 2} = 2$

$$\left(\sqrt{x^2 + x - 2}\right)^2 = (2)^2$$
$$x^2 + x - 2 = 4$$
$$x^2 + x - 6 = 0$$
$$(x + 3)(x - 2) = 0$$

$$x + 3 = 0 \quad \text{or} \quad x - 2 = 0$$
$$x = -3 \quad \text{or} \quad x = 2$$

check:

$x = -3$	$x = 2$
$\sqrt{(-3)^2 + (-3) - 2} = 2$	$\sqrt{2^2 + 2 - 2} = 2$
$\sqrt{9 - 3 - 2} = 2$	$\sqrt{4 + 2 - 2} = 2$
$\sqrt{4} = 2$	$\sqrt{4} = 2$
$2 = 2$	$2 = 2$
True	True

$\{-3, 2\}$

28. $\sqrt{x + 15} - x = 9$

$$\sqrt{x + 15} = x + 9$$
$$\left(\sqrt{x + 15}\right)^2 = (x + 9)^2$$
$$x + 15 = x^2 + 18x + 81$$
$$0 = x^2 + 17x + 66$$
$$0 = (x + 11)(x + 6)$$

$$x + 11 = 0 \quad \text{or} \quad x + 6 = 0$$
$$x = -11 \quad \text{or} \quad x = -6$$

check:

$x = -11$	$x = -6$
$\sqrt{-11 + 15} - (-11) = 9$	$\sqrt{-6 + 15} - (-6) = 9$
$\sqrt{4} + 11 = 9$	$\sqrt{9} + 6 = 9$
$2 + 11 = 9$	$3 + 6 = 9$
$13 = 9$	$9 = 9$
False	True

$\{-6\}$

29. $$1 + \frac{11}{x} + \frac{30}{x^2} = 0$$

$$x^2 \cdot \left(1 + \frac{11}{x} + \frac{30}{x^2}\right) = x^2 \cdot 0$$

$$x^2 + 11x + 30 = 0$$
$$(x + 6)(x + 5) = 0$$

$$x + 6 = 0 \quad \text{or} \quad x + 5 = 0$$
$$x = -6 \quad \text{or} \quad x = -5$$

Neither of these values makes a denominator of the original equation 0.

$\{-6, -5\}$

30.

$$\frac{x}{x-3}-\frac{7}{x-2}=\frac{3}{x^2-5x+6}$$

$$\frac{x}{x-3}-\frac{7}{x-2}=\frac{3}{(x-3)(x-2)}$$

$$(x-3)(x-2)\left(\frac{x}{x-3}-\frac{7}{x-2}\right)=(x-3)(x-2)\cdot\frac{3}{(x-3)(x-2)}$$

$$x(x-2)-7(x-3)=3$$

$$x^2-2x-7x+21=3$$

$$x^2-9x+18=0$$

$$(x-6)(x-3)=0$$

$$x-6=0 \quad\text{or}\quad x-3=0$$
$$x=6 \quad\text{or}\quad x=3$$

$x=3$ makes the term $\dfrac{x}{x-3}$ undefined, therefore it cannot be a solution.

$\{6\}$

31. $x^2-19x+90=0$
$(x-9)(x-10)=0$

$x-9=0 \quad\text{or}\quad x-10=0$
$x=9 \quad\text{or}\quad x=10$
$\{9,10\}$

32. $x^4+5x^2-36=0$

Let $u=x^2$ and $u^2=x^4$.
$$u^2+5u-36=0$$
$$(u-4)(u+9)=0$$

$u-4=0 \quad\text{or}\quad u+9=0$
$u=4 \quad\text{or}\quad u=-9$
$x^2=4 \quad\text{or}\quad x^2=-9$
$\sqrt{x^2}=\pm\sqrt{4} \quad\text{or}\quad \sqrt{x^2}=\pm\sqrt{-9}$
$x=\pm2 \quad\text{or}\quad x=\pm3i$
$\{-2,2,-3i,3i\}$

33. $(x+6)^2+11=3$
$$(x+6)^2=-8$$
$$\sqrt{(x+6)^2}=\pm\sqrt{-8}$$
$$x+6=\pm2i\sqrt{2}$$
$$x=-6\pm2i\sqrt{2}$$
$\left\{-6-2i\sqrt{2},-6+2i\sqrt{2}\right\}$

34. $x^2+4x+16=0$

$$x=\frac{-4\pm\sqrt{4^2-4(1)(16)}}{2(1)}$$

$$x=\frac{-4\pm\sqrt{-48}}{2}$$

$$x=\frac{-4\pm4i\sqrt{3}}{2}$$

$$x=-2\pm2i\sqrt{3}$$

$\left\{-2-2i\sqrt{3},-2+2i\sqrt{3}\right\}$

35. $x^2+10x=0$
$x(x+10)=0$

$x=0 \quad\text{or}\quad x+10=0$
$x=0 \quad\text{or}\quad x=-10$
$\{0,-10\}$

36. $x^2-6x-20=0$

$$x=\frac{-(-6)\pm\sqrt{(-6)^2-4(1)(-20)}}{2(1)}$$

$$x=\frac{6\pm\sqrt{116}}{2}$$

$$x=\frac{6\pm2\sqrt{29}}{2}$$

$$x=3\pm\sqrt{29}$$

$\left\{3-\sqrt{29},3+\sqrt{29}\right\}$

37.
$$\sqrt{3x+49} = x+3$$
$$\left(\sqrt{3x+49}\right)^2 = (x+3)^2$$
$$3x+49 = x^2+6x+9$$
$$0 = x^2+3x-40$$
$$0 = (x+8)(x-5)$$
$$x+8=0 \quad \text{or} \quad x-5=0$$
$$x=-8 \quad \text{or} \quad x=5$$

check:

$x=-8$	$x=5$
$\sqrt{3(-8)+49} = (-8)+3$	$\sqrt{3(5)+49} = (5)+3$
$\sqrt{25} = -5$	$\sqrt{64} = 8$
$5 = -5$	$8 = 8$
False	True

$\{5\}$

38.
$$\sqrt{2x+9} = \sqrt{4x+1}-2$$
$$\left(\sqrt{2x+9}\right)^2 = \left(\sqrt{4x+1}-2\right)^2$$
$$2x+9 = \left(\sqrt{4x+1}-2\right)\left(\sqrt{4x+1}-2\right)$$
$$2x+9 = \left(\sqrt{4x+1}\right)^2 - 2\sqrt{4x+1} - 2\sqrt{4x+1} + 4$$
$$2x+9 = 4x+1 - 4\sqrt{4x+1} + 4$$
$$-2x+4 = -4\sqrt{4x+1}$$
$$(-2x+4)^2 = \left(-4\sqrt{4x+1}\right)^2$$
$$4x^2 - 16x+16 = 16(4x+1)$$
$$4x^2 - 16x+16 = 64x+16$$
$$4x^2 - 80x = 0$$
$$4x(x-20) = 0$$
$$4x=0 \quad \text{or} \quad x-20=0$$
$$x=0 \quad \text{or} \quad x=20$$

check:

$x=0$	$x=20$
$\sqrt{2(0)+9} = \sqrt{4(0)+1}-2$	$\sqrt{2(20)+9} = \sqrt{4(20)+1}-2$
$\sqrt{9} = \sqrt{1}-2$	$\sqrt{49} = \sqrt{81}-2$
$3 = 1-2$	$7 = 9-2$
$3 = -1$	$7 = 7$
False	True

$\{20\}$

39.
$$x-1-\frac{20}{x} = 0$$
$$x\cdot\left(x-1-\frac{20}{x}\right) = x\cdot 0$$
$$x^2-x-20 = 0$$
$$(x+4)(x-5) = 0$$
$$x+4=0 \quad \text{or} \quad x-5=0$$
$$x=-4 \quad \text{or} \quad x=5$$
$$\{-4,5\}$$

40. $3x^2 - 8x + 2 = 0$

$$x = \frac{-(-8) \pm \sqrt{(-8)^2 - 4(3)(2)}}{2(3)}$$

$$x = \frac{8 \pm \sqrt{40}}{6}$$

$$x = \frac{8 \pm 2\sqrt{10}}{6}$$

$$x = \frac{4 \pm \sqrt{10}}{3}$$

$$\left\{ \frac{4 - \sqrt{10}}{3}, \frac{4 + \sqrt{10}}{3} \right\}$$

41. $x + 5x^{1/2} - 24 = 0$

Let $u = x^{1/2}$ and $u^2 = x$.

$$u^2 + 5u - 24 = 0$$
$$(u + 8)(u - 3) = 0$$

$$
\begin{array}{lcl}
u + 8 = 0 & \text{or} & u - 3 = 0 \\
u = -8 & \text{or} & u = 3 \\
x^{1/2} = -8 & \text{or} & x^{1/2} = 3 \\
\left(x^{1/2}\right)^2 = (-8)^2 & \text{or} & \left(x^{1/2}\right)^2 = 3^2 \\
x = 64 & \text{or} & x = 9
\end{array}
$$

check:

$x = 64$	$x = 9$
$64 + 5(64)^{1/2} - 24 = 0$	$9 + 5(9)^{1/2} - 24 = 0$
$64 + 5(8) - 24 = 0$	$9 + 5(3) - 24 = 0$
$64 + 40 - 24 = 0$	$9 + 15 - 24 = 0$
$80 = 0$	$0 = 0$
False	True

$\{9\}$

42. $(x + 9)^2 + 64 = 0$

$$(x + 9)^2 = -64$$
$$\sqrt{(x + 9)^2} = \pm\sqrt{-64}$$
$$x + 9 = \pm 8i$$
$$x = -9 \pm 8i$$
$$\{-9 - 8i, -9 + 8i\}$$

43. $f(x) = (x + 12)^2 - 16 = 0$

$$(x + 12)^2 = 16$$
$$\sqrt{(x + 12)^2} = \pm\sqrt{16}$$
$$x + 12 = \pm 4$$
$$x = -12 \pm 4$$

$$
\begin{array}{lcl}
x = -12 - 4 & \text{or} & x = -12 + 4 \\
x = -16 & \text{or} & x = -8
\end{array}
$$

$\{-16, -8\}$

44. $f(x) = x^2 - 7x - 20 = 0$

$$x = \frac{-(-7) \pm \sqrt{(-7)^2 - 4(1)(-20)}}{2(1)}$$

$$x = \frac{7 \pm \sqrt{129}}{2}$$

$$\left\{ \frac{7 - \sqrt{129}}{2}, \frac{7 + \sqrt{129}}{2} \right\}$$

45.
$$
\begin{array}{lcl}
x = 4 & \text{or} & x = -9 \\
x - 4 = 0 & \text{or} & x + 9 = 0
\end{array}
$$

$$(x - 4)(x + 9) = 0$$
$$x^2 + 9x - 4x + 36 = 0$$
$$x^2 + 5x - 36 = 0$$

46.
$$
\begin{array}{lcl}
x = 4 & \text{or} & x = 4 \\
x - 4 = 0 & \text{or} & x - 4 = 0
\end{array}
$$

$$(x - 4)(x - 4) = 0$$
$$x^2 - 4x - 4x + 16 = 0$$
$$x^2 - 8x + 16 = 0$$

47.
$$
\begin{array}{lcl}
x = \dfrac{4}{3} & \text{or} & x = \dfrac{13}{2} \\
3x = 4 & \text{or} & 2x = 13 \\
3x - 4 = 0 & \text{or} & 2x - 13 = 0
\end{array}
$$

$$(3x - 4)(2x - 13) = 0$$
$$6x^2 - 39x - 8x + 52 = 0$$
$$6x^2 - 47x + 52 = 0$$

48.
$$
\begin{array}{lcl}
x = 2\sqrt{2} & \text{or} & x = -2\sqrt{2} \\
x - 2\sqrt{2} = 0 & \text{or} & x + 2\sqrt{2} = 0
\end{array}
$$

$$(x - 2\sqrt{2})(x + 2\sqrt{2}) = 0$$
$$x^2 + 2\sqrt{2}x - 2\sqrt{2}x - 4 \cdot 2 = 0$$
$$x^2 - 4 \cdot 2 = 0$$
$$x^2 - 8 = 0$$

49.
$$
\begin{array}{lcl}
x = 5i & \text{or} & x = -5i \\
x - 5i = 0 & \text{or} & x + 5i = 0
\end{array}
$$

$$(x - 5i)(x + 5i) = 0$$
$$x^2 + 5ix - 5ix - 25i^2 = 0$$
$$x^2 - 25(-1) = 0$$
$$x^2 + 25 = 0$$

50.
$$
\begin{array}{lcl}
x = 8i & \text{or} & x = -8i \\
x - 8i = 0 & \text{or} & x + 8i = 0
\end{array}
$$

$$(x - 8i)(x + 8i) = 0$$
$$x^2 + 8ix - 8ix - 64i^2 = 0$$
$$x^2 - 64(-1) = 0$$
$$x^2 + 64 = 0$$

51. $y = x^2 - 8x - 20$

vertex:

$$x = \frac{-b}{2a} = \frac{-(-8)}{2(1)} = 4$$

$$y = 4^2 - 8(4) - 20 = 16 - 32 - 20 = -36$$

$(4, -36)$

axis of symmetry: $x = 4$

52. $y = x^2 + 5x - 16$

vertex:

$$x = \frac{-b}{2a} = \frac{-5}{2(1)} = -\frac{5}{2}$$

$$y = \left(-\frac{5}{2}\right)^2 + 5\left(-\frac{5}{2}\right) - 16$$

$$= \frac{25}{4} - \frac{25}{2} - \frac{16}{1}$$

$$= \frac{25}{4} - \frac{50}{4} - \frac{64}{4}$$

$$= -\frac{89}{4}$$

$$\left(-\frac{5}{2}, -\frac{89}{4}\right)$$

axis of symmetry: $x = -\frac{5}{2}$

53. $y = -x^2 - 2x + 35$

vertex:

$$x = \frac{-b}{2a} = \frac{-(-2)}{2(-1)} = -1$$

$$y = -(-1)^2 - 2(-1) + 35 = -1 + 2 + 35 = 36$$

$(-1, 36)$

axis of symmetry: $x = -1$

54. $y = (x - 3)^2 - 4$

The equation is in the form $y = a(x - h)^2 + k$.

vertex: $(h, k) = (3, -4)$

axis of symmetry: $x = h$, or $x = 3$

55. $y = x^2 + 6x + 11$

y-intercept: Let $x = 0$.

$$y = (0)^2 + 6(0) + 11 = 11$$

$(0, 11)$

x-intercepts: Let $y = 0$.

$$0 = x^2 + 6x + 11$$

$$x = \frac{-6 \pm \sqrt{6^2 - 4(1)(11)}}{2(1)}$$

$$x = \frac{-6 \pm \sqrt{-8}}{2}$$

No x-intercepts

56. $y = x^2 - 4x - 3$

y-intercept: Let $x = 0$.

$$y = (0)^2 - 4(0) - 3 = -3$$

$(0, -3)$

x-intercepts: Let $y = 0$.

$$0 = x^2 - 4x - 3$$

$$x = \frac{-(-4) \pm \sqrt{(-4)^2 - 4(1)(-3)}}{2(1)}$$

$$x = \frac{4 \pm \sqrt{28}}{2}$$

$$x = \frac{4 \pm 2\sqrt{7}}{2}$$

$$x = 2 \pm \sqrt{7}$$

$$\left(2 - \sqrt{7}, 0\right), \left(2 + \sqrt{7}, 0\right)$$

57. $y = -x^2 + 10x - 25$

y-intercept: Let $x = 0$.

$$y = -(0)^2 + 10(0) - 25 = -25$$

$(0, -25)$

x-intercepts: Let $y = 0$.

$$0 = -x^2 + 10x - 25$$

$$x^2 - 10x + 25 = 0$$

$$(x - 5)(x - 5) = 0$$

$$x - 5 = 0$$

$$x = 5$$

$(5, 0)$

58. $y = (x + 4)^2 - 8$

y-intercept: Let $x = 0$.

$$y = (0 + 4)^2 - 8 = 16 - 8 = 8$$

$(0, 8)$

x-intercepts: Let $y = 0$.

$$0 = (x + 4)^2 - 8$$

$$8 = (x + 4)^2$$

$$\pm\sqrt{8} = \sqrt{(x + 4)^2}$$

$$\pm 2\sqrt{2} = x + 4$$

$$-4 \pm 2\sqrt{2} = x$$

$$\left(-4 - 2\sqrt{2}, 0\right), \left(-4 + 2\sqrt{2}, 0\right)$$

59. $y = x^2 + 8x + 15$

vertex:

$$x = \frac{-b}{2a} = \frac{-8}{2(1)} = -4$$

$$y = (-4)^2 + 8(-4) + 15 = 16 - 32 + 15 = -1$$
$$(-4, -1)$$

y-intercept: Let $x = 0$.
$$y = (0)^2 + 8(0) + 15 = 15$$
$$(0, 15)$$

x-intercepts: Let $y = 0$.
$$0 = x^2 + 8x + 15$$
$$0 = (x + 5)(x + 3)$$
$$x + 5 = 0 \quad \text{or} \quad x + 3 = 0$$
$$x = -5 \quad \text{or} \quad x = -3$$
$$(-5, 0), (-3, 0)$$

axis of symmetry: $x = -4$
point symmetric to the y-intercept: $(-8, 15)$

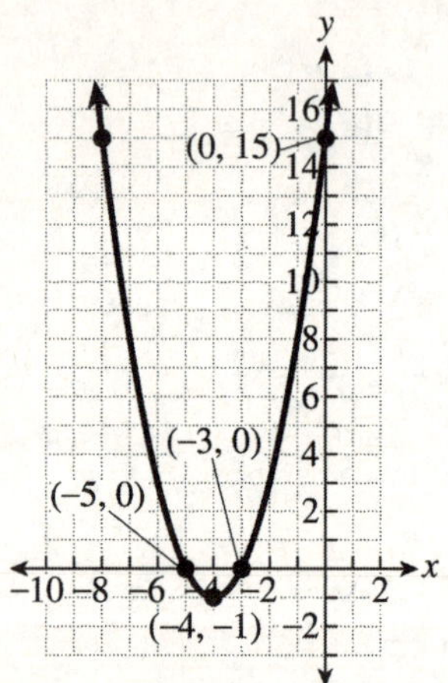

60. $y = -x^2 + 2x + 3$

vertex:

$$x = \frac{-b}{2a} = \frac{-2}{2(-1)} = 1$$

$$y = -(1)^2 + 2(1) + 3 = -1 + 2 + 3 = 4$$
$$(1, 4)$$

y-intercept: Let $x = 0$.
$$y = -(0)^2 + 2(0) + 3 = 3$$
$$(0, 3)$$

x-intercepts: Let $y = 0$.
$$0 = -x^2 + 2x + 3$$
$$x^2 - 2x - 3 = 0$$
$$(x + 1)(x - 3) = 0$$
$$x + 1 = 0 \quad \text{or} \quad x - 3 = 0$$
$$x = -1 \quad \text{or} \quad x = 3$$
$$(-1, 0), (3, 0)$$

axis of symmetry: $x = 1$
point symmetric to the y-intercept: $(2, 3)$

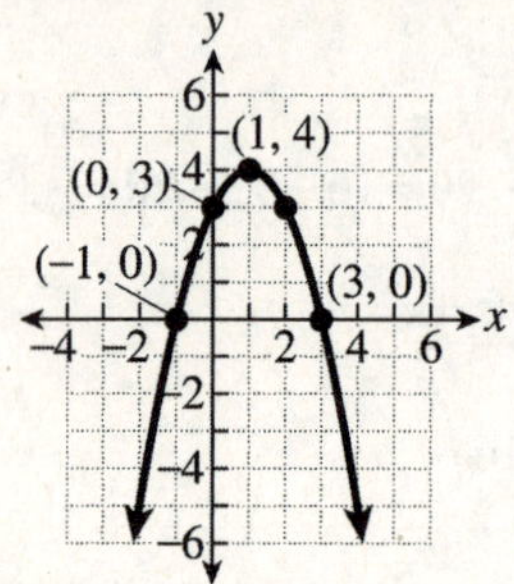

61. $y = x^2 + 4x - 2$

vertex:

$$x = \frac{-b}{2a} = \frac{-4}{2(1)} = -2$$

$$y = (-2)^2 + 4(-2) - 2 = 4 - 8 - 2 = -6$$
$$(-2, -6)$$

y-intercept: Let $x = 0$
$$y = (0)^2 + 4(0) - 2 = -2$$
$$(0, -2)$$

x-intercepts: Let $y = 0$.
$$0 = x^2 + 4x - 2$$
$$x = \frac{-4 \pm \sqrt{4^2 - 4(1)(-2)}}{2(1)}$$
$$x = \frac{-4 \pm \sqrt{24}}{2}$$
$$x = \frac{-4 \pm 2\sqrt{6}}{2}$$
$$x = -2 \pm \sqrt{6}$$
$$\left(-2 - \sqrt{6}, 0\right), \left(-2 + \sqrt{6}, 0\right)$$

axis of symmetry: $x = -2$
point symmetric to the y-intercept: $(-4, -2)$

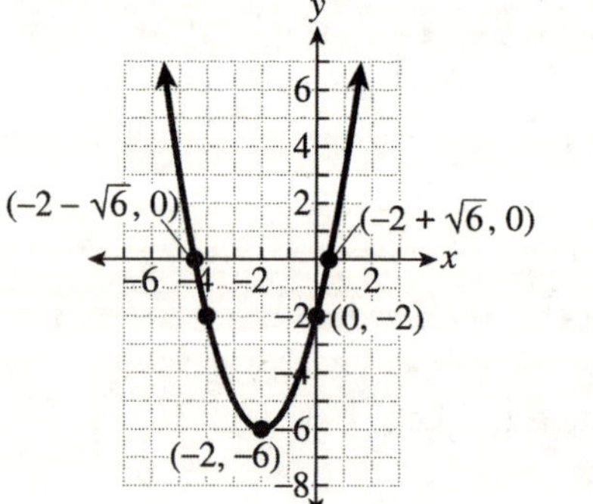

62. $y = x^2 + 4x + 8$

vertex:

$x = \dfrac{-b}{2a} = \dfrac{-4}{2(1)} = -2$

$y = (-2)^2 + 4(-2) + 8 = 4 - 8 + 8 = 4$
$(-2, 4)$

y-intercept: Let $x = 0$.
$y = (0)^2 + 4(0) + 8 = 8$
$(0, 8)$

x-intercepts: The vertex is above the x-axis and the parabola opens upward. No x-intercepts.

axis of symmetry: $x = -2$
point symmetric to the y-intercept: $(-4, 8)$

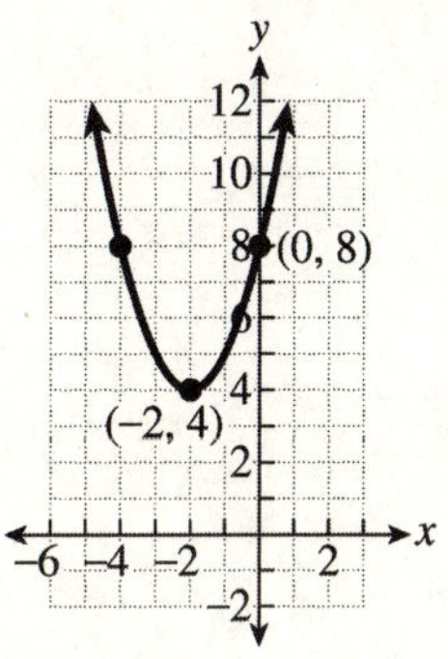

63. $y = -(x + 3)^2 + 4$

Use the formula $y = a(x - h)^2 + k$,

where (h, k) is the vertex.

vertex: $(-3, 4)$

y-intercept: Let $x = 0$
$y = -(0 + 3)^2 + 4 = -9 + 4 = -5$
$(0, -5)$

x-intercepts: Let $y = 0$.
$0 = -(x + 3)^2 + 4$
$(x + 3)^2 = 4$
$\sqrt{(x + 3)^2} = \pm\sqrt{4}$
$x + 3 = \pm 2$
$x = -3 \pm 2$
$x = -3 - 2 \quad$ or $\quad x = -3 + 2$
$x = -5 \quad\quad$ or $\quad x = -1$
$(-5, 0), (-1, 0)$

axis of symmetry: $x = -3$
point symmetric to the y-intercept: $(-6, -5)$

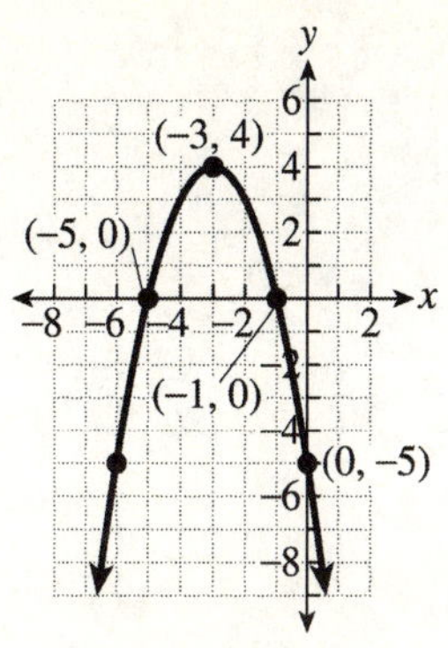

64. $y = (x - 2)^2 - 9$

Use the formula $y = a(x - h)^2 + k$,

where (h, k) is the vertex.

vertex: $(2, -9)$

y-intercept: Let $x = 0$.
$y = (0 - 2)^2 - 9 = 4 - 9 = -5$
$(0, -5)$

x-intercepts: Let $y = 0$.
$0 = (x - 2)^2 - 9$
$(x - 2)^2 = 9$
$\sqrt{(x - 2)^2} = \pm\sqrt{9}$
$x - 2 = \pm 3$
$x = 2 \pm 3$
$x = 2 - 3 \quad$ or $\quad x = 2 + 3$
$x = -1 \quad\;\;$ or $\quad x = 5$
$(-1, 0), (5, 0)$

axis of symmetry: $x = 2$
point symmetric to the y-intercept: $(4, -5)$

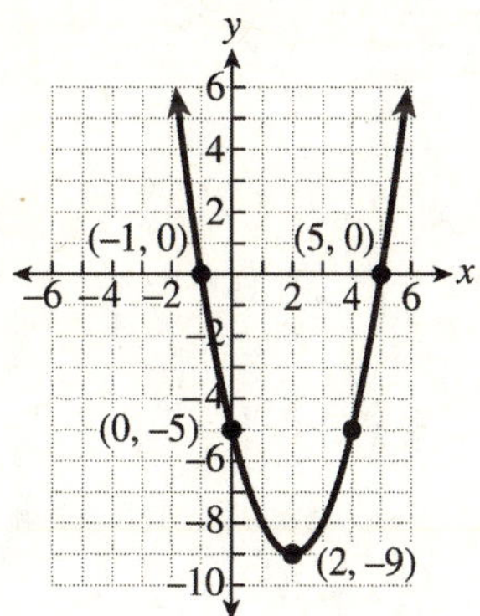

65. $x^2 - 9x + 14 \le 0$
$(x-2)(x-7) \le 0$

Find the zeros:
$(x-2)(x-7) = 0$
$x - 2 = 0$ or $x - 7 = 0$
$\quad x = 2$ or $\quad x = 7$

Test Points	$x-2$	$x-7$	Product
0	$-$	$-$	$+$
4	$+$	$-$	$-$
8	$+$	$+$	$+$

$[2, 7]$

66. $x^2 + 9x + 8 > 0$
$(x+8)(x+1) > 0$

Find the zeros:
$(x+8)(x+1) = 0$
$x + 8 = 0$ or $x + 1 = 0$
$\quad x = -8$ or $\quad x = -1$

Test Points	$x+8$	$x+1$	Product
-10	$-$	$-$	$+$
-4	$+$	$-$	$-$
0	$+$	$+$	$+$

$(-\infty, -8) \cup (-1, \infty)$

67. $x^2 + 4x - 11 > 0$

Find the zeros:
$x^2 + 4x - 11 = 0$
$$x = \frac{-4 \pm \sqrt{4^2 - 4(1)(-11)}}{2(1)}$$
$$x = \frac{-4 \pm \sqrt{60}}{2}$$
$$x = \frac{-4 \pm 2\sqrt{15}}{2}$$
$$x = -2 \pm \sqrt{15}$$
$x \approx -5.87$ or $x \approx 1.87$

Test Points	$x^2 + 4x - 11$	Sign
-6	$(-6)^2 + 4(-6) - 11 = 1$	$+$
0	$(0)^2 + 4(0) - 11 = -11$	$-$
3	$(3)^2 + 4(3) - 11 = 10$	$+$

$\left(-\infty, -2 - \sqrt{15}\right) \cup \left(-2 + \sqrt{15}, \infty\right)$

68. $x^2 + 3x + 40 \ge 0$
Find the zeros:
$x^2 + 3x + 40 = 0$
$$x = \frac{-3 \pm \sqrt{3^2 - 4(1)(40)}}{2(1)}$$
$$x = \frac{-3 \pm \sqrt{-151}}{2}$$
No real zeros.

Test Point	$x^2 + 3x + 40$	Sign
0	$(0)^2 + 3(0) + 40 = 40$	$+$

$(-\infty, \infty)$

69. $\dfrac{x^2 - 15x + 54}{x - 7} \le 0$

$\dfrac{(x-9)(x-6)}{x-7} \le 0$

Numerator is zero:
$x - 9 = 0$ or $x - 6 = 0$
$\quad x = 9$ or $\quad x = 6$

Denominator is zero:
$x - 7 = 0$
$\quad x = 7$

Test Points	$x-9$	$x-6$	$x-7$	$\dfrac{(x-9)(x-6)}{x-7}$
0	$-$	$-$	$-$	$-$
$6\frac{1}{2}$	$-$	$+$	$-$	$+$
8	$-$	$+$	$+$	$-$
10	$+$	$+$	$+$	$+$

$(-\infty, 6] \cup (7, 9]$

70. $\dfrac{x^2-4}{x^2+2x-80} \geq 0$

$\dfrac{(x+2)(x-2)}{(x+10)(x-8)} \geq 0$

Numerator is zero:
$x+2=0$ or $x-2=0$
$x=-2$ or $x=2$

Denominator is zero:
$x+10=0$ or $x-8=0$
$x=-10$ or $x=8$

Test Points	$x+2$	$x-2$	$x+10$	$x-8$	$\dfrac{(x+2)(x-2)}{(x+10)(x-8)}$
-11	$-$	$-$	$-$	$-$	$+$
-4	$-$	$-$	$+$	$-$	$-$
0	$+$	$-$	$+$	$-$	$+$
4	$+$	$+$	$+$	$-$	$-$
10	$+$	$+$	$+$	$+$	$+$

$(-\infty,-10)\cup[-2,2]\cup(8,\infty)$

71. $f(x)=x^2+15x+50\geq 0$
$(x+5)(x+10)\geq 0$

Find the zeros:
$(x+5)(x+10)=0$
$x+5=0$ or $x+10=0$
$x=-5$ or $x=-10$

Test Points	$x+5$	$x+10$	Product
-11	$-$	$-$	$+$
-8	$-$	$+$	$-$
0	$+$	$+$	$+$

$(-\infty,-10]\cup[-5,\infty)$

72. $\dfrac{x^2-16}{x^2-10x+9} \geq 0$

$\dfrac{(x+4)(x-4)}{(x-9)(x-1)} \geq 0$

Numerator is zero:
$x+4=0$ or $x-4=0$
$x=-4$ or $x=4$

Denominator is zero:
$x-9=0$ or $x-1=0$
$x=9$ or $x=1$

Test Points	$x+4$	$x-4$	$x-9$	$x-1$	$\dfrac{(x+4)(x-4)}{(x-9)(x-1)}$
-5	$-$	$-$	$-$	$-$	$+$
0	$+$	$-$	$-$	$-$	$-$
2	$+$	$-$	$-$	$+$	$+$
6	$+$	$+$	$-$	$+$	$-$
10	$+$	$+$	$+$	$+$	$+$

$(-\infty,-4]\cup(1,4]\cup(9,\infty)$

73. Unknown:
time: t

Knowns:
v_o: 104 ft/sec
s: 0 ft
$h(t)$: 0 ft

$0=-16t^2+104t+0$
$0=-t(16t-104)$

$-t=0$ or $16t-104=0$
$t=0$ or $16t=104$
$t=6.5$

Time: 6.5 seconds

74. Unknown:
time: t

Knowns:
v_o: 66 ft/sec
s: 30 ft
$h(t)$: 0 ft

$0=-16t^2+66t+30$
$16t^2-66t-30=0$

$t=\dfrac{-(-66)\pm\sqrt{(-66)^2-4(16)(-30)}}{2(16)}$

$t=\dfrac{66\pm\sqrt{6276}}{32}$

$t\approx -0.41$ or $t\approx 4.54$

Time: 4.54 seconds

75.

Pipe	Time to Complete the Job Alone	Work-Rate	Time Working	Portion of the Job Completed
Large	$t - 2$ hours	$\dfrac{1}{t-2}$	4 hours	$\dfrac{4}{t-2}$
Small	t hours	$\dfrac{1}{t}$	4 hours	$\dfrac{4}{t}$

$$\frac{4}{t-2} + \frac{4}{t} = 1$$

$$t(t-2)\cdot\left(\frac{4}{t-2} + \frac{4}{t}\right) = t(t-2)\cdot 1$$

$$4t + 4(t-2) = t(t-2)$$

$$4t + 4t - 8 = t^2 - 2t$$

$$0 = t^2 - 10t + 8$$

$$t = \frac{-(-10) \pm \sqrt{(-10)^2 - 4(1)(8)}}{2(1)}$$

$$t = \frac{10 \pm \sqrt{68}}{2}$$

$$t \approx 0.877 \text{ or } t \approx 9.1$$

The value $t \approx 0.877$ is omitted because it gives a negative solution for the larger pipe.

It would take the smaller pipe, working alone, 9.1 hours to fill the tank.

76.

Crews	Time to Complete the Job Alone	Work-Rate	Time Working	Portion of the Job Completed
Fast	$t - 6$ hours	$\dfrac{1}{t-6}$	32 hours	$\dfrac{32}{t-6}$
Slow	t hours	$\dfrac{1}{t}$	32 hours	$\dfrac{32}{t}$

$$\frac{32}{t-6} + \frac{32}{t} = 1$$

$$t(t-6)\cdot\left(\frac{32}{t-6} + \frac{32}{t}\right) = t(t-6)\cdot 1$$

$$32t + 32(t-6) = t(t-6)$$

$$32t + 32t - 192 = t^2 - 6t$$

$$0 = t^2 - 70t + 192$$

$$t = \frac{-(-70) \pm \sqrt{(-70)^2 - 4(1)(192)}}{2(1)}$$

$$t = \frac{70 \pm \sqrt{4132}}{2}$$

$$t \approx 2.9 \text{ or } t \approx 67.1$$

The value $t \approx 2.9$ is omitted because it gives a negative solution for the faster crew.

Faster crew: $t - 6 = 67.1 - 6 = 61.1$ hours

The faster crew, working alone, would take 61.1 hours to trim all the trees on campus.

77. Unknowns: Known:
Length: $x+2$ Area: 288 cm^2
Width: x

$$x(x+2) = 288$$
$$x^2 + 2x = 288$$
$$x^2 + 2x - 288 = 0$$
$$(x+18)(x-16) = 0$$

$x+18 = 0$ or $x-16 = 0$
$x = -18$ or $x = 16$
Width: $x = 16$ centimeters
Length: $x+2 = 16+2 = 18$ centimeters

78. Unknowns: Known:
Length: $x+13$ Area: 500 ft^2
Width: x

$$(x+13)x = 500$$
$$x^2 + 13x = 500$$
$$x^2 + 13x - 500 = 0$$

$$x = \frac{-13 \pm \sqrt{13^2 - 4(1)(-500)}}{2(1)}$$

$$x = \frac{-13 \pm \sqrt{2169}}{2}$$

$x \approx -29.8$ or $x \approx 16.8$
Width: $x \approx 16.8$ feet
Length: $x+13 \approx 16.8+13 = 29.8$ feet

79. $(1000)^2 + (800)^2 = d^2$
$$1,640,000 = d^2$$
$$\pm\sqrt{1,640,000} = \sqrt{d^2}$$
$$1280.6 \approx d$$
1280.6 miles

80. Unknowns: Known:
Length: x Diagonal: 12 ft
Width: $x-3$

$$x^2 + (x-3)^2 = 12^2$$
$$x^2 + x^2 - 6x + 9 = 144$$
$$2x^2 - 6x - 135 = 0$$

$$x = \frac{-(-6) \pm \sqrt{(-6)^2 - 4(2)(-135)}}{2(2)}$$

$$x = \frac{6 \pm \sqrt{1116}}{4}$$

$x \approx -6.9$ or $x \approx 9.9$
Length: $x \approx 9.9$ feet
Width: $x-3 \approx 9.9-3 = 6.9$ feet

1. $x^2 + 9x + 18 = 0$
$$(x+6)(x+3) = 0$$
$x+6 = 0$ or $x+3 = 0$
$x = -6$ or $x = -3$
$\{-6, -3\}$

2. $(3x-8)^2 - 6 = 58$

$$(3x-8)^2 = 64$$
$$\sqrt{(3x-8)^2} = \pm\sqrt{64}$$
$$3x - 8 = \pm 8$$
$$3x = 8 \pm 8$$
$$x = \frac{8 \pm 8}{3}$$

$x = 0$ or $x = \dfrac{16}{3}$

$\left\{0, \dfrac{16}{3}\right\}$

3. $x^2 - 14x - 3 = 0$
$$x^2 - 14x = 3$$
$$x^2 - 14x + \left(-\frac{14}{2}\right)^2 = 3 + \left(-\frac{14}{2}\right)^2$$
$$x^2 - 14x + 49 = 3 + 49$$
$$(x-7)^2 = 52$$
$$\sqrt{(x-7)^2} = \pm\sqrt{52}$$
$$x - 7 = \pm 2\sqrt{13}$$
$$x = 7 \pm 2\sqrt{13}$$
$\left\{7 - 2\sqrt{13}, 7 + 2\sqrt{13}\right\}$

4. $x^2 - 6x + 12 = 0$

$$x = \frac{-(-6) \pm \sqrt{(-6)^2 - 4(1)(12)}}{2(1)}$$

$$x = \frac{6 \pm \sqrt{-12}}{2}$$

$$x = \frac{6 \pm 2i\sqrt{3}}{2}$$

$$x = 3 \pm i\sqrt{3}$$
$\left\{3 - i\sqrt{3}, 3 + i\sqrt{3}\right\}$

5. $x^4 - 8x^2 - 48 = 0$

Let $u = x^2$ and $u^2 = x^4$.

$u^2 - 8u - 48 = 0$

$(u - 12)(u + 4) = 0$

$u - 12 = 0 \quad$ or $\quad u + 4 = 0$

$\quad u = 12 \quad$ or $\quad u = -4$

$\quad x^2 = 12 \quad$ or $\quad x^2 = -4$

$\sqrt{x^2} = \pm\sqrt{12} \quad$ or $\quad \sqrt{x^2} = \pm\sqrt{-4}$

$\quad x = \pm 2\sqrt{3} \quad$ or $\quad x = \pm 2i$

$\left\{-2\sqrt{3}, 2\sqrt{3}, -2i, 2i\right\}$

6. $x + 8x^{1/2} - 9 = 0$

Let $u = x^{1/2}$ and $u^2 = x$.

$u^2 + 8u - 9 = 0$

$(u + 9)(u - 1) = 0$

$u + 9 = 0 \quad$ or $\quad u - 1 = 0$

$\quad u = -9 \quad$ or $\quad u = 1$

$\quad x^{1/2} = -9 \quad$ or $\quad x^{1/2} = 1$

$\left(x^{1/2}\right)^2 = (-9)^2 \quad$ or $\quad \left(x^{1/2}\right)^2 = 1^2$

$\quad x = 81 \quad$ or $\quad x = 1$

check:

$x = 81$	$x = 1$
$(81) + 8(81)^{1/2} - 9 = 0$	$(1) + 8(1)^{1/2} - 9 = 0$
$81 + 8(9) - 9 = 0$	$1 + 8(1) - 9 = 0$
$81 + 72 - 9 = 0$	$1 + 8 - 9 = 0$
$144 = 0$	$0 = 0$
False	True

$\{1\}$

7. $(4x - 1)^2 - 13 = 12$

$(4x - 1)^2 = 25$

$\sqrt{(4x - 1)^2} = \pm\sqrt{25}$

$4x - 1 = \pm 5$

$4x = 1 \pm 5$

$x = \dfrac{1 \pm 5}{4}$

$x = \dfrac{-4}{4} \quad$ or $\quad x = \dfrac{6}{4}$

$x = -1 \quad$ or $\quad x = \dfrac{3}{2}$

$\left\{-1, \dfrac{3}{2}\right\}$

8. $2x^2 - 15x - 27 = 0$

$(2x + 3)(x - 9) = 0$

$2x + 3 = 0 \quad$ or $\quad x - 9 = 0$

$\quad 2x = -3 \quad$ or $\quad x = 9$

$\quad x = -\dfrac{3}{2} \quad$ or $\quad x = 9$

$\left\{-\dfrac{3}{2}, 9\right\}$

9. $x^2 - 14x + 40 = 0$

$(x - 4)(x - 10) = 0$

$x - 4 = 0 \quad$ or $\quad x - 10 = 0$

$\quad x = 4 \quad$ or $\quad x = 10$

$\{4, 10\}$

10.

$$\dfrac{x}{x + 8} - \dfrac{6}{x - 4} = \dfrac{8}{x^2 + 4x - 32}$$

$$\dfrac{x}{x + 8} - \dfrac{6}{x - 4} = \dfrac{8}{(x + 8)(x - 4)}$$

$$(x + 8)(x - 4)\left(\dfrac{x}{x + 8} - \dfrac{6}{x - 4}\right) = (x + 8)(x - 4) \cdot \dfrac{8}{(x + 8)(x - 4)}$$

$$x(x - 4) - 6(x + 8) = 8$$

$$x^2 - 4x - 6x - 48 = 8$$

$$x^2 - 10x - 56 = 0$$

$$(x + 4)(x - 14) = 0$$

$x + 4 = 0 \quad$ or $\quad x - 14 = 0$

$\quad x = -4 \quad$ or $\quad x = 14$

Neither of these values makes a denominator of the original equation 0.

$\{-4, 14\}$

11. $x^2 - 10x + 34 = 0$

$$x = \frac{-(-10) \pm \sqrt{(-10)^2 - 4(1)(34)}}{2(1)}$$

$$x = \frac{10 \pm \sqrt{-36}}{2}$$

$$x = \frac{10 \pm 6i}{2}$$

$$x = 5 \pm 3i$$

$$\{5 - 3i, 5 + 3i\}$$

12. $x = 5$ or $x = \dfrac{4}{7}$

$x - 5 = 0$ or $7x = 4$
$x - 5 = 0$ or $7x - 4 = 0$

$$(x - 5)(7x - 4) = 0$$
$$7x^2 - 4x - 35x + 20 = 0$$
$$7x^2 - 39x + 20 = 0$$

13. $x = 4i$ or $x = -4i$
$x - 4i = 0$ or $x + 4i = 0$

$$(x - 4i)(x + 4i) = 0$$
$$x^2 + 4ix - 4ix - 16i^2 = 0$$
$$x^2 - 16(-1) = 0$$
$$x^2 + 16 = 0$$

14. $y = x^2 + 8x - 39$

vertex:
$$x = \frac{-b}{2a} = \frac{-8}{2(1)} = \frac{-8}{2} = -4$$

$$y = (-4)^2 + 8(-4) - 39 = 16 - 32 - 39 = -55$$
$$(-4, -55)$$

15. $y = (x - 7)^2 + 20$

The equation is in the form $y = a(x - h)^2 + k$.
vertex: $(h, k) = (7, 20)$

16. $y = x^2 - 10x + 20$

y-intercept: Let $x = 0$
$y = (0)^2 - 10(0) + 20 = 20$
$(0, 20)$

x-intercepts: Let $y = 0$
$0 = x^2 - 10x + 20$

$$x = \frac{-(-10) \pm \sqrt{(-10)^2 - 4(1)(20)}}{2(1)}$$

$$x = \frac{10 \pm \sqrt{20}}{2}$$

$$x = \frac{10 \pm 2\sqrt{5}}{2}$$

$x = 5 \pm \sqrt{5}$
$x \approx 2.8$ or $x \approx 7.2$
$(2.8, 0), (7.2, 0)$

17. $y = -(x + 1)^2 + 4$

Vertex: using the formula $y = a(x - h)^2 + k$,
where (h, k) is the vertex.
vertex: $(-1, 4)$

y-intercept: Let $x = 0$.
$y = -(0 + 1)^2 + 4 = -1 + 4 = 3$
$(0, 3)$

x-intercepts: Let $y = 0$.
$$0 = -(x + 1)^2 + 4$$
$$(x + 1)^2 = 4$$
$$\sqrt{(x + 1)^2} = \pm\sqrt{4}$$
$$x + 1 = \pm 2$$
$$x = -1 \pm 2$$
$x = -1 - 2$ or $x = -1 + 2$
$x = -3$ or $x = 1$
$(-3, 0), (1, 0)$

axis of symmetry: $x = -1$
point symmetric to the y-intercept: $(-2, 3)$

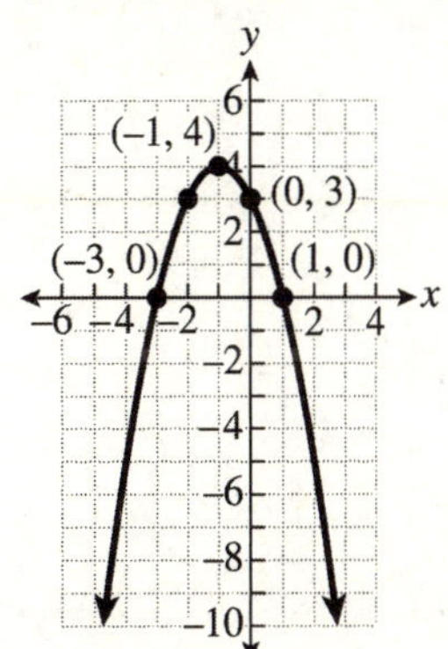

18. $y = x^2 - 10x + 13$

vertex:

$$x = \frac{-b}{2a} = \frac{-(-10)}{2(1)} = 5$$

$$y = 5^2 - 10(5) + 13 = 25 - 50 + 13 = -12$$
$$(5, -12)$$

y-intercept:
The equation is in standard form.
$(0, 13)$

x-intercepts: Let $y = 0$.
$$0 = x^2 - 10x + 13$$
$$x = \frac{-(-10) \pm \sqrt{(-10)^2 - 4(1)(13)}}{2(1)}$$
$$x = \frac{10 \pm \sqrt{48}}{2}$$
$$x = \frac{10 \pm 4\sqrt{3}}{2}$$
$$x = 5 \pm 2\sqrt{3}$$
$$\left(5 - 2\sqrt{3}, 0\right), \left(5 + 2\sqrt{3}, 0\right)$$

axis of symmetry: $x = 5$
point symmetric to the y-intercept: $(10, 13)$

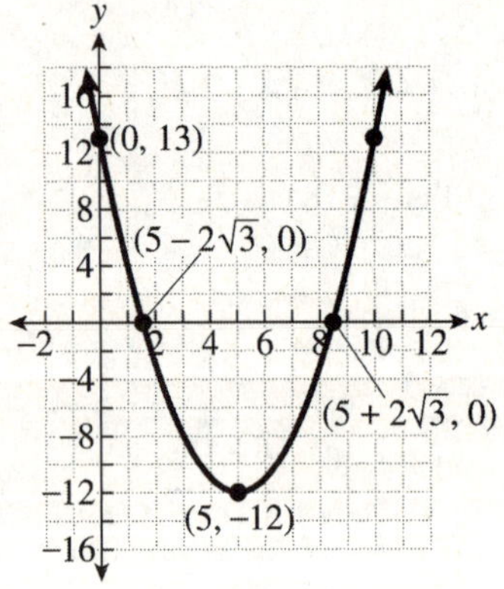

19. $x^2 + 3x - 28 > 0$
$(x + 7)(x - 4) > 0$

Find the zeros:
$(x + 7)(x - 4) = 0$
$x + 7 = 0$ or $x - 4 = 0$
    $x = -7$ or $x = 4$

Test Points	$x + 7$	$x - 4$	Product
-8	$-$	$-$	$+$
0	$+$	$-$	$-$
6	$+$	$+$	$+$

$(-\infty, -7) \cup (4, \infty)$

20. $\dfrac{x^2 + 11x + 24}{x^2 - 5x - 14} \le 0$

$$\frac{(x + 3)(x + 8)}{(x - 7)(x + 2)} \le 0$$

Numerator is zero:
$x + 3 = 0$ or $x + 8 = 0$
    $x = -3$ or $x = -8$

Denominator is zero:
$x - 7 = 0$ or $x + 2 = 0$
    $x = 7$ or $x = -2$

Test Points	$x + 3$	$x + 8$	$x - 7$	$x + 2$	$\dfrac{(x+3)(x+8)}{(x-7)(x+2)}$
-9	$-$	$-$	$-$	$-$	$+$
-6	$-$	$+$	$-$	$-$	$-$
$-2\frac{1}{2}$	$+$	$+$	$-$	$-$	$+$
0	$+$	$+$	$-$	$+$	$-$
8	$+$	$+$	$+$	$+$	$+$

$[-8, -3] \cup (-2, 7)$

21.

Unknowns:	Known:
Length: x	Area: 3600 yd^2
Width: $x - 60$	

$$x(x - 60) = 3600$$
$$x^2 - 60x = 3600$$
$$x^2 - 60x - 3600 = 0$$
$$x = \frac{-(-60) \pm \sqrt{(-60)^2 - 4(1)(-3600)}}{2(1)}$$
$$x = \frac{60 \pm \sqrt{18,000}}{2}$$

$x \approx -37.1$ or $x \approx 97.1$
Length: $x \approx 97.1$ yards
Width: $x - 60 \approx 97.1 - 60 = 37.1$ yards

22.
$$(1500)^2 + (800)^2 = d^2$$
$$2,890,000 = d^2$$
$$\pm\sqrt{2,890,000} = \sqrt{d^2}$$
$$\pm 1700 = d$$
The resort is 1700 miles away.

CHAPTER 11 FUNCTIONS

11.1 QUICK CHECK

1. a) Yes.

  b) No.

2. No.

3. No.

4. $f(-5) = (-5)^2 - 3(-5) - 20$
  $= 25 + 15 - 20$
  $= 40 - 20$
  $= 20$

5. $f(3m - 9) = -4(3m - 9) + 21$
  $= -12m + 36 + 21$
  $= -12m + 57$

11.1 REVIEW OF FUNCTIONS

1. function

3. evaluating

5. Yes. A college graduate will only have one starting salary.

7. Yes. Each person will only have a certain number of jobs in their lifetime.

9. No. More than one city may have the same high temperature last Tuesday.

11. Function.
Domain:
$\{1960, 1964, 1968, 1972, 1976, 1980, 1984\}$
Range:
$\{$Kennedy, Johnson, Nixon, Carter, Reagan$\}$

13. Function.
Domain: $\{5, 3, 1, -1, -3, -5\}$
Range: $\{-5, -3, -1, 1, 3, 5\}$

15. Function.
Domain: $\{-6, -3, 0, 3, 6\}$
Range: $\{5\}$

17. Yes.

19. No.

21. No.

23. No.

25. $f(-6) = 2(-6) - 9 = -12 - 9 = -21$

27. $f\left(\dfrac{3}{5}\right) = -5\left(\dfrac{3}{5}\right) - 2 = -3 - 2 = -5$

29. $f(9) = -\dfrac{2}{3}(9) + 10 = -6 + 10 = 4$

31. $f(a + 6) = 7(a + 6) - 4 = 7a + 42 - 4 = 7a + 38$

33. $f(5n + 2) = -2(5n + 2) - 17$
  $= -10n - 4 - 17$
  $= -10n - 21$

35. $f(5) = (5)^2 + 7(5) - 8 = 25 + 35 - 8 = 52$

37. $f(-7) = (-7)^2 - 9(-7) + 16$
  $= 49 + 63 + 16$
  $= 128$

39. $f(3) = (3 + 6)^2 + 37$
  $= (9)^2 + 37$
  $= 81 + 37$
  $= 118$

41. $f(-4) = |3(-4) - 2| + 17$
  $= |-12 - 2| + 17$
  $= |-14| + 17$
  $= 14 + 17$
  $= 31$

43. Domain: $(-\infty, \infty)$; range: $[-1, \infty)$;
x-intercepts: $(2, 0), (4, 0)$; y-intercept: $(0, 8)$

45. Domain: $[-4, \infty)$; range: $[2, \infty)$;
x-intercept: none; y-intercept: $(0, 4)$

47. $f(-3) = 4$

49. $f(-2) = 8$

51. When $x = 5$, $f(5) = -4$, so $x = 5$.

53. When $x = -1$, $f(-1) = -9$, and
when $x = 9$, $f(9) = -9$, so $x = -1, 9$.

55. Answers will vary. Example:

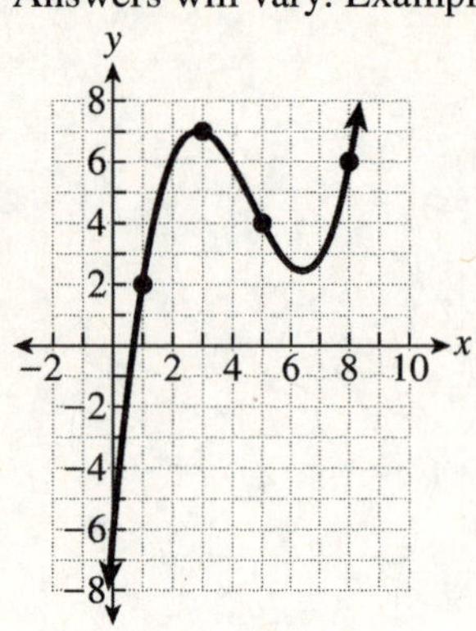

57. Answers will vary. Example:

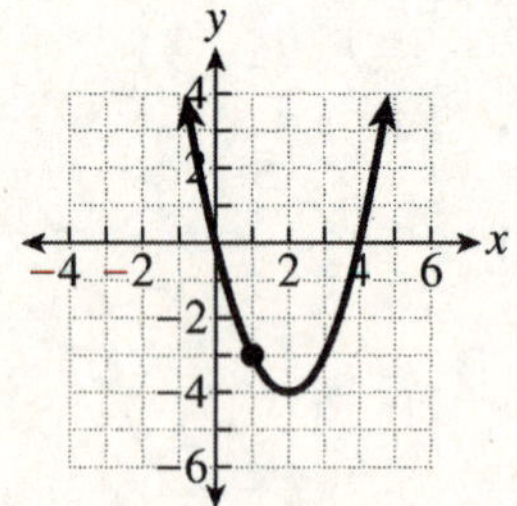

59. Answers will vary. Example:
This does not represent a function since this would mean that a single x-value would "match" with more than one y-value.

61. Answers will vary. Example:
Look for points of the graph whose y value is 2. The corresponding x-coordinate is the desired value.

11.2 **QUICK CHECK**

1. $f(x) = \dfrac{3}{4}x - 6$

y-intercept: Let $x = 0$.

$$f(0) = \frac{3}{4}(0) - 6 = -6$$

$(0, -6)$

x-intercept: Let $f(x) = 0$.

$$0 = \frac{3}{4}x - 6$$

$$6 = \frac{3}{4}x$$

$$\frac{4}{3} \cdot 6 = \frac{4}{3} \cdot \frac{3}{4}x$$

$$8 = x$$

$(8, 0)$

Third Point: $m = \dfrac{3}{4}$

From the y-intercept $(0, -6)$ move up 3, right 4 to the point $(4, -3)$.

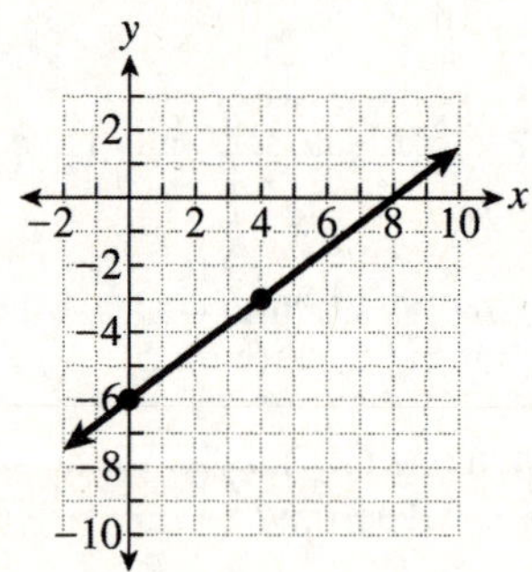

2. $f(x) = -5$

The graph is a horizontal line with the y-intercept of $(0, -5)$.

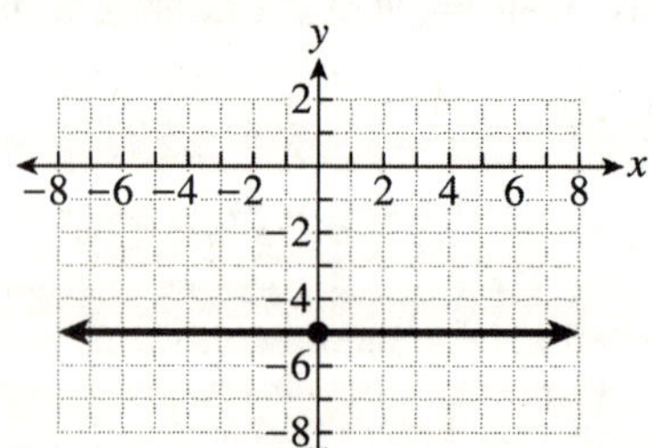

3. Use the points $(0,-6)$ and $(3,0)$.

$$m = \frac{-6-0}{0-3} = \frac{-6}{-3} = 2$$

The function is of the form $f(x) = 2x + b$.

$$f(0) = -6$$
$$2(0) + b = -6$$
$$b = -6$$
$$f(x) = 2x - 6$$

4. $f(150) = 3500$ and $f(80) = 2100$

$$m = \frac{3500 - 2100}{150 - 80} = \frac{1400}{70} = 20$$

The function is of the form $f(x) = 20x + b$.

$$f(80) = 2100$$
$$20(80) + b = 2100$$
$$1600 + b = 2100$$
$$b = 500$$
$$f(x) = 20x + 500$$

5. a) $C(x) = 50 + 1.25x$
$R(x) = 9.95x$
$$\begin{aligned} P(x) &= R(x) - C(x) \\ &= 9.95x - (50 + 1.25x) \\ &= 9.95x - 50 - 1.25x \\ &= 8.70x - 50 \end{aligned}$$

b)
$$P(x) = 1000$$
$$8.70x - 50 = 1000$$
$$8.70x = 1050$$
$$x \approx 121$$

Ross will have to sell 121 calendars to make a $1000 profit.

11.2 Linear Functions

1. linear

3. revenue

5. $f(x) = x - 7$

y-intercept: Let $x = 0$.

$$f(0) = (0) - 7 = -7$$
$$(0, -7)$$

x-intercept: Let $f(x) = 0$.
$$x - 7 = 0$$
$$x = 7$$
$$(7, 0)$$

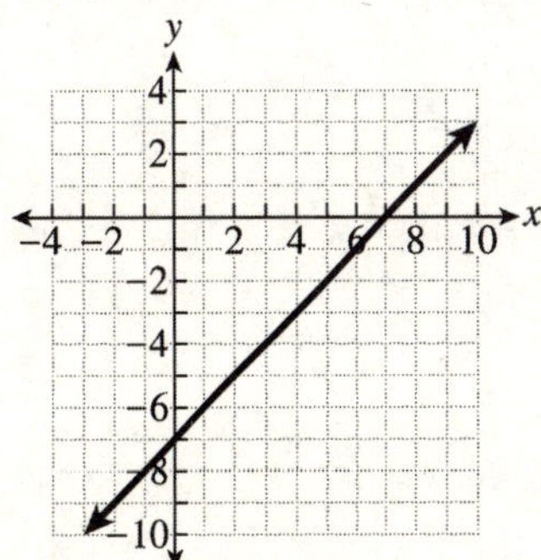

7. $f(x) = -x + 3$

y-intercept: Let $x = 0$.

$$f(0) = -(0) + 3 = 3$$
$$(0, 3)$$

x-intercept: Let $f(x) = 0$.
$$-x + 3 = 0$$
$$3 = x$$
$$(3, 0)$$

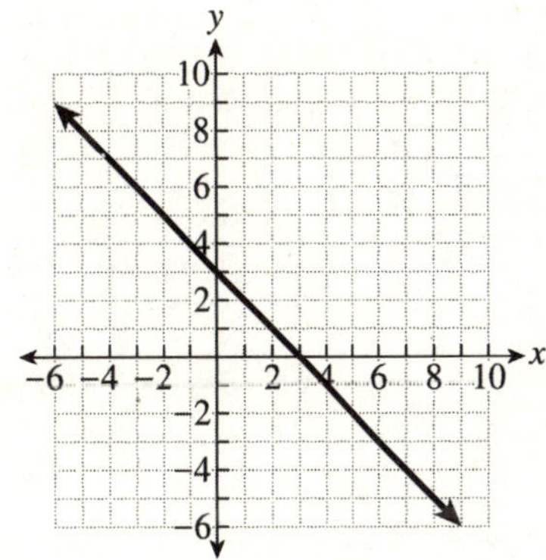

9. $f(x) = 2x + 4$

y-intercept: Let $x = 0$.

$f(0) = 2(0) + 4 = 4$
$(0, 4)$

x-intercept: Let $f(x) = 0$.
$2x + 4 = 0$
$\quad 2x = -4$
$\quad\ \ x = -2$
$(-2, 0)$

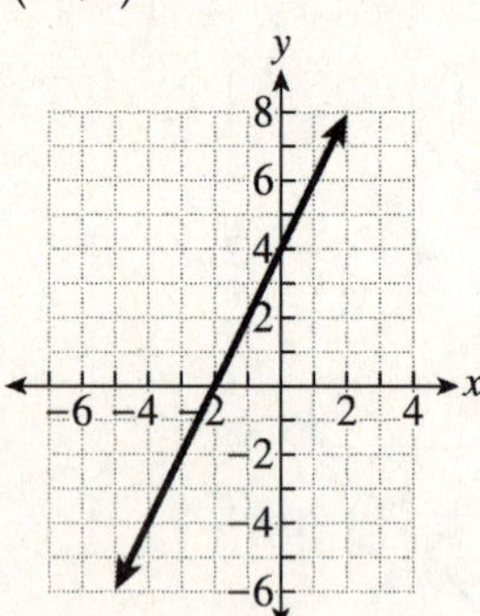

11. $f(x) = -3x + 3$

y-intercept: Let $x = 0$.

$f(0) = -3(0) + 3 = 3$
$(0, 3)$

x-intercept: Let $f(x) = 0$.
$-3x + 3 = 0$
$\quad\ 3 = 3x$
$\quad\ 1 = x$
$(1, 0)$

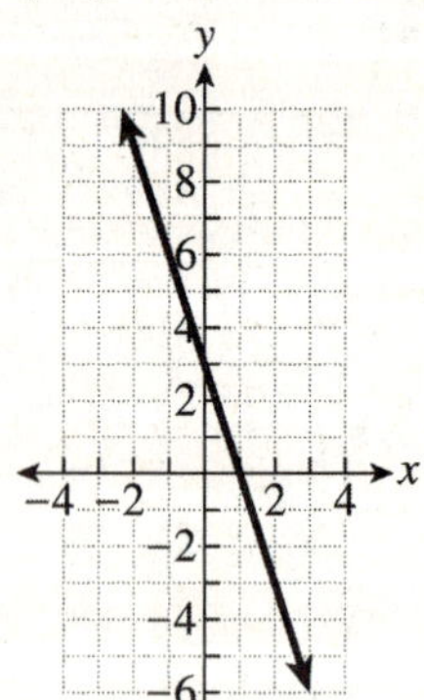

13. $g(x) = 4x - 6$

y-intercept: Let $x = 0$.

$g(0) = 4(0) - 6 = -6$
$(0, -6)$

x-intercept: Let $g(x) = 0$.
$4x - 6 = 0$
$\quad 4x = 6$
$\quad\ \ x = \dfrac{6}{4}$
$\quad\ \ x = \dfrac{3}{2}$
$\left(\dfrac{3}{2}, 0\right)$

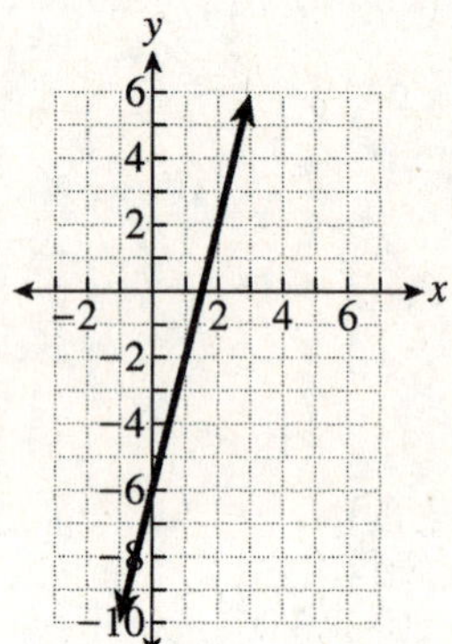

15. $f(x) = \dfrac{2}{3}x - 4$

y-intercept: Let $x = 0$.

$f(0) = \dfrac{2}{3}(0) - 4 = -4$
$(0, -4)$

x-intercept: Let $f(x) = 0$.

$\dfrac{2}{3}x - 4 = 0$

$\quad \dfrac{2}{3}x = 4$

$\dfrac{3}{2} \cdot \dfrac{2}{3}x = \dfrac{3}{2} \cdot 4$

$\quad\ \ x = 6$
$(6, 0)$

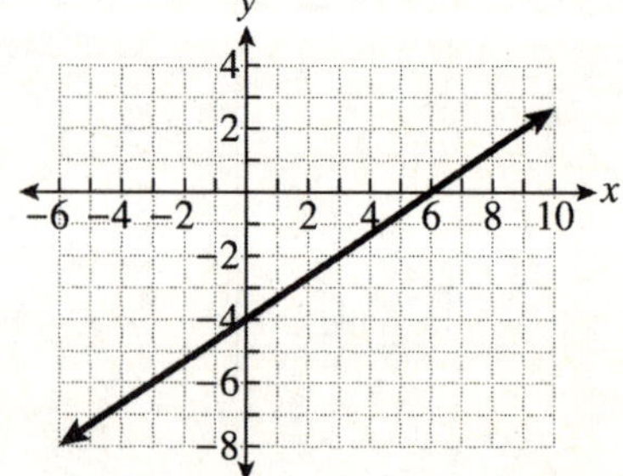

17. $f(x) = -\dfrac{3}{4}x + 6$

y-intercept: Let $x = 0$.

$f(0) = -\dfrac{3}{4}(0) + 6 = 6$

$(0, 6)$

x-intercept: Let $f(x) = 0$.

$-\dfrac{3}{4}x + 6 = 0$

$-\dfrac{3}{4}x = -6$

$x = -6\left(-\dfrac{4}{3}\right)$

$x = 8$

$(8, 0)$

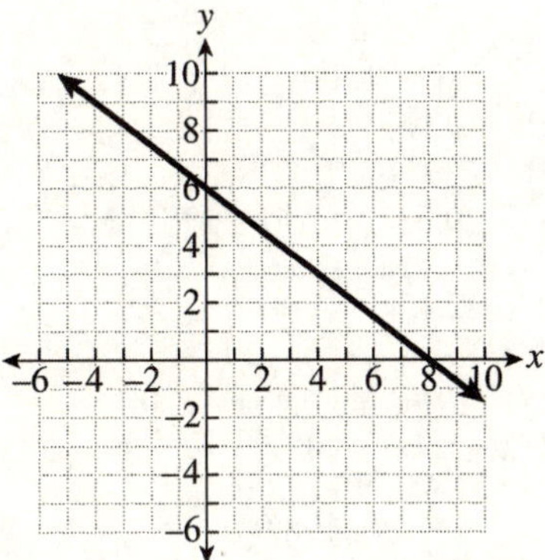

19. $f(x) = \dfrac{3}{7}x$

y-intercept: Let $x = 0$.

$f(0) = \dfrac{3}{7}(0) = 0$

$(0, 0)$

x-intercept: $(0, 0)$

Use slope of $\dfrac{3}{7}$ (3 up, 7 right)

to get the point $(7, 3)$.

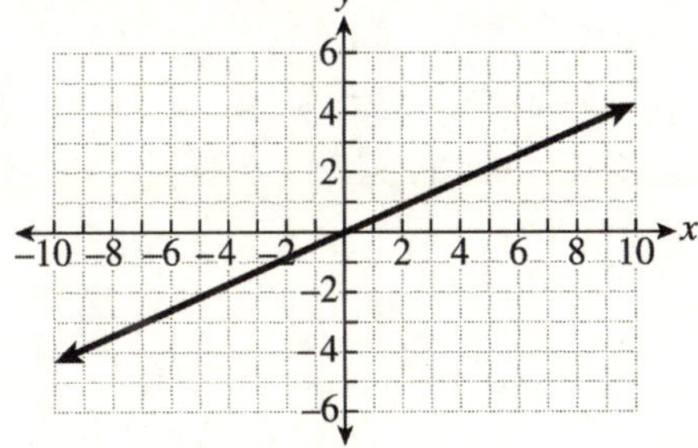

21. $f(x) = 7$

The graph is a horizontal line with the y-intercept of $(0, 7)$.

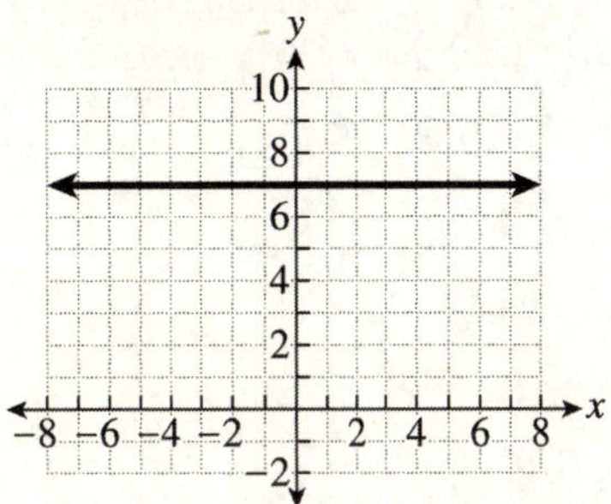

23. Use the points $(-2, 1)$ and $(-1, 4)$.

$m = \dfrac{4 - 1}{-1 - (-2)} = \dfrac{3}{1}$

The function is of the form $f(x) = 3x + b$.

$f(-2) = 1$
$3(-2) + b = 1$
$-6 + b = 1$
$b = 7$
$f(x) = 3x + 7$

25. Use the points $(1, 1)$ and $(4, -3)$.

$m = \dfrac{-3 - 1}{4 - 1} = \dfrac{-4}{3}$

The function is of the form $f(x) = -\dfrac{4}{3}x + b$.

$f(1) = 1$

$-\dfrac{4}{3}(1) + b = 1$

$b = 1 + \dfrac{4}{3}$

$b = \dfrac{3}{3} + \dfrac{4}{3}$

$b = \dfrac{7}{3}$

$f(x) = -\dfrac{4}{3}x + \dfrac{7}{3}$

27. Use the points $(0, 0)$ and $(1, 3)$.

$m = \dfrac{3 - 0}{1 - 0} = \dfrac{3}{1}$

The function is of the form $f(x) = 3x + b$.

$f(0) = 0$
$3(0) + b = 0$
$b = 0$
$f(x) = 3x$

29. The graph is a horizontal line, so it must be a constant function with the y-intercept of $(0, -6)$. $f(x) = -6$

31. a) $f(10) = 0.69$ and $f(30) = 1.29$

$$m = \frac{1.29 - 0.69}{30 - 10} = \frac{0.6}{20} = 0.03$$

The function is of the form
$$f(x) = 0.03x + b.$$
$$f(10) = 0.69$$
$$0.03(10) + b = 0.69$$
$$0.3 + b = 0.69$$
$$b = 0.39$$
$$f(x) = 0.03x + 0.39$$

b) Slope is the cost per minute, $\dfrac{\$0.03}{\text{minute}}$.

y-intercept is the connection fee, $0.39.

c) $f(44) = 0.03(44) + 0.39$
$$= 1.32 + 0.39$$
$$= 1.71$$
$1.71

33. a) $f(100) = 35$ and $f(150) = 42.50$

$$m = \frac{42.50 - 35}{150 - 100} = \frac{7.5}{50} = 0.15$$

The function is of the form
$$f(x) = 0.15x + b.$$
$$f(100) = 35$$
$$0.15(100) + b = 35$$
$$15 + b = 35$$
$$b = 20$$
$$f(x) = 0.15x + 20$$

b) Slope is the price per mile, $\dfrac{\$0.15}{\text{mile}}$.

y-intercept is the base fee, $20.

c) $f(260) = 0.15(260) + 20$
$$= 39 + 20$$
$$= 59$$
$59

35. a) $C(x) = 50 + 75 = 125$
$$R(x) = 4x$$
$$P(x) = R(x) - C(x) = 4x - 125$$

b) $P(75) = 4(75) - 125 = 300 - 125 = 175$
$175

37. a) $C(x) = 12 + 0.40x$
$$R(x) = 3x$$
$$P(x) = R(x) - C(x)$$
$$= 3x - (12 + 0.40x)$$
$$= 3x - 12 - 0.40x$$
$$= 2.6x - 12$$

b) $P(60) = 2.6(60) - 12 = 156 - 12 = 144$
$144

39. a) $C(x) = 7.5x + 1200$
$$R(x) = 29.95x$$
$$P(x) = R(x) - C(x)$$
$$= 29.95x - (7.5x + 1200)$$
$$= 29.95x - 7.5x - 1200$$
$$= 22.45x - 1200$$

b)
$$P(x) = 22.45x - 1200$$
$$22.45x - 1200 = 10,000$$
$$22.45x = 11,200$$
$$x \approx 498.9$$
At least 499 baskets.

41. If $f(x) = mx + 7$ and $f(5) = 22$
$$f(5) = m(5) + 7 = 22$$
$$5m + 7 = 22$$
$$5m = 15$$
$$m = 3$$

43. If $f(x) = mx - 13$ and $f(-6) = -25$
$$f(-6) = m(-6) - 13 = -25$$
$$-6m - 13 = -25$$
$$-6m = -12$$
$$m = 2$$

45. Answers will vary. Example:
To graph a linear function, first find the y-intercept by setting $x = 0$ and simplifying. Then find the x-intercept by setting $f(x) = 0$ and solving for x. Finally, graph the line using these two points.

$f(x) = 3x - 2$

y-intercept: Let $x = 0$.

$f(0) = 3(0) - 2 = -2$
$(0, -2)$

x-intercept: Let $f(x) = 0$.
$0 = 3x - 2$
$2 = 3x$

$\dfrac{2}{3} = x$

$\left(\dfrac{2}{3}, 0\right)$

47. Answers will vary. Example:
Johnny is starting a lemonade stand. Johnny's dad spent \$200 to build a stand and Johnny's mom spent \$38.50 for lemons, cups, and sugar. How many cups of lemonade will he need to sell to break even, if he sells the lemonade for \$0.10 a cup?

1. $f(x) = (x + 2)^2 - 1$

The vertex is $(-2, -1)$.
Shift the basic parabola 2 units to the left and 1 unit down.
y-intercept:

$f(0) = (0 + 2)^2 - 1 = 4 - 1 = 3$
$(0, 3)$

x-intercepts: Let $f(x) = 0$.

$$(x + 2)^2 - 1 = 0$$
$$(x + 2)^2 = 1$$
$$\sqrt{(x + 2)^2} = \pm\sqrt{1}$$
$$x + 2 = \pm 1$$
$$x = -2 \pm 1$$

$x = -2 - 1 = -3$ or $x = -2 + 1 = -1$
$(-3, 0), (-1, 0)$

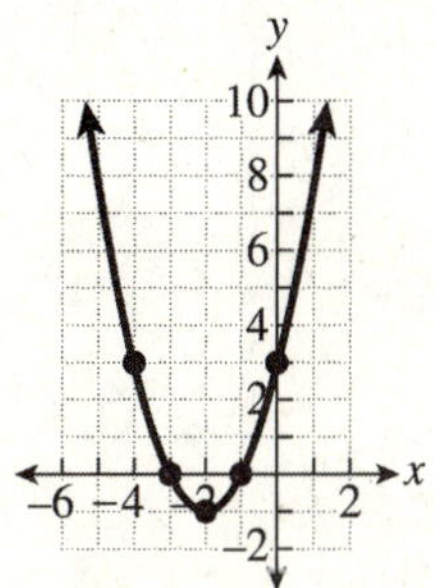

Domain: $(-\infty, \infty)$; range: $[-1, \infty)$

2. $f(x) = (x - 1)^2 + 4$

The vertex is $(1, 4)$.
Shift the basic parabola 1 unit to the right and 4 units up.

y-intercept: Let $x = 0$.

$f(0) = (0 - 1)^2 + 4 = 1 + 4 = 5$
$(0, 5)$

x-intercepts:
The vertex is above the x-axis and the parabola opens upward. There are no x-intercepts.

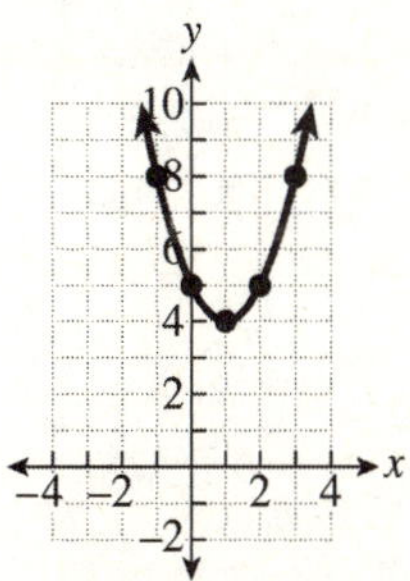

Domain: $(-\infty, \infty)$; range: $[4, \infty)$

3. $f(x) = -(x-2)^2 + 4$
The vertex is $(2,4)$.

Because $a < 0$, the parabola opens downward.
Rotate $f(x) = x^2$ about the x-axis.
Shift the basic parabola 2 units to the right and 4 units up.

y-intercept: Let $x = 0$.

$f(0) = -(0-2)^2 + 4 = -4 + 4 = 0$
$(0,0)$

x-intercepts: Let $f(x) = 0$.

$$0 = -(x-2)^2 + 4$$
$$(x-2)^2 = 4$$
$$\sqrt{(x-2)^2} = \pm\sqrt{4}$$
$$x - 2 = \pm 2$$
$$x = 2 \pm 2$$
$$x = 2 - 2 = 0 \text{ or } x = 2 + 2 = 4$$
$$(0,0), (4,0)$$

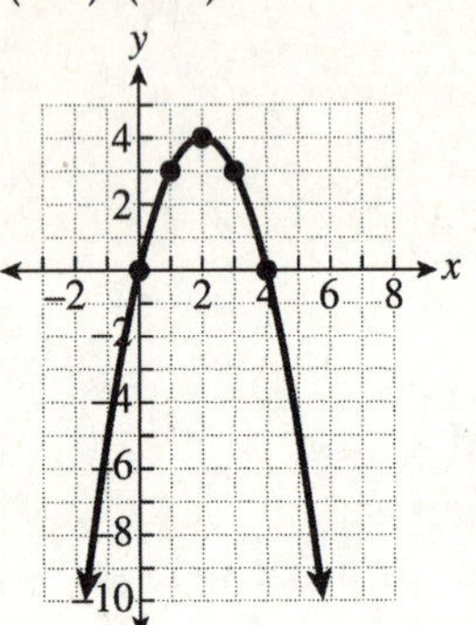

Domain: $(-\infty, \infty)$; range: $(-\infty, 4]$

4. a) $f(x) = (x+13)^2 - 7$
$a > 0$, vertex: $(-13, -7)$
No rotation about the x-axis
Left 13 units, down 7 units

b) $f(x) = -(x-4)^2 + 30$
$a < 0$, vertex: $(4, 30)$
Rotate about the x-axis.
Right 4 units, up 30 units

5. a) $f(x) = -x^2 + 10x - 35$ is a parabola

opening downward, so it has a maximum at the vertex.
vertex:
$$x = -\frac{b}{2a} = -\frac{10}{2(-1)} = 5$$
$$f(5) = -(5)^2 + 10(5) - 35$$
$$= -25 + 50 - 35$$
$$= -10$$
Maximum value is -10.

b) $f(x) = 5(x-2)^2 + 47$ is a parabola

opening upward so it has a minimum at the vertex.

Vertex: $(2, 47)$.

Minimum value is 47.

6. Unknowns:
#1: x
#2: $62 - x$

$$f(x) = x(62-x) = 62x - x^2 = -x^2 + 62x$$

Parabola opens downward, so there is a maximum at the vertex.
vertex:
$$x = -\frac{b}{2a} = -\frac{62}{2(-1)} = 31$$

$$f(31) = -(31)^2 + 62(31) = -961 + 1922 = 961$$
Maximum product is 961.

7. Unknowns:
length: x

$$w = \frac{P}{2} - l = \frac{264}{2} - x = 132 - x$$
width: $132 - x$

$$A(x) = x(132-x) = 132x - x^2 = -x^2 + 132x$$

Parabola opens downward, so there is a maximum at the vertex.
vertex:
$$x = -\frac{b}{2a} = -\frac{132}{2(-1)} = 66$$

$$A(66) = -(66)^2 + 132(66)$$
$$= -4356 + 8712$$
$$= 4356$$
Maximum area is 4356 square meters.

8. $\dfrac{f(x+h) - f(x)}{h}$

$$= \frac{(x+h)^2 + 3(x+h) + 317 - \left(x^2 + 3x + 317\right)}{h}$$

$$= \frac{x^2 + 2xh + h^2 + 3x + 3h + 317 - x^2 - 3x - 317}{h}$$

$$= \frac{2xh + h^2 + 3h}{h}$$

$$= \frac{\cancel{h}(2x + h + 3)}{\cancel{h}}$$

$$= 2x + h + 3$$

11.3 QUADRATIC FUNCTIONS

1. parabola

3. y-intercept

5. horizontal

7. minimum

9. $f(x) = (x+6)^2$
$a > 0$, vertex: $(-6,0)$
No rotation about the x-axis
Left 6 units

11. $f(x) = x^2 + 9$
$a > 0$, vertex: $(0,9)$
No rotation about the x-axis
Up 9 units

13. $f(x) = (x-5)^2 - 9$
$a > 0$, vertex: $(5,-9)$
No rotation about the x-axis
Right 5 units, down 9 units

15. $f(x) = (x+10)^2 - 40$
$a > 0$, vertex: $(-10,-40)$
No rotation about the x-axis
Left 10 units, down 40 units

17. $f(x) = -(x-2)^2 + 13$
$a < 0$, vertex: $(2,13)$
Rotate about the x-axis.
Right 2 units, up 13 units

19. $f(x) = x^2 + 7$
The vertex is $(0,7)$.
Shift the basic parabola 7 units up.
y-intercept:
$(0,7)$

x-intercepts:
The vertex is above the x-axis and the parabola opens upward. There are no x-intercepts.

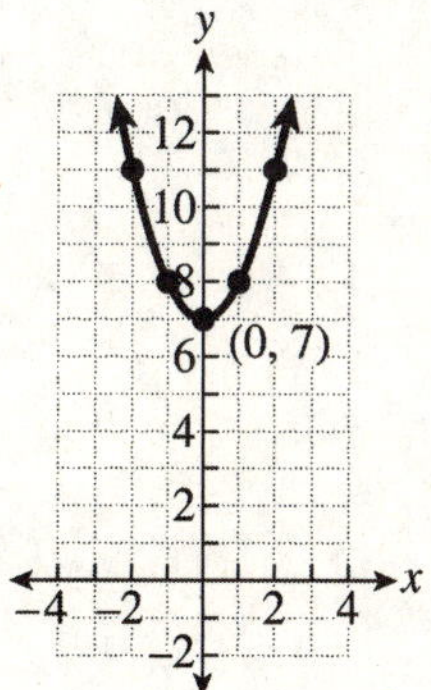

Domain: $(-\infty, \infty)$; range: $[7, \infty)$

21. $f(x) = (x-3)^2$
The vertex is $(3,0)$.
Shift the basic parabola 3 units to the right.
y-intercept: Let $x = 0$.
$f(0) = (0-3)^2 = 9$
$(0,9)$

x-intercepts: Let $f(x) = 0$.

$$(x-3)^2 = 0$$
$$\sqrt{(x-3)^2} = 0$$
$$x - 3 = 0$$
$$x = 3$$
$(3,0)$

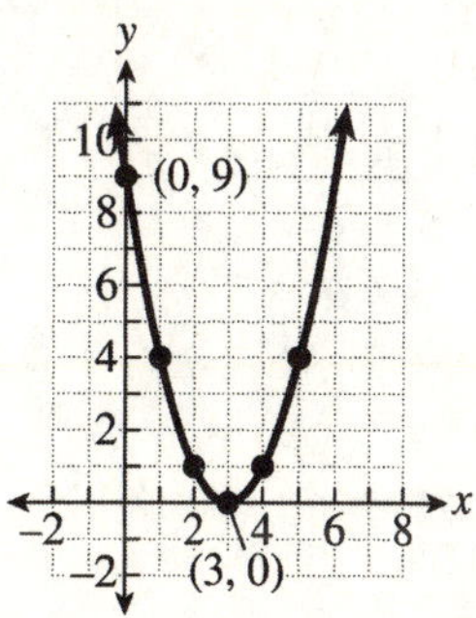

Domain: $(-\infty, \infty)$; range: $[0, \infty)$

23. $f(x) = (x-4)^2 - 1$

The vertex is $(4,-1)$.
Shift the basic parabola 4 units to the right and 1 unit down.
y-intercept: Let $x = 0$.

$f(0) = (0-4)^2 - 1 = 16 - 1 = 15$
$(0,15)$

x-intercepts: Let $f(x) = 0$.

$$(x-4)^2 - 1 = 0$$
$$(x-4)^2 = 1$$
$$\sqrt{(x-4)^2} = \pm\sqrt{1}$$
$$x - 4 = \pm 1$$
$$x = 4 \pm 1$$
$$x = 4 - 1 = 3 \text{ or } x = 4 + 1 = 5$$
$$(3,0), (5,0)$$

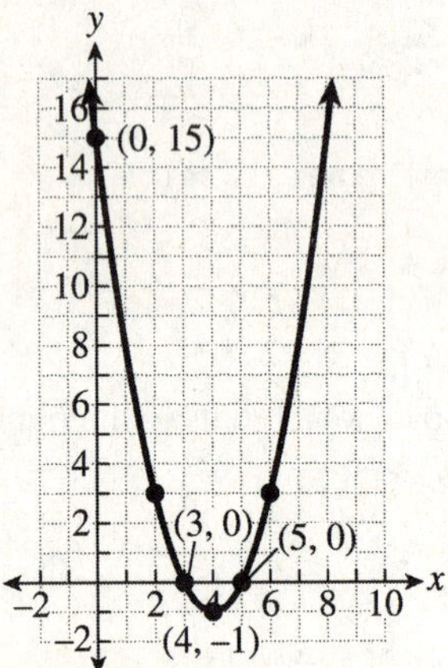

Domain: $(-\infty, \infty)$; range: $[-1, \infty)$

25. $f(x) = (x-1)^2 - 4$

The vertex is $(1,-4)$.
Shift the basic parabola 1 unit to the right and 4 units down.
y-intercept: Let $x = 0$.

$f(0) = (0-1)^2 - 4 = 1 - 4 = -3$
$(0,-3)$

x-intercepts: Let $f(x) = 0$.

$$(x-1)^2 - 4 = 0$$
$$(x-1)^2 = 4$$
$$\sqrt{(x-1)^2} = \pm\sqrt{4}$$
$$x - 1 = \pm 2$$
$$x = 1 \pm 2$$
$$x = 1 - 2 = -1 \text{ or } x = 1 + 2 = 3$$
$$(-1,0), (3,0)$$

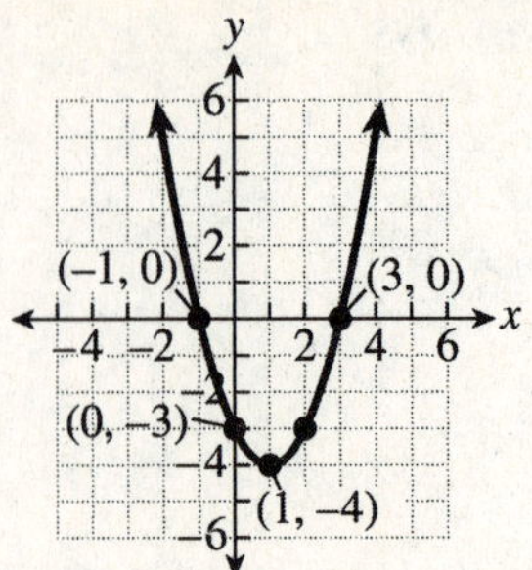

Domain: $(-\infty, \infty)$; range: $[-4, \infty)$

27. $f(x) = (x+3)^2 - 5$

The vertex is $(-3,-5)$.
Shift the basic parabola 3 units to the left and 5 units down.
y-intercept: Let $x = 0$.

$f(0) = (0+3)^2 - 5 = 9 - 5 = 4$
$(0,4)$

x-intercepts: Let $f(x) = 0$.

$$(x+3)^2 - 5 = 0$$
$$(x+3)^2 = 5$$
$$\sqrt{(x+3)^2} = \pm\sqrt{5}$$
$$x + 3 = \pm\sqrt{5}$$
$$x = -3 \pm \sqrt{5}$$
$$x = -3 - \sqrt{5} \approx -5.2 \text{ or } x = -3 + \sqrt{5} \approx -0.8$$
$$(-5.2, 0), (-0.8, 0)$$

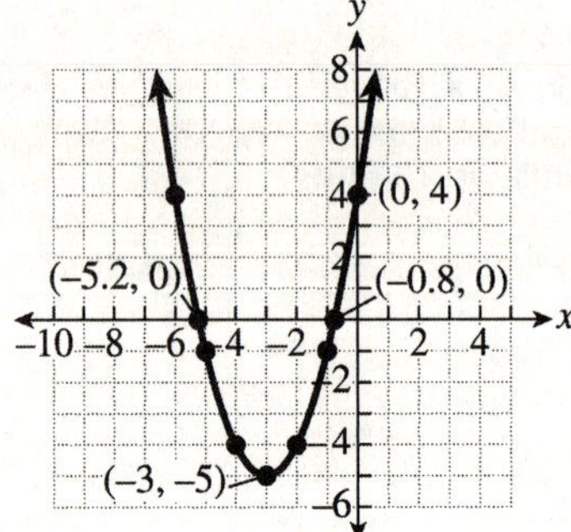

Domain: $(-\infty, \infty)$; range: $[-5, \infty)$

29. $f(x) = (x+1)^2 - 8$

The vertex is $(-1, -8)$.
Shift the basic parabola 1 unit to the left and 8 units down.

y-intercept: Let $x = 0$.

$f(0) = (0+1)^2 - 8 = 1 - 8 = -7$
$(0, -7)$

x-intercepts: Let $f(x) = 0$.

$$(x+1)^2 - 8 = 0$$
$$(x+1)^2 = 8$$
$$\sqrt{(x+1)^2} = \pm\sqrt{8}$$
$$x + 1 = \pm 2\sqrt{2}$$
$$x = -1 \pm 2\sqrt{2}$$

$x = -1 - 2\sqrt{2} \approx -3.8$ or $x = -1 + 2\sqrt{2} \approx 1.8$
$(-3.8, 0), (1.8, 0)$

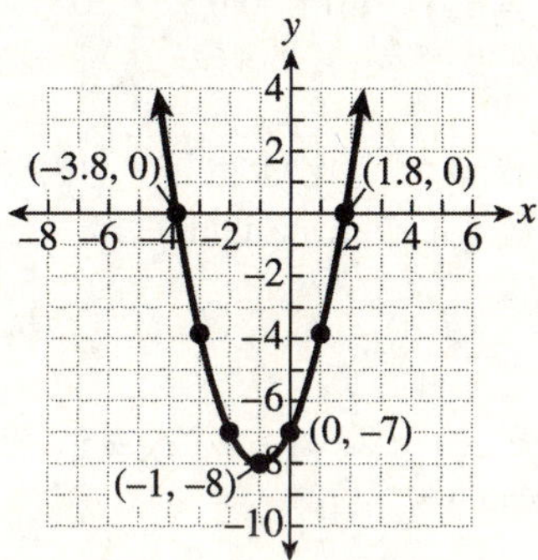

Domain: $(-\infty, \infty)$; range: $[-8, \infty)$

31. $f(x) = (x-2)^2 + 4$

The vertex is $(2, 4)$.
Shift the basic parabola 2 units to the right and 4 units up.

y-intercept: Let $x = 0$.

$f(0) = (0-2)^2 + 4 = 4 + 4 = 8$
$(0, 8)$

x-intercepts:

The vertex is above the x-axis and the parabola opens upward. There are no x-intercepts.

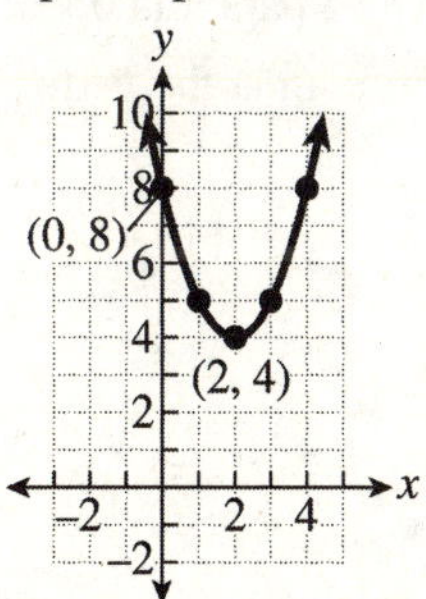

Domain: $(-\infty, \infty)$; range: $[4, \infty)$

33. $f(x) = -(x+4)^2 + 9$
The vertex is $(-4, 9)$.

Because $a < 0$, the parabola opens downward.
Rotate $f(x) = x^2$ about the x-axis.
Shift the basic parabola 4 units to the left and 9 units up.

y-intercept: Let $x = 0$.

$f(0) = -(0+4)^2 + 9 = -16 + 9 = -7$
$(0, -7)$

x-intercepts: Let $f(x) = 0$.

$$0 = -(x+4)^2 + 9$$
$$(x+4)^2 = 9$$
$$\sqrt{(x+4)^2} = \pm\sqrt{9}$$
$$x + 4 = \pm 3$$
$$x = -4 \pm 3$$

$x = -4 - 3 = -7$ or $x = -4 + 3 = -1$
$(-7, 0), (-1, 0)$

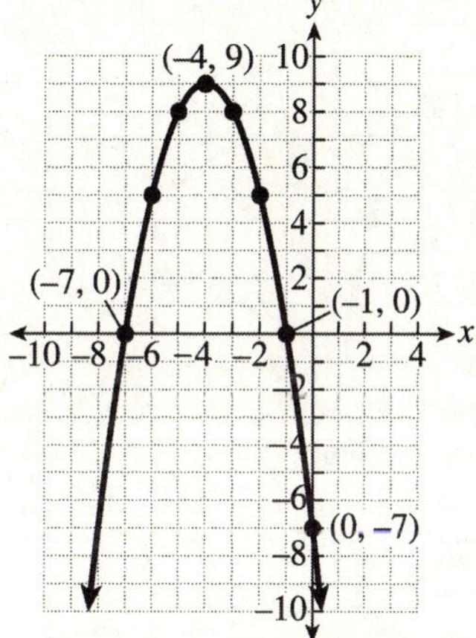

Domain: $(-\infty, \infty)$; range: $(-\infty, 9]$

35. $f(x) = -(x-3)^2 + 7$
The vertex is $(3, 7)$.

Because $a < 0$, the parabola opens downward.
Rotate $f(x) = x^2$ about the x-axis.
Shift the basic parabola 3 units to the right and 7 units up.

y-intercept: Let $x = 0$.

$f(0) = -(0-3)^2 + 7 = -9 + 7 = -2$
$(0, -2)$

x-intercepts: Let $f(x) = 0$.

$$0 = -(x-3)^2 + 7$$
$$(x-3)^2 = 7$$
$$\sqrt{(x-3)^2} = \pm\sqrt{7}$$
$$x - 3 = \pm\sqrt{7}$$
$$x = 3 \pm \sqrt{7}$$
$$x = 3 - \sqrt{7} \approx 0.4 \text{ or } x = 3 + \sqrt{7} \approx 5.6$$
$$(0.4, 0), (5.6, 0)$$

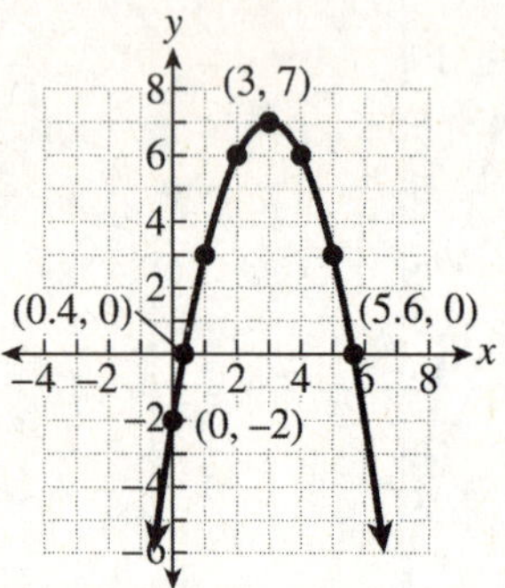

Domain: $(-\infty, \infty)$; range: $(-\infty, 7]$

37. $f(x) = -(x+2)^2 - 3$
The vertex is $(-2, -3)$.

Because $a < 0$, the parabola opens downward.
Rotate $f(x) = x^2$ about the x-axis.
Shift the basic parabola 2 units to the left and 3 units down.

y-intercept: Let $x = 0$.

$f(0) = -(0+2)^2 - 3 = -4 - 3 = -7$
$(0, -7)$

x-intercepts:
The vertex is below the x-axis and the parabola opens downward. There are no x-intercepts.

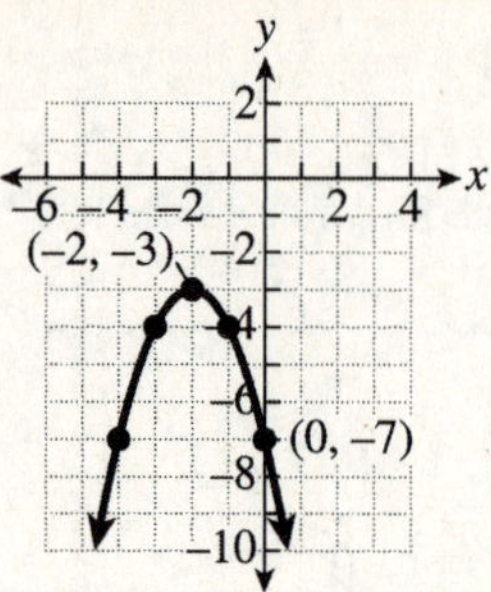

Domain: $(-\infty, \infty)$; range: $(-\infty, -3]$

39. $f(x) = x^2 - 8x + 20$ is a parabola opening upward, so it has a minimum at the vertex.
vertex:

$$x = -\frac{b}{2a} = -\frac{(-8)}{2(1)} = 4$$
$$f(4) = (4)^2 - 8(4) + 20 = 16 - 32 + 20 = 4$$
Minimum value is 4.

41. $f(x) = -x^2 + 4x - 30$ is a parabola opening downward, so it has a maximum at the vertex.
vertex:

$$x = -\frac{b}{2a} = -\frac{4}{2(-1)} = 2$$
$$f(2) = -(2)^2 + 4(2) - 30 = -4 + 8 - 30 = -26$$
Maximum value is -26.

43. $f(x) = (x-7)^2 + 18$ is a parabola opening upward, so it has a minimum at the vertex.
vertex: $(7, 18)$

Minimum value is 18.

45. $f(x) = -(x+16)^2 - 33$ is a parabola opening downward, so it has a maximum at the vertex.
vertex: $(-16, -33)$

Maximum value is -33.

47. $f(x) = 7(x-4)^2 + 46$ is a parabola opening upward, so it has a minimum at the vertex.
vertex: $(4, 46)$

Minimum value is 46.

49. $f(x) = 3x^2 - 20x + 121$ is a parabola opening upward, so it has a minimum at the vertex.

vertex:

$$x = -\frac{b}{2a} = -\frac{(-20)}{2(3)} = \frac{10}{3}$$

$$f\left(\frac{10}{3}\right) = 3\left(\frac{10}{3}\right)^2 - 20\left(\frac{10}{3}\right) + 121$$

$$= 3\left(\frac{100}{9}\right) - 20\left(\frac{10}{3}\right) + \frac{121}{1}$$

$$= \frac{100}{3} - \frac{200}{3} + \frac{363}{3}$$

$$= \frac{263}{3}$$

Minimum value is $\dfrac{263}{3}$.

51. $h(t) = -16t^2 + 32t + 5$ is a parabola opening downward, so it has a maximum at the vertex.

vertex:

$$t = -\frac{b}{2a} = -\frac{32}{2(-16)} = 1$$

$$h(1) = -16(1)^2 + 32(1) + 5 = -16 + 32 + 5 = 21$$

Maximum height of the rock is 21 feet.

53. $h(t) = -4.9t^2 + 2.1t + 1.5$ is a parabola opening downward, so it has a maximum at the vertex.

vertex:

$$t = -\frac{b}{2a} = -\frac{2.1}{2(-4.9)} = \frac{2.1}{9.8} \approx 0.214$$

$$h(0.214) = -4.9(0.214)^2 + 2.1(0.214) + 1.5$$

$$\approx -0.224 + 0.449 + 1.5$$

$$= 1.725$$

Maximum height of the coin is 1.725 meters.

55. Unknowns:
#1: x
#2: $24 - x$

$$f(x) = x(24 - x)$$

$$= 24x - x^2$$

$$= -x^2 + 24x$$

Parabola opens downward, so it has a maximum at the vertex.

vertex:

$$x = -\frac{b}{2a} = -\frac{24}{2(-1)} = 12$$

$$f(12) = -(12)^2 + 24(12)$$

$$= -144 + 288$$

$$= 144$$

The maximum product is 144.

57. Unknowns:
length: x

$$w = \frac{P}{2} - l = \frac{300}{2} - x = 150 - x$$

width: $150 - x$

$$A(x) = x(150 - x)$$

$$= 150x - x^2$$

$$= -x^2 + 150x$$

Parabola opens downward, so it has a maximum at the vertex.

vertex:

$$x = -\frac{b}{2a} = -\frac{150}{2(-1)} = 75$$

length: $x = 75$ ft
width: $150 - x = 150 - 75 = 75$ ft

$$A(75) = -(75)^2 + 150(75)$$

$$= -5625 + 11,250$$

$$= 5625 \text{ sq ft}$$

75 feet by 75 feet, area is 5625 square feet.

59. $f(x) = 0.000625x^2 - 0.0875x + 6.5$

$$x = -\frac{b}{2a} = -\frac{(-0.0875)}{2(0.000625)} = 70 \text{ (thousand)}$$

$$f(70) = 0.000625(70)^2 - 0.0875(70) + 6.5$$

$$= 3.0625 - 6.125 + 6.5$$

$$= 3.4375$$

70,000 pans, \$3.4375/pan

61. $r(x) = -250(x - 3)^2 + 37,750$
vertex: $(3, 37{,}750)$
\$3 increase, \$37,750 revenue

63. New vertex: $(6, 0)$
$$f(x) = (x - 6)^2 + 0$$
$$f(x) = (x - 6)^2$$

65. New vertex: $(-3, -11)$
$$f(x) = \left[x - (-3)\right]^2 - 11$$
$$f(x) = (x + 3)^2 - 11$$

67. New vertex: $(7, 13)$
$$f(x) = -(x - 7)^2 + 13$$

69.
$$\frac{f(x+h)-f(x)}{h} = \frac{(x+h)^2 - 4 - (x^2 - 4)}{h}$$
$$= \frac{x^2 + 2xh + h^2 - 4 - x^2 + 4}{h}$$
$$= \frac{2xh + h^2}{h}$$
$$= \frac{h(2x+h)}{h}$$
$$= 2x + h$$

71.
$$\frac{f(x+h)-f(x)}{h} = \frac{(x+h)^2 - 9(x+h) + 17 - (x^2 - 9x + 17)}{h}$$
$$= \frac{x^2 + 2xh + h^2 - 9x - 9h + 17 - x^2 + 9x - 17}{h}$$
$$= \frac{2xh + h^2 - 9h}{h}$$
$$= \frac{h(2x+h-9)}{h}$$
$$= 2x + h - 9$$

73.
$$\frac{f(x+h)-f(x)}{h} = \frac{3(x+h)^2 - 11(x+h) + 49 - (3x^2 - 11x + 49)}{h}$$
$$= \frac{3(x^2 + 2xh + h^2) - 11(x+h) + 49 - (3x^2 - 11x + 49)}{h}$$
$$= \frac{3x^2 + 6xh + 3h^2 - 11x - 11h + 49 - 3x^2 + 11x - 49}{h}$$
$$= \frac{6xh + 3h^2 - 11h}{h}$$
$$= \frac{h(6x + 3h - 11)}{h}$$
$$= 6x + 3h - 11$$

75. Answers will vary. Example:
Determine whether the parabola opens upward or downward. Find the vertex. If the vertex lies above the x-axis and the parabola opens upward, then there are no x-intercepts. If the vertex lies below the x-axis and the parabola opens downward, then there are no x-intercepts.

77. Answers will vary.

11.4 QUICK CHECK

1. $f(x) = |x+1| + 5$
$h = -1, k = 5$
Shift the graph of $f(x) = |x|$ by 1 unit to the left and by 5 units up.

y-intercept: Let $x = 0$.

$f(0) = |0+1| + 5 = 1 + 5 = 6$
$(0, 6)$

x-intercepts:

From the shifted graph, there are no x-intercepts.

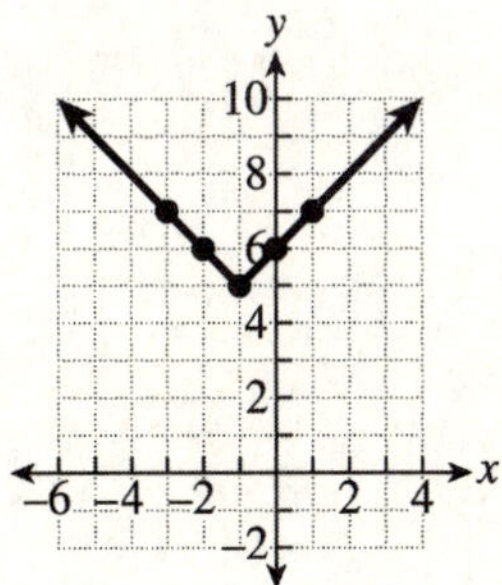

Domain: $(-\infty, \infty)$; range: $[5, \infty)$

2. $f(x) = |x-4| - 2$
$h = 4, k = -2$
Shift the graph of $f(x) = |x|$ by 4 units to the right and by 2 units down.

y-intercept: Let $x = 0$.

$f(0) = |0-4| - 2 = 4 - 2 = 2$
$(0, 2)$

x-intercepts: Let $f(x) = 0$.

$|x-4| - 2 = 0$
$\quad |x-4| = 2$

$x - 4 = -2 \quad \text{or} \quad x - 4 = 2$
$\quad\quad x = 2 \quad \text{or} \quad\quad x = 6$
$(2, 0), (6, 0)$

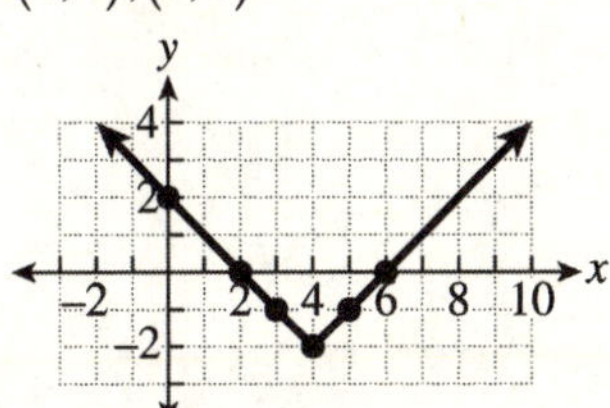

Domain: $(-\infty, \infty)$; range: $[-2, \infty)$

3. $f(x) = -|x-2| - 3$
$h = 2, k = -3$
After rotating the graph of $f(x) = |x|$ about the x-axis, shift the rotated graph by 2 units to the right and by 3 units down.

y-intercept: Let $x = 0$.

$f(0) = -|0-2| - 3 = -2 - 3 = -5$
$(0, -5)$

x-intercepts:

From the shifted graph, there are no x-intercepts.

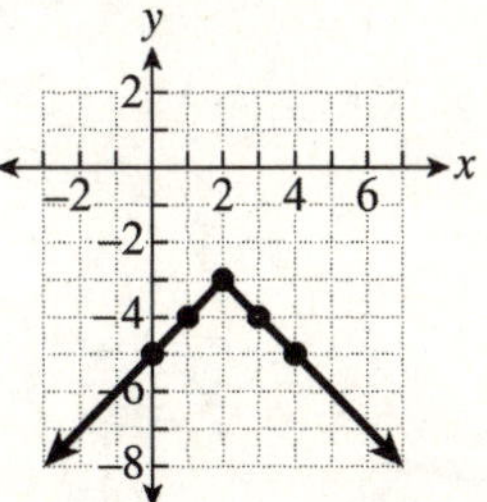

Domain: $(-\infty, \infty)$; range: $(-\infty, -3]$

4. $f(x) = \sqrt{x-2} + 6$
$h = 2, k = 6$
Shift the graph of $f(x) = \sqrt{x}$ by 2 units to the right and by 6 units up.

The graph begins at $(2, 6)$ and moves only up and to the right from there.

y-intercept:
None
x-intercepts:
None

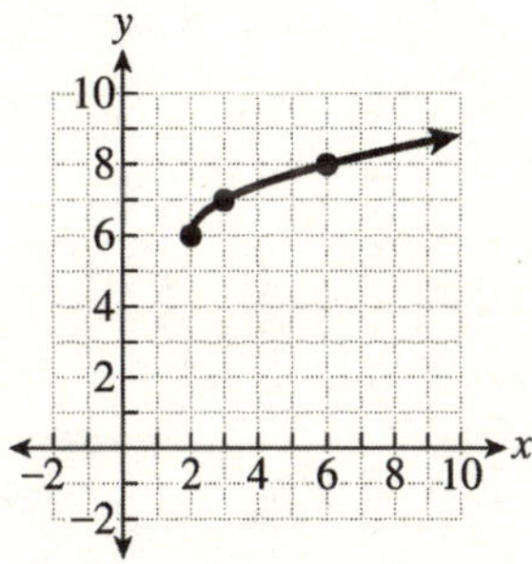

Domain: $[2, \infty)$; range: $[6, \infty)$

5. $f(x) = \sqrt{x+1} - 2$
$h = -1, k = -2$
Shift the graph of $f(x) = \sqrt{x}$ by 1 unit to the left and by 2 units down.

y-intercept: Let $x = 0$.
$f(0) = \sqrt{0+1} - 2 = 1 - 2 = -1$
$(0, -1)$

x-intercept: Let $f(x) = 0$.
$$\sqrt{x+1} - 2 = 0$$
$$\sqrt{x+1} = 2$$
$$\left(\sqrt{x+1}\right)^2 = 2^2$$
$$x + 1 = 4$$
$$x = 3$$
$(3, 0)$

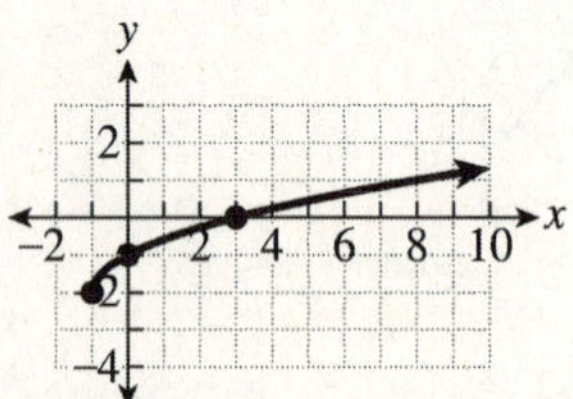

Domain: $[-1, \infty)$; range: $[-2, \infty)$

6. $f(x) = -\sqrt{x+1} - 5$
$h = -1, k = -5$
After rotating the graph of $f(x) = \sqrt{x}$ about the x-axis, shift the rotated graph by 1 unit to the left and by 5 units down.

y-intercept: Let $x = 0$.
$f(0) = -\sqrt{0+1} - 5 = -1 - 5 = -6$
$(0, -6)$

x-intercept:
From the shifted graph, there are no x-intercepts.

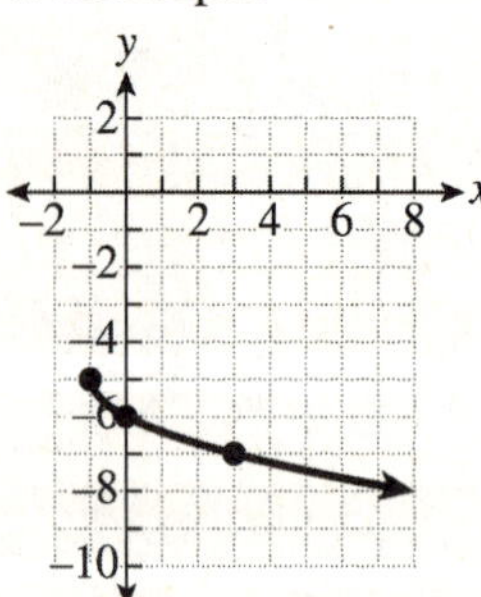

Domain: $[-1, \infty)$; range: $(-\infty, -5]$

7. $f(x) = \sqrt{-(x+4)} - 2$
$h = -4, k = -2$
After rotating the graph of $f(x) = \sqrt{x}$ about the y-axis, shift the rotated graph by 4 units to the left and by 2 units down.

y-intercept:
From the shifted graph, there is no y-intercept.

x-intercept: Let $f(x) = 0$.
$$\sqrt{-(x+4)} - 2 = 0$$
$$\sqrt{-(x+4)} = 2$$
$$\left(\sqrt{-(x+4)}\right)^2 = 2^2$$
$$-(x+4) = 4$$
$$-x - 4 = 4$$
$$-x = 8$$
$$x = -8$$
$(-8, 0)$

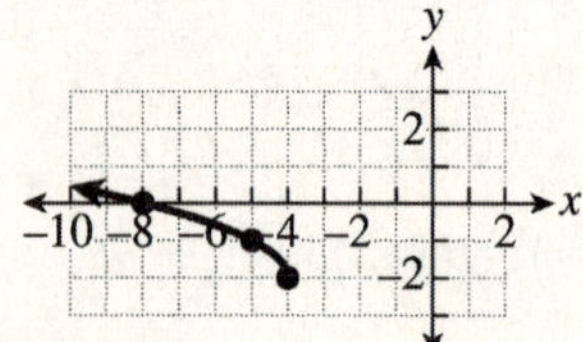

Domain: $(-\infty, -4]$; range: $[-2, \infty)$

8. $f(x) = (x-1)^3 + 1$
$h = 1, k = 1$
Shift the graph of $f(x) = x^3$ by 1 unit to the right and by 1 unit up

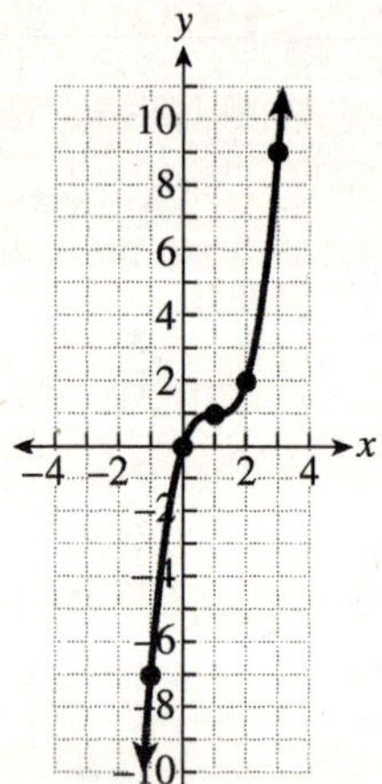

Domain: $(-\infty, \infty)$; range: $(-\infty, \infty)$

9. $f(x) = -(x+3)^3$
$h = -3, k = 0$

After rotating the graph of $f(x) = x^3$ about the x-axis, shift the rotated graph by 3 units to the left.

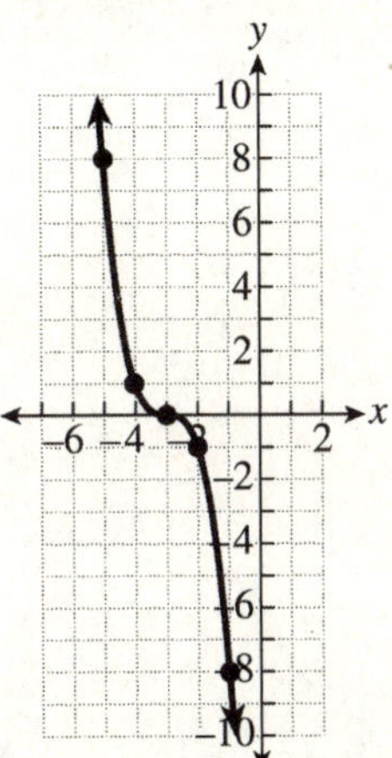

Domain: $(-\infty, \infty)$; range: $(-\infty, \infty)$

10. a) $f(x)$ is a function of the form $f(x) = \sqrt{x}$ that has been shifted by 2 units to the right and by 5 units up, so $h = 2$ and $k = 5$.
The function is $f(x) = \sqrt{x-2} + 5$.

b) $f(x)$ is function of the form $f(x) = |x|$ that has been shifted by 4 units to the left and by 2 units down, so $h = -4$ and $k = -2$.
The function is $f(x) = |x+4| - 2$.

11.4 OTHER FUNCTIONS AND THEIR GRAPHS

1. radicand

3. $f(x) = |x-2| + 6$
$h = 2, k = 6$
Shift the graph of $f(x) = |x|$ by 2 units to the right and by 6 units up.

y-intercept: Let $x = 0$.
$f(0) = |0-2| + 6 = 2 + 6 = 8$
$(0, 8)$

x-intercepts:
From the shifted graph, there are no x-intercepts.

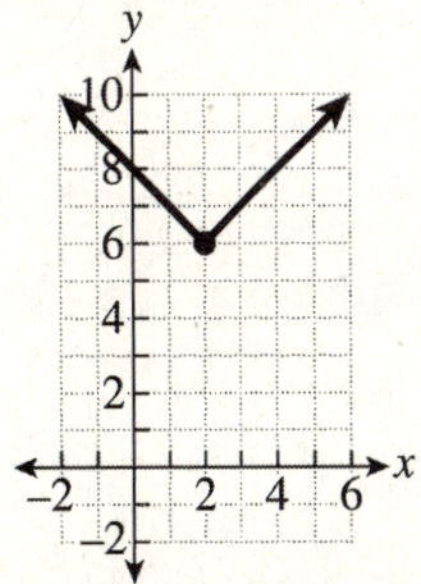

Domain: $(-\infty, \infty)$; range: $[6, \infty)$

5. $f(x) = |x+1| - 5$
$h = -1, k = -5$
Shift the graph of $f(x) = |x|$ by 1 unit to the left and by 5 units down.

y-intercept: Let $x = 0$.
$f(0) = |0+1| - 5 = 1 - 5 = -4$
$(0, -4)$

x-intercepts: Let $f(x) = 0$.

$|x+1| - 5 = 0$
$|x+1| = 5$

$x + 1 = -5$ or $x + 1 = 5$
$x = -6$ or $x = 4$
$(-6, 0), (4, 0)$

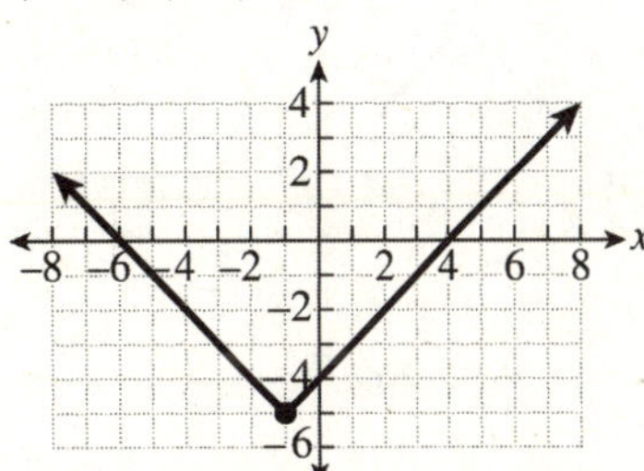

Domain: $(-\infty, \infty)$; range: $[-5, \infty)$

7. $f(x) = -|x+6| + 3$
$h = -6, k = 3$
After rotating the graph of $f(x) = |x|$ about the x-axis, shift the rotated graph by 6 units to the left and by 3 units up.

y-intercept: Let $x = 0$.
$f(0) = -|0+6| + 3 = -6 + 3 = -3$
$(0, -3)$

x-intercepts: Let $f(x) = 0$.
$$-|x+6| + 3 = 0$$
$$|x+6| = 3$$
$$x+6 = -3 \quad \text{or} \quad x+6 = 3$$
$$x = -9 \quad \text{or} \quad x = -3$$
$(-9, 0), (-3, 0)$

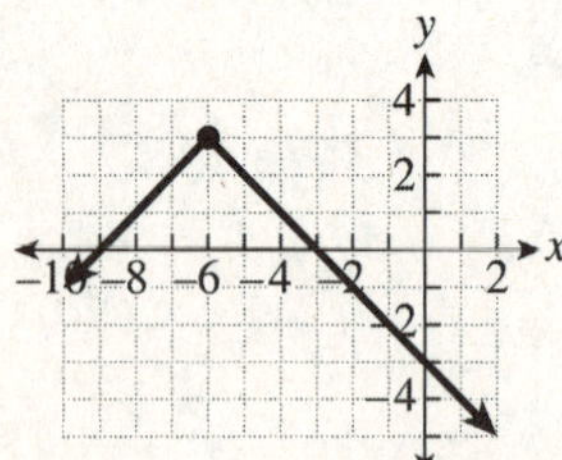

Domain: $(-\infty, \infty)$; range: $(-\infty, 3]$

9. $f(x) = -|x-7| - 2$
$h = 7, k = -2$
After rotating the graph of $f(x) = |x|$ about the x-axis, shift the rotated graph by 7 units to the right and by 2 units down.

y-intercept: Let $x = 0$.
$f(0) = -|0-7| - 2 = -7 - 2 = -9$
$(0, -9)$

x-intercepts:
From the shifted graph, there are no x-intercepts.

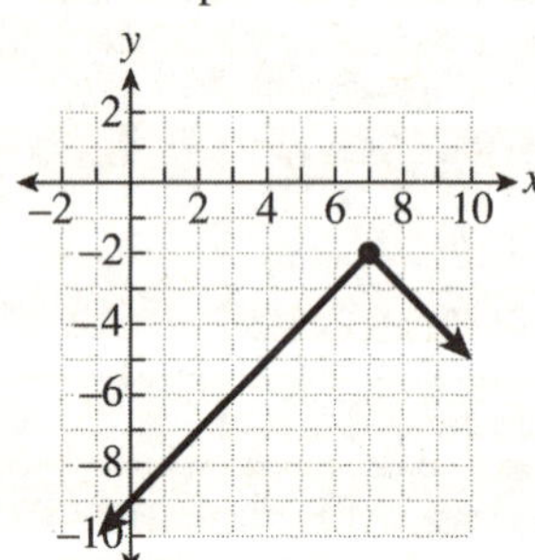

Domain: $(-\infty, \infty)$; range: $(-\infty, -2]$

11. $f(x) = \sqrt{x+2}$
$h = -2, k = 0$
Shift the graph of $f(x) = \sqrt{x}$ by 2 units to the left.

y-intercept: Let $x = 0$.
$f(0) = \sqrt{0+2} = \sqrt{2} \approx 1.4$
$(0, 1.4)$

x-intercept: Let $f(x) = 0$.
$$\sqrt{x+2} = 0$$
$$\left(\sqrt{x+2}\right)^2 = 0^2$$
$$x+2 = 0$$
$$x = -2$$
$(-2, 0)$

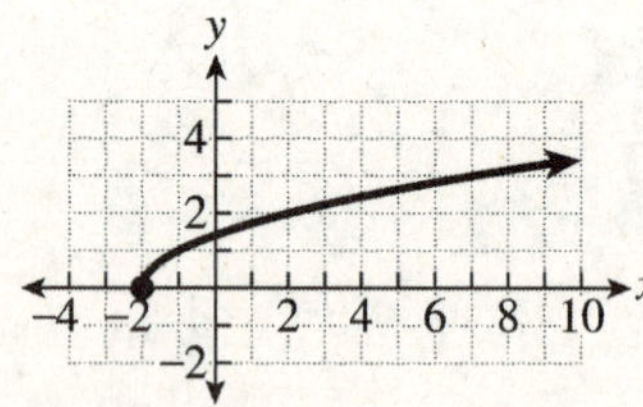

Domain: $[-2, \infty)$; range: $[0, \infty)$

13. $f(x) = \sqrt{x+4} + 6$
$h = -4, k = 6$
Shift the graph of $f(x) = \sqrt{x}$ by 4 units to the left and by 6 units up.

y-intercept: Let $x = 0$.
$f(0) = \sqrt{0+4} + 6 = 2 + 6 = 8$
$(0, 8)$

x-intercept:
From the shifted graph, there are no x-intercepts.

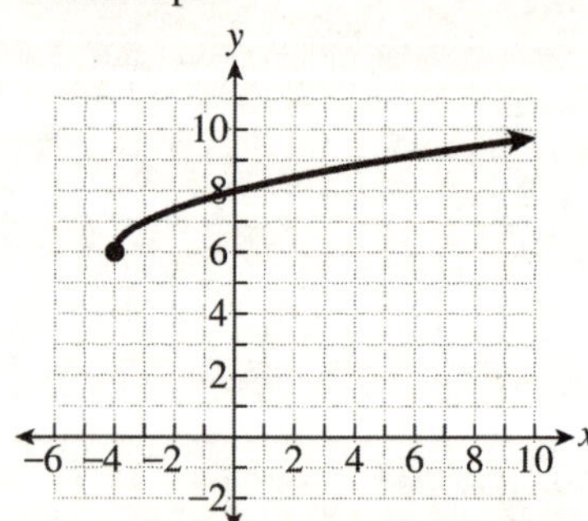

Domain: $[-4, \infty)$; range: $[6, \infty)$

15. $f(x) = \sqrt{x-5} - 2$

$h = 5, k = -2$

Shift the graph of $f(x) = \sqrt{x}$ by 5 units to the right and by 2 units down.

y-intercept:

From the shifted graph, there is no y-intercept.

x-intercept: Let $f(x) = 0$.

$$\sqrt{x-5} - 2 = 0$$
$$\sqrt{x-5} = 2$$
$$\left(\sqrt{x-5}\right)^2 = 2^2$$
$$x - 5 = 4$$
$$x = 9$$

$(9, 0)$

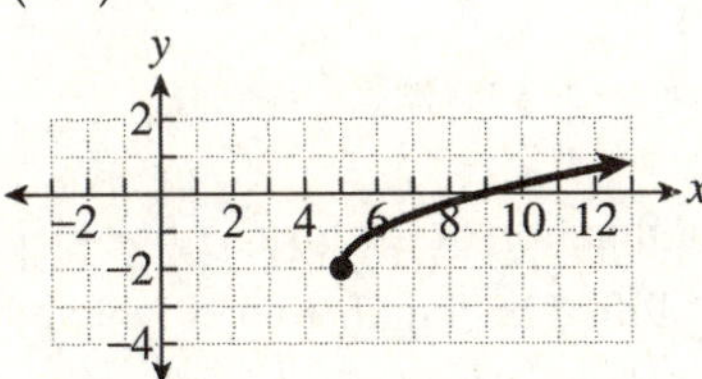

Domain: $[5, \infty)$; range: $[-2, \infty)$

17. $f(x) = -\sqrt{x-2} + 3$

$h = 2, k = 3$

After rotating the graph of $f(x) = \sqrt{x}$ about the x-axis, shift the rotated graph by 2 units to the right and by 3 units up.

y-intercept:

From the shifted graph, there is no y-intercept.

x-intercept: Let $f(x) = 0$.

$$-\sqrt{x-2} + 3 = 0$$
$$\sqrt{x-2} = 3$$
$$\left(\sqrt{x-2}\right)^2 = 3^2$$
$$x - 2 = 9$$
$$x = 11$$

$(11, 0)$

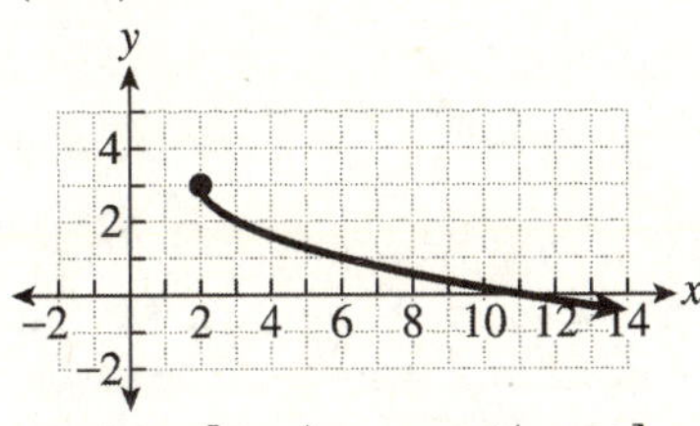

Domain: $[2, \infty)$; range: $(-\infty, 3]$

19. $f(x) = \sqrt{-(x-3)} + 1$

$h = 3, k = 1$

After rotating the graph of $f(x) = \sqrt{x}$ about the y-axis, shift the rotated graph by 3 units to the right and by 1 unit up.

y-intercept: Let $x = 0$.

$$f(0) = \sqrt{-(0-3)} + 1 = \sqrt{3} + 1 \approx 2.7$$
$$(0, 2.7)$$

x-intercept:

From the shifted graph, there is no x-intercept.

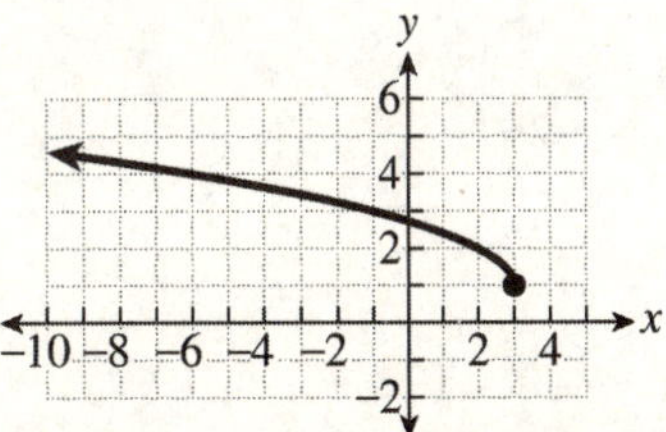

Domain: $(-\infty, 3]$; range: $[1, \infty)$

21. $f(x) = x^3 - 1$

$h = 0, k = -1$

Shift the graph of $f(x) = x^3$ by 1 unit down.

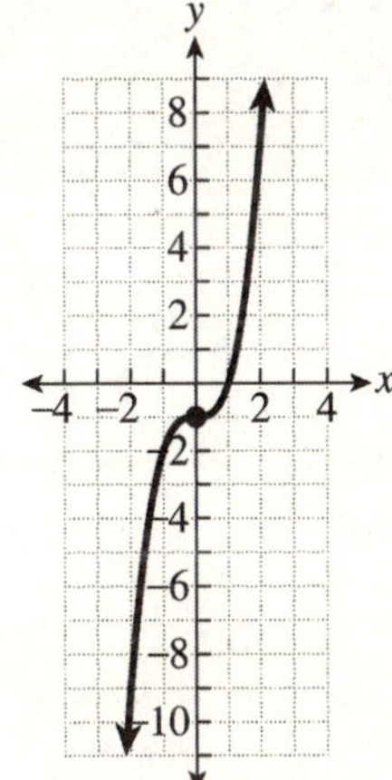

Domain: $(-\infty, \infty)$; range: $(-\infty, \infty)$

23. $f(x) = (x+4)^3 + 1$

$h = -4, k = 1$

Shift the graph of $f(x) = x^3$ by 4 units to the left and by 1 unit up.

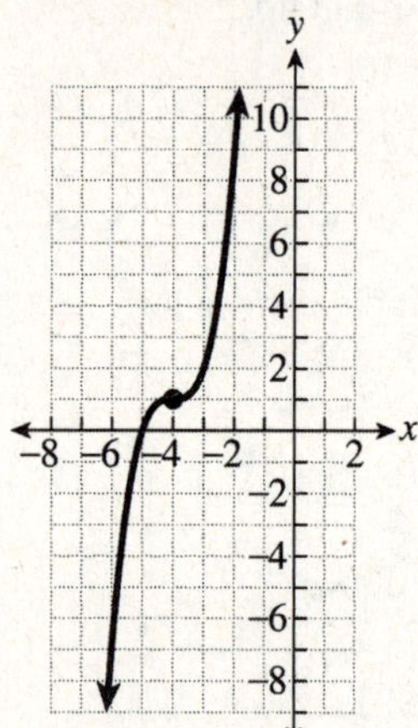

Domain: $(-\infty, \infty)$; range: $(-\infty, \infty)$

25. $f(x) = (x+2)^3 + 7$

$h = -2, k = 7$

Shift the graph of $f(x) = x^3$ by 2 units to the left and by 7 units up.

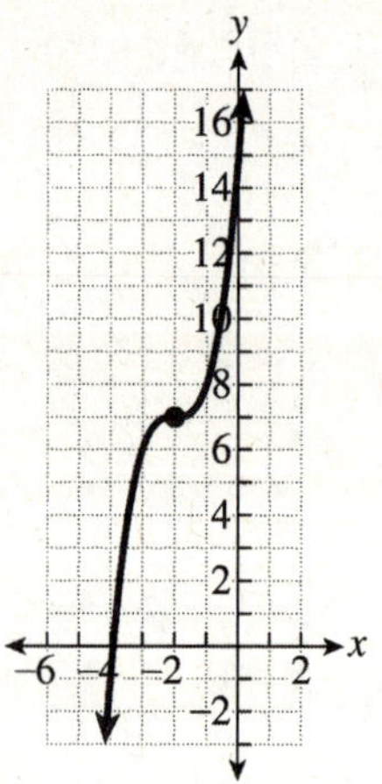

Domain: $(-\infty, \infty)$; range: $(-\infty, \infty)$

27. $f(x) = -(x-1)^3 + 4$

$h = 1, k = 4$

After rotating the graph of $f(x) = x^3$ about the x-axis, shift the rotated graph by 1 unit to the right and by 4 units up.

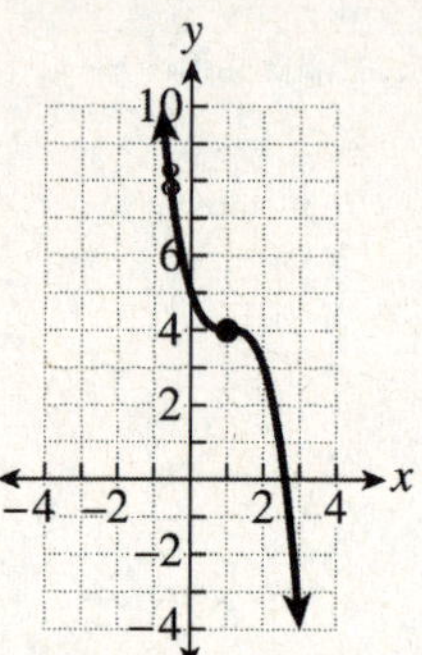

Domain: $(-\infty, \infty)$; range: $(-\infty, \infty)$

29. $f(x)$ is a function of the form $f(x) = |x|$ that has been shifted by 4 units to the right and by 5 units down, so $h = 4$ and $k = -5$.

The function is $f(x) = |x-4| - 5$.

31. $f(x)$ is a function of the form $f(x) = \sqrt{x}$ that has been shifted by 4 units to the right and by 3 units up, so $h = 4$ and $k = 3$.

The function is $f(x) = \sqrt{x-4} + 3$.

33. $f(x)$ is a function of the form $f(x) = \sqrt{x}$ that has been rotated about the y-axis and shifted by 3 units to the right and by 1 unit down, so $h = 3$ and $k = -1$.

The function is $f(x) = \sqrt{-(x-3)} - 1$.

35. $f(x)$ is a function of the form $f(x) = x^3$ that has been shifted by 5 units up, so $h = 0$ and $k = 5$.

The function is $f(x) = x^3 + 5$.

37. Answers will vary. Example:
The graph of a quadratic function is a parabola opening upward or downward. The graph of a square-root function is half of a parabola opening to the left or right.

11.5 QUICK CHECK

1. a)
$$(f+g)(x) = f(x) + g(x)$$
$$= \left(x^2 - 9x + 20\right) + (3x - 44)$$
$$= x^2 - 6x - 24$$

b)
$$(f-g)(x) = f(x) - g(x)$$
$$= \left(x^2 - 9x + 20\right) - (3x - 44)$$
$$= x^2 - 9x + 20 - 3x + 44$$
$$= x^2 - 12x + 64$$

c)
$$(f+g)(7) = (7)^2 - 6(7) - 24$$
$$= 49 - 42 - 24$$
$$= 49 - 66$$
$$= -17$$

d)
$$(f-g)(-2) = (-2)^2 - 12(-2) + 64$$
$$= 4 + 24 + 64$$
$$= 92$$

2.
$$(f \cdot g)(x) = f(x) \cdot g(x)$$
$$= (3x + 4)(4x - 7)$$
$$= 12x^2 - 21x + 16x - 28$$
$$= 12x^2 - 5x - 28$$

3.
$$\left(\frac{f}{g}\right)(x) = \frac{f(x)}{g(x)}$$
$$= \frac{x^2 + 3x - 18}{x^2 + 13x + 42}$$
$$= \frac{(x+6)(x-3)}{(x+6)(x+7)}$$
$$= \frac{x-3}{x+7}, \quad x \neq -6, -7$$

4. a)
$$(f+g)(x) = f(x) + g(x)$$
$$= (111,916x + 9,528,000)$$
$$\quad + (43,973x + 2,591,000)$$
$$= 155,889x + 12,119,000$$

Total number of students enrolled in a public or private college x years after 1990.

b)
$$(f-g)(x) = f(x) - g(x)$$
$$= (111,916x + 9,528,000)$$
$$\quad - (43,973x + 2,591,000)$$
$$= 111,916x + 9,528,000$$
$$\quad - 43,973x - 2,591,000$$
$$= 67,943x + 6,937,000$$

This function tells us how many more students are enrolled at public colleges than at private colleges x years after 1990.

5.
$$(f \circ g)(x) = f\big(g(x)\big)$$
$$= f(2x + 7)$$
$$= 4(2x + 7) - 9$$
$$= 8x + 28 - 9$$
$$= 8x + 19$$

6. a)
$$(f \circ g)(x) = f\big(g(x)\big)$$
$$= f\left(x^2 - 3x - 28\right)$$
$$= \left(x^2 - 3x - 28\right) + 3$$
$$= x^2 - 3x - 25$$

b)
$$(g \circ f)(x) = g\big(f(x)\big)$$
$$= g(x + 3)$$
$$= (x+3)^2 - 3(x+3) - 28$$
$$= (x+3)(x+3) - 3(x+3) - 28$$
$$= x^2 + 3x + 3x + 9 - 3x - 9 - 28$$
$$= x^2 + 3x - 28$$

7.
$$(f \circ g)(x) = f\big(g(x)\big)$$
$$= f(x - 4)$$
$$= \frac{2(x-4) + 5}{(x-4) + 7}$$
$$= \frac{2x - 8 + 5}{x - 4 + 7}$$
$$= \frac{2x - 3}{x + 3}$$

Domain: All real numbers except $x = -3$.

11.5 THE ALGEBRA OF FUNCTIONS

1. sum

3. $(f \cdot g)(x)$

5. composition

7. a)
$$(f+g)(x) = f(x) + g(x)$$
$$= (4x + 11) + (3x - 17)$$
$$= 7x - 6$$

b) $(f+g)(8) = 7(8) - 6 = 56 - 6 = 50$

c) $(f+g)(-3) = 7(-3) - 6 = -21 - 6 = -27$

9. a)
$$(f+g)(x) = f(x)+g(x)$$
$$= (5x+8)+\left(x^2-x-19\right)$$
$$= x^2+4x-11$$

b)
$$(f+g)(8) = (8)^2+4(8)-11$$
$$= 64+32-11$$
$$= 85$$

c)
$$(f+g)(-3) = (-3)^2+4(-3)-11$$
$$= 9-12-11$$
$$= -14$$

11. a)
$$(f-g)(x) = f(x)-g(x)$$
$$= (6x-11)-(-2x-5)$$
$$= 6x-11+2x+5$$
$$= 8x-6$$

b) $(f-g)(6) = 8(6)-6 = 48-6 = 42$

c) $(f-g)(-5) = 8(-5)-6 = -40-6 = -46$

13. a)
$$(f-g)(x) = f(x)-g(x)$$
$$= (10x+37)-\left(x^2-5x-12\right)$$
$$= 10x+37-x^2+5x+12$$
$$= -x^2+15x+49$$

b)
$$(f-g)(6) = -(6)^2+15(6)+49$$
$$= -36+90+49$$
$$= 103$$

c)
$$(f-g)(-5) = -(-5)^2+15(-5)+49$$
$$= -25-75+49$$
$$= -51$$

15. a)
$$(f\cdot g)(x) = f(x)\cdot g(x)$$
$$= (2x-3)(x+4)$$
$$= 2x^2+8x-3x-12$$
$$= 2x^2+5x-12$$

b)
$$(f\cdot g)(5) = 2(5)^2+5(5)-12$$
$$= 2(25)+25-12$$
$$= 50+13$$
$$= 63$$

c)
$$(f\cdot g)(-3) = 2(-3)^2+5(-3)-12$$
$$= 2(9)-15-12$$
$$= 18-27$$
$$= -9$$

17. a)
$$(f\cdot g)(x) = f(x)\cdot g(x)$$
$$= (x-6)\left(x^2+3x-4\right)$$
$$= x^3+3x^2-4x-6x^2-18x+24$$
$$= x^3-3x^2-22x+24$$

b)
$$(f\cdot g)(5) = (5)^3-3(5)^2-22(5)+24$$
$$= 125-3(25)-110+24$$
$$= 125-75-86$$
$$= -36$$

c)
$$(f\cdot g)(-3) = (-3)^3-3(-3)^2-22(-3)+24$$
$$= -27-3(9)+66+24$$
$$= -27-27+90$$
$$= 36$$

19. a)
$$\left(\frac{f}{g}\right)(x) = \frac{f(x)}{g(x)} = \frac{3x+1}{x+7}$$

b)
$$\left(\frac{f}{g}\right)(7) = \frac{3(7)+1}{(7)+7} = \frac{22}{14} = \frac{11}{7}$$

c)
$$\left(\frac{f}{g}\right)(-2) = \frac{3(-2)+1}{(-2)+7} = \frac{-5}{5} = -1$$

21. a)
$$\left(\frac{f}{g}\right)(x) = \frac{f(x)}{g(x)}$$
$$= \frac{x^2-9}{x^2+7x+12}$$
$$= \frac{(x+3)(x-3)}{(x+3)(x+4)}$$
$$= \frac{\cancel{(x+3)}^{1}(x-3)}{\cancel{(x+3)}_{1}(x+4)}$$
$$= \frac{x-3}{x+4}$$

b)
$$\left(\frac{f}{g}\right)(7) = \frac{(7)-3}{(7)+4} = \frac{4}{11}$$

c)
$$\left(\frac{f}{g}\right)(-2) = \frac{(-2)-3}{(-2)+4} = \frac{-5}{2} = -\frac{5}{2}$$

23.
$$(f+g)(x) = f(x)+g(x)$$
$$= 2x+9+5x-1$$
$$= 7x+8$$
$$(f+g)(8) = 7(8)+8 = 56+8 = 64$$

25. $(f-g)(x) = f(x) - g(x)$
$= 2x + 9 - (5x - 1)$
$= 2x + 9 - 5x + 1$
$= -3x + 10$
$(f-g)(-4) = -3(-4) + 10 = 12 + 10 = 22$

27. $(f \cdot g)(x) = f(x) \cdot g(x)$
$= (2x + 9)(5x - 1)$
$= 10x^2 - 2x + 45x - 9$
$= 10x^2 + 43x - 9$

$(f \cdot g)(10) = 10(10)^2 + 43(10) - 9$
$= 10(100) + 430 - 9$
$= 1000 + 421$
$= 1421$

29. $\left(\dfrac{f}{g}\right)(x) = \dfrac{f(x)}{g(x)} = \dfrac{2x + 9}{5x - 1}$

$\left(\dfrac{f}{g}\right)(-2) = \dfrac{2(-2) + 9}{5(-2) - 1} = \dfrac{-4 + 9}{-10 - 1} = \dfrac{5}{-11} = -\dfrac{5}{11}$

31. $(f+g)(x) = f(x) + g(x)$
$= (5x - 9) + (2x - 17)$
$= 7x - 26$

33. $(f \cdot g)(x) = f(x) \cdot g(x)$
$= (5x - 9)(2x - 17)$
$= 10x^2 - 85x - 18x + 153$
$= 10x^2 - 103x + 153$

35. $(f-g)(x) = f(x) - g(x)$
$= (x - 10) - (x^2 - 8x - 20)$
$= x - 10 - x^2 + 8x + 20$
$= -x^2 + 9x + 10$

37. $\left(\dfrac{f}{g}\right)(x) = \dfrac{f(x)}{g(x)}$

$= \dfrac{x - 10}{x^2 - 8x - 20}$

$= \dfrac{x - 10}{(x - 10)(x + 2)}$

$= \dfrac{\overset{1}{\cancel{x - 10}}}{\underset{1}{\cancel{(x - 10)}}(x + 2)}$

$= \dfrac{1}{x + 2}$

39. a) $(f+g)(x) = f(x) + g(x)$
$= (10.4x + 398) + (9.1x + 49)$
$= 19.5x + 447$

$(f+g)(x)$ is the total number of doctors, in thousands, in the U.S. x years after 1980.

b) $(f+g)(40) = 19.5(40) + 447$
$= 780 + 447$
$= 1227$

There will be 1,227,000 doctors in the U.S. in the year 2020.

41. a) $(f-g)(x) = f(x) - g(x)$
$= (9.2x + 181.8) - (3.4x + 160.9)$
$= 9.2x + 181.8 - 3.4x - 160.9$
$= 5.8x + 20.9$

$(f-g)(x)$ tells how many more females than males, in thousands, will earn a master's degree in the U.S. x years after 1990.

b) $(f-g)(30) = 5.8(30) + 20.9$
$= 174 + 20.9$
$= 194.9$

There will be 194,900 more master's degrees earned by females than by males in the U.S. in the year 2020.

43. $(f \circ g)(5) = f(g(5))$
$= f(5 - 3)$
$= f(2)$
$= 4(2) + 7$
$= 8 + 7$
$= 15$

45. $(g \circ f)(-1) = g(f(-1))$
$= g(4(-1) + 7)$
$= g(-4 + 7)$
$= g(3)$
$= 3 - 3$
$= 0$

47. a) $(f \circ g)(x) = f(g(x))$
$= f(x^2 + 5x - 6)$
$= 3(x^2 + 5x - 6) + 2$
$= 3x^2 + 15x - 18 + 2$
$= 3x^2 + 15x - 16$

b) $(f \circ g)(5) = 3(5)^2 + 15(5) - 16$
$= 3(25) + 75 - 16$
$= 75 + 59$
$= 134$

49. a)
$$(f \circ g)(x) = f(g(x))$$
$$= f(x+8)$$
$$= (x+8)^2 - 7(x+8) + 12$$
$$= x^2 + 16x + 64 - 7x - 56 + 12$$
$$= x^2 + 9x + 20$$

b)
$$(f \circ g)(5) = (5)^2 + 9(5) + 20$$
$$= 25 + 45 + 20$$
$$= 90$$

51. a)
$$(f \circ g)(x) = f(g(x))$$
$$= f(2x+4)$$
$$= 3(2x+4) + 5$$
$$= 6x + 12 + 5$$
$$= 6x + 17$$

b)
$$(g \circ f)(x) = g(f(x))$$
$$= g(3x+5)$$
$$= 2(3x+5) + 4$$
$$= 6x + 10 + 4$$
$$= 6x + 14$$

53. a)
$$(f \circ g)(x) = f(g(x))$$
$$= f(x^2 + 3x - 40)$$
$$= (x^2 + 3x - 40) + 4$$
$$= x^2 + 3x - 36$$

b)
$$(g \circ f)(x) = g(f(x))$$
$$= g(x+4)$$
$$= (x+4)^2 + 3(x+4) - 40$$
$$= (x+4)(x+4) + 3(x+4) - 40$$
$$= x^2 + 4x + 4x + 16 + 3x + 12 - 40$$
$$= x^2 + 11x - 12$$

55.
$$(f \circ g)(x) = f(g(x))$$
$$= f(2x+13)$$
$$= \frac{5(2x+13)}{(2x+13) + 7}$$
$$= \frac{10x + 65}{2x + 20}$$
Domain: All real numbers except $x = -10$.

57.
$$(f \circ g)(x) = f(g(x))$$
$$= f(x+11)$$
$$= \sqrt{2(x+11) - 10}$$
$$= \sqrt{2x + 22 - 10}$$
$$= \sqrt{2x + 12}$$
Find Domain:
$$2x + 12 \geq 0$$
$$2x \geq -12$$
$$x \geq -6$$
Domain: $[-6, \infty)$

59.
$$(f \circ f)(x) = f(f(x))$$
$$= f(3x - 10)$$
$$= 3(3x - 10) - 10$$
$$= 9x - 30 - 10$$
$$= 9x - 40$$

61.
$$(f \circ f)(x) = f(f(x))$$
$$= f(x^2 + 6x + 12)$$
$$= (x^2 + 6x + 12)^2 + 6(x^2 + 6x + 12) + 12$$
$$= (x^2 + 6x + 12)(x^2 + 6x + 12) + 6(x^2 + 6x + 12) + 12$$
$$= x^4 + 6x^3 + 12x^2 + 6x^3 + 36x^2 + 72x + 12x^2 + 72x + 144 + 6x^2 + 36x + 72 + 12$$
$$= x^4 + 12x^3 + 66x^2 + 180x + 228$$

63.
$$\begin{aligned}
(f \circ g)(x) &= f\big(g(x)\big) \\
&= f(x+5) \\
&= (x+5)^2 + 4(x+5) + 3 \\
&= (x+5)(x+5) + 4(x+5) + 3 \\
&= x^2 + 10x + 25 + 4x + 20 + 3 \\
&= x^2 + 14x + 48
\end{aligned}$$
$$\begin{aligned}
(g \circ f)(x) &= g\big(f(x)\big) \\
&= g\big(x^2 + 4x + 3\big) \\
&= \big(x^2 + 4x + 3\big) + 5 \\
&= x^2 + 4x + 8
\end{aligned}$$
$$\begin{aligned}
(f \circ g)(x) &= (g \circ f)(x) \\
x^2 + 14x + 48 &= x^2 + 4x + 8 \\
14x + 48 &= 4x + 8 \\
10x + 48 &= 8 \\
10x &= -40 \\
x &= -4
\end{aligned}$$

65.
$$\begin{aligned}
(f \circ g)(x) &= f\big(g(x)\big) \\
&= f\big(x^2 - 25\big) \\
&= \big(x^2 - 25\big) + 7 \\
&= x^2 - 18
\end{aligned}$$
$$\begin{aligned}
(g \circ f)(x) &= g\big(f(x)\big) \\
&= g(x+7) \\
&= (x+7)^2 - 25 \\
&= x^2 + 14x + 49 - 25 \\
&= x^2 + 14x + 24
\end{aligned}$$
$$\begin{aligned}
(f \circ g)(x) &= (g \circ f)(x) \\
x^2 - 18 &= x^2 + 14x + 24 \\
-18 &= 14x + 24 \\
-42 &= 14x \\
-3 &= x
\end{aligned}$$

67. If $f(x) = x - 8$,
$$\begin{aligned}
(f \circ g)(x) &= f\big(g(x)\big) \\
&= f(x+8) \\
&= (x+8) - 8 \\
&= x
\end{aligned}$$
so $f(x) = x - 8$.

69. If $f(x) = \dfrac{x-8}{3}$,
$$\begin{aligned}
(f \circ g)(x) &= f\big(g(x)\big) \\
&= f(3x+8) \\
&= \frac{(3x+8) - 8}{3} \\
&= \frac{3x}{3} \\
&= x
\end{aligned}$$
so $f(x) = \dfrac{x-8}{3}$.

71. Yes, explanations will vary. Example: $(f+g)(x) = f(x) + g(x)$ and $(g+f)(x) = g(x) + f(x)$. By the commutative property of addition $(f+g)(x) = (g+f)(x)$.

11.6 QUICK CHECK

1. No, not one-to-one, since the y-coordinate
-7 corresponds to both $x = -2$ and $x = 2$.

2. No, not one-to-one, since it does not pass the
horizontal line test.

3.

$(f \circ g)(x) = x$	$(g \circ f)(x) = x$
$(f \circ g)(x)$	$(g \circ f)(x)$
$= f(g(x))$	$= g(f(x))$
$= f(4x+7)$	$= g\left(\dfrac{x-7}{4}\right)$
$= \dfrac{(4x+7)-7}{4}$	$= 4\left(\dfrac{x-7}{4}\right)+7$
$= \dfrac{4x}{4}$	$= x-7+7$
$= x$	$= x$

Since $(f \circ g)(x) = x$ and $(g \circ f)(x) = x$, these
two functions are inverses.

4.

$(f \circ g)(x) = x$	$(g \circ f)(x) = x$
$(f \circ g)(x)$	$(g \circ f)(x)$
$= f(g(x))$	$= g(f(x))$
$= f(3x-12)$	$= g\left(\dfrac{1}{3}x+4\right)$
$= \dfrac{1}{3}(3x-12)+4$	$= 3\left(\dfrac{1}{3}x+4\right)-12$
$= x-4+4$	$= x+12-12$
$= x$	$= x$

Since $(f \circ g)(x) = x$ and $(g \circ f)(x) = x$, these
two functions are inverses.

5. $f(x) = \dfrac{1}{2}x + 10$

$$y = \dfrac{1}{2}x + 10$$

Interchange x and y.

$$x = \dfrac{1}{2}y + 10$$
$$x - 10 = \dfrac{1}{2}y$$
$$2(x-10) = 2 \cdot \dfrac{1}{2}y$$
$$2x - 20 = y$$
$$f^{-1}(x) = 2x - 20$$

6. $f(x) = \dfrac{3+5x}{x}$

$$y = \dfrac{3+5x}{x}$$

Interchange x and y.

$$x = \dfrac{3+5y}{y}$$
$$y \cdot x = y \cdot \dfrac{3+5y}{y}$$
$$yx = 3 + 5y$$
$$yx - 5y = 3$$
$$y(x-5) = 3$$
$$y = \dfrac{3}{x-5}$$
$$f^{-1}(x) = \dfrac{3}{x-5}, \quad x \neq 5$$

7. $f(x) = \sqrt{x-9} + 8$

$$y = \sqrt{x-9} + 8$$

Interchange x and y.

$$x = \sqrt{y-9} + 8$$
$$x - 8 = \sqrt{y-9}$$
$$(x-8)^2 = \left(\sqrt{y-9}\right)^2$$
$$x^2 - 16x + 64 = y - 9$$
$$x^2 - 16x + 73 = y$$
$$f^{-1}(x) = x^2 - 16x + 73$$
Domain: $[8, \infty)$

8. Coordinates of $f(x)$:

$(3, -10), (4, -6), (7, -4)$

Coordinates of $f^{-1}(x)$:

$(-10, 3), (-6, 4), (-4, 7)$

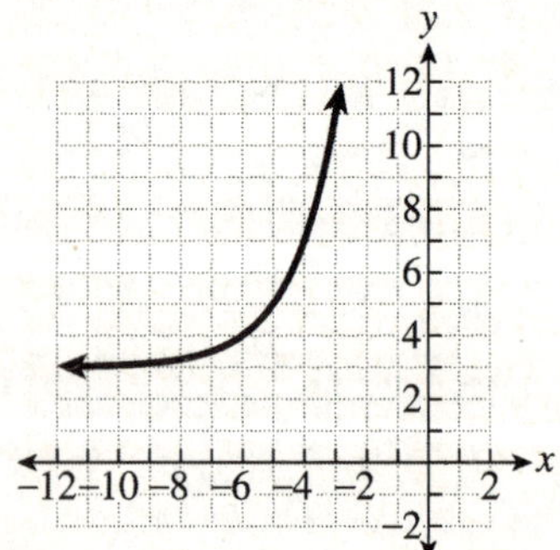

11.6 INVERSE FUNCTIONS

1. one-to-one

3. inverse

5. $f^{-1}(x)$

7. Yes.

9. No, since the coordinate $f(x) = 4$ corresponds to both $x = -2$ and $x = 2$ and $f(x) = 16$ corresponds to both $x = -4$ and $x = 4$.

11. No, since the y-coordinate -3 corresponds to both $x = -9$ and $x = 10$.

13. No, since the y-coordinate 3 corresponds to both $x = -5$ and $x = 4$.

15. No. Since a mother may have more than one child this function will not be one-to-one.

17. Yes, since there is only one governor per state.

19. No.

21. Yes.

23. No.

25. $(f \circ g)(x) = f(g(x))$
$\qquad = f(5 + x)$
$\qquad = 5 + x + 5$
$\qquad = x + 10$
$(f \circ g)(x) \ne x$
$f(x)$ and $g(x)$ are not inverses.

27.

$(f \circ g)(x) = x$	$(g \circ f)(x) = x$
$(f \circ g)(x) = f(g(x))$	$(g \circ f)(x) = g(f(x))$
$\quad = f\left(\dfrac{x}{4}\right)$	$= g(4x)$
$\quad = 4\left(\dfrac{x}{4}\right)$	$= \dfrac{4x}{4}$
$\quad = x$	$= x$

$f(x)$ and $g(x)$ are inverses.

29.

$(f \circ g)(x) = x$	$(g \circ f)(x) = x$
$(f \circ g)(x) = f(g(x))$	$(g \circ f)(x) = g(f(x))$
$\quad = f\left(\dfrac{x+4}{3}\right)$	$= g(3x - 4)$
$\quad = 3\left(\dfrac{x+4}{3}\right) - 4$	$= \dfrac{(3x-4)+4}{3}$
$\quad = x + 4 - 4$	$= \dfrac{3x}{3}$
$\quad = x$	$= x$

$f(x)$ and $g(x)$ are inverses.

31. $(f \circ g)(x) = f(g(x))$
$\qquad = f(x - 13)$
$\qquad = -(x - 13) + 13$
$\qquad = -x + 13 + 13$
$\qquad = -x + 26$
$(f \circ g)(x) \ne x$
$f(x)$ and $g(x)$ are not inverses.

33. $f(x) = x + 9$
$\quad y = x + 9$
Interchange x and y.
$\quad x = y + 9$
$\quad x - 9 = y$
$\quad f^{-1}(x) = x - 9$

35. $f(x) = \dfrac{x}{3}$
$\quad y = \dfrac{x}{3}$
Interchange x and y.
$\quad x = \dfrac{y}{3}$
$\quad 3x = y$
$\quad f^{-1}(x) = 3x$

37. $f(x) = 3x + 14$
$\quad y = 3x + 14$
Interchange x and y.
$\quad x = 3y + 14$
$\quad x - 14 = 3y$
$\quad \dfrac{x - 14}{3} = y$
$\quad f^{-1}(x) = \dfrac{x - 14}{3}$

39. $f(x) = 7x - 6$
$y = 7x - 6$
Interchange x and y.
$x = 7y - 6$
$x + 6 = 7y$
$\dfrac{x+6}{7} = y$
$f^{-1}(x) = \dfrac{x+6}{7}$

41. $f(x) = -\dfrac{3}{4}x + 6$
$y = -\dfrac{3}{4}x + 6$
Interchange x and y.
$x = -\dfrac{3}{4}y + 6$
$x - 6 = -\dfrac{3}{4}y$
$-\dfrac{4}{3}(x-6) = y$
$-\dfrac{4x-24}{3} = y$
$f^{-1}(x) = -\dfrac{4x-24}{3}$

43. $f(x) = mx$
$y = mx$
Interchange x and y.
$x = my$
$\dfrac{x}{m} = y$
$f^{-1}(x) = \dfrac{x}{m}$

45. $f(x) = \dfrac{1}{x} + 3$
$y = \dfrac{1}{x} + 3$
Interchange x and y.
$x = \dfrac{1}{y} + 3$
$y \cdot x = y \cdot \left(\dfrac{1}{y} + 3 \right)$
$xy = 1 + 3y$
$xy - 3y = 1$
$y(x-3) = 1$
$y = \dfrac{1}{x-3}$
$f^{-1}(x) = \dfrac{1}{x-3}$

47. $f(x) = \dfrac{1}{x+3}$
$y = \dfrac{1}{x+3}$
Interchange x and y.
$x = \dfrac{1}{y+3}$
$(y+3) \cdot x = (y+3) \cdot \left(\dfrac{1}{y+3} \right)$
$xy + 3x = 1$
$xy = 1 - 3x$
$y = \dfrac{1-3x}{x}$
$f^{-1}(x) = \dfrac{1-3x}{x}$

49. $f(x) = \dfrac{5-6x}{x}$
$y = \dfrac{5-6x}{x}$
Interchange x and y.
$x = \dfrac{5-6y}{y}$
$y \cdot x = y \cdot \left(\dfrac{5-6y}{y} \right)$
$xy = 5 - 6y$
$xy + 6y = 5$
$y(x+6) = 5$
$y = \dfrac{5}{x+6}$
$f^{-1}(x) = \dfrac{5}{x+6}$

51. $f(x) = \dfrac{8+4x}{3x}$
$y = \dfrac{8+4x}{3x}$
Interchange x and y.
$x = \dfrac{8+4y}{3y}$
$3y \cdot x = 3y \cdot \left(\dfrac{8+4y}{3y} \right)$
$3xy = 8 + 4y$
$3xy - 4y = 8$
$y(3x-4) = 8$
$y = \dfrac{8}{3x-4}$
$f^{-1}(x) = \dfrac{8}{3x-4}$

53. $f(x) = \sqrt{x}$
$$y = \sqrt{x}$$
Interchange x and y.
$$x = \sqrt{y}$$
$$x^2 = \left(\sqrt{y}\right)^2$$
$$x^2 = y$$
$$f^{-1}(x) = x^2$$
Domain of $f^{-1}(x)$: $x \geq 0$

55. $f(x) = \sqrt{x} - 7$
$$y = \sqrt{x} - 7$$
Interchange x and y.
$$x = \sqrt{y} - 7$$
$$x + 7 = \sqrt{y}$$
$$(x+7)^2 = \left(\sqrt{y}\right)^2$$
$$x^2 + 14x + 49 = y$$
$$f^{-1}(x) = x^2 + 14x + 49$$
Domain of $f^{-1}(x)$: $x \geq -7$

57. $f(x) = \sqrt{x-8} + 3$
$$y = \sqrt{x-8} + 3$$
Interchange x and y.
$$x = \sqrt{y-8} + 3$$
$$x - 3 = \sqrt{y-8}$$
$$(x-3)^2 = \left(\sqrt{y-8}\right)^2$$
$$x^2 - 6x + 9 = y - 8$$
$$x^2 - 6x + 17 = y$$
$$f^{-1}(x) = x^2 - 6x + 17$$
Domain of $f^{-1}(x)$: $x \geq 3$

59. $f(x) = x^2 - 9 \quad (x \geq 0)$
$$y = x^2 - 9$$
Interchange x and y.
$$x = y^2 - 9$$
$$x + 9 = y^2$$
$$\sqrt{x+9} = \sqrt{y^2}$$
$$\sqrt{x+9} = y$$
$$f^{-1}(x) = \sqrt{x+9}$$
Domain of $f^{-1}(x)$: $x \geq -9$

61. $f(x) = x^2 - 6x + 8 \quad (x \geq 3)$
$$y = x^2 - 6x + 8$$
$$y - 8 = x^2 - 6x$$
$$y - 8 + \left(-\frac{6}{2}\right)^2 = x^2 - 6x + \left(-\frac{6}{2}\right)^2$$
$$y - 8 + 9 = x^2 - 6x + 9$$
$$y + 1 = (x-3)^2$$
Interchange x and y.
$$x + 1 = (y-3)^2$$
$$\sqrt{x+1} = \sqrt{(y-3)^2}$$
$$\sqrt{x+1} = y - 3$$
$$\sqrt{x+1} + 3 = y$$
$$f^{-1}(x) = \sqrt{x+1} + 3$$
Domain of $f^{-1}(x)$: $x \geq -1$

63. $\{(-17,-5),(-9,-1),(-5,1),(1,4),(7,7)\}$

65. Not one-to-one since the y coordinate 13 corresponds to both $x = 0$ and $x = 2$.

67. Coordinates of $f(x)$:
$$(0,6),(2,3),(4,0),(6,-3)$$
Coordinates of $f^{-1}(x)$:
$$(6,0),\ (3,2),\ (0,4),(-3,6)$$

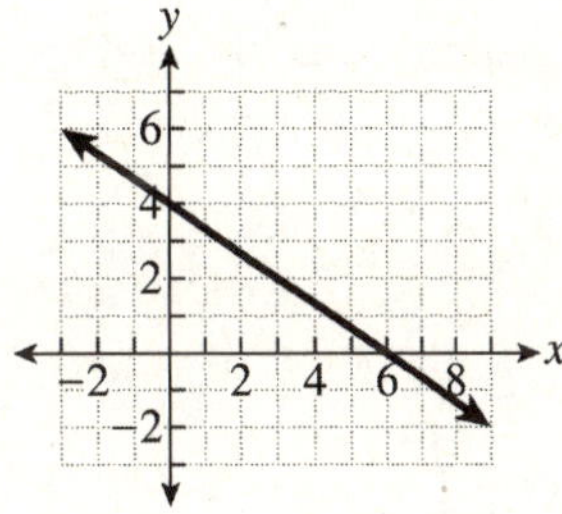

69. Coordinates of $f(x)$:

$(-6,-3),(-5,-2),(-2,-1),(3,0)$

Coordinates of $f^{-1}(x)$:

$(-3,-6),(-2,-5),(-1,-2),(0,3)$

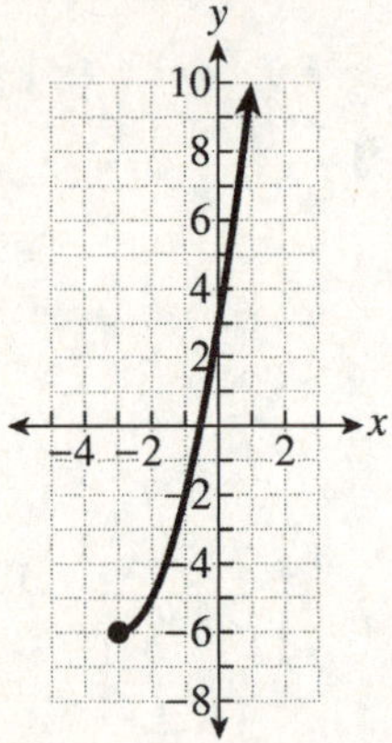

71. Points on the line $f(x)$: $(4,9),(-2,5)$

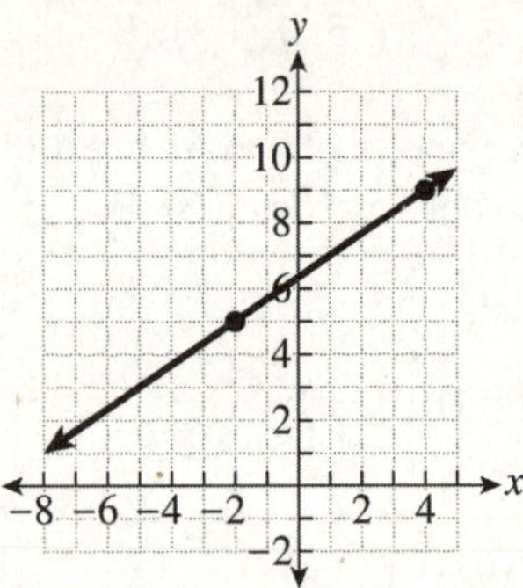

73. Quadratic functions are parabolic in shape but the domain is $[3,\infty)$ with the points:

$(4,0),(5,3),(6,8),$ and $(3,-1)$

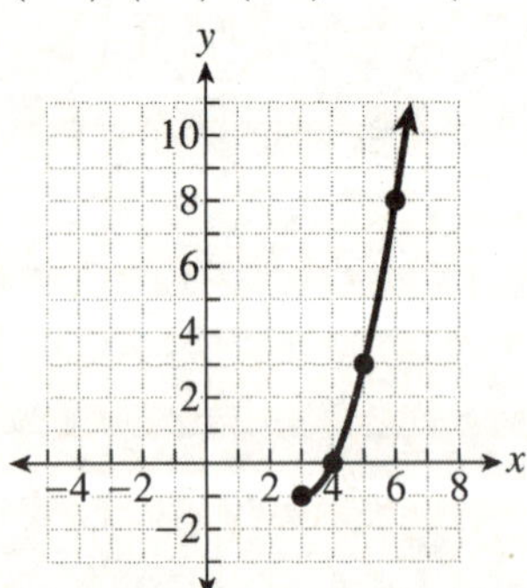

75. Answers will vary. Example:
The horizontal line test shows whether or not each y-coordinate corresponds to exactly one x-coordinate. Thus it shows whether or not the function is one-to-one.

77. Answers will vary.

CHAPTER 11 REVIEW

1. Yes.

2. No.

3. Yes.

4. No.

5. $f(-5)=2(-5)=-10$

6. $f(4)=3(4)-7=12-7=5$

7. $f(3b-8)=7(3b-8)-11$
$=21b-56-11$
$=21b-67$

8. $f(7)=(7)^2-7(7)+32=49-49+32=32$

9. $f(-3)=(-3)^2-9(-3)-36=9+27-36=0$

10. $f(2n+5)$
$=(2n+5)^2+10(2n+5)-25$
$=(2n+5)(2n+5)+10(2n+5)-25$
$=4n^2+10n+10n+25+20n+50-25$
$=4n^2+40n+50$

11. $f(-9)=\left|(-9)+2\right|-17$
$=\left|-7\right|-17$
$=7-17$
$=-10$

12. $f(10)=\sqrt{2(10)+16}+27$
$=\sqrt{20+16}+27$
$=\sqrt{36}+27$
$=6+27$
$=33$

13. Domain: $[-\infty,\infty)$; range: $[-7,\infty)$

14. Domain: $[1,\infty)$; range: $[5,\infty)$

15. $f(4)=-3$

16. $f(-1)=2$

17. When $f(x)=3, x=-4, x=2.$

18. When $f(x)=-5, x=-2,\ x=0.$

19. $f(x) = -3x - 6$

y-intercept: Let $x = 0$.
$f(0) = -3(0) - 6 = -6$
$(0, -6)$

x-intercept: Let $f(x) = 0$.
$-3x - 6 = 0$
$-3x = 6$
$x = -2$
$(-2, 0)$

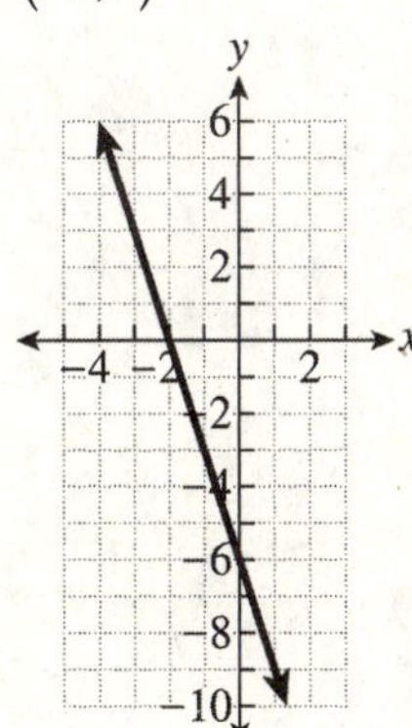

20. $f(x) = 2x + 4$

y-intercept: Let $x = 0$.
$f(0) = 2(0) + 4 = 4$
$(0, 4)$

x-intercept: Let $f(x) = 0$.
$2x + 4 = 0$
$2x = -4$
$x = -2$
$(-2, 0)$

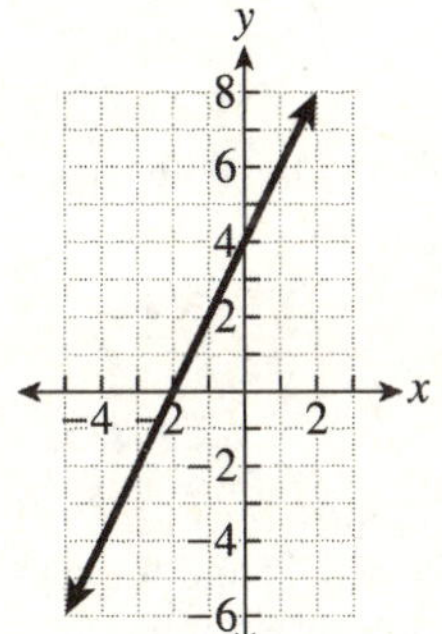

21. $f(x) = \dfrac{2}{5}x - 5$

y-intercept: Let $x = 0$.
$f(0) = \dfrac{2}{5}(0) - 5 = -5$
$(0, -5)$

x-intercept: Let $f(x) = 0$.
$\dfrac{2}{5}x - 5 = 0$

$\dfrac{2}{5}x = 5$

$\dfrac{5}{2} \cdot \dfrac{2}{5}x = \dfrac{5}{2} \cdot 5$

$x = \dfrac{25}{2}$

$\left(\dfrac{25}{2}, 0\right)$

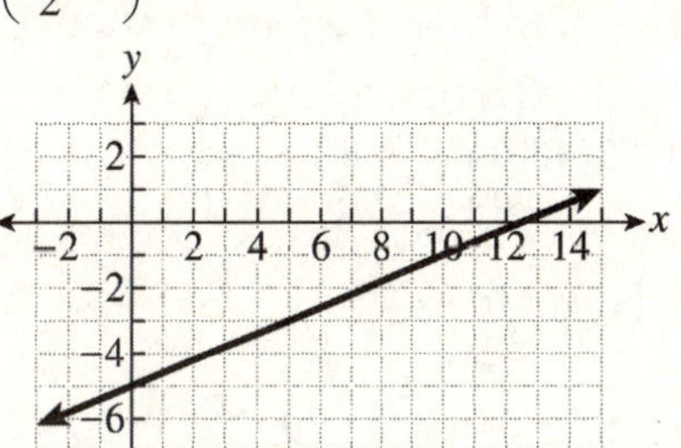

22. $f(x) = -\dfrac{3}{4}x + 6$

y-intercept: Let $x = 0$.
$f(0) = -\dfrac{3}{4}(0) + 6 = 6$
$(0, 6)$

x-intercept: Let $f(x) = 0$.
$-\dfrac{3}{4}x + 6 = 0$

$-\dfrac{3}{4}x = -6$

$-\dfrac{4}{3}\left(-\dfrac{3}{4}\right)x = \left(-\dfrac{4}{3}\right)(-6)$

$x = 8$

$(8, 0)$

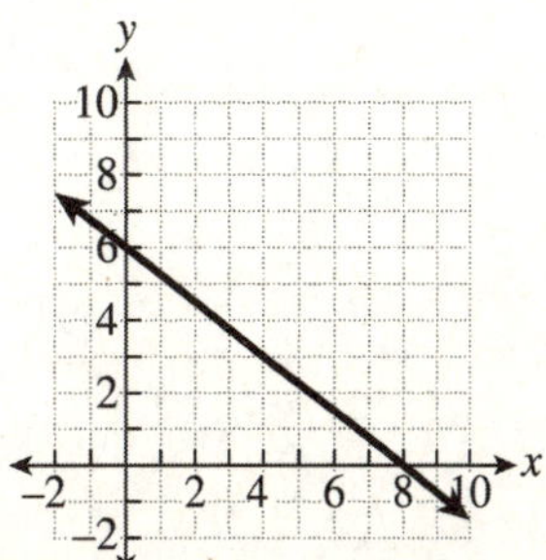

23. y-intercept: $(0,6)$

x-intercept: $(-3,0)$

$$m = \frac{6-0}{0-(-3)} = \frac{6}{3} = 2$$

$$f(x) = 2x + 6$$

24. y-intercept: $(0,0)$

additional point: $(4,-1)$

$$m = \frac{-1-0}{4-0} = -\frac{1}{4}$$

$$f(x) = -\frac{1}{4}x$$

25. a) Use the points $(250, 42.50)$ and $(400, 50)$.

$$m = \frac{50 - 42.50}{400 - 250} = \frac{7.50}{150} = 0.05$$

The function has the form $f(x) = 0.05x + b$.

$$f(400) = 50$$
$$0.05(400) + b = 50$$
$$20 + b = 50$$
$$b = 30$$
$$f(x) = 0.05x + 30$$

b) $f(75) = 0.05(75) + 30 = 3.75 + 30 = 33.75$

$33.75

26. a) $C(x) = 40 + 60 + 0.10x$

$C(x) = 0.10x + 100$

b) $R(x) = 1.50x$

c) $P(x) = R(x) - C(x)$
$$= 1.50x - (0.10x + 100)$$
$$= 1.50x - 0.10x - 100$$
$$= 1.40x - 100$$

d) $P(400) = 1.40(400) - 100$
$$= 560 - 100$$
$$= 460$$

$460

27. $f(x) = (x-2)^2 - 9$

The vertex is $(2,-9)$.

Shift the basic parabola 2 units to the right and 9 units down.

y-intercept: Let $x = 0$.

$$f(0) = (0-2)^2 - 9 = 4 - 9 = -5$$
$$(0,-5)$$

x-intercept: Let $f(x) = 0$.

$$(x-2)^2 - 9 = 0$$
$$(x-2)^2 = 9$$
$$\sqrt{(x-2)^2} = \pm\sqrt{9}$$
$$x - 2 = \pm 3$$
$$x = 2 \pm 3$$
$$x = 2 - 3 = -1 \text{ or } x = 2 + 3 = 5$$
$$(-1,0),(5,0)$$

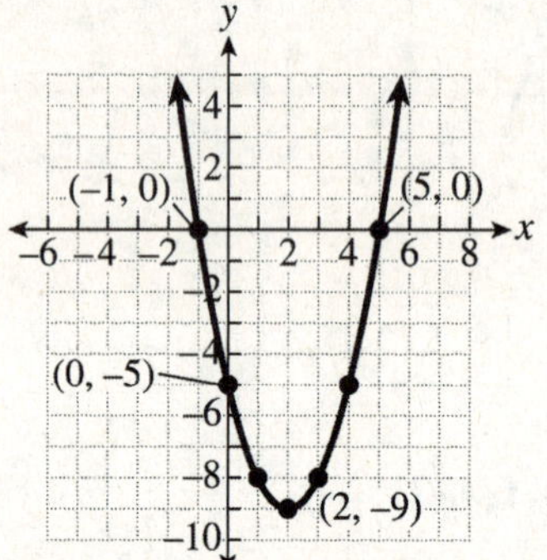

Domain: $(-\infty, \infty)$; range: $[-9, \infty)$

28. $f(x) = (x+1)^2 + 3$

The vertex is $(-1,3)$.

Shift the basic parabola 1 unit to the left and 3 units up.

y-intercept: Let $x = 0$.

$$f(0) = (0+1)^2 + 3 = 1 + 3 = 4$$
$$(0,4)$$

x-intercepts:

The vertex is above the x-axis and the parabola opens upward. There are no x-intercepts.

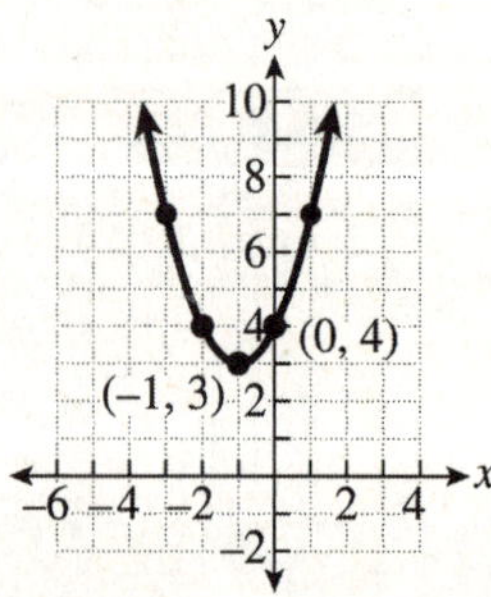

Domain: $(-\infty, \infty)$; range: $[3, \infty)$

29. $f(x) = (x-3)^2 - 7$

The vertex is $(3, -7)$.
Shift the basic parabola 3 units to the right and 7 units down.
y-intercept: Let $x = 0$.
$f(0) = (0-3)^2 - 7 = 9 - 7 = 2$
$(0, 2)$

x-intercept: Let $f(x) = 0$.
$(x-3)^2 - 7 = 0$
$(x-3)^2 = 7$
$\sqrt{(x-3)^2} = \pm\sqrt{7}$
$x - 3 = \pm\sqrt{7}$
$x = 3 \pm \sqrt{7}$
$x = 3 - \sqrt{7} \approx 0.4$ or $x = 3 + \sqrt{7} \approx 5.6$
$(0.4, 0), (5.6, 0)$

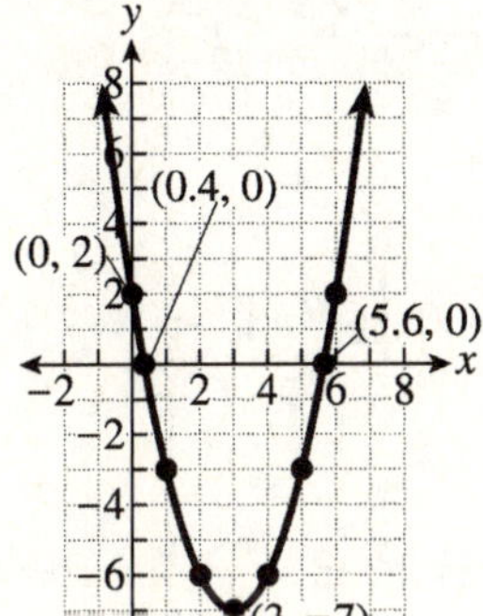

Domain: $(-\infty, \infty)$; range: $[-7, \infty)$

30. $f(x) = -(x+2)^2 + 4$

The vertex is $(-2, 4)$.
Because $a < 0$, the parabola opens downward.
Rotate $f(x) = x^2$ about the x-axis.
Shift the basic parabola 2 units to the left and 4 units up.
y-intercept: Let $x = 0$.
$f(0) = -(0+2)^2 + 4 = -4 + 4 = 0$
$(0, 0)$

x-intercept: Let $f(x) = 0$.
$-(x+2)^2 + 4 = 0$
$(x+2)^2 = 4$
$\sqrt{(x+2)^2} = \pm\sqrt{4}$
$x + 2 = \pm 2$
$x = -2 \pm 2$
$x = -2 - 2 = -4$ or $x = -2 + 2 = 0$
$(-4, 0), (0, 0)$

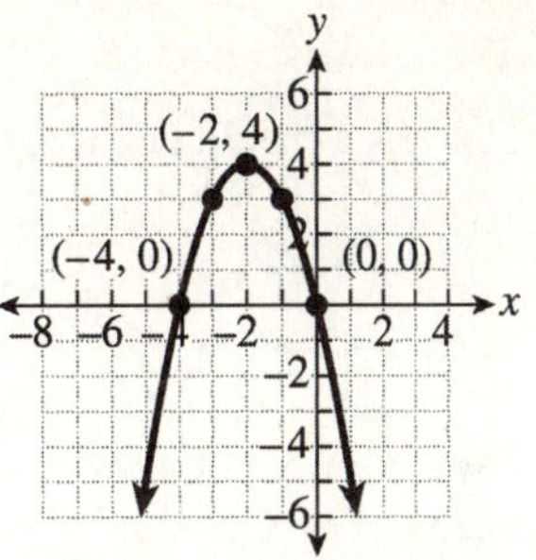

Domain: $(-\infty, \infty)$; range: $(-\infty, 4]$

31. $f(x) = x^2 - 12x + 60$ is a parabola opening up $(a > 0)$, so it has a minimum at the vertex.
vertex:
$x = -\dfrac{b}{2a} = -\dfrac{(-12)}{2(1)} = 6$

$f(6) = (6)^2 - 12(6) + 60 = 36 - 72 + 60 = 24$
Minimum: 24

32. $f(x) = -x^2 + 16x - 70$ is a parabola opening down $(a < 0)$, so it has a maximum at the vertex.
vertex:
$x = -\dfrac{b}{2a} = -\dfrac{16}{2(-1)} = 8$

$f(8) = -(8)^2 + 16(8) - 70$
$\quad\quad = -64 + 128 - 70$
$\quad\quad = -6$

Maximum: -6

33. $f(x) = (x+9)^2 - 25$ is a parabola opening up $(a > 0)$, so it has a minimum at the vertex.
vertex: $(-9, -25)$
Minimum: -25

34. $f(x) = -(x+8)^2 - 39$ is a parabola opening down $(a < 0)$, so it has a maximum at the vertex.
vertex: $(-8, -39)$
Maximum: -39

35. $h(t) = -16t^2 + 176t + 42$ is a parabola opening down $(a < 0)$, so it has a maximum at the vertex.

vertex:

$$t = -\frac{b}{2a} = -\frac{176}{2(-16)} = 5.5$$

$$h(5.5) = -16(5.5)^2 + 176(5.5) + 42$$
$$= -16(30.25) + 176(5.5) + 42$$
$$= -484 + 968 + 42$$
$$= 526$$

526 feet

36. Unknowns:

length: x

$$w = \frac{P}{2} - l = \frac{144}{2} - x = 72 - x$$

width: $72 - x$

$$A(x) = x(72 - x) = 72x - x^2 = -x^2 + 72x$$

Parabola opens down $(a < 0)$, so it has a maximum at the vertex.

$$x = -\frac{b}{2a} = -\frac{72}{2(-1)} = 36$$

$$A(36) = -(36)^2 + 72(36)$$
$$= -1296 + 2592$$
$$= 1296$$

36 feet by 36 feet, area is 1296 square feet

37.
$$\frac{f(x+h) - f(x)}{h} = \frac{\left((x+h)^2 + 13\right) - \left(x^2 + 13\right)}{h}$$
$$= \frac{x^2 + 2xh + h^2 + 13 - x^2 - 13}{h}$$
$$= \frac{2xh + h^2}{h}$$
$$= \frac{\cancel{h}(2x + h)}{\cancel{h}}$$
$$= 2x + h$$

38.
$$\frac{f(x+h) - f(x)}{h}$$
$$= \frac{\left((x+h)^2 - 6(x+h) - 33\right) - \left(x^2 - 6x - 33\right)}{h}$$
$$= \frac{x^2 + 2xh + h^2 - 6x - 6h - 33 - x^2 + 6x + 33}{h}$$
$$= \frac{2xh + h^2 - 6h}{h}$$
$$= \frac{\cancel{h}(2x + h - 6)}{\cancel{h}}$$
$$= 2x + h - 6$$

39. $f(x) = |x - 2| - 4$

$h = 2, k = -4$

Shift the graph of $f(x) = |x|$ by 2 units to the right and by 4 units down.

y-intercept: Let $x = 0$.

$$f(0) = |0 - 2| - 4 = 2 - 4 = -2$$
$$(0, -2)$$

x-intercepts: Let $f(x) = 0$.

$$|x - 2| - 4 = 0$$
$$|x - 2| = 4$$

$x - 2 = -4$ or $x - 2 = 4$

   $x = -2$ or $x = 6$

$(-2, 0), (6, 0)$

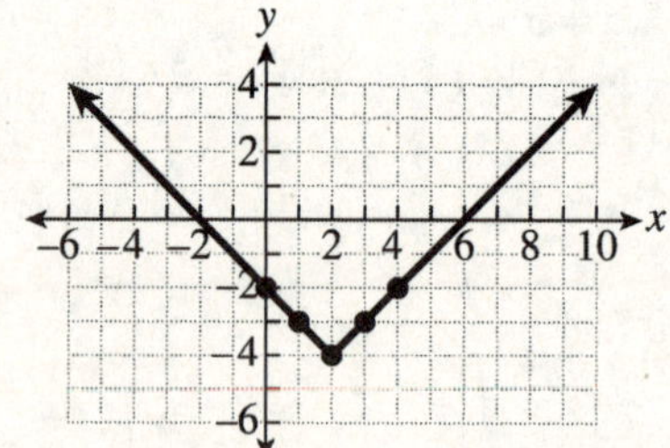

Domain: $(-\infty, \infty)$; range: $[-4, \infty)$

40. $f(x) = -|x + 3| - 2$

$h = -3, k = -2$

After rotating the graph of $f(x) = |x|$ about the x-axis, shift the rotated graph by 3 units to the left and by 2 units down.

y-intercept: Let $x = 0$.

$$f(0) = -|0 + 3| - 2 = -3 - 2 = -5$$
$$(0, -5)$$

x-intercepts:

From the shifted graph, there are no x-intercepts.

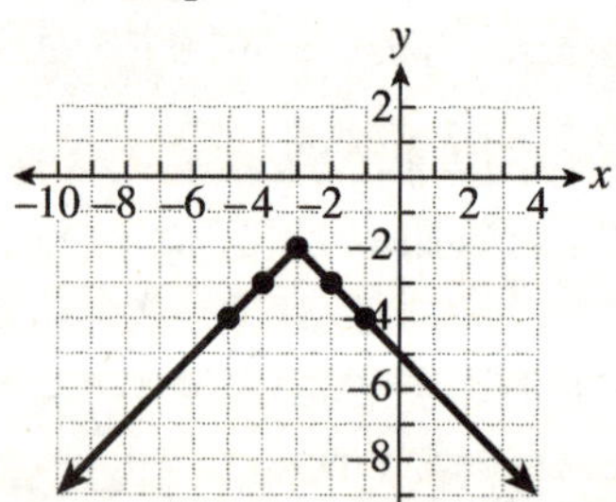

Domain: $(-\infty, \infty)$; range: $(-\infty, -2]$

41. $f(x) = \sqrt{x-2} + 7$
$h = 2, k = 7$
Shift the graph of $f(x) = \sqrt{x}$ by 2 units to the right and by 7 units up.

y-intercept:
From the shifted graph, there is no *y*-intercept.

x-intercept:
From the shifted graph, there is no *x*-intercept.

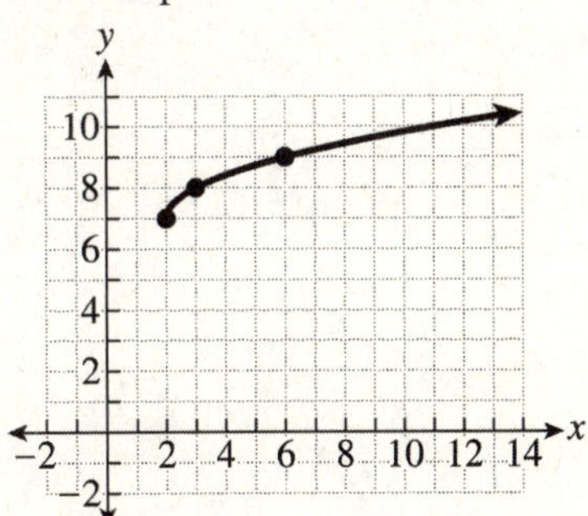

Domain: $[2, \infty)$; range: $[7, \infty)$

42. $f(x) = \sqrt{x+5} - 3$
$h = -5, k = -3$
Shift the graph of $f(x) = \sqrt{x}$ by 5 units to the left and by 3 units down.

y-intercept: Let $x = 0$.
$$f(0) = \sqrt{0+5} - 3 = \sqrt{5} - 3 \approx -0.8$$
$(0, -0.8)$

x-intercept: Let $f(x) = 0$.
$$\sqrt{x+5} - 3 = 0$$
$$\sqrt{x+5} = 3$$
$$\left(\sqrt{x+5}\right)^2 = 3^2$$
$$x + 5 = 9$$
$$x = 4$$
$(4, 0)$

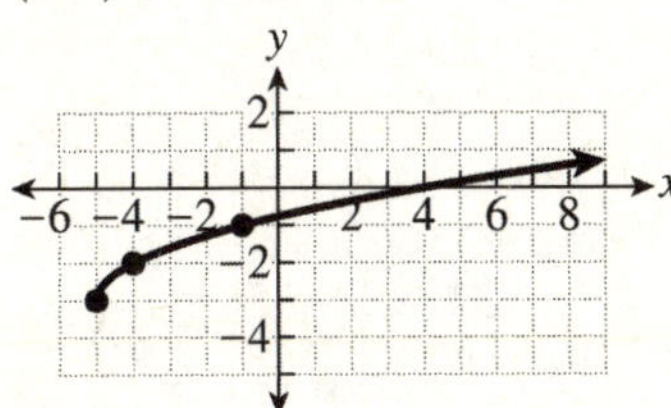

Domain: $[-5, \infty)$; range: $[-3, \infty)$

43. $f(x) = -\sqrt{x+1} + 2$
$h = -1, k = 2$
After rotating the graph of $f(x) = \sqrt{x}$ about the *x*-axis, shift the rotated graph by 1 unit to the left and by 2 units up.

y-intercept: Let $x = 0$.
$$f(0) = -\sqrt{0+1} + 2 = -\sqrt{1} + 2 = -1 + 2 = 1$$
$(0, 1)$

x-intercept: Let $f(x) = 0$.
$$-\sqrt{x+1} + 2 = 0$$
$$\sqrt{x+1} = 2$$
$$\left(\sqrt{x+1}\right)^2 = 2^2$$
$$x + 1 = 4$$
$$x = 3$$
$(3, 0)$

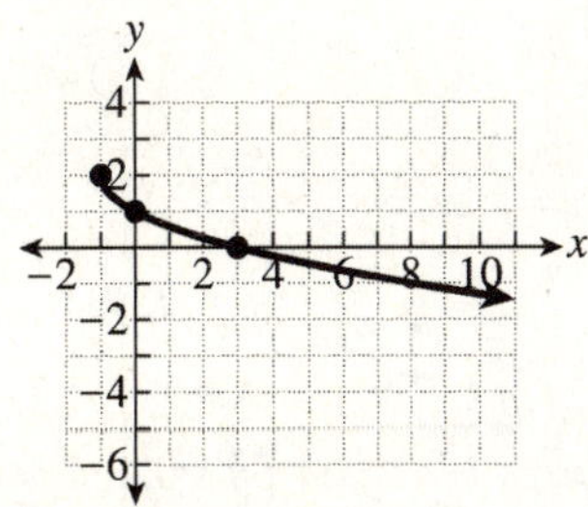

Domain: $[-1, \infty)$; range: $(-\infty, 2]$

44. $f(x) = \sqrt{-(x-2)} + 5$
$h = 2, k = 5$
After rotating the graph of $f(x) = \sqrt{x}$ about the *y*-axis, shift the rotated graph by 2 units to the right and by 5 units up.

y-intercept: Let $x = 0$.
$$f(0) = \sqrt{-(0-2)} + 5 = \sqrt{2} + 5 \approx 6.4$$
$(0, 6.4)$

x-intercept:
From the shifted graph, there is no *x*-intercept.

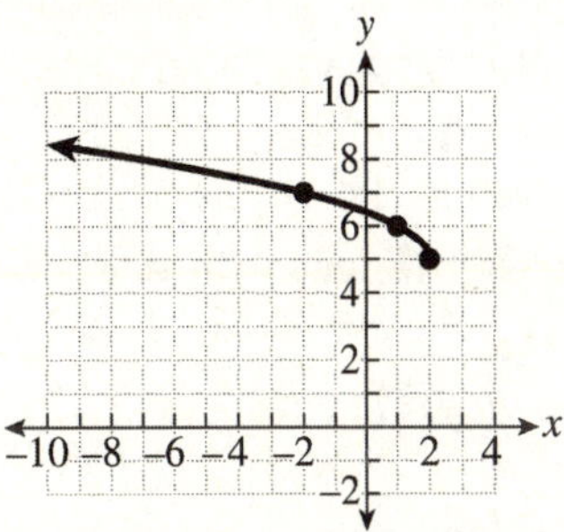

Domain: $(-\infty, 2]$; range: $[5, \infty)$

45. $f(x) = (x-2)^3$
$h = 2, k = 0$
Shift the graph of $f(x) = x^3$ by 2 units to the right.

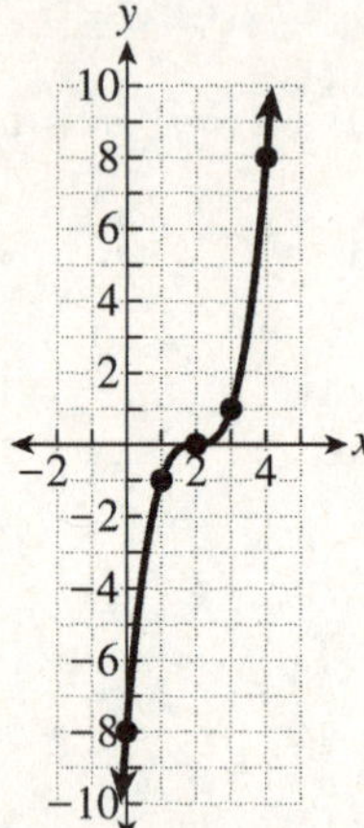

Domain: $(-\infty, \infty)$; range: $(-\infty, \infty)$

46. $f(x) = x^3 - 1$
$h = 0, k = -1$
Shift the graph of $f(x) = x^3$ by 1 unit down.

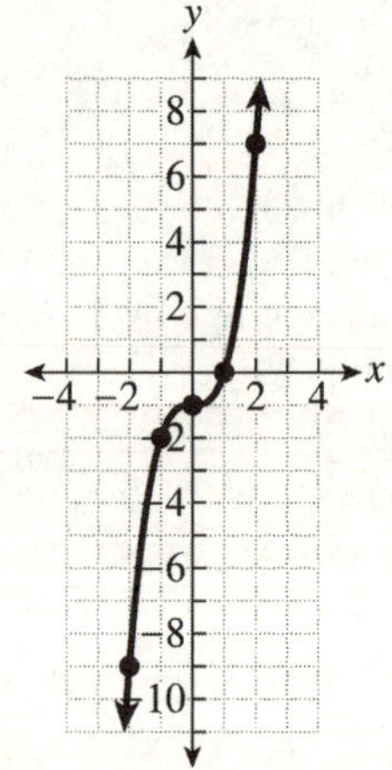

Domain: $(-\infty, \infty)$; range: $(-\infty, \infty)$

47. $f(x)$ is a function of the form $f(x) = \sqrt{x}$ that has been shifted by 1 unit to the right and by 6 units up, so $h = 1$ and $k = 6$.
The function is $f(x) = \sqrt{x-1} + 6$.

48. $f(x)$ is a function of the form $f(x) = x^3$ that has been shifted by 6 units to the right, so $h = 6$ and $k = 0$.
The function is $f(x) = (x-6)^3$.

49. $f(x)$ is a function of the form $f(x) = |x|$ that has been shifted by 6 units to the left and by 3 units up, so $h = -6$ and $k = 3$.
The function is $f(x) = |x+6| + 3$.

50. $(f+g)(-5) = f(-5) + g(-5)$
$\qquad = \left((-5)^2 + 9(-5) - 22\right) + \left((-5) + 11\right)$
$\qquad = (25 - 45 - 22) + (-5 + 11)$
$\qquad = -42 + 6$
$\qquad = -36$

51. $(f-g)(15) = f(15) - g(15)$
$\qquad = \left((15)^2 + 9(15) - 22\right) - \left((15) + 11\right)$
$\qquad = (225 + 135 - 22) - (15 + 11)$
$\qquad = 338 - 26$
$\qquad = 312$

52. $(f \cdot g)(8) = f(8) \cdot g(8)$
$\qquad = \left((8)^2 + 9(8) - 22\right)\left((8) + 11\right)$
$\qquad = (64 + 72 - 22)(8 + 11)$
$\qquad = (114)(19)$
$\qquad = 2166$

53. $\left(\dfrac{f}{g}\right)(-1) = \dfrac{f(-1)}{g(-1)}$
$\qquad = \dfrac{(-1)^2 + 9(-1) - 22}{(-1) + 11}$
$\qquad = \dfrac{1 - 9 - 22}{-1 + 11}$
$\qquad = \dfrac{-30}{10}$
$\qquad = -3$

54. $(f+g)(x) = f(x) + g(x)$
$\qquad = (6x - 11) + (4x + 35)$
$\qquad = 10x + 24$

55. $(f-g)(x) = f(x) - g(x)$
$\qquad = (8x - 33) - (-8x + 55)$
$\qquad = 8x - 33 + 8x - 55$
$\qquad = 16x - 88$

56. $(f \cdot g)(x) = f(x) \cdot g(x)$
$\qquad = (x - 9)(x + 4)$
$\qquad = x^2 + 4x - 9x - 36$
$\qquad = x^2 - 5x - 36$

57.
$$\left(\frac{f}{g}\right)(x) = \frac{f(x)}{g(x)}$$
$$= \frac{x^2 - x - 6}{x^2 - 10x + 21}$$
$$= \frac{(x-3)(x+2)}{(x-3)(x-7)}$$
$$= \frac{x+2}{x-7}$$

58.
$$(f \circ g)(8) = f(g(8))$$
$$= f\left((8)^2 - 14(8) + 48\right)$$
$$= f(64 - 112 + 48)$$
$$= f(0)$$
$$= (0) + 4$$
$$= 4$$

59.
$$(g \circ f)(9) = g(f(9))$$
$$= g(9 + 4)$$
$$= g(13)$$
$$= (13)^2 - 14(13) + 48$$
$$= 169 - 182 + 48$$
$$= 35$$

60.
$$(f \circ g)(x) = f(g(x))$$
$$= f(3x + 20)$$
$$= 6(3x + 20) - 17$$
$$= 18x + 120 - 17$$
$$= 18x + 103$$
$$(g \circ f)(x) = g(f(x))$$
$$= g(6x - 17)$$
$$= 3(6x - 17) + 20$$
$$= 18x - 51 + 20$$
$$= 18x - 31$$

61.
$$(f \circ g)(x) = f(g(x))$$
$$= f(-4x + 9)$$
$$= 2(-4x + 9) + 15$$
$$= -8x + 18 + 15$$
$$= -8x + 33$$
$$(g \circ f)(x) = g(f(x))$$
$$= g(2x + 15)$$
$$= -4(2x + 15) + 9$$
$$= -8x - 60 + 9$$
$$= -8x - 51$$

62.
$$(f \circ g)(x) = f(g(x))$$
$$= f\left(x^2 + 7x - 56\right)$$
$$= \left(x^2 + 7x - 56\right) - 8$$
$$= x^2 + 7x - 64$$
$$(g \circ f)(x) = g(f(x))$$
$$= g(x - 8)$$
$$= (x - 8)^2 + 7(x - 8) - 56$$
$$= x^2 - 16x + 64 + 7x - 56 - 56$$
$$= x^2 - 9x - 48$$

63.
$$(f \circ g)(x) = f(g(x))$$
$$= f\left(x^2 - 9x + 41\right)$$
$$= 3\left(x^2 - 9x + 41\right) + 7$$
$$= 3x^2 - 27x + 123 + 7$$
$$= 3x^2 - 27x + 130$$
$$(g \circ f)(x)$$
$$= g(f(x))$$
$$= g(3x + 7)$$
$$= (3x + 7)^2 - 9(3x + 7) + 41$$
$$= (3x + 7)(3x + 7) - 9(3x + 7) + 41$$
$$= 9x^2 + 21x + 21x + 49 - 27x - 63 + 41$$
$$= 9x^2 + 15x + 27$$

64. No, there may be more than one person who has the same favorite fast food restaurant.

65. Yes, each player has a unique number.

66. Yes.

67. No.

68.
$$(f \circ g)(x) = f(g(x))$$
$$= f(4x - 3)$$
$$= 3(4x - 3) + 4$$
$$= 12x - 9 + 4$$
$$= 12x - 5$$
No, the functions are not inverses;
$$(f \circ g)(x) \neq x.$$

69.

$(f \circ g)(x) = x$	$(g \circ f)(x) = x$
$(f \circ g)(x)$	$(g \circ f)(x)$
$= f(g(x))$	$= g(f(x))$
$= f\left(\dfrac{x+10}{2}\right)$	$= g(2x-10)$
$= 2\left(\dfrac{x+10}{2}\right)-10$	$= \dfrac{(2x-10)+10}{2}$
$= x+10-10$	$= \dfrac{2x}{2}$
$= x$	$= x$

Yes, $f(x)$ and $g(x)$ are inverses.

70. $(f \circ g)(x) = f(g(x))$

$$= f\left(\frac{1}{4}x+\frac{9}{2}\right)$$

$$= 4\left(\frac{1}{4}x+\frac{9}{2}\right)+18$$

$$= 4\cdot\frac{1}{4}x+4\cdot\frac{9}{2}+18$$

$$= x+18+18$$
$$= x+36$$

No, the functions are not inverses, since $(f \circ g)(x) \neq x$.

71.

$(f \circ g)(x) = x$	$(g \circ f)(x) = x$
$(f \circ g)(x) = f(g(x))$	$(g \circ f)(x) = g(f(x))$
$= f\left(\dfrac{9-x}{5}\right)$	$= g(-5x+9)$
$= -5\left(\dfrac{9-x}{5}\right)+9$	$= \dfrac{9-(-5x+9)}{5}$
$= -9+x+9$	$= \dfrac{9+5x-9}{5}$
$= x$	$= \dfrac{5x}{5}$
	$= x$

Yes, $f(x)$ and $g(x)$ are inverses.

72. $f(x) = x-10$
$y = x-10$
Interchange x and y.
$x = y-10$
$x+10 = y$
$f^{-1}(x) = x+10$
Domain of $f^{-1}(x)$: $(-\infty, \infty)$

73. $f(x) = 2x-15$
$y = 2x-15$
Interchange x and y.
$x = 2y-15$
$x+15 = 2y$
$\dfrac{x+15}{2} = y$
$f^{-1}(x) = \dfrac{x+15}{2}$
Domain of $f^{-1}(x)$: $(-\infty, \infty)$

74. $f(x) = -x+19$
$y = -x+19$
Interchange x and y.
$x = -y+19$
$x+y = 19$
$y = -x+19$
$f^{-1}(x) = -x+19$
Domain of $f^{-1}(x)$: $(-\infty, \infty)$

75. $f(x) = -8x+27$
$y = -8x+27$
Interchange x and y.
$x = -8y+27$
$8y = -x+27$
$y = \dfrac{-x+27}{8}$
$f^{-1}(x) = \dfrac{-x+27}{8}$
Domain of $f^{-1}(x)$: $(-\infty, \infty)$

76. $f(x) = \dfrac{9x+2}{3x}$

$y = \dfrac{9x+2}{3x}$

Interchange x and y.

$x = \dfrac{9y+2}{3y}$

$3y \cdot x = 3y \cdot \left(\dfrac{9y+2}{3y}\right)$

$3xy = 9y+2$
$3xy - 9y = 2$
$y(3x-9) = 2$

$y = \dfrac{2}{3x-9}$

$f^{-1}(x) = \dfrac{2}{3x-9}$

Domain of $f^{-1}(x)$: all real numbers except 3.

77. $f(x) = \dfrac{7x}{4x+21}$

$$y = \dfrac{7x}{4x+21}$$

Interchange x and y.

$$x = \dfrac{7y}{4y+21}$$

$$(4y+21)\cdot x = (4y+21)\cdot\left(\dfrac{7y}{4y+21}\right)$$

$$4xy + 21x = 7y$$
$$21x = 7y - 4xy$$
$$21x = y(7-4x)$$

$$\dfrac{21x}{7-4x} = y$$

$$f^{-1}(x) = \dfrac{21x}{7-4x}$$

Domain of $f^{-1}(x)$: all real numbers except $\dfrac{7}{4}$.

78. Coordinates of $f(x)$:

$(-1,1),(-2,2),(-3,5),(-4,10)$

Coordinates of $f^{-1}(x)$:

$(1,-1),(2,-2),(5,-3),(10,-4)$

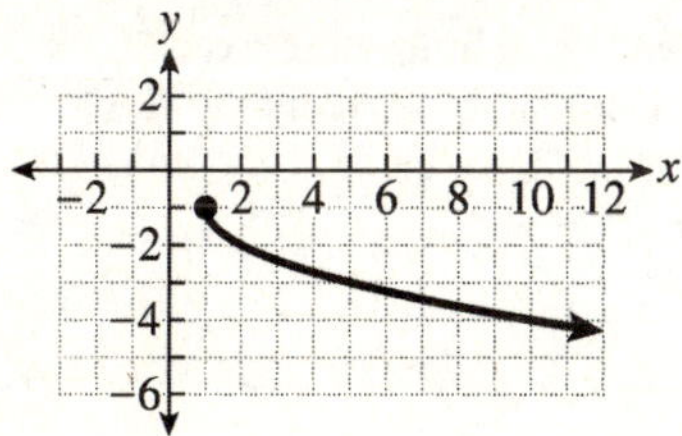

79. Coordinates of $f(x)$:

$(-7,-3),(-3,-5),(2,-6),(9,-7)$

Coordinates of $f^{-1}(x)$:

$(-3,-7),(-5,-3),(-6,2),(-7,9)$

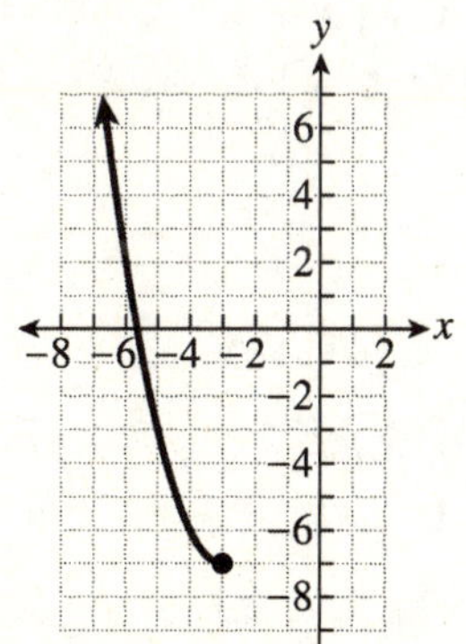

CHAPTER 11 TEST

1. Yes.

2. No.

3. $f(-8) = -2(-8) + 35 = 16 + 35 = 51$

4. $f(6a-5) = 5(6a-5)+16$
$$= 30a - 25 + 16$$
$$= 30a - 9$$

5. $f(-9) = (-9)^2 - 15(-9) + 54$
$$= 81 + 135 + 54$$
$$= 270$$

6. Domain: $(-\infty, \infty)$; range: $(-\infty, 5]$

7. $f(x) = -4x + 6$
y-intercept: Let $x = 0$.
$f(0) = -4(0) + 6 = 6$
$(0,6)$

x-intercept: Let $f(x) = 0$.
$-4x + 6 = 0$
$$-4x = -6$$
$$x = \dfrac{-6}{-4}$$
$$x - \dfrac{3}{2}$$
$\left(\dfrac{3}{2}, 0\right)$

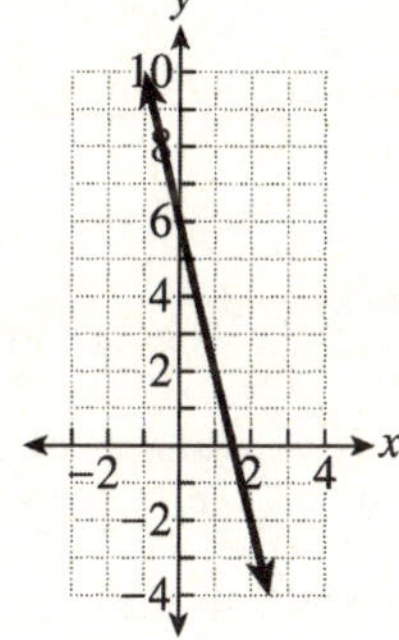

8. $f(x) = \dfrac{2}{3}x - 4$

y-intercept: Let $x = 0$.

$f(0) = \dfrac{2}{3}(0) - 4 = -4$

$(0, -4)$

x-intercept: Let $f(x) = 0$.

$\dfrac{2}{3}x - 4 = 0$

$\dfrac{2}{3}x = 4$

$\dfrac{3}{2} \cdot \dfrac{2}{3}x = \dfrac{3}{2} \cdot 4$

$x = 6$

$(6, 0)$

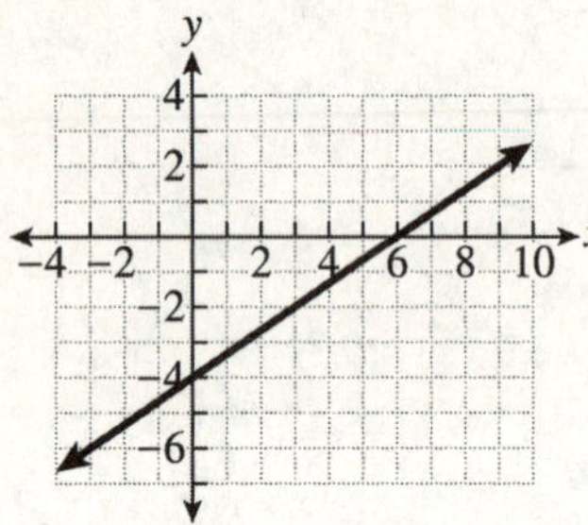

9. Use the points $(10, 105)$ and $(25, 210)$.

$m = \dfrac{210 - 105}{25 - 10} = \dfrac{105}{15} = 7$

The function has the form

$f(x) = 7x + b.$

$f(10) = 105$

$7(10) + b = 105$

$70 + b = 105$

$b = 35$

so $f(x) = 7x + 35$

10. $f(x) = (x + 4)^2 - 6$

The vertex is $(-4, -6)$.
Shift the basic parabola 4 units to the left and 6 units down.

y-intercept: Let $x = 0$.

$f(0) = (0 + 4)^2 - 6 = 16 - 6 = 10$

$(0, 10)$

x-intercepts: Let $f(x) = 0$.

$(x + 4)^2 - 6 = 0$

$(x + 4)^2 = 6$

$\sqrt{(x + 4)^2} = \pm\sqrt{6}$

$x + 4 = \pm\sqrt{6}$

$x = -4 \pm \sqrt{6}$

$x = -4 - \sqrt{6} \approx -6.4$ or $x = -4 + \sqrt{6} \approx -1.6$

$(-6.4, 0), (-1.6, 0)$

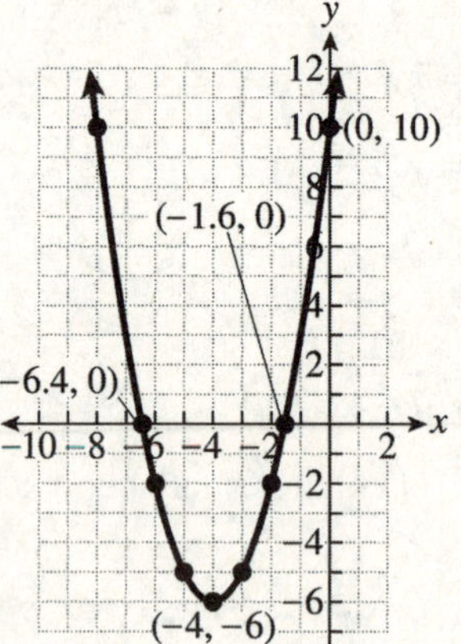

11. $f(x) = -(x + 3)^2 + 4$

The vertex is $(-3, 4)$.

Because $a < 0$, the parabola opens downward.
Rotate $f(x) = x^2$ about the x-axis.
Shift the basic parabola 3 units to the left and 4 units up.

y-intercept: Let $x = 0$.

$f(0) = -(0 + 3)^2 + 4 = -9 + 4 = -5$

$(0, -5)$

x-intercepts: Let $f(x) = 0$.

$-(x + 3)^2 + 4 = 0$

$(x + 3)^2 = 4$

$\sqrt{(x + 3)^2} = \pm\sqrt{4}$

$x + 3 = \pm 2$

$x = -3 \pm 2$

$x = -3 - 2 = -5$ or $x = -3 + 2 = -1$

$(-5, 0), (-1, 0)$

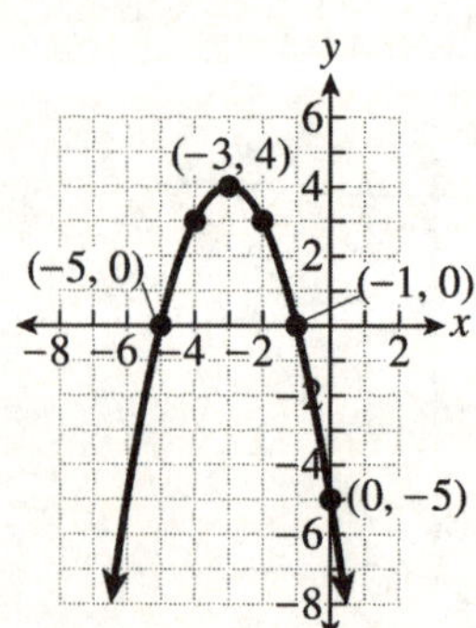

12. $h(t) = -16t^2 + 20t + 3$

vertex:

$$t = -\frac{b}{2a} = -\frac{20}{2(-16)} = \frac{20}{32} = \frac{5}{8}$$

$$h\left(\frac{5}{8}\right) = -16\left(\frac{5}{8}\right)^2 + 20\left(\frac{5}{8}\right) + 3$$

$$= -16\left(\frac{25}{64}\right) + 20\left(\frac{5}{8}\right) + 3$$

$$= -\frac{25}{4} + \frac{50}{4} + \frac{12}{4}$$

$$= \frac{37}{4}$$

$$= 9.25$$

The maximum height is 9.25 feet.

13. $f(x) = |x+4| - 5$

$h = -4, k = -5$

Shift the graph of $f(x) = |x|$ by 4 units to the left and by 5 units down.

y-intercept: Let $x = 0$.

$$f(0) = |0+4| - 5 = 4 - 5 = -1$$
$$(0, -1)$$

x-intercepts: Let $f(x) = 0$.

$$|x+4| - 5 = 0$$
$$|x+4| = 5$$

$$x + 4 = -5 \quad \text{or} \quad x + 4 = 5$$
$$x = -9 \quad \text{or} \qquad x = 1$$
$$(-9, 0), (1, 0)$$

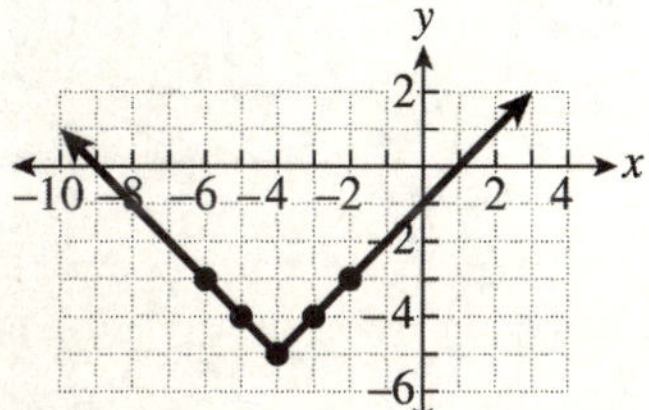

Domain: $(-\infty, \infty)$; range: $[-5, \infty)$

14. $f(x) = \sqrt{x+1} - 3$

$h = -1, k = -3$

Shift the graph of $f(x) = \sqrt{x}$ by 1 unit to the left and by 3 units down.

y-intercept: Let $x = 0$.

$$f(0) = \sqrt{0+1} - 3 = \sqrt{1} - 3 = 1 - 3 = -2$$
$$(0, -2)$$

x-intercept: Let $f(x) = 0$.

$$\sqrt{x+1} - 3 = 0$$
$$\sqrt{x+1} = 3$$
$$\left(\sqrt{x+1}\right)^2 = 3^2$$
$$x + 1 = 9$$
$$x = 8$$
$$(8, 0)$$

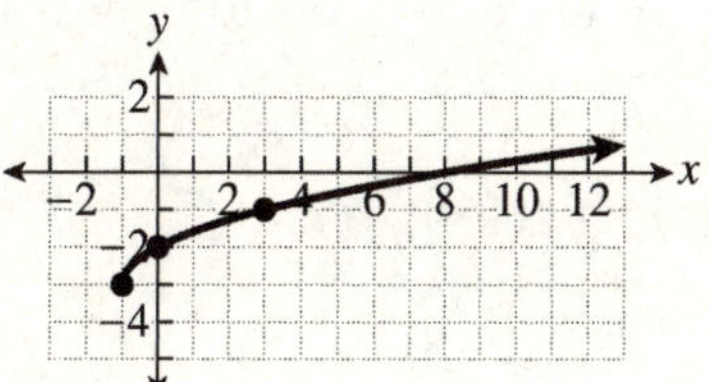

Domain: $[-1, \infty)$; range: $[-3, \infty)$

15. $f(x) = (x+1)^3 + 8$

$h = -1, k = 8$

Shift the graph of $f(x) = x^3$ by 1 unit to the left and by 8 units up.

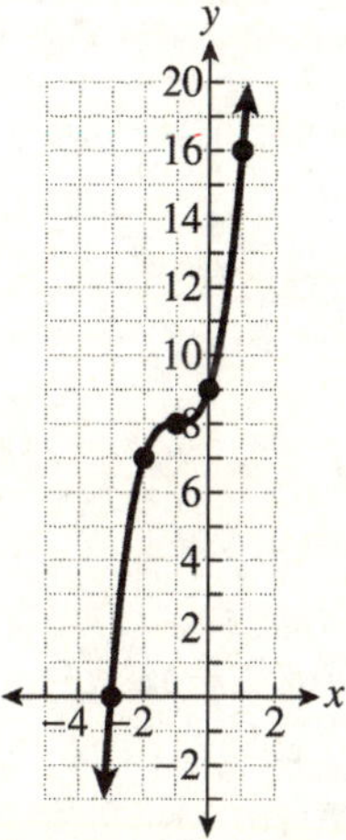

Domain: $(-\infty, \infty)$; range: $(-\infty, \infty)$

16. $f(x)$ is a function of the form $f(x) = \sqrt{x}$ that has been shifted by 6 units to the left, so $h = -6$ and $k = 0$.
The function is $f(x) = \sqrt{x+6}$.

17.
$$
\begin{aligned}
(f+g)(-2) &= f(-2) + g(-2) \\
&= \left(5(-2)-2\right) + \left((-2)^2 - 4(-2) - 13\right) \\
&= (-10-2) + (4+8-13) \\
&= (-12) + (-1) \\
&= -13
\end{aligned}
$$

18.
$$
\begin{aligned}
(f \cdot g)(x) &= f(x) \cdot g(x) \\
&= (x-6)\left(x^2 + 6x + 36\right) \\
&= x^3 + 6x^2 + 36x - 6x^2 - 36x - 216 \\
&= x^3 - 216
\end{aligned}
$$

19.
$$
\begin{aligned}
(f \circ g)(6) &= f\left(g(6)\right) \\
&= f\left(2(6) - 8\right) \\
&= f(12 - 8) \\
&= f(4) \\
&= 4 + 7 \\
&= 11
\end{aligned}
$$

20.
$$
\begin{aligned}
(g \circ f)(6) &= g\left(f(6)\right) \\
&= g(6+7) \\
&= g(13) \\
&= 2(13) - 8 \\
&= 26 - 8 \\
&= 18
\end{aligned}
$$

21.
$$
\begin{aligned}
(f \circ g)(x) &= f\left(g(x)\right) \\
&= f(-2x + 23) \\
&= 4(-2x + 23) + 13 \\
&= -8x + 92 + 13 \\
&= -8x + 105
\end{aligned}
$$

22.
$$
\begin{aligned}
(f \circ g)(x) &= f\left(g(x)\right) \\
&= f(x + 12) \\
&= (x+12)^2 - 9(x+12) - 30 \\
&= x^2 + 24x + 144 - 9x - 108 - 30 \\
&= x^2 + 15x + 6
\end{aligned}
$$

23. No.

24.

$(f \circ g)(x) = x$	$(g \circ f)(x) = x$
$(f \circ g)(x)$	$(g \circ f)(x)$
$= f\left(g(x)\right)$	$= g\left(f(x)\right)$
$= f\left(\dfrac{1}{2}x + 8\right)$	$= g(2x - 16)$
$= 2\left(\dfrac{1}{2}x + 8\right) - 16$	$= \dfrac{1}{2}(2x - 16) + 8$
$= x + 16 - 16$	$= x - 8 + 8$
$= x$	$= x$

Yes, $f(x)$ and $g(x)$ are inverses.

25.
$$
\begin{aligned}
f(x) &= -4x + 20 \\
y &= -4x + 20
\end{aligned}
$$
Interchange x and y.
$$
\begin{aligned}
x &= -4y + 20 \\
4y &= -x + 20 \\
y &= \frac{x - 20}{-4} \\
y &= \frac{-x + 20}{4} \\
f^{-1}(x) &= \frac{-x + 20}{4}
\end{aligned}
$$

26.
$$
f(x) = \frac{x - 9}{4x}
$$
$$
y = \frac{x - 9}{4x}
$$
Interchange x and y.
$$
x = \frac{y - 9}{4y}
$$
$$
4y \cdot x = 4y \cdot \frac{y - 9}{4y}
$$
$$
4xy = y - 9
$$
$$
4xy - y = -9
$$
$$
y(4x - 1) = -9
$$
$$
y = \frac{-9}{4x - 1}
$$
$$
f^{-1}(x) = \frac{-9}{4x - 1}, \quad x \neq \frac{1}{4}
$$

CUMULATIVE REVIEW CHAPTERS 8–11

1. $|x+3|-8=-2$

$$|x+3|=6$$

$x+3=-6 \quad$ or $\quad x+3=6$

$x=-9 \quad$ or $\quad\quad x=3$

$\{-9,3\}$

2. $-2x-9 \geq -17$

$\quad -2x \geq -8$

$$\dfrac{-2x}{-2} \leq \dfrac{-8}{-2}$$

$\quad\quad x \leq 4$

$(-\infty, 4]$

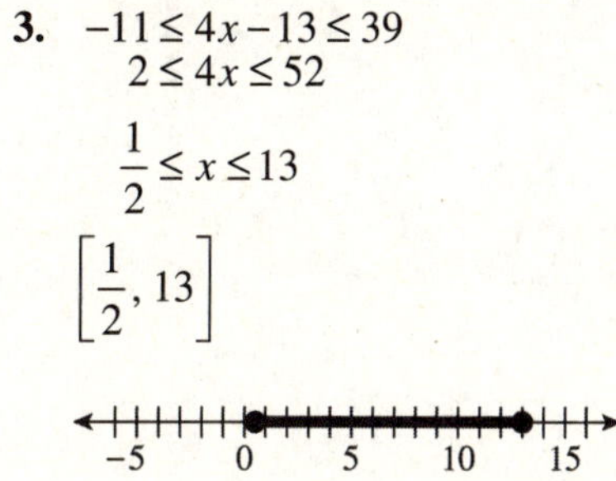

3. $-11 \leq 4x-13 \leq 39$

$\quad 2 \leq 4x \leq 52$

$$\dfrac{1}{2} \leq x \leq 13$$

$\left[\dfrac{1}{2}, 13\right]$

4. $|2x+1| > 5$

$2x+1 < -5 \quad$ or $\quad 2x+1 > 5$

$\quad 2x < -6 \quad$ or $\quad\quad 2x > 4$

$\quad\quad x < -3 \quad$ or $\quad\quad\quad x > 2$

$(-\infty, -3) \cup (2, \infty)$

5. $|x|-8 < -3$

$\quad\quad |x| < 5$

$-5 < x < 5$

$(-5, 5)$

6. $x^2 - 3x - 54 = (x-9)(x+6)$

7. $4x^2 - 36x + 32 = 4\left(x^2 - 9x + 8\right)$

$\quad\quad\quad\quad\quad\quad = 4(x-1)(x-8)$

8. $x^2 + 11x + 30 = (x+5)(x+6)$

9. $49x^2 - 64 = (7x)^2 - (8)^2 = (7x+8)(7x-8)$

10. $3x - 2y = 22$

$7x + 4y = 34$

$\quad 6x - 4y = 44 \quad (2 \cdot \text{Eq. 1})$

$\underline{+ \ 7x + 4y = 34}$

$\quad\quad 13x = 78$

$\quad\quad\quad x = 6$

Substitute $x = 6$ into Eq. 1.

$3(6) - 2y = 22$

$\quad 18 - 2y = 22$

$\quad\quad -2y = 4$

$\quad\quad\quad y = -2$

$(6, -2)$

11. $\quad x + y + z = 6 \quad\quad (\text{Eq. 1})$

$\quad x + 3y - z = 20 \quad (\text{Eq. 2})$

$2x - 4y + z = -16 \quad (\text{Eq. 3})$

$\quad\quad x + y + z = 6 \quad\quad (\text{Eq. 1})$

$\underline{+ \ x + 3y - z = 20 \quad (\text{Eq. 2})}$

$\quad\quad 2x + 4y = 26 \quad (\text{Eq. 4})$

$\quad -x - y - z = -6 \quad\quad (-1 \cdot \text{Eq. 1})$

$\underline{+ \ 2x - 4y + z = -16 \quad\quad (\text{Eq. 3})}$

$\quad\quad x - 5y = -22 \quad\quad (\text{Eq. 5})$

$\quad 2x + 4y = 26 \quad\quad (\text{Eq. 4})$

$\underline{-2x + 10y = 44 \quad (-2 \cdot \text{Eq. 5})}$

$\quad\quad 14y = 70$

$\quad\quad\quad y = 5$

Substitute $y = 5$ into Eq. 5.

$x - 5(5) = -22$

$\quad x - 25 = -22$

$\quad\quad\quad x = 3$

Substitute $x = 3$, $y = 5$ into Eq. 1.

$3 + 5 + z = 6$

$\quad 8 + z = 6$

$\quad\quad z = -2$

$(3, 5, -2)$

12. $f(19) = \sqrt{6(19) - 58}$

$\quad\quad\quad = \sqrt{114 - 58}$

$\quad\quad\quad = \sqrt{56}$

$\quad\quad\quad \approx 7.483$

13. $\left(16x^{10}\right)^{\frac{3}{2}} = \left(\left(16x^{10}\right)^{\frac{1}{2}}\right)^3$

$\quad\quad\quad = \left(\sqrt{16x^{10}}\right)^3$

$\quad\quad\quad = \left(4x^5\right)^3$

$\quad\quad\quad = 64x^{15}$

14. $\left(x^{6/7}\right)^{3/4} = x^{\frac{6}{7} \cdot \frac{3}{4}} = x^{9/14}$

15. $\sqrt{75r^{12}s^{10}t^9} = \sqrt{25\cdot3\cdot\left(r^6\right)^2\cdot\left(s^5\right)^2\cdot\left(t^4\right)^2\cdot t}$

$\qquad = 5r^6s^5t^4\sqrt{3t}$

16. $7\sqrt{108} - 2\sqrt{48} = 7\sqrt{36\cdot3} - 2\sqrt{16\cdot3}$

$\qquad = 7\cdot6\sqrt{3} - 2\cdot4\sqrt{3}$

$\qquad = 42\sqrt{3} - 8\sqrt{3}$

$\qquad = 34\sqrt{3}$

17. $\left(4\sqrt{3} - 2\sqrt{5}\right)\left(5\sqrt{3} + 4\sqrt{5}\right)$

$\qquad = 20\cdot3 + 16\sqrt{15} - 10\sqrt{15} - 8\cdot5$

$\qquad = 60 + 6\sqrt{15} - 40$

$\qquad = 20 + 6\sqrt{15}$

18. $\dfrac{ab^3c^2}{\sqrt{a^3b^8c}} = \dfrac{ab^3c^2}{ab^4\sqrt{ac}}$

$\qquad = \dfrac{ab^3c^2}{ab^4\sqrt{ac}} \cdot \dfrac{\sqrt{ac}}{\sqrt{ac}}$

$\qquad = \dfrac{ab^3c^2\sqrt{ac}}{ab^4(ac)}$

$\qquad = \dfrac{ab^3c^2\sqrt{ac}}{a^2b^4c}$

$\qquad = \dfrac{c\sqrt{ac}}{ab}$

19. $\dfrac{5\sqrt{5} + 6\sqrt{2}}{4\sqrt{5} - 3\sqrt{2}} = \dfrac{\left(5\sqrt{5} + 6\sqrt{2}\right)}{\left(4\sqrt{5} - 3\sqrt{2}\right)} \cdot \dfrac{\left(4\sqrt{5} + 3\sqrt{2}\right)}{\left(4\sqrt{5} + 3\sqrt{2}\right)}$

$\qquad = \dfrac{20\cdot5 + 15\sqrt{10} + 24\sqrt{10} + 18\cdot2}{16\cdot5 + 12\sqrt{10} - 12\sqrt{10} - 9\cdot2}$

$\qquad = \dfrac{100 + 39\sqrt{10} + 36}{80 - 18}$

$\qquad = \dfrac{136 + 39\sqrt{10}}{62}$

20. $\sqrt{2x+7} - 3 = 4$

$\qquad \sqrt{2x+7} = 7$

$\qquad \left(\sqrt{2x+7}\right)^2 = (7)^2$

$\qquad 2x+7 = 49$

$\qquad 2x = 42$

$\qquad x = 21$

check:

$\sqrt{2(21)+7} - 3 = 4$

$\qquad \sqrt{42+7} = 7$

$\qquad \sqrt{49} = 7$

$\qquad 7 = 7$

True

$\{21\}$

21. $\sqrt{x+6} = x$

$\qquad \left(\sqrt{x+6}\right)^2 = (x)^2$

$\qquad x+6 = x^2$

$x^2 - x - 6 = 0$

$(x-3)(x+2) = 0$

$x-3 = 0 \ \text{ or } \ x+2 = 0$

$\quad x = 3 \ \text{ or } \qquad x = -2$

check:

$x = 3$	$x = -2$
$\sqrt{3+6} = 3$	$\sqrt{-2+6} = -2$
$\sqrt{9} = 3$	$\sqrt{4} = -2$
$3 = 3$	$2 = -2$
True	False

$\{3\}$

22. $f(x) = \sqrt{2x-9} = 5$

$\qquad \left(\sqrt{2x-9}\right)^2 = (5)^2$

$\qquad 2x-9 = 25$

$\qquad 2x = 34$

$\qquad x = 17$

23. $\sqrt{-360} = \sqrt{-1\cdot36\cdot10} = 6i\sqrt{10}$

24. $(2+5i)(9-2i) = 18 - 4i + 45i - 10i^2$

$\qquad = 18 + 41i - 10(-1)$

$\qquad = 18 + 41i + 10$

$\qquad = 28 + 41i$

25. $\dfrac{5i}{3+4i} = \dfrac{5i}{3+4i} \cdot \dfrac{3-4i}{3-4i}$

$\qquad = \dfrac{15i - 20i^2}{9 - 12i + 12i - 16i^2}$

$\qquad = \dfrac{15i - 20(-1)}{9 - 16(-1)}$

$\qquad = \dfrac{15i + 20}{9 + 16}$

$\qquad = \dfrac{20 + 15i}{25}$

$\qquad = \dfrac{\cancel{5}^{\,1}(4+3i)}{\cancel{25}_{\,5}}$

$\qquad = \dfrac{4}{5} + \dfrac{3}{5}i$

26. $(x+2)^2 + 41 = 23$

$$(x+2)^2 = -18$$
$$\sqrt{(x+2)^2} = \pm\sqrt{-18}$$
$$x+2 = \pm 3i\sqrt{2}$$
$$x = -2 \pm 3i\sqrt{2}$$
$$\left\{-2 - 3i\sqrt{2}, -2 + 3i\sqrt{2}\right\}$$

27. $3x^2 - 10x - 21 = 0$

$$x = \frac{-(-10) \pm \sqrt{(-10)^2 - 4(3)(-21)}}{2(3)}$$
$$x = \frac{10 \pm \sqrt{352}}{6}$$
$$x = \frac{10 \pm \sqrt{16 \cdot 22}}{6}$$
$$x = \frac{10 \pm 4\sqrt{22}}{6}$$
$$x = \frac{2\left(5 \pm 2\sqrt{22}\right)}{6}$$
$$x = \frac{5 \pm 2\sqrt{22}}{3}$$
$$\left\{\frac{5 - 2\sqrt{22}}{3}, \frac{5 + 2\sqrt{22}}{3}\right\}$$

28. $x^4 - 11x^2 + 18 = 0$

Let $u = x^2$ and $u^2 = x^4$.
$$u^2 - 11u + 18 = 0$$
$$(u-2)(u-9) = 0$$

$$
\begin{array}{lll}
u - 2 = 0 & \text{or} & u - 9 = 0 \\
u = 2 & \text{or} & u = 9 \\
x^2 = 2 & \text{or} & x^2 = 9 \\
\sqrt{x^2} = \pm\sqrt{2} & \text{or} & \sqrt{x^2} = \pm\sqrt{9} \\
x = \pm\sqrt{2} & \text{or} & x = \pm 3
\end{array}
$$

$$\left\{-\sqrt{2}, \sqrt{2}, -3, 3\right\}$$

29. $x + 3\sqrt{x} - 10 = 0$

Let $u = \sqrt{x}$ and $u^2 = x$.
$$u^2 + 3u - 10 = 0$$
$$(u+5)(u-2) = 0$$

$$
\begin{array}{lll}
u + 5 = 0 & \text{or} & u - 2 = 0 \\
u = -5 & \text{or} & u = 2 \\
\sqrt{x} = -5 & \text{or} & \sqrt{x} = 2 \\
\left(\sqrt{x}\right)^2 = (-5)^2 & \text{or} & \left(\sqrt{x}\right)^2 = (2)^2 \\
x = 25 & \text{or} & x = 4
\end{array}
$$

check:

$x = 25$	$x = 4$
$25 + 3\sqrt{25} - 10 = 0$	$4 + 3\sqrt{4} - 10 = 0$
$25 + 3(5) - 10 = 0$	$4 + 3(2) - 10 = 0$
$25 + 15 - 10 = 0$	$4 + 6 - 10 = 0$
$30 = 0$	$0 = 0$
False	True

$\{4\}$

30. $\sqrt{x+23} - x = 3$

$$\sqrt{x+23} = x + 3$$
$$\left(\sqrt{x+23}\right)^2 = (x+3)^2$$
$$x + 23 = (x+3)(x+3)$$
$$x + 23 = x^2 + 3x + 3x + 9$$
$$x + 23 = x^2 + 6x + 9$$
$$0 = x^2 + 5x - 14$$
$$0 = (x+7)(x-2)$$

$$
\begin{array}{lll}
x + 7 = 0 & \text{or} & x - 2 = 0 \\
x = -7 & \text{or} & x = 2
\end{array}
$$

check:

$x = -7$	$x = 2$
$\sqrt{-7+23} - (-7) = 3$	$\sqrt{2+23} - 2 = 0$
$\sqrt{16} + 7 = 3$	$\sqrt{25} - 2 = 3$
$4 + 7 = 3$	$5 - 2 = 3$
$11 = 3$	$3 = 3$
False	True

$\{2\}$

31. $x^2 - 12x + 27 = 0$

$$(x-3)(x-9) = 0$$

$$
\begin{array}{lll}
x - 3 = 0 & \text{or} & x - 9 = 0 \\
x = 3 & \text{or} & x = 9
\end{array}
$$

$\{3, 9\}$

32. $y = -x^2 + 8x + 6$

vertex:
$$x = -\frac{b}{2a} = -\frac{8}{2(-1)} = 4$$

$$y = -(4)^2 + 8(4) + 6 = -16 + 32 + 6 = 22$$
$(4, 22)$

y-intercept: Let $x = 0$.
$$y = -(0)^2 + 8(0) + 6 = 6$$
$(0, 6)$

x-intercept: Let $y = 0$.
$$-x^2 + 8x + 6 = 0$$
$$x^2 - 8x - 6 = 0$$
$$x = \frac{-(-8) \pm \sqrt{(-8)^2 - 4(1)(-6)}}{2(1)}$$

$$x = \frac{8 \pm \sqrt{88}}{2}$$

$$x = \frac{8 \pm 2\sqrt{22}}{2}$$

$$x = 4 \pm \sqrt{22}$$
$\left(4 - \sqrt{22}, 0\right), \left(4 + \sqrt{22}, 0\right)$

axis of symmetry: $x = 4$

point symmetric to the y-intercept: $(8, 6)$

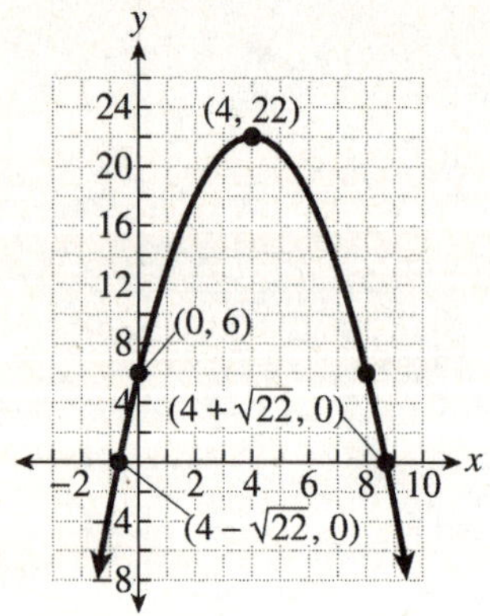

33. $y = x^2 - 4x + 7$

vertex:
$$x = -\frac{b}{2a} = -\frac{-4}{2(1)} = 2$$

$$y = (2)^2 - 4(2) + 7 = 4 - 8 + 7 = 3$$
$(2, 3)$

y-intercept: Let $x = 0$.
$$y = (0)^2 - 4(0) + 7 = 7$$
$(0, 7)$

x-intercept: If we plot these two points, we see that the vertex is above the x-axis, and the parabola moves only in an upward direction. This parabola does not have any x-intercepts.

axis of symmetry: $x = 2$

point symmetric to the y-intercept: $(4, 7)$

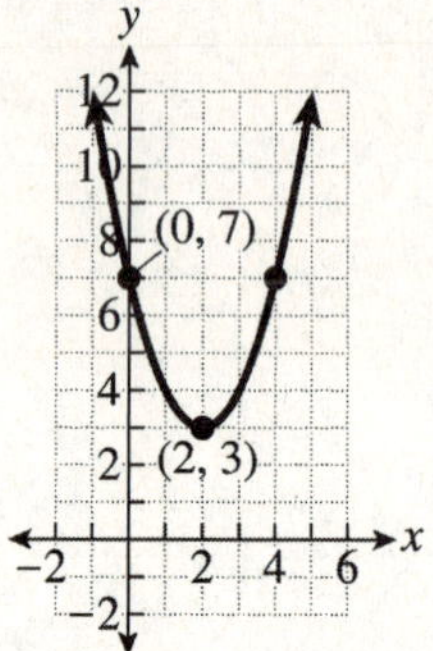

34. $x^2 - 10x + 24 \le 0$
$(x - 4)(x - 6) \le 0$

$$\begin{array}{lll} x - 4 = 0 & \text{or} & x - 6 = 0 \\ x = 4 & \text{or} & x = 6 \end{array}$$

Test Points	$x - 4$	$x - 6$	Product
0	$-$	$-$	$+$
5	$+$	$-$	$-$
8	$+$	$+$	$+$

$[4, 6]$

35. $\dfrac{x^2 - 2x - 48}{x + 5} \le 0$

$\dfrac{(x-8)(x+6)}{x+5} \le 0$

Values of x that make the numerator equal to 0: $x = 8, x = -6$

Values of x that make the denominator equal to 0: $x = -5$

Test Points	$x-8$	$x+6$	$x+5$	$\dfrac{(x-8)(x+6)}{x+5}$
-7	$-$	$-$	$-$	$-$
$-5\frac{1}{2}$	$-$	$+$	$-$	$+$
0	$-$	$+$	$+$	$-$
10	$+$	$+$	$+$	$+$

$(-\infty, -6] \cup (-5, 8]$

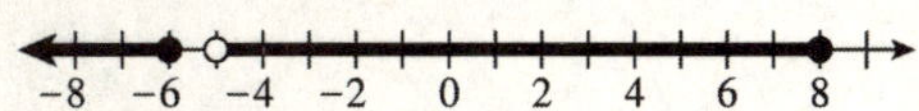

36. Unknowns: Known:
length: $x + 5$ Area: 980 ft^2
width: x

$$x(x+5) = 980$$
$$x^2 + 5x = 980$$
$$x^2 + 5x - 980 = 0$$

$$x = \frac{-5 \pm \sqrt{5^2 - 4(1)(-980)}}{2(1)}$$

$$x = \frac{-5 \pm \sqrt{3945}}{2}$$

$x \approx 28.9$ or $x \approx -33.9$

width: $x \approx 28.9$ feet
length: $x + 5 \approx 28.9 + 5 = 33.9$ feet

37. $h(t) = -16t^2 + 80t + 12$

$$0 = -16t^2 + 80t + 12$$
$$16t^2 - 80t - 12 = 0$$

$$t = \frac{-(-80) \pm \sqrt{(-80)^2 - 4(16)(-12)}}{2(16)}$$

$$t = \frac{80 \pm \sqrt{7168}}{32}$$

$t \approx 5.15$ or $t \approx -0.15$

5.15 seconds

38.

Pipe	Time to Complete the Job Alone	Work Rate	Time Working	Portion of the Job Completed
Large	$t - 3$	$\dfrac{1}{t-3}$	5	$\dfrac{5}{t-3}$
Small	t	$\dfrac{1}{t}$	5	$\dfrac{5}{t}$

$$\frac{5}{t-3} + \frac{5}{t} = 1$$

$$t(t-3) \cdot \left(\frac{5}{t-3} + \frac{5}{t} \right) = t(t-3) \cdot 1$$

$$5t + 5(t-3) = t(t-3)$$
$$5t + 5t - 15 = t^2 - 3t$$
$$10t - 15 = t^2 - 3t$$
$$0 = t^2 - 13t + 15$$

$$t = \frac{-(-13) \pm \sqrt{(-13)^2 - 4(1)(15)}}{2(1)}$$

$$t = \frac{13 \pm \sqrt{109}}{2}$$

$t \approx 11.7$ or $t \approx 1.3$
The value $t \approx 1.3$ is omitted because it gives a negative solution for the large pipe.
11.7 hours

39. $f(4n - 11) = 3(4n - 11) - 10$
$\qquad\qquad\quad = 12n - 33 - 10$
$\qquad\qquad\quad = 12n - 43$

40. $f(-5) = (-5)^2 - 10(-5) + 51$
$\qquad\quad = 25 + 50 + 51$
$\qquad\quad = 126$

41. $f(x) = -\dfrac{2}{3}x - 4$

y-intercept: $(0, -4)$

x-intercept: Let $f(x) = 0$.

$$-\frac{2}{3}x - 4 = 0$$

$$-\frac{2}{3}x = 4$$

$$-\frac{3}{2} \cdot \left(-\frac{2}{3}x\right) = -\frac{3}{2} \cdot 4$$

$$x = -6$$

$(-6, 0)$

$m = -\dfrac{2}{3}$. From the y-intercept $(0, -4)$,

move down 2, right 3 to the point $(3, -6)$.

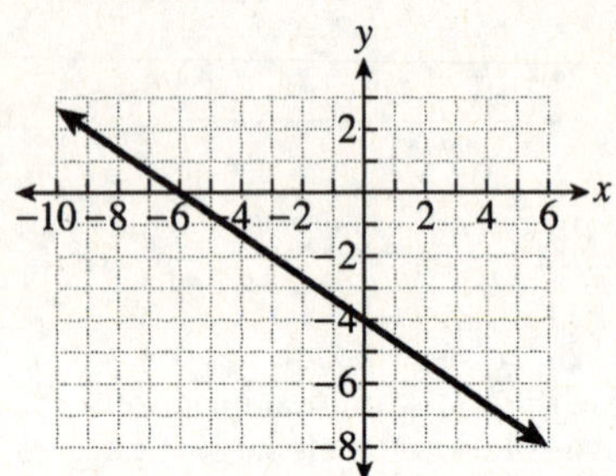

42. a) $m = \dfrac{44 - 35}{160 - 100} = \dfrac{9}{60} = 0.15$

$f(100) = 0.15(100) + b$

$0.15(100) + b = 35$

$15 + b = 35$

$b = 20$

$f(x) = 0.15x + 20$

b) Let $x = 300$.

$f(300) = 0.15(300) + 20 = 45 + 20 = 65$

$\$65$

43. $f(x) = x^2 - 30x + 129$ is a parabola opening

up; the parabola has a minimum at the vertex.

$$x = -\frac{b}{2a} = -\frac{-30}{2(1)} = 15$$

$$f(15) = (15)^2 - 30(15) + 129$$
$$= 225 - 450 + 129$$
$$= -96$$

Minimum value is -96.

44. $h(t) = -16t^2 + 144t + 50$

$$t = -\frac{b}{2a} = -\frac{144}{2(-16)} = 4.5$$

$$h(4.5) = -16(4.5)^2 + 144(4.5) + 50$$
$$= -16(20.25) + 648 + 50$$
$$= -324 + 648 + 50$$
$$= 374$$

374 feet

45. $f(x) = (x - 3)^2 - 4$

The vertex is $(3, -4)$.
Shift the basic parabola 3 units to the right and
4 units down.

y-intercept: Let $x = 0$.

$f(0) = (0 - 3)^2 - 4 = 9 - 4 = 5$

$(0, 5)$

x-intercepts: Let $f(x) = 0$.

$$(x - 3)^2 - 4 = 0$$

$$(x - 3)^2 = 4$$

$$\sqrt{(x - 3)^2} = \pm\sqrt{4}$$

$$x - 3 = \pm 2$$

$$x = 3 \pm 2$$

$x = 3 - 2 = 1$ or $x = 3 + 2 = 5$

$(1, 0), (5, 0)$

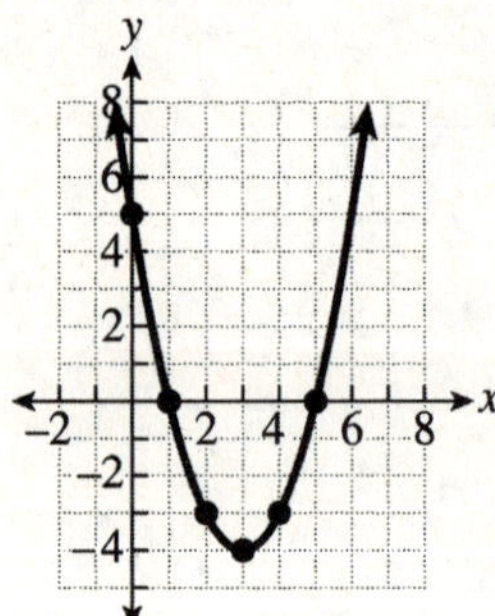

Domain: $(-\infty, \infty)$; range: $[-4, \infty)$

46. $f(x) = |x - 2| - 5$
$h = 2, k = -5$
Shift the graph of $f(x) = |x|$ by 2 units to the right and by 5 units down.

y-intercept: Let $x = 0$.
$f(0) = |0 - 2| - 5 = 2 - 5 = -3$
$(0, -3)$

x-intercepts: Let $f(x) = 0$.

$|x - 2| - 5 = 0$
$\quad |x - 2| = 5$

$x - 2 = -5 \quad$ or $\quad x - 2 = 5$
$\quad\quad x = -3 \quad$ or $\quad\quad x = 7$
$(-3, 0), (7, 0)$

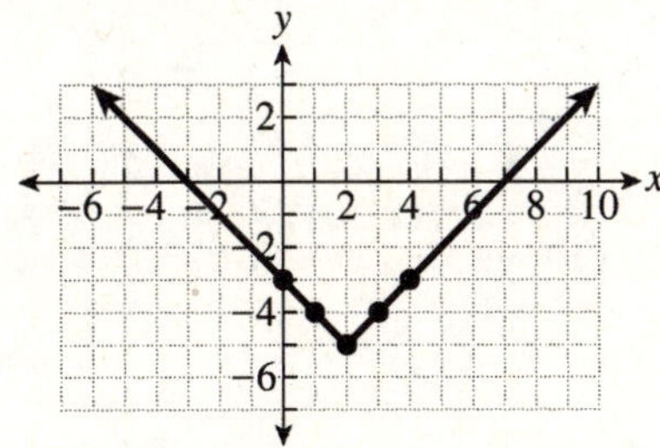

Domain: $(-\infty, \infty)$; range: $[-5, \infty)$

47. $f(x) = \sqrt{x + 4} - 1$
$h = -4, k = -1$
Shift the graph of $f(x) = \sqrt{x}$ by 4 units to the left and by 1 unit down.

y-intercept: Let $x = 0$.
$f(0) = \sqrt{0 + 4} - 1 = 2 - 1 = 1$
$(0, 1)$

x-intercept: Let $f(x) = 0$.
$\sqrt{x + 4} - 1 = 0$
$\quad \sqrt{x + 4} = 1$
$\left(\sqrt{x + 4}\right)^2 = 1^2$
$\quad\quad x + 4 = 1$
$\quad\quad\quad x = -3$
$(-3, 0)$

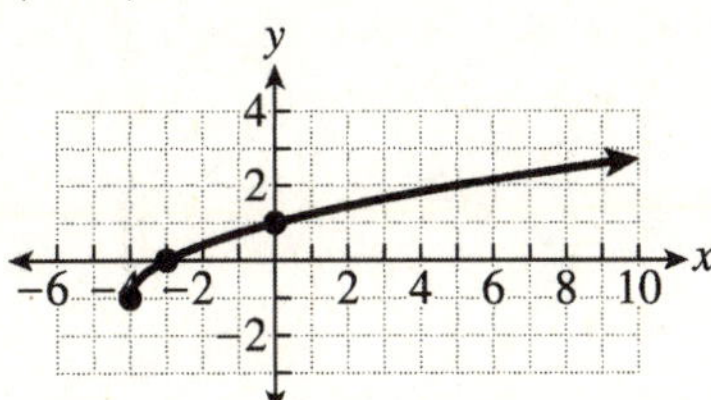

Domain: $[-4, \infty)$; range: $[-1, \infty)$

48. $f(x) = (x + 1)^3$
$h = -1, k = 0$
Shift the graph of $f(x) = x^3$ by 1 unit to the left.

Domain: $(-\infty, \infty)$; range: $(-\infty, \infty)$

49. $(f + g)(x) = f(x) + g(x)$
$\quad\quad\quad = (7x + 11) + (3x - 42)$
$\quad\quad\quad = 10x - 31$

50. $(f \circ g)(x) = f(g(x))$
$\quad\quad\quad = f\left(x^2 - 2x + 12\right)$
$\quad\quad\quad = \left(x^2 - 2x + 12\right) + 6$
$\quad\quad\quad = x^2 - 2x + 18$
$(g \circ f)(x) = g(f(x))$
$\quad\quad\quad = g(x + 6)$
$\quad\quad\quad = (x + 6)^2 - 2(x + 6) + 12$
$\quad\quad\quad = x^2 + 12x + 36 - 2x - 12 + 12$
$\quad\quad\quad = x^2 + 10x + 36$

51.

$(f \circ g)(x) = x$	$(g \circ f)(x) = x$
$(f \circ g)(x)$	$(g \circ f)(x)$
$= f(g(x))$	$= g(f(x))$
$= f\left(\dfrac{x - 8}{3}\right)$	$= g(3x + 8)$
$= 3\left(\dfrac{x - 8}{3}\right) + 8$	$= \dfrac{(3x + 8) - 8}{3}$
$= x - 8 + 8$	$= \dfrac{3x}{3}$
$= x$	$= x$

Yes, $f(x)$ and $g(x)$ are inverses.

52. $f(x) = 3x - 19$
$$y = 3x - 19$$
Interchange x and y.
$$x = 3y - 19$$
$$x + 19 = 3y$$
$$\frac{x + 19}{3} = y$$
$$f^{-1}(x) = \frac{x + 19}{3}$$

53. $f(x) = \dfrac{3}{x - 9}$
$$y = \frac{3}{x - 9}$$
Interchange x and y.
$$x = \frac{3}{y - 9}$$
$$(y - 9) \cdot x = (y - 9) \cdot \frac{3}{y - 9}$$
$$xy - 9x = 3$$
$$xy = 3 + 9x$$
$$y = \frac{3 + 9x}{x}$$
$$f^{-1}(x) = \frac{3 + 9x}{x}, x \neq 0$$

12.1 QUICK CHECK

1. $f(x) = 4^x$

$f(5) = 4^5 = 1024$

$f(-3) = 4^{-3} = \dfrac{1}{4^3} = \dfrac{1}{64}$

2. $f(x) = 2 \cdot 7^{x+3} - 19$

$f(-1) = 2 \cdot 7^{-1+3} - 19$

$\quad\quad = 2 \cdot 7^2 - 19$
$\quad\quad = 2 \cdot 49 - 19$
$\quad\quad = 98 - 19$
$\quad\quad = 79$

3.

x	$f(x) = 5^x$
-1	$5^{-1} = \dfrac{1}{5^1} = \dfrac{1}{5}$
0	$5^0 = 1$
1	$5^1 = 5$

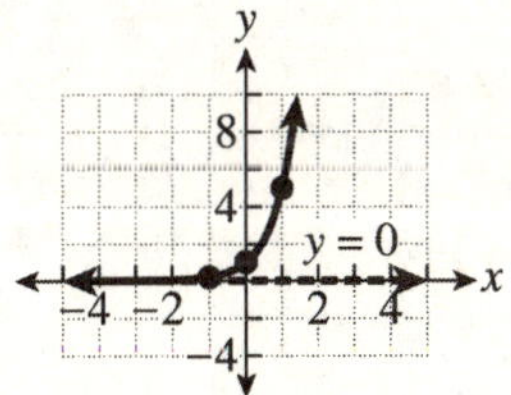

Domain: $(-\infty, \infty)$; range: $(0, \infty)$

4.

x	$f(x) = \left(\dfrac{1}{2}\right)^x$
-1	$\left(\dfrac{1}{2}\right)^{-1} = 2^1 = 2$
0	$\left(\dfrac{1}{2}\right)^0 = 1$
1	$\left(\dfrac{1}{2}\right)^1 = \dfrac{1}{2}$

Domain: $(-\infty, \infty)$; range: $(0, \infty)$

5. $f(x) = e^x$

$f(-1) = e^{-1} = \dfrac{1}{e} \approx 0.368$

$f(4.3) = e^{4.3} \approx 73.700$

6. a) $4^x = 256$
$\quad\quad 4^x = 4^4$
$\quad\quad\ x = 4$
$\quad\quad \{4\}$

b) $9^{2x-17} = 9$
$\quad\ 9^{2x-17} = 9^1$
$\quad\ 2x - 17 = 1$
$\quad\quad\quad 2x = 18$
$\quad\quad\quad\ x = 9$
$\quad \{9\}$

7. The year 2018 corresponds to $x = 26$.

$f(26) = 7.4 \cdot 1.08^{26} \approx 54.7$
Approximately \$54.7 billion

12.1 EXPONENTIAL FUNCTIONS

1. exponential

3. increasing

5. $r = s$

7. $f(4) = 2^4 = 16$

9. $f(0) = 2^0 = 1$

11. $f(2) = \left(\dfrac{2}{5}\right)^2 = \dfrac{4}{25}$

13. $f(-4) = \left(\dfrac{2}{5}\right)^{-4} = \left(\dfrac{5}{2}\right)^4 = \dfrac{625}{16}$

15. $f(5) = 4^{5-3} = 4^2 = 16$

17. $f(4) = 3^{4+2} - 19 = 3^6 - 19 = 729 - 19 = 710$

19. $f(2) = 5^{2(2)+1} + 400$
$= 5^{4+1} + 400$
$= 5^5 + 400$
$= 3125 + 400$
$= 3525$

21. $f(2) = \left(\dfrac{1}{5}\right)^{2+4} = \left(\dfrac{1}{5}\right)^6 = \dfrac{1}{5^6} = \dfrac{1}{15,625}$

23. $f(-2) = 4^{-(-2)} + 18 = 4^2 + 18 = 16 + 18 = 34$

25. $f(x) = 4^x$
horizontal asymptote $y = 0$

x	$f(x) = 4^x$
-1	$4^{-1} = \dfrac{1}{4}$
0	$4^0 = 1$
1	$4^1 = 4$

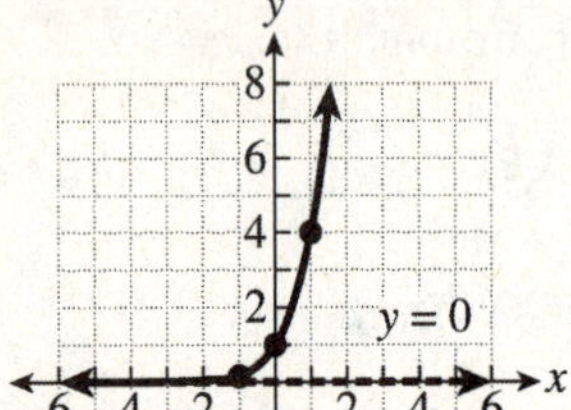

Domain: $(-\infty, \infty)$; range: $(0, \infty)$

27. $f(x) = \left(\dfrac{1}{4}\right)^x$

horizontal asymptote $y = 0$

x	$f(x) = \left(\dfrac{1}{4}\right)^x$
-1	$\left(\dfrac{1}{4}\right)^{-1} = 4$
0	$\left(\dfrac{1}{4}\right)^0 = 1$
1	$\left(\dfrac{1}{4}\right)^1 = \dfrac{1}{4}$

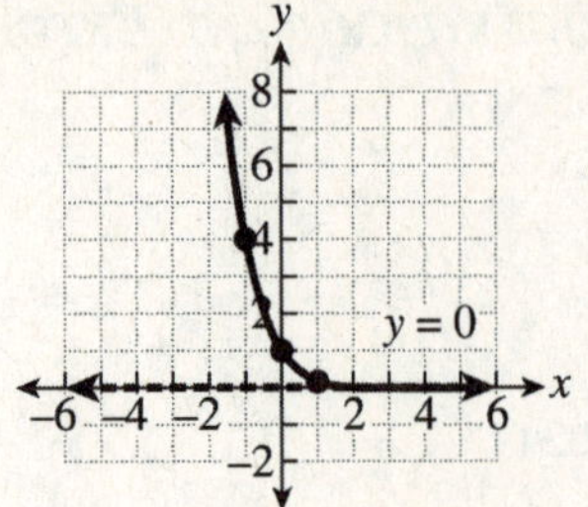

Domain: $(-\infty, \infty)$; range: $(0, \infty)$

29. $f(x) = 7^x$
horizontal asymptote $y = 0$

x	$f(x) = 7^x$
-1	$7^{-1} = \dfrac{1}{7}$
0	$7^0 = 1$
1	$7^1 = 7$

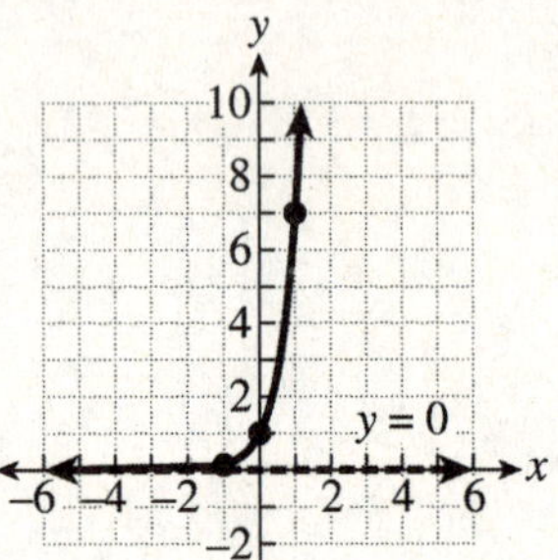

Domain: $(-\infty, \infty)$; range: $(0, \infty)$

31. $f(x) = 6^{-x}$

horizontal asymptote $y = 0$

x	$f(x) = 6^{-x}$
-1	$6^{-(-1)} = 6^1 = 6$
0	$6^{-(0)} = 6^0 = 1$
1	$6^{-1} = \dfrac{1}{6}$

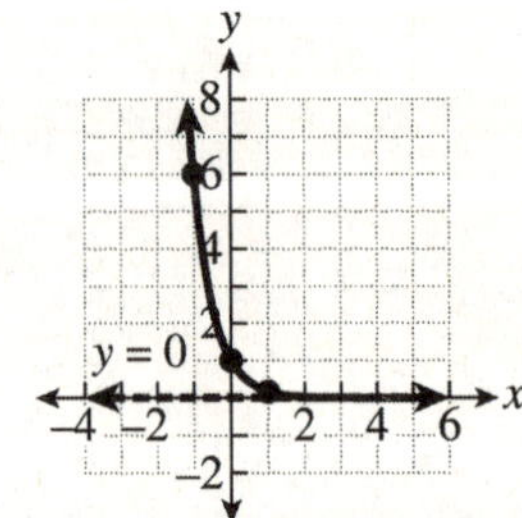

Domain: $(-\infty, \infty)$; range: $(0, \infty)$

33. $f(2) = e^2 \approx 7.389$

35. $f(-5.2) = e^{-5.2} \approx 0.006$

37. $f(-1) = e^{-1+4} = e^3 \approx 20.086$

39. $f(-2) = e^{3(-2)-1} = e^{-6-1} = e^{-7} \approx 0.001$

41. $3^x = 81$
$3^x = 3^4$
$x = 4$
$\{4\}$

43. $5^x = 125$
$5^x = 5^3$
$x = 3$
$\{3\}$

45. $6^x = \dfrac{1}{6}$
$6^x = 6^{-1}$
$x = -1$
$\{-1\}$

47. $2^{x-6} = 32$
$2^{x-6} = 2^5$
$x - 6 = 5$
$x = 11$
$\{11\}$

49. $5^{x+4} = \dfrac{1}{25}$
$5^{x+4} = 5^{-2}$
$x + 4 = -2$
$x = -6$
$\{-6\}$

51. $\left(\dfrac{1}{3}\right)^{x+1} = \dfrac{1}{729}$
$\left(\dfrac{1}{3}\right)^{x+1} = \left(\dfrac{1}{3}\right)^{6}$
$x + 1 = 6$
$x = 5$
$\{5\}$

53. $6^{2x-13} = 216$
$6^{2x-13} = 6^3$
$2x - 13 = 3$
$2x = 16$
$x = 8$
$\{8\}$

55. $3^{4x+3} = \dfrac{1}{243}$
$3^{4x+3} = 3^{-5}$
$4x + 3 = -5$
$4x = -8$
$x = -2$
$\{-2\}$

57. $9^{3x-12} = 1$
$9^{3x-12} = 9^0$
$3x - 12 = 0$
$3x = 12$
$x = 4$
$\{4\}$

59. $2^{x^2+3x-6} = 16$
$2^{x^2+3x-6} = 2^4$
$x^2 + 3x - 6 = 4$
$x^2 + 3x - 10 = 0$
$(x+5)(x-2) = 0$
$x + 5 = 0 \quad$ or $\quad x - 2 = 0$
$\quad x = -5 \quad$ or $\quad\quad x = 2$
$\{-5, 2\}$

61. Let $t = 40$.
$f(40) = 5000(1.13)^{40} \approx 663,907.76$
Account balance is $663,907.76.

63. Let $t = 5$.
$f(5) = 10,000(1.0225)^{4(5)} \approx 15,605.09$
Account balance is $15,605.09.

65. The year 2017 corresponds to $t = 22$.
$f(22) = 8e^{0.1145(22)} \approx 99.33$
The value of the card in 2017 will be $99.33.

67. The year 2016 corresponds to $t = 16$.
$f(16) = 2.8e^{0.1767(16)} \approx 47.3$
The net revenues in 2016 will be $47.3 billion.

69. Let $t = 60$.

$$f(60) = 4 + 76e^{-0.014(60)} \approx 36.8$$

After 60 minutes, the temperature will be $36.8° \text{C}$.

71. The year 2020 corresponds to $t = 27$.

$$f(27) = 0.74 \cdot 1.033^{27} \approx 1.78$$

The cost of a loaf in 2020 will be $1.78.

73. Answers will vary. Example:

The graph of $f(x) = b^x$ for $b > 1$ will be an increasing graph.

The graph of $f(x) = b^x$ for $0 < b < 1$ will be a decreasing graph.

75. Answers will vary. Example:

$$2^{x+1} = 32$$
$$2^{x+1} = 2^5 \qquad \text{Convert to like bases.}$$
$$x + 1 = 5 \qquad \text{Set exponents equal.}$$
$$x = 4 \qquad \text{Solve for } x.$$
$$\{4\}$$

12.2 QUICK CHECK

1. a) Since $5^2 = 25$, $\log_5 25 = 2$.

b) Since $3^{-3} = \dfrac{1}{3^3} = \dfrac{1}{27}$, $\log_3 \dfrac{1}{27} = -3$.

c) Since $12^1 = 12$, $\log_{12} 12 = 1$.

d) Since $6^0 = 1$, $\log_6 1 = 0$.

2. $f\left(\dfrac{1}{256}\right) = \log_4\left(\dfrac{1}{256}\right) = -4$,

since $4^{-4} = \dfrac{1}{4^4} = \dfrac{1}{256}$

3. a) $\log 10,000 = 4$

b) $\log 75 \approx 1.875$

4. a) $\ln e^{-3} = -3$

b) $\ln 605 \approx 6.405$

5. $7^x = 20$ in logarithmic form: $\log_7 20 = x$

6. $\log_3 x = 6$ in exponential form: $3^6 = x$

7. $\log_4 x = 5$

$$4^5 = x \qquad \text{Rewrite in exponential form.}$$
$$1024 = x$$
$$\{1024\}$$

8. $f(x) = \log_3(x + 6)$

$$\log_3(x + 6) = 2$$
$$3^2 = x + 6 \qquad \text{Rewrite in}$$
$$9 = x + 6 \qquad \text{exponential form.}$$
$$3 = x$$
$$\{3\}$$

9. a) $f(x) = \log_5 x$

$b = 5$

Plot points at:
$(1, 0)$
$(b, 1) = (5, 1)$
$\left(\dfrac{1}{b}, -1\right) = \left(\dfrac{1}{5}, -1\right)$

Vertical asymptote at $x = 0$.

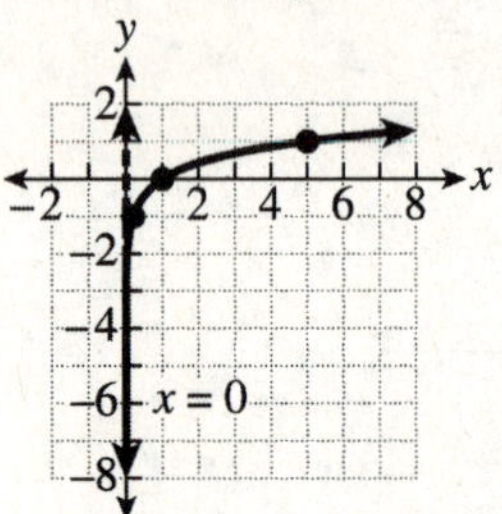

Domain: $(0, \infty)$; range: $(-\infty, \infty)$

b) $f(x) = \log_{1/4} x$

$b = \dfrac{1}{4}$

Plot points at:
$(1, 0)$
$(b, 1) = \left(\dfrac{1}{4}, 1\right)$
$\left(\dfrac{1}{b}, -1\right) = \left(\dfrac{1}{\frac{1}{4}}, -1\right) = (4, -1)$

Vertical asymptote at $x = 0$.

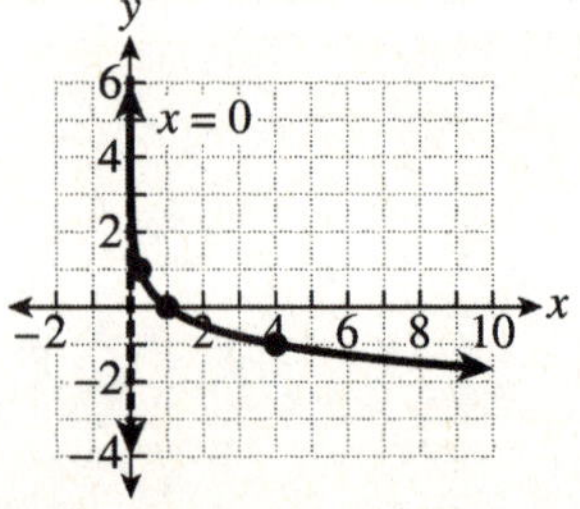

Domain: $(0, \infty)$; range: $(-\infty, \infty)$

10. The year 2020 corresponds to $x = 25$.

$$f(25) = 7.95\ln(25) + 20.95 \approx 46.54$$

Annual Sales in 2020 will be approximately $46.54 billion.

11. Use the formula

$R = \log I,$ where $I = 2500$
$R = \log 2500$
$R \approx 3.4$

The magnitude is 3.4 on the Richter scale.

12.2 LOGARITHMIC FUNCTIONS

1. $\log_b x$

3. common

5. exponential

7. vertical

9. Since $2^2 = 4,\ \log_2 4 = 2$.

11. Since $3^3 = 27,\ \log_3 27 = 3$.

13. Since $4^3 = 64,\ \log_4 64 = 3$.

15. Since $8^0 = 1,\ \log_8 1 = 0$.

17. Since $3^{-2} = \dfrac{1}{3^2} = \dfrac{1}{9},\ \log_3\left(\dfrac{1}{9}\right) = -2$.

19. Since $8^{-3} = \dfrac{1}{8^3} = \dfrac{1}{512},\ \log_8\left(\dfrac{1}{512}\right) = -3$.

21. a) $f(9) = \log_3 9 = 2$

b) $f(243) = \log_3 243 = 5$

23. a) $f(8) = \log_2(8 - 7) = \log_2(1) = 0$

b) $f(135) = \log_2(135 - 7) = \log_2(128) = 7$

25. $\log 1000 = 3$

27. $\log 0.01 = -2$

29. $\log 112 \approx 2.049$

31. $\log 7.42 \approx 0.870$

33. $\ln e = 1$

35. $\ln\left(\dfrac{1}{e^6}\right) = -6$

37. $\ln 32 \approx 3.466$

39. $\ln 16.23 \approx 2.787$

41. $4^3 = 64$ in logarithmic form: $\log_4 64 = 3$

43. $8^{-2} = \dfrac{1}{64}$ in logarithmic form: $\log_8 \dfrac{1}{64} = -2$

45. $5^x = 42$ in logarithmic form: $\log_5 42 = x$

47. $10^x = 321$ in logarithmic form: $\log_{10} 321 = x$, which can be written as $\log 321 = x$

49. $\log_2 128 = 7$ in exponential form: $2^7 = 128$

51. $\log_9 \dfrac{1}{6561} = -4$ in exponential form:

$$9^{-4} = \dfrac{1}{6561}$$

53. $\ln x = 6$ in exponential form: $e^6 = x$

55. $\log x + 5 = 12$ is equivalent to $\log x = 7$.

$\log x = 7$ in exponential form: $10^7 = x$

57. $\log_2 x = 5$

$2^5 = x$ Rewrite in exponential form.
$32 = x$
$\{32\}$

59. $\log_3 x = -4$

$3^{-4} = x$ Rewrite in exponential form.
$\dfrac{1}{3^4} = x$
$\dfrac{1}{81} = x$
$\left\{\dfrac{1}{81}\right\}$

61. $\log x = 4$

$10^4 = x$ Rewrite in exponential form.
$10,000 = x$
$\{10,000\}$

63. $\ln x = 2$

$\quad e^2 = x$ Rewrite in exponential form.

$\quad 7.389 \approx x$

$\quad \{7.389\}$

65. $\log_2 x + 5 = 8$

$\quad \log_2 x = 3$

$\quad\quad 2^3 = x$ Rewrite in exponential form.

$\quad\quad\; 8 = x$

$\quad \{8\}$

67. $\ln(x-3) + 6 = 10$

$\quad\quad \ln(x-3) = 4$

$\quad\quad\quad e^4 = x - 3$ Rewrite in

$\quad\quad\quad e^4 + 3 = x$ exponential form.

$\quad\quad\quad 57.598 \approx x$

$\quad \{57.598\}$

69. $f(x) = \log_3(2x-1)$

$\quad \log_3(2x-1) = 4$

$\quad\quad\quad 3^4 = 2x - 1$ Rewrite in

$\quad\quad\quad 81 = 2x - 1$ exponential form.

$\quad\quad\quad 82 = 2x$

$\quad\quad\quad 41 = x$

$\quad \{41\}$

71. $f(x) = \log(x-8) + 5$

$\quad \log(x-8) + 5 = 7$

$\quad\quad \log(x-8) = 2$

$\quad\quad\quad 10^2 = x - 8$ Rewrite in

$\quad\quad\quad 100 = x - 8$ exponential form.

$\quad\quad\quad 108 = x$

$\quad \{108\}$

73. $\log_5 125 = x$

$\quad\quad 5^x = 125$ Rewrite in exponential form.

$\quad\quad 5^x = 5^3$

$\quad\quad\; x = 3$

$\quad \{3\}$

75. $\log 100,000 = x$

$\quad\quad 10^x = 100,000$ Rewrite in

$\quad\quad 10^x = 10^5$ exponential form.

$\quad\quad\; x = 5$

$\quad \{5\}$

77. $\log_9 27 = x$

$\quad\quad 9^x = 27$ Rewrite in exponential form.

$\quad\quad \left(3^2\right)^x = 3^3$

$\quad\quad 3^{2x} = 3^3$

$\quad\quad 2x = 3$

$\quad\quad x = \dfrac{3}{2}$

$\quad \left\{\dfrac{3}{2}\right\}$

79. $f(x) = \log_3 x$

$b = 3$

Plot points at:

$(1,0)$

$(b,1) = (3,1)$

$\left(\dfrac{1}{b}, -1\right) = \left(\dfrac{1}{3}, -1\right)$

Vertical asymptote at $x = 0$.

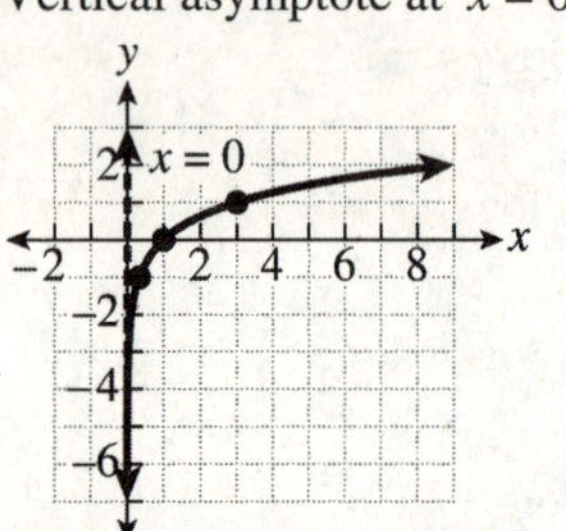

Domain: $(0, \infty)$; range: $(-\infty, \infty)$

81. $f(x) = \log x$

$b = 10$

Plot points at:

$(1,0)$

$(b,1) = (10,1)$

$\left(\dfrac{1}{b}, -1\right) = \left(\dfrac{1}{10}, -1\right)$

Vertical asymptote at $x = 0$.

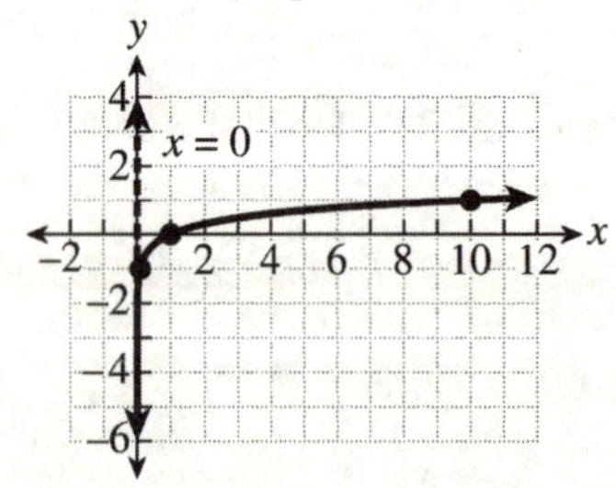

Domain: $(0, \infty)$; range: $(-\infty, \infty)$

83. $f(x) = \log_{1/5} x$

$b = \dfrac{1}{5}$

Plot points at:

$(1, 0)$

$(b, 1) = \left(\dfrac{1}{5}, 1\right)$

$\left(\dfrac{1}{b}, -1\right) = \left(\dfrac{1}{\frac{1}{5}}, -1\right) = (5, -1)$

Vertical asymptote at $x = 0$.

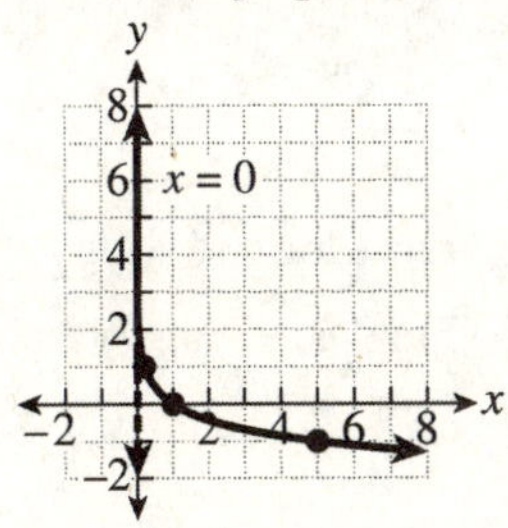

Domain: $(0, \infty)$; range: $(-\infty, \infty)$

85. The year 2030 corresponds to $x = 51$.

$f(51) = 14{,}650 + 1344\ln 51 \approx 19{,}934$

There will be 19,934 airports in 2030.

87. $f(24) = 18.6 + 4.78\ln 24 \approx 33.8$

The average height of girls who are 24 months old is 33.8 inches.

89. $R = \log 16{,}000$

$R \approx 4.2$

The magnitude is 4.2 on the Richter scale.

91. $8.8 = \log I$

in exponential form:

$$10^{8.8} = I$$

$630{,}957{,}345 \approx I$

The earthquake is 630,957,345 times larger.

93. Answers will vary. Example:

A logarithm is a function. We define $\log_b x$ to be the exponent that b must be raised to so that the result is x.

1. $\log 5 + \log 7 + \log x = \log(5 \cdot 7) + \log x$

$\qquad\qquad = \log(5 \cdot 7 \cdot x)$

$\qquad\qquad = \log(35x)$

2. $\log_2(5b) = \log_2 5 + \log_2 b$

3. $\log_2 60 - \log_2 15 = \log_2\left(\dfrac{60}{15}\right) = \log_2 4 = 2$

4. $\log_b\left(\dfrac{x}{b}\right) = \log_b x - \log_b b = \log_b x - 1$

5. $\log_3\left(9x^2 y^4\right) = \log_3 9 + \log_3 x^2 + \log_3 y^4$

$\qquad\qquad = 2 + 2\log_3 x + 4\log_3 y$

6. $\ln \sqrt[7]{x} = \ln x^{1/7} = \dfrac{1}{7}\ln x$

7. $8\log_8 x = \log_8 x^8$

8. $\ln e^{-4} = -4\ln e = -4 \cdot 1 = -4$

9. $e^{\ln 16} = 16$

10. $\log_2 a - \log_2 b + \log_2 c + \log_2 d$

$= \log_2\left(\dfrac{a}{b}\right) + \log_2 c + \log_2 d$

$= \log_2\left(\dfrac{ac}{b}\right) + \log_2 d$

$= \log_2\left(\dfrac{acd}{b}\right)$

11. $6\log a + \dfrac{1}{5}\log b - 9\log c - 12\log d$

$= \log a^6 + \log b^{1/5} - \log c^9 - \log d^{12}$

$= \log a^6 + \log b^{1/5} - \left(\log c^9 + \log d^{12}\right)$

$= \log\left(a^6\sqrt[5]{b}\right) - \log\left(c^9 d^{12}\right)$

$= \log\left(\dfrac{a^6\sqrt[5]{b}}{c^9 d^{12}}\right)$

12.
$$\log_5\left(\frac{z^3}{x^9 y^4}\right) = \log_5 z^3 - \log_5\left(x^9 y^4\right)$$
$$= 3\log_5 z - \left(\log_5 x^9 + \log_5 y^4\right)$$
$$= 3\log_5 z - \log_5 x^9 - \log_5 y^4$$
$$= 3\log_5 z - 9\log_5 x - 4\log_5 y$$

13. $\log_3 125 = \dfrac{\log 125}{\log 3} \approx 4.395$

12.3 PROPERTIES OF LOGARITHMS

1. $\log_b (xy)$

3. $r \cdot \log_b x$

5. $\log_4 3 + \log_4 a = \log_4 (3a)$

7. $\log x + \log y = \log (xy)$

9. $\log_9 3 + \log_9 27 = \log_9 (3 \cdot 27) = \log_9 81 = 2$

11. $\log_3 (10x) = \log_3 10 + \log_3 x$

13. $\ln (3e) = \ln 3 + \ln e = \ln 3 + 1$

15. $\log_4 (abc) = \log_4 a + \log_4 b + \log_4 c$

17. $\log_3 56 - \log_3 8 = \log_3\left(\dfrac{56}{8}\right) = \log_3 7$

19. $\ln x - \ln 3 = \ln\left(\dfrac{x}{3}\right)$

21.
$$\log_4 2 - \log_4 128 = \log_4\left(\frac{2}{128}\right)$$
$$= \log_4\left(\frac{1}{64}\right)$$
$$= -3$$

23.
$$\log_8\left(\frac{10a}{b}\right) = \log_8 (10a) - \log_8 b$$
$$= \log_8 10 + \log_8 a - \log_8 b$$

25.
$$\log_4\left(\frac{6xyz}{w}\right)$$
$$= \log_4 (6xyz) - \log_4 w$$
$$= \log_4 6 + \log_4 x + \log_4 y + \log_4 z - \log_4 w$$

27.
$$\log_5\left(\frac{xy}{25z}\right) = \log_5 (xy) - \log_5 (25z)$$
$$= \log_5 x + \log_5 y - \left(\log_5 25 + \log_5 z\right)$$
$$= \log_5 x + \log_5 y - \log_5 25 - \log_5 z$$
$$= \log_5 x + \log_5 y - 2 - \log_5 z$$

29. $\log_2 x^7 = 7\log_2 x$

31.
$$\log_5\left(x^4 y^7\right) = \log_5 x^4 + \log_5 y^7$$
$$= 4\log_5 x + 7\log_5 y$$

33. $\log_3 \sqrt{x} = \log_3 x^{1/2} = \dfrac{1}{2}\log_3 x$

35.
$$\log_2\left(16x^5 y^7\right) = \log_2 16 + \log_2 x^5 + \log_2 y^7$$
$$= 4 + 5\log_2 x + 7\log_2 y$$

37. $8\log_5 x = \log_5 x^8$

39. $\dfrac{1}{3}\ln x = \ln x^{1/3} = \ln \sqrt[3]{x}$

41. $5\log_3 x^2 = \log_3\left(x^2\right)^5 = \log_3 x^{10}$

43. $\log_5 5^{12} = 12$

45. $\ln e^9 = 9$

47. $2^{\log_2 7} = 7$

49. $3^{\log_3 b} = b$

51.
$$\log_7 4 + \log_7 5 - \log_7 3 = \log_7 20 - \log_7 3$$
$$= \log_7\left(\frac{20}{3}\right)$$

53.
$$4\log_9 2 + 3\log_9 5 = \log_9 2^4 + \log_9 5^3$$
$$= \log_9 16 + \log_9 125$$
$$= \log_9 (16 \cdot 125)$$
$$= \log_9 2000$$

55.
$$\frac{1}{2}\ln 16 - \frac{1}{3}\ln 125 = \ln 16^{1/2} - \ln 125^{1/3}$$
$$= \ln \sqrt{16} - \ln \sqrt[3]{125}$$
$$= \ln 4 - \ln 5$$
$$= \ln\left(\frac{4}{5}\right)$$

57. $\log_2 x - \log_2 y - \log_2 z - \log_2 w$
$= \log_2 x - \left(\log_2 y + \log_2 z + \log_2 w\right)$
$= \log_2 x - \log_2\left(yzw\right)$
$= \log_2\left(\dfrac{x}{yzw}\right)$

59. $4\ln x + 9\ln y - 13\ln z = \ln x^4 + \ln y^9 - \ln z^{13}$
$= \ln\left(x^4 y^9\right) - \ln z^{13}$
$= \ln\left(\dfrac{x^4 y^9}{z^{13}}\right)$

61. $\log_b\left(x+6\right) + \log_b\left(x-8\right)$
$= \log_b\left(\left(x+6\right)\left(x-8\right)\right)$
$= \log_b\left(x^2 - 8x + 6x - 48\right)$
$= \log_b\left(x^2 - 2x - 48\right)$

63. $\log\left(3x-4\right) + \log\left(x+10\right)$
$= \log\left(\left(3x-4\right)\left(x+10\right)\right)$
$= \log\left(3x^2 + 30x - 4x - 40\right)$
$= \log\left(3x^2 + 26x - 40\right)$

65. $3\ln a + \dfrac{1}{3}\ln b - 10\ln c = \ln a^3 + \ln b^{1/3} - \ln c^{10}$
$= \ln a^3 + \ln \sqrt[3]{b} - \ln c^{10}$
$= \ln\left(a^3 \sqrt[3]{b}\right) - \ln c^{10}$
$= \ln\left(\dfrac{a^3 \sqrt[3]{b}}{c^{10}}\right)$

67. $\log_b 6 = \log_b\left(2\cdot 3\right) = \log_b 2 + \log_b 3 = A + B$

69. $\log_b 35 = \log_b\left(5\cdot 7\right) = \log_b 5 + \log_b 7 = C + D$

71. $\log_b 8 = \log_b 2^3 = 3\log_b 2 = 3A$

73. $\log_b 125 = \log_b 5^3 = 3\log_b 5 = 3C$

75. $\log_3\left(\dfrac{ab}{cd}\right) = \log_3 ab - \log_3 cd$
$= \log_3 a + \log_3 b - \left(\log_3 c + \log_3 d\right)$
$= \log_3 a + \log_3 b - \log_3 c - \log_3 d$

77. $\log_2\left(\dfrac{a^3}{8b^4}\right) = \log_2 a^3 - \log_2\left(8b^4\right)$
$= 3\log_2 a - \left(\log_2 8 + \log_2 b^4\right)$
$= 3\log_2 a - \left(3 + 4\log_2 b\right)$
$= 3\log_2 a - 3 - 4\log_2 b$

79. $\ln\left(\dfrac{a^6 \sqrt{b}}{c^7}\right) = \ln a^6 \sqrt{b} - \ln c^7$
$= \ln a^6 + \ln \sqrt{b} - \ln c^7$
$= 6\ln a + \dfrac{1}{2}\ln b - 7\ln c$

81. $-\ln\left(\dfrac{a^3 d^2}{bc^7}\right) = -\left(\ln\left(a^3 d^2\right) - \ln\left(bc^7\right)\right)$
$= -\ln\left(a^3 d^2\right) + \ln\left(bc^7\right)$
$= -\left(\ln a^3 + \ln d^2\right) + \ln b + \ln c^7$
$= -3\ln a - 2\ln d + \ln b + 7\ln c$

83. $\log_2 11 = \dfrac{\log 11}{\log 2} \approx 3.459$

85. $\log_{15} 30 = \dfrac{\log 30}{\log 15} \approx 1.256$

87. $\log_3 12{,}695 = \dfrac{\log 12{,}695}{\log 3} \approx 8.601$

89. $\log_6 4 = \dfrac{\log 4}{\log 6} \approx 0.774$

91. $\log_8 0.13 = \dfrac{\log 0.13}{\log 8} \approx -0.981$

93. $5^? = 19$
Write in exponential form.
$? = \log_5 19$
$? = \dfrac{\log 19}{\log 5}$
$? = 1.829$

95. $15^? = 100$
Write in exponential form.
$? = \log_{15} 100$
$? = \dfrac{\log 100}{\log 15}$
$? = 1.701$

97. Examples will vary. Example:
Let $x = 3, y = 4, b = 10, n = 2$.

$$\log_b (x+y) = \log_b x + \log_b y$$
$$\log (3+4) = \log 3 + \log 4$$
$$\log 7 = \log 3 + \log 4$$
$$0.845 = 0.477 + 0.602$$
$$0.845 = 1.079$$

False.

$$\log_b (x-y) = \log_b x - \log_b y$$
$$\log (3-4) = \log 3 - \log 4$$
$$\log (-1) = 0.477 - 0.602$$

False, because $\log(-1)$ is undefined.

$$\log_b x^n = \left(\log_b x\right)^n$$
$$\log 3^2 = \left(\log 3\right)^2$$
$$\log 9 = \left(\log 3\right)\left(\log 3\right)$$
$$0.954 = (0.477)(0.477)$$
$$0.954 = 0.228$$

False.

12.4 QUICK CHECK

1. a) $3^{3x-2} = 81$

$$3^{3x-2} = 3^4$$
$$3x - 2 = 4$$
$$3x = 6$$
$$x = 2$$
$$\{2\}$$

b) $4^{3x-1} = 32^{x+2}$

$$\left(2^2\right)^{3x-1} = \left(2^5\right)^{x+2}$$
$$2^{6x-2} = 2^{5x+10}$$
$$6x - 2 = 5x + 10$$
$$x - 2 = 10$$
$$x = 12$$
$$\{12\}$$

2. $4^{x-5} - 9 = 3$

$$4^{x-5} = 12$$
$$\log 4^{x-5} = \log 12$$
$$(x-5)\log 4 = \log 12$$
$$x - 5 = \frac{\log 12}{\log 4}$$
$$x = \frac{\log 12}{\log 4} + 5$$
$$x \approx 6.792$$
$$\left\{\frac{\log 12}{\log 4} + 5\right\}$$

3. $f(x) = 9^{x+3} - 17$

$$9^{x+3} - 17 = -6$$
$$9^{x+3} = 11$$
$$\log 9^{x+3} = \log 11$$
$$(x+3)\log 9 = \log 11$$
$$x + 3 = \frac{\log 11}{\log 9}$$
$$x = \frac{\log 11}{\log 9} - 3$$
$$x \approx -1.909$$
$$\left\{\frac{\log 11}{\log 9} - 3\right\}$$

4. $e^{x+2} - 10 = 18$

$$e^{x+2} = 28$$
$$\ln e^{x+2} = \ln 28$$
$$x + 2 = \ln 28$$
$$x = \ln 28 - 2$$
$$x \approx 1.332$$
$$\{\ln 28 - 2\}$$

5. $\log\left(x^2 + 15x - 30\right) = \log 2x$

$$x^2 + 15x - 30 = 2x$$
$$x^2 + 13x - 30 = 0$$
$$(x+15)(x-2) = 0$$
$$x = -15 \quad \text{or} \quad x = 2$$

check: $x = -15$
$$\log\left((-15)^2 + 15(15) - 30\right) = \log 2(-15)$$
$$\log(-30) = \log(-30)$$
Undefined

check: $x = 2$
$$\log\left((2)^2 + 15(2) - 30\right) = \log 2(2)$$
$$\log 4 = \log 4$$
True
$$\{2\}$$

6. $\log_4 (2x - 7) = \log_4 (3x + 2)$
$$2x - 7 = 3x + 2$$
$$-7 = x + 2$$
$$-9 = x$$

check: $x = -9$
$$\log_4 (2(-9) - 7) = \log_4 (3(-9) + 2)$$
$$\log_4 (-25) = \log_4 (-25)$$
Undefined
$$\varnothing$$

7. $\log_7\left(x^2-3x-3\right)=1$

Convert to exponential form.

$7^1 = x^2 - 3x - 3$
$0 = x^2 - 3x - 10$
$0 = (x+2)(x-5)$
$x = -2 \ \text{ or } \ x = 5$
$\{-2, 5\}$

8. $f(x) = \log(3x-14) + 9 = 15$
$\qquad \log(3x-14) = 6$

Convert to exponential form.

$10^6 = 3x - 14$
$1,000,000 = 3x - 14$
$1,000,014 = 3x$
$333,338 = x$
$\{333,338\}$

9. a) $\log_2 x + \log_2(x+4) = 5$
$\qquad \log_2 x(x+4) = 5$

Convert to exponential form.

$x(x+4) = 2^5$
$x^2 + 4x = 32$
$x^2 + 4x - 32 = 0$
$(x+8)(x-4) = 0$
$x = -8 \ \text{ or } \ x = 4$

check: $x = -8$
$\log_2(-8) + \log_2(-8+4) = 5$
$\qquad \log_2(-8) + \log_2(-4) = 5$
Undefined

check: $x = 4$
$\log_2(4) + \log_2(4+4) = 5$
$\qquad\qquad 2 + 3 = 5$
$\qquad\qquad\qquad 5 = 5$

True
$\{4\}$

b) $\log_3(41x-42) - \log_3(x-2) = 4$
$\qquad \log_3\left(\dfrac{41x-42}{x-2}\right) = 4$

Convert to exponential form.

$\dfrac{41x-42}{x-2} = 3^4$

$41x - 42 = 81(x-2)$
$41x - 42 = 81x - 162$
$-40x - 42 = -162$
$\qquad -40x = -120$
$\qquad\qquad x = 3$

check: $x = 3$
$\log_3(41(3)-42) - \log_3(3-2) = 4$
$\qquad \log_3(81) - \log_3 1 = 4$
$\qquad\qquad\qquad 4 - 0 = 4$
$\qquad\qquad\qquad\qquad 4 = 4$
True
$\{3\}$

10. $\log_2(x-3) + \log_2(x-7) = \log_2(3x-1)$
$\qquad \log_2\big((x-3)(x-7)\big) = \log_2(3x-1)$
$\qquad\qquad (x-3)(x-7) = (3x-1)$
$\qquad x^2 - 7x - 3x + 21 = 3x - 1$
$\qquad\qquad x^2 - 10x + 21 = 3x - 1$
$\qquad\qquad x^2 - 13x + 22 = 0$
$\qquad\qquad (x-2)(x-11) = 0$
$\qquad\qquad x = 2 \ \text{ or } \ x = 11$

check: $x = 2$
$\log_2(2-3) + \log_2(2-7) = \log_2(3(2)-1)$
$\qquad \log_2(-1) + \log_2(-5) = \log_2(5)$
Undefined

check: $x = 11$
$\log_2(11-3) + \log_2(11-7) = \log_2(3(11)-1)$
$\qquad \log_2(8) + \log_2(4) = \log_2(32)$
$\qquad\qquad\qquad 3 + 2 = 5$
$\qquad\qquad\qquad\qquad 5 = 5$

True
$\{11\}$

11. $f(x) = e^{x+6} + 9$
$\qquad y = e^{x+6} + 9$
Exchange x and y.

$\qquad x = e^{y+6} + 9$
$\qquad x - 9 = e^{y+6}$
$\ln(x-9) = \ln e^{y+6}$
$\ln(x-9) = (y+6)\ln e$
$\ln(x-9) = (y+6)\cdot 1$
$\ln(x-9) = y + 6$
$\ln(x-9) - 6 = y$
$\qquad f^{-1}(x) = \ln(x-9) - 6$

12. $f(x) = \ln(x+10) - 18$
$\qquad y = \ln(x+10) - 18$
Exchange x and y.

$\qquad x = \ln(y+10) - 18$
$\qquad x + 18 = \ln(y+10)$
Convert to exponential form.
$\qquad e^{x+18} = y + 10$
$\qquad e^{x+18} - 10 = y$
$\qquad f^{-1}(x) = e^{x+18} - 10$

12.4 *EXPONENTIAL AND LOGARITHMIC EQUATIONS*

1. extraneous

3. $3^x = 27$
$3^x = 3^3$
$x = 3$
$\{3\}$

5. $2^{x-4} = 32$
$2^{x-4} = 2^5$
$x - 4 = 5$
$x = 9$
$\{9\}$

7. $7^{3x-25} = 49$
$7^{3x-25} = 7^2$
$3x - 25 = 2$
$3x = 27$
$x = 9$
$\{9\}$

9. $9^x = 27$
$\left(3^2\right)^x = 3^3$
$3^{2x} = 3^3$
$2x = 3$
$x = \dfrac{3}{2}$
$\left\{\dfrac{3}{2}\right\}$

11. $36^{x-3} = 216$
$\left(6^2\right)^{x-3} = 6^3$
$6^{2x-6} = 6^3$
$2x - 6 = 3$
$2x = 9$
$x = \dfrac{9}{2}$
$\left\{\dfrac{9}{2}\right\}$

13. $100^{x-6} = 1000^x$
$\left(10^2\right)^{x-6} = \left(10^3\right)^x$
$10^{2x-12} = 10^{3x}$
$2x - 12 = 3x$
$-12 = x$
$\{-12\}$

15. $4^x = 17$
Rewrite in logarithmic form.
$\log_4 17 = x$
$x = \dfrac{\log 17}{\log 4}$
$x \approx 2.044$
$\{2.044\}$

17. $10^x = 129$
Rewrite in logarithmic form.
$\log 129 = x$
$x \approx 2.111$
$\{2.111\}$

19. $2^{x-5} = 27$
Rewrite in logarithmic form.
$\log_2 27 = x - 5$
$\log_2 27 + 5 = x$
$x = \dfrac{\log 27}{\log 2} + 5$
$x \approx 9.755$
$\{9.755\}$

21. $7^{x+4} - 8 = 192$
$7^{x+4} = 200$
Rewrite in logarithmic form.
$\log_7 200 = x + 4$
$\log_7 200 - 4 = x$
$\dfrac{\log 200}{\log 7} - 4 = x$
$-1.277 \approx x$
$\{-1.277\}$

23. $6^{2x} = 45$
Rewrite in logarithmic form.
$\log_6 45 = 2x$
$\dfrac{\log 45}{\log 6} = 2x$
$\dfrac{\log 45}{2\log 6} = x$
$1.062 \approx x$
$\{1.062\}$

25. $e^{2x-7} = 987$
Rewrite in logarithmic form.
$$\ln 987 = 2x - 7$$
$$\ln 987 + 7 = 2x$$
$$\frac{\ln 987 + 7}{2} = x$$
$$6.947 \approx x$$
$$\{6.947\}$$

27. $8^{2x+9} - 17 = 1233$
$$8^{2x+9} = 1250$$
Rewrite in logarithmic form.
$$\log_8 1250 = 2x + 9$$
$$\frac{\log 1250}{\log 8} = 2x + 9$$
$$\frac{\log 1250}{\log 8} - 9 = 2x$$
$$\frac{\log 1250}{2\log 8} - \frac{9}{2} = x$$
$$-2.785 \approx x$$
$$\{-2.785\}$$

29. $\ln(3x+1) = \ln(5x-7)$
$$3x + 1 = 5x - 7$$
$$-2x + 1 = -7$$
$$-2x = -8$$
$$x = 4$$
Substituting 4 for x in the original equation produces a true statement. The solution checks.
$$\{4\}$$

31. $\log_8(3x+13) = \log_8(7x-23)$
$$3x + 13 = 7x - 23$$
$$-4x + 13 = -23$$
$$-4x = -36$$
$$x = 9$$
Substituting 9 for x in the original equation produces a true statement. The solution checks.
$$\{9\}$$

33. $\log_4(4x-1) = \log_4(5x+3)$
$$4x - 1 = 5x + 3$$
$$-x - 1 = 3$$
$$-x = 4$$
$$x = -4$$
check: $x = -4$
$$\log_4\big(4(-4)-1\big) = \log_4\big(5(-4)+3\big)$$
$$\log_4(-17) = \log_4(-17)$$
Undefined
$\varnothing$

35. $\log_4\big(x^2 - 5x + 30\big) = \log_4 44$
$$x^2 - 5x + 30 = 44$$
$$x^2 - 5x - 14 = 0$$
$$(x+2)(x-7) = 0$$
$$x = -2 \quad \text{or} \quad x = 7$$
Substituting -2 or 7 for x in the original equation produces true statements.
The solutions check.
$$\{-2, 7\}$$

37. $\log_7\big(x^2 + 7x - 29\big) = \log_7(3x+3)$
$$x^2 + 7x - 29 = 3x + 3$$
$$x^2 + 4x - 32 = 0$$
$$(x+8)(x-4) = 0$$
$$x = -8 \quad \text{or} \quad x = 4$$
check: $x = -8$
$$\log_7\big((-8)^2 + 7(-8) - 29\big) = \log_7\big(3(-8)+3\big)$$
$$\log_7(-21) = \log_7(-21)$$
Undefined
check: $x = 4$
$$\log_7\big((4)^2 + 7(4) - 29\big) = \log_7\big(3(4)+3\big)$$
$$\log_7 15 = \log_7 15$$
True
$$\{4\}$$

39. $\log_3 x = 5$
Rewrite in exponential form.
$$x = 3^5$$
$$x = 243$$
$$\{243\}$$

41. $\log_5 x = -4$
Rewrite in exponential form.
$$x = 5^{-4}$$
$$x = \frac{1}{5^4}$$
$$x = \frac{1}{625}$$
$$\left\{\frac{1}{625}\right\}$$

43. $\ln(x-4) = 6$
Rewrite in exponential form.
$$x - 4 = e^6$$
$$x = e^6 + 4$$
$$x \approx 407.429$$
$$\{407.429\}$$

45. $\log_2(x+8) - 7 = -2$

$\log_2(x+8) = 5$

Rewrite in exponential form.

$$x + 8 = 2^5$$
$$x + 8 = 32$$
$$x = 24$$

$\{24\}$

47. $\log_5(x^2 + 5x - 11) = 2$

Rewrite in exponential form.

$$x^2 + 5x - 11 = 5^2$$
$$x^2 + 5x - 11 = 25$$
$$x^2 + 5x - 36 = 0$$
$$(x + 9)(x - 4) = 0$$

$x + 9 = 0$ or $x - 4 = 0$
$x = -9$ or $x = 4$

$\{-9, 4\}$

49. $\log(x^2 - 9x + 30) = 1$

Rewrite in exponential form.

$$x^2 - 9x + 30 = 10^1$$
$$x^2 - 9x + 30 = 10$$
$$x^2 - 9x + 20 = 0$$
$$(x - 4)(x - 5) = 0$$

$x - 4 = 0$ or $x - 5 = 0$
$x = 4$ or $x = 5$

$\{4, 5\}$

51. $\log_4\left(\dfrac{3x+8}{x-6}\right) = 2$

Rewrite in exponential form.

$$\frac{3x+8}{x-6} = 4^2$$
$$\frac{3x+8}{x-6} = 16$$
$$\overset{1}{\cancel{(x-6)}}\left(\frac{3x+8}{\cancel{x-6}_1}\right) = (x-6)\cdot 16$$
$$3x + 8 = 16x - 96$$
$$-13x + 8 = -96$$
$$-13x = -104$$
$$x = 8$$

$\{8\}$

53. $\log_4(x+5) + \log_4(x+7) = \log_4 3$

$$\log_4\big((x+5)(x+7)\big) = \log_4 3$$
$$(x+5)(x+7) = 3$$
$$x^2 + 7x + 5x + 35 = 3$$
$$x^2 + 12x + 32 = 0$$
$$(x+8)(x+4) = 0$$

$x + 8 = 0$ or $x + 4 = 0$
$x = -8$ or $x = -4$

check: $x = -8$

$\log_4\big((-8)+5\big) + \log_4\big((-8)+7\big) = \log_4 3$
$\log_4(-3) + \log_4(-1) = \log_4 3$

Undefined

check: $x = -4$

$\log_4\big((-4)+5\big) + \log_4\big((-4)+7\big) = \log_4 3$
$\log_4 1 + \log_4 3 = \log_4 3$
$0 + \log_4 3 = \log_4 3$
$\log_4 3 = \log_4 3$

True

$\{-4\}$

55. $\log(3x+8) - \log(x-7) = \log 4$

$$\log\left(\frac{3x+8}{x-7}\right) = \log 4$$
$$\frac{3x+8}{x-7} = 4$$
$$\overset{1}{\cancel{(x-7)}}\left(\frac{3x+8}{\cancel{x-7}_1}\right) = (x-7)\cdot 4$$
$$3x + 8 = 4x - 28$$
$$-x + 8 = -28$$
$$-x = -36$$
$$x = 36$$

Substituting 36 for x in the original equation produces a true statement. The solution checks.

$\{36\}$

57.
$$\log_2(x+2) + \log_2(x+6) = 5$$
$$\log_2((x+2)(x+6)) = 5$$
$$(x+2)(x+6) = 2^5$$
$$x^2 + 6x + 2x + 12 = 32$$
$$x^2 + 8x - 20 = 0$$
$$(x+10)(x-2) = 0$$

$$x + 10 = 0 \quad \text{or} \quad x - 2 = 0$$
$$x = -10 \quad \text{or} \quad x = 2$$

check: $x = -10$
$$\log_2((-10)+2) + \log_2((-10)+6) = 5$$
$$\log_2(-8) + \log_2(-4) = 5$$
Undefined

check: $x = 2$
$$\log_2((2)+2) + \log_2((2)+6) = 5$$
$$\log_2 4 + \log_2 8 = 5$$
$$2 + 3 = 5$$
$$5 = 5$$
True
$\{2\}$

59.
$$\log_2 x + \log_2(x-4) = 5$$
$$\log_2 x(x-4) = 5$$
$$x(x-4) = 2^5$$
$$x^2 - 4x = 32$$
$$x^2 - 4x - 32 = 0$$
$$(x+4)(x-8) = 0$$

$$x + 4 = 0 \quad \text{or} \quad x - 8 = 0$$
$$x = -4 \quad \text{or} \quad x = 8$$

check: $x = -4$
$$\log_2(-4) + \log_2((-4)-4) = 5$$
$$\log_2(-4) + \log_2(-8) = 5$$
Undefined

check: $x = 8$
$$\log_2(8) + \log_2((8)-4) = 5$$
$$\log_2 8 + \log_2 4 = 5$$
$$3 + 2 = 5$$
$$5 = 5$$
True
$\{8\}$

61.
$$\log_4(3x-1) + \log_4(x+5) = 3$$
$$\log_4(3x-1)(x+5) = 3$$
$$(3x-1)(x+5) = 4^3$$
$$3x^2 + 15x - x - 5 = 64$$
$$3x^2 + 14x - 69 = 0$$
$$(3x+23)(x-3) = 0$$

$$3x + 23 = 0 \quad \text{or} \quad x - 3 = 0$$
$$3x = -23 \quad \text{or} \quad x = 3$$
$$x = -\frac{23}{3} \quad \text{or} \quad x = 3$$

check: $x = -\dfrac{23}{3}$
$$\log_4\left(3\left(-\frac{23}{3}\right)-1\right) + \log_4\left(\left(-\frac{23}{3}\right)+5\right) = 3$$
$$\log_4(-24) + \log_4\left(-\frac{8}{3}\right) = 3$$
Undefined

check: $x = 3$
$$\log_4(3(3)-1) + \log_4((3)+5) = 3$$
$$\log_4 8 + \log_4 8 = 3$$
$$\log_4(8 \cdot 8) = 3$$
$$\log_4 64 = 3$$
$$3 = 3$$
True
$\{3\}$

63.
$$\log_6(7x+11) - \log_6(2x-9) = 1$$
$$\log_6\left(\frac{7x+11}{2x-9}\right) = 1$$
$$\frac{7x+11}{2x-9} = 6^1$$
$$(2x-9)\left(\frac{7x+11}{2x-9}\right) = (2x-9)\cdot 6$$
$$7x + 11 = 12x - 54$$
$$-5x + 11 = -54$$
$$-5x = -65$$
$$x = 13$$

Substituting 13 for x in the original equation produces a true statement. The solution checks.
$\{13\}$

65.
$$f(x) = 3^{2x-15} + 4 = 31$$
$$3^{2x-15} = 27$$
$$3^{2x-15} = 3^3$$
$$2x - 15 = 3$$
$$2x = 18$$
$$x = 9$$
$\{9\}$

67.
$$f(x) = \log_5\left(x^2 + 3x - 45\right) + 6 = 8$$
$$\log_5\left(x^2 + 3x - 45\right) = 2$$
$$x^2 + 3x - 45 = 5^2$$
$$x^2 + 3x - 45 = 25$$
$$x^2 + 3x - 70 = 0$$
$$(x + 10)(x - 7) = 0$$
$$x + 10 = 0 \quad \text{or} \quad x - 7 = 0$$
$$x = -10 \quad \text{or} \quad x = 7$$
$$\{-10, 7\}$$

69.
$$\log_8\left(x^2 - 10x - 20\right) = \log_8\left(3x + 10\right)$$
$$x^2 - 10x - 20 = 3x + 10$$
$$x^2 - 13x - 30 = 0$$
$$(x + 2)(x - 15) = 0$$
$$x + 2 = 0 \quad \text{or} \quad x - 15 = 0$$
$$x = -2 \quad \text{or} \quad x = 15$$
$$\{-2, 15\}$$

71.
$$5^{x+9} = 407$$
$$\log 5^{x+9} = \log 407$$
$$(x + 9)\log 5 = \log 407$$
$$x + 9 = \frac{\log 407}{\log 5}$$
$$x = \frac{\log 407}{\log 5} - 9$$
$$x \approx -5.267$$
$$\{-5.267\}$$

73.
$$9^{x-4} = 32$$
$$\log 9^{x-4} = \log 32$$
$$(x - 4)\log 9 = \log 32$$
$$x - 4 = \frac{\log 32}{\log 9}$$
$$x = \frac{\log 32}{\log 9} + 4$$
$$x \approx 5.577$$
$$\{5.577\}$$

75.
$$2^{7x+30} = \frac{1}{32}$$
$$2^{7x+30} = 2^{-5}$$
$$7x + 30 = -5$$
$$7x = -35$$
$$x = -5$$
$$\{-5\}$$

77.
$$\log\left(3x - 41\right) = 3$$
$$3x - 41 = 10^3$$
$$3x - 41 = 1000$$
$$3x = 1041$$
$$x = 347$$
$$\{347\}$$

79.
$$2^{6x+25} = 128$$
$$2^{6x+25} = 2^7$$
$$6x + 25 = 7$$
$$6x = -18$$
$$x = -3$$
$$\{-3\}$$

81.
$$7^{x-8} = 56$$
$$\log 7^{x-8} = \log 56$$
$$(x - 8)\log 7 = \log 56$$
$$x - 8 = \frac{\log 56}{\log 7}$$
$$x = \frac{\log 56}{\log 7} + 8$$
$$x \approx 10.069$$
$$\{10.069\}$$

83.
$$\log_{12}\left(9x - 11\right) = \log_{12} 4$$
$$9x - 11 = 4$$
$$9x = 15$$
$$x = \frac{15}{9}$$
$$x = \frac{5}{3}$$
$$\left\{\frac{5}{3}\right\}$$

85. $\log_2 x + \log_2(x+4) = 5$
$\log_2 x(x+4) = 5$

$x(x+4) = 2^5$
$x^2 + 4x = 32$
$x^2 + 4x - 32 = 0$
$(x+8)(x-4) = 0$

$x + 8 = 0 \quad \text{or} \quad x - 4 = 0$
$x = -8 \quad \text{or} \qquad x = 4$
$\{-8, 4\}$

check: $x = -8$
$\log_2(-8) + \log_2((-8)+4) = 5$
$\log_2(-8) + \log_2(-4) = 5$
undefined
check: $x = 4$
$\log_2(4) + \log_2((4)+4) = 5$
$\log_2 4 + \log_2 8 = 5$
$2 + 3 = 5$
$5 = 5$
$\{4\}$

87. $\log_7(4x-19) - \log_7(x+6) = \log_7 3$

$\log_7\left(\dfrac{4x-19}{x+6}\right) = \log_7 3$

$\dfrac{4x-19}{x+6} = 3$

$\overset{1}{\cancel{(x+6)}}\left(\dfrac{4x-19}{\underset{1}{\cancel{x+6}}}\right) = (x+6)\cdot 3$

$4x - 19 = 3x + 18$
$x - 19 = 18$
$x = 37$

$\{37\}$

89. $\log_8(5x-43) + 4 = 7$
$\log_8(5x-43) = 3$
$5x - 43 = 8^3$
$5x - 43 = 512$
$5x = 555$
$x = 111$

$\{111\}$

91. $\log_4(x^2+4x+16) = \log_4(x^2+x-5)$
$x^2 + 4x + 16 = x^2 + x - 5$
$4x + 16 = x - 5$
$3x + 16 = -5$
$3x = -21$
$x = -7$
$\{-7\}$

93. $\log_5(x-7) + \log_5(x+3) = \log_5(5x+15)$
$\log_5(x-7)(x+3) = \log_5(5x+15)$
$(x-7)(x+3) = 5x+15$
$x^2 + 3x - 7x - 21 = 5x + 15$
$x^2 - 4x - 21 = 5x + 15$
$x^2 - 9x - 36 = 0$
$(x+3)(x-12) = 0$

$x + 3 = 0 \quad \text{or} \quad x - 12 = 0$
$x = -3 \quad \text{or} \qquad x = 12$

check: $x = -3$
$\log_5((-3)-7) + \log_5((-3)+3) = \log_5(5(-3)+15)$
$\log_5(-10) + \log_5 0 = \log_5 0$

Undefined

check: $x = 12$
$\log_5((12)-7) + \log_5((12)+3) = \log_5(5(12)+15)$
$\log_5 5 + \log_5 15 = \log_5 75$
$\log_5 75 = \log_5 75$

True
$\{12\}$

95. $\ln(2x+5) = -2$
$2x + 5 = e^{-2}$
$2x = e^{-2} - 5$
$x = \dfrac{e^{-2} - 5}{2}$
$x \approx -2.432$
$\{-2.432\}$

97. $2^{x^2+12x+42} = 64$
$2^{x^2+12x+42} = 2^6$
$x^2 + 12x + 42 = 6$
$x^2 + 12x + 36 = 0$
$(x+6)(x+6) = 0$

$x + 6 = 0$
$x = -6$
$\{-6\}$

99. $\log_2(\log_3(2x-1)) = 2$
$\log_3(2x-1) = 2^2$
$\log_3(2x-1) = 4$

$2x - 1 = 3^4$
$2x - 1 = 81$
$2x = 82$
$x = 41$

$\{41\}$

101.
$$\log_3\left(\log_3\left(x^2-4x-5\right)\right)=1$$
$$\log_3\left(x^2-4x-5\right)=3^1$$
$$\log_3\left(x^2-4x-5\right)=3$$
$$x^2-4x-5=3^3$$
$$x^2-4x-5=27$$
$$x^2-4x-32=0$$
$$(x+4)(x-8)=0$$
$$x+4=0 \quad \text{or} \quad x-8=0$$
$$x=-4 \quad \text{or} \quad x=8$$
$$\{-4,8\}$$

103.
$$\left(\log_2 x\right)^2-3\log_2 x-10=0$$
Let $\log_2 x=u$.
$$u^2-3u-10=0$$
$$(u+2)(u-5)=0$$
$$u+2=0 \quad \text{or} \quad u-5=0$$
$$u=-2 \quad \text{or} \quad x=5$$
$$\log_2 x=-2 \quad \text{or} \quad \log_2 x=5$$
$$x=2^{-2} \quad \text{or} \quad x=2^5$$
$$x=\frac{1}{4} \quad \text{or} \quad x=32$$
$$\left\{\frac{1}{4},32\right\}$$

105.
$$\left(\log_3 x\right)^2+\log_3 x^2=8$$
$$\left(\log_3 x\right)^2+2\log_3 x-8=0$$
Let $\log_3 x=u$.
$$u^2+2u-8=0$$
$$(u+4)(u-2)=0$$
$$u+4=0 \quad \text{or} \quad u-2=0$$
$$u=-4 \quad \text{or} \quad x=2$$
$$\log_3 x=-4 \quad \text{or} \quad \log_3 x=2$$
$$x=3^{-4} \quad \text{or} \quad x=3^2$$
$$x=\frac{1}{81} \quad \text{or} \quad x=9$$
$$\left\{\frac{1}{81},9\right\}$$

107.
$$f(x)=e^{x-2}$$
$$y=e^{x-2}$$
Exchange x and y.
$$x=e^{y-2}$$
$$\ln x=\ln e^{y-2}$$
$$\ln x=(y-2)\ln e$$
$$\ln x=y-2$$
$$\ln x+2=y$$
$$f^{-1}(x)=\ln x+2$$

109.
$$f(x)=e^{x+5}+4$$
$$y=e^{x+5}+4$$
Exchange x and y.
$$x=e^{y+5}+4$$
$$x-4=e^{y+5}$$
$$\ln(x-4)=\ln e^{y+5}$$
$$\ln(x-4)=y+5$$
$$\ln(x-4)-5=y$$
$$f^{-1}(x)=\ln(x-4)-5$$

111.
$$f(x)=2^{x-6}-9$$
$$y=2^{x-6}-9$$
Exchange x and y.
$$x=2^{y-6}-9$$
$$x+9=2^{y-6}$$
$$\log_2(x+9)=\log_2 2^{y-6}$$
$$\log_2(x+9)=y-6$$
$$\log_2(x+9)+6=y$$
$$f^{-1}(x)=\log_2(x+9)+6$$

113.
$$f(x)=\log(x+6)$$
$$y=\log(x+6)$$
Exchange x and y.
$$x=\log(y+6)$$
$$10^x=y+6$$
$$10^x-6=y$$
$$f^{-1}(x)=10^x-6$$

115.
$$f(x)=\ln(x-4)-9$$
$$y=\ln(x-4)-9$$
Exchange x and y.
$$x=\ln(y-4)-9$$
$$x+9=\ln(y-4)$$
$$e^{x+9}=y-4$$
$$e^{x+9}+4=y$$
$$f^{-1}(x)=e^{x+9}+4$$

117.
$$f(x)=\log_3(x+1)+8$$
$$y=\log_3(x+1)+8$$
Exchange x and y.
$$x=\log_3(y+1)+8$$
$$x-8=\log_3(y+1)$$
$$3^{x-8}=y+1$$
$$3^{x-8}-1=y$$
$$f^{-1}(x)=3^{x-8}-1$$

119. Answers will vary. Example:
We must check logarithmic equations for extraneous roots because they have restricted domains. Exponential equations have a domain of all real numbers, so they do not need to be checked.

121. Answers will vary. Example:
To find the inverse of an exponential function, exchange x and y and solve for y by taking the log of both sides.

$$y = e^{x+1} - 3$$

Exchange x and y.

$$x = e^{y+1} - 3$$
$$x + 3 = e^{y+1}$$
$$\ln(x+3) = \ln e^{y+1}$$
$$\ln(x+3) = y + 1$$
$$\ln(x+3) - 1 = y$$
$$f^{-1}(x) = \ln(x+3) - 1$$

To find the inverse of a logarithmic function, exchange x and y and solve for y by converting to an exponential function.

$$y = \log(x-1) + 2$$

Exchange x and y.

$$x = \log(y-1) + 2$$
$$x - 2 = \log(y-1)$$
$$10^{x-2} = y - 1$$
$$10^{x-2} + 1 = y$$
$$f^{-1}(x) = 10^{x-2} + 1$$

12.5 Quick Check

1.
$$A = P\left(1 + \frac{r}{n}\right)^{nt}$$
$$A = 400\left(1 + \frac{0.03}{12}\right)^{12 \cdot 5}$$
$$A = 400(1.0025)^{60}$$
$$A \approx 464.65$$
The balance will be \$464.65.

2.
$$A = P\left(1 + \frac{r}{n}\right)^{nt}$$
$$100{,}000 = 25{,}000\left(1 + \frac{0.10}{4}\right)^{4t}$$
$$4 = 1.025^{4t}$$
$$\log 4 = \log 1.025^{4t}$$
$$\log 4 = 4t \log 1.025$$
$$\frac{\log 4}{4 \log 1.025} = \frac{4t \log 1.025}{4 \log 1.025}$$
$$\frac{\log 4}{4 \log 1.025} = t$$
$$14.0 \approx t$$
Approximately 14.0 years

3.
$$A = Pe^{rt}$$
$$500{,}000 = 50{,}000 e^{0.15t}$$
$$10 = e^{0.15t}$$
$$\ln 10 = \ln e^{0.15t}$$
$$\ln 10 = 0.15t$$
$$\frac{\ln 10}{0.15} = t$$
$$15.35 \approx t$$
Approximately 15.35 years

4. $P = P_0 e^{kt}$

P	P_0	k	t
32	10	unknown	20

$$32 = 10e^{k(20)}$$

$$\frac{32}{10} = e^{20k}$$

$$\ln\left(\frac{32}{10}\right) = \ln e^{20k}$$

$$\ln\left(\frac{32}{10}\right) = 20k$$

$$\frac{\ln\left(\frac{32}{10}\right)}{20} = k$$

$$0.058158 \approx k$$

P	P_0	k	t
unknown	10	0.058158	46

$$P = 10e^{0.058158(46)}$$

$$P = 10e^{2.675268}$$

$$P \approx 145$$

Approximately 145 cents or \$1.45

5. $P = P_0 e^{kt}$

P	P_0	k	t
25	50	unknown	50.6

$$25 = 50e^{k(50.6)}$$

$$0.5 = e^{50.6k}$$

$$\ln 0.5 = \ln e^{50.6k}$$

$$\ln 0.5 = 50.6k$$

$$\frac{\ln 0.5}{50.6} = k$$

$$-0.013699 \approx k$$

P	P_0	k	t
10	50	-0.013699	unknown

$$10 = 50e^{-0.013699t}$$

$$0.2 = e^{-0.013699t}$$

$$\ln 0.2 = \ln e^{-0.013699t}$$

$$\ln 0.2 = -0.013699t$$

$$\frac{\ln 0.2}{-0.013699} = t$$

$$117.5 \approx t$$

Approximately 117.5 days

6. $f(x) = 13,400 \cdot 1.055^x = 125,000$

$$1.055^x = \frac{125,000}{13,400}$$

$$\log 1.055^x = \log \frac{625}{67}$$

$$x \log 1.055 = \log \frac{625}{67}$$

$$x = \frac{\log \dfrac{625}{67}}{\log 1.055}$$

$$x \approx 42$$

$1975 + 42 = 2017$
In 2017

7. $\text{pH} = -\log\left[\text{H}^+\right]$

$$\text{pH} = -\log\left[1.58 \times 10^{-3}\right]$$

$\text{pH} = 2.8$
Since $\text{pH} = 2.8 < 7$, it is an acid.

8. $\text{pH} = -\log\left[\text{H}^+\right]$

$$0.3 = -\log\left[\text{H}^+\right]$$

$$-0.3 = \log\left[\text{H}^+\right]$$

$$10^{-0.3} = \left[\text{H}^+\right]$$

$$\text{H}^+ \approx 5.01 \times 10^{-1} \text{ moles/liter}$$

9. $L = 10\log\left(\dfrac{I}{10^{-12}}\right)$

$$L = 10\log\left(\frac{10^{-6}}{10^{-12}}\right)$$

$$L = 10\log\left(10^{-6-(-12)}\right)$$

$L = 10\log 10^6$
$L = 10 \cdot 6$
$L = 60$
60 db

10.
$$L = 10\log\left(\frac{I}{10^{-12}}\right)$$
$$80 = 10\log\left(\frac{I}{10^{-12}}\right)$$
$$8 = \log\left(\frac{I}{10^{-12}}\right)$$
$$\frac{I}{10^{-12}} = 10^{8}$$
$$10^{-12} \cdot \frac{I}{10^{-12}} = 10^{-12} \cdot 10^{8}$$
$$I = 10^{-4}$$
10^{-4} watts per square meter

12.5 *Applications of Exponential and Logarithmic Functions*

1. compound

3. continuously

5. negative

7. $A = P\left(1 + \frac{r}{n}\right)^{nt}$

$$A = 7500\left(1 + \frac{0.03}{4}\right)^{4(4)}$$
$$= 7500(1.0075)^{16}$$
$$\approx 8452.44$$
The balance will be \$8452.44.

9. $A = P\left(1 + \frac{r}{n}\right)^{nt}$

$$A = 12,413\left(1 + \frac{0.08}{2}\right)^{2(10)}$$
$$= 12,413(1.04)^{20}$$
$$\approx 27,198.41$$
The balance will be \$27,198.41.

11.
$$A = P\left(1 + \frac{r}{n}\right)^{nt}$$
$$6000 = 4000\left(1 + \frac{0.09}{12}\right)^{12t}$$
$$\frac{6000}{4000} = \left(1 + \frac{0.09}{12}\right)^{12t}$$
$$1.5 = 1.0075^{12t}$$
$$\log 1.5 = \log 1.0075^{12t}$$
$$\log 1.5 = 12t\log 1.0075$$
$$\frac{\log 1.5}{12\log 1.0075} = \frac{12t\log 1.0075}{12\log 1.0075}$$
$$\frac{\log 1.5}{12\log 1.0075} = t$$
$$4.5 \approx t$$
4.5 years

13.
$$A = P\left(1 + \frac{r}{n}\right)^{nt}$$
$$2(240) = 240\left(1 + \frac{0.03}{12}\right)^{12t}$$
$$480 = 240\left(1 + \frac{0.03}{12}\right)^{12t}$$
$$2 = 1.0025^{12t}$$
$$\log 2 = \log 1.0025^{12t}$$
$$\log 2 = 12t\log 1.0025$$
$$\frac{\log 2}{12\log 1.0025} = \frac{12t\log 1.0025}{12\log 1.0025}$$
$$\frac{\log 2}{12\log 1.0025} = t$$
$$23.1 \approx t$$
23.1 years

15. $A = Pe^{rt}$

$A = 2000e^{0.043(6)} \approx 2588.68$
In 6 years, the balance will be \$2588.68.

17.
$$A = Pe^{rt}$$
$$6750 = 4500e^{0.07t}$$
$$\frac{6750}{4500} = e^{0.07t}$$
$$1.5 = e^{0.07t}$$
$$\ln 1.5 = \ln e^{0.07t}$$
$$\ln 1.5 = 0.07t$$
$$\frac{\ln 1.5}{0.07} = t$$
$$5.8 \approx t$$
It will take 5.8 years.

19. $P = P_0 e^{kt}$

P	P_0	k	t
1532	1134	unknown	2

$$1532 = 1134e^{k \cdot 2}$$
$$\frac{1532}{1134} = e^{2k}$$
$$\ln\left(\frac{1532}{1134}\right) = \ln e^{2k}$$
$$\ln\left(\frac{1532}{1134}\right) = 2k$$
$$\frac{\ln\left(\frac{1532}{1134}\right)}{2} = k$$
$$0.150411 \approx k$$

P	P_0	k	t
unknown	1134	0.150411	16

$$P = 1134e^{0.150411(16)}$$
$$P \approx 12,583$$
There will be 12,583 stores in 2016.

21. $P = P_0 e^{kt}$

P	P_0	k	t
6	4	unknown	25

$$6 = 4e^{k \cdot 25}$$
$$\frac{6}{4} = e^{25k}$$
$$\ln 1.5 = \ln e^{25k}$$
$$\ln 1.5 = 25k$$
$$\frac{\ln 1.5}{25} = k$$
$$0.016219 \approx k$$

P	P_0	k	t
unknown	4	0.016219	76

$$P = 4e^{0.016219(76)}$$
$$P \approx 13.7$$
The population in 2050 will be 13.7 billion.

23. $P = P_0 e^{kt}$

P	P_0	k	t
48.40	39.43	unknown	4

$$48.40 = 39.43e^{k \cdot 4}$$
$$\frac{48.40}{39.43} = e^{4k}$$
$$\ln\left(\frac{48.40}{39.43}\right) = \ln e^{4k}$$
$$\ln\left(\frac{48.40}{39.43}\right) = 4k$$
$$\frac{\ln\left(\frac{48.40}{39.43}\right)}{4} = k$$
$$0.051243 \approx k$$

P	P_0	k	t
100	39.43	0.051243	unknown

$$100 = 39.43e^{0.051243t}$$
$$\frac{100}{39.43} = e^{0.051243t}$$
$$\ln\left(\frac{100}{39.43}\right) = \ln e^{0.051243t}$$
$$\ln\left(\frac{100}{39.43}\right) = 0.051243t$$
$$\frac{\ln\left(\frac{100}{39.43}\right)}{0.051243} = t$$
$$18 \approx t$$
$$1998 + 18 = 2016$$
The bill will reach \$100 in 2016.

25. $P = P_0 e^{kt}$

P	P_0	k	t
40,000	5000	unknown	60

$$40,000 = 5000 e^{k \cdot 60}$$

$$\frac{40,000}{5000} = e^{60k}$$

$$8 = e^{60k}$$

$$\ln 8 = \ln e^{60k}$$

$$\ln 8 = 60k$$

$$\frac{\ln 8}{60} = k$$

$$0.034657 \approx k$$

P	P_0	k	t
100,000	5000	0.034657	unknown

$$100,000 = 5000 e^{0.034657t}$$

$$20 = e^{0.034657t}$$

$$\ln 20 = \ln e^{0.034657t}$$

$$\ln 20 = 0.034657t$$

$$\frac{\ln 20}{0.034657} = t$$

$$86.4 \approx t$$

It will take 86.4 minutes.

27. $P = P_0 e^{kt}$

P	P_0	k	t
474	593	unknown	8

$$474 = 593 e^{k \cdot 8}$$

$$\frac{474}{593} = e^{8k}$$

$$\ln\left(\frac{474}{593}\right) = \ln e^{8k}$$

$$\ln\left(\frac{474}{593}\right) = 8k$$

$$\frac{\ln\left(\frac{474}{593}\right)}{8} = k$$

$$-0.027998 \approx k$$

P	P_0	k	t
unknown	593	−0.027998	27

$$P = 593 e^{-0.027998(27)}$$
$$P \approx 278$$

In 2020, there will be 278 drive-in theaters.

29. $P = P_0 e^{kt}$

P	P_0	k	t
50	100	unknown	112

$$50 = 100 e^{k \cdot 112}$$
$$0.5 = e^{112k}$$
$$\ln 0.5 = \ln e^{112k}$$
$$\ln 0.5 = 112k$$

$$\frac{\ln 0.5}{112} = k$$

$$-0.006189 \approx k$$

P	P_0	k	t
unknown	100	−0.006189	240

$$P = 100 e^{-0.006189(240)}$$
$$P \approx 22.6$$
After 240 minutes, 22.6 milligrams
will remain.

31. $P = P_0 e^{kt}$

P	P_0	k	t
$0.5(50\%)$	$1(100\%)$	unknown	5730

$$0.5 = 1 e^{k \cdot 5730}$$
$$0.5 = e^{5730k}$$
$$\ln 0.5 = \ln e^{5730k}$$
$$\ln 0.5 = 5730k$$

$$\frac{\ln 0.5}{5730} = k$$

$$-0.000121 \approx k$$

P	P_0	k	t
$0.98(98\%)$	$1(100\%)$	−0.000121	unknown

$$0.98 = 1 e^{-0.000121t}$$
$$\ln 0.98 = \ln e^{-0.000121t}$$
$$\ln 0.98 = -0.000121t$$

$$\frac{\ln 0.98}{-0.000121} = t$$

$$167 \approx t$$

The bone is 167 years old.

33. $P = P_0 e^{kt}$

P	P_0	k	t
2.5	5	unknown	100,000

$$2.5 = 5e^{k \cdot 100,000}$$
$$0.5 = e^{100,000k}$$
$$\ln 0.5 = \ln e^{100,000k}$$
$$\ln 0.5 = 100,000k$$
$$\frac{\ln 0.5}{100,000} = k$$
$$-0.000007 \approx k$$

P	P_0	k	t
1	5	−0.000007	unknown

$$1 = 5e^{-0.000007t}$$
$$0.2 = e^{-0.000007t}$$
$$\ln 0.2 = \ln e^{-0.000007t}$$
$$\ln 0.2 = -0.000007t$$
$$\frac{\ln 0.2}{-0.000007} = t$$
$$229,919.7 \approx t$$

It will take 229,919.7 years.

35. $P = P_0 e^{kt}$

P	P_0	k	t
1457	1878	unknown	62

$$1457 = 1878e^{k \cdot 62}$$
$$\frac{1457}{1878} = e^{62k}$$
$$\ln\left(\frac{1457}{1878}\right) = \ln e^{62k}$$
$$\ln\left(\frac{1457}{1878}\right) = 62k$$
$$\frac{\ln\left(\frac{1457}{1878}\right)}{62} = k$$
$$-0.004094 \approx k$$

P	P_0	k	t
1000	1878	−0.004094	unknown

$$1000 = 1878e^{-0.004094t}$$
$$\frac{1000}{1878} = e^{-0.004094t}$$
$$\ln\left(\frac{1000}{1878}\right) = \ln e^{-0.004094t}$$
$$\ln\left(\frac{1000}{1878}\right) = -0.004094t$$
$$\frac{\ln\left(\frac{1000}{1878}\right)}{-0.004094} = t$$
$$154 \approx t$$

$1940 + 154 = 2094$

In 2094, there will be only 1000 daily newspapers.

37.

P	P_0	k	t
24,000	30,000	unknown	1

$$24,000 = 30,000e^{k \cdot 1}$$
$$\frac{24,000}{30,000} = e^{1k}$$
$$0.8 = e^{k}$$
$$\ln 0.8 = \ln e^{k}$$
$$\ln 0.8 = k$$
$$-0.223144 \approx t$$

P	P_0	k	t
5000	30,000	−0.223144	unknown

$$5000 = 30,000e^{-0.223144t}$$
$$\frac{5000}{30,000} = e^{-0.223144t}$$
$$\frac{1}{6} = e^{-0.223144t}$$
$$\ln\left(\frac{1}{6}\right) = \ln e^{-0.223144t}$$
$$\ln\left(\frac{1}{6}\right) = -0.223144t$$
$$\frac{\ln\left(\frac{1}{6}\right)}{-0.223144} = t$$
$$8 \approx t$$

$2008 + 8 = 2016$

The car will be worth \$5000 in 2016.

39. In 2016, $x = 15$.
$f(15) = 4794 \cdot 1.21^{15} \approx 83{,}652$
There will be 83,652 stores in 2016.

41. $f(x) = 410.57 \cdot 1.035^x = 1000$

$$1.035^x = \frac{1000}{410.57}$$

$$\log 1.035^x = \log\left(\frac{1000}{410.57}\right)$$

$$x \cdot \log 1.035 = \log\left(\frac{1000}{410.57}\right)$$

$$x = \frac{\log\left(\frac{1000}{410.57}\right)}{\log 1.035}$$

$$x \approx 25.9$$

$1994 + 25.9 = 2019.9$
In 2020, the average premium will be $1000.

43. $f(x) = 6830 \cdot 1.038^x = 100{,}000$

$$1.038^x = \frac{100{,}000}{6830}$$

$$\log 1.038^x = \log\left(\frac{100{,}000}{6830}\right)$$

$$x \cdot \log 1.038 = \log\left(\frac{100{,}000}{6830}\right)$$

$$x = \frac{\log\left(\frac{100{,}000}{6830}\right)}{\log 1.038}$$

$$x \approx 72$$

$1975 + 72 = 2047$
In 2047, the average annual income of a person without a high school diploma will be $100,000.

45. $f(x) = 24{,}000 \cdot 0.8^x = 5000$

$$0.8^x = \frac{5000}{24{,}000}$$

$$\log 0.8^x = \log\left(\frac{5000}{24{,}000}\right)$$

$$x \cdot \log 0.8 = \log\left(\frac{5000}{24{,}000}\right)$$

$$x = \frac{\log\left(\frac{5000}{24{,}000}\right)}{\log 0.8}$$

$$x \approx 7$$

In 7 years, the value of the car will be $5000.

47. a) $f(6) = 35 + 61e^{-0.0165(6)} \approx 90.3° \text{F}$

b) $f(t) = 35 + 61e^{-0.0165t} = 98.6$

$$61e^{-0.0165t} = 63.6$$

$$e^{-0.0165t} = \frac{63.6}{61}$$

$$\ln e^{-0.0165t} = \ln\left(\frac{63.6}{61}\right)$$

$$-0.0165t = \ln\left(\frac{63.6}{61}\right)$$

$$t = \frac{\ln\left(\frac{63.6}{61}\right)}{-0.0165}$$

$$t \approx -2.5$$

The person died 2.5 hours before the body was found.

49. $f(x) = 32{,}178 \cdot 1.027^x = 70{,}000$

$$1.027^x = \frac{70{,}000}{32{,}178}$$

$$\log 1.027^x = \log\left(\frac{70{,}000}{32{,}178}\right)$$

$$x \cdot \log 1.027 = \log\left(\frac{70{,}000}{32{,}178}\right)$$

$$x = \frac{\log\left(\frac{70{,}000}{32{,}178}\right)}{\log 1.027}$$

$$x \approx 29.2$$

$1989 + 29.2 = 2018.2$
In 2018, the average annual salary will be $70,000.

51. $\text{pH} = -\log\left[\text{H}^+\right]$

$\text{pH} = -\log\left[2.51 \times 10^{-13}\right] = 12.6$
Since $\text{pH} = 12.6 > 7$, it is a base.

53. $\text{pH} = -\log\left[\text{H}^+\right]$

$\text{pH} = -\log\left[5.01 \times 10^{-3}\right] = 2.3$
Since $\text{pH} = 2.3 < 7$, it is an acid.

55.
$$\text{pH} = -\log\left[\text{H}^+\right]$$
$$5.0 = -\log\left[\text{H}^+\right]$$
$$-5.0 = \log\left[\text{H}^+\right]$$
$$10^{-5.0} = \left[\text{H}^+\right]$$

$\text{H}^+ = 1.0 \times 10^{-5}$ moles per liter

57.
$$\mathrm{pH} = -\log\left[\mathrm{H}^+\right]$$
$$10.2 = -\log\left[\mathrm{H}^+\right]$$
$$-10.2 = \log\left[\mathrm{H}^+\right]$$
$$10^{-10.2} = \left[\mathrm{H}^+\right]$$
$$\mathrm{H}^+ \approx 6.31 \times 10^{-11} \text{ moles per liter}$$

59.
$$L = 10\log\left(\frac{I}{10^{-12}}\right)$$
$$L = 10\log\left(\frac{10^{-4}}{10^{-12}}\right)$$
$$= 10\log\left(10^{-4-(-12)}\right)$$
$$= 10\log 10^8$$
$$= 10 \cdot 8$$
$$= 80$$
80 db

61.
$$L = 10\log\left(\frac{I}{10^{-12}}\right)$$
$$L = 10\log\left(\frac{10^{-1}}{10^{-12}}\right)$$
$$= 10\log\left(10^{-1-(-12)}\right)$$
$$= 10\log 10^{11}$$
$$= 10 \cdot 11$$
$$= 110$$
110 db

63.
$$L = 10\log\left(\frac{I}{10^{-12}}\right)$$
$$110 = 10\log\left(\frac{I}{10^{-12}}\right)$$
$$11 = \log\left(\frac{I}{10^{-12}}\right)$$
$$\frac{I}{10^{-12}} = 10^{11}$$
$$10^{-12} \cdot \left(\frac{I}{10^{-12}}\right) = 10^{-12} \cdot 10^{11}$$
$$I = 10^{-1}$$
10^{-1} watts per square meter

65.
$$L = 10\log\left(\frac{I}{10^{-12}}\right)$$
$$95 = 10\log\left(\frac{I}{10^{-12}}\right)$$
$$9.5 = \log\left(\frac{I}{10^{-12}}\right)$$
$$\frac{I}{10^{-12}} = 10^{9.5}$$
$$10^{-12} \cdot \left(\frac{I}{10^{-12}}\right) = 10^{-12} \cdot 10^{9.5}$$
$$I = 10^{-2.5}$$
$10^{-2.5}$ watts per square meter

67. Since $2025 - 1900 = 125$, $x = 125$.
$f(125) = 11.03 + 14.29\ln(125) \approx 80$
The life expectancy of Americans
born in 2025 is 80 years.

69. $f(x) = 2.648 + 0.181\ln x = 3.2$
$$0.181\ln x = 0.552$$
$$\ln x = \frac{0.552}{0.181}$$
$$x = e^{\frac{0.552}{0.181}}$$
$$x \approx 21.1$$
$1995 + 21.1 = 2016.1$
In 2016, there will be 3.2 million teachers.

71. Answers will vary. Example:
Gretchen invests $10,000 in an account with an
interest rate of 6% compounded annually.
When will Gretchen have $26,927.73 in the
account?

73. Answers will vary. Example:
The half-life of a radioactive element is the
amount of time it takes for half of the substance
to decay.

12.6 QUICK CHECK

1. $f(x) = 3^{x-1} - 9$
$h = 1,\ k = -9$
Shift the graph of $f(x) = 3^x$ by 1 unit to the right and by 9 units down.
The horizontal asymptote is at $y = -9$.
Shift three points. (Add 1 to each x-coordinate and subtract 9 from each y-coordinate.)

$f(x) = 3^x$	$\left(-1, \frac{1}{3}\right)$	$(0,1)$	$(1,3)$
$f(x) = 3^{x-1} - 9$	$\left(0, -8\frac{2}{3}\right)$	$(1,-8)$	$(2,-6)$

y-intercept: By shifting, $\left(0, -8\frac{2}{3}\right)$.

x-intercept: Set $f(x) = 0$ and solve for x.

$$3^{x-1} - 9 = 0$$
$$3^{x-1} = 9$$
$$3^{x-1} = 3^2$$
$$x - 1 = 2$$
$$x = 3$$

$(3, 0)$

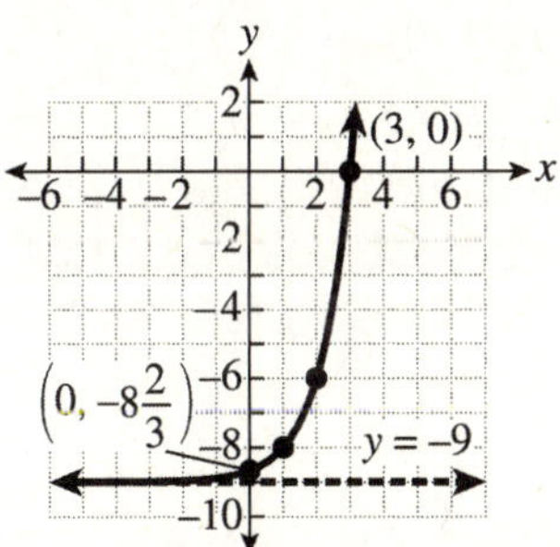

Domain: $(-\infty, \infty)$; range: $(-9, \infty)$

2. $f(x) = 4^{x-2} - 6$
$h = 2,\ k = -6$
Shift the graph of $f(x) = 4^x$ by 2 units to the right and by 6 units down.
The horizontal asymptote is at $y = -6$.
Shift three points. (Add 2 to each x-coordinate and subtract 6 from each y-coordinate.)

$f(x) = 4^x$	$\left(-1, \frac{1}{4}\right)$	$(0,1)$	$(1,4)$
$f(x) = 4^{x-2} - 6$	$\left(1, -5\frac{3}{4}\right)$	$(2,-5)$	$(3,-2)$

y-intercept: Find $f(0)$.

$$f(0) = 4^{0-2} - 6 = \frac{1}{16} - \frac{96}{16} = -\frac{95}{16} \approx -5.9$$
$(0, -5.9)$

x-intercept: Set $f(x) = 0$ and solve for x.

$$0 = 4^{x-2} - 6$$
$$4^{x-2} = 6$$
$$\log\left(4^{x-2}\right) = \log 6$$
$$(x-2)\log 4 = \log 6$$
$$x - 2 = \frac{\log 6}{\log 4}$$
$$x = \frac{\log 6}{\log 4} + 2$$
$$x \approx 3.3$$

$(3.3, 0)$

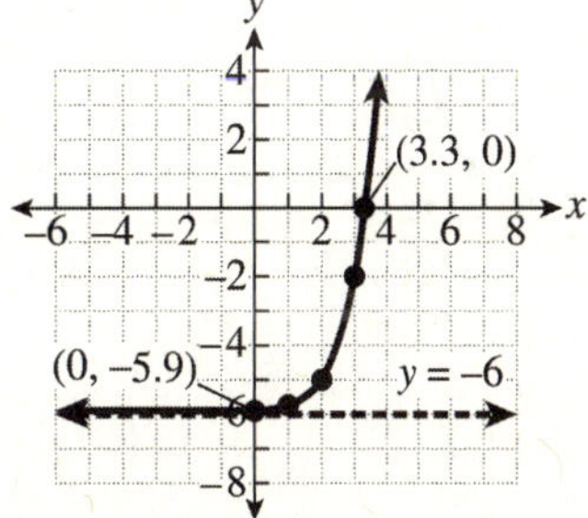

Domain: $(-\infty, \infty)$; range: $(-6, \infty)$

3. $f(x) = e^{x-3} - 9$
$h = 3, \; k = -9$
Shift the graph of $f(x) = e^x$ by 3 units to the right and by 9 units down.
The horizontal asymptote is at $y = -9$.
Shift three points. (Add 3 to each x-coordinate and subtract 9 from each y-coordinate.)

$f(x) = e^x$	$(-1, 0.4)$	$(0, 1)$	$(1, 2.7)$
$f(x) = e^{x-3} - 9$	$(2, -8.6)$	$(3, -8)$	$(4, -6.3)$

y-intercept: Find $f(0)$.

$$f(0) = e^{0-3} - 9 = e^{-3} - 9 \approx -8.95$$

$(0, -8.95)$

x-intercept: Set $f(x) = 0$ and solve for x.

$$e^{x-3} - 9 = 0$$
$$e^{x-3} = 9$$
$$\ln e^{x-3} = \ln 9$$
$$x - 3 = \ln 9$$
$$x = \ln 9 + 3$$
$$x \approx 5.2$$

$(5.2, 0)$

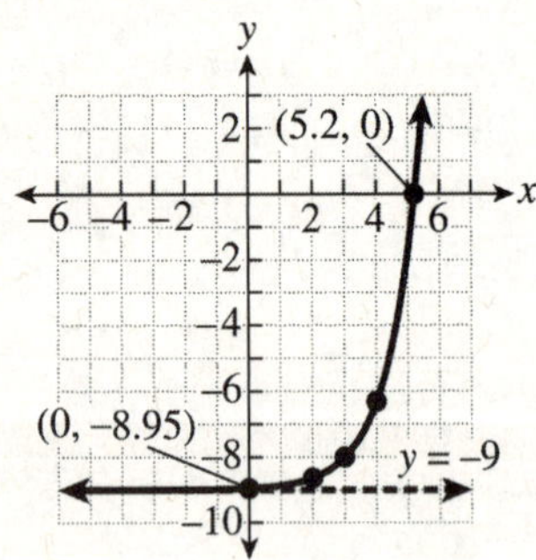

Domain: $(-\infty, \infty)$; range: $(-9, \infty)$

4. $f(x) = \log_3(x - 2) - 1$
$h = 2, \; k = -1$
Shift the graph of $f(x) = \log_3 x$ by 2 units to the right and by 1 unit down.
The vertical asymptote is at $x = 2$.
Shift three points. (Add 2 to each x-coordinate and subtract 1 from each y-coordinate.)

$f(x) = \log_3 x$	$\left(\dfrac{1}{3}, -1\right)$	$(1, 0)$	$(3, 1)$
$f(x) = \log_3(x - 2) - 1$	$\left(\dfrac{7}{3}, -2\right)$	$(3, -1)$	$(5, 0)$

y-intercept: From the graph, there is no y-intercept.

x-intercept: By shifting, $(5, 0)$.

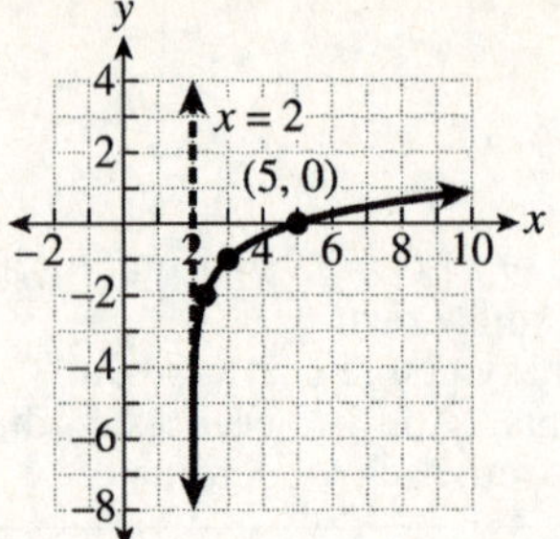

Domain: $(2, \infty)$; range: $(-\infty, \infty)$

5. $f(x) = \log_3(x + 7) - 2$
$h = -7, \; k = -2$
Shift the graph of $f(x) = \log_3 x$ by 7 units to the left and by 2 units down.
The vertical asymptote is at $x = -7$.
Shift three points. (Subtract 7 from each x-coordinate and subtract 2 from each y-coordinate.)

$f(x) = \log_3 x$	$\left(\dfrac{1}{3}, -1\right)$	$(1, 0)$	$(3, 1)$
$f(x) = \log_3(x + 7) - 2$	$\left(-\dfrac{20}{3}, -3\right)$	$(-6, -2)$	$(-4, -1)$

y-intercept: Find $f(0)$.

$$\begin{aligned}
f(0) &= \log_3(0 + 7) - 2 \\
&= \log_3 7 - 2 \\
&= \frac{\log 7}{\log 3} - 2 \\
&\approx -0.2
\end{aligned}$$

$(0, -0.2)$

x-intercept: Set $f(x) = 0$ and solve for x.

$$\log_3(x + 7) - 2 = 0$$
$$\log_3(x + 7) = 2$$
$$x + 7 = 3^2$$
$$x + 7 = 9$$
$$x = 2$$

$(2, 0)$

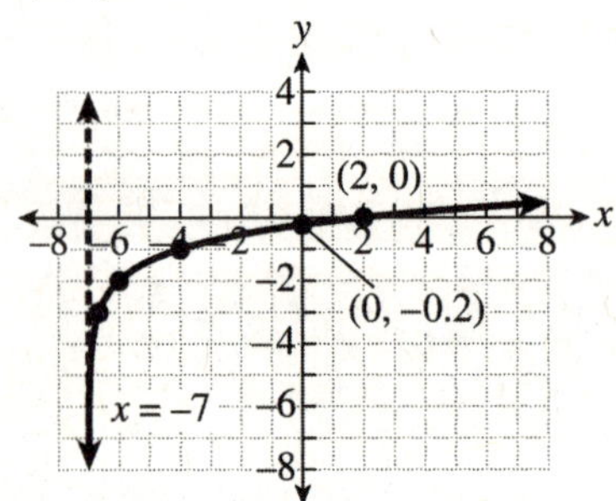

Domain: $(-7, \infty)$; range: $(-\infty, \infty)$

6. $f(x) = \ln(x-1) - 2$
$h = 1,\ k = -2$
Shift the graph of $f(x) = \ln x$ by 1 unit to the right and by 2 units down.
The vertical asymptote is at $x = 1$.
Shift three points. (Add 1 to each x-coordinate and subtract 2 from each y-coordinate.)

$f(x) = \ln x$	$(0.4, -1)$	$(1, 0)$	$(2.7, 1)$
$f(x) = \ln(x-1) - 2$	$(1.4, -3)$	$(2, -2)$	$(3.7, -1)$

y-intercept: From the graph, there is no y-intercept.

x-intercept: Set $f(x) = 0$ and solve for x.

$$\ln(x-1) - 2 = 0$$
$$\ln(x-1) = 2$$
$$x - 1 = e^2$$
$$x = e^2 + 1$$
$$x \approx 8.4$$

$(8.4, 0)$

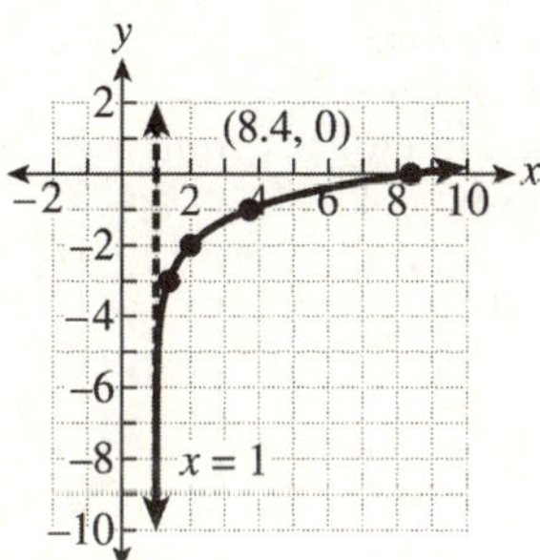

Domain: $(1, \infty)$; range: $(-\infty, \infty)$

12.6 *GRAPHING EXPONENTIAL AND LOGARITHMIC FUNCTIONS*

1. horizontal

3. exponential

5. y-

7. $f(x) = 2^{x+4} + 6$
x-intercept:
$$2^{x+4} + 6 = 0$$
$$2^{x+4} = -6$$
$$\log 2^{x+4} = \log(-6)$$
undefined; none

y-intercept:
$$f(0) = 2^{0+4} + 6 = 2^4 + 6 = 22$$
$(0, 22)$

9. $f(x) = 6^{x+2} - 21$

x-intercept:
$$6^{x+2} - 21 = 0$$
$$6^{x+2} = 21$$
$$\log 6^{x+2} = \log 21$$
$$(x+2)\log 6 = \log 21$$
$$x + 2 = \frac{\log 21}{\log 6}$$
$$x = \frac{\log 21}{\log 6} - 2$$
$$x \approx -0.3$$
$(-0.3, 0)$

y-intercept:
$$f(0) = 6^{0+2} - 21 = 6^2 - 21 = 15$$
$(0, 15)$

11. $f(x) = e^{x+3} - 10$
x-intercept:
$$e^{x+3} - 10 = 0$$
$$e^{x+3} = 10$$
$$\ln e^{x+3} = \ln 10$$
$$x + 3 = \ln 10$$
$$x = \ln 10 - 3$$
$$x \approx -0.7$$
$(-0.7, 0)$

y-intercept:
$$f(0) = e^{0+3} - 10 = e^3 - 10 \approx 10.1$$
$(0, 10.1)$

13. $f(x) = \log_2(x-5) - 3$
x-intercept:
$$\log_2(x-5) - 3 = 0$$
$$\log_2(x-5) = 3$$
$$x - 5 = 2^3$$
$$x - 5 = 8$$
$$x = 13$$
$(13, 0)$

y-intercept:
$$f(0) = \log_2(0-5) - 3 = \log_2(-5) - 3$$
undefined; none

15. $f(x) = \ln(x+3) - 4$
x-intercept:
$\ln(x+3) - 4 = 0$
$\ln(x+3) = 4$
$x + 3 = e^4$
$x = e^4 - 3$
$x \approx 51.6$
$(51.6, 0)$

y-intercept:
$f(0) = \ln(0+3) - 4 = \ln 3 - 4 \approx -2.9$
$(0, -2.9)$

17. $f(x) = 3^{x-5} + 8$
Asymptote: $y = 8$
Domain: $(-\infty, \infty)$
Range: $(8, \infty)$

19. $f(x) = 4^{x-12} - 9$
Asymptote: $y = -9$
Domain: $(-\infty, \infty)$
Range: $(-9, \infty)$

21. $f(x) = \log_7(x-5) - 9$
The argument of $\log_7(x-5)$ is 0 when $x = 5$.
Asymptote: $x = 5$
Domain: $(5, \infty)$
Range: $(-\infty, \infty)$

23. $f(x) = \log(2x-1) + 16$
The argument of $\log(2x-1)$ is 0 when $x = \dfrac{1}{2}$.
Asymptote: $x = \dfrac{1}{2}$
Domain: $\left(\dfrac{1}{2}, \infty\right)$
Range: $(-\infty, \infty)$

25. $f(x) = 2^{x+2} + 4$
$h = -2,\ k = 4$
Shift the graph of $f(x) = 2^x$ by 2 units to the left and by 4 units up.
The horizontal asymptote is at $y = 4$.
Shift three points. (Subtract 2 from each x-coordinate and add 4 to each y-coordinate.)

$f(x) = 2^x$	$\left(-1, \dfrac{1}{2}\right)$	$(0,1)$	$(1,2)$
$f(x) = 2^{x+2} + 4$	$\left(-3, \dfrac{9}{2}\right)$	$(-2,5)$	$(-1,6)$

y-intercept: Find $f(0)$.
$f(0) = 2^{0+2} + 4 = 4 + 4 = 8$
$(0, 8)$
x-intercept: From the graph, there is no x-intercept.

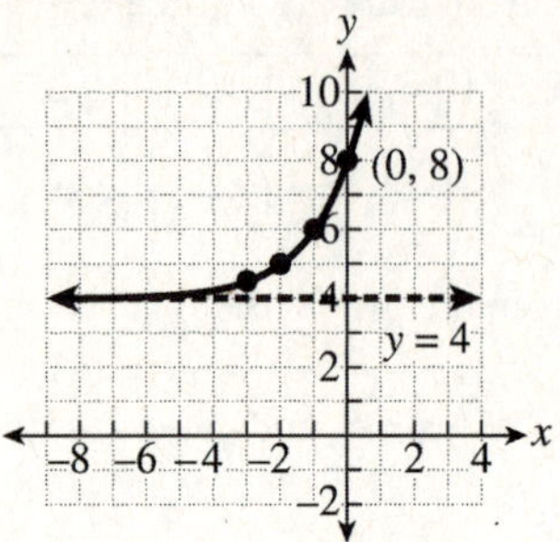

Domain: $(-\infty, \infty)$; range: $(4, \infty)$

27. $f(x) = 3^{x+1} - 9$
$h = -1,\ k = -9$
Shift the graph of $f(x) = 3^x$ by 1 unit to the left and by 9 units down.
The horizontal asymptote is at $y = -9$.
Shift three points. (Subtract 1 from each x-coordinate and subtract 9 from each y-coordinate.)

$f(x) = 3^x$	$\left(-1, \dfrac{1}{3}\right)$	$(0,1)$	$(1,3)$
$f(x) = 3^{x+1} - 9$	$\left(-2, -\dfrac{26}{3}\right)$	$(-1,-8)$	$(0,-6)$

y-intercept: By shifting, $(0, -6)$.

x-intercept: Set $f(x) = 0$ and solve for x.

$3^{x+1} - 9 = 0$
$3^{x+1} = 9$
$3^{x+1} = 3^2$
$x + 1 = 2$
$x = 1$
$(1, 0)$

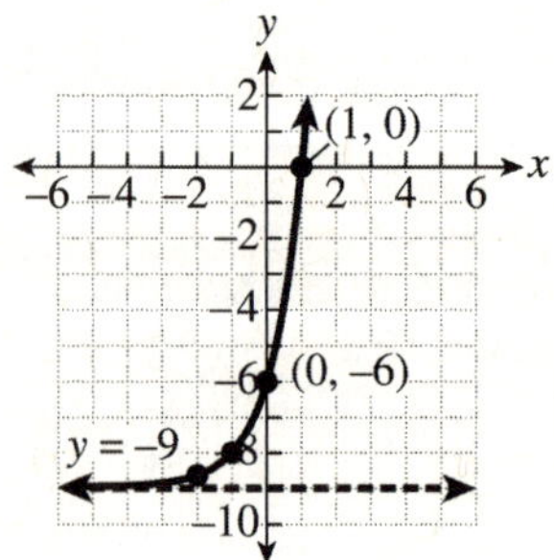

Domain: $(-\infty, \infty)$; range: $(-9, \infty)$

29. $f(x) = \left(\dfrac{1}{2}\right)^{x+4} - 4$

$h = -4,\ k = -4$

Shift the graph of $f(x) = \left(\dfrac{1}{2}\right)^{x}$ by 4 units to the left and by 4 units down.
The horizontal asymptote is at $y = -4$.
Shift three points. (Subtract 4 from each x-coordinate and subtract 4 from each y-coordinate.)

$f(x) = \left(\dfrac{1}{2}\right)^{x}$	$(-1, 2)$	$(0, 1)$	$\left(1, \dfrac{1}{2}\right)$
$f(x) = \left(\dfrac{1}{2}\right)^{x+4} - 4$	$(-5, -2)$	$(-4, -3)$	$\left(-3, -\dfrac{7}{2}\right)$

y-intercept: Find $f(0)$.

$$f(0) = \left(\dfrac{1}{2}\right)^{0+4} - 4 = \dfrac{1}{16} - 4 \approx -3.9$$
$(0, -3.9)$

x-intercept: Set $f(x) = 0$ and solve for x.

$$\left(\dfrac{1}{2}\right)^{x+4} - 4 = 0$$
$$\left(\dfrac{1}{2}\right)^{x+4} = 4$$
$$\left(\dfrac{1}{2}\right)^{x+4} = \left(\dfrac{1}{2}\right)^{-2}$$
$$x + 4 = -2$$
$$x = -6$$
$(-6, 0)$

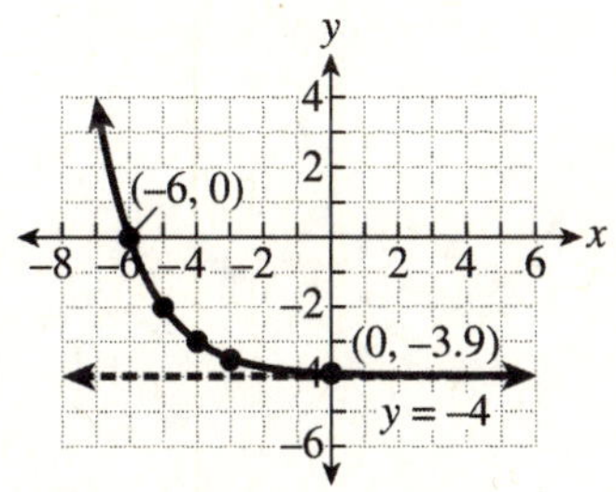

Domain: $(-\infty, \infty)$; range: $(-4, \infty)$

31. $f(x) = 5^{x+1} - 6$
$h = -1,\ k = -6$

Shift the graph of $f(x) = 5^{x}$ by 1 unit to the left and by 6 units down.
The horizontal asymptote is at $y = -6$.
Shift three points. (Subtract 1 from each x-coordinate and subtract 6 from each y-coordinate.)

$f(x) = 5^{x}$	$\left(-1, \dfrac{1}{5}\right)$	$(0, 1)$	$(1, 5)$
$f(x) = 5^{x+1} - 6$	$\left(-2, -\dfrac{29}{5}\right)$	$(-1, -5)$	$(0, -1)$

y-intercept: By shifting, $(0, -1)$.

x-intercept: Set $f(x) = 0$ and solve for x.

$$5^{x+1} - 6 = 0$$
$$5^{x+1} = 6$$
$$\log_5\left(5^{x+1}\right) = \log_5 6$$
$$x + 1 = \dfrac{\log 6}{\log 5}$$
$$x = \dfrac{\log 6}{\log 5} - 1$$
$$x \approx 0.1$$
$(0.1, 0)$

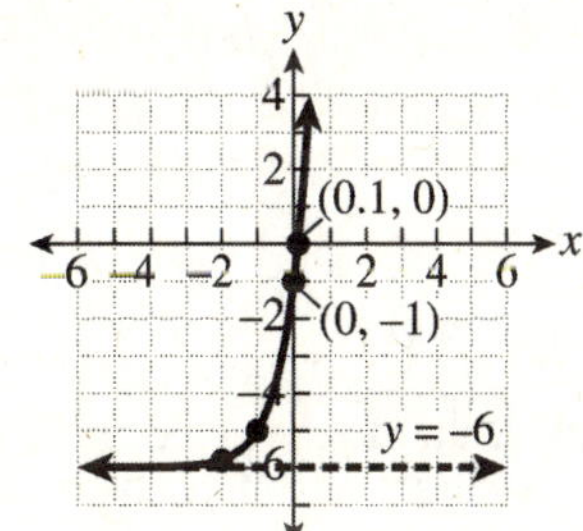

Domain: $(-\infty, \infty)$; range: $(-6, \infty)$

33. $f(x) = e^{x-2} + 3$
$h = 2,\ k = 3$
Shift the graph of $f(x) = e^x$ by 2 units to the right and by 3 units up.
The horizontal asymptote is at $y = 3$.
Shift three points. (Add 2 to each x-coordinate and add 3 to each y-coordinate.)

$f(x) = e^x$	$(-1, 0.4)$	$(0,1)$	$(1, 2.7)$
$f(x) = e^{x-2} + 3$	$(1, 3.4)$	$(2, 4)$	$(3, 5.7)$

y-intercept: Find $f(0)$.

$$f(0) = e^{0-2} + 3 = e^{-2} + 3 \approx 3.1$$

$(0, 3.1)$

x-intercept: From the graph, there is no x-intercept.

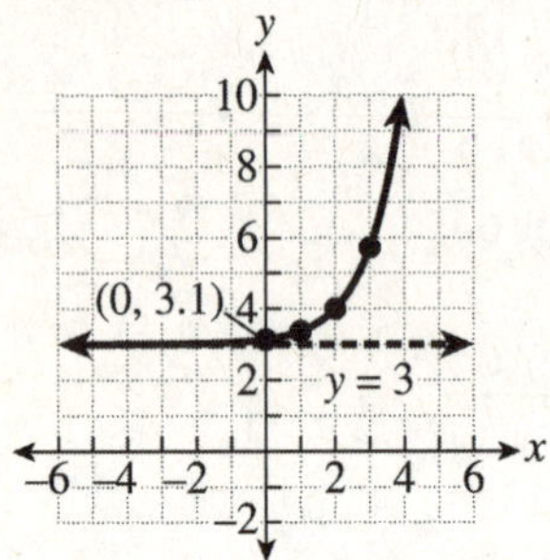

Domain: $(-\infty, \infty)$; range: $(3, \infty)$

35. $f(x) = e^{x+1} - 7$
$h = -1,\ k = -7$
Shift the graph of $f(x) = e^x$ by 1 unit to the left and by 7 units down.
The horizontal asymptote is at $y = -7$.
Shift three points. (Subtract 1 from each x-coordinate and subtract 7 from each y-coordinate.)

$f(x) = e^x$	$(-1, 0.4)$	$(0,1)$	$(1, 2.7)$
$f(x) = e^{x+1} - 7$	$(-2, -6.6)$	$(-1, -6)$	$(0, -4.3)$

y-intercept: By shifting, $(0, -4.3)$.

x-intercept: Set $f(x) = 0$ and solve for x.

$$e^{x+1} - 7 = 0$$
$$e^{x+1} = 7$$
$$\ln\left(e^{x+1}\right) = \ln 7$$
$$x + 1 = \ln 7$$
$$x = \ln 7 - 1$$
$$x \approx 0.9$$

$(0.9, 0)$

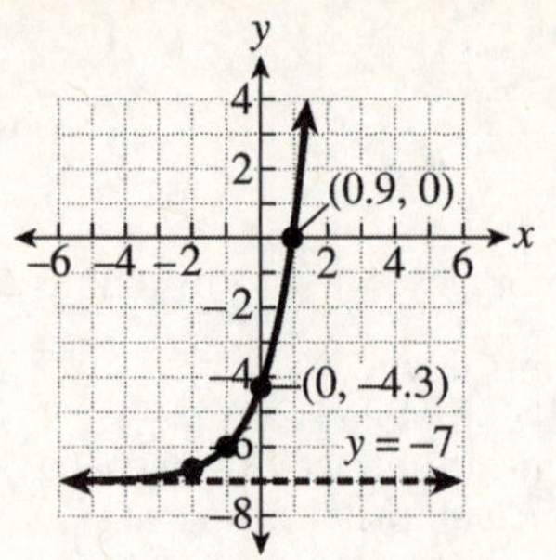

Domain: $(-\infty, \infty)$; range: $(-7, \infty)$

37. $f(x) = \log_4 (x - 2) - 1$
$h = 2,\ k = -1$
Shift the graph of $f(x) = \log_4 x$ by 2 units to the right and by 1 unit down.
The vertical asymptote is at $x = 2$.
Shift three points. (Add 2 to each x-coordinate and subtract 1 from each y-coordinate.)

$f(x) = \log_4 x$	$\left(\dfrac{1}{4}, -1\right)$	$(1, 0)$	$(4, 1)$
$f(x) = \log_4 (x - 2) - 1$	$\left(\dfrac{9}{4}, -2\right)$	$(3, -1)$	$(6, 0)$

y-intercept: From the graph, there is no y-intercept.

x-intercept: By shifting, $(6, 0)$.

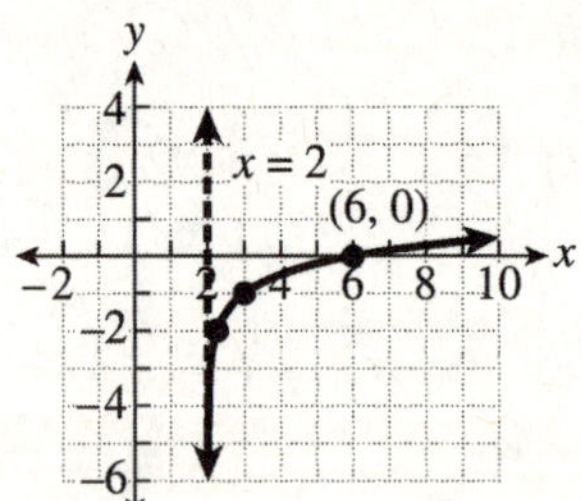

Domain: $(2, \infty)$; range: $(-\infty, \infty)$

39. $f(x) = \log_3(x+8) + 2$
$h = -8,\ k = 2$
Shift the graph of $f(x) = \log_3 x$ by 8 units to the left and by 2 units up.
The vertical asymptote is at $x = -8$.
Shift three points. (Subtract 8 from each x-coordinate and add 2 to each y-coordinate.)

$f(x) = \log_3 x$	$\left(\dfrac{1}{3}, -1\right)$	$(1,0)$	$(3,1)$
$f(x) = \log_3(x+8) + 2$	$\left(-\dfrac{23}{3}, 1\right)$	$(-7, 2)$	$(-5, 3)$

y-intercept: Find $f(0)$.

$$f(0) = \log_3(0+8) + 2 = \frac{\log 8}{\log 3} + 2 \approx 3.9$$

$(0, 3.9)$

x-intercept: Set $f(x) = 0$ and solve for x.

$$\log_3(x+8) + 2 = 0$$
$$\log_3(x+8) = -2$$
$$x + 8 = 3^{-2}$$
$$x + 8 = \frac{1}{9}$$
$$x \approx -7.9$$

$(-7.9, 0)$

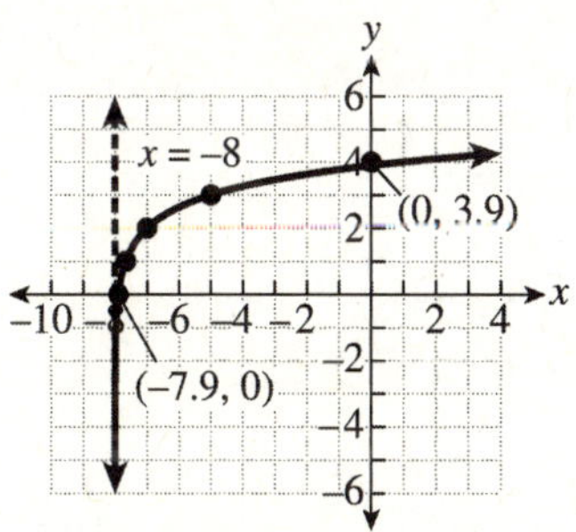

Domain: $(-8, \infty)$; range: $(-\infty, \infty)$

41. $f(x) = \log_2(x+5) - 3$
$h = -5,\ k = -3$
Shift the graph of $f(x) = \log_2 x$ by 5 units to the left and by 3 units down.
The vertical asymptote is at $x = -5$.
Shift three points. (Subtract 5 from each x-coordinate and subtract 3 from each y-coordinate.)

$f(x) = \log_2 x$	$\left(\dfrac{1}{2}, -1\right)$	$(1,0)$	$(2,1)$
$f(x) = \log_2(x+5) - 3$	$\left(-\dfrac{9}{2}, -4\right)$	$(-4, -3)$	$(-3, -2)$

y-intercept: Find $f(0)$.

$$f(0) = \log_2(0+5) - 3 = \frac{\log 5}{\log 2} - 3 \approx -0.7$$

$(0, -0.7)$

x-intercept: Set $f(x) = 0$ and solve for x.

$$\log_2(x+5) - 3 = 0$$
$$\log_2(x+5) = 3$$
$$x + 5 = 2^3$$
$$x + 5 = 8$$
$$x = 3$$

$(3, 0)$

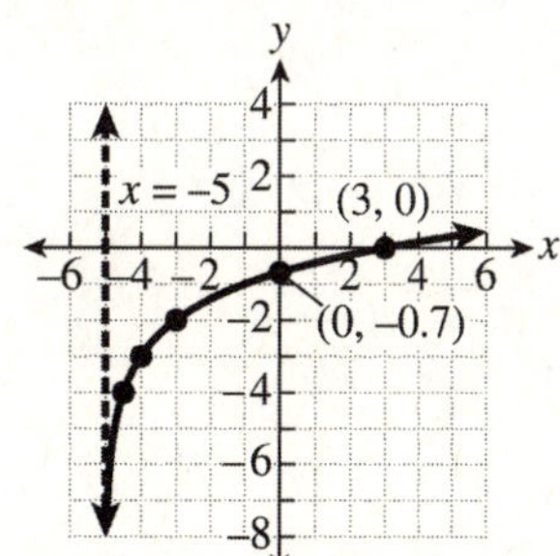

Domain: $(-5, \infty)$; range: $(-\infty, \infty)$

43. $f(x) = \ln(x-3) + 1$
$h = 3,\ k = 1$
Shift the graph of $f(x) = \ln x$ by 3 units to the right and by 1 unit up.
The vertical asymptote is at $x = 3$.
Shift three points. (Add 3 to each x-coordinate and add 1 to each y-coordinate.)

$f(x) = \ln x$	$(0.4, -1)$	$(1,0)$	$(2.7,1)$
$f(x) = \ln(x-3) + 1$	$(3.4, 0)$	$(4, 1)$	$(5.7, 2)$

y-intercept: From the graph, there is no y-intercept.
x-intercept: By shifting, $(3.4, 0)$.

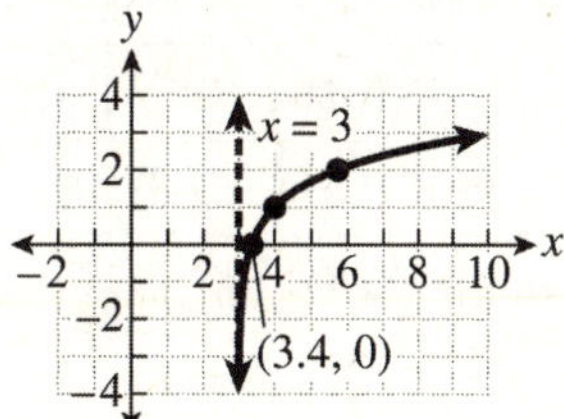

Domain: $(3, \infty)$; range: $(-\infty, \infty)$

45. $f(x) = \ln(x+1) - 2$
$h = -1,\ k = -2$
Shift the graph of $f(x) = \ln x$ by 1 unit to the left and by 2 units down.
The vertical asymptote is at $x = -1$.
Shift three points. (Subtract 1 from each x-coordinate and subtract 2 from each y-coordinate.)

$f(x) = \ln x$	$(0.4, -1)$	$(1, 0)$	$(2.7, 1)$
$f(x) = \ln(x+1) - 2$	$(-0.6, -3)$	$(0, -2)$	$(1.7, -1)$

y-intercept: By shifting, $(0, -2)$.

x-intercept: Set $f(x) = 0$ and solve for x.

$$\ln(x+1) - 2 = 0$$
$$\ln(x+1) = 2$$
$$x + 1 = e^2$$
$$x = e^2 - 1$$
$$x \approx 6.4$$

$(6.4, 0)$

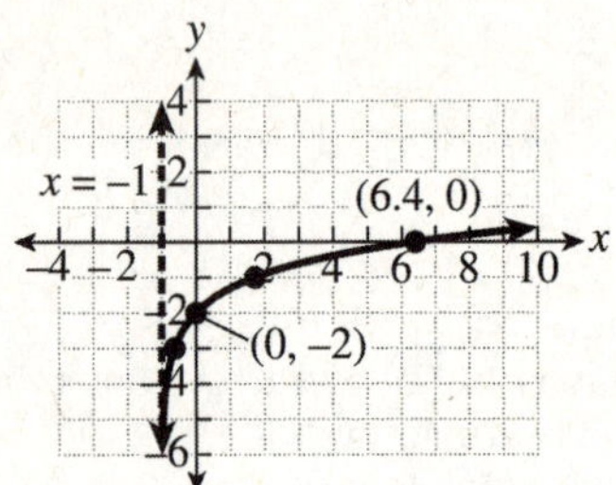

Domain: $(-1, \infty)$; range: $(-\infty, \infty)$

47. $f(x) = e^{x-1} - 4$
$h = 1,\ k = -4$
Shift the graph of $f(x) = e^x$ by 1 unit to the right and by 4 units down.
The horizontal asymptote is at $y = -4$.
Shift three points. (Add 1 to each x-coordinate and subtract 4 from each y-coordinate.)

$f(x) = e^x$	$(-1, 0.4)$	$(0, 1)$	$(1, 2.7)$
$f(x) = e^{x-1} - 4$	$(0, -3.6)$	$(1, -3)$	$(2, -1.3)$

y-intercept: By shifting, $(0, -3.6)$.

x-intercept: Set $f(x) = 0$ and solve for x.

$$e^{x-1} - 4 = 0$$
$$e^{x-1} = 4$$
$$\ln\left(e^{x-1}\right) = \ln 4$$
$$x - 1 = \ln 4$$
$$x = \ln 4 + 1$$
$$x \approx 2.4$$

$(2.4, 0)$

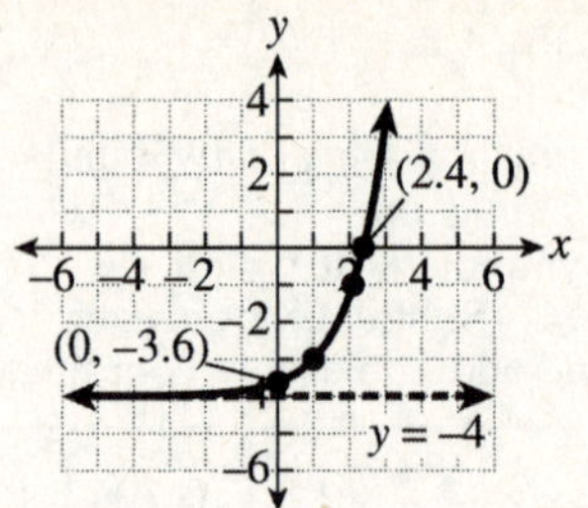

Domain: $(-\infty, \infty)$; range: $(-4, \infty)$

49. $f(x) = \log_3(x+4) - 2$
$h = -4,\ k = -2$
Shift the graph of $f(x) = \log_3 x$ by 4 units to the left and by 2 units down.
The vertical asymptote is at $x = -4$.
Shift three points. (Subtract 4 from each x-coordinate and subtract 2 from each y-coordinate.)

$f(x) = \log_3 x$	$\left(\dfrac{1}{3}, -1\right)$	$(1, 0)$	$(3, 1)$
$f(x) = \log_3(x+4) - 2$	$\left(-\dfrac{11}{3}, -3\right)$	$(-3, -2)$	$(-1, -1)$

y-intercept: Find $f(0)$.

$$f(0) = \log_3(0+4) - 2 = \frac{\log 4}{\log 3} - 2 \approx -0.7$$

$(0, -0.7)$

x-intercept: Set $f(x) = 0$ and solve for x.

$$\log_3(x+4) - 2 = 0$$
$$\log_3(x+4) = 2$$
$$x + 4 = 3^2$$
$$x + 4 = 9$$
$$x \approx 5$$

$(5, 0)$

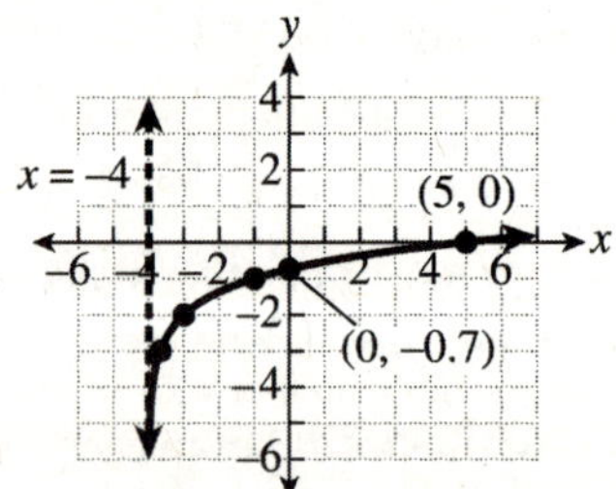

Domain: $(-4, \infty)$; range: $(-\infty, \infty)$

51. $f(x) = \ln(x+6)$
$h = -6,\ k = 0$
Shift the graph of $f(x) = \ln x$ by 6 units to the left.
The vertical asymptote is at $x = -6$.
Shift three points. (Subtract 6 from each x-coordinate.)

$f(x) = \ln x$	$(0.4, -1)$	$(1, 0)$	$(2.7, 1)$
$f(x) = \ln(x+6)$	$(-5.6, -1)$	$(-5, 0)$	$(-3.3, 1)$

y-intercept: Find $f(0)$.

$f(0) = \ln(0+6) = \ln 6 \approx 1.8$

$(0, 1.8)$

x-intercept: By shifting, $(-5, 0)$.

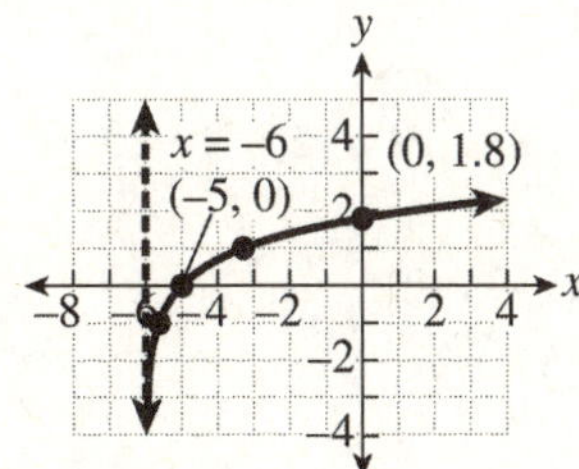

Domain: $(-6, \infty)$; range: $(-\infty, \infty)$

53. $f(x) = 2^{x-1} - 8$
$h = 1,\ k = -8$
Shift the graph of $f(x) = 2^x$ by 1 unit to the right and by 8 units down.
The horizontal asymptote is at $y = -8$.
Shift three points. (Add 1 to each x-coordinate and subtract 8 from each y-coordinate.)

$f(x) = 2^x$	$\left(-1, \dfrac{1}{2}\right)$	$(0, 1)$	$(1, 2)$
$f(x) = 2^{x-1} - 8$	$\left(0, -\dfrac{15}{2}\right)$	$(1, -7)$	$(2, -6)$

y-intercept: By shifting, $\left(0, -\dfrac{15}{2}\right) = (0, -7.5)$.

x-intercept: Set $f(x) = 0$ and solve for x.

$$2^{x-1} - 8 = 0$$
$$2^{x-1} = 8$$
$$2^{x-1} = 2^3$$
$$x - 1 = 3$$
$$x = 4$$

$(4, 0)$

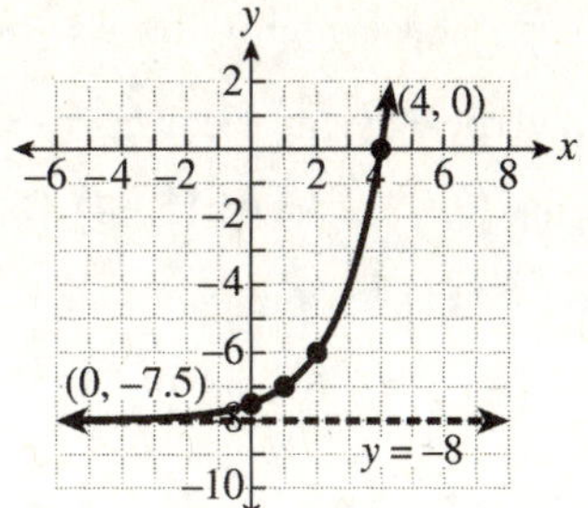

Domain: $(-\infty, \infty)$; range: $(-8, \infty)$

55. $f(x) = \left(\dfrac{1}{3}\right)^x - 3$

$h = 0,\ k = -3$

Shift the graph of $f(x) = \left(\dfrac{1}{3}\right)^x$ by 3 units down.

The horizontal asymptote is at $y = -3$.
Shift three points. (Subtract 3 from each y-coordinate.)

$f(x) = \left(\dfrac{1}{3}\right)^x$	$(-1, 3)$	$(0, 1)$	$\left(1, \dfrac{1}{3}\right)$
$f(x) = \left(\dfrac{1}{3}\right)^x - 3$	$(-1, 0)$	$(0, -2)$	$\left(1, -\dfrac{8}{3}\right)$

y-intercept: By shifting, $(0, -2)$.

x-intercept: By shifting, $(-1, 0)$.

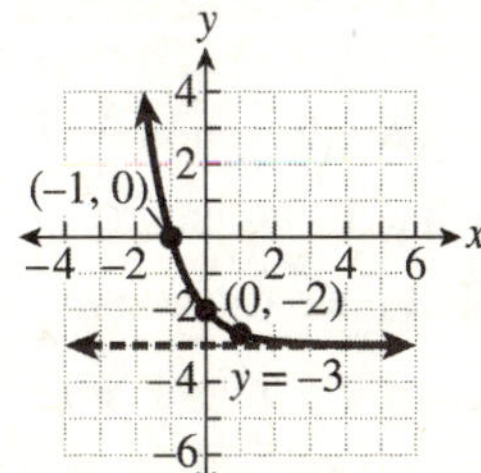

Domain: $(-\infty, \infty)$; range: $(-3, \infty)$

57. $f(x)$ has a vertical asymptote at $x = 4$, so
$f^{-1}(x)$ has a horizontal asymptote at $y = 4$.
$f(x)$ has points $(5, 0), (6, 1)$ and $(8, 2)$, so
$f^{-1}(x)$ has points $(0, 5), (1, 6)$ and $(2, 8)$.

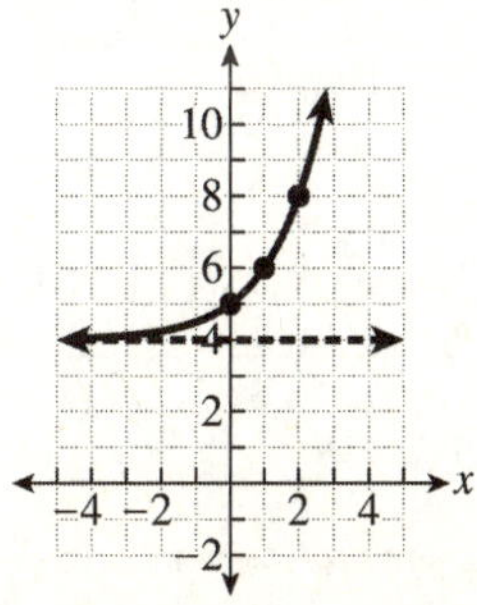

59. $f(x)$ has a horizontal asymptote at $y = -8$, so $f^{-1}(x)$ has a vertical asymptote at $x = -8$.

$f(x)$ has points $(3,-7),(4,-6),(5,-4),(6,0)$ and $(7,8)$, so $f^{-1}(x)$ has points $(-7,3),(-6,4),(-4,5),(0,6)$ and $(8,7)$.

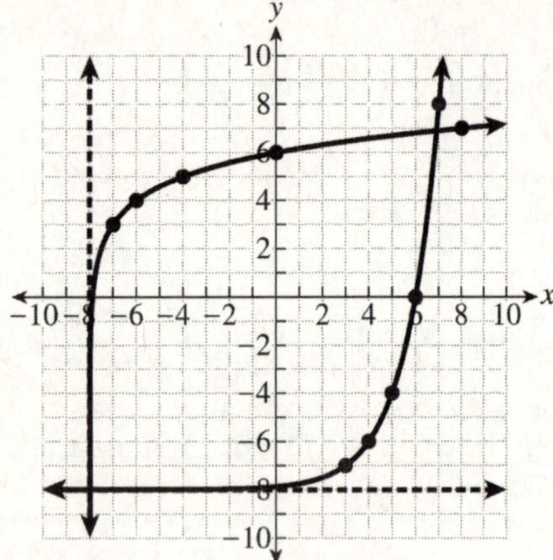

61. b

63. h

65. a

67. f

69. Answers will vary. Example:
First, put the exponential function in the form $f(x) = b^{x-h} + k$. Graph the horizontal asymptote, $y = k$. Shift the basic graph $f(x) = b^x$ using h and k. Find the x-intercept by setting the function equal to zero and solving for x. Find the y-intercept by substituting 0 for x and solving for y.

$f(x) = 3^{x-1} + 2$
$h = 1,\ k = 2$
Shift the graph of $f(x) = 3^x$ by 1 unit to the right and by 2 units up.
The horizontal asymptote is at $y = 2$.
Shift three points. (add 1 to each x-coordinate and add 2 to each y-coordinate.)

$f(x) = 3^x$	$\left(-1,\dfrac{1}{3}\right)$	$(0,1)$	$(1,3)$
$f(x) = 3^{x-1} + 2$	$\left(0,\dfrac{7}{3}\right)$	$(1,3)$	$(2,5)$

y-intercept: By shifting, $\left(0,\dfrac{7}{3}\right)$.

x-intercept: From the graph, there is no x-intercept.

1. $f(5) = 4^5 - 9 = 1024 - 9 = 1015$

2. $f(11) = e^{11-8} - 13 = e^3 - 13 \approx 7.086$

3. Since $6^{-2} = \dfrac{1}{6^2} = \dfrac{1}{36}$, $\log_6\left(\dfrac{1}{36}\right) = -2$.

4. Since $5^5 = 3125$, $\log_5 3125 = 5$.

5. Since $9^0 = 1$, $\log_9 1 = 0$.

6. $\ln e^{12} = 12$

7. $f(52) = \ln(52 - 9) + 20 = \ln(43) + 20 \approx 23.761$

8. $f(1729) = \log(1729 + 23) - 6$
$ = \log(1752) - 6$
$ \approx -2.756$

9. $4^x = 1024$ in logarithmic form: $\log_4 1024 = x$

10. $e^x = 20$ in logarithmic form: $\ln 20 = x$

11. $\log x = 3.2$ in exponential form: $10^{3.2} = x$

12. $\ln x - 3 = 8$ is equivalent to $\ln x = 11$
$\ln x = 11$ in exponential form: $e^{11} = x$

13. $\log_2 13 + \log_2 8 = \log_2(13 \cdot 8) = \log_2 104$

14. $2 \log a - 5 \log b = \log a^2 - \log b^5 = \log\left(\dfrac{a^2}{b^5}\right)$

15. $4\log_2 x - 3\log_2 y + 9\log_2 z$
$= \log_2 x^4 - \log_2 y^3 + \log_2 z^9$
$= \log_2\left(\dfrac{x^4}{y^3}\right) + \log_2 z^9$
$= \log_2\left(\dfrac{x^4 z^9}{y^3}\right)$

16. $-5\log_b 3 - 2\log_b 2 = \log_b 3^{-5} - \log_b 2^2$

$$= \log_b\left(\frac{3^{-5}}{2^2}\right)$$

$$= \log_b\left(\frac{1}{3^5 \cdot 2^2}\right)$$

$$= \log_b\left(\frac{1}{972}\right)$$

17. $\log_8 64x = \log_8 64 + \log_8 x = 2 + \log_8 x$

18. $\log_3\left(\dfrac{n}{27}\right) = \log_3 n - \log_3 27 = \log_3 n - 3$

19. $\log_b\left(\dfrac{a^6 c^3}{b^2}\right) = \log_b\left(a^6 c^3\right) - \log_b b^2$

$$= \log_b a^6 + \log_b c^3 - \log_b b^2$$
$$= 6\log_b a + 3\log_b c - 2$$

20. $\log_2\left(\dfrac{8x^9}{yz^7}\right)$

$$= \log_2 8x^9 - \log_2\left(yz^7\right)$$

$$= \log_2 8 + \log_2 x^9 - \left(\log_2 y + \log_2 z^7\right)$$

$$= 3 + 9\log_2 x - \log_2 y - 7\log_2 z$$

21. $\log_4 25 = \dfrac{\log 25}{\log 4} \approx 2.322$

22. $\log_{1/2} 295 = \dfrac{\log 295}{\log\left(\dfrac{1}{2}\right)} \approx -8.205$

23. $\log_2 1597 = \dfrac{\log 1597}{\log 2} \approx 10.641$

24. $\log_{37} 6,403,200 = \dfrac{\log 6,403,200}{\log 37} \approx 4.340$

25. $10^x = 1000$
$10^x = 10^3$
$x = 3$
$\{3\}$

26. $8^{x-5} = 64$
$8^{x-5} = 8^2$
$x - 5 = 2$
$x = 7$
$\{7\}$

27. $16^x = 32$
$\left(2^4\right)^x = 2^5$
$2^{4x} = 2^5$
$4x = 5$
$x = \dfrac{5}{4}$
$\left\{\dfrac{5}{4}\right\}$

28. $9^{x+6} = 27^x$
$\left(3^2\right)^{x+6} = \left(3^3\right)^x$
$3^{2x+12} = 3^{3x}$
$2x + 12 = 3x$
$12 = x$
$\{12\}$

29. $7^x = 11$
$\log 7^x = \log 11$
$x\log 7 = \log 11$
$x = \dfrac{\log 11}{\log 7}$
$x \approx 1.232$
$\{1.232\}$

30. $6^{x+4} = 3000$
$\log 6^{x+4} = \log 3000$
$(x+4)\log 6 = \log 3000$
$x + 4 = \dfrac{\log 3000}{\log 6}$
$x = \dfrac{\log 3000}{\log 6} - 4$
$x \approx 0.468$
$\{0.468\}$

31. $e^{x-8} = 62$
$\ln e^{x-8} = \ln 62$
$x - 8 = \ln 62$
$x = \ln 62 + 8$
$x \approx 12.127$
$\{12.127\}$

32.
$$5^{2x-9} = 67$$
$$\log 5^{2x-9} = \log 67$$
$$(2x-9)\log 5 = \log 67$$
$$2x-9 = \frac{\log 67}{\log 5}$$
$$2x = \frac{\log 67}{\log 5} + 9$$
$$x = \frac{\frac{\log 67}{\log 5} + 9}{2}$$
$$x \approx 5.806$$
$$\{5.806\}$$

33.
$$\log_4(x+3) = \log_4 17$$
$$x+3 = 17$$
$$x = 14$$
$$\{14\}$$

34.
$$\log\left(x^2+12x+6\right) = \log(3x+16)$$
$$x^2+12x+6 = 3x+16$$
$$x^2+9x-10 = 0$$
$$(x+10)(x-1) = 0$$
$$x+10 = 0 \quad \text{or} \quad x-1 = 0$$
$$x = -10 \quad \text{or} \quad x = 1$$
check: $x = -10$
$$\log\left((-10)^2+12(-10)+6\right) = \log(3(-10)+16)$$
$$\log(100-120-6) = \log(-30+16)$$
$$\log(-14) = \log(-14)$$
Undefined
check: $x = 1$
$$\log\left((1)^2+12(1)+6\right) = \log(3(1)+16)$$
$$\log(1+12+6) = \log(3+16)$$
$$\log(19) = \log(19)$$
True
$$\{1\}$$

35.
$$\ln x = 5$$
$$x = e^5$$
$$x \approx 148.413$$
$$\{148.413\}$$

36.
$$\log_3\left(x^2-13x+57\right)-1 = 2$$
$$\log_3\left(x^2-13x+57\right) = 3$$
$$x^2-13x+57 = 3^3$$
$$x^2-13x+57 = 27$$
$$x^2-13x+30 = 0$$
$$(x-3)(x-10) = 0$$
$$x-3 = 0 \quad \text{or} \quad x-10 = 0$$
$$x = 3 \quad \text{or} \quad x = 10$$
check: $x = 3$
$$\log_3\left(3^2-13(3)+57\right)-1 = 2$$
$$\log_3(9-39+57)-1 = 2$$
$$\log_3 27-1 = 2$$
$$3-1 = 2$$
$$2 = 2$$
True
check: $x = 10$
$$\log_3\left(10^2-13(10)+57\right)-1 = 2$$
$$\log_3(100-130+57)-1 = 2$$
$$\log_3 27-1 = 2$$
$$3-1 = 2$$
$$2 = 2$$
True
$$\{3,10\}$$

37.
$$\log_9 x + \log_9(x+8) = \log_9 20$$
$$\log_9 x(x+8) = \log_9 20$$
$$x(x+8) = 20$$
$$x^2+8x = 20$$
$$x^2+8x-20 = 0$$
$$(x+10)(x-2) = 0$$
$$x+10 = 0 \quad \text{or} \quad x-2 = 0$$
$$x = -10 \quad \text{or} \quad x = 2$$
check: $x = -10$
$$\log_9(-10) + \log_9((-10)+8) = \log_9 20$$
Undefined
check: $x = 2$
$$\log_9(2) + \log_9((2)+8) = \log_9 20$$
$$\log_9(2) + \log_9(10) = \log_9 20$$
$$\log_9 20 = \log_9 20$$
True
$$\{2\}$$

38. $\ln(x+10)-\ln(x-6)=\ln 5$

$$\ln\left(\frac{x+10}{x-6}\right)=\ln 5$$
$$\frac{x+10}{x-6}=5$$
$$x+10=5(x-6)$$
$$x+10=5x-30$$
$$10=4x-30$$
$$40=4x$$
$$10=x$$

check: $x=10$
$$\ln(10+10)-\ln(10-6)=\ln 5$$
$$\ln 20-\ln 4=\ln 5$$
$$\ln\frac{20}{4}=\ln 5$$
$$\ln 5=\ln 5$$
True
$\{10\}$

39. $\log_3(x+5)+\log_3(x-1)=3$
$$\log_3(x+5)(x-1)=3$$
$$(x+5)(x-1)=3^3$$
$$x^2-x+5x-5=27$$
$$x^2+4x-32=0$$
$$(x+8)(x-4)=0$$
$$x+8=0 \quad\text{or}\quad x-4=0$$
$$x=-8 \quad\text{or}\quad x=4$$

check: $x=-8$
$$\log_3((-8)+5)+\log_3((-8)-1)=3$$
$$\log_3(-3)+\log_3(-9)=3$$
Undefined
check: $x=4$
$$\log_3((4)+5)+\log_3((4)-1)=3$$
$$\log_3 9+\log_3 3=3$$
$$2+1=3$$
$$3=3$$
True
$\{4\}$

40. $\log x+\log(x+3)+5=6$
$$\log x(x+3)=1$$
$$x(x+3)=10^1$$
$$x^2+3x-10=0$$
$$(x+5)(x-2)=0$$
$$x+5=0 \quad\text{or}\quad x-2=0$$
$$x=-5 \quad\text{or}\quad x=2$$

check: $x=-5$
$$\log(-5)x+\log((-5)+3)+5=6$$
$$\log(-5)x+\log(-2)+5=6$$
Undefined
check: $x=2$
$$\log(2)+\log((2)+3)+5=6$$
$$\log(2)+\log(5)+5=6$$
$$\log 10+5=6$$
$$1+5=6$$
$$6=6$$
True
$\{2\}$

41. $f(x)=2^{3x-7}+8=13$
$$2^{3x-7}=5$$
$$\log 2^{3x-7}=\log 5$$
$$(3x-7)\log 2=\log 5$$
$$3x-7=\frac{\log 5}{\log 2}$$
$$3x=\frac{\log 5}{\log 2}+7$$
$$x=\frac{\frac{\log 5}{\log 2}+7}{3}$$
$$x\approx 3.107$$
$\{3.107\}$

42. $f(x)=\ln x+4=8$
$$\ln x=4$$
$$x=e^4$$
$$x\approx 54.598$$
$\{54.598\}$

43. $f(x)=e^x+4$
$$y=e^x+4$$
Interchange x and y.
$$x=e^y+4$$
$$x-4=e^y$$
$$\ln(x-4)=\ln e^y$$
$$\ln(x-4)=y$$
$$f^{-1}(x)=\ln(x-4)$$

44.
$$f(x) = 2^{x-5} - 1$$
$$y = 2^{x-5} - 1$$
Interchange x and y.
$$x = 2^{y-5} - 1$$
$$x + 1 = 2^{y-5}$$
$$\log_2(x+1) = \log_2 2^{y-5}$$
$$\log_2(x+1) = y - 5$$
$$\log_2(x+1) + 5 = y$$
$$f^{-1}(x) = \log_2(x+1) + 5$$

45.
$$f(x) = \ln(x+8)$$
$$y = \ln(x+8)$$
Interchange x and y.
$$x = \ln(y+8)$$
$$y + 8 = e^x$$
$$y = e^x - 8$$
$$f^{-1}(x) = e^x - 8$$

46.
$$f(x) = \log_3(x+6) - 10$$
$$y = \log_3(x+6) - 10$$
Interchange x and y.
$$x = \log_3(y+6) - 10$$
$$x + 10 = \log_3(y+6)$$
$$y + 6 = 3^{x+10}$$
$$y = 3^{x+10} - 6$$
$$f^{-1}(x) = 3^{x+10} - 6$$

47.
$$A = P\left(1 + \frac{r}{n}\right)^{nt}$$

$$10,000 = 3500\left(1 + \frac{0.08}{4}\right)^{4t}$$

$$\frac{10,000}{3500} = \left(1 + \frac{0.08}{4}\right)^{4t}$$

$$\frac{20}{7} = 1.02^{4t}$$

$$\log\left(\frac{20}{7}\right) = \log 1.02^{4t}$$

$$\log\left(\frac{20}{7}\right) = 4t \log 1.02$$

$$\frac{\log\left(\frac{20}{7}\right)}{4\log 1.02} = \frac{4t \log 1.02}{4 \log 1.02}$$

$$\frac{\log\left(\frac{20}{7}\right)}{4\log 1.02} = t$$

$$13.3 \approx t$$
13.3 years

48.
$$P = P_0 e^{kt}$$

P	P_0	k	t
250,000	200,000	unknown	10

$$250,000 = 200,000 e^{k \cdot 10}$$
$$\frac{250,000}{200,000} = e^{10k}$$
$$\ln\left(\frac{5}{4}\right) = \ln e^{10k}$$
$$\ln\left(\frac{5}{4}\right) = 10k$$
$$\frac{\ln\left(\frac{5}{4}\right)}{10} = k$$
$$0.022314 \approx k$$

P	P_0	k	t
unknown	200,000	0.022314	23

$$P = 200,000 e^{0.022314(23)}$$
$$P \approx 334,133$$
In 2015, the population will be 334,133.

49.
$$P = P_0 e^{kt}$$

P	P_0	k	t
12.5	25	unknown	5

$$12.5 = 25 e^{k \cdot 5}$$
$$0.5 = e^{5k}$$
$$\ln 0.5 = \ln e^{5k}$$
$$\ln 0.5 = 5k$$
$$\frac{\ln 0.5}{5} = k$$
$$-0.138629 \approx k$$

P	P_0	k	t
unknown	25	−0.138629	10

$$P = 25 e^{-0.138629(10)}$$
$$P \approx 6.25$$
6.25 grams

50. $P = P_0 e^{kt}$

P	P_0	k	t
$0.5\,(50\%)$	$1\,(100\%)$	unknown	5730

$$0.5 = 1e^{k \cdot 5730}$$
$$\ln 0.5 = \ln e^{5730k}$$
$$\ln 0.5 = 5730k$$
$$\frac{\ln 0.5}{5730} = k$$
$$-0.000121 \approx k$$

P	P_0	k	t
$0.98\,(98\%)$	$1\,(100\%)$	-0.000121	unknown

$$0.98 = 1e^{-0.000121(t)}$$
$$\ln 0.98 = \ln e^{-0.000121t}$$
$$\ln 0.98 = -0.000121t$$
$$\frac{\ln 0.98}{-0.000121} = t$$
$$167 \approx t$$

167 years

51. $f(x) = 11.5 \cdot 1.057^x = 30$

$$1.057^x = \frac{30}{11.5}$$
$$\log 1.057^x = \log\left(\frac{30}{11.5}\right)$$
$$x \cdot \log 1.057 = \log\left(\frac{30}{11.5}\right)$$
$$t = \frac{\log\left(\dfrac{30}{11.5}\right)}{\log 1.057}$$
$$t \approx 17.3$$

$1990 + 17.3 = 2007.3$

In 2007, 30% of Americans were obese.

52. $f(x) = 75 + 12e^{-0.206t}$

$$f(6) = 75 + 12e^{-0.206(6)} \approx 78.5$$

The body's temperature 6 hours
after it was found was $78.5°$F.

53. $R = \log I$
$R = \log 20{,}000 \approx 4.3$
The magnitude is 4.3 on the Richter scale.

54. $$\text{pH} = -\log\left[\text{H}^+\right]$$
$$3.5 = -\log\left[\text{H}^+\right]$$
$$-3.5 = \log\left[\text{H}^+\right]$$
$$\text{H}^+ = 10^{-3.5}$$
$$\text{H}^+ \approx 3.16 \times 10^{-4} \text{ moles/liter}$$

55. $f(x) = 3^x + 2$
$h = 0,\ k = 2$
Shift the graph of $f(x) = 3^x$ by 2 units up.
The horizontal asymptote is at $y = 2$.
Shift three points. (Add 2 to each y-coordinate.)

$f(x) = 3^x$	$\left(-1, \dfrac{1}{3}\right)$	$(0,1)$	$(1,3)$
$f(x) = 3^x + 2$	$\left(-1, \dfrac{7}{3}\right)$	$(0,3)$	$(1,5)$

y-intercept: By shifting, $(0,3)$.

x-intercept: From the graph, there is no
x-intercept.

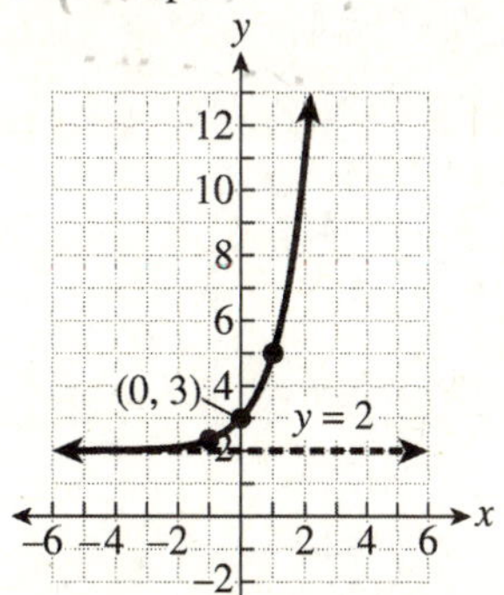

Domain: $(-\infty, \infty)$; range: $(2, \infty)$

56. $f(x) = 2^{x+2} - 8$
$h = -2, \; k = -8$
Shift the graph of $f(x) = 2^x$ by 2 units to the left and by 8 units down.
The horizontal asymptote is at $y = -8$.
Shift three points. (Subtract 2 from each x-coordinate and subtract 8 from each y-coordinate.)

$f(x) = 2^x$	$\left(-1, \dfrac{1}{2}\right)$	$(0,1)$	$(1,2)$
$f(x) = 2^{x+2} - 8$	$\left(-3, -\dfrac{15}{2}\right)$	$(-2,-7)$	$(-1,-6)$

y-intercept: Find $f(0)$.

$f(0) = 2^{0+2} - 8 = 4 - 8 = -4$

$(0,-4)$

x-intercept: Set $f(x) = 0$ and solve for x.

$$2^{x+2} - 8 = 0$$
$$2^{x+2} = 8$$
$$2^{x+2} = 2^3$$
$$x + 2 = 3$$
$$x = 1$$

$(1,0)$

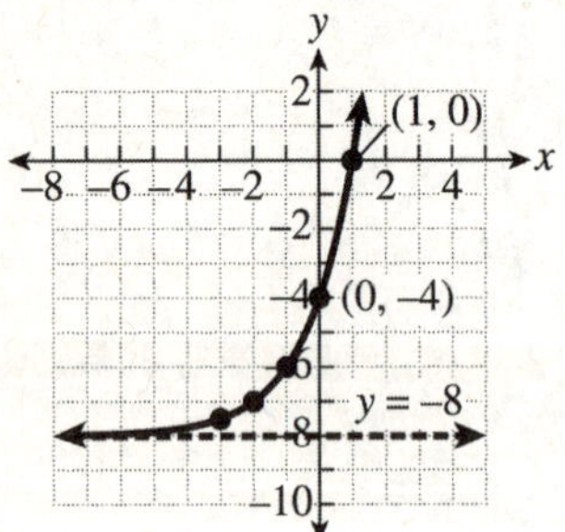

Domain: $(-\infty, \infty)$; range: $(-8, \infty)$

57. $f(x) = e^{x+1} - 6$
$h = -1, \; k = -6$
Shift the graph of $f(x) = e^x$ by 1 unit to the left and by 6 units down.
The horizontal asymptote is at $y = -6$.
Shift three points. (Subtract 1 from each x-coordinate and subtract 6 from each y-coordinate.)

$f(x) = e^x$	$(-1, 0.4)$	$(0,1)$	$(1, 2.7)$
$f(x) = e^{x+1} - 6$	$(-2, -5.6)$	$(-1, -5)$	$(0, -3.3)$

y-intercept: By shifting, $(0, -3.3)$.

x-intercept: Set $f(x) = 0$ and solve for x.

$$e^{x+1} - 6 = 0$$
$$e^{x+1} = 6$$
$$\ln e^{x+1} = \ln 6$$
$$x + 1 = \ln 6$$
$$x = \ln 6 - 1$$
$$x \approx 0.8$$

$(0.8, 0)$

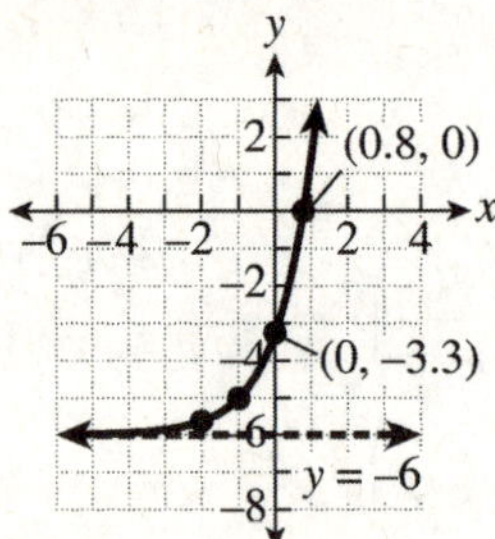

Domain: $(-\infty, \infty)$; range: $(-6, \infty)$

58. $f(x) = \log_2(x-5)$
$h = 5,\ k = 0$
Shift the graph of $f(x) = \log_2 x$ by 5 units to the right.
The vertical asymptote is at $x = 5$.
Shift three points. (Add 5 to each x-coordinate.)

$f(x) = \log_2 x$	$\left(\dfrac{1}{2}, -1\right)$	$(1,0)$	$(2,1)$
$f(x) = \log_2(x-5)$	$\left(\dfrac{11}{2}, -1\right)$	$(6,0)$	$(7,1)$

y-intercept: From the graph, there is no y-intercept.

x-intercept: By shifting, $(6,0)$.

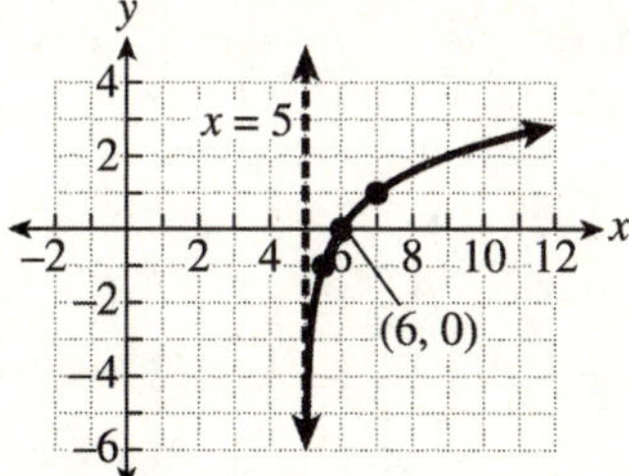

Domain: $(5, \infty)$; range: $(-\infty, \infty)$

59. $f(x) = \log_3(x+9) - 2$
$h = -9,\ k = -2$
Shift the graph of $f(x) = \log_3 x$ by 9 units to the left and by 2 units down.
The vertical asymptote is at $x = -9$.
Shift three points. (Subtract 9 from each x-coordinate and subtract 2 from each y-coordinate.)

$f(x) = \log_3 x$	$\left(\dfrac{1}{3}, -1\right)$	$(1,0)$	$(3,1)$
$f(x) = \log_3(x+9) - 2$	$\left(-\dfrac{26}{3}, -3\right)$	$(-8,-2)$	$(-6,-1)$

y-intercept: Find $f(0)$.

$$f(0) = \log_3(0+9) - 2 = 2 - 2 = 0$$

$(0,0)$

x-intercept: From the y-intercept, $(0,0)$.

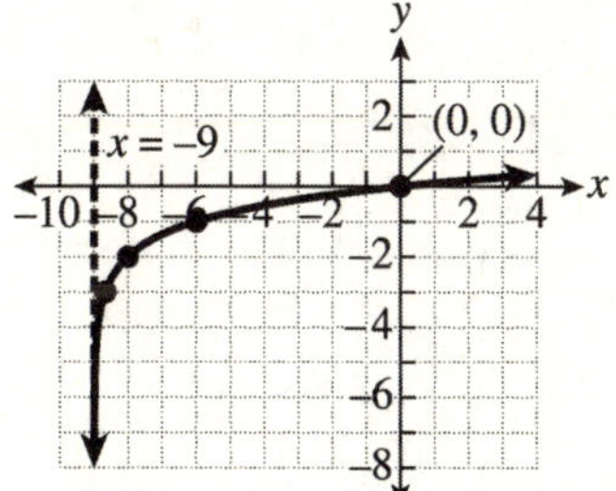

Domain: $(-9, \infty)$; range: $(-\infty, \infty)$

60. $f(x) = \ln(x+8) + 1$
$h = -8,\ k = 1$
Shift the graph of $f(x) = \ln x$ by 8 units to the left and by 1 unit up.
The vertical asymptote is at $x = -8$.
Shift three points. (Subtract 8 from each x-coordinate and add 1 to each y-coordinate.)

$f(x) = \ln x$	$(0.4, -1)$	$(1,0)$	$(2.7,1)$
$f(x) = \ln(x+8) + 1$	$(-7.6, 0)$	$(-7,1)$	$(-5.3, 2)$

y-intercept: Find $f(0)$.

$$f(0) = \ln(0+8) + 1 = \ln 8 + 1 \approx 3.1$$

$(0, 3.1)$

x-intercept: From the graph, $(-7.6, 0)$.

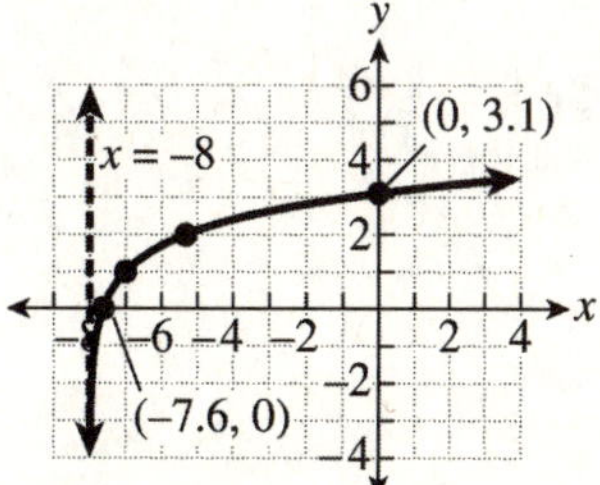

Domain: $(-8, \infty)$; range: $(-\infty, \infty)$

CHAPTER 12 TEST

1. $f(6) = e^{6-4} - 10 = e^2 - 10 \approx -2.611$

2. $f(11) = \ln(11+7) + 4 = \ln(18) + 4 \approx 6.890$

3. Since $12^2 = 144$, $\log_{12} 144 = 2$.

4. $\ln e^{-3} = -3$

5. $\log_4 3136 - \log_4 49 = \log_4 \left(\dfrac{3136}{49}\right)$
$$= \log_4(64)$$
$$= 3$$

6. $2\log_b x + 5\log_b y - 3\log_b z$
$= \log_b x^2 + \log_b y^5 - \log_b z^3$
$= \log_b \left(x^2 y^5\right) - \log_b z^3$
$= \log_b \left(\dfrac{x^2 y^5}{z^3}\right)$

7. $\ln\left(\dfrac{a^4 b^6}{c^8 d}\right) = \ln\left(a^4 b^6\right) - \ln\left(c^8 d\right)$
$$= \ln a^4 + \ln b^6 - \left(\ln c^8 + \ln d\right)$$
$$= 4\ln a + 6\ln b - 8\ln c - \ln d$$

8. $\log_9 62 = \dfrac{\log 62}{\log 9} \approx 1.878$

9. $6^{x+4} = \dfrac{1}{36}$
$6^{x+4} = 6^{-2}$
$x + 4 = -2$
$x = -6$
$\{-6\}$

10. $\qquad 4^{x-9} = 76$
$\log 4^{x-9} = \log 76$
$(x-9)\log 4 = \log 76$
$$x - 9 = \dfrac{\log 76}{\log 4}$$
$$x = \dfrac{\log 76}{\log 4} + 9$$
$$x \approx 12.124$$
$\{12.124\}$

11. $e^{x+2} - 6 = 490$
$e^{x+2} = 496$
$\ln e^{x+2} = \ln 496$
$x + 2 = \ln 496$
$x = \ln 496 - 2$
$x \approx 4.207$
$\{4.207\}$

12. $\log_3(2x - 11) = \log_3 5$
$2x - 11 = 5$
$2x = 16$
$x = 8$
$\{8\}$

13. $\log_5(x + 65) + 5 = 8$
$\log_5(x + 65) = 3$
$x + 65 = 5^3$
$x + 65 = 125$
$x = 60$
$\{60\}$

14. $\log_2 x + \log_2(x - 4) = 5$
$\log_2 x(x - 4) = 5$
$x(x - 4) = 2^5$
$x^2 - 4x = 32$
$x^2 - 4x - 32 = 0$
$(x + 4)(x - 8) = 0$
$x + 4 = 0 \quad \text{or} \quad x - 8 = 0$
$x = -4 \quad \text{or} \qquad x = 8$
check:
$x = -4$
$\log_2(-4) + \log_2(-4 - 4) = 5$
$\log_2(-4) + \log_2(-8) = 5$
Undefined
check:
$x = 8$
$\log_2(8) + \log_2(8 - 4) = 5$
$\log_2(8) + \log_2(4) = 5$
$3 + 2 = 5$
$5 = 5$
True
$\{8\}$

15. $f(x) = e^{x+9} - 17$
$y = e^{x+9} - 17$
Interchange x and y.
$x = e^{y+9} - 17$
$x + 17 = e^{y+9}$
$\ln(x + 17) = \ln e^{y+9}$
$\ln(x + 17) = y + 9$
$\ln(x + 17) - 9 = y$
$f^{-1}(x) = \ln(x + 17) - 9$

16.

$$A = P\left(1+\frac{r}{n}\right)^{nt}$$

$$6000 = 4200\left(1+\frac{0.06}{12}\right)^{12t}$$

$$\frac{6000}{4200} = \left(1+\frac{0.06}{12}\right)^{12t}$$

$$\frac{10}{7} = 1.005^{12t}$$

$$\log\left(\frac{10}{7}\right) = \log 1.005^{12t}$$

$$\log\left(\frac{10}{7}\right) = 12t\log 1.005$$

$$\frac{\log\left(\frac{10}{7}\right)}{12\log 1.005} = \frac{12t\log 1.005}{12\log 1.005}$$

$$\frac{\log\left(\frac{10}{7}\right)}{12\log 1.005} = t$$

$$6 \approx t$$

6 years

17. $P = P_0 e^{kt}$

P	P_0	k	t
25,000	15,000	unknown	10

$$25,000 = 15,000 e^{k\cdot 10}$$

$$\frac{25,000}{15,000} = e^{10k}$$

$$\ln\left(\frac{5}{3}\right) = \ln e^{10k}$$

$$\ln\left(\frac{5}{3}\right) = 10k$$

$$\frac{\ln\left(\frac{5}{3}\right)}{10} = k$$

$$0.051083 \approx k$$

P	P_0	k	t
unknown	15,000	0.051083	33

$P = 15,000 e^{0.051083(33)}$
$P \approx 80,946.56$
In 2025, the painting will be worth
$80,946.56.

18. $P = P_0 e^{kt}$

P	P_0	k	t
0.5(50%)	1(100%)	unknown	5730

$$0.5 = 1e^{k\cdot 5730}$$

$$\ln 0.5 = \ln e^{5730k}$$
$$\ln 0.5 = 5730k$$

$$\frac{\ln 0.5}{5730} = k$$

$$-0.000121 \approx k$$

P	P_0	k	t
0.9(90%)	1(100)	−0.000121	unknown

$$0.9 = 1e^{-0.000121t}$$

$$\ln 0.9 = \ln e^{-0.000121t}$$
$$\ln 0.9 = -0.000121t$$

$$\frac{\ln 0.9}{-0.000121} = t$$

$$870.7 \approx t$$
870.7 years

19. $f(x) = e^{x-2} - 9$
$h = 2,\ k = -9$
Shift the graph of $f(x) = e^x$ by 2 units to the right and by 9 units down.
The horizontal asymptote is at $y = -9$.
Shift three points. (Add 2 to each x-coordinate and subtract 9 from each y-coordinate.)

$f(x) = e^x$	$(-1, 0.4)$	$(0, 1)$	$(1, 2.7)$
$f(x) = e^{x-2} - 9$	$(1, -8.6)$	$(2, -8)$	$(3, -6.3)$

y-intercept: Find $f(0)$.

$f(0) = e^{0-2} - 9 = e^{-2} - 9 \approx -8.9$

$(0, -8.9)$

x-intercept: Set $f(x) = 0$ and solve for x.

$e^{x-2} - 9 = 0$
$e^{x-2} = 9$
$\ln\left(e^{x-2}\right) = \ln 9$
$x - 2 = \ln 9$
$x = \ln 9 + 2$
$x \approx 4.2$

$(4.2, 0)$

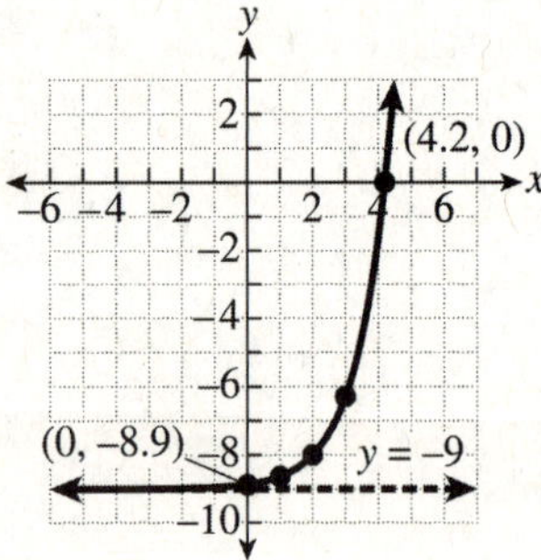

Domain: $(-\infty, \infty)$; range: $(-9, \infty)$

20. $f(x) = \log_3(x+4) + 1$
$h = -4,\ k = 1$
Shift the graph of $f(x) = \log_3 x$ by 4 units to the left and by 1 unit up.
The vertical asymptote is at $x = -4$.
Shift three points. (Subtract 4 from each x-coordinate and add 1 to each y-coordinate.)

$f(x) = \log_3 x$	$\left(\dfrac{1}{3}, -1\right)$	$(1, 0)$	$(3, 1)$
$f(x) = \log_3(x+4) + 1$	$\left(-\dfrac{11}{3}, 0\right)$	$(-3, 1)$	$(-1, 2)$

y-intercept: Find $f(0)$.

$f(0) = \log_3(0+4) + 1 = \dfrac{\log 4}{\log 3} + 1 \approx 2.3$

$(0, 2.3)$

x-intercept: By shifting, $\left(-\dfrac{11}{3}, 0\right) \approx (-3.7, 0)$.

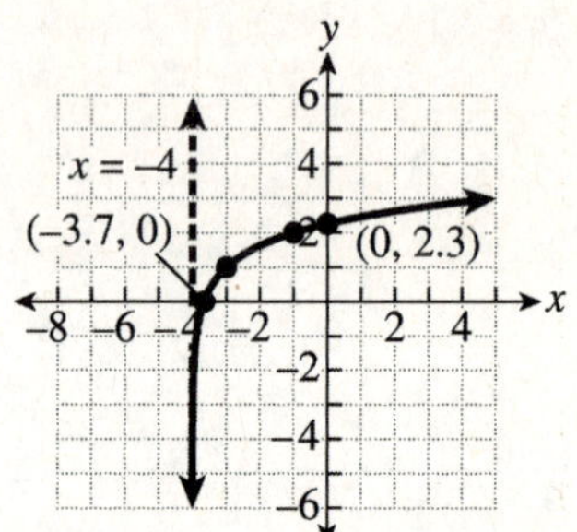

Domain: $(-4, \infty)$; range: $(-\infty, \infty)$

CHAPTER 13   CONIC SECTIONS

13.1 QUICK CHECK

1. $y = -x^2 + 2x + 15$

Use the form $y = ax^2 + bx + c.$

$a < 0,$ parabola opens downward.

vertex:

$$x = \frac{-b}{2a} = \frac{-2}{2(-1)} = 1$$

$$y = -(1)^2 + 2(1) + 15 = -1 + 2 + 15 = 16$$
$$(1,16)$$

y-intercept:
$$y = -(0)^2 + 2(0) + 15 = 15$$
$$(0,15)$$

x-intercept:
$$0 = -x^2 + 2x + 15$$
$$0 = x^2 - 2x - 15$$
$$0 = (x+3)(x-5)$$
$$x = -3 \text{ or } x = 5$$
$$(-3,0),\ (5,0)$$

axis of symmetry: $x = 1$

point symmetric to the y-intercept: $(2,15)$

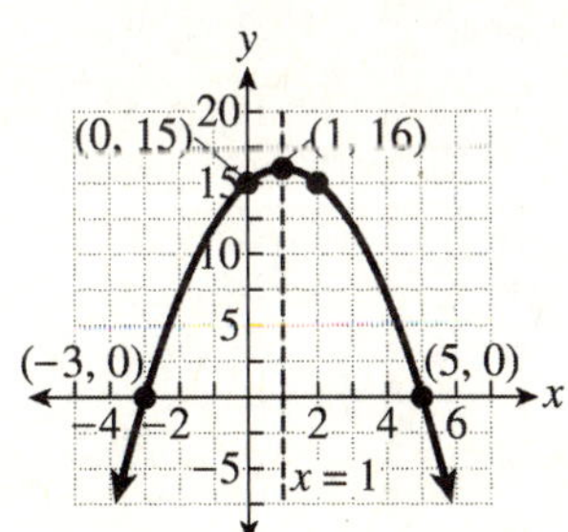

2. $y = (x+4)^2 + 3$ or $y = \left(x-(-4)\right)^2 + 3$

Use the form $y = a(x-h)^2 + k.$

$a > 0,$ parabola opens upward.

vertex: $(-4,3)$

Shift the graph of $y = x^2$ by 4 units to the left and by 3 units up.

y-intercept:
$$y = (0+4)^2 + 3 = 16 + 3 = 19$$
$$(0,19)$$

x-intercept:
The vertex is above the x-axis and the parabola opens upward. There are no x-intercepts.

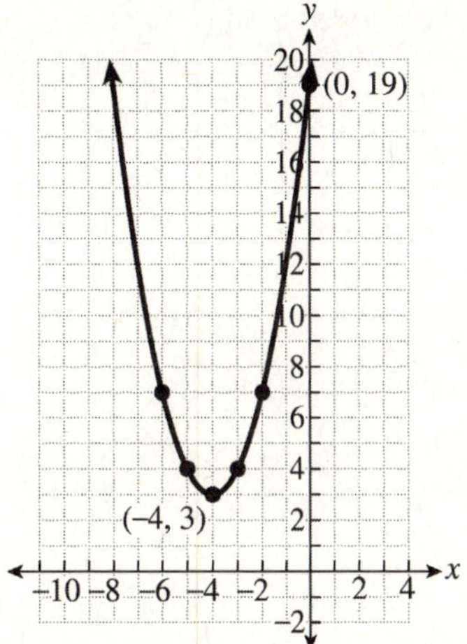

3. $x = y^2 + 4y - 2$

Use the form $x = ay^2 + by + c.$

$a > 0,$ parabola opens to the right.

vertex:

$$y = \frac{-b}{2a} = \frac{-4}{2(1)} = -2$$

$$x = (-2)^2 + 4(-2) - 2 = 4 - 8 - 2 = -6$$
$$(-6,-2)$$

x-intercept:
$$x = (0)^2 + 4(0) - 2 = -2$$
$$(-2,0)$$

y-intercept:
$$0 = y^2 + 4y - 2$$

$$y = \frac{-(4) \pm \sqrt{(4)^2 - 4(1)(-2)}}{2(1)}$$

$$y = \frac{-4 \pm \sqrt{24}}{2}$$

$$y = \frac{-4 \pm 2\sqrt{6}}{2}$$

$$y = -2 \pm \sqrt{6}$$

$$\left(0, -2 - \sqrt{6}\right), \left(0, -2 + \sqrt{6}\right)$$

axis of symmetry: $y = -2$

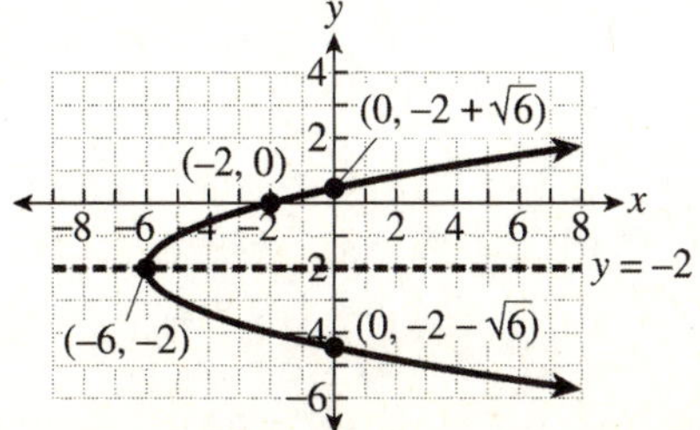

4. $x = -y^2 + 8y - 12$

Use the form $x = ay^2 + by + c$.

$a < 0,$ parabola opens to the left.

vertex:

$$y = \frac{-b}{2a} = \frac{-8}{2(-1)} = 4$$

$x = -(4)^2 + 8(4) - 12 = -16 + 32 - 12 = 4$

$(4, 4)$

x-intercept:

$x = -(0)^2 + 8(0) - 12 = -12$

$(-12, 0)$

y-intercept:

$0 = -y^2 + 8y - 12$

$0 = y^2 - 8y + 12$

$0 = (y - 2)(y - 6)$

$y = 2$ or $y = 6$

$(0, 2), (0, 6)$

axis of symmetry: $y = 4$

point symmetric to the x-intercept: $(-12, 8)$

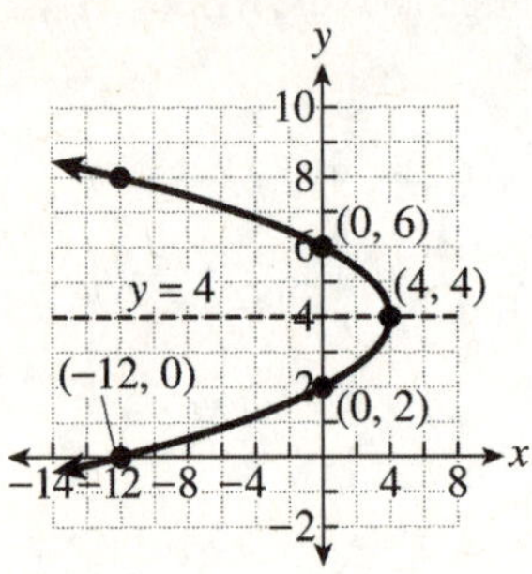

5. $x = (y - 3)^2 - 1$

Use the form $x = a(y - k)^2 + h$.

$a > 0,$ parabola opens to the right.

vertex: $(-1, 3)$

Shift the graph of $x = y^2$ by 1 unit to the left and by 3 units up.

x-intercept:

$x = (0 - 3)^2 - 1 = 9 - 1 = 8$

$(8, 0)$

y-intercept:

$0 = (y - 3)^2 - 1$

$1 = (y - 3)^2$

$\pm 1 = y - 3$

$3 \pm 1 = y$

$y = 4$ or $y = 2$

$(0, 4), (0, 2)$

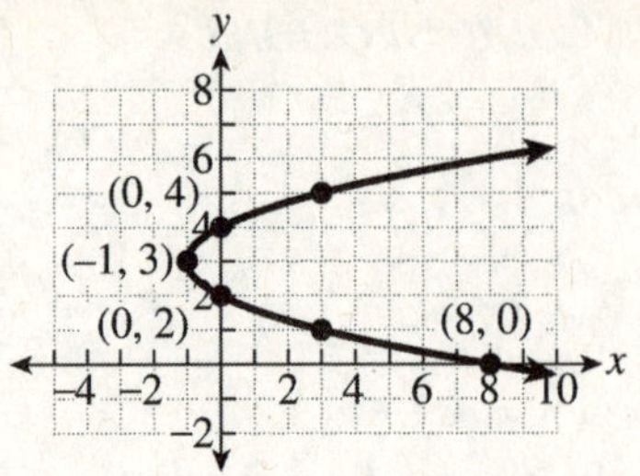

6. $x = -(y - 4)^2 - 3$

Use the form $x = a(y - k)^2 + h$.

$a < 0,$ parabola opens to the left, so rotate the graph of $x = y^2$ about the y-axis.

vertex: $(-3, 4)$

Shift the rotated graph by 3 units to the left and by 4 units up.

x-intercept:

$x = -(0 - 4)^2 - 3 = -16 - 3 = -19$

$(-19, 0)$

y-intercept:

The vertex is to the left of the y-axis and the parabola opens to the left. There are no y-intercepts.

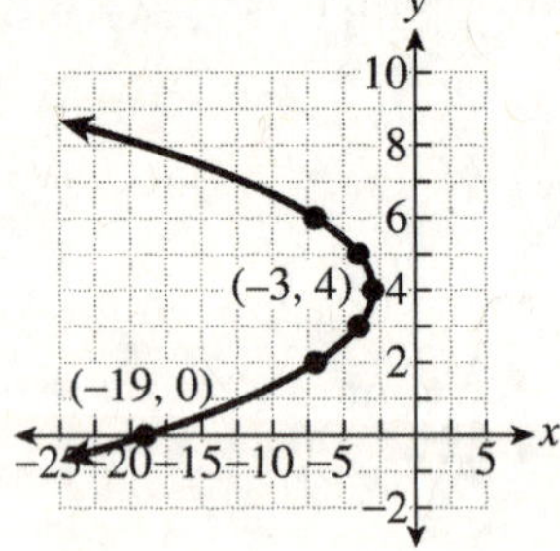

7. a)
$$y = x^2 + 4x + 20$$
$$y - 20 = x^2 + 4x$$
$$y - 20 + \left(\frac{4}{2}\right)^2 = x^2 + 4x + \left(\frac{4}{2}\right)^2$$
$$y - 20 + 4 = x^2 + 4x + 4$$
$$y - 16 = (x+2)^2$$
$$y = (x+2)^2 + 16$$
vertex: $(-2,16)$

b)
$$x = -y^2 - 6y + 17$$
$$x - 17 = -\left(y^2 + 6y\right)$$
$$x - 17 - \left(\frac{6}{2}\right)^2 = -\left(y^2 + 6y + \left(\frac{6}{2}\right)^2\right)$$
$$x - 17 - 9 = -\left(y^2 + 6y + 9\right)$$
$$x - 26 = -(y+3)^2$$
$$x = -(y+3)^2 + 26$$
vertex: $(26,-3)$

8. Since the parabola opens upward, the equation is of the form $y = ax^2 + bx + c$, where $a > 0$. The x-intercepts are $(2,0)$ and $(4,0)$, so $x - 2$ and $x - 4$ are factors.
$$y = a(x-2)(x-4)$$
$$y = a\left(x^2 - 6x + 8\right)$$
Use y-intercept $(0,8)$
$$8 = a\left((0)^2 - 6(0) + 8\right)$$
$$8 = 8a$$
$$1 = a$$
so,
$$y = 1\left(x^2 - 6x + 8\right)$$
$$y = x^2 - 6x + 8$$

1. upward; downward

3. right; left

5. (h,k)

7. $y = x^2 + 8x + 8$

Use the form $y = ax^2 + bx + c$.

$a > 0$, parabola opens upward.

vertex:
$$x = \frac{-b}{2a} = \frac{-8}{2(1)} = -4$$
$$y = (-4)^2 + 8(-4) + 8 = 16 - 32 + 8 = -8$$
$$(-4,-8)$$

y-intercept:
$$y = (0)^2 + 8(0) + 8 = 8$$
$$(0,8)$$

x-intercept:
$$0 = x^2 + 8x + 8$$
$$x = \frac{-(8) \pm \sqrt{(8)^2 - 4(1)(8)}}{2(1)}$$
$$x = \frac{-8 \pm \sqrt{32}}{2}$$
$$x = \frac{-8 \pm 4\sqrt{2}}{2}$$
$$x = -4 \pm 2\sqrt{2}$$
$$\left(0, -4 - 2\sqrt{2}\right), \left(0, -4 + 2\sqrt{2}\right)$$

axis of symmetry: $x = -4$

point symmetric to the y-intercept: $(-8,8)$

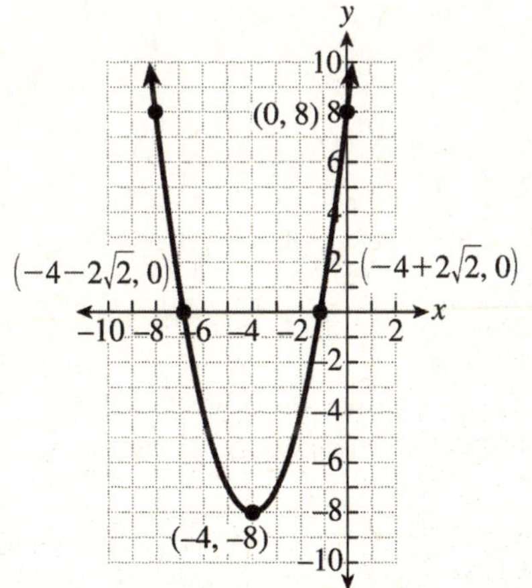

9. $y = -x^2 + 2x + 15$

Use the form $y = ax^2 + bx + c$.

$a < 0$, parabola opens downward.

vertex:

$x = \dfrac{-b}{2a} = \dfrac{-2}{2(-1)} = 1$

$y = -(1)^2 + 2(1) + 15 = -1 + 2 + 15 = 16$
$(1, 16)$

y-intercept:
$y = -(0)^2 + 2(0) + 15 = 15$
$(0, 15)$

x-intercept:
$0 = -x^2 + 2x + 15$
$0 = x^2 - 2x - 15$
$0 = (x - 5)(x + 3)$
$x = 5$ or $x = -3$
$(5, 0), (-3, 0)$

axis of symmetry: $x = 1$

point symmetric to the y-intercept: $(2, 15)$

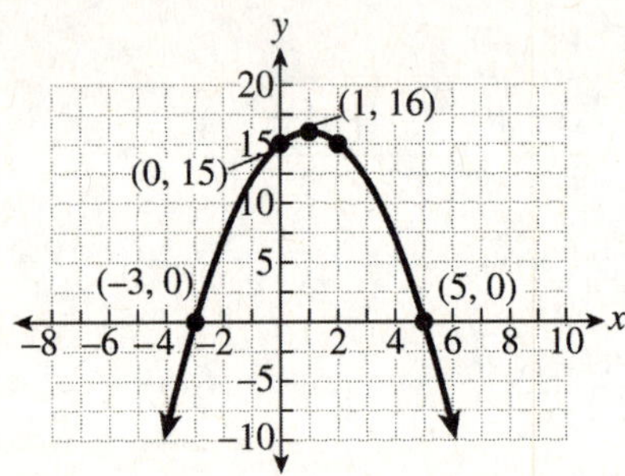

11. $y = (x + 5)^2 + 5$ or $y = (x - (-5))^2 + 5$

Use the form $y = a(x - h)^2 + k$.

$a > 0$, parabola opens upward.

vertex: $(-5, 5)$

Shift the graph of $y = x^2$ by 5 units to the left
and by 5 units up.

y-intercept:
$y = (0 + 5)^2 + 5 = 25 + 5 = 30$
$(0, 30)$

x-intercept:
The vertex is above the x-axis and the parabola
opens upward. There are no x-intercepts.

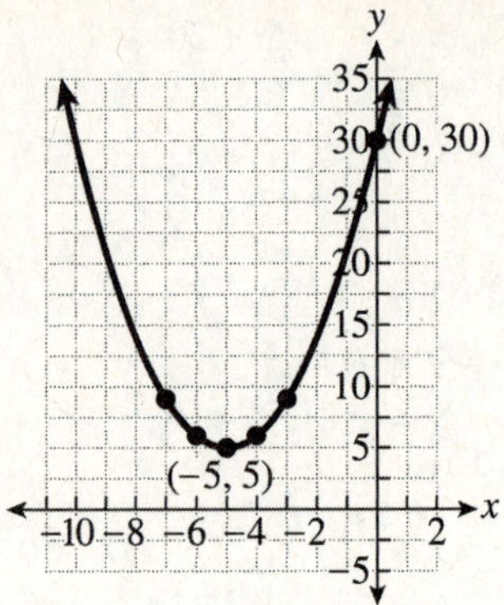

13. $y = -(x + 2)^2 + 9$ or $y = -(x - (-2))^2 + 9$

Use the form $y = a(x - h)^2 + k$.

$a < 0$, parabola opens downward, so rotate the
graph of $y = x^2$ about the x-axis.

vertex: $(-2, 9)$

Shift the rotated graph by 2 units to the left and
by 9 units up.

y-intercept:
$y = -(0 + 2)^2 + 9 = -4 + 9 = 5$
$(0, 5)$

x-intercept:
$0 = -(x + 2)^2 + 9$
$(x + 2)^2 = 9$
$x + 2 = \pm\sqrt{9}$
$x = -2 \pm 3$
$x = -5$ or $x = 1$
$(-5, 0), (1, 0)$

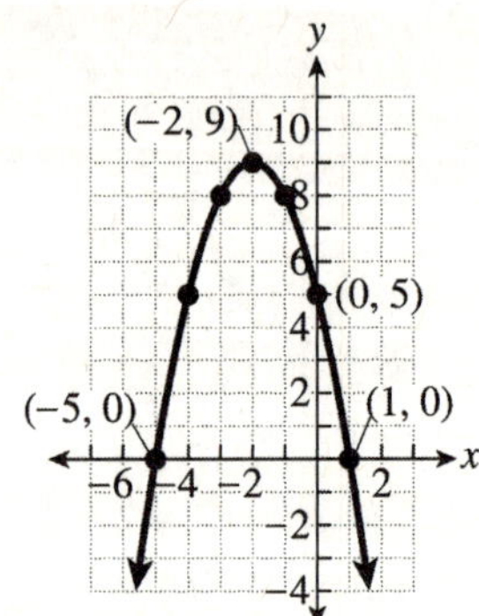

15. $x = y^2 + 4y - 5$

Use the form $x = ay^2 + by + c$.

$a > 0$, parabola opens to the right.

vertex:

$$y = \frac{-b}{2a} = \frac{-4}{2(1)} = -2$$

$x = (-2)^2 + 4(-2) - 5 = 4 - 8 - 5 = -9$
$(-9, -2)$

x-intercept:
$x = (0)^2 + 4(0) - 5 = -5$
$(-5, 0)$

y-intercept:
$0 = y^2 + 4y - 5$
$0 = (y + 5)(y - 1)$
$y = -5$ or $y = 1$
$(0, -5), (0, 1)$

axis of symmetry: $y = -2$

point symmetric to the x-intercept: $(-5, -4)$

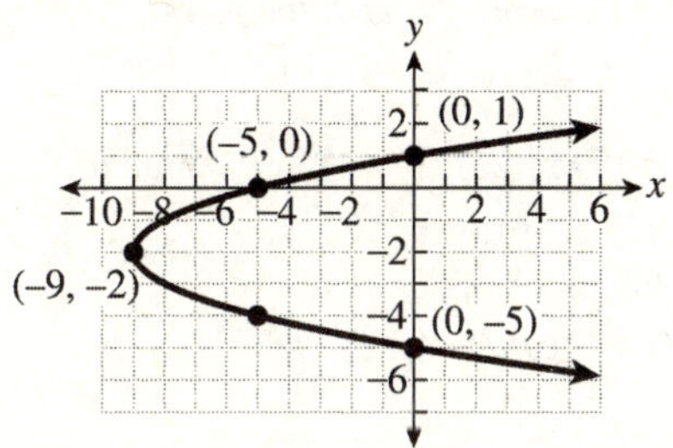

17. $x = y^2 - 4y + 6$

Use the form $x = ay^2 + by + c$.

$a > 0$, parabola opens to the right.

vertex:

$$y = \frac{-b}{2a} = \frac{-(-4)}{2(1)} = 2$$

$x = (2)^2 - 4(2) + 6 = 4 - 8 + 6 = 2$

$(2, 2)$

x-intercept:
$x = (0)^2 - 4(0) + 6 = 6$
$(6, 0)$

y-intercept:
The vertex is to the right of the y-axis and the parabola opens to the right. There are no y-intercepts.

axis of symmetry: $y = 2$

point symmetric to the x-intercept: $(6, 4)$

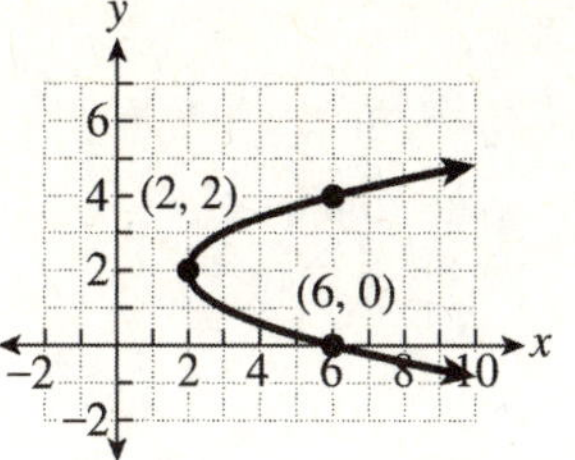

19. $x = -y^2 + 6y - 8$

Use the form $x = ay^2 + by + c$.

$a < 0$, parabola opens to the left.

vertex:

$$y = \frac{-b}{2a} = \frac{-6}{2(-1)} = 3$$

$x = -(3)^2 + 6(3) - 8 = -9 + 18 - 8 = 1$
$(1, 3)$

x-intercept:
$x = -(0)^2 + 6(0) - 8 = -8$
$(-8, 0)$

y-intercept:
$0 = -y^2 + 6y - 8$
$0 = y^2 - 6y + 8$
$0 = (y - 2)(y - 4)$
$y = 2$ or $y = 4$
$(0, 2), (0, 4)$

axis of symmetry: $y - 3$

point symmetric to the x-intercept: $(-8, 6)$

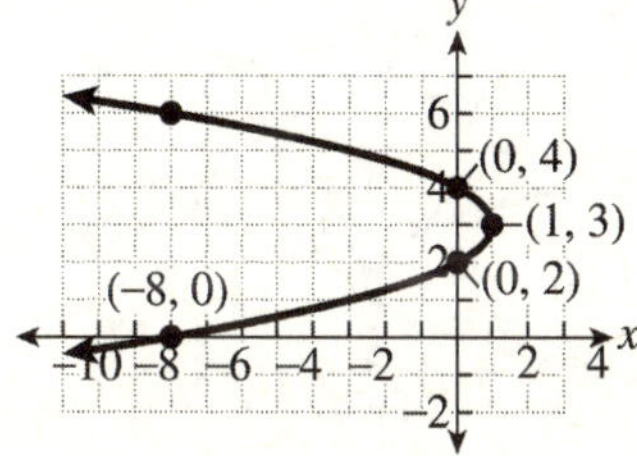

21. $x = -y^2 + 5y + 1$

Use the form $x = ay^2 + by + c.$

$a < 0,$ parabola opens to the left.

vertex:
$$y = \frac{-b}{2a} = \frac{-5}{2(-1)} = \frac{5}{2}$$

$$x = -\left(\frac{5}{2}\right)^2 + 5\left(\frac{5}{2}\right) + 1 = -\frac{25}{4} + \frac{50}{4} + \frac{4}{4} = \frac{29}{4}$$

$$\left(\frac{29}{4}, \frac{5}{2}\right)$$

x-intercept:
$$x = -(0)^2 + 5(0) + 1 = 1$$
$$(1, 0)$$

y-intercept:
$$0 = -y^2 + 5y + 1$$
$$0 = y^2 - 5y - 1$$

$$y = \frac{-(-5) \pm \sqrt{(-5)^2 - 4(1)(-1)}}{2(1)}$$

$$y = \frac{5 \pm \sqrt{29}}{2}$$

$$\left(0, \frac{5 + \sqrt{29}}{2}\right), \left(0, \frac{5 - \sqrt{29}}{2}\right)$$

axis of symmetry: $y = \dfrac{5}{2}$

point symmetric to the x-intercept: $(1, 5)$

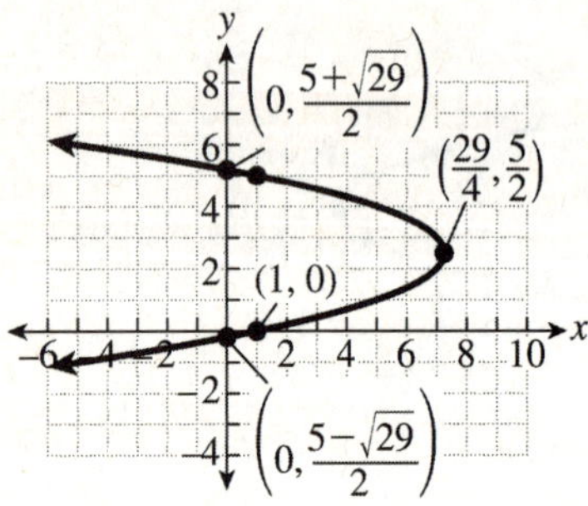

23. $x = (y + 3)^2 - 1$ or $x = \left(y - (-3)\right)^2 - 1$

Use the form $x = a(y - k)^2 + h.$

$a > 0,$ parabola opens to the right.

vertex: $(-1, -3)$

Shift the graph of $x = y^2$ by 1 unit to the left and by 3 units down.

x-intercept:
$$x = (0 + 3)^2 - 1 = 9 - 1 = 8$$
$$(8, 0)$$

y-intercept:
$$0 = (y + 3)^2 - 1$$
$$1 = (y + 3)^2$$
$$\pm 1 = y + 3$$
$$-3 \pm 1 = y$$
$$y = -4 \text{ or } y = -2$$
$$(0, -4), (0, -2)$$

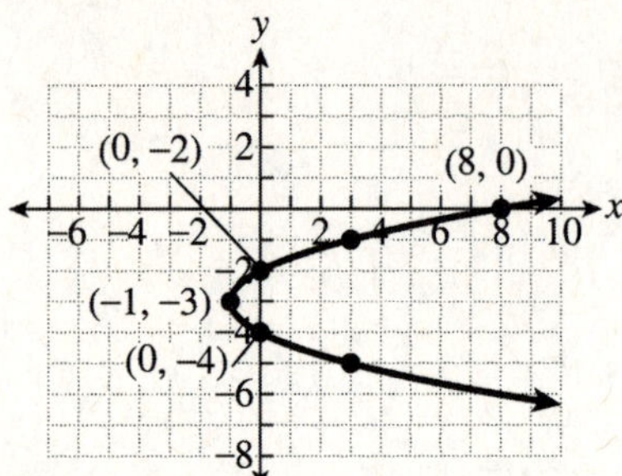

25. $x = (y + 2)^2 + 2$ or $x = \left(y - (-2)\right)^2 + 2$

Use the form $x = a(y - k)^2 + h.$

$a > 0,$ parabola opens to the right.

vertex: $(2, -2)$

Shift the graph of $x = y^2$ by 2 units to the right and by 2 units down.

x-intercept:
$$x = (0 + 2)^2 + 2 = 4 + 2 = 6$$
$$(6, 0)$$

y-intercept:
The vertex is to the right of the y-axis and the parabola opens to the right. There are no y-intercepts.

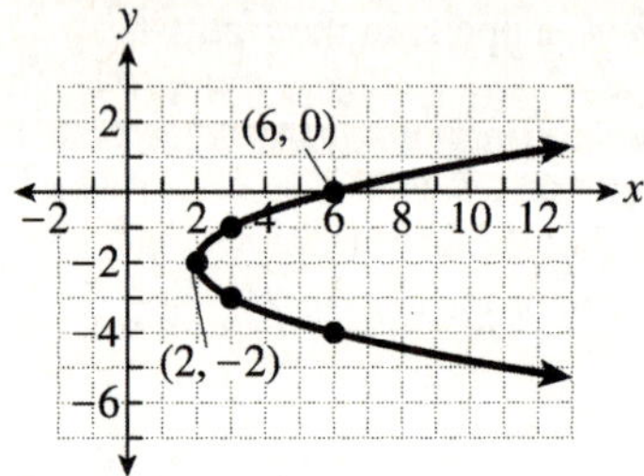

27. $x = -(y-2)^2 + 4$

Use the form $x = a(y-k)^2 + h$.

$a < 0$, parabola opens to the left, so rotate the graph of $x = y^2$ about the y-axis.

vertex: $(4,2)$

Shift the rotated graph by 4 units to the right and by 2 units up.

x-intercept:
$x = -(0-2)^2 + 4 = -4 + 4 = 0$
$(0,0)$

y-intercept:
$$0 = -(y-2)^2 + 4$$
$$(y-2)^2 = 4$$
$$y - 2 = \pm 2$$
$$y = 2 \pm 2$$
$y = 0$ or $y = 4$
$(0,0), (0,4)$

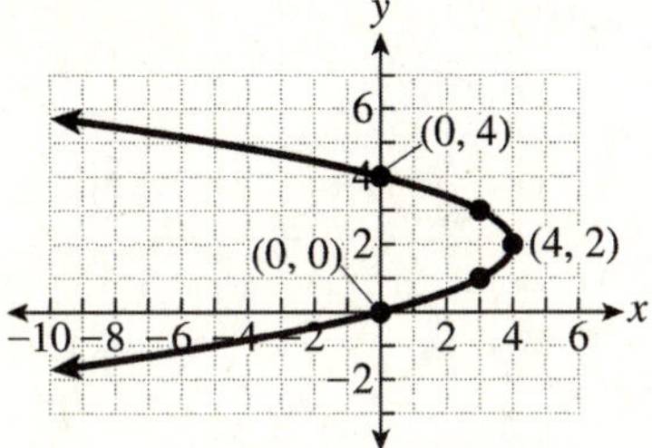

29. $x = -y^2 - 6y - 15$

Use the form $x = ay^2 + by + c$.

$a < 0$, parabola opens to the left.

vertex:
$$y = \frac{-b}{2a} = \frac{-(-6)}{2(-1)} = -3$$

$x = -(-3)^2 - 6(-3) - 15 = -9 + 18 - 15 = -6$
$(-6,-3)$

x-intercept:
$x = -(0)^2 - 6(0) - 15 = -15$
$(-15,0)$

y-intercept:
The vertex is to the left of the y-axis and the parabola opens to the left. There are no y-intercepts.

axis of symmetry: $y = -3$

point symmetric to the x-intercept: $(-15,-6)$

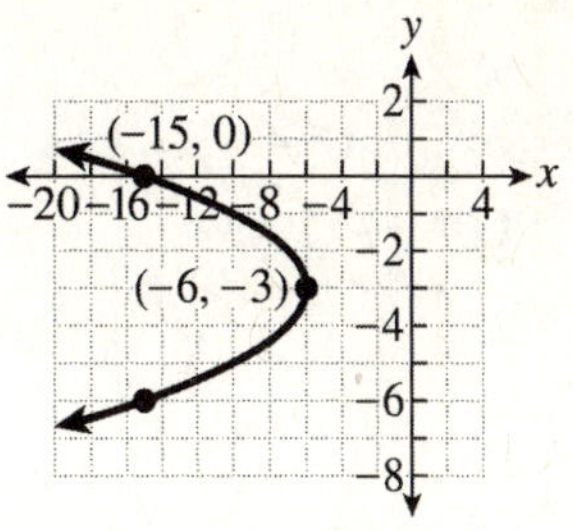

31. $x = (y-1)^2 - 7$

Use the form $x = a(y-k)^2 + h$.

$a > 0$, parabola opens to the right.

vertex: $(-7,1)$

Shift the graph of $x = y^2$ by 7 units to the left and by 1 unit up.

x-intercept:
$x = (0-1)^2 - 7 = 1 - 7 = -6$
$(-6,0)$

y-intercept:
$$0 = (y-1)^2 - 7$$
$$7 = (y-1)^2$$
$$\pm\sqrt{7} = y - 1$$
$$1 \pm \sqrt{7} = y$$
$\left(0, 1-\sqrt{7}\right), \left(0, 1+\sqrt{7}\right)$

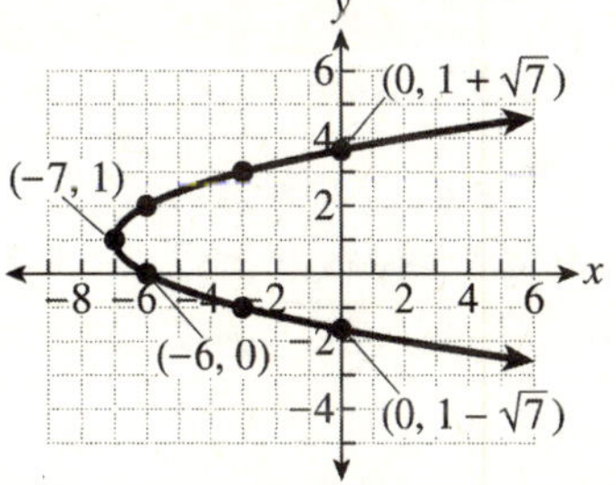

33.　$y = -x^2 - 4x - 7$

Use the form $y = ax^2 + bx + c$.

$a < 0$, parabola opens downward.

vertex:

$$x = \frac{-b}{2a} = \frac{-(-4)}{2(-1)} = -2$$

$$y = -(-2)^2 - 4(-2) - 7 = -4 + 8 - 7 = -3$$
$$(-2, -3)$$

y-intercept:

$$y = -(0)^2 - 4(0) - 7 = -7$$
$$(0, -7)$$

x-intercept:
The vertex is below the x-axis and the parabola opens downward. There are no x-intercepts.

axis of symmetry: $x = -2$

point symmetric to the y-intercept: $(-4, -7)$

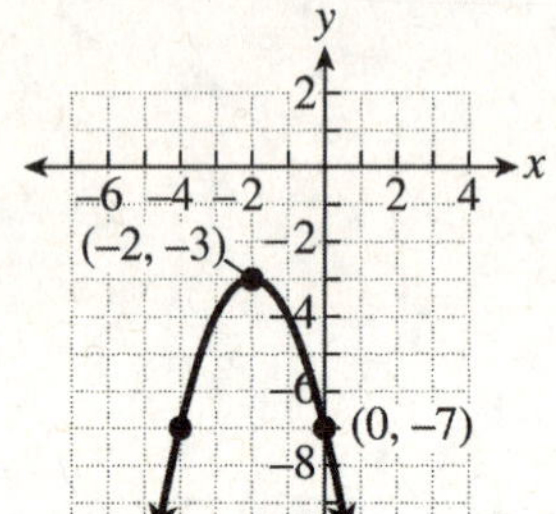

35.　$y = -(x-1)^2 + 9$

Use the form $y = a(x-h)^2 + k$.

$a < 0$, parabola opens downward, so rotate the graph of $y = x^2$ about the x-axis.

vertex: $(1, 9)$

Shift the rotated graph by 1 unit to the right and by 9 units up.

y-intercept:

$$y = -(0-1)^2 + 9 = -1 + 9 = 8$$
$$(0, 8)$$

x-intercept:

$$0 = -(x-1)^2 + 9$$
$$(x-1)^2 = 9$$
$$x - 1 = \pm\sqrt{9}$$
$$x = 1 \pm 3$$
$$x = -2 \text{ or } x = 4$$
$$(-2, 0), (4, 0)$$

axis of symmetry: $x = 1$

point symmetric to the y-intercept: $(2, 8)$

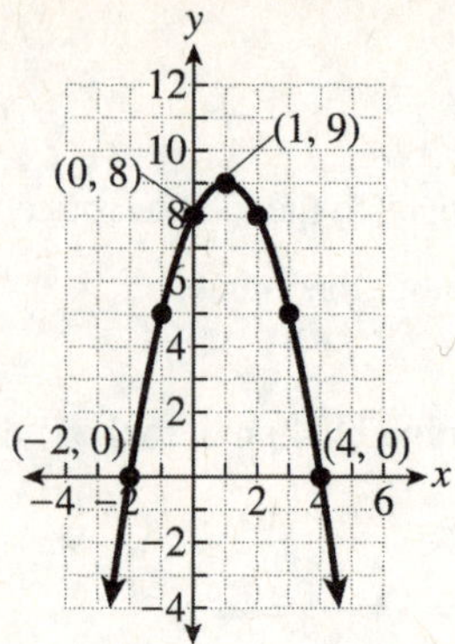

37.　$x = (y+2)^2 - 4$ or $x = (y - (-2))^2 - 4$

Use the form $x = a(y-k)^2 + h$.

$a > 0$, parabola opens to the right.

vertex: $(-4, -2)$

Shift the graph of $x = y^2$ by 4 units to the left and by 2 units down.

x-intercept:

$$x = (0+2)^2 - 4 = 4 - 4 = 0$$
$$(0, 0)$$

y-intercept:

$$(y+2)^2 - 4 = 0$$
$$(y+2)^2 = 4$$
$$y + 2 = \pm\sqrt{4}$$
$$y = -2 \pm 2$$
$$y = -4 \text{ or } y = 0$$
$$(0, -4), (0, 0)$$

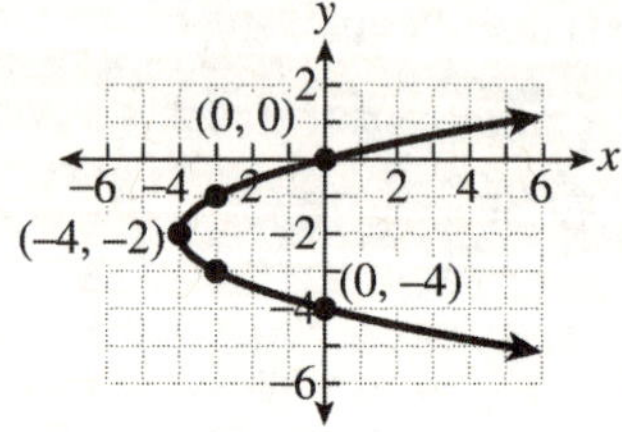

39. $x = -(y+1)^2 - 3$ or $x = -(y-(-1))^2 - 3$

Use the form $x = a(y-k)^2 + h$.

$a < 0$, parabola opens to the left, so rotate the graph of $x = y^2$ about the y-axis.

vertex: $(-3,-1)$

Shift the rotated graph by 3 units to the left and by 1 unit down.

x-intercept:

$x = -(0+1)^2 - 3 = -1 - 3 = -4$
$(-4,0)$

y-intercept:
The vertex is to the left of the y-axis and the parabola opens to the left. There are no y-intercepts.

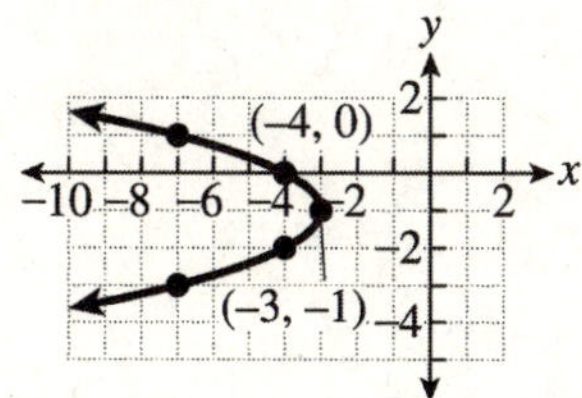

41.
$$y = x^2 - 2x + 15$$
$$y - 15 = x^2 - 2x$$
$$y - 15 + \left(-\frac{2}{2}\right)^2 = x^2 - 2x + \left(-\frac{2}{2}\right)^2$$
$$y - 15 + 1 = x^2 - 2x + 1$$
$$y - 14 = (x-1)^2$$
$$y = (x-1)^2 + 14$$

vertex: $(1,14)$

43.
$$y = -x^2 + 14x + 76$$
$$y - 76 = -\left(x^2 - 14x\right)$$
$$y - 76 - \left(-\frac{14}{2}\right)^2 = -\left(x^2 - 14x + \left(-\frac{14}{2}\right)^2\right)$$
$$y - 76 - 49 = -\left(x^2 - 14x + 49\right)$$
$$y - 125 = -(x-7)^2$$
$$y = -(x-7)^2 + 125$$

vertex: $(7,125)$

45.
$$x = y^2 + 6y - 40$$
$$x + 40 = y^2 + 6y$$
$$x + 40 + \left(\frac{6}{2}\right)^2 = y^2 + 6y + \left(\frac{6}{2}\right)^2$$
$$x + 40 + 9 = y^2 + 6y + 9$$
$$x + 49 = (y+3)^2$$
$$x = (y+3)^2 - 49$$

vertex: $(-49,-3)$

47.
$$x = -y^2 + 4y - 51$$
$$x + 51 = -\left(y^2 - 4y\right)$$
$$x + 51 - \left(-\frac{4}{2}\right)^2 = -\left(y^2 - 4y + \left(-\frac{4}{2}\right)^2\right)$$
$$x + 51 - 4 = -\left(y^2 - 4y + 4\right)$$
$$x + 47 = -(y-2)^2$$
$$x = -(y-2)^2 - 47$$

vertex: $(-47,2)$

49. Since the parabola opens upward, the equation is of the form $y = ax^2 + bx + c$, where $a > 0$.

The x-intercepts are $(1,0)$ and $(5,0)$, so we know that $x - 1$ and $x - 5$ are factors.

$$y = a(x-1)(x-5)$$
$$y = a\left(x^2 - 6x + 5\right)$$

Use the y-intercept $(0,5)$.

$$5 = a\left((0)^2 - 6(0) + 5\right)$$
$$5 = 5a$$
$$1 = a$$

so,

$$y = 1\left(x^2 - 6x + 5\right)$$
$$y = x^2 - 6x + 5$$

51. Since the parabola opens to the right, the equation is of the form $x = ay^2 + by + c$, where $a > 0$. The y-intercepts are $(0, 5)$ and $(0, -1)$, so we know that $y - 5$ and $y + 1$ are factors.

$$x = a(y - 5)(y + 1)$$
$$x = a(y^2 - 4y - 5)$$

Use the x-intercept $(-5, 0)$.

$$-5 = a((0)^2 - 4(0) - 5)$$
$$-5 = -5a$$
$$1 = a$$

so,

$$x = 1(y^2 - 4y - 5)$$
$$x = y^2 - 4y - 5$$

53. a) Since the parabola opens upward, the equation is of the form $y = ax^2 + bx + c$, where $a > 0$. The x-intercepts are $(-50, 0)$ and $(50, 0)$, so we know that $x + 50$ and $x - 50$ are factors.

$$y = a(x - 50)(x + 50)$$
$$y = a(x^2 - 2500)$$

Use the y-intercept $(0, -5)$.

$$-5 = a((0)^2 - 2500)$$
$$-5 = -2500a$$
$$\frac{-5}{-2500} = a$$
$$\frac{1}{500} = a$$

so,

$$y = \frac{1}{500}(x^2 - 2500)$$
$$y = \frac{1}{500}x^2 - 5$$

b) Walking 30 feet in from $x = 50$, makes $x = 20$.

$$y = \frac{1}{500}(20)^2 - 5$$
$$= \frac{1}{500}(400) - 5$$
$$= \frac{4}{5} - \frac{25}{5}$$
$$= -\frac{21}{5}$$
$$= -4.2$$

The person has dropped 4.2 feet in elevation.

55. b

57. h

59. g

61. c

63. Answers will vary. Example:

If the equation has an x^2 term the parabola will either open upward or downward. If the equation has a y^2 term the parabola will either open to the left or to the right.

13.2 QUICK CHECK

1. a) Let $(1,2)$ be the point (x_1, y_1) and $(9,8)$ be the point (x_2, y_2).

$$d = \sqrt{(9-1)^2 + (8-2)^2}$$
$$d = \sqrt{8^2 + 6^2}$$
$$d = \sqrt{64 + 36}$$
$$d = \sqrt{100}$$
$$d = 10$$

b) Let $(3,-6)$ be the point (x_1, y_1) and $(-1,6)$ be the point (x_2, y_2).

$$d = \sqrt{(-1-3)^2 + (6-(-6))^2}$$
$$d = \sqrt{(-4)^2 + (12)^2}$$
$$d = \sqrt{16 + 144}$$
$$d = \sqrt{160}$$
$$d \approx 12.6$$

2. Let $(-4,-3)$ be the point (x_1, y_1) and $(5,7)$ be the point (x_2, y_2).

x-coordinate: $\dfrac{x_1 + x_2}{2} = \dfrac{-4 + 5}{2} = \dfrac{1}{2}$

y-coordinate: $\dfrac{y_1 + y_2}{2} = \dfrac{-3 + 7}{2} = \dfrac{4}{2} = 2$

$$\left(\frac{1}{2}, 2\right)$$

3. $x^2 + y^2 = 12$
$$x^2 + y^2 = \left(\sqrt{12}\right)^2$$
$$x^2 + y^2 = \left(2\sqrt{3}\right)^2$$

Use the form $x^2 + y^2 = r^2$.

Center: $(0,0)$; radius: $r = 2\sqrt{3} \approx 3.5$

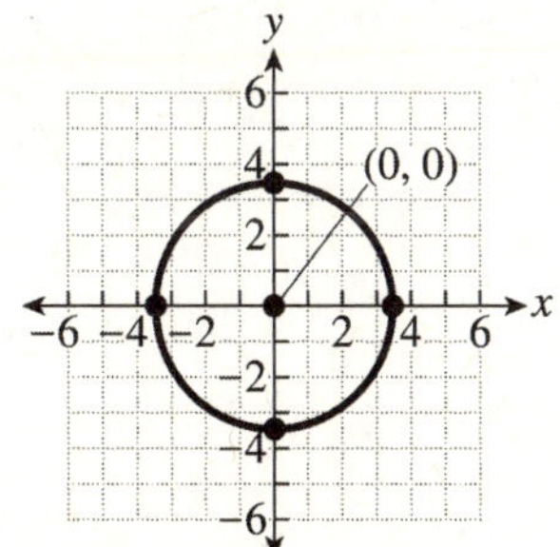

4. $(x-6)^2 + (y-5)^2 = 9$
$$(x-6)^2 + (y-5)^2 = 3^2$$

Use the form $(x-h)^2 + (y-k)^2 = r^2$.

$h = 6, k = 5, r = 3$

Center: $(6,5)$; radius: $r = 3$

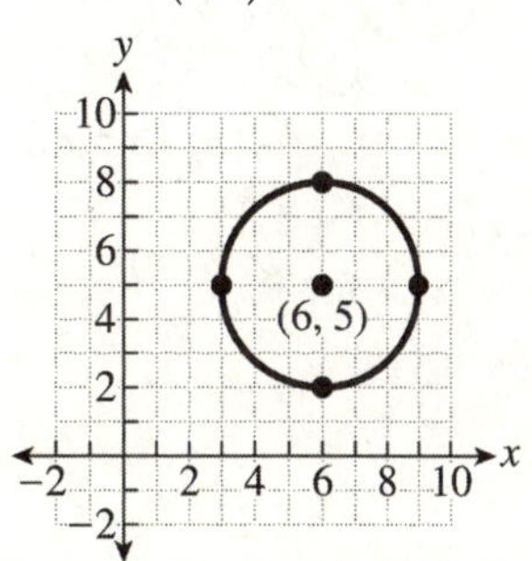

5. a) $(x-1)^2 + (y+6)^2 = 16$
$$(x-1)^2 + (y-(-6))^2 = 4^2$$

Use the form $(x-h)^2 + (y-k)^2 = r^2$.

$h = 1, k = -6, r = 4$

Center: $(1,-6)$; radius: $r = 4$

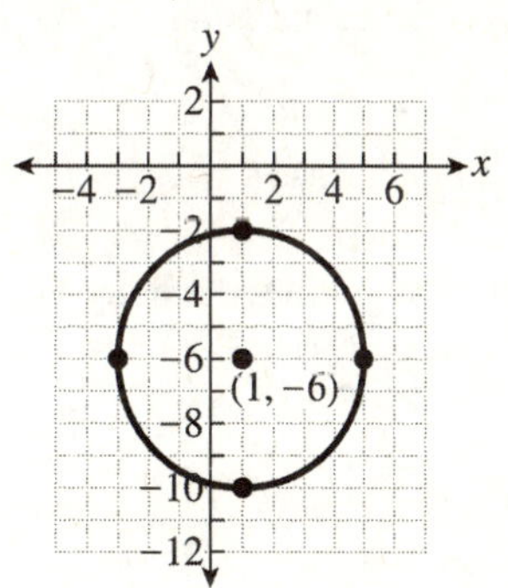

b) $(x+2)^2 + (y-3)^2 = 64$
$$(x-(-2))^2 + (y-3)^2 = 8^2$$

Use the form $(x-h)^2 + (y-k)^2 = r^2$.

$h = -2, k = 3, r = 8$

Center: $(-2,3)$; radius: $r = 8$

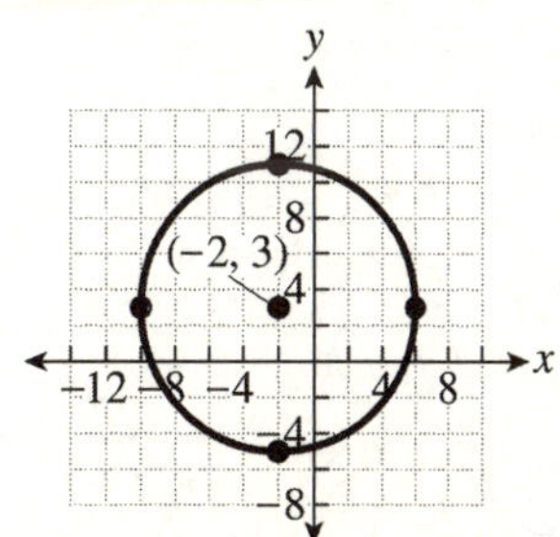

6.
$$x^2 + y^2 + 14x - 6y + 57 = 0$$
$$\left(x^2 + 14x\right) + \left(y^2 - 6y\right) = -57$$
$$\left(x^2 + 14x + \left(\frac{14}{2}\right)^2\right) + \left(y^2 - 6y + \left(-\frac{6}{2}\right)^2\right) = -57 + \left(\frac{14}{2}\right)^2 + \left(-\frac{6}{2}\right)^2$$
$$\left(x^2 + 14x + 49\right) + \left(y^2 - 6y + 9\right) = -57 + 49 + 9$$
$$(x + 7)^2 + (y - 3)^2 = 1$$
$$(x - (-7))^2 + (y - 3)^2 = 1^2$$

Center: $(-7, 3)$; radius: $r = 1$

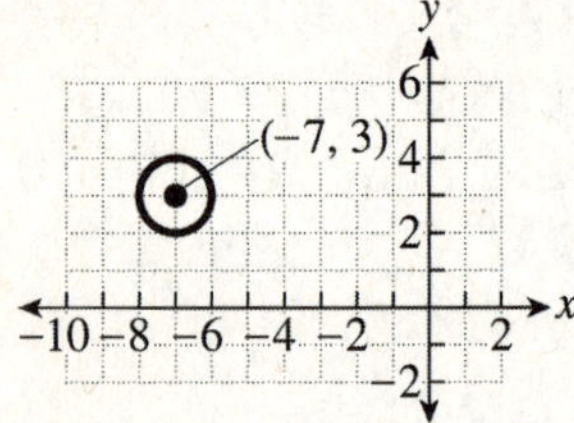

7. Center: $(-3, 5)$; radius: $r = 4$

Use the form $(x - h)^2 + (y - k)^2 = r^2$

where (h, k) is the center and r is the radius.

$$(x - (-3))^2 + (y - 5)^2 = 4^2$$
$$(x + 3)^2 + (y - 5)^2 = 16$$

8. Find the center by finding the midpoint.
$$x = \frac{x_1 + x_2}{2} = \frac{4 + (-2)}{2} = 1$$
$$y = \frac{y_1 + y_2}{2} = \frac{-7 + 3}{2} = -2$$
$$(1, -2)$$

Use the point $(-2, 3)$ and the center $(1, -2)$.

$$d = \sqrt{(x_2 - x_1)^2 + (y_2 - y_1)^2}$$
$$d = \sqrt{(1 - (-2))^2 + (-2 - 3)^2}$$
$$d = \sqrt{9 + 25}$$
$$d = \sqrt{34}$$

Radius: $r = \sqrt{34}$

$$(x - h)^2 + (y - k)^2 = r^2$$
$$(x - 1)^2 + (y - (-2))^2 = \left(\sqrt{34}\right)^2$$
$$(x - 1)^2 + (y + 2)^2 = 34$$

13.2 CIRCLES

1. $d = \sqrt{(x_2 - x_1)^2 + (y_2 - y_1)^2}$

3. circle

5. radius

7. $x^2 + y^2 = r^2$

9. $d = \sqrt{(x_2 - x_1)^2 + (y_2 - y_1)^2}$
$$d = \sqrt{(4 - 7)^2 + (2 - 6)^2}$$
$$d = \sqrt{9 + 16}$$
$$d = \sqrt{25}$$
$$d = 5$$

11. $d = \sqrt{(x_2 - x_1)^2 + (y_2 - y_1)^2}$
$$d = \sqrt{(54 - 15)^2 + (27 - (-53))^2}$$
$$d = \sqrt{1521 + 6400}$$
$$d = \sqrt{7921}$$
$$d = 89$$

13. $d = \sqrt{(x_2 - x_1)^2 + (y_2 - y_1)^2}$

$d = \sqrt{(-4 - (-4))^2 + (5 - (-6))^2}$

$d = \sqrt{0 + 121}$

$d = \sqrt{121}$

$d = 11$

15. $d = \sqrt{(x_2 - x_1)^2 + (y_2 - y_1)^2}$

$d = \sqrt{(-5 - 2)^2 + (4 - (-3))^2}$

$d = \sqrt{49 + 49}$

$d = \sqrt{98}$

$d \approx 9.9$

17. $d = \sqrt{(x_2 - x_1)^2 + (y_2 - y_1)^2}$

$d = \sqrt{(-18 - (-13))^2 + (1 - (-9))^2}$

$d = \sqrt{25 + 100}$

$d = \sqrt{125}$

$d \approx 11.2$

19. x-coordinate: $\dfrac{x_1 + x_2}{2} = \dfrac{0 + 6}{2} = 3$

y-coordinate: $\dfrac{y_1 + y_2}{2} = \dfrac{0 + 4}{2} = 2$

$(3, 2)$

21. x-coordinate: $\dfrac{x_1 + x_2}{2} = \dfrac{6 + 8}{2} = 7$

y-coordinate: $\dfrac{y_1 + y_2}{2} = \dfrac{-9 + (-25)}{2} = -17$

$(7, -17)$

23. $x^2 + y^2 = 16$

$x^2 + y^2 = 4^2$

Center: $(0,0)$; radius: $r = 4$

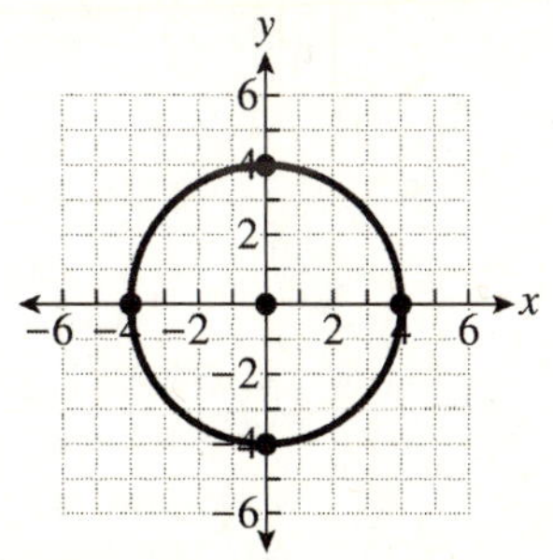

25. $x^2 + y^2 = 100$

$x^2 + y^2 = 10^2$

Use the form $x^2 + y^2 = r^2$.

Center: $(0,0)$; radius: $r = 10$

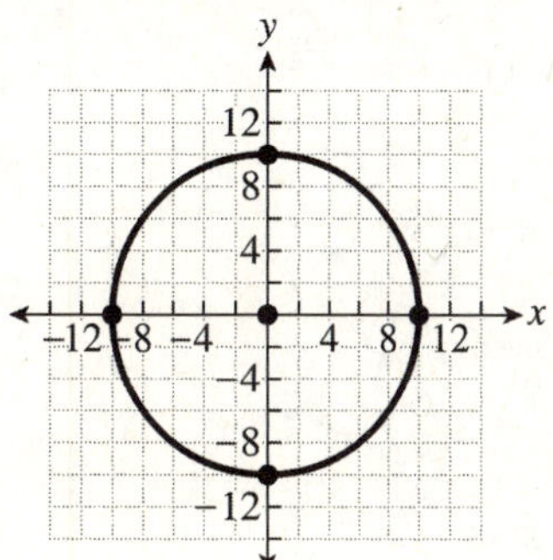

27. $x^2 + y^2 = 6$

$x^2 + y^2 = \left(\sqrt{6}\right)^2$

Use the form $x^2 + y^2 = r^2$.

Center: $(0,0)$; radius: $r = \sqrt{6}$

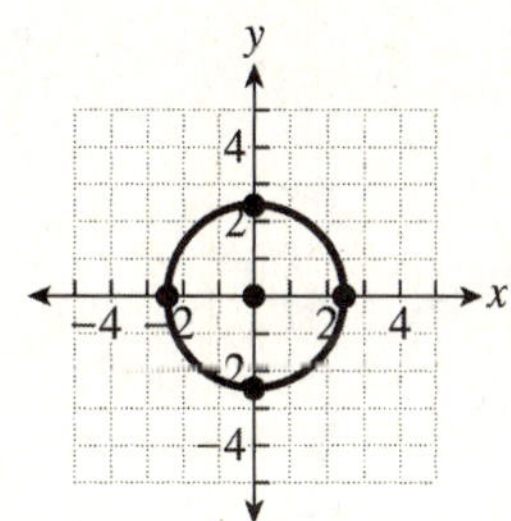

29. $(x - 2)^2 + (y - 7)^2 = 16$

$(x - 2)^2 + (y - 7)^2 = 4^2$

Use the form $(x - h)^2 + (y - k)^2 = r^2$.

$h = 2, \ k = 7, \ r = 4$

Center: $(2, 7)$; radius: $r = 4$

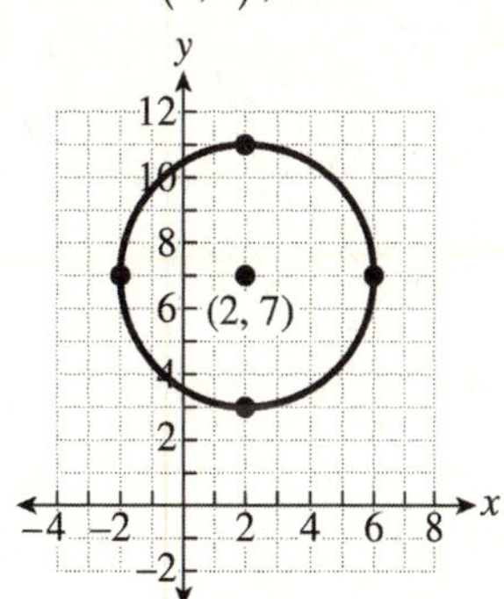

31.
$$(x+5)^2 + (y-1)^2 = 25$$
$$(x-(-5))^2 + (y-1)^2 = 5^2$$

Use the form $(x-h)^2 + (y-k)^2 = r^2$.

$h = -5$, $k = 1$, $r = 5$

Center: $(-5,1)$; radius: $r = 5$

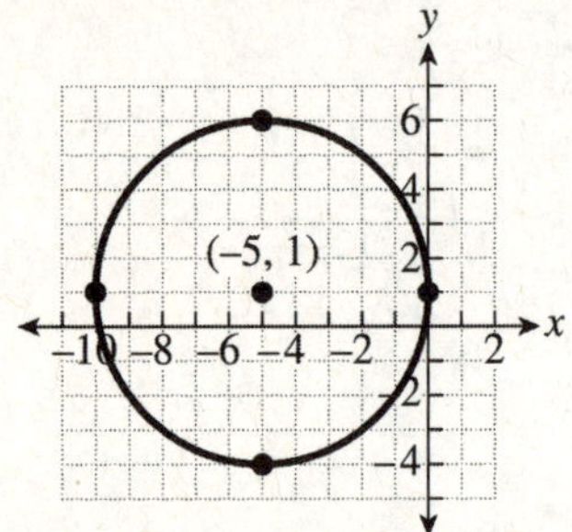

33.
$$(x+2)^2 + (y-1)^2 = 49$$
$$(x-(-2))^2 + (y-1)^2 = 7^2$$

Use the form $(x-h)^2 + (y-k)^2 = r^2$.

$h = -2$, $k = 1$, $r = 7$

Center: $(-2,1)$; radius: $r = 7$

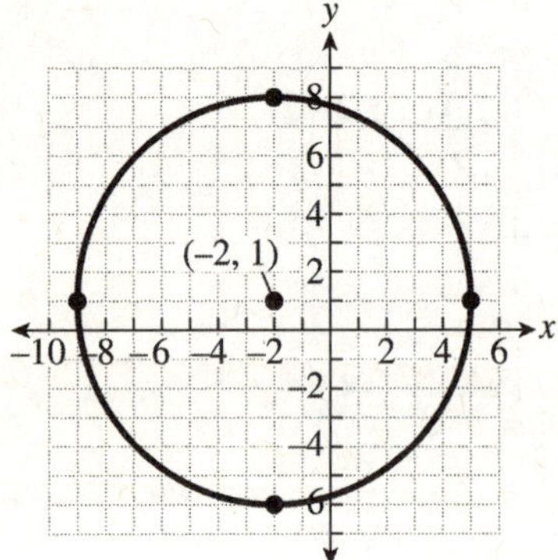

35.
$$(x+5)^2 + y^2 = 25$$
$$(x-(-5))^2 + (y-0)^2 = 5^2$$

Use the form $(x-h)^2 + (y-k)^2 = r^2$.

$h = -5$, $k = 0$, $r = 5$

Center: $(-5,0)$; radius: $r = 5$

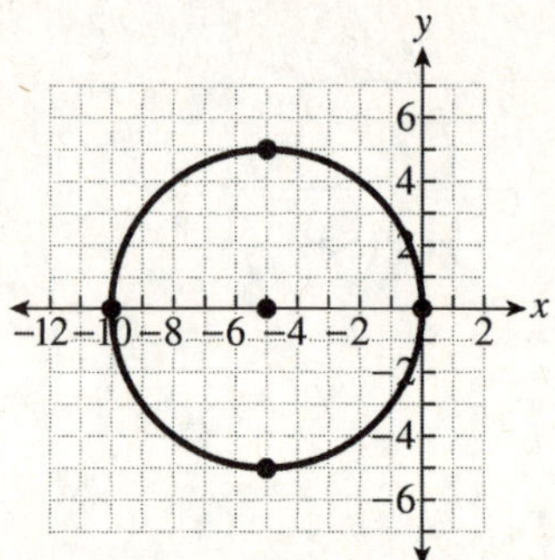

37.
$$(x-2)^2 + (y+6)^2 = 18$$
$$(x-2)^2 + (y-(-6))^2 = \left(\sqrt{18}\right)^2$$
$$(x-2)^2 + (y-(-6))^2 = \left(3\sqrt{2}\right)^2$$

Use the form $(x-h)^2 + (y-k)^2 = r^2$.

$h = 2$, $k = -6$, $r = 3\sqrt{2}$

Center: $(2,-6)$; radius: $r = 3\sqrt{2}$

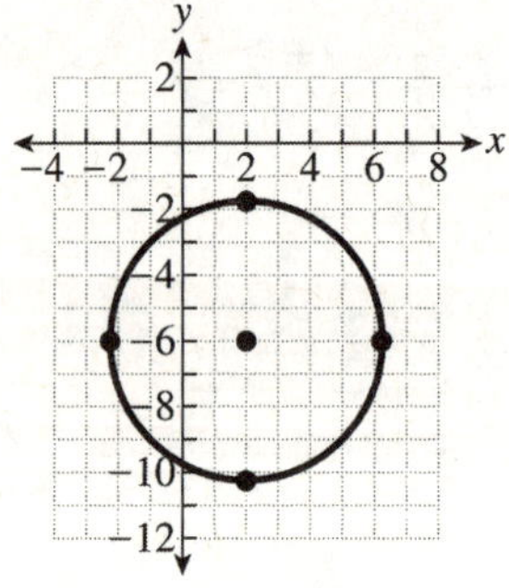

39.
$$x^2 + y^2 + 4x + 12y - 41 = 0$$
$$\left(x^2 + 4x\right) + \left(y^2 + 12y\right) = 41$$
$$\left(x^2 + 4x + \left(\frac{4}{2}\right)^2\right) + \left(y^2 + 12y + \left(\frac{12}{2}\right)^2\right) = 41 + \left(\frac{4}{2}\right)^2 + \left(\frac{12}{2}\right)^2$$
$$\left(x^2 + 4x + 4\right) + \left(y^2 + 12y + 36\right) = 41 + 4 + 36$$
$$(x + 2)^2 + (y + 6)^2 = 81$$
$$\left(x - (-2)\right)^2 + \left(y - (-6)\right)^2 = 9^2$$

Center: $(-2, -6)$; radius: $r = 9$

41.
$$x^2 + y^2 - 10x + 2y - 23 = 0$$
$$\left(x^2 - 10x\right) + \left(y^2 + 2y\right) = 23$$
$$\left(x^2 - 10x + \left(-\frac{10}{2}\right)^2\right) + \left(y^2 + 2y + \left(\frac{2}{2}\right)^2\right) = 23 + \left(-\frac{10}{2}\right)^2 + \left(\frac{2}{2}\right)^2$$
$$\left(x^2 - 10x + 25\right) + \left(y^2 + 2y + 1\right) = 23 + 25 + 1$$
$$(x - 5)^2 + (y + 1)^2 = 49$$
$$(x - 5)^2 + \left(y - (-1)\right)^2 = 7^2$$

Center: $(5, -1)$; radius: $r = 7$

43.
$$x^2 + y^2 + 16x - 57 = 0$$
$$\left(x^2 + 16x\right) + y^2 = 57$$
$$\left(x^2 + 16x + \left(\frac{16}{2}\right)^2\right) + y^2 = 57 + \left(\frac{16}{2}\right)^2$$
$$\left(x^2 + 16x + 64\right) + y^2 = 57 + 64$$
$$(x + 8)^2 + y^2 = 121$$
$$\left(x - (-8)\right)^2 + (y - 0)^2 = 11^2$$

Center: $(-8, 0)$; radius: $r = 11$

45.
$$x^2 + y^2 + 4x + 14y - 27 = 0$$
$$\left(x^2 + 4x\right) + \left(y^2 + 14y\right) = 27$$
$$\left(x^2 + 4x + \left(\frac{4}{2}\right)^2\right) + \left(y^2 + 14y + \left(\frac{14}{2}\right)^2\right) = 27 + \left(\frac{4}{2}\right)^2 + \left(\frac{14}{2}\right)^2$$
$$\left(x^2 + 4x + 4\right) + \left(y^2 + 14y + 49\right) = 27 + 4 + 49$$
$$(x + 2)^2 + (y + 7)^2 = \left(\sqrt{80}\right)^2$$
$$\left(x - (-2)\right)^2 + \left(y - (-7)\right)^2 = \left(4\sqrt{5}\right)^2$$

Center: $(-2, -7)$; radius: $r = 4\sqrt{5}$

47. Center: $(0,0)$; radius: $r = 6$

Use the form $(x - h)^2 + (y - k)^2 = r^2$

where (h, k) is the center and r is the radius.

$$(x - 0)^2 + (y - 0)^2 = 6^2$$
$$x^2 + y^2 = 36$$

49. Center: $(4, 2)$; radius: $r = 7$

Use the form $(x - h)^2 + (y - k)^2 = r^2$

where (h, k) is the center and r is the radius.

$$(x - 4)^2 + (y - 2)^2 = 7^2$$
$$(x - 4)^2 + (y - 2)^2 = 49$$

51. Center: $(-9, -10)$; radius: $r = 5$

Use the form $(x - h)^2 + (y - k)^2 = r^2$

where (h, k) is the center and r is the   radius.

$$(x - (-9))^2 + (y - (-10))^2 = 5^2$$
$$(x + 9)^2 + (y + 10)^2 = 25$$

53. Center: $(0, 0)$; radius: $r = 8$

Use the form $(x - h)^2 + (y - k)^2 = r^2$

where (h, k) is the center and r is the radius.

$$(x - 0)^2 + (y - 0)^2 = 8^2$$
$$x^2 + y^2 = 64$$

55. Center: $(1, 5)$; radius: $r = 4$

Use the form $(x - h)^2 + (y - k)^2 = r^2$

where (h, k) is the center and r is the radius.

$$(x - 1)^2 + (y - 5)^2 = 4^2$$
$$(x - 1)^2 + (y - 5)^2 = 16$$

57. Center: $(-1, 1)$; radius: $r = 7$

Use the form $(x - h)^2 + (y - k)^2 = r^2$

where (h, k) is the center and r is the   radius.

$$(x - (-1))^2 + (y - 1)^2 = 7^2$$
$$(x + 1)^2 + (y - 1)^2 = 49$$

59. Find the center by finding the midpoint.

$$x = \frac{x_1 + x_2}{2} = \frac{1 + 9}{2} = 5$$

$$y = \frac{y_1 + y_2}{2} = \frac{7 + 7}{2} = 7$$

Center: $(5, 7)$

Use the point $(1, 7)$ and the center $(5, 7)$.

$$d = \sqrt{(x_2 - x_1)^2 + (y_2 - y_1)^2}$$
$$d = \sqrt{(5 - 1)^2 + (7 - 7)^2}$$
$$d = \sqrt{16 + 0}$$
$$d = \sqrt{16}$$
$$d = 4$$

Radius: $r = 4$

$$(x - h)^2 + (y - k)^2 = r^2$$
$$(x - 5)^2 + (y - 7)^2 = 4^2$$
$$(x - 5)^2 + (y - 7)^2 = 16$$

61. Find the center by finding the midpoint.

$$x = \frac{x_1 + x_2}{2} = \frac{-2 + 6}{2} = 2$$

$$y = \frac{y_1 + y_2}{2} = \frac{10 + 4}{2} = 7$$

Center: $(2, 7)$

Use the point $(6, 4)$ and the center $(2, 7)$.

$$d = \sqrt{(x_2 - x_1)^2 + (y_2 - y_1)^2}$$
$$d = \sqrt{(2 - 6)^2 + (7 - 4)^2}$$
$$d = \sqrt{16 + 9}$$
$$d = \sqrt{25}$$
$$d = 5$$

Radius: $r = 5$

$$(x - h)^2 + (y - k)^2 = r^2$$
$$(x - 2)^2 + (y - 7)^2 = 5^2$$
$$(x - 2)^2 + (y - 7)^2 = 25$$

63. Find the center by finding the midpoint.

$$x = \frac{x_1 + x_2}{2} = \frac{4 + 12}{2} = 8$$

$$y = \frac{y_1 + y_2}{2} = \frac{1 + (-3)}{2} = -1$$

Center: $(8, -1)$

Use the point $(4, 1)$ and the center $(8, -1)$.

$$d = \sqrt{(x_2 - x_1)^2 + (y_2 - y_1)^2}$$

$$d = \sqrt{(8 - 4)^2 + (-1 - 1)^2}$$

$$d = \sqrt{16 + 4}$$

$$d = \sqrt{20}$$

Radius: $r = \sqrt{20} = 2\sqrt{5}$

$$(x - h)^2 + (y - k)^2 = r^2$$

$$(x - 8)^2 + (y - (-1))^2 = \left(\sqrt{20}\right)^2$$

$$(x - 8)^2 + (y + 1)^2 = 20$$

65.

$$x^2 + (y + 6)^2 = 16$$

$$(x - 0)^2 + (y - (-6))^2 = 4^2$$

Use the form $(x - h)^2 + (y - k)^2 = r^2$.

$h = 0$, $k = -6$, $r = 4$

Center: $(0, -6)$; radius: $r = 4$

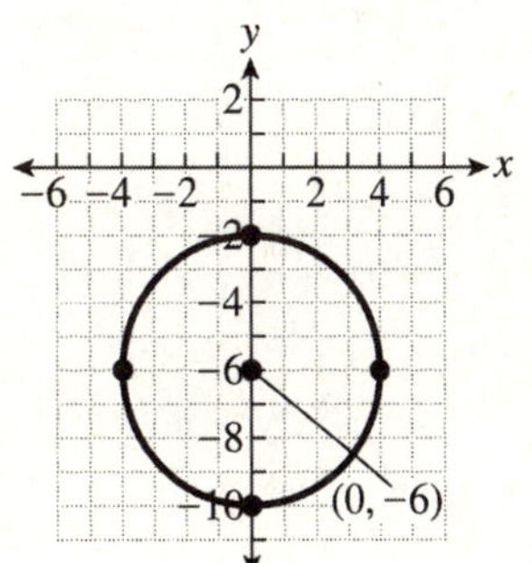

67.

$$x^2 + y^2 + 18x - 6y - 10 = 0$$

$$\left(x^2 + 18x\right) + \left(y^2 - 6y\right) = 10$$

$$\left(x^2 + 18x + \left(\frac{18}{2}\right)^2\right) + \left(y^2 - 6y + \left(-\frac{6}{2}\right)^2\right) = 10 + \left(\frac{18}{2}\right)^2 + \left(-\frac{6}{2}\right)^2$$

$$\left(x^2 + 18x + 81\right) + \left(y^2 - 6y + 9\right) = 10 + 81 + 9$$

$$(x + 9)^2 + (y - 3)^2 = 100$$

$$(x - (-9))^2 + (y - 3)^2 = 10^2$$

Use the form $(x - h)^2 + (y - k)^2 = r^2$.

$h = -9$, $k = 3$, $r = 10$

Center: $(-9, 3)$; radius: $r = 10$

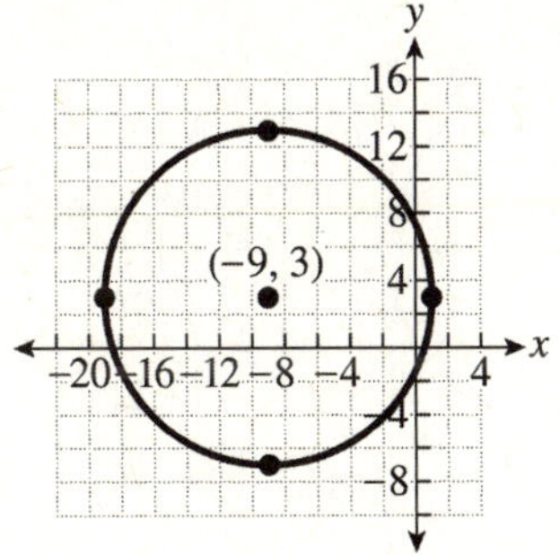

69.
$$y = x^2 + 6x - 7$$
$$y + 7 = x^2 + 6x$$
$$y + 7 + \left(\frac{6}{2}\right)^2 = x^2 + 6x + \left(\frac{6}{2}\right)^2$$
$$y + 7 + 9 = x^2 + 6x + 9$$
$$y + 16 = (x + 3)^2$$
$$y = (x + 3)^2 - 16$$

Use the form $y = a(x - h)^2 + k$.

$a = 1$, $h = -3$, $k = -16$

The parabola opens upward.

vertex: $(-3, -16)$

Shift the graph of $y = x^2$ by 3 units to the left and by 16 units down.

y-intercept:
$$y = (0 + 3)^2 - 16$$
$$y = 9 - 16$$
$$y = -7$$
$$(0, -7)$$

x-intercept:
$$0 = (x + 3)^2 - 16$$
$$16 = (x + 3)^2$$
$$\pm\sqrt{16} = x + 3$$
$$-3 \pm 4 = x$$
$$x = -7 \text{ or } x = 1$$
$$(-7, 0), (1, 0)$$

axis of symmetry: $x = -3$

point symmetric to the y-intercept: $(-6, -7)$

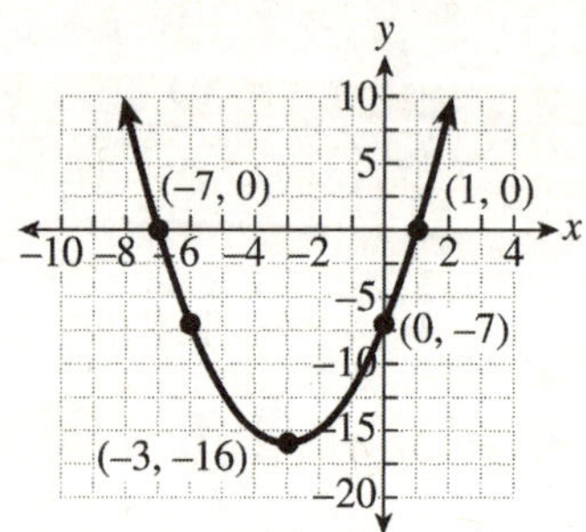

71. $x = -(y + 1)^2 - 5$

Use the form $x = a(y - k)^2 + h$.

$a = -1$, $h = -5$, $k = -1$

The parabola opens to the left, so rotate the graph of $x = y^2$ about the y-axis.

vertex: $(-5, -1)$

Shift the rotated graph by 5 units to the left and

by 1 unit down.

x-intercept:
$$x = -(0 + 1)^2 - 5 = -1 - 5 = -6$$
$$(-6, 0)$$

y-intercept:

The vertex is to the left of the y-axis and the parabola opens to the left. There are no y-intercepts.

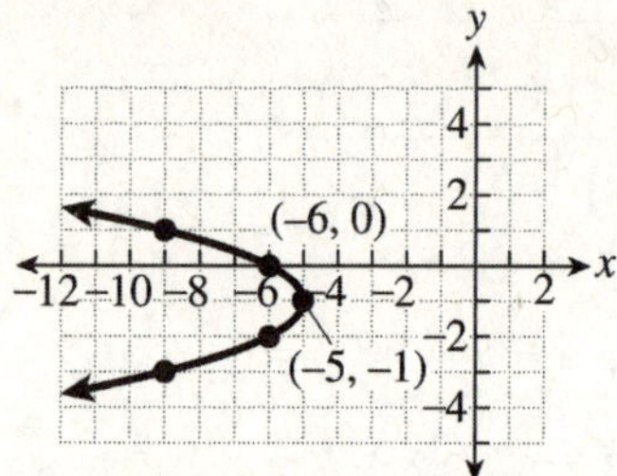

73.
$$(x + 8)^2 + y^2 = 1$$
$$(x - (-8))^2 + (y - 0)^2 = 1^2$$

Use the form $(x - h)^2 + (y - k)^2 = r^2$.

$h = -8$, $k = 0$, $r = 1$

Center: $(-8, 0)$; radius: $r = 1$

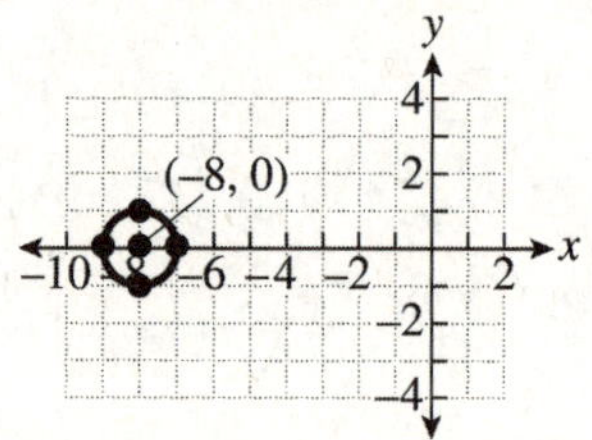

75.
$$x^2 + (y - 7)^2 = 25$$
$$(x - 0)^2 + (y - 7)^2 = 5^2$$

Use the form $(x - h)^2 + (y - k)^2 = r^2$.

$h = 0$, $k = 7$, $r = 5$

Center: $(0, 7)$; radius: $r = 5$

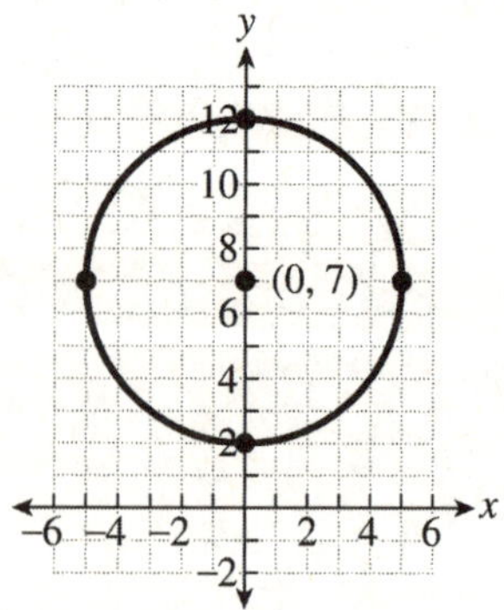

77. a) Place the origin at the center of the barbeque area. The coordinates of the diagonal corners are $(25,10)$ and $(-25,-10)$.

The center of the circle is $(0,0)$.

Use the point $(25,10)$ and the center $(0,0)$.

$$d = \sqrt{(x_2 - x_1)^2 + (y_2 - y_1)^2}$$
$$d = \sqrt{(25 - 0)^2 + (10 - 0)^2}$$
$$d = \sqrt{625 + 100}$$
$$d = \sqrt{725}$$

Radius: $r = \sqrt{725}$

$$(x - h)^2 + (y - k)^2 = r^2$$
$$(x - 0)^2 + (y - 0)^2 = \left(\sqrt{725}\right)^2$$
$$x^2 + y^2 = 725$$

b) Circle area $= \pi r^2$
$$= \pi \left(\sqrt{725}\right)^2$$
$$= 725\pi$$

Rectangle area $= (50)(20) = 1000$

Sod area $=$ Circle area $-$ Rectangle area
$$= 725\pi - 1000 \approx 1277.7$$
The area of the sod is 1277.7 square feet.

79. Answers will vary. Example:
Parabolas that open up or down have the form
$$y = a(x - h)^2 + k.$$
If $a > 0$, the parabola opens up.

Example: $y = 2(x - 3)^2 + 4$

If $a < 0$, the parabola opens down.

Example: $y = -(x + 4)^2 - 1$

Parabolas that open left or right have the form
$$x = a(y - k)^2 + h.$$
If $a > 0$, the parabola opens to the right.

Example: $x = (y + 2)^2 - 3$

If $a < 0$, the parabola opens to the left.

Example: $x = -(y - 2)^2 + 3$

Circles can be written in the form
$$(x - h)^2 + (y - k)^2 = r^2.$$

Example: $(x + 3)^2 + (y - 2)^2 = 100.$

13.3 QUICK CHECK

1. a)
$$\frac{x^2}{25} + \frac{y^2}{9} = 1$$
$$\frac{x^2}{5^2} + \frac{y^2}{3^2} = 1$$

Use the form $\dfrac{x^2}{a^2} + \dfrac{y^2}{b^2} = 1$.

Center: $(0,0)$

$a = 5$ and $b = 3$

x-intercepts: $(-5,0),(5,0)$
y-intercepts: $(0,-3),(0,3)$

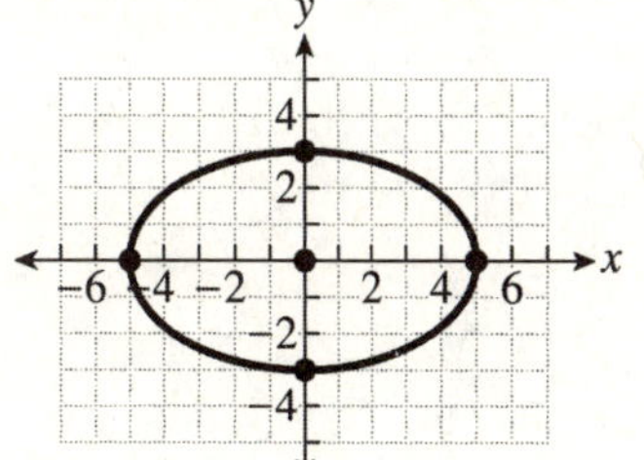

b)
$$\frac{x^2}{9} + \frac{y^2}{16} = 1$$
$$\frac{x^2}{3^2} + \frac{y^2}{4^2} = 1$$

Use the form $\dfrac{x^2}{a^2} + \dfrac{y^2}{b^2} = 1$.

Center: $(0,0)$

$a = 3$ and $b = 4$

x-intercepts: $(-3,0),(3,0)$
y-intercepts: $(0,-4),(0,4)$

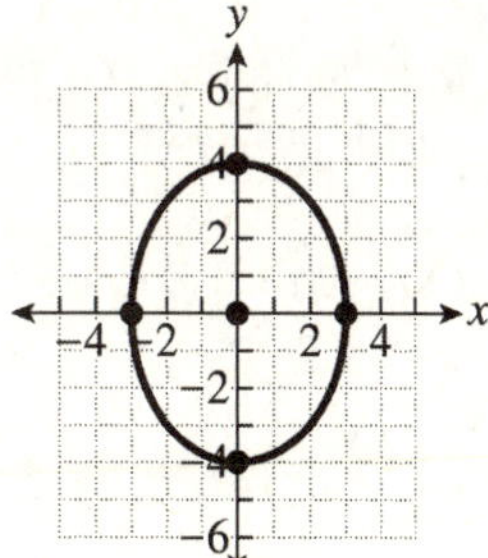

2. $4x^2 + 36y^2 = 144$

$$\frac{4x^2 + 36y^2}{144} = \frac{144}{144}$$

$$\frac{4x^2}{144} + \frac{36y^2}{144} = 1$$

$$\frac{x^2}{36} + \frac{y^2}{4} = 1$$

$$\frac{x^2}{6^2} + \frac{y^2}{2^2} = 1$$

Use the form $\dfrac{x^2}{a^2} + \dfrac{y^2}{b^2} = 1$.

Center: $(0,0)$

$a = 6$ and $b = 2$

x-intercepts: $(-6,0),(6,0)$
y-intercepts: $(0,-2),(0,2)$

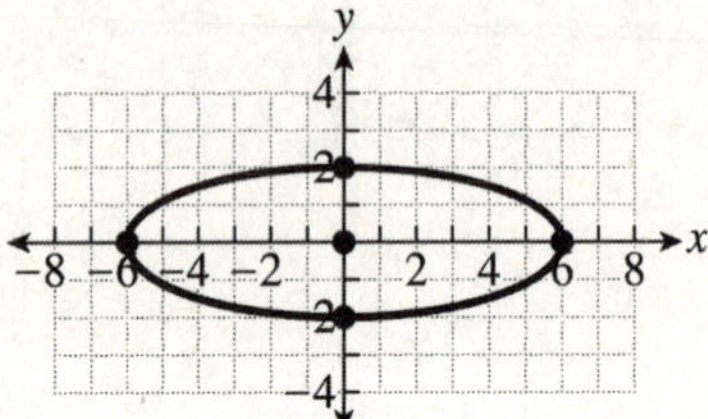

3. $\dfrac{(x-8)^2}{49} + \dfrac{(y-7)^2}{25} = 1$

$$\frac{(x-8)^2}{7^2} + \frac{(y-7)^2}{5^2} = 1$$

Use the form $\dfrac{(x-h)^2}{a^2} + \dfrac{(y-k)^2}{b^2} = 1$.

Center: $(8,7)$

$a = 7$ and $b = 5$

The endpoints of the horizontal axis are 7 units to the left and right of the center.

$(1,7)$ and $(15,7)$

The endpoints of the vertical axis are 5 units above and below the center.

$(8,12)$ and $(8,2)$

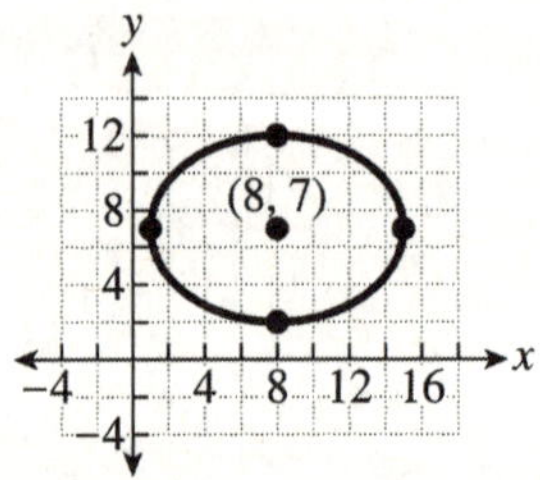

4. $\dfrac{(x-4)^2}{20} + (y+5)^2 = 1$

$$\frac{(x-4)^2}{\left(2\sqrt{5}\right)^2} + \frac{\left(y-(-5)\right)^2}{1^2} = 1$$

Use the form $\dfrac{(x-h)^2}{a^2} + \dfrac{(y-k)^2}{b^2} = 1$.

Center: $(4,-5)$

The endpoints of the horizontal axis are $2\sqrt{5}$ units to the left and right of the center.

$\left(4 - 2\sqrt{5}, -5\right)$ and $\left(4 + 2\sqrt{5}, -5\right)$

The endpoints of the vertical axis are 1 unit above and below the center.

$(4,-4)$ and $(4,-6)$

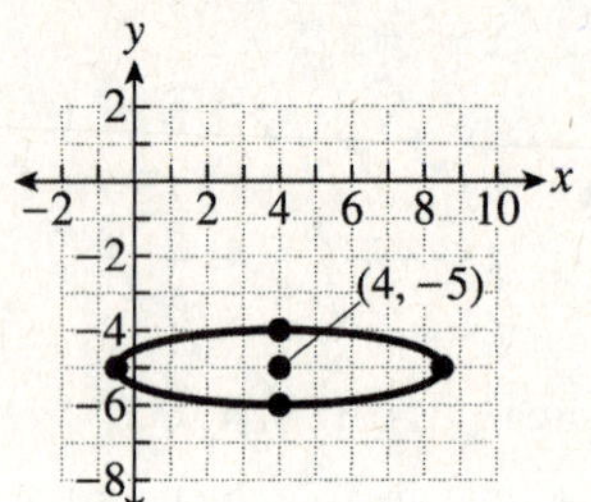

5.
$$9x^2 + 16y^2 - 36x + 160y + 292 = 0$$
$$\left(9x^2 - 36x\right) + \left(16y^2 + 160y\right) = -292$$
$$9\left(x^2 - 4x\right) + 16\left(y^2 + 10y\right) = -292$$
$$9\left(x^2 - 4x + \left(-\frac{4}{2}\right)^2\right) + 16\left(y^2 + 10y + \left(\frac{10}{2}\right)^2\right) = -292 + 9\left(-\frac{4}{2}\right)^2 + 16\left(\frac{10}{2}\right)^2$$
$$9\left(x^2 - 4x + 4\right) + 16\left(y^2 + 10y + 25\right) = -292 + 36 + 400$$
$$9(x-2)^2 + 16(y+5)^2 = 144$$
$$\frac{9(x-2)^2}{144} + \frac{16(y+5)^2}{144} = \frac{144}{144}$$
$$\frac{(x-2)^2}{16} + \frac{(y+5)^2}{9} = 1$$
$$\frac{(x-2)^2}{4^2} + \frac{(y-(-5))^2}{3^2} = 1$$

Use the form $\dfrac{(x-h)^2}{a^2} + \dfrac{(y-k)^2}{b^2} = 1$.

Center: $(2,-5)$

$a = 4$ and $b = 3$

The endpoints of the horizontal axis are 4 units to the left and right of the center.
$(-2,-5)$ and $(6,-5)$

The endpoints of the vertical axis are 3 units above and below the center.
$(2,-2)$ and $(2,-8)$

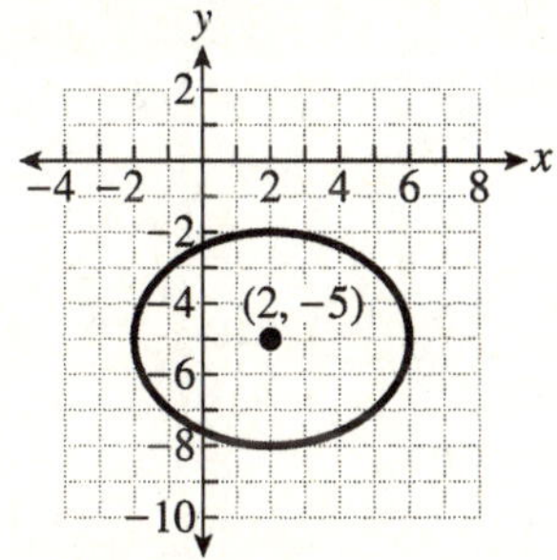

6. Center: $(0,5)$

The distance from the center to an endpoint of the horizontal axis is 7 units, so $a = 7$. The distance from the center to an endpoint of the vertical axis is 3 units, so $b = 3$.

In standard form:
$$\frac{(x-0)^2}{7^2} + \frac{(y-5)^2}{3^2} = 1$$
$$\frac{x^2}{49} + \frac{(y-5)^2}{9} = 1$$

13.3 ELLIPSES

1. ellipse

3. center

5. vertices

7. $\dfrac{x^2}{a^2}+\dfrac{y^2}{b^2}=1$

9. $\dfrac{x^2}{4}+\dfrac{y^2}{9}=1$

$\dfrac{x^2}{2^2}+\dfrac{y^2}{3^2}=1$

Use the form $\dfrac{x^2}{a^2}+\dfrac{y^2}{b^2}=1$.

Center: $(0,0)$

$a=2$ and $b=3$

x-intercepts: $(-2,0),(2,0)$
y-intercepts: $(0,-3),(0,3)$

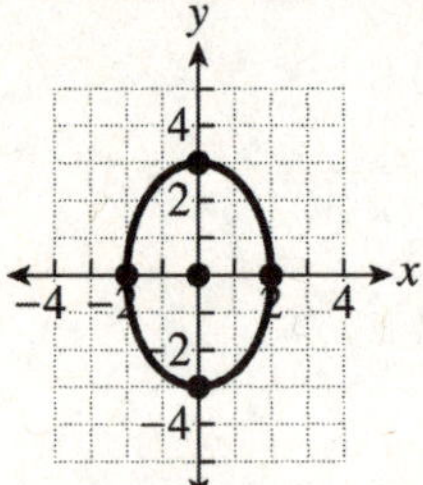

11. $\dfrac{x^2}{25}+\dfrac{y^2}{4}=1$

$\dfrac{x^2}{5^2}+\dfrac{y^2}{2^2}=1$

Use the form $\dfrac{x^2}{a^2}+\dfrac{y^2}{b^2}=1$.

Center: $(0,0)$

$a=5$ and $b=2$

x-intercepts: $(-5,0),(5,0)$
y-intercepts: $(0,-2),(0,2)$

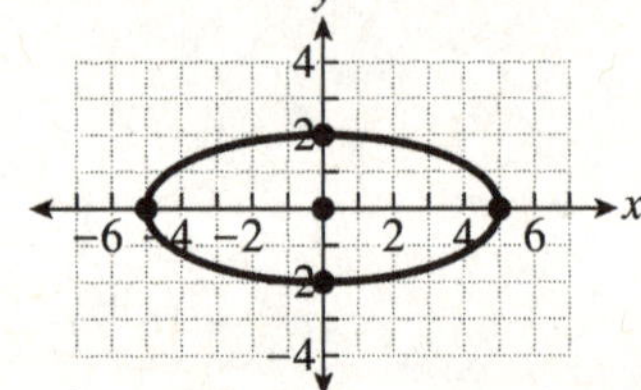

13. $\dfrac{x^2}{36}+\dfrac{y^2}{12}=1$

$\dfrac{x^2}{6^2}+\dfrac{y^2}{\left(\sqrt{12}\right)^2}=1$

$\dfrac{x^2}{6^2}+\dfrac{y^2}{\left(2\sqrt{3}\right)^2}=1$

Use the form $\dfrac{x^2}{a^2}+\dfrac{y^2}{b^2}=1$.

Center: $(0,0)$

$a=6$ and $b=2\sqrt{3}$

x-intercepts: $(-6,0),(6,0)$
y-intercepts: $\left(0,2\sqrt{3}\right),\left(0,-2\sqrt{3}\right)$

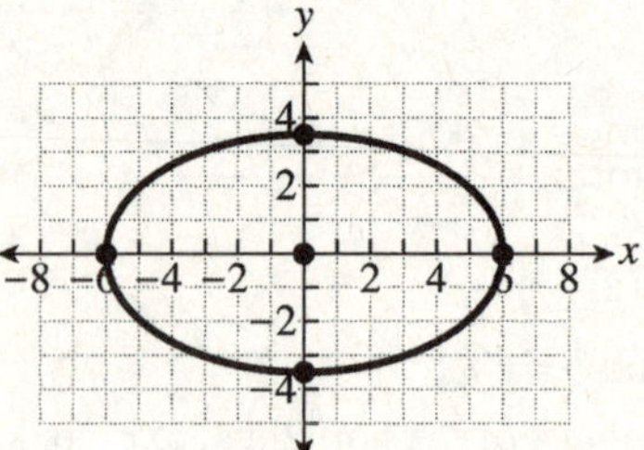

15. $81x^2+4y^2=324$

$\dfrac{81x^2}{324}+\dfrac{4y^2}{324}=\dfrac{324}{324}$

$\dfrac{x^2}{4}+\dfrac{y^2}{81}=1$

$\dfrac{x^2}{2^2}+\dfrac{y^2}{9^2}=1$

Use the form $\dfrac{x^2}{a^2}+\dfrac{y^2}{b^2}=1$.

Center: $(0,0)$

$a=2$ and $b=9$

x-intercepts: $(-2,0),(2,0)$
y-intercepts: $(0,-9),(0,9)$

Continued on next page.

15. Continued.

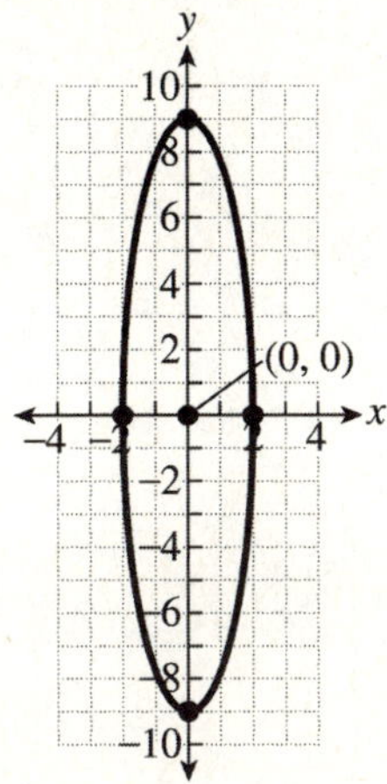

17. $18x^2 + 32y^2 = 288$

$$\frac{18x^2}{288} + \frac{32y^2}{288} = \frac{288}{288}$$

$$\frac{x^2}{16} + \frac{y^2}{9} = 1$$

$$\frac{x^2}{4^2} + \frac{y^2}{3^2} = 1$$

Use the form $\dfrac{x^2}{a^2} + \dfrac{y^2}{b^2} = 1$.

Center: $(0,0)$

$a = 4$ and $b = 3$

x-intercepts: $(-4,0),(4,0)$
y-intercepts: $(0,-3),(0,3)$

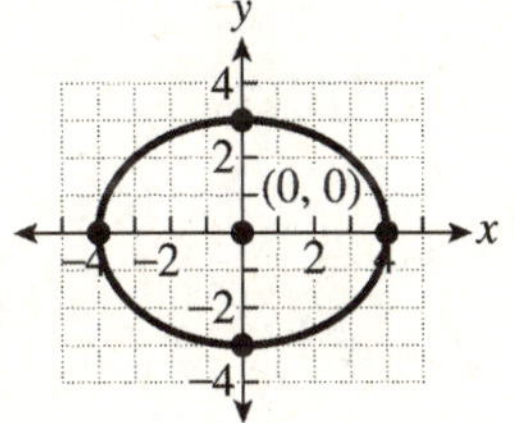

19. $\dfrac{(x-6)^2}{25} + \dfrac{(y-8)^2}{36} = 1$

$$\frac{(x-6)^2}{5^2} + \frac{(y-8)^2}{6^2} = 1$$

Use the form $\dfrac{(x-h)^2}{a^2} + \dfrac{(y-k)^2}{b^2} = 1$.

Center: $(6,8)$

$a = 5$ and $b = 6$

The endpoints of the horizontal axis are 5 units to the left and right of the center.

$(1,8)$ and $(11,8)$

The endpoints of the vertical axis are 6 units above and below the center.

$(6,14)$ and $(6,2)$

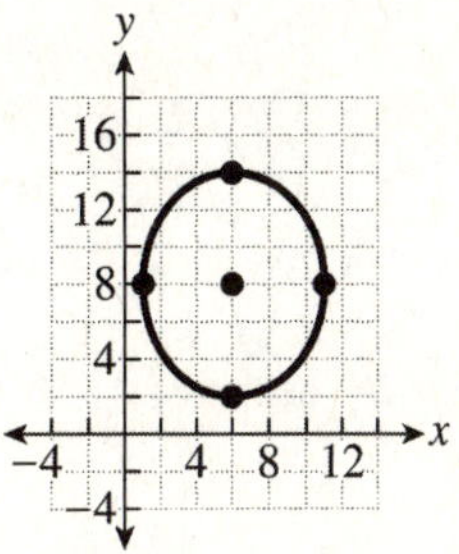

21. $\dfrac{(x+3)^2}{36} + \dfrac{(y-5)^2}{4} = 1$

$$\frac{(x-(-3))^2}{6^2} + \frac{(y-5)^2}{2^2} = 1$$

Use the form $\dfrac{(x-h)^2}{a^2} + \dfrac{(y-k)^2}{b^2} = 1$.

Center: $(-3,5)$

$a = 6$ and $b = 2$

The endpoints of the horizontal axis are 6 units to the left and right of the center.

$(-9,5)$ and $(3,5)$

The endpoints of the vertical axis are 2 units above and below the center.

$(-3,3)$ and $(-3,7)$

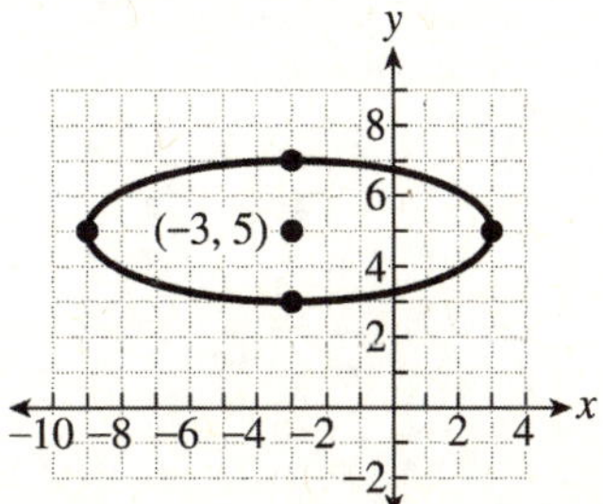

23.
$$\frac{(x+5)^2}{4}+\frac{y^2}{36}=1$$

$$\frac{(x-(-5))^2}{2^2}+\frac{(y-0)^2}{6^2}=1$$

Use the form $\dfrac{(x-h)^2}{a^2}+\dfrac{(y-k)^2}{b^2}=1$.

Center: $(-5,0)$

$a=2$ and $b=6$

The endpoints of the horizontal axis are 2 units to the left and right of the center.

$(-7,0)$ and $(-3,0)$

The endpoints of the vertical axis are 6 units above and below the center.

$(-5,6)$ and $(-5,-6)$

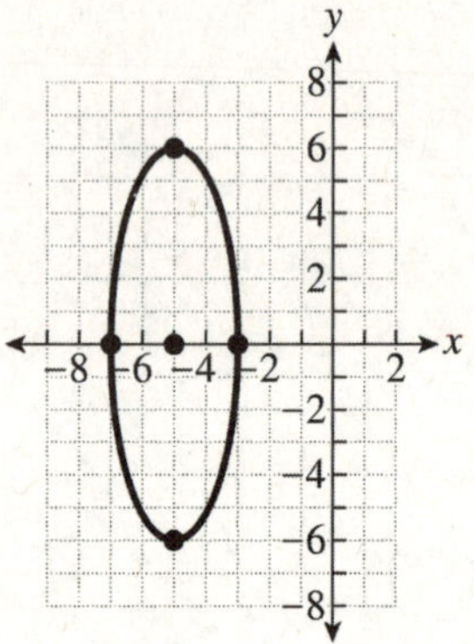

25.
$$\frac{(x+2)^2}{25}+(y-5)^2=1$$

$$\frac{(x-(-2))^2}{5^2}+\frac{(y-5)^2}{1^2}=1$$

Use the form $\dfrac{(x-h)^2}{a^2}+\dfrac{(y-k)^2}{b^2}=1$.

Center: $(-2,5)$

$a=5$ and $b=1$

The endpoints of the horizontal axis are 5 units to the left and right of the center.

$(-7,5)$ and $(3,5)$

The endpoints of the vertical axis are 1 unit above and below the center.

$(-2,4)$ and $(-2,6)$

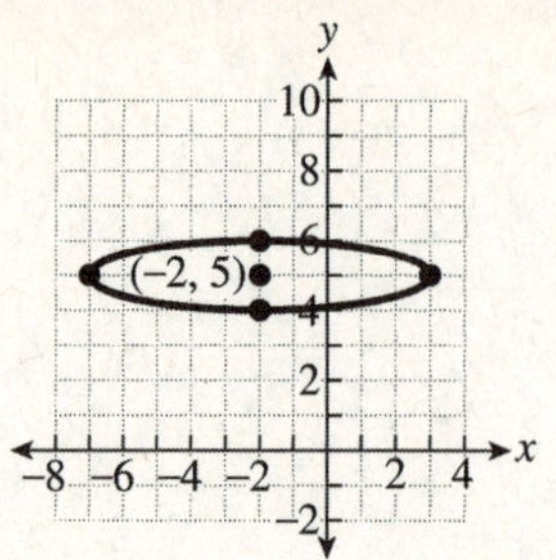

27.
$$\frac{(x-6)^2}{20}+\frac{(y+6)^2}{4}=1$$

$$\frac{(x-6)^2}{\left(\sqrt{20}\right)^2}+\frac{(y-(-6))^2}{2^2}=1$$

$$\frac{(x-6)^2}{\left(2\sqrt{5}\right)^2}+\frac{(y-(-6))^2}{2^2}=1$$

Use the form $\dfrac{(x-h)^2}{a^2}+\dfrac{(y-k)^2}{b^2}=1$.

Center: $(6,-6)$

$a=2\sqrt{5}$ and $b=2$

The endpoints of the horizontal axis are $2\sqrt{5}$ units to the left and right of the center.

$\left(6-2\sqrt{5},-6\right)$ and $\left(6+2\sqrt{5},-6\right)$

The endpoints of the vertical axis are 2 units above and below the center.

$(6,-4)$ and $(6,-8)$

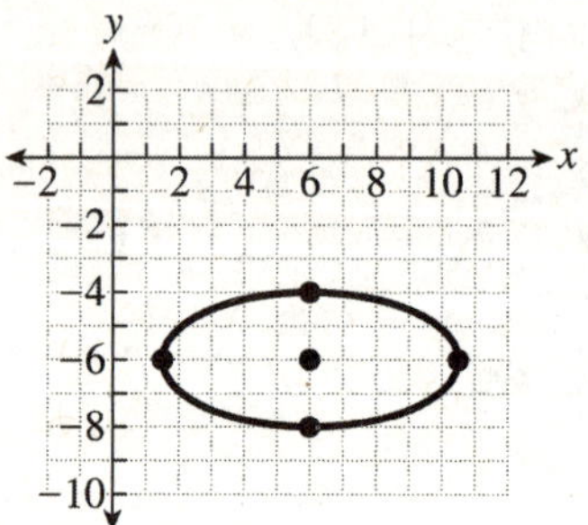

29.

$$4x^2 + 9y^2 - 24x - 90y + 225 = 0$$

$$\left(4x^2 - 24x\right) + \left(9y^2 - 90y\right) = -225$$

$$4\left(x^2 - 6x\right) + 9\left(y^2 - 10y\right) = -225$$

$$4\left(x^2 - 6x + \left(-\frac{6}{2}\right)^2\right) + 9\left(y^2 - 10y + \left(-\frac{10}{2}\right)^2\right) = -225 + 4\left(-\frac{6}{2}\right)^2 + 9\left(-\frac{10}{2}\right)^2$$

$$4\left(x^2 - 6x + 9\right) + 9\left(y^2 - 10y + 25\right) = -225 + 36 + 225$$

$$4(x-3)^2 + 9(y-5)^2 = 36$$

$$\frac{4(x-3)^2}{36} + \frac{9(y-5)^2}{36} = \frac{36}{36}$$

$$\frac{(x-3)^2}{9} + \frac{(y-5)^2}{4} = 1$$

$$\frac{(x-3)^2}{3^2} + \frac{(y-5)^2}{2^2} = 1$$

Use the form $\dfrac{(x-h)^2}{a^2} + \dfrac{(y-k)^2}{b^2} = 1$.

Center: $(3,5)$

$a = 3$ and $b = 2$

The endpoints of the horizontal axis are 3 units to the left and right of the center.

$(0,5)$ and $(6,5)$

The endpoints of the vertical axis are 2 units above and below the center.

$(3,7)$ and $(3,3)$

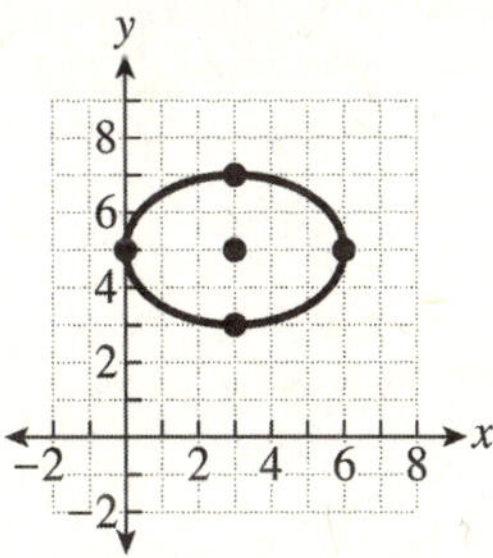

31.

$$2x^2 + 8y^2 + 16x - 64y + 88 = 0$$

$$\left(2x^2 + 16x\right) + \left(8y^2 - 64y\right) = -88$$

$$2\left(x^2 + 8x\right) + 8\left(y^2 - 8y\right) = -88$$

$$2\left(x^2 + 8x + \left(\frac{8}{2}\right)^2\right) + 8\left(y^2 - 8y + \left(-\frac{8}{2}\right)^2\right) = -88 + 2\left(\frac{8}{2}\right)^2 + 8\left(-\frac{8}{2}\right)^2$$

$$2\left(x^2 + 8x + 16\right) + 8\left(y^2 - 8y + 16\right) = -88 + 32 + 128$$

$$2(x+4)^2 + 8(y-4)^2 = 72$$

$$\frac{2(x+4)^2}{72} + \frac{8(y-4)^2}{72} = \frac{72}{72}$$

$$\frac{(x+4)^2}{36} + \frac{(y-4)^2}{9} = 1$$

$$\frac{(x-(-4))^2}{6^2} + \frac{(y-4)^2}{3^2} = 1$$

Use the form $\dfrac{(x-h)^2}{a^2} + \dfrac{(y-k)^2}{b^2} = 1$.

Center: $(-4, 4)$

$a = 6$ and $b = 3$

The endpoints of the horizontal axis are 6 units to the left and right of the center.

$(-10, 4)$ and $(2, 4)$

The endpoints of the vertical axis are 3 units above and below the center.

$(-4, 7)$ and $(-4, 1)$

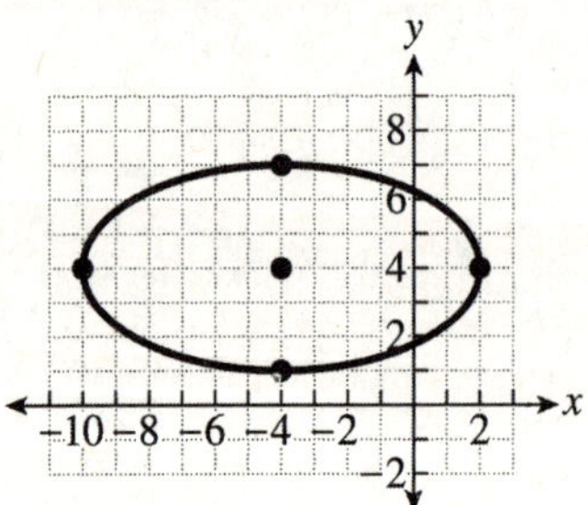

33. Center: $(0, 0)$

The endpoints of the horizontal axis are $(-5, 0)$ and $(5, 0)$; $a = 5$.

The endpoints of the vertical axis are $(0, -3)$ and $(0, 3)$; $b = 3$.

$$\frac{x^2}{5^2} + \frac{y^2}{3^2} = 1$$

$$\frac{x^2}{25} + \frac{y^2}{9} = 1$$

35. Center: $(4,0)$

The endpoints of the horizontal axis are $(-1,0)$ and $(9,0)$; $a = 5$.

The endpoints of the vertical axis are $(4,-2)$ and $(4,2)$; $b = 2$.

$$\frac{(x-4)^2}{5^2} + \frac{(y-0)^2}{2^2} = 1$$

$$\frac{(x-4)^2}{25} + \frac{y^2}{4} = 1$$

37. Center: $(0,-4)$

The endpoints of the horizontal axis are $(-3,-4)$ and $(3,-4)$; $a = 3$.

The endpoints of the vertical axis are $(0,-9)$ and $(0,1)$; $b = 5$.

$$\frac{(x-0)^2}{3^2} + \frac{(y-(-4))^2}{5^2} = 1$$

$$\frac{x^2}{9} + \frac{(y+4)^2}{25} = 1$$

39. Center: $(3,-4)$

The endpoints of the horizontal axis are $(-3,-4)$ and $(9,-4)$; $a = 6$.

The endpoints of the vertical axis are $(3,-9)$ and $(3,1)$; $b = 5$.

$$\frac{(x-3)^2}{6^2} + \frac{(y-(-4))^2}{5^2} = 1$$

$$\frac{(x-3)^2}{36} + \frac{(y+4)^2}{25} = 1$$

41.
$$(x+4)^2 + (y-6)^2 = 9$$
$$(x-(-4))^2 + (y-6)^2 = 3^2$$

A circle in standard form is

$$(x-h)^2 + (y-k)^2 = r^2;$$

$h = -4$, $k = 6$, and $r = 3$.

Center: $(-4,6)$ and radius: $r = 3$

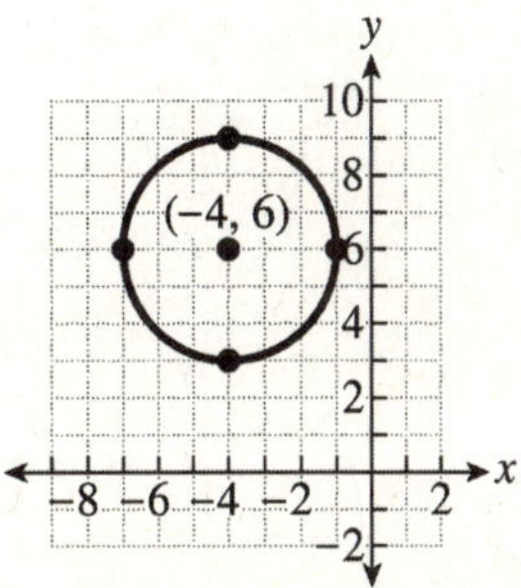

43. $x = y^2 + 6y + 10$

The parabola has the form $x = ay^2 + by + c$. Since $a > 0$, the parabola opens to the right.

vertex:

$$y = \frac{-b}{2a} = \frac{-6}{2(1)} = -3$$

$$x = (-3)^2 + 6(-3) + 10 = 9 - 18 + 10 = 1$$
$$(1,-3)$$

x-intercept:
$$x = (0)^2 + 6(0) + 10 = 10$$
$$(10,0)$$

y-intercept:
The vertex is to the right of the y-axis and the parabola opens to the right. There are no y-intercepts.

axis of symmetry: $y = -3$

point symmetric to the x-intercept: $(10,-6)$

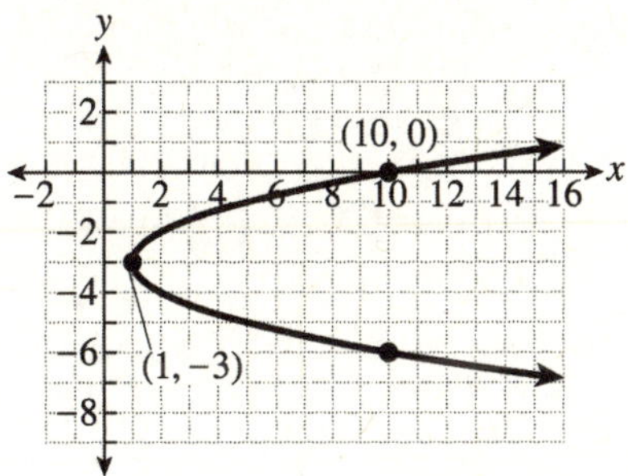

45.

$$3x^2 + 5y^2 = 45$$

$$\frac{3x^2}{45} + \frac{5y^2}{45} = \frac{45}{45}$$

$$\frac{x^2}{15} + \frac{y^2}{9} = 1$$

$$\frac{x^2}{\left(\sqrt{15}\right)^2} + \frac{y^2}{3^2} = 1$$

An ellipse in standard form is

$$\frac{(x-h)^2}{a^2} + \frac{(y-k)^2}{b^2} = 1.$$

Center: $(0,0)$

$a = \sqrt{15}$ and $b = 3$

x-intercepts: $\left(-\sqrt{15},0\right), \left(\sqrt{15},0\right)$

y-intercepts: $(0,-3), (0,3)$

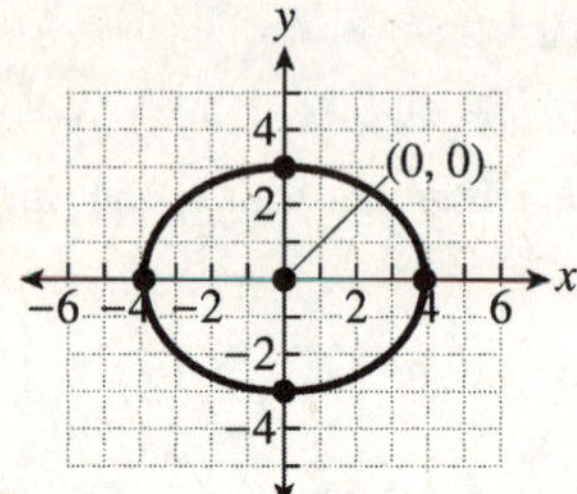

47.

$$\frac{(x-1)^2}{49} + \frac{(y+3)^2}{36} = 1$$

$$\frac{(x-1)^2}{7^2} + \frac{(y-(-3))^2}{6^2} = 1$$

An ellipse in standard form is

$$\frac{(x-h)^2}{a^2} + \frac{(y-k)^2}{b^2} = 1.$$

Center: $(1,-3)$

$a = 7$ and $b = 6$

The endpoints of the horizontal axis are 7 units to the left and right of the center.

$(-6,-3)$ and $(8,-3)$

The endpoints of the vertical axis are 6 units above and below the center.

$(1,-9)$ and $(1,3)$

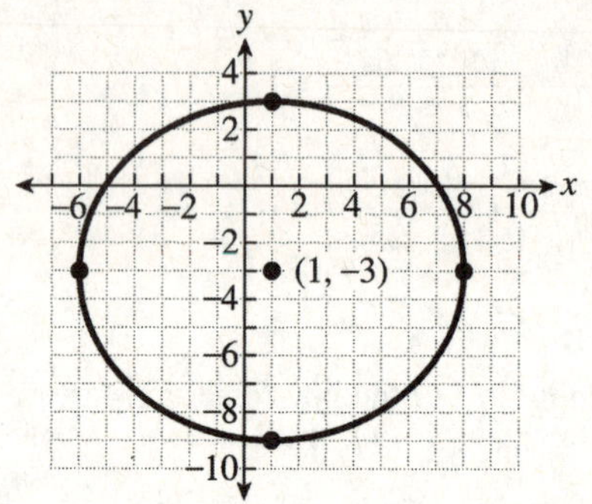

49.
$$9x^2 + 25y^2 + 54x - 100y - 44 = 0$$
$$\left(9x^2 + 54x\right) + \left(25y^2 - 100y\right) = 44$$
$$9\left(x^2 + 6x\right) + 25\left(y^2 - 4y\right) = 44$$
$$9\left(x^2 + 6x + \left(\frac{6}{2}\right)^2\right) + 25\left(y^2 - 4y + \left(-\frac{4}{2}\right)^2\right) = 44 + 9\left(\frac{6}{2}\right)^2 + 25\left(-\frac{4}{2}\right)^2$$
$$9\left(x^2 + 6x + 9\right) + 25\left(y^2 - 4y + 4\right) = 44 + 81 + 100$$
$$9(x+3)^2 + 25(y-2)^2 = 225$$
$$\frac{9(x+3)^2}{225} + \frac{25(y-2)^2}{225} = \frac{225}{225}$$
$$\frac{(x+3)^2}{25} + \frac{(y-2)^2}{9} = 1$$
$$\frac{(x-(-3))^2}{5^2} + \frac{(y-2)^2}{3^2} = 1$$

An ellipse in standard form is $\dfrac{(x-h)^2}{a^2} + \dfrac{(y-k)^2}{b^2} = 1$.

Center: $(-3, 2)$

$a = 5$ and $b = 3$
The endpoints of the horizontal axis are 5 units
to the left and right of the center.
$(-8, 2)$ and $(2, 2)$

The endpoints of the vertical axis are 3 units
above and below the center.
$(-3, 5)$ and $(-3, -1)$

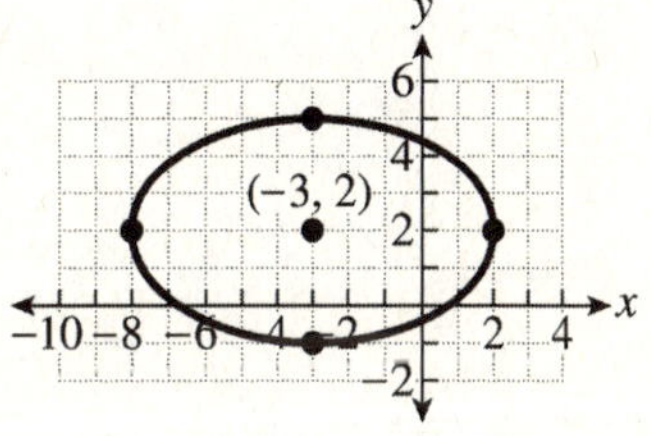

51.
$$x^2 + y^2 - 12x - 14y - 59 = 0$$
$$\left(x^2 - 12x\right) + \left(y^2 - 14y\right) = 59$$
$$\left(x^2 - 12x + \left(-\frac{12}{2}\right)^2\right) + \left(y^2 - 14y + \left(-\frac{14}{2}\right)^2\right) = 59 + \left(-\frac{12}{2}\right)^2 + \left(-\frac{14}{2}\right)^2$$
$$\left(x^2 - 12x + 36\right) + \left(y^2 - 14y + 49\right) = 59 + 36 + 49$$
$$(x-6)^2 + (y-7)^2 = 144$$
$$(x-6)^2 + (y-7)^2 = 12^2$$

A circle in standard form is
$(x-h)^2 + (y-k)^2 = r^2;$

$h = 6$, $k = 7$, and $r = 12$.

Center: $(6, 7)$ and radius: $r = 12$

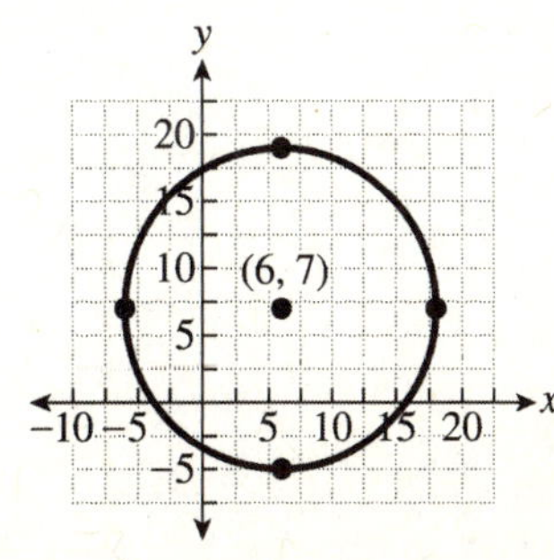

53. $x^2 + y^2 = 32$

$x^2 + y^2 = \left(\sqrt{32}\right)^2$

$x^2 + y^2 = \left(4\sqrt{2}\right)^2$

A circle in standard form is

$(x-h)^2 + (y-k)^2 = r^2;$

$h = 0$, $k = 0$, and $r = 4\sqrt{2}$.

Center: $(0,0)$ and radius: $r = 4\sqrt{2}$

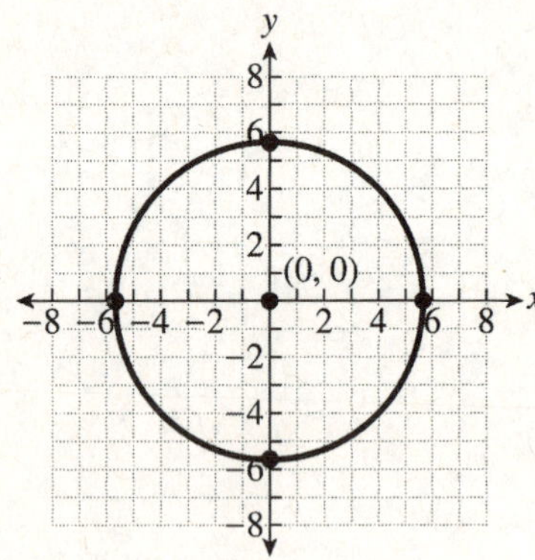

55. $x = -(y+4)^2 - 9$

$x = -(y-(-4))^2 - 9$

The parabola has the form $x = a(y-k)^2 + h;$

$a = -1$, $h = -9$, and $k = -4$.

Since $a < 0$, the parabola opens to the left.

vertex: $(-9,-4)$

x-intercept:

$x = -(0+4)^2 - 9 = -16 - 9 = -25$

$(-25,0)$

y-intercept:

The vertex is to the left of the y-axis and the parabola opens to the left. There are no y-intercepts.

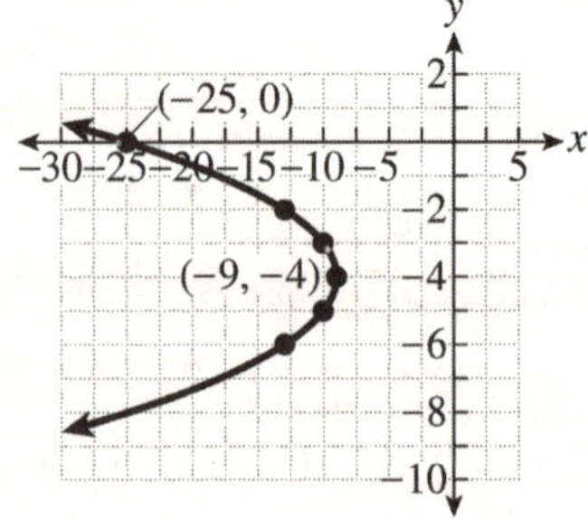

57.

$$\frac{(x+6)^2}{25} + \frac{(y+1)^2}{4} = 1$$

$$\frac{(x-(-6))^2}{5^2} + \frac{(y-(-1))^2}{2^2} = 1$$

An ellipse in standard form is

$$\frac{(x-h)^2}{a^2} + \frac{(y-k)^2}{b^2} = 1.$$

Center: $(-6,-1)$

$a = 5$ and $b = 2$

The endpoints of the horizontal axis are 5 units to the left and right of the center.

$(-11,-1)$ and $(-1,-1)$

The endpoints of the vertical axis are 2 units above and below the center.

$(-6,1)$ and $(-6,-3)$

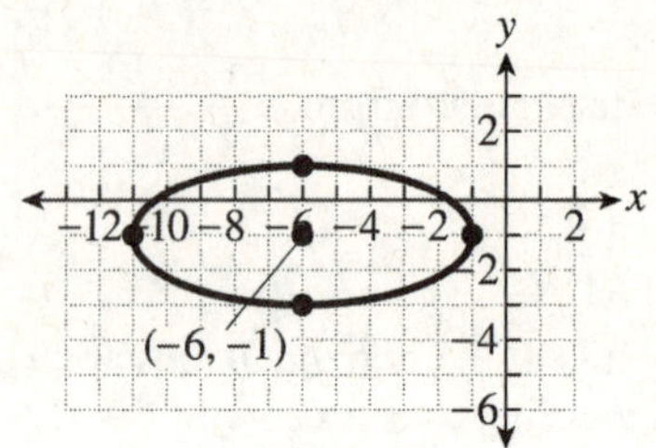

59. a)

$$1089x^2 + 2500y^2 = 2,722,500$$

$$\frac{1089x^2}{2,722,500} + \frac{2500y^2}{2,722,500} = \frac{2,722,500}{2,722,500}$$

$$\frac{x^2}{2500} + \frac{y^2}{1089} = 1$$

$$\frac{x^2}{50^2} + \frac{y^2}{33^2} = 1$$

The width of the track from west to east is the length of the horizontal axis. Since $a = 50$, the length of the horizontal axis is 100. The width of the track from east to west is 100 feet.

b) The width of the track from north to south is the length of the vertical axis. Since $b = 33$, the length of the vertical axis is 66. The width of the track from north to south is 66 feet.

61. Consider the major axis of the ellipse to be horizontal with length 30 feet. Then the distance from the endpoint to the center is 15 feet; $a = 15$. The length of the minor axis is 20 feet. Therefore, the difference from the endpoint to the center is 10 feet; $b = 10$. The area is found using the formula $A = \pi ab$.

$$A = \pi(15)(10) = 150\pi \approx 471.2$$

The area of the pool is 471.2 square feet.

63. Answers will vary. Example:
A circle and ellipse differ graphically because the horizontal and vertical axes of a circle are of equal length. Thus, the values of a and b are the same. However, they are similar in the fact that they both can be described by the equation $\dfrac{(x-h)^2}{a^2} + \dfrac{(y-k)^2}{b^2} = 1$.

13.4 QUICK CHECK

1. $\dfrac{x^2}{16} - \dfrac{y^2}{9} = 1$

$\dfrac{x^2}{4^2} - \dfrac{y^2}{3^2} = 1$

The hyperbola is in the form $\dfrac{x^2}{a^2} - \dfrac{y^2}{b^2} = 1$.

The hyperbola opens to the left and to the right where $a = 4$, $b = 3$.

Center: $(0,0)$

Vertices: $(-4,0), (4,0)$

Asymptotes: $y = \pm\dfrac{3}{4}x$

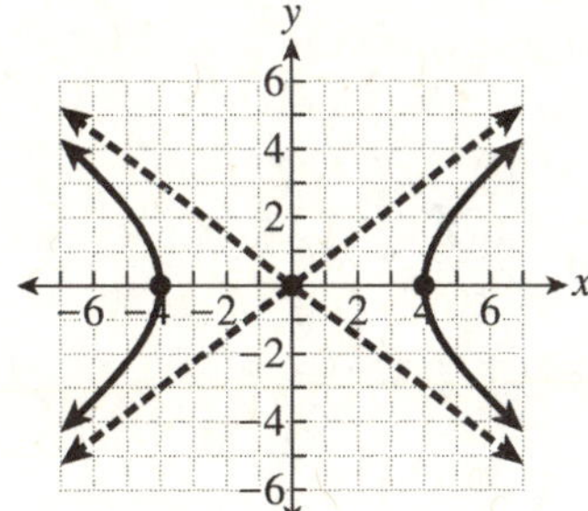

2. $\dfrac{y^2}{4} - \dfrac{x^2}{36} = 1$

$\dfrac{y^2}{2^2} - \dfrac{x^2}{6^2} = 1$

The hyperbola is in the form $\dfrac{y^2}{b^2} - \dfrac{x^2}{a^2} = 1$.

The hyperbola opens upward and downward where $a = 6$, $b = 2$.

Center: $(0,0)$

Vertices: $(0,-2), (0,2)$

Asymptotes: $y = \pm\dfrac{2}{6}x$

$y = \pm\dfrac{1}{3}x$

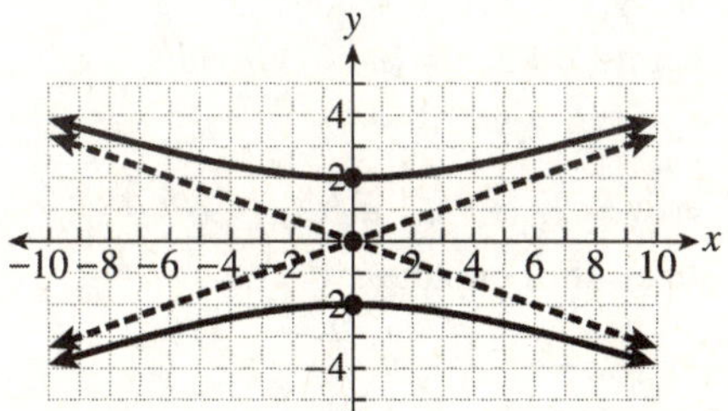

3. $\dfrac{(x-3)^2}{4} - \dfrac{(y-5)^2}{16} = 1$

$\dfrac{(x-3)^2}{2^2} - \dfrac{(y-5)^2}{4^2} = 1$

The hyperbola is in the form $\dfrac{(x-h)^2}{a^2} - \dfrac{(y-k)^2}{b^2} = 1$.

The hyperbola opens to the left and to the right where $h = 3, k = 5$, $a = 2$, and $b = 4$.

Center: $(3,5)$

Vertices: $(1,5), (5,5)$

Slopes of asymptotes:

$$\dfrac{b}{a} = \dfrac{4}{2} = 2, \quad \dfrac{-b}{a} = \dfrac{-4}{2} = -2$$

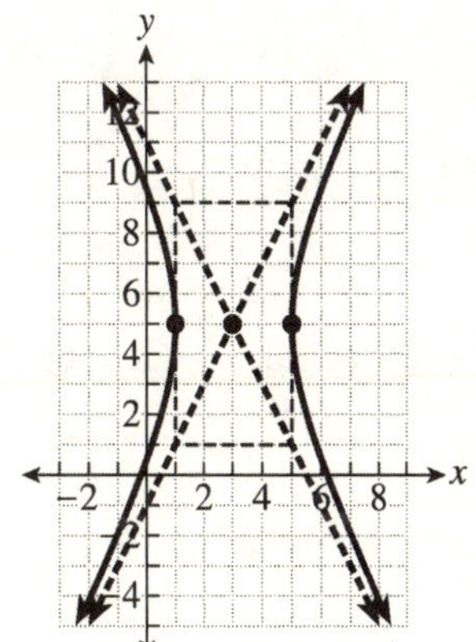

4.
$$\frac{(y-2)^2}{49} - \frac{(x+3)^2}{9} = 1$$

$$\frac{(y-2)^2}{7^2} - \frac{(x-(-3))^2}{3^2} = 1$$

The hyperbola is in the form
$$\frac{(y-k)^2}{b^2} - \frac{(x-h)^2}{a^2} = 1.$$

The hyperbola opens upward and downward where $h = -3$, $k = 2$, $a = 3$, and $b = 7$.

Center: $(-3, 2)$

Vertices: $(-3, 9)$, $(-3, -5)$

Slopes of asymptotes:
$$\frac{b}{a} = \frac{7}{3}, \quad \frac{-b}{a} = \frac{-7}{3} = -\frac{7}{3}$$

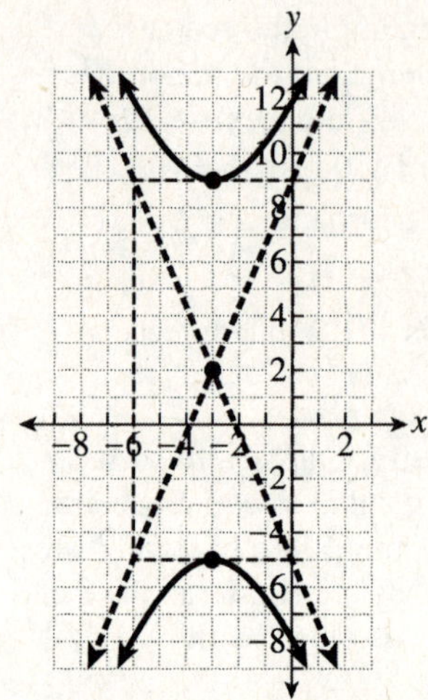

5.
$$16x^2 - 9y^2 - 64x - 54y - 161 = 0$$

$$\left(16x^2 - 64x\right) - 9y^2 - 54y = 161$$

$$16\left(x^2 - 4x\right) - 9\left(y^2 + 6y\right) = 161$$

$$16\left(x^2 - 4x + \left(-\frac{4}{2}\right)^2\right) - 9\left(y^2 + 6y + \left(\frac{6}{2}\right)^2\right) = 161 + 16\left(-\frac{4}{2}\right)^2 - 9\left(\frac{6}{2}\right)^2$$

$$16\left(x^2 - 4x + 4\right) - 9\left(y^2 + 6y + 9\right) = 161 + 64 - 81$$

$$16(x-2)^2 - 9(y+3)^2 = 144$$

$$\frac{16(x-2)^2}{144} - \frac{9(y+3)^2}{144} = \frac{144}{144}$$

$$\frac{(x-2)^2}{9} - \frac{(y+3)^2}{16} = 1$$

$$\frac{(x-2)^2}{3^2} - \frac{(y+3)^2}{4^2} = 1$$

The hyperbola is in the form
$$\frac{(x-h)^2}{a^2} - \frac{(y-k)^2}{b^2} = 1.$$

The hyperbola opens to the left and to the right. where $h = 2, k = -3, a = 3$, and $b = 4$.

Center: $(2, -3)$

Vertices: $(-1, -3)$, $(5, -3)$

Asymptotes: $y = \pm\dfrac{b}{a}(x-h) + k$

$$y = \pm\frac{4}{3}(x-2) + (-3)$$

$$y = \pm\frac{4}{3}(x-2) - 3$$

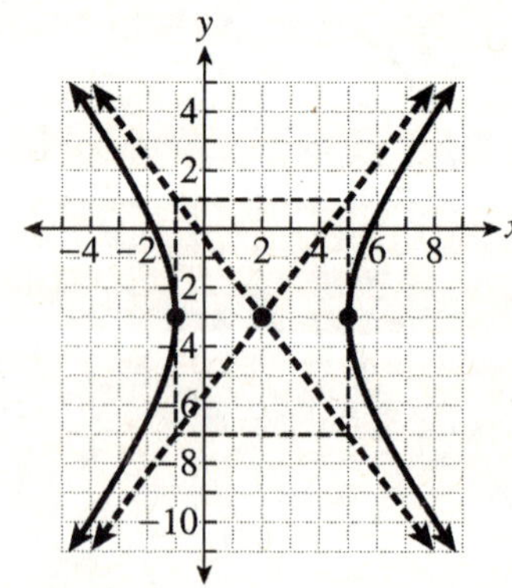

6. The hyperbola is in the form
$$\frac{(y-k)^2}{b^2} - \frac{(x-h)^2}{a^2} = 1.$$
The hyperbola opens upward and downward.

Center: $(-3,4)$

The distance from the center to one of the vertices is 2, so $b = 2$.

The horizontal distance from a vertex to one of the asymptotes is 5, so $a = 5$.

$$\frac{(y-4)^2}{2^2} - \frac{(x-(-3))^2}{5^2} = 1$$

$$\frac{(y-4)^2}{4} - \frac{(x+3)^2}{25} = 1$$

13.4 HYPERBOLAS

1. hyperbola

3. center

5. transverse axis

7. $\dfrac{x^2}{a^2} - \dfrac{y^2}{b^2} = 1$; $\dfrac{y^2}{b^2} - \dfrac{x^2}{a^2} = 1$

9. $\dfrac{x^2}{9} - \dfrac{y^2}{4} = 1$

$$\frac{x^2}{3^2} - \frac{y^2}{2^2} = 1$$

The hyperbola is in the form $\dfrac{x^2}{a^2} - \dfrac{y^2}{b^2} = 1.$

The hyperbola opens to the left and to the right where $a = 3, b = 2$.

Center: $(0,0)$

Vertices: $(-3,0),(3,0)$

Asymptotes: $y = \pm \dfrac{2}{3}x$

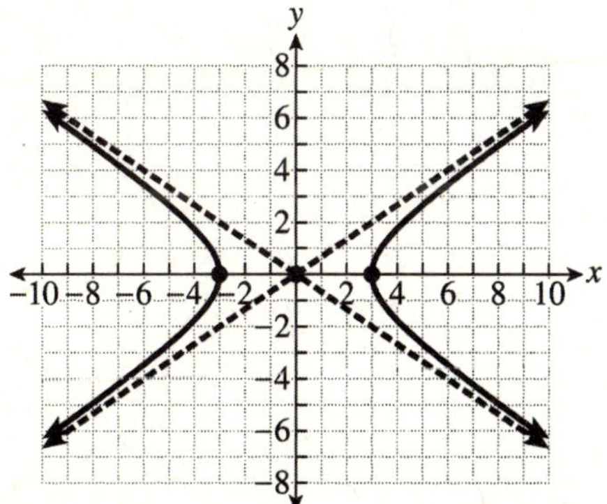

11. $\dfrac{y^2}{49} - \dfrac{x^2}{9} = 1$

$$\frac{y^2}{7^2} - \frac{x^2}{3^2} = 1$$

The hyperbola is in the form $\dfrac{y^2}{b^2} - \dfrac{x^2}{a^2} = 1.$

The hyperbola opens upward and downward where $a = 3, b = 7$.

Center: $(0,0)$

Vertices: $(0,-7),(0,7)$

Asymptotes: $y = \pm \dfrac{7}{3}x$

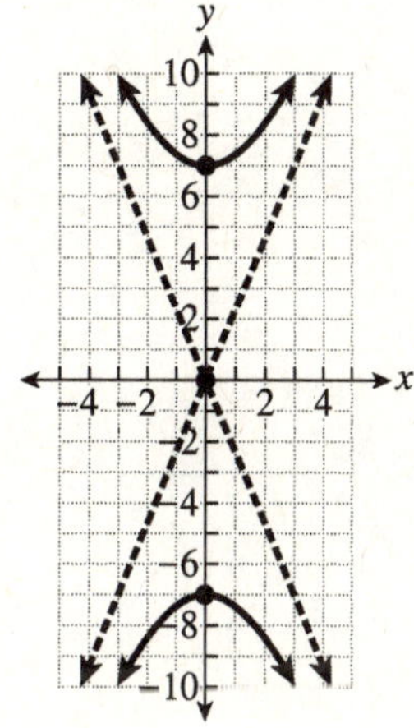

13. $\dfrac{y^2}{25} - x^2 = 1$

$\dfrac{y^2}{5^2} - \dfrac{x^2}{1^2} = 1$

The hyperbola is in the form $\dfrac{y^2}{b^2} - \dfrac{x^2}{a^2} = 1$.

The hyperbola opens upward and downward where $a = 1, b = 5$.

Center: $(0,0)$

Vertices: $(0,-5),(0,5)$

Asymptotes: $y = \pm\dfrac{5}{1}x$

$\qquad\qquad y = \pm 5x$

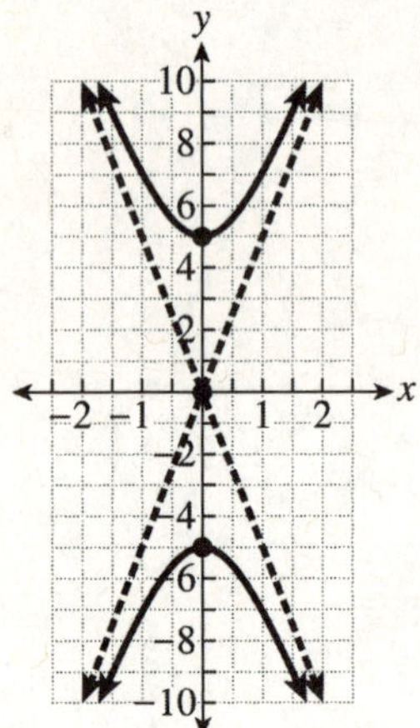

15. $\dfrac{x^2}{12} - \dfrac{y^2}{9} = 1$

$\dfrac{x^2}{\left(\sqrt{12}\right)^2} - \dfrac{y^2}{3^2} = 1$

The hyperbola is in the form $\dfrac{x^2}{a^2} - \dfrac{y^2}{b^2} = 1$.

The hyperbola opens to the left and to the right where $a = \sqrt{12} = 2\sqrt{3}, b = 3$.

Center: $(0,0)$

Vertices: $\left(-2\sqrt{3},0\right),\left(2\sqrt{3},0\right)$

Asymptotes: $y = \pm\dfrac{3}{2\sqrt{3}}x$

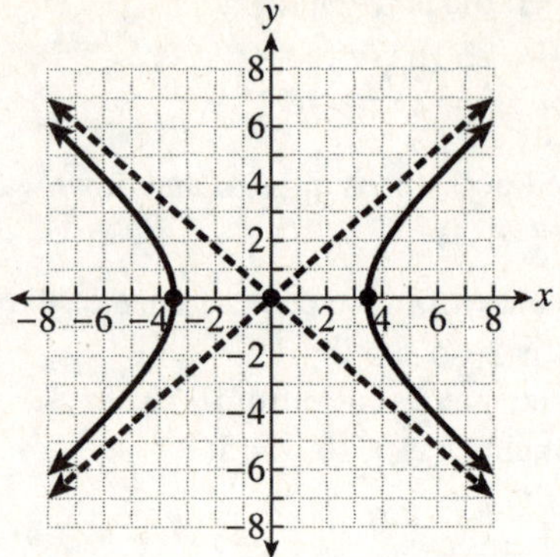

17. $49x^2 - 4y^2 = 196$

$\dfrac{49x^2}{196} - \dfrac{4y^2}{196} = \dfrac{196}{196}$

$\dfrac{x^2}{4} - \dfrac{y^2}{49} = 1$

$\dfrac{x^2}{2^2} - \dfrac{y^2}{7^2} = 1$

The hyperbola is in the form $\dfrac{x^2}{a^2} - \dfrac{y^2}{b^2} = 1$.

The hyperbola opens to the left and to the right where $a = 2, b = 7$.

Center: $(0,0)$

Vertices: $(-2,0),(2,0)$

Asymptotes: $y = \pm\dfrac{7}{2}x$

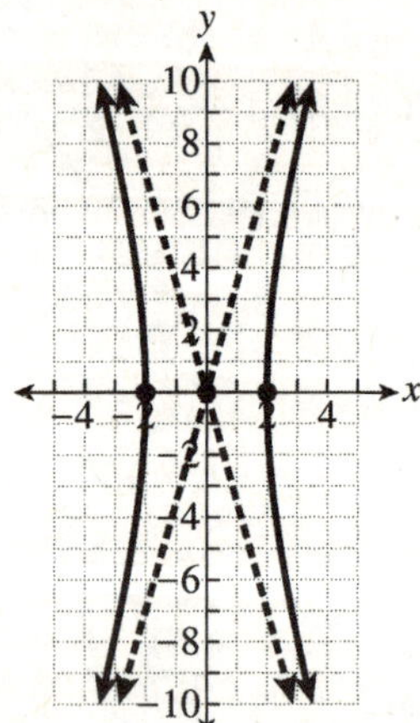

19. $y^2 - 9x^2 = 9$

$$\frac{y^2}{9} - \frac{9x^2}{9} = \frac{9}{9}$$

$$\frac{y^2}{9} - x^2 = 1$$

$$\frac{y^2}{3^2} - \frac{x^2}{1^2} = 1$$

The hyperbola is in the form $\dfrac{y^2}{b^2} - \dfrac{x^2}{a^2} = 1$.

The hyperbola opens upward and downward where $a = 1, b = 3$.

Center: $(0,0)$

Vertices: $(0,-3),(0,3)$

Asymptotes: $y = \pm\dfrac{3}{1}x$

$$y = \pm 3x$$

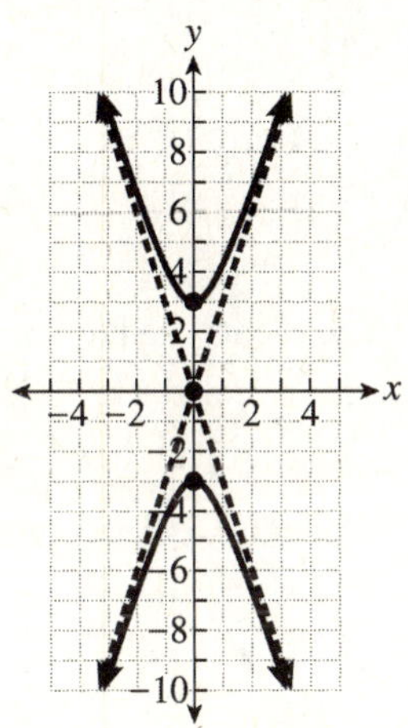

21. $\dfrac{(y-5)^2}{16} - \dfrac{(x-2)^2}{9} = 1$

$$\frac{(y-5)^2}{4^2} - \frac{(x-2)^2}{3^2} = 1$$

The hyperbola is in the form

$$\frac{(y-k)^2}{b^2} - \frac{(x-h)^2}{a^2} = 1.$$

The hyperbola opens upward and downward where $h = 2, k = 5$, $a = 3$, and $b = 4$.

Center: $(h,k) = (2,5)$

Vertices: $(2,1), (2,9)$

Asymptotes: $y = \pm\dfrac{b}{a}(x-h) + k$

$$y = \pm\frac{4}{3}(x-2) + 5$$

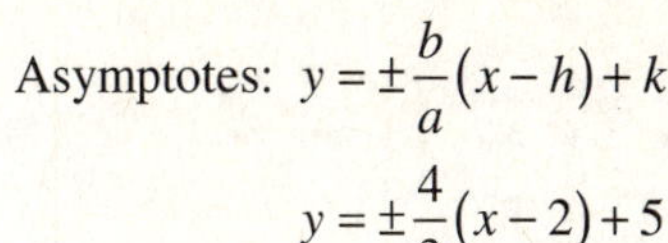

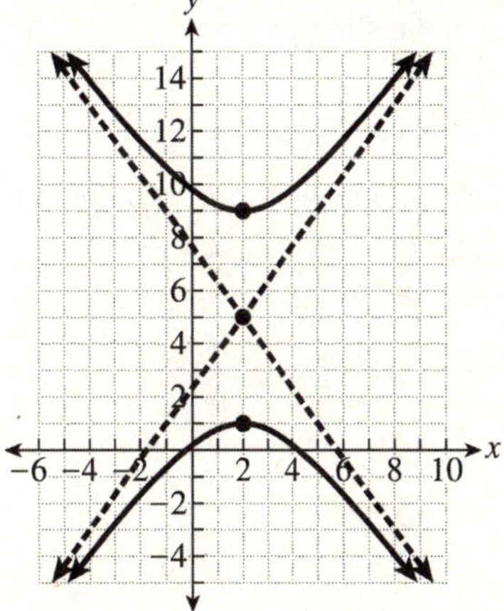

23. $\dfrac{(x+1)^2}{9} - \dfrac{(y+4)^2}{4} = 1$

$$\frac{(x-(-1))^2}{3^2} - \frac{(y-(-4))^2}{2^2} = 1$$

The hyperbola is in the form

$$\frac{(x-h)^2}{a^2} - \frac{(y-k)^2}{b^2} = 1.$$

The hyperbola opens to the left and to the right where $h = -1, k = -4$, $a = 3$, and $b = 2$.

Center: $(h,k) = (-1,-4)$

Vertices: $(2,-4), (-4,-4)$

Asymptotes: $y = \pm\dfrac{b}{a}(x-h) + k$

$$y = \pm\frac{2}{3}(x+1) - 4$$

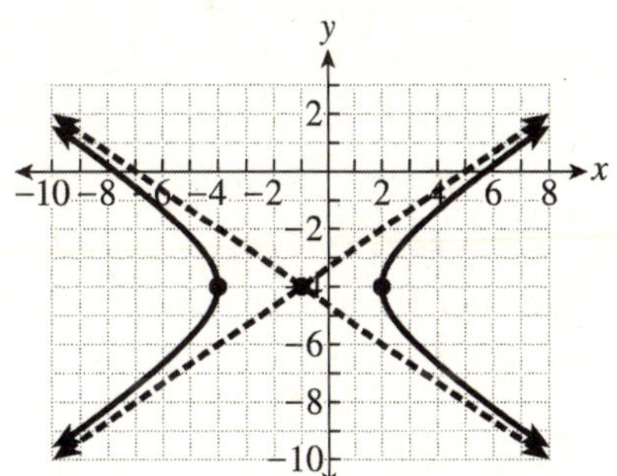

25.
$$\frac{(y-1)^2}{4} - \frac{x^2}{4} = 1$$
$$\frac{(y-1)^2}{2^2} - \frac{(x-0)^2}{2^2} = 1$$

The hyperbola is in the form
$$\frac{(y-k)^2}{b^2} - \frac{(x-h)^2}{a^2} = 1.$$

The hyperbola opens upward and downward where $h=0, k=1, a=2,$ and $b=2$.

Center: $(h,k) = (0,1)$

Vertices: $(0,-1), (0,3)$

Asymptotes: $y = \pm\frac{b}{a}(x-h)+k$
$$y = \pm\frac{2}{2}(x-0)+1$$
$$y = \pm x+1$$

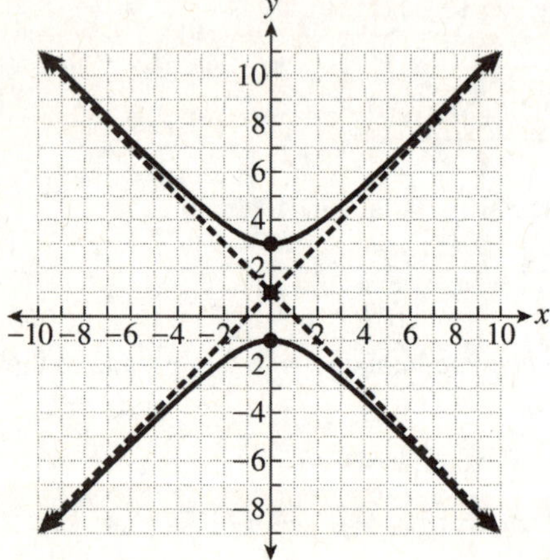

27.
$$\frac{(x-4)^2}{49} - \frac{(y+2)^2}{16} = 1$$
$$\frac{(x-4)^2}{7^2} - \frac{(y-(-2))^2}{4^2} = 1$$

The hyperbola is in the form
$$\frac{(x-h)^2}{a^2} - \frac{(y-k)^2}{b^2} = 1.$$

The hyperbola opens to the left and to the right where $h=4, k=-2, a=7,$ and $b=4$.

Center: $(h,k) = (4,-2)$

Vertices: $(11,-2), (-3,-2)$

Asymptotes: $y = \pm\frac{b}{a}(x-h)+k$
$$y = \pm\frac{4}{7}(x-4)-2$$

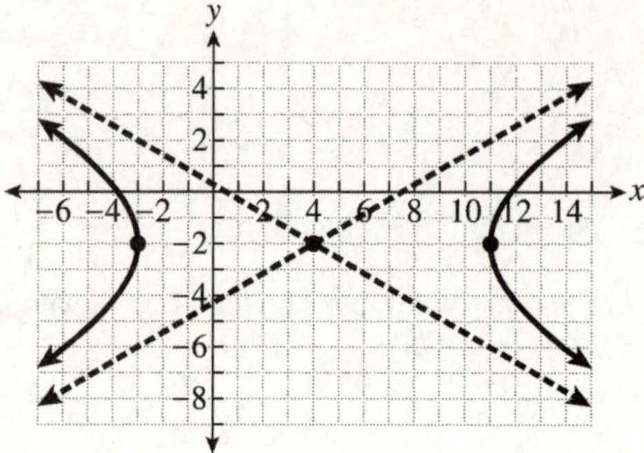

29.
$$\frac{(x+2)^2}{25} - \frac{(y+3)^2}{28} = 1$$
$$\frac{(x-(-2))^2}{5^2} - \frac{(y-(-3))^2}{(2\sqrt{7})^2} = 1$$

The hyperbola is in the form
$$\frac{(x-h)^2}{a^2} - \frac{(y-k)^2}{b^2} = 1.$$

The hyperbola opens to the left and to the right where $h=-2, k=-3, a=5,$ and $b=2\sqrt{7}$.

Center: $(h,k) = (-2,-3)$

Vertices: $(-7,-3), (3,-3)$

Asymptotes: $y = \pm\frac{b}{a}(x-h)+k$
$$y = \pm\frac{2\sqrt{7}}{5}(x+2)-3$$

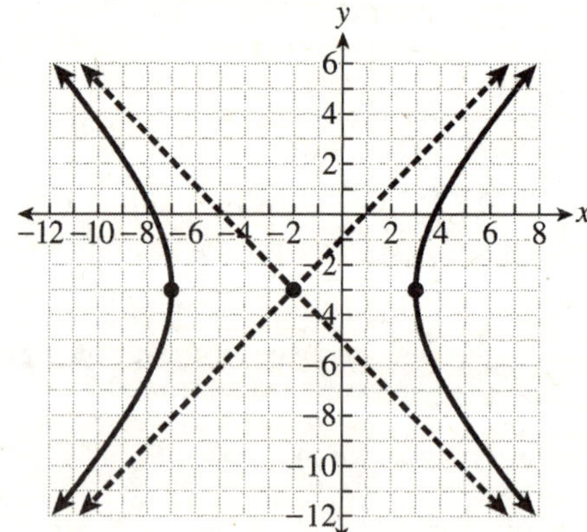

31.

$$9x^2 - 4y^2 + 18x - 24y - 63 = 0$$

$$9x^2 + 18x - 4y^2 - 24y = 63$$

$$9\left(x^2 + 2x\right) - 4\left(y^2 + 6y\right) = 63$$

$$9\left(x^2 + 2x + \left(\frac{2}{2}\right)^2\right) - 4\left(y^2 + 6y + \left(\frac{6}{2}\right)^2\right) = 63 + 9\left(\frac{2}{2}\right)^2 - 4\left(\frac{6}{2}\right)^2$$

$$9\left(x^2 + 2x + 1\right) - 4\left(y^2 + 6y + 9\right) = 63 + 9 - 36$$

$$9(x+1)^2 - 4(y+3)^2 = 36$$

$$\frac{9(x+1)^2}{36} - \frac{4(y+3)^2}{36} = \frac{36}{36}$$

$$\frac{(x+1)^2}{4} - \frac{(y+3)^2}{9} = 1$$

$$\frac{(x-(-1))^2}{2^2} - \frac{(y-(-3))^2}{3^2} = 1$$

The hyperbola is in the form $\dfrac{(x-h)^2}{a^2} - \dfrac{(y-k)^2}{b^2} = 1$.

The hyperbola opens to the left and to the right where $h = -1, k = -3$, $a = 2$, and $b = 3$.

Center: $(h,k) = (-1,-3)$

Vertices: $(-3,-3)$, $(1,-3)$

Asymptotes: $y = \pm\dfrac{b}{a}(x-h) + k$

$$y = \pm\frac{3}{2}(x+1) - 3$$

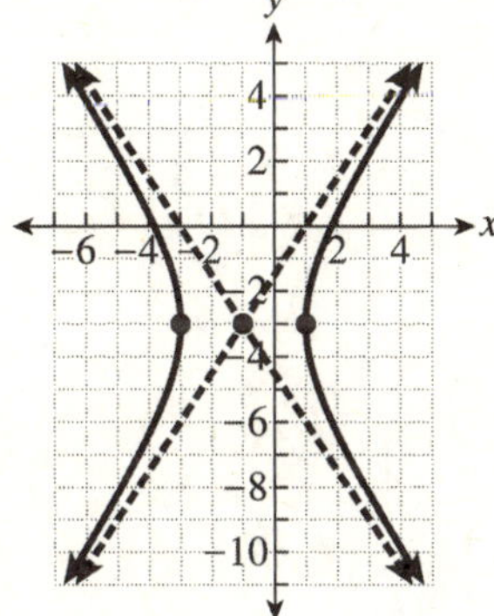

33.

$$-25x^2 + 16y^2 - 100x - 224y + 284 = 0$$

$$16y^2 - 224y - 25x^2 - 100x = -284$$

$$16\left(y^2 - 14y\right) - 25\left(x^2 + 4x\right) = -284$$

$$16\left(y^2 - 14y + \left(-\frac{14}{2}\right)^2\right) - 25\left(x^2 + 4x + \left(\frac{4}{2}\right)^2\right) = -284 + 16\left(-\frac{14}{2}\right)^2 - 25\left(\frac{4}{2}\right)^2$$

$$16\left(y^2 - 14y - 49\right) - 25\left(x^2 + 4x + 4\right) = -284 + 784 - 100$$

$$16(y - 7)^2 - 25(x + 2)^2 = 400$$

$$\frac{16(y - 7)^2}{400} - \frac{25(x + 2)^2}{400} = \frac{400}{400}$$

$$\frac{(y - 7)^2}{25} - \frac{(x + 2)^2}{16} = 1$$

$$\frac{(y - 7)^2}{5^2} - \frac{(x - (-2))^2}{4^2} = 1$$

The hyperbola is in the form $\dfrac{(y - k)^2}{b^2} - \dfrac{(x - h)^2}{a^2} = 1$.

The hyperbola opens upward and downward where $h = -2, k = 7, a = 4,$ and $b = 5$.

Center: $(h, k) = (-2, 7)$

Vertices: $(-2, 2), (-2, 12)$

Asymptotes: $y = \pm\dfrac{b}{a}(x - h) + k$

$$y = \pm\frac{5}{4}(x + 2) + 7$$

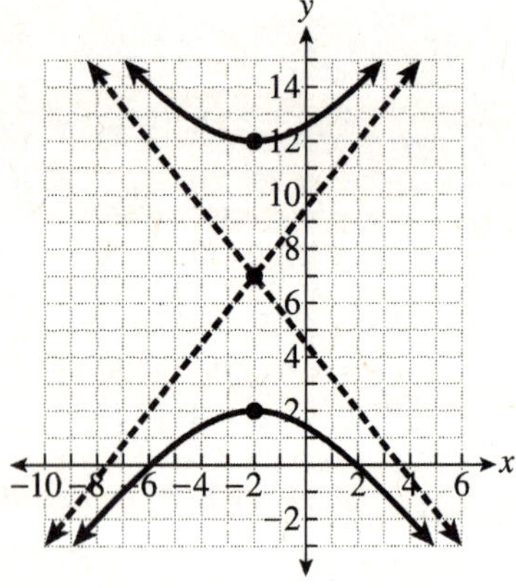

35. Center: $(0,0)$; $h = 0$, $k = 0$

Vertices: $(0,-5),(0,5)$, $b = 5$

The horizontal distance from a vertex to one of the asymptotes is 3, so $a = 3$.

$$\frac{(y-0)^2}{5^2} - \frac{(x-0)^2}{3^2} = 1$$

$$\frac{y^2}{25} - \frac{x^2}{9} = 1$$

37. Center: $(3,2)$; $h = 3$, $k = 2$

Vertices: $(3,-3),(3,7)$, $b = 5$

The horizontal distance from a vertex to one of the asymptotes is 2, so $a = 2$.

$$\frac{(y-2)^2}{5^2} - \frac{(x-3)^2}{2^2} = 1$$

$$\frac{(y-2)^2}{25} - \frac{(x-3)^2}{4} = 1$$

39. Center: $(-1,-5)$; $h = -1$, $k = -5$

Vertices: $(-6,-5),(4,-5)$, $a = 5$

The vertical distance from a vertex to one of the asymptotes is 3, so $b = 3$.

$$\frac{(x-(-1))^2}{5^2} - \frac{(y-(-5))^2}{3^2} = 1$$

$$\frac{(x+1)^2}{25} - \frac{(y+5)^2}{9} = 1$$

41. $y = -x^2 + 6x - 17$

The equation is of the form $y = ax^2 + bx + c$ where $a < 0$ so the parabola opens downward.

vertex:

$$x = \frac{-b}{2a} = \frac{-6}{2(-1)} = 3$$

$$y = -(3)^2 + 6(3) - 17 = -9 + 18 - 17 = -8$$
$$(3,-8)$$

y-intercept:
$$y = -(0)^2 + 6(0) - 17 = -17$$
$$(0,-17)$$

x-intercept: none

axis of symmetry: $x = 3$

point symmetric to the y-intercept: $(6,-17)$

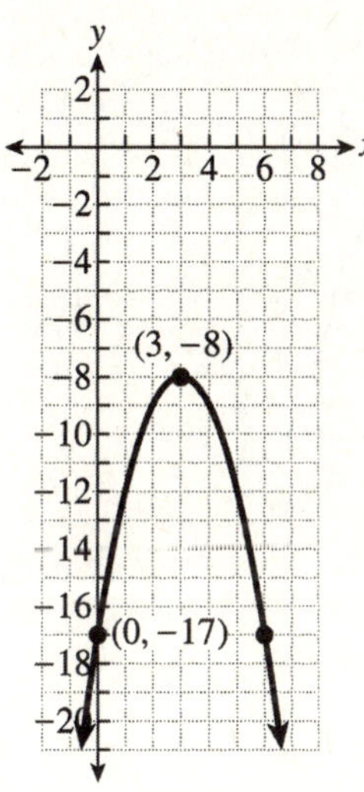

43.
$$\frac{(x-6)^2}{9}+\frac{y^2}{49}=1$$
$$\frac{(x-6)^2}{3^2}+\frac{(y-0)^2}{7^2}=1$$

The ellipse is in the form
$$\frac{(x-h)^2}{a^2}+\frac{(y-k)^2}{b^2}=1,$$

where $h=6, k=0, a=3,$ and $b=7.$

Center: $(6,0)$

The endpoints of the horizontal axis are 3 units to the left and to the right of the center.

$(3,0)$ and $(9,0)$

The endpoints of the vertical axis are 7 units above and below the center.

$(6,7)$ and $(6,-7)$

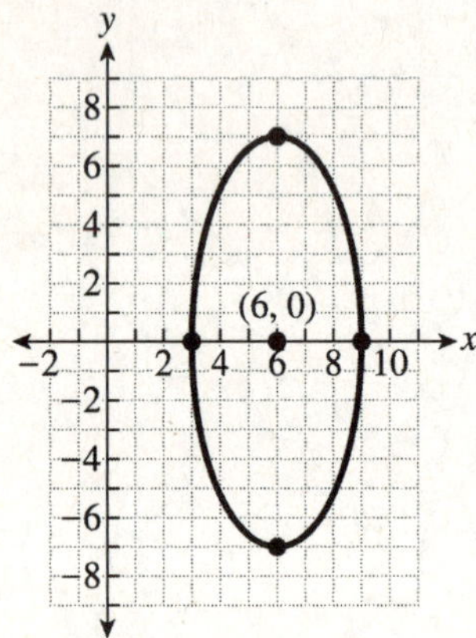

45.
$$\frac{(x+2)^2}{25}+\frac{(y-5)^2}{9}=1$$
$$\frac{(x-(-2))^2}{5^2}+\frac{(y-5)^2}{3^2}=1$$

The ellipse is in the form
$$\frac{(x-h)^2}{a^2}+\frac{(y-k)^2}{b^2}=1,\ \text{where}$$

$h=-2, k=5, a=5,$ and $b=3.$

Center: $(-2,5)$

The endpoints of the horizontal axis are 5 units left and right of the center.

$(-7,5)$ and $(3,5)$

The endpoints of the vertical axis are 3 units above and below the center.

$(-2,8)$ and $(-2,2)$

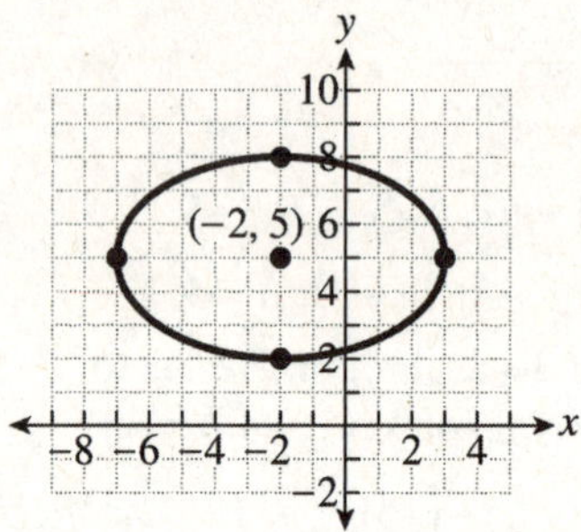

47. $x^2+y^2=64$
$$x^2+y^2=8^2$$

The circle is in the form $x^2+y^2=r^2.$

Center: $(0,0)$ and radius: $r=8$

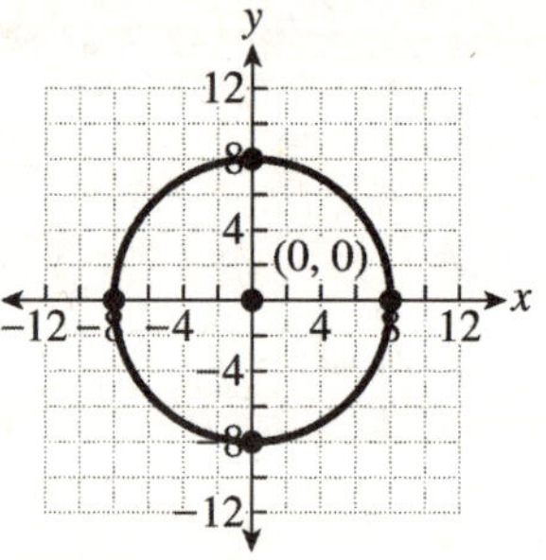

49. $4x^2 - 9y^2 + 40x + 54y - 17 = 0$

$$4x^2 + 40x - 9y^2 + 54y = 17$$

$$4\left(x^2 + 10x\right) - 9\left(y^2 - 6y\right) = 17$$

$$4\left(x^2 + 10x + \left(\frac{10}{2}\right)^2\right) - 9\left(y^2 - 6y + \left(-\frac{6}{2}\right)^2\right) = 17 + 4\left(\frac{10}{2}\right)^2 - 9\left(-\frac{6}{2}\right)^2$$

$$4\left(x^2 + 10x + 25\right) - 9\left(y^2 - 6y + 9\right) = 17 + 100 - 81$$

$$4(x+5)^2 - 9(y-3)^2 = 36$$

$$\frac{4(x+5)^2}{36} - \frac{9(y-3)^2}{36} = \frac{36}{36}$$

$$\frac{(x+5)^2}{9} - \frac{(y-3)^2}{4} = 1$$

$$\frac{(x-(-5))^2}{3^2} - \frac{(y-3)^2}{2^2} = 1$$

The hyperbola is in the form $\dfrac{(x-h)^2}{a^2} - \dfrac{(y-k)^2}{b^2} = 1$.

The hyperbola opens to the left and to the right where $h = -5, k = 3$, $a = 3$, and $b = 2$.

Center: $(h,k) = (-5,3)$

Vertices: $(-8,3),\ (-2,3)$

Asymptotes: $y = \pm\dfrac{b}{a}(x-h) + k$

$$y = \pm\frac{2}{3}(x+5) + 3$$

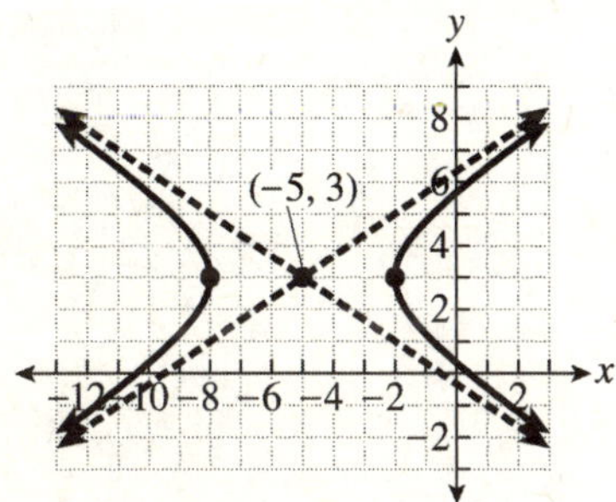

51.

$$x^2 + y^2 - 8x + 18y + 48 = 0$$
$$x^2 - 8x + y^2 + 18y = -48$$
$$\left(x^2 - 8x + \left(-\frac{8}{2}\right)^2\right) + \left(y^2 + 18y + \left(\frac{18}{2}\right)^2\right) = -48 + \left(-\frac{8}{2}\right)^2 + \left(\frac{18}{2}\right)^2$$
$$\left(x^2 - 8x + 16\right) + \left(y^2 + 18y + 81\right) = -48 + 16 + 81$$
$$(x-4)^2 + (y+9)^2 = 49$$
$$(x-4)^2 + (y-(-9))^2 = 7^2$$

The circle is in the form $(x-h)^2 + (y-k)^2 = r^2$ where $h = 4, k = -9$ and $r = 7$.

Center: $(4,-9)$ and radius: $r = 7$

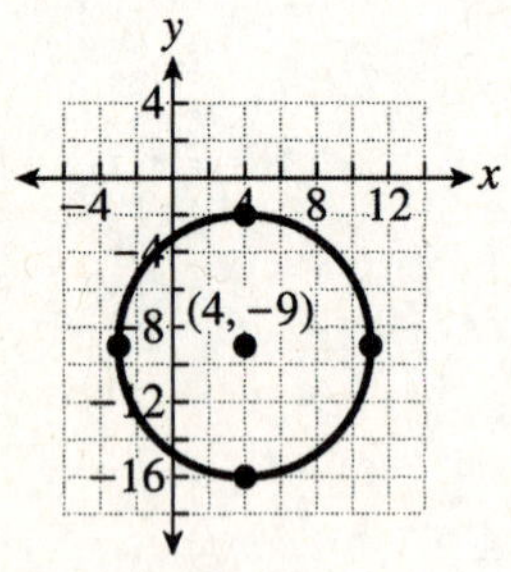

53.

$$\frac{(x+6)^2}{9} - \frac{(y+2)^2}{16} = 1$$
$$\frac{(x-(-6))^2}{3^2} - \frac{(y-(-2))^2}{4^2} = 1$$

The hyperbola is in the form $\dfrac{(x-h)^2}{a^2} - \dfrac{(y-k)^2}{b^2} = 1.$

The hyperbola opens to the left and to the right where $h = -6, k = -2, a = 3,$ and $b = 4.$

Center: $(h,k) = (-6,-2)$

Vertices: $(-9,-2), (-3,-2)$

Asymptotes: $y = \pm\dfrac{b}{a}(x-h) + k$

$$y = \pm\frac{4}{3}(x+6) - 2$$

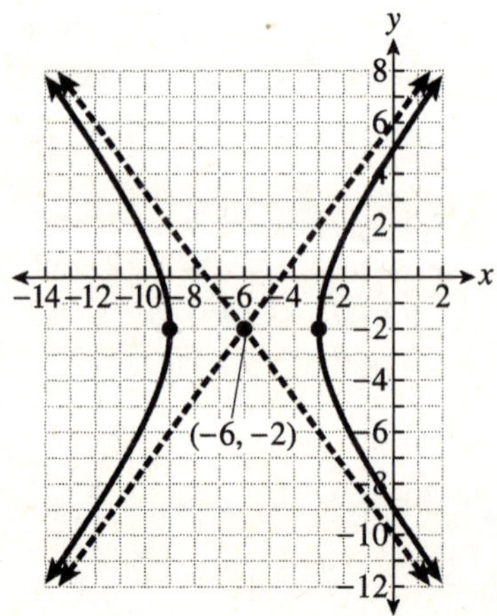

55. e

57. c

59. b

61. Answers will vary. Example:
If the equation is of the form

$$\frac{(x-h)^2}{a^2} - \frac{(y-k)^2}{b^2} = 1, \text{ the hyperbola opens to}$$

the left and to the right.
If the equation is of the form

$$\frac{(y-k)^2}{b^2} - \frac{(x-h)^2}{a^2} = 1, \text{ the hyperbola opens}$$

upward and downward.

13.5 QUICK CHECK

1. $4x^2 + y^2 = 41$ (Eq. 1)
$x + 2y = -8$ (Eq. 2)

Solve for x in Eq. 2.
$x = -2y - 8$
Substitute $x = -2y - 8$ into Eq. 1.

$$4(-2y-8)^2 + y^2 = 41$$
$$4(4y^2 + 32y + 64) + y^2 = 41$$
$$16y^2 + 128y + 256 + y^2 = 41$$
$$17y^2 + 128y + 215 = 0$$
$$(17y + 43)(y + 5) = 0$$

$17y + 43 = 0$ or $y + 5 = 0$
$17y = -43$ or $y = -5$

$$y = -\frac{43}{17} \quad \text{or} \quad y = -5$$

Find the x-coordinates using $x = -2y - 8$.

$y = -\dfrac{43}{17}$ $y = -5$

$x = -2\left(-\dfrac{43}{17}\right) - 8$ $x = -2(-5) - 8$

$x = \dfrac{86}{17} - \dfrac{136}{17}$ $x = 10 - 8$

$x = -\dfrac{50}{17}$ $x = 2$

$\left(-\dfrac{50}{17}, -\dfrac{43}{17}\right)$ $(2, -5)$

2. $y = 2x - 8$ (Eq. 1)
$x^2 + y^2 = 9$ (Eq. 2)

Substitute $y = 2x - 8$ into Eq. 2

$$x^2 + (2x - 8)^2 = 9$$
$$x^2 + 4x^2 - 32x + 64 = 9$$
$$5x^2 - 32x + 55 = 0$$

$$x = \frac{-(-32) \pm \sqrt{(-32)^2 - 4(5)(55)}}{2(5)}$$

$$x = \frac{32 \pm \sqrt{-76}}{10}$$

undefined
$\varnothing$

3. $y = x^2$ (Eq. 1)
$5x^2 + y^2 = 24$ (Eq. 2)

Substitute $y = x^2$ in Eq. 2.

$$5(y) + y^2 = 24$$
$$y^2 + 5y - 24 = 0$$
$$(y + 8)(y - 3) = 0$$

$y = -8$ or $y = 3$

Find the x-coordinates using $y = x^2$.

$y = -8$ $y = 3$
$-8 = x^2$ $3 = x^2$
$x = \pm\sqrt{-8}$ $x = +\sqrt{3}$
undefined

$\left(-\sqrt{3}, 3\right), \left(\sqrt{3}, 3\right)$

4. $4x^2 + 7y^2 = 211$ (Eq. 1)
$x^2 + y^2 = 34$ (Eq. 2)

$4x^2 + 7y^2 = 211$ (Eq. 1)
$+\ \ \underline{-4x^2 - 4y^2 = -136}$ $(-4 \cdot \text{Eq. 2})$
$3y^2 = 75$
$y^2 = 25$
$y = \pm 5$

Find the x-coordinates using $x^2 + y^2 = 34$.

$y = -5$ $y = 5$
$x^2 + (-5)^2 = 34$ $x^2 + (5)^2 = 34$
$x + 25 = 34$ $x + 25 = 34$
$x^2 = 9$ $x^2 = 9$
$x = \pm 3$ $x = \pm 3$

$(-3, -5), (3, -5), (-3, 5), (3, 5)$

5. Unknowns:
Length: x
Width: y

$2x + 2y = 30$ (Perimeter)
$x \cdot y = 36$ (Area)

$2x = 30 - 2y$
$x = 15 - y$

Substitute into the area equation.

$(15 - y)y = 36$
$15y - y^2 = 36$
$0 = y^2 - 15y + 36$
$0 = (y - 3)(y - 12)$
$y = 3$ or $y = 12$

Find the x-coordinates using $x = 15 - y$.

$y = 3$ $y = 12$
$x = 15 - 3$ $x = 15 - 12$
$x = 12$ $x = 3$

Dimensions are 3 feet by 12 feet.

6. Unknowns:
Length: x
Width: y

$2x + 2y = 42$ (Perimeter)
$x = 21 - y$

Use Pythagorean's theorem to find an equation using the diagonal.

$x^2 + y^2 = 15^2$
$x^2 + y^2 = 225$

Substitute the perimeter equation into the Pythagorean Theorem.

$(21 - y)^2 + y^2 = 225$
$441 - 42y + y^2 + y^2 = 225$
$2y^2 - 42y + 216 = 0$
$2(y^2 - 21y + 108) = 0$
$2(y - 9)(y - 12) = 0$
$y = 9$ or $y = 12$

Find the x-coordinates using $x = 21 - y$.

$y = 12$ $y = 9$
$x = 21 - 12$ $x = 21 - 9$
$x = 9$ $x = 12$

Dimensions are 9 inches by 12 inches.

7. Use the function for the height of an object:

$h(t) = -16t^2 + v_0 t + s$

$y = -16t^2 + 40t$ (Eq. 1)
$y = -16t^2 + 35t + 10$ (Eq. 2)

Substitute $y = -16t^2 + 40t$ into Eq. 2.
$-16t^2 + 40t = -16t^2 + 35t + 10$
$40t = 35t + 10$
$5t = 10$
$t = 2$

2 seconds

13.5 NONLINEAR SYSTEMS OF EQUATIONS

1. nonlinear

3. $x^2 + y^2 = 100$ (Eq. 1)
$x - 7y = 50$ (Eq. 2)

Solve for x in Eq. 2.
$x = 7y + 50$

Substitute $x = 7y + 50$ into Eq. 1.
$(7y + 50)^2 + y^2 = 100$
$49y^2 + 700y + 2500 + y^2 = 100$
$50y^2 + 700y + 2400 = 0$
$50(y^2 + 14y + 48) = 0$
$50(y + 8)(y + 6) = 0$
$y = -8$ or $y = -6$

Find the x-coordinates using $x = 7y + 50$.

$y = -8$ $y = -6$
$x = 7(-8) + 50$ $x = 7(-6) + 50$
$x = -56 + 50$ $x = -42 + 50$
$x = -6$ $x = 8$
$(-6, -8), (8, -6)$

5. $x^2 + y^2 = 85$ (Eq. 1)
$3x + y = -29$ (Eq. 2)

Solve for y in Eq. 2.
$y = -3x - 29$

Substitute $y = -3x - 29$ into Eq. 1.
$$x^2 + (-3x - 29)^2 = 85$$
$$x^2 + 9x^2 + 174x + 841 = 85$$
$$10x^2 + 174x + 756 = 0$$
$$2(5x^2 + 87x + 378) = 0$$
$$2(5x + 42)(x + 9) = 0$$
$5x + 42 = 0$ or $x + 9 = 0$
$x = -\dfrac{42}{5}$ or $x = -9$

Find the y-coordinates using $y = -3x - 29$.

$x = -\dfrac{42}{5}$ $x = -9$

$y = -3\left(-\dfrac{42}{5}\right) - 29$ $y = -3(-9) - 29$

$y = \dfrac{126}{5} - \dfrac{145}{5}$ $y = 27 - 29$

$y = -\dfrac{19}{5}$ $y = -2$

$\left(-\dfrac{42}{5}, -\dfrac{19}{5}\right), (-9, -2)$

7. $4x^2 + y^2 = 16$ (Eq. 1)
$y = 2x - 4$ (Eq. 2)

Substitute $y = 2x - 4$ into Eq. 1.
$$4x^2 + (2x - 4)^2 = 16$$
$$4x^2 + 4x^2 - 16x + 16 = 16$$
$$8x^2 - 16x = 0$$
$$8x(x - 2) = 0$$
$x = 0$ or $x = 2$

Find the y-coordinates using $y = 2x - 4$.
$x = 0$ $x = 2$
$y = 2(0) - 4$ $y = 2(2) - 4$
$y = -4$ $y = 0$
$(0, -4), (2, 0)$

9. $y = x^2 - 3x + 8$ (Eq. 1)
$-2x + y = 4$ (Eq. 2)

Substitute $y = x^2 - 3x + 8$ into Eq. 2.
$$-2x + (x^2 - 3x + 8) = 4$$
$$x^2 - 5x + 8 = 4$$
$$x^2 - 5x + 4 = 0$$
$$(x - 4)(x - 1) = 0$$
$x = 4$ or $x = 1$

Find the y-coordinates using
$y = x^2 - 3x + 8$.
$x = 4$ $x = 1$
$y = (4)^2 - 3(4) + 8$ $y = (1)^2 - 3(1) + 8$
$y = 12$ $y = 6$
$(4, 12), (1, 6)$

11. $x^2 - y^2 = 8$ (Eq. 1)
$y = x - 2$ (Eq. 2)

Substitute $y = x - 2$ into Eq. 1.
$$x^2 - (x - 2)^2 = 8$$
$$x^2 - (x^2 - 4x + 4) = 8$$
$$x^2 - x^2 + 4x - 4 = 8$$
$$4x = 12$$
$$x = 3$$
Find the y-coordinate using $y = x - 2$.
$y = 3 - 2 = 1$
$(3, 1)$

13. $y = 3x - 9$ (Eq. 1)
$x^2 + y^2 = 4$ (Eq. 2)

Substitute $y = 3x - 9$ into Eq. 2
$$x^2 + (3x - 9)^2 = 4$$
$$x^2 + 9x^2 - 54x + 81 = 4$$
$$10x^2 - 54x + 77 = 0$$

$$x = \frac{-(-54) \pm \sqrt{(-54)^2 - 4(10)(77)}}{2(10)}$$

$$x = \frac{54 \pm \sqrt{-164}}{20}$$
undefined
$\varnothing$

15. $x^2 + y^2 = 25$ (Eq. 1)
$y = x^2 - 5$ (Eq. 2)

Solve for x^2 in Eq. 2.
$x^2 = y + 5$

Substitute $x^2 = y + 5$ into Eq. 1.
$$(y + 5) + y^2 = 25$$
$$y^2 + y - 20 = 0$$
$$(y + 5)(y - 4) = 0$$
$y = -5$ or $y = 4$

Find the x-coordinates using $x^2 = y + 5$.
$y = -5$ $y = 4$
$x^2 = (-5) + 5$ $x^2 = (4) + 5$
$x^2 = 0$ $x^2 = 9$
$x = 0$ $x = \pm 3$
$(0, -5), (-3, 4), (3, 4)$

17. $x^2 + y^2 = 30$ (Eq. 1)
$x = y^2$ (Eq. 2)

Substitute $y^2 = x$ into Eq. 1.
$$x^2 + x = 30$$
$$x^2 + x - 30 = 0$$
$$(x+6)(x-5) = 0$$
$$x = -6 \text{ or } x = 5$$

Find the y-coordinates using $x = y^2$.
$$x = -6 \qquad\qquad x = 5$$
$$-6 = y^2 \qquad\qquad 5 = y^2$$
$$\pm\sqrt{-6} = y \qquad\quad \pm\sqrt{5} = y$$
undefined
$$\left(5, -\sqrt{5}\right), \left(5, \sqrt{5}\right)$$

19. $x^2 + 4y^2 = 4$ (Eq. 1)
$y = x^2 + 1$ (Eq. 2)

Solve for x^2 in Eq. 2.
$$x^2 = y - 1$$

Substitute $x^2 = y - 1$ into Eq. 1.
$$(y-1) + 4y^2 = 4$$
$$4y^2 + y - 5 = 0$$
$$(4y+5)(y-1) = 0$$
$$4y + 5 = 0 \quad \text{or} \quad y - 1 = 0$$
$$4y = -5 \quad \text{or} \quad y = 1$$
$$y = -\frac{5}{4}$$

Find the x-coordinates using $x^2 = y - 1$.
$$y = -\frac{5}{4} \qquad\qquad y = 1$$
$$x^2 = -\frac{5}{4} - 1 \qquad x^2 = (1) - 1$$
$$x^2 = -\frac{9}{4} \qquad\qquad x^2 = 0$$
$$x = \pm\sqrt{-\frac{9}{4}} \qquad x = 0$$
undefined
$$(0, 1)$$

21. $x^2 - y^2 = 1$ (Eq. 1)
$x = y^2 + 11$ (Eq. 2)

Solve for y^2 in Eq. 2.
$$y^2 = x - 11$$

Substitute $y^2 = x - 11$ into Eq. 1.
$$x^2 - (x-11) = 1$$
$$x^2 - x + 11 = 1$$
$$x^2 - x + 10 = 0$$
$$x = \frac{-(-1) \pm \sqrt{(-1)^2 - 4(1)(10)}}{2(1)}$$
$$x = \frac{1 \pm \sqrt{-39}}{2}$$
undefined
$$\varnothing$$

23. $x^2 - 4y^2 = 16$ (Eq. 1)
$x = y^2 - 11$ (Eq. 2)

Solve for y^2 in Eq. 2.
$$y^2 = x + 11$$

Substitute $y^2 = x + 11$ into Eq. 1.
$$x^2 - 4(x+11) = 16$$
$$x^2 - 4x - 44 = 16$$
$$x^2 - 4x - 60 = 0$$
$$(x+6)(x-10) = 0$$
$$x = -6 \text{ or } x = 10$$

Find the y-coordinates using $y^2 = x + 11$.
$$x = -6 \qquad\qquad x = 10$$
$$y^2 = (-6) + 11 \qquad y^2 = (10) + 11$$
$$y^2 = 5 \qquad\qquad y^2 = 21$$
$$y = \pm\sqrt{5} \qquad\qquad y = \pm\sqrt{21}$$
$$\left(-6, -\sqrt{5}\right), \left(-6, \sqrt{5}\right), \left(10, -\sqrt{21}\right), \left(10, \sqrt{21}\right)$$

25. $y = x^2 - 7$ (Eq. 1)
$y = -x^2 + 11$ (Eq. 2)

Substitute $y = -x^2 + 11$ into Eq. 1.
$$-x^2 + 11 = x^2 - 7$$
$$0 = 2x^2 - 18$$
$$0 = 2\left(x^2 - 9\right)$$
$$0 = 2(x+3)(x-3)$$
$$x = -3 \text{ or } x = 3$$

Find the y-coordinates using $y = x^2 - 7$.
$$x = -3 \qquad\qquad x = 3$$
$$y = (-3)^2 - 7 \qquad y = (3)^2 - 7$$
$$y = 2 \qquad\qquad y = 2$$
$$\left(-3, 2\right), (3, 2)$$

27. $y = x^2 - 6$ (Eq. 1)
$x^2 + (y-4)^2 = 8$ (Eq. 2)

Solve for x^2 in Eq. 1.
$x^2 = y + 6$

Substitute $x^2 = y + 6$ into Eq. 2.
$$y + 6 + (y-4)^2 = 8$$
$$y + 6 + y^2 - 8y + 16 = 8$$
$$y^2 - 7y + 14 = 0$$
$$y = \frac{-(-7) \pm \sqrt{(-7)^2 - 4(1)(14)}}{2(1)}$$
$$y = \frac{7 \pm \sqrt{-7}}{2}$$
undefined
$\varnothing$

29. $x^2 + y^2 = 5$ (Eq. 1)
$x^2 - y^2 = 3$ (Eq. 2)

$$\begin{array}{ll} x^2 + y^2 = 5 & \text{(Eq. 1)} \\ + \; x^2 - y^2 = 3 & \text{(Eq. 2)} \\ \hline 2x^2 = 8 \\ x^2 = 4 \\ x = \pm 2 \end{array}$$

Find the y-coordinates using $x^2 + y^2 = 5$.

$$\begin{array}{ll} x = -2 & x = 2 \\ (-2)^2 + y^2 = 5 & (2)^2 + y^2 = 5 \\ 4 + y^2 = 5 & 4 + y^2 = 5 \\ y^2 = 1 & y^2 = 1 \\ y = \pm 1 & y = \pm 1 \end{array}$$
$(-2,-1), (-2,1), (2,-1), (2,1)$

31. $3x^2 + y^2 = 14$ (Eq. 1)
$x^2 - y^2 = 2$ (Eq. 2)

$$\begin{array}{ll} 3x^2 + y^2 = 14 & \text{(Eq. 1)} \\ + \; x^2 - y^2 = 2 & \text{(Eq. 2)} \\ \hline 4x^2 = 16 \\ x^2 = 4 \\ x = \pm 2 \end{array}$$

Find the y-coordinates using
$3x^2 + y^2 = 14$.

$$\begin{array}{ll} x = -2 & x = 2 \\ 3(-2)^2 + y^2 = 14 & 3(2)^2 + y^2 = 14 \\ 12 + y^2 = 14 & 12 + y^2 = 14 \\ y^2 = 2 & y^2 = 2 \\ y = \pm\sqrt{2} & y = \pm\sqrt{2} \end{array}$$
$\left(-2,-\sqrt{2}\right), \left(-2,\sqrt{2}\right), \left(2,-\sqrt{2}\right), \left(2,\sqrt{2}\right)$

33. $x^2 + 2y^2 = 18$ (Eq. 1)
$y^2 - x^2 = 6$ (Eq. 2)

$$\begin{array}{ll} x^2 + 2y^2 = 18 & \text{(Eq. 1)} \\ + \; -x^2 + y^2 = 6 & \text{(Eq. 2)} \\ \hline 3y^2 = 24 \\ y^2 = 8 \\ y = \pm\sqrt{8} \\ y = \pm 2\sqrt{2} \end{array}$$

Find the x-coordinates using $x^2 + y^2 = 18$.

$$\begin{array}{ll} y = -2\sqrt{2} & y = 2\sqrt{2} \\ x^2 + 2\left(-2\sqrt{2}\right)^2 = 18 & x^2 + 2\left(2\sqrt{2}\right)^2 = 18 \\ x^2 + 16 = 18 & x^2 + 16 = 18 \\ x^2 = 2 & x^2 = 2 \\ x = \pm\sqrt{2} & x = \pm\sqrt{2} \end{array}$$
$\left(-\sqrt{2},-2\sqrt{2}\right), \left(\sqrt{2},-2\sqrt{2}\right), \left(-\sqrt{2},2\sqrt{2}\right),$
$\left(\sqrt{2},2\sqrt{2}\right)$

35. $9x^2 + 4y^2 = 87$ (Eq. 1)
$x^2 - 2y^2 = 6$ (Eq. 2)

$$\begin{array}{ll} 9x^2 + 4y^2 = 87 & \text{(Eq. 1)} \\ + \; 2x^2 - 4y^2 = 12 & (2 \cdot \text{Eq. 2}) \\ \hline 11x^2 = 99 \\ x^2 = 9 \\ x = \pm 3 \end{array}$$

Find the y-coordinates using
$9x^2 + 4y^2 = 87$.

$$\begin{array}{ll} x = -3 & x = 3 \\ 9(-3)^2 + 4y^2 = 87 & 9(3)^2 + 4y^2 = 87 \\ 81 + 4y^2 = 87 & 81 + 4y^2 = 87 \\ 4y^2 = 6 & 4y^2 = 6 \\ y^2 = \dfrac{6}{4} & y^2 = \dfrac{6}{4} \\ y^2 = \dfrac{3}{2} & y^2 = \dfrac{3}{2} \\ y = \pm\sqrt{\dfrac{3}{2}} & y = \pm\sqrt{\dfrac{3}{2}} \\ y = \pm\dfrac{\sqrt{3}}{\sqrt{2}} & y = \pm\dfrac{\sqrt{3}}{\sqrt{2}} \\ y = \pm\dfrac{\sqrt{6}}{2} & y = \pm\dfrac{\sqrt{6}}{2} \end{array}$$
$\left(-3,-\dfrac{\sqrt{6}}{2}\right), \left(-3,\dfrac{\sqrt{6}}{2}\right), \left(3,-\dfrac{\sqrt{6}}{2}\right), \left(3,\dfrac{\sqrt{6}}{2}\right)$

37. $x^2 + y^2 = 24$ (Eq. 1)
$x^2 + 5y^2 = 60$ (Eq. 2)

$$
\begin{aligned}
x^2 + y^2 &= 24 \quad &&\text{(Eq. 1)} \\
+\; -x^2 - 5y^2 &= -60 \quad &&(-1 \cdot \text{Eq. 2}) \\
\hline
-4y^2 &= -36 \\
y^2 &= 9 \\
y &= \pm 3
\end{aligned}
$$

Find the x-coordinates using $x^2 + y^2 = 24$.

$y = -3$ $\qquad\qquad$ $y = 3$
$x^2 + (-3)^2 = 24$ $\qquad$ $x^2 + (3)^2 = 24$
$\qquad x^2 = 15$ $\qquad\qquad\quad x^2 = 15$
$\qquad\quad x = \pm\sqrt{15}$ $\qquad\qquad x = \pm\sqrt{15}$

$\left(-\sqrt{15}, -3\right), \left(\sqrt{15}, -3\right), \left(-\sqrt{15}, 3\right), \left(\sqrt{15}, 3\right)$

39. $x^2 + y^2 = 6$ (Eq. 1)
$4x^2 + 9y^2 = 39$ (Eq. 2)

$$
\begin{aligned}
-4x^2 - 4y^2 &= -24 \quad &&(-4 \cdot \text{Eq. 1}) \\
+\; 4x^2 + 9y^2 &= 39 \quad &&\text{(Eq. 2)} \\
\hline
5y^2 &= 15 \\
y^2 &= 3 \\
y &= \pm\sqrt{3}
\end{aligned}
$$

Find the x-coordinates using $x^2 + y^2 = 6$.

$y = -\sqrt{3}$ $\qquad\qquad$ $y = \sqrt{3}$
$x^2 + \left(-\sqrt{3}\right)^2 = 6$ $\qquad$ $x^2 + \left(\sqrt{3}\right)^2 = 6$
$\qquad x^2 = 3$ $\qquad\qquad\qquad x^2 = 3$
$\qquad\quad x = \pm\sqrt{3}$ $\qquad\qquad\quad x = \pm\sqrt{3}$

$\left(\sqrt{3}, -\sqrt{3}\right), \left(-\sqrt{3}, -\sqrt{3}\right), \left(\sqrt{3}, \sqrt{3}\right), \left(-\sqrt{3}, \sqrt{3}\right)$

41. $x^2 + 12y^2 = 117$ (Eq. 1)
$x^2 + 4y^2 = 45$ (Eq. 2)

$$
\begin{aligned}
x^2 + 12y^2 &= 117 \quad &&\text{(Eq. 1)} \\
+\; -x^2 - 4y^2 &= -45 \quad &&(-1 \cdot \text{Eq. 2}) \\
\hline
8y^2 &= 72 \\
y^2 &= 9 \\
y &= \pm 3
\end{aligned}
$$

Find the x-coordinates using $x^2 + 4y^2 = 45$.

$y = -3$ $\qquad\qquad$ $y = 3$
$x^2 + 4(-3)^2 = 45$ $\qquad$ $x^2 + 4(3)^2 = 45$
$\qquad\quad x^2 = 9$ $\qquad\qquad\qquad x^2 = 9$
$\qquad\quad\; x = \pm 3$ $\qquad\qquad\quad x = \pm 3$

$(-3, -3), (3, -3), (-3, 3), (3, 3)$

43. $3x^2 + 2y^2 = 21$ (Eq. 1)
$15x^2 + 4y^2 = 51$ (Eq. 2)

$$
\begin{aligned}
-6x^2 - 4y^2 &= -42 \quad &&(-2 \cdot \text{Eq. 1}) \\
+\; 15x^2 + 4y^2 &= 51 \quad &&\text{(Eq. 2)} \\
\hline
9x^2 &= 9 \\
x^2 &= 1 \\
x &= \pm 1
\end{aligned}
$$

Find the y-coordinates using $3x^2 + 2y^2 = 21$.

$x = -1$ $\qquad\qquad$ $x = 1$
$3(-1)^2 + 2y^2 = 21$ $\qquad$ $3(1)^2 + 2y^2 = 21$
$\qquad\quad 2y^2 = 18$ $\qquad\qquad\quad 2y^2 = 18$
$\qquad\quad\; y^2 = 9$ $\qquad\qquad\qquad y^2 = 9$
$\qquad\qquad y = \pm 3$ $\qquad\qquad\quad y = \pm 3$

$(-1, -3), (-1, 3), (1, -3), (1, 3)$

45. Unknowns :
Length: x
Width: y

$2x + 2y = 44$ (Perimeter)
$x \cdot y = 105$ (Area)

Solve the perimeter equation for y.
$2y = -2x + 44$
$y = -x + 22$

Substitute into the area equation.
$x(-x + 22) = 105$
$-x^2 + 22x = 105$
$\qquad 0 = x^2 - 22x + 105$
$\qquad 0 = (x - 7)(x - 15)$
$x = 7$ or $x = 15$

Solve for the width (y) using $y = -x + 22$.
$x = 7$ $\qquad\qquad$ $x = 15$
$y = -(7) + 22$ $\qquad$ $y = -(15) + 22$
$y = 15$ $\qquad\qquad$ $y = 7$
Dimensions are 7 feet by 15 feet.

47. Unknowns:
Length: x
Width: y

$2x + 2y = 82$ (Perimeter)
$x \cdot y = 400$ (Area)

Solve the perimeter equation for y.
$2y = -2x + 82$
$y = -x + 41$

Substitute into the area equation.

$x(-x + 41) = 400$
$-x^2 + 41x = 400$
$0 = x^2 - 41x + 400$
$0 = (x - 16)(x - 25)$
$x = 16$ or $x = 25$

Solve for the width (y) using $y = -x + 41$.

$x = 16$ $x = 25$
$y = -(16) + 41$ $y = -(25) + 41$
$y = 25$ $y = 16$
Dimensions are 16 feet by 25 feet.

49. Unknowns:
Length: x
Width: y

$2x + 2y - 4 = 60$ (Perimeter)
$x \cdot y = 192$ (Area)

Solve the perimeter equation for x.
$2x + 2y = 64$
$2x = -2y + 64$
$x = -y + 32$

Substitute into the area equation.

$(-y + 32)y = 192$
$-y^2 + 32y = 192$
$0 = y^2 - 32y + 192$
$0 = (y - 8)(y - 24)$
$y = 8$ or $y = 24$

Solve for the length (x) using $x = -y + 32$.

$y = 8$ $y = 24$
$x = -(8) + 32$ $x = -(24) + 32$
$x = 24$ $x = 8$
Dimensions are 8 feet by 24 feet.

51. Unknowns:
Length: x
Width: y

$2x + 2y = 210$ (Perimeter)
$x^2 + y^2 = 75^2$ (Diagonal)

Solve the perimeter equation for x.
$2x + 2y = 210$
$2x = -2y + 210$
$x = -y + 105$

Substitute into the diagonal equation.

$(-y + 105)^2 + y^2 = 75^2$
$y^2 - 210y + 11{,}025 + y^2 = 5625$
$2y^2 - 210y + 5400 = 0$
$2(y^2 - 105y + 2700) = 0$
$2(y - 45)(y - 60) = 0$
$y = 45$ or $y = 60$

Solve for the length (x) using $x = -y + 105$.

$y = 45$ $y = 60$
$x = -(45) + 105$ $x = -(60) + 105$
$x = 60$ $x = 45$
Dimensions are 45 feet by 60 feet.

53. Unknowns:
Length: x
Width: y

$x \cdot y = 800$ (Area)
$x^2 + y^2 = \left(20\sqrt{5}\right)^2$ (Diagonal)

Solve the area equation for x.

$x = \dfrac{800}{y}$

Substitute into the diagonal equation.

$$\left(\frac{800}{y}\right)^2 + y^2 = \left(20\sqrt{5}\right)^2$$

$$\frac{640{,}000}{y^2} + y^2 = 2000$$

$$y^2 \cdot \left(\frac{640{,}000}{y^2} + y^2\right) = y^2 \cdot 2000$$

$$640{,}000 + y^4 = 2000y^2$$

$y^4 - 2000y^2 + 640{,}000 = 0$
$\left(y^2 - 400\right)\left(y^2 - 1600\right) = 0$

$y^2 - 400 = 0$ or $y^2 - 1600 = 0$
$y^2 = 400$ or $y^2 = 1600$
$y = \pm 20$ or $y = \pm 40$

Since the width and the length
cannot be negative, $y = 20$ or $y = 40$.

Solve for the length (x) using $x = \dfrac{800}{y}$.

$y = 20$ $y = 40$
$x = \dfrac{800}{20}$ $x = \dfrac{800}{40}$
$x = 40$ $x = 20$
Dimensions are 20 feet by 40 feet.

55. $h(t) = -16t^2 + v_o t + s$

$h(t) = -16t^2 + 120t$

$h(t) = -16t^2 + 540$

$-16t^2 + 120t = -16t^2 + 540$

$120t = 540$

$t = 4.5$

4.5 seconds

57. a) $h(t) = -16t^2 + v_o t + s$

$h(t) = -16t^2 + 60t$

$h(t) = -16t^2 + 32t + 84$

$-16t^2 + 60t = -16t^2 + 32t + 84$

$60t = 32t + 84$

$28t = 84$

$t = 3$

3 seconds

b) In 3 seconds, $t = 3$.

$h(t) = -16t^2 + 60t$

$h(3) = -16(3)^2 + 60(3)$

$\quad = -16(9) + 60(3)$

$\quad = -144 + 180$

$\quad = 36$

36 feet high

59. Answers will vary. Example:

Find the dimensions of a rectangle whose area is 1500 square feet and whose perimeter is 160 feet.

CHAPTER 13 REVIEW

1. $y = x^2 - 8x - 9$

Use the form $y = ax^2 + bx + c$.

$a > 0,$ parabola opens upward.

vertex:

$x = \dfrac{-b}{2a} = -\dfrac{-8}{2(1)} = 4$

$y = (4)^2 - 8(4) - 9 = 16 - 32 - 9 = -25$

$(4, -25)$

y-intercept:

$y = (0)^2 - 8(0) - 9 = -9$

$(0, -9)$

x-intercept:

$0 = x^2 - 8x - 9$

$0 = (x + 1)(x - 9)$

$x = -1$ or $x = 9$

$(-1, 0), (9, 0)$

axis of symmetry: $x = 4$

point symmetric to the y-intercept: $(8, -9)$

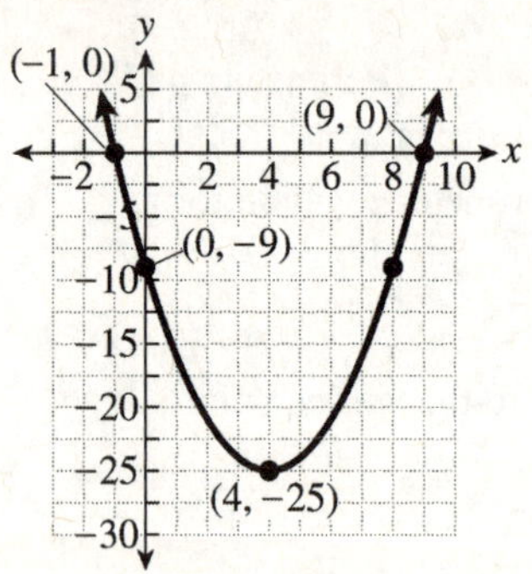

2. $y = -x^2 + 6x - 3$

Use the form $y = ax^2 + bx + c$.

$a < 0,$ parabola opens downward.

vertex:

$x = \dfrac{-b}{2a} = \dfrac{-6}{2(-1)} = 3$

$y = -(3)^2 + 6(3) - 3 = -9 + 18 - 3 = 6$

$(3, 6)$

y-intercept:

$y = -(0)^2 + 6(0) - 3 = -3$

$(0, -3)$

Continued on next page.

2. Continued.

x-intercept:

$0 = -x^2 + 6x - 3$

$x^2 - 6x + 3 = 0$

$x = \dfrac{-(-6) \pm \sqrt{(-6)^2 - 4(1)(3)}}{2(1)}$

$x = \dfrac{6 \pm \sqrt{24}}{2}$

$x = \dfrac{6 \pm 2\sqrt{6}}{2}$

$x = 3 \pm \sqrt{6}$

$\left(3 - \sqrt{6}, 0\right), \left(3 + \sqrt{6}, 0\right)$

axis of symmetry: $x = 3$

point symmetric to the y-intercept: $(6, -3)$

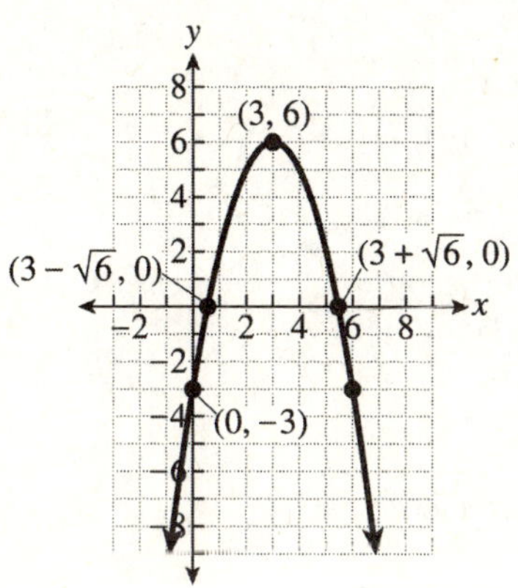

3. $y = (x - 4)^2 - 1$

Use the form $y = a(x - h)^2 + k$.

$a > 0$, parabola opens upward.

vertex: $(4, -1)$

Shift the graph of $y = x^2$ by 4 units to the right and by 1 unit down.

y-intercept:

$y = (0 - 4)^2 - 1 = 16 - 1 = 15$

$(0, 15)$

x-intercept:

$0 = (x - 4)^2 - 1$

$1 = (x - 4)^2$

$\pm\sqrt{1} = x - 4$

$4 \pm 1 = x$

$x = 3$ or $x = 5$

$(3, 0), (5, 0)$

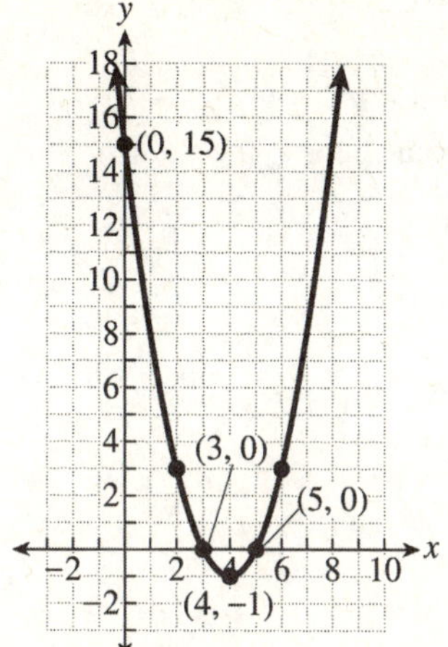

4. $y = -(x + 3)^2 - 4$ or $y = -\left(x - (-3)\right)^2 - 4$

Use the form $y = a(x - h)^2 + k$.

$a < 0$, parabola opens downward, so rotate the graph of $y = x^2$ about the x-axis.

vertex: $(-3, -4)$

Shift the rotated graph by 3 units to the left and by 4 units down.

y-intercept:

$y = -(0 + 3)^2 - 4 = -9 - 4 = -13$

$(0, -13)$

x-intercept:

The vertex is below the x-axis and the parabola opens downward. There are no x-intercepts.

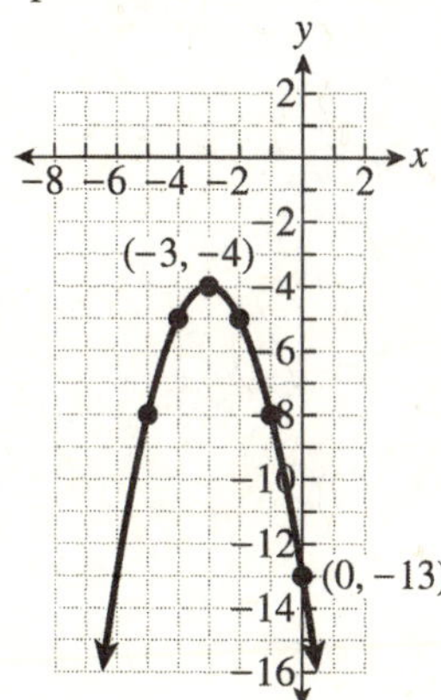

5. $x = y^2 - 10y + 15$

Use the form $x = ay^2 + by + c$.

$a > 0$, parabola opens to the right.

vertex:

$$y = \frac{-b}{2a} = \frac{-(-10)}{2(1)} = 5$$

$$x = (5)^2 - 10(5) + 15 = 25 - 50 + 15 = -10$$
$(-10, 5)$

x-intercept:
$$x = (0)^2 - 10(0) + 15 = 15$$
$(15, 0)$

y-intercept:
$$0 = y^2 - 10y + 15$$

$$y = \frac{-(-10) \pm \sqrt{(-10)^2 - 4(1)(15)}}{2(1)}$$

$$y = \frac{10 \pm \sqrt{40}}{2}$$

$$y = \frac{10 \pm 2\sqrt{10}}{2}$$

$$y = 5 \pm \sqrt{10}$$
$\left(0, 5 - \sqrt{10}\right), \left(0, 5 + \sqrt{10}\right)$

axis of symmetry: $y = 5$

point symmetric to the x-intercept: $(15, 10)$

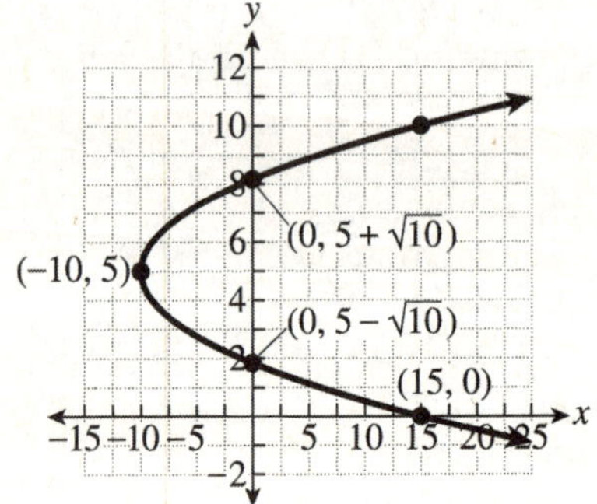

6. $x = y^2 + 3y - 28$

Use the form $x = ay^2 + by + c$.

$a > 0$, parabola opens to the right.

vertex:

$$y = \frac{-b}{2a} = \frac{-3}{2(1)} = -\frac{3}{2}$$

$$x = \left(-\frac{3}{2}\right)^2 + 3\left(-\frac{3}{2}\right) - 28$$

$$= \frac{9}{4} - \frac{9}{2} - \frac{28}{1}$$

$$= \frac{9}{4} - \frac{18}{4} - \frac{112}{4}$$

$$= -\frac{121}{4}$$

$\left(-\frac{121}{4}, -\frac{3}{2}\right)$

x-intercept:
$$x = (0)^2 + 3(0) - 28 = -28$$
$(-28, 0)$

y-intercept:
$$0 = y^2 + 3y - 28$$
$$0 = (y + 7)(y - 4)$$
$$y = -7 \quad \text{or} \quad y = 4$$
$(0, -7), (0, 4)$

axis of symmetry: $y = -\frac{3}{2}$

point symmetric to the x-intercept: $(-28, -3)$

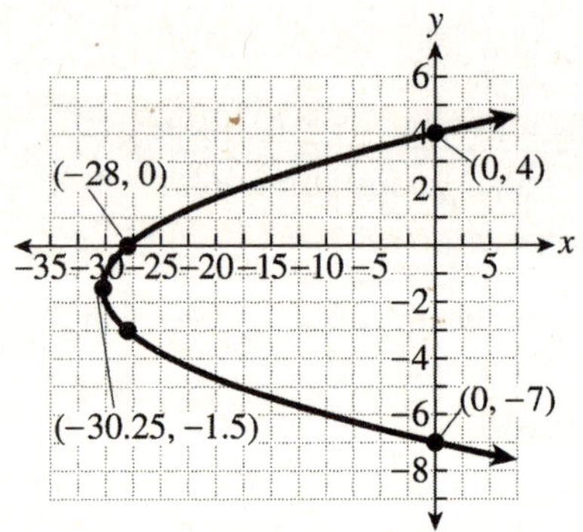

7. $x = -y^2 + y + 20$

Use the form $x = ay^2 + by + c$.

$a < 0,$ parabola opens to the left.

vertex:

$$y = \frac{-b}{2a} = \frac{-1}{2(-1)} = \frac{1}{2}$$

$$x = -\left(\frac{1}{2}\right)^2 + \frac{1}{2} + 20$$

$$= -\frac{1}{4} + \frac{1}{2} + 20$$

$$= -\frac{1}{4} + \frac{2}{4} + \frac{80}{4}$$

$$= \frac{81}{4}$$

$$\left(\frac{81}{4}, \frac{1}{2}\right)$$

x-intercept:

$x = -(0)^2 + (0) + 20 = 20$
$(20, 0)$

y-intercept:

$0 = -y^2 + y + 20$
$y^2 - y - 20 = 0$
$(y + 4)(y - 5) = 0$
$y = -4$ or $y = 5$
$(0, -4), (0, 5)$

axis of symmetry: $y = \dfrac{1}{2}$

point symmetric to the x-intercept: $(20, 1)$

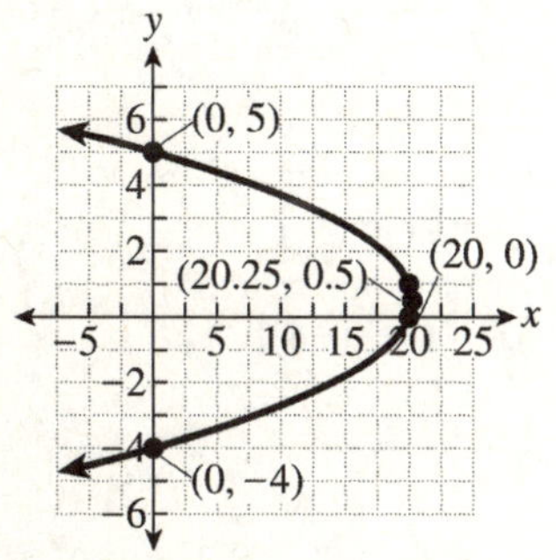

8. $x = -y^2 - 6y - 14$

Use the form $x = ay^2 + by + c$.

$a < 0,$ parabola opens to the left.

vertex:

$$y = \frac{-b}{2a} = -\frac{(-6)}{2(-1)} = -3$$

$x = -(-3)^2 - 6(-3) - 14 = -9 + 18 - 14 = -5$
$(-5, -3)$

x-intercept:

$x = -(0)^2 - 6(0) - 14 = -14$
$(-14, 0)$

y-intercept:
The vertex is to the left of the y-axis and the parabola opens to the left. There are no y-intercepts.

axis of symmetry: $y = -3$

point symmetric to the x-intercept: $(-14, -6)$

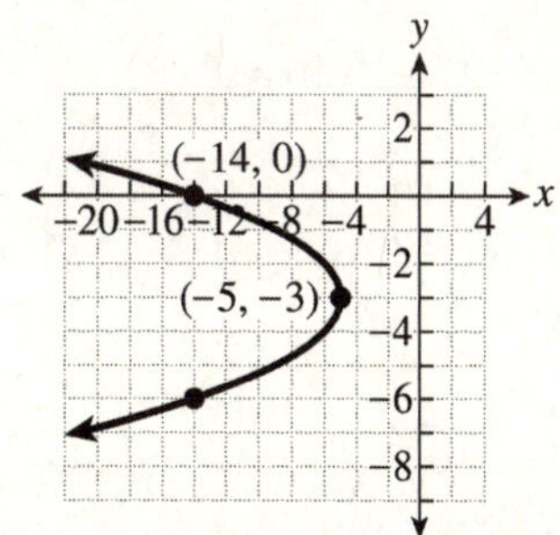

9. $x = (y + 5)^2 - 4$ or $x = \left(y - (-5)\right)^2 - 4$

Use the form $x = a(y - k)^2 + h$.

$a > 0,$ parabola opens to the right.

vertex: $(-4, -5)$

Shift the graph of $x = y^2$ by 4 units to the left and by 5 units down.

x-intercept:

$x = (0 + 5)^2 - 4 = 25 - 4 = 21$
$(21, 0)$

y-intercept:

$$0 = (y + 5)^2 - 4$$
$$4 = (y + 5)^2$$
$$\pm 2 = y + 5$$
$$-5 \pm 2 = y$$
$y = -7$ or $y = -3$
$(0, -7)(0, -3)$

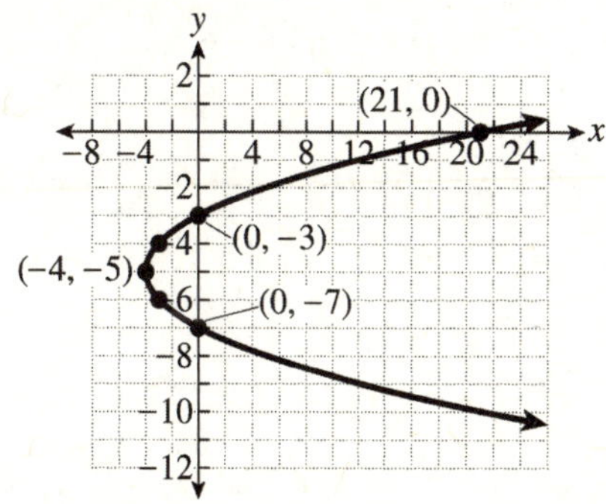

10. $x = -(y+1)^2 + 10$ or $x = -(y-(-1))^2 + 10$

Use the form $x = a(y-k)^2 + h$.

$a < 0$, parabola opens to the left, so rotate the graph of $x = y^2$ about the y-axis.

vertex: $(10, -1)$

Shift the rotated graph by 10 units to the right and by 1 unit down.

x-intercept:

$x = -(0+1)^2 + 10 = -1 + 10 = 9$

$(9, 0)$

y-intercept:

$$0 = -(y+1)^2 + 10$$
$$(y+1)^2 = 10$$
$$y + 1 = \pm\sqrt{10}$$
$$y = -1 \pm \sqrt{10}$$
$$\left(0, -1-\sqrt{10}\right), \left(0, -1+\sqrt{10}\right)$$

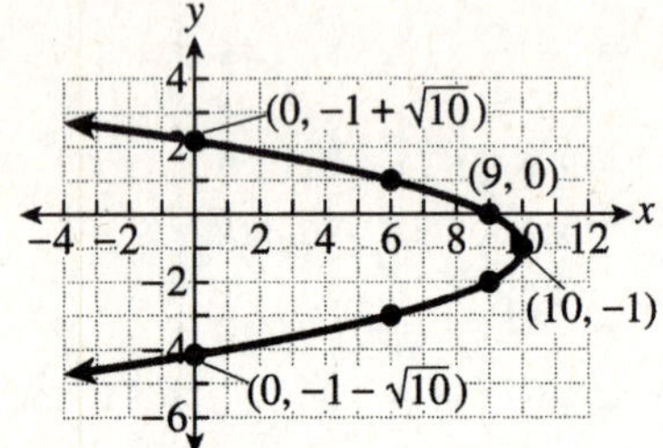

11. $x = -(y-4)^2 + 9$

Use the form $x = a(y-k)^2 + h$.

$a < 0$, parabola opens to the left, so rotate the graph of $x = y^2$ about the y-axis.

vertex: $(9, 4)$

Shift the rotated graph by 9 units to the right and by 4 units up.

x-intercept:

$x = -(0-4)^2 + 9 = -16 + 9 = -7$

$(-7, 0)$

y-intercept:

$$0 = -(y-4)^2 + 9$$
$$9 = (y-4)^2$$
$$\pm 3 = y - 4$$
$$4 \pm 3 = y$$
$$y = 1 \text{ or } y = 7$$
$$(0, 1), (0, 7)$$

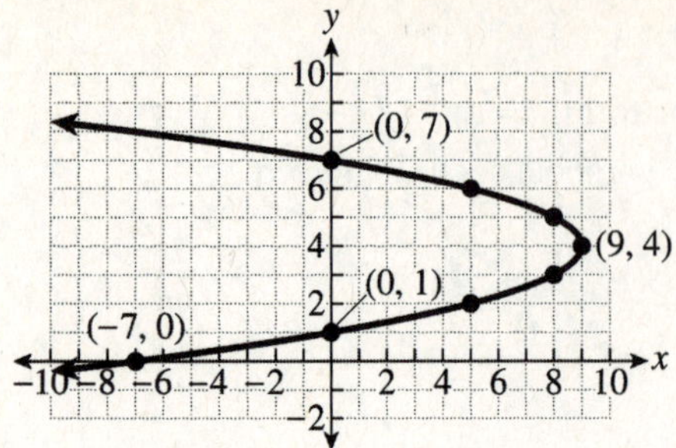

12. $x = (y-3)^2 + 6$

Use the form $x = a(y-k)^2 + h$.

$a > 0$, parabola opens to the right.

vertex: $(6, 3)$

Shift the graph of $x = y^2$ by 6 units to the right and by 3 units up.

x-intercept:

$x = (0-3)^2 + 6 = 9 + 6 = 15$

$(15, 0)$

y-intercept:

The vertex is to the right of the y-axis and the parabola opens to the right. There are no y-intercepts.

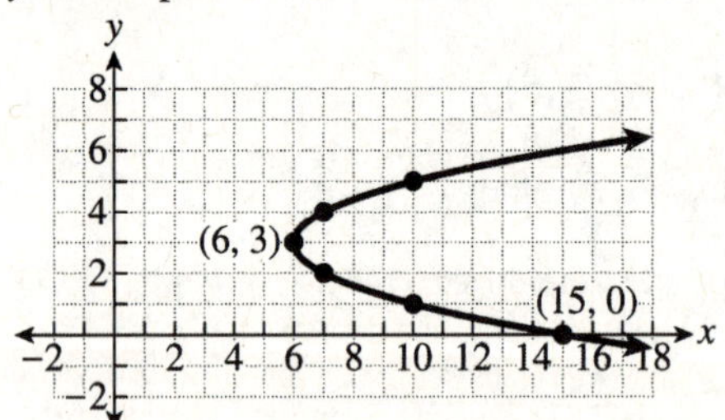

13. $d = \sqrt{(x_2 - x_1)^2 + (y_2 - y_1)^2}$

$d = \sqrt{(1-(-7))^2 + (7-(-8))^2}$

$d = \sqrt{64 + 225}$

$d = \sqrt{289}$

$d = 17$

14. $d = \sqrt{(x_2 - x_1)^2 + (y_2 - y_1)^2}$

$d = \sqrt{(-6-3)^2 + (-2-(-5))^2}$

$d = \sqrt{81 + 9}$

$d = \sqrt{90}$

$d \approx 9.5$

15. $x^2 + y^2 = 36$
$x^2 + y^2 = 6^2$

Use the form $x^2 + y^2 = r^2$.

Center: $(0,0)$; radius: $r = 6$

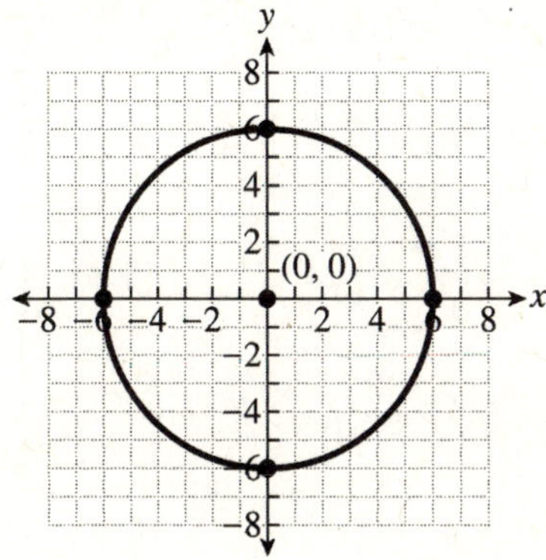

16. $x^2 + y^2 = 20$
$x^2 + y^2 = \left(\sqrt{20}\right)^2$

Use the form $x^2 + y^2 = r^2$.

Center: $(0,0)$; radius: $r = \sqrt{20} = 2\sqrt{5}$

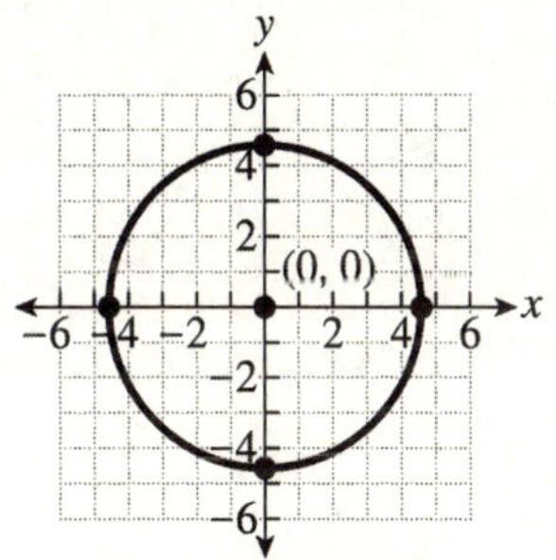

17. $(x+3)^2 + (y-4)^2 = 4$
$(x-(-3))^2 + (y-4)^2 = 2^2$

Use the form $(x-h)^2 + (y-k)^2 = r^2$.

$h = -3$, $k = 4$, and $r = 2$.

Center: $(-3,4)$ and radius: $r = 2$

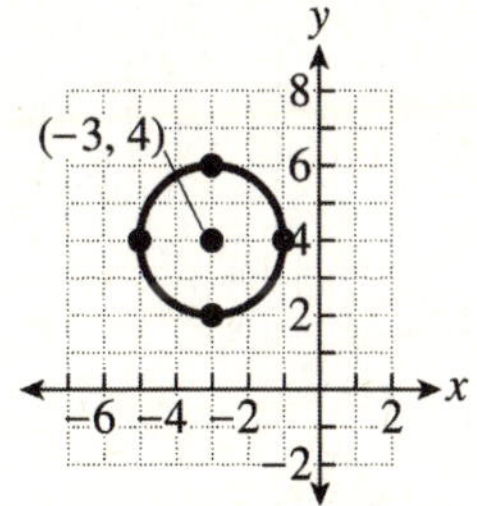

18. $(x-8)^2 + (y+3)^2 = 9$
$(x-8)^2 + (y-(-3))^2 = 3^2$

Use the form $(x-h)^2 + (y-k)^2 = r^2$.

$h = 8$, $k = -3$, and $r = 3$.

Center: $(8,-3)$; radius: $r = 3$

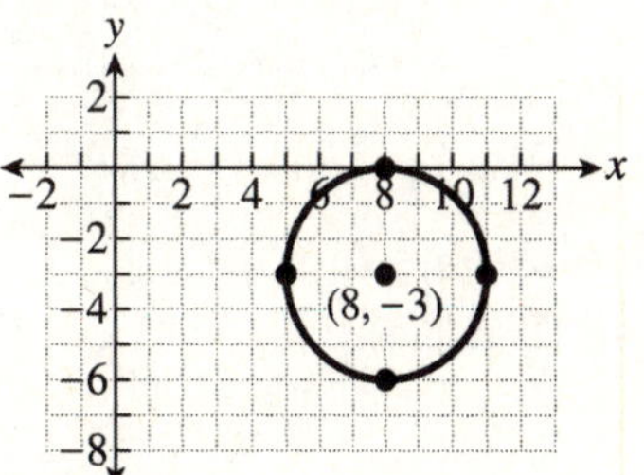

19. $(x+4)^2 + (y-5)^2 = 25$
$(x-(-4))^2 + (y-5)^2 = 5^2$

Use the form $(x-h)^2 + (y-k)^2 = r^2$.

$h = -4$, $k = 5$, and $r = 5$.

Center: $(-4,5)$; radius: $r = 5$

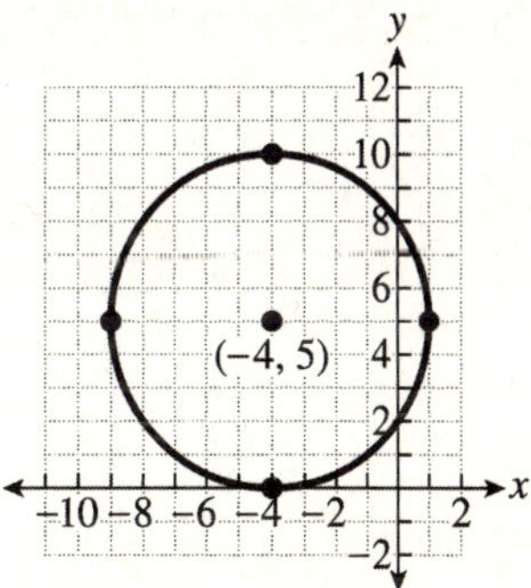

20. $(x-2)^2 + y^2 = 16$
$(x-2)^2 + (y-0)^2 = 4^2$

Use the form $(x-h)^2 + (y-k)^2 = r^2$.

$h = 2$, $k = 0$, and $r = 4$.

Center: $(2,0)$; radius: $r = 4$

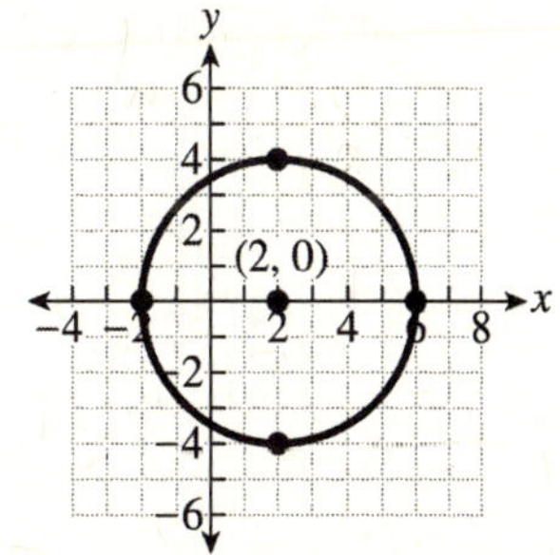

21.
$$x^2 + y^2 + 12x - 2y - 12 = 0$$
$$x^2 + 12x + y^2 - 2y = 12$$
$$\left(x^2 + 12x + \left(\frac{12}{2}\right)^2\right) + \left(y^2 - 2y + \left(-\frac{2}{2}\right)^2\right) = 12 + \left(\frac{12}{2}\right)^2 + \left(-\frac{2}{2}\right)^2$$
$$\left(x^2 + 12x + 36\right) + \left(y^2 - 2y + +1\right) = 12 + 36 + 1$$
$$(x + 6)^2 + (y - 1)^2 = 49$$
$$(x - (-6))^2 + (y - 1)^2 = 7^2$$

Use the form $(x - h)^2 + (y - k)^2 = r^2$.

$h = -6$, $k = 1$, and $r = 7$.

Center: $(-6, 1)$; radius: $r = 7$

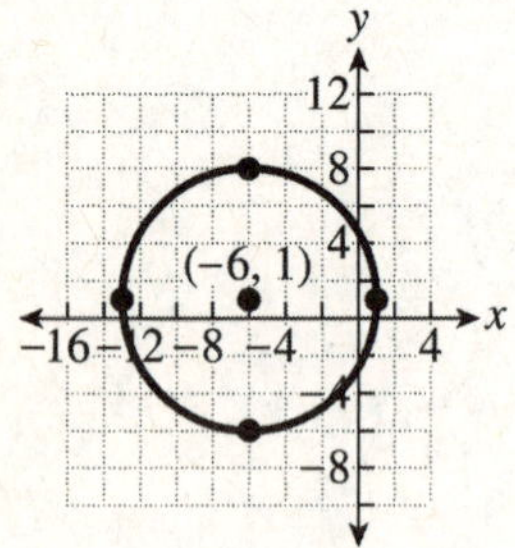

22.
$$x^2 + y^2 + 6x - 18y - 10 = 0$$
$$x^2 + 6x + y^2 - 18y = 10$$
$$\left(x^2 + 6x + \left(\frac{6}{2}\right)^2\right) + \left(y^2 - 18y + \left(-\frac{18}{2}\right)^2\right) = 10 + \left(\frac{6}{2}\right)^2 + \left(-\frac{18}{2}\right)^2$$
$$\left(x^2 + 6x + 9\right) + \left(y^2 - 18y + 81\right) = 10 + 9 + 81$$
$$(x + 3)^2 + (y - 9)^2 = 100$$
$$(x - (-3))^2 + (y - 9)^2 = 10^2$$

Use the form $(x - h)^2 + (y - k)^2 = r^2$.

$h = -3$, $k = 9$, and $r = 10$.

Center: $(-3, 9)$; radius: $r = 10$

23.
$$x^2 + y^2 - 8x - 20y + 99 = 0$$
$$x^2 - 8x + y^2 - 20y = -99$$
$$\left(x^2 - 8x + \left(-\frac{8}{2}\right)^2\right) + \left(y^2 - 20y + \left(-\frac{20}{2}\right)^2\right) = -99 + \left(-\frac{8}{2}\right)^2 + \left(-\frac{20}{2}\right)^2$$
$$\left(x^2 - 8x + 16\right) + \left(y^2 - 20y + 100\right) = -99 + 16 + 100$$
$$(x - 4)^2 + (y - 10)^2 = 17$$
$$(x - 4)^2 + (y - 10)^2 = \left(\sqrt{17}\right)^2$$

Use the form $(x - h)^2 + (y - k)^2 = r^2$.

$h = 4$, $k = 10$, and $r = \sqrt{17}$.

Center: $(4, 10)$; radius: $r = \sqrt{17}$

24. Center: $(0,0)$; radius: $r = 3$

$$x^2 + y^2 = 3^2$$
$$x^2 + y^2 = 9$$

25. Center: $(-2,4)$; radius: $r = 5$

$$\left(x-(-2)\right)^2 + (y-4)^2 = 5^2$$
$$(x+2)^2 + (y-4)^2 = 25$$

26. $\dfrac{x^2}{25} + \dfrac{y^2}{9} = 1$

$$\dfrac{x^2}{5^2} + \dfrac{y^2}{3^2} = 1$$

Use the form $\dfrac{x^2}{a^2} + \dfrac{y^2}{b^2} = 1$.

Center: $(0,0)$

$a = 5$ and $b = 3$

x-intercepts: $(-5,0),(5,0)$
y-intercepts: $(0,-3),(0,3)$

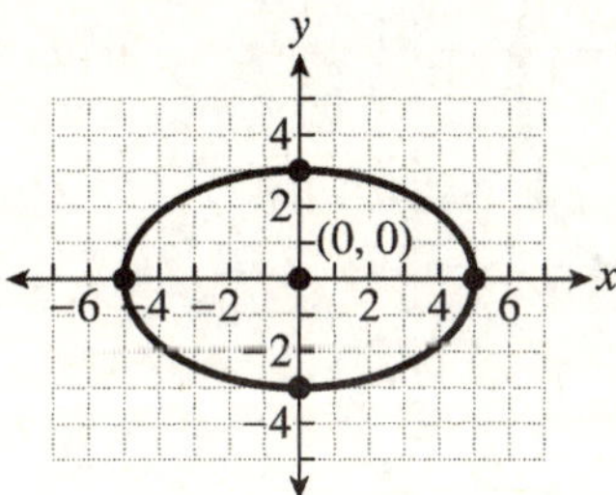

27. $x^2 + \dfrac{y^2}{16} = 1$

$$\dfrac{x^2}{1^2} + \dfrac{y^2}{4^2} = 1$$

Use the form $\dfrac{x^2}{a^2} + \dfrac{y^2}{b^2} = 1$.

Center: $(0,0)$

$a = 1$ and $b = 4$

The ellipse has a vertical major axis.
Center: $(0,0)$

x-intercepts: $(-1,0),(1,0)$
y-intercepts: $(0,-4),(0,4)$

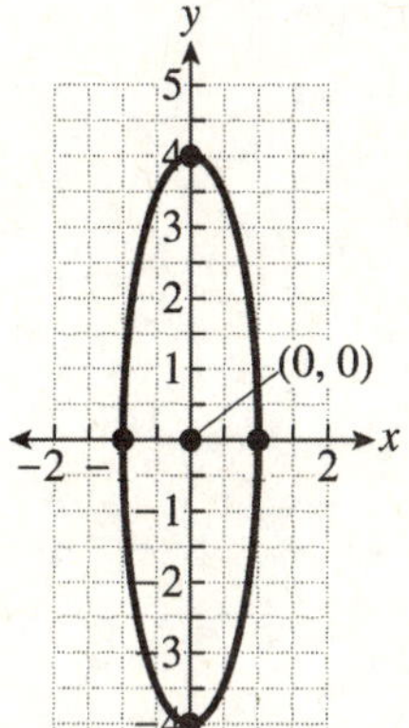

28. $\dfrac{(x+3)^2}{9}+\dfrac{(y-4)^2}{16}=1$

$\dfrac{(x-(-3))^2}{3^2}+\dfrac{(y-4)^2}{4^2}=1$

Use the form $\dfrac{(x-h)^2}{a^2}+\dfrac{(y-k)^2}{b^2}=1$.

Center: $(-3,4)$

$a=3$ and $b=4$

The endpoints of the horizontal axis are 3 units to the left and right of the center.

$(-6,4),(0,4)$

The endpoints of the vertical axis are 4 units above and below the center.

$(-3,0),(-3,8)$

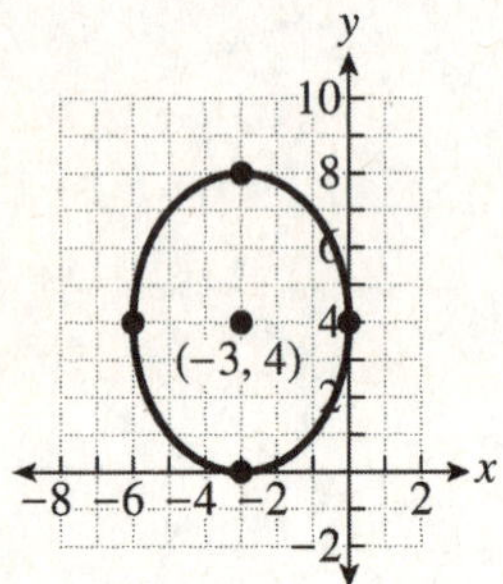

30. $\dfrac{(x+6)^2}{9}+\dfrac{y^2}{36}=1$

$\dfrac{(x-(-6))^2}{3^2}+\dfrac{(y-0)^2}{6^2}=1$

Use the form $\dfrac{(x-h)^2}{a^2}+\dfrac{(y-k)^2}{b^2}=1$.

Center: $(-6,0)$

$a=3$ and $b=6$

The endpoints of the horizontal axis are 3 units to the left and right of the center.

$(-9,0),(-3,0)$

The endpoints of the vertical axis are 6 units above and below the center.

$(-6,-6),(-6,6)$

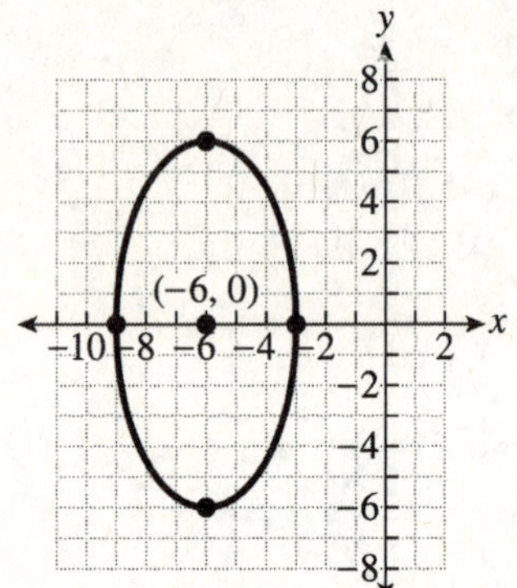

29. $\dfrac{(x-8)^2}{25}+\dfrac{(y-1)^2}{4}=1$

$\dfrac{(x-8)^2}{5^2}+\dfrac{(y-1)^2}{2^2}=1$

Use the form $\dfrac{(x-h)^2}{a^2}+\dfrac{(y-k)^2}{b^2}=1$.

Center: $(8,1)$

$a=5$ and $b=2$

The endpoints of the horizontal axis are 5 units to the left and right of the center.

$(3,1),(13,1)$

The endpoints of the vertical axis are 2 units above and below the center.

$(8,-1),(8,3)$

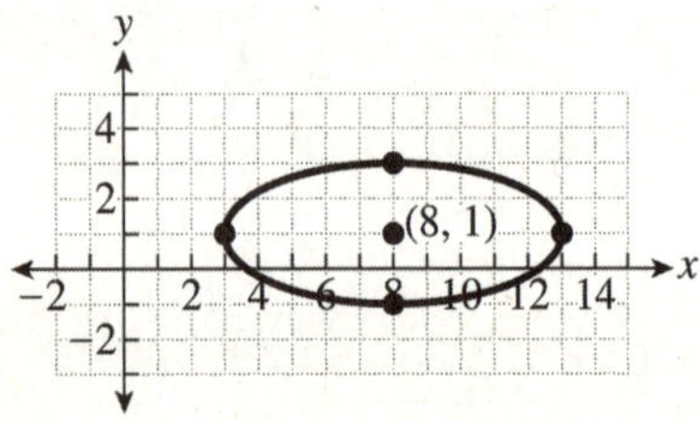

31.
$$9x^2 + 16y^2 + 90x - 64y + 145 = 0$$
$$9x^2 + 90x + 16y^2 - 64y = -145$$
$$9\left(x^2 + 10x\right) + 16\left(y^2 - 4y\right) = -145$$
$$9\left(x^2 + 10x + \left(\frac{10}{2}\right)^2\right) + 16\left(y^2 - 4y + \left(-\frac{4}{2}\right)^2\right) = -145 + 9\left(\frac{10}{2}\right)^2 + 16\left(-\frac{4}{2}\right)^2$$
$$9\left(x^2 + 10x + 25\right) + 16\left(y^2 - 4y + 4\right) = -145 + 225 + 64$$
$$9(x+5)^2 + 16(y-2)^2 = 144$$
$$\frac{9(x+5)^2}{144} + \frac{16(y-2)^2}{144} = \frac{144}{144}$$
$$\frac{(x+5)^2}{16} + \frac{(y-2)^2}{9} = 1$$
$$\frac{\left(x-(-5)\right)^2}{4^2} + \frac{(y-2)^2}{3^2} = 1$$

Use the form $\dfrac{(x-h)^2}{a^2} + \dfrac{(y-k)^2}{b^2} = 1$.

Center: $(-5, 2)$

$a = 4$ and $b = 3$

The endpoints of the horizontal axis are 4 units to the left and right of the center.

$(-9, 2), (-1, 2)$

The endpoints of the vertical axis are 3 units above and below the center.

$(-5, -1), (-5, 5)$

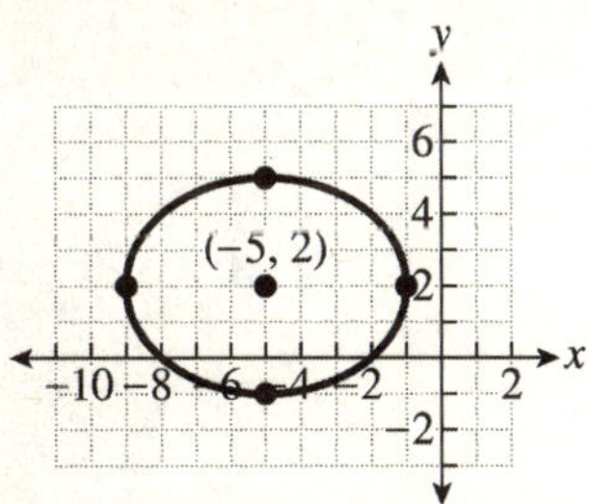

32.

$$4x^2 + 49y^2 - 48x - 98y - 3 = 0$$

$$4x^2 - 48x + 49y^2 - 98y = 3$$

$$4\left(x^2 - 12x\right) + 49\left(y^2 - 2y\right) = 3$$

$$4\left(x^2 - 12x + \left(-\frac{12}{2}\right)^2\right) + 49\left(y^2 - 2y + \left(-\frac{2}{2}\right)^2\right) = 3 + 4\left(-\frac{12}{2}\right)^2 + 49\left(-\frac{2}{2}\right)^2$$

$$4\left(x^2 - 12x + 36\right) + 49\left(y^2 - 2y + 1\right) = 3 + 144 + 49$$

$$4(x-6)^2 + 49(y-1)^2 = 196$$

$$\frac{4(x-6)^2}{196} + \frac{49(y-1)^2}{196} = \frac{196}{196}$$

$$\frac{(x-6)^2}{49} + \frac{(y-1)^2}{4} = 1$$

$$\frac{(x-6)^2}{7^2} + \frac{(y-1)^2}{2^2} = 1$$

Use the form $\dfrac{(x-h)^2}{a^2} + \dfrac{(y-k)^2}{b^2} = 1$.

Center: $(6,1)$; $a = 7$ and $b = 2$

33.

$$81x^2 + 16y^2 + 324x + 128y - 716 = 0$$

$$81x^2 + 324x + 16y^2 + 128y = 716$$

$$81\left(x^2 + 4x\right) + 16\left(y^2 + 8y\right) = 716$$

$$81\left(x^2 + 4x + \left(\frac{4}{2}\right)^2\right) + 16\left(y^2 + 8y + \left(\frac{8}{2}\right)^2\right) = 716 + 81\left(\frac{4}{2}\right)^2 + 16\left(\frac{8}{2}\right)^2$$

$$81\left(x^2 + 4x + 4\right) + 16\left(y^2 + 8y + 16\right) = 716 + 324 + 256$$

$$81(x+2)^2 + 16(y+4)^2 = 1296$$

$$\frac{81(x+2)^2}{1296} + \frac{16(y+4)^2}{1296} = \frac{1296}{1296}$$

$$\frac{(x+2)^2}{16} + \frac{(y+4)^2}{81} = 1$$

$$\frac{(x-(-2))^2}{4^2} - \frac{(y-(-4))^2}{9^2} = 1$$

Use the form $\dfrac{(x-h)^2}{a^2} + \dfrac{(y-k)^2}{b^2} = 1$.

Center: $(-2,-4)$; $a = 4$ and $b = 9$

34. Center: $(0,0)$ with $a = 2$, and $b = 5$

$$\frac{x^2}{2^2} + \frac{y^2}{5^2} = 1$$

$$\frac{x^2}{4} + \frac{y^2}{25} = 1$$

35. Center: $(-1,4)$ with $a = 8$, and $b = 3$

$$\frac{(x-(-1))^2}{8^2} + \frac{(y-4)^2}{3^2} = 1$$

$$\frac{(x+1)^2}{64} + \frac{(y-4)^2}{9} = 1$$

36. $\dfrac{x^2}{9} - \dfrac{y^2}{25} = 1$

$$\frac{x^2}{3^2} - \frac{y^2}{5^2} = 1$$

The hyperbola is in the form $\dfrac{x^2}{a^2} - \dfrac{y^2}{b^2} = 1$.

The hyperbola opens to the left and to the right where $a = 3$, $b = 5$.

Center: $(0,0)$

Vertices: $(-3,0)$, $(3,0)$

Asymptotes: $y = \pm\dfrac{b}{a}x$

$$y = \pm\frac{5}{3}x$$

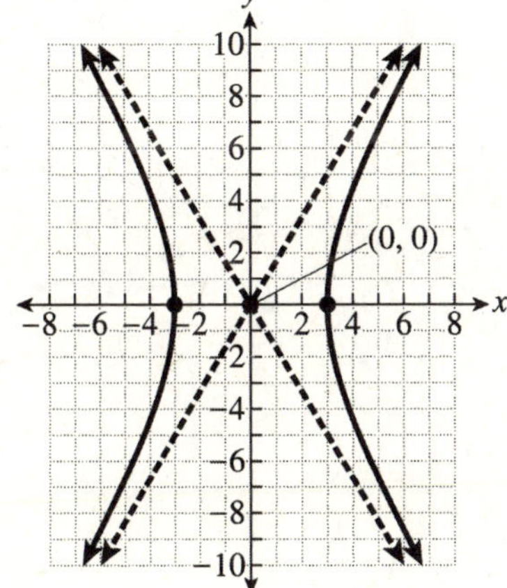

37. $\dfrac{y^2}{16} - \dfrac{x^2}{16} = 1$

$$\frac{y^2}{4^2} - \frac{x^2}{4^2} = 1$$

The hyperbola is in the form $\dfrac{y^2}{b^2} - \dfrac{x^2}{a^2} = 1$.

The hyperbola opens upward and downward where $a = 4$, $b = 4$.

Center: $(0,0)$

Vertices: $(-4,0)$, $(4,0)$

Asymptotes: $y = \pm\dfrac{b}{a}x$

$$y = \pm\frac{4}{4}x$$

$$y = \pm x$$

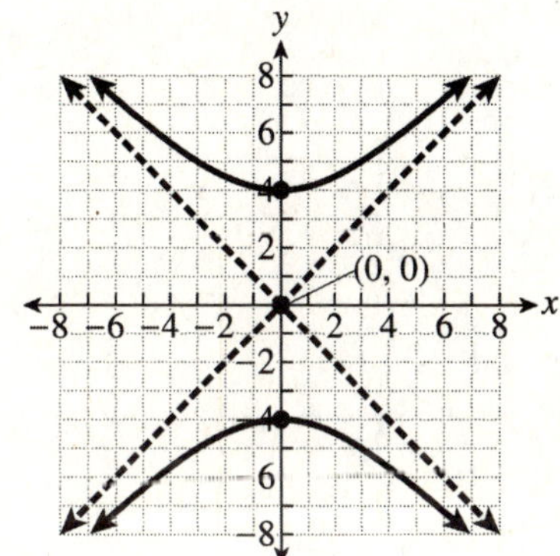

38.
$$\frac{(x-1)^2}{4} - \frac{(y-3)^2}{9} = 1$$

$$\frac{(x-1)^2}{2^2} - \frac{(y-3)^2}{3^2} = 1$$

The hyperbola is in the form

$$\frac{(x-h)^2}{a^2} - \frac{(y-k)^2}{b^2} = 1.$$

The hyperbola opens to the left and to the right where $h = 1$, $k = 3$, $a = 2$, and $b = 3$.

Center: $(1, 3)$

Vertices: $(-1, 3)$, $(3, 3)$

Asymptotes: $y = \pm \dfrac{b}{a}(x-h) + k$

$$y = \pm \frac{3}{2}(x-1) + 3$$

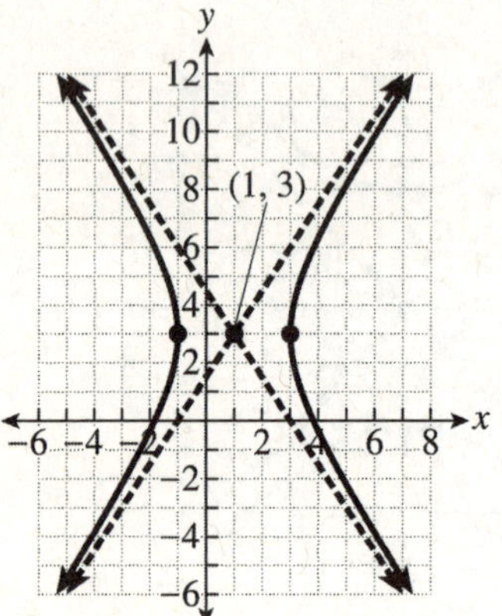

39.
$$\frac{(x+4)^2}{25} - \frac{(y-2)^2}{36} = 1$$

$$\frac{(x-(-4))^2}{5^2} - \frac{(y-2)^2}{6^2} = 1$$

The hyperbola is in the form

$$\frac{(x-h)^2}{a^2} - \frac{(y-k)^2}{b^2} = 1.$$

The hyperbola opens to the left and to the right where $h = -4$, $k = 2$, $a = 5$, and $b = 6$.

Center: $(-4, 2)$

Vertices: $(-9, 2)$, $(1, 2)$

Asymptotes: $y = \pm \dfrac{b}{a}(x-h) + k$

$$y = \pm \frac{6}{5}(x+4) + 2$$

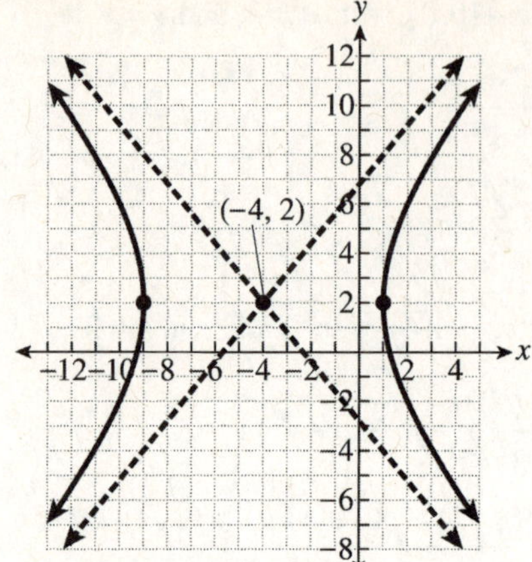

40.
$$\frac{(y+3)^2}{4} - \frac{(x+2)^2}{81} = 1$$

$$\frac{(y-(-3))^2}{2^2} - \frac{(x-(-2))^2}{9^2} = 1$$

The hyperbola is in the form

$$\frac{(y-k)^2}{b^2} - \frac{(x-h)^2}{a^2} = 1.$$

The hyperbola opens upward and downward where $h = -2$, $k = -3$, $a = 9$, and $b = 2$.

Center: $(-2, -3)$

Vertices: $(-2, -5)$, $(-2, -1)$

Asymptotes: $y = \pm \dfrac{b}{a}(x-h) + k$

$$y = \pm \frac{2}{9}(x+2) - 3$$

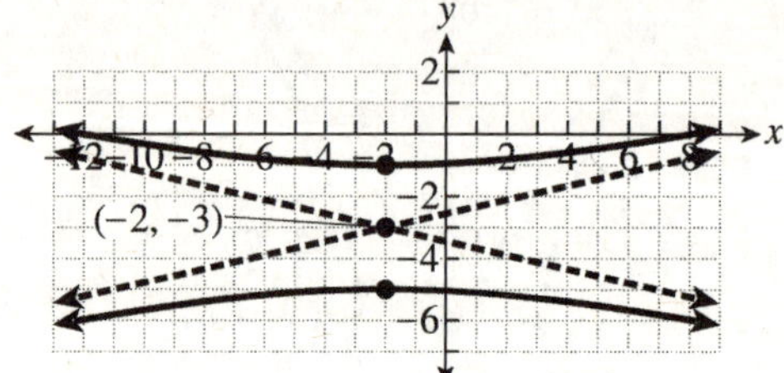

41.

$$-4x^2 + 25y^2 - 24x + 300y + 764 = 0$$

$$25y^2 + 300y - 4x^2 - 24x = -764$$

$$25\left(y^2 + 12y\right) - 4\left(x^2 + 6x\right) = -764$$

$$25\left(y^2 + 12y + \left(\frac{12}{2}\right)^2\right) - 4\left(x^2 + 6x + \left(\frac{6}{2}\right)^2\right) = -764 + 25\left(\frac{12}{2}\right)^2 - 4\left(\frac{6}{2}\right)^2$$

$$25\left(y^2 + 12y + 36\right) - 4\left(x^2 + 6x + 9\right) = -764 + 900 - 36$$

$$25(y+6)^2 - 4(x+3)^2 = 100$$

$$\frac{25(x+6)^2}{100} - \frac{4(x+3)^2}{100} = \frac{100}{100}$$

$$\frac{(y+6)^2}{4} - \frac{(x+3)^2}{25} = 1$$

$$\frac{\left(y-(-6)\right)^2}{2^2} - \frac{\left(x-(-3)\right)^2}{5^2} = 1$$

The hyperbola is in the form $\dfrac{(y-k)^2}{b^2} - \dfrac{(x-h)^2}{a^2} = 1$.

The hyperbola opens upward and downward where $h = -3$, $k = -6$, $a = 5$, and $b = 2$.

Center: $(-3, -6)$

Vertices: $(-3, -8)$, $(-3, -4)$

Asymptotes: $y = \pm\dfrac{b}{a}(x-h) + k$

$$y = \pm\frac{2}{5}(x+3) - 6$$

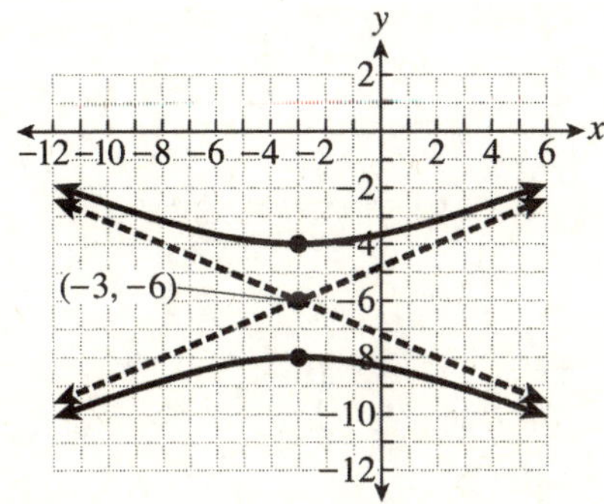

42.

$$-9x^2 + 4y^2 + 90x - 56y - 65 = 0$$

$$4y^2 - 56y - 9x^2 + 90x = 65$$

$$4\left(y^2 - 14y\right) - 9\left(x^2 - 10x\right) = 65$$

$$4\left(y^2 - 14y + \left(-\frac{14}{2}\right)^2\right) - 9\left(x^2 - 10x + \left(-\frac{10}{2}\right)^2\right) = 65 + 4\left(-\frac{14}{2}\right)^2 - 9\left(-\frac{10}{2}\right)^2$$

$$4\left(y^2 - 14y + 49\right) - 9\left(x^2 - 10x + 25\right) = 65 + 196 - 225$$

$$4(y - 7)^2 - 9(x - 5)^2 = 36$$

$$\frac{4(y - 7)^2}{36} - \frac{9(x - 5)^2}{36} = \frac{36}{36}$$

$$\frac{(y - 7)^2}{9} - \frac{(x - 5)^2}{4} = 1$$

$$\frac{(y - 7)^2}{3^2} - \frac{(x - 5)^2}{2^2} = 1$$

The hyperbola is in the form $\dfrac{(y - k)^2}{b^2} - \dfrac{(x - h)^2}{a^2} = 1$.

The hyperbola opens upward and downward where $h = 5$, $k = 7$, $a = 2$, and $b = 3$.

Center: $(5, 7)$

Transverse axis: $2b = 2(3) = 6$

Conjugate axis: $2a = 2(2) = 4$

43.

$$49x^2 - 36y^2 + 784x + 288y + 796 = 0$$

$$49x^2 + 784x - 36y^2 + 288y = -796$$

$$49\left(x^2 + 16x\right) - 36\left(y^2 - 8y\right) = -796$$

$$49\left(x^2 + 16x + \left(\frac{16}{2}\right)^2\right) - 36\left(y^2 - 8y + \left(-\frac{8}{2}\right)^2\right) = -796 + 49\left(\frac{16}{2}\right)^2 - 36\left(-\frac{8}{2}\right)^2$$

$$49\left(x^2 + 16x + 64\right) - 36\left(y^2 - 8y + 16\right) = -796 + 3136 - 576$$

$$49(x + 8)^2 - 36(y - 4)^2 = 1764$$

$$\frac{49(x + 8)^2}{1764} - \frac{36(y - 4)^2}{1764} = \frac{1764}{1764}$$

$$\frac{(x + 8)^2}{36} - \frac{(y - 4)^2}{49} = 1$$

$$\frac{(x - (-8))^2}{6^2} - \frac{(y - 4)^2}{7^2} = 1$$

The hyperbola is in the form $\dfrac{(x - h)^2}{a^2} - \dfrac{(y - k)^2}{b^2} = 1$.

The hyperbola opens to the left and to the right where $h = -8$, $k = 4$, $a = 6$, and $b = 7$.

Center: $(h, k) = (-8, 4)$

Transverse Axis: $2a = 2(6) = 12$

Conjugate Axis: $2b = 2(7) = 14$

44. Center: $(0,0)$ with $a = 2$, and $b = 5$

$$\frac{x^2}{2^2} - \frac{y^2}{5^2} = 1$$

$$\frac{x^2}{4} - \frac{y^2}{25} = 1$$

45. Center: $(-1,2)$ with $a = 4$, and $b = 3$

$$\frac{(y-2)^2}{3^2} - \frac{(x-(-1))^2}{4^2} = 1$$

$$\frac{(y-2)^2}{9} - \frac{(x+1)^2}{16} = 1$$

46. $x^2 + y^2 = 50$ (Eq. 1)
$x + 2y = 15$ (Eq. 2)

Solve for x in Eq. 2.
$x = -2y + 15$

Substitute $x = -2y + 15$ into Eq. 1.
$$(-2y+15)^2 + y^2 = 50$$
$$4y^2 - 60y + 225 + y^2 = 50$$
$$5y^2 - 60y + 175 = 0$$
$$5(y^2 - 12y + 35) = 0$$
$$5(y-5)(y-7) = 0$$
$y = 5$ or $y = 7$

Find the x-coordinates using $x = -2y + 15$.

$y = 5$ $\qquad\qquad$ $y = 7$
$x = -2(5) + 15$ $\qquad$ $x = -2(7) + 15$
$x = 5$ $\qquad\qquad$ $x = 1$
$(5,5), (1,7)$

47. $4x^2 + 3y^2 = 48$ (Eq. 1)
$y = 2x - 4$ (Eq. 2)

Substitute $y = 2x - 4$ into Eq. 1.
$$4x^2 + 3(2x-4)^2 = 48$$
$$4x^2 + 3(4x^2 - 16x + 16) = 48$$
$$4x^2 + 12x^2 - 48x + 48 = 48$$
$$16x^2 - 48x = 0$$
$$16x(x-3) = 0$$
$x = 0$ or $x = 3$

Find the y-coordinates using $y = 2x - 4$.

$x = 0$ $\qquad\qquad$ $x = 3$
$y = 2(0) - 4$ $\qquad$ $y = 2(3) - 4$
$y = -4$ $\qquad\qquad$ $y = 2$
$(0,-4), (3,2)$

48. $y = x^2 - 8x + 23$ (Eq. 1)
$2x - y = 1$ (Eq. 2)

Substitute $y = x^2 - 8x + 23$ into Eq. 2.
$$2x - (x^2 - 8x + 23) = 1$$
$$2x - x^2 + 8x - 23 = 1$$
$$-x^2 + 10x - 23 = 1$$
$$0 = x^2 - 10x + 24$$
$$0 = (x-4)(x-6)$$
$x = 4$ or $x = 6$

Find the y-coordinates using $y = x^2 - 8x + 23$.

$x = 4$ $\qquad\qquad$ $x = 6$
$y = (4)^2 - 8(4) + 23$ $\qquad$ $y = (6)^2 - 8(6) + 23$
$y = 7$ $\qquad\qquad$ $y = 11$
$(4,7), (6,11)$

49. $4y^2 - x^2 = 96$ (Eq. 1)
$6y - x = 32$ (Eq. 2)

Solve for x in Eq. 2.
$-x = -6y + 32$
$x = 6y - 32$

Substitute $x = 6y - 32$ into Eq. 1.
$$4y^2 - (6y-32)^2 = 96$$
$$4y^2 - (36y^2 - 384y + 1024) = 96$$
$$4y^2 - 36y^2 + 384y - 1024 = 96$$
$$-32y^2 + 384y - 1120 = 0$$
$$-32(y^2 - 12y + 35) = 0$$
$$-32(y-5)(y-7) = 0$$
$y = 5$ or $y = 7$

Find the x-coordinates using $x = 6y - 32$.

$y = 5$ $\qquad\qquad$ $y = 7$
$x = 6(5) - 32$ $\qquad$ $x = 6(7) - 32$
$x = -2$ $\qquad\qquad$ $x = 10$
$(-2,5), (10,7)$

50. $x^2 + y^2 = 17$ (Eq. 1)
$y = x^2 + 3$ (Eq. 2)

Solve for x^2 in Eq. 2.
$x^2 = y - 3$

Substitute $x^2 = y - 3$ into Eq. 1.
$(y - 3) + y^2 = 17$
$y^2 + y - 20 = 0$
$(y + 5)(y - 4) = 0$
$y = -5$ or $y = 4$

Find the x-coordinates using $x^2 = y - 3$.
$y = -5$ $\qquad$ $y = 4$
$x^2 = (-5) - 3$ $\qquad$ $x^2 = 4 - 3$
$x^2 = -8$ $\qquad$ $x^2 = 1$
$x = \pm\sqrt{-8}$ $\qquad$ $x = \pm\sqrt{1}$
undefined $\qquad$ $x = \pm 1$
$(-1, 4), (1, 4)$

51. $3x^2 + 5y^2 = 72$ (Eq. 1)
$y = x^2 - 6$ (Eq. 2)

Solve for x^2 in Eq. 2.
$x^2 = y + 6$

Substitute $x^2 = y + 6$ into Eq. 1.
$3(y + 6) + 5y^2 = 72$
$3y + 18 + 5y^2 = 72$
$5y^2 + 3y - 54 = 0$
$(5y + 18)(y - 3) = 0$
$5y + 18 = 0$ $\quad$ or $\quad$ $y - 3 = 0$
$y = -\dfrac{18}{5}$ $\quad$ or $\quad$ $y = 3$

Find the x-coordinates using $x^2 = y + 6$.
$y = -\dfrac{18}{5}$ $\qquad$ $y = 3$
$x^2 = \left(-\dfrac{18}{5}\right) + 6$ $\qquad$ $x^2 = 3 + 6$
$x^2 = \dfrac{12}{5}$ $\qquad$ $x^2 = 9$
$x = \pm\sqrt{\dfrac{12}{5}}$ $\qquad$ $x = \pm\sqrt{9}$
$x = \pm\dfrac{2\sqrt{15}}{5}$ $\qquad$ $x = \pm 3$
$\left(-\dfrac{2\sqrt{15}}{5}, -\dfrac{18}{5}\right), \left(\dfrac{2\sqrt{15}}{5}, -\dfrac{18}{5}\right), (-3, 3), (3, 3)$

52. $9x^2 - y^2 = 9$ (Eq. 1)
$y = x^2 - 1$ (Eq. 2)

Solve for x^2 in Eq. 2.
$x^2 = y + 1$

Substitute $x^2 = y + 1$ into Eq. 1.
$9(y + 1) - y^2 = 9$
$9y + 9 - y^2 = 9$
$0 = y^2 - 9y$
$0 = y(y - 9)$
$y = 0$ or $y = 9$

Find the x-coordinates using $x^2 = y + 1$.
$y = 0$ $\qquad$ $y = 9$
$x^2 = 0 + 1$ $\qquad$ $x^2 = 9 + 1$
$x^2 = 1$ $\qquad$ $x^2 = 10$
$x = \pm\sqrt{1}$ $\qquad$ $x = \pm\sqrt{10}$
$x = \pm 1$
$(-1, 0), (1, 0), \left(-\sqrt{10}, 9\right), \left(\sqrt{10}, 9\right)$

53. $y = -x^2 + 2x + 12$ (Eq. 1)
$y = x^2$ (Eq. 2)

Substitute $y = x^2$ into Eq. 1.
$x^2 = -x^2 + 2x + 12$
$2x^2 - 2x - 12 = 0$
$2\left(x^2 - x - 6\right) = 0$
$2(x - 3)(x + 2) = 0$
$x = 3$ or $x = -2$

Find the y-coordinates using $y = x^2$.
$x = 3$ $\qquad$ $x = -2$
$y = (3)^2$ $\qquad$ $y = (-2)^2$
$y = 9$ $\qquad$ $y = 4$
$(3, 9), (-2, 4)$

54. $x^2 + y^2 = 100$ (Eq. 1)
$y^2 - x^2 = 62$ (Eq. 2)

$\quad x^2 + y^2 = 100$ (Eq. 1)
$+\quad \underline{-x^2 + y^2 = 62}$ (Eq. 2)
$ 2y^2 = 162$
$ y^2 = 81$
$ y = \pm 9$

Find the x-coordinates using $x^2 + y^2 = 100$.
$y = -9$ $\qquad$ $y = 9$
$x^2 + (-9)^2 = 100$ $\qquad$ $x^2 + (9)^2 = 100$
$x^2 + 81 = 100$ $\qquad$ $x^2 + 81 = 100$
$x^2 = 19$ $\qquad$ $x^2 = 19$
$x = \pm\sqrt{19}$ $\qquad$ $x = \pm\sqrt{19}$
$\left(-\sqrt{19}, -9\right), \left(\sqrt{19}, -9\right), \left(-\sqrt{19}, 9\right), \left(\sqrt{19}, 9\right)$

55. $3x^2 + 4y^2 = 31$ (Eq. 1)
$x^2 - 2y^2 = 7$ (Eq. 2)

$$\begin{array}{rl} 3x^2 + 4y^2 = 31 & \text{(Eq. 1)} \\ + \ 2x^2 - 4y^2 = 14 & (2 \cdot \text{Eq. 2}) \\ \hline 5x^2 = 45 & \\ x^2 = 9 & \\ x = \pm 3 & \end{array}$$

Find the y-coordinates using $x^2 - 2y^2 = 7$.

$$\begin{array}{ll} x = -3 & x = 3 \\ (-3)^2 - 2y^2 = 7 & (3)^2 - 2y^2 = 7 \\ 9 - 2y^2 = 7 & 9 - 2y^2 = 7 \\ -2y^2 = -2 & -2y^2 = -2 \\ y^2 = 1 & y^2 = 1 \\ y = \pm 1 & y = \pm 1 \end{array}$$

$(-3, -1), (-3, 1), (3, -1), (3, 1)$

56. $x^2 + y^2 = 8$ (Eq. 1)
$2x^2 + 15y^2 = 68$ (Eq. 2)

$$\begin{array}{rl} -2x^2 - 2y^2 = -16 & (-2 \cdot \text{Eq. 1}) \\ + \ 2x^2 + 15y^2 = 68 & \text{(Eq. 2)} \\ \hline 13y^2 = 52 & \\ y^2 = 4 & \\ y = \pm 2 & \end{array}$$

Find the x-coordinates using $x^2 + y^2 = 8$.

$$\begin{array}{ll} y = -2 & y = 2 \\ x^2 + (-2)^2 = 8 & x^2 + (2)^2 = 8 \\ x^2 = 4 & x^2 = 4 \\ x = \pm 2 & x = \pm 2 \end{array}$$

$(-2, -2), (2, -2), (-2, 2), (2, 2)$

57. $6x^2 + y^2 = 159$ (Eq. 1)
$2x^2 + 9y^2 = 131$ (Eq. 2)

$$\begin{array}{rl} 6x^2 + y^2 = 159 & \text{(Eq. 1)} \\ + \ -6x^2 - 27y^2 = -393 & (-3 \cdot \text{Eq. 2}) \\ \hline -26y^2 = -234 & \\ y^2 = 9 & \\ y = \pm 3 & \end{array}$$

Find the x-coordinates using $2x^2 + 9y^2 = 131$.

$$\begin{array}{ll} y = -3 & y = 3 \\ 2x^2 + 9(-3)^2 = 131 & 2x^2 + 9(3)^2 = 131 \\ 2x^2 = 50 & 2x^2 = 50 \\ x^2 = 25 & x^2 = 25 \\ x = \pm 5 & x = \pm 5 \end{array}$$

$(-5, -3), (5, -3), (-5, 3), (5, 3)$

58. Unknowns :
Length: x
Width: y

$2x + 2y = 70$ (Perimeter)
$x \cdot y = 300$ (Area)

Solve the perimeter equation for x.
$2x = -2y + 70$
$x = -y + 35$

Substitute into the area equation.

$$\begin{aligned} (-y + 35)y &= 300 \\ -y^2 + 35y &= 300 \\ 0 &= y^2 - 35y + 300 \\ 0 &= (y - 15)(y - 20) \\ y = 15 \ &\text{or} \ y = 20 \end{aligned}$$

Solve for the length (x) using $x = -y + 35$.

$$\begin{array}{ll} y = 15 & y = 20 \\ x = -(15) + 35 & x = -(20) + 35 \\ x = 20 & x = 15 \end{array}$$

Dimensions are 15 feet by 20 feet.

59. Unknowns :
Length: x
Width: y

$2x + 2y = 42$ (Perimeter)
$x^2 + y^2 = 15^2$ (Diagonal)

Solve the perimeter equation for x.
$2x = -2y + 42$
$x = -y + 21$

Substitute into the diagonal equation.

$$\begin{aligned} (-y + 21)^2 + y^2 &= 15^2 \\ y^2 - 42y + 441 + y^2 &= 225 \\ 2y^2 - 42y + 441 &= 225 \\ 2y^2 - 42y + 216 &= 0 \\ 2(y^2 - 21y + 108) &= 0 \\ 2(y - 9)(y - 12) &= 0 \\ y = 9 \ \text{or} \ y &= 12 \end{aligned}$$

Solve for the length (x) using $x = -y + 21$.

$$\begin{array}{ll} y = 9 & y = 12 \\ x = -(9) + 21 & x = -(12) + 21 \\ x = 12 & x = 9 \end{array}$$

Dimensions are 9 inches by 12 inches.

60. a)
$$h(t) = -16t^2 + v_0 t + s$$
$$h(t) = -16t^2 + 76t$$
$$h(t) = -16t^2 + 41t + 140$$

$$-16t^2 + 76t = -16t^2 + 41t + 140$$
$$76t = 41t + 140$$
$$35t = 140$$
$$t = 4$$

4 seconds

b) In 4 seconds, $t = 4$.
$$h(t) = -16t^2 + 76t$$
$$h(4) = -16(4)^2 + 76(4)$$
$$= -16(16) + 304$$
$$= -256 + 304$$
$$= 48$$

48 feet high

CHAPTER 13 TEST

1. $y = x^2 - 4x + 7$

The equation is of the form

$y = ax^2 + bx + c$ where $a > 0$

so the parabola opens upward.

vertex:
$$x = \frac{-b}{2a} = \frac{-(-4)}{2(1)} = 2$$

$$y = (2)^2 - 4(2) + 7 = 4 - 8 + 7 = 3$$
$$(2, 3)$$

y-intercept:
$$y = (0)^2 - 4(0) + 7 = 7$$
$$(0, 7)$$

x-intercept:
The vertex is above the x-axis and the parabola opens upward. There are no x-intercepts.

axis of symmetry: $x = 2$

point symmetric to the y-intercept: $(4, 7)$

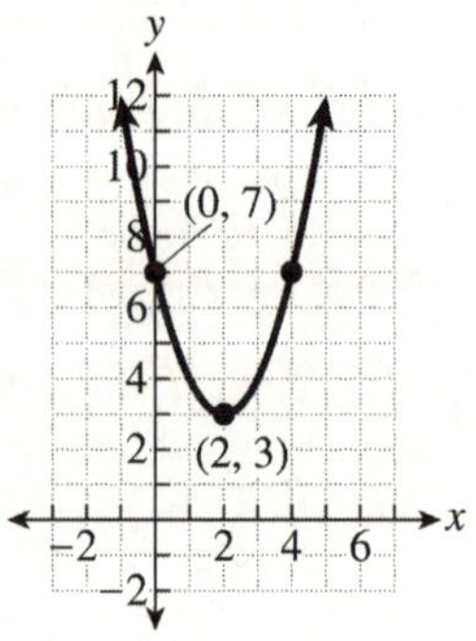

2. $y = -(x + 3)^2 + 9$

The equation is of the form

$y = a(x - h)^2 + k$ where $a < 0$

so to parabola opens downward.

vertex:
$$(-3, 9)$$

y-intercept:
$$y = -(0 + 3)^2 + 9 = -9 + 9 = 0$$
$$(0, 0)$$

x-intercept:
$$0 = -(x + 3)^2 + 9$$
$$(x + 3)^2 = 9$$
$$x + 3 = \pm 3$$
$$x = -3 \pm 3$$
$$x = -6 \text{ or } x = 0$$
$$(-6, 0), (0, 0)$$

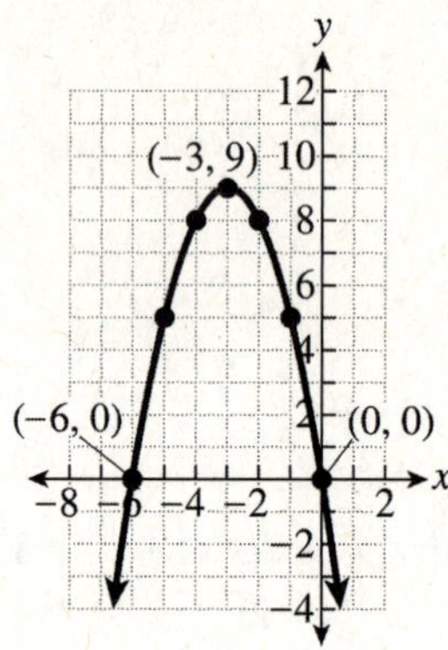

3. $x = -y^2 + 9y - 14$

The equation is of the form

$x = ay^2 + by + c$ where $a < 0$

so the parabola opens to the left.

vertex:

$$y = \frac{-b}{2a} = \frac{-9}{2(-1)} = \frac{9}{2} = 4.5$$

$$x = -\left(\frac{9}{2}\right)^2 + 9\left(\frac{9}{2}\right) - 14$$

$$= -\frac{81}{4} + \frac{81}{2} - 14$$

$$= -\frac{81}{4} + \frac{162}{4} - \frac{56}{4}$$

$$= \frac{25}{4} = 6.25$$

$$\left(\frac{25}{4}, \frac{9}{2}\right) \text{ or } (6.25, 4.5)$$

x-intercept:

$x = -(0)^2 + 9(0) - 14 = -14$

$(-14, 0)$

y-intercept:

$0 = -y^2 + 9y - 14$

$y^2 - 9y + 14 = 0$

$(y - 2)(y - 7) = 0$

$y = 2$ or $y = 7$

$(0, 2), (0, 7)$

axis of symmetry: $y = \dfrac{9}{2}$

point symmetric to the x-intercept: $(-14, 9)$

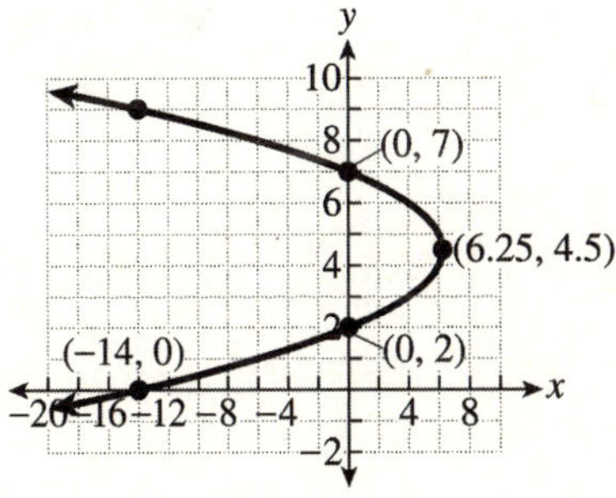

4. $x = (y + 1)^2 - 8$

The equation is of the form

$x = a(y - k)^2 + h$ where $a > 0$

so the parabola opens to the right.

vertex: $(-8, -1)$

x-intercept:

$x = (0 + 1)^2 - 8 = 1 - 8 = -7$

$(-7, 0)$

y-intercept:

$$0 = (y + 1)^2 - 8$$

$$8 = (y + 1)^2$$

$$\pm\sqrt{8} = y + 1$$

$$-1 \pm 2\sqrt{2} = y$$

$$\left(0, -1 - 2\sqrt{2}\right)\left(0, -1 + 2\sqrt{2}\right)$$

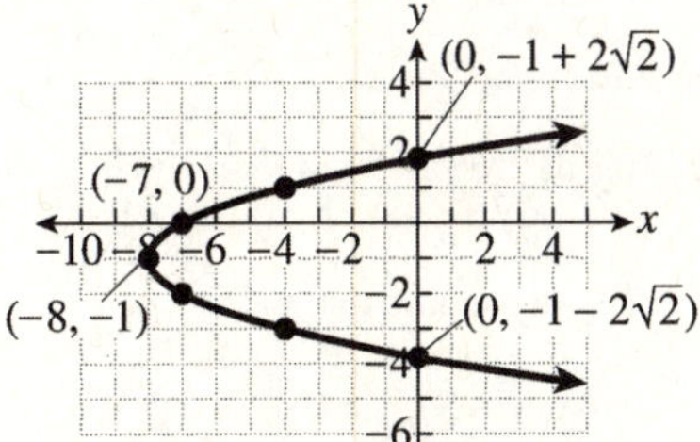

5. $d = \sqrt{(x_2 - x_1)^2 + (y_2 - y_1)^2}$

$d = \sqrt{(3 - 7)^2 + (5 - (-3))^2}$

$d = \sqrt{16 + 64}$

$d = \sqrt{80}$

$d \approx 8.9$

6. $(x + 4)^2 + (y - 2)^2 = 25$

$(x - (-4))^2 + (y - 2)^2 = 5^2$

Use the form $(x - h)^2 + (y - k)^2 = r^2$.

$h = -4$, $k = 2$, and $r = 5$.

Center: $(-4, 2)$; radius: $r = 5$

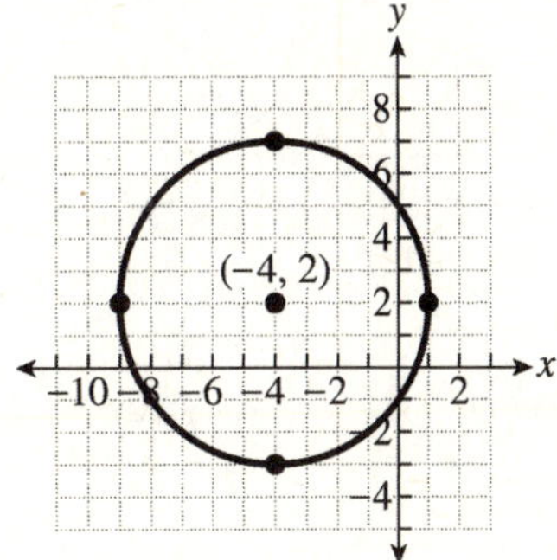

7.
$$x^2 + y^2 - 12x - 6y + 13 = 0$$
$$x^2 - 12x + y^2 - 6y = -13$$
$$\left(x^2 - 12x + \left(-\frac{12}{2}\right)^2\right) + \left(y^2 - 6y + \left(-\frac{6}{2}\right)^2\right) = -13 + \left(-\frac{12}{2}\right)^2 + \left(-\frac{6}{2}\right)^2$$
$$\left(x^2 - 12x + 36\right) + \left(y^2 - 6y + 9\right) = -13 + 36 + 9$$
$$(x-6)^2 + (y-3)^2 = 32$$
$$(x-6)^2 + (y-3)^2 = \left(4\sqrt{2}\right)^2$$

Use the form $(x-h)^2 + (y-k)^2 = r^2$.

$h = 6$, $k = 3$, and $r = 4\sqrt{2}$.

Center: $(6,3)$; radius: $r = 4\sqrt{2}$

8. Center: $(3,-8)$; radius: $r = 16$

$$(x-3)^2 + (y-(-8))^2 = 16^2$$
$$(x-3)^2 + (y+8)^2 = 256$$

9.
$$\frac{(x-4)^2}{16} + \frac{(y+7)^2}{9} = 1$$
$$\frac{(x-4)^2}{4^2} + \frac{(y-(-7))^2}{3^2} = 1$$

The ellipse is in the form $\dfrac{(x-h)^2}{a^2} + \dfrac{(y-k)^2}{b^2} = 1$,

where $h = 4, k = -7$, $a = 4$, and $b = 3$.

Center: $(4,-7)$

The endpoints of the horizontal axis are 4 units to the left and to the right of the center.
$(0,-7), (8,-7)$

The endpoints of the vertical axis are 3 units above and below the center.
$(4,-10), (4,-4)$

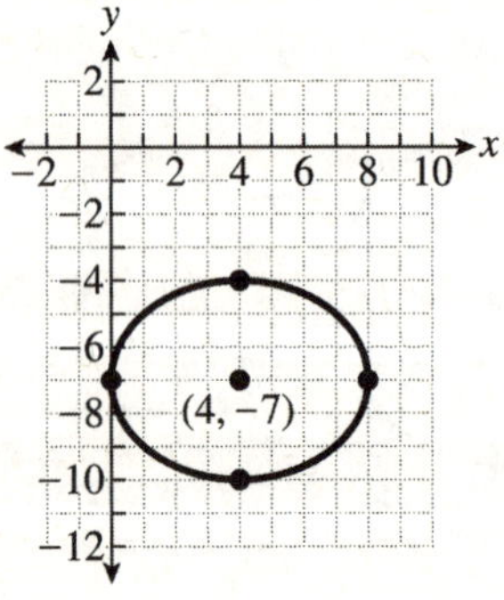

10.

$$4x^2 + 25y^2 - 64x + 100y + 256 = 0$$

$$4x^2 - 64x + 25y^2 + 100y = -256$$

$$4\left(x^2 - 16x\right) + 25\left(y^2 + 4y\right) = -256$$

$$4\left(x^2 - 16x + \left(-\frac{16}{2}\right)^2\right) + 25\left(y^2 + 4y + \left(\frac{4}{2}\right)^2\right) = -256 + 4\left(-\frac{16}{2}\right)^2 + 25\left(\frac{4}{2}\right)^2$$

$$4\left(x^2 - 16x + 64\right) + 25\left(y^2 + 4y + 4\right) = -256 + 256 + 100$$

$$4(x-8)^2 + 25(y+2)^2 = 100$$

$$\frac{4(x-8)^2}{100} + \frac{25(y+2)^2}{100} = \frac{100}{100}$$

$$\frac{(x-8)^2}{25} + \frac{(y+2)^2}{4} = 1$$

$$\frac{(x-8)^2}{5^2} + \frac{(y-(-2))^2}{2^2} = 1$$

The ellipse is in the form $\dfrac{(x-h)^2}{a^2} + \dfrac{(y-k)^2}{b^2} = 1,$ where $h = 8, k = -2, a = 5,$ and $b = 2.$

Center: $(8,-2)$

11. $\dfrac{x^2}{16} - y^2 = 1$

$$\frac{x^2}{4^2} - \frac{y^2}{1^2} = 1$$

The hyperbola is in the form

$$\frac{(x-h)^2}{a^2} - \frac{(y-k)^2}{b^2} = 1.$$

The hyperbola opens to the left and to the right where $a = 4,$ and $b = 1.$

Center: $(h,k) = (0,0)$

Vertices: $(-4,0), (4,0)$

Asymptotes: $y = \pm \dfrac{b}{a} x$

$$y = \pm \frac{1}{4} x$$

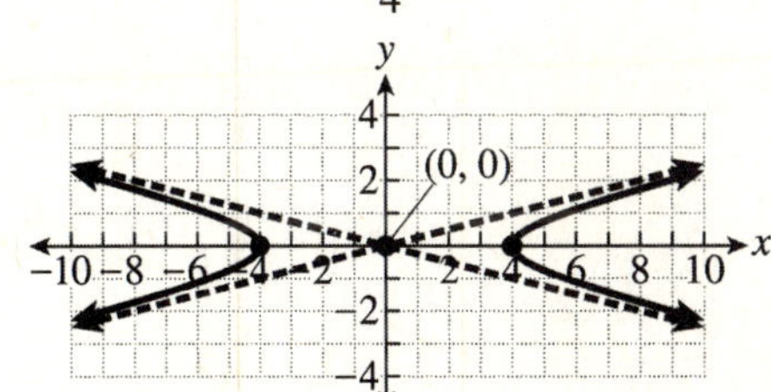

12. Center: $(3,2)$; radius: $r = 5$

$$(x-3)^2 + (y-2)^2 = 5^2$$
$$(x-3)^2 + (y-2)^2 = 25$$

13. Center: $(5,-1)$ with $a = 2,$ and $b = 6$

$$\frac{(x-5)^2}{2^2} + \frac{(y-(-1))^2}{6^2} = 1$$

$$\frac{(x-5)^2}{4} + \frac{(y+1)^2}{36} = 1$$

14. $x^2 + y^2 = 100 \quad (\text{Eq. 1})$
$x + 3y = 10 \quad\;\; (\text{Eq. 2})$

Solve for x in Eq. 2.
$x = -3y + 10$

Substitute $x = -3y + 10$ into Eq. 1.

$$(-3y+10)^2 + y^2 = 100$$
$$9y^2 - 60y + 100 + y^2 = 100$$
$$10y^2 - 60y = 0$$
$$10y(y-6) = 0$$

$y = 0$ or $y = 6$

Find the x-coordinates using $x = -3y + 10.$

$y = 0 \qquad\qquad y = 6$
$x = -3(0) + 10 \qquad x = -3(6) + 10$
$x = 10 \qquad\qquad x = -8$
$(10,0), (-8,6)$

15. $x^2 + 6y^2 = 49$ (Eq. 1)
$x - 2y = 1$ (Eq. 2)

Solve for x in Eq. 2.
$x = 2y + 1$

Substitute $x = 2y + 1$ into Eq. 1.
$(2y+1)^2 + 6y^2 = 49$
$4y^2 + 4y + 1 + 6y^2 = 49$
$10y^2 + 4y - 48 = 0$
$2(5y^2 + 2y - 24) = 0$
$2(5y + 12)(y - 2) = 0$
$5y + 12 = 0$ or $y - 2 = 0$
$5y = -12$ or $y = 2$
$y = -\dfrac{12}{5}$ or $y = 2$

Find the x-coordinates using $x = 2y + 1$.

$y = -\dfrac{12}{5}$ $\qquad\qquad$ $y = 2$

$x = 2\left(-\dfrac{12}{5}\right) + 1$ $\qquad$ $x = 2(2) + 1$

$x = -\dfrac{24}{5} + \dfrac{5}{5}$ $\qquad$ $x = 4 + 1$

$x = -\dfrac{19}{5}$ $\qquad\qquad$ $x = 5$

$\left(-\dfrac{19}{5}, -\dfrac{12}{5}\right), (5, 2)$

16. $y^2 - x^2 = 10$ (Eq. 1)
$y = x^2 - 32$ (Eq. 2)

Solve for x^2 in Eq. 2.
$x^2 = y + 32$

Substitute $x^2 = y + 32$ into Eq. 1.
$y^2 - (y + 32) = 10$
$y^2 - y - 32 = 10$
$y^2 - y - 42 = 0$
$(y + 6)(y - 7) = 0$
$y = -6$ or $y = 7$

Find the x-coordinates using $x^2 = y + 32$.
$y = -6$ $\qquad\qquad$ $y = 7$
$x^2 = -6 + 32$ $\qquad$ $x^2 = 7 + 32$
$x^2 = 26$ $\qquad\qquad$ $x^2 = 39$
$x^2 = \pm\sqrt{26}$ $\qquad$ $x^2 = \pm\sqrt{39}$
$\left(-\sqrt{26}, -6\right), \left(\sqrt{26}, -6\right), \left(-\sqrt{39}, 7\right), \left(\sqrt{39}, 7\right)$

17. $x^2 + 4y^2 = 85$ (Eq. 1)
$3x^2 + 2y^2 = 165$ (Eq. 2)

$\;-3x^2 - 12y^2 = -255$ $(-3 \cdot$ Eq. 1$)$
$+\;\underline{3x^2 + 2y^2 = 165}$ (Eq. 2)
$-10y^2 = -90$
$y^2 = 9$
$y = \pm 3$

Find the x-coordinates using $x^2 + 4y^2 = 85$.
$y = -3$ $\qquad\qquad$ $y = 3$
$x^2 + 4(-3)^2 = 85$ $\qquad$ $x^2 + 4(3)^2 = 85$
$x^2 + 36 = 85$ $\qquad\qquad$ $x^2 + 36 = 85$
$x^2 = 49$ $\qquad\qquad$ $x^2 = 49$
$x = \pm 7$ $\qquad\qquad$ $x = \pm 7$

$(-7, -3), (7, -3), (-7, 3), (7, 3)$

18. <u>Unknowns</u>:
Length: x
Width: y

$2x + 2y = 340$ (Perimeter)
$x \cdot y = 6000$ (Area)

Solve the perimeter equation for x.
$2x = -2y + 340$
$x = -y + 170$

Substitute into the area equation.

$(-y + 170)y = 6000$
$-y^2 + 170y = 6000$
$0 = y^2 - 170y + 6000$
$0 = (y - 50)(y - 120)$
$y = 50$ or $y = 120$

Solve for the length (x) using $x = -y + 170$.

$y = 50$ $\qquad\qquad$ $y = 120$
$x = -(50) + 170$ $\qquad$ $x = -(120) + 170$
$x = 120$ $\qquad\qquad$ $x = 50$
Dimensions are 50 feet by 120 feet.

CHAPTER 14   SEQUENCES, SERIES, AND THE BINOMIAL THEOREM

14.1 QUICK CHECK

1. $a_n = \dfrac{1}{3n+4}$

$a_1 = \dfrac{1}{3(1)+4} = \dfrac{1}{7}$

$a_2 = \dfrac{1}{3(2)+4} = \dfrac{1}{10}$

$a_3 = \dfrac{1}{3(3)+4} = \dfrac{1}{13}$

$a_4 = \dfrac{1}{3(4)+4} = \dfrac{1}{16}$

$\dfrac{1}{7}, \dfrac{1}{10}, \dfrac{1}{13}, \dfrac{1}{16}$

2. $a_n = \dfrac{(-1)^{n+1}}{2n}$

$a_1 = \dfrac{(-1)^{1+1}}{2(1)} = \dfrac{1}{2}$

$a_2 = \dfrac{(-1)^{2+1}}{2(2)} = -\dfrac{1}{4}$

$a_3 = \dfrac{(-1)^{3+1}}{2(3)} = \dfrac{1}{6}$

$a_4 = \dfrac{(-1)^{4+1}}{2(4)} = -\dfrac{1}{8}$

$\dfrac{1}{2}, -\dfrac{1}{4}, \dfrac{1}{6}, -\dfrac{1}{8}$

3. If we disregard the signs, the sequence is $7, 13, 19, 25, 31, \ldots$, an increase of 6 between each term. The terms of the sequence have alternating signs. The next three terms are $-37, 43, -49$.

$a_n = (-1)^{n+1}(6n+1)$

4. $a_n = (a_{n-1})^2 + 1$

$a_1 = 2$

$a_2 = (a_{2-1})^2 + 1 = (a_1)^2 + 1 = (2)^2 + 1 = 5$

$a_3 = (a_{3-1})^2 + 1 = (a_2)^2 + 1 = (5)^2 + 1 = 26$

$a_4 = (a_{4-1})^2 + 1 = (a_3)^2 + 1 = (26)^2 + 1 = 677$

$a_5 = (a_{5-1})^2 + 1 = (a_4)^2 + 1 = (677)^2 + 1 = 458330$

$2, 5, 26, 677, 458330$

5. $a_n = n^2 + 3n$

$a_1 = (1)^2 + 3(1) = 4$

$a_2 = (2)^2 + 3(2) = 10$

$a_3 = (3)^2 + 3(3) = 18$

$a_4 = (4)^2 + 3(4) = 28$

$\begin{aligned} s_4 &= a_1 + a_2 + a_3 + a_4 \\ &= 4 + 10 + 18 + 28 \\ &= 60 \end{aligned}$

6. $a_n = 3a_{n-1} + 4$

$a_1 = 1$

$a_2 = 3a_{2-1} + 4 = 3a_1 + 4 = 3(1) + 4 = 7$

$a_3 = 3a_{3-1} + 4 = 3a_2 + 4 = 3(7) + 4 = 25$

$a_4 = 3a_{4-1} + 4 = 3a_3 + 4 = 3(25) + 4 = 79$

$\begin{aligned} s_4 &= a_1 + a_2 + a_3 + a_4 \\ &= 1 + 7 + 25 + 79 \\ &= 112 \end{aligned}$

7. $\displaystyle\sum_{i=1}^{4} 6i = 6(1) + 6(2) + 6(3) + 6(4)$

$\qquad = 6 + 12 + 18 + 24$

$\qquad = 60$

8. The original value of the copy machine was $8000.

Find the value after one year:

$(0.20)(8000) = 1600 \,(\text{decrease})$

$a_1 = 8000 - 1600 = 6400$

Find the value after two years:

$(0.20)(6400) = 1280 \,(\text{decrease})$

$a_2 = 6400 - 1280 = 5120$

Find the value after three years:

$(0.20)(5120) = 1024 \,(\text{decrease})$

$a_3 = 5120 - 1024 = 4096$

Find the value after four years:

$(0.20)(4096) = 819.20 \,(\text{decrease})$

$a_4 = 4096 - 819.20 = 3276.80$

$6400, \$5120, \$4096, \$3276.80$

14.1 SEQUENCES AND SERIES

1. sequence

3. general

5. series

7. finite

9. $a_n = 3n - 4$
$a_5 = 3(5) - 4$
$a_5 = 11$

11. $a_n = (-1)^n \cdot 3^{n+1}$
$a_6 = (-1)^6 \cdot 3^{6+1}$
$a_6 = 1 \cdot 3^7$
$a_6 = 2187$

13. $a_n = 4n$
$a_1 = 4(1) = 4$
$a_2 = 4(2) = 8$
$a_3 = 4(3) = 12$
$a_4 = 4(4) = 16$
$a_5 = 4(5) = 20$
$4, 8, 12, 16, 20$

15. $a_n = 2n + 9$
$a_1 = 2(1) + 9 = 2 + 9 = 11$
$a_2 = 2(2) + 9 = 4 + 9 = 13$
$a_3 = 2(3) + 9 = 6 + 9 = 15$
$a_4 = 2(4) + 9 = 8 + 9 = 17$
$a_5 = 2(5) + 9 = 10 + 9 = 19$
$11, 13, 15, 17, 19$

17. $a_n = (-1)^n \cdot (2n + 3)$
$a_1 = (-1)^1 \cdot (2(1) + 3) = -5$
$a_2 = (-1)^2 \cdot (2(2) + 3) = 7$
$a_3 = (-1)^3 \cdot (2(3) + 3) = -9$
$a_4 = (-1)^4 \cdot (2(4) + 3) = 11$
$a_5 = (-1)^5 \cdot (2(5) + 3) = -13$
$-5, 7, -9, 11, -13$

19. $a_n = 3 \cdot 4^n$
$a_1 = 3 \cdot 4^1 = 3 \cdot 4 = 12$
$a_2 = 3 \cdot 4^2 = 3 \cdot 16 = 48$
$a_3 = 3 \cdot 4^3 = 3 \cdot 64 = 192$
$a_4 = 3 \cdot 4^4 = 3 \cdot 256 = 768$
$a_5 = 3 \cdot 4^5 = 3 \cdot 1024 = 3072$
$12, 48, 192, 768, 3072$

21. $5, 10, 15, 20, 25, \ldots$
Each successive term increases by 5.
The next three terms are: 30, 35, 40.
$a_n = 5n$

23. $1, 3, 9, 27, 81, \ldots$
Each successive term increases by a factor of 3.
The next three terms are: 243, 729, 2187.
$a_n = 3^{n-1}$

25. $-\dfrac{1}{5}, \dfrac{1}{10}, -\dfrac{1}{15}, \dfrac{1}{20}, -\dfrac{1}{25}, \ldots$
The sign alternates and each successive denominator increases by 5.
The next three terms are: $\dfrac{1}{30}, -\dfrac{1}{35}, \dfrac{1}{40}$.
$a_n = \dfrac{(-1)^n}{5n}$

27. $1, 8, 27, 64, 125, \ldots$
Each term is perfect cube.
The next three terms are: 216, 343, 512.
$a_n = n^3$

29. $a_n = a_{n-1} + 3$
$a_1 = 6$
$a_2 = a_{2-1} + 3 = 6 + 3 = 9$
$a_3 = a_{3-1} + 3 = 9 + 3 = 12$
$a_4 = a_{4-1} + 3 = 12 + 3 = 15$
$a_5 = a_{5-1} + 3 = 15 + 3 = 18$
$6, 9, 12, 15, 18$

31. $a_n = 3a_{n-1} + 8$
$a_1 = -5$
$a_2 = 3a_{2-1} + 8 = 3(-5) + 8 = -7$
$a_3 = 3a_{3-1} + 8 = 3(-7) + 8 = -13$
$a_4 = 3a_{4-1} + 8 = 3(-13) + 8 = -31$
$a_5 = 3a_{5-1} + 8 = 3(-31) + 8 = -85$
$-5, -7, -13, -31, -85$

33. $a_n = -3a_{n-1}$

$a_1 = 16$
$a_2 = -3a_{2-1} = -3(16) = -48$
$a_3 = -3a_{3-1} = -3(-48) = 144$
$a_4 = -3a_{4-1} = -3(144) = -432$
$a_5 = -3a_{5-1} = -3(-432) = 1296$
$16, -48, 144, -432, 1296$

35. Given the sequence $13, 20, 27, 34, \ldots$

$s_8 = 13 + 20 + 27 + 34 + 41 + 48 + 55 + 62 = 300$

37. Given the sequence $-18, -31, -44, -57, \ldots$

$s_7 = -18 + (-31) + (-44) + (-57)$
$\quad + (-70) + (-83) + (-96)$
$\quad = -399$

39. Given the sequence $7, 14, 28, 56, \ldots$

$s_6 = 7 + 14 + 28 + 56 + 112 + 224 = 441$

41. Given the sequence $25, -36, 49, -64, \ldots$

$s_6 = 25 + (-36) + 49 + (-64) + 81 + (-100)$
$\quad = -45$

43. $a_n = 3n + 10$

$a_1 = 3(1) + 10 = 13$
$a_2 = 3(2) + 10 = 16$
$a_3 = 3(3) + 10 = 19$
$a_4 = 3(4) + 10 = 22$
$a_5 = 3(5) + 10 = 25$

$s_5 = 13 + 16 + 19 + 22 + 25 = 95$

45. $a_n = -7n$

$a_1 = -7(1) = -7$
$a_2 = -7(2) = -14$
$a_3 = -7(3) = -21$
$a_4 = -7(4) = -28$
$a_5 = -7(5) = -35$
$a_6 = -7(6) = -42$

$s_6 = -7 + (-14) + (-21) + (-28) + (-35) + (-42)$
$\quad = -147$

47. $a_n = n^2$

$a_1 = (1)^2 = 1$
$a_2 = (2)^2 = 4$
$a_3 = (3)^2 = 9$
$a_4 = (4)^2 = 16$
$a_5 = (5)^2 = 25$

$s_5 = 1 + 4 + 9 + 16 + 25 = 55$

49. $a_n = (-1)^n \cdot (5n)$

$a_1 = (-1)^1 \cdot (5 \cdot 1) = -5$
$a_2 = (-1)^2 \cdot (5 \cdot 2) = 10$
$a_3 = (-1)^3 \cdot (5 \cdot 3) = -15$
$a_4 = (-1)^4 \cdot (5 \cdot 4) = 20$
$a_5 = (-1)^5 \cdot (5 \cdot 5) = -25$
$a_6 = (-1)^6 \cdot (5 \cdot 6) = 30$
$a_7 = (-1)^7 \cdot (5 \cdot 7) = -35$
$a_8 = (-1)^8 \cdot (5 \cdot 8) = 40$
$a_9 = (-1)^9 \cdot (5 \cdot 9) = -45$
$a_{10} = (-1)^{10} \cdot (5 \cdot 10) = 50$

$s_{10} = -5 + 10 + (-15) + 20 + (-25)$
$\quad + 30 + (-35) + 40 + (-45) + 50$
$\quad = 25$

51. $a_n = a_{n-1} + 7$

$a_1 = 2$
$a_2 = a_{2-1} + 7 = (2) + 7 = 9$
$a_3 = a_{3-1} + 7 = (9) + 7 = 16$
$a_4 = a_{4-1} + 7 = (16) + 7 = 23$
$a_5 = a_{5-1} + 7 = (23) + 7 = 30$
$a_6 = a_{6-1} + 7 = (30) + 7 = 37$

$s_6 = 2 + 9 + 16 + 23 + 30 + 37 = 117$

53. $a_n = a_{n-1} - 9$

$a_1 = 5$
$a_2 = a_{2-1} - 9 = (5) - 9 = -4$
$a_3 = a_{3-1} - 9 = (-4) - 9 = -13$
$a_4 = a_{4-1} - 9 = (-13) - 9 = -22$
$a_5 = a_{5-1} - 9 = (-22) - 9 = -31$
$a_6 = a_{6-1} - 9 = (-31) - 9 = -40$
$a_7 = a_{7-1} - 9 = (-40) - 9 = -49$
$a_8 = a_{8-1} - 9 = (-49) - 9 = -58$

$s_8 = 5 + (-4) + (-13) + (-22) + (-31)$
$\quad + (-40) + (-49) + (-58)$
$\quad = -212$

55. $a_n = 20a_{n-1}$

$a_1 = 3$
$a_2 = 20a_{2-1} = 20(3) = 60$
$a_3 = 20a_{3-1} = 20(60) = 1200$
$a_4 = 20a_{4-1} = 20(1200) = 24,000$

$s_4 = 3 + 60 + 1200 + 24,000 = 25,263$

57. $\displaystyle\sum_{i=1}^{15} i = 1 + 2 + 3 + 4 + 5 + 6 + 7 + 8 + 9 + 10 + 11$
$\quad + 12 + 13 + 14 + 15$
$\quad = 120$

59. $\displaystyle\sum_{i=1}^{8}(2i-1)=(2\cdot1-1)+(2\cdot2-1)+(2\cdot3-1)$
$$+(2\cdot4-1)+(2\cdot5-1)+(2\cdot6-1)$$
$$+(2\cdot7-1)+(2\cdot8-1)$$
$$=1+3+5+7+9+11+13+15$$
$$=64$$

61. $\displaystyle\sum_{i=1}^{10}i^2=1^2+2^2+3^2+4^2+5^2+6^2+7^2+8^2$
$$+9^2+10^2$$
$$=1+4+9+16+25+36+49+64+81$$
$$+100$$
$$=385$$

63. $\displaystyle\sum_{i=1}^{15}\left(\frac{1}{i}-\frac{1}{i+1}\right)=\left(\frac{1}{1}-\frac{1}{1+1}\right)+\left(\frac{1}{2}-\frac{1}{2+1}\right)$
$$+\left(\frac{1}{3}-\frac{1}{3+1}\right)+\left(\frac{1}{4}-\frac{1}{4+1}\right)$$
$$+\left(\frac{1}{5}-\frac{1}{5+1}\right)+\left(\frac{1}{6}-\frac{1}{6+1}\right)$$
$$+\left(\frac{1}{7}-\frac{1}{7+1}\right)+\left(\frac{1}{8}-\frac{1}{8+1}\right)$$
$$+\left(\frac{1}{9}-\frac{1}{9+1}\right)+\left(\frac{1}{10}-\frac{1}{10+1}\right)$$
$$+\left(\frac{1}{11}-\frac{1}{11+1}\right)+\left(\frac{1}{12}-\frac{1}{12+1}\right)$$
$$+\left(\frac{1}{13}-\frac{1}{13+1}\right)+\left(\frac{1}{14}-\frac{1}{14+1}\right)$$
$$+\left(\frac{1}{15}-\frac{1}{15+1}\right)$$
$$=1-\frac{1}{2}+\frac{1}{2}-\frac{1}{3}+\frac{1}{3}-\frac{1}{4}+\frac{1}{4}-\frac{1}{5}$$
$$+\frac{1}{5}-\frac{1}{6}+\frac{1}{6}-\frac{1}{7}+\frac{1}{7}-\frac{1}{8}+\frac{1}{8}$$
$$-\frac{1}{9}+\frac{1}{9}-\frac{1}{10}+\frac{1}{10}-\frac{1}{11}+\frac{1}{11}$$
$$-\frac{1}{12}+\frac{1}{12}-\frac{1}{13}+\frac{1}{13}-\frac{1}{14}+\frac{1}{14}$$
$$-\frac{1}{15}+\frac{1}{15}-\frac{1}{16}$$
$$=1-\frac{1}{16}$$
$$=\frac{15}{16}$$

65. $\displaystyle 7+14+21+28=\sum_{i=1}^{4}7i$

67. $5+(-10)+15+(-20)+25+(-30)$
$$=\sum_{i=1}^{6}(-1)^{i+1}\cdot5i$$

69. $\displaystyle 1+4+9+16+25+36+49=\sum_{i=1}^{7}i^2$

71. Find the value after one year:
$$(0.20)(30,000)=6000\,(\text{decrease})$$
$$a_1=30,000-6000=24,000$$
Find the value after 2 years:
$$(0.20)(24,000)=4800\,(\text{decrease})$$
$$a_2=24,000-4800=19,200$$
Find the value after 3 years:
$$(0.20)(19,200)=3840\,(\text{decrease})$$
$$a_2=19,200-3840=15,360$$
Find the value after four years:
$$(0.20)(15,360)=3072\,(\text{decrease})$$
$$a_2=15,360-3072=12,288$$
$$\$24,000,\ \$19,200,\ \$15,360,\ \$12,288$$

73. a) Find the value after one year:
$$a_1=50,000$$
Find the value after two years:
$$(0.04)(50,000)=2000\,(\text{increase})$$
$$a_2=50,000+2000=52,000$$
Find the value after three years:
$$(0.04)(52,000)=2080\,(\text{increase})$$
$$a_3=52,000+2080=54,080$$
Find the value after four years:
$$(0.04)(54,080)=2163.2\,(\text{increase})$$
$$a_4=54,080+2163.2=56,243.2$$
Find the value after five years:
$$(0.04)(56243.2)=2249.728\,(\text{increase})$$
$$a_5=56,243.2+2249.728=58,492.928$$
$$\$50,000,\ \$52,000,\ \$54,080,\ \$56,243.20,$$
$$\$58,492.93$$

b) $s_5=\$50,000+\$52,000+\$54,080$
$$+\$56,243.20+\$58,492.93$$
$$=\$270,816.13$$

75. Answers will vary. Example:
One real-world sequence is a function whose input is the number of years that you have a mortgage and whose output is the balance of the mortgage at the end of the year.

14.2 QUICK CHECK

1. $a_1 = -8$
$d = 3$
$a_2 = a_1 + d = -8 + 3 = -5$
$a_3 = a_1 + 2d = -8 + 2(3) = -2$
$a_4 = a_1 + 3d = -8 + 3(3) = 1$
$a_5 = a_1 + 4d = -8 + 4(3) = 4$
$a_6 = a_1 + 5d = -8 + 5(3) = 7$
$-8, -5, -2, 1, 4, 7$

2. $2, 14, 26, 38, 50, \ldots$
$a_1 = 2$
$d = 14 - 2 = 12$
$a_n = a_1 + (n-1)d$
$a_n = 2 + (n-1)12$
$a_n = 2 + 12n - 12$
$a_n = 12n - 10$

3. $57, 59, 61, 63, 65, \ldots$
$a_1 = 57$
$d = 59 - 57 = 2$
$a_n = a_1 + (n-1)d$
$a_n = 57 + (n-1)2$
$a_n = 57 + 2n - 2$
$a_n = 2n + 55$
$a_{61} = 2(61) + 55 = 177$

4. $20, 26, 32, 38, 44, \ldots$
$d = 6$
The 6th, 7th 8th and 9th terms are 50, 56, 62, 68.
$s_9 = 20 + 26 + 32 + 38 + 44 + 50 + 56 + 62 + 68$
$\quad = 396$

5. $690, 671, 652, 633, 614, \ldots$
$a_1 = 690$
$d = 671 - 690 = -19$
$a_n = a_1 + (n-1)d$
$a_n = 690 + (n-1)(-19)$
$a_n = 690 - 19n + 19$
$a_n = -19n + 709$
Since $n = 36$
$a_{36} = -19(36) + 709 = -684 + 709 = 25$
$s_n = \dfrac{n}{2}(a_1 + a_n)$
$s_{36} = \dfrac{36}{2}(690 + 25) = 18(715) = 12{,}870$

6. $2, 4, 6, 8, \ldots$
$a_1 = 2$
$d = 4 - 2 = 2$
$a_n = a_1 + (n-1)d$
$a_n = 2 + (n-1)2$
$a_n = 2 + 2n - 2$
$a_n = 2n$
Since $n = 62$
$a_{62} = 2(62) = 124$
$s_n = \dfrac{n}{2}(a_1 + a_n)$
$s_{62} = \dfrac{62}{2}(2 + 124) = 31(126) = 3906$

14.2 ARITHMETIC SEQUENCES AND SERIES

1. arithmetic

3. $a_n = a_{n-1} + d$

5. arithmetic series

7. $d = 10 - 6 = 4$

9. $d = \dfrac{29}{6} - \dfrac{7}{2} = \dfrac{29}{6} - \dfrac{21}{6} = \dfrac{8}{6} = \dfrac{4}{3}$

11. $d = -10 - (-13) = 3$

13. $d = \dfrac{41}{4} - \dfrac{9}{4} = \dfrac{32}{4} = 8$

15. Since $d = 5$, each subsequent term increases by 5.
$1, 6, 11, 16, 21$

17. Since $d = 12$, each subsequent term increases by 12.
$-9, 3, 15, 27, 39$

19. Since $d = -21$, each subsequent term decreases by 21.
$-10, -31, -52, -73, -94$

21. $18, 29, 40, 51, 62, \ldots$
$a_1 = 18$
$d = 29 - 18 = 11$
$a_n = a_1 + (n-1)d$
$a_n = 18 + (n-1)11 = 18 + 11n - 11 = 11n + 7$

23. $3, 1, -1, -3, -5, \ldots$
$a_1 = 3$
$d = 1 - 3 = -2$

$a_n = a_1 + (n-1)d$
$a_n = 3 + (n-1)(-2) = 3 - 2n + 2 = -2n + 5$

25. $-7, 13, 33, 53, 73, , \ldots$
$a_1 = -7$
$d = 13 - (-7) = 20$

$a_n = a_1 + (n-1)d$
$a_n = -7 + (n-1)20 = -7 + 20n - 20 = 20n - 27$

27. $-\dfrac{3}{2}, \dfrac{5}{2}, \dfrac{13}{2}, \dfrac{21}{2}, \dfrac{29}{2}, \ldots$

$a_1 = -\dfrac{3}{2}$

$d = \dfrac{5}{2} - \left(-\dfrac{3}{2}\right) = \dfrac{8}{2} = 4$

$a_n = a_1 + (n-1)d$

$a_n = -\dfrac{3}{2} + (n-1)4 = -\dfrac{3}{2} + 4n - 4 = 4n - \dfrac{11}{2}$

29. $16, 21, 26, 31, 36, \ldots$
$a_1 = 16$
$d = 21 - 16 = 5$
$a_n = a_1 + (n-1)d$
$a_n = 16 + (n-1)5 = 16 + 5n - 5 = 5n + 11$

$a_{19} = 5(19) + 11 = 106$

31. $23, 29, 35, 41, 47, \ldots$
$a_1 = 23$
$d = 29 - 23 = 6$
$a_n = a_1 + (n-1)d$
$a_n = 23 + (n-1)6 = 23 + 6n - 6 = 6n + 17$

$\qquad 389 = 6n + 17$
$389 - 17 = 6n + 17 - 17$
$\qquad 372 = 6n$

$\qquad \dfrac{372}{6} = n$

$\qquad 62 = n$
$a_{62} = 389,$ the 62nd term

33. $611, 603, 595, 587, 79, \ldots$
$a_1 = 611$
$d = 603 - 611 = -8$
$a_n = a_1 + (n-1)d$
$a_n = 611 + (n-1)(-8)$
$\qquad = 611 - 8n + 8$
$\qquad = -8n + 619$

$\qquad -365 = -8n + 619$
$-365 - 619 = -8n + 619 - 619$
$\qquad -984 = -8n$

$\qquad \dfrac{-984}{-8} = n$

$\qquad 123 = n$
$a_{123} = -365,$ the 123rd term

35. $a_1 = 8$ and $d = 5$
$a_n = a_1 + (n-1)d$
$a_n = 8 + (n-1)5 = 8 + 5n - 5 = 5n + 3$
$a_7 = 5(7) + 3 = 38$

$s_n = \dfrac{n}{2}(a_1 + a_n)$

$s_7 = \dfrac{7}{2}(a_1 + a_7)$

$s_7 = \dfrac{7}{2}(8 + 38) = 161$

37. $a_1 = -6$ and $d = 11$
$a_n = a_1 + (n-1)d$
$a_n = -6 + (n-1)11 = -6 + 11n - 11 = 11n - 17$
$a_{15} = 11(15) - 17 = 148$

$s_n = \dfrac{n}{2}(a_1 + a_n)$

$s_{15} = \dfrac{15}{2}(a_1 + a_{15})$

$s_{15} = \dfrac{15}{2}(-6 + 148) = 1065$

39. $a_1 = 65$ and $d = -8$
$a_n = a_1 + (n-1)d$
$a_n = 65 + (n-1)(-8) = 65 - 8n + 8 = -8n + 73$
$a_{36} = -8(36) + 73 = -215$

$s_n = \dfrac{n}{2}(a_1 + a_n)$

$s_{36} = \dfrac{36}{2}(a_1 + a_{36})$

$s_{36} = 18(65 + (-215)) = -2700$

41. $a_1 = 109$ and $d = 99$
$a_n = a_1 + (n-1)d$
$a_n = 109 + (n-1)99$
$\quad = 109 + 99n - 99$
$\quad = 99n + 10$
$a_{32} = 99(32) + 10 = 3178$

$s_n = \dfrac{n}{2}(a_1 + a_n)$

$s_{32} = \dfrac{32}{2}(a_1 + a_{32})$

$s_{32} = 16(109 + 3178) = 52,592$

43. $1, 21, 41, 61, \ldots$

$a_1 = 1$ and $d = 21 - 1 = 20$
$a_n = 1 + (n-1)20 = 1 + 20n - 20 = 20n - 19$
$a_{10} = 20(10) - 19 = 181$

$s_n = \dfrac{n}{2}(a_1 + a_n)$

$s_{10} = \dfrac{10}{2}(a_1 + a_{10})$

$s_{10} = \dfrac{10}{2}(1 + 181) = 910$

45. $-17, -1, 15, 31, \ldots$

$a_1 = -17$ and $d = -1 - (-17) = 16$
$a_n = -17 + (n-1)16 = -17 + 16n - 16 = 16n - 33$
$a_{17} = 16(17) - 33 = 239$

$s_n = \dfrac{n}{2}(a_1 + a_n)$

$s_{17} = \dfrac{17}{2}(a_1 + a_{17})$

$s_{17} = \dfrac{17}{2}(-17 + 239) = 1887$

47. $86, 47, 8, -31, \ldots$

$a_1 = 86$ and $d = 47 - 86 = -39$
$a_n = 86 + (n-1)(-39)$
$\quad = 86 - 39n + 39$
$\quad = -39n + 125$
$a_{20} = -39(20) + 125 = -655$

$s_n = \dfrac{n}{2}(a_1 + a_n)$

$s_{20} = \dfrac{20}{2}(a_1 + a_{20})$

$s_{20} = \dfrac{20}{2}(86 + (-655)) = -5690$

49. $1, 2, 3, 4, 5, \ldots$

Since $a_1 = 1$ and $d = 2 - 1 = 1$,
$a_n = 1 + (n-1)1 = 1 + n - 1 = n$
$a_{1000} = 1000$

$s_n = \dfrac{n}{2}(a_1 + a_n)$

$s_{1000} = \dfrac{1000}{2}(a_1 + a_{1000})$

$s_{1000} = \dfrac{1000}{2}(1 + 1000) = 500,500$

51. $1, 3, 5, 7, 9, \ldots$

$a_1 = 1$ and $d = 3 - 1 = 2$
$a_n = 1 + (n-1)2 = 1 + 2n - 2 = 2n - 1$
$a_{67} = 2(67) - 1 = 133$

$s_n = \dfrac{n}{2}(a_1 + a_n)$

$s_{67} = \dfrac{67}{2}(a_1 + a_{67})$

$s_{67} = \dfrac{67}{2}(1 + 133)$

$s_{67} = 4489$

53. $2, 4, 6, 8, 10, \ldots$

$a_1 = 2$ and $d = 4 - 2 = 2$
$a_n = 2 + (n-1)2 = 2 + 2n - 2 = 2n$
$a_{200} = 2(200) = 400$

$s_n = \dfrac{n}{2}(a_1 + a_n)$

$s_{200} = \dfrac{200}{2}(a_1 + a_{200})$

$s_{200} = \dfrac{200}{2}(2 + 400) = 40,200$

55. a) Since $a_1 = 10$ and $d = 1$, the sequence is:
$\quad$ \$10, \$11, \$12, \$13, \$14.

b) $a_n = a_1 + (n-1)d$
$\quad a_n = 10 + (n-1)1 = 10 + n - 1 = n + 9$

57. a) Since $a_1 = 38,500$ and $d = 750,$ the sequence is: \$38,500, \$39,250, \$40,000, \$40,750, \$41,500, \$42,250.

b) $a_n = a_1 + (n-1)d$
$a_n = 38,500 + (n-1)750$
$\quad = 38,500 + 750n - 750$
$\quad = 750n + 37,750$

c) Find his salary at year 15.
$a_{15} = 750(15) + 37,750 = 49,000$

Add his salary for the first 15 years.

$s_{15} = \dfrac{15}{2}(a_1 + a_{15})$

$s_{15} = \dfrac{15}{2}(38,500 + 49,000) = 656,250$

\$656,250

59. Answers will vary. Example:
A gym membership starts at \$300 per year and increases \$25 each year.

61. Answers will vary.

14.3 QUICK CHECK

1. $256, 64, 16, 4, 1, \ldots$

$r = \dfrac{64}{256} = \dfrac{1}{4}$

2. $-120, 60, -30, 15, -\dfrac{15}{2}, \ldots$

$r = \dfrac{60}{-120} = -\dfrac{1}{2}$

3. $a_1 = 5$ and $r = -\dfrac{3}{4}$

$a_2 = -\dfrac{3}{4}(5) = -\dfrac{15}{4}$

$a_3 = -\dfrac{3}{4}\left(-\dfrac{15}{4}\right) = \dfrac{45}{16}$

$a_4 = -\dfrac{3}{4}\left(\dfrac{45}{16}\right) = -\dfrac{135}{64}$

$a_5 = -\dfrac{3}{4}\left(-\dfrac{135}{64}\right) = \dfrac{405}{256}$

$5, -\dfrac{15}{4}, \dfrac{45}{16}, -\dfrac{135}{64}, \dfrac{405}{256}$

4. $9, 18, 36, 72, \ldots$
$a_1 = 9$

$r = \dfrac{18}{9} = 2$

$a_n = a_1 \cdot r^{n-1}$
$a_n = 9 \cdot 2^{n-1}$

5. $11, -55, 275, -1375, \ldots$
$a_1 = 11$

$r = \dfrac{-55}{11} = -5$

$a_n = a_1 \cdot r^{n-1}$
$a_n = 11 \cdot (-5)^{n-1}$

6. $12, 36, 108, 324, \ldots$

$r = \dfrac{36}{12} = 3$

$a_5 = 3(324) = 972$
$a_6 = 3(972) = 2916$
$a_7 = 3(2916) = 8748$
$a_8 = 3(8748) = 26,244$

$s_8 = 12 + 36 + 108 + 324 + 972 + 2916 + 8748$
$\qquad + 26,244$
$\quad = 39,360$

7. $1, -3, 9, -27, \ldots$

$a_1 = 1$ and $r = \dfrac{-3}{1} = -3$

$s_n = a_1 \cdot \dfrac{1 - r^n}{1 - r}$

$s_{10} = 1 \cdot \dfrac{1 - (-3)^{10}}{1 - (-3)} = -14,762$

8. $3, 2, \dfrac{4}{3}, \dfrac{8}{9}, \ldots$

$r = \dfrac{2}{3}$

$|r| = \left|\dfrac{2}{3}\right| = \dfrac{2}{3} < 1$

Yes, since $|r| < 1,$ the infinite series has a limit.

$s = \dfrac{a_1}{1 - r}$

$s = \dfrac{3}{1 - \dfrac{2}{3}} = \dfrac{3}{\dfrac{1}{3}} = 9$

9. $-2, 8, -32, 128, \ldots$

$$r = \frac{8}{-2} = -4$$

$$|r| = |-4| = 4 > 1$$

No, since $|r| > 1$, the infinite series has no limit.

10. a) The value of the car decreases by 30% each year. Thus, the value of the car is 70% of the value the previous year.

$$a_1 = 0.7(25,000) = 17,500$$
$$a_2 = 0.7(17,500) = 12,250$$
$$a_3 = 0.7(12,250) = 8575$$
$$a_4 = 0.7(8575) = 6002.50$$
$$a_5 = 0.7(6002.50) = 4201.75$$
$$a_6 = 0.7(4201.75) = 2941.23$$

$17,500, \$12,250, \$8575, \$6002.50, \$4201.75, \$2941.23

b) $a_1 = 17,500$ and $r = \dfrac{12,250}{17,500} = 0.7$

$$a_n = a_1 \cdot r^{n-1}$$
$$a_{10} = 17,500 \cdot (0.7)^{10-1} \approx 706.19$$

$706.19

14.3 *GEOMETRIC SEQUENCES AND SERIES*

1. geometric sequence

3. $a_n = r \cdot a_{n-1}$

5. $s_n = a_1 \cdot \dfrac{1 - r^n}{1 - r}$

7. $1, 8, 64, 512, 4096, \ldots$

$$r = \frac{8}{1} = 8$$

9. $16, 4, 1, \dfrac{1}{4}, \dfrac{1}{16}, \ldots$

$$r = \frac{4}{16} = \frac{1}{4}$$

11. $-3, 21, -147, 1029, -7203, \ldots$

$$r = \frac{21}{-3} = -7$$

13. $-6, 8, -\dfrac{32}{3}, \dfrac{128}{9}, -\dfrac{512}{27}, \ldots$

$$r = \frac{8}{-6} = -\frac{4}{3}$$

15. $a_1 = 1$ and $r = 10$

$$a_2 = 1(10) = 10$$
$$a_3 = 10(10) = 100$$
$$a_4 = 100(10) = 1000$$
$$a_5 = 1000(10) = 10,000$$

$1, 10, 100, 1000, 10000$

17. $a_1 = 4$ and $r = -2$

$$a_2 = 4(-2) = -8$$
$$a_3 = -8(-2) = 16$$
$$a_4 = 16(-2) = -32$$
$$a_5 = -32(-2) = 64$$

$4, -8, 16, -32, 64$

19. $a_1 = 8$ and $r = \dfrac{3}{4}$

$$a_2 = 8\left(\frac{3}{4}\right) = 6$$
$$a_3 = 6\left(\frac{3}{4}\right) = \frac{9}{2}$$
$$a_4 = \frac{9}{2}\left(\frac{3}{4}\right) = \frac{27}{8}$$
$$a_5 = \frac{27}{8}\left(\frac{3}{4}\right) = \frac{81}{32}$$

$8, 6, \dfrac{9}{2}, \dfrac{27}{8}, \dfrac{81}{32}$

21. $a_1 = -20$ and $r = -\dfrac{7}{4}$

$$a_2 = -20\left(-\frac{7}{4}\right) = 35$$
$$a_3 = 35\left(-\frac{7}{4}\right) = -\frac{245}{4}$$
$$a_4 = -\frac{245}{4}\left(-\frac{7}{4}\right) = \frac{1715}{16}$$
$$a_5 = \frac{1715}{16}\left(-\frac{7}{4}\right) = -\frac{12,005}{64}$$

$-20, 35, -\dfrac{245}{4}, \dfrac{1715}{16}, -\dfrac{12,005}{64}$

23. $1, \dfrac{2}{3}, \dfrac{4}{9}, \dfrac{8}{27}, \ldots$

$a_1 = 1$

$r = \dfrac{\frac{2}{3}}{1} = \dfrac{2}{3}$

$a_n = a_1 \cdot r^{n-1}$

$a_n = 1 \cdot \left(\dfrac{2}{3}\right)^{n-1} = \left(\dfrac{2}{3}\right)^{n-1}$

25. $-1, 4, -16, 64 \ldots$

$a_1 = -1$

$r = \dfrac{4}{-1} = -4$

$a_n = a_1 \cdot r^{n-1}$

$a_n = -1 \cdot (-4)^{n-1}$

27. $2, 14, 98, 686, \ldots$

$a_1 = 2$

$r = \dfrac{14}{2} = 7$

$a_n = a_1 \cdot r^{n-1}$

$a_n = 2 \cdot 7^{n-1}$

29. $16, -12, 9, -\dfrac{27}{4}, \ldots$

$a_1 = 16$

$r = \dfrac{-12}{16} = -\dfrac{3}{4}$

$a_n = a_1 \cdot r^{n-1}$

$a_n = 16 \cdot \left(-\dfrac{3}{4}\right)^{n-1}$

31. $a_1 = 5, r = 4$

$s_n = a_1 \cdot \dfrac{1 - r^n}{1 - r}$

$s_7 = 5 \cdot \dfrac{1 - 4^7}{1 - 4} = 27,305$

33. $a_1 = 48, r = -2$

$s_n = a_1 \cdot \dfrac{1 - r^n}{1 - r}$

$s_{20} = 48 \cdot \dfrac{1 - (-2)^{20}}{1 - (-2)} = -16,777,200$

35. $a_1 = 256, r = \dfrac{1}{2}$

$s_n = a_1 \cdot \dfrac{1 - r^n}{1 - r}$

$s_9 = 256 \cdot \dfrac{1 - \left(\frac{1}{2}\right)^9}{1 - \left(\frac{1}{2}\right)} = 511$

37. $1, -8, 64, -512, \ldots$

$a_1 = 1, \ r = \dfrac{-8}{1} = -8$

$s_n = a_1 \cdot \dfrac{1 - r^n}{1 - r}$

$s_6 = 1 \cdot \dfrac{1 - (-8)^6}{1 - (-8)} = -29,127$

39. $4, 12, 36, 108, \ldots$

$a_1 = 4, \ r = \dfrac{12}{4} = 3$

$s_n = a_1 \cdot \dfrac{1 - r^n}{1 - r}$

$s_7 = 4 \cdot \dfrac{1 - 3^7}{1 - 3} = 4372$

41. $13, 156, 1872, 22,464, \ldots$

$a_1 = 13, \ r = \dfrac{156}{13} = 12$

$s_n = a_1 \cdot \dfrac{1 - r^n}{1 - r}$

$s_6 = 13 \cdot \dfrac{1 - 12^6}{1 - 12} = 3,528,889$

43. $30, 10, \dfrac{10}{3}, \dfrac{10}{9}, \ldots$

$a_1 = 30, \ r = \dfrac{10}{30} = \dfrac{1}{3}$

$|r| = \left|\dfrac{1}{3}\right| = \dfrac{1}{3} < 1$

Yes, since $|r| < 1,$ the infinite series has a limit.

$s = \dfrac{a_1}{1 - r}$

$s = \dfrac{30}{1 - \frac{1}{3}} = \dfrac{30}{\frac{2}{3}} = 45$

45. $1, -\dfrac{4}{5}, \dfrac{16}{25}, -\dfrac{64}{125}, \ldots$

$a_1 = 1, \; r = \dfrac{-\dfrac{4}{5}}{1} = -\dfrac{4}{5}$

$|r| = \left|-\dfrac{4}{5}\right| = \dfrac{4}{5} < 1$

Yes, since $|r| < 1$, the infinite series has a limit.

$s = \dfrac{a_1}{1-r}$

$s = \dfrac{1}{1-\left(-\dfrac{4}{5}\right)} = \dfrac{1}{\dfrac{9}{5}} = \dfrac{5}{9}$

47. $\dfrac{81}{100}, \dfrac{243}{200}, \dfrac{729}{400}, \dfrac{2187}{800}, \ldots$

$a_1 = \dfrac{81}{100}, \; r = \dfrac{\dfrac{243}{200}}{\dfrac{81}{100}} = \dfrac{243}{200} \cdot \dfrac{100}{81} = \dfrac{3}{2}$

$|r| = \left|\dfrac{3}{2}\right| = \dfrac{3}{2} > 1$

No, since $|r| > 1$, the infinite series has no limit.

49. $\dfrac{1}{2}, -\dfrac{1}{4}, \dfrac{1}{8}, -\dfrac{1}{16}, \ldots$

$a_1 = \dfrac{1}{2}, \; r = \dfrac{-\dfrac{1}{4}}{\dfrac{1}{2}} = -\dfrac{1}{2}$

$|r| = \left|-\dfrac{1}{2}\right| = \dfrac{1}{2} < 1$

Yes, since $|r| < 1$, the infinite series has a limit.

$s = \dfrac{a_1}{1-r}$

$s = \dfrac{\dfrac{1}{2}}{1-\left(-\dfrac{1}{2}\right)} = \dfrac{\dfrac{1}{2}}{\dfrac{3}{2}} = \dfrac{1}{3}$

51. $a_n = 5 \cdot \left(\dfrac{5}{6}\right)^{n-1}$

$a_1 = 5, \; r = \dfrac{5}{6}$

$|r| = \left|\dfrac{5}{6}\right| = \dfrac{5}{6} < 1$

Yes, since $|r| < 1$, the infinite series has a limit.

$s = \dfrac{a_1}{1-r}$

$s = \dfrac{5}{1-\dfrac{5}{6}} = \dfrac{5}{\dfrac{1}{6}} = 30$

53. $a_n = -8 \cdot \left(\dfrac{3}{10}\right)^{n-1}$

$a_1 = -8, \; r = \dfrac{3}{10}$

$|r| = \left|\dfrac{3}{10}\right| = \dfrac{3}{10} < 1$

Yes, since $|r| < 1$, the infinite series has a limit.

$s = \dfrac{a_1}{1-r}$

$s = \dfrac{-8}{1-\dfrac{3}{10}} = \dfrac{-8}{\dfrac{7}{10}} = -\dfrac{80}{7} = -11\dfrac{3}{7}$

55. $a_n = \dfrac{5}{7} \cdot \left(\dfrac{11}{8}\right)^{n-1}$

$a_1 = \dfrac{5}{7}, \; r = \dfrac{11}{8}$

$|r| = \left|\dfrac{11}{8}\right| = \dfrac{11}{8} > 1$

No, since $|r| > 1$, the infinite series has no limit.

57. a) The copy machine decreases in value by 20%, so it retains 80% of its previous value. The value at the end of each year:

$a_1 = 0.8(8000) = 6400$
$a_2 = 0.8(6400) = 5120$
$a_3 = 0.8(5120) = 4096$

\$6400, \$5120, \$4096

b) $a_1 = 6400$ and $r = \dfrac{5120}{6400} = 0.8$

$a_n = a_1 \cdot r^{n-1}$
$a_8 = 6400 \cdot (0.8)^{8-1} \approx 1342.18$

\$1342.18

59. The value decreases by 50% each year, so the actual value is found by multiplying 0.5 times the value of the previous year.

$$a_n = 40,000 \cdot (0.5)^n$$

Or,

$$a_0 = 40,000 \text{ and } r = 0.5$$
$$a_1 = a_0 \cdot r$$
$$a_1 = 40,000 \cdot 0.5 = 20,000$$
$$a_n = a_1 \cdot r^{n-1}$$
$$a_n = 20,000 \cdot (0.5)^{n-1}$$
$$a_n = 20,000 \cdot (0.5)^{n-1} \text{ or } 40,000 \cdot (0.5)^n$$

61. $a_1 = 10,000,000$ and $r = 1.1$

$$s_n = a_1 \cdot \frac{1-r^n}{1-r}$$

$$s_{10} = 10,000,000 \cdot \frac{1-1.1^{10}}{1-1.1} = 159,374,246$$

$$\$159,374,246$$

63. Answers will vary. Example:
A copy machine purchased for \$4000 maintains 90% of its value each year.

65. Answers will vary. Example:
The difference between terms in an arithmetic sequence is constant. For a geometric sequence, each term is a multiple of the previous term.

67. Answers will vary.

14.4 QUICK CHECK

1. $9! = 1 \cdot 2 \cdot 3 \cdot 4 \cdot 5 \cdot 6 \cdot 7 \cdot 8 \cdot 9 = 362,880$

2. $\displaystyle {}_nC_r = \frac{n!}{r! \cdot (n-r)!}$

$$\displaystyle {}_6C_2 = \frac{6!}{2! \cdot (6-2)!} = \frac{6!}{2! \cdot 4!} = \frac{720}{2 \cdot 24} = \frac{720}{48} = 15$$

3. $\displaystyle {}_nC_n = \frac{n!}{n! \cdot 0!}$

$$\displaystyle {}_{20}C_{20} = \frac{20!}{20! \cdot 0!} = \frac{20!}{20! \cdot 1} = 1$$

4. Use the properties: $\quad {}_9C_0 = {}_9C_9 = 1$
$$\qquad {}_9C_1 = {}_9C_8 = 9$$
$$\qquad {}_9C_2 = {}_9C_7$$
$$\qquad {}_9C_3 = {}_9C_6$$
$$\qquad {}_9C_4 = {}_9C_5$$

To complete the table: $\quad {}_9C_0 = 1$
$$\qquad {}_9C_1 = 9$$
$$\qquad {}_9C_2 = 36$$
$$\qquad {}_9C_3 = 84$$
$$\qquad {}_9C_4 = 126$$
$$\qquad {}_9C_5 = 126$$
$$\qquad {}_9C_6 = 84$$
$$\qquad {}_9C_7 = 36$$
$$\qquad {}_9C_8 = 9$$
$$\qquad {}_9C_9 = 1$$

5. $(x+y)^3 = {}_3C_0 \cdot x^3 y^0 + {}_3C_1 \cdot x^{3-1}y^1 + {}_3C_2 \cdot x^{3-2}y^2 + {}_3C_3 \cdot x^0 y^3$

$$= x^3 + \frac{3!}{1! \cdot (3-1)!} x^2 y + \frac{3!}{2! \cdot (3-2)!} xy^2 + y^3$$

$$= x^3 + 3x^2 y + 3xy^2 + y^3$$

6. $(x-4)^4 = {}_4C_0 \cdot x^4 (-4)^0 + {}_4C_1 \cdot x^{4-1}(-4)^1 + {}_4C_2 \cdot x^{4-2}(-4)^2 + {}_4C_3 \cdot x^{4-3}(-4)^3 + {}_4C_4 \cdot x^0 (-4)^4$

$$= x^4 + \frac{4!}{1! \cdot (4-1)!} x^3 (-4) + \frac{4!}{2! \cdot (4-2)!} x^2 (16) + \frac{4!}{3! \cdot (4-3)!} x(-64) + 256$$

$$= x^4 - 16x^3 + 96x^2 - 256x + 256$$

7. Apply the binomial theorem with $n = 7$. Pascal's triangle gives the coefficients $1, 7, 21, 35, 35, 21, 7, 1$.

$$(x+y)^7 = x^7 + 7x^6 y + 21x^5 y^2 + 35x^4 y^3 + 35x^3 y^4 + 21x^2 y^5 + 7xy^6 + y^7$$

14.4 THE BINOMIAL THEOREM

1. n factorial; $n!$

3. binomial

5. $6! = 1 \cdot 2 \cdot 3 \cdot 4 \cdot 5 \cdot 6 = 720$

7. $8! = 1 \cdot 2 \cdot 3 \cdot 4 \cdot 5 \cdot 6 \cdot 7 \cdot 8 = 40,320$

9. $3! = 1 \cdot 2 \cdot 3 = 6$

11. $1! = 1$

13. $1 \cdot 2 \cdot 3 \ldots \cdot 17 = 17!$

15. $1 \cdot 2 \cdot 3 \ldots \cdot 33 = 33!$

17. $\dfrac{11!}{4!} = \dfrac{\cancel{1} \cdot \cancel{2} \cdot \cancel{3} \cdot \cancel{4} \ldots \cdot 11}{\cancel{1} \cdot \cancel{2} \cdot \cancel{3} \cdot \cancel{4}}$
$= 5 \cdot 6 \cdot 7 \cdot 8 \cdot 9 \cdot 10 \cdot 11$
$= 1,663,200$

19. $\dfrac{9!}{3! \cdot 6!} = \dfrac{\cancel{1} \cdot \cancel{2} \cdot \cancel{3} \ldots \cdot 9}{(1 \cdot 2 \cdot 3)(\cancel{1} \cdot \cancel{2} \cdot \cancel{3} \ldots \cancel{6})} = \dfrac{7 \cdot 8 \cdot 9}{1 \cdot 2 \cdot 3} = 84$

21. $8! + 7! = 40,320 + 5040 = 45,360$

23. $_nC_r = \dfrac{n!}{r! \cdot (n-r)!}$
$_7C_2 = \dfrac{7!}{2! \cdot (7-2)!} = \dfrac{7!}{2! \cdot 5!} = \dfrac{7 \cdot \cancel{6}^{3} \cdot \cancel{5!}^{1}}{\cancel{2}_1 \cdot \cancel{5!}_1} = 21$

25. $_nC_r = \dfrac{n!}{r! \cdot (n-r)!}$
$_8C_3 = \dfrac{8!}{3! \cdot (8-3)!}$
$= \dfrac{8!}{3! \cdot 5!}$
$= \dfrac{8 \cdot 7 \cdot \cancel{6}^{1} \cdot \cancel{5!}^{1}}{\cancel{6}_1 \cdot \cancel{5!}_1}$
$= 56$

27. $_nC_r = \dfrac{n!}{r! \cdot (n-r)!}$
$_{12}C_{11} = \dfrac{12!}{11! \cdot (12-11)!} = \dfrac{12!}{11! \cdot 1!} = \dfrac{12 \cdot \cancel{11!}^{1}}{\cancel{11!}_1} = 12$

29. $_nC_r = \dfrac{n!}{r! \cdot (n-r)!}$
$_{13}C_4 = \dfrac{13!}{4! \cdot (13-4)!}$
$= \dfrac{13!}{4! \cdot 9!}$
$= \dfrac{13 \cdot \cancel{12}^{1} \cdot 11 \cdot 10 \cdot \cancel{9!}^{1}}{\cancel{24}_2 \cdot \cancel{9!}_1}$
$= \dfrac{13 \cdot 11 \cdot \cancel{10}^{5}}{\cancel{2}_1}$
$= 715$

31. $_nC_r = \dfrac{n!}{r! \cdot (n-r)!}$
$_8C_0 = \dfrac{8!}{0! \cdot (8-0)!} = \dfrac{\cancel{8!}^{1}}{1 \cdot \cancel{8!}_1} = 1$

33. Use the properties:
$_{11}C_0 = {}_{11}C_{11} = 1$
$_{11}C_1 = {}_{11}C_{10} = 11$
$_{11}C_2 = {}_{11}C_9$
$_{11}C_3 = {}_{11}C_8$
etc.

To complete the table:
$_{11}C_0 = 1$
$_{11}C_1 = 11$
$_{11}C_2 = 55$
$_{11}C_3 = 165$
$_{11}C_4 = 330$
$_{11}C_5 = 462$
$_{11}C_6 = 462$
$_{11}C_7 = 330$
$_{11}C_8 = 165$
$_{11}C_9 = 55$
$_{11}C_{10} = 11$
$_{11}C_{11} = 1$

35. Use the binomial theorem and its properties.

$$(x+y)^3 = {}_3C_0 \cdot x^3 y^0 + {}_3C_1 \cdot x^{3-1} y^1 + {}_3C_2 \cdot x^{3-2} y^2 + {}_3C_3 \cdot x^0 y^3$$
$$= x^3 + \frac{3!}{1! \cdot (3-1)!} x^2 y + \frac{3!}{2! \cdot (3-2)!} xy^2 + y^3$$
$$= x^3 + 3x^2 y + 3xy^2 + y^3$$

37. Use the binomial theorem and its properties.

$$(x+2)^3 = {}_3C_0 \cdot x^3 2^0 + {}_3C_1 \cdot x^{3-1} 2^1 + {}_3C_2 \cdot x^{3-2} 2^2 + {}_3C_3 \cdot x^0 2^3$$
$$= x^3 + \frac{3!}{1! \cdot (3-1)!} x^2 \cdot 2 + \frac{3!}{2! \cdot (3-2)!} x \cdot 4 + 8$$
$$= x^3 + 6x^2 + 12x + 8$$

39. Use the binomial theorem and its properties.

$$(x+6)^4 = {}_4C_0 \cdot x^4 6^0 + {}_4C_1 \cdot x^{4-1} 6^1 + {}_4C_2 \cdot x^{4-2} 6^2 + {}_4C_3 \cdot x^{4-3} 6^3 + {}_4C_4 \cdot x^0 6^4$$
$$= x^4 + \frac{4!}{1! \cdot (4-1)!} x^3 \cdot 6 + \frac{4!}{2! \cdot (4-2)!} x^2 \cdot 36 + \frac{4!}{3! \cdot (4-3)!} x \cdot 216 + 1296$$
$$= x^4 + 24x^3 + 216x^2 + 864x + 1296$$

41. Use the binomial theorem and its properties.

$$(x-y)^5 = {}_5C_0 \cdot x^{5-0} (-y)^0 + {}_5C_1 \cdot x^{5-1} (-y)^1 + {}_5C_2 \cdot x^{5-2} (-y)^2 + {}_5C_3 \cdot x^{5-3} (-y)^3 + {}_5C_4 \cdot x^{5-4} (-y)^4$$
$$+ {}_5C_5 \cdot x^{5-5} (-y)^5$$
$$= x^5 - 5x^4 y + 10x^3 y^2 - 10x^2 y^3 + 5xy^4 - y^5$$

43. Use the binomial theorem and its properties.

$$(x-3)^4 = {}_4C_0 \cdot x^{4-0} (-3)^0 + {}_4C_1 \cdot x^{4-1} (-3)^1 + {}_4C_2 \cdot x^{4-2} (-3)^2 + {}_4C_3 \cdot x^{4-3} (-3)^3 + {}_4C_4 \cdot x^{4-4} (-3)^4$$
$$= x^4 + 4x^3 (-3) + 6x^2 (9) + 4x(-27) + 81$$
$$= x^4 - 12x^3 + 54x^2 - 108x + 81$$

45. Use the binomial theorem and its properties.

$$(x+3y)^3 = {}_3C_0 \cdot x^{3-0} (3y)^0 + {}_3C_1 \cdot x^{3-1} (3y)^1 + {}_3C_2 \cdot x^{3-2} (3y)^2 + {}_3C_3 \cdot x^{3-3} (3y)^3$$
$$= x^3 + 3x^2 (3y) + 3x(9y^2) + 27y^3$$
$$= x^3 + 9x^2 y + 27xy^2 + 27y^3$$

47. Use the binomial theorem and its properties.

$$(4x-y)^3 = {}_3C_0 \cdot (4x)^{3-0} (-y)^0 + {}_3C_1 \cdot (4x)^{3-1} (-y)^1 + {}_3C_2 \cdot (4x)^{3-2} (-y)^2 + {}_3C_3 \cdot (4x)^{3-3} (-y)^3$$
$$= 64x^3 + 3(16x^2)(-y) + 3(4x) y^2 + (-y^3)$$
$$= 64x^3 - 48x^2 y + 12xy^2 - y^3$$

49. Start with row $n = 7$ from Pascal's Triangle and fill in rows 8 through 10.

n											
7		1	7	21	35	35	21	7	1		
8	1	8	28	56	70	56	28	8	1		
9	1	9	36	84	126	126	84	36	9	1	
10	1	10	45	120	210	252	210	120	45	10	1

1, 10, 45, 120, 210, 252, 210, 120, 45, 10, 1

51. Sum of the binomial coefficients:

n		Sum
1	$1+1=2$	2
2	$1+2+1=4$	4
3	$1+3+3+1=8$	8
4	$1+4+6+4+1=16$	16
5	$1+5+10+10+5+1=32$	32
6	$1+6+15+20+15+6+1=64$	64

53. Use the binomial theorem and Pascal's triangle $(n=5)$.

$$(x+y)^5 = 1x^{5-0}y^0 + 5x^{5-1}y^1 + 10x^{5-2}y^2 + 10x^{5-3}y^3 + 5x^{5-4}y^4 + 1x^{5-5}y^5$$
$$= x^5 + 5x^4y + 10x^3y^2 + 10x^2y^3 + 5xy^4 + y^5$$

55. Use the binomial theorem and Pascal's triangle $(n=7)$.

$$(x+2)^7 = 1x^{7-0}(2)^0 + 7x^{7-1}(2)^1 + 21x^{7-2}(2)^2 + 35x^{7-3}(2)^3 + 35x^{7-4}(2)^4 + 21x^{7-5}(2)^5 + 7x^{7-6}(2)^6 + 1x^{7-7}(2)^7$$
$$= x^7 + 14x^6 + 84x^5 + 280x^4 + 560x^3 + 672x^2 + 448x + 128$$

57. Use the binomial theorem and Pascal's triangle $(n=3)$.

$$(x-y)^3 = 1x^{3-0}(-y)^0 + 3x^{3-1}(-y)^1 + 3x^{3-2}(-y)^2 + 1x^{3-3}(-y)^3$$
$$= x^3 - 3x^2y + 3xy^2 - y^3$$

59. Use the binomial theorem and Pascal's triangle $(n=4)$.

$$(x-7)^4 = 1x^{4-0}(-7)^0 + 4x^{4-1}(-7)^1 + 6x^{4-2}(-7)^2 + 4x^{4-3}(-7)^3 + 1^{4-4}(-7)^4$$
$$= x^4 - 28x^3 + 294x^2 - 1372x + 2401$$

61. Answers will vary. Example:

$$_nC_a = \frac{n!}{a!\cdot(n-a)!} \quad \text{and} \quad _nC_b = \frac{n!}{b!\cdot(n-b)!}$$

If $a+b=n$ then $a=n-b$ and/or $b=n-a$.

Substituting,

$$_nC_a = \frac{n!}{(n-b)!\cdot b!} \quad \text{and} \quad _nC_b = \frac{n!}{(n-a)!\cdot a!}$$

Therefore, $_nC_a = {}_nC_b$.

CHAPTER 14 REVIEW

1. $a_n = n + 7$
    $a_1 = 1 + 7 = 8$
    $a_2 = 2 + 7 = 9$
    $a_3 = 3 + 7 = 10$
    $a_4 = 4 + 7 = 11$
    $a_5 = 5 + 7 = 12$
    $8, 9, 10, 11, 12$

2. $a_n = 5n - 19$
    $a_1 = 5(1) - 19 = -14$
    $a_2 = 5(2) - 19 = -9$
    $a_3 = 5(3) - 19 = -4$
    $a_4 = 5(4) - 19 = 1$
    $a_5 = 5(5) - 19 = 6$
    $-14, -9, -4, 1, 6$

3. $13, 26, 39, 52, 65, \ldots$

    Each successive term increases by 13.
    The next three terms are: $78, 91, 104$.

    $a_n = 13n$

4. $12, 19, 26, 33, 40, \ldots$

    Each successive term increases by 7.
    The next three terms are: $47, 54, 61$.

    $a_n = 7n + 5$

5. $a_n = (a_{n-1}) + 17$
    $a_1 = -9$
    $a_2 = (-9) + 17 = 8$
    $a_3 = (8) + 17 = 25$
    $a_4 = (25) + 17 = 42$
    $a_5 = (42) + 17 = 59$
    $-9, 8, 25, 42, 59$

6. $a_n = 2(a_{n-1}) - 23$
    $a_1 = -4$
    $a_2 = 2(-4) - 23 = -31$
    $a_3 = 2(-31) - 23 = -85$
    $a_4 = 2(-85) - 23 = -193$
    $a_5 = 2(-193) - 23 = -409$
    $-4, -31, -85, -193, -409$

7. $3, 17, 31, 45, 59, \ldots$

    The difference between each term is 14.
    $a_n = a_{n-1} + 14$

8. $5, -10, 20, -40, 80, \ldots$

    The terms alternate signs. Neglecting sign, each
    successive terms doubles in value.
    $a_n = -2a_{n-1}$

9. $4, 14, 24, 34, \ldots$

    The difference between each term is 10, so
    a_5 and a_6 are 44 and 54.

    $s_6 = 4 + 14 + 24 + 34 + 44 + 54 = 174$

10. $2, 12, 72, 432, \ldots$

    Each successive factor differs by a factor of 6.
    $a_5 = 6(432) = 2592$
    $a_6 = 6(2592) = 15,552$
    $a_7 = 6(15,552) = 93,312$
    $a_8 = 6(93,312) = 559,872$
    $a_9 = 6(559,872) = 3,359,232$

    $s_9 = 2 + 12 + 72 + 432 + 2592 + \ldots + 3,329,232$
    $ = 4,031,078$

11. $a_n = 2n - 5$
    $a_1 = 2(1) - 5 = -3$
    $a_2 = 2(2) - 5 = -1$
    $a_3 = 2(3) - 5 = 1$
    $a_4 = 2(4) - 5 = 3$
    $a_5 = 2(5) - 5 = 5$
    $a_6 = 2(6) - 5 = 7$
    $a_7 = 2(7) - 5 = 9$

    $s_7 = -3 + (-1) + 1 + 3 + 5 + 7 + 9 = 21$

12. $a_n = 5n + 6$
    $a_1 = 5(1) + 6 = 11$
    $a_2 = 5(2) + 6 = 16$
    $a_3 = 5(3) + 6 = 21$
    $a_4 = 5(4) + 6 = 26$
    $a_5 = 5(5) + 6 = 31$
    $a_6 = 5(6) + 6 = 36$
    $a_7 = 5(7) + 6 = 41$
    $a_8 = 5(8) + 6 = 46$

    $s_8 = 11 + 16 + 21 + 26 + 31 + 36 + 41 + 46 = 228$

13. $a_n = a_{n-1} - 8$
    $a_1 = 50$
    $a_2 = 50 - 8 = 42$
    $a_3 = 42 - 8 = 34$
    $a_4 = 34 - 8 = 26$
    $a_5 = 26 - 8 = 18$
    $a_6 = 18 - 8 = 10$
    $a_7 = 10 - 8 = 2$

    $s_7 = 50 + 42 + 34 + 26 + 18 + 10 + 2 = 182$

14. $a_n = 4a_{n-1} + 5$
$a_1 = 20$
$a_2 = 4(20) + 5 = 85$
$a_3 = 4(85) + 5 = 345$
$a_4 = 4(345) + 5 = 1385$
$a_5 = 4(1385) + 5 = 5545$

$s_5 = 20 + 85 + 345 + 1385 + 5545 = 7380$

15. $\displaystyle\sum_{i=1}^{12} i = 1 + 2 + 3 + 4 + \ldots + 11 + 12 = 78$

16. $\displaystyle\sum_{i=1}^{11} (5i + 6) = (5(1) + 6) + (5(2) + 6) \ldots$
$\qquad\qquad + (5(11) + 6)$
$\qquad\quad = 11 + 16 + 21 + 26 + 31 + 36 + 41$
$\qquad\qquad + 46 + 51 + 56 + 61$
$\qquad\quad = 396$

17. $7, 31, 55, 79, 103, \ldots$
$d = 31 - 7 = 24$

18. $-3, -9, -15, -21, -27, \ldots$
$d = -9 - (-3) = -6$

19. $a_1 = 16$ and $d = -9$

$a_2 = a_1 + d = 16 + (-9) = 7$
$a_3 = a_1 + 2d = 16 + 2(-9) = -2$
$a_4 = a_1 + 3d = 16 + 3(-9) = -11$
$a_5 = a_1 + 4d = 16 + 4(-9) = -20$
$16, 7, -2, -11, -20$

20. $a_1 = -18$ and $d = 11$

$a_2 = a_1 + d = -18 + 11 = -7$
$a_3 = a_1 + 2d = -18 + 2(11) = 4$
$a_4 = a_1 + 3d = -18 + 3(11) = 15$
$a_5 = a_1 + 4d = -18 + 4(11) = 26$
$-18, -7, 4, 15, 26$

21. $4, 11, 18, 25, 32, \ldots$

$a_1 = 4$
$d = 11 - 4 = 7$

$a_n = a_1 + (n-1)d$
$a_n = 4 + (n-1)7 = 4 + 7n - 7 = 7n - 3$

22. $8, 1, -6, -13, -20, \ldots$

$a_1 = 8$
$d = 1 - 8 = -7$

$a_n = a_1 + (n-1)d$
$a_n = 8 + (n-1)(-7) = 8 - 7n + 7 = -7n + 15$

23. $-15, -13, -11, -9, -7, \ldots$

$a_1 = -15$
$d = -13 - (-15) = 2$

$a_n = a_1 + (n-1)d$
$a_n = -15 + (n-1)2 = -15 + 2n - 2 = 2n - 17$

24. $\dfrac{5}{6}, \dfrac{19}{12}, \dfrac{7}{3}, \dfrac{37}{12}, \dfrac{23}{6}, \ldots$

$a_1 = \dfrac{5}{6}$

$d = \dfrac{19}{12} - \dfrac{5}{6} = \dfrac{19}{12} - \dfrac{10}{12} = \dfrac{9}{12} = \dfrac{3}{4}$

$a_n = a_1 + (n-1)d$

$a_n = \dfrac{5}{6} + (n-1)\dfrac{3}{4} = \dfrac{5}{6} + \dfrac{3}{4}n - \dfrac{3}{4} = \dfrac{3}{4}n + \dfrac{1}{12}$

25. $25, 46, 67, 88, 109, \ldots$

$a_1 = 25$
$d = 46 - 25 = 21$

$a_n = a_1 + (n-1)d$
$a_n = 25 + (n-1)21 = 25 + 21n - 21 = 21n + 4$
$a_{39} = 21(39) + 4 = 823$

26. $-9, -32, -55, -78, -101, \ldots$

$a_1 = -9$
$d = -32 - (-9) = -23$

$a_n = a_1 + (n-1)d$
$a_n = -9 + (n-1)(-23)$
$\qquad = -9 - 23n + 23$
$\qquad = -23n + 14$
$a_{27} = -23(27) + 14 = -607$

27. $-12, -5, 2, 9, 16, \ldots$

$a_1 = -12$
$d = -5 - (-12) = 7$

$a_n = a_1 + (n-1)d$
$a_n = -12 + (n-1)7 = -12 + 7n - 7 = 7n - 19$

$1633 = 7n - 19$
$1652 = 7n$
$236 = n$
The 236th term is 1633.

28. $40, 21, 2, -17, -36, \ldots$

$a_1 = 40$
$d = 21 - 40 = -19$

$a_n = a_1 + (n-1)d$
$a_n = 40 + (n-1)(-19)$
$\quad = 40 - 19n + 19$
$\quad = -19n + 59$
$-9346 = -19n + 59$
$-9405 = -19n$
$\quad 495 = n$
The 495th term is -9346.

29. $10, 17, 24, 31, \ldots$

$a_1 = 10$
$d = 17 - 10 = 7$

$a_n = a_1 + (n-1)d$
$a_n = 10 + (n-1)7 = 10 + 7n - 7 = 7n + 3$
$a_8 = 7(8) + 3 = 59$

$s_n = \dfrac{n}{2}(a_1 + a_n)$

$s_8 = \dfrac{8}{2}(a_1 + a_8)$

$s_8 = \dfrac{8}{2}(10 + 59) = 276$

30. $-102, -84, -66, -48, \ldots$

$a_1 = -102$
$d = -84 - (-102) = 18$

$a_n = a_1 + (n-1)d$
$a_n = -102 + (n-1)18$
$\quad = -102 + 18n - 18$
$\quad = 18n - 120$
$a_{15} = 18(15) - 120 = 150$

$s_n = \dfrac{n}{2}(a_1 + a_n)$

$s_{15} = \dfrac{15}{2}(a_1 + a_{15})$

$s_{15} = \dfrac{15}{2}(-102 + 150) = 360$

31. $16, 13, 10, 7, \ldots$

$a_1 = 16$
$d = 13 - 16 = -3$

$a_n = a_1 + (n-1)d$
$a_n = 16 + (n-1)(-3) = 16 - 3n + 3 = -3n + 19$
$a_{24} = -3(24) + 19 = -53$

$s_n = \dfrac{n}{2}(a_1 + a_n)$

$s_{24} = \dfrac{24}{2}(a_1 + a_{24})$

$s_{24} = \dfrac{24}{2}(16 + (-53)) = -444$

32. $-10, -3, 4, , 11, \ldots$

$a_1 = -10$
$d = -3 - (-10) = 7$

$a_n = a_1 + (n-1)d$
$a_n = -10 + (n-1)7 = -10 + 7n - 7 = 7n - 17$
$a_{30} = 7(30) - 17 = 193$

$s_n = \dfrac{n}{2}(a_1 + a_n)$

$s_{30} = \dfrac{30}{2}(a_1 + a_{30})$

$s_{30} = \dfrac{30}{2}(-10 + 193) = 2745$

33. Positive integers: $1, 2, 3, 4, \ldots$

$a_1 = 1$ and $d = 2 - 1 = 1$
$a_n = a_1 + (n-1)d$
$a_n = 1 + (n-1)1 = 1 + n - 1 = n$
$a_{340} = 340$

$s_n = \dfrac{n}{2}(a_1 + a_n)$

$s_{340} = \dfrac{340}{2}(a_1 + a_{340})$

$s_{340} = \dfrac{340}{2}(1 + 340) = 57{,}970$

34. a) Each successive term in the sequence increases by 2,000,000.

$\$20,000,000, \$22,000,000, \$24,000,000,$
$\$26,000,000, \$28,000,000$

b) $a_1 = 20,000,000$
$d = 2,000,000$

$a_n = a_1 + (n-1)d$
$a_n = 20,000,000 + (n-1)2,000,000$
$a_n = 20,000,000 + 2,000,000n - 2,000,000$
$\quad = 2,000,000n + 18,000,000$

c) $a_{10} = 2,000,000(10) + 18,000,000$
$\quad\quad = 38,000,000$

$s_n = \dfrac{n}{2}(a_1 + a_n)$

$s_{10} = \dfrac{10}{2}(a_1 + a_{10})$

$s_{10} = \dfrac{10}{2}(20,000,000 + 38,000,000)$
$\quad\quad = 290,000,000$
$\$290,000,000$

35. $1, 10, 100, 1000, 10000, \ldots$

$r = \dfrac{10}{1} = 10$

36. $5, 30, 180, 1080, 6480, \ldots$

$$r = \frac{30}{5} = 6$$

37. $1, 12, 144, 1728, \ldots$

$$a_1 = 1$$
$$r = \frac{12}{1} = 12$$
$$a_n = a_1 \cdot r^{n-1}$$
$$a_n = 1 \cdot 12^{n-1}$$

38. $8, 6, \dfrac{9}{2}, \dfrac{27}{8}, \ldots$

$$a_1 = 8$$
$$r = \frac{6}{8} = \frac{3}{4}$$
$$a_n = a_1 \cdot r^{n-1}$$
$$a_n = 8 \cdot \left(\frac{3}{4}\right)^{n-1}$$

39. $-4, -12, -36, -108, \ldots$

$$a_1 = -4$$
$$r = \frac{-12}{-4} = 3$$
$$a_n = a_1 \cdot r^{n-1}$$
$$a_n = -4 \cdot 3^{n-1}$$

40. $-24, 8, -\dfrac{8}{3}, \dfrac{8}{9}, \ldots$

$$a_1 = -24$$
$$r = \frac{8}{-24} = -\frac{1}{3}$$
$$a_n = a_1 \cdot r^{n-1}$$
$$a_n = -24 \cdot \left(-\frac{1}{3}\right)^{n-1}$$

41. $9, 18, 36, 72, \ldots$

$$a_1 = 9$$
$$r = \frac{18}{9} = 2$$
$$a_n = a_1 \cdot r^{n-1}$$
$$a_n = 9 \cdot 2^{n-1}$$
$$a_8 = 9 \cdot 2^{8-1} = 1152$$

42. $a_1 = 1$

$$r = 3$$
$$s_n = a_1 \cdot \frac{1-r^n}{1-r}$$
$$s_9 = 1 \cdot \frac{1-3^9}{1-3} = 9841$$

43. $a_1 = 10$

$$r = -2$$
$$s_n = a_1 \cdot \frac{1-r^n}{1-r}$$
$$s_{11} = 10 \cdot \frac{1-(-2)^{11}}{1-(-2)} = 6830$$

44. $1, 6, 36, 216, \ldots$

$$a_1 = 1$$
$$r = \frac{6}{1} = 6$$
$$s_n = a_1 \cdot \frac{1-r^n}{1-r}$$
$$s_7 = 1 \cdot \frac{1-6^7}{1-6} = 55{,}987$$

45. $\dfrac{3}{2}, 12, 96, 768, \ldots$

$$a_1 = \frac{3}{2}$$
$$r = \frac{12}{\frac{3}{2}} = 8$$
$$s_n = a_1 \cdot \frac{1-r^n}{1-r}$$
$$s_8 = \frac{3}{2} \cdot \frac{1-8^8}{1-8} = 3{,}595{,}117.5$$

46. $320, 160, 80, 40, \ldots$

$$a_1 = 320$$
$$r = \frac{160}{320} = \frac{1}{2}$$
$$s_n = a_1 \cdot \frac{1-r^n}{1-r}$$
$$s_{10} = 320 \cdot \frac{1-\left(\frac{1}{2}\right)^{10}}{1-\frac{1}{2}} = 639.375$$

47. $35, -140, 560, -2240, \ldots$

$a_1 = 35$

$r = \dfrac{-140}{35} = -4$

$s_n = a_1 \cdot \dfrac{1 - r^n}{1 - r}$

$s_9 = 35 \cdot \dfrac{1 - (-4)^9}{1 - (-4)} = 1{,}835{,}015$

48. $64, 80, 100, 125, \ldots$

$a_1 = 64$

$r = \dfrac{80}{64} = \dfrac{5}{4}$

$|r| = \left|\dfrac{5}{4}\right| = \dfrac{5}{4} > 1$

No, since $|r| > 1$, the infinite series has no limit.

49. $\dfrac{36}{7}, \dfrac{24}{7}, \dfrac{16}{7}, \dfrac{32}{21}, \ldots$

$a_1 = \dfrac{36}{7}$

$r = \dfrac{\frac{24}{7}}{\frac{36}{7}} = \dfrac{2}{3}$

since $|r| = \left|\dfrac{2}{3}\right| = \dfrac{2}{3} < 1$

Yes, since $|r| < 1$, the infinite series has a limit.

$s = \dfrac{a_1}{1 - r}$

$s = \dfrac{\frac{36}{7}}{1 - \frac{2}{3}} = \dfrac{\frac{36}{7}}{\frac{1}{3}} = \dfrac{108}{7} = 15\dfrac{3}{7}$

50. $a_n = -21 \cdot \left(\dfrac{2}{7}\right)^{n-1}$

$a_1 = -21$

$r = \dfrac{2}{7}$

$|r| = \left|\dfrac{2}{7}\right| = \dfrac{2}{7} < 1$

Yes, since $|r| < 1$, the infinite series has a limit.

$s = \dfrac{a_1}{1 - r}$

$s = \dfrac{-21}{1 - \frac{2}{7}} = \dfrac{-21}{\frac{5}{7}} = -\dfrac{147}{5} = -29\dfrac{2}{5}$

51. $a_n = 45 \cdot \left(-\dfrac{7}{8}\right)^{n-1}$

$a_1 = 45$

$r = -\dfrac{7}{8}$

$|r| = \left|-\dfrac{7}{8}\right| = \dfrac{7}{8} < 1$

Yes, since $|r| < 1$, the infinite series has a limit.

$s = \dfrac{a_1}{1 - r}$

$s = \dfrac{45}{1 - \left(-\frac{7}{8}\right)} = \dfrac{45}{\frac{15}{8}} = 24$

52. The budget increases by 10% per year, so the budget value
in each successive year differs by a factor of 1.1.

$a_1 = 150{,}000$

$r = 1.1$

$s_n = a_1 \cdot \dfrac{1 - r^n}{1 - r}$

$s_8 = 150{,}000 \cdot \dfrac{1 - 1.1^8}{1 - 1.1} \approx 1{,}715{,}383.22$

$\$1{,}715{,}383.22$

53. $9! = 1 \cdot 2 \cdot 3 \cdot 4 \cdot 5 \cdot 6 \cdot 7 \cdot 8 \cdot 9 = 362{,}880$

54. $7! = 1 \cdot 2 \cdot 3 \cdot 4 \cdot 5 \cdot 6 \cdot 7 = 5040$

55. $_nC_r = \dfrac{n!}{r! \cdot (n - r)!}$

$_8C_3 = \dfrac{8!}{3! \cdot (8 - 3)!} = 56$

56. $_nC_r = \dfrac{n!}{r! \cdot (n - r)!}$

$_{11}C_8 = \dfrac{11!}{8! \cdot (11 - 8)!} = 165$

57. Use the binomial theorem and its properties.

$$(x+y)^6 = {}_6C_0 \cdot x^{6-0}y^0 + {}_6C_1 \cdot x^{6-1}y^1 + {}_6C_2 \cdot x^{6-2}y^2 + {}_6C_3 \cdot x^{6-3}y^3 + {}_6C_4 \cdot x^{6-4}y^4 + {}_6C_5 \cdot x^{6-5}y^5 + {}_6C_6 \cdot x^{6-6}y^6$$

$$= x^6 + \frac{6!}{1! \cdot (6-1)!}x^5 y + \frac{6!}{2! \cdot (6-2)!}x^4 y^2 + \frac{6!}{3! \cdot (6-3)!}x^3 y^3 + \frac{6!}{4! \cdot (6-4)!}x^2 y^4 + \frac{6!}{5! \cdot (6-5)!}xy^5 + y^6$$

$$= x^6 + 6x^5 y + 15x^4 y^2 + 20x^3 y^3 + 15x^2 y^4 + 6xy^5 + y^6$$

58. Use the binomial theorem and its properties.

$$(x+6)^4 = {}_4C_0 \cdot x^{4-0}6^0 + {}_4C_1 \cdot x^{4-1}6^1 + {}_4C_2 \cdot x^{4-2}6^2 + {}_4C_3 \cdot x^{4-3}6^3 + {}_4C_4 \cdot x^{4-4}6^4$$

$$= x^4 + 4x^3 \cdot 6 + 6x^2 \cdot 36 + 4x \cdot 216 + 1296$$

$$= x^4 + 24x^3 + 216x^2 + 864x + 1296$$

59. Use the binomial theorem and its properties.

$$(x-9)^3 = {}_3C_0 \cdot x^{3-0}(-9)^0 + {}_3C_1 \cdot x^{3-1}(-9)^1 + {}_3C_2 \cdot x^{3-2}(-9)^2 + {}_3C_3 \cdot x^{3-3}(-9)^3$$

$$= x^3 + 3x^2(-9) + 3x(81) - 729$$

$$= x^3 - 27x^2 + 243x - 729$$

60. Use the binomial theorem and Pascal's triangle $(n = 9)$.

$$(x+y)^9 = x^9 + 9x^8 y + 36x^7 y^2 + 84x^6 y^3 + 126x^5 y^4 + 126x^4 y^5 + 84x^3 y^6 + 36x^2 y^7 + 9xy^8 + y^9$$

CHAPTER 14 TEST

1. $11, 26, 41, 56, 71, \ldots$

Each successive term increases by 15.
The next three terms are: 86, 101, 116.

$$a_n = 15n - 4$$

2. $\displaystyle\sum_{i=1}^{9}(4i+11)$

$$= (4(1)+11) + (4(2)+11) \ldots + (4(9)+11)$$

$$= 15 + 19 + 23 + 27 + 31 + 35 + 39 + 43 + 47$$

$$= 279$$

3. $a_1 = 9$ and $d = 8$

$$a_2 = a_1 + d = 9 + 8 = 17$$
$$a_3 = a_1 + 2d = 9 + 2(8) = 25$$
$$a_4 = a_1 + 3d = 9 + 3(8) = 33$$
$$a_5 = a_1 + 4d = 9 + 4(8) = 41$$

$$9, 17, 25, 33, 41$$

4. $20, 13, 6, -1, -8, \ldots$

$$a_1 = 20$$
$$d = 13 - 20 = -7$$
$$a_n = a_1 + (n-1)d$$
$$a_n = 20 + (n-1)(-7) = 20 - 7n + 7 = -7n + 27$$

5. $14, \dfrac{35}{2}, 21, \dfrac{49}{2}, \ldots$

$$a_1 = 14$$

$$d = \frac{35}{2} - 14 = \frac{35}{2} - \frac{28}{2} = \frac{7}{2}$$

$$a_n = a_1 + (n-1)d$$

$$a_n = 14 + (n-1)\frac{7}{2} = 14 + \frac{7}{2}n - \frac{7}{2} = \frac{7}{2}n + \frac{21}{2}$$

$$a_{15} = \frac{7}{2}(15) + \frac{21}{2} = 63$$

$$s_n = \frac{n}{2}(a_1 + a_n)$$

$$s_{15} = \frac{15}{2}(a_1 + a_{15})$$

$$s_{15} = \frac{15}{2}(14 + 63) = 577.5$$

6. $2, 4, 6, 8, \ldots$

$a_1 = 2$
$d = 4 - 2 = 2$

$a_n = a_1 + (n-1)d$
$a_n = 2 + (n-1)2 = 2 + 2n - 2 = 2n$
$a_{200} = 2(200) = 400$

$s_n = \dfrac{n}{2}(a_1 + a_n)$

$s_{200} = \dfrac{200}{2}(a_1 + a_{200})$

$s_{200} = \dfrac{200}{2}(2 + 400) = 40,200$

7. $a_1 = 5$ and $r = 8$
$a_2 = a_1 \cdot r^{2-1} = 5 \cdot 8^1 = 40$
$a_3 = a_1 \cdot r^{3-1} = 5 \cdot 8^2 = 320$
$a_4 = a_1 \cdot r^{4-1} = 5 \cdot 8^3 = 2560$
$a_5 = a_1 \cdot r^{5-1} = 5 \cdot 8^4 = 20,480$
$5, 40, 320, 2560, 20480$

8. $4, 28, 196, 1372, \ldots$

$a_1 = 4$

$r = \dfrac{28}{4} = 7$

$a_n = a_1 \cdot r^{n-1}$

$a_n = 4 \cdot 7^{n-1}$

9. $3, 15, 75, 375, \ldots$

$a_1 = 3$

$r = \dfrac{15}{3} = 5$

$s_n = a_1 \cdot \dfrac{1 - r^n}{1 - r}$

$s_{10} = 3 \cdot \dfrac{1 - 5^{10}}{1 - 5} = 7,324,218$

10. $10, -6, \dfrac{18}{5}, -\dfrac{54}{25}, \ldots$

$a_1 = 10$

$r = \dfrac{-6}{10} = -\dfrac{3}{5}$

$|r| = \left|-\dfrac{3}{5}\right| = \dfrac{3}{5} < 1$

Yes, since $|r| < 1$, the infinite series has a limit.

$s = \dfrac{a_1}{1 - r}$

$s = \dfrac{10}{1 - \left(-\dfrac{3}{5}\right)} = \dfrac{10}{\dfrac{8}{5}} = \dfrac{25}{4} = 6\dfrac{1}{4}$

11. a) Each successive term in the sequence increases by \$500.

$\$2500, \$3000, \$3500, \$4000, \$4500, \$5000$

b) $a_1 = 2500$
$d = 3000 - 2500 = 500$

$a_n = a_1 + (n-1)d$
$a_n = 2500 + (n-1)500$
$\quad = 2500 + 500n - 500$
$\quad = 500n + 2000$

c) $a_{18} = 500(18) + 2000 = 11,000$

$s_n = \dfrac{n}{2}(a_1 + a_n)$

$s_{18} = \dfrac{18}{2}(a_1 + a_{18})$

$s_{18} = \dfrac{18}{2}(2500 + 11,000) = 121,500$

$\$121,500$

12. a) A new car decreases in value by 25% each year, so its value at the end of a year is 0.75 times the value of the previous year.

$a_1 = 18,000(0.75) = 13,500$
$a_2 = 13,500(0.75) = 10,125$
$a_3 = 10,125(.075) = 7593.75$
$a_4 = 7593.75(0.75) \approx 5695.31$

$\$13,500, \$10,125, \$7593.75, \$5695.31$

b) $a_n = 18,000(0.75)^{n-1}$

13. $_nC_r = \dfrac{n!}{r! \cdot (n-r)!}$

$_{12}C_5 = \dfrac{12!}{5! \cdot (12-5)!} = 792$

14. Use the binomial theorem and its properties:

$$(x+1)^4 = {}_4C_0 \cdot x^{4-0} \cdot 1^0 + {}_4C_1 \cdot x^{4-1} \cdot 1^1 + {}_4C_2 \cdot x^{4-2} \cdot 1^2 + {}_4C_3 \cdot x^{4-3} \cdot 1^3 + {}_4C_4 \cdot x^{4-4} \cdot 1^4$$
$$= x^4 + 4x^3 + 6x^2 + 4x + 1$$

15. $(4x+y)^7 = 1(4x)^7 y^0 + 7(4x)^6 y^1 + 21(4x)^5 y^2 + 35(4x)^4 y^3 + 35(4x)^3 y^4 + 21(4x)^2 y^5 + 7(4x)^1 y^6$
$$+ 1 \cdot (4x)^0 y^7$$
$$= 16{,}384x^7 + 28{,}672x^6 y + 21{,}504x^5 y^2 + 8960x^4 y^3 + 2240x^3 y^4 + 336x^2 y^5 + 28xy^6 + y^7$$

CUMULATIVE REVIEW CHAPTERS 12–14

1. $f(2) = e^{2+3} - 21 = e^5 - 21 \approx 127.413$

2. $f(25) = \ln(25+7) + 32$
$$= \ln(32) + 32$$
$$\approx 35.466$$

3. $\log_3 81 = \log_3 3^4 = 4$

4. $\ln e^8 = 8$

5. $2\ln x - 4\ln y + 5\ln z = \ln x^2 - \ln y^4 + \ln z^5$
$$= \ln \frac{x^2}{y^4} + \ln z^5$$
$$= \ln\left(\frac{x^2 z^5}{y^4}\right)$$

6. $\log\left(\frac{b^4 c^5}{a}\right) = \log\left(b^4 c^5\right) - \log a$
$$= \log b^4 + \log c^5 - \log a$$
$$= 4\log b + 5\log c - \log a$$

7. $\log_{13} 200 = \dfrac{\log 200}{\log 13} \approx 2.066$

8. $\log_9 3224 = \dfrac{\log 3224}{\log 9} \approx 3.677$

9. $5^{x-6} = 25$
$5^{x-6} = 5^2$
$x - 6 = 2$
$x = 8$
$\{8\}$

10. $7^x = 219$
$\log 7^x = \log 219$
$x \cdot \log 7 = \log 219$
$x = \dfrac{\log 219}{\log 7}$
$x \approx 2.769$
$\{2.769\}$

11. $e^{x+7} = 6205$
$\ln e^{x+7} = \ln 6205$
$x + 7 = \ln 6205$
$x = \ln 6205 - 7$
$x \approx 1.733$
$\{1.733\}$

12. $\log\left(x^2 + 7x - 9\right) = \log(3x + 23)$
$x^2 + 7x - 9 = 3x + 23$
$x^2 + 4x - 32 = 0$
$(x+8)(x-4) = 0$
$x + 8 = 0$ or $x - 4 = 0$
$x = -8$ or $x = 4$
check: $x = -8$
$\log\left((-8)^2 + 7(-8) - 9\right) = \log(3(-8) + 23)$
$\log(-1) = \log(-1)$
Undefined
check: $x = 4$
$\log\left(4^2 + 7(4) - 9\right) = \log(3(4) + 23)$
$\log(35) = \log(35)$
True
$\{4\}$

13. $\log_3(x - 9) = 4$
Convert to exponential form.
$x - 9 = 3^4$
$x - 9 = 81$
$x = 90$
$\{90\}$

14.
$$\log_2(x+3) + \log_2(x+9) = 4$$
$$\log_2(x+3)(x+9) = 4$$
Convert to exponential form.
$$(x+3)(x+9) = 2^4$$
$$x^2 + 12x + 27 = 16$$
$$x^2 + 12x + 11 = 0$$
$$(x+11)(x+1) = 0$$
$$x = -11 \quad \text{or} \quad x = -1$$

check: $x = -11$
$$\log_2(-11+3) + \log_2(-11+9) = 4$$
$$\log_2(-8) + \log_2(-2) = 4$$
Undefined

check: $x = -1$
$$\log_2(-1+3) + \log_2(-1+9) = 4$$
$$\log_2(2) + \log_2(8) = 4$$
$$1 + 3 = 4$$
$$4 = 4$$

True
$\{-1\}$

15.
$$f(x) = e^{x-2} + 3$$
$$y = e^{x-2} + 3$$
Exchange x and y.
$$x = e^{y-2} + 3$$
$$x - 3 = e^{y-2}$$
$$\ln(x-3) = \ln e^{y-2}$$
$$\ln(x-3) = y - 2$$
$$\ln(x-3) + 2 = y$$
$$f^{-1}(x) = \ln(x-3) + 2$$

16.
$$f(x) = \ln(x+9) - 16$$
$$y = \ln(x+9) - 16$$
Exchange x and y.
$$x = \ln(y+9) - 16$$
$$x + 16 = \ln(y+9)$$
Convert to exponential form.
$$y + 9 = e^{x+16}$$
$$y = e^{x+16} - 9$$
$$f^{-1}(x) = e^{x+16} - 9$$

17.
$$A = P\left(1 + \frac{r}{n}\right)^{nt}$$
$$100{,}000 = 2000\left(1 + \frac{0.06}{4}\right)^{4t}$$
$$5 = (1.015)^{4t}$$
$$\log 5 = \log(1.015)^{4t}$$
$$\log 5 = 4t\log(1.015)$$
$$\frac{\log 5}{4\log(1.015)} = \frac{4t\log(1.015)}{4\log(1.015)}$$
$$\frac{\log 5}{4\log(1.015)} = t$$
$$27.02 \approx t$$
It will take 27 years.

18.
$$P = P_0 e^{kt}$$

P	P_0	k	t
150,000	120,000	unknown	10

$$150{,}000 = 120{,}000 e^{k(10)}$$
$$\frac{150{,}000}{120{,}000} = e^{10k}$$
$$\ln\frac{5}{4} = \ln e^{10k}$$
$$\ln\frac{5}{4} = 10k$$
$$\frac{\ln\frac{5}{4}}{10} = k$$
$$0.022314 \approx k$$

P	P_0	k	t
unknown	120,000	0.022314	30

$$P = 120{,}000 e^{0.022314(30)}$$
$$P \approx 234{,}373$$
Approximately $234{,}373$ people

19. $f(x) = 2^x - 1$
$h = 0,\ k = -1$
Shift the graph of $f(x) = 2^x$ by 1 unit down.
The horizontal asymptote is at $y = -1$.
Shift three points. (*Subtract 1 from each*
y-coordinate.)

$f(x) = 2^x$	$\left(-1, \dfrac{1}{2}\right)$	$(0,1)$	$(1,2)$
$f(x) = 2^x - 1$	$\left(-1, -\dfrac{1}{2}\right)$	$(0,0)$	$(1,1)$

y-intercept: By shifting, $(0,0)$.

x-intercept: By shifting, $(0,0)$.

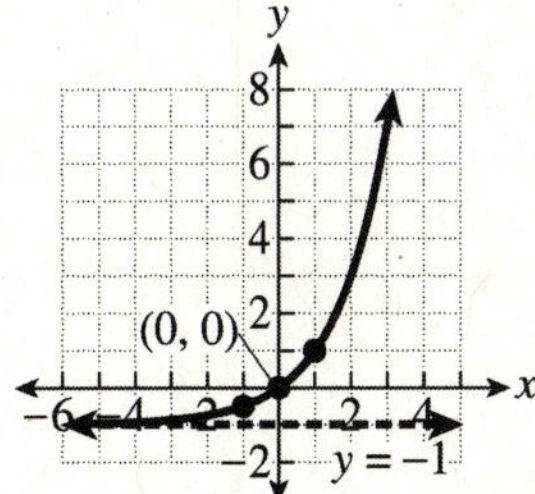

Domain: $(-\infty, \infty)$; range: $(-1, \infty)$

20. $f(x) = \ln(x+5) - 3$
$h = -5,\ k = -3$
Shift the graph of $f(x) = \ln x$ by 5 units to
the left and by 3 units down.
The vertical asymptote is at $x = -5$.
Shift three points. (Subtract 5 from each
x-coordinate and subtract 3 from each
y-coordinate.)

$f(x) = \ln x$	$(0.4, -1)$	$(1,0)$	$(2.7,1)$
$f(x) = \ln(x+5) - 3$	$(-4.6, -4)$	$(-4, -3)$	$(-2.3, -2)$

y-intercept: Find $f(0)$.

$f(0) = \ln(0+5) - 3 = \ln 5 - 3 \approx -1.39$

$(0, -1.39)$

x-intercept: Set $f(x) = 0$ and solve for *x*.

$\ln(x+5) - 3 = 0$
$\quad \ln(x+5) = 3$
$\qquad\quad x + 5 = e^3$
$\qquad\qquad x = e^3 - 5$
$\qquad\qquad x \approx 15.09$

$(15.09, 0)$

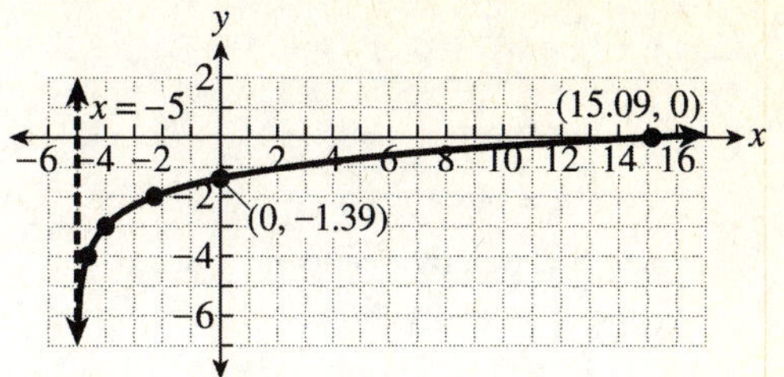

Domain: $(-5, \infty)$; range: $(-\infty, \infty)$

21. $y = x^2 + 6x - 8$

The equation is of the form $y = ax^2 + bx + c$
where $a > 0$, so the parabola opens upward.
vertex:
$$x = \frac{-b}{2a} = \frac{-6}{2(1)} = -3$$

$$y = (-3)^2 + 6(-3) - 8 = 9 - 18 - 8 = -17$$
$(-3, -17)$

y-intercept:
$$y = (0)^2 + 6(0) - 8$$
$$y = -8$$
$(0, -8)$

x-intercept:
$$0 = x^2 + 6x - 8$$

$$x = \frac{-(6) \pm \sqrt{(6)^2 - 4(1)(-8)}}{2(1)}$$

$$x = \frac{-6 \pm \sqrt{68}}{2}$$

$$x = \frac{-6 \pm 2\sqrt{17}}{2}$$

$$x = -3 \pm \sqrt{17}$$

$\left(-3 - \sqrt{17}, 0\right),\ \left(-3 + \sqrt{17}, 0\right)$

axis of symmetry: $x = -3$

point symmetric to the *y*-intercept: $(-6, -8)$

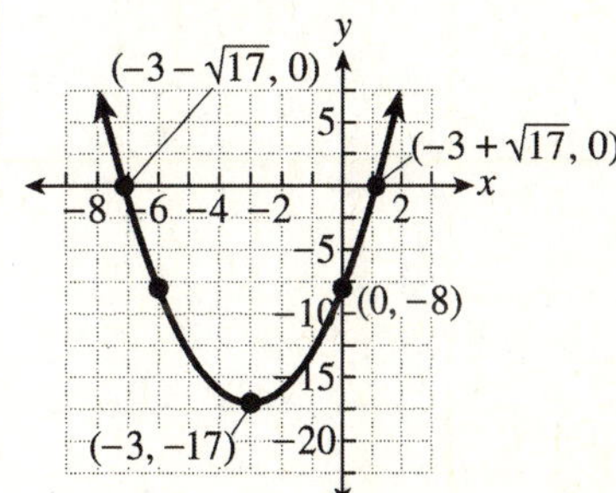

22. $y = (x+3)^2 - 4$ or $y = (x-(-3))^2 - 4$

Use the form $y = a(x-h)^2 + k$.

$a = 1, h = -3, k = -4$

$a > 0$, the parabola opens upward.

vertex: $(-3, -4)$

Shift the graph of $y = x^2$ by 3 units to the left and by 4 units down.

y-intercept:
$y = (0+3)^2 - 4 = 9 - 4 = 5$
$(0, 5)$

x-intercept:
$$0 = (x+3)^2 - 4$$
$$4 = (x+3)^2$$
$$\pm\sqrt{4} = x + 3$$
$$-3 \pm 2 = x$$
$$x = -5 \text{ or } x = -1$$
$$(-5, 0), (-1, 0)$$

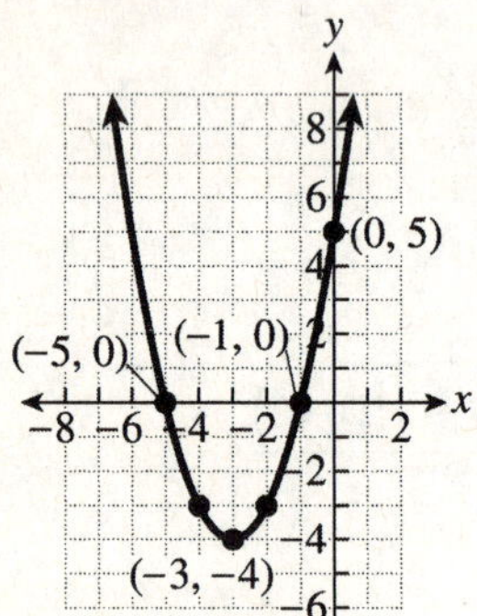

23. $x = -y^2 + y + 6$

Use the form $x = ay^2 + by + c$.

$a < 0$, parabola opens to the left.

vertex:
$$y = \frac{-b}{2a} = \frac{-1}{2(-1)} = \frac{1}{2}$$

$$x = -\left(\frac{1}{2}\right)^2 + \left(\frac{1}{2}\right) + 6 = -\frac{1}{4} + \frac{2}{4} + \frac{24}{4} = \frac{25}{4}$$

$$\left(\frac{25}{4}, \frac{1}{2}\right)$$

x-intercept:
$x = -(0)^2 + (0) + 6 = 6$
$(6, 0)$

y-intercept:
$$0 = -y^2 + y + 6$$
$$y^2 - y - 6 = 0$$
$$(y+2)(y-3) = 0$$
$$y = -2 \text{ or } y = 3$$
$$(0, -2), (0, 3)$$

axis of symmetry: $y = \dfrac{1}{2}$

point symmetric to the x-intercept: $(6, 1)$

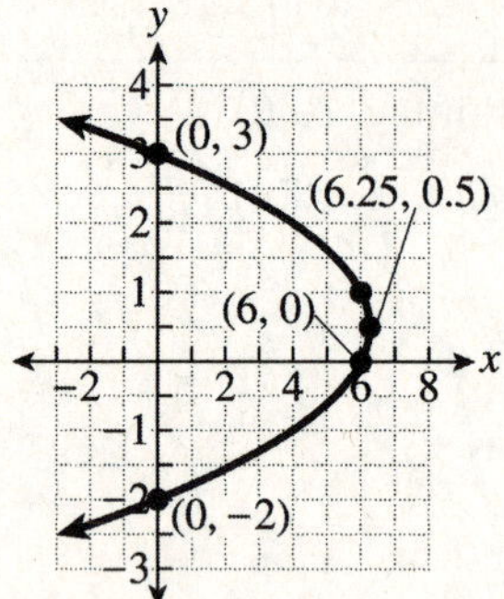

24. $x = (y-2)^2 + 3$

Use the form $x = a(y-k)^2 + h$.

$a = 1, h = 3, k = 2$

$a > 0$, parabola opens to the right.

vertex: $(3, 2)$

Shift the graph of $x = y^2$ by 3 units to the right and by 2 units up.

x-intercept:
$x = (0-2)^2 + 3 = 4 + 3 = 7$
$(7, 0)$

y-intercept:
The vertex is to the right f the y-axis and the parabola opens to the right. There are no y-intercepts.

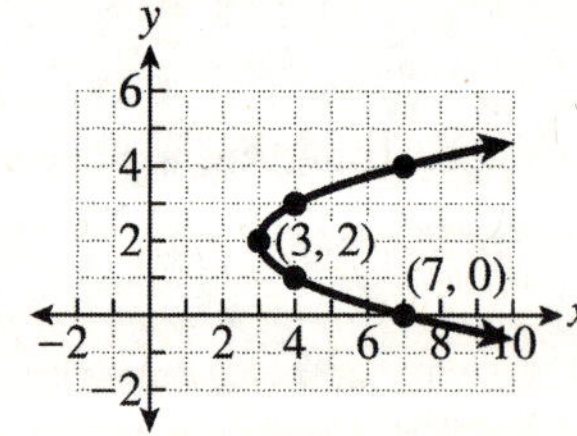

25. $x^2 + y^2 = 64$
$x^2 + y^2 = 8^2$

Use the form $x^2 + y^2 = r^2$.

Center: $(0,0)$; radius: $r = 8$

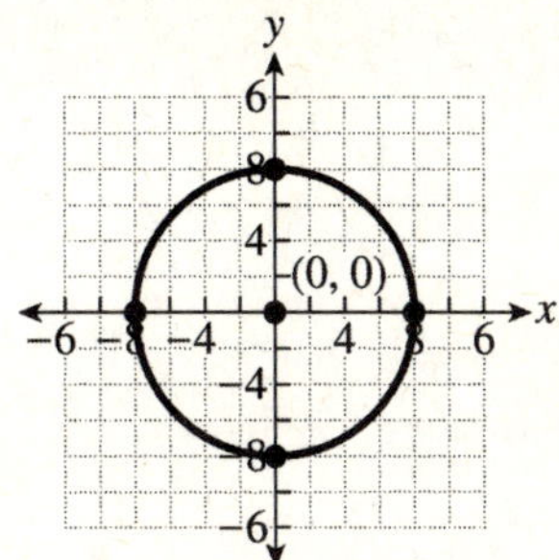

26. $(x+3)^2 + (y-2)^2 = 25$
$(x-(-3))^2 + (y-2)^2 = 5^2$

Use the form $(x-h)^2 + (y-k)^2 = r^2$.

$h = -3$, $k = 2$, and $r = 5$

Center: $(-3, 2)$; radius: $r = 5$

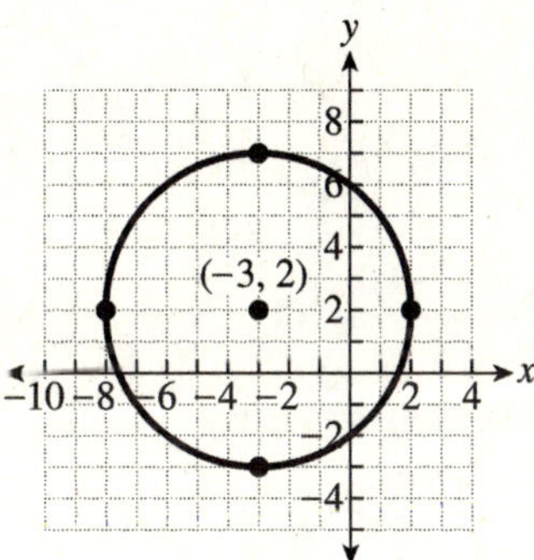

27. $\dfrac{x^2}{9} + \dfrac{y^2}{49} = 1$

$\dfrac{x^2}{3^2} + \dfrac{y^2}{7^2} = 1$

Use the form $\dfrac{x^2}{a^2} + \dfrac{y^2}{b^2} = 1$.

Center: $(0,0)$

$a = 3$, and $b = 7$.

x-intercepts: $(-3,0)$, $(3,0)$
y-intercepts: $(0,-7)$, $(0,7)$

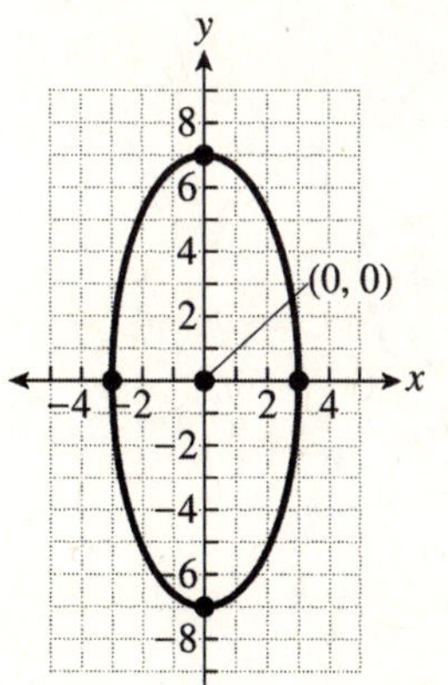

28.
$$36x^2 + 25y^2 - 288x - 300y + 576 = 0$$
$$36x^2 - 288x + 25y^2 - 300y = -576$$
$$36(x^2 - 8x) + 25(y^2 - 12y) = -576$$
$$36\left(x^2 - 8x + \left(-\frac{8}{2}\right)^2\right) + 25\left(y^2 - 12y + \left(-\frac{12}{2}\right)^2\right) = -576 + 36\left(-\frac{8}{2}\right)^2 + 25\left(-\frac{12}{2}\right)^2$$
$$36(x^2 - 8x + 16) + 25(y^2 - 12y + 36) = -576 + 576 + 900$$
$$36(x-4)^2 + 25(y-6)^2 = 900$$
$$\frac{36(x-4)^2}{900} + \frac{25(y-6)^2}{900} = \frac{900}{900}$$
$$\frac{(x-4)^2}{25} + \frac{(y-6)^2}{36} = 1$$
$$\frac{(x-4)^2}{5^2} + \frac{(y-6)^2}{6^2} = 1$$

Continued on next page.

28. Continued.

Use the form $\dfrac{(x-h)^2}{a^2}+\dfrac{(y-k)^2}{b^2}=1$.

Center: $(4,6)$

$a=5$ and $b=6$

The endpoints of the horizontal axis are 5 units to the left and right of the center.

$(-1,6)$ and $(9,6)$

The endpoints of the vertical axis are 6 units above and below the center.

$(4,0),(4,12)$

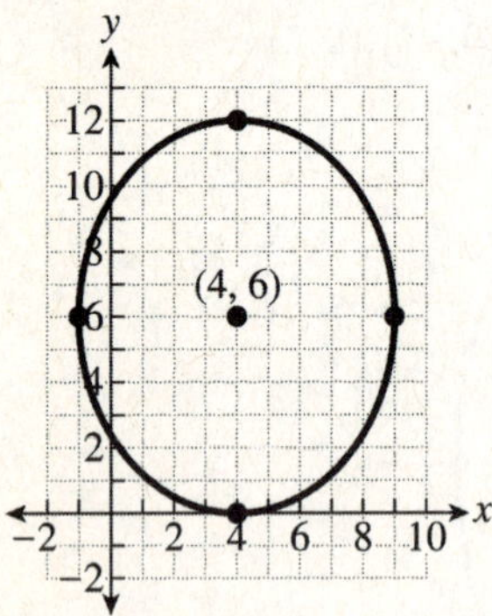

29. $\dfrac{x^2}{25}-\dfrac{y^2}{4}=1$

$\dfrac{x^2}{5^2}-\dfrac{y^2}{2^2}=1$

The hyperbola is in the form $\dfrac{x^2}{a^2}-\dfrac{y^2}{b^2}=1$.

The hyperbola opens to the left and to the right where $a=5,$ and $b=2.$

Center: $(0,0)$

Vertices: $(-5,0),\ (5,0)$

Asymptotes: $y=\pm\dfrac{b}{a}x$

$y=\pm\dfrac{2}{5}x$

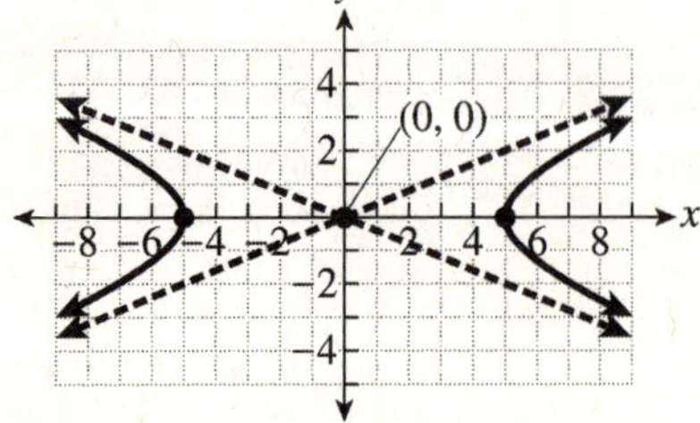

30. $(y-2)^2-\dfrac{(x+4)^2}{9}=1$

$\dfrac{(y-2)^2}{1^2}-\dfrac{(x-(-4))^2}{3^2}=1$

The hyperbola is in the form

$\dfrac{(y-k)^2}{b^2}-\dfrac{(x-h)^2}{a^2}=1.$

The hyperbola opens upward and downward where $h=-4, k=2,\ a=3,$ and $b=1.$

Center: $(-4,2)$

Vertices: $(-4,1),\ (-4,3)$

Asymptotes: $y=\pm\dfrac{b}{a}(x-h)+k$

$y=\pm\dfrac{1}{3}(x+4)+2$

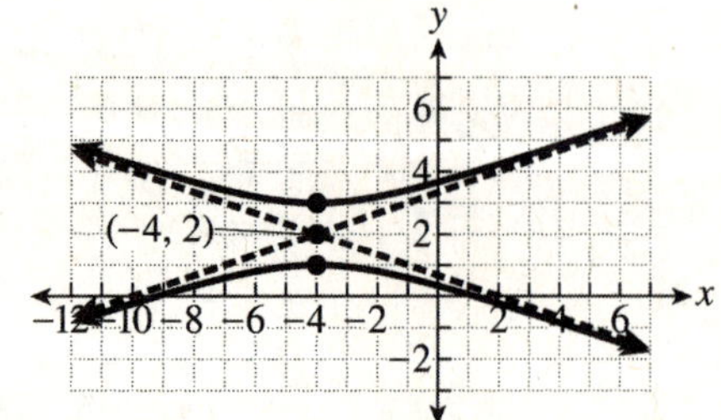

31. Center: $(3,-2)$; radius: 6

Use the form $(x-h)^2+(y-k)^2=r^2$

where (h,k) is the center and r is the radius.

$$(x-3)^2+(y-(-2))^2=6^2$$
$$(x-3)^2+(y+2)^2=36$$

32. Center: $(0,1)$

The endpoints of the horizontal axis are $(-4,1)$ and $(4,1)$; $a=4$.

The endpoints of the vertical axis are $(0,-6)$ and $(0,8)$; $b=7$.

$$\frac{(x-0)^2}{4^2}+\frac{(y-1)^2}{7^2}=1$$
$$\frac{x^2}{16}+\frac{(y-1)^2}{49}=1$$

33. $x^2+y^2=40$ (Eq. 1)
$2x+y=10$ (Eq. 2)

Solve for y in Eq. 2.
$y=-2x+10$

Substitute $y=-2x+10$ into Eq. 1.
$$x^2+(-2x+10)^2=40$$
$$x^2+4x^2-40x+100=40$$
$$5x^2-40x+60=0$$
$$5(x^2-8x+12)=0$$
$$5(x-2)(x-6)=0$$
$x=2$ or $x=6$

Find the y-coordinates using $y=-2x+10$.

$x=2$	$x=6$
$y=-2(2)+10$	$y=-2(6)+10$
$y=-4+10$	$y=-12+10$
$y=6$	$y=-2$

$(2,6),(6,-2)$

34. $x^2+2y^2=18$ (Eq. 1)
$y=x-3$ (Eq. 2)

Substitute $y=x-3$ into Eq. 1.
$$x^2+2(x-3)^2=18$$
$$x^2+2(x^2-6x+9)=18$$
$$x^2+2x^2-12x+18=18$$
$$3x^2-12x=0$$
$$3x(x-4)=0$$
$x=0$ or $x=4$

Find the y-coordinates using $y=x-3$.

$x=0$	$x=4$
$y=(0)-3$	$y=(4)-3$
$y=-3$	$y=1$

$(0,-3),(4,1)$

35. $x^2+y^2=100$ (Eq. 1)
$x^2-y^2=28$ (Eq. 2)

$$\begin{array}{ll} x^2+y^2=100 & \text{(Eq. 1)} \\ +\ \ x^2-y^2=28 & \text{(Eq. 2)} \\ \hline 2x^2=128 & \\ x^2=64 & \\ x=\pm8 & \end{array}$$

Find the y-coordinates using $x^2+y^2=100$.

$x=-8$	$x=8$
$(-8)^2+y^2=100$	$(8)^2+y^2=100$
$64+y^2=100$	$64+y^2=100$
$y^2=36$	$y^2=36$
$y=\pm6$	$y=\pm6$

$(-8,-6),(-8,6),(8,-6),(8,6)$

36. $2x^2+5y^2=38$ (Eq. 1)
$4x^2-3y^2=24$ (Eq. 2)

$$\begin{array}{ll} -4x^2-10y^2=-76 & (-2\cdot\text{Eq. 1}) \\ +\ \ 4x^2-3y^2=24 & \text{(Eq. 2)} \\ \hline -13y^2=-52 & \\ y^2=4 & \\ y=\pm2 & \end{array}$$

Find the x-coordinates using $2x^2+5y^2=38$.

$y=-2$	$y=2$
$2x^2+5(-2)^2=38$	$2x^2+5(2)^2=38$
$2x^2+20=38$	$2x^2+20=38$
$2x^2=18$	$2x^2=18$
$x^2=9$	$x^2=9$
$x=\pm3$	$x=\pm3$

$(-3,-2),(3,-2),(-3,2),(3,2)$

37. Unknowns:
Length: x
Width: y

$2x + 2y = 100$ (Perimeter)
$x \cdot y = 600$ (Area)

Solve the perimeter equation for x, the length.
$2x = -2y + 100$
$x = -y + 50$

Substitute into the area equation.

$(-y + 50)y = 600$
$-y^2 + 50y = 600$
$0 = y^2 - 50y + 600$
$0 = (y - 30)(y - 20)$
$y = 30$ or $y = 20$

Solve for the length (x) using $x = -y + 50$.

$y = 30$ $\qquad$ $y = 20$
$x = -(30) + 50$ $\qquad$ $x = -(20) + 50$
$x = 20$ $\qquad$ $x = 30$
Dimensions are 20 feet by 30 feet.

38. $-4, 9, 22, 35, 48, \ldots$

Each successive term increases by 13.
The next three terms are: 61, 74, 87.
$a_n = 13n - 17$

39. $a_n = 2(a_{n-1}) + 5$

$a_1 = 13$
$a_2 = 2(a_{2-1}) + 5 = 2(13) + 5 = 31$
$a_3 = 2(a_{3-1}) + 5 = 2(31) + 5 = 67$
$a_4 = 2(a_{4-1}) + 5 = 2(67) + 5 = 139$
$a_5 = 2(a_{5-1}) + 5 = 2(139) + 5 = 283$

$13, 31, 67, 139, 283$

40. $3, 12, 21, 30, 39, \ldots$
$a_1 = 3$
$d = 12 - 3 = 9$

$a_n = a_1 + (n-1)d$
$a_n = 3 + (n-1)9 = 3 + 9n - 9 = 9n - 6$

41. $-6, -1, 4, 9, \ldots$

$a_1 = -6$
$d = -1 - (-6) = 5$

$a_n = n + (n-1)d$
$a_n = -6 + (n-1)5 = -6 + 5n - 5 = 5n - 11$
$a_{12} = 5(12) - 11 = 49$

$s_n = \dfrac{n}{2}(a_1 + a_n)$

$s_{12} = \dfrac{12}{2}(a_1 + a_{12})$

$s_{12} = \dfrac{12}{2}(-6 + 49) = 258$

42. $1, 2, 3, 4, \ldots$

$a_1 = 1$
$d = 2 - 1 = 1$

$a_n = 1 + (n-1)1 = 1 + n - 1 = n$
$a_{200} = 200$

$s_n = \dfrac{n}{2}(a_1 + a_n)$

$s_{200} = \dfrac{200}{2}(a_1 + a_{200})$

$s_{200} = \dfrac{200}{2}(1 + 200) = 20,100$

43. $12, 9, \dfrac{27}{4}, \dfrac{81}{16}, \ldots$

$r = \dfrac{9}{12} = \dfrac{3}{4}$

$a_n = a_1 \cdot r^{n-1}$

$a_n = 12 \cdot \left(\dfrac{3}{4}\right)^{n-1}$

44. $a_1 = 1$
$r = 4$

$s_n = a_1 \cdot \dfrac{1 - r^n}{1 - r}$

$s_8 = 1 \cdot \dfrac{1 - 4^8}{1 - 4} = 21,845$

45. $2, 14, 98, 686, \ldots$

$a_1 = 2$

$r = \dfrac{14}{2} = 7$

$s_n = a_1 \cdot \dfrac{1 - r^n}{1 - r}$

$s_7 = 2 \cdot \dfrac{1 - 7^7}{1 - 7} = 274,514$

46. $a_n = 6 \cdot \left(\dfrac{3}{8}\right)^{n-1}$

$a_1 = 6$ and $r = \dfrac{3}{8}$

$|r| = \left|\dfrac{3}{8}\right| = \dfrac{3}{8} < 1$

Yes, since $|r| < 1$, the infinite series has a limit.

$s = \dfrac{a_1}{1 - r}$

$s = \dfrac{6}{1 - \dfrac{3}{8}} = \dfrac{6}{\dfrac{5}{8}} = \dfrac{48}{5} = 9\dfrac{3}{5}$

47. $\displaystyle {}_nC_r = \frac{n!}{r! \cdot (n-r)!}$

$\displaystyle {}_{10}C_4 = \frac{10!}{4! \cdot (10-4)!} = \frac{10!}{4! \cdot 6!} = 210$

48. Use the binomial theorem and its properties.

$$(x+y)^4 = {}_4C_0 \cdot x^{4-0}y^0 + {}_4C_1 \cdot x^{4-1}y^1 + {}_4C_2 \cdot x^{4-2}y^2 + {}_4C_3 \cdot x^{4-3}y^3 + {}_4C_4 \cdot x^{4-4}y^4$$

$$= x^4 + \frac{4!}{1! \cdot (4-1)!}x^3 y + \frac{4!}{2! \cdot (4-2)!}x^2 y^2 + \frac{4!}{3! \cdot (4-3)!}xy^3 + y^4$$

$$= x^4 + 4x^3 y + 6x^2 y^2 + 4xy^3 + y^4$$

49. Use the binomial theorem and its properties:

$$(x-y)^6 = {}_6C_0 \cdot x^{6-0}(-y)^0 + {}_6C_1 \cdot x^{6-1}(-y)^1 + {}_6C_2 \cdot x^{6-2}(-y)^2 + {}_6C_3 \cdot x^{6-3}(-y)^3 + {}_6C_4 \cdot x^{6-4}(-y)^4$$

$$+ {}_6C_5 \cdot x^{6-5}(-y)^5 + {}_6C_6 \cdot x^{6-6}(-y)^6$$

$$= x^6 + \frac{6!}{1! \cdot (6-1)!}x^5(-y) + \frac{6!}{2! \cdot (6-2)!}x^4(-y)^2 + \frac{6!}{3! \cdot (6-3)!}x^3(-y)^3 + \frac{6!}{4! \cdot (6-4)!}x^2(-y)^4$$

$$+ \frac{6!}{5! \cdot (6-5)!}x(-y)^5 + (-y)^6$$

$$= x^6 + 6x^5(-y) + 15x^4(-y)^2 + 20x^3(-y)^3 + 15x^2(-y)^4 + 6x(-y)^5 + (-y)^6$$

$$= x^6 - 6x^5 y + 15x^4 y^2 - 20x^3 y^3 + 15x^2 y^4 - 6xy^5 + y^6$$

50. $(x+y)^8 = x^8 + 8x^7 y + 28x^6 y^2 + 56x^5 y^3 + 70x^4 y^4 + 56x^3 y^5 + 28x^2 y^6 + 8xy^7 + y^8$